The Haplochromine Fishes of the East African Lakes

Collected papers on their taxonomy, biology and evolution (with an introduction and species index)

by

P. H. Greenwood
*Department of Zoology,
British Museum (Natural History)*

Cornell University Press
Ithaca, New York

© Trustees of the British Museum (Natural History), 1981
Copyright in 'Two new species of Haplochromis (Pisces, Cichlidae) from Lake Victoria', Taylor and Francis Ltd., 1966

All rights reserved. Except for brief quotations in a review, this book, or parts thereof, must not be reproduced in any form without permission in writing from the publisher. For information address Cornell University Press, 124 Roberts Place, Ithaca, New York 14850.

First published 1981 by Cornell University Press by Permission of the copyright holders.

This edition is not for sale in the United Kingdom, the British Commonwealth (excluding Canada), Continental Europe and Japan.

International Standard Book Number 0-8014-1346-X
Library of Congress Catalog Card Number 80-69736

Reproduced from copies in the British Museum (Natural History)

Printed in the Federal Republic of Germany

CONTENTS

Since the title pages (and *verso*) of several papers are not reproduced in these volumes, a full bibliographical reference is given for each paper. Whenever possible a precise date of publication is cited, and where that cannot be done, the month of publication is given.

Where a paper has joint authorship, the names of both authors are given; otherwise, the author is P. H. Greenwood.

Introduction	i
The monotypic genera of cichlid fishes in Lake Victoria. *Bull. Br. Mus. nat. Hist.* (Zool.), **3** (7): 295–333 (1956: February)	1
A revision of the Lake Victoria *Haplochromis* species (Pisces, Cichlidae), Part I. *Bull. Br. Mus. nat. Hist.* (Zool.), **4** (5): 223–44 (1956: November)	39
A revision of the Lake Victoria *Haplochromis* species (Pisces, Cichlidae), Part II. *Bull. Br. Mus. nat. Hist.* (Zool.), **5** (4): 73–97 (1957: October)	59
The monotypic genera of cichlid fishes in Lake Victoria, Part II. *Bull. Br. Mus. nat. Hist.* (Zool.), **5** (7): 163–77 (1959: February)	83
A revision of the Lake Victoria *Haplochromis* species (Pisces, Cichlidae), Part III. *Bull. Br. Mus. nat. Hist.* (Zool.), **5** (7): 179–218 (1959: February)	97
A revision of the Lake Victoria *Haplochromis* species (Pisces, Cichlidae), Part IV. *Bull. Br. Mus. nat. Hist.* (Zool.), **6** (4): 227–81 (1960: March)	137
A revision of the Lake Victoria *Haplochromis* species (Pisces, Cichlidae), Part V. *Bull. Br. Mus. nat. Hist.* (Zool.), **9** (4): 139–214 (1962: November)	191
Two new species of *Haplochromis* (Pisces, Cichlidae) from Lake Victoria. *Ann. Mag. nat. Hist.* (13) **8**: 303–18 (May, 1965, actually published 8th March, 1966)	265
A revision of the Lake Victoria *Haplochromis* species (Pisces, Cichlidae), Part VI. *Bull. Br. Mus. nat. Hist.* (Zool.), **15** (2): 29–119 (1967: 13th January)	281
Greenwood, P. H. & Gee, J. M. A revision of the Lake Victoria *Haplochromis* species (Pisces, Cichlidae), Part VII. *Bull. Br. nat. Hist.* (Zool.), **18** (1): 1–65 (1969: 29th April)	371
Greenwood, P. H. & Barel, C. D. N. A revision of the Lake Victoria *Haplochromis* species (Pisces, Cichlidae), Part VIII. *Bull. Br. Mus. nat. Hist.* (Zool.), **33** (2): 141–92 (1978: 23rd February)	435

The cichlid fishes of Lake Nabugabo, Uganda. *Bull. Br. Mus. nat. Hist.* (Zool.), **12** (9): 315–57 (1965: August) — 487

A revision of the *Haplochromis* and related species (Pisces, Cichlidae) from Lake George, Uganda. *Bull. Br. Mus. nat. Hist.* (Zool.), **25** (5): 139–242 (1973: 27th June) — 531

The *Haplochromis* species (Pisces: Cichlidae) of Lake Rudolf, East Africa. *Bull. Br. Mus. nat. Hist.* (Zool.), **27** (3): 139–65 (1974: 19th July) — 633

Towards a phyletic classification of the 'genus' *Haplochromis* (Pisces, Cichlidae) and related taxa. Part I. *Bull. Br. Mus. nat. Hist.* (Zool.), **35** (4): 265–322 (1979: 31st May) — 659

Towards a phyletic classification of the 'genus' *Haplochromis* (Pisces, Cichlidae) and related taxa. Part II: the species from Lakes Victoria, Nabugabo, Edward, George and Kivu. *Bull. Br. Mus. nat. Hist.* (Zool.), **39** (1): 1–101 (1980: 30th October) — 717

Species flocks and explosive evolution. *In* Greenwood, P. H. & Forey, P. L. (Eds), pp. 61–74, *Chance, Change and Challenge – The evolving biosphere.* Cambridge University Press & British Museum (Nat. Hist.) London. (1981). — 819

INTRODUCTION

The Pocket Oxford Dictionary defines an 'introduction' as something containing 'preliminary matter', but it describes a 'prologue' as a 'preliminary discourse'. In those terms this essay is both an introduction and a prologue. First, then, the preliminary matter.

Fourteen of the papers reprinted here cover a species-level revision of the numerous cichlid fishes occurring in Lakes Victoria, Edward, George, Nabugabo and Turkana (formerly Rudolf). When these papers were first published, most of the species were referred to a single, geographically widespread genus, *Haplochromis*; most were, and still are, endemic to the lakes in which they occur.

Two papers are concerned with a reappraisal of the *Haplochromis* generic concept. They represent a first attempt to split that polyphyletic 'genus' into a number of monophyletic lineages. Because of the constraints imposed by the Linnaean categorical hierarchy, and because it is impossible to demonstrate that *Haplochromis* as previously conceived is monophyletic, each of those lineages is now accorded generic rank.

As a result, the total number of *Haplochromis* species in Africa is reduced from over 300 to a mere 5, but the number of *Haplochromis*-like genera has increased from 6 to 19 (see Greenwood, 1979a; p. 717 below). These genera can be referred to, conveniently, as the 'haplochromines', it being understood that that name carries with it no phylogenetic implications.

The majority of new haplochromine genera are at present known only from Lake Victoria and from Lakes Nabugabo, Edward, George and Kivu which are historically related to Victoria (see Greenwood, 1980). That position may, however, change when the haplochromines of Lake Malawi are revised at a generic level. The haplochromine species from Lake Turkana are members of a genus widespread in the Nile and in the Zaire river system (see Greenwood, 1979a; p. 684 below).

One of the two non-endemic genera in Lakes Victoria, Edward, George, Nabugabo and Kivu, *Astatotilapia*, is widespread in Africa (Greenwood, 1979, a; p. 675 below). All its member species are trophically generalised, and none can be considered in any way anatomically specialised, at least when compared with other haplochromines. Unfortunately it is impossible to be more specific about their phyletic relationships with the species (and genera) of the lakes, nor is it possible to establish relationships between *Astatotilapia* species within and without the lake basins. Thus it is no longer strictly correct to refer to the lake haplochromines as members of a 'species flock', because that term implies a close relationship between all its members. Nevertheless, like the word 'haplochromine', it is a convenient term and I shall continue to use it, again without phylogenetic overtones, to describe the endemic taxa of Lake Victoria. Incidentally, about 99% of the haplochromine species from that lake are strictly endemic, with only a few penetrating the affluent and effluent rivers of the lake, and then for no great distances. A similar level of endemicity characterises the species from Lakes Edward, George and Kivu; the figure is lower in Nabugabo, but the species of that lake are outstanding for the recency of their origin (see p. 523).

Perhaps at this point it should be noted that the papers describing species from Lake Victoria (and the other lakes too) were written before I had investigated fully the possible phyletic relationships of the various species. Nor at that time had I been able to test the validity of the flock's presumed mono- or oligophyletic origin. Such an origin had been suggested by previous workers, and I had accepted it at the outset of my own research. So, whenever phrases with the word 'related' (or its derivatives and synonyms) are used in the descriptive papers, any implied phyletic relationship should be discounted. Rather, the words should be taken to mean 'resemblance' or 'similarity' in a strictly phenotypic sense.

A great deal of work remains to be done on the intra- and interrelationships of the haplochromines from Victoria, George, Edward and Kivu, especially on their relationships with the hundred or more '*Haplochromis*' species from Lake Malawi. Many of the latter resemble taxa in the other lakes, both anatomically and in their ecological requirements. As far as I can determine, however, the genus *Haplochromis* (*sensu stricto*) is not represented in Lake Malawi, and many of the anatomical resemblances between taxa in that and the other lakes may only be superficial or convergent similarities.

Lake Malawi, like Lake Victoria, has provided the site for the evolution of an ecologically and taxonomically complex species flock. In both lakes the majority of haplochromine species are endemic, and in each lake the species of a particular lineage appear to be more closely related to one another than to species occurring outside the lake (Lake Victoria in this instance being taken to include Lakes Kioga, Nabugabo, Edward and George, and Lake Kivu: see p. 718 and Greenwood, 1980).

The species revision of the haplochromines from Lake Victoria was undertaken as an essential preliminary to a broad study of the flock in that lake, a fascinating example of trophic and habitudinal specialisation and, since the lake is only about three-quarters of a million years old, explosive speciation as well. An interim review of this research was published a few years ago (Greenwood, 1974). Since that paper is not reprinted here, the prologue part of this introduction takes the form of a brief account of the Lake Victoria flock seen from an evolutionary and ecological viewpoint. Further thoughts on the evolution of the flock, and on the broader evolutionary implications of species flocks in general are contained in a recent paper (Greenwood, 1981; p. 819 below).

A recent estimate of the number of haplochromine taxa in Lake Victoria indicates that there are about 170–200 endemic species present. Some of these have been described, others await description. The great majority of recently discovered species, however, appear to be members of already known lineages and trophic groups. Taxonomically speaking, the undescribed material increases the existing complexity of known lineages, but does not seem to add to the number of genera already recognised.

Lake Victoria's haplochromines together occupy most habitats in the lake and utilise most, if not all, available food sources. They coexist, or at least did before man's interference in the late 1950's by the introduction of exotic species, with a mere 38 species of non-cichlid fishes (representing 11 families).

Both in terms of fish biomass (some 83%) and the actual number of haplochromine species present, Lake Victoria can be considered a 'cichlid lake', as it also can from the way in which the cichlid species dominate its overall bioeconomy.

Haplochromine species live at all depths, from the shallow, inshore areas to the deepest parts of the lake (c 100 m). They occupy calm and sheltered habitats, live over rocky or sandy, wave-washed shores, and are found amongst marginal grass-swamps and water-lily stands. Indeed, they are seemingly absent only from the poorly oxygenated papyrus swamps.

Most species appear to be somewhat restricted in their habitats, but with some having a wider habitat tolerance than others, particularly with regard to the nature of the substrate. Thus the species occupying a sheltered bay with a sandy bottom will, on the whole, differ from those in a superficially similar bay in which the bottom is muddy. In fact, the species composition of a particular beach will change if persistent onshore winds result in the sand being covered by mud washed in from off-shore areas.

Another obvious division is between species living in deeper water (i.e. more than 30 m) and those from shallower regions. Species from the shallower zone seem, in most cases, to have a fairly wide depth tolerance within that zone, although some are restricted to the littoral and immediately sublittoral regions (possibly because of their food preferences, especially in the case of vegetarians). Deeper water species, on the other hand, seem to have a much narrower depth range, but that comment needs further corroboration by more extensive sampling in the deeper waters of the lake, a region which has only been studied in the last decade or so, and then not throughout the whole lake.

Some, indeed most shallow-water species may have a lake-wide distribution, occurring wherever suitable habitats and substrata are present. The picture is not so clear for deep-water species. At present it would seem that many have a much more restricted intralake distribution than do their inshore counterparts.

In all these various ways, some more obvious than others, the haplochromines differ from the non-cichlid species which co-exist with them in all the habitats so far sampled. Unlike the haplochromines, the non-cichlids do appear to have a wider range of habitat tolerance, and some have a much greater depth range as well. An outstanding difference between the two groups lies in the number of haplochromine species present in any one habitat: as many as thirty, compared with the usual three or four non-cichlid species (which are often members of different families). Furthermore, several haplochromine species will be tapping the same food source, whereas each of the non-cichlid species will usually be members of different trophic groups (whose food, however, will be the same as that of particular cichlids in the same habitat). I say usually because non-cichlid piscivores all tend to tap the same general food source, namely haplochromines (which are the chief food of piscivorous haplochromine species as well).

The greatest contrast between haplochromine and non-cichlid taxa lies in the wide range of trophic specialisations shown by the former. Amongst the Lake Victoria representatives of the Cichlidae, particularly the haplochromines, there are species which exploit every available food source in the lake. That fact must be seen in contrast to the 38 non-cichlid species from 11 different families which, only when taken together, almost cover the same trophic spectrum as the cichlids.

The contrast is further emphasised when one recalls that most of the 170-odd haplochromine species were, until recently, placed in a single genus, *Haplochromis* (Greenwood, 1979a & 1980; p. 659 & p. 717 below). Ignoring for the moment the phylogenetic arguments which can be levelled against the old *Haplochromis* generic concept (Greenwood, 1979a; p. 660 below), the fact remains that superficially the various haplochromine species of the lake (together with those from Lakes Edward, George and Kivu) do resemble one another very closely indeed. The differences only become apparent when attention is focussed on the finer details of jaw and skull anatomy, and or dental morphology.

Trophic specialisations encompassed by the haplochromines include insectivory, mollusc-eating, phytophagy, piscivory in various forms, feeding on elements of the zooplankton, and even lepidophagy (see Greenwood, 1974).

Within each of these broad trophic divisions there are further subdivisions of specialisation. The mollusc eaters may crush their prey between hypertrophied pharyngeal bones and teeth, or between the jaws, or they may lever the snail's body from its shell before ingesting the soft parts. Phytophages may graze algae from plants or rock faces, or browse on rooted plants. Insectivorous species may feed on larval or pupal stages living in or on the bottom, or the prey may be taken in midwater; some insectivores are equipped, anatomically, to extract insect pupae from their burrows in rock or submerged wood. One group of piscivores has specialised in feeding on haplochromine embryos and larvae removed from the buccal cavity of brooding female fishes; the great majority of piscivores, however, are predators on free-living haplochromines.

Each trophic specialisation is, of course, represented by several species, and in one group, the piscivores, by as many as forty. Generally there are no apparent interspecific differences in the feeding methods involved, nor in the prey organisms consumed by any particular group of trophic specialists. But the apparent degree of anatomical specialisation manifest by the various taxa in a phyletic lineage can be arranged along a polarised morphocline leading from the least to the most derived forms. In other words, within many lineages one can still trace, in co-existing species, the anatomical stages involved in the evolution of an increasingly derived character or suite of characters.

Since one is dealing here with biologically distinct, coeval sister taxa, the phenomenon could well be described as cladistic gradualism (as contrasted with the supposed phyletic gradualism, beloved by many palaeontologists [see Greenwood, 1979b & 1981; p. 830 below]). The Lake Victoria haplochromines would seem to be a good example of punctuational evolution (Gould and Eldredge, 1977) at a relatively early (and rather messy) stage of its overall ontogeny.

Lest it be thought that each lineage is a straightforward morphocline composed of single species which can be arranged serially, it must be stressed that most levels of derivation in a morphocline are represented by more than one species. The species clustered at any one level differ from each other not in the principal features characterising the cline, but in slight morphometric and meristic details and, especially, in the coloration of breeding males. Species specific male coloration is probably the major (or at least a major) factor in the prevention of interspecific crosses.

As might be expected, the anatomical features involved are chiefly those concerned with food gathering and preparation. That is the shape and disposition of the jaw teeth, the morphology and correlated functional aspects of the jaw elements, the shape of the skull (especially the relative proportions of its pre- and postorbital parts), and the form of the pharyngeal bones and dentition. Body shape, too, is involved but to a much lesser degree; it is only markedly different in certain members of some piscivorous lineages.

When all these characters are analysed for the flock as a whole, it is apparent that the changes involved in their derivation from the primitive 'bauplan' (as manifest, it is thought, by species of the genus *Astatotilapia*) are basically ones of differential growth (see Greenwood, 1974). The presumed simplicity of the ontogenetic modifications required to effect such changes, coupled with the suitability of the basic haplochromine 'bauplan' as a substrate for modifications of that kind, may in part explain the seemingly explosive trophic radiation which the flock has undergone since its inception less than a million years ago.

The causal factors underlying the explosive speciation which initiated these various lines of differentiation are much debated and still far from resolved (see Greenwood, 1974 & 1981, p. 830 below; Fryer, 1977), and at the genetical level, not even explored.

The geomorphological history of the lake basin would seem to provide an ideal physical situation for allopatric speciation, but there are grounds for seriously considering that some type of sympatric speciation is also implicated, especially in the origin of deep-water species. These problems are discussed at length in my 1974 paper, and are reviewed in a broader context in an essay dealing with species-flocks and evolution (see p. 819). Likewise, the ecological implications of multiple species replication in a trophic group are also considered in those papers, especially with regard to their apparent negation of the competitive exclusion principle. This particular problem is further complicated by the species also having, or so it would seem, identical demands on the habitat other than just the trophic ones.

So far the evolutionary story outlined above has centred on the haplochromines of present-day Lake Victoria, a shallow but expansive water-body occupying some 69,000 km^2 of Africa just south of the equator. Recent research has corroborated earlier ideas that the cichlids of Victoria are closely related to those of the neighbouring but much smaller lakes, Edward, George and Kivu. Indeed it seems that the haplochromines of these four lakes should be considered as elements of one evolutionary super-unit (Greenwood, 1980). It is even possible that, historically, the lakes were once part of a single water-body; at the species level, however, each lake now has its own endemic taxa. Ecologically, the flock inhabiting the interconnected Lakes Edward and George, and that inhabiting Lake Kivu, are much less complex than the Victoria flock, partly no doubt because of the fewer species involved in the former lakes (c 40 in Edward and George, only c 10 in Kivu).

The broader phyletic relationships of the Victoria–Edward–Kivu unit are as yet unknown. For the moment it can be said only that the taxa do not have any obvious close relationships with species from the Nile, or with those from Lakes Turkana and Albert. Species from those lakes, and from the Nile, however, do seem to share a recent common ancestry (Greenwood, 1979a; p. 708 below).

Much is left to be learned about the haplochromines of Lake Victoria. The work embodied in the papers reprinted here is very much a preliminary study.

The truth of that remark is underlined by recent field research carried out by my colleagues from Leiden University, drs Franz and Els Witte, and drs Martien van Oijen. During the last three years they have been involved in a detailed ecological and taxonomic survey of the haplochromine species in the southern (Tanzanian) waters of the lake, and with very interesting results.

Apart from collecting many new species recognisable on the basis of trenchant anatomical and meristic differences, they also have evidence suggesting that some of the 'species' I described or redescribed are in fact polyspecific aggregates. In these instances the true species are virtually identical in their anatomy and general morphology, but can be recognised in the field by their specific and distinctive male breeding coloration.

Thus, the complexity of the Lake Victoria haplochromine flock grows, the while further justifying its high rank amongst the known examples of explosive speciation (see Greenwood, 1981; p. 819 below).

References

FRYER, G. 1977. Evolution of species flocks of cichlid fishes in African lakes. *Z. Zool. Syst. Evolforsch.*, **15**: 141–65.

GOULD, S. J. & ELDREDGE, N. 1977. Punctuated equilibria: the tempo and mode of evolution reconsidered. *Paleobiol.*, **3**: 115–51.

GREENWOOD, P. H. 1974. Cichlid fishes of Lake Victoria, East Africa: the biology and evolution of a species flock. *Bull. Br. Mus. nat Hist.* (Zool.) **Suppl. 6**: 1–134.

—— 1979a. Towards a phyletic classification of the 'genus' *Haplochromis* (Pisces, Cichlidae) and related taxa. Part I. *Bull. Br. Mus. nat. Hist.* (Zool.) **35**: 263–322.

—— 1979b. Macroevolution – myth or reality? *J. Linn. Soc. (Biol.)* **12** (4): 293–304.

—— 1980. Towards a phyletic classification of the 'genus' *Haplochromis* (Pisces, Cichlidae), and related taxa. Part II. The species from Lakes Victoria, Nabugabo, Edward, George and Kivu. *Bull. Br. Mus. nat. Hist.* (Zool.) **39**: 1–101.

—— 1981. Species flocks and explosive evolution. *In* Greenwood, P. H. & Forey, P. L. (Eds), pp. 61–74, *Chance, Change and Challenge – The evolving biosphere*. Cambridge University Press & British Museum (Nat. Hist.) London.

ERRATA

Asterisks in the margins of the text draw the reader's attention to the list of errata.

p. 51, *for* (M=32.4) *read* (M=23.4).
p. 65, *delete* specimen number 201.
p. 338, *for* in fishes < 100 mm . . . per cent of head.
read 30.8 – 43.0 (M=35.7) per cent of head; eye diameter in fishes < 100 mm S. L., 26.0 – 32.1 (M=29.1) per cent of head, in larger individuals 20.6 – 25.0, M=23.3 per cent.
p. 413, *for* (M=18.2) *read* (M=28.2)

THE MONOTYPIC GENERA OF CICHLID FISHES IN LAKE VICTORIA

By P. H. GREENWOOD, B.Sc.

CONTENTS

	Page
INTRODUCTION	298
SYNOPSIS OF GENERA OF THE *Haplochromis* GROUP OCCURRING IN LAKE VICTORIA	299
Genus *Macropleurodus*	299
Generic characters and synonymy	299
Diagnosis	301
Macropleurodus bicolor (Boulenger)	301
Synonymy	301
Description	304
Syncranium and associated musculature	305
Coloration and polychromatism	308
Ecology	310
Affinities	311
Study material and distribution records	311
Genus *Platytaeniodus*	312
Generic characters and synonymy	312
Diagnosis	314
Platytaeniodus degeni Boulenger	315
Synonymy	315
Description	315
Syncranium	316
Ecology	317
Affinities	318
Study material and distribution records	318
Genus *Hoplotilapia*	319
Generic characters and synonymy	319
Diagnosis	319
Hoplotilapia retrodens Hilgendorf	321
Description	322
Syncranium	322
Coloration	323
Ecology	324
Affinities	324
Study material and distribution records	326
Genus *Paralabidochromis* nov.	327
Diagnosis	327
Paralabidochromis victoriae sp. nov.	328
Description	328
Affinities	329
DISCUSSION	329
SUMMARY	332
ACKNOWLEDGMENTS	333
REFERENCES	333

INTRODUCTION

In his revision of the Lake Victoria Cichlidae, Regan (1922) recognized four endemic monotypic genera, the species being *Astatoreochromis alluaudi* Pellegrin, 1904, *Hoplotilapia retrodens* Hilgendorf, 1888, *Platytaeniodus degeni* Boulenger, 1906, and *Macropleurodus bicolor* (Boulenger), 1906. Subsequent collections made by the Cambridge Expedition (1930–1931) extended the range of *A. alluaudi* to include the Lake Edward system and Lakes Kachira and Nakavali (Trewavas, 1933), but provided no further distributional data for the other three genera. Since *Astatoreochromis alluaudi* occurs beyond the Victoria system, it is preferable to delay revision of this species until numerous specimens from the different localities can be examined. Some notes on the osteology and possible phyletic relationships of *Astatoreochromis* have been published already (Greenwood, 1954).

A fifth genus which, but for its geographical separation, would not have been distinguished from *Labidochromis* of Lake Nyasa, is recorded for the first time.

The present study is based on specimens collected by and for the East African Fisheries Research Organization during a field study of the Lake Victoria Cichlidae, and on material in the collections of the British Museum (Nat. Hist.), Muséum National d'Histoire naturelle, Paris, and the Museo Civico di Storia Naturale, Genoa. It forms the first part of a revision of the Lake Victoria *Haplochromis* species flock.

Notes on counts and measurements

The counts and measurements used are as defined by Trewavas (1935), except that "length of head" is measured directly from the posterior margin of the operculum to the premaxillary symphysis. This method has been found to yield more consistent results than measurements taken between verticals through the posterior tip of the operculum and the *level* of the tip of the snout, along a line parallel to the longitudinal axis of the body. Likewise, length of snout is measured directly.

Other measurements and counts are:

Depth of preorbital: measured from about the middle of the orbital rim of the preorbital bone along a line which continues the radius of the eye at this point; the line approximately bisects the bone.

Interorbital width: the least width of the roofing part of the frontal bones, between the eyes.

Lower jaw: measured directly from the angle to the symphysis.

Lateral line scale series: After last upper lateral line scale, proceed to the scale of the lower lateral line next behind the transverse row that includes the last scale of the upper lateral line and slopes downwards and forwards from it.

In addition to those defined by Trewavas. Depth of cheek: the greatest depth measured vertically from the lower orbital margin to the lower edge of the *adductor mandibulae* muscles. In most specimens this is virtually a measurement of the depth of the scaled portion of the cheek.

Post-ocular part of the head : measured directly from the posterior orbital margin to the posterior tip of the operculum.

SYNOPSIS OF GENERA OF THE HAPLOCHROMIS GROUP OCCURRING IN LAKE VICTORIA

1. Anal spines three . . . . 2.
   Anal spines four or more . . . . *Astatoreochromis*
2. Teeth of upper jaw in two or more series anteriorly, but only in a single (rarely double) series laterally . . . . 3.
   Teeth of upper jaw in 2–5 series both anteriorly and laterally . . . . 4.
3. Anterior outer teeth disproportionately longer than the adjacent lateral teeth, slender, unicuspid and procurrent. . . *Paralabidochromis* gen. nov., p. 327
   Anterior outer teeth forming a graded series with the lateral teeth, not forwardly directed . . . . *Haplochromis*
4. Outer teeth enlarged and stout, with obliquely truncated and inwardly directed crowns . . . . *Macropleurodus*, p. 299
   Outer teeth of both jaws small, bi- or unicuspid; inner teeth in broad bands anteriorly and laterally . . . . 5.
5. Lower jaw broad and flat, tooth bands in both jaws of uniform breadth antero-posteriorly; those of lower jaw continued posteriorly on to the ascending part of the dentary . . . . *Hoplotilapia*, p. 319
   Lower jaw stout, rounded anteriorly; teeth aggregated anteriorly into two pyriform bands, contiguous at the symphysis . . . . . *Platytaeniodus*, p. 312

Genus **MACROPLEURODUS** Regan, 1922

Bayonia Boulenger, 1911 (*nec.* Bocage 1865), type species *Bayonia xenodonta* Blgr.
Macropleurodus Regan, 1922, Proc. zool. Soc., Lond. 189; type species: *Haplochromis bicolor* Boulenger, 1906, Ann. Mag. nat. Hist. (7) **17**, 444.
Haplochromis (part)
Paratilapia (part) } For references see synonymy under species.
Hemitilapia (part)

Generic characters and synonymy

Prior to Regan's revision of the Lake Victoria Cichlidae (Regan, 1922) specimens of the genus here recognized as *Macropleurodus* had been described as belonging to several genera and species. When Regan (*op. cit.*) defined the apparently new genus. *Macropleurodus* he did not have at his disposal specimens covering the wide range necessary to determine ontogenetic changes in certain characters. Consequently he failed to recognize that his genus was identical with Boulenger's *Bayonia* (Boulenger, 1911), and incorrectly assigned the type specimens of this and one other synonymous "species" to *Haplochromis*.

Regan's diagnosis of *Macropleurodus* gave particular emphasis to the posterior premaxillary teeth:

" . . . several inner series of small teeth anteriorly and three or four series of enlarged teeth laterally, which are exposed when the mouth is shut" (Regan, 1922).

The two specimens on which this description was based have the posterolateral inner teeth not only enlarged, but also similar in form to adjacent outer teeth. It is now known, however, that in small individuals these inner teeth are only slightly enlarged, if at all, and do not closely resemble the outer teeth. Specimens available indicate that most fish between 80 and 90 mm. standard length have a dentition intermediate between juvenile and adult types. In these, many of the typically juvenile, stoutly bi- and unicuspid inner teeth are replaced laterally by teeth differing only slightly in size and form from those of the outer series. Individuals below 80 mm. lack enlarged inner teeth.

Thus, although Regan's diagnosis is incisive for fishes over 90 mm. S.L., it is not sufficiently comprehensive to include smaller individuals.

On the other hand, the present collection shows that the morphology of the outer teeth is equally diagnostic and, moreover, is little affected by the size of the individual. Basically, the outer series is composed of stout teeth, having enlarged and obliquely truncated anterior cusps and disproportionately smaller posterior cusps. In the upper jaw these teeth are implanted obliquely to the long axis of the premaxilla, so that their crowns lie at an angle to it and the anterior cusp points inwards. Teeth of this type are present in all the specimens examined and are not known to occur in any *Haplochromis* or related species.

The peculiar form of the outer teeth in *Macropleurodus* at once suggests affinity with *Bayonia xenodonta* Boulenger (1911), in which the outer teeth are described as having " . . . very large compressed crowns, with long anterior cusp directed inwards and very short or indistinct posterior cusp . . . ". In this respect, the published figure of *B. xenodonta* (Boulenger, 1911 and 1914) is somewhat misleading since the teeth are stouter than depicted.

Through the courtesy of Dr. D. Guiglia, re-examination of the type specimen of *B. xenodonta* has been possible and has confirmed that the outer teeth are alike in *Bayonia* and *Macropleurodus*. Further, as Regan and Trewavas (1928) first observed, Boulenger's description of the inner teeth is inaccurate. Posteriorly these teeth form a double series, with six teeth on either side nearly as large as the outer, and in this respect conform to the dental pattern of young *Macropleurodus*. Also, the more anterior teeth of the two inner series are not " minute and conical " but are in fact tricuspid. Thickening of the buccal mucosa (probably a fixation artefact) has buried the inner teeth so that only their major cusps protrude. Two types of tricuspid teeth are present: one, the usual small tricuspid tooth found in many species of *Haplochromis*; the other, stout and with the minor cusps displaced so as to form a triangular crown having the major cusp at its apex. Similar trigonid teeth are also found in *Macropleurodus*, both young and adult, but have not been observed in *Haplochromis*.

There is also agreement in morphometric characters and, although *Bayonia xenodonta* has only twenty-four outer teeth in the upper jaw, this number is within the known lower range for *Macropleurodus* (see below).

There seems little doubt, therefore, that in all salient features the type and unique specimen of *Bayonia xenodonta* is identical with small specimens of *Macropleurodus bicolor*, and that the two are conspecific.

One other name has been given to cichlid fishes from Lake Victoria with teeth of the *Macropleurodus* type, namely *Hemitilapia materfamilias* Pellegrin, 1913. I have examined the holotype (no. 12.278 in the collection of the Muséum d'Histoire naturelle, Paris) and found it to agree closely with *Macropleurodus* of similar size. The teeth and dental pattern of the type and three other specimens represent an advanced stage in the transition from juvenile to adult condition; that is, the postero-lateral premaxillary teeth of the first inner series are nearly comparable in size and form with the adjacent outer teeth. The remaining inner teeth are small and unequally bicuspid.

Regan (1922) tentatively synonymized *H. materfamilias* with *Haplochromis obliquidens* Hilgendorf; apparently he did not examine the type of *H. materfamilias* (*op. cit.*, pp. 157 and 158) and was misled by Pellegrin's description. Comparison of the holotype with specimens of *H. obliquidens* at once reveals the existence of fundamental differences in the dentition of the two species. In *H. obliquidens* the movably implanted outer teeth are fine and numerous (50–70 in upper jaw); the posterior cusp is wanting, except very occasionally in small fish, whilst the anterior cusp, although obliquely truncate, is compressed and slender. In contradistinction, the immovable outer teeth of *H. materfamilias* are coarse and less numerous (24–40); with few exceptions a posterior cusp is present and the stout anterior cusp is circular in cross section.

Diagnosis

Cichlid fishes of the *Haplochromis* group as defined by Regan (1920, 1922) but differing from *Haplochromis* in having stout outer teeth with inwardly directed and obliquely truncated crowns; anterior cusp long, slightly decurved and not compressed, the posterior cusp small and indistinct. Fishes over 90 mm. S.L. have one or more inner premaxillary tooth-series composed laterally of enlarged teeth similar in form to the adjacent outer teeth. Consequent upon the enlargement of the lateral teeth, the dentigerous surface of the premaxilla is broader laterally than anteriorly. In small individuals, where the inner teeth are small and bi- or unicuspid throughout the series, the outer teeth are already characteristic. Teeth usually exposed laterally, even when the mouth is shut.

Macropleurodus bicolor (Boulenger) 1906

(Fig. 1)

Haplochromis bicolor (part) Blgr. 1906, Ann. Mag. nat. Hist. (7) **17**, 444 (type only).
Paratilapia bicolor (part) Blgr. 1907, Fish. Nile 479, pl. lxxxix, fig. 1; *Idem*, 1911, Ann. Mus. Genova (3) **5**, 68; *Idem*, 1915, Cat. Afr. Fish. **3**, 346, fig. 234.
Bayonia xenodonta Blgr. 1911, Ann. Mus. Genova (3) **5**, 70; *Idem*, 1915, *op. cit.*, 488, fig. 338.
Hemitilapia materfamilias Pellegrin, 1913, Bull. Soc. zool. France, **37**, 313; Boulenger, 1915, *op. cit.*, 492.
Haplochromis obliquidens (part), Regan 1922, Proc. zool. Soc., London, 188.
Macropleurodus bicolor, Regan, 1922, *op. cit.*, 189.

FIG. 1.—*Macropleurodus bicolor*, holotype. 7/8 N.S. (From Boulenger, *Fishes of the Nile*.)

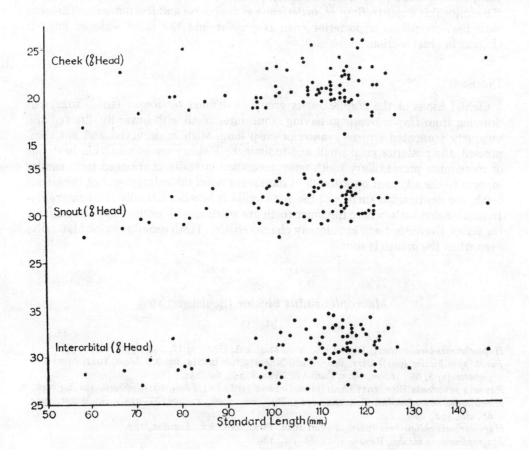

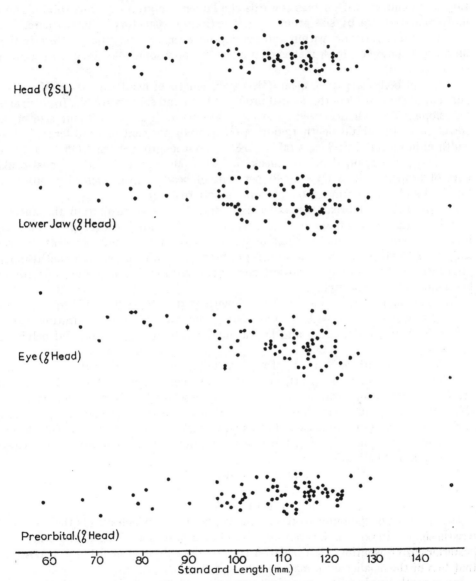

Fig. 2.—Scatter-diagram to show individual variation and allometry of the characters indicated. The isometric relationship of head length and standard length is usual, as is the allometry of head length and the inter-related snout, eye and preorbital measurements. The negative allometry between length of lower jaw and head length is unusual and seems to be related to the stronger jaws and enlarged teeth of larger specimens. These diagrams show scatter but not frequency. Within the size-range 100 to 125 mm. S.L., each dot in the denser aggregates represents at least two specimens.

Description

The selectivity of sampling gear used has resulted in a very unequal size distribution of specimens, with a bias towards the larger size groups. Thus subdivision of morphometric data by size groups is not entirely satisfactory; furthermore, there is considerable variation within and overlap between various groups. Despite these limitations, however, most metric characters do show some allometry with standard length (Fig. 2).

Depth of body 34·4–40·8, mean (M)=37·6, length of head 29·0–35·3 (M = 31·7) per cent of standard length. Dorsal profile of head and snout variable, from straight but sloping through decurved to strongly decurved (Fig. 3), without size or sex correlation. Preorbital depth 13·6–18·3 (M = 16·0) per cent of head length; least width of interorbital 25·8–34·5 (M = 30·8); snout length 26·5–34·2 (M = 31·5), eye 21·4–32·0 (M = 27·0), depth of cheek 16·7–26·0 (M = 21·3), length of post-ocular part of head 41·0–50·0 (M = 47·8) per cent of head length. Described from one hundred and one specimens, 60–150 mm. standard length.

Mouth short and broad; maxilla extending to the vertical from the anterior margin of the orbit, or, more frequently, to anterior third of eye. Jaws unequal, lower jaw somewhat shorter than upper, 28·0–36·0 (M = 31·9) per cent of head length, its length/breadth ratio from 1·3 to broader than long, with a mode at unity.

Gill rakers short and stout, lowest one or two reduced; 7–8 (rarely 9 or 10) on lower limb of anterior arch.

Scales ctenoid; lateral line interrupted, with 31 (f.7), 32 (f.36), 33 (f.29), 34 (f.26), or 35 (f.3) scales. Cheek with three or four (rarely two) rows of imbricating scales; 6–8 scales between dorsal fin and lateral line, 7–9 between pectoral and pelvic fin insertions.

Fins. Dorsal with 23 (f.1), 24 (f.13), 25 (f.67), 26 (f.18), or 27 (f.2) rays; anal with 11 (f.7), 12 (f.82), 13 (f.11), or 14 (f.1) rays comprising XIV–XVII 8–11 and III 8–11 spinous and soft rays for the fins respectively. Pectoral fin 25·2–32·0 (M = 28·8) per cent of standard length. Pelvics with the first soft ray produced, variable in its posterior extension, but usually reaching to vent in immature fishes, and as far as the spinous part of the anal fin in adults. Caudal sub-truncate, scaled on the proximal half only.

Teeth. Little remains to be added to the description given above. There are 24–40 outer teeth in the upper jaw (mode 34); the number of teeth shows a weak positive correlation with size, especially in fish less than 100 mm. S.L.

Two variants of the outer teeth are known, neither of which affects their characteristic shape. In one there is developed a double posterior cusp; in the other, the posterior cusp is wanting. As this latter type occurs only in large fishes, it is possible that loss of the smaller cusp may be due to attrition.

Inner tooth bands with 2–4 series anteriorly and laterally, narrowing to 1 or 2 series posteriorly. In contrast to the outer series, inner teeth show considerable variation in form, being tricuspid, or variously bicuspid. Tricuspid teeth with the cusps arranged in triangular outline, occur at all sizes; the type of *Bayonia xenodonta* is exceptional in possessing inner teeth mainly of this type.

There is noticeable asymmetry in the degree to which lateral teeth of all series are enlarged. Most specimens examined show a distinct tendency for the dextral tooth band to be wider and its teeth larger than on the left side. Only a few specimens show symmetrical or sinistral hypertrophy. In both fresh and preserved material the outermost premaxillary teeth are usually exposed laterally even when the mouth is closed, although there is marked variation in the symmetry and degree of exposure. Hypertrophy of lateral teeth on one side is usually associated with a greater exposure of the teeth on that side.

Pharyngeal bone. Triangular, short and broad; teeth small and cuspidate, the median series often enlarged.

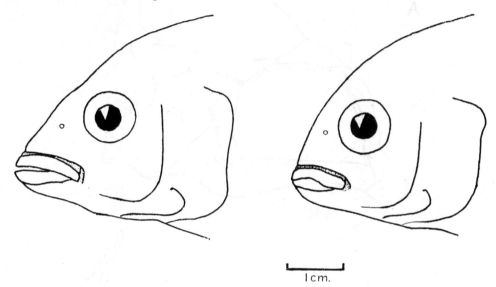

Fig. 3.—*Macropleurodus bicolor.* Individual variability of head profile.

Syncranium and associated musculature

A general impression gained from the syncranium of *Macropleurodus* is one of antero-posterior compression. This is due to the short lower jaw, steeply aligned ethmo-vomerine complex and deep concavity in the entopterygoid, which accommodates part of the short and broad *adductor-mandibulae* muscles.

Both the neurocranium and premaxilla warrant description (Fig. 4). Unpublished observations on skulls of Lake Victoria *Haplochromis* show clearly that the neurocranium of *Macropleurodus* is atypical, although closely paralleled by that of *Haplochromis prodromus* Trewavas, 1935 (= *H. annectens* Regan, 1922, nec *Cyrtocara annectens* Regan, 1921). From the generalised, but common *Haplochromis* skull-type, it differs principally in having a shortened and strongly curved preorbital face, with the anterior profile ascending almost vertically. In generalized *Haplochromis* skulls (Fig. 5) and more particularly in skulls of elongate species, this part of the neurocranium lies parallel to a straight and gently sloping line connecting the foremost part of the ethmoid with the anterior extremity of the supra-occipital (Fig.

5, A). Posteriorly, the neurocranium does not differ significantly from the general type, except that the pharyngeal apophysis is not greatly depressed below parasphenoid, and its pro-otic buttress is broad and bullate.

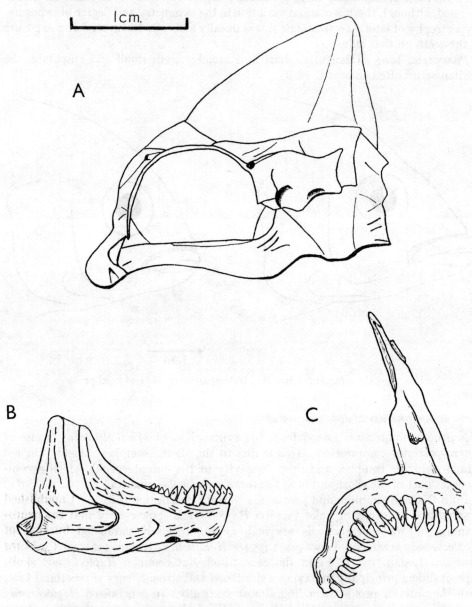

FIG. 4.—*Macropleurodus bicolor*. (A) Neurocranium in left lateral view; (B) dentary, (C) premaxilla, both in right lateral view. Skeleton prepared from a specimen of 115 mm. S.L.

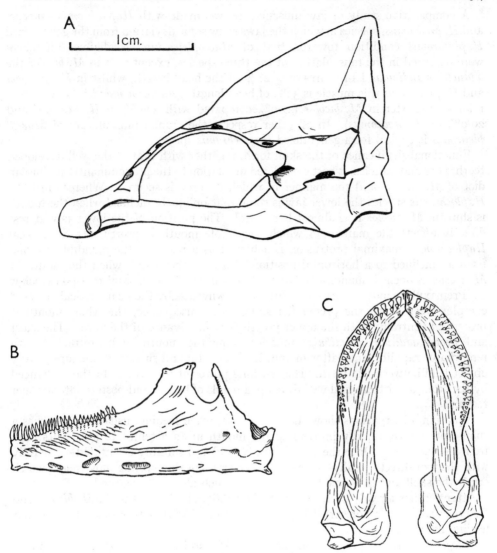

Fig. 5.—*Haplochromis michaeli*. (A) Neurocranium, (B) dentary, left lateral view; (C) dentary, occlusal view; to show the skull and jaws of a generalized *Haplochromis* from Lake Victoria. Skeleton prepared from a specimen of 115 mm. S.L.

The premaxilla shows considerable departure from the basic type found in most *Haplochromis* species. Whereas in *Haplochromis* there is slight ventral arching of the horizontal limb of this bone, in *Macropleurodus* the arch is greatly exaggerated, with its point of maximum curvature at the broadest part of the tooth band (Fig. 4, B). Individuals with a marked dextrally developed tooth pattern have a corresponding degree of asymmetry in the two halves of the premaxilla, which is then more acutely arched on the right side.

A comparative study of jaw musculature was made with *Haplochromis sauvagei* and *H. prodromus*, species in which the jaws show some deviation from the generalized *Haplochromis* condition towards that of *Macropleurodus*. Negligible differences were observed in the musculature of the three species, except that in *M. bicolor* the *adductor mandibulae* I measures only 28% of the head length, whilst in *H. sauvagei* and *H. prodromus* this muscle is 33% of head length. *Adductor mandibulae* II is also relatively shorter in *M. bicolor* (21·0% compared with 33·6% in *H. sauvagei* and 26·9% in *H. prodromus*). In all other respects the cranial musculature of *Macropleurodus* is typical for a generalized *Haplochromis* species.

Functional significance of the skull form, together with that of the well developed teeth, jaws and muscles, is best considered in relation to the predominantly molluscan diet of *M. bicolor*, and the manner in which its prey is secured. Whereas in most *Haplochromis* species the lower jaw is somewhat obliquely inclined when the mouth is shut, in *M. bicolor* it is almost horizontal. The position of the lower jaw at rest directly affects the manner in which the whole mouth is protruded; in typical *Haplochromis*, maximal protrusion is achieved as a result of the mandible moving from an inclined to a horizontal position. On the other hand, when the mouth of *M. bicolor* is opened, mandibular movement is from the horizontal to a point below it. Premaxillary movement is also directed downwards by the near vertical ethmoid complex over which the premaxilla slides. In consequence, the whole mouth is protruded ventrally, with the upper jaw slightly in advance of the lower. The short and broad *adductor mandibulae* muscles allow the mouth to be retracted with remarkable rapidity, thereafter mounting a powerful and sustained pressure on any object held between the teeth. The crushing power of the jaws is further enhanced by the stout outer teeth and well developed bands of lateral and postero-lateral inner teeth.

Aquarium observations show that *M. bicolor*, when feeding, usually approaches a snail from above, rapidly protruding the mouth in an attempt to snatch its prey from the substrate. Once the snail is firmly held and suitably orientated—generally with the foot directly orally—there follows a series of short biting movements which crush the shell and thus free the soft parts, which alone are ingested.

Attention has already been drawn to the similar skull structure in *M. bicolor* and *H. prodromus*; it is not surprising therefore to find that both species have similar feeding habits.

Ontogeny. Alizarin preparations of larval *M. bicolor*, *H. macrops* (a generalized species), and *H. prodromus* have been compared. These specimens reveal no fundamental differences in osteology or dentition of the three species when compared at morphologically equivalent developmental stages. For example, at the latest stage examined (9·0 mm. total length; yolk sac almost completely resorbed) the small conical outer teeth are morphologically and numerically identical in all three species.

From these admittedly few observations it would seem that characteristic adult skull form and outer teeth must develop during post-larval ontogeny.

Coloration and polychromatism

Coloration in life. Adult females with greenish-yellow ground colour, becoming lighter or silver ventrally. Dorsal and anal fins yellow-green, the former with or without two irregular, dark, longitudinal stripes, the latter with two to four ill-defined yellow spots on the posterior part. Caudal and ventral fins generally colourless, though the latter may sometimes be slightly dusky. Young of both sexes and sexually inactive males have similar coloration.

Adult males (breeding coloration). Dark slate-blue ground colour, lighter, sometimes silver, ventrally. Chest, branchiostegal membrane, lower jaw and ventral aspects of the cheek, black. Operculum and flank sometimes with a faint scarlet flush. Dorsal fin dusky, with deep red spots and streaks between the rays, especially intense on the soft part. Anal dusky, with well marked scarlet ocelli. Ventral fins black. Intensity of male coloration is correlated with sexual state ; inter-grades are known between the coloration described above and that of typical female or juvenile coloration.

Besides normal sexually dimorphic coloration, certain fish exhibit a third colour pattern, in the form of an individually variable piebald, black on a yellow-green ground. The holotype is such a specimen (Fig. 1). The *bicolor* pattern is clearly composed of vertically arranged irregular and often interrupted dark bands, which are generally continued across the body on to the vertical and paired fins. Although some are more intensely blotched than others, no intergrades are known between normal female coloration and *bicolor* variants.

With two exceptions all *bicolor* individuals examined were females. The colour-pattern and degree of pigmentation differ in the two exceptional male fishes. Since protandry might be suspected, the gonads were sectioned and examined microscopically. In both fishes, however, there was evidence only of testicular tissue. In one fish the pattern is typical ; in the other the pattern is less intense and occurs on a darker ground than is typical for female fishes.

Accurate frequency-estimates for *bicolor* individuals are difficult to obtain, since collectors show marked sampling bias in favour of these strikingly coloured fishes. However, in more rigorously controlled collections from one area, *bicolor* frequency amongst female fish in the 105 mm. to 125 mm. size class is approximately 30%, an incidence sufficiently high to justify regarding the phenomenon as being due to polymorphism and not to the maintenance of an atypical phenotype by recurrent mutation.

If the two female colour patterns are accepted as an example of polymorphism, it is necessary, *ex hypothesi*, to consider the selective balance which must exist between the two forms. This question is further complicated by the apparently almost completely sex-limited polymorphism in *M. bicolor*.

Since the genetical basis of polymorphism and sex determination is unknown for *Macropleurodus*, some hypothesis at least is desirable before considering the question of selective values for the two colour patterns.

If, as in many fishes, the female is the heterogametic sex, then a possible (and doubtless oversimplified) explanation for this sex-limited polymorphism is that the gene or gene complex underlying development of a *bicolor* pattern may lie in a sex chromosome, be recessive to the gene or genes for normal colour, and be linked with a recessive lethal gene. Thus full expression of *bicolor* pattern could only be manifest in the heterogametic sex. Males carrying the double complement of recessive *bicolor* genes necessary for phenotypic expression in that sex, would, on the linkage supposition, die as a result of simultaneously receiving the two recessive lethal genes. Since linkage is sometimes broken, a small percentage of male *bicolor* individuals might well be expected and the two *bicolor* male fishes in this collection are possibly such individuals.

By this reasoning, either selection in favour of polymorph genes must be sufficient to compensate for loss of males and consequent unbalance of the sex-ratio or, alternatively, the unbalanced sex ratio may be the factor preventing spread of *bicolor* genes throughout the species, should these have a selective value slightly higher than "normal".

Two possible advantages associated with *bicolor* patterns, or genotypes, present themselves. Firstly, a *bicolor* pattern is, in effect, a disruptive one and may thus provide some protection against the attacks of predators. Studies on fish-eating birds (Cott, 1952) and on piscivorous fishes such as *Bagrus* and *Clarias* (personal observation) neither support nor negate this possibility, since *M. bicolor* has not been found among the prey of these animals. Secondly, there is the possibility, also unproven, that a female *bicolor* genotype, or a male heterozygous for *bicolor*, may possess some physiological advantage over other genotypes.

Regrettably, then, insufficient positive evidence is available at present to warrant further discussion on the evolutionary aspects of polymorphism in *Macropleurodus*.

Ecology

Habitat. *M. bicolor* is widely distributed within Lake Victoria, occurring most frequently in littoral and sublittoral regions, especially where the bottom is hard (sand, rock or shingle), but only rarely over mud. Depth distribution is fairly restricted, with a maximum of between 30 and 40 feet (see also Graham, 1929).

Food. From gut analyses of numerous specimens (throughout the size range 60–150 mm.) it is apparent that snails and insect larvae are the predominant food organisms (see also Graham, 1929). Shell fragments are rarely found in the stomach or intestine although opercula are usually present. Aquarium observations confirm that almost the entire body of the snail is removed from its shell before ingestion takes place (*vide* p. 325), although small snails and, on occasion, thin-shelled species, may be crushed intra-orally before being swallowed. As a result of this feeding mechanism snail remains are so fragmentary as to preclude accurate identification; remains of *Gabbia* sp. have, however, been positively identified on several occasions.

The insects most commonly recorded from the pabulum of *M. bicolor* are larvae of the boring may-fly, *Povilla adusta* Navás, with other larval Ephemeroptera, and larval Chironomidae occurring less frequently. The proportion of insect to molluscan food eaten is difficult to determine and is probably related to local and cyclic abundance of these organisms.

In the light of numerous gut analyses which are now available for this species, Graham's record (1929) of fish and cichlid eggs from the stomach of *M. bicolor* requires some comment. Apart from this record no other instances of piscivorous habits are known for *M. bicolor*. Unfortunately, Graham does not give a detailed analysis of the gut contents, particularly of the number of specimens from which his data are derived. Bearing in mind this limitation, it is suggested that the fish-remains and eggs were from the stomachs of different individuals and consequent upon brooding female fishes swallowing their own young, a relatively common occurrence when the fishes are caught in gill-nets.

Breeding. The exact spawning grounds of *M. bicolor* are unknown : eggs and larvae at all stages of development have been found in the mouths of various female fishes and it is presumed that the species is a mouth-brooder. Brooding females have been caught in all parts of the species range. The smallest sexually active fish examined was a female 96 mm. long (*ex* Kisumu). The habitat of post-larval fishes is unknown.

Affinities

Unlike the other monotypic genera *M. bicolor* can apparently be related to an extant species of *Haplochromis*. Similarity in skull architecture of *M. bicolor* and *H. prodromus* has already been noted. There are additional similarities in the short and stoutly constructed jaws of both species, besides a marked resemblance in general facies. Against these resemblances must be set the very different tooth form and dental pattern of *H. prodromus*, although in this species the teeth are stout and the inner series well developed. The structure of the head and dental patterns suggest that, at a functional level, the condition represented by *H. prodromus* might well be considered pre-adaptive to the development of a relatively massive dentition, such as that of *M. bicolor*.

Study material and distribution records

Museum and Reg. No.	S.L. (mm.).	Locality.	Collector.
British Museum (N.H.) :			
1906.5.30.414 (type of *H. bicolor*)	125	Bunjako (Uganda)	Degen
1906.5.30.378	115	Ditto	,,
1928.5.24.493–503	115–125	See below	M. Graham
1928.5.24.1–3	120–125	Ditto	Ditto
Paris Museum :			
12.278 (holotype of *H. materfamilias*	109	Port Florence, Kenya	Alluaud and Jeannel
12.279–281 (paratypes of *H. materfamilias*; proportions not included in description above)	91–97	Ditto	Ditto
Genoa Museum :			
(Type of *Bayonia xenodonta* Blgr.)	73	Jinja	Bayon
B.M.(N.H.) :			
1955.2.10.5–11	78–110	Kisumu	E.A.F.R.O.
1955.2.10.50–57	101–123	Jinja, Napoleon Gulf	Ditto
1955.2.10.43–46	79–104	Beach nr. Nasu Point, Buvuma Channel	,,

Museum and Reg. No.	S.L. (mm.)	Locality	Collector
B.M.(N.H.):—cont.			
1955.2.10.22–23	82 and 97	Grant Bay (Uganda)	E.A.F.R.O.
1955.2.10.21	102	Dagusi Island	Ditto
1955.2.10.65–73	105–120	Off southern tip of Buvuma Is.	,,
1955.2.10.24–42 and 74–83	90–122	Harbour at Entebbe	,,
1955.2.10.58–62	58–67	Bugonga Beach	,,
1955.2.10.12, 16–20 and 48	112–130	Busongwe Bay (Kagera River mouth)	,,
1955.2.10.63–64	110 and 147	Majita Beach (Tanganyika Terr.)	,,
1955.2.10.4, 49 and 84	115–119	Mwanza	,,

Graham (1929) records the occurrence of *M. bicolor* as follows :

Kenya : Kavirondo Gulf : Off Sukuri Island.
Off Ulambwi bay.
Mbita Passage.
Near Nzoia River.
Kadimu Bay.
Tanganyika Territory : Mwanza.
Smith Sound.

Genus *PLATYTAENIODUS* Boulenger, 1906

Platytaeniodus Boulenger, 1906, Ann. Mag. nat. Hist. (7) **17**, 451 ; *Idem*, 1907, Fish. Nile, 493 ; *Idem*, 1915, Cat. Afr. Fish. **3**, 426, fig. 292 ; Regan, 1922, Proc. zool. Soc., Lond. 190. Type species : *Platytaeniodus degeni* Blgr. 1906.

Generic synonyms :
Astatotilapia (part) ⎱ For references see synonymy under species.
Haplochromis (part) ⎰

Generic characters and synonomy

Both Boulenger and Regan considered the premaxilla and its tooth pattern diagnostic. Boulenger (1914) states :

" . . . the alveolar surface of the premaxilla widening towards the pharynx, the band of teeth in the upper jaw horseshoe shaped . . . "

Additional material shows, however that in fishes below 100 mm. standard length the posterior premaxillary dentigerous (alveolar) surfaces are not always expanded medially. Nevertheless, even in small specimens the premaxilla is stouter and its dentigerous surface wider than in *Haplochromis* ; posteriorly the teeth are arranged in several rows, so that the premaxillary tooth band is always clearly U-shaped, with the arms at least as broad as the medial part. Broadening of the posterior alveolar surfaces is gradual and shows positive allometry with standard length ; in some large individuals the left and right surfaces are closely apposed in the mid-line (Fig. 6, c).

Tooth bands in the lower jaw are more readily diagnostic and less subject to variation with absolute size than those of the premaxilla. In *Platytaeniodus* the mandibular teeth are confined to the anterior and antero-lateral portions of the dentary and are grouped into two broad and roughly pyriform patches, contiguous at the symphysis; posteriorly there is a short, single row of four to seven teeth

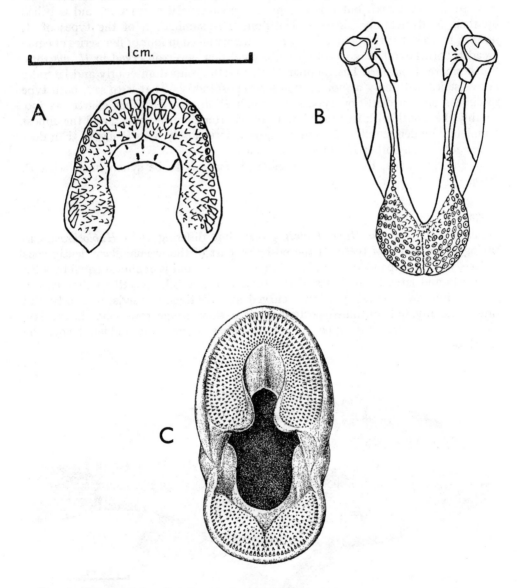

Fig. 6.—*Platytaeniodus degeni*. (A) Premaxilla, (B) dentary, both in occlusal view. Skeleton prepared from a specimen 80 mm. S.L. (C) Mouth of the holotype, *ca.* × 4. [(C) from Boulenger, *Fishes of the Nile*.]

lying between the ascending part of the ramus and the anterior tooth bands (Fig. 6, B and C).

When provisionally referring *Astatotilapia jeanneli* Pellegrin to *Haplochromis macrops* Blgr., Regan (1922) was apparently misled by the large eye and shallow preorbital of *A. jeanneli*. There is undoubtedly some resemblance between *H. macrops* and *P. degeni*, but this is confined to superficial characters, and is belied by their fundamentally different dentition. Re-examination of the types of *A. jeanneli* reveals that the premaxillary teeth are arranged in four or five series of equal breadth both laterally and posteriorly, a condition never observed in *H. macrops*. Furthermore *A. jeanneli* has the mandibular teeth grouped anteriorly and laterally in five series, with only a short, single series posteriorly. That is to say, both type specimens have a dentition typical for small *P. degeni*. In other characters too, notably the mouth with its broad lower jaw, shorter than the upper, and the almost completely hidden maxilla, *A. jeanneli* agrees more closely with *P. degeni* than does *H. macrops* or any other *Haplochromis* species.

On these grounds, therefore, I consider *A. jeanneli* to be synonomous with *P. degeni*.

Diagnosis

Cichlid fishes of the *Haplochromis* group, but differing from *Haplochromis* in having broad bands of teeth on the posterior part of the premaxillary dentigerous surface, which is expanded medially in large specimens but is of almost equal breadth anteriorly and laterally in fishes of less than 100 mm. S.L. Teeth on the dentary grouped into two, broad, pyriform, curved and contiguous bands anteriorly and antero-laterally, but continued posteriorly as a short single row only. Lower jaw usually shorter than the upper; maxilla almost completely hidden below the preorbital.

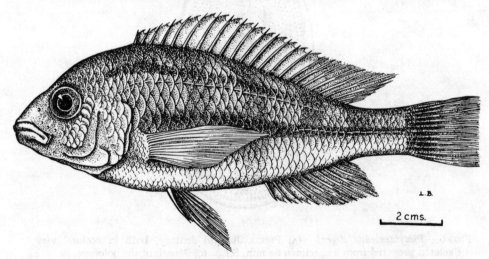

FIG. 7.—*Platytaeniodus degeni*, ♀. Drawn by Miss L. Buswell.

Platytaeniodus degeni Boulenger, 1906
(Fig. 7)

Platytaeniodus degeni Boulenger, 1906, *l.c.*; 1907, *l.c.* and pl. xci, fig. 1; 1915, *l.c.* and fig. 292; Regan, 1922, 190, fig. 14.
Astatotilapia jeanneli Pellegrin, 1913, Bull. Soc. zool. France, **37**, 313.
Haplochromis jeannelli (Pellegrin), Blgr. 1915, Cat. Afr. Fish. **3**, 291.
Haplochromis macrops (part) Regan, 1922, Proc. zool. Soc. London, 166.

Description

Thirty-six specimens (size range 67 to 154 mm.) comprising the type and other specimens in the British Museum (Nat. Hist.) (including material newly collected by E.A.F.R.O.), as well as the types of *A. jeanneli* are considered in this description.

Since most characters tabulated below show some allometry with standard length these data are grouped into two size classes. In some characters intra-group variability is high, but further subdivision into smaller groups is impracticable.

In the table of proportions, head length, depth of body and length of caudal peduncle are expressed as percentages of standard length; all other characters are expressed as percentages of head length. Range and mean are given for each character.

TABLE I.

Standard Length.	67–93 mm. (23 specimens).	98–154 mm. (13 specimens).
Depth of body	32·5–40·5 $M = 35·6$	34·78–43·0 $M = 38·4$
Length of head	30·0–36·5 $M = 32·4$	31·5–34·7 $M = 32·7$
Depth of preorbital	12·0–16·0 $M = 14·2$	13·7–17·4 $M = 15·2$
Least interorbital width	25·0–30·8 $M = 27·1$	28·2–34·8 $M = 31·2$
Length of snout	25·0–32·0 $M = 30·0$	31·75–37·4 $M = 34·8$
Diameter of eye	27·4–33·4 $M = 30·8$	22·1–26·0 $M = 24·8$
Depth of cheek	18·5–26·1 $M = 19·7$	20·0–26·6 $M = 23·4$
Length of lower jaw	32·1–39·6 $M = 37·0$	35·0–38·1 $M = 36·0$
Caudal peduncle	13·9–21·5 $M = 16·5$	14·0–18·5 $M = 16·7$

Dorsal profile of head and snout gently (in a single specimen somewhat strongly) decurved; mouth horizontal; lower jaw equal to or more usually shorter than upper, its length/breadth ratio 1·16–1·68: in some specimens the lower jaw, including lips, is slightly broader than the upper. Lips well developed and somewhat thickened; maxilla almost completely hidden beneath the preorbital, with only its posteroventral tip exposed and extending to below the anterior orbital margin, or slightly beyond.

The holotype, a male of 115·0 mm. S.L., is figured by Boulenger (1907 and 1915). This fish is both somewhat atypical and slightly distorted in preservation and the impression given of a deep ventral profile and a slightly oblique mouth is not characteristic.

Gill rakers. Short and stout, 7–8 on lower limb of the anterior arch, the lowest one or two usually reduced.

Dentition. Premaxilla with 4–8 rows of teeth ; dentary with 4–6 rows.

With few exceptions, specimens over 100 mm. S.L. have the posterior part of the premaxillary dentigerous surface expanded medially, so that the upper tooth bands are broader posteriorly than anteriorly. In small individuals, although there are several series of teeth posteriorly, the tooth band is either of equal width at all points, or medial expansion of the premaxilla may have begun, causing the band to be very slightly wider posteriorly (Fig. 6, A). Expansion of the posterior surface is not correlated with an increase in the number of tooth rows borne on it, which are, in fact, equal to or slightly fewer than those on the anterior part of the premaxilla.

Mandibular dental pattern as described for the genus ; pyriform band from half to two thirds as broad as long.

Teeth are variable in form, those of the outer series slenderly conical, with or without an admixture of unequally bicuspid teeth. Teeth of the inner series all unicuspid in most specimens above 100 mm. S.L. and in some below that size ; otherwise the outermost teeth unicuspid and the remainder tricuspid or exceptionally, bicuspid.

Scales ctenoid ; lateral line interrupted, with 31 (f.2), 32 (f.5), 33 (f.14), 34 (f.10), 35 (f.2), or 36 (f.3) scales. Cheek with 3–4 series of imbricating scales ; 7–9 scales between lateral line and origin of dorsal fin ; 7–9 (rarely 10) between the pectoral and pelvic fins.

Fins. Dorsal with 24 (f.3), 25 (f.23) or 26 (f.10) rays, anal with 11 (f.6), 12 (f.23) or 13 (f.7) rays, comprising XV–XVII 8–10 and III 8–10 spinous and soft rays. Pectoral fin 22·6–31·8 (M = 28·0) per cent of standard length. Caudal truncate, scaled on its proximal half only. Pelvics with the first ray produced, extending to the vent in a few specimens and to the spinous part of the anal in most.

Syncranium. Since the form of premaxilla and dentary in *P. degeni* is correlated with the well-developed tooth pattern, both these bones depart very strikingly from the typical *Haplochromis* condition.

In small individuals of *P. degeni*, the premaxilla bears a superficial resemblance to that of *Hoplotilapia*, particularly with regard to the dental pattern, but in large fishes it is unique. On the other hand, the dental pattern and morphology of the dentary are comparable in both large and small individuals. The dentary is characterized by its broad and laterally expanded anterior tooth-bearing portion, which imparts to this bone an appearance unique amongst the Lake Victoria cichlids.

The neurocranium of *P. degeni* is intermediate between the generalized *Haplochromis* type and that of *M. bicolor*. It is strictly comparable with the neurocrania of species of the *H. crassilabris* group. Here the skull is characterized by a somewhat shortened and steeply inclined ethmo-vomer complex ; in consequence, the anterior skull profile is also steep. In a typical *Haplochromis* skull the ethmo-vomer is longer and rises less steeply, meeting the downward sloping frontals at a wide angle. As a result, the anterior profile is shallower and also more acute than in the " *crassilabris* " type skull (Fig. 5).

Jaw musculature in *P. degeni* dos not differ greatly from that of a generalized *Haplochromis* species.

Coloration. Preserved material. Males: dusky to dark grey ; dorsal and anal fins dark ; pelvics black, caudal colourless. *Females and immature individuals:* silver-grey or light brown ; fins colourless.

Transverse and longitudinal banding sometimes occurs, being most clearly marked in females and young individuals ; when present, there is a well marked median longitudinal stripe, a fainter and interrupted band running slightly below the dorsal fin, and eight or nine narrow transverse stripes on the flank and caudal peduncle. Faint lachrymal and interocular stripes may also be present. The presence and intensity of these markings is apparently related to the emotional state of the fish or may only appear after death.

Coloration in life. Sexually active males: ground colour light blue-grey, lips iridescent blue. Chest and branchiostegal membrane black. Fins ; dorsal sooty, lappets and spots on soft part red ; caudal with red flush, most intense along margin ; anal with dusky pink flush and several yellow ocelli. *Females:* ground colour golden fawn ; all fins neutral, dorsal with orange lappets and spots, especially on the soft part ; caudal with orange margin and maculae ; anal with faint or well marked yellow ocelli.

Ecology

P. degeni is recorded from several areas in Lake Victoria (see below), but as so few specimens are known it is not possible to generalize on habitat preferences. From the scanty data available it appears that the species is probably restricted to littoral and sub-littoral regions where the water is less than fifty feet deep. Specimens have been caught in nets set over both hard and soft substrates, but the greater number came from stations having a sand or shingle bottom.

Food. The distinctive dentition of this species suggests a highly specialized diet. Tantalizingly few fish, however, have yielded ingested material. Twelve specimens have been examined, all of which were caught in nets set overnight or in seines operated during varied daylight hours. According to the substrate over which they were living, ten fishes had either sand grains or organic mud in the stomach and intestines, together with fairly dense aggregations of mucus. Two fishes, caught on different occasions at a station near the southern tip of Ramafuta Island (Buvuma Channel), had the entire alimentary tract filled with the diatom *Melosira*. Diatoms from the stomachs of these fishes showed only slight signs of digestion, but samples taken from the mid-intestines and recta were almost completely digested. Animal remains, occurring sporadically, included insect larvae, Hydracarina, fragments of Copepoda, Ostracoda, and in two specimens shell fragments of Pelecypoda (Sphaeriidae).

Most guts also contained some diatoms and blue-green algae, the former apparently digested, the latter intact. The very small quantity of ingested material in any one individual is striking and no particular organism, or group of organisms, occurs with sufficient frequency to indicate what the food of *P. degeni* may be. Since sand and bottom debris is significant in the majority of specimens, it is possible that the species may feed on the micro-fauna and flora living on and within the substrate. Thus, broad bands of jaw teeth may serve to rasp and loosen food from the surrounding sand.

Breeding. Spawning sites are unknown; only two females, both from beaches in the Mwanza area, have been found with eggs in the mouth. The smallest individual with demonstrably active gonads was a female, 71·0 mm. long.

Affinities

There is no obvious relationship between *P. degeni* and any known Lake Victoria species or species-group of *Haplochromis*, with which genus the species shows fundamental affinities. The peculiar premaxillary and mandibular tooth patterns serve to set *P. degeni* apart from even those *Haplochromis* with several series of inner teeth. Regan (1922) considered *P. degeni* as being " very near " to *H. prodromus*, which species he believed " shows a slight departure from the normal *Haplochromis* dentition towards the *Platytaeniodus* type ". His opinion was based on the holotype and then unique specimen of *H. prodromus*. Summarizing unpublished data on *H. prodromus*, it is clear that the type specimen has an aberrant dental pattern and that its resemblance to *P. degeni* is purely superficial. Whereas in large *Platytaeniodus* there is an actual expansion of tooth-bearing surfaces, in *H. prodromus* only the tooth-band is apparently expanded; its increased breadth is actually due to the posterior teeth being more widely separated from one another than are the anterior teeth. In no specimen of *H. prodromus* is the upper tooth band as broad posteriorly as anteriorly, yet this is the usual condition in *P. degeni*. Further, the dentary of *P. degeni* differs considerably from that of *H. prodromus*. The evolution of a wholly multi-seriate dentition has probably occurred more than once within the Lake Victoria species-flock, as for instance in the *H. sauvagei* group and again in the monotypic genera. Thus any apparent relationship between *P. degeni* and *H. prodromus* should be considered as consequent upon convergent evolutionary trends, the ultimate expressions of which are achieved by manifestly dissimilar means.

Study material and distribution records

Museum and Reg. No.	S.L. (mm.).	Locality.	Collector.
British Museum (Nat. Hist.) :			
1906.5.30.511 (holotype)	114	Bunjako (Uganda)	Degen
1909.3.29.10	98	Sesse Is. (Uganda)	
1928.5.24	93	Mbita Passage (Kenya)	M. Graham
Paris Museum :			
12.262 (holotype of *A. jeanneli*)	72	Port Florence (Kenya)	Alluaud and Jeannel
12.262 (paratype of *A. jeanneli*)	67	Ditto	Ditto

Museum and Reg. No.	S.L. (mm.).	Locality.	Collector.
B.M. (N.H.):			
1955.2.10.91–94	73–79	Kisumu (Kenya)	E.A.F.R.O.
1955.2.10.105–106	78–90	Kamarenga (Kenya)	Ditto
1955.2.10.115	74	Kendu (Kenya)	,,
1955.2.10.88	74	Likungu (Kenya)	,,

Genus *HOPLOTILAPIA* Hilgendorf, 1888

Hoplotilapia Hilgendorf, 1888, S.B. Ges. naturf. Fr. Berlin, 76–77 (type species (*Paratilapia* ?) *retrodens* Hilgendorf, l.c.) ; Regan, 1922, Proc. zool. Soc., Lond. 190.
Cnestrostoma Regan, 1920, Ann. Mag. nat. Hist. (9) 5, footnote p. 45 (type species *Paratilapia polyodon*, Blgr.).
Haplochromis (part)
Paratilapia (part) } For references see synonymy under species below.
Hemichromis (part)

Generic characters and synonymy

The holotype of *Hoplotilapia retrodens* (in the collections of the Zoologisches Museum der Humboldt-Universität, Berlin) has not been examined by me, nor can it be definitely established whether this specimen is still in existence. Dr. Kurt Deckert of the Zoologisches Museum has, on two occasions, kindly attempted to locate several specimens, including the type of *H. retrodens*. Of these he writes : " Ich muss Ihnen leider mittheilen, dass unser Suchen nach den verlangten Typen ohne Erfolg geblieben ist, obwohl ich mit grosser Sicherheit anneheme, dass sie nicht verloren-gegangen sind."

Hilgendorf's original description of (*Paratilapia?*) *retrodens* (1888), although brief and lacking detail, nevertheless stresses characters which clearly separate this species from others of the *Haplochromis* group : viz. a multi-seriate dental pattern with stout and enlarged posterior teeth. Pfeffer's redescription (1896) of the same specimen confirms and extends this account. In the material at my disposal, however, the posterior teeth are clearly enlarged only in the upper jaw, and not in both, as stated for the type. Specimens described below agree closely with the type in the other characters described by Hilgendorf and Pfeffer ; slight differences in scale numbers can probably be attributed to different methods of making these counts.

In addition to the four specimens of *H. retrodens* in the British Museum (Nat. Hist.) it has been possible, through the kindness of Dr. Delfa Guiglia, to examine the type of *Paratilapia polyodon* Blgr. and one other specimen (Museo Civico di Storia Naturale, Genoa, reg. no. G.E. 12.994) determined by Boulenger as *P. polyodon*, and to confirm Regan's view (1922) that these are conspecific with *H. retrodens*.

Diagnosis

Differing from *Haplochromis* as defined by Regan (1920 and 1922*b*) in having broad bands of teeth in both jaws, well developed and usually of almost uniform breadth throughout or very slightly narrower posteriorly. Posterior teeth of the upper jaw enlarged and stout, those of the lower jaw slightly, if at all, enlarged, but the tooth-band continued posteriorly on to the steep ascending contour of the dentary. Lower jaw wide and flat, almost square in anterior outline, slightly shorter than the upper.

Since only three of the collected specimens of this genus are small it is not possible to generalize on differences which apparently exist between the dentition of adult

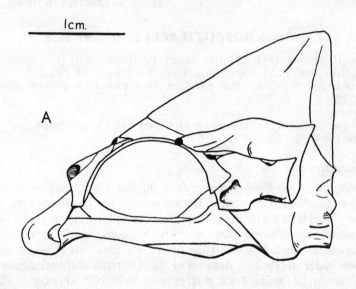

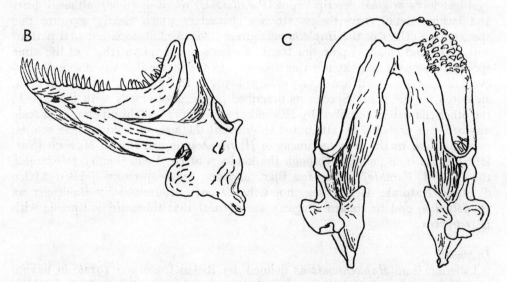

FIG. 8.—*Hoplotilapia retrodens*. (A) Neurocranium, (B) dentary in lateral view, (C) dentary, occlusal view. Dentition only part indicated. From a specimen of 125 mm. S.L.

and juvenile fishes. In these three specimens (74·0, 76·0 and 55·0 mm. S.L.) the tooth bands are broad anteriorly, being composed of 5, 3, and 3 series respectively. Laterally, however, they are reduced to two series, whilst postero-laterally only the outer series persists. In none is the dentition continued onto the ascending part of the dentary, although the shape of the lower jaw is as in the adult.

Two adults of 96 and 110 mm. S.L., collected in a single seine haul at Bukakata, retain these presumedly juvenile dental characters. They form a graded morphological series with a third specimen (134·0 mm. S.L., from the same station) which exhibits only slight departure from the " typical " condition.

The broad and shallow lower jaw (Fig. 8, B and C) of typical individuals is unique amongst Lake Victoria Cichlidae. A few specimens of *H. retrodens* have the dentary, at least in external appearance, similar to that of *Haplochromis*, although in every case the dental pattern is typical for *Hoplotilapia*.

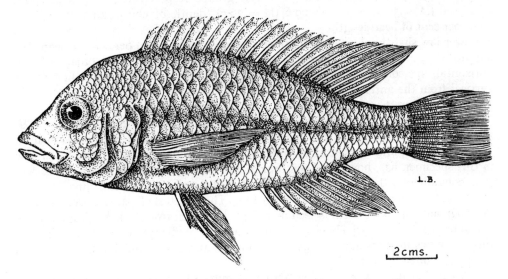

FIG. 9.—*Hoplotilapia retrodens*, ♀. Drawn by Miss L. Buswell.

Hoplotilapia retrodens Hilgendorf, 1888

FIG. 9.

(*Paratilapia* ?) *retrodens* Hilgendorf, 1888, S.B. Ges. naturf. Fr. Berlin, 76.
Hoplotilapia retrodens Hilgendorf, t.c., 77.
Hemichromis retrodens, Pfeffer, 1896, Thierw. O. Afr. Fische, 19.
Haplochromis bicolor (part) Boulenger, 1906, Ann. Mag. nat. Hist. (7) **17**, 444 ; two specimens B.M. no. 1906.5.30.417 and 418.
Paratilapia bicolor (part) Boulenger 1915, Cat. Afr. Fish. **3**, 346.
Paratilapia polyodon Boulenger, 1909, Ann. Mus. Genova (3) **4**, 306, fig. ; *Idem*, 1911, *ibid.* **5**, 68 ; *Idem* 1915, Cat. Afr. Fish. **3**, 349, fig. 236.
Cnestrostoma polyodon (Boulenger), Regan, 1920, Ann. Mag. nat. Hist. (9) **5**, footnote p. 45.

Description

From available material it would appear that only the interorbital width shows marked allometry with standard length. It must, however, be borne in mind that paucity of specimens within the smaller and larger size groups may obscure such relationships, especialy since, in those size-groups which are well represented, individual variability is high. With the exception therefore of interorbital width, ranges and means are given for the sample as a whole. For interorbital width the range and mean are given for three size groups, 74–115 mm. (N = 21), 116–130 mm. (N = 32), and 131–144 mm. (N = 11).

Depth of body 31·6–41·6 (M = 38·3), length of head 30·1–34·8 (M = 32·0) per cent of standard length. Dorsal profile of head and snout straight or slightly concave, steeply sloping. Preorbital depth 12·5–18·8 (M = 16·4) per cent of head length; least width of interorbital 24·2–33·8 (M = 28·7), 28·6–35·1 (M = 31·5) and 31·1–33·4 (M = 31·6) per cent for the three size groups respectively; length of snout 29·0–36·8 (M = 32·5), eye 23·8–29·6 (M = 26·9), depth of cheek 19·5–28·2 (M = 23·9) per cent of head length.

Lower jaw slightly shorter than upper 33·7–40·8 (M = 38·2) per cent of head, the length/breadth ratio from broader than long to 1·33 times as long as broad. Mouth horizontal, lips somewhat thickened. Posterior tip of the maxilla extending to the vertical from the anterior margin of the orbit or as far as the pupil.

Described from 64 specimens, 55–144 mm. standard length.

With few exceptions, there is remarkable uniformity in the general facies of *H. retrodens*. In this respect the figure of *Paratilapia polyodon* type specimen in Boulenger (1909 and 1915) can be considered fairly representative. The lower jaw in this specimen has, however, been broken and subsequently distorted in preservation, consequently the ventral head profile of the figured specimen is inclined upward and is not horizontal, as it would be in life. The greatest departure from typical physiognomy and body form is seen in a single specimen from Bukoba (Tanganyika) which has the body relatively elongate and the head profile strongly curved. Despite this aberrance in gross morphology, the dentition and other fundamental characters of this specimen are typical.

Teeth and dental pattern. Fishes above 90 mm. S.L. have 5 to 8 series of teeth anteriorly and 4 to 5 series posteriorly in the upper jaw; 5 to 8 (rarely 9 or 10) series anteriorly and 3 to 5 series posteriorly in the lower jaw. From 40 to 68 teeth in outer series of the upper jaw. Teeth small, those of the outermost series largest, variable in form, but usually unicuspid; some bi- and tricuspid teeth occur in the inner series. Two specimens from Majita and Mwanza (Tanganyika Territory) have markedly aberrant, stoutly conical or nearly molariform teeth in all series of both jaws. In both specimens the dental pattern is otherwise typical.

Syncranium. The premaxilla is comparable in form with that of *Haplochromis*, except that it is stouter, especially posteriorly, and its dentigerous surface broader.

As noted above, the broad and greatly flattened dentary is unique amongst Lake Victoria cichlids (Fig. 8).

The neurocranium agrees closely with that of *Platytaeniodus* and with *Haplochromis* of the " *crassilabris* " species-group (see p. 316).

Pharyngeal bone with a triangular and approximately equilateral dentigerous surface; pharyngeal teeth cuspidate, the median series often enlarged.

Scales ctenoid, lateral line interrupted, with 31 (f.1), 32 (f.10), 33 (f.38), 34 (f.14) or 35 (f.1) scales. (Hilgendorf (1888) gives L.S. 30 [probably on a mid-lateral series] for the holotype.) Cheek with three or four series of imbricating scales. (Hilgendorf [*loc. cit.*] gives 4–5 rows); 7–8 scales—rarely 6—between lateral line and origin of dorsal fin; 8–9, less commonly 7, between pectoral and pelvic fin bases.

Fins. Dorsal with 24 (f.4), 25 (f.45) or 26 (f.13) rays, comprising XV–XVII 8–10 spinous and soft rays. Two specimens have the formulae XIV 8 and XVII 6, but there are indications that these are the results of wounding and subsequent irregular healing. Anal with 11 (f.6), 12 (f.54) or 13 (f.4) rays, i.e. III, 8–10. Caudal fin truncate, scaled on proximal half only. Pectoral fin 23·9–33·0 (M = 28·4) per cent of standard length. Pelvics with first ray produced in both sexes but proportionately longer in adult males.

Coloration. Colours of preserved specimens are variable. *Adult males*: dark grey to black, median fins dark, caudal colourless or maculate; pelvics black. *Females and immature males*: light olive brown to silver, all fins yellow or colourless. Sometimes faint traces of transverse bars, an interrupted band below insertion of dorsal fin and a well-marked mid-lateral stripe.

Colours in life. Adult males: ground colour dark grey-green (darkening to deep slate-grey immediately after death), chest and branchiostegal membrane black. Dorsal fin dark grey-green, lappets red; red spots, often coalescing into streaks, between the rays. Caudal with proximal third to half black, distal part red. Anal with dark base, remainder red; ocelli yellow. Pectoral colourless; pelvics black, with faint red flush along median rays. *Adult females*: ground colour light olivaceous dorsally, shading through sulphur-yellow to pearly-white ventrally; usually two longitudinal bands as described above, of variable intensity. Lips yellow. Dorsal and caudal fins olivaceous; anal colourless or with slight yellow flush; small and ill-defined orange spots sometimes occur in the position of the male ocelli. Pelvics and pectoral fins yellow.

Three specimens (2 ♀, 1 ♂) from Bukakata, which were noted when discussing atypical dental patterns (p. 321), also exhibit aberrant coloration. This is known only from preserved material, in which it has the form of three large and elongate black spots, arranged mid-laterally, on a light ground colour.

Almost completely sex-limited polychromatism occurs in this species and, as in *M. bicolor*, it is the female which usually exhibits atypical coloration. Besides the black and yellow piebald, as described for *M. bicolor*, there is a second and more colourful, if less distinctive, pattern. Any attempt to describe this pattern must perforce be imprecise, since intensity and detail show a remarkable range of individual variability. The ground colour is invariably a light sandy-yellow, with a superimposed orange flush, usually most intense on the head. Dorsally there are a number of irregular and ill-defined dark blotches, separate or confluent, which occasionally extend on to the flank. In some individuals this dorsal pigmentation is comparable with the clearly defined blotches of a typical *bicolor* pattern; in others it is more diffuse and individual elements are only faintly discernible. The whole body, including the fins, is also peppered with small melanophores, particularly on the head and opercula. Fin coloration is extremely variable, but the caudal and anal are usually flushed with bright orange.

Sampling bias in favour of atypically-coloured individuals undoubtedly occurs and precludes the accurate estimation of frequency. Of thirty-five females examined, thirteen were " *bicolor* " and eleven of the other pattern. Polychromatism is known in populations from most areas in which the species has been collected.

A single male (119·0 mm. S.L., Busongwe, Kagera River area) had an incipient *bicolor* pattern, resembling in appearance and degree of pigmentation one of the " *bicolor* " males of *M. bicolor* (p. 309).

The genetical basis of, and selection factors maintaining polychromatism in *H. retrodens* remain undetermined, as previously explained in the case of *M. bicolor*.

Ecology

H. retrodens is widespread throughout Lake Victoria (see below). Sampling in many habitats shows the species to be restricted to littoral and sub-littoral areas, especially where the substrate is hard, and usually where there is submerged vegetation. The majority of specimens was caught in water from twelve to twenty feet deep, with a few from slightly deeper water.

Food. The gut contents of sixty-five individuals (size-range 75·0–144·0 mm. S.L.) have been examined; of these, only seven were empty. Mollusca (particularly Lamellibranchiata) are the predominating food, being recorded from thirty-six fishes. The majority of Mollusca is represented by finely broken shells. Insecta (f.8)—especially Ephemeroptera, Trichoptera and Chironomidae—together with Crustacea (f.3) and Hydracarina (f.2) occur less frequently. In fifteen fishes the ingested material consisted almost entirely of fragmented plant epidermis, whilst eleven others contained only sand grains or organic mud.

Observations made on fishes living in aquaria show that *H. retrodens*, when feeding, repeatedly makes short, darting movements into or over the substrate, the broad and horizontal mouth serving as a scoop or shovel. Much bottom material is spilled from the mouth or with the exhalent current. Snails were retained and apparently crushed within the jaws before being passed back to the pharynx. The presence of macerated plant epidermis in several fishes may be explained either as ingested bottom debris, or as the result of the fish actively scraping the leaves and stems of submerged plants, for which purpose the multi-seriate dentition would seem adapted. It may be significant to note that the frustules of epiphytic diatoms found in the stomachs of these fishes were empty, as were the epidermal cells which had been ruptured.

Breeding. No information is available on the breeding habits of *H. retrodens*. The smallest sexually mature individual was a female of 96 mm. S.L. (Bukakata, Uganda).

Affinities

Morphologically, *Hoplotilapia retrodens*, like *P. degeni* is relatively far removed from any Lake Victoria *Haplochromis* species or species group. Neither is it closely related to the other monotypic genera of that lake. Regan's remark (1922, p. 159) that " A remarkable group of three species includes *Haplochromis sauvagei* and the monotypic genera *Macropleurodus* and *Hoplotilapia*, which scarcely differ from each other except for the considerable differences in dentition, . . . " is difficult to endorse, unless the various structures whose forms are modified in association with the dentition and feeding habits are included in the term " dentition ". *Macropleurodus* and *Hoplotilapia* have manifestly dissimilar jaw morphology and also differ in the form of the neurocranium. These structural differences impart a characteristic physiognomy to the species, which allows them to be recognized without reference to the dental form and pattern. The occurrence of " bicolor " patterns in both *M. bicolor* and *H. retrodens* is suggestive, but at present little importance can be attached to this character since it occurs in at least three widely divergent species of *Haplochromis* as well as in *M. bicolor*. If neurocranial form can be considered as being of phylogenetic value, then the affinities of *Hoplotilapia* lie with the *H. crassilabris* species group, although *H. retrodens* has departed considerably from this complex in the form of its jaws and dental pattern.

Hoplotilapia and *Macropleurodus* show interesting trophic parallels, both between the two genera and in relation to the three mollusc-eating cichlids of Lake Victoria, viz. *Astatoreochromis alluaudi*, *Haplochromis pharyngomylus* and *H. ishmaeli*. Ecologically, *Hoplotilapia* and *Macropleurodus* occur together with *A. alluaudi* and *H. pharyngomylus*, but the diet of *Hoplotilapia* and especially of *Macropleurodus*, unlike that of the two last-named species, includes a substantial number of insects.

The parallelism between *Hoplotilapia* and *Macropleurodus* extends to the method of feeding, and particularly the manner in which the hard-shelled prey is crushed. It is this characteristic which most clearly emphasizes the morphological disparity between the two genera under discussion on the one hand and the three mollusc-eating species on the other ; these latter species, although including a monotypic genus, have deviated less markedly from the generalized *Haplochromis* anatomy. Whereas in *Astatoreochromis alluaudi*, *H. ishmaeli* and *H. pharyngomylus* the food is crushed entirely by means of the hypertrophied pharyngeal bones and teeth, in *Hoplotilapia* and *Macropleurodus* the food is broken mainly by the peculiarly developed jaws and oral dentition, although the relatively poorly-developed pharyngeal mill doubtless continues the process.

Despite functional similarity in the jaws and dentition of *Hoplotilapia* and *Macropleurodus*, there is considerable divergence in the detailed morphology of these elements. On the one hand, in *Macropleurodus* the jaws are short and stout, with a narrow gape ; associated with the stout supporting skeleton, the teeth are strong. On the other hand, the dentary and premaxilla of *Hoplotilapia*, although broad and encompassing a wide gape, appear relatively fragile. The dentition of *Hoplotilapia*, when compared with *Macropleurodus*, is seen to be composed of small and slender teeth which are arranged in bands broader both anteriorly and posteriorly than the corresponding teeth of *Macropleurodus*. Since the shells of Gastropoda in Lake Victoria are stouter than those of Lamellibranchiata, it would not be unreasonable to suppose that the strong, laterally concentrated and enlarged teeth of *Macropleurodus*, and the uniformly multi-seriate, finely-pointed teeth of *Hoplotilapia*, are associated with the predominantly gastropod diet of the former species, and the predominantly lamellibranch diet of the latter.

Study material and distribution records

Museum and Reg. No.	S.L. (mm.)	Locality.	Collector.
British Museum (Nat. Hist.) :			
1906.5.30.417–418	90 and 106	Buganga (Uganda)	Degen
1909.5.4.16	112	Sesse Is. (Uganda)	Bayon
1911.3.3.34	144	Jinja, Ripon falls (Uganda)	,,
1928.5.24.489–492	125–132	Lake Victoria	Graham
Genoa Museum :			
C.E.12.995 (holotype of *P. polyodon*)	135	—	—
C.E.12.994	112		
B.M. (N.H.) :			
1955.2.10.141, 147–149	128–142	Rusinga Is. (Kenya)	E.A.F.R.O.
1955.2.10.145	129	Homa Bay (Kenya)	Ditto
1955.2.10.142–144, 146	105–135	Kamaringa (Kenya)	,,
1955.2.10.137–140	104–117	Kisumu (Kenya)	,,
1955.2.10.116–123, 180	76–138	Jinja (Uganda)	,,
1955.2.10.177–178	55	Beach nr. Nasu Point. (Buvuma Channel, Uganda)	,,
1955.2.10.170–171	105 and 107	Pilkington Bay (Uganda)	,,
1955.2.10.176	114	Ramafuta Is. (Uganda)	,,
1955.2.10.168–169	155	Yempita Is. (Rosebery Channel, Uganda)	,,
1955.2.10.124–132	117–124	Harbour, Entebbe	,,
1955.2.10.133	109	Bugonga beach, Entebbe	,,
1955.2.10.172–175	96–142	Old Bukakata (Uganda)	,,
1955.2.10.150–163	104–132	Busongwe (Kagera R. mouth, Uganda)	,,
1955.2.10.179	127	Beach south of Bukoba (Tanganyika)	,,
1955.2.10.164–167	106–117	Majita beach (Tanganyika)	,,
1955.2.10.134	86	Harbour, Mwanza (Tanganyika)	,,
1955.2.10.135–136	74 and 139	Capri Bay, Mwanza (Tanganyika)	,,

Graham (1929) lists the distribution of *H. retrodens* as follows :

Kenya Colony : Mbita passage ; Kavirondo Gulf ; Kadimu Bay.
Tanganyika Territory : Mussonya Bay (Ukerewe Is.); trawl near Bukoba.

PARALABIDOCHROMIS gen. nov.

Diagnosis

Cichlid fishes of the *Haplochromis* group, but differing from that genus in having the anterior teeth in both jaws procurrent and disproportionately longer than the adjacent lateral teeth. Jaws narrowing at the symphysis; lips thickened. Known only from Lake Victoria.

Type species: *Paralabidochromis victoriae* sp. nov.

The single specimen of *Paralabidochromis* available provides an interesting taxonomic and phylogenetic problem. No characters have been found which will distinguish this fish generically from specimens of the genus *Labidochromis* Trewavas; a genus otherwise known only from Lake Nyasa. Unfortunately comparisons must be limited to characters apparent in preserved material and then only to the few specimens available. Nothing is known of the coloration in life of adult males in either genus. This is regrettable since coloration might well provide a reliable indication of the affinities of the two genera, both in relation to one another, and to the species flocks of Victoria and Nyasa (*vide* Regan, 1921, 686). The presence of a dark sub-marginal band on the dorsal fin of *Labidochromis vellicans*, in contradistinction to its absence in *Paralabidochromis* is probably of some importance. A sub-marginal band is not known in any Lake Victoria *Haplochromis* species, but is present in most species of the group of Nyasa genera to which *L. vellicans* is apparently related (*vide* Trewavas, 1935, p. 71).

Although on purely morphological grounds it might seem advisable to include the Lake Victoria species within the genus *Labidochromis*, such a decision would imply phyletic relationships between the Victoria and Nyasa species closer than those between either species and others of its own lake. To avoid this I have given greater weight to the difference in colour-pattern than would perhaps have been justified if both inhabited the same lake.

Apart from the presence of the pan-African genera *Tilapia* and *Haplochromis*, there is no obvious relationship between the Cichlidae of Lakes Nyasa and Victoria. Superficial resemblances between individual species, or genera, in the two lakes have been associated with differences which point to their being examples of convergent evolution (Regan, 1922, p. 159), although it would perhaps be preferable to consider this convergence of morphological characters as parallel evolution since the phenomenon occurs between species within a group of related genera.

It would seem most probable, therefore, that *Paralabidochromis* represents a remarkable example of exact and detailed parallel evolution with *Labidochromis*. Apart from the enlarged anterior teeth, neither *Labidochromis* nor *Paralabidochromis* departs greatly from the generalized *Haplochromis* type, as represented in the rivers of East and Central Africa. Thus, it is possible that the two genera were independently evolved from different parental *Haplochromis* species, which, however, shared the generalized facies of fluviatile species.

Paralabidochromis victoriae sp. nov.

FIG. 10.

Description

Depth of body 33·0; length of head 31·6 per cent of standard length. Dorsal profile of head and snout slightly curved and sloping moderately steeply. Preorbital depth 16·7 per cent of head length; interorbital width 25·0, snout length 33·4, diameter of eye 29·2 and depth of cheek 20·8 per cent of head length. Caudal peduncle 1·33 times as long as deep, its length 15·8 per cent of standard length.

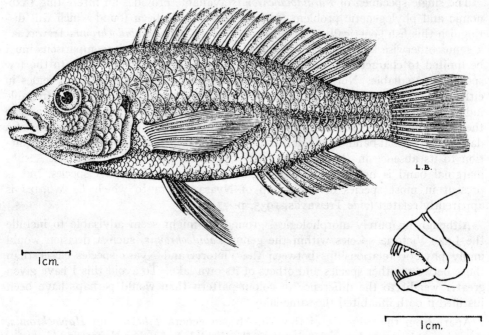

FIG. 10.—*Paralabidochromis victoriae*, ♂, holotype. Drawn by Miss L. Buswell.

Mouth almost horizontal, lips well developed. Posterior tip of the premaxilla extending to a point nearer the nostril than the anterior orbital margin. The angle between the rami of each jaw is acute giving a beak-like appearance, which is enhanced by the peculiar dentition.

Dentition. Thirty-eight teeth in the outer series of the premaxilla; the six anterior teeth in both jaws procurrent, greatly elongate, slender, slightly recurved and dagger-like; movable (Fig. 10). Postero-lateral outer teeth small and unicuspid, becoming progressively larger and weakly bicuspid laterally. Inner series in both jaws composed of tricuspid and compressed teeth, of which there are three rows in the upper and four in the lower jaw; outermost row of inner teeth in both jaws, somewhat enlarged and standing slightly apart from the remaining rows.

Lower pharyngeal bone triangular, the teeth small and cuspidate, with only the median series slightly enlarged.

Gill rakers short ; seven on lower limb of first arch.

Scales ctenoid ; lateral line interrupted, with 32 scales. Cheek with 2 series of imbricating scales ; 6 scales between origin of the dorsal fin and the upper lateral line ; 7 between pectoral and pelvic fins. Nuchal scales not exceptionally small.

Fins. Dorsal, XV 8 ; anal, III 9. Pectoral fin 25·0% of standard length ; pelvics with the first soft ray produced and extending posteriorly to the third anal spine. Caudal truncate, scaled on proximal half only.

Coloration. Preserved specimen light brown, with indications of seven faint transverse bands on the flanks, and a dark lachrymal stripe. Pelvic fins black, all other fins colourless.

Type locality. Sandy littoral, near Nasu Point, Buvuma Channel.

Described from a single specimen, a male 76·0 mm. S.L. (91·0 mm. T.L.), collected by the author whilst seine-netting at night, 29th May, 1951.

Affinities

Paralabidochromis victoriae is closely related morphologically, to *Labidochromis vellicans* of Lake Nyasa, from which it differs principally in possessing large scales on the chest, cheek and nape. The dorsal profile of the head and snout in *L. vellicans* apparently differs from that of *P. victoriae*, in being straighter and more steeply sloping.

Amongst the *Haplochromis* species of Lake Victoria, *P. victoriae*, has some structural affinity with *Haplochromis chilotes*. Both species have short and narrow jaws, thickened lips and a certain similarity of dentition. For example, the teeth of *H. chilotes* are slender and elongate anteriorly, few in number and arranged in an acute dental arch. However, the disproportionately long and procurrent anterior teeth of *P. victoriae*, and the hypertrophied lips of *H. chilotes*, immediately serve as diagnostic characters. The phylogeny of *H. chilotes*, is at present, obscure, but the species would seem more closely related to the *H. crassilabris* species group in Lake Victoria, than to *Paralabidochromis*.

DISCUSSION

It is clear from Regan's analysis of the Lake Victoria Cichlidae, that he did not consider the monotypic genera far removed phylogenetically from certain *Haplochromis* species, a view which is strengthened by the additional data now available. Although the monotypic genera are readily defined by trenchant characters they retain fundamental affinities with the *Haplochromis* species of Lake Victoria. But the morphological differentiation which these genera have undergone creates an impression of greater divergence than is shown by other adaptive groups within the Lake Victoria species flock. Analysis of the diagnostic characters of the genera described here, shows that, in any one genus, the anatomical characters of the head are functionally related to the dentition. Further, the ontogenetic basis for most of these characters is probably attributable to differential growth and not to any large-scale qualitative change. Thus the first evolutionary steps involved may well have been relatively simple and similar, in the earliest stages, to those which initiated the often slight differences characterizing the greater number of *Haplochromis* species.

Despite fundamental inter-relationship, three of the four monotypic genera exhibit considerable inter-generic divergence in the characters affected by the peculiar development of their dentition. The degree of divergence and the means by which it has been effected in *Macropleurodus*, *Platytaeniodus* and *Hoplotilapia* are such that the genera cannot be related *inter se*. That is to say, these genera represent three independent evolutionary offshoots from the basic *Haplochromis* stock. The fourth genus, *Paralabidochromis*, shows least departure from the generalized *Haplochromis* condition, but must also be looked upon as having evolved independently.

A broad outline of the probable phyletic relationships within the *Haplochromis* species flock is perhaps necessary before attempting further discussion of the monotypic genera.

There are some Lake Victoria *Haplochromis* species which, when seen in isolation or when known only from a few specimens, might seem almost as distinctive as the monotypic genera. It is possible, however, to relate these outstanding species to others more typical of their particular adaptive group (mollusc eaters, piscivores, epiphytic-algal grazers, etc.). Furthermore, when larger series of specimens are examined, intra-specific variation is sufficiently high to reduce considerably any apparent inter-specific gap.

With few exceptions, specific differences amongst the *Haplochromis* are quantitative and generally consequent upon the differential growth of various syncranial parts. Qualitative differences, on the other hand, are usually those which can be related to, and are used in defining, adaptational groups. At both inter- and intra-group levels, qualitative and quantitative differences tend to be small and intergrading.

It is in both qualitative and quantitative characters that the monotypic genera depart most markedly from *Haplochromis* but, unlike the inter-group differences within the *Haplochromis* flock, the morphological gap is clear cut, and remains so even when large series of specimens are examined.

On this interpretation, *Hoplotilapia* and *Platytaeniodus* probably examplify the phenomenon of " quantum evolution " as described by Simpson (1944 and 1953). It is difficult to suggest any ancestry for these genera nearer than a present-day species group, *viz*. the *H. crassilabris* complex. Even at this level relationship is extremely tenuous, and based only on similarity of skull form, in itself probably an adaptational character and therefore of doubtful phylogenetic significance. The jaws and dental pattern are so dissimilar in *Hoplotilapia* and *Platytaeniodus* that one must consider the genera separately and not as elements of a single lineage.

Macropleurodus is less readily regarded as being a product of quantum evolution, and will be discussed later.

Simpson (1944) has noted that quantum evolution is usually associated with a shift from one adaptive zone to another, and that the interzonal populations or species would be relatively ill-adapted, unstable and short-lived. Thus, morphological discontinuity is generally observed between the parental and divergent lineages.

The morphological discontinuity existing between *Platytaeniodus* and *Hoplotilapia*, and between these two genera and *Haplochromis* has been demonstrated ; it is the more regrettable then that pertinent ecological data for *P. degeni* are both inadequate and confusing, so that it is difficult to equate the apparent morphological specialization of this species with any particular ecological niche. The food of *P. degeni* is virtually unknown, although inorganic material and scant but varied organic remains suggest bottom-feeding habits. A multi-serial and concentrated dentition, such as that of *P. degeni* may possibly have adaptive value, particularly if food must be freed from the substrate or if it requires trituration before digestion can be effected. (It should be noted that no *Haplochromis* has consistently yielded such baffling residua in the gut.)

Hoplotilapia retrodens is somewhat better understood. In this species food is predominantly Molluscan. The functional significance of the jaw structure and dental pattern in *Hoplotilapia*, particularly with regard to its observed diet of thin-shelled bivalves, has been discussed above (p. 325). From these characters and the predominance of Mollusca over the other food organisms it would seem that the essential elements of quantum evolution are fairly well defined in *H. retrodens*. That is to say, the species shows both morphological discontinuity and entry into an adaptive niche different from that of the presumed parental stock.

There is, of course, the possibility that *Platytaeniodus* and *Hoplotilapia* represent an early stage in quantum evolution and that these species may be relatively ill-adapted to existing ecological conditions. Their further evolutionary development or, alternatively, extinction, will therefore depend upon environment change effecting, or failing to effect, the realization of characters which at present could only be considered prospective adaptations (*sensu* Simpson, 1953, p. 188).

The slight and often indeterminable adaptive differences between *Haplochromis* species occupying the same ecological zone, together with numerous instances of inter-specific overlap in feeding habits, would seem to suggest that there is, and has been, only slight selection-pressure acting through competition for food. Thus the continued existence of *Platytaeniodus* and *Hoplotilapia*, like that of many closely related *Haplochromis* species, could be attributed to a period of decreased selection pressure.

Although *Macropleurodus*, like *Platytaeniodus* and *Hoplotilapia*, is separated from *Haplochromis* by a clearly defined morphological gap, the gap is of lesser degree. Whereas *P. degeni* and *H. retrodens* exhibit unique jaw morphology and dental patterns without departure from *Haplochromis* in tooth form, the basic dental pattern of *M. bicolor* is foreshadowed in two *Haplochromis* species, as is the form of the dentary. Only in tooth form and shape of the premaxilla does *M. bicolor* show great differentiation from *H. prodromus*. The latter species is morphologically closely related to *H. sauvagei*, a smaller species, which in turn shows departure from the basic *Haplochromis* type towards *H. prodromus*. In *H. sauvagei* and *H. prodromus* the outer teeth are relatively stout and, anteriorly, there are several rows of inner teeth; the neurocranium and dentary of these species also approach the *M. bicolor* condition. Both species feed on Mollusca and Insecta, the very slight differences in the feeding habits of *H. prodromus* and *H. sauvagei* being attributable to the smaller size of *H. sauvagei*.

It is tempting, therefore, to consider the members of the series *H. sauvagei-H. prodromus-M. bicolor* as representing stages of a lineage, although the possibility of independent evolution of the same adaptive characters, cannot be ignored. Until more critical evidence is available for the phyletic relationship of these species, they may be regarded either as separate end points of different lineages, or as a " stufenreihe", or single phyletic line. The three species do illustrate the gross anatomical and functional stages through which *Macropleurodus* could have evolved, although the transition in shape of teeth from a typical *Haplochromis* to *Macropleurodus* is not represented in any living species.

The isolated position of *Paralabidochromis* in relation to the *Haplochromis* species of Lake Victoria was commented upon above.

Little information is available on the evolutionary relationships of *Paralabidochromis*. Although taxonomically isolated from *Haplochromis* the genus is nearer the *Haplochromis* stem than either *Hoplotilapia*, *Platytaeniodus* or *Macropleurodus*. In Lake Nyasa, Trewavas (1935 and 1949) considers *Labidochromis vellicans* as belonging to a group of nine genera (which group excludes *Haplochromis*), that, although lacking an absolute character to distinguish them, are more closely related to each other than to any other genus. Such grouping is impossible for the monotypic genera of Lake Victoria. Within this lake the divergent genera of the *Haplochromis* group must be considered as being distinct from one another, as well as from the parental stock.

SUMMARY

(1) The genera *Macropleurodus* Regan, 1922, *Platytaeniodus* Boulenger, 1906, and *Hoplotilapia* Hilgendorf, 1888, are redefined on the basis of new and fairly extensive collections. Similarly, the species *M. bicolor* (Blgr.) 1906, *P. degeni* Boulenger, 1906, and *H. retrodens* Hilgendorf, 1888, are redescribed.

(2) Generic and specific characters are discussed, with particular regard to ontogenetic changes. Information gained from small specimens shows that three species previously considered as *Haplochromis* must be added to the synonymies of *M. bicolor* and *P. degeni*.

(3) Comparative anatomical and osteological studies of the head indicate that *Hoplotilapia* and *Platytaeniodus* are not closely related to any extant *Haplochromis* species in Lake Victoria. Morphological stages leading to the syncranial type found in *Macropleurodus* are, however, represented by two endemic *Haplochromis* species. On the basis of syncranial morphology, it is clear that the monotypic genera are not closely related *inter se*.

(4) Apparently sex-limited polychromatism occurring in *Macropleurodus* and *Hoplotilapia* is described and discussed.

(5) A fifth monotypic genus, *Paralabidochromis victoriae* is described. This genus exhibits remarkably close morphological parallelism with *Labidochromis vellicans* from Lake Nyasa.

(6) Locality lists and notes on the ecology of the genera are given, together with observations on the feeding habits of *Macropleurodus* and *Hoplotilapia*.

(7) The evolutionary relationships of the four genera are discussed. It is suggested that **Hoplotilapia** and **Platytaeniodus** may represent the products of low-level **quantum evolution.**

ACKNOWLEDGMENTS

I wish to acknowledge my gratitude and thanks to the Trustees of the British Museum (Natural History) for facilities afforded me during the tenure of a Colonial Fisheries Research Studentship ; to Professor L. Bertin of the Muséum National d'Histoire naturelle, Paris, for allowing me to study type specimens of Lake Victoria Cichlidae described by Pellegrin ; to Dr. Delfa Guiglia of the Museo Civico di Storia Naturale, Genoa, for the courtesies mentioned in the text ; to Dr. Kurt Deckert, Zoologisches Museum, Berlin, for attempting to locate the type specimen of *Hoplotilapia retrodens* ; to the officers of the Lake Victoria Fisheries Service who collected much valuable material ; to my colleagues Miss R. H. Lowe and Dr. P. S. Corbet for many discussions and to the latter for his invaluable identifications of material from gut-contents ; and to Mr. Denys W. Tucker, for his very helpful criticism of the manuscript. I am especially indebted to Dr. Ethelwynn Trewavas for the interest she has shown in my work, and for much helpful information and advice.

REFERENCES

This list includes only those references which occur in the general sections of the text. All others appear under the corresponding species headings.

COTT, H. B. 1952. In *East African Fisheries Research Organization Annual Report*. East Africa High Commission, Nairobi (p. 21).

GRAHAM, M. 1929. *A report on the Fishing Survey of Lake Victoria, 1927–1928, and appendices.* Crown Agents, London.

GREENWOOD, P. H. 1954. On two cichlid fishes from the Malagarazi river (Tanganyika), with notes on the pharyngeal apophysis in species of the *Haplochromis* group. *Ann. Mag. nat. Hist.* (12) **7** : 401.

REGAN, C. T. 1920. The classification of the fishes of the family Cichlidae. I. The Tanganyika genera. *Ann. Mag. nat. Hist.* (9) **5** : 33.

—— 1921. The cichlid fishes of Lake Nyasa. *Proc. zool. Soc. Lond.* 675.

—— 1922a. The cichlid fishes of Lake Victoria. *Proc. zool. Soc. Lond.* 157.

—— 1922b. The classification of the fishes of the family Cichlidae. II. On African and Syrian genera not restricted to the Great Lakes. *Ann. Mag. nat. Hist.* (9) **10** : 249.

SIMPSON, G. G. 1944. *Tempo and Mode in Evolution.* New York.

—— 1953. *The Major Features of Evolution.* New York.

TREWAVAS, E. 1933. Scientific results of the Cambridge expedition to the East African lakes, 1930–1. II. The cichlid fishes. *J. linn. Soc. (Zool.)* **38** : 309.

—— 1935. A synopsis of the cichlid fishes of Lake Nyasa. *Ann. Mag. nat. Hist.* (10) **16** : 65.

—— 1949. The origin and evolution of the cichlid fishes of the great African lakes, with special reference to Lake Nyasa. *Rapp.* 13e *Congrès Intern. Zool.*, Section 5B, 365. Paris.

A REVISION OF THE LAKE VICTORIA *HAPLOCHROMIS* SPECIES (PISCES, CICHLIDAE) PART I: *H. OBLIQUIDENS* HILGEND., *H. NIGRICANS* (BLGR.), *H. NUCHISQUAMULATUS* (HILGEND.) AND *H. LIVIDUS*, SP. N.

By P. H. GREENWOOD.

CONTENTS

	Page
INTRODUCTION	226
Haplochromis obliquidens Hilgend.	226
Synonymy and description	227
Distribution	229
Ecology	229
Affinities and taxonomic status of the species	230
Study material and distribution records	232
Haplochromis lividus sp. nov.	232
Synonymy and description	232
Distribution	234
Ecology	234
Diagnosis	236
Study material and distribution records	237
Haplochromis nigricans (Blgr.)	237
Synonymy and description	237
Distribution	239
Ecology	239
Diagnosis	240
Study material and distribution records	240
Haplochromis nuchisquamulatus (Hilgend.)	241
Synonymy and description	241
Distribution	242
Ecology	242
Diagnosis	242
Study material and distribution records	243
SUMMARY	243
ACKNOWLEDGMENTS	243
REFERENCES	243

INTRODUCTION

The species described in this paper form a well-defined ecological group within the *Haplochromis* species flock of Lake Victoria. All feed principally by grazing on epiphytic and epilithic algae.

As a group and severally they show obvious morphological adaptations to this particular feeding habit. Adaptation is most clearly seen in tooth-form and arrangement, which depart from those common to the majority of *Haplochromis* species.

Several other ecologico-morphological groups have evolved in Lake Victoria. Their existence raises questions regarding the possibility of providing a realistic basis for subdividing the present phylogenetically amorphous arrangement of the species.

There are, however, certain difficulties inherent in this procedure. A strictly morphological approach to sub-division is unworkable. Intergradation, rather than discreteness, of morphological group-characters might be said to typify this species flock. Such a situation is, however, not unexpected in a large group of oligophyletic origin which has undergone intense adaptive radiation during a short period of geological time (Regan, 1922; Greenwood, 1951).

Also, although some morphologically distinct species-complexes occupy equally distinctive ecological niches, there are other morphologically-homogeneous groups which cut across any attempted ecological classification.

Furthermore, in any ecologically-defined group there are grades of anatomical specialization such that the most and least specialized species are only with difficulty included in a supra-specific category defined by morphological criteria alone. The species described here typify this situation. Tooth-form in *H. obliquidens* is unlike that of most species at present included in the genus *Haplochromis*. Yet three algal-grazing species are known, which partially bridge this morphological gap. At the opposite extreme *H. nuchisquamulatus* exhibits incipient dental adaptation only slightly removed from a generalized *Haplochromis* type.

In Lake Victoria, then, there exist several nascent supra-specific groups which are more readily identified by ecological than morphological criteria. Since conventional taxonomic characters are, so to speak, also nascent, formal recognition of these categories is impossible. I propose, therefore, to recognize their biological and evolutionary significance only by drawing attention to their existence.

Haplochromis obliquidens Hilgendorf, 1888

Chromis (*Haplochromis*) *obliquidens* Hilgendorf, 1888, *S. B. Ges. naturf. Fr. Berlin*, 76.
Ctenochromis obliquidens, Pfeffer, 1897, *Arch. f. Naturg.*, **63**, 60.
Tilapia obliquidens, Boulenger, 1898, *Trans. zool. Soc., Lond.*, **15**, 5.
Hemitilapia bayoni Boulenger, 1908, *Ann. Mus. Genova* (3) **4**, 6; Idem, 1911, *Ibid.* (3) **5**, 69; Idem, 1915, *Cat. Afr. Fish.*, **3**, 491, fig. 340.
Haplochromis nuchisquamulatus (part), Boulenger, 1915, *op. cit.*, 290.
Clinodon bayoni (Blgr.), Regan, 1920, *Ann. Mag. nat. Hist.* (9), **5**, 33.
Haplochromis obliquidens (part), Regan, 1922, *Proc. zool. Soc., Lond.* 188.

The holotype of *Haplochromis obliquidens* could not be examined; it is amongst those specimens, once housed in the Berlin Museum, and which cannot be located at present. However, the characters noted in Hilgendorf's original description are diagnostic.

Through the courtesy of Dr. D. Guiglia (Museo Civico di Storia Naturale, Genoa) I was able to study the holotype of *Hemitilapia bayoni* Boulenger, and thus to confirm Regan's synonymy of this species with *H. obliquidens*.

On the other hand, I cannot agree with Regan's tentative synonymy of *Hemitilapia materfamilias* Pellegrin, 1913, and *Haplochromis obliquidens* (Regan, 1922). Re-examination of *H. materfamilias* type specimen revealed that Pellegrin's original description is misleading, particularly in respect of the dentition, and that the species should be referred to *Macropleurodus bicolor* (Blgr.) (Greenwood 1956).

Description. Based on fifty-seven fishes (size range 48–89 mm. standard length) including the holotype of *Hemitilapia bayoni* and five specimens in the British Museum (Natural History). Three other British Museum (Nat. Hist.) specimens were examined, but are not included in the morphometric data.

Since no marked allometry with size was determined for any character examined, measurements are given for the collection as a whole, with the exception of the smallest specimen, which is treated separately.

Depth of body 33·4–41·2, mean (M) 37·5, length of head 29·4–34·0 (M = 32·3) per cent of standard length. Dorsal profile of head and snout straight; fairly steeply sloping in most fishes but decurved in a few individuals. Preorbital depth 12·5–17·4 (M = 15·2) per cent of head length; least interorbital width 27·8–34·7 (M = 31·8); snout as broad as or somewhat broader than long, rarely longer than broad, its length 26·6–33·3 (M = 29·2) per cent head length. Eye 29·1–33·3 (M = 31·4); depth of cheek 19·0–25·0 (M = 21·5) per cent head length.

Caudal peduncle about $1\frac{1}{2}$ times as long as deep; 13·2–16·4 (M = 15·0) per cent of standard length.

Corresponding ratios for the smallest individual (48 mm. S. L.)—not included in the mean values given above—are: Head 32·3; preorbital 11·1; interorbital 27·8; snout 27·8; eye 27·8; cheek 16·7; and caudal peduncle 16·6 per cent.

Mouth short and horizontal or very slightly oblique; posterior maxillary tip extending to the vertical from the anterior orbital margin, or almost so. Jaws equal anteriorly, the lower 31·6–41·6 (M = 37·2) per cent of head length; its length/breadth ratio from 1·1–1·7 (mode 1·4).

Gill rakers short; 8 or 9, rarely 7 or 10, on the lower limb of the anterior arch.

Scales ctenoid; lateral line interrupted, with 30 (f.6), 31 (f.31) or 32 (f.18) scales. Cheek with 3 (rarely 2 or 4) series of imbricating scales. 5 or 6 scales between the dorsal fin origin and the lateral line; 5–7 between pectoral and pelvic fin insertions.

Fins. Dorsal with 24 (f.25), 25 (f. 31) or 26 (f.1) rays, anal 10 (f.1), 11 (f.11), 12 (f.43) or 13 (f.2), comprising XIV–XVI, 8–10 and III, 7–10 spines and soft rays for the fins respectively. First pelvic ray slightly produced and variable in its posterior extension, usually reaching the spinous anal fin in adults and occasionally to the soft part in ripe males. Caudal sub-truncate.

Teeth. Teeth forming the outer series are movably implanted and have slender necks with undivided, expanded, compressed and obliquely truncate crowns (Text-fig. 1). In many specimens a few postero-lateral teeth in both jaws are bicuspid but otherwise retain almost the same crown form as the more anterior teeth; in some the second cusp is incipient, but in others it is clearly differentiated. There is no correlation between length of fish and the presence or number of undifferentiated postero-lateral teeth.

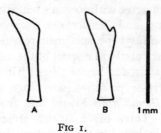

FIG 1.

A weak positive correlation exists between the number of teeth in the outer series of the upper jaw, and standard length.

S.L. (mm.)	48	55–64	65–74	75–87
Tooth number	50	42–60	50–70	58–70
Mean	50	54	59	64
N	1	20	21	12

The inner teeth are mostly tricuspid and arranged in 2–4 rows, with a distinct interspace separating them from the outer series. Some obtusely tricuspid teeth, and others differing only in their smaller size from those in the outer series, frequently occur in the first inner row. These obliquely truncate teeth are larger than their tricuspid associates.

Alizarin preparations of two larvae (10 and 11 mm. total length) obtained from the mouth of a brooding female, show larval dentition to be comparable with that of *H. macrops* (Blgr.) *H. prodromus* Trewavas and *Macropleurodus bicolor* at an equivalent developmental stage. The teeth of *H. obliquidens* larvae differ considerably from the adult condition, being slender and setiform, with slightly recurved unicuspid crowns; 14–16 outer teeth, aggregated medially, are present in the upper jaw.

In certain fishes from Kisumu, it was noticed that the crowns of all teeth were coarse and irregular, and that their typical golden-brown coloration was replaced by black. Similar structural differences and discoloration have been observed in specimens of *H. michaeli* Trewavas from various localities. The cause of this aberrancy is unknown.

Lower pharyngeal bone sub-equilaterally triangular, its dentigerous surface broader than long. The numerous teeth are fine, compressed and directed posteriorly, and have truncated crowns; a small anterior cusp is present in all.

Skeleton. Differs in no important respect from that of generalized *Haplochromis* species. Vertebrae : 14 + 17, 13 + 17, 13 + 16 or 12 + 17.

Coloration in life : Breeding males. Ground colour bright yellow-green, becoming yellower ventrally ; chest and branchiostegal membrane blackish ; lips slightly irridescent. Dorsal fin yellow-grey, lappets red ; orange-red spots and streaks on the posterior spinous and entire soft parts. Anal with a pinkish flush ; 3 or 4 yellow ocelli. Pelvics with black outer and clear or faint pink inner half. Non-breeding adult males are similarly coloured except that the body is more nearly olivaceous and the chest not darkened. *Females and juveniles of both sexes.* Ground colour silvery-yellow. Dorsal and caudal fins neutral ; anal and pelvic fins pale yellow. Darker, almost olivaceous females are known.

Transverse banding occurs in both sexes, but is rarely apparent in life.

Preserved material : Adult males. Dusky, the vertical bars partly or completely obscured. Dorsal fin sooty, lappets black, the posterior spinous and entire soft part maculate ; anal colourless ; pelvics dark laterally, pale mesially ; caudal maculate. *Females and juveniles.* Grey to brown, with or without six to ten narrow transverse bars on the flanks ; less frequently a faint mid-lateral stripe and a fainter stripe approximately following the upper lateral line. Fins colourless and immaculate.

Distribution. *Haplochromis obliquidens* has been collected from many localities in Lake Victoria. It is also known from the Victoria Nile.

Recently, Miss R. H. Lowe obtained a small sample of *H. obliquidens* from Lake Bunyoni (Uganda). Earlier reports (Worthington, 1932) indicated the probability that no *Haplochromis* were then present in this lake. It is presumed that those now occurring there were accidentally introduced on occasions when the lake has been stocked with *Tilapia* species ; that one of the two species now recorded is probably *H. nigripinnis* Regan (otherwise endemic to Lake Edward) and the other is *H. obliquidens* supports this assumption, since *Tilapia* have been introduced from both Lakes Edward and Victoria.

The nine Lake Bunyoni *H. obliquidens* (size range 63–85 mm. S.L.) differ slightly from the Lake Victoria population in the following characters : body more slender, 30·2–34·5 (M = 32·4) per cent of standard length ; the preserved coloration of sexually mature males is apparently more melanic ; in three specimens the outer series of teeth is entirely composed of bicuspids similar to the undifferentiated postero-lateral teeth of Lake Victoria fishes. In all other observed morphological characters the two populations are identical.

Ecology : Habitat. Shallow littoral zone, particularly in the vicinity of emergent vegetation ; less commonly in the water-lily zone, over exposed sandy beaches and at the margin of papyrus swamps. There are indications, both from fishing and direct observation, that *H. obliquidens* may frequent rocky shore-lines, where the substrate is largely composed of broken rocks and boulders. The species has often been collected and seen around rock foundations of piers.

Food. The intestine of *H. obliquidens* is long and much coiled ($2\frac{1}{2}$–3 times S.L.) ; stomach large and distensible. Stomach and intestinal contents of fifty-three individuals (size range 48–89 mm. S.L.) from various localities, have been examined.

Diatoms comprised the main digested contents in forty-four individuals ; the genera principally recorded were : *Melosira, Suririella, Gomphonema, Rhopalodia, Navicula* and *Cyclotella*.

Small fragments of plant epidermis occurred in the stomachs of twenty-nine fishes. The quantity ingested by individuals varied considerably. It was observed that, unless ruptured, most epidermal cells were apparently undigested.

Blue-green algae, especially *Rivularia* and *Microcystis*, and less frequently *Anabaena* and *Oscillatoria*, were recorded from nineteen stomachs; none of these plants showed signs of digestion.

Filamentous green algae, chiefly *Spirogyra* and to a lesser extent *Oedogonium*, occurred in sixteen stomachs. No digestion was noted.

The stomach contents of one individual comprised only partly digested fragments of Ephemeroptera larvae, probably taken at the time of their emergence. Fragmentary remains of both adult and larval insects were found in the intestines of three other fishes.

The frequent occurrence of epiphytic algae and epidermal fragments of phanerogams suggests that *H. obliquidens* feeds partly by scraping the surface of submerged leaves and stems. This supposition is confirmed by observations made on the feeding behaviour of these fishes in the lake; the peculiar dentition of *H. obliquidens* would seem to be highly adapted for such habits.

On the other hand, sand grains and bottom debris were also found in many stomachs; indeed, it was often difficult to determine whether ingested plant fragments were the partly digested remains of epidermis scraped from living plants or whether they were derived from the semi-decayed debris which accumulates near dense plant stands. Probably *H. obliquidens* feeds both by grazing on plants and by utilizing plant material contained in the bottom detritus. In either eventuality it is clear that diatoms are the principal food organisms utilized, and that much ingested plant material is voided undigested.

Although rather infrequent, the occurrence of insects in the pabulum could indicate that the species is partly facultative in its feeding habits and may utilize temporary and seasonal abundances of animal food.

Breeding. Breeding behaviour and spawning sites of *H. obliquidens* are unknown. However, females carrying young in the buccal cavity have been obtained from most localities.

The smallest sexually active fish was a female 61 mm. long; above 68 mm. S.L., most individuals were found to be mature.

Affinities and taxonomic status of the species

Particular interest attaches to *H. obliquidens*, since although it is the type species of the genus its dental morphology is unique amongst the very numerous species of *Haplochromis*. Throughout this discussion the generic diagnosis is taken to be that prepared by Regan (1920) in which particular emphasis was laid on neurocranial osteology. Subsequently this definition has been modified by the recognition of several related genera distinguished from *Haplochromis* by their divergent dentition (Regan, 1922; Trewavas, 1938; Greenwood, 1956).

Within the genus thus defined two types of outer teeth predominate, a unicuspid, conical form and a bicuspid compressed type. The common dental pattern is a single outer series distinctly separated from the inner series, usually comprising two or three rows anteriorly and a single row postero-laterally.

Tooth form and pattern have played an important part in species discrimination and in the actual or attempted delimitation of supra-specific groups amongst Lake Victoria species. Extreme dental specialization, associated with osteological changes, characterises four of the five monotypic cichlid genera in this lake (Regan, 1922; Greenwood, 1956), whilst less obvious dental characters were used in an attempt to subdivide the endemic *Haplochromis* into five genera (Regan, 1920). Two years later, Regan abandoned this concept, reducing some of his genera to subgeneric rank and discarding others (*idem*, 1922).

Because the dental morphology of *H. obliquidens* does not conform with that usual for *Haplochromis*, there might appear to be grounds, as Regan suggested (*op. cit.*), for recognizing at least one sub-genus to accommodate those species with unequally bicuspid or conical outer teeth. The sub-genus *Ctenochromis* (Pfeffer, 1893) would be available for such species (Regan, *loc. cit.*). In that paper Regan first indicated *H. astatodon* Regan of Lake Kivu as providing a dental type, which although invariably bicuspid, linked the "*obliquidens*" tooth form with that of the commonly occurring *Ctenochromis* type. The teeth of *H. astatodon* exhibit some diversity in the degree to which they approach the "*obliquidens*" condition, but the greatest number of individuals has teeth approximating more closely to this type than to "*Ctenochromis*". The common tooth form in *H. astatodon* may be likened to a bicuspid variant of typical *H. obliquidens* teeth; indeed, similar teeth frequently occur postero-laterally in both jaws of *H. obliquidens*.

Two other annectent species have since been found: *H. annectidens* Trewavas from Lake Nabugabo and a new species (described below) from Lake Victoria. Intra-specific variation in the tooth form of this latter species is as great as that of *H. astatodon*, but most individuals possess teeth similar to the undifferentiated postero-lateral teeth of *H. obliquidens*.

It is clear, then, that although the teeth of *H. obliquidens* may represent an extreme form, intermediates linking them to the usual bicuspid *Haplochromis* type are found as the characteristic dentition in three extant species. The gap separating the most "*obliquidens*"-like teeth of *H. astatodon* and the new species from those of *H. obliquidens* is relatively slight; it represents no more than the loss of a small cusp from an expansive, compressed and obliquely truncate crown. Less modified crown structure as seen in some teeth of these two species, grades through the condition found in *H. nuchisquamulatus*, into the more usual, acutely bicuspid form.

Thus, the case for recognizing at least two sub-genera of *Haplochromis* on the basis of dental morphology (Regan, 1920 and 1922) is weakened. As was mentioned earlier, several ecologically defined groups, each comprising apparently related species, are known from Lake Victoria. In every case, the group shows certain morphological divergence from the generalized *Haplochromis* type, but no clear-cut gap has evolved which would allow for its formal recognition as a sub-genus.

Study material and distribution records

Museum and Reg. No.	Locality.	Collector.
Genoa Museum. Holotype of *Hemitilapia bayoni*	Sesse Islands	Bayon.
British Museum (N.H.) 1908, 10.19.6 (paratype of *H. bayoni*)	Sesse Islands	Bayon.
British Museum (N.H.) 1911, 3.3.80	Jinja (Ripon Falls)	Bayon.
British Museum (N.H.) 1913, 9.30.13–18	Lake Victoria	Bayon.
British Museum (N.H.) 1956, 7.9.1–16	Jinja (Pier)	E.A.F.R.O.
,, ,, ,, ,, ,, ,, 17–20	Beach near Nasu Point (Buvuma Channel)	,,
,, ,, ,, ,, ,, ,, 21–27	Grant Bay (Buvuma Channel)	,,
,, ,, ,, ,, ,, ,, 28	Napoleon Gulf, near Bugungu (opp. Jinja)	,,
,, ,, ,, ,, ,, ,, 29–32	Entebbe Harbour	,,
,, ,, ,, ,, ,, ,, 33–45	Kisumu, Kavirondo Gulf	,,
,, ,, ,, ,, ,, ,, 46–55	Mwanza, Capri Bay	,,
,, ,, ,, ,, ,, ,, 56–57	Godziba Island	,,
,, ,, ,, ,, ,, ,, 169–170	Kalagala, Victoria Nile	,,

Haplochromis lividus sp. nov.

Haplochromis nuchisquamulatus (part), Blgr., 1915, *Cat. Afr. Fish.*, **3**, 290, Fig. 197.
Haplochromis desfontainesii (part), Blgr., 1915, *op. cit.*, 302.
Haplochromis nubilus (part), Regan, 1922, *Proc. zool. Soc., London*, 164.

Type specimen. A male 90 + 21 mm. from Bugungu (near Jinja), Uganda.

Description. Based on seventy-seven fishes (size range 46–90 mm. S.L.) from Lake Victoria. Five specimens from Lake Kyoga are considered separately.

Within the size range of individuals studied no character showed marked allometry with standard length or length of head; measurements are therefore given for the whole collection with the exception of the smallest fish, which was not included when determining means.

Depth of body 33·3–41·2 (M = 36·5); length of head 31·0–35·0 (M = 32·7) per cent of standard length. Dorsal head profile straight and moderately steeply sloping (*ca.* 45°), rarely somewhat curved. Preorbital depth 12·0–16·7 (M = 14·7) per cent head length; least interorbital width 26·2–33·3 (M = 29·7); snout as broad as long, its length 26·0–32·0 (M = 28·8) per cent of head. Eye 28·0–36·0 (M = 31·4); depth of cheek 17·0–24·1 (M = 20·1) per cent head length.

Caudal peduncle 1·1–1·7 (M = 1·4) times as long as deep; 12·2–18·5 (M = 15·5) per cent standard length.

Corresponding ratios for the smallest fish (46 mm. S.L.) are: Depth 39·0, head 39·0 per cent of standard length. Preorbital 12·8, interorbital 23·2, snout 27·8 and cheek 16·7 per cent of head-length. Caudal peduncle 15·2 per cent of S.L.

Mouth horizontal or slightly oblique; posterior maxillary tip reaching the vertical to the anterior orbital margin or nearly so, and to the eye in some. Lips slightly thickened. Lower jaw 33·3–41·0 (M = 37·2) per cent of head, its length/breadth ratio 1·3–2·0 (mode 1·6).

Gill rakers short, 8 or 9 (less frequently 7 or 10) on the lower part of the first arch.

Scales ctenoid, lateral-line interrupted, with 30 (f.7), 31 (f.16), 32 (f.48), 33 (f.5), or 34 (f.1) scales. Cheek with 2 or 3 (rarely 4) series. 5 or 6 scales between dorsal fin origin and the lateral line; 5 or 6 between pectoral and pelvic fin insertions.

Fins. Dorsal with 23 (f.1), 24 (f.34), 25 (f.40) or 26 (f.2); anal 11 (f.18), 12 (f.55), and 13 (f.4), rays, comprising XV–XVI 8–10 and III, 8–10 spinous and soft rays for the fins respectively. First pelvic ray produced, variable in its posterior extension, but reaching the spinous anal in most adults. Pectoral fins as long as, or slightly shorter than the head. Caudal sub-truncate.

Teeth. In the form and pattern of its teeth, *H. lividus* departs from the generality of *Haplochromis* species. The anterior and antero-lateral teeth in the outer series are movably implanted and have slender necks (somewhat stouter than in *H. obliquidens*) with compressed, expanded and obliquely truncated, unequally bicuspid crowns. The posterior cusp shows some variation in size, but it is always smaller than the anterior, from which it is narrowly separated (Text-fig. 2). It should be

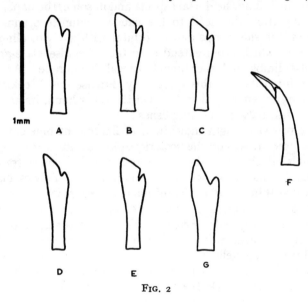

FIG. 2

noted that these teeth bear a striking resemblance to the undifferentiated postero-lateral teeth of *H. obliquidens*. Posterolateral teeth in *H. lividus* are either similar to the anterior teeth or indistinguishable from the generalized acutely bicuspid type (Text-fig. 2G). Less frequently, unicuspid teeth occur in this position. A weak positive correlation exists between the number of teeth in the outer series of the upper jaw and standard length.

S.L. (mm.)	46	56–65	66–75	76–85	86–91
Tooth number	36	42–58	38–66	44–75	52–66
Mean	36	50	54	61	57
N	1	8	27	33	5

Teeth forming the inner rows are invariably tricuspid and small. No obtusely cuspidate teeth, or teeth similar to those of the outer series, have been observed (*cf. H. obliquidens* in which such teeth are frequently encountered).

There is considerable variation in the number and disposition of inner rows. From 2–5 and from 2–4 series occur in the upper and lower jaws; individuals with more than three rows usually have the interspace between outer and inner teeth greatly reduced or even absent, particularly in the upper jaw.

Lower pharyngeal bone sub-equilaterally triangular; dentigerous surface slightly broader than long. Pharyngeal teeth similar to those described for *H. obliquidens*.

Osteology. That of a typical generalized *Haplochromis* species; vertebrae 14 + 16 (in two specimens).

Coloration. Breeding colours of male *H. lividus* are perhaps the most distinctive morphological characteristic of the species and are not repeated or even approached in any other Lake Victoria *Haplochromis*. In preserved material their brilliance is lost. *Sexually active males*. Ground colour light olive-green shading to slate-grey ventrally; flanks (including the dorsal aspects and in some, the nape) with a golden-red flush extending from the head to the caudal peduncle origin. Inter-orbital region of the head, the snout, lips and preorbital with a vivid, almost fluorescent, blue sheen, traces of which often extend onto the otherwise slate-grey lower jaw, lower preopercular limbs and the branchiostegal membrane. As far as can be determined, the intensity of this peculiarly intense head coloration is little influenced by the fishes' emotional state. On the other hand, its greatest extension is apparently manifest only in breeding fishes.

Dorsal fin grey to sooty, slight indications of fluorescent blue can be detected in some individuals; red streaks on the posterior spinous and entire soft part; lappets orange-red. Caudal dark, with ill-defined red maculae concentrated proximally on the upper half. Anal dark, with 2–4 yellow ocelli. Pelvics black, becoming lighter on the medial third. Coloration of *immature males* is similar except that the blue head colour is less concentrated and intense, or it may even be absent. The flanks are also less intensely red. *Females*. Ground colour light grey-green, becoming silver ventrally. Dorsal, caudal and pectoral fins colourless or faintly yellow-grey. Anal and pelvic fins yellow.

Preserved material: Males. Ground colour variable, usually grey. Dorsal caudal and anal fins clear or dark, the two former maculate as in life; pelvics black. The blue head coloration is lost, but in most individuals it is faintly represented by a dead-white or ashen colour (at least in formalin fixed material preserved in spirit for five years). *Females and immature males*. Ground colour as above. All fins clear. From 5–7 transverse bars on the flanks; the posterior pair rarely extend below the level of the lower lateral-line and are often joined by a short longitudinal stripe. Faint indications of a mid-lateral stripe are present in some individuals. Banding and striping are sometimes apparent in living fishes, but are intensified after death.

Distribution. *H. lividus* is known from several localities in Lake Victoria. (See below.)

Ecology: Habitat. Shallow littoral zone, especially in the vicinity of emergent and submerged vegetation, less frequently in the water-lily zone and at the margin of papyrus swamps; the species is commonly encountered over rock foundations of piers. Thus, the habitat preferences of *H. lividus* are similar to those of *H. obliquidens* with which species it is usually captured. There are, however, indications that *H. lividus* may inhabit the deeper littoral zone where *H. obliquidens* are relatively scarce.

Food. The intestine is long ($2-2\frac{1}{4}$ times standard length) and much coiled; the stomach large and distensible. The stomach contents of sixty-two individuals (size range 56–90 mm. S.L.) from most localities have been examined. In general the food of *H. lividus* is similar to that of *H. obliquidens*.

Diatoms of the genera *Melosira* and *Rhopalodia* comprised the predominating digested contents in the stomachs of forty-five fishes, and were significant in twelve others.

Fragments of plant epidermis were found in thirty-three stomachs; as in *H. obliquidens* the amount and fragment-size showed considerable variation.

Filamentous green-algae, represented by *Spirogyra*, were recorded from only four fishes; in none was there any indication of digestion. Blue-green algae (especially *Rivularia* and *Microcystis*) were found in twenty-four stomachs. Again, the algae were apparently not digested.

Very fragmentary animal remains (Ostracoda, Crustacea [Decapoda] and Insecta [larval Chironomidae]) were recorded from sixteen individuals.

The occurrence of Insecta (winged Hymenoptera) as the main stomach contents in twelve fishes collected contemporaneously at one station, is of particular interest. Besides insect remains, diatoms were well represented in these stomachs. This observation suggests that *H. lividus* may feed facultatively on animal food at times of local abundance. Insects also comprised the main contents of two other specimens, both from different localities.

Feeding habits of *H. lividus* are probably similar to those of *H. obliquidens*. Direct observation shows the species to be a grazer on submerged plants and stones, whilst the occurrence of sand-grains and bottom debris in some stomach contents indicates occasional benthic feeding.

Breeding. Breeding habits and sites are imperfectly known. Three females carrying young in the buccal cavity have been collected; one from an exposed beach flanked by dense emergent vegetation and two from an off-shore water-lily stand. From one of these fishes twenty-five larvae of 12 mm. total length were recovered; the other females had jettisoned the greater part of their broods.

Affinities. Disregarding for the moment its peculiar male breeding coloration, *H. lividus* shows marked affinity with *H. astatodon*, *H. annectidens* and *H. obliquidens*, especially with regard to dental characteristics. Save *H. annectidens*, for which no information is available, the food of these species is similar, and composed mainly of epiphytic algae and plant debris (Poll and Damas, 1939, for food of *H. astatodon*).

In fact, *H. lividus*, *H. astatodon* and *H. annectidens* seem to provide examples of herbivorous intermediates linking generalized and usually insectivorous *Haplochromis* species with the specialized algal-grazer, *H. obliquidens*.

Within the species flock of Lake Victoria *H. lividus* shows some morphological relationship with *H. nuchisquamulatus*. Anatomical and dental characteristics might entitle *H. nuchisquamulatus* to consideration as the extant representative of an annectant form between *H. lividus* and the generalized species typified in Lake Victoria by *H. nubilus* (Blgr.) and *H. macrops* (Blgr.).

The present distribution of species having "*lividus*"-like teeth requires little comment. *Haplochromis astatodon* is endemic to Lake Kivu, whose *Haplochromis* species flock has long been recognized as having well-defined Victorian affinities (Regan, 1921), although the possibility of convergent evolution in the two lakes cannot be entirely discounted. *Haplochromis annectidens* is endemic to Lake Nabugabo and is part of a small species group which could only have been derived from that of Lake Victoria (Trewavas, 1933).

The presence within Lake Victoria of both *H. lividus* and its morphological derivative *H. obliquidens* is suggestive of an ancestor-descendant relationship. Before accepting this apparent phylogeny, due regard must be paid to the unique male breeding coloration of *H. lividus*. Baerends and Baerends van Roon (1950) expressed the opinion that male coloration plays an important part in species recognition amongst cichlids. Thus, we may assume the importance of male coloration as a barrier to interspecific mating. Field observations on the *Haplochromis* of Lake Victoria lend weight to this hypothesis. Although male colours and colour-patterns are broadly repeated in several species, no instance has yet been recorded of related species with identical or near identical male coloration breeding in the same habitat.

Therefore, although the distinctive coloration of *H. lividus* might be used in argument against close relationship with *H. obliquidens*, it might equally well be interpreted as resulting from selection strengthening mating barriers between species which occupy similar habitats, especially if the species are closely related and of recent origin.

Diagnosis. *Haplochromis lividus* differs from other *Haplochromis* in Lake Victoria in having distally compressed and expanded teeth whose crowns are unequally bicuspid and obliquely truncated. Dentition serves to distinguish this species from the fluviatile *Haplochromis* of East Africa. In life male coloration is the most obvious diagnostic character.

From species with similar dental morphology *H. lividus* may be differentiated as follows : from *H. astatodon* by its larger eye/cheek ratio ; from *H. annectidens* by its slightly wider interorbital region and somewhat stouter, shorter teeth. In life coloration distinguishes *H. lividus* and *H. astatodon* ; live colours are unknown for *H. annectidens*.

Five specimens from Lake Kyoga (*Tilapia nubila* B.M. (N.H.) reg. nos. 1911.3.3. 141–145 ; 60–66 mm. S.L.) have teeth and dental patterns of the *H. lividus* type, but differ from Lake Victoria specimens in the following characters : dorsal head profile steeper ; body deeper ; and greater depth of cheek (23·3–25·2, mean 24·1 per cent head length). Should further collections from Lake Kyoga show that these fishes have *H. lividus* coloration (as is suggested in the preserved material) and should they also maintain the observed differences in morphology, then it will be necessary to recognize a distinct sub-species in that lake.

Study material and distribution records

Museum and reg. no.	Locality.	Collector.
British Museum (N.H.) 1906.5.30.318–320	. Entebbe	. Degen.
British Museum (N.H.) 1956.7.9.63–65 .	. Jinja Pier	. E.A.F.R.O.
,, ,, ,, ,, ,, ,, 58–62 .	. Beach near Jinja	. ,,
,, ,, ,, ,, ,, ,, 66–74 . (Type and paratypes)	. Napoleon Gulf, near Bugungu (opp. Jinja)	. ,,
,, ,, ,, ,, ,, ,, 75–83 .	. Kirenia (near Jinja)	. ,,
,, ,, ,, ,, ,, ,, 84–94 .	. Entebbe Harbour	. ,,
,, ,, ,, ,, ,, ,, 95–99 .	. Beach near Nasu Point, Buvuma Channel	. ,,
,, ,, ,, ,, ,, ,, 100–102	. Hannington Bay (Uganda)	. ,,
,, ,, ,, ,, ,, ,, 103–124	. Grant Bay (Uganda)	. ,,
,, ,, ,, ,, ,, ,, 125 .	. Mwanza, Capri Bay	. ,,
,, ,, ,, ,, ,, ,, 126–128	. Majita (Tanganyika Territory)	. ,,

Haplochromis nigricans (Blgr.) 1906

Tilapia nigricans (part) Blgr., 1906, *Ann. Mag. nat. Hist.* (7) **17**, 448 ; *Idem*, 1907, *Fish. Nile*, 518 ; *Idem*, 1911, *Ann. Mus. Genova* (3) **5**, 75 ; *Idem*, 1915, *Cat. Afr. Fish.*, **3**, 241, fig. 160.
Tilapia simotes Blgr., 1911, *Ann. Mus. Genova* (3) **5**, 75 ; *Idem.*, 1915, *op. cit.*, 242, fig. 161.
Neochromis nigricans (Blgr.), Regan, 1920, *Ann. Mag. nat. Hist.* (9) **5**, 33.
Haplochromis (Neochromis) nigricans (Blgr.), Regan, 1922, *Proc. zool. Soc., London*, 163.

As Regan (1922) first showed, Boulenger's figure of *Tilapia nigricans* is misleading. It was prepared from a specimen distorted in preservation and consequently the head profile differs considerably from that of *T. simotes*. However, in the important characters of dental pattern and morphology both species are identical. If the head of the figured specimen is restored to its natural position the characteristically decurved profile of *Haplochromis nigricans* is apparent.

Description. Based on fourty-four specimens (size range 49–94 mm. standard length) including holotypes of *T. nigricans* and *T. simotes*. Other specimens in the British Museum (Nat. Hist.) collections were examined but are not included in the morphometric data. The paratype of *T. nigricans* is clearly not referable to this species and should probably be placed in *H. lividus*. Its small size permits only tentative identification. Both skeletons in the British Museum (Nat. Hist.) are of *H. nigricans*.

Depth of body 34·5–40·0 (mean 36·9), length of head 28·0–33·3 (M = 31·2) per cent of standard length. Dorsal head profile strongly decurved. Preorbital depth 11·8–16·7 (M = 14·6) per cent of head length ; least interorbital width 25·0–31·5 (M = 28·8) ; snout broader than long in most specimens of more than 65 mm. S.L., and as long as broad in smaller fishes, its length 26·3–35·2 (M = 30·4) per cent of head. Eye 25·9–33·3 (M = 30·0), depth of cheek 19·4–27·3 (M = 32·4) per cent of head.

Caudal peduncle from 1·1–1·8 (mode 1·3) times as long as deep, its length 11·4–17·6 (M = 15·4) per cent of standard length.

Mouth horizontal; posterior maxillary tip reaching the vertical from the anterior orbital margin or extending somewhat beyond. Jaws equal anteriorly, the lower short and broad, from 30·0–37·8 (M = 35·6) per cent of head length, its length/breadth ratio 1·0–1·4 (mode 1·2). *H. simotes* holotype is unusual in having its lower jaw only 28 per cent of the head length.

Gill rakers short, 8 or 9 (rarely 10) on the lower part of the first arch.

Scales ctenoid; lateral line interrupted, with 30 (f.1), 31 (f.12) 32 (f.28), or 33 (f.3) scales. Cheek with 2 or 3 (rarely 4) series of scales; 6 or 7 (less frequently 5 or 5½) between dorsal fin origin and the lateral line; 7 or 8 (rarely 6 or 9) between pectoral and pelvic fin insertions.

Fins. Dorsal with 24 (f.7), 25 (f.33) or 26 (f.4) rays, anal 11 (f.7), 12 (f.32) or 13 (f.5), comprising XIV–XVIII, 8–10 and III, 8–10 spinous and soft rays. First pelvic ray produced, extending to the vent or even to the soft anal; its posterior extension not correlated with sex or maturity. Pectoral fin shorter than the head. Caudal sub-truncate or feebly rounded; scaled on the proximal half to two-thirds.

Teeth. The outer series is composed of close set, movably implanted bicuspid teeth, with long, slender necks and expanded crowns. Cusp size in some individuals is markedly disparate, whilst in others the cusps are sub-equal. In the upper jaw, teeth situated postero-laterally are either tri- or unicuspid.

A weak positive correlation exists between the number of teeth in the outer series of the upper jaw and standard length.

S.L. (mm.)	49–58	59–68	69–78	79–93
Tooth number	40–50	46–60	46–60	54–70
Mean	46	52	52	62
N	19	6	14	4

The inner series is composed of small tricuspid teeth; 3–7 (mode 4) rows in each jaw. Compared with other Lake Victoria *Haplochromis* (except some individuals of *H. nuchisquamulatus* and *H. lividus*) the space separating inner and outer tooth series is greatly reduced in *H. nigricans*; it is non-existent in 30 per cent of the specimens examined.

Lower pharyngeal bone sub-equilaterally triangular; dentigerous surface somewhat broader than long; teeth numerous, and similar to those in *H. obliquidens* and *H. lividus*.

Cranial skeleton. The short and strongly decurved snout is reflected in the neurocranial shape. This differs slightly from that of generalized *Haplochromis* by having a more steeply sloping ethmo-vomer complex. Also, the dentary is relatively stouter and more massive in *H. nigricans*.

Coloration in life: Breeding males. Ground colour black, shot with metallic blue; snout, lips, interorbital region and to a lesser degree, cheeks and opercula, bluish. Dorsal fin black, lappets and maculae on the soft part deep crimson; anal dusky crimson, ocelli yellow; caudal crimson, pelvics black. *Adult females and juveniles.* Ground colour olivaceous; a faint golden-yellow flush over the opercula and branchiostegal membrane. Dorsal and anal fins dark yellow; caudal grey-green; pelvics dusky yellow.

Preserved material: Adult males. Black or slate grey; seven or eight transverse bars visible on the flanks of light coloured fishes. Dorsal fin black, with pale margin and maculae; caudal black proximally, pale distally; anal pale; pelvics black. *Females and juveniles.* Ground colour greyish-brown, with seven or eight dark transverse bars on the flank; a pronounced lachrymal stripe. All fins hyaline or slightly darkened.

Particular interest attaches to a single adult female with black and yellow piebald coloration similar to that described in *Macropleurodus bicolor* (Blgr.) and *Hoplotilapia retrodens* Hilgen. (Greenwood, 1956). The significance of this atypical individual is difficult to assess. In other characters *H. nigricans* does not manifest any apparent relationship with the monotypic genera, nor with *H. sauvagei* (Pfeffer), another species exhibiting sex-limited polychromism. It is probably the result of independent but parallel mutation occurring in *H. nigricans*, and therefore of no phyletic value. Such a phenomenon might be expected amongst members of a recently evolved and oligophyletic species flock.

Distribution. Lake Victoria and the Victoria Nile. Although most localities represented in the present collection are in Uganda, this should not be taken to indicate that *H. nigricans* is confined to, or more abundant in, these waters. The species has been seen in many areas, but its lithophilic habits render capture difficult except by unconventional or specialized gear.

Numerous specimens have been caught at Godziba Island (1° 29′ S., 32° 36′ E.). This small, rocky outcrop lies slightly south and west of the centre of Lake Victoria and is distant from either the mainland or other off-shore islands. Because there is no indication of *H. nigricans* ever occurring in deep or sub-littoral waters, one is led to suppose that Godziba fishes are at present isolated from coastal populations, and have been isolated for some considerable time. With this in mind, the Godziba sample was carefully compared with others from the mainland, but no phenotypic peculiarities could be detected.

Ecology: Habitat. *H. nigricans* is apparently confined to rocky and shallow areas of the littoral zone. Since rock exposures are not infrequent in the exposed littoral, its habitat, broadly speaking, overlaps that occupied by other algal-grazing *Haplochromis* species. No data are available for populations living in the Nile.

Food. The intestine is long (*ca.* $2\frac{1}{2}$–3 times S.L.) and coiled. Observations on fishes in the lake indicate that *H. nigricans* feeds by grazing on algae from rock surfaces, a conclusion which is supported by stomach content analyses.

Ingested material from thirty-two stomachs showed a preponderance of diatoms over all other material. Specific identification of these plants was impossible, but the genera represented (chiefly *Navicula*, *Synedra*, *Rhopalodia* and *Gomphonema*) are typically epilithic or epiphytic in Lake Victoria (Ross, 1954). The absence, except from two stomachs, of fragmentary phanerogam tissue (an important element in stomach contents of other algal grazing species) was noteworthy, but explicable if *H. nigricans* graze from rock surfaces.

Filamentous green algae (*Spirogyra* and *Oedogonium*) and blue-green algae occurred less frequently, and were apparently undigested.

Very fine, sand-grain-like particles were recorded from thirteen stomachs. That these might have been fragments derived from rock surfaces and not the bottom seems likely in the absence of bottom debris typically associated with a sand substrate.

Breeding. Spawning sites are unknown. Courtship activity has been observed amongst fishes living over rocks near the Ripon Falls, but actual spawning was not seen. Two females have been found with embryos and larvae in the buccal cavity ; it is assumed that *H. nigricans*, like the generality of *Haplochromis* species, is a mouth-brooder.

The smallest adult fishes recorded were a female 51 mm. and a male 55mm. in standard length. Males apparently reach a larger size than females since no female greater than 70 mm. S.L. has been captured.

Affinities. *Haplochromis nigricans* is closely related to *H. serridens* Regan of Lake Edward (*vide* Trewavas, 1933). Both species have almost identical dental morphology and pattern, as well as similarity in general facies and preserved coloration. No clear-cut quantitative characters can be found to separate the species. There is, however, a subtle difference in their gross morphology, probably attributable to the more rounded physiognomy of *H. serridens*. Also, the inner tooth bands of this species are usually broader and possess more teeth than those of H. *nigricans*.

Tooth form, and less obviously the dental pattern, in one other Lake Edward species, *H. fuscus* Regan, is similar to that of *H. nigricans* ; but the species are readily distinguished by the smaller nuchal and thoracic scales in *H. fuscus* and also by its thicker lips and more abruptly declivous dorsal head profile.

Amongst Lake Victoria species *H. nigricans* is probably related to, and derived from a species resembling *H. nuchisquamulatus*.

Diagnosis. *H. nigricans* is distinguished from other Lake Victoria *Haplochromis* with bicuspid outer teeth by the following combination of characters : a short and broad lower jaw (modal length/breadth ratio 1 : 2) ; slender, movably implanted outer teeth narrowly separated, if at all, from the broad bands of inner teeth ; a strongly decurved dorsal head profile ; a long and convoluted intestine.

Study material and distribution records

Museum and reg. no.	Locality.	Collector.
British Museum (N.H.) 1906.5.30.469 (Holotype of *Tilapia nigricans*)	Entebbe	Degen.
Genoa Museum (Holotype of *Tilapia simotes*)	Kakindu (Victoria Nile)	Bayon.
British Museum (N.H.) 1911.3.3.160–163 (Paratypes of *T. simotes*)	Jinja (Ripon Falls)	Bayon
British Museum (N.H.) 1911.3.3.156–158, plus one additional specimen (Paratypes of *T. simotes*)	Kakindu	Bayon
British Museum (N.H.) 1956.7.9.129–136	Napoleon Gulf, near Ripon Falls	E.A.F.R.O.
,, ,, ,, ,, ,, ,, 137–148	Jinja Pier	,,
,, ,, ,, ,, ,, ,, 149–150	Napoleon Gulf, near Jinja	,,
,, ,, ,, ,, ,, ,, 151	Beach near Nasu Point, Buvuma Channel	,,
,, ,, ,, ,, ,, ,, 152	Buka Bay (Uganda)	,,
,, ,, ,, ,, ,, ,, 153–165	Godziba Island	,,

Haplochromis nuchisquamulatus (Hilgendorf) 1888

Chromis nuchisquamulatus Hilgend., 1888, *S. B. Ges. naturf. Fr. Berlin*, 76.
Ctenochromis nuchisquamulatus (Hilgend.), Pfeffer, 1896, *Thierw. O. Afr. Fische*, 14.
Tilapia nigricans (part), Blgr., 1915, *Cat. Afr. Fish.*, 3, 241.
Haplochromis nuchisquamulatus (part), *idem, ibid.*, 290.
Haplochromis (Neochromis) nuchisquamulatus (Hilgend.), Regan, 1922, *Proc. zool. Soc., London*, 163.

The holotype of *H. nuchisquamulatus* is amongst those specimens, once housed in the Berlin Museum, which cannot be located at the present time. It is thus the more regrettable that Hilgendorf's original description is totally inadequate for modern taxonomic purposes.

As a basis for comparison I have therefore relied upon Regan's identification of two British Museum (Nat. Hist.) specimens. From Regan's paper (1922) it is clear that he, too, was unable to study the type specimen, but he apparently gained sufficient information from photographs and data supplied by Dr. Pappenheim to identify his material. Of this I have located only one specimen (British Museum (N.H.) reg. no. 1911.3.3.155, from Kakindu, Victoria Nile). In its general morphology this fish agrees closely with a photograph of the type. Further, with the aid of a binocular microscope it has proved possible to check certain other characters visible in this remarkably clear photograph.[1]

Although this species is represented in my study-material by only six specimens, I have little doubt as to its biological validity. Morphologically *H. nuchisquamulatus* is intermediate between *H. lividus* and *H. nigricans*: it may well represent the stock from which these species diverged. The tooth form of *H. nuchisquamulatus* is less specialized than that of *H. lividus* and is nearer *H. nigricans*. That is to say, the outer teeth are slender, bicuspid and movable, whilst those of the inner series show a tendency towards an increase in the number of rows and a decrease in the space separating them from the outer series. The lower jaw is more slender than in *H. nigricans* and is similar to the dentary in *H. lividus* and *H. obliquidens*, and in other, more generalized *Haplochromis*.

None of the dental and associated characters considered above lies within the known range of intra-specific variability for *H. nigricans* or *H. lividus*. Neither is there any indication by analogy with well-defined *Haplochromis* species that the "*nuchisquamulatus*" character-complex is an extreme variant of some other species

Description. The principal morphometric characters for each of the six specimens examined are tabulated below. All are adult males.

S.L.	Depth.*	Head.*	Po. %	Io. %	Snt. %	Eye. %	Ch. %	Lj. %	C.P.*
83·0	38·0	31·3	15·4	30·8	30·8	30·8	23·0	36·5	15·7
86·0	37·2	32·5	14·3	28·6	28·6	32·2	23·2	35·7	15·2
93·0	38·7	32·3	14·5	29·0	32·2	29·0	24·2	38·6	15·1
98·0	38·8	31·6	16·2	30·0	32·2	25·8	22·6	38·7	15·3
99·0	37·4	32·8	15·4	30·8	30·8	30·8	24·6	40·0	15·2
113·0	37·0	30·3	17·4	26·5	31·8	29·0	26·5	36·2	18·6

* Percentage standard length.
% Percentage head-length.

[1] To be reproduced in a later part of this series.

Dorsal head profile curved and sloping. Mouth horizontal; posterior maxillary tip extending to the vertical from the anterior orbital margin or slightly beyond. Jaws equal anteriorly, the length/breadth ratio of the lower 1·4–1·7 (mode 1·5).

Teeth. Outer teeth unequally bicuspid; a few slender and unicuspid teeth occur posteriorly in the upper jaw, in which there are from 50–70 teeth. Although relatively fine, the neck in these teeth is stouter and less clearly demarcated from the expanded crown, than in *H. lividus* or *H. obliquidens*.

Inner teeth small and tricuspid, occurring in 4–8 and 3–6 rows in the upper and lower jaws respectively; the space separating inner and outer series is reduced.

Lower pharyngeal bone sub-equilaterally triangular, its dentigerous surface slightly broader than long. Teeth fine and numerous; in the three larger specimens, the median teeth are enlarged.

Gill rakers short, 8–10 on the lower part of the first arch.

Scales ctenoid: lateral line interrupted, with 31 (f.2), 32 (f.2) or 33 (f.2) scales. Cheek with 2 or 3 series. 6–8 scales between dorsal fin origin and the lateral line; 6–8 scales between pectoral and pelvic fin insertions.

Fins. Dorsal with 24 (f.1), 25 (f.2) or 26 (f.3) rays, anal 12 (f.4) or 13 (f.2), comprising XV–XVII, 9 or 10 and III, 9 or 10 spinous and soft rays. First pelvic ray produced, extending to the second anal ray. Pectoral fins slightly shorter than the head. Caudal sub-truncate.

Skeleton. That of a generalized *Haplochromis*.

Coloration. Unknown in life and known only for preserved males. Ground colour dark greyish-brown, the dorsal and ventral surfaces darker than the flanks, across which seven transverse bars are visible; in two specimens the chest is black. Well-defined, narrow lachrymal and two interorbital stripes; two broad bands across the nape, one immediately post-ocular in position, the other slightly more posterior.

Ecology. Of the six specimens studied, five were caught in exposed littoral zones of Lake Victoria, and one in the Victoria Nile. The type specimen is frcm Lake Victoria, but no precise locality is given.

Food. Fragments of plant tissue and numerous epiphytic algae were recorded from four of the five stomachs examined, whilst the fifth contained filaments of *Oedogonium* and some fragmentary plant tissue.

Diagnosis. *H. nuchisquamulatus* is distinguished from other Lake Victoria *Haplochromis* with bicuspid outer teeth by the following combination of characters: long and convoluted intestine (*ca.* 3 × S.L.); relatively slender, movably implanted and numerous outer teeth; increased number of inner tooth rows (3–8) narrowly separated from the outer series. From *H. nigricans* it is recognized by the narrower lower jaw and less strongly decurved dorsal head profile; outer teeth in *H. nuchisquamulatus* are also somewhat stouter than those of *H. nigricans*. These acutely cuspidate teeth serve to separate *H. nuchisquamulatus* from *H. lividus*.

The diagnostic character used by Hilgendorf (small nuchal scales whose exposed surface is less than half that of flank scales) cannot be considered valid. In most *Haplochromis* nuchal scales are smaller than those on the flank, and furthermore are subject to quite considerable intra-specific size-variation.

Study material and distribution records

Museum and reg. no.	Locality.	Collector.
British Museum (N.H.) 1911.3.3.154	Kakindu (Victoria Nile)	Bayon.
British Museum (N.H.) 1906.5.30.316–317	Entebbe	Degen.
British Museum (N.H.) 1956.7.9.166–167	Beach near Nasu Point, Buvuma Channel	E.A.F.R.O.
,, ,, ,, ,, ,, ,, 168	Godziba Island	E.A.F.R.O.

SUMMARY

1. The algal-grazing species *Haplochromis obliquidens* Hilgendorf 1888, *H. nigricans* (Boulenger) 1906, and *H. nuchisquamulatus* (Hilgendorf) 1888, are redescribed on the basis of new and more extensive collections.

2. A new species, *H. lividus*, apparently related to *H. obliquidens* is described.

3. Data on the food and ecology of these species are given.

4. Consideration is given to the possibility of recognizing a number of supra-specific groups of *Haplochromis* in Lake Victoria. At present, although such groups may be determined, it is impossible to give them formal taxonomic status.

ACKNOWLEDGMENTS

I wish to acknowledge my gratitude and thanks to the Trustees of the British Museum (Natural History) for facilities afforded me during the tenure of a Colonial Fisheries Research Studentship; to Professor L. Bertin of the Muséum National d'Histoire naturelle, Paris, for allowing me to study type specimens of Lake Victoria Cichlidae described by Pellegrin; to Dr. Delfa Guiglia of the Museo Civico di Storia Naturale, Genoa, for the courtesies mentioned in the text; to my colleague Dr. Philip S. Corbet for identifying some of the material from gut contents; and to Mr. Denys W. Tucker, for his very helpful criticism of the manuscript. I am especially indebted to Dr. Ethelwynn Trewavas for much helpful information and advice.

REFERENCES

(Other than those given in full in the text)

BAERENDS, G. P. & BAERENDS VAN ROON, J. M. 1950. An introduction to the study of the ethology of cichlid fishes. *Behaviour*, Supplement 1, pp. 1–242. Leiden.

GREENWOOD, P. H. 1951. Evolution of the African cichlid fishes; the *Haplochromis* species flock in Lake Victoria. *Nature*, London, **167** : 19.

—— 1956. The monotypic genera of cichlid fishes in Lake Victoria. *Bull. Br. Mus. nat. Hist., Zool.* **3**, no. 7.

PFEFFER, G. 1893. Ostafrikanische Fische gesammelt von Herrn Dr. F. Stuhlmann. *Jahrb. Hamburg. wiss. Anst.* **10** (2), 129–177, 3 pls. (1–49 sep. pag.).

POLL, M. & DAMAS, H. 1939. Poissons. *Exploration du Parc National Albert, mission H. Damas* (1935–1936), fasc. 6, pp. 1–73.

REGAN, C. T. 1920. The classification of the fishes of the family Cichlidae. I. The Tanganyika genera. *Ann. Mag. nat. Hist.* (9) **5** : 33.

REGAN, C. T. 1921. The cichlid fishes of Lakes Albert Edward and Kivu. *Ann. Mag. nat. Hist.* (9) **8** : 632.
—— 1922. The cichlid fishes of Lake Victoria. *Proc. zool. Soc. Lond.* 157.
Ross, R. 1954. The algae of the East African Great Lakes. *Proc. International Assoc. : Theoretical and Applied Limnology.*
TREWAVAS, E. 1933. Scientific results of the Cambridge expedition to the East African lakes, 1930–1. II. The cichlid fishes. *J. Linn. Soc. (Zool.)* **38** : 309.
—— 1935. A synopsis of the cichlid fishes of Lake Nyasa. *Ann. Mag. nat. Hist.* (10) **16** : 65.
WORTHINGTON, E. B. 1932. *A Report on the Fisheries of Uganda.* Crown Agents, London.

A REVISION OF THE LAKE VICTORIA *HAPLOCHROMIS* SPECIES (PISCES, CICHLIDAE) PART II[1]: *H. SAUVAGEI* (PFEFFER), *H. PRODROMUS* TREWAVAS, *H. GRANTI* BLGR. AND *H. XENOGNATHUS*, SP. N.

By P. H. GREENWOOD

CONTENTS

	Page
INTRODUCTION	76
Haplochromis sauvagei (Pfeffer)	76
Synonymy and description	77
Distribution	80
Ecology	80
Diagnosis and affinities	80
Study material and distribution records	81
Haplochromis prodromus Trewavas	82
Synonymy and description	82
Distribution	85
Ecology	85
Study material and distribution records	85
Diagnosis and affinities	86
Haplochromis granti Boulenger	86
Synonymy and description	86
Distribution	89
Ecology	89
Diagnosis and affinities	89
Study material and distribution records	90
Haplochromis xenognathus sp. nov.	90
Synonymy and description	91
Distribution	94
Ecology	94
Diagnosis and affinities	94
Study material and distribution records	95
DISCUSSION	95
SUMMARY	97
ACKNOWLEDGMENTS	97
REFERENCES	97

[1] Part I, see Greenwood, 1956b.

INTRODUCTION

MORPHOLOGICALLY, the *H. sauvagei* complex stands apart from any other species-group in Lake Victoria. The principal group character is that of the dentition which combines recurved outer teeth with multiseriate inner tooth-bands (Text-fig. 3).

Furthermore, the shape of the neurocranium, although differing intra-specifically within the group, is unlike that of other *Haplochromis*. This character is probably associated with the multiseriate dentition and relatively powerful jaw musculature. Indeed, amongst the non-piscivorous predators such marked divergence in cranial anatomy is otherwise only found in mollusc-eating species with hyper-developed pharyngeal bones and musculature. Two specifically constant forms of neurocranium occur in the "sauvagei" group, but neither can be correlated with the type of dental pattern present.

Trophically, members of the group may be classed as mollusc eaters, although available data indicate that other food organisms do contribute to their diet, usually in a subsidiary capacity. Unlike other mollusc-eating *Haplochromis* in this lake, species of the "sauvagei" group do not swallow the shells of their prey, but remove the soft parts before ingestion takes place. In this respect the feeding method is like that of *Macropleurodus bicolor* (Blgr.), a monotypic genus apparently derived from this group.

Haplochromis sauvagei (Pfeffer), 1896
(Text-fig. 1 and Pl. 1 upper fig.)

Ctenochromis sauvagei Pfeffer, 1896, *Thier. Afr. Fische*, 15.
Haplochromis nuchisquamulatus (part), Boulenger, 1915, *Cat. Afr. Fish.*, **3**, 290.
Paratilapia granti (part), Boulenger, 1915, *op. cit.*, 342.
Paratilapia bicolor (part) Boulenger, 1915, *op. cit.*, 346.
Paratilapia retrodens (part), Boulenger, 1915, *op. cit.*, 235.
Haplochromis sauvagei (part), Regan, 1922, *Proc. zool. Soc., Lond.*, 167.
? *Paratilapia crassilabris* (part), Boulenger, 1915, *op. cit.*, 345.

I was unable to examine the holotype of *H. sauvagei* which was mislaid during the 1939–45 war ; at present the Berlin Museum authorities cannot confirm whether this specimen has been lost. Pending more definite information, no neotype can be selected, but, should such a step become necessary, I suggest that the specimen B.M. (N.H.) Reg. No. 1956.9.17.1, a male from Entebbe (Text-fig. 1) be given neotypical status.

Fortunately, Pfeffer's original description of *Ctenochromis sauvagei* is comprehensive, and, when coupled with a photograph of the type, clearly indicates to which *Haplochromis* species his specimen should be referred. The photograph, preserved in the British Museum (Natural History), is reproduced in Plate 1.

Additional material discloses only one important discrepancy with the original description, in which the mouth and lower-jaw profile are described as rising steeply : " . . . ; das untere Profil der Unterkinnlade steigt viel stärker. Die von dicken und breiten Lippen umgebene kurze Mundspalte steigt nach vorn sehr steil auf."

In most specimens the ventral head profile is almost horizontal, or, at most, slightly oblique. A possible explanation for this discrepancy may lie in the fact that Pfeffer's description was taken from a fish preserved with its mouth open. From the photograph it is clear that, if the jaws were restored to their natural position, the cleft and lower jaw profile would be slightly oblique.

The present synonymy for *H. sauvagei* is essentially that prepared by Regan (1922), but some of the specimens which he referred to this species are now placed in others. In this connection, reference should be made to the list of study material.

Paratilapia crassilabris part (Boulenger, 1915) is tentatively retained in the synonymy on the basis of a single specimen (B.M. (N.H.) Reg. no. 1911.3.3.32). This individual cannot be identified with certainty, but it is nearer *H. sauvagei* than any other species with thickened lips and dentition not of the generalized type.

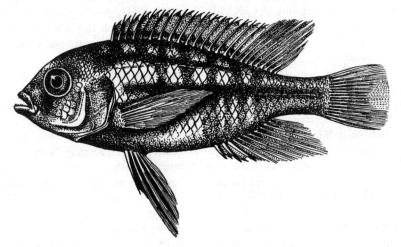

Fig. 1. *Haplochromis sauvagei*, ♂, B.M. (N.H.) 1956.9.17.1. Drawn by Miss L. Buswell.

Description. Based on 85 specimens, 58–105 mm. standard length. Of the measurements made, only cheek depth clearly shows allometry with standard length.

Depth of body 30·4–41·8, mean (M) = 35·6 ; length of head 29·6–34·5 (M = 31·9) per cent of standard length. Dorsal head profile varying from decurved to straight, but strongly sloping, the former shape occurring more frequently.

Preorbital depth 15·4–20·2 (M = 17·3) per cent of head length, least interorbital width 23·0–31·2 (M = 27·0) per cent. Snout as broad as long, its length 27·2–35·5 (M = 30·8) per cent of head ; eye diameter 25·7–33·4 (M = 28·9) per cent. Cheek becoming relatively deeper with increasing standard length ; four size-groups are recognized, 58–69 mm. S.L. (N = 13), 70–80 (N = 21), 81–90 (N = 27) and 91–105 (N = 24), for which the cheek depth is 21·0–25·0 (M = 23·2), 20·4–26·0 (M = 23·7), 22·0–26·9 (M = 24·4) and 24·1–26·6 (M = 25·1).

Caudal peduncle 13·9–19·3 (M = 16·4) per cent of standard length; its length 1·1–1·6, times its depth.

Mouth horizontal or slightly oblique; posterior maxillary tip reaching or almost reaching the vertical to the anterior orbital margin. Lips thickened; the depth of the upper lip, measured mid-laterally, is contained 4–4½ times in the eye-diameter. Jaws equal anteriorly, the lower 30·6–37·7 (M = 34·5) per cent of head length and 1·0–1·5 (mode 1·3) times as long as broad.

Gill rakers short, 7–9 (rarely 10) on the lower limb of the first arch.

Scales ctenoid; lateral line with 31 (f.7), 32 (f.29), 33 (f.39), 34 (f.9) or 35 (f.1) scales; cheek with 3–4 (rarely 2) series. 7–9 (less frequently 6) scales between dorsal fin origin and upper lateral line; 7 or 8 (less commonly 6 or 9) between pectoral and pelvic fin-bases.

Fins. Dorsal with 24 (f.11), 25 (f.56) or 26 (f.18) rays, anal 10 (f.1), 11 (f.10), 12 (f.71) or 13 (f.3), comprising XV–XVII, 8–10, and III, 7–10 spinous and branched rays for the fins respectively. Pectoral slightly shorter than the head, or occasionally of equal length. Pelvic with the first ray produced, variable in its posterior extension but longer in adult males than females. Caudal sub-truncate.

Lower pharyngeal bone triangular, its dentigerous area $1\frac{1}{3}$–$1\frac{1}{4}$ times as broad as long; pharyngeal teeth slender and cuspidate. In some specimens, teeth in the median rows are slightly enlarged, but retain their bicuspid crowns.

Teeth. In the outermost series of both jaws, the teeth have strongly recurved tips and are unequally bicuspid or unicuspid. The predominant tooth form is apparently correlated with length. Fishes less than 80 mm. S.L. have mainly bicuspid teeth, those in the range 80–90 mm. have either unicuspids or an admixture of uni- and bicuspids, whilst larger individuals possess mainly unicuspid teeth. When both types of teeth are present, the unicuspid form usually occurs anteriorly and laterally. There are 32–56 (mode 42) outer teeth in the upper jaw.

Inner teeth are either tri- or unicuspid; as in the outer series, unicuspid teeth are commoner in fishes above 80 mm. S.L. Antero-medially, the teeth are arranged in a broad band comprising 3–8 (mode 4) and 2–6 (modes 3 and 4) rows in the upper and lower jaws respectively. Laterally, the band narrows to a single series. A distinct inter-space separates the inner and outer series.

Syncranium and associated musculature. Neurocranial form in *H. sauvagei* departs quite considerably from the generalized *Haplochromis* type, and approaches that of *Macropleurodus bicolor* (Greenwood, 1956a). Essentially the same points of difference with the generalized type occur in both species. The skull has a fore-shortened appearance due to the strongly decurved and almost vertically disposed ethmovomerine region. This curvature affects the morphology of the entire pre-orbital skull which is less gently curved than in the generalized neurocranium.

On the other hand, the jaws do not exhibit such radical departure from the generalized condition. The premaxilla, apart from a slight broadening of its dentigerous area, compares closely with that of other *Haplochromis*; the dentary is somewhat shorter, more massive and has a wider median dentigerous area than is common in generalized species. Consequent upon these modifications slight differences are apparent in the suspensorium.

Muscle disposition and form are similar to those of basic *Haplochromis* species. However, the *adductor mandibulae* I is slightly longer (38–44 per cent head length compared with 36–39 per cent) and broader (length/breadth ratio 3·0–3·7 cf. 4·4–5·5).

The syncranium, its musculature and the dentition all foreshadow the condition found in *H. prodromus*, and therefore that which reaches its ultimate expression in the genus *Macropleurodus* Regan (Greenwood, *op. cit.*). It is perhaps significant that variability in the degree to which the syncranium departs from the basic type is greater in *H. sauvagei* than in *H. prodromus*.

Causal factors responsible for the characteristic preorbital face in both *H. sauvagei* and *H. prodromus* are not readily determined. From an examination of larval fishes it is manifest that, as in *M. bicolor*, this form develops during post-larval ontogeny. It is probably effected by differential growth of various syncranial parts, especially since the ethmovomerine region is not directly affected by the moulding influence of muscle insertions. However, the premaxilla, which is closely associated both anatomically and functionally with the dentary, could exert considerable influence over this region. In *H. sauvagei* the dentary is short in relation to the head and also in comparison with other *Haplochromis* species of comparable size. If, during post-larval ontogeny, this bone increased in length more slowly than the neurocranium, and if it is to remain functionally integrated with the upper jaw, then there can be two morphological results : either a skull of the *H. sauvagei* type, or one in which the upper jaw projects anteriorly beyond the lower. A third possibility, that the suspensorium be rotated anteriorly, cannot be considered in this case, since its almost vertical alignment in *H. sauvagei* is typically that of the basic type.

H. sauvagei includes Gastropoda as a substantial part of its diet. As in *H. prodromus* and *M. bicolor* the soft parts alone are ingested. The feeding habits of *M. bicolor* have been described elsewhere (Greenwood *op. cit.*) : when feeding on snails, aquarium-kept *H. sauvagei* follow the same general pattern, except that after grasping the foot of the snail between its jaws the fish then uses the shell as a fulcrum to lever out the soft parts. Only rarely is the shell crushed by the jaws.

Coloration in life : Breeding males. Ground colour dark grey-green or blue-grey, lighter or yellowish ventrally ; a suffused coppery sheen on the flanks and ventral aspects of the operculum. Dorsal fin black basally, becoming slate-coloured distally ; lappets red ; red spots, often coalesced, between the soft rays. Anal dark with a red flush ; ocelli yellow. Caudal dark grey proximally, lighter distally, and with an overall orange-red flush. Pelvics black laterally, orange-red medially. *Non-breeding males* have similar coloration except that the copper flush is absent and other bright colours are less intense. *Females and immature males.* Ground colour golden-green, shading to pearly-white ventrally. All fins yellow-green.

In both sexes there may develop after death a dark longitudinal band running mid-laterally from the eye to the dorsal fin base, a second band running dorso-laterally approximately along the upper lateral-line, and 6–10 narrow transverse bars across the flanks. In life these markings are rarely discernible.

Amongst females a second type of coloration is known. This takes the form of irregular black blotches on a yellow ground and is identical with the *bicolor* pattern described for certain female *Macropleurodus bicolor*, *Hoplotilapia retrodens*, and *Haplochromis nigricans* (Greenwood, 1956a and b).

Since collectors show some predilection for fishes with a striking colour pattern, it is difficult to obtain accurate frequency estimates for the *bicolor* pattern. In the present sample, 25 per cent of females are *bicolor*. As most specimens were obtained by collectors aware of possible biasing factors, this figure may be accepted as fairly reliable. No male *bicolor* variants have yet been recorded. Thus, the incidence of *bicolor* variants seems sufficiently high to recognize the phenomenon as sex-limited polychromatism, and not merely the maintenance of an atypical genotype by recurrent mutation. Aberrantly coloured females were found in most localities. None exhibits a pattern intergrading with that usual for females.

Sex-limited polychromatism involving the same phenotypic expression was observed in *M. bicolor* and *Hoplotilapia retrodens* (Greenwood, 1956a). It seems probable that hypotheses regarding its genic basis and evolutionary significance in these species are also applicable to *H. sauvagei*. The possible significance of *bicolor* females as indicating phyletic relationship amongst the various species in which they occur has also been discussed (Greenwood, *op. cit.*). It was concluded that, in general, no reliability could be placed on this character, and that its repeated appearance was probably attributable to the oligophyletic origin of the Lake Victoria species-flock. Nevertheless, it is suggestive that both *H. sauvagei* and *M. bicolor* exhibit " *bicolor* " polychromatism as well as an apparent similarity in fundamental syncranial morphology.

Colour in preserved material : Adult males. Slate-grey to sooty, the longitudinal and transverse banding often obscured. Spinous dorsal fin grey, soft part hyaline but maculate. Anal and caudal hyaline. Pelvics black on the outer half, hyaline mesially. A dark lachrymal stripe and two bars across the snout are often present. *Females and immature males.* Ground colour variable, from silver-grey to brownish. Banding, as described above, usually developed. All fins hyaline, the soft dorsal and upper half of the caudal, maculate.

Distribution. Known only from Lake Victoria.

Ecology : Habitat. Restricted to littoral zones where the bottom is hard (sand or shingle) ; the species is especially common over exposed sandy beaches.

Food. The gut contents of forty-five fishes from various localities indicate that *H. sauvagei* feed mainly on Gastropoda (f.19), bottom deposits, which included insect larvae, Copepoda and diatoms (f.19), and Insecta (chiefly larval boring may-flies, *Povilla adusta* Navás) (f.4). No fragments of snail shell were observed, although opercula occurred frequently in the stomach and intestine (see also p. 79).

Breeding. Spawning sites and behaviour are unknown. In many localities, sexually active and quiescent fishes, and brooding females occur together.

The smallest adult fish was a female 72 mm. S.L. All specimens over 80 mm. were adult. No difference was detected in the sizes of adult males and females.

Diagnosis. *H. sauvagei* is distinguished from other *Haplochromis* in Lake Victoria by combinations of the following characters : lips thickened ; outer teeth with strongly recurved tips ; usually more than three inner rows of teeth in the upper jaw (mode 4). The species closely resembles *H. prodromus*, from which it may be separated by its slightly thinner lips and smaller adult size. In life, male breeding coloration serves to separate the two species.

Affinities. Similarity in the skull architecture and the dentition of *H. sauvagei* and *H. prodromus* suggest a phyletic relationship between the species. Consequent upon these anatomical similarities, the species show a close parallel in their feeding habits and food preferences, although in this respect *H. sauvagei* may be considered less specialized than *H. prodromus*.

Study material and distribution records

Museum and Reg. No.	Locality.	Collector.
Uganda		
B.M. (N.H.) 1908.5.30.365–366 (as *Paratilapia granti*)	Bunjako	Degen.
B.M. (N.H.) 1906.5.30.371–372 (as *P. granti*)	Bugonga (Entebbe)	,,
B.M. (N.H.) 1906.5.30.413	Sesse Is.	Bayon.
,, ,, 1911.3.3.27	,,	,,
,, ,, 1909.3.29.9 (all as *P. bicolor*)	,,	,,
B.M. (N.H.) 1909.5.11.11	,,	,,
,, ,, 1906.5.30.374–377	Bunjako	Degen.
,, ,, 1956.9.17.1 (See Text-fig. 1)	Entebbe, Airport beach	E.A.F.R.O.
,, ,, 1956.10.9.1–25	,, ,, ,,	,,
,, ,, ,, ,, ,, 26–30	Entebbe, harbour	,,
,, ,, ,, ,, ,, 31–34	Bugungu (Napoleon Gulf)	,,
,, ,, ,, ,, ,, 35–36, 201	Jinja pier	,,
,, ,, ,, ,, ,, 37–40	Shore opposite Kirinya Point (Napoleon Gulf)	,,
,, ,, ,, ,, ,, 41	Kirinya Point	,,
,, ,, ,, ,, ,, 42	Old Bukakata	,,
,, ,, ,, ,, ,, 43	Katebo	,,
Tanganyika Territory		
,, ,, ,, ,, ,, 44	Mwanza	,,
,, ,, ,, ,, ,, 45–72	Majita	,,
,, ,, ,, ,, ,, 73	Ukerewe Is.	,,
,, ,, ,, ,, ,, 74–75	Bukoba	,,
Kenya		
,, ,, ,, ,, ,, 76	Kisumu	,,
,, ,, ,, ,, ,, 77–80	Kamaringa (Kavirondo Gulf)	,,
,, ,, ,, ,, ,, 81	Kach Bay (Kavirondo Gulf)	,,
,, ,, ,, ,, ,, 82	Open water 5 miles N. of Kendu (Kavirondo Gulf)	,,
,, ,, ,, ,, ,, 83	Rusinga Island	,,

Haplochromis prodromus Trewavas, 1935
(Text-figs. 2 and 3)

Paratilapia retrodens (part), Boulenger, 1915, *Cat. Afr. Fish.*, **3**, 235.
Haplochromis ishmaeli (part), Boulenger, 1915, *op. cit.*, 293.
Haplochromis annectens Regan 1922 (nec. *Cyrtocara annectens* Regan, 1921), *Proc. zool. Soc. Lond.*, 167, fig. 2.

Description. Based on sixty-two specimens (including the holotype), 68–130 mm. S.L. None of the morphometric characters studied shows allometry with standard length.

In its general appearance *H. prodromus* closely resembles *H. sauvagei*, from which species it is distinguished by its thicker lips, slightly deeper cheek and larger adult size.

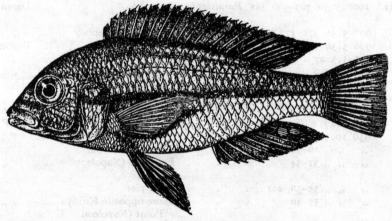

FIG. 2. *Haplochromis prodromus*, ♂, holotype (from Regan, the cichlid fishes of Lake Victoria, *Proc. Zool. Soc.*, 1922, 168, fig. 2).

Depth of body 32·8–40·0 (M = 36.2); length of head 29·4–33·6 (M = 31·5) per cent of standard length. Dorsal head profile somewhat variable, but always curved; strongly decurved in some large individuals, less so in smaller fishes (70–75 mm. S.L.).

Preorbital depth 14·0–19·1 (M = 15·8) per cent head length; least interorbital width 24·0–31·3 (M = 28·1) per cent. Snout as broad as or slightly broader than long, rarely longer than broad, its length 27·5–36·8 (M = 32·7) per cent of head; eye diameter 25·8–33.3 (M = 27·8); cheek 22·0–30·5 (M = 26·7) per cent.

Caudal peduncle 12·6–18·1 per cent of standard length, its length 1·0–1·7 (mode 1·3) times its depth.

Mouth horizontal; posterior maxillary tip reaching or almost reaching the vertical to the anterior orbital margin. Lips thickened; the depth of the upper lip, measured mid-laterally, contained 3–3⅓ times in eye diameter. Jaws equal anteriorly, or infrequently the lower very slightly shorter; lower jaw 30·5–37·8 (M = 34·3) per cent of head length, up to 1·3 (mode 1·1) times as long as broad.

Gill rakers short, 7–9 on the lower limb of the anterior arch.

Scales ctenoid ; lateral line with 30 (f.1), 31 (f.7), 32 (f.16), 33 (f.35) or 34 (f.2) scales ; cheek with 3 or 4 series. 7 or 8 (rarely $6\frac{1}{2}$ or 9) scales between origin of dorsal fin and the lateral line, 7 or 8 (less frequently 9) between pectoral and pelvic fin bases.

Fins. Dorsal with 24 (f.7), 25 (f.40) or 26 (f.15) rays, anal with 11 (f.7), 12 (f.47) or 13 (f.8), comprising XV–XVII, 8–10 and III, 8–10 spinous and branched rays for the fins respectively. Pectoral shorter than the head. Pelvic fins with the first ray produced and of variable posterior extension, but reaching the anal fin in most adult fishes. Caudal sub-truncate.

Lower pharyngeal bone triangular, its dentigerous surface about $1\frac{1}{2}$ times as broad as long ; pharyngeal teeth slender and cuspidate ; those of the median series sometimes enlarged.

Teeth. The dental pattern and tooth form in *H. prodromus* closely resemble those of *H. sauvagei*.

In the outer series of both jaws the teeth have strongly recurved tips and are unequally bicuspid and unicuspid. Bicuspid and weakly bicuspid teeth are the predominating forms in fishes less than 100 mm. S.L. Above this size most teeth are unicuspid. 26–56 (mode 40) outer teeth occur in the upper jaw.

Inner teeth are either tri- or unicuspid, the tricuspid form occurring most frequently in fishes less than 90 mm. S.L. Antero-medially the teeth are arranged in 3–7 (modes 4 and 5) and 3–6 (modes 3 and 4) series in the upper and lower jaws respectively. The posterior medial margin of the upper tooth-band is straight or slightly curved, that of the lower band is distinctly curved (Text-fig. 3).

The dental pattern of the holotype must be considered aberrant ; it is not repeated in any of the sixty-one additional specimens. In the type, some posterolateral inner teeth are displaced medially from their series, thereby giving a spurious impression of a tooth band widened at that point. There is no increase in the width of the underlying premaxillary alveolar surface, nor is there an increase in the number of tooth rows (see fig. 14 in Regan, 1922). In all other respects the dentition of this specimen agrees closely with those described above.

Syncranium and associated musculature. The neurocranium and premaxilla of *H. prodromus* are virtually identical with those of *H. sauvagei*. The dentary, however, is relatively more massive and the mental profile is almost vertical.

Likewise, the jaw musculature compares closely with that of *H. sauvagei*, except that the *adductor mandibulae* I is somewhat shorter (36–39 per cent head length).

Observations made on the feeding methods of *H. prodromus* kept in aquaria, indicate that snails are removed from their shells in a manner similar to that employed by *Macropleurodus bicolor*. That is, the shell is crushed free by the jaws before ingestion takes place. The species was not seen to lever out the soft parts as is usual with *H. sauvagei*.

Coloration in life : Adult males. Ground colour slatey blue-grey ; a peacock-blue sheen on the belly and ventral flanks. Chest and branchiostegal membrane black, operculum with a golden flush. Very faint indications of a dark mid-lateral stripe and seven transverse bars ; also a faint lachrymal stripe. Dorsal dark, with a deep red flush between both spinous and soft rays ; lappets orange. Anal sooty, ocelli deep yellow. Caudal sooty, with a faint orange flush along its posterior margin. Pelvics black. *Females and immature males.* Ground colour silver-grey above the mid-lateral stripe and silver below, with a faint peacock-blue flush on the flanks. Transverse barring is indistinct. Dorsal fin dark. Caudal and anal hyaline. Pelvics faintly yellow.

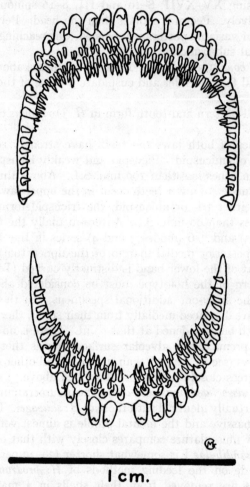

Fig. 3. The premaxillary and mandibular tooth bands in *H. prodromus*.

Colour in preserved material: *Adult males.* Ground colour dark grey; in some, faint traces of transverse and longitudinal banding. Chest and branchiostegal membrane black. Dorsal, caudal and anal dark, the soft dorsal maculate. Pelvics black. *Females and immature males.* Pale, banding variable but usually a distinct mid-lateral stripe and a faint, more dorsal band running along the upper lateral line; five to nine transverse bars across the flank. All fins hyaline.

Distribution. Known only from Lake Victoria.

Ecology : Habitat. Restricted to littoral zones, particularly where the substrate is hard (sand or shingle) and occurring less frequently over mud. Thus, the habitat of *H. prodromus* broadly overlaps that of *H. sauvagei*. Nevertheless, although biasing factors are introduced by the size selectivity of sampling gear and the limitations imposed by the habitat on the use of certain gear, it seems that *H. sauvagei* occur most frequently over shallow exposed beaches—where *H. prodromus* are less common—and that *H. prodromus* are more abundant in off-shore to deeper waters. This assumption is supported by results obtained when such non-selective collecting methods as explosives were used in both habitats.

Study material and distribution records

Museum and Reg. No.	Locality.	Collector.
Uganda		
B.M. (N.H.) 1907.5.7.78 (holotype *H. prodromus*)	Buddu coast	Simon.
,, ,, 1906.5.30.379 (as *P. retrodens*)	Bunjako	Degen.
B.M. (N.H.) 1956.10.9 84–97	Jinja	E.A.F.R.O.
,, ,, ,, ,, ,, 98–99	Shore opposite Kirinya Point (Napoleon Gulf)	,,
,, ,, ,, ,, ,, 100–105	Beach near Nasu Point (Buvuma Channel)	,,
,, ,, ,, ,, ,, 106	Pilkington Bay	,,
,, ,, ,, ,, ,, 107	Hannington Bay	,,
,, ,, ,, ,, ,, 108–125	Entebbe harbour	,,
,, ,, ,, ,, ,, 126	Katebo	,,
,, ,, ,, ,, ,, 127–129	Busungwe Bay (Kagera river mouth)	,,
,, ,, ,, ,, ,, 130	Dagusi Island	,,
Tanganyika Territory		
,, ,, ,, ,, ,, 131–133	Mwanza, Capri Bay	,,
,, ,, ,, ,, ,, 134–135	Godziba Island	,,
,, ,, ,, ,, ,, 197–199	Majita	,,
Kenya		
,, ,, ,, ,, ,, 136–137, 196	Kamaringa (Kavirondo Gulf)	,,
,, ,, ,, ,, ,, 138	Kisumu	,,

Food. Stomach and intestinal contents of seventy-four fishes were examined. Of these, eleven were empty, fifty-seven contained only the remains of Gastropoda, three contained Gastropoda and Insecta, and three yielded unidentifiable sludge. Due to their very fragmentary nature the specific identification of molluscan remains was difficult ; where identification was possible the genus *Bellamya* predominated. As many as twenty-two snail opercula were recorded from the intestine of a single fish, although the modal estimated number of snails per individual was about four.

Breeding. Sexually active and quiescent individuals were associated in all localities, but no data were collected on breeding sites or spawning behaviour. Only one female was found carrying larvae in the buccal cavity. There is apparently no sex-correlated adult size difference in this species; the smallest sexually active individual was a male 102 mm. long.

Diagnosis. The same character complex serves to separate *H. sauvagei* and *H. prodromus* from the other *Haplochromis* of Lake Victoria. *H. prodromus* is distinguished from *H. sauvagei* by its larger adult size, thicker lips, slightly deeper cheek, and, in life, by male breeding coloration.

Affinities. The apparent phyletic relationship between *H. prodromus* and *H. sauvagei* on the one hand, and the more specialized *Macropleurodus bicolor* on the other, has been discussed above and elsewhere (Greenwood 1956a). In the latter paper, it was shown that Regan's suggested relationship between *H. prodromus* and *Platytaeniodus degeni* Blgr. can no longer be considered valid. Regan's views were based on the type and then unique specimen of *H. prodromus* whose dental pattern is aberrant. In any case, the posterior widening of the premaxillary dental surface is apparent and not actual in this fish, whereas in *P. degeni* the premaxilla undergoes a localized but distinct broadening during post-larval ontogeny.

Haplochromis granti Boulenger, 1906
(Text-figs. 4 and 5)

Paratilapia granti (part), Boulenger, 1915, *Cat. Afr. Fish.*, **3**, 342, Fig. 231.
Haplochromis sauvagei (part), Regan, 1922, *Proc. zool. Soc., Lond.*, 167.

In Regan's revision of the Lake Victoria Cichlidae (*ibid.*, 1922), *H. granti* was treated as a synonym of *H. sauvagei*. After comparing the type with other specimens now available, I conclude that the two species should be regarded as distinct. This conclusion is supported by field observations. Both species have in common the "*sauvagei*" group characters of broad inner tooth bands, outer teeth with strongly recurved tips, and thickened lips. But they differ considerably in gross morphology and in certain details of dental pattern. The holotype of *H. granti* (figured in Boulenger, 1915) does not present a specifically typical appearance. However, its dental pattern indicates conspecificity with the specimens here described as *H. granti*. Furthermore, in the type, characters which contribute to gross morphology, for instance the form of the dentary and the head shape, intergrade with those of other specimens possessing a more typical facies.

One rather damaged specimen (B.M. (N.H.) Reg. No. 1911.3.3.28), identified by Boulenger as *Paratilapia retrodens* and later by Regan as *H. sauvagei*, should probably be referred to *H. granti*. Because of this uncertainty *P. retrodens* is not included in the revised synonymy of *H. granti*.

Description.—Based on the type, two paratypes and twenty-six additional specimens in the size range 70–122 mm. S.L. No clear-cut allometry with standard length was observed in the morphometric characters listed below.

Depth of body 32·7–39·3 ($M = 35·4$); length of head 28·8–33·3 ($M = 31·5$) per cent of standard length. Dorsal head profile slightly curved, or, less frequently, straight and gently to steeply sloping.

Preorbital depth 15·3–19·0 (13·3 in the smallest specimen) (M = 17·1) per cent of head length ; least interorbital width 25·0–32·8 (M = 28·6) per cent. Snout as broad as or slightly broader than long, its length 29·0–36·0 (M = 31·6) per cent of head ; eye diameter 25·0–31·0 (M = 27·5) ; depth of cheek 22·0–30·6 (M = 26·8) per cent.

Caudal peduncle 13·6–19·0 per cent of standard length, 1·2–1·7 times as long as deep.

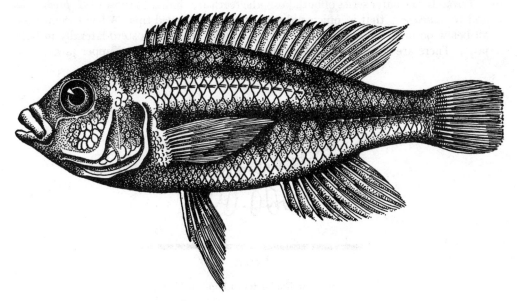

Fig. 4. *Haplochromis granti*, ♀, B.M. (N.H.) 1956.9.17.2. Drawn by Miss L. Buswell.

Mouth usually somewhat oblique ; posterior maxillary tip almost reaching the vertical to the anterior orbital margin, or occasionally reaching this line. Lips thick, sub-equally developed in a few specimens (e.g. the type), but the upper lip clearly thicker than the lower in most. Jaws equal anteriorly, or the lower jaw slightly projecting, its length 22·2–30·6 (M = 26·8) per cent of head length and 1·0–1·5 (mode 1·3) times its width.

The oblique mouth and unequally thickened lips give an appearance of deformity to many specimens. This impression is apparently misleading since there is no indication of any impairment to the efficiency of the jaw mechanism, either as a mechanical unit or in relation to feeding habits.

Gill-rakers short, 7–9 on the lower limb of the first arch.

Scales ctenoid, lateral line with 32 (f.9), 33 (f.11), or 34 (f.9) scales ; cheek with 3 or 4 (in one specimen 2) series ; 7 or 8 scales between origin of dorsal fin and lateral line ; 7 or 8 (rarely 9) between pectoral and pelvic fin bases.

Fins. Dorsal with 25 (f.11), 26 (f.17) or 27 (f.1) rays; anal 11 (f.7) or 12 (f.21), comprising XV–XVII, 9 or 10 and III, 8 or 9 spinous and branched rays for the fins respectively. In one specimen the anal fin had been damaged and subsequently healed irregularly, giving II, 10 rays. Pectoral shorter than the head, except in two specimens where it is of the same length. Pelvic fins extending to the vent in immature fishes and to the anal fin in adults; the first ray is proportionately more produced in sexually active males. Caudal fin truncate or sub-truncate.

Lower pharyngeal bone triangular, its dentigerous surface $1\frac{1}{3}$–$1\frac{1}{4}$ times as broad as long; pharyngeal teeth similar to those of *H. prodromus*.

Teeth. In the outer series of both jaws, the teeth are similar to those of *H. prodromus* and *H. sauvagei*; that is, unicuspid with strongly recurved tips. A few specimens—all below 90 mm. S.L.—have some bicuspid teeth situated postero-laterally in both jaws. There are 28–46 (mode, ill defined : 36) outer teeth in the upper jaw.

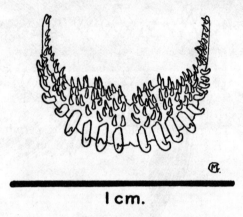

Fig. 5. Mandibular tooth band in *H. granti*.

The inner series are composed of tricuspid teeth in most fishes below 90 mm. and of unicuspid teeth in larger specimens. An admixture of both types is known from three fishes. The teeth are arranged in 2–6 (mode 4) rows in both jaws, but narrow to single series laterally. In many specimens the lower tooth band is wider than the upper; antero-medially, the posterior margin of this band is straight or very gently curved, thus contrasting with the lower series in *H. sauvagei* and *H. prodromus*, where the margin is clearly curved (Text-fig. 5).

Syncranium and associated musculature. The preorbital face of the neurocranium is intermediate in form between that of *H. sauvagei* and the generalized *Haplochromis* type. Greatest departure from the condition observed in *H. sauvagei* and *H. prodromus* is seen in the maxilla, which in *H. granti* is shorter and more bowed in its long axis. Also, the inner face of the posterior limb is markedly concave, which results in the outer face appearing more bullate than in other members of the " *sauvagei* " group. The dentary resembles that of *H. sauvagei* but differs in its less rounded, more angular, anterior outline.

Shortage of material allowed only two dissections of head musculature to be made. The major muscles are distributed as in *H. prodromus* and *H. sauvagei* but the origin of the *adductor mandibulae* I is deeper and more fan-shaped in *H. granti*. In the two specimens dissected, the length of this muscle (42 and 43 per cent of head) is somewhat greater than in *H. prodromus* but equal to that in *H. sauvagei*.

Coloration in life: Adult males. Ground colour light blue-grey; branchiostegal membrane dusky, especially between the rami of the lower jaw. Dorsal fin blue-grey, darkest on the proximal third; lappets orange-red, as are the spots and streaks between the soft rays. Caudal blue-grey, darker on the proximal half; margin outlined in red; orange-red spots between the rays. Anal dusky blue-grey, with an overall pink flush; ocelli yellow. Pelvics black, faint pink mesially. *Females and immature males.* Coloration in life unknown.

Colour in preserved material: Adult males. Ground colour grey or brown; branchiostegal membrane dark grey. An intense black mid-lateral stripe and often traces of a lachrymal stripe and 5–7 vertical bars across the flanks. Dorsal, caudal and anal fins hyaline or dusky; pelvics black. *Females and immature males.* Ground colour silver-white, darkest dorsally. An intense mid-lateral stripe and often faint indications of an interrupted upper band running between the dorsal fin base and the upper lateral line. Seven to nine faint transverse bars are usually present on the flanks and caudal peduncle; no lachrymal stripe. All fins hyaline.

Distribution. Confined to Lake Victoria.

Ecology: Habitat. Too few records are available to permit generalization on the habitat preferences of *H. granti*. The twenty-six specimens whose habitat had been recorded were caught in littoral zones and in water less than forty feet deep. Most localities represented in the collection can be classified either as sandy beaches on exposed shores or as exposed coastlines with a hard substrate. The few remaining localities are sheltered bays where the bottom is composed of organic mud.

Food. Twelve of the twenty-six fishes examined contained ingested material in the stomach or intestine. In each case only the soft parts of Gastropoda were found, except for some Lamellibranchiata shell fragments in one individual. From these admittedly few observations it is inferred that *H. granti* feed in a manner similar to that observed for *H. prodromus* and *H. sauvagei*.

Breeding. There is no information on any aspect of the breeding behaviour in this species; all specimens below 90 mm. S.L. were immature.

Diagnosis. Haplochromis granti differs from other Lake Victoria *Haplochromis* in possessing broad bands of inner teeth (2–6, mode 4, series) in both jaws and by its unequally thickened lips, the upper usually thicker than the lower. This latter character, together with the oblique mouth and straight posterior margin to the inner tooth band of the lower jaw, serves to distinguish *H. granti* from *H. prodromus* and *H. sauvagei*.

Affinities. By virtue of its dentition, *H. granti* must be included in the *H. sauvagei-H. prodromus* species-group. Other characters probably associated with dentition, such as the shape of the premaxilla and dentary, are closely similar in all three species. But, despite resemblances in these dental and osteological characters, and in the associated musculature, the neurocranial morphology of *H. granti* has not departed so radically from the generalized *Haplochromis* type. Morphologically speaking, the relationship between *H. sauvagei* and *H. prodromus* is directly linear, whilst that of *H. granti* is somewhat divergent but with a parallel trophic trend.

Study material and distribution records

Museum and Reg. No.	Locality.	Collector.
Uganda		
B.M. (N.H.) 1903.5.30.367 (holotype *P. granti*)	Bunjako	Degen.
B.M. (N.H.) 1903.5.30.368–369 (paratypes *P. granti*)	,,	,,
B.M. (N.H.) 1956.10.9.139	Bay opposite Kirinya Point (Napoleon Gulf)	E.A.F.R.O.
,, ,, ,, ,, ,, 140–141	Bugungu (Napoleon Gulf)	,,
,, ,, ,, ,, ,, 142–144	Beach nr. Nasu Point (Buvuma Channel)	,,
E. African Fisheries Res. Lab. Jinja	Pilkington Bay	,,
B.M.(N.H.) 1956.10.9.145–147	Thruston Bay	,,
,, ,, 1956.9.17.2	Ekunu Bay	,,
,, ,, 1956.10.9.148–152	Entebbe, harbour	,,
,, ,, ,, ,, ,, 153	Near Busungwe Is.	,,
,, ,, ,, ,, ,, 154	Busungwe Bay (Kagera river mouth)	,,
,, ,, ,, ,, ,, 155–157	Beach near Grant Bay	,,
,, ,, ,, ,, ,, 200	Buka Bay	,,
Kenya		
,, ,, ,, ,, ,, 158	Kisumu	,,
Tanganyika		
,, ,, ,, ,, ,, 159	Ukerewe Is.	,,
,, ,, ,, ,, ,, 160–163	Majita	,,

Haplochromis xenognathus sp. nov.
(Text-figs. 6 and 7)

The high intra-specific variability of *H. xenognathus* makes this species of particular interest when considering the evolution of monotypic cichlid genera. Some of the more aberrant specimens, if studied in isolation, might well be given a status equal with the monotypic genera recognized at present. Less extreme individuals, on the other hand are not immediately distinguishable from *H. sauvagei*.

The modal type tooth-pattern and the usual arrangement of the jaws are, however, unlike those of other species in the "*sauvagei*" group (Text fig. 7). I am led to include *H. xenognathus* in this group because of its "*sauvagei*"-like tooth form and the multiseriate dental pattern.

The sample provides sufficient intra-specific variation to indicate morphological stages through which the typical specific facies may have passed in its evolution from a form similar to the extant *H. sauvagei*.

Type specimen. A male, 91 + 19 mm.; from Entebbe harbour.

Description. Based on thirty-five specimens 80–113 mm. S.L.

Depth of body 31·2–38·0 (M = 34·8); length of head 29·2–35·4 (M = 33·1) per cent of standard length. Dorsal head profile usually straight and somewhat steeply sloping; curved in a few specimens.

Preorbital depth 16·0–20·7 (M = 17·7) per cent head length; least interorbital width 23·5–29·0 (M = 26·8) per cent. Snout from $1\frac{1}{5}$–$1\frac{1}{3}$ longer than broad, its length 31·8–37·8 (M = 35.2) per cent of head; eye 23·2–28·7 (M = 26·0); depth of cheek 23·2–28·7 (M = 26·0) per cent.

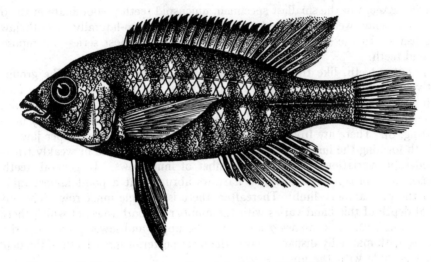

Fig. 6. *Haplochromis xenognathus*, ♂, holotype, B.M. (N.H.) 1956.9.17.3. Drawn by Miss L. Buswell.

Caudal peduncle 13·8–19·0 (M = 15·9) per cent of standard length; 1·1–1·7 (mode 1·4) times as long as deep.

Mouth horizontal, the posterior maxillary tip reaching, or almost reaching, the vertical to the anterior orbital margin; lips slightly thickened. The lower jaw is clearly shorter than the upper in 74 per cent of the specimens examined and subequal to the upper in 26 per cent. Even in this latter group the outermost teeth of the lower jaw occlude behind the equivalent upper jaw series. Lower jaw 32·0–38·0 (M = 34·5) per cent of head, and 1·1–1·8 (mode 1·4) times as long as broad.

Gill rakers short, 7–9 (rarely 10) on the lower limb of the first arch.

Scales ctenoid; lateral line with 31 (f.4), 32 (f.14), 33 (f.11), 34 (f.4) or 35 (f.2) scales; cheek with 3 or 4 (rarely 2 or 5) series; 6 or 7 (less frequently 8) scales between origin of dorsal fin and lateral line; 7 or 8 (rarely 6 or 9) between pectoral and pelvic fin bases.

Fins. Dorsal with 24 (f.13), 25 (f.19), or 26 (f.3) rays; anal with 11 (f.9), 12 (f.24) or 13 (f.1). One specimen has only two spines, giving a total count of 10 rays. The spinous and branched ray counts for the fins are XV–XVII, 8–10 and III, 8–10. Pectoral fins shorter than the head. Pelvic fins with first ray produced, extending to the anterior part of the anal fin in females and more posteriorly in adult males. Caudal truncate.

Lower pharyngeal bone triangular, its dentigerous surface $1\frac{1}{3}$–$1\frac{1}{4}$ times as broad as long. The pharyngeal teeth are slender and bicuspid, those of the median series not noticeably enlarged. In one specimen, the lower pharyngeal bone is stout, the toothed surface slightly longer than broad and the median teeth enlarged and molariform; this fish also shows a somewhat atypical oral dentition, in that the teeth are bluntly cuspidate.

Teeth. Except in the smallest specimen, unicuspid teeth predominate in the outer series, but some weakly bicuspid teeth do occur postero-laterally in both jaws of large fishes. In the smallest specimen, the entire outermost series is composed of bicuspid teeth.

The outer teeth, like those of *H. sauvagei* and other species of the group, are relatively stout and have strongly recurved tips. In the lower jaw, the anterior teeth are implanted at an acute angle, so that their necks lie almost horizontally; but recurvature of the crown is such that the tip points almost vertically upwards (Text-fig. 8). There are from 32–52 (mode 44) outer teeth in the upper jaw.

Teeth forming the inner series are small and either unicuspid or weakly tricuspid. Considerable variation exists in the number of inner rows. In general, teeth are disposed in a broad crescent which narrows abruptly at a point almost mid-way along the premaxillary limb. Thereafter, there is a single inner row. The antero-medial depth of this band varies with the number of tooth rows, of which there are from 3–9 in both jaws (modes 7 and 5 for the upper and lower jaws respectively). In fishes with markedly disparate jaws, the most posterior inner teeth of the dentary do not occlude with the upper series.

The toothed surface of the dentary is often slightly convex, so that when viewed laterally several points on the inner band are higher than the crowns of the outer teeth (Text-fig. 8).

Fishes with narrow tooth bands in both jaws have a dental pattern closely resembling that of *H. sauvagei*; the resemblance to this species is enhanced by the sub-equal jaws of these specimens. In contradistinction, other *H. xenognathus* with sub-equal jaws have a broad and specifically typical tooth pattern.

Syncranium and associated musculature. The neurocranium of *H. xenognathus* is identical with that of *H. granti*. The shape of the premaxilla varies slightly in relation to the number of inner rooth rows, but is otherwise comparable with the premaxilla of *H. sauvagei*. Likewise the dentary is similar in the two species, except that the dentigerous surface is inclined forwards and downwards in *H. xenognathus*. Perhaps the most characteristic appearance of this bone is imparted by the almost horizontally implanted anterior teeth.

A syncranial skeleton prepared from a specimen with *H. sauvagei*-like facies did

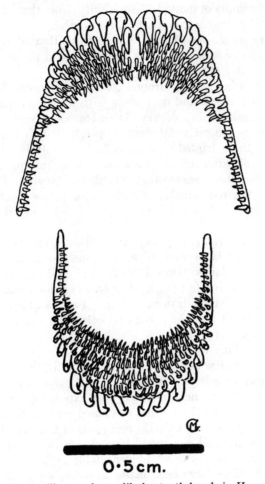

Fig. 7. Premaxillary and mandibular tooth bands in *H. xenognathus*.

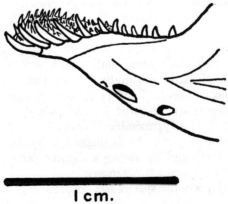

Fig. 8. Lateral view of the anterior part of the dentary of *H. xenognathus*.

not differ, beyond the limits of individual variability, from that of a typical specimen and was clearly distinguishable from that of *H. sauvagei*.

Head musculature in *H. xenognathus* is similar to that of *H. sauvagei* and *H. prodromus*. The *adductor mandibulae* I is shorter than in the former species, but is equal to that of the latter (33·2–39·4 per cent of head).

Coloration in life: Adult males. Ground colour dark bronze dorsally, shading to grey-bronze ventrally: cheek and operculum with a distinct bronze sheen; chest and branchiostegal membrane bluish-grey. Dorsal fin sooty, with red streaks between the soft rays; lappets red. Caudal dark, upper half with red spots, lower half flushed with red. Anal clouded; ocelli yellow. Ventrals sooty, the first ray bluish-white. Live coloration of *immature males* is unknown. *Females.* Ground colour as in males, but the branchiostegal membrane greyish. Dorsal fin with red lappets but lacking the red streaks. Caudal dark yellow ventrally, lighter and maculate above. Anal and ventral fins dark olive-yellow, the first pelvic ray bluish-white.

Colour in preserved material: Breeding males. Dark grey, the flank with a faint coppery sheen; a distinct lachrymal stripe. In some specimens five transverse bars may be discerned on the flanks. Dorsal, caudal and anal fins grey, the upper third of the caudal maculate. Pelvic fins black. *Non-breeding and immature males.* Ground colour silver-grey; 7–9 distinct transverse bars. Fins as above. *Females.* As for non-breeding males, except that the pelvics are colourless. In some specimens there is a fairly distinct mid-lateral stripe.

Distribution. Known only from Lake Victoria.

Ecology: Habitat. The few and scattered records indicate that *H. xenognathus* is confined to littoral zones where the substrate is hard. Most specimens in the collection were obtained from seine nets operated over exposed sandy beaches.

Food. Seventeen of twenty-three fishes examined contained ingested material in the stomach or intestine: in ten, only the soft parts of Gastropoda were found; in three, Insecta (chironomid and ephemerid larvae); in two, bottom detritus (sand-grains and plant tissue); and in three, unidentifiable sludge.

Although no observations have been made on living fishes, the presence of gastropod remains without shells suggests that the feeding methods of *H. xenognathus* are like those of *H. sauvagei*, *H. prodromus* and *H. granti*.

Breeding. No information is available. The smallest sexually mature fish was a female 87 mm. S.L.

Diagnosis. Haplochromis xenognathus may be distinguished from other species of the genus by the following characters: outer teeth with strongly recurved tips, those of the lower jaw implanted horizontally or almost so; inner teeth in the upper jaw arranged in a broad, crescentic band (3–7, mode 5 series); lower jaw usually much shorter than the upper and also with broad bands of teeth (3–9, mode 5, inner series). Some individuals closely resemble *H. sauvagei* both in gross morphology and in details of dentition, but may be distinguished by the peculiar implantation of their anterior lower teeth and by having a slightly narrower and longer snout; in life the coloration of both sexes is diagnostic.

Affinities. In many respects the species has departed considerably from the basic

" *sauvagei* " type as represented by the nominate species. Yet, it is apparently with *H. sauvagei* that *H. xenognathus* shows greatest morphological affinity. The resemblance is most clearly seen in the least typical members of *H. xenognathus*, but is obscured in other forms. These latter individuals seem to indicate that if future evolution in *H. xenognathus* is continued along such lines and is coupled with a reduction in morphological variation, then the species could acquire a status equivalent to the monotypic genera recognized at present.

Study material and distribution records

Museum and Reg. No.	Locality.	Collector.
B.M. (N.H.) 1956.10.9.164–167	Jinja	E.A.F.R.O.
„ „ „ „ „ 168	Bay opposite Kirinya Point (Napoleon Gulf)	„
„ „ „ „ „ 169	Beach nr. Nasu Point (Buvuma Channel)	„
„ „ „ „ „ 170	Hannington Bay	„
„ „ 1956.9.17.3 (type)	Entebbe, harbour	„
„ „ 1956.10.9.171–173	„ „	„
„ „ „ „ „ 174	Entebbe, Airport beach	„
„ „ „ „ „ 175	Bugonga, Entebbe peninsula	„
„ „ „ „ „ 176–177	Katebo	„
„ „ „ „ „ 178–182	Beach south of Busungwe (Kagera river mouth)	„
Tanganyika		
„ „ „ „ „ 183	Bukoba	„
„ „ „ „ „ 184–191	Majita	„
„ „ „ „ „ 192–195	Mwanza, Capri Bay	„

DISCUSSION

In an adaptively multi-radiate species-flock, the differentiation of true phyletic relationship from parallel trends is difficult, particularly when the flock, like that of Lake Victoria, is oligophyletic in origin. It has been noticed, however, that in many adaptive sub-groups some species show greater morphological affinity with one another than with other members. This I interpret as indicating that both lineal descent and parallel evolution have contributed to the origins of the groups.

The remarkable uniformity of the distinctively shaped outer teeth in all four members of the " *sauvagei* " complex probably indicates a monophyletic origin for the group. Their multiseriate dentition, on the other hand, cannot be considered of value in indicating phylogeny. Broad tooth-bands have evolved in several other and unrelated species, for example, the algal-grazers *H. nigricans* and *H. nuchisquamulatus*, and in certain monotypic genera.

As an ecologically defined group, the " *sauvagei* " complex exhibits considerable variation in species morphology. There are three distinctive forms, represented by

H. sauvagei and *H. prodromus*, *H. granti*, and *H. xenognathus*. The two first-mentioned species must be considered nearer the generalized type and the two latter as showing progressive but independent divergence.

When discussing the evolution of *Macropleurodus bicolor*, I drew attention to the possible relationship between this species and *Haplochromis prodromus* (Greenwood, 1956a). No conclusion can be reached at present, but the prospective adaptational significance of a " *prodromus* " type cranial anatomy in the evolution of *M. bicolor* cannot be disregarded. That both species should have almost identical feeding habits would seem to lend additional weight to this argument.

Morphological and ecological differences between *H. prodromus* and *H. sauvagei* are of the slightest order. If their lineal relationship can be accepted, one is tempted to consider the species in an ancestor-descendant category. In all probability, present-day *H. sauvagei* differ genotypically and even phenotypically from the presumed ancestral type, yet, despite these limitations, it is difficult to imagine a species more similar to *H. prodromus*. Several other such *Haplochromis* species-pairs are known from Lake Victoria. Each, except for their temporal coexistence, would fulfil the palaeontological requirements for ancester-descendant relationship. Indeed, coexistence, even within one habitat, of species which could be lineally derived appears to be a major feature of the Lake Victoria *Haplochromis* species-flock. The Lake's geological history provides a possible explanation of this phenomenon (Brooks, 1950 ; Greenwood, 1951). During the inter-Pluvial periods of the Pleistocene, Lake Victoria was probably reduced to a series of small lakes and swamps. Under such conditions a species would be isolated into several discontinuous groups. If some surviving populations underwent genic reorganization as a result of isolation, or of increased selection pressure, it is possible that they might retain their discreteness if brought into contact with the parental stocks when the lakes were joined during the succeeding Pluvial period. That the derived and parental species were able to coexist even when their ecological requirements were similar, seems to indicate drastically reduced selection pressure.

The third member of the " *sauvagei* " group, *H. granti*, is unlike either *H. sauvagei* or *H. prodromus*. However, the differences lie in characters which could be derived by heterogonic growth of certain cranial parts from a species less differentiated than *H. sauvagei*, but possessing the group dental characters.

From an evolutionary viewpoint, *H. xenognathus* is undoubtedly the most interesting species. When the first few specimens came to my notice, I considered them to be members of a distinct and apparently monotypic genus. The degree of morphological differentiation of this supposedly new genus was at least equal to that of *Hoplotilapia retrodens* Hilg. When more specimens were collected, however, it was obvious that the species was extremely variable and should be retained within the genus *Haplochromis*. The less typical specimens differed only slightly from *H. sauvagei*, whilst the typical fishes were clearly distinct from that species. Thus, it seems legitimate to look upon *H. xenognathus* as an example of a stage through which species might pass in the evolution of a genus.

Ecologically, the " *sauvagei* " group has entered an adaptive zone unique for *Haplochromis*, but occupied by two monotypic genera, *Macropleurodus bicolor* and

Hoplotilapia retrodens. All other known predominantly mollusc-eating *Haplochromis* species crush their prey by means of hypertrophied pharyngeal bones and dentition.

If, in conclusion, one considers the morphological and consequent ecological adaptations of the *H. sauvagei* group, the impression is gained of a species-complex partially advanced on the path of trophic specialization. From its present peak it could supply, and may even have supplied, raw material for further specialization.

SUMMARY

1. *Haplochromis sauvagei* (Pfeffer) 1896, and *H. prodromus* Trewavas 1935, are re-described.
2. The species *H. granti* Blgr. 1906, previously synonymized with *H. sauvagei*, is reinstated.
3. A new species, *Haplochromis xenognathus*, is described.
4. Data are given on the ecology of all four species.
5. Sex-limited polychromatism, involving a piebald female coloration, is described for *H. sauvagei*.
6. The evolutionary status of the species is discussed.

REFERENCES

BROOKS, J. L. 1950. Speciation in ancient lakes. *Quart. Rev. Biol.* **25** : 131.
GREENWOOD, P. H. 1951. Evolution of the African cichlid fishes ; the *Haplochromis* species-flock in Lake Victoria. *Nature*, London, **167** : 19.
—— 1956*a*. The monotypic genera of cichlid fishes in Lake Victoria. *Bull. Br. Mus. (Nat. Hist.) Zool.* **3** : 297.
—— 1956*b*. A revision of the Lake Victoria *Haplochromis* species (Pisces, Cichlidae) Part I. *Ibid.* **4** : 225.
REGAN, C. T. 1922. The cichlid fishes of Lake Victoria. *Proc. zool. Soc. Lond.* 157.

ACKNOWLEDGMENTS

I wish to express my gratitude to the Trustees of the British Museum (Natural History) for the numerous facilities afforded me in their museum ; to Dr. Ethelwynn Trewavas for her constant help and for reading the manuscript of this paper ; and to my colleague Dr. Philip S. Corbet who identified the insects recorded from stomach contents.

PLATE 4

The types of *Haplochromis sauvagei* (Pfeffer) upper and *H. nuchisquamulatus* (Hilgendorf) lower (see Greenwood, 1956, *Bull B.M. (N.H.) Zool.* **4**, 241), from photographs made in the Berlin Museum in 1921. Both specimens may be lost.

THE MONOTYPIC GENERA OF CICHLID FISHES IN LAKE VICTORIA, PART II[1]

By P. H. GREENWOOD[2]

British Museum (Natural History), London

CONTENTS

	Page
GENERIC DIAGNOSIS AND DISCUSSION	165
Astatoreochromis alluaudi Pellegrin	167
Description	167
Osteology	169
Affinities	173
Description and diagnosis of *A. a. alluaudi* Pellegrin and *A. a. occidentalis* subsp. nov.	174
Ecology	174
SUMMARY	175
ACKNOWLEDGMENTS	177
REFERENCES	177

A REVISION of the four endemic monotypic cichlid genera of Lake Victoria, *Macropleurodus bicolor* (Blgr.), *Platytaeniodus degeni* Blgr., *Hoplotilapia retrodens* Hilg., and *Paralabidochromis victoriae* Greenwood has already been published (Greenwood, 1956). These species differ from *Haplochromis* in various dental characters. Unlike the other monotypic genera, *Astatoreochromis alluaudi* is not confined to the Lake Victoria basin; its range includes Lakes Edward, George, Nakavali and Kachira (Trewavas, 1933). Furthermore, *Astatoreochromis* differs from *Haplochromis* only in having an increased number of spines in the anal fin; the oral dentition is typically that of a non-piscivorous *Haplochromis*.

Genus *ASTATOREOCHROMIS* Pellegrin, 1903

Astatoreochromis Pellegrin, 1903, *Mém. Soc. zool. France*, **16**, 385; *Idem*, 1905, *ibid*. **17**, 185, pl. XVI, fig. 2; *Idem*, 1910, *ibid*. **22**, 297; Regan, 1922, *Proc. zool. Soc., London*, 188; Fowler, 1936, *Proc. Acad. nat. Sci. Philad*. **88**, 333, fig. 138 (mis-spelt *Astatore*); Poll, 1939, *Explor. Parc. Nat. Albert, mission* H. Damas (1935–36), fasc. 6, 1–73. *Haplochromis* (part) Boulenger, 1907, *Fish, Nile*, 505 pl. XC, fig. 4; *Idem*, 1911, *Ann. Mus. Genova* (3), **5**, 71; *Idem*, 1915, *Cat. Afr. Fish*. **3**, 305, fig. 206.

Type species. Astatoreochromis alluaudi Pellegrin, 1903.

Diagnosis. Astatoreochromis differs from *Haplochromis* only in having four or more spines in the anal fin. From other genera in the *Haplochromis* group with more than four anal fin spines, *Astatoreochromis* is distinguished by the absence of a marked antero-posterior differentiation in the form of the premaxillary teeth.

[1] Part I was published in *Bull. Br. Mus. nat. Hist., Zool*. **3**, No 7, 1956.
[2] Formerly East African Fisheries Research Organization, Jinja, Uganda.

In comparison with the *Haplochromis* of Lakes Victoria, Edward, and Kachira, *Astatoreochromis* shows an increased ratio of spinous to branched rays in the dorsal and anal fins. From other *Haplochromis*-like genera in these lakes, *Astatoreochromis* differs both in having more anal fin spines and in the nature of its oral dentition.

Discussion. As Boulenger (1907) pointed out, the principal diagnostic character for *Astatoreochromis* cannot be considered trenchant because some four-spined specimens of normally three-spined *Haplochromis* species have been recorded. He cites as an example an aberrant *H. desfontainesi* from Tunisia. Nevertheless, throughout the very numerous species of *Haplochromis* it is very exceptional to find an individual with more (or less) than three anal fin spines and as yet no specimens of *Astatoreochromis* with less than four anal spines have been found. It cannot be denied that *Astatoreochromis* and *Haplochromis* are closely related (as are *Haplochromis* and the other monotypic genera of Lakes Edward and Victoria) and it might seem that little is to be gained from recognizing *Astatoreochromis* as a distinct genus.

However, *Astatoreochromis* differs from the *Haplochromis* of Lakes Victoria and Edward in four other characters which, if taken together, may indicate that it has a different lineage from these species. In an earlier paper (Greenwood, 1954) I drew attention to the form of the pharyngeal apophysis in *Astatoreochromis* and showed that it resembles the apophysis occurring in *Haplochromis vanderhorsti* Greenwood (Malagarasi River) and *H. mahagiensis* David & Poll (Lake Albert). The other Victoria species with enlarged pharyngeal bones (*H. ishmaeli* Blgr. and *H. pharyngomylus* Trewavas) have a different apophyseal form. A summary of these observations is given in Text-fig. 2.

Contrary to my earlier views, I now consider that, taken by itself, apophyseal form is of doubtful value as an indicator of phyletic relationship. For example, both the *H. mahagiensis-H. vanderhorsti* and the *H. ishmaeli-H. pharyngomylus* types of apophysis are found in Lake Nyasa *Haplochromis* with enlarged pharyngeal bones; *Haplochromis placodon* Regan (a species with hypertrophied pharyngeals) has the " *ishmaeli* " type whilst *H. sphaerodon* Regan, *H. latristriga* (Günther) and *H. selenurus* (Regan) (species with less massive pharyngeals) have the " *mahagiensis* " type. There is no evidence to suggest that Nyasa fishes with " *mahagiensis* "-like apophyses are more closely related to one another than to *H. placodon*, or that they represent an exotic element within the Nyasa flock. Certainly there is no indication of their being related to the *H. mahagiensis-H. vanderhorsti* species group. Thus, one must conclude that similarity of apophyseal form is yet another example of convergent evolution, at least at an inter-group level.

Considering *Astatoreochromis* in relation to the cichlid species flocks of Lakes Victoria and Edward it is clear that this genus does not conform to the general morphological pattern of the endemic species and genera. Three characters, the shape of the caudal fin, the coloration and the high number of anal ocelli, set *Astatoreochromis* apart. Excepting *H. melanopterus* (a species of doubtful validity, see Greenwood, p. 192) all the endemic *Haplochromis* of Lake Victoria have a truncate or subtruncate caudal fin; the caudal of *Astatoreochromis* is distinctly rounded.

A considerable variety of colour and colour patterns is exhibited by the endemic

Haplochromis, but all can be broken down into various combinations of several basic types. The golden-green ground colour of *Astatoreochromis* does not occur in any endemic species. The third outstanding characteristic of *Astatoreochromis* is the high number of ocelli on the anal fin of male fishes. Not only are the ocelli more numerous than in *Haplochromis*, but they are arranged in three or four horizontal rows; it is extremely rare to find more than two rows in any *Haplochromis* from Lake Victoria or Edward.

In all these characters, *Astatoreochromis* resembles *H. vanderhorsti*. There is also one other point of close inter-specific resemblance; both species show only slight dimorphism in the coloration of the two sexes. In contrast the coloration of Lake Victoria *Haplochromis* is markedly dimorphic.

Thus, although the form of the pharyngeal apophysis alone is of doubtful value in showing phyletic relationships, I consider that the additional evidence supports my original conclusion that *Astatoreochromis* was derived from an *H. vanderhorsti*-like stem. The two other Victoria species with enlarged pharyngeal bones and dentition (*H. ishmaeli* and *H. pharyngomylus*) are apparently related to one another. Their origin was probably by way of two forms represented in the present lake by a generalized species formerly confused with *H. michaeli* [see Greenwood, 1954 and 1956a], but now known to be an undescribed species and a species partly advanced towards extreme hypertrophy of the pharyngeal mill (*H. obtusidens*).

The apparently distinct origin of *Astatoreochromis alluaudi* in relation to the rest of the Victoria-Edward *Haplochromis* species flock is a further and perhaps more fundamental reason for maintaining the species as a distinct genus.

Astatoreochromis alluaudi Pellegrin, 1903
(Text-fig. 1)

For synonymy see under genus.

Lectotype. A female 122 mm. S.L. from the Kavirondo Gulf, Lake Victoria; Reg. No. 04, 137 of the Paris Museum.

Description. From the available material it seems that only two characters (length of the caudal fin and the extent to which the lower pharyngeal bones are hypertrophied) show clear-cut differences between populations inhabiting the various lakes. These two characters will be treated separately but all others are given for the species as a whole.

The general species description is based on the following material: Lakes Victoria and Kyoga (including the Victoria Nile), 77 specimens, 20–163 mm. S.L. (of which 40, including the four syntypes, were used in obtaining proportional measurements): Lakes Edward and George 11 specimens 24·0–80·0 mm. S.L.; Lake Nakavali, 18 specimens, 50–137 mm. S.L. (of which 11 were used for proportional measurements); Lake Kachira, three specimens 66–78 mm. S.L.

Depth of body 33·8–43·3 per cent of standard length, length of head 32·1–40·0, mean (M) = 35 per cent. Dorsal head profile fairly steeply sloping, straight or somewhat decurved, becoming concave in larger individuals.

Preorbital depth, showing slight positive allometry with standard length, 11·1–17·5 (M = 15·0) per cent of head length, least interorbital width 25·2–31·7 (M = 28·3) per cent. Snout as broad as long, its length 25·0–33·3 (M = 29·2) per cent of head. Eye diameter shows negative allometry with standard length, being 31·5–23·2 (M = 27·2) per cent of head in fishes 20–80 mm. S.L. and 24·3–18·8 (M = 22·1) per cent in larger individuals. Depth of cheek positively allometric with standard length; 12·8–26·0 (M = 21·3) and 20·0–27·9 (M = 24·2) per cent of head in the two size groups mentioned above.

Caudal peduncle 11·0–15·2 per cent of standard length, its length/depth ratio 1·0–1·4 (modal range 1·0–1·1) or, rarely, deeper than long.

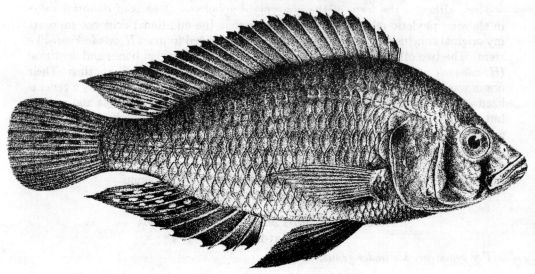

FIG. 1. *Astatoreochromis alluaudi alluaudi* (from Boulenger, *Fishes of the Nile*).

Mouth horizontal or slightly oblique. Jaws equal anteriorly or, occasionally, lower somewhat projecting; posterior tip of the maxilla reaching or almost reaching the vertical to the anterior orbital margin. Lower jaw 35·0–45·3 (M = 40·0) per cent of head length and 1·3–2·0 (rarely) times as long as broad (modal range 1·5–1·6).

Gill rakers short and stout; 8 or 9 (occasionally 10, rarely 7) on the lower limb of the first gill-arch.

Scales ctenoid; lateral line with 30 (f.12), 31 (f.21), 32 (f.20) or 33 (f.2) scales; cheek with 3 or 4 (occasionally 5) series; 4 or 5 (occasionally 6) scales between the origin of the dorsal fin and the lateral line; 4–6 (rarely 7) between the pectoral and pelvic fin bases.

Fins. Dorsal with 23 (f.2), 24 (f.4), 25 (f.15), 26 (f.68), 27 (f.11) or 28 (f.1) rays, comprising 16 (f.5), 17 (f.16), 18 (f.59), 19 (f.20) or 20 (f.1) spinous and 7 or 8 (rarely 9) branched rays. Anal fin with 11 (f.3), 12 (f.67), 13 (f.30) or 14 (f.2) rays comprising

4 (f.28), 5 (f.63) or 6 (f.11) spinous and 7 or 8 (rarely 6 or 9) branched rays. Pectoral fin shorter than the head, 22·3–29·4 per cent of standard length.

Caudal fin rounded, longer in fishes from Lakes Nakavali, Edward and George than in those from Lake Victoria; namely: length of caudal fin in Victoria specimens (N = 41) 21·4–28·5 (Mean 24·3) per cent of standard length; in Lake Nakavali fishes (N = 4) 24·0–31·6 (M = 27·4) per cent, and in Lake Edward fishes (N = 9), 24·0–31·6 (M = 27·0). This fin was damaged in two of the three specimens from Lake Kachira.

Pelvic fin with the first ray produced and extending to beyond the vent or as far as the spinous part of the anal fin.

Teeth. Even in the smallest specimen examined, the most posterior teeth in the upper jaw were unicuspid. In fishes less than 100 mm. S.L., the anterior and lateral teeth of the upper jaw and the entire outer series of teeth in the lower jaw are unequally bicuspid and relatively stout. In larger specimens, the dentition is a mixture of weakly bicuspid and unicuspid teeth; fishes over 140 mm. S.L. (and some smaller individuals) have only stout, unicuspid teeth in the outer series of both jaws. There are 28–56 (modal range 40–46) outer teeth in the upper jaw.

The small, tricuspid or unicuspid inner teeth are arranged in one or two rows.

Osteology. Vertebrae: 15 + 14 in the single specimen examined B.M. (N.H.) Reg. No. 1911.3.3.111, from Kakindu, Victoria Nile.

Neurocranial apophysis for the upper pharyngeal bones. The form of this apophysis was mentioned in the discussion on generic characters. Since the apophysis is of importance in defining cichlid genera, its variation and the probable factors influencing its variability in *Astatoreochromis* will be outlined briefly.

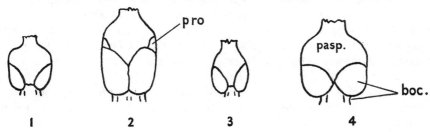

Fig. 2. Semi-diagrammatic representation of the shape and proportions of elements contributing to the upper pharyngeal apophysis in: (1) young *Astatoreochromis a. alluaudi*; (2) adult *A. a. alluaudi*; (3) adult *Haplochromis vanderhorsti*; (4) adult *Haplochromis ishmaeli*. Scale constant.

Although the shape and proportions of elements contributing to the apophysis are affected by the relative size of the pharyngeal bones, the characteristic group facies (see p. 170) is developed even in the absence of markedly hypertrophied pharyngeals (Text-fig. 2, (i)). In *A. alluaudi* it appears that the extent to which the basioccipital facets are enlarged and expanded depends primarily on the relative hypertrophy of the pharyngeals, and secondarily on the size of the fish. Thus, in two specimens from Lake Victoria, one, 73 mm. S.L. with weakly developed pharyngeals, has proportionately smaller basioccipital facets than the other, 63 mm. S.L. and with enlarged pharyngeal bones and teeth (cf. Text-fig. 2 (i) and 2 (ii)). Likewise, fishes 70 mm., 76 mm., and 80 mm. S.L., from Lake Edward, and two specimens 71 mm. and 82 mm. S.L. from Lake Nakavali all have weakly developed pharyngeals, and apophyses comparable with the 73 mm. fish mentioned above. In this size-range it would appear that the size of the pharyngeal bones is exerting full influence on apophyseal form.

The effect of overall size is demonstrated in a fish 125 mm. S.L. from Lake Nakavali. In this specimen the pharyngeal bones are weak in comparison with those of a comparable sized fish from Lake Victoria (cf. Text-fig. 3, lower row, left and right). Yet, the apophyseal form is similar in the two specimens except for a slightly smaller surface area in the Nakavali fish.

Lower pharyngeal bone triangular. The form of this bone (which depends on the degree to which it is hypertrophied) and the nature of its teeth show a marked difference between fishes from Lake Victoria (including Kyoga) and those from the other lakes (see Text-fig. 3). When specimens of equal sizes from different lakes are compared it is immediately obvious that those from Lake Victoria have more massive bones with a greater proportion of molariform teeth. As far as can be determined from available material there is a little geographical variation of this character in fishes from Lakes Edward, George, Nakavali and Kachira. In all these populations the bone is clearly less massive than in Lake Victoria fishes and there are fewer molariform teeth. When present, such teeth are generally confined to the two median rows ; any enlarged teeth in the lateral series are usually cuspidate.

The difference in pharyngeal bone size can be expressed quantitatively by using the ratio of head length to pharyngeal bone width (measured from tip to tip of the upper arms) ; it is, however, less impressive an indication of disparity in massiveness than an actual comparison of individual bones. The ratio for specimens from the various lakes is : *Victoria* (including Kyoga) ; 2·4–3·1 (Mean 2·7 ; 32 specimens examined) ; *Nakavali* : 2·6–3·6 (Mean 3·1 ; 16 specimens) ; *Edward and George* : 2·8–3·6 (Mean 3·0 ; 10 specimens) ; *Kachira* : 2·7–3·1 (Mean 3·0 ; three specimens).

As specimens of *A. alluaudi* from Lake Victoria cover a sufficiently wide size-range it is possible to determine ontogenetic changes in tooth form and in the proportions of the bone. In the smallest specimen (20 mm. S.L.) the two median tooth-rows are composed of enlarged but cuspidate teeth and the bone is relatively coarse (Text-fig. 3 top row, left). With increasing size, the bone becomes proportionately stouter and the median teeth larger and blunter (Text-fig. 3 middle row, left), as do some of the teeth in the lateral rows. In the great majority of fishes over 60 mm. S.L., only the most lateral series of teeth, and those in the upper corners of the bone, remain slender and cuspidate. The number of such non-molariform teeth is even further reduced in fishes greater than 120 mm. S.L. Only seven of the 78 fishes examined had pharyngeal bones and dentition less hypertrophied than the modal condition for their respective size-groups.

Ontogenetic changes are less marked in *A. alluaudi* from the western lakes. The impression gained from these specimens is that the pharyngeal bones, apart

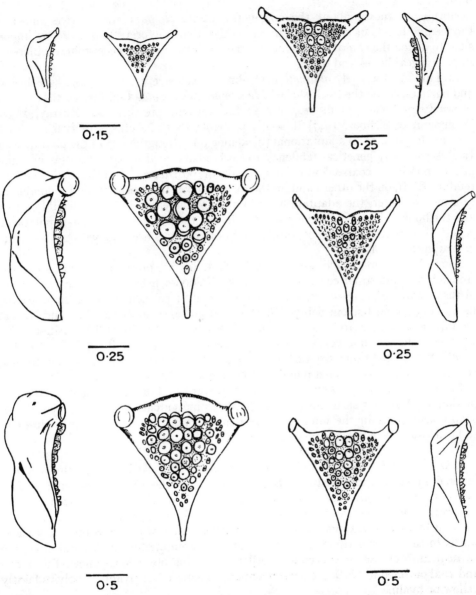

FIG. 3. Lower pharyngeal bones and teeth (lateral and occlusal views) of: Top row, left *Astatoreochromis a. alluaudi* 20 mm. S.L.; right, *A. a. alluaudi* 48 mm. S.L. Middle row, left, *A. a. alluaudi* 60 mm. S.L.; right, *A. a. occidentalis* (Lake Nakavali) 63 mm. S.L. Bottom row, left, *A. a. alluaudi* 120 mm. right, *A. a. occidentalis* (Lake Nakavali) 123 mm. S.L. Scale in centimetres.

from their greater size, may be compared with those of 20–30 mm. *A. alluaudi* from Lake Victoria.

Nothing is known about the epigenetics of *A. alluaudi* and little is known of the feeding habits of populations in lakes other than Victoria. It is therefore impossible to define the causal factors for the marked intra-specific, geographical difference in pharyngeal bones and teeth.

In Lake Victoria, *A. alluaudi* feed almost exclusively on Mollusca (see below) and particularly on the thick-shelled *Melanoides tuberculata*. Considering the extreme plasticity of bone and its response to intermittent pressure (see Murray, 1932; Weinmann & Sichner, 1947) it seems probable that the effects of crushing such prey might produce an adaptational thickening and strengthening of the pharyngeals. In this way, any genetic tendency towards pharyngeal hypertrophy (as manifest in the relatively coarse lower pharyngeals of post-larval *A. alluaudi*) would be reinforced. If, on the other hand, in the western lakes the species is not predominantly a mollusc eater, the adaptational stimulus for increased bone size would be less, and the bones might be relatively weak. Finally, the possibility of inter-populational genetic differences cannot be discounted, especially since the various lakes are geographically isolated.

Some data seem to add weight to the first, i.e. adaptational, hypothesis. The stomach and intestinal contents of 13 Lake Nakavali fishes have been examined; of these, two were empty. Five of the remaining 11 fishes had fed on small cichlid fishes, and six on bottom dèbris (plant tissue) and insects (both adult and larval). Despite a careful search, no remains of Mollusca were identified. Admittedly, 13 specimens do not constitute an adequate sample, but, if 13 Lake Victoria *A. alluaudi* in the same size-range were examined, every specimen with intestinal contents would have yielded remains of Mollusca.

Likewise, in four *A. alluaudi* from Lake Edward and one from Lake George, the predominant food was insects, although three individuals had scanty remains of small Gastropoda in the intestines. The snails could not be identified, except in so far as they were not *Melanoides* sp.

Coloration in life (known only from Lake Victoria). Sexual dimorphism is less marked in this species than in *Haplochromis* and the other monotypic genera. *Females and immature males.* Ground colour golden, overlain with olivaceous green, becoming yellow ventrally; a dark band runs obliquely downwards through the eye and becomes continuous with the lachrymal stripe, which runs obliquely backwards to the anterior tip of the preoperculum; often another dark band along the vertical limb of the preoperculum. All median fins olivaceous-yellow, the dorsal and anal outlined in black; caudal maculate. Pectoral fins hyaline; pelvics faintly yellow or hyaline.

Breeding males. Coloration essentially that of females except that the spinous dorsal is suffused with maroon, as is the entire anal fin, and the soft dorsal is densely spotted with maroon maculae. Anal fin with numerous yellow ocelli arranged in three or four vertical and the same number of horizontal rows. Pelvic fins black, the first ray pearly. Cephalic markings usually more intense than in females.

Preserved material : Adult males. Ground colour greyish-brown to brown, lighter ventrally ; five or six dark transverse bars, often interrupted ventrally, on the flanks ; occasionally an interrupted mid-lateral stripe. Cephalic markings as described above. Soft dorsal fin and entire caudal maculate ; lappets of spinous dorsal, margin of soft dorsal and entire margin of anal fin black. Pelvics black laterally, the first ray pearly. Ocelli on anal fin dark grey. *Females and immature males.* Ground colour as in males but lighter. Soft dorsal and entire caudal weakly maculate or immaculate. Cephalic markings fainter than in males. Anal fin without ocelli, but in some individuals a few, small, light spots occur in the position of the ocelli. Pectoral and pelvic fins hyaline.

Affinities. The relationship of *Astatoreochromis alluaudi* to the other monotypic genera of Lake Victoria and to certain species of *Haplochromis* was discussed above. It only remains to consider Regan's suggestion that *A. alluaudi* is " Near *H. gestri*, especially distinguished by the increased number of dorsal and anal spines and the large blunt pharyngeal teeth ". (*Haplochromis gestri* is a synonym of *H. obesus* (Blgr.) (see p. 182).

With the information now available on the anatomy and ecology of both species, it is clear that *A. alluaudi* and *H. obesus* are not closely related. *Haplochromis obesus* belongs to a group of endemic Lake Victoria species which has developed the highly specialized habit of feeding on the embryos and larvae of other cichlid fishes (p. 187.) *Astatoreochromis*, on the other hand, possesses the potentialities for developing into a highly specialized mollusc-eater, although one subspecies is apparently a generalized bottom feeder. Besides the morphological differences noted by Regan, there are marked dissimilarities in the dentition and jaws of the two species. On the scale of divergence found in the *Haplochromis* and related species occurring in Lake Victoria, *A. a. alluaudi* and *H. obesus* must be placed in very distinct lineages.

Differences in caudal fin length and the form of the pharyngeal bones are sufficiently well-marked to warrant the recognition of two subspecies of *Astatoreochromis*, one occurring in Lakes Victoria and Kyoga (including the Victoria Nile), and the other in Lakes Edward, George, Nakavali and Kachira, and in the Semliki River.

Admittedly one of the characters distinguishing the two groups could be considered a response to environmental differences (see p. 172). On the other hand, the importance of geographical isolation must be recognized. At present, and probably for a considerable period in the past, the western group of Lakes (Edward, Nakavali and Kachira) have been isolated from Lake Victoria by extensive papyrus-swamp divides on the interconnecting river systems (see Worthington, 1932). Likewise, Lakes Kachira and Nakavali are isolated from Lake Edward by intervening papyrus-swamps. Thus, although *Astatoreochromis* is relatively tolerant of papyrus-swamp conditions (see p. 174) the existence of such extensive swamp divides must considerably reduce any gene flow between the different lakes. Unfortunately, there is insufficient material from Lakes Edward, Kachira and Nakavali to determine whether a distinct subspecies occurs in each lake. At present, therefore, only two subspecies can be recognized.

Astatoreochromis alluaudi alluaudi Pellegrin

Diagnosis. *Astatoreochromis a. alluaudi* differs from the other subspecies in having a more massive lower pharyngeal bone with a greater number of molariform teeth, see Text-fig. 3 (ratio of head length to width of lower pharyngeal bone 2·4–3·1, Mean 2·7), and in having a shorter caudal fin (21·4–28·5 [Mean 24·3] per cent of standard length).

Other, ecological differences will be discussed below.

Distribution. Lakes Victoria, Kyoga and the Victoria Nile.

Astatoreochromis alluaudi occidentalis subsp. nov.

Type specimen. A male, 125 + 35·0 mm. long, B.M. (N.H.) Reg. No. 1933.2.23. 146, collected by Worthington from Lake Nakavali.

Diagnosis. Differs from the nominate subspecies in having a finer lower pharyngeal bone with fewer molariform teeth, see Text-fig. 3 (ratio of head length to width of lower pharyngeal bone 2·6–3·6, Mean 3·0) and in having a longer caudal fin (24·0–31·6, Mean 27·2 per cent of standard length).

Distribution. Lakes Edward, George, Nakavali and Kachira; the Semliki River above the rapids.

Ecology. Habitat. *A. a. alluaudi*, unlike the majority of *Haplochromis* species in Lake Victoria, is not confined to any particular type of substrate. Indeed, in this lake the subspecies is ubiquitous in all areas where the water is less than 60 feet deep. There are also indications that in Lake Victoria *A. a. alluaudi* may extend into deeper water. Graham collected one specimen in surface nets set over 193 feet of water some distance off-shore (Station 71; 0° 20¾' S., 33° 1½' E.; in the collections of E.A.F.R.O. there is one other specimen caught by nets set on the bottom at *ca.* 180 feet (0° 4' S., 33° 14' E.).

During rainy seasons, post-larval *A. a. alluaudi* have been found in pools and streams some distance inside papyrus-swamps. Larger young (40–50 mm. S.L.) enter small temporary streams when these are flowing into the lake. Neither the papyrus-swamp habitat nor that of temporary streams is occupied by endemic *Haplochromis* or related species. Young and adults of the widely-distributed, fluviatile-lacustrine species *H. nubilus* (Blgr.) and *H. multicolor* (Schoeller) do, however, live in such habitats.

No habitat data are available for *A. a. alluaudi* in the Victoria Nile and Lake Kyoga, nor for *A. a. occidentalis* in any lake. Specimens of the latter have been collected from the Semliki River near its source in Lake Edward.

Food. *Astatoreochromis a. alluaudi* (Lake Victoria). The stomach and intestinal contents of 40 fishes (48–163 mm. S.L.) from different localities clearly indicate that *A. a. alluaudi* feeds almost exclusively on Mollusca, especially Gastropoda. In most of the specimens examined, some insect larvae were also found; but, both in volume and numbers, these represented only a small fraction of the ingested material. The very fragmentary nature of the shells found in the alimentary tract precluded accurate identification of the mollusc species eaten. However, it seems most probable that the principal gastropod prey is *Melanoides tuberculata* (Müller), and the chief lamellibranch, *Corbicula* sp.

Astatoreochromis a. occidentalis. Lake Nakavali. Thirteen specimens 50–137 mm. S.L. were examined ; two were empty. In the largest fish, the entire alimentary tract was filled with plant debris ; five specimens (79–123 mm. S.L.) each contained fragmentary remains of small cichlid fishes (probably *Haplochromis*), with, in two, a little plant debris and some insect remains. The five smaller fishes (50–72 mm.) contained fragmentary insect remains (especially larval and adult Diptera) and plant debris.

Lake Edward. Only four specimens (62–76 mm. S.L.) were available for gut analysis ; three contained a few unidentifiable fragments of mollusc shells together with bottom debris and the fourth (71 mm. S.L.), mostly adult insects (Diptera) and the very fragmentary remains of a small fish. Although the mollusc fragments could not be identified positively they were not derived from *Melanoides*.

Lake George. The alimentary tract of the single fish available (80 mm. S.L.) contained fragments of adult insects.

Lake Kachira. The three specimens examined (66–78 mm. S.L.) were all from one station and contained only bottom debris and plant remains (including water-lily seeds) ; a few fragments of insects were found in the intestine of one individual.

Breeding. Both subspecies of *Astatoreochromis alluaudi* are female mouth-brooders; exact spawning sites are not known. In Lake Victoria, males of *A. a. alluaudi* less than 100 mm. S.L. are immature but females are mature at about 95 mm. S.L. The three specimens of *A. a. occidentalis* from Lake Kachira (66–78 mm. S.L. 1 ♂ and 2 ♀) are all sexually active, thus suggesting that in this lake the subspecies reaches maturity at a smaller size than *A. a. alluaudi* in Lake Victoria. Little information was obtained on the size of sexually mature *A. a. occidentalis* in other lakes ; a brooding female 57 mm. S.L. from Lake Nakavali and a ripe female 62 mm. long from Lake Edward seem to indicate that in these lakes female *A. a. occidentalis* also mature at a smaller size than do the females of *A. a. alluaudi* in Lake Victoria. It is possible that differences in the feeding habits of the two subspecies may be primarily responsible for the smaller adult size of *A. a. occidentalis*.

A marked disparity was noticed in the sex ratio of *A. a. alluaudi* from Lake Victoria and *A. a. occidentalis* from Lake Nakavali ; there is insufficient material to determine the sex ratio in other localities. Using only those specimens whose sex could be ascertained with certainty, the ratio is 16 ♀ : 46 ♂ in Lake Victoria, and 1 ♀ : 7 ♂ in Lake Nakavali. Reasons for this discrepancy are obscure but at least any bias introduced by collectors selecting brightly coloured males can be discounted ; both sexes are remarkably similar in colour. Furthermore, collections from Lake Victoria were made so as to eliminate this bias.

SUMMARY

1. The monotypic genus *Astatoreochromis alluaudi* is redescribed.
2. The generic characters are discussed, particularly from the phylogenetic viewpoint. It is thought that *A. a. alluaudi* was not derived from the same stem as other

Victoria and Edward species with hypertrophied pharyngeal bones and teeth. By the same tokens, *Astatoreochromis* is not closely related to the other and endemic monotypic genera of the two lakes. The genus is apparently related to such fluviatile species as *Haplochromis vanderhorsti* (Malagarasi River system) and *H. straeleni* (Congo system).

3. Two subspecific groups may be recognized, one from the Lake Victoria system and the other from lakes in western Uganda. These groups are given subspecific status, namely: *Astatoreochromis a. alluaudi* from Lakes Victoria and Kyoga, and the Victoria Nile; and *A. a. occidentalis* from Lakes Edward, George, Nakavali and Kachira, and the Semliki River.

4. The feeding habits of the two subspecies are described.

Study Material and Distribution Records.

Astatoreochromis a. alluaudi

Museum and Reg. No.	Locality	Collector
	Kenya	
Paris Museum 04,137 (Lectotype)	Kavirondo Bay	Alluaud
04,138–9 (Paratypes)	,, ,,	,,
B.M. (N.H.).—1904.6.281 (Paratype, presented by Paris Museum)	,, ,,	,,
B.M. (N.H.).—1958.7.9.2	Kisumu Harbour	E.A.F.R.O.
	Uganda	
B.M. (N.H.).—1906.5.30.506–9	Entebbe	Degen
,, 1906.5.30.505	Bunjako	,,
,, 1907.5.7.73–76	Buddu Coast	Simon
,, 1911.3.3.112–3113	Jinja, Ripon Falls	Bayon
,, 1958.7.9.3–5	Grant Bay	E.A.F.R.O.
,, 1958.7.9.6	Karinya (near Jinja)	,,
,, 1958.7.9.7–16	Jinja	,,
,, 1958.7.9.18–21	Pilkington Bay	,,
,, 1958.7.9.22	Thruston Bay	,,
,, 1958.7.9.23	0° 4′ S., 33° 14′ E.	,,
,, 1958.7.9.24–37	Entebbe Harbour	,,
,, 1958.7.9.38	Beach nr. Nasu Point	,,
,, 1958.7.9.39–40	Stream at Bugungu, Napoleon Gulf	,,
,, 1958.7.9.50	Ekunu Bay	,,
,, 1958.7.9.51–58	Ramafuta Island	,,
	Tanganyika	
,, 1958.7.9.1	Mwanza	,,
,, 1958.7.9.17	Majita	,,
,, 1958.7.9.41–49	Godziba Is.	,,
	Lake Victoria, Locality Unknown	
,, 1908.5.19.51	—	D. Radcliffe
,, 1928.5.24.370–372	—	M. Graham

Museum and Reg. No.		Locality	Collector
		Lake Kyoga and the Victoria Nile	
,,	1911.3.27.21	Between Lake Kyoga and the Murchison Falls	F. Melland
,,	1911.3.3.108	Bululo, Lake Kyoga	Bayon
,,	1911.3.3.109–110	Kakindu, Victoria Nile	,,
		Astatoreochromis a. occidentalis	
		Lake Kachira	
B.M. (N.H.).—1933.2.23.160–162			E. B. Worthington
		Lake Edward	
,,	1933.2.23.137–140		,,
		Lake George	
,,	1933.2.23.141		,,
		Lake Nakavali	
,,	1933.2.23.142–159		,,

ACKNOWLEDGMENTS

It is with great pleasure that I acknowledge my gratitude to Dr. Ethelwynn Trewavas for her helpful advice and criticism ; to the authorities of the Muséum National d'Histoire naturelle, Paris for allowing me to examine Pellegrin's type specimens ; to Dr. M. Poll of the Museé Royal du Congo Belge, Tervueren, who placed at my disposal several specimens from the Semliki River, and to Dr. Denys W. Tucker for his helpful criticism of the manuscript.

REFERENCES

(Other than those given in full in the synonymy)

GREENWOOD, P. H. 1954. On two cichlid fishes from the Malagarazi River (Tanganyika) etc. *Ann. Mag. nat. Hist.* (12) **7** : 401–414.

—— 1956. The monotypic genera of cichlid fishes in Lake Victoria. *Bull. Br. Mus. nat. Hist., Zool.* **3**, No. 7.

MURRAY, P. D. F. 1936. *Bones.* Cambridge.

REGAN, C. T. 1922. The cichlid fishes of Lake Victoria. *Proc. zool. Soc. Lond.* : 157–191.

TREWAVAS, E. 1933. Scientific results of the Cambridge expedition to the East African lakes, 1930–1. II. The cichlid fishes. *J. Linn. Soc. (Zool.)* **38** : 308–341.

WEINMANN, J. P. & SICHER, H. 1947. *Bone and Bones.* Henry Kimpton, London.

WORTHINGTON, E. B. 1932. *A Report on the Fisheries of Uganda.* Crown Agents, London.

A REVISION OF THE LAKE VICTORIA *HAPLOCHROMIS* SPECIES (PISCES, CICHLIDAE), PART III

By P. H. GREENWOOD
British Museum (Natural History) London

CONTENTS

	Page
INTRODUCTION	179
Haplochromis cronus sp. nov.	180
Haplochromis obesus (Boulenger)	182
Haplochromis maxillaris Trewavas	189
Haplochromis melanopterus Trewavas	192
Haplochromis parvidens (Boulenger)	194
Haplochromis cryptodon sp. nov.	198
Haplochromis microdon (Boulenger)	200
DISCUSSION OF THE SEVEN FOREGOING SPECIES	203
Haplochromis plagiodon Regan & Trewavas	205
Haplochromis chilotes (Boulenger)	207
Haplochromis chromogynos sp. nov.	212
Haplochromis aelocephalus sp. nov.	214
SUMMARY	218
ACKNOWLEDGMENTS	218
REFERENCES	218

INTRODUCTION

FIVE of the seven species described in the first part of this paper are known to feed almost exclusively on the embryos and larvae of other cichlid fishes, especially *Haplochromis*. Data on the food of the sixth species are very inadequate but are nevertheless indicative of similar habits. The seventh species is known from only a few specimens, but various morphological similarities between it and two other species of this group suggest embryo and larval fish-eating habits.

If the species on which these fishes prey are mouth-brooders, it can be said that none of the young found in the stomachs of the predators was of a size at which it would normally have left the parental mouth.

Despite identical feeding habits, the members of this species group are morphologically heterogeneous and exhibit convergence only in a tendency for the teeth to be deeply embedded in the oral mucosa and in having capacious mouths. Furthermore, in most species there is a marked intra-specific variability in gross morphology, especially of the head. It seems that the group is of polyphyletic origin.

The four species dealt with in the second part of the paper are, morphologically speaking, somewhat isolated from the other *Haplochromis* of Lake Victoria. All are insectivores.

Haplochromis cronus sp. nov.
(Text-fig. 1)

Holotype. A female, 135 mm. standard length, from Buka Bay, Uganda.

Description, based on eight specimens, including the holotype, 114–135 mm. standard length.

Depth of body 39·5–43·5 per cent of standard length, length of head 30·3–34·6 per cent. Dorsal head profile strongly curved, with a well-defined but localized

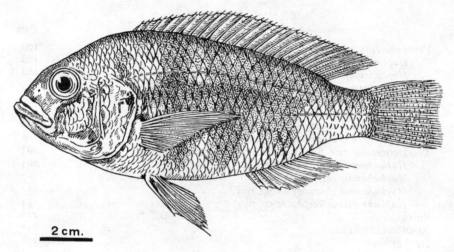

Fig. 1. *Haplochromis cronus*; holotype. Drawn by Miss D. Fitchew.

swelling above the anterior part of the eye. Preorbital depth 16·7–18·2, mean (M) 17·5 per cent of head length; least interorbital width 31·9–35·5 (M = 33·1) per cent. Snout slightly broader than long, its length 31·6–35·7 (M = 33·7) per cent of head; eye diameter 23·1–26·3 (M = 25·2), depth of cheek 29·3–34·2 (M = 30·1) per cent.

Caudal peduncle 15·2–17·3 per cent of standard length, 1·1–1·3 (mode 1·3) times as long as deep.

Jaws equal anteriorly; lips thickened; posterior tip of the maxilla not bullate and almost completely hidden beneath the preorbital, extending to the vertical through the anterior part of the eye. Lower jaw stout and deep, its length 29·3–34·2 (M = 30·1) per cent of the head, 1·2–1·4 times as long as broad.

Gill rakers stout; 8–10 on the lower limb of the first arch.

Scales ctenoid; lateral line with 32 (f.5), 33 (f.1) or 34 (f.2) scales. Cheek with four or five series. Five to 8 scales between the dorsal fin origin and the lateral line; 8 or 9 between the pectoral and pelvic fin bases.

Fins. Dorsal with 24 (f.7) or 25 (f.1) rays, anal (damaged in one specimen) with 12 (f.7), comprising XV or XVI, 9–10 and III, 9 spinous and branched rays for the fins respectively. Pectoral fin shorter than the head; pelvic fins with the first soft ray produced and extending to the vent in females and to the anal fin in males. Caudal truncate, the rays noticeably coarse; densely scaled over about four-fifths of its length (a most unusual character in Lake Victoria *Haplochromis* species).

Teeth. The outer row in both jaws is composed of unicuspid, fairly stout teeth, implanted vertically and not hidden by thickened oral mucosa; there are 40–56 teeth in this series of the upper jaw. The inner teeth are small and unicuspid, arranged in two rows (three in one specimen) in both jaws, and are separated from the outer series by a distinct space.

Lower pharyngeal bone triangular, the dentigerous area about 1·3 times as broad as long; the teeth are slender and cuspidate.

Syncranium. The syncranium is noticeable for its short and deep neurocranium (comparable with *H. obesus*; see p. 185) and for the stout but otherwise unspecialized dentary. These characters were determined from a radiograph B.M. (N.H.) Reg. No. 957 and the partial dissection of one specimen.

Vertebrae: 13 + 16 in the single specimen radiographed.

Coloration of preserved material: *Adult females and sexually quiescent males.* Ground colour dark golden above, lighter below, with traces of a golden-yellow flush on the operculum: a broad, mid-lateral stripe of variable depth and intensity crossed by four or five broad but faint transverse bars on the flanks; a well-defined lachrymal stripe. Dorsal fin hyaline, with dark spots and bars on the soft part (probably deep red in life); caudal hyaline (densely maculate in males); proximal two-thirds of anal fin dark, remainder light; pelvic fins dark (black laterally in males).

One of the three females available has a typical "*bicolor*" (piebald black and yellow) coloration, similar to that described in several other and apparently unrelated *Haplochromis* species and in two monotypic genera (Greenwood, 1957, and p. 213).

Sexually active male. Dark brown above, sooty-grey below; transverse and lateral stripes faint except at their junction mid-laterally. Dorsal fin dusky, the soft part maculate; caudal dusky and densely maculate; anal dark, except for its extreme tip and two colourless ocelli. Pelvic fins black on the lateral half and dusky mesially.

Distribution. Known only from Lake Victoria.

Ecology: Habitat. Five of the eight specimens are from an exposed beach habitat, one is from the sandy littoral of a sheltered gulf, one from the mud-bottom sublittoral of a sheltered bay and one from shallow water near a reed bed (no other data available). In no locality was the water more than 20 feet deep.

Food. Four specimens contained food in the stomach; in each, only larval cichlid fishes were found (in three fishes these were identified as *Haplochromis*); the number of larvae in each fish was: 127 (*ca.* 11 mm. long); 50 (*ca.* 11 mm.) and 41 (*ca.* 11 mm.). The remains found in the fourth fish were too fragmentary to allow even an estimate of numbers.

Breeding. Two females were found with, in one, larvae and in the other, newly hatched embryos in the buccal cavity. Since the ovarian condition of these fishes was clearly " spent " it can be assumed that the young were the fishes' own brood and not prey.

Affinities. Haplochromis cronus belongs to the small group of deep-bodied, broad-headed *Haplochromis* whose adult size is ca. 100–140 mm. S.L. The specialized mollusc-eating species *H. pharyngomylus* Trewavas and *H. ishmaeli* Boulenger may be cited as examples of this morphotype. *H. cronus* differs from all other members of the group in having a densely and extensively scaled caudal fin. The species shows some affinity with *H. obesus* and *H. maxillaris*, forms which may have evolved independently from an *H. cronus*-like ancestor.

Diagnosis. From other species with a similar gross morphology, *H. cronus* can be distinguished, primarily, by its almost completely scaled caudal fin (four-fifths scaled in *H. cronus* cf. two-thirds scaled in other species). The relatively large, completely exposed, caniniform and recurved teeth of *H. cronus*, together with an unmodified lower pharyngeal dentition and the presence of a supra-orbital swelling, also serve to distinguish *H. cronus* from other morphologically similar species.

Study material and distribution records

Museum and Reg. No.	Locality	Collector
	Uganda	
B.M. (N.H.).—1958.1.16.85 (Holotype)	Buka Bay	E.A.F.R.O.
B.M. (N.H.).—1958.1.16.86–89	,, ,,	,,
,, 1958.1.16.90	Napoleon Gulf near Jinja	,,
,, 1958.1.16.91	Pilkington Bay	,,
	Kenya	
,, 1928.5.24.408	Port Victoria (Graham's St. No. 84)	M. Graham

Haplochromis obesus (Boulenger) 1906
(Text-figs. 2 and 3)

Pelmatochromis obesus (part) Boulenger, 1906, *Ann. Mag. nat. Hist.* (17) **17**, 447 (type specimen, by restriction [specimen figured in *Fish. Nile*], only) ; *Idem*, 1907, *Fish. Nile*, 491, pl. LXXXIX fig. 5 ; *Idem*, 1915, *Cat. Afr. Fish.* **3**, 414, fig. 283.
Lipochromis obesus (Boulenger), Regan, 1920, *Ann. Mag. nat. Hist.* (9) **5**, 45 (foot-note).
Haplochromis obesus (Boulenger), Regan, 1922, *Proc. zool. Soc. London*, 170.
Paratilapia gestri Boulenger, 1911, *Ann. Mus. Genova* (3), **5**, 67, pl. I, fig. 3.
Paratilapia gestri (part, holotype only). Boulenger, 1915, *Cat. Afr. Fish.* **3**, 318, fig. 211.
Haplochromis gestri (part, holotype of *P. gestri* only), Regan, 1922, *Proc. zool. Soc. London*, 170.

The union of *H. obesus* and *H. gestri* might be questioned if only the type specimens of the two species were available ; indeed, for a long time I thought that the species were distinct. However, after examining a large series of *H. gestri*-like specimens, I am forced to conclude that the type and unique specimen of *H. obesus* is merely an aberrant individual from a species whose modal morphotype is " *gestri* "-like.

A REVISION OF THE LAKE VICTORIA *HAPLOCHROMIS* SPECIES

In my opinion, the critical specific character-complex is the broad and stout lower jaw, combined with a relatively fine premaxilla and a stout, posteriorly bullate maxilla. These characters are easily verified in the type specimens of *H. obesus* and *H. gestri*, and have been further confirmed in a radiograph of the former.

In its gross morphology and partly in its physiognomy, the type of *H. obesus* differs from all except one of the 46 specimens now referred to this species; these differences may be partly attributable to *post-mortem* changes and poor preservation. In Boulenger's figure (reproduced here as Text-fig. 2) the mouth is shown as it appeared when closed artifically, with the result that the gape is very oblique. With the passage of time, the specimen has softened and it is now possible to close the mouth more easily. If this is done, it will be seen that the angle of the mouth is only slightly

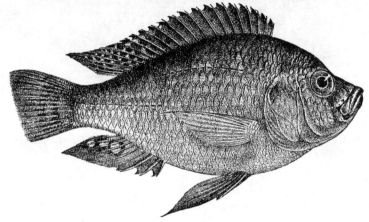

FIG. 2. *Haplochromis obesus*; holotype (from Boulenger, *Fishes of the Nile*).

oblique and that the dorsal head profile, although sloping steeply, is not so markedly concave as it appears in the figure. The lower jaw closes within the upper and only the anterior part of the maxilla is covered by the preorbital. Whether or not this is due to a natural deformity or to *post-mortem* distortion, I cannot say.

Although, in appearance, the majority of specimens resemble the holotype of *H. gestri*, there are several others which depart from that mode but still retain the diagnostic dentary, upper jaw elements and dentition of the species.

Description, based on 48 fishes (71–170 mm. S.L.) including the type, and the holotype of *H. gestri*.

Depth of body 33·6–47·3 per cent of standard length; length of head 30·3–35·9 per cent. Physiognomy variable, the dorsal head profile straight or very slightly concave in the interorbital region, sloping steeply (Text-fig. 4); most fishes resemble the figured specimen. Preorbital depth 12·5–17·3 (M = 15·4) per cent of head length, least interorbital width 27·3–37·0 (M = 32·2) per cent. Snout 1·20–1·33 times as broad as long, its length 28·0–39·5 (M = 33·5) per cent of head; eye diameter shows negative allometry with standard length, being 23·7–30·4 (M = 27·6) per cent in

23 fishes less than 125 mm. S.L. and 20·5–27·9 (M = 23·6) per cent in 25 larger individuals; depth of cheek 21·8–31·8 (M = 27·6) per cent.

Caudal peduncle 12·8–17·9 per cent of standard length, its length 1·0–1·6 (modal range 1·0–1·2) times its depth.

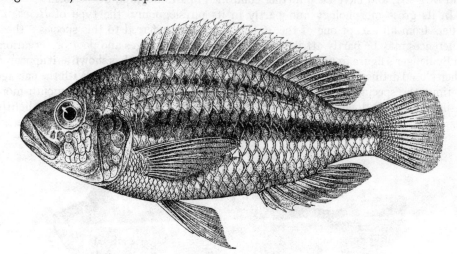

Fig. 3. *Haplochromis obesus*; typical form (holotype of *Paratilapia gestri*). From Boulenger, *Ann. Mus. Genova*, 1911.

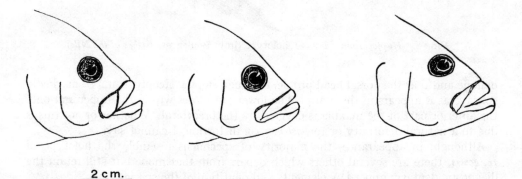

Fig. 4. *Haplochromis obesus*; individual variability of head profile.

Mouth slightly oblique, maxilla bullate posteriorly and only partly covered by the preorbital, reaching the vertical through some part of anterior half of eye. Lips somewhat thickened; jaws equal anteriorly, the length of the lower showing a positive but widely scattered allometry with standard length, 40·0–54·5 per cent of head length. Length/breadth ratio of the lower jaw 1·0–1·6 (mode 1·3).

Gill rakers short and stout, 9 or 10 (rarely 8 or 11) on the lower part of the first arch.

Scales ctenoid; lateral line with 29 (f.1), 31 (f.11), 32 (f.18), 33 (f.15) or 34 (f.2) scales; cheek with 3 or 4 (rarely 2) series. Six or 7 (rarely 8) scales between the dorsal fin origin and the lateral line; 6–8 between pectoral and pelvic fin bases.

Fins. Dorsal with 24 (f.21), 25 (f.26) or 26 (f.1) rays, anal with 10 (f.2), 11 (f.23), 12 (f.21) or 13 (f.1), comprising XV–XVI (rarely XVII), 8–10 and III, 8 or 9 (rarely 10) spinous and branched rays for the fins respectively. Pectoral shorter than the head. Pelvic fins with the first soft ray produced, more particularly so in adult males. Caudal subtruncate or less commonly, obliquely truncate.

Teeth. Both the inner and outer series of teeth are deeply embedded in the oral mucosa, so that only the tips protrude.

Except for the smallest specimens, the outer teeth in both jaws are relatively stout and unicuspid with conical crowns. In small specimens most teeth are unequally bicuspid, or there may be an admixture of uni- and bicuspid forms. There are 34–52 teeth in the outer series of the upper jaw.

The shape of the teeth is variable; in most specimens the anterior and some lateral outer teeth in the lower jaw have the crown bent so that its tip is directed anteriorly. In the upper jaw there is an admixture of such teeth with the more usual recurved and conical types. Teeth with anteriorly directed crowns are known only in *H. obesus*, *H. maxillaris* and *H. melanopterus*.

The inner teeth are unicuspid and slenderly conical in fishes over 100 mm. S.L.; in smaller individuals there is a combination of unicuspid and weakly tricuspid teeth. Anteriorly in both jaws the inner teeth are arranged in one or, less commonly, two series; the interspace between inner and outer teeth is greatly reduced or even absent.

Lower pharyngeal bone short and broad, the dentigerous surface 1·2–1·6 times as broad as long; pharyngeal teeth cuspidate and laterally compressed.

Syncranium. The most outstanding skeletal characteristic of *H. obesus* is the broad and stout lower jaw (Text-fig. 5A). The "*obesus*"-type dentary is unique amongst the *Haplochromis* of Lake Victoria. When compared with one of the larger but generalized species the dentary of *H. obesus* is noticeably bullate in the region where each ramus divides into ascending and horizontal rami.

Departure from a generalized *Haplochromis* condition is also seen in the maxilla, which is deeper and more bullate posteriorly. The neurocranium closely resembles that of *Hoplotilapia retrodens* Hilgendorf and *Platytaeniodus degeni* Boulenger, since the preorbital face is short and the supra-occipital crest deep (Greenwood, 1956). This intergeneric convergence is probably associated with the relatively massive lower jaw of all three species.

When compared with other *Haplochromis* (e.g. *H. cronus* and *H. pharyngomylus*) having approximately the same adult size and similar body-form, it is obvious that the mouth of *H. obesus* is more distensible and more protrusible. These factors may be associated with the specialized feeding habits of the species (see p. 204).

Vertebrae: 13 + 16 and 13 + 15 (type *H. obesus*).

Coloration in life: Adult males. Ground colour dark malachite green shading to silvery-blue ventrally; a coppery sheen on the operculum, chest and belly; a distinct dark mid-lateral stripe. Dorsal fin dark, with an overall pinkish flush; caudal and anal fins dark, the latter with four or five orange-yellow ocelli arranged in one or, more frequently, two rows. Sexually quiescent males have a similar but less intense coloration. *Adult females.* Olivaceous-silver shading to silver ventrally; a distinct, dark mid-lateral stripe. All fins hyaline; in some individuals there are small yellow spots in the position of the anal ocelli in males. Some females show a typical "*bicolor*" black and silver (or yellow) piebald coloration. No estimate of the frequency of "*bicolor*" individuals can be made from the data available; such females have, so far, only been found in the Napoleon Gulf, near Jinja, Uganda.

A second atypical colour-form is also known. Fishes showing the extreme expression of this coloration are uniformly black, but lighter (sooty) ventral coloration is

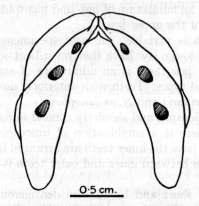

Fig. 5 (A). *Haplochromis obesus*; outline of dentary, ventral view.

more usual. Unlike the "*bicolor*" pattern, the dark form is not sex-limited and is known to occur in several different areas of the lake. Furthermore, it shows some intergradation with the usual coloration, at least in males.

Colour in preserved material: Both sexes. Ground colour golden to dark brown (adult males generally darkest); a well defined, dark, mid-lateral stripe and an ill-defined dorsal stripe following the contour of the upper lateral-line; 5–9 vertical bars on the flanks and caudal peduncle; often faint indications of a lachrymal stripe. Pelvic fins black in adult males, otherwise colourless, as are all other fins; soft dorsal and the entire caudal weakly maculate.

Distribution. Lake Victoria and Lake Kwania (Kyoga system).

Ecology: Habitat. Haplochromis obesus is apparently restricted to water less than 50 feet deep in the littoral and sublittoral zones of Lake Victoria. Most of the specimens were caught over a hard substrate (sand, shingle or rock) but two were caught over a soft mud bottom. In all probability, the distribution of *H. obesus* is closely linked with the spawning and brooding areas of the cichlid fishes on whose embryos and larvae it preys.

Food. Of the 73 specimens examined, 18 had food in the stomach. In every one of these fishes, only fish embryos or larvae were found; with one exception (a small cyprinid fish) the prey could be identified as Cichlidae. A hundred embryos at the same stage of development were recorded from one stomach and in many others the embryos or larvae were all at the same developmental stage; embryos at different ontogenetic stages were, however found in some individuals.

The possibility that these stomach contents did not represent food but rather the fishes' own young accidently swallowed, can be overruled by the following considerations: a mixture of early and advanced ontogenetic stages was found in one stomach; embryos and larvae were found in the stomachs of both male and female fishes and it is unknown amongst the Lake Victoria cichlids for both parents to share brooding duties; early embryos were identified in the stomach contents of

Fig. 5 (B). *Haplochromis parvidens* ; outline of dentary, ventral view.

an immature female; and finally, personal observations show that it is unusual for a brooding female to swallow her brood when she is captured; generally, the young are jettisoned.

It is not known how *H. obesus* or the other larval fish-eating species obtain their prey. The question is complicated because the principal source of food for these species is the young of other cichlid fishes. Both species of *Tilapia* in Lake Victoria and all species of *Haplochromis* whose breeding habits are known are female mouth-brooders. Although late larval cichlids do leave the parental mouth, the earlier, non-free-swimming stages do not, except when the parent is so harrassed that it jettisons the brood. Unless a number of *Haplochromis* are not mouth-brooders, it seems that the larval and embryo fish-eating species employ some means of forcing the parent fish to abandon its brood. It may be added that there is no evidence to indicate that any Lake Victoria *Haplochromis* are not mouth-brooders.

Breeding. A single brooding female was recorded: young removed from the buccal cavity were in the germ-ring stage of development. There is no sex-correlated size difference in adult fishes and sexual maturity is reached at a standard length of about 85 mm.

Affinities. Regan (1922) suggested that *H. gestri* (= *H. obesus*) was near *Astatoreochromis alluaudi* Pellegrin. With the additional information now available on both species, this opinion is no longer tenable. *A. alluaudi* is a specialized mollusc-eater with hypertrophied pharyngeal bones and the consequent modifications to the syncranial architecture (see Greenwood, 1954). Although *H. obesus* has a markedly modified lower jaw and somewhat atypical upper jaw features, it is more closely related to the generalized *Haplochromis* species. The relationships of *A. alluaudi* lie, apparently, with some of the semi-specialized fluviatile *Haplochromis* of the Malagarasi and Congo rivers (Greenwood op. cit. and p. 167). Any resemblance between *A. alluaudi* and *H. obesus* is entirely superficial and attributable to the stout bodies and broad heads of the two species.

Perhaps the closest relatives of *H. obesus* are *H. cronus* and *H. maxillaris*, with which species it not only shows certain similarity in gross and detailed morphology, but it also shares the same food requirements.

Diagnosis. The shape of the lower jaw (in which there is usually a predominance of unicuspid outer teeth with anteriorly directed crowns) is the most trenchantly diagnostic character. The deeply embedded teeth, together with certain morphometric characters of the head, serve to distinguish *H. obesus* from other Lake Victoria *Haplochromis* species.

Study material and distribution records

Museum and Reg. No.	Locality	Collector
	Uganda	
B.M. (N.H.).—1906.5.30.311 (Holotype *Pelmatochromis obesus*)	Bunjako	Degen.
Genoa Museum (Holotype *Paratilapia gestri*)	Jinja	Bayon.
B.M. (N.H.).—1958.1.16.140	Ekunu Bay	E.A.F.R.O.
„ 1958.1.16.141	Buka Bay	„
„ 1958.1.16.142	Channel between Yempita and Busiri Isles, Buvuma Channel	„
„ 1958.1.16.143–150	Beach near Nasu Point, Buvuma Channel	„
„ 1958.1.16.154–156	SE. tip of Ramafuta Is., Buvuma Channel	„
„ 1958.1.16.157	Karinya, Napoleon Gulf	„
„ 1958.1.16.158–161	Entebbe Harbour	„
	Kenya	
„ 1958.1.16.151–153	Kisumu Harbour	„
	Lake Victoria, Locality Unknown	
„ 1958.1.16.162–164	—	„
„ 1928.5.24.341–2	—	M. Graham.
	Lake Kwania	
„ 1929.1.24.509	—	E. B. Worthington.

106

Haplochromis maxillaris Trewavas, 1928
(Text-fig. 6)

Pelmatochromis microdon (part) Boulenger, 1906, *Ann. Mag. nat. Hist.* (7) **17**, 441 ; *Idem*, 1915, *Cat. Afr. Fish.* **3**, 412.
Haplochromis microdon (Boulenger), (part), Regan, 1922, *Proc. zool. Soc. London*, 173.
Haplochromis maxillaris Trewavas, 1928, *Ann. Mag. nat. Hist.* (10) **2**, 94.

Lectotype. A male 114 mm. standard length (B.M. (N.H.) Reg. No. 1928.5.24.486) from Emin Pasha Gulf, Tanganyika Territory (2° 31½′ S., 31° 43½′ E.), Michael Graham's station 227.

Description, based on 58 specimens (including the types) 90–160 mm. S.L.

Depth of body 32·0–42·8 per cent of standard length, length of head 30·0–34·8 per cent. Physiognomy variable, its shape partly dependent on the angle of the mouth

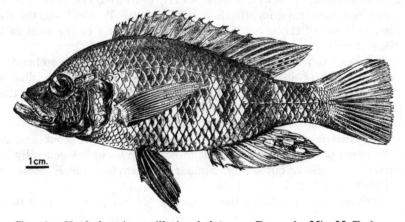

Fig. 6. *Haplochromis maxillaris* ; holotype. Drawn by Miss M. Fasken.

and whether the lower jaw protrudes or not ; dorsal head profile concave (markedly so in a few specimens) and sloping at an angle of 40°–50°. A few specimens bear a superficial resemblance to *H. obesus*, but despite this variability in gross morphology there is a distinct modal specific facies (see Text-fig. 6).

Preorbital depth 11·4–16·3 (M = 13·9) per cent of head length, least interorbital width 22·6–31·3 (M = 26·5) per cent. Snout slightly broader than long, rarely as long as broad, its length 25·8–34·0 (M = 30·3) per cent of head. Eye diameter shows negative allometry with standard length : in 15 fishes 60–100 mm. S.L. it is 30·0–38·0 (M = 33·2) per cent of head and in 43 larger individuals it is 25·0–31·4 (M = 27·3) per cent. Depth of cheek positively allometric with standard length : for the two size groups as above, 18·8–24·2 (M = 20·7) and 21·6–27·8 (M = 26·4) per cent head length.

Caudal peduncle 12·6–18·5 per cent of standard length, 1·1–1·7 (modal range 1·2–1·3) times as long as deep.

Mouth distensible and usually somewhat oblique when closed, but horizontal in a few specimens. Maxilla partially hidden by the preorbital, its posterior tip bullate and reaching the vertical to the anterior part of the eye or even as far as the pupil. Lips thickened. Lower jaw usually projecting, but in a few fishes the jaws are equal anteriorly. Length of lower jaw shows positive allometry with standard length, in fishes 60–100 mm. S.L. it is 39·0–47·5 (M = 44·8) per cent of head length, and in larger fishes 46·5–56·0 (M = 50·0) per cent. Breadth of lower jaw contained 1·3–2·2 (modal range 1·5–1·8) times in its length.

Gill rakers short, 10 or 11 (rarely 9 or 12) on the lower part of the first gill-arch, the lower one or two rakers often greatly reduced.

Scales ctenoid : lateral line with 29 (f.2), 30 (f.7), 31 (f.22), 32 (f.20), 33 (f.4) or 34 (f.1) scales ; cheek with 2 or 3 series ; $5\frac{1}{2}$–7 scales between the dorsal fin origin and the lateral line, 5–7 (rarely 8) between the pectoral and pelvic fin bases.

Fins. Dorsal with 24 (f.19), 25 (f.36) or 26 (f.3) rays, anal with 11 (f.18), 12 (f.37) or 13 (f.1), comprising XV–XVI (rarely XVII), 8–10 and III, 8 or 9 (rarely 10) spinous and branched rays for the fins respectively. Pectoral shorter than the head. First soft ray of the pelvic fin produced, extending to the vent in females and to the soft part of the anal fin in adult males.

Teeth. The inner and outer rows of teeth in both jaws are deeply embedded in the thickened oral mucosa ; in many specimens the inner series are invisible without dissection. Furthermore, the outer teeth of the upper jaw are covered by the thickened and inwardly curved margin of the lip.

Fishes less than 80 mm. S.L. have small, weakly and unequally bicuspid outer teeth in both jaws. In larger fishes these teeth are also small, but stout and conical ; those in the upper jaw are recurved, whilst those in the lower jaw generally have the crown curved anteriorly or outwardly. Similar teeth are found in *H. obesus*, but are not the predominant form in that species.

In the three skeletons examined there were 34, 36 and 40 outer teeth in the premaxilla.

The inner teeth are weakly tricuspid in small fishes and unicuspid in larger individuals ; arranged in one or, rarely, two series and separated from the outer row by a small interspace. Inner teeth in the upper jaw are slightly recurved and implanted so as to slope posteriorly ; those of the lower jaw are vertical or directed anteriorly.

Lower pharyngeal bone triangular, the dentigerous area 1·0–1·4 times as broad as long ; teeth slender and cuspidate, those of the two median rows sometimes coarser.

Syncranium. The dentary of *H. maxillaris* departs slightly from the generalized type. As in the dentary of *H. obesus* there is a pronounced lateral bullation of the area surrounding the bifurcation into ascending and horizontal rami.

The premaxilla and maxilla are similar to those of *H. obesus*, except that the premaxillary teeth are restricted to the anterior and antero-lateral areas of the bone. The neurocranium is of a generalized *Haplochromis* type.

Vertebrae : 13 + 16 and 12 + 16 in two skeletons.

Coloration in life : Adult males. Ground colour dark blue-grey, lighter ventrally, with faint indications of darker transverse bars on the flanks. Dorsal fin dusky, with maroon spots between the rays of the soft part; lappets orange-red. Caudal and anal fins smoky-grey, the latter with three to five yellow ocelli arranged in either one or two rows. Pelvic fins dusky. Coloration in life of *immature males* unknown. *Females.* Silver-grey ground colour. Dorsal fin greyish; anal and caudal fins similar but with a yellowish flush and, on the caudal, ill-defined, dark spots. Pelvic fins very faint yellow. Several dark, but faint, transverse bars may appear on the flanks immediately after death.

Colour of preserved material: Males. Dark, some with an underlying silvery ground colour, others almost black. Seven or more transverse bands are sometimes visible on the flanks. Dorsal and caudal fins hyaline and maculate, or dusky; anal hyaline or dusky. Pelvic fins black. *Females.* Ground colour yellowish-silver to brown; some with seven or more transverse bars. All fins hyaline or somewhat dusky; anal and caudal weakly maculate.

Distribution. Known only from Lake Victoria.

Ecology : Habitat. Haplochromis maxillaris is apparently restricted to water less than 30 feet deep, and particularly to the littoral and sublittoral zones of the lake. Most specimens were caught over a hard substrate (sand or shingle), but a few were recorded from mud substrates. Thus the habitat preferences of *H. maxillaris* are almost identical with those of *H. obesus*, the two species frequently being caught in the same gear.

Food. Forty of the 118 individuals examined had identifiable food in the stomach. The smallest specimen (44 mm. S.L.) proved exceptional in that the stomach was filled with Copepoda and blue-green algae. All the remaining 39 fishes had eaten cichlid embryos or larvae. In some individuals, the entire stomach contents were of prey at uniform developmental stage, whilst in other fishes two or more stages (often as widely different as early cleavage embryos and late larval fishes) were present. Both sexes were represented amongst the fishes examined, which came from numerous localities.

The remarkable similarity between the food of *H. maxillaris* and that of *H. obesus* is noteworthy. Again, it is difficult to imagine how the food is obtained if the species preys on mouth-brooding cichlids.

Breeding. There is no information on any aspect of breeding behaviour in this species. All fishes below 100 mm. S.L. were immature; it seems probable that sexual maturity is reached at a length of about 105 mm. Males and females attain the same adult size.

Affinities. The distensible mouth, stout and posteriorly bullate maxilla, and the thickened lips of *H. maxillaris* all suggest affinity with *H. obesus*. Furthermore, conical outer teeth in which the crown is directed anteriorly or laterally are known only in these two species (and *H. melanopterus*, see below, p. 194). Certain specimens of both species show convergence in gross morphology, but the characteristic lower jaw of *H. obesus* usually allows for immediate identification. Apart from these few convergent individuals, the two species differ in certain morphometric details of the head and each has a distinctive modal facies. It is difficult to assess the phyletic significance of the resemblances and differences between *H. obesus* and *H. maxillaris*. The species could be equally well derived from a common stem or from unrelated ancestors within the Lake Victoria species flock.

Study material and distribution records

Museum and Reg. No.	Locality	Collector
	Tanganyika	
B.M. (N.H.).—1928.5.24.486 (Lectotype, *H. maxillaris*)	Emin Pasha Gulf	M. Graham.
„ 1958.1.16.182–184	Majita	E.A.F.R.O.
„ 1958.1.16.188	Mwanza, Capri Bay	„
	Uganda	
„ 1958.1.16.165–171	Beach near Nasu Point, Buvuma Channel	„
„ 1958.1.16.172–179	Ramafuta Is., Buvuma Channel	„
„ 1958.1.16.180	Njoga, Williams Bay	„
„ 1958.1.16.181	Buka Bay	„
„ 1958.1.16.185–187	Beach near Hannington Bay	„
„ 1958.1.16.189	Pilkington Bay	„
„ 1958.1.16.190–193	Bukafu Bay	„
„ 1958.1.16.194–198	Entebbe Harbour	„
„ 1906.5.30.310	Entebbe	Degen.
„ 1958.1.16.199	Ekunu Bay	E.A.F.R.O.
„ 1958.1.16.205–214	Napoleon Gulf, near Jinja	„
	Kenya	
„ 1958.1.16.200–204	Kisumu Harbour	„
	Lake Victoria, Locality Unknown	
„ 1928.5.24.480–485 (Syntypes, *H. maxillaris*)	—	M. Graham.

Haplochromis melanopterus Trewavas 1928
(Text-fig. 7)

H. melanopterus Trewavas, 1928, *Ann. Mag. nat. Hist.* (10) **2**, 94.

Discussion. This problematical species is known from a single specimen which had suffered some *post-mortem* distortion before preservation. Its status is, therefore, all the more difficult to decide.

Superficially, *H. melanopterus* is most distinctive. The lower jaw (which closes entirely within the upper) is short, narrow and pointed anteriorly. The preorbital is very shallow so that the greater part of the maxilla is exposed when the mouth is shut. This latter character, together with the peculiar arrangement of the lower jaw in relation to the upper, may be an artefact of preservation and *post-mortem* distortion. The shallow preorbital and the short, pointed lower jaw cannot, however, be attributed to these causes.

The dentition closely resembles that of *H. maxillaris* both in the form of the teeth and their restricted distribution on the premaxilla (see p. 190). The immediate question raised is, are not perhaps the shallow preorbital and the lower jaw size and shape the result of some ontogenetic disturbance in the development of an individual *H. maxillaris*? The apparent distortion of the upper jaw might then be considered teratological rather than the result of *post-mortem* distortion.

In general appearance, *H. melanopterus* is unlike *H. maxillaris* but as Trewavas noted in her original description of the species, it is nearer the *H. maxillaris-obesus* complex than any other species group. Thus, it is impossible to give an adequate

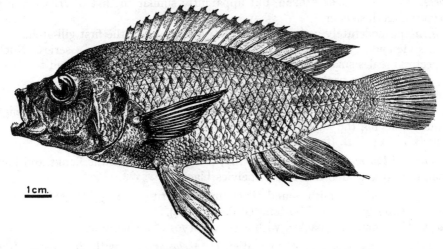

Fig. 7. *Haplochromis melanopterus*; holotype. Drawn by Miss M. Fasken.

answer to the question posed above. I have decided to treat *H. melanopterus* as a distinct species mainly on the grounds that it is difficult to determine whether or not the peculiar jaws are a teratological feature, and because the nuchal and pectoral squamation of the type is manifestly smaller than in either *H. obesus* or *H. maxillaris*. Also, the rounded caudal fin is a most unusual feature in a Lake Victoria *Haplochromis*.

Description, based on the holotype, an adult male 127 mm. S.L.

Depth*	Head*	Po. %	Io. %	Snt. %	Eye %	Cheek %	L.j. %	C.P.*
35·5	33·5	11·8	27·0	30·6	28·2	23·5	37·6	14·2

* Percentage standard length.
% Percentage head-length.

Dorsal head profile very concave. Mouth probably oblique; maxilla stout and bullate posteriorly, about three-quarters exposed even when the mouth is shut. In this specimen the mouth is open and the mandible lies horizontally, but the lateral limbs of the premaxilla and the maxillae are almost vertical. Lower jaw narrowing rapidly at a point almost half-way along its length; greatest width contained 1·3 times in the length; the entire lower jaw closing within the upper; lips thickened.

Dentition very similar to that of *H. maxillaris*. Outer teeth conical, those of the upper jaw recurved and restricted to the anterior and antero-lateral aspects of the premaxilla. Teeth in the lower jaw have the crown directed anteriorly or laterally; the anterior teeth of this series are somewhat stouter than the equivalent teeth in *H. maxillaris*. Teeth of the inner series small and unicuspid, arranged in two irregular rows in each jaw.

The oral mucosa appears to have shrunk; consequently the outer teeth are more exposed than those of *H. obesus* or *H. maxillaris*, but the inner teeth are deeply embedded.

Lower pharyngeal bone broken, but apparently similar to that of *H. maxillaris*; pharyngeal teeth slender.

Gill rakers moderately coarse; ten on the lower part of the first gill-arch.

Scales ctenoid; 33 scales in the lateral line; cheek with 3 or 4 series. Nuchal and pectoral scales small. Seven scales between the dorsal fin origin and the lateral line; 9 between the pectoral and pelvic fin bases.

Fins. Dorsal with XV, 8 rays, anal with III, 8. Pectoral very slightly shorter than the head; pelvic fins with the first and second soft rays of about equal length, not quite reaching the anal fin. Caudal rounded.

Vertebrae: 14 + 16 (from a radiograph, B.M. (N.H.) Reg. No. 955A).

Colour: *Adult male*. Brownish dorsally, brownish-silver on the flanks and belly. Dorsal, caudal and anal fins dusky, pelvics black.

Ecology: *Habitat*. Smith Sound, Tanganyika Territory (2° 33′ S., 32° 50′ E.) in 12 feet of water over a mud bottom (Graham, 1929).

Food. The stomach is packed with early embryos of a cichlid fish.

Affinities. Trewavas (1929) compared *H. melanopterus* with *H. obesus* (then known only from the holotype). Now that more material of *H. obesus* is available, the resemblance is found to be less marked. In some respects the morphology of the types is similar, but in the detailed structure of the head and dentition, *H. melanopterus* would seem to be more closely allied with *H. maxillaris*. It may yet prove to be merely a teratological specimen of that species.

Haplochromis parvidens (Boulenger) 1911
(Text-fig. 8)

Paratilapia parvidens Boulenger, 1911, *Ann. Mus. Genova* (3), **5**, 65, pl. I, fig. 1; *Idem*, 1915, *Cat. Afr. Fish*. **3**, 322 fig. 215.

Haplochromis nigrescens (Pellegrin) (part, holotype of *P. parvidens* only), Regan 1922, *Proc. zool. Soc. London*, 172.

Regan (1922) considered *H. parvidens* to be a synonym of *H. nigrescens* (Pellegrin) 1909. I have re-examined the holotypes of both species and find that, although at first sight the species do resemble one another, the dentition and form of the lower jaw in *H. parvidens* is most distinctive. Additional material now available confirms and emphasizes these differences. The two species also differ in their feeding habits; *H. nigrescens* is an insectivore and predator on small fishes, whilst *H. parvidens* is a specialized predator on embryo and larval fishes.

Haplochromis parvidens differs from the other larval and embryo fish-eating species in having a more slender and acutely pointed head, characters which typify the less specialized piscivorous predatory *Haplochromis* in Lake Victoria. The shape of the lower jaw is, however, unlike that of any predatory *Haplochromis* species (see Text-fig. 5B).

Description, based on 32 fishes (including the holotype) 63–163 mm. S.L.

Depth of body 33·3–38·2 per cent of standard length, length of head 33·3–37·5 per cent. Physiognomy relatively constant, the dorsal head profile straight or gently concave, sloping at an angle of 30°–35°. Preorbital depth 15·9–20·5 (M = 18·6) per cent of head length; least interorbital width 22·2–28·0 (M = 25·1) per cent. Snout length 1·2–1·33 times its breadth and 32·0–41·3 (M = 37·7) per cent of the head; eye diameter 20·3–27·2 (M = 23·0), depth of cheek 19·7–27·0 (M = 24·0) per cent.

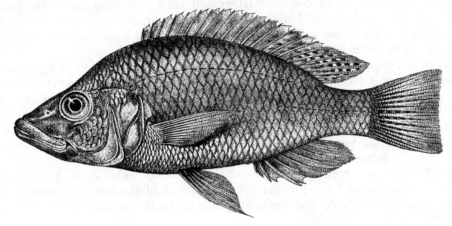

Fig. 8. *Haplochromis parvidens*; holotype (from Boulenger, *Ann. Mus. Genova*, 1911).

Caudal peduncle 13·6–16·8 per cent of standard length, 1·1–1·5 times as long as deep (modal range 1·3–1·4 times).

Mouth widely distensible and protractile, slightly oblique or horizontal when closed. Lips thickened. Lower jaw of a characteristic shape (Text-fig. 5B), somewhat rounded in cross-section and narrowing rapidly from a point about half-way along its length; consequently, the anterior part closes within the upper jaw. Length of lower jaw 43·3–55·5 (M = 48·0) per cent of head length (showing a weak positive allometry with standard length) and 1·5–2·5 (modal range 1·9–2·1) times its width. Posterior tip of the maxilla bullate and almost completely hidden beneath the preorbital, usually not reaching the vertical to the anterior orbital margin, but extending to below the anterior part of the eye in a few specimens.

Gill rakers. Nine to 11 (rarely 8 or 12, mode 10) on the lower part of the first gill-arch.

Scales ctenoid; lateral line with 30 (f.2), 31 (f.14), 32 (f.15) or 33 (f.1) scales. Cheek with 3 or 4 series; 6 or 7 (rarely 5) scales between the origin of the dorsal fin and the lateral line; 6 or 7 (less frequently 5) between the pectoral and pelvic fin bases.

Fins. Dorsal with 23 (f.1), 24 (f.10), 25 (f.19) or 26 (f.2) rays, anal with 11 (f.9), 12 (f.22) or 13 (f.1), comprising XV–XVI (rarely XIV), 9 or 10 (rarely 8) and III, 8 or 9 (rarely 10) spinous and branched rays for the fins respectively. Pectoral fin shorter than the head. First soft ray of the pelvic fin produced and extending to the anal fin; proportionately longer in adult males.

Teeth. Both the inner and outer rows of teeth are deeply embedded in the oral epithelium, with the inner series often completely hidden. The outer teeth are mainly biscuspid in fishes 63–110 mm. S.L., with some unicuspids present in larger individuals. In fishes above this size, the outer teeth are predominantly unicuspid, relatively slender and recurved; laterally placed teeth point inwards. In the three skeletons available, there are 50, 54 and 62 outer teeth in the upper jaw.

Teeth in the inner series are small, slender and weakly tricuspid in fishes less than 115 mm. S.L., but are unicuspid in larger individuals. In most fishes, the inner teeth are implanted almost horizontally, so that their crowns point backwards. One or two (rarely three) series of inner teeth occur in each jaw and are separated from the outer teeth by a small but distinct interspace.

Lower pharyngeal bone with a triangular dentigerous surface, 1·1–1·2 times as broad as long or, rarely, somewhat wider. Lower pharyngeal teeth fine and cuspidate.

Syncranium. The premaxilla and the dentary of *H. parvidens* are outstanding osteological characters. In combination they are diagnostic of the species.

In *H. parvidens*, the premaxillary pedicels are as long as, or longer than the dentigerous limb of the bone, whereas in the majority of Lake Victoria *Haplochromis* (including the large-mouthed species) the pedicels are shorter.

The mandible has been described above; its skeleton clearly shows the marked anterior narrowing and the peculiar lateral bullation of the area surrounding the bifurcation into ascending and horizontal rami (a similar swelling is also seen in *H. obesus* and *H. maxillaris*).

The maxilla is strictly comparable with that of *H. obesus* and *H. maxillaris*, but the neurocranium differs in having an elongate preorbital face. The preorbital part of the skull is about one-third of the basilar length as compared with one-fifth to one-quarter in generalized *Haplochromis*. In this respect the neurocranium of *H. parvidens* resembles that of a small predatory *Haplochromis* such as *H. nigrescens*.

Vertebrae: 13 + 16 in two skeletons examined.

Coloration in life: Adult males. Ground colour dark blue-black dorsally, silvery-blue ventrally. Dorsal fin sooty, with orange-red lappets; caudal sooty, the dorsal and ventral tips orange-red. Anal fin deep maroon, with two or three red ocelli. Pelvics black. Coloration of *immature males* unknown. *Females.* Ground colour an overall olivaceous-green, with faint indications of five to nine dark transverse bars. All fins olivaceous.

Colour in preserved material: Adult and immature males. Dark brown, some adults almost black; faint traces of five to nine transverse bars may be visible on the flanks and caudal peduncle. Dorsal fin dark, the soft part faintly maculate; pelvics black; anal and caudal fins dusky. *Females.* Golden-brown ground coloration, some faintly barred. All fins hyaline, the dorsal and caudal fins sometimes maculate.

Distribution. Lake Victoria and Lake Salisbury (a single specimen in the collections of the Uganda Game and Fisheries Department, Entebbe).

Ecology : Habitat. Like *H. maxillaris* and *H. obesus*, *Haplochromis parvidens* is apparently confined to littoral and sublittoral zones where the water is less than 50 feet deep. Unlike the former species, however, *H. parvidens* is less closely restricted to a particular substrate. Although all three species have been caught in the same habitat, the available data suggest that *H. parvidens* may be the only member of this trophic group to occur commonly over a mud bottom.

Food. Seventeen of the 60 specimens examined had food in the stomach ; of these, 15 had eaten cichlid embryos or larvae. The stomachs of the two other fishes were filled with a fatty, yellow, yolk-like substance.

Breeding. There is no information on the breeding habits of *H. parvidens*. Fishes less than 105 mm. S.L. are immature ; there is no apparent correlation between sex and maximum adult size.

Affinities. Despite the deeply embedded teeth, long premaxillary pedicels and unusual lower jaw, there is an overall similarity between *H. parvidens* and some of the structurally less-specialized predatory *Haplochromis*. *Haplochromis nigrescens*, with which *H. parvidens* was previously synonymized, exemplifies this apparent relationship. There is also some similarity between *H. parvidens* and *H. cryptodon*, and more particularly with *H. microdon*. *Haplochromis parvidens* could have evolved from either an *H. nigrescens*-like stem or from a species resembling *H. cryptodon*.

Study material and distribution records

Museum and Reg. No.	Locality	Collector
	Uganda	
Genoa Museum (Holotype)	Ripon Falls, Jinja	Bayon.
B.M. (N.H.).—1911.3.3.33	,, ,,	,,
,, 1958.1.16.95	Kaianje	E.A.F.R.O.
,, 1958.1.16.96–98	Entebbe Harbour	,,
,, 1958.1.99	Busungwe Bay	,,
,, 1958.1.16.100	Ekunu Bay	,,
,, 1958.1.16.101	Napoleon Gulf, near Jinja	,,
,, 1958.1.16.108	,, ,,	,,
,, 1958.1.16.130–139	,, ,,	,,
,, 1958.1.16.107	Macdonald Bay	,,
,, 1958.1.16.109–113	Pilkington Bay	,,
,, 1958.1.16.114–115	Njoga, Williams Bay	,,
,, 1958.1.16.116–129	Beach near Nasu Point, Buvuma Channel	,,
	Kenya	
,, 1958.1.16.104–106	Kisumu Harbour	,,
	Tanganyiha	
,, 1958.1.16.92–94	Mwanza, Capri Bay	,,

Museum and Reg. No.	Locality	Collector
Lake Victoria, Locality Unknown		
„ 1958.1.16.102–103	—	„
„ 1928.5.24.112	—	M. Graham.
„ 1928.5.24.399–400	—	„
„ 1928.5.24.401–402	—	„

Haplochromis cryptodon sp. nov.

Holotype. A male, 123 mm. standard length, from a beach near Nasu Point, Buvuma Channel, Uganda.

Description, based on 31 specimens (including the holotype) 92–130 mm. standard length.

Depth of body 27·5–35·6 per cent of standard length, length of head 30·3–34·9 per cent. Physiognomy relatively uniform, the dorsal head profile straight and sloping at an angle of 35°–40°.

Preorbital depth 12·5–17·6 (M = 15·4) per cent of head length, least interorbital width 21·2–25·7 (M = 23·6) per cent. Snout as broad as long or very slightly broader, its length 27·5–34·2 (M = 31·3) per cent of head; eye 23·1–29·4 (M = 25·8), depth of cheek 17·7–25·7 (M = 23·6) per cent.

Caudal peduncle 14·7–17·7 per cent of standard length, 1·3–1·7 (modal range 1·3–1·5) times as long as deep.

(Four specimens [two from near Nasu Point, Buvuma Channel, Uganda, and two from Majita, Tanganyika Territory] differ in being noticeably more slender [depth 27·5–31·0 per cent S.L.] and in having less steeply sloping heads. The two Uganda specimens also have a somewhat longer lower jaw [46·0 per cent head length] than is modal. In all other characters these specimens agree with the generality of individuals. Since they are amongst the five smallest specimens available, it is possible that their divergent characters may be " juvenile ").

Mouth slightly oblique and moderately distensible; lips slightly thickened. Posterior tip of the maxilla somewhat bullate and reaching or almost reaching the vertical to the anterior orbital margin. Lower jaw with a tendency to narrow rather abruptly at about its mid-point, but not narrowing so markedly as in *H. parvidens*; in some specimens (particularly individuals less than 100 mm. S.L.) this character is only visible after dissection. Length of lower jaw 39·2–46·5 (M = 42·3) per cent of head length, 1·3–1·9 (modal range 1·5–1·6) times as long as broad.

Gill rakers moderately slender, 10 or 11 (less frequently 9), on the lower limb of the anterior arch.

Scales ctenoid; lateral line with 30 (f.2), 31 (f.7), 32 (f.17), 33 (f.2) or 34 (f.2) scales. Cheek with 2 or 3 (rarely 4) series; 5–7 scales between the lateral line and the origin of the dorsal fin; 6–8 (rarely 9) between the pectoral and pelvic fin insertions.

Fins. Dorsal with 23 (f.1), 24 (f.14) or 25 (f.16) rays, anal with 11 (f.7), 12 (f.23) or 13 (f.1) comprising XV–XVI, 9 or 10 (rarely 8) and III, 8 or 9 (rarely 10) spinous and soft rays for the fins respectively. Pectoral fins shorter than the head. First pelvic ray produced and extending to the spinous part of the anal.

Teeth. The inner and outer series of teeth are deeply embedded in the oral epithelium, so that only the tips of the outer teeth are visible. In specimens less than 100 mm. S.L. the outer teeth are weakly and unequally bicuspid. In larger fishes this row is composed of small, unicuspid and slightly recurved teeth.

Teeth in the inner series are either unicuspid or weakly tricuspid, and are arranged in one or two rows. In fresh material it is usually impossible to see these teeth unless the oral mucosa is dissected away.

Lower pharyngeal bone triangular, its dentigerous area 1·0–1·5 times as broad as long; pharyngeal teeth slender and cuspidate.

Syncranium. The syncranium of *H. cryptodon* resembles the generalized *Haplochromis* type, except that the maxilla is somewhat stouter and the dentary shows an incipient departure from the generalized condition towards that of *H. parvidens* (see p. 196 and text fig. 5B).

Vertebrae: 14 + 16 in the single specimen examined (Radiograph, B.M. (N.H.) Reg. No. 958).

Coloration in life: Adult and immature females. Ground colour dark green-brown shading to light gold ventrally. All fins hyaline. The live coloration of *males* is unknown.

Colour of preserved material: Adult males. Ground colour dark gun-metal dorsally, shading to greyish-green on the flanks and ventral surfaces; chest and branchiostegal membrane dusky; faint traces of a coppery sheen on the operculum and flanks. Five to seven faint but dark transverse bars are visible on the flanks and caudal peduncle. Dorsal fin dusky; caudal and anal fins hyaline, the latter with two to four dead-white ocelli; pelvic fins black. *Females*. Ground colour light golden-yellow, slightly darker dorsally; in some individuals there are faint traces of about five, broad, transverse bands on the flanks. One adult is of particular interest since it displays incipient male coloration; the pelvic fins are dusky as are the chest and branchiostegal membrane. In addition there are traces of a coppery sheen on the operculum.

Distribution. Habitat. H. cryptodon has been recorded from only three localities, namely, the Napoleon Gulf near Jinja, a beach near Nasu Point (Buvuma Channel) and a beach at Majita, Tanganyika Territory. The apparent absence of *H. crytodon* from other localities is difficult to explain since it was one of the more abundant species at the Nasu Point station and formed a regular element of the seine-net catches there. Perhaps it is significant that the majority of *H. cryptodon* caught at Nasu Point were brooding or " ripe " females, thus suggesting that the area is used as a breeding ground. If this is so, the species may normally occur in some other habitat which could not be fished by conventional gear.

Food. Only one fish, a female from Nasu Point, had ingested material in the stomach and intestine. The stomach was packed with recently fertilized cichlid ova, whilst numerous small fish vertebrae were found in the posterior intestine. The stomach contents may have been the female's own brood, but the presence of larval fish vertebrae in the faeces cannot be explained on the same grounds.

With such little positive evidence it is impossible to generalize on the feeding habits of *H. cryptodon*. But the single record of gut-contents, taken together with the jaw structure of this species, suggests a diet of embryo and larval fishes.

Breeding. *H. cryptodon* is a female mouth-brooder. The smallest individual caught, a female 92 mm. S.L., was sexually mature. Males and females reach the same adult size.

Affinities. With the exception of *H. cronus*, *H. cryptodon* is the most generalized of the species referred to this trophic group. In structure and proportions it shows greater affinity with *H. microdon* and *H. parvidens* than with the *H. obesus-H. maxillaris-H. cronus* section of the group. *H. cryptodon* was probably evolved from the complex of piscivorous-insectivorous species which are not markedly differentiated (except for their larger size) from the generalized *Haplochromis* stock in Lake Victoria.

Diagnosis. From the generality of Lake Victoria *Haplochromis* species, *H. cryptodon* may be distinguished by its distensible mouth and almost completely hidden dentition. From other species showing these characters, it is distinguished by the shape of the head (and particularly of the lower jaw) and an absence of teeth with anteriorly directed crowns.

Study material and distribution records

Museum and Reg. No.	Locality	Collector
	Uganda	
B.M. (N.H.).—1958.1.16.31 (Holotype, *H. cryptodon*)	Beach near Nasu Point, Buvuma Channel	E.A.F.R.O.
„ 1958.1.26.32–33	„	„
„ 1958.1.16.37–62	„	„
„ 1958.1.16.34	Napoleon Gulf, near Jinja	„
	Tanganyika	
„ 1958.1.16.35–36	Majita	„

Haplochromis microdon (Boulenger) 1906
(Text-fig. 9)

Pelmatochromis microdon (part; holotype only) Blgr., 1906, *Ann. Mag. nat. Hist.* (7) **17**, 441; *Idem*, 1913, *Cat. Afr. Fish.* **3**, 412, fig. 282.
Haplochromis microdon (Blgr.), (part), Regan, 1922, *Proc. Zool. Soc., London*, 173.

When redefining *H. microdon* Regan (1922) noted his belief that the type specimen had a malformed lower jaw (which did not bite against the upper) and that its small teeth were due to this malformation. In the light of additional specimens, I am unable to agree with Regan, and conclude that the shape of the lower jaw and its small teeth are, indeed, some of the diagnostic characters of the species. Consequently, I find that the other species which Regan referred to *H. microdon* can no longer be considered conspecific; they will be dealt with in a subsequent paper.

The peculiar lower jaw of *H. microdon* closely resembles that of *H. parvidens* but the two species differ in other osteological characters.

A REVISION OF THE LAKE VICTORIA *HAPLOCHROMIS* SPECIES

Description, based on eight specimens (including the holotype) 114–148 mm. S.L.

Depth of body 33·1–37·6 per cent standard length, length of head 31·2–34·8 per cent. Physiognomy variable and dependent on whether the dorsal head profile is moderately or strongly concave. Preorbital depth 14·6–18·8 (M = 17·1) per cent of head length, least interorbital width 24·6–29·6 (M = 26·6) per cent. Snout as long as broad or very slightly longer; its length 30·8–34·5 (M = 32·5) per cent of head, eye diameter 24·0–28·2 (M = 25·5), depth of cheek 19·2–26·0 (M = 23·8) per cent.

Caudal peduncle 15·3–17·0 per cent of standard length, 1·3–1·5 times as long as deep.

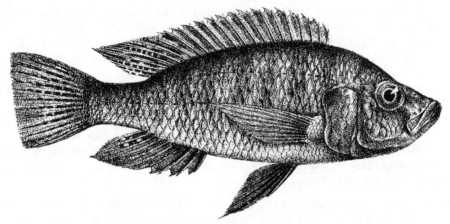

FIG. 9. *Haplochromis microdon*; holotype (from Boulenger, *Fishes of the Nile*).

Mouth oblique, distensible and moderately protractile. Jaws equal anteriorly or the lower very slightly shorter; lower jaw always closing within the upper, its length 43·5–48·0 (M = 46·2) per cent of head length, 1·6–2·3 (modal range 1·9–2·2) times as long as broad. Premaxillary pedicels shorter than the dentigerous limb. Posterior tip of the maxilla slightly bullate, partly hidden by the preorbital and extending to the vertical to the anterior orbital margin, or almost so.

[One specimen, an adult female 114·0 mm. S.L., from Pilkington Bay, is not included in the general description given above. Although it differs sufficiently from the other specimens to raise doubts as to its identity, I do not consider it to be a distinct species. The possibility of this fish being a hybrid between *H. microdon* and some other species (? *H. parvidens*) cannot, however, be excluded.

The dentition and lower jaw are of the "*microdon-parvidens*" type, but the dentary is narrower anteriorly and it is shorter than even the extreme specimens of either species. These characters, coupled with the large eye and short snout give this fish an unusual appearance which departs from both the "*microdon*" and the "*parvidens*" types. The mouth is oblique and the premaxillary pedicels short, so that the sum of characters places the specimen nearest *H. microdon*.

The principal morphometric characters are :

Depth*	Head*	Po. %	Io. %	Snt. %	Eye %	Cheek %	L.j. %	C.P.*
33·3	34·2	18·0	25·2	29·5	30·8	20·5	41·0	15·8

* Percentage S.L.
% Percentage head-length.]

Gill-rakers slender, 9–11 (mode 11) on the lower part of the anterior arch.

Scales ctenoid ; lateral line with 30 (f.2), 32 (f.4) or 33 (f.2) scales.

Fins. Dorsal fin with 24 (f.4) or 25 (f.4) rays, anal with 10 (f.1), 11 (f.3) or 12 (f.4), comprising XV–XVI, 9 or 10 and III, 7–9 spinous and branched rays for the fins respectively. Pectoral fin shorter than the head ; first soft pelvic ray produced, reaching to the vent in females and to the spinous anal fin in males.

Teeth. All the teeth are so deeply embedded in the oral epithelium that they are invisible in fresh material. The teeth in the outer series of both jaws are small, unicuspid and slightly recurved. The single or rarely double row of unicuspid inner teeth is implanted at an acute angle. A distinct interspace separates the inner and outer series of teeth.

Lower pharyngeal bone triangular, the dentigerous area slightly broader than long ; lower pharyngeal teeth slender and cuspidate.

Syncranium (from a radiograph, B.M. (N.H.) Reg. No. 961, and a partial dissection). The dentary is almost identical with that of *H. parvidens*, but is more slender. The premaxilla is of a generalized type and does not have the elongated pedicels which characterize *H. parvidens*. The neurocranium is apparently intermediate between that of *H. cryptodon* and that of *H. parvidens*.

Colour of preserved specimens : Adult males. Ground colour very dark brown, the ventral abdominal region lighter in one specimen ; five or six, broad and dark transverse bars visible on the flanks of some individuals. Branchiostegal membrane black. Dorsal, caudal, pelvic and anal fins dark, almost black, the anal with four, whitish ocelli.

Adult females. Ground colour greyish-brown, darker dorsally ; very faint indications of five or six broad transverse bars on the flanks. Dorsal, anal, and caudal fins hyaline, the last weakly maculate especially on the upper half ; pelvic fins black.

Distribution. Lake Victoria.

Ecology : Habitat. Seven of the eight fishes examined were from littoral zones of the lake, and were taken over both firm and soft substrates ; two were caught at a depth of 30–36 feet and the others in water 6–12 feet deep. The habitat of the eighth specimen is not known.

Food unknown. Since the dentition and jaw structure of *H. microdon* so closely resemble those of *H. parvidens*, the feeding habits of the two species may well be similar.

Breeding. No data are available ; the eight specimens are all adults.

Affinities. The most obvious relative of *H. microdon* is *H. parvidens* ; the lower jaw in both species is of a type otherwise unrepresented in the Lake Victoria *Haplochromis* species flock. The dentary of *H. microdon* is, however, a more extreme modification of the generalized type than is the dentary of *H. parvidens*. On the other hand the elongated premaxillary pedicels of *H. parvidens* do not occur in *H. microdon* which retains a premaxilla of the generalized type. Consequently the mouth is less protrusible in this species. *H. parvidens*, in its gross morphology, and particularly in the shape of its neurocranium, shows strong affinity with some of the piscivorous predators (e.g. *H. nigrescens*). In corresponding characters, *H. microdon* is near *H. cryptodon*, from which it could be derived by further specialization of the lower jaw. Thus, the resemblance between *H. microdon* and *H. parvidens* may be the result of convergent evolutionary trends.

Diagnosis. Haplochromis microdon may be distinguished from other species with deeply embedded teeth by the shape of the lower jaw, the concave dorsal head profile and the oblique mouth. It may be further distinguished from *H. parvidens* by its having premaxillary pedicels which are shorter than the horizontal (dentigerous) limb of this bone.

Study material and distribution records

Museum and Reg. No.	Locality	Collector
	Uganda	
B.M. (N.H.).—1906.5.30.309 (Holotype *Pelmatochromis microdon*)	Bunjako	Degen.
,, 1958.1.16.24	Pilkington Bay	E.A.F.R.O.
,, 1958.1.16.25	Entebbe Harbour	,,
,, 1958.1.16.26	Ekunu Bay	,,
,, 1958.1.16.29–30	Beach near Nasu Point, Buvuma Channel	,,
	Lake Victoria, Locality Unknown	
,, 1958.1.16.27–28	—	,,

DISCUSSION

The embryo-larval fish-eating habits of this species-group were briefly mentioned in the introduction. As far as I can determine from published accounts, no other African cichlids have occupied this particular niché.

That the group preys almost exclusively on the embryos and larvae of other cichlids is probably due to the fact that only the Cichlidae breed continuously; the non-cichlid fishes spawn biannually, at the periods of maximum rainfall. On the other hand, the mouth-brooding habits of most Lake Victoria *Haplochromis*, and both *Tilapia* species, would seem to provide very little opportunity for these predators unless they have evolved a method of forcing the parent to jettison its brood. Since embryos at all stages of development have been found in the stomach contents, it is clear that the species do not obtain their food solely by preying on newly fertilized eggs before these are picked up by the brooding parent. The method of attack is unknown, but from the large number of embryos or larvae taken it must be highly efficient.

Two anatomical features (a distensible and somewhat protrusible mouth, and a weak to moderately developed dentition deeply embedded in the thickened oral epithelium) characterize six of the seven species in this trophic group. It is not known if either of the characters has any functional significance in connection with the feeding habits. The seventh species, *H. cronus*, has strong and fully-exposed teeth, and the mouth is not noticeably distensible or protractile. Unfortunately, only a few specimens of *H. cronus* had food in the stomach. However, it may be significant that these fishes had fed on larger (*ca.* 11 mm.) late larvae, whilst members of the other species had fed on embryos and early larvae. That is to say, *H. cronus* had taken young of an age when they frequently leave the parent's mouth for short periods. Those species with distensible, protractile mouths and hidden teeth had taken mainly intra-oral young. Perhaps these larger mouthed species engulf the head or mouth of a brooding female and in this way force the parent to abandon its young?

As in most other trophically defined species-groups there is evidence of both phyletic and convergent relationships between the species. But, unlike most others, this group shows greater divergence and more tenuous relationships within the apparently phyletic lines.

Haplochromis cronus is the least specialized species, but it differs from the generality of Lake Victoria *Haplochromis* in having the caudal fin almost completely scaled (a characteristic of Lake Nyasa *Haplochromis*).

Haplochromis obesus and *H. maxillaris* seem to be much specialized derivatives of a form resembling *H. cronus*, but neither of these species has a scaled caudal fin. The dentition is similar in *H. obesus* and *H. maxillaris* and quite unlike that of *H. cronus*; in other characters (especially the shape of the lower jaw) the two species are markedly different. It is, in fact, impossible to decide whether the species are of the same lineage or the descendants of distinct but related ancestral stocks.

A similar state of affairs exists when *H. cryptodon* and *H. microdon* are considered. A further complication is introduced by the resemblance between *H. microdon* and *H. parvidens*. In this case, however, it is possible that *H. microdon* was derived from an *H. cryptodon*-like ancestor, and *H. parvidens* from one of the less-specialized piscivorous predators.

In all these species, anatomical differences between members of possible lineages are certainly greater than those encountered between species in the algal-grazing and mollusc-"shelling" groups (Greenwood, 1956*b* and 1957).

Two of the species described above, *H. obesus* and *H. maxillaris*, clearly demonstrate a phenomenon which is common amongst the Lake Victoria *Haplochromis*, namely, the intraspecific constancy of certain osteological and dental characters contrasting with the variability of other and often anatomically related characters. For example, the lower jaw and dentition of *H. obesus* is readily diagnostic, whilst the physiognomy is so variable that difficulty would be experienced in identifying some specimens were it not for the characteristic lower jaw. Likewise, there is marked variation in the gross morphology of *H. maxillaris*, yet the dentary and the dentition of both jaws are relatively constant. *Haplochromis parvidens*, however, shows only slight variation in its gross morphology.

In certain characters, *Haplochromis taurinus* Trewavas, of Lake Edward, resembles fishes of the *H. maxillaris-H. obesus* complex. The shape of the head approaches that of *H. maxillaris*, particularly with regard to the stout jaws and thickened lips; also, the outer teeth in the upper jaw are hidden by a fold of lip-tissue. The likeness does not extend to the dentition, which is a critical character in this group. The teeth of *H. taurinus* are large, distinct and biscuspid (at least in the holotype, a fish 164 mm. total length) and are of a form rarely encountered in any Lake Victoria *Haplochromis*. There can be little doubt therefore, that the resemblance between *H. taurinus* and *H. maxillaris* or *H. obesus* is superficial and of little phyletic significance.

Haplochromis plagiodon Regan & Trewavas 1928

(Text-fig. 10)

Haplochromis crassilabris Blgr. (part), 1906, *Ann. Mag. nat. Hist.* (7) **17**, 445.
Paratilapia crassilabris (Blgr.), part, Blgr., 1915, *Cat. Afr. Fish.* **3**, 345.
Haplochromis crassilabris Blgr. (part), Regan, 1922, *Proc. zool. Soc. London*, 167.
Hemitilapia bayoni Blgr. (part), Blgr., 1908, *Ann. Mus. Genova* (3), **4**, 6; *Idem*, 1915, *Cat. Afr. Fish.* **3**, 491.
Clinodon bayoni (Blgr.) (part), Regan, 1920, *Ann. Mag. nat. Hist.* (9) **5**, 45 (footnote).
Haplochromis obliquidens Hilgendorf (part), Regan, 1922, *Proc. zool. Soc. London*, 188.
Haplochromis plagiodon Regan & Trewavas, 1928, *Ann. Mag. nat. Hist.* (10), **2**, 224.

Description, based on five specimens (including the holotype), 56–85 mm. standard length.

H. plagiodon is a generalized species, except that it has teeth of an unusual form, resembling stout and somewhat modified versions of the teeth found in *H. lividus* Greenwood. Although *H. plagiodon* is represented by only five specimens, the form of the teeth is sufficiently distinctive and constant to warrant the assumption that the species is biologically valid.

The principal morphometric characters for each of the five specimens are tabulated below. All are males.

S.L.	Depth*	Head*	Po. %	Io. %	Snt. %	Eye %	Cheek %	L.j. %	C.P.*
56·0	35·7	31·2	13·1	25·7	28·6	30·2	21·7	34·2	17·0
72·0	36·0	30·5	15·0	27·2	27·2	29·2	22·7	36·3	16·7
76·0	36·8	31·6	15·9	25·0	29·2	31·3	23·0	37·5	15·2
†81·0	32·7	29·6	16·7	28·4	29·2	28·4	20·8	34·6	18·0
85·0	36·5	30·3	15·6	26·9	26·9	30·8	23·0	34·6	16·5

* Percentage standard length.
% Percentage head length.
† Holotype.

Dorsal head profile straight and steeply sloping (*ca.* 50°). Mouth horizontal; posterior tip of the maxilla extending to the vertical to the anterior orbital margin, or slightly beyond. Jaws equal anteriorly, the lower 1·3–1·6 times as long as broad; lips not thickened.

Teeth. Outer teeth stout, erect and bicuspid, with the major cusp obliquely truncate and somewhat compressed, the minor cusp conical. In one specimen

(72 mm. S.L.) the teeth are very worn, so that the minor cusp is indicated merely as a faint groove on the labial aspect of the tooth. There are 30–38 teeth in the outer series of the upper jaw.

In most respects, the shape of these teeth closely resembles one of the variants occurring in *H. lividus* (see Text-fig. 2B in Greenwood, 1956b), except that in *H. plagiodon* the teeth are not recurved, are stouter and do not have a distinct neck.

The inner teeth are tricuspid and arranged in two or three rows in each jaw and are separated from the outer row by a distinct interspace.

Lower pharyngeal bone triangular, fairly stout (as compared with, for example, fishes of the *H. lividus-H. nuchisquamulatus* group); dentigerous area 1·2–1·5

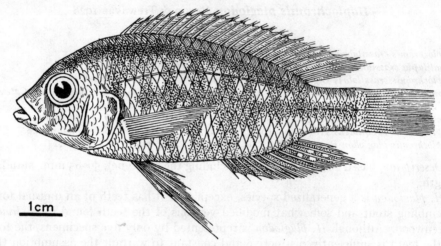

Fig. 10. *Haplochromis plagiodon*. Drawn by Miss D. Fitchew.

times as broad as long. Lower pharyngeal teeth cuspidate, those of the two median series enlarged; conical in three specimens but cuspidate in two others.

Gill rakers short and stout; 7 or 8 on the lower limb of the anterior arch.

Scales ctenoid; lateral line with 31 (f.1), 32 (f.3) or 33 (f.1) scales. Cheek with 3 or 4 (rarely 2) series; 6 or 7 scales between the dorsal fin origin and the lateral line, 8 or 9 between the pelvic and pectoral fin bases.

Fins. Dorsal with 24 (f.1), 25 (f.2) or 26 (f.2) rays, anal with 11 (f.1), 12 (f.3) or 13 (f.1) comprising XV–XVI, 9 or 10 and III, 8–10 spinous and branched rays for the fins respectively. First soft pelvic ray produced and extending to the spinous part of the anal fin. Pectoral fin slightly shorter than the head.

Coloration. Unknown in life and known only for preserved males. Ground colour silver-grey to brownish-grey; eight to ten transverse bars on the flanks and caudal peduncle, a fairly well-defined mid-lateral stripe, with indications of an interrupted band running slightly above the upper lateral line. Faint traces of two interocular bands and a lachrymal stripe. Dorsal, caudal and anal fins pale and immaculate;

six to eight small ocelli arranged in two rows on the anal fin. Pelvic fins dusky on the outer two-thirds in two specimens and entirely pale in the remainder.

Ecology. No ecological data are available for the type and one other specimen; the three other fishes were caught in a seine-net operated in shallow water over an exposed, sandy beach at Entebbe, Uganda, In two of these specimens, remains of larval Diptera and Ephemeroptera (together with many fine sand grains) were found in the intestines.

Diagnosis and affinities. Haplochromis plagiodon may be distinguished from other species in Lake Victoria by its peculiar teeth. The relatively coarse lower pharyngeal bone and the enlarged median pharyngeal teeth, together with the stout, firmly implanted and few (30–36) outer teeth, separate *H. plagiodon* from the two other species (*H. lividus* and *H. nuchisquamulatus*) with obliquely truncated, biscuspid outer teeth.

Regan & Trewavas (1928) compared the teeth of *H. plagiodon* holotype with those of the type of *Bayonia xenodonta* Blgr. (now considered a synonym of *Macropleurodus bicolor* (Blgr.), Greenwood, 1956a). They emphasized the differences existing between the two species, even though there appeared to be some resemblance in the shape of the teeth. Now that additional specimens of comparable sizes are available for both species, it is clearer than ever that *H. plagiodon* is not closely related to *Macropleurodus bicolor*. Regan & Trewavas also suggested that *H. plagiodon* might be related to *H. humilior* (Blgr.). Although both these species have somewhat enlarged lower pharyngeal bones and median pharyngeal teeth, the resemblance in gross morphology is less marked and ceases when the oral dentition is compared. Likewise, although the oral teeth of *H. plagiodon* resemble those of *H. lividus*, the latter species has a fine lower pharyngeal bone with slender, cuspidate median teeth, and finer, more numerous oral teeth.

Thus, it is only possible to suggest that *H. plagiodon* represents an independent offshoot from the generalized *Haplochromis* stem.

Study material and distribution records

Museum and Reg. No.	Locality	Collector
	Uganda	
B.M. (N.H.).—1909.5.4.29 (Holotype *H. plagiodon*)	Sesse Isles	Bayon.
„ 1906.5.30.427	Entebbe	Degen.
„ 1958.1.16.245–247	Entebbe, Bugonga Beach	E.A.F.R.O.

Haplochromis chilotes (Blgr.) 1911
(Text-fig. 11)

Paratilapia chilotes Blgr., 1911, *Ann. Mus. Genova* (3), **5**, 68, pl. II, fig. 2 ; *Idem, Cat. Afr. Fish.* **3**, 338, fig. 228.
Haplochromis chilotes (Blgr.), Regan, 1922, *Proc. zool. Soc. London*, 170.

As defined below, *H. chilotes* exhibits a high degree of individual variability. The species may be divided into two morphotypes : first, those individuals with hypertrophied and lobed lips and secondly, those in which the lips are thickened but not produced into well-defined lobes. Seventeen of the 25 specimens available fall into the first category and eight into the second. The division is not sharp, however, since the lips of some individuals in the second category do have a lobe-like, median swelling.

Certain other characters are apparently correlated with the extent of lip development. For example, in most specimens with strongly lobed lips, the upper dental arcade is narrow and acutely rounded anteriorly. Also, in these fishes, the lower jaw is usually longer. The correlation is not complete and some lobe-lipped fishes

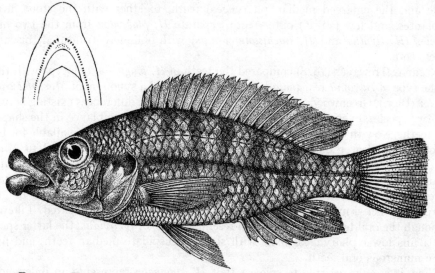

FIG. 11. *Haplochromis chilotes* ; holotype (from Boulenger, *Fishes of the Nile*).

have a mixture of these characters. It is because the sample as a whole shows a complete intergradation of lip, jaw and dental characters that I consider the material to represent a single species. Furthermore, I can discover no ecological differences between the two morphotypes, nor is there any obvious difference in the breeding coloration of the males. Nevertheless, the present arrangement should probably be considered tentative until more material and further field observations are available.

Description, based on 25 specimens (including the holotype) 70–148 mm. standard length.

Depth of body 32·5–40·8 per cent of standard length ; length of head (excluding the lips) apparently correlated with the degree of lip hypertrophy, 31·4–34·7 (M = 32·7) per cent of standard length in fishes without clearly lobed lips and 32·0–38·2 (M = 35·5) per cent in those with lobed lips. Dorsal head profile straight and gently sloping, or slightly decurved. Preorbital depth 15·0–18·2 (M = 16·8) per cent of head ; least interorbital width 19·3–27·2 (M = 23·8) ; snout longer than broad in lobe-lipped fishes and as broad as long in others, its length 30·8–38·4 (M = 34·0) per cent of head; eye 28·6–21·8 (M = 25·4), depth of cheek 18·1–25·4 (M = 20·5) per cent.

Caudal peduncle 12·5–17·2 (M = 15·3) per cent of standard length.

Mouth horizontal, posterior tip of the maxilla reaching or almost reaching the vertical to the anterior orbital margin. Lips thickened, grossly so in some specimens in which each lip is produced medially into a tongue-shaped or globose lobe; in others there may be an incipient lobe or even no indication of any such development. Jaws equal anteriorly, the lower proportionately longer (36·0–49·0, M = 39·6 per cent of head) in fishes with lobed lips than in the others (30·0–36·6, M = 33·2 per cent).

Gill-rakers short and stout, 7–9 on the lower limb of the anterior arch.

Scales ctenoid; lateral line with 31 (f.3), 32 (f.10), 33 (f.9), 34 (f.2) or 35 (f.1) scales. Cheek with 3 (rarely 2 or 4) series. Seven or 8 (rarely 6) scales between the origin of the dorsal fin and the lateral line; 8 or 9 (less commonly 6, 7 or 10) between the pectoral and pelvic fin bases.

Fins. Dorsal with 24 (f.3), 25 (f.18) or 26 (f.4) rays, anal with 11 (f.3) or 12 (f.22), comprising XV–XVI, 9 or 10 and III, 8 or 9 spinous and branched rays for the fins respectively. Pectoral fin shorter than the head; first soft pelvic ray produced, variable in its posterior extension but proportionately longer in adult males than females. Caudal subtruncate.

Teeth. In most specimens with hypertrophied and lobed lips the outer teeth in the upper jaw are arranged in an acutely rounded arcade, a pattern not found in any other *Haplochromis* from Lake Victoria. Specimens with thickened but non-lobed lips have more broadly rounded dental arcades, which are, nevertheless, more acutely rounded anteriorly than those of most other species. Complete intergradation exists between the various arcade shapes; the correlation between lip development and dental pattern is not complete since some fishes with lobed lips have a broad arcade.

In fishes 70–108 mm. standard length the outer tooth row is composed of unicuspid and weakly bicuspid, slender and slightly recurved teeth. In larger individuals these teeth are always unicuspid, are stouter and very slightly, if at all, curved. From 16 to 46 teeth may occur in the outer series of the upper jaw; there is apparently some correlation between lip development and the number of teeth, with a tendency for lobe-lipped fishes to have more teeth.

The inner series are composed of tricuspid and weakly tricuspid teeth in fishes less than 110 mm. standard length, and of predominantly unicuspid teeth in larger individuals. These teeth are arranged in two or three rows (less commonly four or one) anteriorly in each jaw.

Lower pharyngeal bone triangular, the dentigerous area 1·0–1·5 (modal range 1·2–1·3) times as broad as long. With one exception, all specimens have the median teeth (particularly the posterior few pairs) enlarged but still clearly cuspidate. In the exceptional specimen, the median teeth are slender.

Coloration in life: Breeding males. Ground colour greyish-black or black; lips and branchiostegal membrane black. Dorsal fin black, lappets and margin of soft part red, as are the maculae between the branched rays. Caudal black basally, but light yellow or hyaline distally ; anal yellow or hyaline, with three or four reddish-yellow ocelli ; pelvics black. *Non-breeding males* : Ground colour variable, usually greyish-black ; a dark mid-lateral band and a less intense, interrupted, wavy dorsal band, are generally visible. Dorsal, caudal and anal fins greyish ; pelvics black.

Females and immature males. Ground colour greyish-silver, creamy-white on the chest and ventral surface of the head. Lips grey or cream ; six to nine dark transverse bars on the flanks and caudal peduncle, which are broadest at the points of intersection with the broad, mid-lateral stripe and the narrower, interrupted dorsal band. Dorsal caudal and anal fins greyish ; pelvics hyaline.

One of the eight females available was a piebald, black and yellow " *bicolor* " variant (p. 212) ; it was of the group with hypertrophied and lobed lips.

Preserved material : Sexually active males. Ground colour blackish, lower jaw, branchiostegal membrane and chest dusky ; lips usually pale but dusky in a few specimens. A dark, mid-lateral stripe and a less intense dorsal band, together with six or seven broad transverse bars, are usually visible on the flanks. Dorsal fin black, with light lappets and margin to the soft part. Anal fin pale orange-yellow, with three or four white ocelli arranged in a single row. Caudal fin black on the basal third to half, orange-yellow distally. Pelvic fins black, somewhat lighter medially. *Sexually inactive males* have a similar coloration except that the ground colour is lighter, as is the lower jaw and the branchiostegal membrane, whilst the chest is silvery. *Females and juvenile males* are brownish-fawn, lightest ventrally ; the banding and barring is more obvious than in adult males. All fins hyaline, with the soft part of the dorsal fin and the caudal fin maculate.

Distribution. Known from Lake Victoria and probably also from the Victoria Nile since the type locality is given as " Victoria Nile at Ripon Falls " (Boulenger, 1911).

Ecology : Habitat. H. chilotes is apparently confined to the littoral and sublittoral zones of the lake, where the depth of water is less than 50 feet and to localities with a hard substrate (sand, shingle and rocks). Only two specimens were caught over a mud bottom. Furthermore, it seems that the species may be confined to sheltered bays and gulfs ; with two exceptions, no *H. chilotes* have been recorded from exposed habitats. The exceptional fishes were caught near islands lying some distance from the mainland.

Food. Eight of the 23 specimens contained food in the stomach and intestines ; these fishes represent five different localities, four of which are geographically distant from one another.

One fish had fed almost exclusively on prawns (*Caridina nilotica* Roux) and the others on insect larvae. Larvae of the boring may-fly (*Povilla adusta* Navás) formed the main insect prey, and it was noticed that the silk case of the larva had also been ingested. Other insects eaten included Trichoptera and Diptera larvae. From the amount of plant debris and sand occurring in the stomach it would seem that *H. chilotes* feeds from the bottom.

Breeding habits. Females 70 mm. S.L. are sexually active, but the smallest adult male in this sample was 97 mm. S.L.; from the available data it seems probable that males reach a greater adult size than females. One female was found with three late larvae in the buccal cavity; since the condition of this fish was clearly " spent " it is assumed that the larvae were part of a larger brood which was lost when the female was captured.

Morphology of late larval H. chilotes. The three young fishes (all *ca*. 10 mm. S.L.) referred to above are indistinguishable from the larvae of other and unrelated *Haplochromis* species. Although the female parent had hypertrophied and lobed lips, no trace of these characters was visible in the larvae.

Affinities. *Haplochromis chilotes* was probably derived from an *H. chromogynos*-like ancestor by further development of the lips, narrowing of the head and the consequent effects on the dental pattern. *H. chilotes* without lobed lips resemble *H. chromogynos* more closely than do specimens with lobed lips. But the similarity between the two species, even at its closest, is less marked than for example, that between *H. sauvagei* and *H. prodromus* (Greenwood, 1957).

There is a striking superficial resemblance between *H. chilotes* and *Lobochilotes labiatus* (Blgr.) of Lake Tanganyika. The range of lip development in *L. labiatus* is about equal to that of *H. chilotes* but with this difference; it is positively correlated with size in *L. labiatus*. Small fishes have weakly lobed lips whilst, in larger individuals the lobes are well developed. Tooth form in *Lobochilotes* is quite unlike that of *H. chilotes*, but there is a tendency for the dental arcade to be acutely rounded anteriorly.

Lobed lips are also developed in *Haplochromis lobochilotes* of Lake Nyasa and thus there is some resemblance between this species and *H. chilotes*. In this instance, however, the similarity is less marked because the general proportions of the two species are somewhat different; again, the form of the teeth is dissimilar.

Of the lobe-lipped species occurring outside Lake Victoria, *Melanochromis labrosus* Trewavas, of Lake Nyasa shows the greatest overall and detailed likeness with *H. chilotes*. The dentition of *M. labrosus* and *H. chilotes* of a comparable size is similar, as is the gross and finer morphology of the head. The nuchal and pectoral squamation of *M. labrosus* is, however, markedly smaller than that of *H. chilotes*. Unfortunately, *M. labrosus* is known only from one specimen so a detailed comparison of the two species cannot be made.

Diagnosis. *H. chilotes* with hypertrophied and lobed lips are immediately distinguishable from other Lake Victoria species on this character alone; from *H. lobochilotes* of Lake Nyasa and *Lobochilotes labiatus* of Lake Tanganyika, it is distinguished by differences in the shape of the teeth and by various morphometric characters *Haplochromis chilotes* with moderately developed and weakly or non-lobed lips may be confused with *H. chromogynos* or with species of the *H. sauvagei* complex. They are immediately distinguishable from the latter group in having fewer rows of inner teeth and by the shape of the outer teeth, which do not have strongly recurved tips. From *H. chromogynos*, *H. chilotes* is distinguished by its narrower interorbital region and longer snout.

Study material and distribution records

Museum and Reg. No.	Locality	Collector
	Uganda	
Genoa Museum	Jinja, Ripon Falls	Bayon.
(Type *Paratilapia chilotes*)		
B.M. (N.H.).—1911.3.3.33 .	,, ,, .	,,
(Paratype *P. chilotes*)		
,, 1958.1.16.3 .	Ramafuta Is. .	E.A.F.R.O.
,, 1958.1.16.4 .	Pilkington Bay .	,,
,, 1958.1.16.5 .	Lukula Is. .	,,
,, 1958.1.16.6 .	Channel between Dagusi Is. and mainland .	,,
,, 1958.1.16.9 .	Ekunu Bay .	,,
,, 1958.1.16.10–16 .	Off south tip of Buvuma Is. .	,,
,, 1958.1.16.17–23 .	Napoleon Gulf, near Jinja .	,,
	Tanganyika	
,, 1958.1.16.1 .	Kaseiraji Is. .	,,
,, 1958.1.16.8 .	Godziba Is. .	,,
	Kenya	
,, 1958.1.16.2 .	Kisumu Harbour .	,,
	Lake Victoria, Locality Unknown	
,, 1958.1.16.7 .	— .	,,

Haplochromis chromogynos sp. nov.

Haplochromis bicolor Blgr. (part), 1906, *Ann. Mag. nat. Hist.* (7) **17**, 444.
Paratilapia bicolor (Blgr.) (part), Blgr., 1915, *Cat. Afr. Fish.* **3**, 346.
Haplochromis sauvagei (Pfeff.), (part), Regan, 1922, *Proc. zool. Soc. London*, 167.

Haplochromis chromogynos is unique amongst the *Haplochromis* of Lake Victoria (and probably the other lakes as well) since a high percentage, if not all, of the females have a piebald, black and yellow " *bicolor* " coloration. " *Bicolor* " female variants are known to occur in several *Haplochromis* species, but in none does the frequency of the variants exceed about 30 per cent. All 22 specimens of female *H. chromogynos* are " *bicolor* ". These fishes were collected from several different parts of the lake and include fishes at various stages of sexual development.

Holotype. A female, 79 mm. standard length, from the Napoleon Gulf, near Jinja, Uganda.

Description, based on 29 specimens (including the holotype) 50–110 mm. standard length.

Depth of body 32·5–42·3 (M = 35·0) per cent of standard length ; length of head 30·4–37·3 (M = 33·2) per cent. Dorsal head profile curved and moderately declivous. Preorbital depth 13·2–17·0 (M = 15·4) per cent of head length, least interorbital width 22·6–31·4 (M = 27·5) per cent. Snout slightly longer than broad, or less commonly, broader than long, its length 26·3–33·3 (M = 30·7) per cent of head ; eye 25·7–32·7 (M = 28·6) ; depth of cheek 17·9–24·6 (M = 21·6) per cent of head.

Caudal peduncle 13·6–18·5 per cent of standard length, 1·1–1·6 times as long as deep.

Mouth horizontal; jaws equal anteriorly, the lower 1·1–1·5 (mode 1·3) times as long as broad, its length 30·0–34·4 (M = 32·5) per cent of head length. Lips thickened; posterior tip of the maxilla extending to the vertical through the anterior orbital margin or slightly beyond.

Gill rakers short, 8 (less frequently 9 and rarely 6 or 7) on the lower limb of the anterior arch.

Scales ctenoid; lateral line with 31 (f.3), 32 (f.10), 33 (f.13) or 34 (f.3) scales. Cheek with 3 (less frequently 2 or 4) series. Six to 8 scales between the origin of the dorsal fin and the lateral line; 8 or 9 between the pectoral and pelvic fin bases; chest scales small.

Fins. Dorsal with 24 (f.3), 25 (f.24) or 26 (f.2) rays, anal with 11 or 12, comprising XV–XVI, 8–10 and III, 8 or 9 spinous and branched rays for the fins respectively. Pectoral fin slightly shorter than the head; first pelvic ray produced, variable in its posterior extension but usually reaching the spinous part of the anal fin.

Teeth. In the outer row of both jaws, the teeth are slender and gently recurved. Fishes less than 65 mm. S.L., have only unequally bicuspid teeth; individuals 65–95 mm. S.L. have an admixture of bi- and unicuspid teeth in which either type may predominate. Fishes more than 95 mm. S.L. have only unicuspid teeth. There are 24–42 (modal range 30–32) outer teeth in the upper jaw.

The inner teeth are tricuspid in fishes less than 95 mm. S.L. and unicuspid in larger individuals; an admixture of both types may occur. These teeth are arranged in three rows (less frequently two or four) in both jaws.

Lower pharyngeal bone triangular, the dentigerous area 1·1–1·4 (mode 1·2) times as broad as long. Occasionally the median series of teeth are enlarged and submolariform; more frequently, only the posterior few pairs are markedly enlarged. In a few specimens, no median teeth are enlarged.

Coloration. As mentioned above, *H. chromogynos* is unique in apparently having only " *bicolor* " females. The colour patterns of these fishes are variable, but are within the range known for other species with " *bicolor* " females. In preserved material, the yellowish-silver ground colour appears yellowish-white, silver or brown. *The colours of live males* are unknown.

Coloration of preserved males. Ground colour greyish-brown to grey; lips, lower jaw and the anterior part of the branchiostegal membrane, lighter; six or seven faint transverse bars visible on the flanks and caudal peduncle; a faint lachrymal stripe is often present. Dorsal fin with the spinous part dusky, lappets lighter; soft part orange-yellow. Anal dusky on the basal half, orange-yellow distally, with one to three white ocelli arranged in a single row. Caudal fin dark, but with a broad, orange-yellow margin. Pelvic fins black on the outer half, orange mesially.

Ecology : Habitat. H. chromogynos is probably confined to the littoral zone and to water less than 20 feet deep; it has only been caught over a firm substrate (rock, sand or shingle).

Food. One record of stomach and intestinal contents is available; the main contents were the remains of Trichoptera larvae and sand-grain cases, but a few larval chironomids and baétids were also identified.

Breeding. One female carrying embryos in the buccal cavity was recorded. Sexual maturity is reached at a standard length of 90–100 mm. in both sexes.

Affinities. H. chromogynos may be related to H. chilotes (see p. 211). The similarity between these species is most pronounced when the non-lobed lip forms of H. chilotes are compared with H. chromogynos. Superficially, H. chromogynos resembles H. crassilabris (Blgr.) but the dentition of the two species is markedly different.

The available specimens of H. paucidens Regan, from Lake Kivu, indicate a very close relationship between the two species; the most marked difference is the shallower cheek of H. chromogynos (mean depth 21·6, cf. 27·6 per cent for H. paucidens). Unfortunately there is no information on the coloration of female H. paucidens or on the breeding colours of male fishes of either species.

Study material and distribution records

Museum and Reg. No.	Locality	Collector
	Uganda	
B.M. (N.H.).—1958.1.16.83 (Holotype)	Napoleon Gulf, near Jinja	E.A.F.R.O.
„ 1958.1.16.71–75	Napoleon Gulf, near Jinja golf course	„
„ 1958.1.16.76–81	Napoleon Gulf, Jinja pier	„
„ 1958.1.16.82	Napoleon Gulf, bay opposite Jinja pier	„
„ 1958.1.16.84	Entebbe Harbour	„
„ 1958.1.16.69	Grant Bay	„
„ 1958.1.16.67–68	Ramafuta Is., Buvuma Channel	„
„ 1906.5.30.415–416	Bugangu	Degen.
„ 1906.5.30.407–412	Bunjako	„
	Tanganyika	
„ 1958.1.16.63–66	Mwanza, Capri Bay	E.A.F.R.O.
	Kenya	
„ 1958.1.16.70	Kisumu Harbour	„

Haplochromis aelocephalus sp. nov.
(Text-fig. 12)

Holotype. A male 96 mm. S.L., from Igwe Island.

An interesting feature of this species is its wide range of variation in head shape; the more extreme individuals might well be considered distinct species were it not for the presence of annectent forms (Text-fig. 13). This variation is not correlated with sex or size. The most constant specific characters are the multiseriate dentition, the small scales on the pectoral region, and the thickened lips.

Description. Based on the holotype and 21 other specimens, 63–120 mm. standard length.

Depth of body 31·3–38·4 per cent of standard length; length of head 33·0–38·6 per cent. Dorsal head profile straight or very slightly concave, sloping gently; physiognomy variable. Preorbital depth 14·7–19·4 (M = 17·2) per cent of head length, least interorbital width 21·9–26·4 (M = 24·8) per cent. Snout 1·1–1·2 times as long as broad, except in a few extreme individuals where it is 1·25–1·30 times as long as broad; snout length 32·0–39·0 (M = 35·3) per cent of head. Diameter of eye shows fairly clear-cut negative allometry with standard length, 25·0–30·8 (M = 27·8) per cent of head in fishes 62–100 mm. S.L. and 23·2–25·7 (M = 24·5) per cent in larger individuals; depth of cheek 17·3–24·7 (M = 20·5) per cent.

Caudal peduncle 12·8–18·5 (M = 16·2) per cent of standard length, 1·1–1·6 (modal range 1·4–1·5) times as long as deep.

Mouth horizontal, lower jaw projecting slightly; posterior tip of the maxilla not quite reaching the vertical through the anterior margin of the orbit, except in one specimen. Lips thickened and variable; in a few fishes there are faint indications of a lobe-like swelling on the lower lip. In all specimens there is a pronounced sub-

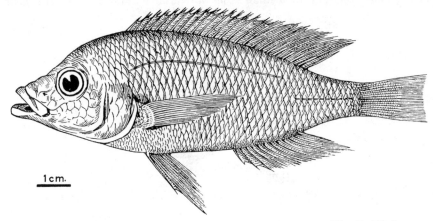

Fig. 12. *Haplochromis aelocephalus*; holotype. Drawn by Miss D. Fitchew.

mental thickening which extends posteriorly for a short distance. Lower jaw length apparently correlated with head shape, being greatest in the more extreme individuals; namely, in seven "extreme" specimens (Text-fig. 13) 42·5–48·5 (M = 45·1) per cent of head and in the remaining specimens 37·0–46·9 (M = 41·3) per cent. The length/breadth ratio of the lower jaw 1·6–2·6 (modal range 1·8–2·0).

Gill rakers short, 7–9 on the lower limb of the anterior arch.

Scales ctenoid; lateral line with 31 (f.1), 32 (f.5), 33 (f.10) or 34 (f.6) scales. Cheek with 3 or 4 (rarely 2) series. Six to 8 scales between the origin of the dorsal fin and the lateral line; 8 or 9 (rarely 7 or 10) scales between the pectoral and pelvic fin bases; chest scales small.

Fins. Dorsal with 24 (f.3), 25 (f.17) or 26 (f.2) rays, anal with 11 (f.1), 12 (f.16) or 13 (f.5), comprising XV or XVI, 9 or 10 and III, 8–10 spinous and branched rays for the fins respectively. Pelvic fins with the first ray produced. Pectoral fin shorter than the head. Caudal truncate or subtruncate.

Teeth. The outer row in both jaws of fishes less than 65 mm. S.L. is composed of slender and slightly recurved bicuspid teeth; specimens 65–95 mm. S.L. have an admixture of bi- and unicuspids in which either form may predominate, whilst in larger fishes, all the outer teeth are unicuspid. There are 24–42 (mode 32) teeth in the outer row of the upper jaw. The dental arcade in the lower jaw is narrow anteriorly; in a few specimens it is rather acutely pointed and resembles that of lobe-lipped *H. chilotes* (see p. 209).

Teeth in the inner series of both jaws are generally tricuspid in fishes less than 95 mm. S. L. and unicuspid in larger individuals; a mixture of both types is found in specimens of an intermediate size. There are three to five, rarely two (mode five), series of inner teeth anteriorly in the lower jaw and three to six (mode five) in the upper. The innermost series of the lower jaw is usually implanted so as to lie almost horizontally.

Lower pharyngeal bone triangular, the dentigerous area 1·0–1·4 (mode 1·2) times as broad as long. Teeth in the median series are variable in form. In most specimens

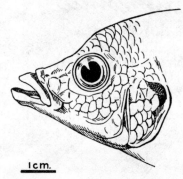

Fig. 13. *Haplochromis aelocephalus*; individual variant of head profile (extreme form).

these teeth (especially the upper three or four pairs) are somewhat enlarged and weakly cuspidate; the next most common variant has these teeth slightly enlarged and clearly cuspidate. Finally, in a few fishes the median teeth are unmodified and resemble the other teeth. Two exceptional fishes had the entire median series enlarged and molariform and the pharyngeal bone noticeably stouter.

Coloration. The colours of living fishes are unknown.

Preserved material: *Sexually active males*. Ground colour grey-black, chest and branchiostegal membrane black; faint indications of a coppery sheen on the operculum and flanks. Dorsal fin black except for the light lappets and a colourless band outlining the soft part of the fin; caudal black basally, light (? orange) distally; anal fin dark on the basal half and light (? orange) distally, with three or four hyaline ocelli arranged in a single row. Pelvics black. *Females, quiescent and juvenile males*. Ground colour greyish-silver (in sexually quiescent males there is a faint trace of coppery sheen on the operculum and the chest is dusky) with, in some, an interrupted or continuous, dark, mid-lateral stripe and five or six transverse bars on the flanks. Dorsal and anal fins yellowish, slightly dusky on the proximal half in quiet males, but hyaline and faintly maculate in females and immature males. Pelvic fins yellowish or hyaline in females and immature males, dusky in quiet males.

Distribution. Known only from Lake Victoria.

Ecology : Habitat. The species has been found in relatively few localities and only in the Uganda waters of the lake. However, the available data suggest that *H. aelocephalus* is restricted to water less than 40 feet deep and to areas where the substrate is firm (sand and rock).

Food. Sixteen of the 20 specimens examined had food in the stomach or intestines ; from these it would seem that *H. aelocephalus* preys on a variety of invertebrate animals, and possibly even small fishes.

In ten specimens the predominant food organisms were insects (particularly dipterous larvae, but also Ephemeroptera [*Povilla adusta*] and Trichoptera larvae). The non-insect food identified was : in two fishes, oligochaet worms ; in one, the remains of a prawn (*Caridina nilotica* Roux) ; in another, fragments of plant-tissue and a few Ostracoda ; and in two others, numerous fragments of lamellibranch and gastropod shells. One exceptional individual contained the remains of a small cichlid fish.

The presence of sand grains in the stomach and intestines of many individuals suggests that the species may be a bottom feeder.

Breeding. No data are available.

Affinities. Haplochromis aelocephalus shows no special affinity with any other *Haplochromis* species in Lake Victoria ; the less extreme individuals resemble members of the *H. nigrescens* species-complex of piscivorous-insectivorous predators. The multiseriate dentition, however, disqualifies *H. aelocephalus* from a place in this complex, but suggests relationship with species of the *H. sauvagei-H. prodromus* group, and particularly *H. xenognathus*. In shape, the teeth of *H. aelocephalus* are unlike those of *H. xenognathus* which have characteristically recurved tips (Greenwood, 1957). In certain cephalic characters, especially the shape of the lower jaw, the narrow lower dental arcade and the semi-lobate lips, *H. aelocephalus* approaches *H. chilotes* but in all other characters there is no obviously close relationship between the two species.

Diagnosis. H. aelocephalus may be distinguished by the following combination of characters : proportions of the head ; a multiseriate dentition with the outer teeth slender and gently recurved ; lips somewhat thickened.

Study material and distribution records

Museum and Reg. No.	Locality	Collector
	Uganda	
B.M. (N.H.).—1958.1.16.244 (Holotype)	Igwe Isl.	E.A.F.R.O.
,, 1958.1.16.215	Ekunu Bay	,,
,, 1958.1.16.216	Entebbe, Bugonga Beach	,,
,, 1958.1.16.217	Beach near Nasu Point, Buvuma Channel	,,
,, 1958.1.16.218–224	Igwe Isl.	,,
,, 1958.1.16.225–228	Bay opposite Jinja, Napoleon Gulf	,,
,, 1958.1.16.232–233	Pilkington Bay	,,
,, 1958.1.16.234–235	Buka Bay	,,
,, 1958.1.16.236–242	Napoleon Gulf, near Jinja	,,
,, 1958.1.16.243	Unknown	,,

SUMMARY

1. Seven species, which feed almost exclusively on the embryos and larvae of other Cichlidae, are discussed. *Haplochromis obesus* (Blgr.), *H. maxillaris* Trewavas, *H. melanopterus* Trewavas, *H. parvidens* (Blgr.) and *H. microdon* (Blgr.) are redescribed on the basis of new and more extensive collections. Two new species, *H. cronus* and *H. cryptodon* are described.

2. Notes on the ecology and feeding habits of these species are given.

3. The relationships of these species are discussed and it is concluded that the group has a polyphyletic origin.

4. Four other species are considered. These are all insectivorous and do not appear to be closely related to the other species of *Haplochromis* in Lake Victoria. *Haplochromis plagiodon* Regan & Trewavas and *H. chilotes* (Blgr.) are redescribed, and two new species, *H. chromogynos* and *H. aelocephalus* are described.

5. *H. chromogynos* is of particular interest since the normal female coloration is apparently the "*bicolor*" piebald which occurs as an infrequent and sex-limited mutant amongst the females of other and unrelated species.

6. Both *H. chilotes* and *H. aelocephalus* are noteworthy for the wide range of individual variability which they show.

ACKNOWLEDGMENTS

I wish to acknowledge my gratitude and thanks to the Trustees of the British Museum (Natural History) for the facilities afforded me ; to the authorities of the Muséum National d'Histoire naturelle, Paris and of the Museo Civico di Storia Naturale, Genoa, who graciously allowed me to study type-material in their collections ; and to Mr. A. C. Wheeler of the Zoology Department, British Museum (Natural History) who was responsible for making several radiographs used in this study. I am especially indebted to Dr. Ethelwynn Trewavas for her most helpful advice and criticism.

REFERENCES

GRAHAM, M. 1929. *A Report on the Fishing Survey of Lake Victoria, 1927–1928, and Appendices.* Crown Agents, London.

GREENWOOD, P. H. 1954. On two cichlid fishes from the Malagarazi River (Tanganyika) with notes on the pharyngeal apophysis in species of the *Haplochromis* group. *Ann. Mag. nat. Hist.* (12) **7** : 401.

—— 1956a. The monotypic genera of cichlid fishes in Lake Victoria. *Bull. Br. Mus. nat. Hist., Zool.* **3** : No. 7.

—— 1956b. A revision of the Lake Victoria *Haplochromis* species (Pisces, Cichlidae). Part I. *Ibid.* **4** : No. 5.

—— 1957. A revision of the Lake Victoria *Haplochromis* species (Pisces, Cichlidae). Part II. *Ibid.* **5** : No. 4.

REGAN, C. T. 1922. The cichlid fishes of Lake Victoria. *Proc. zool. Soc. Lond.* 157.

REGAN, C. T. & TREWAVAS, E. 1928. Four new cichlid fishes from Lake Victoria. *Ann. Mag. nat. Hist.* (10) **2** : 224.

TREWAVAS, E. 1928. Descriptions of five new cichlid fishes of the genus *Haplochromis* from Lake Victoria. *Ibid.* (10) **2** : 93.

A REVISION OF THE LAKE VICTORIA *HAPLOCHROMIS* SPECIES (PISCES, CICHLIDAE) PART IV

By P. H. GREENWOOD

Department of Zoology, British Museum (Natural History)

CONTENTS

	Page
INTRODUCTION	229
Haplochromis lacrimosus (Blgr.)	230
Haplochromis pallidus (Blgr.)	233
Haplochromis macrops (Blgr.)	236
Haplochromis cinereus (Blgr.)	239
Haplochromis niloticus nom. nov.	243
Haplochromis martini (Blgr.)	245
Haplochromis humilior (Blgr.)	248
Haplochromis riponianus (Blgr.)	252
Haplochromis saxicola sp. nov.	256
Haplochromis theliodon sp. nov.	260
Haplochromis empodisma sp. nov.	262
Haplochromis obtusidens Trewavas	266
Haplochromis pharyngomylus Regan	270
Haplochromis ishmaeli Blgr.	275
SUMMARY	280
ACKNOWLEDGMENTS	280
REFERENCES	280

INTRODUCTION

IN previous parts of this series I have revised species-groups possessing common or related structural peculiarities and, in most cases, similar feeding habits. The present paper deals with a greater variety of structural and trophic types and many of the species show no obvious relationship to one another. With one exception, the species described below fall into three groups, namely, structurally generalized insectivorous species, specialized mollusc-eaters and species showing various degrees of structural and adaptational intermediacy between the other two groups. The exceptional fish, *Haplochromis martini* is a piscivorous predator; it is included simply because of its overall resemblance to one of the insectivorous species described here.

Formerly, some of the generalized species reviewed in this paper were synonymized with one of the mollusc-eaters (*H. ishmaeli*); the others are included because of various resemblances to species now resurrected and redefined. One such is *Haplochromis cinereus*, a species previously considered to be the extant representative of the ancestral type from which at least part of the present flock had evolved. These views on the central evolutionary position of *H. cinereus* are no longer tenable since the "species" thought to be *H. cinereus* was a complex of several distinct species, some more generalized than the others. *Haplochromis cinereus, sensu stricto* is, in fact, an anatomically somewhat specialized derivative from an even more generalized form.

The most outstanding structural character in many of the species described below is an increase in the strength and size of the pharyngeal bones and musculature. As might be expected, these changes are reflected in the diet of the species, which usually include Mollusca as an important element in their food. Two species, *H. ishmaeli* and *H. pharyngomylus*, feed almost entirely on snails and bivalves.

Those species with the pharyngeal mill in an intermediate stage of hypertrophy are able to deal with small molluscs and also with the tubicolous larvae of certain Trichoptera, an otherwise infrequent element in the food of insectivorous *Haplochromis* without strengthened pharyngeals.

By crushing their molluscan prey within the pharynx these species stand in sharp contradistinction to the other groups of mollusc-eating *Haplochromis* in Lake Victoria. Species in this latter group remove the snail from its shell by holding the foot between the jaws and then levering the soft parts free before ingestion takes place (Greenwood, 1956a and 1957).

Haplochromis lacrimosus (Blgr.) 1906
(Text-fig. 1)

Tilapia lacrimosa (part) Boulenger, 1906, *Ann. Mag. nat. Hist.* (7) **17**, 450; Idem, 1907, *Fish. Nile*, 515; Idem, 1915, *Cat. Afr. Fish.* **3**, 234, fig. 154.
Haplochromis cinereus (part) Regan, 1922, *Proc. zool. Soc. London*, 166.

Lectotype. An adult male 76 mm. standard length (B.M. [N.H.] 1906.5.30.471) from Entebbe.

Description, based on 36 specimens (including the lectotype and 10 paratypes) 66·0–97·0 mm. S.L.

Depth of body 31·8–38·7 (M = 35·5) per cent of standard length, length of head 30·8–35·5 (M = 33·5) per cent. Dorsal head profile straight or slightly curved, sloping moderately steeply. Preorbital depth 13·6–18·0 (M = 15·5) per cent of head length, showing weak positive allometry with standard length; least interorbital width 20·8–26·9 (M = 23·5), snout length 26·6–32·2 (M = 29·6) per cent. Diameter of eye 26·1–32·6 (M = 30·4) per cent of head, ratio of eye diameter to preorbital depth 1·5–2·3 (mode 2·0); depth of cheek 17·6–23·5 (M = 20·8) per cent. Caudal peduncle 15·0–19·2 (M = 17·2) per cent of standard length, 1·2–1·8 (mode 1·5) times as long as deep.

Mouth horizontal or almost so; jaws equal anteriorly, the lower 31·4–41·3 (M = 37·1) per cent of head and 1·2–2·0 (mode 1·6) times as long as broad. Posterior tip of the maxilla extending to the vertical to the anterior orbital margin or slightly beyond, rarely not quite reaching the anterior orbital margin.

Gill rakers variable, from moderately stout to slender; 7–9 (mode 8), rarely 6 on the lower part of the first gill arch, the lowermost two or three reduced.

138

Scales ctenoid; lateral line with 31 (f.5), 32 (f.21) or 33 (f.8) scales; cheek with 2 or 3 (rarely 4) series; 6 or 7 (rarely 5½) between the lateral line and the dorsal fin origin, 7 or 8 (rarely 6) between the pectoral and pelvic fin bases. Scales of the pectoral region small or moderate.

Fins. Dorsal with 24 (f.19) or 25 (f.16) rays, anal with 11 (f.19), 12 (f.16) or 13 (f.1), comprising XV–XVI, 8–10 and III, 8–10 spinous and branched rays for the fins respectively. Pectoral 82·5–100·0 (M = 88·5) per cent of head. Caudal truncate or subtruncate.

Teeth. The outer row in both jaws is composed of unequally bicuspid, relatively slender and sometimes slightly recurved teeth. Rarely, a few unicuspid teeth may occur anteriorly in this row; likewise a few posterolateral teeth in the upper jaw

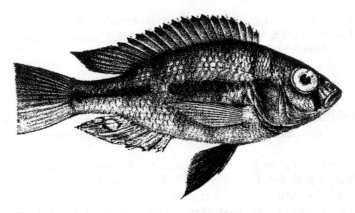

FIG. 1. *Haplochromis lacrimosus*; lectotype (from Boulenger, *Fishes of the Nile*).

may be unicuspid. There are 40–60 (mode 54, modal range 48–54) teeth in the upper jaw.

Teeth in the inner series are tricuspid and implanted at a slight, posteriorly directed angle (cf. *H. cinereus* where the inner teeth lie almost horizontally); there are 2 or 3 (rarely 4) rows of inner teeth in the upper jaw and 2 (less commonly 3, rarely 4) in the lower.

Lower pharyngeal bone triangular and slender; a few slightly enlarged but bicuspid teeth may occur in the median tooth-rows.

Coloration. The colours of live fishes are unknown.

Preserved material: Sexually active males. Ground colour yellowish-silver, chest dusky; a dark lachrymal stripe is always present and in some specimens it may extend obliquely upwards through the eye and on to the nape; there are usually two transverse bands across the snout.

Two common patterns of body markings are known. (i) A large mid-lateral blotch situated slightly posterior to the pelvic fin insertion and a mid-lateral stripe running from a point above the second anal ray on to the caudal fin ; a faint transverse bar is visible immediately posterior to the edge of the operculum and two others lie between the mid-lateral blotch and the origin of the posterior stripe. The blotch itself appears to be the intensified mid-portion of a vertical bar.

(ii) Nine, close-set and ventrally ill-defined transverse bars on the flanks ; ventrally, the bars tend to run into one another so that the lower region of the flank is steely-grey.

All fins, except the pelvics, hyaline, the upper part of the caudal sometimes weakly maculate, the mid-part dark ; anal with two or three ocelli ; pelvic fins black.

Females and quiescent males. Ground colour greyish-silver, brownish above. Seven to nine faint transverse bars on the flanks, not reaching the ventral or dorsal outlines of the body. All fins hyaline.

Distribution. At present, *H. lacrimosus* is known with certainty from Lake Victoria ; Pappenheim & Boulenger (1914) recorded a specimen from Lake Edward, but I have not been able to examine their material.

Ecology : Habitat. No precise details are available for fishes already in the collections of the B.M. (Nat. Hist.) ; specimens collected by E.A.F.R.O. come from only two localities, both exposed, sandy beaches with the water depth less than 20 feet. Thus, it is impossible to generalize on the habitat preferences of *H. lacrimosus*.

Food. The stomach and intestinal contents of twenty fishes were examined ; with one exception (a fish from Entebbe) these specimens were caught at one time and at a single locality (Majita, Tanganyika Territory). The gut contents of the sixteen specimens containing food were varied. Twelve fishes contained fine sand-grains, bottom detritus (including fragments of plant epidermis and diatom frustules) and some Cladocera ; five contained remains of insect larvae (probably Diptera), one an adult dipteran, one the remains of a larval *Povilla adusta* Navás (Ephemeroptera) and one an insect egg-mass. Two fishes yielded, besides insect fragments, the remains of an oligochaet worm, whilst two others each contained the foot and soft parts of a snail. From these scanty and topographically restricted data, *H. lacrimosus* should perhaps be classified as a bottom-feeding omnivore.

Breeding. The breeding habits are unknown. Two of the smallest fishes (male and female, both 66·0 mm. S.L.) are adult. It seems possible that adult males reach a larger maximum size than do females.

Affinities. *Haplochromis lacrimosus* is one of the structurally and ecologically generalized species of Lake Victoria. Its most striking and apparently diagnostic character is the markings of male fishes ; but, it must be stressed that coloration is known only from preserved material. In general appearance *H. lacrimosus* resembles *H. pallidus* (see p. 233) but the two species differ in several characters besides male coloration.

Study material and distribution records

Museum and Reg. No.	Locality	Collector
	Uganda	
B.M. (N.H.).—1906.5.30.471 (Lectotype *Tilapia lacrimosa*)	Entebbe	Degen
B.M. (N.H.).—1906.5.30.472–478	,,	,,
,, 1906.5.30.483–484	Bunjako	,,
,, 1906.5.30.488–489	Buganga	,,
,, 1907.5.7.81–82	Buddu Coast	Simon
,, 1908.10.19.2–5	Sesse Islands	Bayon
,, 1959.4.28.24	Entebbe Harbour	E.A.F.R.O.
	Tanganyika	
,, 1959.4.28.1–23	Majita	,,

Haplochromis pallidus (Blgr.) 1911
(Text-figs. 2 and 3)

Tilapia pallida (part) Boulenger, 1911, *Ann. Mus. Genova* (3), **5**, 74; Idem, 1915, *Cat. Afr. Fish.* **3**, 231–2.
Labrochromis pallidus Regan, 1920, *Ann. Mag. nat. Hist.* (9), **5**, 45 (footnote).
Haplochromis cinereus (part), Regan, 1922, *Proc. zool. Soc. London*, 166.

This synonymy is tentative, as I have been unable to locate three specimens of *T. pallida* which Regan (1922) referred to *Haplochromis guiarti* (Pellegrin). Regan's genus *Labrochromis*, based on a skeleton wrongly identified as *T. pallida*, is discussed on page 275.

Description, based on twenty specimens (including the holotype and four paratypes) 43–74 mm. S.L.

Depth of body 33·3–38·8 (M = 35·4) per cent of standard length, length of head 32·3–35·3 (M = 34·1) per cent. Dorsal head profile straight or slightly curved, sloping at about 30°–40°. Preorbital depth 13·2–18·2 (M = 16·5) per cent of head; least interorbital width 21·0–30·0 (M = 24·7) per cent. Snout as long as broad or slightly longer; its length 29·1–33·4 (M = 31·2) per cent of head; diameter of eye 26·1–33·4 (M = 29·5) per cent, ratio of eye diameter to preorbital depth 1·5–2·3 (mode 1·7); depth of cheek 19·0–25·0 (M = 21·7) per cent of head. Caudal peduncle 14·2–18·5 (M = 16·2) per cent of standard length, 1·2–1·7 (mode 1·2) times as long as deep.

Mouth horizontal; jaws equal anteriorly, the lower 32·2–40·9 (M = 37·6) per cent of head and 1·4–1·8 (mode 1·6) times as long as broad. Posterior tip of the maxilla extending to the vertical through the anterior orbital margin or slightly beyond.

Gill rakers short and stout (slender in one specimen), 7–9 (mode 9), rarely 10, on the lower part of the first arch, the lowermost 1–4 (or even 5) rakers greatly reduced.

Scales ctenoid; lateral line with 31 (f.6), 32 (f.13) or 33 (f.1) scales; cheek with 2 or 3 (rarely 4) series; 6 or 7 (occasionally 5) between the lateral line and the dorsal fin origin, 6 or 7 (occasionally 8) between the pectoral and pelvic fin bases. Scales on the chest rather small.

Fins. Dorsal with 24 (f.5) or 25 (f.15) rays, anal with 11 (f.3) or 12 (f.17), comprising XV–XVI, 8–10 and III, 8 or 9 spinous and branched rays for the fins respectively. Caudal truncate; pectoral fin 78·0–87·0 (M = 81·0) per cent of head.

Teeth. The outer row in both jaws is composed of unequally bicuspid, moderately stout teeth implanted erectly. In most fishes more than 67 mm. S.L. some unicuspid teeth occur postero-laterally in the upper jaw. The holotype (the largest specimen examined) has only weakly bicuspid teeth in the outer row. Three small specimens from near the Ripon Falls have somewhat more slender outer teeth than other specimens. There are 36–48 teeth in the upper outer series; no clear-cut mode can be determined from the sample studied.

The small and tricuspid inner teeth are implanted at a very slight angle and arranged in 3 (rarely 2) rows in the upper jaw and 2 or 3 rows in the lower.

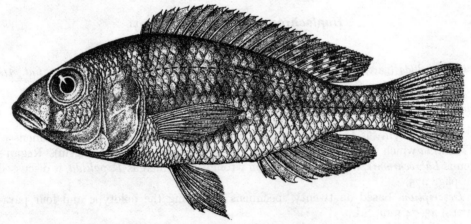

FIG. 2. *Haplochromis pallidus*; holotype (from Boulenger, *Ann. Mus. Genova*).

Lower pharyngeal bone triangular, usually slender but slightly thickened in three specimens. Most fishes have the two median rows of teeth slightly enlarged (especially the most posterior one or two pairs); in a few, all the pharyngeal teeth are slender.

Coloration. The colours of live fishes are unknown.

Preserved material. Adult males. Ground colour greyish, with faint traces of up to seven dark transverse bars on the flank and caudal peduncle; branchiostegal membrane greyish. A distinct, vertical lachrymal stripe, continued at an angle, runs through the centre of the eye; a very faint stripe runs from the posterior orbital margin to the angle of the preoperculum. Dorsal fin greyish, with dark lappets and a dark band along the basal two-thirds of the fin anteriorly, narrowing to the basal third or quarter posteriorly. Caudal dark proximally and along the mid-line. Anal hyaline, with two large, dead-white ocelli. Pelvics black, darkest laterally.

Females and immature males. Ground colour greyish-silver, six or seven faint transverse bars on the flanks and, in some, two bars on the caudal peduncle; a very faint lachrymal stripe. All fins hyaline.

Boulenger's (1911) account of preserved coloration differs somewhat from that given above, but as his material represented at least two and possibly three species, the discrepancies are not considered significant.

Distribution. Known only from Lake Victoria, unless the locality " Jinja, Ripon Falls " for specimens nos. 1911.3.3.127–130 implies that these fishes were caught *below* the falls in the Victoria Nile.

Ecology. Habitat. The only precise bionomic data available are for those specimens collected by E.A.F.R.O. All these were from one locality, an exposed, shallow and sandy beach near Entebbe Airport.

Food. Thirteen of the sixteen fishes examined had ingested matter in the stomach and intestine ; all these specimens were collected at one locality and at the same

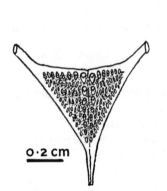 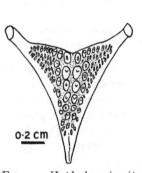 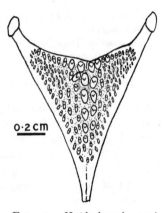

FIG. 3. *Haplochromis pallidus;* lower pharyngeal bone, occlusal view.

FIG. 4. *Haplochromis riponianus;* lower pharyngeal bone, occlusal view.

FIG. 5. *Haplochromis saxicola;* lower pharyngeal bone, occlusal view.

time. Except for three fishes, all contained moderately large quantities of bottom debris (sand grains, diatom frustules and fragments of plant tissue) together with fragmentary insect larvae (? Diptera). The exceptional specimens contained only bottom detritus.

Breeding. Haplochromis pallidus is a female mouth brooder ; the two smallest fishes examined (a male 54 mm. S.L. and a female 58 mm. S.L.) are both adult.

Affinities. Haplochromis pallidus must be considered one of the many small and generalized species in Lake Victoria. Within this group it is extremely difficult to suggest phyletic relationships because the degree of inter-specific differentiation is so slight. In general appearance *H. pallidus* perhaps comes nearest to *H. lacrimosus*, from which species it is distinguished by a higher modal number of gill rakers (9 cf. 8), fewer and somewhat stouter teeth, a lower modal eye/preorbital ratio (1·7 cf. 2·0) and particularly, by differences in the preserved coloration of male fishes. The nature of both oral and pharyngeal dentition suggests that a "*pallidus*"-like species could have been ancestral to the adaptational grade at present represented by *Haplochromis humilior* (see p. 252).

Study material and distribution records

Museum and Reg. No.	Locality	Collector
	Uganda	
Genoa Museum (C.E. 12912)	Jinja	Bayon
B.M. (N.H.).—1911.3.3.127–130	Jinja, Ripon Falls	,,
,, 1959.4.28.25–40	Entebbe, Airport beach	E.A.F.R.O.

Haplochromis macrops (Blgr.) 1911
(Text-fig. 6)

Haplochromis stanleyi (part) Boulenger, 1914, *Cat. Afr. Fish.* **3**, 295.
Haplochromis ishmaeli (part) Boulenger, 1914, *tom. cit.* 293.
Tilapia macrops Boulenger, 1911, *Ann. Mus. Genova* (3), **5**, 73. pl. III, fig. 1 ; Idem, 1914, *Cat. Afr. Fish.* **3**, 238.
Haplochromis macrops (part, i.e. the species as described but excluding the tentative synonymy of *Astatotilapia jeanneli* Pellegrin), Regan, 1922, *Proc. zool. Soc. London*, 166.

Description, based on forty specimens from Lake Victoria (including the holotype [Genoa Museum] and one of the paratypes) 66–91 mm. S.L.

Depth of body 32·5–38·2 (M = 35·8) per cent of standard length, length of head 31·0–35·1 (M = 33·0) per cent. Dorsal head profile straight or slightly curved, sloping at a moderate angle (*ca.* 35°–40°). Preorbital depth 11·5–16·3 (M = 14·2) per cent of head, least width of interorbital 26·6–32·2 (M = 29·7) per cent. Snout as broad as long or slightly broader, its length 26·6–31·0 (M = 29·0) per cent of head, diameter of eye 28·6–35·4 (M = 33·0) per cent, ratio of eye diameter to preorbital depth 2·0–2·9 (mode 2·3) ; depth of cheek 17·8–24·2 (M = 21·1) per cent of head. Caudal peduncle 14·1–18·4 (M = 16·8) per cent of standard length, 1·2–1·6 (mode 1·4) times as long as deep.

Mouth slightly oblique, posterior tip of the maxilla extending to the vertical to the anterior orbital margin or even as far as the pupil. Lower jaw 38·0–42·5 (M = 39·5) per cent of head, 1·4–2·2 (modal range 1·7–1·8) times as long as broad.

Gill rakers slender or, occasionally, relatively stout ; 8–11 (mode 9) on the lower part of the first arch, the lower 1–4 rakers reduced.

Scales ctenoid ; lateral line with 30 (f.6), 31 (f.17), 32 (f.13) or 33 (f.3) scales ; cheek with 2 or 3 series. Five or 6 (occasionally 7) scales between the lateral line and the dorsal fin origin ; 6 or 7 (occasionally 5, rarely 8) between the pectoral and pelvic fin bases. Scales on the pectoral region (relative to those on the ventral abdominal region) moderate to large.

Fins. Dorsal with 24 (f.6), 25 (f.30) or 26 (f.4) rays, anal with 11 (f.9), 12 (f.30) or 13 (f.1), comprising XV–XVII, 8–10 and III, 8–10 spinous and branched rays for the fins respectively. Pectoral 68·0–96·0 (M = 84·0) per cent of head. Caudal truncate.

Teeth. The outer row of teeth in both jaws (except those situated postero-laterally in the upper) is composed mainly of bicuspid, moderately stout teeth. In certain fishes over 80 mm. S.L. some weakly cuspidate teeth may occur. *H. macrops* is, however, unusual in that the postero-lateral teeth in the upper jaw are generally tricuspid in fishes less than 85 mm. S.L. ; in larger individuals these teeth may be unicuspid. One aberrant individual has only tricuspid teeth in the upper, outer row ; the corresponding row in the lower jaw has a mixture of bi- and tricuspids, with the former predominating. There are 46–66 (mode 60, modal range 56–60) teeth in the upper, outer row.

The inner tooth-rows are made up of tri- and unicuspid teeth arranged in 2 or 3 (rarely 4) rows in the upper jaw and 1 or 2 (less commonly 3) rows in the lower. Inner teeth are implanted so as to stand erect or with a very slight, posteriorly directed slope.

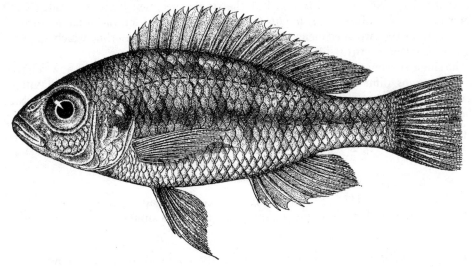

Fig. 6. *Haplochromis macrops* ; holotype (from Boulenger, *Ann. Mus. Genova*).

Lower pharyngeal bone triangular, not enlarged ; teeth fine and cuspidate.

Coloration in life. Sexually active males. Ground colour dusky to intensely black. Dorsal fin black, lappets and margin of the soft part red, as are the spots and blotches between the branched rays. Anal dusky with a diffuse red flush becoming more intense distally : ocelli yellow. Caudal dusky, ventral half with a red flush. Pelvic fins dusky yellow. *Quiescent males* have a female-type coloration, but with yellow anal ocelli and red spots on the soft dorsal. *Females.* Ground colour greyish-yellow to silver-grey. Dorsal and anal fins light yellow, caudal yellowish.

Colour of preserved material. Adult males. Dark blackish-brown, somewhat lighter, except in sexually active fishes, on the chest and branchiostegal membrane ; faint indications of six transverse bars on the flanks (generally not reaching the dorsal and ventral outlines) ; a dark but faint lachrymal stripe is visible in some specimens, as is a faint dark bar along the preoperculum. Dorsal fin dark (especially along the basal third), lappets black. Caudal fin dark. Anal dark on the basal third to half, pale distally, with two or three ocelli. Pelvic fins black. *Females.* Brownish-yellow, some with eight or nine faint transverse bars on the flank and caudal peduncle. All fins hyaline but the caudal somewhat darker.

Distribution. Definite records of *Haplochromis macrops* are available only from Lake Victoria although there is an indication that the species may also occur in the Lake Edward basin. I have examined material identified as *H. macrops* from Lake Edward (one specimen B.M. (N.H.) Reg. No. 1933.2.23.397; see Trewavas, 1933) and from rivers flowing into Lake Edward (see Poll, 1939 and Poll & Damas, 1939). Of these latter specimens (twelve in all) only one (R. G. Mus. Congo 31095, det. David, 1936), from Rutshuru, compares closely with the Victoria population of *H. macrops*. I hesitate to identify the remaining Congo Museum specimens from Rutshuru (Reg. Nos. 64888–64899), but the single specimen from the B.M. (N.H.) seems most closely allied to *H. lividus* Greenwood of Lake Victoria. I have not been able to study the two specimens from Lake Edward (now in Berlin) which Pappenheim & Boulenger (1914) identified as *H. macrops*.

The single specimen (R. G. Mus. Congo 31095) from the Edward basin now referred to *H. macrops* differs slightly from the generality of Victoria fishes in having a somewhat larger eye and longer lower jaw; it is an adult female, 73·0 mm. S.L. The principal morphometric characters of this fish are:

D*	H*	Po.%	Io.%	Eye%	Snt.%	Ck.%	Lj.%	C.P.*	Eye/Po.
32·8	31·4	15·2	30·4	37·0	30·4	21·7	43·4	15·0	2·4

* Per cent standard length.
% Per cent head length.

Dorsal XV, 9; anal III, 9; pectoral 91·3 per cent of head.

In characters of dentition and squamation this fish is similar to those from Lake Victoria.

Ecology. Habitat. The species is apparently confined to the shallow, sandy regions of the lake.

Food. The predominating food organisms in the stomachs and intestines of twenty-four fishes (mainly from one locality, but caught on different occasions) are sub-imaginal Ephemeroptera; two fishes had, however, fed almost exclusively on winged termites (Isoptera) and colonial blue-green algae. Typical bottom debris and sand-grains usually found in the guts of other insectivorous *Haplochromis* were not recorded. The occurrence of sub-imaginal or adult insects and planktonic blue-green algae, together with the absence of bottom debris, suggests that *H. macrops* feeds at the surface or in mid-water. Since the algae were not digested and did not constitute a major proportion of the ingested matter, they may be taken accidentally as the fishes dart after insect prey.

Breeding. Haplochromis macrops is a female mouth-brooder. The smallest sexually active fish is a female 73 mm. S.L., the smallest adult male is 78 mm. S.L. It appears that males attain a larger maximum adult size than do females.

Affinities. As with most of the structurally and trophically generalized *Haplochromis* of Lake Victoria, the detailed affinities of *H. macrops* are impossible to determine. The similarities existing between *H. macrops* and *H. cinereus* are discussed elsewhere (p. 242). The two species differ in several characters, particularly in their dentition and the larger eye/preorbital ratio of *H. macrops* (2·0–2·9, mode 2·3, cf. 1·5–1·9, mode 1·8 for *H. cinereus*). Also, the gill rakers of *H. macrops* are finer and more numerous (mode 9) than in *H. cinereus* (mode 7).

The large eye and shallow preorbital of *H. macrops* serve to distinguish the species from most other members of the " generalized species " group.

Study material and distribution records

Museum and Reg. No.	Locality	Collector
	Uganda	
Genoa Museum (C.E. 12928) (Holotype)	Bussu	Bayon
B.M. (N.H.).—1911.3.3.137 (Paratype)	Bussu	Bayon
B.M. (N.H.).—1911.3.3.114–115	Jinja, Ripon Falls	,,
,, 1959.4.28.51–78	Beach near Nasu Point	E.A.F.R.O.
,, 1959.4.28.79–84	Buka Bay	,,
	Tanganyika	
,, 1959.4.28.85	Mwanza, Capri Bay	,,
,, 1959.4.28.86	Majita	,,
,, 1959.4.28.87	Beach near Majita	,,

Haplochromis cinereus (Blgr.) 1906
(Text-fig. 7)

Paratilapia cinerea Boulenger, 1906, *Ann. Mag. nat. Hist.* (7), **17**, 439; Idem, 1907, *Fish. Nile*, 478; Idem, 1915, *Cat. Afr. Fish.* **3**, 344.
Haplochromis stanleyi (part), Boulenger, 1915, *tom. cit.*, 295.
Tilapia lacrimosa (part) Boulenger, 1906, *Ann. Mag. nat. Hist.* (7), **17**, 450; Idem, 1915, *tom. cit.*, 234.
Haplochromis cinereus (part), Regan, 1922, *Proc. zool. Soc. London*, 166.
Haplochromis melanopus (part), Regan, 1922, *op. cit.* 165.

Description, based on twelve specimens (including the holotype) 71–81 mm. S.L.
Depth of body 34·6–39·0 per cent of standard length, length of head 30·8–37·3 (M = 34·7) per cent. Dorsal head profile straight or slightly curved, sloping at *ca.* 40°–50°. Preorbital depth 15·0–18·0 (M = 16·4) per cent of head length, least interorbital width 23·3–28·0 (M = 25·3) per cent. Snout as long as broad or slightly longer, its length 29·2–34·6 (M = 32·2) per cent of head, eye diameter 26·2–32·0 (M = 28·7) per cent, ratio of eye diameter to preorbital depth 1·5–1·9 (mode 1·8). Depth of cheek 20·8–26·0 (M = 23·0) per cent of head. Caudal peduncle 15·7–18·7 per cent of standard length, 1·4–1·6 times as long as deep.

Mouth horizontal or very slightly oblique; jaws equal anteriorly, the lower 34·6–41·3 (M = 37·7) per cent of head and 1·4–1·8 (mode 1·6) times as long as broad. Lips slightly thickened. Posterior tip of the maxilla extending almost to the vertical through the anterior orbital margin or as far as the eye.

Gill rakers moderately stout, 7–9 (mode 7) on the lower part of the first gill-arch, the lowermost two or three rakers reduced.

Scales ctenoid; lateral line with 30 (f.1), 31 (f.1), 32 (f.4) or 33 (f.6) scales; cheek with 3 or 4 series; 5–6½ scales between the lateral line and the dorsal fin origin; 7 or 8 (rarely 6 or 9) between the pectoral and pelvic fin bases. Scales on the pectoral region small or, less frequently, moderate.

Fins. Dorsal with 24 (f.8) or 25 (f.4) rays, anal with 11 (f.4) or 12 (f.8), comprising XV or XVI, 9 or 10 and III,'8 or 9 spinous and branched rays for the fins respectively. Caudal truncate. Pectoral 72·5–92·0 (M = 80·2) per cent of head.

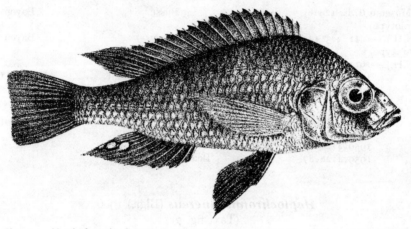

FIG. 7. *Haplochromis cinereus*; holotype (from Boulenger, *Fishes of the Nile*).

Teeth. Teeth in the outer row of both jaws are weakly bicuspid or unicuspid, relatively stout and slightly recurved; in most fishes there is an admixture of both types, with unicuspids predominating. There are 40–54 teeth in the upper, outer row.

The inner series are composed of either unicuspid or bicuspid teeth; less commonly there is a mixture of both types. A characteristic feature of the inner tooth-rows is the way in which the teeth are implanted obliquely so that the crowns point posteriorly. Three or 4 (rarely 2) inner rows occur in the upper jaw and 2 or 3 in the lower.

The dentition of *H. cinereus* is unlike that of other species in the " generalized species " group; it is, indeed, typical of the dentition found throughout the large group of piscivorous predators. However, in other characters (syncranial architecture, feeding habits and body-form) *H. cinereus* is one of the generalized species (see p. 242).

Lower pharyngeal bone triangular, the dentigerous surface slightly broader than long. Teeth in the two median rows are slightly enlarged in five of the specimens examined, but are slender in the remaining seven fishes; the other pharyngeal teeth are slender in all specimens.

Coloration in life unknown. *Colour of preserved fishes: Adult males.* Ground colour greyish-brown; chest and branchiostegal membrane dusky; a faint lachrymal stripe. All fins hyaline, with very faint indications of dark maculae on the soft dorsal; dorsal lappets dark. Anal fin with two or three whitish ocelli (orange surrounded by red in newly preserved material, according to Boulenger). Pelvics black except for a large light area extending over the distal half but not including the first branched ray. *Females.* Brownish, silvery grey ventrally, with eight or nine dark transverse bars on the flanks and caudal peduncle. All fins hyaline, the basal third to half of the caudal weakly maculate.

Distribution. Known only from Lake Victoria.

Ecology. No ecological data (except the locality) are available for the three specimens collected by Degen. The remaining nine fishes were caught in water less than 20 feet deep, over a sandy bottom on both protected and exposed shores.

Food. The stomach and intestinal contents of ten specimens (from four localities) were examined. Of these, three were empty and the remainder yielded sand-grains, bottom detritus (including fragments of plant epidermis) and some larvae of dipterous insects.

Breeding. The species is a female mouth-brooder. The smallest specimens available (a male and a female 71 mm. S.L.) are both sexually mature. As far as can be determined from this inadequate sample, both sexes reach the same adult size.

Discussion of affinities and synonymy. In Regan's revision of the Lake Victoria *Haplochromis* (1922) the definition of *H. cinereus* was expanded to embrace a number of small and generalized or near-generalized forms previously recognized as distinct species. The first attempt to prune this complex was Lohberger's resurrection of *H riponianus* (see p. 252). Now, with more specimens available and some knowledge of *Haplochromis* in nature, it is clear that Regan's definition of *H. cinereus* must be narrowed considerably and that a further two species (*H. lacrimosus* and *H. pallidus*) should be resurrected. Amongst the group of anatomically and trophically unspecialized *Haplochromis* in Lake Victoria, *H. cinereus* is unusual because of its oral dentition (see p. 240). Relatively stout, clearly bicuspid teeth in the outer series of the jaws and erect tricuspid inner teeth are usual in the generalized species. The dentition of *H. cinereus*, on the other hand, shows a marked tendency for slender, unicuspid teeth to predominate in the outer rows; the few bicuspid teeth present are weakly cuspidate. The inner teeth of *H. cinereus* are also atypical for the group in that the usual erect and tricuspid form is largely replaced by slender unicuspids implanted so as to point posteriorly. In fact, the dentition of *H. cinereus* is very like that of many predatory species. With so few specimens of *H. cinereus* known it is impossible to generalize on its feeding habits; however, the gut contents of seven fishes from four different localities do not even hint at the species being a piscivorous predator.

At this point it should perhaps be stressed that my observations are confined to adult fishes 70–90 mm. long ; juveniles have still to be discovered.

Because " *Haplochromis cinereus* " had become something of a dumping ground for any small *Haplochromis* species or specimen, the published information (Graham, 1929) on distribution and habitats can no longer be considered reliable. I have examined " *H. cinereus* " material collected by Graham and find that none of these specimens is referable to *H. cinereus, sensu stricto*. The bulk of this material is of undescribed species and will be dealt with in subsequent papers. Thus, Graham's remark that " the species (*H. cinereus*) is therefore widely distributed except in the deepest part of the lake " and Brooks' (1950, p. 159) elaboration of these data do not apply to *H. cinereus* but rather to the whole species-complex of generalized, bottom feeding *Haplochromis* in the lake.

Haplochromis cinereus has been cited as representing the ancestral type from which the present species-flock could have evolved (Regan, 1922 ; Greenwood, 1951). For the reasons mentioned above this concept must be abandoned ; there are several other *Haplochromis* species still surviving in Lake Victoria which are structurally closer to the ancestral type, for example *H. lacrimosus, H. nubilus* or *H. pallidus*.

The particular affinities of *H. cinereus* are difficult to determine. In gross anatomy and appearance *H. cinereus* does not differ markedly from the majority of small *Haplochromis* ; only when its dentition is considered does the difference appear striking. Like so many members of the generalized group, *H. cinereus* seems to be an independent offshoot from one of the basic stocks.

In overall appearance and perhaps in at least some ecological requirements *Haplochromis macrops* is the one extant species most like *H. cinereus*. The two species are differentiated principally by the larger eye, shallower preorbital and the stouter, more numerous outer teeth of *H. macrops*.

Haplochromis cinereus is not, as Regan suggested, closely related to *H. ishmaeli*. To stress this supposed relationship, Regan noted that six of the *H. ishmaeli* syntypes were actually specimens of *H. cinereus*. As a result of this present revision none of these specimens is still retained in *H. cinereus*. *Haplochromis ishmaeli* belongs to a distinct phyletic line, discussed more fully on pages 269 and 273.

Study material and distribution records

Museum and Reg. No.	Locality	Collector
	Uganda	
B.M. (N.H.).—1906.5.30.292 (Holotype)	Buganga	Degen
B.M. (N.H.).—1906.5.30.350	Entebbe	,,
,, 1906.5.30.482	Bunjako	,,
,, 1959.4.28.41	Entebbe, Harbour	E.A.F.R.O.
,, 1959.4.28.48–50	Entebbe, Airport beach	,,
	Tanganyika	
,, 1959.4.28.42–47	Mwanza, Capri Bay	E.A.F.R.O.

150

A REVISION OF THE LAKE VICTORIA *HAPLOCHROMIS* SPECIES 243

Haplochromis niloticus nom. nov.
(Text-fig. 8)

Tilapia bayoni Boulenger, 1911, *Ann. Mus. Genova* (3), **5**; 72, pl. III, fig. 2 (*nec Paratilapia bayoni* [= *Haplochromis bayoni* (Blgr.), see Regan, 1922, p. 176] Blgr., 1909, *Ann. Mus. Genova* (3), **4**, 304, fig.
Haplochromis humilior (part), Regan, 1922, *Proc. zool. Soc. London*, 169.

Description, based on the holotype (100 mm. S.L.) and two other specimens 96 and 102 mm. S.L.

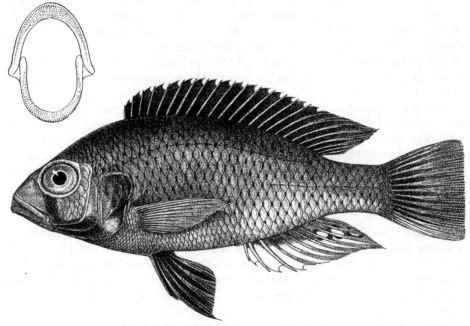

FIG. 8. *Haplochromis niloticus*; holotype (from Boulenger, *Ann. Mus. Genova*).

The principal morphometric characters are given below:

S.L.	Depth*	Head*	Po. %	Io. %	Snt. %	Eye %	Ck. %	Lj. %	C.P.*
96·0	35·4	33·3	17·2	21·8	31·3	31·3	20·4	39·0	18·2
†100·0	33·0	33·0	15·2	24·3	30·3	33·5	24·0	33·3	17·5
102·0	34·3	33·8	17·4	24·7	29·0	30·5	23·2	37·7	17·1

† Holotype.
% Per cent. of head length.
* Per cent. of standard length.

Dorsal head profile sloping rather steeply (*ca.* 45°–50°) and slightly curved. Mouth horizontal; jaws equal anteriorly, the length/breadth ratio of the lower 1·6–1·9. Posterior tip of the maxilla extending to the vertical through the anterior part of the eye. Lips not markedly thickened.

Gill rakers moderately stout; 7 or 8 (the lowermost one or two reduced) on the lower part of the first arch.

Scales ctenoid; 32 in the lateral line; cheek with 2 or 3 series. Six or 7 scales between the lateral line and the dorsal fin origin, 8 or 9 between the pectoral and pelvic fin bases; scales on the pectoral region small.

Fins. Dorsal with 25 (f.2) or 26 (f.1) rays, anal with 12, comprising XVI, 9–10 and III, 9 spinous and branched rays for the fins respectively. Pelvic fins with the first branched ray only slightly produced and extending to the origin of the anal. Caudal sub-truncate, scaled only on its proximal half.

Teeth. The outer row in both jaws is composed of moderately slender, movably implanted and unequally bicuspid teeth; the most posterior five or six teeth in the upper jaw are caniniform and stouter than those situated anteriorly and laterally. There are 65–70 teeth in the upper, outer series.

Teeth in the inner series are small and tricuspid, and are arranged in four or five rows in the upper jaw and four in the lower. The interspace between the inner and outer series is very narrow.

Lower pharyngeal bone triangular, slender or slightly enlarged, the two median rows of teeth relatively coarse in two specimens (including the holotype) and somewhat more enlarged in the third. In the latter fish the next lateral row of teeth is also enlarged and the most posterior teeth of the median rows are sub-molariform.

Osteology. A complete skeleton was prepared from one of the specimens caught at the same time as the three fishes described above. However, since *H. niloticus* is very similar to *H. nuchisquamulatus* (which also occurs in the Victoria Nile) it is difficult to confirm the specific identity of the skeleton. Apparently the sole diagnostic osteological character is the lower pharyngeal bone, which is slender in *H. nuchisquamulatus* and slightly thickened in *H. niloticus*. The lower pharyngeal of the skeleton is that of *H. niloticus* and on this character alone the skeleton is referred to *H. niloticus*. In all other characters, except the oral dentition, the skeleton of *H. niloticus* resembles that of a generalized *Haplochromis* species. There are 14 + 16 vertebrae.

Coloration of live fishes is unknown. The three preserved specimens are all apparently males (judging from the well-defined anal ocelli) and adult. Because most of the coloration is now lost (the fishes are a uniform brownish-grey) I quote the description given by Boulenger (1911) of the then newly preserved specimens. " Back dark olive to blackish, sides brassy yellow to coppery red; a more or less distinct black bar below the eye; dorsal and ventrals brown to black; anal pink, blackish at the base, usually with two or three large orange ocellar spots; caudal brown or blackish the lower third often pink." From this description the coloration of male fishes seems to be remarkably like that of *H. humilior* from Lake Victoria (see p. 250).

Distribution. *Haplochromis niloticus* is known only from the Victoria Nile; no information is available on its habitat or on feeding and breeding habits.

Affinities. *Haplochromis niloticus* has been compared with two *Haplochromis* species from Lake Victoria. In his original description, Boulenger compared the species with *H. martini*, whilst Regan (1922) considered *H. niloticus* to be conspecific with *H. humilior*. In my opinion *H. martini* and *H. humilior* are not closely related to one another and *H. niloticus* is not allied to either. The three species differ in several fundamental characters, especially in the nature of their dentition. Admittedly, the somewhat enlarged median pharyngeal teeth of *H. niloticus* approach the condition found in some specimens of *H. humilior*, but the oral dentition of the two species is very dissimilar. The outer teeth are finer and more numerous in *H. niloticus* and there are more rows of inner teeth barely separated from the outer row. These same characters, together with a somewhat different arrangement of the jaw skeleton, serve to separate *H. niloticus* from *H. martini*.

The nature of the dentition in *H. niloticus* suggests affinity with *H. nigricans* and *H. nuchisquamulatus*, particularly the latter. From *H. nigricans*, *H. niloticus* is distinguished by its more generalized neurocranium and dentary (see Greenwood, 1956b) and its slightly enlarged lower pharyngeal teeth. From *H. nuchisquamulatus*, *H. niloticus* is again distinguished by having somewhat enlarged pharyngeal teeth and by a narrower interorbital region.

The affinities of *H. niloticus* are not especially obvious; the species is probably yet another slightly specialized side branch from the generalized *Haplochromis* stem.

Study material and distribution records

Museum and Reg. No.	Locality	Collector
	Uganda	
Genoa Museum (C.E. 12932) (Holotype *T. bayoni* Blgr. 1911)	Kakindu, Victoria Nile	Bayon
B.M. (N.H.).—1911.3.3.124–5	,, ,,	,,
,, 1911.3.3.126 (skeleton)	,, ,,	,,

Haplochromis martini (Blgr.) 1906
(Text-fig. 9)

Tilapia martini, Boulenger, 1906, *Ann. Mag. nat. Hist.* (7), **17**, 449.
T. martini (part), Idem, 1914, *Cat. Afr. Fish.* **3**, 239, fig. 158.
Haplochromis martini, Regan, 1922, *Proc. zool. Soc. London*, 171.

Regan (1922) based his redescription of *H. martini* on three of the six syntypes, but did not indicate to what species he referred the remaining type specimens. However, amongst the material Regan identified as *H. cinereus* there are three type specimens of *T. martini* (B.M. (N.H.) Reg. No. 1906.5.30.466–468). Presumably it was Regan's intention to include these in the published synonymy of *H. cinereus*. The three specimens are not *H. cinereus* but are, in fact, *Haplochromis martini*.

Lectotype. A brooding female 88 mm. standard length (B.M. (N.H.) Reg. No. 1906.5.30.465) from Bunjako, Uganda.

Description, based on twenty-nine specimens (including the lectotype and four paratypes) 59–104 mm. S.L.

Depth of body 30·8–38·0 (M = 34·4) per cent of standard length, length of head 31·0–38·1 (M = 35·4) per cent. Dorsal head profile very strongly decurved, the snout sloping at an angle of *ca.* 50°–70°. Depth of preorbital 13·0–20·0 (M = 16·6) per cent of head, least interorbital width 20·4–26·8 (M = 24·1) per cent. Snout slightly longer than broad, its length 27·3–34·4 (M = 30·2) per cent of head; diameter of eye 29·4–37·5 (M = 31·7) per cent, depth of cheek 20·4–27·7 (M = 24·6) per cent. Caudal peduncle 15·3–20·6 (M = 17·3) per cent of standard length, 1·2–1·8 (mode 1·6) times as long as deep.

Jaws equal anteriorly, the lower 38·4–45·8 (M = 42·6) per cent of head, 1·6–2·1 (modal range 1·7–2·0) times as long as broad; mouth horizontal or slightly oblique. Posterior tip of the maxilla extending to the vertical through the pupil or, less commonly, only to the anterior part of the eye. Such a marked posterior extension of the maxilla is unusual in Lake Victoria *Haplochromis* and may be considered one of the diagnostic characters of *H. martini*.

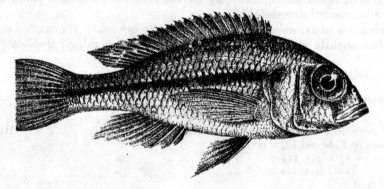

FIG. 9. *Haplochromis martini*; lectotype (from Boulenger, *Fishes of the Nile*).

Gill rakers on the first arch moderately stout (but rather slender in a few specimens), one or two of the uppermost often flattened and bifid; 8 or 9, rarely 7 or 10 (mode 9) gill rakers on the lower part of the arch.

Scales ctenoid; lateral line with 31 (f.5), 32 (f.7), 33 (f.9), 34 (f.7) or 35 (f.1) scales; cheek with 3 or 4 rows. Six to 8 scales between the lateral line and the dorsal fin origin, 7 or 8 (rarely 6) between the pectoral and pelvic fin bases.

Fins. Dorsal with 24 (f.9), 25 (f.19) or 26 (f.1) rays, anal with 11 (f.8), 12 (f.19) or 13 (f.2), comprising XIV–XVI, 8–10 and III, 8–10 spinous and branched rays for the fins respectively. Pectoral fin 73·5–86·5 (M = 82·0) per cent of head; caudal truncate or sub-truncate.

Teeth. In the outer row of both jaws the teeth are slender and mainly unequally bicuspid in fishes less than 85 mm. S.L. Above this size both weakly bicuspid and unicuspid teeth occur together. In fishes of all sizes a few tricuspid teeth are found postero-laterally in the upper jaw, an uncommon character in Lake Victoria *Haplochromis* (see *H. macrops*, p. 237). There are 46–76 teeth in the upper jaw (ill-defined modal range 68–70).

Tricuspid teeth predominate in the inner series although in one large fish (101 mm. S.L.) the inner teeth are all unicuspid. The inner series are arranged in 2 (less commonly 3) rows in the upper jaw and in 1 or 2 (less commonly 3) rows in the lower.

Lower pharyngeal bone triangular, not enlarged and with slender, cuspidate teeth.

Osteology. The neurocranium does not differ greatly from the generalized *Haplochromis* type, but the premaxilla has relatively longer dentigerous arms.

Coloration in life. Sexually active males. Ground colour golden-yellow, shading to silvery-white, with three or four faint black blotches below the insertion of the dorsal fin, and a distinct coppery sheen on the nape and anterior part of the flank. Dorsal fin hyaline, with a pinkish flush. Distal half of the anal scarlet, the proximal half colourless ; anal ocelli orange-red. Caudal fin flushed with scarlet, especially intense on the distal half. Pelvics dark on the outer half, reddish-yellow mesially. *Quiescent males* golden-yellow shading to silvery-white ventrally ; a fairly distinct dark mid-lateral stripe and an interrupted upper stripe slightly below the insertion of the dorsal fin. Dorsal and caudal fins darkish, the anal and pelvics lighter ; anal ocelli yellow. *Females and juvenile males.* Ground colour and banding as above, the upper band usually broken into rather indistinct blotches. All fins light yellow, the caudal somewhat darker.

Preserved material. Both sexes. Ground colour yellowish-silver to brownish, an intense, narrow mid-lateral black line extends from the upper angle of the operculum to the caudal peduncle and, in some specimens even on to the caudal fin ; a fainter, often interrupted black stripe runs mid-way between the upper lateral line and the dorsal fin base. In some fishes there are traces of a very faint interocular band. All fins hyaline ; in males the dorsal has dusky lappets and the pelvics are dark.

Distribution. Haplochromis martini is known only from Lake Victoria. Specimens from Lake Edward once identified as *Tilapia martini* (Boulenger, 1914) were later referred to *H. schubotzi* (Regan, 1921).

Ecology. Habitat. Available records (from eight localities) suggest that *H. martini* is restricted to water less than 40 feet deep, where the species is ubiquitous but nowhere common. The species has been found over both sand and mud bottoms, on exposed shores and in sheltered bays. There are some indications that it may not occur close inshore since the only record of *H. martini* in beach-operated seine nets came from an area (Majita) where the nets were shot about 300 yards off-shore.

Food. Sixteen of the twenty-two specimens examined had ingested material in the stomach and intestines ; in three of these specimens, however, the contents were unidentifiable sludge. Eleven of the remaining thirteen fishes contained, as the exclusive or predominating food, the fragmentary remains of small fishes (identified in two cases as *Haplochromis*) ; one of these individuals had also fed on larval Diptera and another had eaten what appeared to be the foot and other soft parts of a snail. Another fish had fed only on larval Diptera and one was empty except for some small fish bones in the posterior intestine.

Breeding. *Haplochromis martini* is a female mouth-brooder; females reach sexual maturity at *ca.* 80 mm. S.L.; no data are available for males. Both sexes appear to attain the same maximum adult size.

Affinities. Because of its strongly decurved snout, large eye and the marked posterior extension of the premaxilla, *H. martini* is one of the more immediately recognizable species. Yet, despite these characters *H. martini* retains most of the fundamental features of a generalized species such as *H. macrops*. On the other hand, *H. martini* differs from members of this species group (and probably most other Lake Victoria species) in its bright yellow coloration.

When attempting to assess the phyletic affinities of *H. martini* one is faced with these rather contradictory characters and with the fact that, despite its generalized dentition and body-form, *H. martini* can be a piscivorous predator. The majority of piscivorous *Haplochromis* in Lake Victoria are larger than *H. martini* and have elongate bodies and heads; the teeth in these species are usually large and caniniform.

Taking into account the various structural and ecological characters *H. martini* should perhaps be considered a superficially but trophically distinct branch from the generalized stem as represented, perhaps, by an *H. macrops*-like ancestor.

Study material and distribution records

Museum and Reg. No.	Locality	Collector
Uganda		
B.M. (N.H.)—1906.5.30.463–5 (Lectotype and paratypes)	Bunjako	Degen
„ 1906.5.30.466–8 (Paratypes)	„	„
„ 1959.4.28.124–132	Pilkington Bay	E.A.F.R.O.
„ 1959.4.28.138	Old Bukakata Bay	„
„ 1959.4.28.140	Napoleon Gulf, near Jinja	„
Tanganyika		
„ 1959.4.28.116–123	Majita	„
Kenya		
„ 1959.4.28.133–137	Off Port Southby	„
„ 1959.4.28.139	Beach below Usoma Lighthouse	„

Haplochromis humilior (Blgr.) 1911
(Text-figs. 10 and 11)

Tilapia humilior Boulenger, 1911, *Ann. Mus. Genova* (3), **5**, 74, pl. III, fig. 3.
Tilapia lacrimosa (part), Boulenger, 1915, *Cat. Afr. Fish.* **3**, 234.
Haplochromis desfontainesii (part), Boulenger, 1915, *tom. cit.*, 303.
Haplochromis nubilus (part), Regan, 1922, *Proc. zool. Soc. London*, 164.
Haplochromis humilior (part), Regan, 1922, *op. cit.*, 169.
? *Paratilapia granti* (part), Boulenger, 1915, *tom. cit.*, 342.

Lectotype. A male 90 mm. standard length from Kakindu, Victoria Nile, collected by Bayon (now in the collections of the Museo Civico di Storia Naturale, Genoa).

Description, based on thirty specimens 65–90 mm. S.L. (including the lectotype and one paratype; the second paratype [B.M. (N.H.) Reg. No. 1911.3.3.152] is very poorly preserved and although examined, is not included in the description).

Depth of body 29·0–37·5 (M = 34·4) per cent of standard length, length of head 31·6–37·8 (M = 34·7) per cent. Dorsal head profile curved, sloping at an angle of 45°–50°. Preorbital depth 13·6–17·9 (M = 16·3) per cent of head, least interorbital width 21·0–28·6 (M = 24·2) per cent. Snout as long as broad or slightly longer, its length 27·0–34·8 (M = 30·9) per cent of head, diameter of eye 27·0–32·5 (M = 30·3), depth of cheek 18·5–23·2 (M = 21·2) per cent. Caudal peduncle 15·2–19·1 (M = 17·4) per cent of standard length, 1·3–1·8 (mode 1·5) times as long as deep.

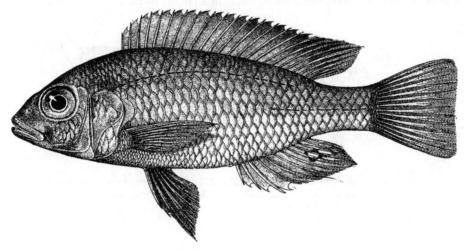

Fig. 10. *Haplochromis humilior*, lectotype (from Boulenger, *Ann. Mus. Genova*).

Mouth horizontal, the lower jaw often slightly shorter than the upper; length of lower jaw 33·4–39·6 (M = 36·6) per cent of head and 1·3–2·2 (modal range 1·5–1·8) times as long as broad. Posterior tip of the maxilla extending to the vertical through the anterior orbital margin or somewhat beyond (to below the pupil in one specimen).

Populations of *H. humilior* from different localities in the lake appear to have characteristic facies which make any one population more or less readily identifiable; unfortunately no means of quantifying these characters could be determined. Also, it has so far proved impossible, through lack of material, to decide whether Lake Victoria *H. humilior* differ from those inhabiting the Victoria Nile.

Gill rakers short and stout, 6–8 (modes 6 and 7), rarely 9 on the lower part of the first arch.

Scales ctenoid; lateral line with 30 (f.7), 31 (f.7), 32 (f.7) or 33 (f.7) scales; cheek with 2 or 3 (rarely 4) series. Six or 7 scales between the lateral line and the origin of the dorsal fin, 6–8 (rarely 5 or 9) between the pectoral and pelvic fin bases. Scales on the pectoral region relatively small.

Fins. Dorsal with 24 (f.6), 25 (f.21) or 26 (f.3) rays, anal with 11 (f.4) or 12 (f.24) comprising XV–XVI, 8–10 and III, 8 or 9 spinous and branched rays for the fins respectively. Pectoral 69·0–92·3 (M = 81·0) per cent of head. Caudal truncate.

Teeth. The outer teeth in both jaws are moderately stout and unequally or, less frequently, sub-equally bicuspid. In fishes over 70 mm. S.L. some weakly bicuspid or even unicuspid teeth may occur; there are 36–52 (modal range 46–48) teeth in the upper, outer series.

Inner teeth are tricuspid and arranged in 2 or 3 (rarely 4) rows in the upper jaw and 1 or 2 (rarely 4) in the lower. A distinct space separates the inner and outer tooth series.

Lower pharyngeal bone and teeth. The lower pharyngeal bone, although relatively stout is less massive than the bone in a specimen of *H. ishmaeli* or *H. pharyngomylus*

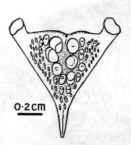 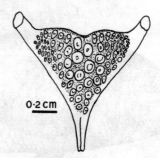

FIG. 11. *Haplochromis humilior*; lower pharyngeal bone, occlusal view, from a Lake Victoria specimen.

FIG. 12. *Haplochromis theliodon*; lower pharyngeal bone, occlusal view.

of the same size. Teeth in the two median rows are always enlarged, with the most posterior one or two pairs largest. There is some variation in the degree of enlargement of the median teeth, which may be bicuspid, conical or even molariform; no clear-cut correlation could be detected between, on the one hand, tooth-size and form and, on the other, the sex and size of the fish. There is, however, a tendency for larger individuals to have the coarsest median pharyngeal teeth. Sagittal sections through a number of bones (from fishes 65–70 mm. S.L.) suggest that both time and chance may influence the nature of the dentition, since unerupted replacement teeth are always molariform irrespective of the nature of the functional teeth which they underlie.

In addition to the two median rows, the next lateral row on each side may also contain a number of molariform teeth.

Coloration in life. Breeding males. Ground colour dark silvery-grey with intense dusky blotches on the head; branchiostegal membrane dull black. A coppery flush extends over the cheek, operculum and flank as far as the origin of the anal fin. Anal and caudal fins light red, the colour becoming more intense along the margins of both fins and the upper and lower posterior angles of the caudal; two or three yellow anal ocelli. Dorsal fin dusky, with an orange-red margin to the soft part and red spots and bars between the rays of the posterior half of the spinous dorsal and over the entire soft part. Pelvics black. *Quiescent males.* General coloration as in females except that the pelvics are somewhat dusky and the unpaired fins have a pinkish flush; anal ocelli are present. *Females.* Ground colour silvery-yellow. Dorsal, pelvic and anal fins pale yellow, the dorsal with red spots distributed as in males.

Colour of preserved material. Males. Greyish, darker dorsally, the chest and branchiostegal membrane sooty; in some specimens there are traces of seven or eight, narrow transverse bars on the flank and caudal peduncle. A lachrymal stripe, a pair of transverse stripes across the snout and, in some fishes, a broad band across the interorbital region are also visible, as are one or two bands on the nape. Dorsal fin variable, from hyaline to dusky. Anal hyaline in quiescent fishes and whitish in active individuals. Pelvics black (darkest in active fishes) but with a whitish overlay on the proximal half. *Females* silvery, some with very faint traces of vertical bars usually most obvious on the mid-lateral aspects of the flanks. All fins hyaline.

Distribution. The species occurs in Lake Victoria and the Victoria Nile.

Ecology. Habitat. In Lake Victoria the species is confined to shallow water over sandy beaches in both exposed and sheltered areas. No data are available for the riverine populations.

Food. The gut contents of thirty-two fishes (from two localities) yielded identifiable material. Of these specimens, twenty-four contained bottom debris (sand grains, plant fragments, diatom frustules and blue-green algae) together with remains of both larval and pupal insects (especially Diptera and Trichoptera, less frequently, Ephemeroptera); four fishes contained only insect remains and one only the diatom *Melosira*. Ten individuals had eaten, in addition to insects, both bivalves (unidentifiable) and gastropods (*Melanoides* sp. and *Bellamya* sp.). These molluscan fragments were too finely divided to allow for any estimate of the number of animals eaten. Nevertheless it does seem, from this sample at least, that Mollusca are not a major element in the food of *H. humilior*.

The large quantities of sand found with the remains of Trichoptera larvae in most fishes is of interest, particularly since the grains are small and of a remarkably uniform size. This suggests that the sand grains could be derived from the sand-grain cases made by certain caddis larvae. The moderately large pharyngeal mill of *H. humilior* may thus serve the dual purpose of crushing mollusc shells and the sand-grain cases of certain larval insects.

Breeding. Nothing is known about the breeding habits of *H. humilior*. The smallest specimens available (a male and a female both 65 mm. S.L.) are adult; the sexes apparently do not differ in maximum adult size attained.

Affinities. Haplochromis humilior differs from the generality of small *Haplochromis* species in several characters, particularly in having a relatively massive lower pharyngeal bone and in having a low modal number of gill rakers (6–7). Another pronounced difference lies in the tendency for the lower jaw to be shorter than the upper. In most other characters, *H. humilior* resembles *H. pallidus*, a species which shows incipient hypertrophy of the lower pharyngeal bone. *Haplochromis humilior* could be a more specialized off-shoot from an *H. pallidus*-like stem. Unlike many other presumed phyletic lines within the Victoria species flock, this one is not continued by one or more extant species. *Haplochromis humilior* does not appear to have any close relationship to the principal group of species with enlarged pharyngeals, namely *H. obtusidens*, *H. ishmaeli* and *H. pharyngomylus*.

Study material and distribution records

Museum and Reg. No.	Locality	Collector
	Uganda	
Genoa Museum (C.E. 12910) (Lectotype)	Kakindu, Victoria Nile	Bayon
B.M. (N.H.).—1911.3.3.152–3 (Paratypes)	,,	,,
,, 1906.5.20.314	Entebbe	Degen
,, 1959.4.28.88–107	Beach near Nasu Point	E.A.F.R.O.
,, 1959.4.28.108	Near Grant Bay	,,
,, 1959.4.28.109–112	Entebbe, Harbour	,,
	Tanganyika	
,, 1959.4.28.113–115	Majita	,,
	Kenya	
,, 1909.11.15.38	Kisumu Bay	Blayney-Percival

Haplochromis riponianus (Blgr.) 1911
(Text-figs. 4 and 13)

Pelmatochromis riponianus (part) Boulenger, 1911, *Ann. Mus. Genova* (3), **5**, 69, pl. II, fig. 3; Idem, 1915, *Cat. Afr. Fish.* **3**, 411, fig. 280.
Haplochromis riponianus, Lohberger, 1929, *Zool. Anz.* **86**, 222.
Paratilapia serranus (part), Boulenger, 1915, *tom. cit.*, 334.
Paratilapia victoriana (part), Boulenger, 1915, *tom. cit.*, 341.
Haplochromis ishmaeli (part) Boulenger, 1906, *Ann. Mag. nat. Hist.* (7), **17**, 446; Idem, 1915, *tom. cit.* 293.
Haplochromis cinereus (part), Regan, 1922, *Proc. zool. Soc. London*, 166.

On the basis of specimens in the Vienna Museum, Lohberger (1929) decided that Regan's views on the conspecificity of *Pelmatochromis riponianus* and *H. cinereus* could not be substantiated; consequently he resurrected the former species as *Haplochromis riponianus*. I have not examined Lohberger's specimens but, from studying considerably more material than was available to either Regan or Lohberger, I can fully endorse the latter's action.

Lectotype. A male 95·5 mm. standard length, from Jinja, Uganda, collected by Bayon (Genoa Museum, C.E. 12996).

A REVISION OF THE LAKE VICTORIA *HAPLOCHROMIS* SPECIES

Description, based on twenty-eight specimens (including the lectotype and two paratypes) 57–104 mm. S.L. One other paratype (B.M. (N.H.) Reg. No. 1911.3.3.37) is not included in the description.

Depth of body 33·3–39·4 (M = 35·7) per cent of standard length, length of head 32·8–37·7 (M = 35·7) per cent. Dorsal head profile straight or very slightly curved, sloping at an angle of *ca.* 35°–45°. Preorbital depth 16·3–19·5 (M = 17·6) per cent of head, least interorbital width 23·0–28·1 (M = 25·1) per cent. Snout slightly longer than broad, its length 30·5–35·4 (M = 33·6) per cent of head; diameter of eye 24·2–31·0 (M = 26·6), depth of cheek 19·2–25·0 (M = 22·7) per cent. Caudal peduncle 14·4–18·4 per cent of standard length, 1·2–1·7 times as long as deep (modal range 1·3–1·5).

FIG. 13. *Haplochromis riponianus*; lectotype (from Boulenger, *Ann. Mus. Genova*).

Mouth horizontal or very slightly oblique, jaws equal anteriorly, the lower 33·4–42·2 (M = 38·5) per cent of head and 1·3–2·0 (modal range 1·6–1·8) times as long as broad. Posterior tip of the maxilla extending to the vertical through the anterior orbital margin or almost so, occasionally to below the anterior quarter of the eye. Lips noticeably thickened but not produced into median lobes.

Gill rakers on the first arch moderately stout in most fishes but slender in a few others, the lowermost one to three reduced and the pair nearest the epi-ceratobranchial angle often flattened and tri- or quadrifid; 6–8 (mode 7) rakers on the lower part of the arch.

Scales ctenoid; lateral line with 30 (f.2), 31 (f.2), 32 (f.11), 33 (f.10) or 34 (f.2) scales, cheek with 3 or 4 series; 5½–7 (rarely 5 or 8) scales between the lateral line and the dorsal fin origin; 7 or 8 (rarely 6) between the pectoral and pelvic fin bases. Scales on the pectoral region moderate.

Fins. Dorsal with 24 (f.3), 25 (f.24) or 26 (f.1) rays, anal with 11 (f.3) or 12 (f.25), comprising XV–XVI, 9–10 and III, 8 or 9 spinous and branched rays for the fins respectively. Pectoral 69·0–88·5 (M = 78·5) per cent of head. Caudal sub-truncate.

Teeth. The outer teeth in both jaws are relatively slender and slightly to strongly recurved; the basic cusp pattern is unequally bicuspid, but the crowns are often so worn that the teeth appear to be weakly cuspidate or even unicuspid and bluntly incisiform. In some fishes over 80 mm. S.L. initially unicuspid teeth occur and may even be the predominating form in fishes more than 100 mm. S.L. The number of teeth in the upper, outer rows shows slight positive allometry with standard length; there are 38–62 teeth in this row.

Unicuspid and weakly tricuspid teeth are found in the inner rows; often both types of teeth occur together, especially in fishes over 80 mm. S.L. The inner teeth are implanted at an angle and may be buried in the thickened oral mucosa (possibly a preservation artefact). In the upper jaw, the inner teeth are arranged in 3 or 4 (rarely 2) rows and in the lower in 2–4 (rarely 1) rows.

Boulenger's description (1911 and 1915) of the inner teeth as " minute " appears to stem from his being misled by the thickened oral epithelium which has hidden all but the tips of these teeth.

Lower pharyngeal bone and teeth. The lower pharyngeal bone is triangular and in most specimens fairly stout. The relative degree to which the bone is enlarged is somewhat greater than that of the lower pharyngeal in *H. humilior* (see p. 250). In a few fishes, however, the bone is slender. This variation in stoutness is not entirely correlated with size since, although the smallest specimens have slender or but slightly thickened bones, some of the larger fishes do not have proportionately enlarged pharyngeals.

The form of the teeth in the four median rows does show correlation with both the size of the individual and the stoutness of the bone. In fishes less than *ca.* 70 mm. S.L. some of the posterior teeth in the two median rows are enlarged; in fishes 70–80 mm. most of the median teeth are enlarged as are some or all of the teeth in the row on each side of the two median rows. Next, in specimens above 80 mm. S.L. (except those with slender pharyngeal bones) the teeth of these four rows are larger still, whilst those in the two median rows tend to be molariform. Finally, in fishes over 98 mm. S.L. some of the more lateral teeth are also molariform. Large individuals with slender lower pharyngeals have only slightly enlarged median teeth, comparable with those of the smaller (< 90 mm.) specimens described above.

Osteology. The shape of the neurocranium departs from the generalized type towards the form found in the elongated, piscivorous predators (as typified by *H. mento*). In *H. riponianus*, the slope from the anterior tip of the vomer to the base of the supraoccipital crest is less steep and more nearly straight than in the generalized type of skull; also, the preorbital region in *H. riponianus* is relatively longer than in, for example, the skull of *H. obliquidens* or *H. macrops*.

Coloration. The colours of live fishes are unknown.

Preserved material. Sexually active males. Ground colour greyish-brown, chest dusky; a dark lachrymal stripe and faint traces of four interrupted transverse bars on the flanks. Dorsal, caudal and anal fins hyaline, the anal with two or three dead-white ocelli, usually arranged in a single row, but occasionally in two rows. Pelvic fins black. *Immature males* similar to females but with darker pelvics and small, distinct ocelli on the anal fin. *Females* silvery-grey shading to silver ventrally; six to ten transverse bars of variable width (narrowest when most numerous) across the flanks and caudal peduncle; sometimes, a very faint lachrymal stripe. All fins hyaline, the caudal occasionally maculate; in a few specimens there are one or two, small, dead-white spots in the same position as the ocelli on the anal fin of males.

Distribution. Lake Victoria and possibly the Victoria Nile. The latter locality is surmised from the data given in the original description of the species. In that paper (Boulenger, 1911), the locality is given only as Jinja, Ripon Falls, but in the introduction Boulenger implies that the entire collection, of which *H. riponianus* formed part, was from the Victoria Nile, that is *below* the Ripon Falls. Later (*Cat. Afr. Fish.* **3**, 1914) the type locality is given more specifically as " Ripon Falls, Victoria Nile."

Ecology. Habitat. The species is apparently confined to sand or rock substrates in the littoral regions of the lake; it has been caught in both exposed and sheltered localities.

Food. Stomach and intestinal contents of thirty-two fishes (from one locality) show that at least this population of *H. riponianus* was mainly insectivorous, although some fishes had also fed on Mollusca. Insect larvae (especially Trichoptera and Ephemeroptera) were found in every fish. As in *H. humilior* (see p. 251) large quantities of uniformly sized sand-grains were found in the intestines of most fishes. Since these grains closely resemble those forming the cases of certain Caddis-fly larvae found intact in some individuals, it is suggested that the sand was derived, at least partly, from crushed cases. In addition to the main contents listed above, eight fishes contained relatively large quantities of crushed bivalve shells (? *Corbicula* sp.); a few fragments of gastropod shells were also found in these individuals.

Breeding. No information is available on the breeding habits of *H. riponianus*. The smallest sexually mature fish is a female 84 mm. S.L.; the smallest adult male is 86 mm. S.L.

Affinities. In combination, the oral and pharyngeal dentition set *H. riponianus* apart from other Lake Victoria species. In appearance and in possessing a similar oral dentition, *H. saxicola* (see p. 257) appears to be the nearest relative of *H. riponianus* although it differs in having a relatively slender pharyngeal bone with only the two median tooth rows slightly enlarged. The two species, which differ somewhat in their ecological relationships, may represent separate adaptive lines derived from a common ancestor.

Haplochromis pallidus seems structurally suited for consideration as the extant representative of the presumed ancestor. (See also p. 235.)

Although the lower pharyngeal bone and its dentition are similar in *H. humilior* and *H. riponianus* and although both species have very similar ecological requirements, the species show quite marked divergence in body-form (especially head shape) and oral dentition. What relationships there are probably lie far back in their phylogenetic history; certainly the two species cannot be considered members of a recently evolved phyletic line.

Boulenger's view (1911) that *H. riponianus* and *H. microdon* are closely related can no longer be held; the two species have very different phylogenies, as witnessed by several anatomical and osteological characters, and equally distinctive habits (Greenwood, 1959). Boulenger was undoubtedly misled by the supposedly small teeth of both species (Greenwood, op. cit., and above) whereas in fact the teeth of *H. riponianus* are not minute and are of a markedly different form from those of *H. microdon*.

Study material and distribution records

Museum and Reg. No.	Locality	Collector
	Uganda	
Genoa Museum (C.E. 12996 (Lectotype)	Jinja, Ripon Falls	Bayon
B.M. (N.H.).—1911.3.3.37–39	,, ,,	,,
,, 1911.3.3.24	,, ,,	,,
,, 1906.5.30.280	Entebbe	Degen
,, 1906.5.30.394 (Paratype *H. ishmaeli*)	Bunjako	,,
,, 1929.8.13.1	Entebbe	Hoare
,, 1959.4.28.141–157	Entebbe, Airport beach	E.A.F.R.O.
,, 1959.4.28.158	Hannington Bay	,,
,, 1959.4.28.159	Buka Bay	,,
,, 1959.4.28.160–162	Entebbe, Harbour	,,
	Lake Victoria, Locality Unknown	
,, 1901.6.24.86–87	—	Sir H. Johnson
,, 1928.5.24.136–138	—	M. Graham

Haplochromis saxicola,, sp. nov.
(Text-figs. 5 and 14)

Pelmatochromis riponianus (part) Boulenger, 1911, *Ann. Mus. Genova* (3), **5**, 69; *Idem*, 1915, *Cat. Afr. Fish.* **3**, 411.
Haplochromis cinereus (part) Regan, 1922, *Proc. zool. Soc. London*, 166.

Holotype. A female 111·0 + 25·0 mm. long (B.M. (N.H.) Reg. No. 1959.4.28.249) from Ramafuta island, Uganda.

Description, based on twenty-seven specimens (including the holotype of the species and two paratypes of *Pelmatochromis riponianus*) 106–123 mm. S.L.

Depth of body 34·8–42·5 (M = 37·8) per cent of standard length, length of head 35·3–42·5 (M = 37·8) per cent. Dorsal head profile straight or gently curved, sloping at *ca.* 30°–40°. Preorbital depth 15·2–19·0 (M = 17·6) per cent of head, least interorbital width 23·9–29·8 (M = 26·9) per cent. Snout slightly longer than broad or, less commonly, as long as broad, its length 34·0–41·8 (M = 37·5) per cent of head ; diameter of eye 20·4–26·8 (M = 24·4), depth of cheek 21·4–26·2 (M = 23·9) per cent. Caudal peduncle 13·1–17·7 (M = 15·4) per cent of standard length, 1·0–1·5 (mode 1·3) times as long as deep.

Mouth horizontal or very slightly oblique ; lips variably thickened but always noticeably enlarged. Lower jaw 39·7–46·5 (M = 43·0) per cent of head and 1·5–1·9 (mode 1·8) times as long as broad. Posterior tip of the maxilla reaching or almost reaching the vertical through the anterior orbital margin, sometimes extending to below the eye.

Gill rakers moderately stout, the uppermost two or three either slender or divided

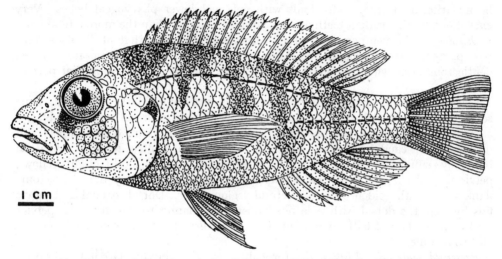

Fig. 14. *Haplochromis saxicola* ; holotype. Drawn by John Norris Wood.

and somewhat flattened, the lowermost one to three usually reduced ; 7–9 (mode 8) rarely 6 on the lower part of the first arch.

Scales ctenoid ; 31 (f.2), 32 (f.15), 33 (f.9) or 34 (f.1) in the lateral line ; cheek with 3 or 4 series. Six or 7 (rarely 5) between the lateral line and the dorsal fin origin ; 6–8 (rarely 5 or 9) between the pectoral and pelvic fin bases. Scales on the pectoral region small.

Fins. Dorsal with 24 (f.4), 25 (f.20) or 26 (f.2) rays, anal with 11 (f.1), 12 (f.24) or 13 (f.1), comprising XIV–XVI, 9 or 10 and III, 8–10 spinous and branched rays for the fins respectively. One specimen has XVI, 6 rays in the dorsal and another II, 8 in the anal. Pectoral 67·8–81·0 (M = 74·5) per cent of head. Caudal truncate or sub-truncate.

Teeth. In the outer series of each jaw the teeth are slender, recurved and generally unicuspid, but a few weakly bicuspid teeth may occur in fishes less than 115 mm. S.L.

The crowns of the teeth are often worn so as to assume a bluntly incisiform shape. There are 52–68 teeth in the outer row of the upper jaw (modal range 60–62).

The inner tooth rows are composed of obliquely implanted and either unicuspid or both uni- and tricuspid teeth arranged in 2 or 3 (rarely 4 or 5) rows in each jaw.

The dentition of *H. saxicola* bears a very strong resemblance to that of *H. riponianus*, the major difference lying in the higher percentage of primarily unicuspid teeth in *H. saxicola*. However, it must be borne in mind that no available specimens of *H. riponianus* are as large as even the smallest *H. saxicola*.

Lower pharyngeal bone and teeth. In ninety per cent of specimens examined, the lower pharyngeal bone is not enlarged; in the remaining ten per cent it is somewhat thickened and resembles the lower pharyngeal of *H. riponianus*.

The two median tooth rows are slightly enlarged in all specimens and may even be molariform in those individuals with enlarged lower pharyngeal bones. Very exceptionally, the median teeth are not noticeably larger than the more lateral ones.

Osteology. The neurocranium of *H. saxicola* is very like that of *H. riponianus* (see p. 254); the premaxilla is, however, distinctive for its noticeably arched dentigerous arms which impart a characteristic peak to the antero-medial part of the bone.

Coloration in life. Sexually active males. Ground colour dark grey-green, some scales on the flank with golden centres; chest and branchiostegal membrane black; a coppery-red flush on the operculum and flanks. Dorsal fin dark, lappets and maculae red; caudal blue-grey with red posterior and ventral margins; anal fin blue-grey, suffused with pink, especially the distal margins, ocelli bright yellow; pelvic fins black. *Quiescent males* pale silvery-blue, almost grey dorsally; some flank scales with golden centres, pectoral region silver. Dorsal, caudal and pelvic fins hyaline, the dorsal with pale red lappets and margin to the soft part, pelvics dusky on the lateral half, hyaline mesially. *Females*. Ground colour golden-grey; all fins hyaline.

Preserved material. Adult males (probably sexually active). Ground coloration dark grey-black, especially dark on the head, dorsal aspects of the flanks and along the ventral body surface. Lips light grey; branchiostegal membrane black. Dorsal fin dark, with light maculae on the soft part. Anal fin black on the basal third, the distal two-thirds yellowish, with two grey ocelli. Caudal greyish ventrally, yellow around the margin.

Females, juvenile and some apparently adult (? quiescent) males. Brownish-yellow ground coloration, darkest on the dorsal surface of the head; four to seven, fairly narrow dark transverse bars on the flanks and caudal peduncle. Dorsal and caudal fins greyish, all other fins colourless, except for the black pelvics in adult males.

Distribution. The species is known definitely from Lake Victoria but may also occur in the Victoria Nile. This locality is suggested by the presumed provenance of two *P. riponianus* paratypes now considered to be *H. saxicola*. For further discussion, see under " Distribution " in *H. riponianus*, p. 255.

Ecology. Habitat. The habitats in which *H. saxicola* has been caught are rather more varied than is common for most Lake Victoria *Haplochromis*. The species occurs in shallow water over exposed sand or shingle beaches, amongst dense plant stands near rocky shores and over shingle and small boulders in water 10–30 feet deep. Finally, one specimen was caught more than a mile off-shore in nets set on a mud bottom in water *ca.* 180 feet deep. The most consistent factors in all these habitats (except the latter) are the hard substrates of sand, rock or shingle.

Food. Twenty-six fishes (from nine different localities) were examined; of these, only twelve contained any ingested material in the stomach or intestine. Sand grains and small pebbles (*ca.* 2 mm. in diameter) were found in eleven individuals; fragmentary insect larvae (Diptera, probably chironomids) formed the principal food in eight fishes and snails in one other. All specimens contained a few fragmentary Ostracoda and one, an almost entire larva of the boring May-fly *Povilla adusta*.

These very unsatisfactory data suggest that *H. saxicola* is a bottom feeder which preys on various invertebrates, particularly insect larvae.

Breeding. No information was obtained on the breeding habits of *H. saxicola*; the smallest specimen (106 mm. S.L.) is sexually mature. Both sexes appear to reach the same maximum adult size.

Affinities. In general appearance, oral dentition, neurocranial morphology and in certain ecological characters, *Haplochromis saxicola* closely resembles *H. riponianus*. The most striking difference between the species is the typically slender lower pharyngeal bone of *H. saxicola* and the stouter pharyngeal bone and teeth of *H. riponianus*. The mean snout and lower jaw proportional lengths for *H. saxicola* are somewhat greater than those for *H. riponianus*. If, however, these characters are taken for both species and plotted against standard length, no marked discontinuity is observed; in all morphometric characters *H. saxicola* could well be large *H. riponianus*.

The differences in the lower pharyngeal bones and dentition, however, are against this interpretation, the more slender bone with but slightly enlarged median teeth characterizing the larger specimens described as *H. saxicola*. In no *Haplochromis* species with enlarged pharyngeals does the bone and its dentition become less coarse with growth. Indeed, the reverse is usual (see Greenwood, 1959 and p. 277).

Study material and distribution records

Museum and Reg. No.	Locality	Collector
	Uganda	
B.M. (N.H.).—1959.4.28.249 (Holotype)	Ramafuta Isl.	E.A.F.R.O.
,, 1911.3.3.39 (Paratype of *Pelmatochromis riponianus*)	Jinja, Ripon Falls	Bayon
Genoa Museum (A paratype of *P. riponianus*)	,, ,,	Bayon
B.M. (N.H.).—1959.4.28.250–255	Ramafuta Island	E.A.F.R.O.
,, 1959.4.28.256–267	,, ,,	,,
,, 1959.4.28.268–270	Beach near Nasu Point	,,
,, 1959.4.28.271–272	Jinja	,,
,, 1959.4.28.273	0° 4′ S., 33° 14′ E.	,,

Haplochromis theliodon sp. nov.
(Text-figs. 12 and 15)

Holotype. A fish 95·0 + 20·0 mm. long (B.M. (N.H.) Reg. No. 1959.4.28.163) from Majita, Tanganyika Territory.

Description, based on seven specimens (including the holotype) 79–95 mm. S.L. The principle morphometric characters are summarized below:

S.L.	Depth*	Head*	Po. %	Io. %	Snt. %	Eye %	Ck. %	L.j. %	C.P.*
75·0	37·3	37·3	17·9	25·0	32·2	25·0	21·4	39·3	15·3
84·0	35·7	36·9	16·2	25·8	32·2	26·8	25·8	38·7	14·3
85·0	36·6	36·6	16·1	22·6	35·5	24·2	22·6	38·7	15·4
86·0	38·4	33·7	17·3	24·1	33·5	24·1	24·1	38·0	15·1
88·0	38·6	36·3	18·1	25·0	34·4	25·0	24·4	37·4	14·8
89·0	38·2	36·5	18·5	24·6	33·8	24·6	27·0	38·4	15·2
95·0	36·8	33·7	17·2	25·0	34·4	26·6	25·0	39·0	13·7

* Per cent. of standard length.
% Per cent. of head length.

Dorsal head profile straight except for a slight concavity above the eye, sloping fairly steeply; snout longer than broad. Mouth slightly oblique; lips, especially the upper, thickened. Lower jaw 1·5–1·8 times as long as broad. Posterior tip of the maxilla almost reaching the vertical to the anterior orbital margin.

Gill rakers variable, from stout (the commonest) to relatively slender; 7 (f.1), 8 (f.5) or 9 (f.1) on the lower part of the first arch. The lowermost two to four rakers may be reduced.

Scales ctenoid; lateral line with 31 (f.2), 32 (f.3) or 33 (f.2) scales; cheek with 3 or 4 series. Six to 9 scales between the lateral line and the dorsal fin origin, 7–9 between the bases of the pectoral and pelvic fins. Scales on the pectoral region small and deeply embedded.

Fins. Dorsal with 25 rays, anal with 12, comprising XV–XVI, 9 or 10 and III, 9 spinous and branched rays for the fins respectively. Pectoral 64·3–72·0 (M = 69·5) per cent of head. Caudal truncate or sub-truncate.

Teeth. In six of the seven specimens examined, the outer row in both jaws is composed of relatively stout, unequally bicuspid teeth, except for the most posterior pair in the upper jaw which are unicuspid. In the seventh specimen, the anterior and postero-lateral teeth are unicuspid and the lateral teeth unequally bicuspid. There are 36–46 outer teeth in the upper jaw.

The inner teeth are small and tricuspid (with a few unicuspids in the exceptional specimen mentioned above) and are arranged in 3 or 4 rows in the upper jaw and 3 or 4 (rarely 2) in the lower. Inner teeth are implanted vertically or somewhat obliquely.

Lower pharyngeal bone and teeth. The lower pharyngeal bone is thickened and moderately stout; that is to say, it is comparable with the lower pharyngeal in most *H. humilior* but finer than that in *H. obtusidens* (see p. 267).

Five of the seven specimens examined have the four median tooth rows composed of large and molariform teeth, whilst the next lateral row of each side contains enlarged but clearly cuspidate teeth. In the exceptional fishes, only the median pair of tooth rows contains molariform elements and only one lateral row on each side has enlarged teeth.

Coloration. The colours of live fishes are unknown.

Preserved material. Quiescent male. Ground colour dusky bronze, with a very faint mid-lateral stripe, six interrupted transverse bars and a distinct lachrymal stripe. Dorsal fin dark, lappets pale ; caudal dark except for its pale distal margin ; anal pale (? orange) with four, small, white ocelli arranged in a single row. Pelvics black on the outer two-thirds, lighter medially. *Females and immature males* uniformly light brown except for the darker dorsal head surface and nape ; a dark

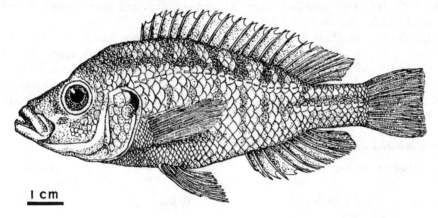

Fig. 15. *Haplochromis theliodon* ; holotype. Drawn by Miss G. Osterritter.

mid-lateral stripe of variable intensity is continuous in some specimens but interrupted in others above and slightly anterior to the anal fin ; when interrupted, the line is thickened to form a black blotch above the pectoral fin. A fainter, continuous dorsal stripe runs above the upper lateral line ; there are traces of seven to nine, variously interrupted transverse bars on the flanks but not on the caudal peduncle. Dorsal and anal fins hyaline, darker between the branched rays. Caudal maculate on the upper half, immaculate below. Pelvic fins hyaline in females but with the outer half dusky in males.

Distribution. Known only from Lake Victoria.

Ecology. Habitat. The few specimens known came from two localities, both exposed, shallow and sandy beaches.

Food. Analysis of stomach and intestinal contents from all seven specimens indicate that *H. theliodon* is a bottom feeder with a varied diet including small fishes (cichlids, 20 25 mm. long), Gastropoda (foot and soft parts only), Lamellibranchiata and insect larvae, e.g. *Povilla adusta* (Ephemeroptera) and Trichoptera (including the cases). Each fish had also ingested fairly large quantities of bottom debris.

Breeding. Few data are available ; one female 75 mm. S.L. is sexually mature whilst another, 88 mm. S.L. appears to be a juvenile maturing for the first time.

Diagnosis and affinities. The nature of its lower pharyngeal bone and dentition places *H. theliodon* on the same level of structural modification as *H. humilior* and *H. riponianus.* In other characters (especially general appearance and oral dentition) *H. theliodon* is unlike both the former species. It differs from *H. humilior* in several morphometric characters (straight and not curved dorsal head profile, deeper preorbital, longer snout and smaller eye) and from *H. riponianus* in the shape of the head and the nature of its oral dentition. *Haplochromis theliodon* also differs from both the other species in having much smaller and more deeply embedded pectoral scales.

The affinities of *H. theliodon* are not easily determined except in so far as the species is clearly a little-modified derivative of the generalized stem. The small pectoral scales of *H. theliodon* are not common in generalized species still extant in Lake Victoria ; but, small chest scales do characterize certain of the generalized and fluviatile species of East Africa, some of which also show an incipient hypertrophy of the pharyngeal bones and dentition (Greenwood, unpublished).

Study material and distribution records

Museum and Reg. No.	Locality	Collector
	Tanganyika	
B.M. (N.H.).—1959.4.28.163 (Holotype)	Majita	E.A.F.R.O.
„ 1959.4.28.164–168	„	„
	Uganda	
„ 1959.4.28.169	Jinja Pier	„

Haplochromis empodisma sp. nov.
(Text-figs. 16 and 17)

Haplochromis ishmaeli (part) Boulenger, 1906, *Ann. Mag. nat. Hist.* (7), **17**, 446 ; *Idem*, 1915, *Cat. Afr. Fish.* **3**, 293 ; Regan, 1922, *Proc. zool. Soc. London*, 169.
Tilapia lacrimosa (part), Boulenger 1906, *op. cit.*, 450 ; *Idem*, 1915, *tom. cit.*, 234.
Haplochromis cinereus (part) Regan, 1922, *op. cit.*, 166.

Note. Because two specimens of *H. empodisma* (collected during Graham's 1927–28 survey of Lake Victoria) were mistaken for syntypes of *H. michaeli* Trewavas, this species was referred to as *H. michaeli* in previous papers (Greenwood, 1954, 1956a). All references to " *H. michaeli* " in these publications should now be corrected to read *Haplochromis empodisma.*

Holotype. An adult male 117 + 23 mm. total length (B.M. (N.H.) Reg. No.

1959.4.28.170), caught on the bottom in 90 feet of water off the southern tip of Kibibi Island (0° 10′ N.; 33° 10′ E.), Uganda.

Description, based on thirty-nine fishes (including the holotype) 65–117 mm. S.L.

Depth of body 33·0–43·8 (M = 39·3) per cent of standard length, length of head 33·3–39·4 (M = 36·7) per cent. Dorsal head profile straight or gently curved, sloping at an angle of 35°–40°. Preorbital depth 15·1–20·5 (M = 18·1) per cent of head; least interorbital width 20·6–28·6 (M = 24·3), length of snout 27·5–37·2 (M = 32·9), diameter of eye 24·4–34·0 (M = 27·3) per cent. Depth of cheek shows a marked positive allometry with standard length, being 21·2–23·0 per cent of head in fishes less than 80 mm. S.L. (N = 4) and 23·5–31·4 (M = 27·9) per cent in larger specimens (N = 35). Caudal peduncle 14·5–20·0 per cent of standard length and 1·3–1·7 (mode 1·4) times as long as deep.

Mouth horizontal or slightly oblique; lips not noticeably thickened. Posterior tip of the maxilla reaching the vertical through the anterior margin of the orbit or to

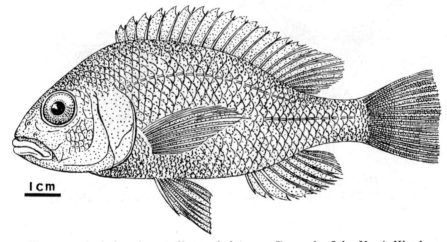

Fig. 16. *Haplochromis empodisma*; holotype. Drawn by John Norris Wood.

below the anterior part of the eye; less frequently, not reaching as far as the orbit. Lower jaw 39·1–48·7 (M = 43·9) per cent of head, 1·5–2·3 (mode, ill-defined, 2·0) times as long as broad. Malformation of the lower jaw is relatively common in this species; individuals so affected have the lower jaw broader than the upper (which closes within the lower) and are distinctly prognathous.

Gill rakers usually slender, although relatively stout rakers also occur; 7–10 (modal range 8–9) on the lower part of the first gill arch, the lowermost 1–3 rakers reduced, occasionally the uppermost 1 or 2 somewhat flattened and bi- or trifid.

Scales ctenoid; lateral line with 30 (f.10), 31 (f.9), 32 (f.15) or 33 (f.4) scales. Cheek with 3 or 4 (occasionally 5) series. Six or 7 scales between the lateral line and the dorsal fin origin, 6–8 between the pectoral and pelvic fin bases.

Fins. Dorsal with 23 (f.8), 24 (f.30) or 25 (f.1) rays, anal with 10 (f.1), 11 (f.22) or 12 (f.16), comprising XIV–XVI, 8 or 9 and III, 7–9 spinous and branched rays for the fins respectively. Pectoral fin 73·0–96·5 (M = 85·8) per cent of head. Caudal truncate or sub-truncate.

Teeth. The teeth in the outer row of both jaws are slender and often gently recurved; in fishes less than 95 mm. S.L. the teeth are unequally bicuspid but in larger individuals are weakly bicuspid or unicuspid; it is usual to find both types of teeth in large fishes. There are 54–82 (modal range 70–72) teeth in the upper jaw.

Teeth forming the inner series are small and usually tricuspid but some unicuspids may occur, particularly in fishes over 100 mm. S.L.; there are 2 or 3 (rarely 4) rows in the upper jaw and 1–3 in the lower. The innermost row, especially in the upper jaw, is implanted obliquely.

A common abnormality affecting both inner and outer teeth is for the crowns to be coarse, slightly swollen and darkly pigmented; in such teeth the crown is globose.

Lower pharyngeal bone and teeth. The lower pharyngeal is fine and rather narrow,

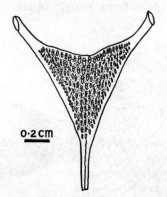

Fig. 17. *Haplochromis empodisma;* lower pharyngeal bone, occlusal view.

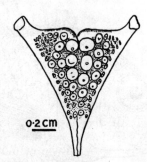

Fig. 18. *Haplochromis obtusidens;* lower pharyngeal bone, occlusal view.

the dentigerous surface having the outline of an isosceles triangle. Usually, none of the pharyngeal teeth is markedly enlarged, but in certain specimens of all sizes a few teeth in the two median rows are slightly coarser. Even when enlarged these teeth retain the same form as their lateral and more slender congeners.

Osteology. The neurocranium and dentary of *H. empodisma* were figured and briefly described in an earlier paper (Greenwood, 1956a, p. 305, fig. 5; the species was then wrongly identified as *H. michaeli*). Both the neurocranium and the jaw elements are directly comparable with those of a generalized species such as *H. macrops*, except that in *H. empodisma* the outer teeth are more numerous.

Coloration in life. Sexually active males. Ground colour dark turquoise on the flanks and dorsal body surface, silver-yellow ventrally; snout and dorsal head surface dark red. Dorsal and caudal fins diffuse red; anal black anteriorly, dark red posteriorly, the ocelli orange-red. Pelvic fins black. *Quiescent males* as for females (see below) but dorsal fin with red maculae between the branched rays. Ripening males show some reddening of the head and snout, whilst the maculae of the dorsal are more intense. *Females*. Ground colour yellowish-silver, darker dorsally. Dorsal and caudal fins dark neutral ; anal yellow ; pelvics very pale yellow.

Preserved material. *Males* (sexually active), dark grey becoming black ventrally ; a distinct lachrymal stripe and two bars across the snout (quiescent males greyish brown, dusky on the belly ; very faint indications of up to five incomplete transverse bars on the flanks ; lachrymal stripe visible but no bars across the snout). Dorsal fin dark, the lappets black ; caudal dark proximally, lighter distally ; basal third of anal black, remainder of fin light, with two or three large white ocelli. Pelvic fins black. *Females* brownish-silver ; in a few specimens there are very faint indications of a lachrymal stripe and of incomplete transverse bars on the flanks. All fins hyaline ; in some fishes the upper half of the caudal is slightly maculate.

Distribution. Known only from Lake Victoria.

Ecology. *Habitat*. It seems that *H. empodisma* is generally restricted to those areas of the lake where the bottom is composed of organic mud. The species has been caught in bottom nets set at depths from 10–90 feet.

Food. The stomach and intestinal contents of thirty-seven fishes (from numerous localities) show that *H. empodisma* is a bottom feeder. The principal food organisms are the larval (and less frequently pupal) stages of dipterous insects, together with diatoms derived from the bottom mud. Only two fishes yielded remains of Mollusca ; in one fish a few fragments of bivalve shell were found and in the other an operculum from a large gastropod. Since no other snail fragments were found in this fish, it is possible that the operculum was accidentally ingested.

Breeding. It is not known whether *H. empodisma* spawns in the habitats described above or whether spawning takes place over a more solid substrate, for example outcrops of rock or sand. Females carrying embryos and larvae have been caught together with non-breeding fishes. The smallest sexually active fish is a female 84 mm. S.L. ; the smallest adult male is 90 mm. S.L. Both sexes reach the same maximum adult size.

Affinities. Structurally and in its feeding habits, *H. empodisma* must be considered a generalized species. It differs from the majority of generalized forms in Lake Victoria only by its larger size, greater number of teeth (possibly a correlate of the larger size) and in the wide range of depths at which it has been caught. The nearest living relative of *H. empodisma* is a small and as yet undescribed species which occurs in the same habitat but is confined to shallow water.

Haplochromis empodisma and *H. obtusidens* are strikingly similar except for one structure, the lower pharyngeal bone. In *H. obtusidens* the lower pharyngeal is thickened and carries a number of enlarged, crushing teeth, whereas in *H. empodisma* this bone is slender and carries numerous fine teeth. *Haplochromis empodisma* could well represent the ancestral type from which *H. obtusidens* was derived by an increase in size of the pharyngeal bones and a correlated change in the pharyngeal dentition. Although the inter-specific differences in the nature of the pharyngeal mill are fairly clear-cut, some individuals deviate from the specific mode in such a way as to indicate a likely transitional condition in the evolution of an " *obtusidens* "-like species.

Since, in turn, *H. obtusidens* provides a structural type basic to the evolution of a specialized crushing pharyngeal mill (as found in *H. ishmaeli* and *H. pharyngomylus*), *H. empodisma* could represent the extant representative of the basal species in the phyletic line which culminated in *H. ishmaeli* and *H. pharyngomylus*.

Study material and distribution records

Museum and Reg. No.	Locality	Collector
	Uganda	
B.M. (N.H.).—1959.4.28.170 (Holotype)	Off Kibibi Island	E.A.F.R.O.
„ 1906.5.30.472	Entebbe	Degen
„ 1906.5.30.402 (Paratype, *H. ishmaeli*)	Bunjako	„
„ 1906.5.30.404 (Paratype, *H. ishmaeli*)	„	„
„ 1959.4.28.180–189	Pilkington Bay	E.A.F.R.O.
„ 1959.4.28.194–195	Thruston Bay	„
„ 1959.4.28.196–197	Ekunu Bay	„
„ 1959.4.28.198	Jinja	„
„ 1959.4.28.199	Napoleon Gulf, near Jinja	„
„ 1959.4.28.202–203	Kibibi Island	„
„ 1959.4.28.358	Off S. tip of Kibibi Island	„
	Tanganyika	
„ 1959.4.28.190–193	Nyamakyamwa	E.A.F.R.O.
	Kenya	
„ 1959.4.28.200–201	Manadu Island	„
„ 1959.4.28.171–179	Off Port Southby	„
	Lake Victoria, Locality Unknown	
„ 1928.6.2.37–38	—	M. Graham

Haplochromis obtusidens Trewavas 1928
(Text-fig. 18)

Haplochromis desfontainesi (part), Boulenger, 1915, *Cat. Afr. Fish.*, **3**, 302.
Tilapia lacrimosa (part), Boulenger, 1915, *tom. cit.*, 234.
Haplochromis cinereus, (part) Regan, 1922, *Proc. zool. Soc. London*, 166.
Haplochromis obtusidens Trewavas, 1928, *Ann. Mag. nat. Hist.* (10), **2**, 95.

Lectotype. An adult male 107·0 + 23·0 mm. total length (B.M. (N.H.) Reg. No. 1928.5.30.21).

Description, based on forty-four specimens (including the lectotype and one paratype) 60–114 mm. S.L.

Certain proportions show fairly well-marked allometry with standard length; for these, ranges and means are given for each of the relevant size-groups.

Depth of body 35·0–44·3 (M = 38·6) per cent of standard length, length of head 32·5–38·2 (M = 35·8) per cent. Dorsal head profile gently curved or, less commonly, straight, sloping at an angle of 40°–50°. Preorbital depth in fishes < 70 mm. S.L.

(N = 4), 13·0–16·7 (M = 14·5) per cent of head, in larger fishes (N = 40) 15·1–20·5 (M = 17·9) ; least interorbital width 21·8–29·0 (M = 24·2) per cent ; snout length in fishes < 85 mm. S.L. (N = 6), 26·0–31·0 (M = 28·5) per cent, in larger individuals (N = 38), 29·0–36·4 (M = 33·3) ; diameter of eye in fishes < 85 mm., 27·6–34·8 (M = 31·2) per cent and in larger fishes 24·3–30·8 (M = 27·2) ; depth of cheek in specimens < 85 mm., 18·6–23·8 (M = 21·5) and in larger individuals 21·2–30·0 (M = 26·7) per cent of head. Caudal peduncle 15·2–19·7 (M = 17·5) per cent of standard length, 1·2–1·8 (mode 1·5) times as long as deep.

Jaws equal anteriorly ; mouth horizontal or slightly oblique ; lips not markedly thickened. Posterior tip of the maxilla extending to the vertical through the anterior orbital margin or to below the anterior part of the eye ; exceptionally, not reaching the orbit. Lower jaw 37·9–45·5 (M = 41·8) per cent of head and 1·3–2·0 (mode 1·6) times as long as broad.

Gill rakers relatively slender, although short and stout rakers are found in a few individuals ; 7–9 (modal range 7–8), rarely 6, on the lower part of the first arch, the upper pair of rakers often flattened.

Scales ctenoid ; 29 (f.1), 30 (f.7), 31 (f.21), 32 (f.11) or 33 (f.4) in the lateral line ; cheek with 3 or 4 (rarely 5) series ; $5\frac{1}{2}$–7 scales between the lateral line and the dorsal fin origin, 5–7 (rarely 9) between the pectoral and pelvic fin bases.

Fins. Dorsal with 23 (f.13), 24 (f.25) or 25 (f.6) rays, anal with 11 (f.24), 12 (f.19) or 13 (f.1), comprising XIV–XVI, 8–10 and III, 8–10 spinous and branched rays for the fins respectively. Pectoral 73·5–103·0 (M = 86·8) per cent of head. Caudal truncate or sub-truncate.

Teeth. In fishes less than 90 mm. S.L., the outer row in both jaws is composed of unequally bicuspid and relatively slender teeth. Larger individuals have an admixture of either unequally bicuspid and unicuspid teeth or of weakly bi- and unicuspids. A few individuals have exclusively bicuspid or unicuspid teeth. There are 40–80 teeth (ill-defined mode at 70 ; modal range 66–70) in the outer row of the upper jaw.

Teeth in the inner series are generally tricuspid and small ; in some fishes over 100 mm. S.L. there may be either a mixture of tri- and unicuspid teeth or, more rarely, only unicuspids. The innermost row of teeth (especially in the upper jaw) is implanted obliquely and in some individuals the whole inner series lie at an oblique angle. There are 2 or 3 (rarely 1 or 4) rows of inner teeth in the upper jaw and 1 to 3 (usually 2, rarely 4) in the lower.

Lower pharyngeal bone and teeth. Despite some individual variability, the lower pharyngeal bone of *H. obtusidens* is always obviously thickened and the two median rows of teeth are enlarged and molariform or sub-molariform. As might be expected, there is a positive correlation between size and the degree to which the pharyngeal bones and teeth are enlarged. In fishes over 90 mm. S.L. as many as six rows of teeth may be composed of molariform elements ; even in specimens with a few molariform rows the remaining teeth are enlarged, except in the upper corners of the bone.

Osteology. In all respects except the form of the pharyngeal apophysis, the neurocranium of *H. obtusidens* resembles that of *H. empodisma* ; the pharyngeal apophysis, however, is stouter and broader, thereby foreshadowing the condition found in *H. ishmaeli* and *H. pharyngomylus*. In both these species the basioccipital facets are more expanded than in *H. obtusidens*.

Coloration in life. Sexually active males. Ground colour light blue-black with, ventrally, a silver patch extending from the isthmus almost to the vent; branchiostegal membrane black. Dorsal fin dark with red lappets and margin to the soft part and red spots or dashes between the branched rays. Caudal dark with a blood-red margin and a diffuse red centre. Anal dark proximally, blood-red distally; ocelli orange or orange-red. Pelvic fins black. *Quiescent males* as for females but lacking the red spots or flush on the dorsal fin. *Females.* Ground colour silver-grey. Dorsal fin dark with traces of a red flush. Caudal dark. Anal and pelvic fins yellowish.

Preserved material. Adult males. Ground colour brownish, dusky ventrally (below the level of the lower lateral line), with faint indications of four transverse stripes originating from the dark area but not reaching the base of the dorsal fin; the intensity of the dusky area varies considerably, from charcoal to coal-black. Dorsal fin hyaline to dusky, the basal region and the lappets black; soft part of the fin sometimes maculate. Caudal dark on its proximal half, light (yellowish) distally. Anal dark on the basal half to third, light distally, with two large white ocelli. Pelvic fins black. *Females* silver-grey to yellowish-brown; in some fishes there are very faint indications of four to six, incomplete transverse bars on the flanks. Dorsal fin dark or hyaline. Caudal dark, the upper half maculate in some individuals. All other fins hyaline.

Note on four atypical individuals. Four specimens, 80–94 mm. S.L. (B.M. (N.H.) Reg. Nos. 1959.4.28.236–239) from Old Bukakata Bay, Uganda are included in the description given above although they differ from the generality of specimens in certain characters. The lower pharyngeal bones in two of these fishes are somewhat more slender than is modal but are typical in the other two specimens. All four fishes have fewer upper outer teeth than is usual (40–52 cf. modal range 66–70) and somewhat shallower cheek than equivalent sized specimens from other areas. In this latter character, however, they resemble the two type specimens. Finally it must be mentioned that these fishes resemble one another in general facies rather more closely than they resemble the other specimens. Such obvious but undefinable and geographically localized facies are fairly common amongst Lake Victoria *Haplochromis* species.

The exact status of the four specimens from Bukakata cannot be determined for want of more material from this locality.

Distribution. Known only from Lake Victoria.

Ecology. Habitat. Haplochromis obtusidens is predominantly a species of shallow water (less than 30 feet), apparently restricted to a substrate of soft, organic mud; a few individuals have, however, been taken over sand and in water about 60 feet deep.

Food. The gut contents of forty-six fishes (from numerous localities) indicate that the principal food organisms of *H. obtusidens* are insects (especially larval Diptera) and molluscs (particularly the bivalve *Corbicula*, although some gastropods

[*Melanoides*] are also eaten). Together with these organisms, the fishes had ingested fairly substantial quantities of bottom mud, which, in the areas inhabited by *H. obtusidens*, is almost entirely composed of living and moribund diatoms. A comparison of stomach and intestinal contents shows that the protoplasm of these diatoms is digested by the fish. Other plants, in this case mostly blue-green algae, are apparently undigested.

Breeding. *Haplochromis obtusidens* is a female mouth-brooder; exact spawning sites were not discovered. A male 83 mm. S.L. is the smallest mature fish in the sample studied; the smallest mature female is 89 mm. S.L. Both sexes reach the same maximum adult size.

Affinities. The relationships of *H. obtusidens* seem to lie with *H. empodisma* which it resembles in body-form and most syncranial characters. *Haplochromis obtusidens* differs in having an enlarged neurocranial apophysis for the upper pharyngeal bones and in the correlated character, enlarged pharyngeal bones and teeth. The species also shows certain fairly marked affinities with *H. ishmaeli* and *H. pharyngomylus* and could well represent the extant version of an ancestral type from which these two species evolved (see also p. 265).

In Lake Edward, *H. malacophagus* Poll has reached a comparable evolutionary stage leading towards extreme pharyngeal hypertrophy, but the two species are not closely related.

Study material and distribution records

Museum and Reg. No.	Locality	Collector
	Uganda	
B.M. (N.H.).—1906.5.30.531	Buganga	Degen
,, 1959.4.28.204–212	Pilkington Bay	E.A.F.R.O.
,, 1959.4.28.213–218	Ekunu Bay	,,
,, 1959.4.28.228–229	Pilkington Bay	,,
,, 1959.4.28.236–239	Old Bukakata Bay	,,
,, 1959.4.28.243–244	Dagusi Island	,,
,, 1959.4.28.245	Buka Bay	,,
,, 1959.4.28.247	Sesse Islands	,,
	Kenya	
,, 1909.7.27.43	Kavirondo Bay	Alluaud
,, 1959.4.28.219–227	Off Port Southby	E.A.F.R.O.
,, 1959.4.28.230–232	Kach Bay (Kavirondo Gulf)	,,
,, 1959.4.28.240–241	Beach below Usoma lighthouse	,,
,, 1959.4.28.242	Off Port Southby	,,
,, 1959.4.28.246	Kavirondo Gulf	,,
,, 1959.4.28.248	Off mouth of Nzoia River	,,
	Tanganyika	
,, 1959.4.28.233–235	Beach near Majita	,,
	Lake Victoria, Locality Unknown	
,, 1928.5.30.21 (Lectotype)	—	M. Graham
B.M. (N.H.).—1928.5.20.20 (Paratype)	—	,,

Haplochromis pharyngomylus Regan 1929
(Text-figs. 19 and 20)

Haplochromis ishmaeli (part) Boulenger, 1906, *Ann. Mag. nat. Hist.* (7), **17**, 446;
Idem, 1915, *Cat. Afr. Fish.*, 3, 293; Regan, 1922, *Proc. zool. Soc. London*, 169.
Haplochromis pharyngomylus Regan, 1929, *Ann. Mag. nat. Hist.* (10) **3**, 388.

Description, based on thirty-eight specimens (including the holotype) 70–126 mm. S.L.

Depth of body 33·8–42·0 (M = 38·5) per cent of standard length, length of head 31·5–36·8 (M = 34·6) per cent. Dorsal head profile straight or somewhat curved (occasionally concave between the eyes), sloping fairly steeply. Preorbital depth 13·8–19·0 (M = 16·8) per cent of head, least interorbital width 23·7–28·5 (M = 26·3), length of snout 27·3–33·3 (M = 30·8), diameter of eye 23·0–31·8 (M = 26·5), depth of cheek 19·7–27·0 (M = 24·1) per cent of head. Caudal peduncle 13·6–18·5 (M = 16·2) per cent of standard length, 1·1–1·6 (modal range 1·3–1·5) times as long as broad.

Mouth horizontal or slightly oblique, jaws equal anteriorly, the lower 35·8–44·0 (M = 38·6) per cent of head, 1·3–2·0 (mode 1·4) times as long as broad. Posterior tip of the maxilla reaching or almost reaching the vertical to the anterior orbital margin.

Gill rakers moderately stout, 6–8 (mode 7), rarely 9, on the lower part of the first arch, the lowermost 1 or 2 rakers reduced.

Scales ctenoid; lateral line with 30 (f.1), 31 (f.6), 32 (f.12), 33 (f.13), 34 (f.4) or 35 (f.1) scales; cheek with 3 or 4 series. Six or 7 (less frequently 8) scales between the lateral line and the dorsal fin, 7 or 8 (less frequently 6 or 9, rarely 10) between the pectoral and pelvic fin bases.

Fins. Dorsal with 24 (f.7), 25 (f.28) or 26 (f.3) rays, anal with 11 (f.5), 12 (f.28) or 13 (f.5), comprising XV–XVI, 8–10 (rarely 11) and III, 8–10 spinous and branched rays for the fins respectively. Pectoral 68·5–91·0 (M = 79·6) per cent of head. Caudal truncate.

Teeth. In fishes 70–90 mm. S.L. only unequally bicuspid, relatively stout teeth occur in the outer row of both jaws. Fishes 90–100 mm. S.L. show some variability in the form of these teeth, which may be unicuspids, a mixture of bi- and weakly bicuspid teeth or both bi- and unicuspids. There are 30–42 (mode 36, modal range 36–40) outer teeth in the upper jaw.

Teeth in the inner rows are usually tricuspid in fishes less than 100 mm. S.L. In larger individuals these teeth may be tricuspid or there can be a mixture of tri- and weakly tricuspids; it is uncommon to find only unicuspid teeth in the inner rows. There are 2 or 3 (rarely 4) rows in the upper jaw and 1 or 2 (rarely 3) in the lower.

Lower pharyngeal bone and teeth. With one exception, the lower pharyngeal bone is massive, even in the smallest specimens. Nevertheless, a slight size correlated increase in relative stoutness can be detected in large fishes. Apart from *H. ishmaeli* (p. 277) and *Astatoreochromis alluaudi*, no other Lake Victoria Cichlidae have such large pharyngeal bones.

A REVISION OF THE LAKE VICTORIA *HAPLOCHROMIS* SPECIES 271

The exceptional specimen mentioned above was caught near Jinja and measured 98 mm. standard length. The lower pharyngeal of this fish can be compared with that of an *H. obtusidens* of a similar size. Since in all other respects this fish resembles *H. pharyngomylus* more closely than *H. obtusidens* it is included in the description of *H. pharyngomylus*. Four other specimens with even finer pharyngeal bones and dentition, but otherwise resembling *H. pharyngomylus*, are not included in the description but are dealt with in a separate appendix (p. 274).

The lower pharyngeal teeth of *H. pharyngomylus* are large and molariform except for a few teeth situated in the upper, lateral angles of the bone. These teeth, although relatively stout, are cuspidate and small; the number of such teeth decreases markedly in the largest individuals. By analogy with other species having

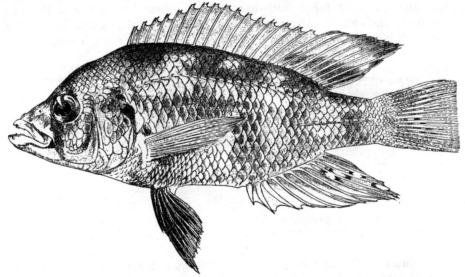

Fig. 19. *Haplochromis pharyngomylus*; holotype. Drawn by Miss M. Fasken.

enlarged pharyngeals and for which a greater size range of specimens is available is seems likely that small *H. pharyngomylus* should have less massive bones and fewer molariform teeth, confined to the median tooth series (see Greenwood, 1959).

Osteology of the neurocranium. Apart from an enlarged and strengthened articular apophysis for the upper pharyngeals, the neurocranium of *H. pharyngomylus* is that of a large, generalized *Haplochromis*. The apophysis has been figured and described previously (Greenwood, 1954). Compared with the apophysis of *H. obtusidens* that of *H. pharyngomylus* has a greater surface area and the prootic buttresses are more obvious. These characters are clearly correlated with the more massive pharyngeal bones and musculature of *H. pharyngomylus*. Only slight difrences exist in the apophyseal region of *H. pharyngomylus* and *H. ishmaeli*; but both species differ from *Astatoreochromis alluaudi* in the form taken by the various elements contributing to the apophysis (Greenwood, 1959. In that paper, I also briefly described the possible effects and interactions of genetical and environmental factors on the development of apophyseal form and size in mollusc-crushing species).

Coloration in live fishes. Adult males. Ground colour blue-grey overlying silver; a distinct coppery sheen on the flanks. Dorsal fin hyaline with pinkish lappets and margin to the soft part. Anal hyaline, ocelli yellow. Caudal hyaline with a pink flush most intense distally and on the ventral half of the fin. Pelvics black. *Females.* Golden-green, becoming silvery-white ventrally. Dorsal and anal fins dark; the caudal yellowish-green, darker proximally.

Preserved material. Adult males. Greyish, darker in sexually active individuals; a dark lachrymal stripe sometimes visible; barring on the flanks variable, but usually consisting of seven to ten narrow vertical bars and a very faint longitudinal stripe following the course of the upper lateral line. Dorsal fin hyaline, dark basally in juvenile and quiescent fishes but almost black in sexually active individuals; lappets dark, the soft dorsal intensely maculate in sexually active fishes. Anal hyaline, dark basally, with five or six ocelli arranged in one or two horizontal rows.

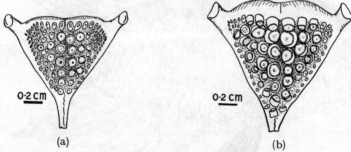

FIG. 20. *Haplochromis pharyngomylus*; lower pharyngeal bone, in occlusal view, of (*a*) a fish 98 mm. S.L., and (*b*) a fish 126 mm. S.L.

Caudal hyaline in quiescent fishes, dusky in active ones. Pelvics hyaline in juveniles, black in adults. *Females*, silvery-grey to light brown; striping variable; when present, as described for males. Dorsal fin hyaline or greyish, the soft part weakly maculate in adults. Caudal hyaline or dark, the upper half faintly maculate in some specimens. Anal and pelvic fins hyaline.

Distribution. Known only from Lake Victoria.

Ecology. Habitat. Haplochromis pharyngomylus is apparently confined to water less than 40 feet deep and to areas of the lake where the bottom is hard (sand, shingle or, less frequently, rock); the species is often found amongst stands of aquatic plants. A few specimens have been caught in the areas of intergradation between sand and mud substrates.

Food. Analyses of stomach and intestinal contents from thirty specimens (representing most localities) show that fishes in the size-range 75–115 mm. S.L. feed mainly or even exclusively on Mollusca; only one fish had eaten mollusca and insect larvae. The sample examined also indicated that both bivalves and gastropods are eaten in approximately equal proportions. The fragmented shells do not permit accurate identification of the prey species; *Sphaerium* sp. and *Corbicula* sp. were recognized amongst the bivalve remains and *Melanoides tuberculata* (the predominant snail), *Bellamya* and less frequently *Biomphalaria* amongst the gastropod fragments.

It is clear from the small size of most shell fragments that the pharyngeal mill in *H. pharyngomylus* is an efficient crushing mechanism. Yet, despite this powerful barrier some shells do pass into the stomach almost undamaged. Since these shells are invariably empty it seems that the digestive enzymes (especially those of the stomach) are capable of breaking down the bodies of snails without preliminary and physical assistance from the pharyngeal teeth. Trewavas (1938) noticed that the gut contents of *H. mahagiensis* (a mollusc eater from Lake Albert) were " impregnated and held in a hard mass by botryoidal aggregates of calcite. Whether these were formed before or after death and fixation is a matter for conjecture." My observations on the gut contents of *H. pharyngomylus* support Trewavas' observations only when the material had been fixed in formol. This would suggest that the calcite aggregates are formed as a result of a chemical reaction between the slightly acid formol and the calcium of the shell. I have never observed aggregates in fresh gut contents, where the various shell fragments could be separated easily.

Breeding. The actual spawning sites and breeding behaviour of *H. pharyngomylus* are unknown, but females carrying embryos and larvae in the buccal cavity occur together with non-breeding fishes in most localities. The smallest adult recorded is a female 90 mm. S.L.; the smallest adult male is 94 mm. S.L. Both sexes reach the same maximum adult size.

Affinities and diagnosis. Extreme hypertrophy of the pharyngeal mill serves to set *H. pharyngomylus* apart from all except one species of Lake Victoria *Haplochromis*. The other species is *H. ishmaeli* which appears to be very closely allied to *H. pharyngomylus*. Although in structural characters the two species are similar an experienced observer can, in most instances, readily distinguish between the two species. But, as is so often the case with *Haplochromis*, the subjective characters used for " field " identifications cannot be quantified or adequately described on paper. When seen alive, adult males of the two species have distinctive coloration.

Haplochromis pharyngomylus is distinguished from *H. ishmaeli* by the following characters: fewer outer teeth in the upper jaw (30–42, modal range 36–40 cf. 38–66, modal range 44–52 in *H. ishmaeli*); shorter pectoral fin (68·0–91·0, M = 79·6 per cent of head cf. 75·0–102·0, M = 88·5); differences in male breeding coloration (see p. 272 and p. 277) and differences in habitat preference (*H. pharyngomylus* is essentially a species of hard substrates whilst *H. ishmaeli* shows a marked preference for muddy areas). With the exception of differences in male breeding colours, none of these characters alone is trenchant; taken together, however, they provide fairly reliable diagnostic features.

The phylogenetic position of *H. pharyngomylus* has been discussed above (p. 269) and in an earlier paper (Greenwood, 1954).

In Lake Edward, *H. placodus* Poll represents the equivalent evolutionary phase in the development of hypertrophied pharyngeals. *Haplochromis placodus* and *H. pharyngomylus* could be derived from a common ancestral stem, but of course, the two species could equally well be examples of convergent evolution.

APPENDIX

Four specimens (115–120 mm. S.L.) from Jinja (B.M. (N.H.) Reg. No. 1959. 4.28.352–355) bear a strong resemblance to *H. pharyngomylus* except that the lower pharyngeal bone in these specimens is barely enlarged and the pharyngeal dentition is correspondingly weak. These fishes cannot be distinguished from *H. pharyngomylus* on proportional measurements, teeth of jaws, fin and scale counts, general appearance or preserved coloration. All four fishes are adult males but unfortunately their live coloration was not recorded.

From *H. pharyngomylus* material described above and from field observations on numerous other specimens it is clear that intraspecific variability in the enlargement of the pharyngeal bones is slight and mainly correlated with size. No specimens were found which could be considered intermediate between the typical condition for the species and that seen in the aberrant individuals. On the other hand, a study of *Astatoreochromis alluaudi* (a *Haplochromis*-like monotypic genus) showed that some populations have enlarged pharyngeals whilst others exhibit only slight hypertrophy of these bones (Greenwood, 1959).

Thus for the moment it is impossible to dismiss the possibility that the four "*pharyngomylus*"-like fishes are indeed aberrant members of that species. I do not propose, however, to include them in this revised description or to describe them as distinct species until further collections dictate one step or the other.

Study material and distribution records

Museum and Reg. No.	Locality	Collector
	Uganda	
B.M. (N.H.).—1906.5.30.383	Entebbe	Degen
„ 1907.5.7.71	Buddu Coast	Simon
„ 1958.12.5.30–33	Entebbe	Pitman
„ 1959.4.28.317–321	Beach near Nasu Point	E.A.F.R.O.
„ 1959.4.28.322–326	Entebbe, Harbour	„
„ 1959.4.28.327–329	Entebbe, Airport beach	„
„ 1959.4.28.333–334	Beach near Grant Bay	„
„ 1959.4.28.335–336	Between Yempita and Busiri Isls., Buvuma Channel	E.A.F.R.O.
„ 1959.4.28.340–343	Beach near Hannington Bay	„
„ 1959.4.28.344	Entebbe, Harbour	„
„ 1959.4.28.345–347	Napoleon Gulf, near Jinja	„
„ 1959.4.28.351	Ramafuta Island	„
„ 1959.4.28.358	Jinja Pier	„
	Kenya	
„ 1959.4.28.348	Kasingiri Gingo (Kavirondo Gulf)	„

Study material and distribution records (cont.)

Museum and Reg. No.	Locality	Collector
	Tanganyika	
,, 1959.4.28.337–339 .	Majita	. ,,
,, 1959.4.28.349 .	Beach near Majita	. ,,
,, 1959.4.28.350 .	Mwanza, Capri Bay	. ,,
	Lake Victoria, Locality Unknown	
,, 1928.5.24.313 . (Holotype)	—	. M. Graham
,, 1959.4.28.330–332 .	—	E.A.F.R.O.

Haplochromis ishmaeli Blgr. 1906
(Text-fig. 21)

Haplochromis ishmaeli (part) Boulenger, 1906, *Ann. Mag. nat. Hist.* (7), **17**, 446; *Idem*, 1915, *Cat. Afr. Fish.* **3**, 293; Regan, 1922, *Proc. zool. Soc. London*, 169.
Tilapia pallida (part) Boulenger, 1911, *Ann. Mus. Genova* (3), **5**, 74; *Idem*, 1915, *tom. cit.*, 231–2.
Labrochromis pallidus Regan, 1920, *Ann. Mag. nat. Hist.* (9), **5**, 45 (footnote).
Tilapia martini (Part) Boulenger, 1915, *tom. cit.*, 239.
Paratilapia victoriana (part) Boulenger, 1915, *tom. cit.*, 341.
Haplochromis macrops (part) Regan, 1922, *Proc. zool. Soc. London*, 166.

In the original description of *H. ishmaeli*, Boulenger listed thirteen type specimens (syntypes) all collected by Degen at Bunjako, Uganda. However, only eleven specimens (ten in spirit and one skeleton) answering this description can be found in the collections of the B.M. (N.H.); furthermore, only eleven such specimens are recorded in the Museum's catalogue of accessions. That the number thirteen was a slip of the pen seems certain because Boulenger (1915) only lists ten types from Bunjako in the third volume of his Catalogue of Fresh-water Fishes of Africa. The only other specimen from this locality is listed in the "Catalogue" as a skeleton and is presumably the eleventh syntype.

These figures agree with the number of specimens I could locate and, more importantly, they agree with the number registered in the Museum's record of accessions. But, the situation is still somewhat obscure because in the "Catalogue" (*tom. cit. loc. cit.*) Boulenger lists as a type a specimen collected from *Entebbe*; since no reference is made to this fish in the original description I am treating its later inclusion in the list of types as erroneous.

Note on the genus Labrochromis, Regan, 1920. This genus, briefly described in a footnote to Regan's paper on the genera of Tanganyika cichlids was apparently based on a single specimen prepared from one of the syntypes of *T. pallida* (B.M. (N.H.) Reg. No. 1911.3.3.132). In his revision of the Lake Victoria Cichlidae, Regan (1922) correctly identified this skeleton as being that of *Haplochromis ishmaeli* and abandoned the genus *Labrochromis* on the grounds that, apart from possessing hypertrophied pharyngeals, *H. ishmaeli* is "nearly identical with *H. cinereus.*" Whilst I do not agree with the latter part of this statement, I fully endorse Regan's action in not maintaining the monotypic genus *Labrochromis* for *Haplochromis ishmaeli*.

Lectotype of Haplochromis ishmaeli. An adult female 104·0 + 23·0 mm. total length (B.M. (N.H.) Reg. No. 1906.5.30.400) collected by Degen at Bunjako, Uganda.

Description, based on thirty-five specimens (including the lectotype and two paratypes) 82–136 mm. S.L., all from Lake Victoria. Two specimens from Lake Edward are described separately on p. 278.

Depth of body 37·0–45·5 (M = 40·1) per cent of standard length, length of head 33·8–37·5 (M = 34·8) per cent. Dorsal head profile slightly curved or straight, sloping fairly steeply. Preorbital depth 15·3–20·5 (M = 17·0) per cent of head, least interorbital width 24·0–32·0 (M = 27·6), length of snout 29·0–36·0 (M = 31·6), diameter of eye 23·0–31·0 (M = 27·7), depth of cheek 20·7–31·0 (M = 25·5) per cent. Caudal peduncle 14·6–18·8 (M = 17·6) per cent of standard length, 1·2–1·6

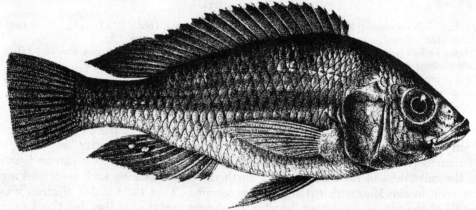

FIG. 21. *Haplochromis ishmaeli*: lectotype (from Boulenger, *Fishes of the Nile*).

(modal range 1·3–1·5) times as long as deep. Jaws equal anteriorly, the lower 35·8–42·5 (M = 39·1) per cent of head and 1·4–2·0 (modal range 1·4–1·6) times as long as broad. Mouth horizontal or very slightly oblique; the posterior tip of the maxilla reaching the vertical through the anterior orbital margin or somewhat beyond, rarely not quite reaching the orbit.

Gill rakers stout, but relatively slender in a few specimens; 6–8 (mode 7), rarely 9, on the lower part of the first arch, the lowermost one or two rakers usually reduced.

Scales ctenoid; lateral line with 30 (f.3), 31 (f.8), 32 (f.15), 33 (f.5) or 34 (f.4) scales; cheek with 3 or 4 rows. Five to 7 (rarely 8) scales between the lateral line and the dorsal fin origin; 8–9 (occasionally 7) between the pectoral and pelvic fin bases.

Fins. Dorsal with 23 (f.1), 24 (f.13) or 25 (f.21) rays, anal with 11 (f.13), 12 (f.21) or 13 (f.1) comprising XV–XVI, 8–10 and III, 8–10 spinous and branched rays for the fins respectively. Pectoral fin 75·0–102·0 (M = 88·5) per cent of head. Caudal truncate or subtruncate.

Teeth. The outer row in both jaws usually contains a mixture of relatively stout, unequally bicuspid and unicuspid teeth; less frequently only unicuspids occur in this row. There is apparently no size-correlated difference in the type or the number of teeth present. There are 38–66 (modal range 44–52) teeth in the outer row of the upper jaw.

Teeth in the inner series are generally tricuspid, but in some fishes the entire inner series are composed of unicuspids. There are 2 or 3 (rarely 1) rows of teeth in the upper jaw and 1 or 2 (less commonly 3) rows in the lower. The inner teeth are implanted somewhat obliquely.

Lower pharyngeal bone and teeth. The lower pharyngeal bone in *H. ishmaeli* is massive and almost all the teeth are molariform; only those situated in the posterior angles of the bone remain small, with pointed crowns. There is some individual variation in the relative enlargement of the bone, but this is slight in comparison with the variation known from such species as *H. obtusidens* and *Astatoreochromis alluaudi*. A slight size-correlated increase in the relative stoutness of the pharyngeal bones was observed in the material studied; likewise, there is an increase in the number of molariform teeth in larger fishes.

In size, shape and dentition, the pharyngeal bones of *H. ishmaeli* are directly comparable with those of *H. pharyngomylus*.

Neurocranial osteology. The neurocranium of *H. ishmaeli* is virtually identical with that of *H. pharyngomylus*.

Coloration of live fishes. Sexually active males. Ground colour light yellow-green dorsally, shading to yellow on the flanks and greyish-white ventrally. Dorsal fin yellow-green, lappets of the anterior spines dusky, the remainder scarlet, as are the spots and dashes between the branched rays. Caudal fin greyish, with red maculae (sometimes coalesced) between the rays. Anal smoky grey with black lappets and an overall scarlet flush; ocelli yellow. Pelvic fins black. *Females* golden-green shading to silvery-white ventrally, the pectoral region faintly blackish. Dorsal fin hyaline, with a narrow red margin. Anal light yellowish-green. Caudal and pelvic fins pale yellow.

Preserved material. Adult males, yellowish-grey, dusky on the chest and branchiostegal membrane (lighter in sexually quiescent fishes); very faint indications of six or seven vertical stripes on the flanks. (These bars are more widely spaced than in *H. pharyngomylus*, see p. 272). A distinct lachrymal stripe. Dorsal fin hyaline in juvenile fishes but darker and with the soft part maculate in mature males; lappets dark. Caudal hyaline, the upper part often maculate. Anal hyaline, the basal third dark in sexually active individuals; three to five large, dead-white ocelli arranged in one or, rarely, two rows. *Females,* brownish to yellowish-silver; sometimes, very faint traces of five to seven transverse bars on the flanks; a weak lachrymal stripe visible in some fishes. All fins hyaline, the dorsal darkest.

Distribution. Lakes Victoria and Edward.

Ecology. The data given in this section relate to fishes from Lake Victoria; nothing is known for fishes from Lake Edward.

Habitat. Haplochromis ishmaeli is essentially a species of inshore regions where the water is less than 30 feet deep and the bottom composed of soft, organic mud.

No specimens have been recorded from depths greater than 60 feet, but some have been caught in nets set over sand and shingle substrates. *Haplochromis ishmaeli* would seem to be the ecological (soft substrate) counterpart of *H. pharyngomylus* (solid substrates).

Because of the difficulty in distinguishing *H. ishmaeli* from *H. pharyngomylus* and especially because the latter species was not recognized until after Graham's collections were brought to England, it is impossible to use Graham's (1929) catch records as an additional source of information on the intralacustrine distribution of *H. ishmaeli*.

Food. The stomach and intestinal contents of nineteen fishes (from several localities) indicate that *H. ishmaeli* in the size-range 84–135 mm. S.L. feed almost exclusively on Mollusca ; the few insect larvae found together with snails in the stomach of one fish suggest that insects may be ingested accidentally.

The number of fishes with food remains is insufficient to determine whether bivalves or gastropods predominate in the diet. Slightly more gastropods (*Melanoides tuberculata*) than bivalves (*Corbicula* sp.) were found in the sample examined.

Breeding. No information is available on the breeding sites or habits of *H. ishmaeli* although the number of females with " spent " ovaries and ventrally distended mouths suggests that the species is probably a female mouth-brooder. The smallest mature females are 97 mm. S.L. (but see Appendix 2) and the smallest adult male is 98 mm. S.L. Females apparently reach a greater maximum size than do males.

Affinities and diagnosis of *H. ishmaeli* are discussed on page 273, with reference to its closest relative, *H. pharyngomylus*.

APPENDIX

(1) *Haplochromis ishmaeli* from Lake Edward. I have been able to examine only two specimens from Lake Edward, one an adult female 120·0 + 28·0 mm. long and the other an adult male 118·0 + 27·0 mm. long (see Trewavas, 1933). In general appearance, in most morphometric characters and in scale and fin ray counts the two fishes are indistinguishable from specimens of a similar size from Lake Victoria. The nature of the pharyngeal bones and dentition is also identical. Nevertheless, the Lake Edward fishes do differ slightly in three characters.

(i) The interorbital is somewhat narrower (24·1 and 25·0 per cent of head) than the mean interorbital width of Victoria fishes (27·6 per cent) although still within the range known from this population. (ii) The number of outer teeth in the upper jaw (40 and 42) is in the lower section of the range for Lake Victoria fishes. (iii) The caudal peduncle is stouter in the majority of Victoria specimens. Now that more specimens are available from Lake Victoria, Trewavas' (op. cit.) observation on the larger eye of the Edward specimens is no longer applicable.

None of the differences commented upon above is so marked as those characterizing the Lake Victoria and Lake Edward populations of *Astatoreochromis alluaudi* (Greenwood, 1959).

Nothing is known about the bionomics of *H. ishmaeli* in Lake Edward ; the two specimens studied were caught in a seine net fished from the eastern shore of the lake near Kisenyi.

(2) Three *H. ishmaeli*-like fishes from Lake Victoria. Three specimens (82, 96 and 100 mm. S.L. (B.M. (N.H.) Reg. Nos. 1959.4.28.304–306)) caught in a trawl off the mouth of the Nzoia river, Kenya, present something of a problem. In appearance and in all standard counts and measurements these fishes are typically *H. ishmaeli*. However, their collector, Mr. S. H. Deathe, recorded the live coloration of these fishes as bright pink. Because all three specimens are adult males, I consider that this striking departure from the usual male coloration of *H. ishmaeli* may be significant, especially since pink is not one of the basic colours in *H. ishmaeli* pigmentation. A further interesting difference is that the smallest specimen is sexually mature whereas the smallest adult *H. ishmaeli* recorded is 98 mm. S.L.

Unfortunately I have only these three specimens and I did not see them when alive. Thus, it is difficult for me to assess fully this seemingly outstanding difference in coloration.

No typical *H. ishmaeli* were reported in the same haul and I do not consider that there is enough evidence to decide whether these peculiar individuals represent an aberrant population of *H. ishmaeli* or a distinct species differing from *H. ishmaeli* in the coloration of its adult males.

Study material and distribution records

Museum and Reg. No.	Locality	Collector
	Uganda	
B.M. (N.H.).—1906.5.30.400 (Lectotype)	Bunjako	Degen
,, 1906.5.30.401–402a (Paratypes)	,,	,,
,, 1906.5.30.396 (Paratype)	,,	,,
,, 1911.3.3.131 (Paratype *Tilapia pallida*)	Jinja	Bayon
,, 1909.5.4.8–10	Sesse Isls.	,,
,, 1906.5.30.275–279	Bunjako	Degen
,, 1959.4.28.281–290	Pilkington Bay	E.A.F.R.O.
,, 1959.4.28.296–300	Ekunu Bay	,,
,, 1959.4.28.301–303	Pilkington Bay	,,
,, 1959.4.28.307–309	Buka Bay	,,
,, 1959.4.28.312–313	Entebbe, Harbour	,,
,, 1959.4.28.314	Macdonald Bay	,,
,, 1959.4.28.316	0° 4′ S., 33° 14′ E.	,,
	Kenya	
,, 1909.11.15.40	Kisumu	A. B. Percival
,, 1959.4.28.274–280	South of Port Southby	E.A.F.R.O.
,, 1959.4.28.356	Kavirondo Gulf	M. Graham
	Tanganyika	
,, 1959.4.28.291–295	Majita	E.A.F.R.O
,, 1959.4.28.310–311	Beach near Majita	,,

SUMMARY

1. Eleven species are redescribed on the basis of new material.
2. In addition, three new species (*Haplochromis theliodon*; *H. empodisma* and *H. saxicola*) are described.
3. Several phyletic lines are represented amongst these fourteen species, which include *H. cinereus*, a species once thought to represent one of the basic types from which the present-day species-flock had evolved. Evidence now available suggests that, anatomically, *H. cinereus* is not sufficiently generalized to retain this distinction.
4. Notes are given on the feeding habits and bionomics of the species.
5. Three species (*H. obtusidens*, *H. pharyngomylus* and *H. ishmaeli*) are largely or entirely mollusc-eaters; three others (*H. humilior*, *H. theliodon* and *H. riponianus*) feed on both insects and molluscs; one, *H. martini*, is a piscivorous predator and seven others are insectivore/omnivores.
6. One species, *H. niloticus* (nom. nov. for *Tilapia bayoni* Blgr. 1911) is known only from the Victoria Nile, whilst *H. ishmaeli* occurs in both Lakes Victoria and Edward; *H. humilior* is found in Lake Victoria and the Victoria Nile.
7. The assumed distribution of *H. macrops* in both Lakes Victoria and Edward is discussed; no definite conclusion can be drawn from the material now available.
8. Two groups of specimens are dealt with in separate appendices because of their uncertain taxonomic position. One group is apparently allied to or even conspecific with *H. pharyngomylus* and the other with *H. ishmaeli*. No conclusion can be reached on the status of these aberrant fishes.

ACKNOWLEDGMENTS

It is with great pleasure that I acknowledge my thanks to Dr. Ethelwynn Trewavas for her advice and criticism and for reading the manuscript of this paper. My thanks are also due to Mr. C. C. Cridland of E.A.F.R.O. Jinja, Uganda, for identifying many of the Mollusca recovered from gut contents.

REFERENCES

BROOKS, J. L. 1950. Speciation in ancient lakes. *Quart. Rev. Biol.* **25** : 131.

GRAHAM, M. 1929. *A report on the Fishing Survey of Lake Victoria, 1927–1928, and Appendices.* Crown Agents, London.

GREENWOOD, P. H. 1951. Evolution of the African Cichlid fishes; the *Haplochromis* species flock in Lake Victoria. *Nature*, London, **167** : 19.

—— 1954. On two cichlid fishes from the Malagarazi River (Tanganyika), etc., *Ann. Mag. nat. Hist.* (12), **7** : 401.

—— 1956a. The monotypic genera of cichlid fishes in Lake Victoria. *Bull. Br. Mus. nat. Hist., Zool.* **3** : No. 7.

—— 1956b. A revision of the Lake Victoria *Haplochromis* species (Pisces, Cichlidae). Part I. *Ibid*, **4** : No. 5.

—— 1957. A revision of the Lake Victoria *Haplochromis* species, *etc.* Part II. *Ibid.* **5** : No. 4.

—— 1959. The monotypic genera of cichlid fishes in Lake Victoria, Part II, and A revision of the Lake Victoria *Haplochromis* species, etc. Part III. *Ibid.* **5** : No. 7.

PAPPENHEIM, P. & BOULENGER, G. A. 1914. Fische. *Wissenschaftliche Ergebnisse der Deutschen Zentral-Afrika Expedition*, 1907–1908, **5** : 225.

POLL, M. 1939. Poissons. *Exploration du Parc National Albert, mission G. F. de Witte* (1933–1935), fasc. **2** : 78.

—— & DAMAS, H. 1939. Poissons. *Exploration du Parc National Albert, mission H. Damas* (1935–1936), fasc. **6** : 73.

REGAN, C. T. 1921. The cichlid fishes of Lakes Albert Edward and Kivu. *Ann. Mag. nat. Hist.* (9), **8** : 632.

TREWAVAS, E. 1933. Scientific results of the Cambridge expedition to the East African lakes, 1930–1931. II. The cichlid fishes. *J. Linn. Soc. (Zool.)*, **38** : 309.

—— 1938. Lake Albert fishes of the genus *Haplochromis*. *Ann. Mag. nat. Hist.* (11), **1** : 435.

A REVISION OF THE LAKE VICTORIA *HAPLOCHROMIS* SPECIES (PISCES, CICHLIDAE) PART V

By P. H. GREENWOOD

CONTENTS

	Page
INTRODUCTION	141
Haplochromis brownae sp. nov.	142
Haplochromis guiarti (Pellegrin)	145
Haplochromis bayoni (Blgr.)	149
Haplochromis serranus (Pfeffer)	152
Haplochromis victorianus (Pellegrin)	156
Haplochromis nyanzae sp. nov.	159
Haplochromis bartoni sp. nov.	161
Haplochromis estor Regan	164
Haplochromis dentex Regan	167
Haplochromis artaxerxes sp. nov.	170
Haplochromis longirostris (Hilgen.)	171
Haplochromis mento Regan	174
Haplochromis mandibularis sp. nov.	178
Haplochromis gowersi Trewavas	180
Haplochromis macrognathus Regan	183
Haplochromis pellegrini Regan	186
Haplochromis percoides (Blgr.)	189
Haplochromis flavipinnis (Blgr.)	192
Haplochromis cavifrons (Hilgen.)	196
Haplochromis plagiostoma Regan	199
Haplochromis michaeli Trewavas	203
DISCUSSION	206
SUMMARY	212
ACKNOWLEDGEMENTS	213
REFERENCES	213

INTRODUCTION

THIS paper is one of two dealing with the piscivorous *Haplochromis* of Lake Victoria. In this part, I have tried to consider a representative sample of species which cover the different morphological types in this trophic group. A species that I consider to represent a morphological type from which the fish-eating species could have evolved, is also described.

Haplochromis brownae sp. nov.

Text-figs. 1 and 25

Haplochromis stanleyi (part) Boulenger, 1915, *Cat. Afr. Fish.* **3**, 295 (one specimen B.M. (Nat. Hist.) Reg. No. 1909.5.4.28 ; collected by Bayon in the Sesse Islands ; apparently this specimen was not considered by Regan in his revision of 1922).
Paratilapia guiarti (part) : Boulenger, 1915, *op. cit.* **3**, 336 (one specimen B.M. (Nat. Hist.) Reg. No. 1906.5.30.354).
Haplochromis guiarti (part) : Regan, 1922, *Proc. zool. Soc. London*, 174 (the same specimen as above).

Holotype : A specimen 104·0 mm. S.L. from Entebbe ; B.M. Reg. No. 1906.5.30.354.

Description. Based on forty-nine specimens (including the holotype) 72–104 mm. standard length.

Depth of body 32·2–39·8 (mean, M, 35·1) per cent of standard length, length of head 30·2–33·4 (M = 31·6) per cent. Dorsal profile of head straight or slightly concave above the eyes, sloping at about 45°.

Preorbital depth 13·8–19·4 (M = 16·3) per cent of head, least interorbital width 26·0–34·0 (M = 29·8) per cent. Snout as long as broad or very slightly broader, its length 28·0–33·3 (M = 30·8) per cent of head ; eye diameter 26·0–31·3 (M = 28·6), depth of cheek 18·9–25·9 (M = 22·7) per cent.

Caudal peduncle length 14·6–20·4 (M = 17·6) per cent of standard length, 1·3–1·8 (mode 1·6) times as long as deep.

Mouth horizontal or slightly oblique, the jaws equal anteriorly, the lips not thickened. Length of lower jaw 38·0–42·9 (M = 40·3) per cent of head, 1·7–2·1 (mode 2·0) times as long as broad. Posterior tip of the maxilla reaching or almost reaching the vertical through the anterior orbital margin.

Gill rakers 9–12 (modal numbers 10 and 11) on the lower part of the first gill arch, the lowermost two to five rakers short and stout, the remainder usually slender (specially in fishes with more than ten rakers) but occasionally stout.

Scales ctenoid ; lateral line with 30 (f.1), 31 (f.5), 32 (f.32) or 33 (f.11) scales. Cheek with 2 or 3 imbricating rows. Six or 7 (rarely 8) scales between the lateral line and the dorsal fin origin, 6–8 (modes 7 and 8) between the pectoral and pelvic fin bases.

Fins. Dorsal with 22 (f.1) 24 (f.4), 25 (f.37) or 26 (f.7) rays, comprising 13 (f.1), 14 (f.1) 15 (f.18) or 16 (f.29) spinous and 8 (f.1), 9 (f.24) or 10 (f.24) branched rays. Anal with 11 (f.4), 12 (f.42) or 13 (f.3) rays, comprising 3 spinous and 8–10 branched rays. Caudal truncate, scaled on the basal half only. Pectoral fin from slightly shorter than, to as long as the head, 24·4–33·4 (M = 28·8) per cent of standard length.

Teeth. In most specimens the outer row in both jaws is composed of relatively stout and bicuspid teeth, but in many fishes (irrespective of size) the posterolateral teeth in the upper jaw are unicuspid or tricuspid. In a few specimens there is an admixture of bi- and unicuspid teeth anteriorly in both jaws. There are 50–66 (M = 56) teeth in the upper, outer series.

Teeth in the *inner rows* are tricuspid and arranged in 2 or 3 (rarely 4) series in the upper jaw and 1 or 2 series in the lower jaw.

Osteology. The neurocranium is that of a generalized *Haplochromis*. The supra-occipital crest slopes fairly steeply and is moderately deep ; the ethmoidal region slopes fairly steeply and the vomer is curved ventrally. The preotic part of the skull is 55–58% of the basal length (as measured from the tip of the vomer to the posterior rim of the basioccipital). The lower *pharyngeal* bone is slender, the teeth cuspidate and fine, without markedly enlarged teeth in the median rows. The dentigerous surface is as broad as it is long and there are 34–40 rows (counted antero-posteriorly) of teeth.

Number of vertebrae : 29 (f.2), 30 (f.9) or 31 (f.3), comprising 12 (f.1), 13 (f.4) 14 (f.8) or 15 (f.1) precaudal and 16 (f.8) or 17 (f.6) caudal elements.

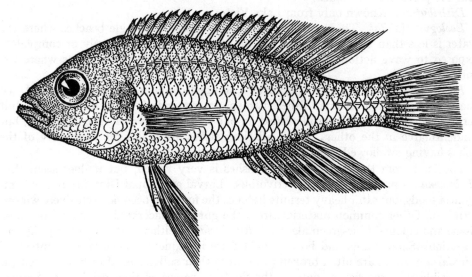

Fig. 1. *Haplochromis brownae* ; natural size. (Drawn by Lavinia Beard.)

Coloration in life. Adult males. Ground colour grey-green dorsally, shading to silver-blue ventrally ; snout dark grey-green. In sexually active fishes a carrot-orange flush develops on the cheek, operculum, flanks and belly ; in such fishes there is also a pronounced lachrymal stripe. Dorsal, caudal and anal fins sooty-grey, the dorsal with three to five horizontal rows of ruddy spots between the spines and rays ; anal ocelli (usually a single row of three or four) orange. Pelvic fins black.

Females have a silver-grey ground coloration (lighter than that of the males) shading to silver ventrally. All fins are hyaline, except the pelvics which are light lemon-yellow. In some individuals there are small orange spots in the position of the anal ocelli in males.

Colour of preserved material. Females. Brownish above, silvery below and on the cheek ; a faint lachrymal stripe is visible as are five transverse bars on the flanks, none of which reaches the ventral profile ; in some specimens there is a short dark, midlateral streak situated above the anal fin. All fins are colourless. In *males* the coloration is more variable, some fishes having a female type coloration (see above) but with a more intense lachrymal stripe and a distinct vertical postorbital bar. All fins, except the pelvics, are hyaline, the dorsal with dusky lappets ; the pelvics are dark laterally but greyish medially.

Other specimens have a similar coloration but the ground colour is darker, the pectoral region sooty and the pelvic fins entirely black. These differences may be associated with the fish's sexual state.

A third varient (undoubtedly attributable to differences in the mode of preservation) almost resembles the live colours. Here the belly, operculum and cheek still retain traces of the carroty flush, the transverse bars are very faint, the lachrymal stripe is intense and the dorsal fin distinctly maculate. The pectoral region is dusky and the pelvic fins are black.

Distribution. Known only from Lake Victoria.

Ecology. Habitat. The species is confined to sandy or shingle beaches where the water is less than thirty feet deep ; such areas are always relatively or completely exposed to wave action. No specimens have been recorded from areas where the substrate is soft.

Food. Twenty of the fifty specimens examined contained ingested material. Of these, three had fed on small fishes (two specimens exclusively so and one on adult termites as well) tentatively identified, in one case, as *Engraulicypris argenteus*, but unidentifiable in the others. Since all three fishes are males the possibility of the fishes having swallowed their own young is eliminated.

The food from the seventeen other fishes is very varied, but it does seem that *H. brownae* is predominantly insectivorous. Larval and pupal Diptera are the commonest foods, but after heavy termite hatches the food can become exclusively winged termites. Other common materials from the gut are macerated fragments of plant tissue and colonial blue-green algae (cf. *Rivularia*). Neither plant shows any signs of digestion. Sand grains and bottom debris are infrequent. Since large numbers of *Rivularia* colonies are often broken away from their substrates after heavy swells, it is possible that the fishes snap up the floating masses as they drift past. Similar concentrations of colonial blue-greens have been found in the guts of other *Haplochromis* whose usual feeding habits do not involve feeding from the bottom. Certainly plants can contribute little to the diet of *Haplochromis brownae* because plant material is apparently indigestible.

Breeding. The species is a female mouth brooder ; the smallest specimen available (72 mm. S.L.) is brooding. Both sexes seem to reach the same maximum adult size.

Affinities. Structurally, and in its diet, *H. brownae* is a generalized lacustrine *Haplochromis*. It differs, however, from the other generalized and insectivorous species of Lake Victoria in having a high number of gill rakers (9–12 cf. 6–9) ; in fact very few Victoria species of any structural or trophic group have more than ten rakers. In most other characters *H. brownae* is very similar to *H. melanopus*, another littoral insectivore with similar ecology. The two species do, however, differ in the shape of the lower pharyngeal bone and its dentigerous surface. In *H. melanopus* the toothed area is clearly broader than long, whereas in *H. brownae* the area is equilateral, or almost so. The species also differ markedly in the colours of the breeding male.

Study material and distribution records

Museum and Reg. No.	Locality	Collector
	Uganda	
B.M. (N.H.).—1906.5.30.354 (Holotype)	Entebbe	Degen
„ 1909.5.4.28	Sesse Isls.	Bayon
„ 1911.3.3.22	Jinja, Ripon falls	„
„ 1962.3.2.96–7	Grant Bay	E.A.F.R.O.
„ 1962.3.2.109–113	Beach near Nasu pt.	„
„ 1962.3.2.114–9	Buka Bay	„
„ 1962.3.2.120–7	Entebbe Harbour	„
„ 1962.3.2.128	Beach near Nasu pt.	„
	Tanganyika	
„ 1962.3.2.76–88	Mwanza, Capri Bay	„
„ 1962.3.2.89–95	Majita Beach	„
„ 1962.3.2.98–108	Beach near Majita	„

Haplochromis guiarti (Pellegrin) 1904

Text-figs. 2 and 25

Tilapia guiarti Pellegrin, 1904, *Bull. Soc. zool. France*, **29**, 186; Idem., 1905, *Mem. Soc. zool. France*, **17**, 184, pl. 16, fig. 1.
Paratilapia guiarti (part): Blgr., 1915, *op. cit.*, **3**, 334 (not the figured specimen).
Paratilapia victoriana (part): Boulenger, 1915, *op. cit.*, 341 (one specimen, B.M. (Nat. Hist.) Reg. No. 1906.5.30.281).
Haplochromis cinereus (part): Regan, 1922, *Proc. zool. Soc. London*, 166 (specimen noted above).
Haplochromis guiarti (part): Regan, 1922, *op. cit.*, 174.
Haplochromis nigroventralis Lohberger, 1929, *Akad. Anz. Wien.*, **66**, 207.

Specimens included in Regan's (1922) synonymy as *Tilapia perrieri* and *Paratilapia longirostris* are no longer considered to be specimens of *H. guiarti*. In his synonymy Regan also included part of *Tilapia pallida* (Boulenger, 1915 Cat. Afr. Fish. **3**, 232). Despite an extensive search, I cannot find any specimens referred to this species which Regan might have examined. For this reason I have not included *T. pallida* (part) in my synonymy.

Holotype. A female 114 mm. S.L. (Paris Museum No. 04 × 150) from the Kavirondo Gulf, Kenya.

Description. Based on fifty-five specimens (83–177 mm. S.L.) including the holotype of the species and that of *H. nigriventralis*. All these fishes came from Lake Victoria (see below, p. 147).

Depth of body 27·3–36·5 (M = 32·3) per cent of standard length, length of head 29·5–33·8 (M = 31·4) per cent. Dorsal profile of head slightly curved and sloping at an angle of *ca* 40°–45°.

Preorbital depth 16·3–21·5 (M = 18·3) per cent of head, least interorbital width 23·4–30·2 (M = 27·4) per cent. Snout longer than broad, or, rarely, as long as broad, its length 31·7–37·5 (M = 34·4) per cent of head. Eye diameter shows slight negative allometry with standard length, being 23·6–29·8 (M = 26·5) per cent in fishes < 115 mm. S.L. (N = 16) and 19·7–25·3 (M = 22·0) in larger individuals (N = 39). Depth of cheek shows very slight positive allometry, 20·0–29·0 (M = 25·9) per cent of head for the whole sample.

Caudal peduncle 16·2–20·8 (M = 18·9) of standard length, 1·4–2·0 (mode 1·5) times as long as deep.

Mouth slightly to moderately oblique, the jaws equal anteriorly. Length of lower jaw 39·2–48·2 (M = 44·4) per cent of head, 1·5–2·3 (mode 2·0) times as long as broad. Lips not thickened, the posterior tip of the maxilla not quite reaching the vertical through the anterior orbital margin in most fishes but reaching this point in a few. The dentigerous arm of the premaxilla shows no medial antero-posterior lengthening.

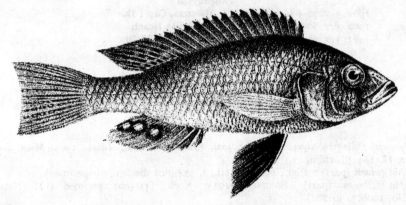

FIG. 2. *Haplochromis guiarti*; about ½ × N.S. (From Boulenger, *Fish. Nile*.)

Gill rakers moderately stout but not stubby; 9–11 (mode 10) on the lower part of the first arch, the lowermost 1–3 rakers reduced.

Scales ctenoid; lateral line with 32 (f.2), 33 (f.26), 34 (f.18), 35 (f.7) or 37 (f.1) scales. Cheek with 3 (less commonly 4 and rarely 2) imbricating rows. Six or 7 (rarely 5, 5½ or 8) scales between the lateral line and the dorsal fin origin; 6 or 7 (rarely 8 or 9) between the pectoral and pelvic fin bases.

Fins. Dorsal with 25 (f.35), 26 (f.19) or 27 (f.1) rays, comprising 15 (f.12), 16 (f.41) or 17 (f.2) spinous and 9 (f.25), 10 (f.29) or 11 (f.1) branched rays. Anal with 11–13 rays, comprising 3 spinous and 8–10 (mode 9) branched rays; in one exceptional specimen there are 4 spines and 9 rays. Caudal truncate, pelvics (particularly in adult males) with the first two branched rays elongated but rarely extending to beyond the first or second branched anal ray (cf. *H. bayoni* where these pelvic rays are greatly produced). Pectoral fin 21·2–27·4 (M = 25·0) per cent of standard length.

Teeth. In fishes <115 mm. S.L. the anterior teeth of the outer row in both jaws are usually unicuspid whilst the lateral teeth are slender and bicuspid. The majority of fishes >115 mm. S.L. have all the outer teeth unicuspid, relatively slender and slightly incurved. However, in several specimens the posterolateral teeth (especially in the upper jaw) are slender and bicuspid or weakly bicuspid. There are 48–74 (M = 62) teeth in the outer row of the upper jaw, the number showing a weak positive correlation with standard length.

In most fishes of all sizes the *inner* teeth are tricuspid but it is not uncommon to find a mixture of tri- and unicuspid teeth, the former predominating. Fishes with the entire inner series composed of unicuspids are rare even amongst specimens >150 mm. S.L. The inner teeth are implanted obliquely and arranged in 3 or 4 (less frequently 2 or 5, rarely 7) rows in the upper jaw and usually in 2 (less commonly 1, 3 or 4) rows in the lower jaw. It seems likely that the number of inner tooth rows is positively correlated with size since there is a tendency for fishes over 125 mm. S.L. to have the higher numbers of rows. Only one fish has seven premaxillary rows (it is a large specimen [170 mm. S.L.] but not the largest); the arrangement of these teeth is very irregular and suggests some ontogenetic disturbance.

Osteology. The neurocranium of *H. guiarti* can be considered as a basic type amongst the piscivorous species. Although it shows several characteristics of the " extreme " predator skull (see p. 209) in an early phase of development, it also shows affinity with the generalized *Haplochromis* skull type as seen, for instance, in *H. brownae*. These points are discussed on pages 207–9.

Vertebrae. Thirty or 31, comprising 13 or 14 precaudal and 16–18 caudals. (Based on six specimens.)

Lower pharyngeal bone fine, its dentigerous surface equilateral or somewhat broader than long. The teeth are cuspidate and slender, the most posterior two or three teeth of the two median rows are generally enlarged; there are 22–28 (mode 26) rows of teeth.

Coloration in life. Adult males. Dorsal surface of head and body intense malachite green shading ventrally to silver. All fins colourless except the pelvics which are black, and the caudal which is dark; anal ocelli (3 or 4 in a single row) bright orange. *Females* have similar coloration except that the pelvics are light yellow and the anal ocelli are absent or represented by small orange spots. After death the colours change rapidly to dark grey-black above and silver below.

Colour in preserved material. Adult males. Ground colour greenish-grey overlying silver, chest and belly sooty; a dark lachrymal stripe is generally present. Dorsal fin sooty, the lappets darker and with a small hyaline area between each lappet; caudal dark on the proximal two-thirds, hyaline distally. Anal fin greyish, the ocelli darker grey; pelvics black. *Sexually quiescent* males have a female type of coloration but with the transverse bars fainter and the anteroventral part of the body sooty. The dorsal fin is sooty, with black lappets; the caudal is maculate, the anal hyaline and the pelvics black. *Adult females* are dark grey dorsally, greyish-silver below and becoming silvery-white on the chest and belly. Often there are traces of six transverse bars of lengths varying from short to almost the full depth of the body; sometimes these bars are connected by a fainter longitudinal stripe. All the fins are hyaline, the caudal often maculate on its proximal half to two-thirds.

Distribution. Lake Victoria and Lake Edward. I have examined three specimens from the latter area. The fishes seem to be referable to *H. guiarti* but I await further material before confirming the occurrence of the species in Lake Edward.

Ecology. Habitat. The species has a lake-wide distribution but is confined to sand or shingle beaches where the water is less than 20 ft. deep. No specimens have been taken over a soft substrate. Information on the distribution and feeding habits of *H. guiarti* given by Graham (1929) must be discounted because several species were confused under this name in his report.

Food. The material found in the alimentary tract of *H. guiarti* may be divided, in order of abundance, into three groups : (i) Fishes (particularly small cichlids) (ii) Insects (particularly winged Termites [Isoptera] and chironomid pupae [Diptera], less frequently larval *Povilla adusta* [the boring may-fly]) (iii) Fragments of phanerogam tissue associated with the colonial blue-green alga *Rivularia*.

Basing the estimate on both frequency of occurrence and on volume, the difference between the amount of fish and the amount of insect food eaten is not great nor is it correlated with the size of the fish. It is difficult to assess the nature of insect-eating habits in this species. At one locality where seine hauls were carried out regularly, it was possible to associate the habit with any high level of insect activity. In this particular case insect-eating might be considered a faculative response to a sudden abundance of a readily available food. The simultaneous occurrence of other species also gorged with the same insects (particularly when the usual diet of these fishes was not primarily insects) seems to indicate that many *Haplochromis* species are " opportunistic " in their feeding habits.

None of the plant matter showed signs of digestion and cannot thus be considered as " food ". As in the case of *H. brownae* (see p. 144) the material may have been accidentally ingested.

In general, the food and feeding habits of *H. guiarti* are similar to those of *H. brownae*. The principal difference lies in the greater proportion of fish eaten by *H. guiarti*.

Breeding. Haplochromis guiarti is a female mouth-brooder ; brooding females are found in the same habitat as non-breeding individuals. Sexual maturity is reached at a length of about 100 mm. although some individuals of 110–113 mm. are still immature. There is no marked sexual dimorphism in adult size.

Affinities. Perhaps the nearest living relative of *H. guiarti* is *H. brownae*. The species differ in many respects but the divergence of *H. guiarti* lies in those anatomical (and associated morphometric) characters which we find in a more exaggerated condition amongst the entirely piscivorous predatory species. The resemblance between *H. guiarti* and *H. brownae* is most apparent in young specimens (80–90 mm. S.L.) of the former and adults of the latter species (70–90 mm. S.L.), but even at this size *H. guiarti* is distinguishable on certain characters (longer jaws and snout) which are part of the " predatory species " character complex. However, it is possible that *H. guiarti*, itself still a relatively generalized species amongst the predators, evolved from an *H. brownae*-like stem.

Amongst the larger predatory species, *H. guiarti* is perhaps related to *H. squamulatus* ; but the overall resemblance is less than that between *H. guiarti* and *H. brownae*. Another species with about the same degree of relationship as *H. squamulatus* is *H. bayoni* (see p. 152).

Study material and distribution records

Museum and Reg. No.	Locality	Collector
Uganda		
B.M. (N.H.).—1906.5.30.210–2	Entebbe	Degen
,, 1906.5.30.213–5	,,	,,
,, 1906.5.30.220–8	,,	,,
,, 1906.5.30.230	Nsonga	,,
,, 1906.5.30.281	Entebbe	,,
,, 1906.5.30.355–361	Entebbe	,,
,, 1909.5.4.6–7	Sesse Isls.	Bayon
,, 1928.1.25.22	Entebbe	Pitman
,, 1962.3.2.246	Entebbe (Airport Beach)	,,
,, 1962.3.2.240–2	Kagera Port	E.A.F.R.O.
,, 1962.3.2.243–5	Ramafuta Isl. (Buvuma Channel)	,,
,, 1962.3.2.260–3	Entebbe Harbour	,,
,, 1962.3.2.253–9	Bufuka Bay	,,
,, 1962.3.2.264–270	Beach near Nasu point	,,
,, 1962.3.2.248	Karenia (near Jinja)	,,
,, 1962.3.2.271–3	Jinja pier	,,
,, 1962.3.2.274–6	Buka Bay	,,
Kenya		
Paris Museum 04 × 150 (Holotype)	Kavirondo Gulf	
B.M. (N.H.).—1962.3.2.247	Near Usoma Lighthouse, Kavirondo Gulf	,,
,, 1962.3.2.250–1	Kasingiri Gingo, Kavirondo Gulf	,,
Tanganyika		
B.M. (N.H.).—1962.3.2.249	Majita Beach	,,
Lake Victoria, Locality Unknown		
B.M. (N.H.).—1901.6.24.89		Sir H. Johnston
,, 1911.3.3.23		Bayon
,, 1962.3.2.252		E.A.F.R.O.

Haplochromis bayoni (Blgr.) 1909

Text-fig. 3

Paratilapia bayoni (part) Blgr., 1909, *Ann. Mus. Genova* (3), **4**, 304, fig. (the figured specimen only; this specimen is now chosen as the lectotype); *Idem*, 1915, *Cat. Afr. Fish.*, **3**, 337, fig. 227 (figured specimen only).

The syntype on which Regan (1922) based his redescription of the species is no longer considered to be *H. bayoni*.

Lectotype. A male, 147 mm. S.L. (collected in the Sesse Islands by Bayon) i.e. the specimen figured by Boulenger (1909), now in the collections of the Museo Civicio distoria Naturale, Genoa (No. C.E. 12976).

Description. Based on twenty-three fishes (including the lectotype) 82–154 mm. S.L.

Depth of body 27·0–35·3 (M = 32·2) per cent of standard length, length of head 35·0–38·0 (M = 36·3) per cent. Dorsal head profile sloping at an angle of *ca.* 30°–35°, its otherwise straight outline broken anteriorly by the prominence of the premaxillary pedicels.

Preorbital depth 17·7–21·6 (M = 19·6) per cent of head length, least interorbital width 20·8–25·9 (M = 23·2) per cent. Snout longer than broad (1·2–1·3 times), its length 35·3–39·5 (M = 37·5) per cent of head. Eye diameter 17·5–21·9 (M = 20·0) per cent, depth of cheek 23·1–28·1 (M = 25·0) per cent.

Caudal peduncle 14·5–17·2 (M = 16·2) per cent of standard length, 1·1–1·6 (modal range 1·3–1·4) times as long as deep.

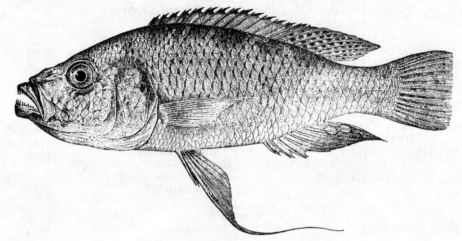

Fig. 3. *Haplochromis bayoni*; lectotype, ·73 × N.S. (From Boulenger, *Ann. Mus. Genova.*)

Mouth slightly oblique, the jaws equal anteriorly or the lower projecting slightly; lower jaw length 42·5–48·0 (M = 45·6) per cent of head, 1·7–2·5 (no definite mode) times as long as deep. Lips slightly thickened, the premaxilla noticeably expanded anteroposteriorly in the midline. Posterior tip of the maxilla not reaching the level of the anterior orbital margin, but generally reaching a point slightly behind the vertical through the posterior tip of the premaxillary pedicels.

Gill rakers usually slender or a mixture of slender and stout, 8–10 (mode 9) on the lower part of the first gill arch, the lowermost 1–3 rakers reduced.

Scales ctenoid; the lateral line with 29 (f.1), 30 (f.1), 31 (f.9), 32 (f.8), 33 (f.3) or 34 (f.1) scales. Cheek with 4 (less frequently 3) series of imbricating scales. Five or 6 (rarely 6½ or 7) scales between the lateral line and the dorsal fin origin, 5 or 6 (less frequently 7) between the pectoral and pelvic fin bases.

Fins. Dorsal with 23 (f.1), 24 (f.8), 25 (f.13) or 26 (f.1) rays, comprising 14 (f.4), 15 (f.18) or 16 (f.1) spinous and 9 (f.8), 10 (f.13) or 11 (f.2) branched rays. Anal with

12 (mode) or 13 rays, comprising 3 spinous and 9 or 10 branched elements. Caudal truncate. Length of pectoral fin 21·2–28·0 (M = 25·5) per cent of standard length. The pelvic fins have the first unbranched ray greatly produced and filamentous, its tip reaching to at least the third branched anal ray and usually to between the third and sixth rays. In some specimens it reaches to beyond the anal base. Both Boulenger (1911) and Regan (1922) imply that the ray is longer in males than in females but from my sample I can find no clear-cut sexual dimorphism. (Boulenger's remarks are probably attributable to his having another species in his study material). This lack of marked dimorphism is very unusual since in all other Victoria *Haplochromis* the adult male has noticeably longer outer pelvic rays. Indeed, the marked hypertrophy of the pelvic rays in both sexes constitutes one of the most reliable diagnostic characters for *H. bayoni*.

Teeth. In every specimen examined the *outer row* of teeth in both jaws is composed of unicuspid, slightly to moderately curved and relatively stout teeth. There are 34–52 (M = 44) outer teeth in the upper jaw, the number perhaps showing a positive correlation with standard length.

The *inner rows* are also composed of unicuspid teeth, and are implanted obliquely; there are 2 (less commonly 3) series in the upper jaw and 2 (less commonly 1) in the lower.

Osteology. The neurocranium is clearly derived from the *H. guiarti* type but is somewhat more advanced towards the " extreme " predator type of *H. mento* and *H. macrognathus* (see p. 209). Compared with *H. guiarti* the neurocranium has the ethmoid-vomer region more strongly decurved, the slope of the preorbital face less steep (*ca.* 35°), its height less (*ca.* 3 times in basal length *cf.* 2·5 times) and a lower supraoccipital crest. The relative length of the preotic portion of the skull is the same in both species (65% of basal length).

Vertebrae. 28–30, comprising 12–13 precaudal and 16 or 17 caudal elements (4 specimens examined).

Lower pharyngeal bone triangular, its dentigerous surface equilateral. The teeth are relatively coarse and are cuspidate; the teeth of the two median rows slightly enlarged. There are 18–20 tooth rows.

Coloration. Data on the live colours of *H. bayoni* are not available. In preserved fishes *females and immature males* are greyish-brown above and silver below, sometimes with very faint traces of four short, transverse bars of irregular outline lying midlaterally; less frequently these bars are joined by an even fainter longitudinal band. All fins are hyaline, the soft dorsal maculate and the proximal two-thirds of the caudal dark with traces of maculae. The only *adult males* available are sexually quiescent and have a coloration like that of the females except for a more intense midlateral stripe and more distinctly maculate dorsal. The pelvics are slightly dusky and there is a single row of three opaque ocelli on the anal fin.

Distribution. Known only from Lake Victoria. Earlier records (Boulenger, 1915) from Lake Kyoga were based on a misidentified specimen.

Ecology. Habitat. The species is apparently confined to water less than thirty feet deep and to hard substrates (sand or shingle). The majority of specimens come from exposed sandy beaches but some are from fairly exposed bays where the bottom was of sand or sand overlain by a thin slick of mud. Since only a few of Graham's specimens were preserved, it is impossible to use his records (Graham, 1929) for distributional purposes. If it be assumed that he identified his specimens correctly, then his locality records confirm my idea of the species' distribution.

Food. Information on the feeding habits of *H. bayoni* is scanty. Seven of the fifteen fishes examined (size range 82–154 mm. S.L.; from six different localities) had food in the stomach and/or intestines. In each case the food consisted entirely of finely macerated fish remains, identifiable in four specimens as being small cyprinid fishes (probably *Barbus* sp.). One fish (153 mm. S.L.) had the remains of two cyprinids (*ca.* 35 mm. S.L.) in the stomach and the remains of at least one other fish in the intestines.

Breeding. Nothing is known about the breeding habits of this species. Both sexes reach maturity at a size between 110 and 125 mm. S.L., and both reach the same maximum adult size.

Affinities. The extreme elongation of the first pelvic ray is unique amongst Lake Victoria *Haplochromis* and serves as a ready diagnostic character. On more fundamental structures, particularly the syncranial architecture, *H. bayoni* shows affinity with *H. guiarti* but the relationship is not especially close. In these same characters *H. bayoni* exhibits a further continuation of the trend leading towards the *H. mento*—*H. macrognathus* level of syncranial organization. Phyletically, *H. bayoni* could be considered as an isolated (but by no means aberrant) offshoot from an *H. guiarti*-like stem.

Study material and distribution records

Museum and Reg. No.	Locality	Collector
	Uganda	
Genoa Museum, No. C.E. 12976 (Lectotype)	Sesse Isls.	Bayon
B.M. (N.H.).—1962.3.2.160–3	Beach near Hannington Bay	E.A.F.R.O.
,, 1962.3.2.170–1	Beach near Nasu point	,,
,, 1962.3.2.172–4	Entebbe, Airport beach	,,
,, 1962.3.2.175–6	Entebbe, Harbour	,,
,, 1962.3.2.177	Kagera port	,,
	Tanganyika	
,, 1962.3.2.164–8	Majita Beach	,,
	Kenya	
,, 1962.3.2.169	Kavirondo Gulf	,,
	Lake Victoria, Locality Unknown	
,, 1962.3.2.178	.	,,

Haplochromis serranus (Pfeffer) 1896

Text-figs. 4, 5 and 25

Hemichromis serranus Pfeffer, 1896. *Thierw. D. ost. Afr., Fische,* 23.
Paratilapia prognatha (part): Blgr., 1915, *Cat. Afr. Fish.*, **3**, 333 (specimen B.M. (N.H.), Reg. No. 1906, 5.30.263).
Haplochromis acutirostris Regan, 1922, *Proc. zool. Soc. Lond.*, 180 (the lectotype [and figured specimen] Reg. No. as above).

I do not consider that any of the specimens used by Boulenger (1915) or Regan (1922) to redescribe the species can be referred to *H. serranus*. My concept of *H. serranus* is based on Pfeffer's original description and on supplementary notes and a figure made from the presumed type specimen by Dr. E. Trewavas. The additional material now available agrees more closely with my idea of the type than do Boulenger's and Regan's fishes.

Description. Based on fifty-two specimens, 101–205 mm. S.L. (including the type of *H. acutirostris*).

Depth of body 32·7–39·2 (M = 36·0) per cent of standard length, length of head 34·8–38·7 (M = 36·3) per cent. Dorsal head profile straight or slightly curved, sloping at an angle of 30°–40°, the premaxillary pedicels very prominent.

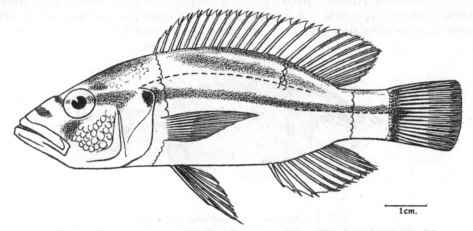

Fig. 4. *Haplochromis serranus*. (Outline drawing of a specimen in the Berlin Museum believed to be the holotype ; made by Dr. E. Trewavas.)

Preorbital depth 14·6–20·0 (M = 17·7) per cent of head, ratio Eye/Preorbital 1·1–1·5, (M = 1·3.) Least interorbital width 20·4–26·8 (M = 23·3) per cent. Snout as long as broad in most specimens, but in some fishes <130 mm. S.L. it is slightly longer than broad ; snout length 30·8–37·0 (M = 34·0) per cent of head, diameter of eye 20·4–26·0 (M = 23·3) per cent, depth of cheek 22·9–31·5 (M = 27·5).

Caudal peduncle length 13·3–19·6 (M = 15·4) per cent of standard length, 1·1–1·5 (mode 1·2) times as long as deep.

Lower jaw moderately oblique, sloping at an angle of 25°–30°, projecting slightly and with a distinct mental bump ; its length 47·7–60·0 (M = 54·3) per cent of head and 1·8–2·5 (mode 2·0) times the breadth. Posterior tip of the maxilla extending to below the eye or to the vertical through the anterior orbital margin in most fishes, but not quite reaching this point in a few.

Gill rakers. Short and stout, 8 or 9 (rarely 7 or 10) on the lower part of the first gill arch, the lowermost one or two rakers reduced.

Scales ctenoid ; lateral line with 30 (f.2), 31 (f.25), 32 (f.18), 33 (f.5) or 34 (f.1) scales, cheek with 4 (less frequently 5, rarely 3 or 6) rows ; 6 or 7 scales between the lateral line and the dorsal fin origin, 7 or 8 (less frequently 9, rarely 6) between the pectoral and pelvic fin bases.

Fins. Dorsal with 24 (f.30), 25 (f.20) or 26 (f.2) rays, comprising 14 (f.2), 15 (f.40) or 16 (f.10) spinous and 8 (f.1), 9 (f.36) or 10 (f.15) branched rays. Anal with 11 (f.6), 12 (f.45) or 13 (f.1) rays, comprising 3 spines and 8–10 branched rays. Pectoral fin length 23·8–33·0 (M = 27·0) per cent of standard length. Caudal truncate (the lower posterior angle somewhat obliquely truncate in a few specimens) scaled on its basal half to two-thirds (rarely). Pelvic fin with the first ray slightly produced and filamentous, proportionately more so in adult males.

Teeth. In all specimens the majority of *outer teeth* are unicuspid but in some fishes <140 mm. S.L. a few bicuspids occur posterolaterally in the upper jaw. The teeth are moderately to strongly curved, the lateral teeth often more so than the anterior ones. There are 44–70 (M = 63) teeth in the upper outer row, the number showing some positive correlation with length in fishes <140 mm. S.L.

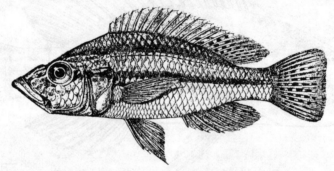

FIG. 5. *Haplochromis serranus*; ·72 × N.S.
(From Regan, *Proc. zool. Soc. Lond.*; the type of *H. acutirostris* Regan.)

All the rows of *inner teeth* (except in some fishes <140 mm. S.L.) are composed of unicuspids.

In the smaller fishes the innermost row (especially in the upper jaw) may be composed of tricuspids or the entire inner series of both jaws may be of tricuspids. The inner teeth are arranged in 2 or 3, rarely 4, rows in the upper jaw and 2, less frequently 1 or 3 rows in the lower jaw.

Osteology. The neurocranium closely resembles that of *H. victorianus* (see p. 157) although the dorsal profile is slightly more curved. It can be considered as a development of the *H. guiarti* type.

Vertebrae. Twenty-eight or 29 (mode 29), comprising 12 (f.1) or 13 (f.6) precaudal and 15 (f.1) or 16 (f.6) caudal elements.

Lower pharyngeal bone triangular, its dentigerous surface very slightly broader than it is long. The teeth are moderately stout and cuspidate, and are arranged in 22–24 rows.

Coloration. The colours of live fishes are unknown.

Preserved material. Adult males. Ground colour light brown, greyish on chest and belly; branchiostegal membrane black or dark grey. In some specimens there is a broad, dark, but faint midlateral band running from behind the operculum to the base of the caudal fin or onto the fin itself, and sometimes an even fainter, interrupted band running slightly below the base of the dorsal fin. The dark lachrymal stripe is very prominent and runs obliquely backwards and downwards from the lower anterior margin of the orbit to the angle of the lower jaw. Dorsal fin brown but with dark streaks between the rays and, in some specimens, with black lappets. Caudal fin darkly maculate. Anal grey-brown, darkest at the base of the spinous part ; ocelli 2–4 in number, greyish and often ill-defined. If there are more than three ocelli, they are arranged in two rows of one and three. Pelvic fins black.

Adult females silvery brown, darker dorsally and apparently without longitudinal bands. All fins are brownish. *Juveniles of both sexes :* ground colour silvery-brown with a broad (sometimes faint) dark midlateral band from the operculum to the basal part of the caudal fin ; lachrymal stripe, if visible, very faint. All fins are greyish, often with traces of maculae on the soft part of the dorsal.

Ecology. Habitat. Most records of *H. serranus* are from sheltered bays and gulfs where the bottom is of soft, organic mud and the depth of water less than 25 feet. A few specimens came from exposed habitats and were caught over sand or shingle, but again the water did not exceed 25 feet in depth. It should be noted that all the latter localities are near muddy areas. No locality is more than half a mile from the shore.

Food. The food of fishes in the size range 100–205 mm. S.L. is exclusively fish and predominantly *Haplochromis* ; no other genus could be identified in the very macerated gut contents of the thirty specimens examined. Some food was, however, too finely divided to even hazard an identification.

Breeding. No information is available on the breeding habits of *H. serranus*. Most individuals <140 mm. S.L. are immature, but one female of 118 mm. S.L. shows early stages of oogenesis. Males and females reach the same adult size.

Distribution. Lake Victoria.

Affinities. *Haplochromis serranus* is closely related to *H. victorianus* and *H. nyanzae*, the three species apparently forming a species group amongst the larger predatory *Haplochromis* of Lake Victoria. In turn, this group is related to *H. spekii* and the relationship will be discussed in another paper. The differences distinguishing *H. serranus* from *H. nyanzae* are outlined on p. 161 ; from *H. victorianus*, it differs in its more oblique and longer lower jaw (47·4–60·0, M = 54·3 per cent of head *cf.* 44·0–51·8, M = 47·1 per cent), shorter pectoral fin (23·8–33·0, M = 27·0 per cent of standard length *cf.* 26·2–32·7, M = 30·4 per cent) and in having fewer teeth in the outer row of the upper jaw (44–70, M = 63, *cf.* 67–86 M = 74).

Study material and distribution records

Museum and Reg. No.	Locality	Collector
	Uganda	
B.M. (N.H.).—1906.5.30.263	Bunjako	Degen
(Type of *H. acutirostris*)		
,, 1962.3.2.27–36	Pilkington Bay	E.A.F.R.O.
,, 1962.3.2.37–51	Ekunu Bay	,,

Museum and Reg. No.	Locality	Collector
	Uganda (continued)	
B.M. (N.H.).—1962.3.2.52–3	Bukafu Bay	E.A.F.R.O.
,, 1962.3.2.54–5	Jinja pier	,,
,, 1962.3.2.56–7	Karenia (near Jinja)	,,
,, 1962.3.2.61–3	Pilkington Bay	,,
,, 1962.3.2.74–5	Williams Bay	,,
,, 1962.3.2.58	Ramafuta Isl.	,,
	Kenya	
,, 1962.3.2.26	Rusinga Isl.	,,
,, 1962.3.2.60	Naia Bay	,,
	Tanganyika	
,, 1962.3.2.59	Ihogororo	,,

Haplochromis victorianus (Pellegrin) 1904

Plate I

Paratilapia victoriana Pellegrin, 1904, *Bull. Soc. zool. France*, **29**, 185 ; Idem, 1905, *Mem. Soc. zool. France*, **17**, 182, pl. 17, fig. 3.

Pelmatochromis spekii (part) Blgr., 1906, *Ann. Mag. nat. Hist.* (7), **17**, 440 ; Idem, 1915, *Cat. Afr. Fish.*, **3**, 416 (one of the types, B.M. (N.H.) Reg. No. 1906.5.30.300).

Regan (1922) tentatively included *Paratilapia victoriana* in the synonymy of *Haplochromis nubilus*. I have examined the types of both species and can find nothing to substantiate this arrangement ; indeed, the two species are only distantly related. None of the specimens identified by Boulenger (1915) as *P. victoriana* can be referred to Pellegrin's species.

Holotype. A fish 120 mm. S.L. (Paris Museum Reg. No. 04 × 148) from Kavirondo Bay, Kenya.

Description. Based on twenty specimens 117–166 mm. S.L., including the holotype.

Depth of body 33·4–41·3 (M = 37·3) per cent of standard length, length of head 33·5–36·0 (M = 34·8) per cent. Dorsal profile of head usually straight (but sometimes with a slight concavity due to more prominent premaxillary pedicels), sloping at ca. 35°–45°.

Preorbital depth 17·9–20·5 (M = 19·2) per cent of head, ratio Eye/Preorbital 1·1–1·3 (mean 1·2), least interorbital width 21·5–24·5 (M = 22·6) per cent. Snout length 31·8–36·0. (M = 34·1) per cent of head, equal to its width ; diameter of eye 21·7–25·5 (M = 23·6), depth of cheek 22·5–26·2 (M = 24·6) per cent.

Caudal peduncle 13·5–19·6 (M = 17·1) per cent of standard length, 1·1–1·8 (mode 1·5) times as long as deep.

Lower jaw horizontal or slightly oblique, anteriorly equal to the upper or projecting slightly, its length 44·0–51·8 (M = 47·1) per cent of head, 1·6–2·2 (modal range 1·8–2·0) as long as broad ; there is always a well-developed mental prominence at the symphysis. The posterior tip of the maxilla reaches the vertical through the anterior orbital margin, or somewhat more posteriorly, in most specimens but does not reach the orbit in a few fishes.

Gill rakers short and moderately stout, 8 or 9 on the lower part of the first gill arch, the lowermost 1–3 rakers reduced.

Scales ctenoid, lateral line with 31 (f.3), 32 (f.6), 33 (f.9) or 34 (f.2) scales, cheek with 3 or 4 series. Six to 8 scales between the lateral line and the dorsal fin origin, 7 or 8 (rarely 6) between the pectoral and pelvic fin bases.

Fins. Dorsal with 24 (f.1), 25 (f.16), or 26 (f.3) rays, comprising 15 (f.4), 16 (f.14) or 17 (f.2) spinous and 9 (f.16) or 10 (f.4) branched rays. Anal with 11 (f.1), 12 (f.16) or 13 (f.3) rays comprising 3 spines and 8 (f.1), 9 (f.16) or 10 (f.3) branched rays. Caudal fin truncate, scaled on its basal half (rarely two-thirds). Pectoral 26·2–32·7 (M = 30·4) per cent of standard length. First pelvic ray slightly produced and filamentous in both sexes (an unusual feature).

Teeth. In fishes less than 125 mm. S.L., the *outer teeth* in both jaws are mostly bicuspids, but some may be weakly so ; fishes between 125 mm. and 130 mm. S.L. show an admixture of unicuspids and weakly bicuspids, or the unicuspids may predominate ; all fishes >135 mm. S.L. have only unicuspid teeth in this series. In all specimens the teeth are relatively slender and slightly curved. There are 64–86 (M = 74) in the outer row of the upper jaw.

The inner teeth are arranged in two or three rows in each jaw ; most specimens <130 mm. S.L. have either a mixture of uni- and tricuspids in apparently equal proportions, or one form may predominate. Fishes >135 mm. S.L. have only unicuspids in the inner series.

Osteology. The neurocranium is very similar to that of *H. serranus*. It is clearly distinct from that of *H. bayoni* (which seems to lead to the *H. mento* type) and is somewhat more substantial than the neurocranium of *H. guiarti* (which is nearest the generalized insectivore type of, say, *H. brownae*).

Vertebrae. Thirty (in the six specimens examined) comprising 13 precaudal and 17 caudal elements.

Lower pharyngeal bone triangular, the dentigerous surface noticeably broader than long. The teeth are cuspidate and slender, and are arranged in 22–24 rows ; the teeth of the two median rows are often enlarged relative to the others.

Coloration. Live fishes. In adult males the ground colour is a silvery-turquoise dorsally, shading to silver on the flanks and belly. Dorsal fin dark neutral, with black lappets and deep red maculae between the branched rays. Caudal fin dark neutral but with blood-red posterior and ventral margins. Anal dark neutral basally, black between the rays, and blood-red proximally ; the ocelli are orange-yellow and numerous. The pelvic fins are black. *Females,* have a similar ground coloration but lack the red areas on the caudal and anal fins, and the black area over the anal spines ; the pelvics are neutral and there are no anal ocelli.

Preserved material. Adult males. Ground colour brownish-grey above, becoming silver-grey on the flanks and belly. The branchiostegal membrane is sooty and there is a faint blackening on the chest and belly. Dorsal fin hyaline but with black lappets and dark maculae on the soft part. Caudal fin colourless but with numerous dark spots and streaks between the rays. Anal fin black between the spines, otherwise sooty-grey, the basal half being darker ; the greyish-white ocelli vary in number from three to seven and are arranged in one or two rows, those of the lower row being much smaller. Pelvic fins mottled black (appearing uniformly black when closed).

Females are brownish silver above and silver to silvery-yellow on the flanks and belly. All fins are colourless, the soft dorsal and the caudal sometimes weakly maculate. In some specimens there is a faint but broad and interrupted midlateral stripe from the posterior margin of the operculum to the base of the caudal fin.

Distribution. Lake Victoria.

Ecology. Habitat. The species has been recorded from only six different localities; in five the bottom is of thick organic mud and in the sixth, shingle (but this place is near an area of mud); all localities are sheltered and the depth of water is between 20 and 40 feet.

Breeding. Very little information is available on the breeding habits of *H. victorianus*. With two exceptions (specimens 128 and 131 mm. S.L.) all the fishes examined were adult and included several individuals smaller than the exceptional juveniles. The largest fishes (166 mm. S.L.) are a male and a female.

Food. Fourteen of the twenty-seven fishes examined had ingested material in the gut. Of these, twelve had fed on fishes and two on organic mud. The fish remains are very finely divided and come from small fishes *ca.* 10–15 mm. S.L. (and this irrespective of the size of the predator). In two cases the fishes are identifiable as post-larval *Haplochromis*. The preponderance of small fishes in the prey is interesting because even in such a small sample most other predatory *Haplochromis* would have yielded remains of much larger prey fishes. There is, of course, usually some correlation between prey size and predator size, but in a sample covering a comparable size-range the prey fishes would be from 30–60 mm. S.L. Perhaps *H. victorianus* has specialized in feeding on post-larval fishes?

The occurrence of mud in two specimens is inexplicable, particularly since there are no indications that the fishes had been feeding on insect larvae, a common alternative food in many otherwise piscivorous species.

Affinities. Haplochromis victorianus is very closely allied to *H. serranus* and *H. nyanzae*, see p. 161. The three species form a fairly well-defined group of deep-bodied and anatomically rather generalized species amongst the piscivorous predators of Lake Victoria.

Study material and distribution records

Museum and Reg. No.	Locality	Collector
	Uganda	
B.M. (N.H.).—1906.5.30.300 (Paratype *P. speki*)	Entebbe	Degen
,, 1962.3.2.478	Ramafuta Isl. (Buvuma channel)	E.A.F.R.O.
,, 1962.3.2.479–82	Karenia, near Jinja	,,
,, 1962.3.2.483–87	Pilkington Bay	,,
,, 1962.3.2.488–494	Ekunu Bay	,,
	Kenya	
Paris Museum 04 × 148 (Holotype)	Kavirondo Bay	,,
B.M. (N.H.).—1962.3.2.477	Naia Bay	,,

REVISION OF THE LAKE VICTORIA *HAPLOCHROMIS* SPECIES 159

Haplochromis nyanzae sp. nov.

Text-fig. 6

Pelmatochromis spekii (part) : Blgr., 1915, *Cat. Afr. Fish.*, (one specimen B.M. (N.H.) Reg. No. 1909.5.4.17).
H. serranus (part) : Regan, 1922 ; *Proc. zool. Soc. Lond.* 174 (specimen as above).

Holotype. A specimen 154·0 mm. S.L. (B.M. (N.H.) Reg. No. 1962.3.2.495) from Jinja.

Description. Based on thirteen specimens (including the holotype) 126–171 mm. S.L.

Depth of body 33·8–38·6 (M = 36·3) per cent of standard length, length of head 33·6–36·7 (M = 35·4) per cent. Dorsal head profile straight or moderately concave, sloping at an angle of 30°–40°.

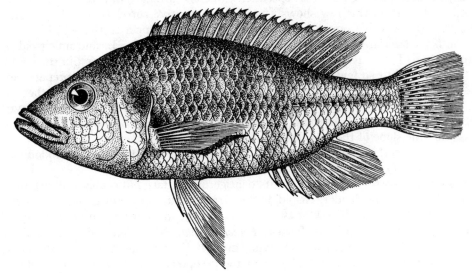

FIG. 6. *Haplochromis nyanzae* ; ·75 × N.S. (Drawn by Lavinia Beard.)

Preorbital depth 17·0–20·2 (M = 18·9) per cent of head, least interorbital width 20·0–24·5 (M = 22·2) per cent. Snout length 33·4–35·8 (M = 34·5) per cent of head, 1·0–1·1 times as long as broad. Eye diameter 19·1–24·0 (M = 22·1) per cent, ratio Eye/Preorbital 1·1–1·3 (M = 1·2), depth of cheek 24·4–27·6 (M = 25·9) per cent.

Caudal peduncle length 14·0–17·6 (M = 15·9) per cent of standard length, 1·1–1·4 (mode 1·4) times as long as deep.

Lower jaw moderately oblique, sloping at an angle of 25°–30°, anteriorly equal to the upper jaw or projecting slightly, its length 45·0–51·6 (M = 48·0) per cent of head, 1·5–2·0 (Mode 1·7) times as long as broad. The anterior outline of the dentary is smoothly curved and lacks a strong mental projection. The posterior tip of the maxilla usually reaches the vertical through the anterior orbital margin, but in some fishes it extends beyond this point or does not reach it.

Gill rakers short and stout (rarely short and slender), 8 or 9 (rarely 7 or 10) on the lower part of the first gill arch, the lowermost one or two rakers reduced.

Scales ctenoid ; lateral line with 31 (f.1), 32 (f.8) or 33 (f.4) scales, cheek with 4 or 5 (rarely 3) rows. Seven or 8 (rarely 6 or 9) scales between the dorsal fin origin and the lateral line, and between the pectoral and pelvic fin bases.

Fins. Dorsal with 24 (f.5) or 25 (f.8) rays, comprising 15 (f.2) or 16 (f.11) spinous and 8 (f.4), 9 (f.8) or 10 (f.1) branched rays. Anal with 12 rays (3 spinous, 9 branched) in all except one specimen which has only 2 spines and 9 rays. Pectoral fin 22·8–28·2 (M = 24·9) per cent of standard length. First pelvic ray very slightly produced in males and less so in females. Caudal truncate, scaled on its basal two-thirds.

Teeth. The *outer* row in both jaws is composed of unicuspid, moderately stout teeth, strongly incurved in most specimens, but less strongly in a few others. The number of teeth in this row shows a very slight positive correlation with standard length ; for the whole sample there are 50–76 (M = 60) upper teeth.

The *inner teeth* are uni- and tricuspid in fishes <135 mm. S.L. and unicuspid in larger fishes. There are 2–4 (usually 3) rows in the upper jaw and 2 or 3 in the lower.

Osteology. With so few specimens available, I have not been able to prepare any skeletal material. However, on comparing radiographs of this species with others of *H. victorianus* I can find no great differences in neurocranial form or general syncranial arrangement. If anything, the supraoccipital crest in *H. nyanzae* is relatively lower and the slope of the dorsal skull profile is a little less steep.

Vertebrae. 29 (f.2), 30 (f.4) or 31 (f.1), comprising 13 (f.5) or 14 (f.2) abdominal and 16 (f.3) or 17 (f.4) caudal elements.

Lower pharyngeal bone triangular, its dentigerous surface noticeably broader than long. The teeth are slender, cuspidate and arranged in 22–24 rows. Except in the most posterior transverse row or two the teeth of the median rows are not enlarged.

Coloration. Unknown in live fishes. *Preserved material. Adult males* have a dark brown ground colour with an overlying greyish tinge ; the belly, ventro-lateral aspects of the flanks, the ventral part of the preoperculum and the branchiostegal membrane are sooty. There is a distinct but narrow black lachrymal stripe, a faint black midlateral band and a fainter dorsolateral band following the upper lateral line. The lateral bands are crossed by five or six faint transverse bars. The dorsal fin is dark brown with darker spots and streaks between the soft rays. Caudal fin dark brown, anal brown with a faint and narrow darker flush along its base. Pelvic fins mottled black. In other males (whose sexual state could not be determined) the general ground coloration is much darker, the lower jaw, snout and ventral aspects of the cheek, preoperculum and operculum are black, the dorsal fin is darker (almost black) and the spots on the soft part are more obvious. The caudal fin too is darker, but the anal is similar. The anal ocelli in all males are difficult to distinguish.

Females are brown above, shading to silver-bronze on the flanks and belly. All fins are yellowish-brown, the dorsal with a grey overtone which is concentrated basally and outlined in black on the soft dorsal so that there appears to be a ventro-caudally curved dark stripe passing across it from the tip of the last spine to the middle of the last branched ray. Usually there is a dark midlateral band on the body, ending as a distinct blotch on the caudal fin base; traces of three or four transverse bands may be visible on the flanks.

Immature and quiescent males are indistinguishable from females, although quiescent males often show a distinct darkening of the pelvic fins and faint indications of ocelli on the anal fin.

Ecology. Habitat. The available specimens of *H. nyanzae* are from five localities, of which four lie within the Napoleon Gulf and the fifth (a small island) in the nearby Buvuma Channel. In all localities the bottom is hard (shingle or rock) and the water less than twenty feet deep. The Napoleon Gulf stations are relatively sheltered but the island is exposed.

Food. Eight of the thirteen fishes examined contained food in the stomach or intestines. All eight had fed on fishes (identifiable in each case as *Haplochromis*) but one had a few insect remains in the intestine. Judging from the size of the scales and vertebrae in the gut contents, the prey fishes must have been between 30 and 60 mm. S.L.

Breeding. No data are available. All the specimens are adult.

Affinities. Haplochromis nyanzae is closely related to both *H. victorianus* and *H. serranus*. It differs from both species in the nature of the coloration in preserved males. From *H. victorianus* it is distinguished by its more oblique lower jaw (which also lacks a pronounced mental bump), shorter pectoral fin (22·8–28·2, M = 24·9% of S.L., *cf.* 26·2–32·7, M = 30·4%) and fewer teeth in the upper jaw (50–76, M = 60; *cf.* 64–86, M = 74). From *H. serranus* it is most readily differentiated by its rounded dentary (i.e. no pronounced mental bump), shorter lower jaw (45·0–51·6 M = 48·0, *cf.* 47·4–60·0, M = 54·3% of head) and shorter pectoral fin (22·8–28·2, M = 24·9, *cf.* 23·8–33·0 M = 27·0% of standard length).

Study material and distribution records

Museum and Reg. No.	Locality	Collector
	Uganda	
B.M. (N.H.).—1962.3.2.495 (Holotype)	Jinja bay	E.A.F.R.O.
,, 1962.3.2.500	Jinja pier	,,
,, 1962.3.2.501–3	Karenia (near Jinja)	,,
,, 1962.3.2.504–7	Jinja (below golf course)	,,
,, 1962.3.2.496–8	Ramafuta Isl. (Buvuma channel)	,,
,, 1909.5.4.17	Sesse Isls.	Bayon

Haplochromis bartoni sp. nov.
Text-fig. 7

Note. The trivial name is given because the species resembles *H. worthingtoni*, a species named in honour of Dr. E. Barton Worthington.

Holotype. A specimen 145·0 mm. S.L. (B.M. [N.H.] Reg. No. 1962.3.2.277) from Ekunu Bay.

Description based on thirty-five specimens (including the holotype) 135–195 mm. S.L.

Depth of body 31·4–37·9 (M = 34·0) per cent of standard length, length of head 36·2–39·7 (M = 37·5) per cent. Dorsal head profile straight or very slightly curved, sloping at about 40°; premaxillary pedicels moderately prominent, sometimes giving the profile a noticeable concavity.

Preorbital depth 17·0–22·4 (M = 20·0) per cent of head, least interorbital width 17·0–21·0 (M = 18·6) per cent. Snout 33·4–40·3 (M = 36·0) per cent of head, and 1·1–1·2 times as long as broad; diameter of eye 20·3–24·1 (M = 22·5), depth of cheek 23·4–30·2 (M = 27·0) per cent.

Caudal peduncle 13·3–17·2 (M = 15·0) per cent of standard length, 1·2–1·5 (mode 1·3) times as long as deep.

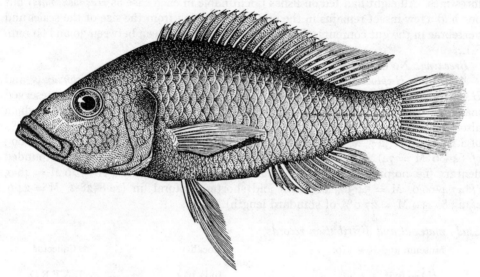

FIG. 7. *Haplochromis bartoni*; ·75 × N.S. (Drawn by Lavinia Beard.)

Mouth slightly oblique; the lower jaw sloping at *ca.* 10°–15° and projecting beyond the upper, usually with a slight mental protuberance. Length of lower jaw 50·8–57·0 (M = 52·5) per cent of head, 1·9–3·0 tines as long as broad. The medial dentigerous part of the premaxilla is not expanded anteroposteriorly. Lips slightly thickened. Posterior tip of the maxilla reaching the vertical through the anterior orbital margin in most fishes (56% of sample) and to a point slightly beyond or slightly anterior to the vertical in the remainder.

Gill rakers short and stout, 8 or 9 (rarely 7) on the lower part of the first arch, the lowermost 1–3 rakers reduced.

Scales ctenoid, lateral line with 31 (f.6), 32 (f.8), 33 (f.19) or 34 (f.2) scales, cheek with 4 (less frequently 3, rarely 5) rows. Six or 7 scales (rarely 7½) between the lateral line and the dorsal origin, 7 or 8 (rarely 6) between the pectoral and pelvic fin bases.

Fins. Dorsal with 23 (f.2), 24 (f.6), 25 (f.18) or 26 (f.8) rays, comprising 14 (f.3), 15 (f.10) or 16 (f.21) spines and 8 (f.2), 9 (f.19), 10 (f.11) or 11 (f.2) branched rays. Anal fin with 11 (f.4), 12 (f.30) or 13 (f.1) rays comprising 3 spines and 8–10 branched rays. Pectoral fin 23·3–27·0 (M = 25·0) per cent of standard length. First pelvic ray slightly elongate in adult males. Caudal with a slightly oblique distal margin in many specimens, but vertically truncate in others; the obliquely truncate type is less oblique than in *H. plagiostoma* (see p. 200). The caudal is scaled on its proximal two-thirds to four-fifths.

Teeth. The *outer row* in both jaws is composed of slender, curved and unicuspid teeth, of which there are 50–80 (M = 62) in the upper jaw. The *inner rows* in most fishes >150 mm. S.L. are composed of unicuspid teeth but in most smaller fishes there is an admixture of uni- and weakly tricuspids in both jaws, or either type of tooth may predominate. In the latter case it is usually the innermost series of the upper jaw which is predominantly tri- or weakly tricuspid, the lower jaw containing mostly unicuspids. There are 2 or 3 (less frequently 4) rows in the upper jaw and 1 or 2 (very rarely 3) in the lower jaw; all inner teeth are implanted obliquely, so that in many specimens the teeth lie horizontally.

Osteology. The *neurocranium* is roughly intermediate between the *H. bayoni*-type (see p. 151) and the *H. mento*-type (see p. 176). The medial toothed part of the premaxilla is not markedly expanded, but it does show some development in that direction. This bone is, however, nearer the generalized *H. guiarti*-type than is the premaxilla of *H. bayoni*.

Vertebrae. Twenty-nine or 30 (mode), comprising 13 abdominal and 16 (f.2) or 17 (f.4) caudal elements.

Lower pharyngeal bone triangular, its dentigerous surface broader than long or equilateral; only rarely is it slightly longer than broad. The teeth are slender (but become coarser in larger fishes), cuspidate and arranged in 20–22 somewhat irregular rows.

Coloration. Live colours are unknown.

Preserved females are brown above the upper lateral line and on the dorsal surface of the head. A dark midlateral stripe runs from behind the operculum to the base of the caudal fin and, in some specimens, can also be seen on the basal half of the caudal fin membrane. Less well-defined, and absent in some fishes, is a dark band from the snout, through the eye and onto the operculum where it becomes continuous with the midlateral band. All fins colourless, the soft dorsal weakly maculate. *Adult males* are uniformly greyish-brown, except for the branchiostegal membrane, chest and belly which are sooty. Dorsal fin grey, with black lappets and dark, often coalesced maculae on the soft part. Anal fin black in the region of the spines and dark along its proximal half; distal part colourless. Anal ocelli whitish-grey, five or six in number and arranged in two rows. Caudal fin brown with dark streaks between the rays; pelvic fins black.

Distribution. Lake Victoria.

Ecology. Habitat. The species has been recorded from several different habitats, whose common features are: a depth of water less than twenty feet and the nearness of dense plant stands. The habitats include a sheltered gulf with a hard substrate, sheltered bays with deep mud bottoms, sheltered beaches with hard and soft substrates, an exposed sandy beach and a rock shelf extending from a partially exposed island. *Haplochromis bartoni* seems to be more abundant in the sheltered habitats and, from its relative scarcity in seine-net catches, it seems to be most abundant some distance offshore (between 200–300 yards).

Food. Nineteen of the thirty-one specimens examined contained food in the stomach or intestines; all had fed exclusively on fishes and the remains could be identified as follows: Cichlidae (undetermined) (f.5), *Haplochromis* spp. (f.11), Cyprinidae (undertermined) (f.1); unidentifiable fish remains (f.2).

Breeding. Little information is available on this species. Sexual maturity is reached at about the same size (145 mm. S.L.) in both sexes and there is no marked dimorphism in the maximum size attained.

Affinities. Some specimens of *H. bartoni* bear a superficial resemblance to the holotype and only specimen of *H. worthingtoni*, a Lake Kyoga species. However, the likeness is purely superficial and the two species differ in several morphological characters. Within the Victoria species-flock, *H. bartoni* is somewhat isolated, not by any outstanding morphological characters but by the sum of several small characters. The neurocranium is nearest that of the *H. mento* complex but it still retains some of the more generalized characters. The premaxilla is of the specialized " beaked " type found in *H. macrognathus*. Another superficial resemblance is to the *H. serranus*-*H. victorianus* complex, but here the likeness is confined to general facies and is not borne out by any deeper-lying details. *Haplochromis bartoni* could represent one of the stages in the evolution of a *H. mento*-type from a *H. guiarti*-like stem, but a stage nearer the " *mento* " than the " *guiarti* " level of organization. Equally. it could link the *H. serranus* group with the *H. mento* complex, but again it would be nearer the " *mento* " than the " *serranus* " condition.

Study material and distribution records

Museum and Reg. No.	Locality	Collector
	Uganda	
B.M. (N.H.).—1962.3.2.277 (Holotype)	Ekunu Bay	E.A.F.R.O.
,, 1962.3.2.278–82	Ekunu Bay	,,
,, 1962.3.2.283–6	Pilkington Bay	,,
,, 1962.3.2.287–92	Ramafuta Isl. (Buvuma ch.)	,,
,, 1962.3.2.293–8	Fielding Bay	,,
,, 1962.3.2.299–300	Karenia (near Jinja)	,,
,, 1962.3.2.301	Old Bukakata	,,
,, 1962.3.2.302	Beach near Nasu point	,,
,, 1962.3.2.303–6	Jinja pier	,,
,, 1962.3.2.307–13	Jinja (below golf course)	,,

Haplochromis estor Regan 1929

Text-figs. 8 and 25

Haplochromis estor Regan, 1929, *Ann. Mag. nat. Hist.*, (10), **3**, 391.

Holotype. A specimen 153 mm. S.L. from an unknown locality in Lake Victoria, collected by M. Graham. B.M. (N.H.) Reg. No. 1959.7.2.1.

Description. Based on the holotype and eight other specimens, 141–170 mm. S.L.

Depth of body 29·6–32·4 (M = 30·3) per cent of standard length, length of head 37·2–38·5 (M = 37·8) per cent. Dorsal head profile sloping fairly steeply (*ca.* 40°), the premaxillary pedicels prominent.

Preorbital depth 19·3–20·6 (M = 19·8) per cent of head, least interorbital width 20·4–23·7 (M = 21·6) per cent. Snout 1·2–1·3 times longer than broad, its length 34·5–37·0 (M = 36·2) per cent of head; diameter of eye 20·4–23·5 (M = 21·7) per cent, depth of cheek 22·8–25·4 (M = 24·5) per cent.

Caudal peduncle 14·3–16·1 (M = 15·8) per cent of standard length, 1·3–1·6 times as long as deep.

Mouth slightly oblique (15°–20°), lower jaw always projecting, sometimes strongly so; its length 54·2–57·5 (M = 55·5) per cent of head, 2·1–2·5 times as long as broad. Lips slightly thickened, the premaxilla with the medial dentigerous surface antero-

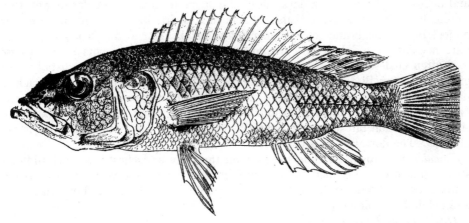

Fig. 8. *Haplochromis estor*; holotype. (Drawn by Miss M. Fasken.)

posteriorly expanded. Posterior tip of the maxilla usually reaching the vertical through the anterior orbital margin, occasionally extending somewhat behind this line.

Gill rakers short and stout, 8 or 9 on the lower part of the first gill arch, the lowermost one to four rakers reduced.

Scales ctenoid, lateral line with 32 (f.3), 33 (f.5) or 34 (f.1) scales, cheek with 4 (rarely 3) series of scales; 6 or 7 (rarely $5\frac{1}{2}$ or 8) between the lateral line and the dorsal fin origin, 7 or 8 (rarely 6) between the pectoral and pelvic fin bases.

Fins. Dorsal with 24 (f.1), 25 (f.5) or 26 (f.3) rays, comprising 15 (f.1) or 16 (f.8) spines and 9 (f.6) or 10 (f.3) branched rays. Anal with 11 or 12 (mode) rays, comprising 3 spines and 8 or 9 branched rays. Pectoral fin 21·8–25·3 (M = 23·6) per cent of standard length. First pelvic ray produced in males. Caudal truncate or very slightly emarginate, scaled on its basal third or half.

Teeth. The *outer row* in both jaws is composed of moderately stout, strongly curved unicuspid teeth, of which there are 52–70 (M = 60) in the upper jaw. The *inner teeth* are tricuspid in the smallest specimen examined (141 mm. S.L.) but are unicuspid in all others. These teeth are implanted obliquely and arranged in 3 or 4 (less frequently 2) rows in the upper jaw and in 2 or an irregular single row in the lower jaw.

Osteology. The neurocranium of *H. estor* is of the *H. mento* type and, because of its more ventrally curved vomer, closely approaches that of *H. dentex* (see p. 168).

Vertebrae : 29–30 comprising 12 or 13 precaudal and 16 or 17 caudal elements in the eight specimens examined.

Lower pharyngeal bone triangular, the breadth of the dentigerous surface slightly greater than its length or, less frequently, equal to its length. The teeth are relatively slender and bicuspid, and are arranged in 18–22 rows.

Coloration unknown in life. *Preserved males* (sexually active). Ground colour dark chocolate-brown above becoming dusky over silver below, especially on the chest and belly ; faint traces of a dark midlateral stripe from the opercular margin to the caudal base are sometimes visible. Lachrymal stripe broad. Lips and lower jaw almost black, the branchiostegal membrane black. Dorsal fin dark grey, the soft part maculate. Caudal fin dark grey and densely maculate, the spots on the proximal half often coalesced so that the fin is dark proximally and lighter distally. Anal greyish with an ill-defined darker band along the distal margin, and a black but narrow band along the basal part ; 2–5 grey-white ocelli are present, arranged in two rows if there are more than four ocelli. Pelvics black. *Quiescent males* have a similar coloration but are much lighter ; consequently the midlateral stripe is more obvious. The dorsal fin lacks spots.

Females are brown, shading to silver on the belly and chest ; a faint midlateral stripe, and in one specimen three very faint and incomplete vertical cross bars are visible. Dorsal fin neutral, the soft part maculate. Caudal dark and maculate. Anal and pelvic fins yellowish.

Distribution. Lake Victoria.

Ecology. Habitat. Four of the localities from which *H. estor* has been obtained are sheltered bays and gulfs where the substrate is mud and the water between 10 and 20 feet deep. No information is available for the two other localities. It can certainly be said that the species (at least when adult) does not occur commonly, if at all, in exposed inshore areas of the lake.

Food. Five of the eight specimens examined had food in the stomach and intestines. Each had fed exclusively on fishes (identifiable as *Haplochromis* in four cases and merely as " fish " in the fifth).

Breeding. No information is available. All the specimens are adult ; the two smallest are females and the rest males.

Affinities. The affinities of *H. estor* are discussed in connection with *H. dentex* (p. 169) *Haplochromis estor* is more advanced towards the extreme *H. macrognathus* type, but is nevertheless more closely allied to *H. dentex* than to *H. macrognathus* or even *H. mento*. Regan (1929) compared *H. estor* with *H. pellegrini*. There is a superficial resemblance and probably an overall phyletic relationship between the species, but it is not, in my opinion, as close as the relationships suggested above.

Study material and distribution records

Museum and Reg. No.	Locality	Collector
	Uganda	
B.M. (N.H.).—1962.3.2.231	Karenia (near Jinja)	E.A.F.R.O.
,, 1962.3.2.232–4	Jinja pier	,,
,, 1962.3.2.235–6	Bugungu (opp. Jinja)	,,
,; 1962.3.2.237	Entebbe	,,
,, 1962.3.2.238	Ekunu Bay	,,
,, 1962.3.2.239	Pilkington Bay	,,
	Lake Victoria, Locality Unknown	
,, 1959.7.2.1 (Holotype)		M. Graham

Haplochromis dentex Regan 1922

Text-figs. 9 and 25

Paratilapia longirostris (part) : Blgr., 1915, *Cat. Afr. Fish.*, **3**, 332.
Haplochromis dentex Regan, 1922, *Proc. zool. Soc. Lond.*, 182, pl. 3, fig. 1.

Holotype. A specimen 127·0 mm. S.L. from the Sesse Islands, B.M. (N.H.) Reg. No. 1909.5.4.1.

Description based on fifteen specimens (91·0–159·0 mm. S.L.) including the holotype.

Depth of body 24·6–29·5 (M = 26·7) per cent of standard length, length of head 33·3–36·2 (M = 34·9) per cent. Dorsal head profile gently curved, the premaxillary pedicels prominent.

Preorbital depth 18·7–24·5 (M = 21·7) per cent of head, least interorbital width 20·0–24·5 (M = 22·5) per cent. Snout longer (1·3–1·5 times) than broad, its length 36·0–41·5 (M = 38·8) per cent of head ; diameter of eye 17·3–25·7 (M = 20·4) per cent, depth of cheek 22·1–27·5 (M = 24·6) per cent.

Caudal peduncle length 17·0–19·8 (M = 18·2) per cent of standard length, 1·6–2·0 times longer than deep.

Angle of mouth variable, from almost horizontal to slightly oblique (15°–20°). Lower jaw projecting slightly, its length 43·8–49·0 (M = 46·0) per cent of head and 2·0–2·5 times its breadth. Lips moderately thickened, the medial dentigerous surface of the premaxilla expanded anteroposteriorly. Posterior tip of the maxilla not reaching the vertical through the anterior orbital margin but always behind a vertical through the posterior tip of the premaxillary pedicels.

Gill rakers moderately slender ; 9 or 10 (mode) on the lower part of the first gill arch, the lowermost one to three rakers reduced.

Scales ctenoid ; lateral line with 33 (f.10) or 34 (f.5) scales (in one specimen there are no lateral line pores on one side). Cheek with 3 or 4 (mode) rows. Five to 7 scales between the lateral line and the dorsal fin origin, 7 or 8 between the pectoral and pelvic fin bases.

Fins. Dorsal with 24 (f.1), 25 (f.12) or 26 (f.2) rays, comprising 15 (f.8) or 16 (f.7) spines and 9 (f.6) or 10 (f.9) branched rays. Anal fin with 11–13 rays, comprising 3 spines and 8–10 (mode 9) branched rays. Pectoral fin 20·7–25·3 (M = 24·6) per cent of standard length. Caudal truncate or weakly emarginate, scaled on its basal two-thirds. First pelvic ray produced in males.

Teeth. The *outer row* in both jaws is composed of large, well-spaced, unicuspid, moderately slender and curved teeth. There are 32–48 (M = 36) in the outer row of the upper jaw.

In all except the two smallest fishes (91 and 100 mm. S.L.) the *inner teeth* are unicuspid. The two small fishes have a mixture of unicuspids, tricuspids and weakly tricuspids, in which the unicuspids predominate. The outermost row in both jaws is composed of teeth only a little smaller than those of the outer row. There are two rows of inner teeth (sometimes irregularly arranged and giving the impression of three rows) in the upper jaw and a single (rarely) double row in the lower jaw.

Osteology. The neurocranium of *H. dentex* clearly belongs to the *H. estor-mento-macrognathus* group. The preotic part is long (65·8% of basal length), the skull is narrow and shallow (neurocranial height 3·4 times in basal length) and the supraoccipital crest is low. It differs, however, from other members of the group in its

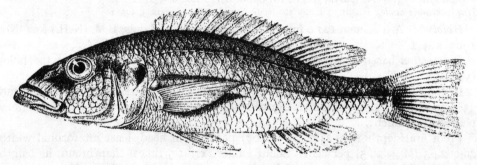

FIG. 9. *Haplochromis dentex*, holotype, ·8 × N.S. (From Regan, *Proc. zool. Soc. Lond.*)

sharply decurved ethmoid-vomer region which slopes at about 40° (the neurocranial roof slopes at *ca.* 25°). In general, the neurocranium of *H. dentex* is nearest that of *H. estor*, differing mainly in having a sharply decurved vomer.

Vertebrae. 30–32 (mode 31 in the ten specimens examined), comprising 13 (f.3) or 14 (f.7) precaudal and 16 (f.1), 17 (f.6) or 18 (f.3) caudal elements.

Lower pharyngeal bone triangular, its dentigerous surface as long as broad or slightly broader than long. The pharyngeal teeth are slender, fine and bicuspid except for the coarser and less obviously cuspidate teeth in the two median and last transverse rows. There are 18–20 rows of teeth.

Coloration. The colours of live fishes are unknown. *In preserved material* there is little difference in the coloration of males and females. The ground colour is dark grey above becoming silvery-grey below (darker in males). In some females there is a faint and narrow midlateral longitudinal stripe from the hind margin of the operculum to the caudal origin. Dorsal and caudal fins are greyish, the former with black lappets and the latter sometimes maculate. The anal and pelvic fins are hyaline in females, whereas in males the anal is greyish with a narrow black basal line running above the spines. In males there are two dead-white anal ocelli, and the pelvics are black.

Distribution. Lake Victoria.

Ecology. Habitat. Since only fifteen specimens are known, it is impossible to generalize on habitat preferences particularly since about half the specimens are from sheltered bays with mud substrates, and the others from exposed, sandy beaches, open off-shore waters with rock and shingle bottoms, and a fairly exposed gulf, also with a hard substate. The depth range extends to at least 25 feet, but the species cannot be considered common in any of the habitats investigated.

Food. Only three of the fifteen specimens available had ingested material in the gut. Two specimens had fed on small *Haplochromis* ; in the intestine of the third was a small quantity of fragmentary plant tissue.

Breeding. The breeding habits of *H. dentex* are unknown. The two smallest fishes (91 and 100 mm. S.L.) are both immature and the next largest (128 mm.) may also be a juvenile. All the other specimens (142–159 mm.) are adult.

Affinities. Superficially, *H. dentex* looks intermediate between the *H. guiarti*-type and the more specialized *H. mento-estor* types. But, closer study shows that it has greater affinity with the latter group, particularly with regard to its neurocranial form and the expanded medial dentigerous part of the premaxilla. *Haplochromis dentex* is perhaps most closely related to *H. estor*, although the latter has progressed further along the *H. mento-H. macrognathus* path of specialization. The supposed relationship between *H. dentex* and *H. estor* is based both on points of overall similarity and on likeness in the neurocrania of the two species. *Haplochromis dentex* is, however, easily distinguished by its fewer and larger teeth (mean number of teeth in the outer row of the upper jaw 37 *cf.* 60) and its shorter lower jaw (43·8–49·0, M = 46·0 per cent of head *cf.* 48·0–57·5, M = 54·9 per cent). The ancestry of *H. dentex* is obscure ; possibly it was derived from a *H. pellegrini*-like stem.

Study material and distribution records

Museum and Reg. No.	Locality	Collector
	Uganda	
B.M. (N.H.).—1909.5.4.1 (Holotype)	Sesse Isls. .	Bayon
,, 1962.3.2.129	Sesse Isls. .	E.A.F.R.O.
,, 1962.3.2.130–1	Macdonald Bay .	,,
,, 1962.3.2.132	Grant Bay .	,,
,, 1962.3.2.133–4	Jinja .	,,
,, 1962.3.2.135–6	Buka Bay .	,,
,, 1962.3.2.137	Kagera port .	,,
,, 1962.3.2.138	Ramafuta Isl. (Buvuma channel) .	,,
,, 1962.3.2.139	Ekunu Bay .	,,
,, 1962.3.2.140	Thruston Bay .	,,
,, 1962.3.2.142	Old Bukakata Bay .	,,
	Kenya	
,, 1962.3.2.141	Kamiriga, Kavirondo Gulf .	,,

Haplochromis artaxerxes sp. nov.

This peculiar species is represented by a single specimen. Its diagnostic characters are such, however, that I have little hesitation in basing the description on one fish. The specific name is derived from Artaxerxes, King of Persia, also known as *Longimanus* and alludes to the extremely long pectoral fins of this species.

Holotype a male 147·0 mm. standard length, from the Napoleon Gulf near Jinja; B.M. (N.H) Reg. No. 1962.3.2.508.

Description. Depth of body 27·9 per cent of standard length, length of head 32·7 per cent. Dorsal head profile gently curved, the premaxillary pedicels not making a prominent projection.

Preorbital depth 17·7 per cent of head, least interorbital width 20·8 per cent. Snout 1·25 times as long as broad, its length 33·3 per cent of head; eye diameter 22·9 per cent, depth of cheek 22·9 per cent.

Caudal peduncle 19·0 per cent of standard length, 1·7 times as long as deep.

Lower jaw rather flat and closing within the upper, the anterior tip projecting. Length of lower jaw 48·0 per cent of head, 2·3 times as long as broad. Median dentigerous area of the premaxilla not expanded anteroposteriorly. Posterior tip of the maxilla reaching the vertical through the anterior orbital margin. Mouth very slightly oblique, sloping at about 10°.

Gill rakers short and stout, 9 on the lower part of the first gill arch, the three lowermost rakers reduced.

Scales ctenoid; lateral line with 34 scales, cheek with 4 series. Seven scales between the lateral line and the dorsal origin, 9 between the pectoral and pelvic fin bases.

Fins. The pectoral fins of this species provide the most readily diagnostic character since they are longer (34·7% of standard length) than in any other Lake Victoria species. Only the third and fourth pectoral rays are produced so that the shape of the fin is also characteristic.

Dorsal with 15 spinous and 10 branched rays, anal with 3 spines and 9 branched rays. The distal margin of the caudal is damaged so its outline cannot be determined. The first branched pelvic ray is elongate and filamentous, in fact, intermediate between the extreme condition found in *H. bayoni* (see p. 151) and that of other species.

Teeth. Most of the outer teeth in both jaws are missing. The few remaining teeth are unicuspid, slender and very strongly curved. The inner teeth are unicuspid and arranged in two series in the upper jaw and a single, irregular row in the lower jaw.

Osteology. Neurocranial shape cannot be determined from a radiograph. There are 31 vertebrae (13 precaudal and 18 caudal). The lower pharyngeal bone is slender and triangular, the dentigerous surface being broader than long. The pharyngeal teeth are slender, fine and cuspidate, and are arranged in about twenty rows.

Coloration. Unknown in life. The *preserved adult male* is dark brown above, rapidly shading to a light brass colour with an overall greyish tinge. Lips, lower jaw, horizontal limb of the preoperculum and the posterior opercular margin black. There is a very broad (*ca.* half diameter of eye) lachrymal band and two, narrow parallel bands across the snout. Dorsal fin yellowish-brown, the soft part faintly marbled. Caudal dark, anal with a narrow, yellow basal band but otherwise sooty, particularly in the area of the spines, except for a narrow and yellow distal margin.

Ecology. The only ecological information is that the fish was caught in *ca.* 10 feet of water over a mud bottom and near a fringing stand of swamp grass. The location is near Jinja and is in the relatively sheltered Napoleon Gulf.

Diagnosis and affinities. The species is characterized by the following combination of characters : pectorals long (35% of standard length and longer than the head), the third and fourth rays greatly produced ; lower jaw flat and closing within the upper jaw.

Because of these particular characters, it is difficult to suggest possible relationships for the species. The elongate pectoral is, of course, a character which could easily and suddenly develop from the pectoral of any *Haplochromis*. The lower jaw, on the other hand, requires a more fundamental change and one rarely encountered in the Victoria species flock. It is otherwise found only in some of the larval fish-eating species (Greenwood, 1959). Certainly none of these could represent the ancestral or descendant condition of a species like *H. artaxerxes*. Without any knowledge of the skeleton in this species (and especially without more information on its dentition) little can be guessed about its phyletic position. Superficially, *H. artaxerxes* does resemble *H. estor* but it is immediately distinguished by the diagnostic characters listed above, and by its shorter head and lower jaw.

Haplochromis longirostris (Hilgen.) 1888

Text-figs. 10 and 11

Paratilapia longirostris Hilgendorf, 1888, *Sitzb. ges. naturf.-Fr. Berlin*, 77 ; Boulenger (part), 1915, *Cat. Afr. Fish.*, **3**, 332.
Haplochromis longirostris (part) ; Regan, 1922, *Proc. zool. Soc. Lond.*, 187, Pl. 4, fig. 2 (two of the three specimens described *loc. cit.*).
Haplochromis gracilicauda Regan, 1922, *op. cit.*, 188, Pl. 4, fig. 2.
Haplochromis tenuis Borodin, 1931. *Proc. New Eng. zool. Club*, **12**, 50.

The characters which Regan used to separate *H. gracilicauda* from *H. longirostris* (snout $1\frac{1}{3}$ to $1\frac{2}{3}$ diameter of eye *cf.* snout $1\frac{2}{3}$ to twice diameter of eye) are really functions of growth, his species *H. gracilicauda* representing juvenile *H. longirostris*. Through the courtesy of Dr. K. Deckert of the Berlin Museum, I have been able to examine the holotype of *Paratilapia longirostris* (Z.M. Berlin Reg. No. 12744) and this confirms the identity of Regan's *H. longirostris* material. Two of these fishes (BM. [N.H.] Reg. No's. 1911.3.3.13 and 1906.5.30.516) are retained in the species but the third (B.M. [N.H.] 1906.5.30.274) is referred to *H. argenteus*.

Description based on twenty-nine specimens (including the holotype, and the syntypes of *H. gracilicauda*) 85–145 mm. S.L.

Depth of body 24·6–30·4 (M = 27·2) per cent of standard length, length of head 29·2–36·2 (M = 33·0) per cent. Dorsal head profile slightly concave or, less commonly, straight ; sloping at an angle of *ca.* 30°. Premaxillary pedicels moderately prominent in large individuals.

Preorbital depth 18·2–22·5 (M = 20·8) per cent of head, least interorbital width 17·1–24·0 (M = 21·2) per cent. Snout length 32·0–38·6 (M = 36·0) per cent of head, 1·2–1·5 times as long as broad, narrowest in large fishes. Diameter of eye 18·8–24·3 (M = 21·7) per cent, depth of cheek 21·8–28·4 (M = 24·3) per cent.

Caudal peduncle 17·2–22·2 (M = 19·2) per cent of standard length, 1·7–2·3 (modal range 1·9–2·0) as long as deep.

Lower jaw markedly oblique, sloping at 40°–50°, its tip projecting slightly beyond the upper jaw in some fishes and level with it in others. Length of lower jaw 42·2–51·4 (M = 46·0) per cent of head, 2·4–3·3 times as long as broad. Lips slightly thickened, the median dentigerous part of the premaxilla not noticeably expanded. The posterior tip of the maxilla usually reaches the vertical from the posterior end of the premaxillary pedicels, or slightly beyond, but never reaches to below the anterior orbital margin.

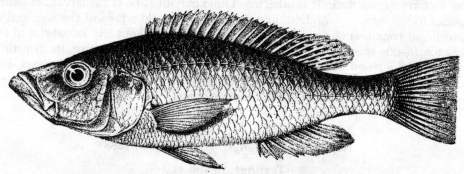

FIG. 10. *Haplochromis longirostris*, × N.S. (From Regan, *Proc. zool. Soc. Lond.*)

Gill rakers variable, from long and slender to short and stout, sometimes flattened and broadly branched ; 9–11 (rarely 8) rakers on the lower part of the first gill arch.

Scales ctenoid, lateral line with 32 (f.6), 33 (f.11) or 34 (f.12) scales, cheek with 3 or 4 rows. Five to 7 (mode 6) scales between the lateral line and the dorsal origin, 5–7 (mode 6), rarely 8, between the pectoral and pelvic fin bases.

Fins. Dorsal with 24 (f.4), 25 (f.19) or 26 (f.5) rays, comprising 15 (f.3), 16 (f.23) or 17 (f.2) spinous and 8 (f.4), 9 (f.18) or 10 (f.6) branched rays. The dorsal fin of the holotype is badly damaged and a count gives *ca.* 14, 9 rays. Anal fin with 11–13 rays comprising 3 spines and 8 or 9 (rarely 10) rays. Caudal truncate, scaled on its proximal two-thirds. Pectoral 20·8–25·2 (M = 23·0) per cent of standard length. First pelvic ray only slightly produced (not filamentous) in adult males.

Teeth. The *outer row* in both jaws is composed of slender and fine, moderately to strongly curved teeth, those situated posterolaterally in the upper jaw are almost hair-like. These outer teeth are unicuspid in most fishes, but in three (85·5 mm. and two of 111 mm. S.L.) there is a mixture of bi- and unicuspids or weakly bicuspids and unicuspids. There are 40–70 (M = 56) teeth in the upper jaw, the number showing no clear-cut correlation with standard length.

The *inner teeth* are either all unicuspid (fishes >125 mm S.L. and a few in the 95–

125 mm. range) or a mixture of uni- and tricuspids, unicuspids and weakly tricuspids or, as in the smallest fish (a syntype of *H. gracilicauda*), all tricuspids. The inner rows are implanted obliquely so that the teeth lie almost horizontally. There are usually 2 inner rows in the upper jaw (occasionally 3, rarely 1) and one or, less frequently, 2 in the lower jaw.

Osteology. The neurocranium barely differs from that of *H. mento* despite the more oblique angle of the lower jaw in *H. longirostris*. This greater jaw angle is apparently brought about by a slight difference in the articulatory surfaces of the quadrate and angular.

Vertebrae. 31 (f.7) or 32 (f.2) comprising 13 (f.1) or 14 (f.8) precaudal and 17 (f.6) or 18 (f.3) caudal elements.

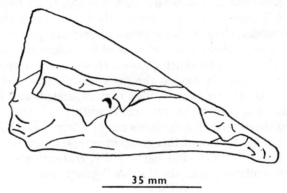

Fig. 11. *Haplochromis longirostris*, neurocranium.

Lower pharyngeal bone triangular, its dentigerous surface broader than long. The teeth are fine, cuspidate and compressed (those of the two median rows sometimes slightly enlarged) and arranged in 24–26 irregular rows.

Coloration. The colours of live fishes are unknown.

Preserved females are grey-brown above, silvery below with, on the flanks and belly, a slight brassy overtone. The lower lip and sometimes the mental area are sooty. All fins are hyaline, the dorsal with dark lappets and the caudal often densely and darkly maculate.

Sexually active males have an overall sooty appearance but with silvery areas showing through, particularly on the anterior flanks and the lateral aspects of the chest and belly ; the ventral parts of the chest and belly, however, are always darker than the dorsal parts of the body. The lower jaw, ventral part of the preoperculum and the branchiostegal membrane are intensely blank. Dorsal fin dark grey with black lappets and dark streaks between the soft rays. Caudal dark, especially on its proximal half. Anal dark, with a black area on the spinous part continued posteriorly as a narrow black band along the fin base ; the ocelli are barely visible as faint grey blotches. Pelvic fins black.

The amount of silver visible on the flanks may be correlated with the fish's state of sexual activity. *Juvenile males* are indistinguishable from females.

Distribution. Lake Victoria.

Ecology. Habitat. The species is known from only a few localities, each of which is a sandy beach either exposed to the open lake or within a sheltered gulf ; apparently *H. longirostris* is nowhere common.

Breeding. One female contains advanced embryos in the buccal cavity ; no other brooding fishes are known. Most individuals >110 mm. S.L. are sexually mature, as are some smaller fishes. Apparently both sexes reach the same maximum adult size.

Food. From the scanty data available, *H. longirostris* appears to feed on both insects (particularly pupal stages) and small fishes. Fourteen of the twenty-two specimens examined contained food. Eight had fed exclusively on insects, one on insects and fishes, and five on fishes only. The fish and insect remains were very fragmentary so identification could not be taken far. Only one fish could be identified (a small cyprinid, probably *Engraulicypris argenteus*) ; the insects are mainly pupal Bäetids.

Affinities. The slender, elongate body form and very oblique jaws of *H. longirostris* are immediate diagnostic characters and ones which isolate the species from all but one other in Lake Victoria. This other species is *H. argenteus*. *Haplochromis argenteus* differs from *H. longirostris* in having the jaws less oblique, the premaxilla more clearly " beaked " (i.e. the median toothed part expanded) and in having the eye diameter noticeably larger than the interorbital width.

On neurocranial characters, *H. longirostris* can be referred to the *H. mento* complex. But it differs from other species of this group in several character combinations and also in its mixed insect-fish diet. Phylogenetically, *H. longirostris* can be considered a somewhat isolated offshoot from the " *mento* "-group stem.

Study material and distribution records

Museum and Reg. No.	Locality	Collector
	Uganda	
B.M. (N.H.).—1906.5.30.262	Bunjako	Degen
(Paratype *H. gracilicauda*)		
„ 1906.5.30.516	Bunjako	„
„ 1906.5.30.268	Entebbe	„
(Lectotype *H. gracilicauda*)		
„ 1911.3.3.13	Jinja, Ripon Falls	Bayon
„ 1962.3.2.10–13	Jinja	E.A.F.R.O.
„ 1962.3.2.1	Grant Bay	„
„ 1962.3.2.5	Karenia (near Jinja)	„
„ 1962.3.2.14–25	Beach near Nasu Point	„
	Tanganyika	
„ 1962.3.2.2	Majita Beach	„
„ 1962.3.2.3–4	Mwanza, Capri Bay	„
	Lake Victoria, Locality Unknown	
„ 1962.3.2.6–9		„

Haplochromis mento Regan 1922

Text-figs. 12 and 25

Paratilapia longirostris (part) : Blgr., 1915, *Cat. Afr. Fish.*, **3**, 332, fig. 223.
Haplochromis mento Regan, 1922, *Proc. zool. Soc. Lond.*, 183.

Holotype. A specimen 174·0 mm. S.L., from Bunjako; B.M. (N.H.) Reg. No. 1906.5.30.258.

Description, based on twenty-five specimens (113–178 mm. S.L.) including the holotype.

Depth of body 24·1–31·5 (M = 28·9) per cent of standard length, length of head 31·5–37·1 (M = 34·5) per cent. There is some variation in the lateral outline of the head; in most specimens the head gives an impression of attenuation and of being pointed, but in others this impression is less marked and the head seems coarser and more bluntly rounded. There are no clear-cut morphometric differences between specimens belonging to either group, and intergrades exist.

Depth of preorbital 18·4–23·6 (M = 21·2) per cent of head, least interorbital width 20·4–24·4 (M. = 22·5) per cent. Snout 1·5–1·8 times as long as broad, its length 36·6–43·4 (M = 39·3) per cent of head; diameter of eye 16·8–22·0 (M = 19·2) per cent, depth of cheek 18·9–27·8 (M = 23·6) per cent.

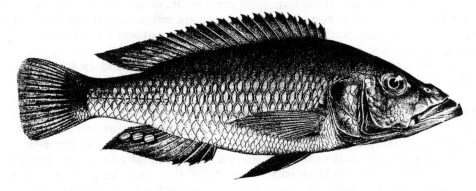

Fig. 12. *Haplochromis mento*; holotype, ·6 × N.S. (From Boulenger, *Fish. Nile.*)

Caudal peduncle length 14·1–20·9 (M = 17·5) per cent of standard length, 1·3–2·2 (modal range 1·7–1·8) times as long as deep.

Lower jaw projecting slightly in fishes <150 mm. S.L. and more markedly prominent in larger fishes; its length 41·8–50·0 (M = 46·8) per cent of head and 2·1–2·8 times as long as broad. Mouth slightly oblique or even horizontal, the medial dentigerous part of the premaxilla expanded anteroposteriorly. Posterior tip of the maxilla generally reaching to a vertical midway between the nostril and the anterior orbital margin, but sometimes extending a little more posteriorly.

Gill rakers short and stout, 8–10 (mode 9) on the lower part of the first arch, the lowermost 1–3 rakers reduced.

Scales ctenoid; lateral line with 32 (f.1), 33 (f.7), 34 (f.14) or 35 (f.3) scales, cheek with 3 or 4 (rarely 5) rows. Six or 7 (rarely 5) scales between the lateral line and the dorsal fin origin, 7 or 8 (rarely 9) between the pectoral and pelvic fin bases.

Fins. Dorsal with 23 (f.1), 24 (f.2), 25 (f.12) or 26 (f.10) rays, comprising 13 (f.1), 15 (f.5), 16 (f.18) or 17 (f.1) spinous and 8 (f.1), 9 (f.11), 10 (f.12) or 11 (f.1) branched rays. Anal with 11, 12 (mode) or 13 rays, comprising 3 (4 in one fish) spines and 8–10 rays. Pectoral fin 19·7–24·7 (M = 23·4) per cent of standard length. First pelvic ray produced in adult males. Caudal truncate, scaled on its basal half to two-thirds.

Teeth. The *outer row* in each jaw is composed of unicuspid, very strongly curved and moderately stout teeth, there being 38–66 (M = 52) in the upper jaw. The teeth of the *inner series* are predominantly unicuspids but in a few specimens there is an admixture of unicuspid and weakly tricuspid teeth. Teeth in the outermost row of the inner series, especially in the upper jaw, are often enlarged. All inner teeth are implanted at a very oblique angle. There are 2 or 3 (rarely 4) rows in the upper jaw and 2 (rarely 1 or 3) in the lower jaw.

Osteology. The neurocranium of *H. mento* presents no outstanding specific characteristics; it is typical of the long, shallow and narrow skull found also in *H. macrognathus*, *H. estor*, *H. gowersi* and *H. dentex*. The premaxilla shows pronounced medial expansion, a character usually associated with the " *mento* " skull type but probably more marked in this species than in the others of the group.

Lower pharyngeal bone triangular, small and fine, its dentigerous surface broader than long. The teeth are cuspidate and generally fine but some of the median series may be coarser. The teeth are rather sparsely distributed and are arranged in 16 (rarely) to 20 (most common) rows.

Vertebrae. 30–32 (mode 30) comprising 13 or 14 precaudal and 17 or 18 caudal elements (7 specimens examined).

Coloration. Live fishes. Adult males have a steely grey-blue ground colour, darker (almost sooty) on the chest and belly, branchiostegal membrane dark grey. Snout, cheek and opercular region have an irridescent sheen. The spinous part of the dorsal is irridescent blue, the soft part dark neutral; lappets dusky. Caudal fin dark neutral. Spinous part of the anal dusky, the soft part dark neutral and bearing the dull orange-red ocelli. Pelvics black on the outer third, the remainder of the fin dusky. The colours of *live females* and *immature males* are unknown but a recently dead female has the dorsal aspects of the body and head bright green and the flanks silver. All fins are yellowish neutral.

Preserved material. Sexually active males. Ground colour brownish with a sooty overlay on the belly, the branchiostegal membrane blackish-grey and a broad but indistinct lachrymal blotch. The dorsal fin is dark brown with black lappets. Caudal dark, densely maculate on its proximal two-thirds. Anal brownish to sooty-grey, the ocelli indistinct and greyish-brown. Pelvics black on the outer third, the remaining rays black but the intervening membrane light grey. *Quiescent males* have a female-type coloration (see below) except that the pelvics are sooty on the outer three rays, the branchiostegal membrane is dark and there are faint traces of grey-white ocelli on the anal fin. The vertical bars on the flank are not always visible. *Females* are brownish-grey above, becoming golden-silver below the level of the lower lateral line. There are about nine faint vertical bars on the flank and caudal peduncle; each bar extends from slightly above the upper lateral line to a point some one to three scales below the level of the lower lateral line. The first five bars may be joined by a midlateral stripe of about the same width as the bars. The dorsal fin is colourless or light brown, the lappets dark. The anal, caudal and pelvic fins are yellowish brown to greyish.

Distribution. Lake Victoria.

Ecology. Habitat. Haplochromis mento is apparently confined to areas where the substrate is sand or sand and rocks, and where the water is not more than twenty feet deep. Most localities from which specimens were obtained are exposed but one is a sheltered gulf; all are within two-hundred yards of the shore, the majority within one hundred yards.

Food. Eleven of the thirty-one specimens examined contained food in the stomach or intestines; of these specimens, one had only the remains of a large dragonfly larva and the others only fish remains. It is regrettable that so few specimens contained food because the prey species are predominantly cyprinid fishes; most other piscivorous *Haplochromis* seem to concentrate on cichlids. Of the ten specimens with fish in the guts, three had fed on Cichlidae (determined as *Haplochromis* in two cases) and seven on Cyprinidae (identified as *Engraulicypris argenteus* in two cases). Admittedly the contents of seven out of eleven stomachs is far too small a sample on which to base generalizations. However, it should be remembered that in samples of a like size from many other species, the identifiable food is entirely of cichlid origin and predominantly *Haplochromis*. Also, the eleven *H. mento* came from five different localities. Possibly, then, *H. mento* has specialized in preying on cyprinids, a group not heavily tapped by other piscivorous fish-predators (see Corbet, 1961).

Breeding. The species is a female mouth brooder. Fishes less than 135 mm. S.L. are immature. Possible sexual dimorphism in the maximum size attained cannot be determined from the sample available since few females are represented; the three fishes of 170 mm. S.L. and over are all males.

Affinities. Structurally, *H. mento* is a specialized predator and, therefore, shows at least group affinity with *H. estor, H. gowersi* and *H. longirostris*. At this level there is also some affinity between *H. mento* and *H. macrognathus*, the latter being a structurally more extreme form of *H. mento*. The affinities of these two species are discussed on p. 174. *Haplochromis estor* and *H. gowersi* both differ from *H. mento* in several ways, but primarily in the shape of the head and the more oblique jaws of the former species. Thus, it is impossible to consider *H. mento* as having any close morpho-relatives, although the species does have phyletically close affinities with several others. Since *H. mento* shows the general characters of its specialized predatory line (particularly syncranial characters) it may represent a fairly basic anatomical state in that line.

Study material and distribution records

Museum and Reg. No.	Locality	Collector
	Uganda	
B.M. (N.H.).—1906.5.30.258 (Holotype)	Bunjako	Degen
,, 1962.3.2.179–191	Beach near Nasu Point	E.A.F.R.O.
,, 1962.3.2.192	Old Bukakata Bay	,,
,, 1962.3.2.193	Between Yempita and Busiri Isls. (Buvuma Channel)	,,
,, 1962.3.2.194	Jinja (below golf course)	,,
,, 1962.3.2.195	Grant Bay	,,
,, 1962.3.2.196-7	Kagera Port	,,

Museum and Reg. No.	Locality	Collector
	Uganda (continued)	
B.M. (N.H.).—1962.3.2.198–200	Entebbe (airport beach)	E.A.F.R.O.
,, 1962.3.2.206–7	Entebbe Harbour	,,
,, 1962.3.2.201–5	Bukafu Bay	,,
,, 1962.3.2.209	Thruston Bay	,,
,, 1962.3.2.210–1	Ramafuta Isl. (Buvuma Channel)	,,
,, 1962.3.2.212–7	Buka Bay	,,
	Tanganyika	
,, 1962.3.2.208	Beach near Majita	,,

Haplochromis mandibularis sp. nov.

Holotype. A specimen 140·0 mm. S.L., from Jinja; B.M. (N.H.) Reg. No. 1962.3.2.222.

Description based on ten specimens (131–174 mm. S.L.) including the holotype. Only one of these specimens is a female.

Depth of body 31·6–34·3 (M = 33·1) per cent of standard length, length of head 38·0–39·3 (M = 38·6) per cent; breadth of head immeditely anterior to orbits 22·4–28·3 (M = 25·9) per cent of head length. Dorsal profile of head sloping at an angle of 30°–40°, slightly curved and becoming concave in large fishes; premaxillary pedicels slightly prominent.

Preorbital depth 17·7–20·7 (M = 19·3) per cent of head, least interorbital width 19·6–22·2 (M = 21·2) per cent. Snout 1·5–1·8 times as long as broad, its length 36·2–39·7 (M = 38·6) per cent of head; diameter of eye 17·2–22·2 (M = 19·4) per cent, depth of cheek 24·5–29·2 (M = 26·1) per cent.

Caudal peduncle 12·2–15·2 (M = 14·2) per cent of standard length, 1·0–1·3 (mode) times as long as deep.

Mouth slightly oblique, lower jaw extension variable, from projecting markedly to no extension beyond the upper jaw. Lips somewhat thickened, the median toothed portion of the premaxilla only slightly expanded anteroposteriorly. The posterior tip of the maxilla reaches a point near the anterior orbital margin in most fishes, but it extends to the level of the orbit (or even slightly beyond) in a few specimens. Lower jaw 47·3–56·8 (M = 51·5) per cent of head, 2·0–2·9 (mode 2·5) times as long as broad.

Gill rakers short and stout, 8 (mode) or 9 on the lower part of the first arch, the lowermost 1 or 2 rakers reduced.

Scales ctenoid, lateral line with 32 (f.4) or 33 (f.6) scales, cheek with 4 (rarely 5) rows. Six or 7 scales between the lateral line and the dorsal fin origin, 6 or 7 (rarely 5) between the pectoral and pelvic bases.

Fins. Dorsal with 24 (f.1), 25 (f.8) or 26 (f.1) rays, comprising 15 (f.1) or 16 (f.9) spinous and 8 (f.1), 9 (f.7) or 10 (f.2) branched rays. Anal with 12 rays, comprising 3 spines and 9 branched rays. Pectoral fin 21·4–24·3 (M = 22·2) per cent of standard length. First pelvic ray somewhat produced and filamentous in adult males. Caudal truncate, the posterior margin running somewhat obliquely forwards and downwards; scaled on the proximal half to two-thirds.

Teeth. The *outer row* in both jaws is composed of fairly stout unicuspid teeth, recurved anteriorly and strongly incurved laterally; there are 72–94 (M = 82) teeth in the upper jaw. The *inner teeth* are all unicuspid and obliquely implanted, being arranged in 4–6 rows in the upper jaw and 2 or 3 (rarely 4) in the lower jaw.

Osteology. The neurocranium closely approaches the *H. mento-macrognathus* type but is somewhat less extreme, particularly with regard to its anterior profile which is steeper. This gives the skull a stouter and more compact appearance. The premaxilla is as beaked as that of *H. macrognathus*.

Lower pharyngeal bone triangular, its dentigerous surface as long as it is broad, but with the posterior margin deeply indented in all specimens except the two largest. The teeth are fine, cuspidate and somewhat irregularly arranged in 20–24 rows.

Coloration. The live colours are unknown. A *preserved female* is brownish-grey above, shading to silvery-brown below. On the flanks are traces of about seven dark transverse bars and there is a very weak spot on the caudal fin base. The snout and upper jaw are dark grey, the lower jaw is paler. Dorsal fin hyaline but with sooty lappets and an oblique, ventrally directed dark bar on the soft part; the base of the soft dorsal is dark. Caudal greyish-black, the basal half densely and darkly maculate. Pelvic and anal fins hyaline, the latter with one, faint, dead-white ocellus.

Quiescent males are brownish-grey above and on the snout and upper jaw, becoming silvery below. There is an interrupted midlateral black band from the posterior opercular margin to the caudal base, where it ends as a faint spot; the spot may extend onto the caudal fin. There are also very faint traces of an interrupted dark band running immediately above the upper lateral line. The dorsal fin is hyaline but with dusky lappets and is densely maculate on the soft part. Anal hyaline, as is the caudal although the latter is densely maculate. Pelvics faintly dusky.

Sexually active males are dark brown above, brassy on the flanks, and black ventrally, particularly on the belly, chest and branchiostegal membrane. The lower part of the preoperculum and the lower third of the operculum are also black. The dorsal fin is dusky with black lappets; the soft part is hyaline on the distal half and densely maculate proximally. The basal third of the caudal is dark brown or black, the distal part yellowish. The anal is yellowish with a narrow black basal band and a faint black area over the last two rays; one or two dark ocelli are present. Pelvic fins are black.

Distribution. Lake Victoria.

Ecology. Habitat. The five localities from which *H. mandibularis* were obtained are all close inshore and have a hard sand substrate; four are in sheltered areas and one is partly exposed. The maximum depth at which the specimens could have been living is between 20 and 25 feet

Food. Five of the ten specimens examined contained food. In each case the gut contents were fishes, identifiable as *Haplochromis* in two specimens and as cichlids in the other three.

Breeding. All the specimens are adult and only one (135 mm. S.L.) is a female. No other data are available.

Affinities and diagnosis. Superficially and in many morphometric characters, *H. mandibularis* resembles *H. macrognathus*. It is distinguishable, however, by the lack of a large, well-defined black spot at the base of the caudal fin, the lack of a black mental spot and by its deeper body. An obvious difference, but one which cannot be quantified, is the shape of the head which is less compressed and less acute in *H. mandibularis*; also in this species, at least in the size-range available, the eye diameter is equal to or is slightly larger than the preorbital depth. In *H. macrognathus* of the same size (131–174 mm.) the preorbital is deeper than the eye. Anatomically and superficially *H. mandibularis* could represent the ancestral condition from which *H. macrognathus* evolved. *Haplochromis mandibularis* also shows relationship with *H. mento* but this is less intimate than that with *H. macrognathus*; again the relationship is of a generalized to a more specialized species.

Study material and distribution records

Museum and Reg. No.	Locality	Collector
	Uganda	
B.M. (N.H.).—1962.3.2.222 (Holotype)	Jinja (below golf course)	E.A.F.R.O.
,, 1962.3.2.223–4	Jinja (below golf course)	,,
,, 1962.3.2.225	Jinja Pier	,,
,, 1962.3.2.226	Bugungu (opp. Jinja)	,,
,, 1962.3.2.227–8	Beach near Nasu Point	,,
	Lake Victoria, Locality Unknown	
,, 1962.3.2.229–230		,,

Haplochromis gowersi Trewavas, 1928

Text-figs. 13 and 14

Haplochromis gowersi Trewavas, 1928, *Ann. Mag. nat. Hist.* (10), **2**, 94.

Lectotype. A specimen 154·0 mm. S.L.; B.M. (N.H.) Reg. No. 1928.5.24.478.

Description based on twenty-two specimens (145–224 mm. S.L.), including the lecto- and paratypes.

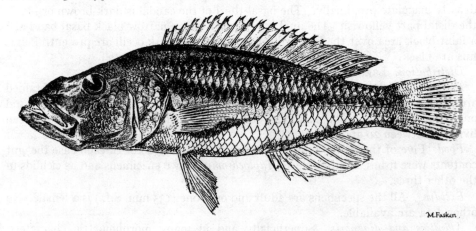

Fig. 13. *Haplochromis gowersi*; type, ·68 × N.S. (Drawn by Miss M. Fasken.)

Depth of body 26·5–33·5 (M = 29·4) per cent of standard length, length of head 35·8–38·4 (M = 37·0) per cent. Dorsal head profile slightly curved, sloping at an angle of 30°–35°; premaxillary pedicels slightly prominent, but not sufficiently prominent to give a pronounced interorbital concavity to the profile.

Preorbital depth 19·3–24·0 (M = 22·0) per cent of head, least interorbital width 16·1–23·3 (M = 19·7) per cent. Snout 1·3–1·5 times as long as broad, its length 36·8–42·2 (M = 39·6) per cent of head; diameter of eye 15·5–19·3 (M = 17·5) per cent, depth of cheek 27·8–33·3 (M = 29·5) per cent.

Caudal peduncle 13·3–17·6 (M = 14·8) per cent of standard length, 1·3–1·6 (mode 1·4) times as long as deep.

Mouth oblique (sloping at *ca.* 30°–35°), lips slightly thickened, the median dentigerous surface of the premaxilla expanded anteroposteriorly; lower jaw always projecting but to a variable extent. Length of lower jaw 49·1–55·1 (M = 52·0) per cent of head,

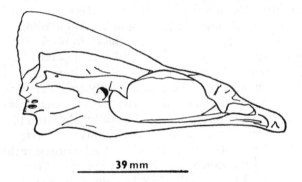

Fig. 14. *Haplochromis gowersi*, neurocranium.

2·1–2·7 times as long as broad. Posterior tip of the maxilla not reaching the vertical through the anterior orbital margin, usually reaching a point about midway between the nostril and the orbit, sometimes a little more posteriorly.

Gill rakers short and stout, 8–10 (mode 9) on the lower part of the first gill arch.

Scales ctenoid; lateral line with 31 (f.3), 32 (f.2), 33 (f.8) or 34 (f.2) scales, cheek with 4 or 5 (rarely 6) rows. Six or 7 (less frequently 8) scales between the lateral line and the dorsal origin, 7 or 8 (rarely 5 or 6) between the pectoral and pelvic fin bases.

Fins. Dorsal with 24 (f.1), 25 (f.10) or 26 (f.9) rays, comprising 15 (f.7) or 16 (f.13) spinous and 9 (f.7), 10 (f.11) or 11 (f.2) branched rays. Anal with 12 or 13 rays, comprising 3 spines and 9 or 10 branched rays. Pectoral fin length very variable, 17·6–27·3 (M = 20·5) per cent of standard length. First pelvic ray slightly produced in adult males. Caudal truncate, scaled on its proximal third to half.

Teeth. The *outer row* in both jaws is composed of unicuspid, moderately stout and slightly to strongly curved teeth. There are 38–52 teeth in the upper jaw, the number showing a very slight positive correlation with standard length. The *inner teeth* are unicuspid and implanted obliquely; there are 3 or 4 (mode), rarely 5 rows in the upper jaw and 2 or 3 (rarely 4 or 5) in the lower jaw.

Coloration. The colours, and particularly the colour pattern, of *H. flavipinnis* are characteristic, even in preserved material.

Live colours. Adult males. Ground colour light silvery-orange dorsally, becoming greyish-black ventrally ; flanks and caudal peduncle crossed by four, broad, and dark bars. the two first becoming continuous with the dark ventral coloration. The snout and interorbital region are olivaceous, the cheek sooty ; anteriorly there is a broad-based, triangular lachrymal stripe. The dorsal fin is light olive-yellow, with dark, red-brown blotches between the soft rays. The caudal and anal fins are yellowish with an overall reddish brown tinge ; the anal ocelli are yellowish-red. Pelvic fins are black. *Adult females* have a light silver-yellow ground colour, but are dark, almost olive on the dorsal surface of the head. The body is crossed by four broad and irregular bars, the two anterior bars reaching further ventrally than the posterior pair. Dorsal fin dark neutral with darker spots on the soft part. The distal two-thirds of the caudal are dark, the proximal part is lighter. Anal fin yellow, the pelvics very light yellow except along the anterior margin where there is a brownish tinge.

Preserved material. Sexual dimorphism is less marked in fixed material, except that the chest and belly of sexually active males are dusky to black, and the entire pelvic fin is black. The ground colour is variable (probably dependent both on the fish's sexual condition if it is a male, and on preservation), from dark pinkish-brown to faintly orange-silver ; four, broad, dark and slightly irregular bars cross the flank and caudal peduncle, the two posterior bars do not extend much below the level of the lower lateral line, the two anterior bars extend (usually) almost to the ventral surface. The first and second bars are joined by a narrow midlateral stripe, as are the third and fourth bars ; from the latter there is often a posteriorly directed tongue which extends onto the base of the caudal fin. In many specimens the first three vertical bars extend upward onto the base of the dorsal fin and even part way up the fin membrane. The lachrymal stripe is a very characteristic feature, being a broad-based triangle which extends over almost half the cheek ; the remainder of the cheek is sometimes very dusky (perhaps a reflection of the fish's emotional state ?). The posterior margin of the preoperculum is outlined by a narrow vertical bar which merges dorsally with an anterior prolongation of the horizontal bar connecting the first two transverse flank bars. Often an obliquely directed, chevron-shaped interocular band is visible on the nape. The dorsal, caudal and anal fins are orange-pink or neutral, the soft dorsal maculate. In males there are three to five dead-white ocelli on the anal ; if there are more than three ocelli, they are usually arranged in two rows, the upper containing the greater number. In both sexes the pelvic spine and the first to third rays are blackish ; in sexually active males the entire fin is black.

Distribution. Lake Victoria.

Ecology. Habitat. It is impossible to generalize on the habitat preferences of *H. flavipinnis* from the data available. Specimens have been obtained from exposed sandy beaches, sheltered bays and gulfs where the substrate is of organic mud, and from rocky shelves running out from exposed islands. The only common factor in each locality has been the depth of water : less than twenty-five feet. Since *H. flavipinnis* is not abundant, nothing further could be determined.

Food. Ten of the eighteen specimens examined had food in the gut. Of these specimens, eight had fed exclusively on fishes and two contained fragments of Ephemeroptera in the stomach. The fish remains were too macerated for accurate identification beyond familial or generic levels; in five guts the fish were identified as cichlids (and, since the scales were ctenoid, probably *Haplochromis*). No specimen had the remains of more than one fish in its stomach. The length of the prey species is estimated as ranging from 25–35 mm. S.L.

Breeding. No data are available on breeding habits. The sexes reach maturity at *ca.* 115 mm. S.L. and there is an indication that males may grow to a larger size than females.

Affinities. The coloration, and particularly the colour pattern, of *H. flavipinnis* serves as an immediate diagnostic character. It is not repeated in other Victoria *Haplochromis*. However, the transverse baring of *H. flavipinnis* is related to that of *H. percoides* (see p. 190). These two species are also related anatomically, although the jaws of *H. flavipinnis* slope more steeply and the caudal fin is obliquely truncated. Osteologically, the species show great similarity, particularly with regard to neurocranial architecture. The few differences (apart from coloration and caudal fin shape) which separate *H. flavipinnis* from *H. percoides* could be attributed to the larger adult size of *H. flavipinnis*, especially since the differences involve characters which often show differential growth rates.

Haplochromis flavipinnis also resembles *H. cavifrons*, at least anatomically, but again there are differences in coloration, in this case more marked than those between *H. percoides* and *H. flavipinnis*. The morphometric differences (e.g. the longer lower jaw and wider interorbital of *H. cavifrons*) are not correlated with size.

Phylogenetically, it seems probable that *H. flavipinnis* was derived from a small, inshore species like the present *H. percoides*, and that besides differences in coloration the evolution involved a differential growth of certain mouth-parts and associated neurocranial areas. That *H. flavipinnis* is less rigidly confined to a definite habitat may suggest that it is in the process of becoming an off-shore, mud-bottom species.

Study material and distribution records

Museum and Reg. No.	Locality	Collector
	Uganda	
B.M. (N.H.).—1906.5.30.308 (Holotype)	Bugonga	Degen
,, 1962.3.2.143–4	Ramafuta Isl. (Buvuma Channel)	E.A.F.R.O.
,, 1962.3.2.145–6	Jinja (below golf course)	,,
,, 1962.3.2.149–50	Jinja Pier	,,
,, 1962.3.2.154–5	Napoleon Gulf, near Jinja	,,
,, 1962.3.2.147	Ekunu Bay	,,
,, 1962.3.2.151–3	Entebbe Harbour	,,
,, 1962.3.2.156–8	Pilkington Bay	,,
	Tanganyika	
,, 1962.3.2.148	Majita Beach	,,
	Lake Victoria, Locality Unknown	
,, 1962.3.2.159		,,

Scales ctenoid; lateral line with 30 (f.2), 31 (2), 32 (f.13) or 33 (f.6) scales, cheek with 3–5 (mode 4) rows. Six or 7 (rarely 8) scales between the lateral line and the dorsal origin, 6–8 between the pectoral and pelvic fin bases.

Fins. Dorsal with 24 (f.8), 25 (f.14) or 26 (f.1) rays, comprising 14 (f.1), 15 (f.19) or 16 (f.13) spines and 8 (f.3), 9 (f.13) or 10 (f.7) branched rays. Anal with 11 (f.3), 12 (f.18) or 13 (f.2) rays, comprising 3 spines and 8–10 branched rays. Pectoral 20·7–25·0 (M = 22·7) per cent of standard length. Pelvics with the first ray moderately produced in adult males. Caudal truncate, scaled on its proximal two-thirds, or, occasionally, as much as four-fifths.

Teeth. The *outer row* of teeth in fishes >100 mm. S.L. is composed entirely of stout, strongly curved unicuspids, but smaller fishes have an admixture of weakly bicuspid and unicuspid teeth. There are 50–100 (M = 80) teeth in the upper outer row. Teeth

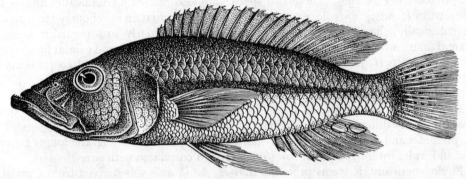

Fig. 15. *Haplochromis macrognathus*, holotype, .66 × N.S. (Drawn by Lavinia Beard.)

of the *inner series* in most fishes >110 mm. S.L. (but also in some between 95 and 100 mm.) are unicuspid; in smaller fishes there is usually a mixture of uni- and tricuspid teeth, sometimes with one type predominating in one jaw but not in the other. All the inner teeth are implanted obliquely so that in the upper jaw their crowns lie almost horizontally. The number of inner tooth rows in the upper jaw shows some positive correlation with the fish's size, there being 3–6 (mode 5) for the size range available. No such correlation exists in the lower jaw where there are 2 or 3 (less commonly 1) rows.

Osteology. The *neurocranium* is almost identical with that of *H. mento*, differing only in being slightly shallower and narrower, and in having the dorsal profile somewhat flatter. The premaxilla shows an extension of the trend seen in *H. mento* in that the median dentigerous section is greatly expanded anteroposteriorly and is almost beak-like.

Vertebrae. Twenty-nine or 30, comprising 13 or 14 precaudal and 16 or 17 caudal elements (5 specimens examined).

Lower pharyngeal bone. The outline is somewhat variable and is correlated with the head-shape of the fish. The dentigerous surface is triangular and usually equilateral but sometimes it is longer than broad, or, less frequently, broader than long.

The teeth are relatively fine (the two median rows somewhat coarser), are cuspidate and are densely crowded in 22 to 24 rows.

Coloration. The colour pattern of preserved females is a salient feature of the species, and is rare amongst Victoria *Haplochromis*. Nevertheless, it only represents the complete manifestation of the basic patterns shown in part by many species.

The colours of *live males* are unknown; *females* are silvery-grey with a distinct midlateral black stripe on the flank and caudal peduncle, ending in a black blotch on the caudal fin near its base; a very faint dark stripe runs slightly above the upper lateral line and there is a dark mental spot on the lower jaw. The chest is yellowish but the branchiostegal membrane is a dark saffron yellow. The dorsal fin is neutral but with dark spots and streaks between the rays. The upper half of the caudal is neutral and darkly maculate, the lower half is yellow. The anal fin is faintly yellow and the pelvics are saffron yellow.

Preserved material. Females. The ground colour is grey-brown dorsally, silver on the flanks and belly. A dark midlateral stripe runs from the eye to the caudal fin origin where it expands into a large blotch on the basal part of the fin; the line is narrow on the head but widens shortly after it passes the posterior margin of the operculum. A second dark stripe runs slightly above the upper lateral line, from a point above the posterior opercular margin to the origin, or slightly beyond the origin of the caudal peduncle. At the base of the dorsal fin there is a series of dark spots which extend slightly upwards onto the fin membrane. There is usually a dark lachrymal spot, and always a dark blotch at the tip of the mandible. All fins, except the pelvics, are colourless although the dorsal and caudal may be darkly maculate. The pelvics are greyish, being darkest along their anterior margins.

Adult males. The ground colour is dark orange-brown. The stripes and other markings are as in females, with the addition of five or six rather faint transverse bars linking the two lateral stripes and extending a little below the lower stripe. The snout, lower jaw, branchiostegal membrane, lower limb of the preoperculum, lower part of the operculum, the chest and belly are black. The dorsal fin is sooty, the caudal dark and maculate. The anal has a black basal stripe which spreads diffusely out over most of the fin except for its distal margin which is yellowish-brown; there are as many as six, irregularly arranged, blackish ocelli. The pelvics are black. Less sexually active males have a similar coloration but the pattern is not so intense, especially with regard to the black antero-ventral parts of the head and body.

Distribution. Lake Victoria.

Ecology. Habitat. The species has been recorded from three different habitats, viz.: sheltered gulfs with a hard substrate; sheltered bays with a mud substrate; and exposed, sandy beaches. From catch records it seems that *H. macrognathus* is commoner in areas where the bottom is hard than in places where it is muddy, and that the species does not occur at depths greater than twenty-five feet.

Food. Only five of the twenty specimens examined had ingested material in the stomach or intestines. Four of the five had fed exclusively on fish, and the fifth contained only fragments of plant tissue with attached colonies of blue-green algae (see also p. 144). The fish remains were identifiable in two cases as *Haplochromis*, in one as Cichlidae and in the fourth were too fragmentary for identification.

Breeding. *Haplochromis macrognathus* is a female mouth brooder. Most fishes <125 mm. S.L. are immature (one male 121 mm. S.L. was classified as " starting ") ; there is no sexual dimorphism in the maximum size attained.

Affinities. Anatomically, *Haplochromis macrognathus* belongs to the *H. mentogowersi-estor* complex; but it is immediately distinguished from these species by the marked compression of the head and its almost beaked premaxillaries. The nearest living relative is, perhaps, *H. mandibularis* which although anatomically somewhat less specialized than *H. mento* shows certain characters (head compression, shape of mouth, coloration) strongly suggestive of the *H. macrognathus* condition.

Study material and distribution records

Museum and Reg. No.	Locality	Collector
	Uganda	
B.M. (N.H.).—1906.5.30.260 (Holotype)	Bunjako	Degen
,, 1962.3.2.420	Ekunu Bay	E.A.F.R.O.
,, 1962.3.2.421–5	Jinja Pier	,,
,, 1962.3.2.431–3	Jinja, below golf course	,,
,, 1962.3.2.434	Fielding Bay	,,
,, 1962.3.2.435	Pilkington Bay	,,
,, 1962.3.2.436	Entebbe, near Buganga	,,
,, 1962.3.2.437	Between Vempita and Busiri Isls. (Buvuma Channel)	,,
	Kenya	
,, 1962.3.2.418–9	Trawl S. of Port Southby	,,
	Tanganyika	
,, 1962.3.2.416–7	Beach near Majita	,,
,, 1962.3.2.426–430	Majita Beach	,,

Haplochromis pellegrini Regan 1922

Text-fig. 16

Paratilapia prognatha (part) : Blgr., 1915, *Cat. Afr. Fish.*, **3**, 333.
Haplochromis pellegrini Regan, 1922, *Proc. zool. Soc. Lond.*, 185, fig. 11.

Lectotype. A specimen 104·0 mm. S.L., from Entebbe, B.M. (N.H.) Reg. No. 1906.5.30.253.

Description based on twenty-five specimens (71–104 mm. S.L.) including the lecto- and paratypes.

Depth of body 29·0–33·6 (M = 31·1) per cent of standard length, length of head 34·6–37·9 (M = 36·3) per cent. Dorsal head profile sloping at *ca.* 30°, somewhat concave above the eyes, the premaxillary pedicels fairly prominent.

Depth of preorbital 15·4–22·2 (M = 17·7) per cent of head, least interorbital width 18·2–24·0 (M = 21·0) per cent. Snout longer than broad, its length 30·8–36·0 (M = 34·0) per cent of head ; diameter of eye 20·5–27·0 (M = 23·6) per cent, depth of cheek 21·4–27·0 (M = 24·3).

Caudal peduncle length 13·2–17·8 (M = 15·4) per cent of standard length, 1·1–1·5 (mode 1·3) times as long as deep.

Mouth moderately oblique ; lower jaw sometimes projecting, its length 42·3–51·5 (M = 46·8) per cent of head, 1·7–2·4 (mode 2·0) times as long as broad. Posterior tip of the maxilla not reaching the vertical through the anterior orbital margin, but approaching this point in a few specimens. The median dentigerous part of the premaxilla slightly expanded anteroposteriorly.

Gill rakers moderately coarse, 7 (f.3), 8 (f.8), 9 (f.12) or 10 (f.1) on the lower part of the first arch, the lowermost 1–3 rakers reduced.

Scales ctenoid ; lateral line with 30 (f.3), 31 (f.3), 32 (f.12) or 33 (f.7) scales, cheek with 3 or 4 (mode) rows ; 6–8 scales between the lateral line and the dorsal fin origin, 6 or 7 (rarely 8) between the pectoral and pelvic fin bases.

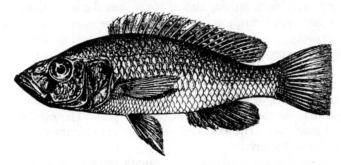

FIG. 16. *Haplochromis pellegrini*, lectotype, ·7 × N.S. (From Regan, *Proc. zool. Soc. Lond.*)

Fins. Dorsal with 23 (f.1), 24 (f.6), 25 (f.16) or 26 (f.2) rays, comprising 14 (f.5), 15 (f.13) or 16 (f.7) spinous and 9 (f.9), 10 (f.15) or 11 (f.1) branched rays. Anal with 3 spines and 8–10 (mode 9) branched rays. Pectoral fin 19·4–25·3 (M = 21·3) per cent of standard length. First ray of pelvic fin slightly produced in both sexes. Caudal truncate or subtruncate, scaled on its basal two-thirds.

Teeth. In most fishes <85 mm. S.L. the *outer row* is composed of weakly bicuspid and unicuspid teeth or only weakly bicuspids ; in a few specimens this row may contain only unicuspids. Fishes >85 mm. have only unicuspid teeth in the outer row. All the outer teeth are slightly curved. The number of teeth in the upper jaw (44–62) shows slight positive correlation with standard length.

In the majority of specimens, all the *inner* teeth are tricuspid, but in some there is an admixture of tricuspids and weakly tricuspids. The inner teeth are arranged in 2 or 3 (rarely 4) rows in the upper jaw and 2 (rarely 1) in the lower.

Osteology. The neurocranium is similar to that of *H. bartoni* and *H. percoides* and does not closely approach the *H. guiarti* type. In other words it is of a type fairly advanced along the " extreme " predator line of evolution.

Lower pharyngeal bone triangular, its dentigerous surface broader than long. The teeth are fine and cuspidate, and are arranged in 22–24 rows.

Vertebrae. Twenty-eight or 29 (mode 28) comprising 13 precaudal and 15 (f.5) or 16 (f.3) caudal elements.

Coloration. Live females have a dark chocolate-brown ground colour, lighter on the chest and belly. All the fins are blackish-brown. The live colours of *males* are unknown.

Preserved material. Females. The ground colour is brownish-silver, grey on the upper surface of the head and above the upper lateral line. There is a distinct (if sometimes faint) lachrymal stripe and, in some specimens, traces of four incomplete and very faint dark vertical bars on the flanks; the two anterior bars are crossed by a midlateral stripe originating behind the operculum but not extending beyond the first pair of bars. Dorsal fin greyish-black, as is the greater part of the caudal except for a narrow chevron of yellow-grey which extends forwards from the posterior margin. Anal greyish, pelvics yellow-brown with a sooty area at their base.

Males are uniformly dark brown, almost black, but the branchiostegal membrane is lighter; the lachrymal stripe is very intense. The dorsal, caudal and anal fins are dark grey to sooty, the anal with two small elongate and dark ocelli. The pelvics are uniformly black except for a lighter innermost ray.

These notes are from specimens fixed in formol and preserved in alcohol. The two type specimens (both females) were fixed in alcohol and are light brown above the upper lateral line and bright silver on the flanks and belly. The dorsal fin is hyaline, with brown maculae on the basal third of the soft part. The caudal is also hyaline but is uniformly maculate. The anal and pelvic fins are colourless.

Distribution. Lake Victoria.

Ecology. Habitat. The material examined came from five localities all of which are relatively exposed sandy beaches near dense stands of submerged and emergent plants. The types came from Entebbe but no further details are known. If it is assumed that they were caught in native fishing gear (and, with regard to the date I think it reasonable to make this assumption) then the chances are greatly in favour of their having come from a similar habitat. Because of its small adult size, I cannot be certain that *H. pellegrini* does not occur in other habitats where fishing was carried out with large meshed gear (e.g. in many areas with a soft mud substrate). However, no specimens were caught in small-mesh trawls fished in these places.

Food. Of the thirty-five specimens examined, twenty-five had ingested material in the gut. In three fishes, this material was an unidentifiable sludge, in one there was sludge plus insect remains; five had fed only on insects (principally Ephemeropteran pupae and larval Diptera) one on insects and small fishes, and fifteen had fed exclusively on fishes. The fish remains were very fragmentary but in most cases were identifiable as cichlids. The size range of the prey is 10–15 mm., that is immediately post-larval fishes. From these records it seems that *H. pellegrini* takes over from those predators which feed on embryo and larval cichlids (Greenwood, 1959). There appears to be some correlation between the diet of *H. pellegrini* and its habitat preference, since it occurs near areas where brooding cichlids congregate (e.g. reed beds).

Breeding. The species is a female mouth brooder. All the specimens seen are adults and both sexes reach the same maximum size.

Affinities. The body-form of *H. pellegrini* is similar to that of the *H. mento-H. gowersi* group and also *H. percoides*. The species is predominantly piscivorous but its small adult size is unusual for this trophic group. Anatomically it belongs to the group of moderately specialized fish-eaters, its neurocranium is fairly well advanced towards the grade of specialization shown by *H. estor* and *H. mento*, and again there is a strong resemblance to *H. percoides*. On skull and body form, *H. pellegrini* must be allocated to the group of moderately specialized predators, but there are no characters which allow one to assign it unequivocally to the *H. estor* complex or to the *H. percoides-flavipinnis* group.

Study material and distribution records

Museum and Reg. No.	Locality	Collector
	Uganda	
B.M. (N.H.).—1906.5.30.253 (Lectotype)	Entebbe	Degen
,, 1906.5.30.254 (Paratype)	Entebbe	Degen
,, 1962.3.2.510–2	Grant Bay	E.A.F.R.O.
,, 1962.3.2.513–5	Beach in Buvuma Channel	,,
,, 1962.3.2.509	Beach near Nasu Point	,,
,, 1962.3.2.516–44	Karenia (near Jinja)	,,

Haplochromis percoides Blgr. 1915

Text-figs. 17 and 25

Haplochromis percoides (part) Blgr., 1915, *Cat. Afr. Fish.*, **3**, 296, fig. 201 (the two syntypes only).
Haplochromis flavipinnis (part) : Regan, 1922, *Proc. zool. Soc. Lond.*, 172 (the two syntypes of *H. percoides* only).

Lectotype. A male 79·0 mm. S.L., from Entebbe, B.M. (N.H.) Reg. No. 1906.5.30.313.

Description based on twenty specimens 67–93 mm. S.L., including the lecto- and paratype.

Depth of body 29·5–33·0 (M = 30·9) per cent of standard length, length of head 34·3–38·1 (M = 35·8) per cent. Dorsal head profile concave, the premaxillary pedicels prominent.

Preorbital depth 16·7–19·5 (M = 17·7) per cent of head, least interorbital width 19·2–22·8 (M = 20·7). Snout slightly longer than broad or as long as broad, its length 31·4–35·0 (M = 34·7) per cent of head ; diameter of eye 21·9–26·6 (M = 24·2) per cent, depth of cheek 21·4–27·4 (M = 25·0).

Caudal peduncle length 12·9–16·4 (M = 15·3) per cent of standard length, 1·1–1·5 (modal range 1·4–1·5) times as long as deep.

Mouth slightly oblique, the jaws equal anteriorly or, more commonly, the lower projecting slightly. Length of lower jaw 42·3–49·2 (M = 45·2) per cent of head, 1·7–2·2 times as long as broad. Posterior tip of the maxilla close to or, occasionally, reaching the vertical through the anterior orbital margin.

Gill rakers usually slender, but in some specimens rather coarse and flat; 7–9 (mode 8) on the lower part of the first arch.

Scales ctenoid; lateral line with 30 (f.2), 31 (f.7), 32 (f.7), 33 (f.2) or 34 (f.1) scales, cheek with 3 or 4 (mode) usually imbricating but occasionally irregular rows. Six to 8 (mode 7) scales between the lateral line and the dorsal origin, 7 or 8 (rarely 6 or 9) between the pectoral and pelvic fin bases.

Fins. Dorsal with 22 (f.1), 23 (f.1), 24 (f.10), 25 (f.7) or 26 (f.1) rays, comprising 14 (f.2), 15 (f.12) or 16 (f.6) spines and 8 (f.4), 9 (f.10) or 10 (f.6) branched rays. Anal with 12 or 13 rays comprising 3 spines and 9 or 10 branched rays. Pectoral fin 22·4–26·6 (M = 24·4) per cent of standard length. First pelvic ray but slightly produced and then only in males. Caudal truncate, about the proximal half covered with small scales (*cf. H. flavipinnis*).

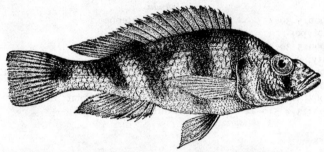

FIG. 17. *Haplochromis percoides*, lectotype ·94 × N.S. (From Boulenger, *Fish. Nile*).

Teeth in the *outer series* are predominantly unicuspid and slightly curved, but in some specimens of all sizes the lateral teeth are bi- or weakly bicuspid; there are 44–56 (mean 50) outer teeth in the upper jaw. Except in one fish (91 mm. S.L.) the *inner series* of both jaws are composed of tricuspid teeth; in the exceptional specimen all the inner teeth are unicuspid. There are 2 or 3 rows in the upper and 1 or 2 in the lower jaw.

Osteology. In its general form, the neurocranium of *H. percoides* is near that of *H. pellegrini*. The preotic face is long (67% of basal length) and its dorsal surface slopes at a gentle angle (*ca* 25°). The most noticeable difference is in the more depressed orbital region of the *H. percoides* neurocranium (reflected in the gentler slope of the dorsal surface). There is also a close resemblance between the neurocranium of *H. percoides* and that of *H. flavipinnis*. Again, the orbital region in *H. percoides* is more depressed.

The *lower pharyngeal bone* is triangular, its dentigerous surface broader than long. The teeth are slender, compressed and cuspidate, albeit weakly so in some of the anterior teeth. There are 18–22 rows of teeth.

Vertebrae. Twenty-nine, comprising 13 and 16 (f.7) or 12 and 17 (f.1) precaudal and caudal elements.

Coloration. The live colours of *H. percoides* are unknown, but judging from the well-marked vertical bands in all preserved specimens these must be conspicuous features in life. No marked sexual dimorphism is shown in the coloration of *preserved specimens*, except that the pelvic fins in males are black. The ground colour varies from silvery to dark brown ; there are five or six dark and broad transverse bands (extending from the dorsal fin base almost to the ventral midline) on the flank and caudal peduncle. Each of the two first and the two last bands are often linked by a narrow midlateral bar ; the last band frequently has a midlateral tongue extending onto the caudal fin base. Usually there are two transverse and parallel bars across the snout ; a lachrymal stripe of variable intensity is always present. The dorsal, caudal and anal fins are yellowish-orange and without other markings. The pectorals and pelvics are yellowish the latter variously dusky in males. Generally in males there is only one, ill-defined greyish blotch in the position of an anal ocellus.

Distribution. Lake Victoria.

Ecology. Habitat. The species is apparently restricted to shallow water over exposed or partly exposed sandy beaches. It is nowhere common but seems to be widely distributed around the lakeshore.

Food. Fourteen of the twenty specimens examined contained food, and all the identifiable remains were of fishes. In eight specimens the fish were small *Haplochromis* (*ca* 12 mm. S.L.), and in three the identification could not be carried beyond that of Cichlidae. In all the remaining specimens, the fish remains were too fragmentary for further identification. It was noticed that in five specimens the stomach contained four or five post-larval fishes of a similar size, thus suggesting that *H. percoides* may prey on shoals of young cichlids.

Breeding. No information is available on the breeding habits of this species. The sample studied consisted mainly of males and the only female showing signs of gonadial activity was a fish 78 mm. S.L.

Affinities. Haplochromis percoides occupies a rather isolated position amongst the Lake Victoria species, both with regard to its gross morphology and its colour pattern. The most closely related species is *H. flavipinnis*. The colour pattern of both species is virtually identical and both show a very concave upper head profile with pronounced nuchal hump. In *H. flavipinnis*, however, the jaws are inclined more steeply and the body is deeper. In details of neurocranial architecture, *H. percoides* shows a strong resemblance to *H. pellegrini* and it is conceivable that the latter species represents a stem type from which *H. percoides* evolved. There is little doubt that *H. flavipinnis* arose from an *H. percoides*-like ancestor.

Study material and distribution records

Museum and Reg. No.	Locality	Collector
	Uganda	
B.M. (N.H.).—1906.5.30.313 (Holotype)	Entebbe	Degen
,, 1962.3.2.345	Beach near Nasu Point	E.A.F.R.O.
,, 1962.3.2.346–53	Entebbe (airport beach)	,,
,, 1962.3.2.337–41	Entebbe Harbour	,,
	Tanganyika	
,, 1962.3.2.342–4	Beach near Majita	,,
,, 1962.3.2.332	Majita Beach	,,

Haplochromis flavipinnis (Blgr.) 1906
Text-figs. 18 and 19

Pelmatochromis flavipinnis Blgr., 1906, *Ann. Mag. nat. Hist.*, (7), **17**, 441 ; Idem, 1907, *Fish. Nile*, 448, pl. 89, fig. 3 ; Idem, 1915, *Cat. Afr. Fish.*, **3**, 418, fig. 286.

Haplochromis flavipinnis (part) : Regan, 1922, *Proc. zool. Soc. Londn.*, 172 (the type specimen only).

Holotype. A specimen 118·0 mm. S.L., from Bugonga ; B.M. (N.H.) Reg. No. 1906.5.30.308.

Description based on nineteen specimens (including the holotype), 69–156 mm. S.L.

Depth of body increasing allometrically with standard length, 30·8–39·2 per cent of the former. Length of head 33·6–38·0 (M = 35·8) per cent of standard length, dorsal profile concave, the concavity becoming more marked with increasing size.

Preorbital depth 16·7–21·8 (M = 19·5) per cent of head, least interorbital width 19·4–23·6 (M = 21·1) per cent. Snout as long as broad in most specimens but occasionally somewhat broader than long in fishes >130 mm. S.L. ; its length 29·2–36·4

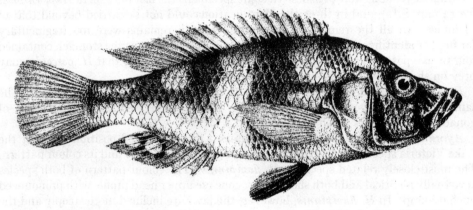

FIG. 18. *Haplochromis flavipinnis*, holotype, ·9 × N.S. (From Boulenger, *Fish. Nile*.)

(M = 33·3) per cent of head. Diameter of eye 19·0–25·0 (M = 21·8) per cent, depth of cheek (showing slight positive allometry) 25·0–32·7 per cent.

Caudal peduncle length 13·8–18·6 (M = 16·0) per cent of standard length, 1·2–1·6 (modal range 1·2–1·4) times as long as deep.

Mouth oblique (sloping at 40°–50°), the lower jaw projecting slightly or, infrequently, the jaws equal anteriorly. Length of lower jaw 43·8–51·0 (M = 47·8) per cent of head, 1·6–2·0 times as long as broad. Posterior tip of the maxilla reaching or nearly reaching the vertical through the anterior orbital margin.

Gill rakers short and stout in fishes <90 mm. S.L., stout and flattened in larger individuals, 8 (mode) or 9, rarely 7 or 10, on the lower part of the first arch.

Scales ctenoid, the lateral line with 30 (f.2), 31 (f.5) 32 (f.10) or 33 (f.2) scales, cheek with 4 (mode) or 5, rarely 6 rows ; 7 or 7½ (less frequently 8) scales between the lateral line and the dorsal origin, 7 or 8 (less frequently 6) between the pectoral and pelvic fin bases.

Fins. Dorsal with 23 (f.1), 24 (f.12) or 25 (f.6) rays, comprising 15 (f.11) or 16 (f.8) spines and 8 (f.3) or 9 (f.16) branched rays. Anal with 11 (f.8) or 12 (f.11) rays, comprising 3 spines and 8 or 9 rays. Pectoral fin 23·2–29·6 (M = 25·6) per cent of standard length. Caudal fin obliquely truncate, the posterior margin sloping downwards and forwards to meet the somewhat curved ventral margin. In outline this fin resembles that of *H. plagiostoma* (see p. 200) but the lower rays do not have the appearance of regenerated structures. The caudal is truncate in the two smallest fishes examined (69 and 70 mm. S.L.) but is slightly oblique in a specimen 86 mm. S.L. At all sizes, the caudal is scaled over about two-thirds to four-fifths of its length, an unusually large area for a Lake Victoria species. The first pelvic ray is produced in adult males.

Teeth. The *outer row* of teeth in both jaws is composed of relatively slender, moderately curved unicuspids; there are 44–66 (M = 55) teeth in the upper jaw, the number showing a very slight positive correlation with size in fishes <120 mm. S.L.

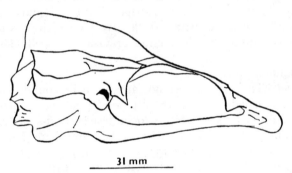

Fig. 19. *Haplochromis flavipinnis*, neurocranium.

The *inner rows* of teeth, in most individuals, are composed of unicuspids but in some fishes <100 mm. S.L. (and even in one of 150 mm.) these rows may contain a mixture of tricuspids, weakly tricuspids and unicuspids. There are 2 or 3 (less commonly 4) rows of teeth in the upper jaw and 1 or 2 rows in the lower.

Osteology. The neurocranium of *H. flavipinnis* resembles that of *H. percoides* in outline except for the shape of the supraoccipital and the points of divergence (slope of preorbital face, height in the interorbital region) are slight; as in *H. percoides*, the neurocranium has a long preotic part (67% of basal length).

Vertebrae. Twenty-nine or 30, comprising 13 and 16 (f.6), 12 and 17 (f.2) or 14 and 16 (f.1) precaudal and caudal elements.

Lower pharyngeal bone triangular, the dentigerous surface slightly broader than long. The teeth are bicuspid and arranged in 18–20 rows, those of the two median rows are somewhat enlarged. In small fishes the teeth are relatively slender but are coarser in large individuals.

Osteology. The *neurocranium* of *H. gowersi* is virtually identical with that of *H. mento* and is thus similar to the neurocranium of *H. estor*.

Vertebrae. Twenty-nine or 30 (mode) in the six specimens examined, comprising 13 or 14 precaudal and 16 or 17 caudal elements.

Lower pharyngeal bone triangular, the dentigerous surface longer than broad, or, less frequently, as long as broad. The teeth are moderately stout and cuspidate, and are arranged in 18–20 slightly irregular rows.

Coloration is unknown in life. *Preserved females* have a silvery ground colour, darker dorsally and on the head. A prominent and fairly broad midlateral stripe runs from the opercular margin to the caudal origin; a fainter, narrower and sometimes partly interrupted stripe runs above the upper lateral line. In some specimens there is a faint lachrymal stripe. All fins are hyaline, the soft dorsal and the caudal are densely maculate and the former has faintly sooty lappets. *Immature males* are coloured like females except that the pelvics are darker. *Sexually active males* have a dark brown ground colour, becoming sooty from below the lateral line, with a faint golden sheen on the flanks and lower part of the operculum. Midlateral and dorsolateral stripes are developed as in females. The branchiostegal membrane is black and there is a faint, dark lachrymal stripe. The dorsal and caudal fins are dark, the anal is yellowish, sometimes with a dusky overlay which does not extend to the distal margin. There are one to three large, greyish anal ocelli. The pelvic fins are black.

Distribution. Lake Victoria.

Ecology. Habitat. The species does not appear to be confined to any particular substrate, since it is caught over both soft (mud) and hard bottoms (sand, rock, shingle); it is found in exposed habitats (open beaches, off-shore islands) as well as in sheltered bays and gulfs. In no locality was the water more than 20 feet deep.

Breeding. One specimen (a female 153 mm. S.L.), from a sandy beach near the Kagera river mouth, had larvae in the buccal cavity. Sexual maturity is reached at a length of about 150 mm.

Food. All the fifteen specimens with food in the stomach or intestines had fed on fishes, but in two there was also a quantity of macerated plant tissue. The fish remains were identified as: *Haplochromis* (f.9), Cichlidae (f.4); unidentifiable (f.2).

Affinities. As Trewavas (1928) first suggested, *H. gowersi* has affinities with *H. mento*. It differs from *H. mento* in having a deeper cheek (mean depth 29·5% of head *cf.* 23·6%), a longer lower jaw (M = 52% of head *cf.* 46·8%) and a more oblique mouth.

Study material and distribution records

Museum and Reg. No.	Locality	Collector
	Uganda	
B.M. (N.H.).—1928.5.24.478 (Lectotype)	Entebbe	Graham
B.M. (N.H.).—1928.5.24.479 (Paratype)	Entebbe	Graham
„ 1962.3.2.392	Williams Bay	E.A.F.R.O.
„ 1962.3.2.393–5	Beach near Nasu Point	„
„ 1962.3.2.396–7	Karenia (near Jinja)	„

Museum and Reg. No.	Locality	Collector
	Uganda (continued)	
B.M. (N.H.).—1962.3.2.398	Thruston Bay	E.A.F.R.O.
,, 1962.3.2.399–400	Kagera Port	,,
,, 1962.3.2.401–3	Buka Bay	,,
,, 1962.3.2.404–5	Bukassa	,,
,, 1962.3.2.406	Pilkington Bay	,,
,, 1962.3.2.407	Ramafuta Isl. (Buvuma Channel)	,,
,, 1962.3.2.409–415	Jinja Pier	,,
	Lake Victoria, Locality Unknown	
,, 1962.3.2.408		,,

Haplochromis macrognathus Regan 1922

Text-figs. 15 and 25

Paratilapia longirostris (part) : Blgr., 1915, *Cat. Afr. Fish.*, **3**, 332.
Haplochromis macrognathus Regan, 1922, *Proc. zool. Soc. London*, 182 ; Pl. 2, fig. 2.

Holotype. A specimen 160·0 mm. S.L.; from Bunjako. B.M. (N.H.) Reg. No. 1906.5.30.260.

Description based on twenty-three specimens (including the holotype) 80–174 mm. S.L.

Depth of body 26·6–33·3 (M = 30·7) per cent of standard length, length of head 33·8–41·4 (M = 38·2) per cent. Head noticeably compressed, its breadth (as measured immediately anterior to the orbital margin) 20·4–26·0 (M = 22·9) per cent of its length. Dorsal head profile straight or slightly curved, sloping at 25°–30° ; premaxillary pedicels moderately prominent and breaking the slope of the head as seen in profile.

Depth of preorbital 18·2–23·3 (M = 21·3) per cent of head, least interorbital width 16·5–22·2 (M = 18·6) per cent. Snout 1·5–2·2 times as long as broad (broadest in smaller fishes ; 1·5–1·6 times), its length 33·3–44·2 (M = 39·0) per cent of head, but with indications of weak positive allometry. Diameter of eye 16·1–24·1 (M = 18·5) per cent of head, slightly greater than, or equal to the preorbital depth in fishes 80–104 mm. S.L., less than the preorbital in larger individuals. Depth of cheek 24·2–31·2 (M = 27·0) per cent.

Caudal peduncle length 13·0–20·0 (M = 15·0) per cent of standard length, 1·1–1·8 (mode 1·4) times as long as deep.

Lower jaw projecting beyond the upper, variable in extent but generally more prominent in larger fishes, its length 48·1–58·8 (M = 54·0) per cent of head and 2·7–3·5 times as long as broad. Mouth moderately oblique, the lower jaw sloping at 25°–30° ; lips slightly thickened. The median dentigerous part of the premaxilla is greatly expanded anteroposteriorly so that in some specimens the premaxilla has a beak-like appearance. The posterior tip of the maxilla rarely reaches the vertical through the anterior orbital margin ; usually it extends to a vertical through the posterior tip of the premaxillary pedicels.

Gill rakers short and stout, 8 (mode) or 9, rarely 10 on the lower part of the first arch, the lowermost 1–3 rakers reduced.

Haplochromis cavifrons (Hilgend.) 1888

Text-figs. 20 and 21

Paratilapia cavifrons Hilgendorf, 1888, *Sitzb. Ges. nat. Fr. Berlin*, 77.
Pelmatochromis cavifrons: Blgr., 1915, *Cat. Afr. Fish.*, **3**, 419, fig. 287.
Haplochromis cavifrons: Regan, 1922, *Proc. zool. Soc. Lond.*, 183.

Description based on forty-one specimens, 108–195 mm. S.L.; I have not examined the type.

Depth of body 31·0–41·5 (M = 35·6) per cent of standard length, length of head 35·4–39·6 (M = 37·3) per cent. Dorsal head profile sloping at an angle of *ca* 30°, its outline sharply broken by the prominent premaxillary pedicels and thus appearing concave.

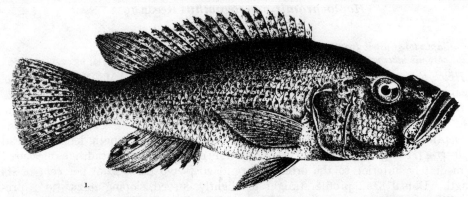

FIG. 20. *Haplochromis cavifrons*, ·75 × N.S. (From Boulenger, *Fish. Nile*.)

Preorbital depth 18·3–21·5 (M = 20·2) per cent of head, least interorbital width 21·5–27·5 (M = 25·4) per cent. Snout as long as broad, or slightly broader than long, occasionally somewhat longer than broad (the latter relationship is found most frequently in fishes <130 mm. S.L.); its length 34·0–40·0 (M = 36·5) per cent of head; diameter of eye 18·8–23·4 (M = 20·3) per cent, depth of cheek 27·2–35·4 (M = 30·2) per cent.

Caudal peduncle length 12·7–17·5 (M = 15·0) per cent of standard length, 1·0–1·4 (mode 1·3) times as long as deep.

Mouth oblique (35°–40°), the jaws equal anteriorly, or the lower projecting slightly. Lower jaw 49·3–60·5 (M = 55·5) per cent of head, its breadth contained 1·8–2·4 (mode 2·0) times in its length in fishes <175 mm. S.L., and 1·4–2·0 (mode 1·6) times in larger fishes. There is usually a broad mental projection developed at the mandibular symphysis. Posterior tip of the maxilla not reaching the vertical through the anterior orbital margin, and usually not covered by the preorbital.

Gill rakers short and stout 7 (rare)–10, mode 8, on the lower part of the first gill arch, the lowermost one or two rakers reduced.

Scales ctenoid; lateral line with 31 (f.6), 32 (f.18), 33 (f.14) or 34 (f.1) scales (in one specimen these scales are very irregularly arranged and give a count of 36); cheek with 4–6 (rarely 3) rows; 8–11 scales between the lateral line and the dorsal origin, 8 or 9 (less frequently 7 or 10) between the pectoral and pelvic fin bases.

Fins. Dorsal with 23 (f.1), 24 (f.14) or 25 (f.25) rays, comprising 15 (f.22) or 16 (f.18) spinous and 8 (f.4), 9 (f.27) or 10 (f.9) branched rays. Anal with 11 or 12 (rarely 13) rays, comprising 3 spines and 8, 9 (mode) or 10 branched rays. Pectoral 21·7–29·3 (M = 24·7) per cent of standard length. First pelvic ray slightly produced in adult males. Caudal truncate, scaled except for the distal quarter.

Teeth. The *outer row* in both jaws is composed of unicuspid, fairly stout and slightly curved teeth. There are 56–74 (M = 63) in this row of the upper jaw.

The *inner rows* in two of the five smallest specimens (108–129 mm. S.L.) are made up of uni- and tricuspid teeth, whilst in the three others and in fishes >130 mm. S.L. the inner teeth are all unicuspids. The inner rows are implanted obliquely, and there are 2–4 (rarely 1) rows in the upper jaw and 1 or 2 (occasionally 3) in the lower.

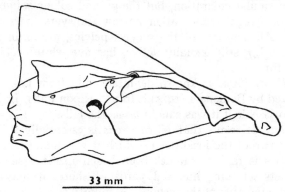

FIG. 21. *Haplochromis cavifrons*, neurocranium.

Osteology. The neurocranium is very similar to that of *H. flavipinnis*; that is, a moderately advanced type with the preorbital part sloping less steeply and the vomer not so markedly curved below the level of the parasphenoid as in the generalized forms.

Vertebrae. Twenty-nine (f.4) or 30 (f.1), comprising 13 precaudal and 16 or 17 caudal elements.

Lower pharyngeal bone triangular, the dentigerous surface as long as broad or slightly broader than long. The pharyngeal teeth are all basically bicuspid, but in larger fishes the minor cusp is vestigial except in the teeth which occupy the dorso-lateral corners of the bone. The teeth are somewhat irregularly arranged in about twenty rows.

Coloration. The densely freckled appearance of *H. cavifrons*, both alive and dead, is very characteristic. *Live colours. Adult males.* Ground colour olive to yellow-brown, shading to silvery below, the head dark brown. Both the head and body are densely speckled with brown spots and blotches. The dorsal fin is muddy yellow, sometimes (? in sexually active fishes) with traces of deep red mottling over its entire length. Caudal muddy-yellow with traces of deep red between the rays. Anal dark on its proximal third to half, muddy-red to pink distally; ocelli yellow. Pelvics are mottled black and pink, becoming blacker in sexually active males. *Females* have a similar coloration except that the pelvic and caudal fins are muddy yellow with dark blotches ; the dorsal is like that of the male. Two aberrantly coloured females were caught near Mwanza (Tanganyika) ; in these fishes the ground colour is a dark olive green shading to light grey-green, the flecks darker green. The dorsal fin is mottled yellow-green, as is the caudal. The anal and pelvic fins are dark yellow with a faint green tinge.

Preserved material. Males. The ground colour is brown, varying from light bronze to a dark blackish-brown (in sexually active adults). The entire body and head are densely peppered with darker spots and blotches. The dorsal fin is dusky brown, marbled with darker pigment, especially on the spinous part. The caudal is brownish and faintly maculate. Anal fin light orange-brown, darker basally ; ocelli faint, represented by one to three whitish spots. The pelvics are black in sexually active fishes, brownish in quiescent individuals.

Females have a similar coloration, but the ground colour is lighter and there are often traces of six, irregular and rather narrow transverse bars on the flanks and caudal peduncle. All fins, except the greyish pelvics, are as in males. One large female (190 mm. S.L.), still sexually active, has five white ocelli arranged in two rows on the anal fin.

Distribution. H. cavifrons is known definitely only from Lake Victoria ; one of the specimens listed by Boulenger (1915) is from " Ripon Falls, Jinja " and may thus be from the Victoria Nile if it was caught *below* the falls.

Ecology. Habitat. It seems that *H. cavifrons* is essentially a " hard substrate " species since only two of the localities in which it was found had a mud substrate. Because the species is rare in catches from beach-operated seines, but relatively common in gill-nets set some hundred yards off-shore, it may be inferred that *H. cavifrons* is not a member of the inshore community as are, for example, *H. guiarti* and *H. bayoni*. No specimens were obtained from water more than 40 feet deep.

Food. Of the twenty-eight fishes (110–190 mm. S.L.) with ingested material in the gut, twenty-two had fed on fishes alone, two on fishes and insects and three on insects only. The insects eaten were : dragonfly larvae, chironomid larvae, and *Povilla* (Ephemeroptera) egg masses and larvae. In the individuals that had fed on fishes, the prey could be identified as follows : *Haplochromis* (f.15) ; Cichlidae of indeterminable genus (f.5) ; *Clarias* (f.1) ; indeterminable (f.1).

Breeding. No information was obtained on the breeding habits of *H. cavifrons*. Sexual maturity is reached in both sexes at lengths between 135 and 155 mm. S.L., and there is no dimorphism in the maximum size attained.

Affinities. No single character or combination of characters gives a clear-cut indication of the phyletic relationship of *H. cavifrons*. The coloration is unique and not easily derived from any other type now existing in the lake. Anatomically, the general suggestion is of relationships with *H. flavipinnis*, particularly with regard to head shape and mouth form. This suggestion is supported by the form of the neurocranium which is nearly identical in the two species. In turn, the same neurocranial characters associate *H. cavifrons* with *H. percoides*, a more generalized species than *H. flavipinnis*. But whereas *H. percoides* and *H. flavipinnis* have a related colour pattern, that of *H. cavifrons* is completely distinct.

It could be argued that similarity in neurocranial architecture is due to functional convergence. Against this it can be shown that similar oblique mouths and large jaws have been evolved in other species (e.g. *H. plagiostoma*) without producing a similar neurocranial shape. In fact, the neurocranium of *H. plagiostoma* is readily distinguished from that of *H. cavifrons*. If on these grounds convergence can be overruled, then *H. cavifrons* is possibly an isolated derivative from the *H. percoides* and *H. flavipinnis* stem.

Study material and distribution records

Museum and Reg. No.	Locality	Collector
	Uganda	
B.M. (N.H.).—1906.5.30.203–7	Bunjako	Degen
,, 1906.5.30.209	Buganga	,,
,, 1911.3.3.35	Jinja, Ripon Falls	Degen
,, 1962.3.2.358–65	Jinja	E.A.F.R.O.
,, 1962.3.2.354–5	Entebbe Harbour	,,
,, 1962.3.2.366–75 and 380–6	Ramafuta Isl. (Buvuma Channel)	,,
,, 1962.3.2.376–9	Pilkington Bay	,,
,, 1962.3.2.391	Bukarra (Sesse Isls.)	,,
	Kenya	
,, 1962.3.2.356	Naia Bay, Kavirondo Gulf	,,
	Tanganyika	
,, 1928.5.24.449	Mazinga Isl.	Graham
,, 1962.3.2.357	Mwanza	,,
	Lake Victoria, Locality Unknown	
,, 1962.3.2.388–90		E.A.F.R.O.

Haplochromis plagiostoma Regan 1922

Text-figs. 22 and 25

Paratilapia longirostris (part): Boulenger, 1915, *Cat. Afr. Fish.*, **3**, 332.
Haplochromis plagiostoma Regan, 1922, *Proc. zool. Soc. Londn.*, 181, Text-fig. 8.

Holotype: a male 113·0 mm. S.L. from Bunjako, Uganda B.M. (N.H.) reg. no. 1906.5.30.261.

Description based on thirty specimens (including the holotype) 69–147 mm. S.L.

Depth of body 32·6–39·0 (M = 36·4) per cent of standard length, length of head 34·0–37·5 (M = 36·0) per cent. Dorsal profile of head gently concave or straight, sloping at an angle of 20°–25°.

Preorbital depth 18·0–21·5 (M = 19·8) per cent of head, least interorbital width 20·6–25·0 (M = 23·4) per cent. Snout slightly broader than long or as long as it is broad, its length 28·2–35·5 (M = 32·5) per cent of the head; depth of cheek 28·0–36·8 (M = 33·0) per cent.

Caudal peduncle 13·2–18·5 (M = 15·8) per cent of standard length and 1·0–1·5 (modal range 1·2–1·3) times as long as deep.

Mouth markedly oblique, directed upwards at an angle of 35°–50° from the horizontal: jaws equal anteriorly or, occasionally, the lower projecting slightly; length of lower jaw 44·0–54·5 (M = 49·2) per cent of head, 1·7–2·4 (mode 2·0) times as long as broad.

Gill rakers stout but occasionally rather slender, 8 or 9 (mode), less frequently 10, on the lower part of the first arch, the lowermost 1–3 rakers reduced.

Scales ctenoid, lateral line with 30 (f.8), 31 (f.5) 32 (f.12) or 33 (f.5) scales, cheek with 4 (mode) or 5, rarely 3, rows of scales; 5–6½ scales between the lateral line and the dorsal origin, 6 or 7 (rarely 5 or 8) between the pectoral and pelvic fin bases.

Fins. Dorsal with 23 (f.4), 24 (f.18) or 25 (f.8) rays, comprising 14 (f.4), 15 (f.19) or 16 (f.7) spinous and 8 (f.6), 9 (f.18) or 10 (f.6) branched rays. Anal with 11, 12 or, rarely, 13 rays comprising 3 spines and 8 (mode), 9 or 10 branched rays. Pectoral fin shorter than the head, 24·6–31·0 (M = 28·1) per cent of standard length. Pelvics with the first branched ray produced, proportionately longer in adult males. The

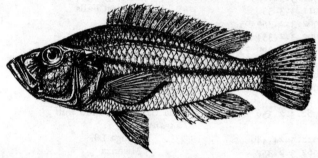

FIG. 22. *Haplochromis plagiostoma*, holotype, ·63 × N.S. (From Regan, *Proc. zool. Soc. Lond.*)

caudal fin of *H. plagiostoma* is obliquely truncate, the posterior margin sloping forward and downwards to meet the upwardly curved ventral margin. The fin shape is identical in both sexes and throughout the size range studied. The lower four or five principal rays in most specimens have the appearance of rays regenerated after damage; commonly only one of the distal branches reaches the fin margin and sometimes neither branch extends much beyond the dichotomy. At present I cannot explain this peculiar fin shape, neither can I suggest whether it is due to some ontogenetic disturbance or to some behavioural trait of the species resulting in damage to the lower fin margin. Occasional specimens of other *Haplochromis* have an obliquely truncate caudal but *H. plagiostoma* and *H. flavipinnis* are the only species in which this peculiarity can be regarded as one of the specific characters.

Teeth. Throughout the size range studied, the *outer row* in both jaws is composed of unicuspid, relatively stout and curved teeth, those of the upper jaw numbering 44–68 (M = 57) and somewhat more curved than the teeth of the lower jaw.

The inner rows are composed of unicuspids in all fishes >120 mm. S.L. and in some specimens in the range 70–120 mm.; other fishes in the latter size group have either all tricuspids or a mixture of uni- and tricuspid teeth. The inner teeth are arranged in 2 (less frequently 1) series in the upper jaw and 1 (less frequently 2) in the lower jaw. All these teeth are implanted obliquely, the angle sometimes approaching the horizontal.

Osteology. The neurocranium of *H. plagiostoma* is of the generalized *H. guiarti* type. Specifically, it differs from *H. guiarti* in having a slightly narrower interorbital distance and a shorter preotic region. Despite the marked difference in jaw angle of the two species (almost horizontal in *H. guiarti* and very oblique in *H. plagiostoma*) the interspecific differences in syncranial proportions and morphology are slight. The steep jaw angle is brought about by several minor changes. For example, the gentler slope of the ethmoid region (over which the premaxillary pedicels slide) allows the dentigerous area of the premaxilla to lie at a greater angle from the horizontal, thus matching the increased slope of the dentary. The latter gains its slope from slight changes in the articular surfaces of the quadrate and angular which allow the dentary to move dorsally through a wider arc than is the case in *H. guiarti*. There is also a compensatory change in neurocranial proportions. The preotic part of the skull in *H. plagiostoma* is relatively shorter than in *H. guiarti*. This shortening of the anterior part of the skull, correlated with a relatively longer dentary and more mobile hinge, allows the dorsal (i.e. most anterior part of the gape) to be moved from a position near the horizontal to one nearer the vertical. The lengthening of the lower jaw permits this change without causing any alteration in the near vertical angle of the hyomandibula.

The *lower pharyngeal bone* is triangular, its dentigerous area as long as broad or slightly longer than broad. The teeth are coarse and cuspidate, those of the two median rows being slightly enlarged; there are 16–20 rows of teeth.

Vertebrae. Twenty-eight (f.1) or 29 (f.9), comprising 12 (f.2) or 13 (f.8) precaudal and 15 (f.1), 16 (f.7) or 17 (f.2) caudal elements.

Coloration in life. Sexually active males are light blue-grey on the flanks and postero-ventral surfaces, smokey grey dorsally and on the chest (which is darker than the back). The head is smokey-grey with lighter blue flecks on the cheek and operculum; the branchiostegal membrane is sooty with a blue overlay. The dorsal fin is dark grey with deep crimson spots between the posterior spines and over the entire soft part; the lappets and the margin of the soft dorsal are deep crimson. The caudal is dark with a crimson margin and spots on the proximal half. The anal is deep red, the ocelli (one to four arranged in a single row) yolk-yellow. The pelvics are black. In *quiescent males* the ground colour is more silvery, the dorsal fin less intensely coloured and the anal is hyaline except for a red flush which is most intense along the margin of the fin.

Females are silver with a faint powder-blue sheen posteriorly on the flanks and faint midlateral and dorsolateral stripes, the latter much shorter than the former.

Colour in preserved material. *Females* are brownish above, silvery below. There is a broad midlateral band, sometimes of irregular thickness, running from behind the operculum to the caudal origin; a distinct but often faint lachrymal stripe is generally present. All the fins are hyaline but the caudal is densely maculate. *Sexually active males* are brownish above shading to a greyish copper on the flanks, and becoming sooty on the chest and belly; the branchiostegal membrane is black. There is often a midlateral stripe but rarely are there any transverse bars; a lachrymal stripe is usually visible. The dorsal, caudal and anal fins are brownish and sometimes faintly dusky. In some specimens there is a narrow black band on the dorsal fin, running obliquely backwards and downwards from the lappet of the last spine to about the middle of the last branched ray. The pelvics are black. The anal ocelli (1–4, usually 2) are opaque and are arranged in a single row.

Distribution. Lake Victoria.

Ecology. Habitat. The species has been caught over both soft (organic mud) and hard (sand and shingle) substrates. Common features of all localities are their sheltered nature (gulfs and bays) and relatively shallow depths (never more than 30 feet).

Food. Although *H. plagiostoma* is one of the commoner species, little is known about its feeding habits. The reason for this is the high frequency of fishes without food in the gut on capture. Those specimens with food (twenty-two in all) are exclusively piscivorous, the prey species being predominantly *Haplochromis* and, less often, small Cyprinidae (probably *Engraulicypris argenteus*). The two smallest fishes examined (81 and 91 mm. S.L.) had fed on fishes; the stomach of the smaller individual contained two post-larval *Haplochromis* (S.L., ca. 8 mm.).

Breeding. A single female (107 mm. S.L.) has been recorded with eggs in the mouth; this specimen is also the smallest sexually mature individual in the collection. One larger fish (115 mm. S.L.) is immature, but all other specimens 112 mm. S.L. and above are adult. Females may reach a larger size than males but I have insufficient material to confirm this point.

Affinities. No close relative of *H. plagiostoma* has been found amongst the present-day *Haplochromis* of Lake Victoria. The species is immediately characterized by its oblique mouth, obliquely truncated caudal fin and lack of distinct transverse bars on the body. Only one other species, *H. flavipinnis*, has an obliquely truncated caudal developed in all individuals, but its coloration is an obvious diagnostic character. An oblique mouth is not uncommon amongst the piscivorous species, but is usually associated with a different type of skull architecture than that of *H. plagiostoma*, which retains the basic *H. guiarti*-type. The nature of the neurocranium, the body-form and the male coloration all suggest that *H. plagiostoma* should be placed with the *H. victorianus*-*H. serranus* species-group, but representing an anatomically rather specialized offshoot. The other members of this group have typical subtruncate caudal fins and horizontal or slightly oblique mouths.

Study material and distribution records

Museum and Reg. No.	Locality	Collector
	Uganda	
B.M. (N.H.).—1906.5.30.261 (Holotype)	Bunjako	Degen
,, 1962.3.2.438–9	Sesse Isls.	E.A.F.R.O.
,, 1962.3.2.440	Jinja (below golf course)	,,
,, 1962.3.2.446	Jinja, Pier	,,
,, 1962.3.2.441	Karenia (near Jinja)	,,
,, 1962.3.2.442	Buvuma Isl.	,,
,, 1962.3.2.443–5	Thruston Bay	,,
,, 1962.3.2.448–64	Pilkington Bay	,,
,, 1962.3.2.465–73	Ekunu Bay	,,
,, 1962.3.2.474–6	Buka Bay	,,
	Lake Victoria, Locality Unknown	
,, 1962.3.2.447		,,

Haplochromis michaeli Trewavas 1928

Text-figs. 23 and 24

Haplochromis michaeli Trewavas, 1928, *Ann. Mag. nat. Hist.*, (10), **2**, 93.

Note. I have already drawn attention (Greenwood, 1960) to a mistake over the identity of the two syntypes of *H. michaeli*. The two specimens which were thought to be the syntypes are actually referrable to a species now named *H. empodisma*. Before this error was realised, these specimens were used as the basis for comparison with other species. Thus, references to *H. michaeli* made in any of my papers published before 1960 should be corrected to read *H. empodisma*. I have no doubt about the identity of the specimens now considered to be the types of *H. michaeli*.

Lectotype. A specimen 145·0 mm. S.L., from Rusinga Island; B.M. (N.H.) Reg. No. 1928.5.24.487.

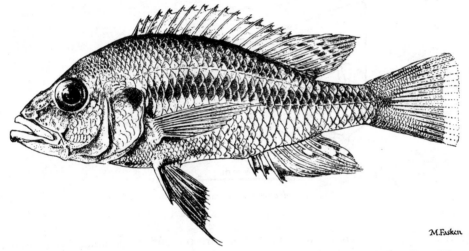

FIG. 23. *Haplochromis michaeli*, lectotype, ·7 × N.S. (Drawn by Miss M. Fasken.)

Description based on twenty-one specimens, 117–145 mm. S.L. (including the lecto- and paratype).

Depth of body 30·8–37·6 (M = 34·3) per cent of standard length, length of head 33·8–38·4 (M = 36·2) per cent. Dorsal head profile straight or slightly curved, if straight often with a gentle interorbital concavity.

Preorbital depth 14·8–19·2 (M = 17·4) per cent of head; least interorbital width 24·8–29·4 (M = 26·6) per cent. Snout length 32·6–38·0 (M = 34·7) per cent of head, 1·0–1·2 times as long as broad; diameter of eye 24·0–29·1 (M = 27·0), depth of cheek 22·9–27·7 (M = 25·8) per cent of head.

Caudal peduncle length 15·8–19·7 (M = 17·3) per cent of standard length, 1·3–1·9 (modal range 1·5–1·6) as long as deep.

Lower jaw slightly oblique, projecting a little in some specimens and always with a pronounced mental bump at the symphysis; its length 43·3–53·5 (M = 48·0) per cent of head, 1·6–2·4 times as long as broad (modal range 1·8–2·0). Lips somewhat thickened, the posterior tip of the maxilla extending to below the pupil in most fishes but only to below the anterior part of the eye in others.

Gill rakers variable, from short but slender to short and stout; 8–10 (mode 9) rakers on the lower part of the first gill arch.

Scales ctenoid; lateral line with 31 (f.1), 32 (f.7), 33 (f.8), 34 (f.3) or 35 (f.2) scales, cheek with 3 or 4 rows. Six or 7 scales between the lateral line and the origin of the dorsal fin, 7 or 8 (rarely 6 or 9) between the pectoral and pelvic fin bases. The scales of the chest are noticeably small.

Fins. Dorsal with 24 (f.10), 25 (f.10) or 26 (f.1) rays, comprising 15 (f.19) or 16 (f.2) spinous and 9 (f.12), 10 (f.8) or 11 (f.1) branched rays. Anal with 11 (f.12), 12 (f.7) or 13 (f.1) rays, comprising 3 spines and 8 (f.12), 9 (f.7) or 10 (f.1) branched

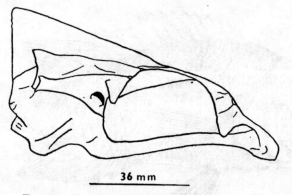

FIG. 24. *Haplochromis michaeli*, neurocranium.

elements. In one specimen, only two spines are present in the anal. The caudal is truncate and scaled on its basal half to two-thirds. The first pelvic ray is slightly produced and filamentous in adult males. Pectoral fin 24·1–30·4 (M = 26·2) per cent of standard length.

Teeth. The *outer row* in each jaw is composed of slender, slightly curved and unicuspid teeth, of which there are 60–82 (M = 66) in the upper jaw.

The *inner teeth* are either all unicuspids or a mixture of uni- and weakly tricuspids; there is apparently no correlation between the nature of the inner teeth and the size of the fish. In one exceptional specimen some of the inner teeth are bicuspid and the remainder are tricuspid. There are 3 (rarely 2 or 4) inner rows in the upper jaw and 2 or 3 (rarely 1) in the lower jaw.

Osteology. The *neurocranium* is of the generalized predator type and compares closely with *H. guiarti* or *H. serranus*. It differs from the generalized non-predator type (e.g. *H. brownae*) only in the somewhat more elongate preotic part of the skull (65% of basal length, *cf.* 55%) and consequently in the less steeply sloping dorsal profile. There is also an associated lengthening of the dentary as compared with *H. brownae*.

Vertebrae. Thirty (in all seven specimens examined), comprising 13 precaudal and 17 caudal elements.

Lower pharyngeal bone triangular, the dentigerous surface slightly to noticeably broader than long. The teeth are small, slender and cuspidate, occasionally with the two median rows composed of stouter teeth. The dental arrangement is irregular; the teeth are spaced and arranged in 22–24 rows.

Coloration. The colours of live fishes are unknown : *Preserved females* are brown above and yellowish below the well-defined, broad midlateral dark stripe. This stripe is often broken into two parts of variable length; it ends posteriorly at the base of the caudal fin. In some fishes there are traces of four, short and faint transverse bars on the flank; a dark lachrymal blotch is often present. All the fins are grey or greyish-yellow, the soft dorsal and the caudal are maculate.

Adult males are brown above the midlateral stripe and on the head; the stripe and occasional transverse bars are as in females, but the lachrymal marking is often in the form of a full stripe from the orbit to the angle of the jaws. Below the midlateral stripe, the colour is brassy-silver or greyish silver with the chest and ventral abdominal surfaces dusky, as is the branchiostegal membrane. All fins (except the pelvics) are sooty-brown, the dorsal with black lappets and darkened membrane between the branched rays. The caudal is maculate, the anal has a black patch across the spinous part and a black band along the distal margin of the fin. The anal ocelli (when present) are small and ill-defined black areas, four to five in number and arranged in a single line. The pelvics are black.

Distribution. Lake Victoria.

Ecology. Habitat. The species has been collected in only five different localities but in each the habitat was similar : a mud substrate in a sheltered gulf or bay, or, in the lee of a large island. The depth of water varied from 10–50 feet; all the specimens were caught in nets set on the bottom and within two-hundred yards of the shore. The exact locality at which the type specimens were collected is not known, nor have I been able to check the identity of specimens listed as *H. michaeli* in Graham's (1929) catch records; apparently these specimens are no longer in the Museum.

Food. Seven of the fourteen specimens examined contained food in the gut. Six of these had fed exclusively on fishes (unidentifiable); the other specimen had ingested large quantities of plant material, none of which showed signs of digestion.

Breeding. No data are available on the breeding habits of this species. The smallest fish (117 mm. S.L.) is immature and the next larger (127 mm.) is adult. The largest male is 138 mm. S.L. and the three largest specimens (141–145 mm.) are females.

Affinities. The nearest relative of *H. michaeli*, at least on anatomical grounds, is *H. martini* (see Greenwood, 1960). The external characters (excluding coloration) which separate the species are : the smaller eye of *H. michaeli* (24·0–29·1 (M = 27·0) of head *cf.* 29·4–37·5 (M = 31·7) per cent), its longer lower jaw (43·3–53·5 (M = 48·0) per cent of head, *cf.* 38·4–45·8 (M = 42·6) per cent), the predominance of unicuspid teeth in the outer row of both jaws (this row in larger *H. martini* [90–100 mm. S.L.] contains both bi- and unicuspids) and a less strongly decurved dorsal head profile. The yellow coloration of live and immediately *postmortem H. martini* is outstanding, and the *post mortem* colours of *H. michaeli* do not suggest any similarity. The preserved colour patterns of the two species are basically similar yet in *H. martini* an upper, interrupted lateral band is always present and the midlateral stripe rarely extends onto the caudal fin base as is usual in *H. michaeli*.

The morphological and anatomical differences between the species are slight and could well be the result of differential growth, particularly since the smallest specimen of *H. michaeli* is 13 mm. longer than the largest *H. martini*. As *H. martini* are sexually mature at a standard length of 80 mm. but *H. michaeli* are still immature at 117 mm. S.L., there can be little reason to suppose that " *H. michaeli* " specimens are merely large individuals of *H. martini*. But I do suggest that the species are very closely related and that *H. michaeli* evolved from an *H. martini*-like stem, one of its divergent characters (and biologically the most important) being the greater size attained. Like *H. martini*, *Haplochromis michaeli* represents a slight deviation from the generalized insectivore type and does not show any great anatomical specialization adapting it for its role as a piscivorous predator. The most obvious specialization *vis a vis* the generalized insectivore, is the larger gape and longer lower jaw, a character also developed in *H. martini*.

The resemblance between *H. michaeli* and *H. empodisma* is entirely superficial. The species belong to different phyletic lines, *H. michaeli* to one with only two extant species and *H. empodisma* to a longer line which includes the extant assemblage of mollusc-eating species (*H. pharyngomylus*, *H. ishmaeli* and *H. obtusidens* ; see Greenwood, 1960).

Study material and distribution records

Museum and Reg. No.	Locality	Collector
	Uganda	
B.M. (N.H.).—1962.3.2.314	Sesse Isls.	E.A.F.R.O.
,, 1962.3.2.315	Buvuma Isl.	,,
,, 1962.3.2.316–7	Fielding Bay	,,
,, 1962.3.2.318	Bugungu Bay (opp. Jinja)	,,
,, 1962.3.2.319	Ekunu Bay	,,
,, 1962.3.2.320–31	Pilkington Bay	,,
	Kenya	
,, 1928.5.24.487 (Lectotype)	Rusinga Isl.	Graham
,, 1928.5.24.488 (Paratype)	Rusinga Isl.	Graham

DISCUSSION

Evolution and Phylogeny

The piscivorous species form a large element of the Lake Victoria species-flock. I estimate that about 40 per cent of the known species are predominantly fish-eating. The specialized predators on embryo and larval fishes (Greenwood, 1959) are included in this estimate.

Like the majority of *Haplochromis* in the Lake, the predators probably evolved from one or two insectivorous species. There is no direct evidence to suggest that they evolved from an originally fluviatile piscivorous stem species. The present-day fluviatile predators are clearly more differentiated from the generalized *Haplochromis* than are the predatory *Haplochromis* of Lake Victoria. In fact, the most widely distributed species are placed in the genus *Serranochromis* which, despite its obviously close relationship to *Haplochromis*, is distinguishable from any predatory *Haplochromis* of Lake Victoria.

Within the Victoria species-flock there is one species, *H. brownae*, which seems to provide annectant feeding habits and certain anatomical details linking the generalized insectivores with the anatomically least specialized piscivores.

The anatomical characters involved in the evolution of the predator group are relatively simple and may be summarized as follows :

(i) An increase in adult size relative to species in other trophic groups ; the modal adult size-range for the species described here is 140–160 mm. S.L., whereas for the majority of non-piscivores it is 80–100 mm. Amongst the fish-eating predators there are exceptions to this generalization ; for example *H. percoides* and *H. pellegrini* with a modal adult length of 100 mm. Species showing the most specialized anatomical traits invariably belong to groups having the largest adult size-ranges.

(ii) A differential lengthening in the neurocranium. It appears that only the preotic region of the skull is involved in this relative growth. The changed proportions of the preotic skull are shown in the following figures. In adult non-piscivores of a generalized type (or even those with specialized feeding habits like algal-grazing) the preotic portion comprises some 55–58% of the neurocranial length. The proportion is probably affected by the adult size of the fish since in certain of the larger species (e.g. *H. ishmaeli*, a mollusc-crusher, or *H. empodisma* an insectivore) the preotic length is from 57 to 62% of neurocranial length. Intraspecifically it appears that there are insignificant ontogenetic changes in preotic proportion, at least over the latter part of the growth range. (e.g. in *Haplochromis mento* the preotic proportion in neurocrania of 25 and 30 mm. length is 66 and 67% respectively).

Amongst the piscivores the preotic proportion is from 62·0–71·0 (mean *ca.* 66·0) per cent, the figures being based on one adult neurocranium for each of the species described above except *H. artaxerxes*. Again, the differences in proportion cannot be entirely due to differences in neurocranial size. For example, the preotic is 71% in a 31 mm. neurocranium of *H. flavipinnis* but only 66% in a 38 mm. skull of *H. dentex*.

There are correlated changes in several syncranial characters. Two obvious changes are the relative lengthening of the lower jaw and deepening of the cheek region. The latter represents a general lengthening of the suspensorium and a broadening of the pterygoid bones ; it is used as a measure of these changes because it is easily determined by direct mensuration.

The lower jaw does not always lengthen in proportion to the lengthening of the preotic region, and the jaw may be relatively longer in one of two species with equal preotic proportions. In such a case, the lower jaw either projects beyond the upper or the suspensorium is greatly lengthened and the angle of the closed mouth becomes very oblique. The latter condition is seen in *H. plagiostoma* where the preotic proportion is 62% of the neurocranium, the lower jaw is 49% of the head length and the

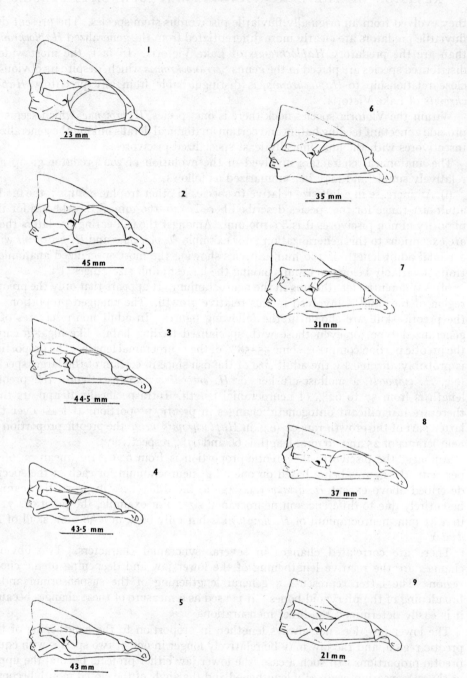

cheek is 33% of the head. These figures may be compared with *H. guiarti* (a species with a near horizontal mouth) where the preotic length is 65·5% of the neurocranium, the lower jaw 44·4% of the head length and the cheek only 25%.

(iii) Other changes in neurocranial proportions are a narrowing of the entire neurocranium and a decrease in its height, particularly of the otic-occipital region and the supraoccipital crest. Correlated with the decrease in height is a decrease in slope of the dorsal surface of the skull. This particular group of characters is not, however, a general one and many predatory species retain a skull whose outline (except for the longer preotic region) differs little from the generalized *Haplochromis* type. The slope of the dorsal surface, especially in the ethmoid region, affects the angle at which the premaxilla lies and hence the angle of the jaws. Species with a gently sloping ethmoid region tend to have the mouth more oblique than those with a steep ethmoid angle. Broadly associated with neurocranial outline is the body shape of the fish. Species with elongate, shallow neurocrania tend to have a slender elongate body shape, whilst the deeper bodied species have taller neurocrania.

(iv) A general character amongst the piscivores is the presence of unicuspid, curved and often strong teeth in the outer row of both jaws. However, in those species showing the greatest anatomical specialization, all or most of the teeth in small (i.e. juvenile) specimens are of the generalized bicuspid type. The pharyngeal teeth are generally few in number, moderately stout and with simple bicuspid or weakly bicuspid crowns. The oro-pharyngeal dentition as a whole is one adapted for gripping and macerating, not, as in many non-cichlid predators, merely for gripping the prey before it is bolted whole.

These, then, are the principal character complexes involved in the anatomical evolution of predatory species from a presumed insectivore stem.

In other morphological characters, such as the number of vertebrae, scales and fin rays, the piscivores do not differ from the generalized species. In this respect the Lake Victoria *Haplochromis* differ from those of Lake Nyasa. There, many species with a predatory facies and with piscivorous habits have more vertebrae and fin rays than do the generalized species. (I am indebted to Dr. E. Trewavas for giving me data on certain Nyasa species and for discussing this point with me).

Fig. 25. Diagram illustrating (in the left-hand column) the main evolutionary trend in neurocranial shape amongst the predatory species, from the generalized predator (1 and 2) to the "extreme" form (4 and 5). In the right-hand column, each neurocranium (except that of *H. guiarti*, No. 6) is a representative of the principal species or species group deviating from the major trend in other syncranial characters. *Haplochromis guiarti* represents a slight modification of the condition found in *H. brownae* (see text, p. 207).

All neurocrania are drawn to the same basal length; the actual length is shown below each specimen.

(1) *H. brownae* (2) *H. serranus* (3) *H. mento*
(4) *H. estor* (5) *H. macrognathus* (6) *H. guiarti*
(7) *H. plagiostoma* (8) *H. dentex* (9) *H. percoides*

Because this paper only deals with a representative sample of the Lake Victoria predators, it is not possible to discuss fully the phyletic relationship of the various species. When I have finished working on the other predatory species some clearer picture may emerge. At present a few supposedly phyletic groups can be recognized. Many species, although showing the broad characterization of either the " generalized predator " type (exemplified by *H. serranus*) or the " extreme predator " type (as seen in *H. macrognathus*), do not exhibit particularly close relationship with other living species. It is certainly more difficult to separate the predators into groups than it is to divide the non-predators into possible phyletic assemblages. One reason for this is, of course, the fact that the non-predators are divisible into phyletic-trophic groups on the basis of anatomical adaptations to a particular feeding habit. Convergence and parallelism do occur but are usually recognizable as affecting whole subgroups and not merely a single species. Amongst the predators the problem is complicated by the relatively few characters with which to build a group facies.

Despite such difficulties it does seem possible to reconstruct a putative ancestral type from which the base of the predator stem could have emerged. *Haplochromis brownae* may show the early stages of evolution leading from an insectivore towards a fully piscivorous predator. Anatomically, there are indications of some prospective adaptations for evolution in that direction. The mixed insect and small-fish diet is another important link. In most respects, however, *H. brownae* is very close to the extant and widespread, usually fluviatile species which we think must resemble the early colonizers of Lake Victoria.

From an *H. brownae*-like stem it is easy to imagine how the *H. serranus*, *H. victorianus*, *H. nyanzae* group was evolved. The species show some new characters (longer preotic skull, unicuspid teeth in the jaws) which may be the result of the larger adult size attained, but which may have evolved independently. Related to the *H. serranus* group is *H. plagiostoma*, an isolated species with very oblique jaws but a neurocranium of the type found in *H. serranus* and its allies.

Another " generalized " predator, but at a slightly higher level of differentiation, is *H. guiarti*. In this species the early stages in the development of an " extreme " predator type of neurocranium are still preserved, the body is more slender than in *H. brownae* and the adult size is greater. Yet, it is sometimes difficult to distinguish between young *H. guiarti* and adult *H. brownae* on the superficial characters of preserved specimens. Live colours, however, provide an immediate means of separating the species. *Haplochromis guiarti* may represent an ancestral type from which the " extreme " predator group (or groups) evolved, mainly by accentuation of the syncranial characters discussed above. The nearest advanced relative of *H. guiarti* appears to be *H. bayoni*, but the skull of this species is further developed towards that of the *H. mento* type.

For the moment it is difficult to suggest the relationships of such species as *H. bartoni*, *H. estor*, *H. dentex*, *H. longirostris*, *H. mento*, and *H. macrognathus*. All bear some resemblance to one another but this is not so close as that found in the non-predator groups (see Greenwood, 1956, 1957, 1960) or amongst the subgroups of larval-embryo fish-eating species. The situation may become more amenable to analysis when the status of species like *H. taeniatus*, *H. macrodon* and *H. prognathus* is fixed. *Haplochromis prognathus* could provide an indication of a stage in the evolution of the " extreme " species. The more basic ancestral stock of this group could, on present evidence, be either the *H. guiarti*-type or the *H. serranus*-type.

Two other predator groups are more easily defined. The smaller group contains only *H. martini* and *H. michaeli*. *Haplochromis martini* retains the appearance and anatomy of a generalized species yet it is partly piscivorous in its habits (see Greenwood, 1960). *Haplochromis michaeli* is little more than an enlarged version of *H. martini* with more definitely piscivorous habits.

The second group contains three species, *H. percoides*, *H. flavipinnis* and *H. cavifrons*, which show (in that order) progressive departure from the generalized stem towards a morphological type of their own. The group is characterized by the species having oblique jaws, concave dorsal head profiles, and unique colouring and colour patterns. The pattern and, to a lesser extent, the colours of *H. percoides* and *H. flavipinnis* are related but that of *H. cavifrons* is most distinctive and unrelated. On all available evidence I consider *H. percoides* and *H. flavipinnis* to be closely related, and *H. cavifrons* to be an independent derivative from the same stem. The neurocranium in these species shows departure from the *H. guiarti* type and resembles the neurocranium of the *H. bartoni-H. longirostris* type. On this character, as well as the general syncranial architecture, the *H. percoides-H. cavifrons* group might have evolved from a species resembling *Haplochromis pellegrini*. This is an intriguing species; it is one of the smallest predominantly piscivorous forms and its syncranium is that of the more " advanced " predators.

None of the species considered in this paper shows close relationship with the larval and embryo fish-eating species (Greenwood, 1959), a group at least diphyletic in origin.

Feeding methods. In general the fish-eating *Haplochromis* macerate their food and do not swallow it whole as do the majority of non-cichlid piscivores. The pharyngeal teeth play a major part in this process, the oral dentition serving to hold the prey during maceration. Exceptions to this generalization are those species preying on larvae and embryos, and, occasionally, those species or individuals that prey on fishes less than 15 mm. S.L.

Because the water of Lake Victoria is so murky, direct observations on feeding habits cannot easily be made. The description which follows is based entirely on aquarium studies. The species involved were *H. gowersi* (two specimens each about 22 cms. total length) as predators and small *Tilapia esculenta* and *Haplochromis* spp. (40–70 mm. total length) as prey. Live food was introduced in the *H. gowersi* tank (4 × 3 × 3 feet) whenever it was available; otherwise the predators were fed on chopped liver or mince-meat, a diet on which they did not thrive. For some minutes after the prey was introduced, the predator remained stationary or slowly aligned itself with the prey. Then it suddenly darted forward, inevitably catching the prey fish by the caudal peduncle. Never once did I see a frontal attack. The prey struggled for a short while but soon became motionless. The predator appeared to make no further attempt to swallow its food, although slight, almost trembling jaw and opercular movements were detectable. After four or five minutes (during which the predator might swim slowly forwards) the prey fish was released. The greater part of its caudal musculature had been grated away so that the terminal vertebrae and hypurals were exposed ; the caudal fin showed less signs of damage. At this stage the prey might attempt to swim off, but usually it remained motionless. The predator then either took hold of its prey from behind, and continued to rasp away the caudal region, or it positioned itself for a frontal attack. The latter course resulted in the head and fore-part of the prey being taken into the mouth. This, of course distorted the predator's mouth and branchial region and it gulped and " chewed " vigorously for some eight to ten minutes. At the end of that time only the prey's caudal fin protruded from the jaws. In all, the process of capture and ingestion took from fifteen to twenty minutes, a lengthy business and one destroying my preconceived notions (based on *H. gowersi's* anatomy and appearance) of a fast-moving rapacious predator.

Regrettably, I do not know whether these feeding methods are general or even natural. The fact that it is unusual to find more than one fish in a predator's stomach, even when the remains of that one individual are in a state of advanced digestion, suggests that the predators feed at fairly protracted intervals. Again, the usual occurrence in any one sample of less than fifty per cent of individuals with food in the stomach seems to reinforce the idea that feeding is not a process immediately triggered off by the digestion of the last meal. For comparison, it should be noted that a comparable sample of, say, mollusc-eating or insectivorous species, would yield about eighty per cent of specimens with food in the stomach.

Prey species. Because the food is so macerated it is difficult to identify the prey species with any degree of refinement. Piecing together what identifiable remains there are, it seems that the predators described in this paper feed mainly on other Cichlidae and to a lesser extent on small Cyprinidae, generally *Engraulicypris argenteus*. The identity of the cichlids is almost impossible to ascertain beyond the generic level. The most frequent stomach contents are of *Haplochromis* ; no definite remains of *Tilapia* are recorded but one cannot be certain that *Tilapia* are not represented amongst the numerous records of " unidentifiable cichlid remains ". Against the possibility of *Tilapia* being an important element in the diet, it must be stressed that predatory *Haplochromis* are rarely encountered in the habitats favoured by *Tilapia* of a size vulnerable to *Haplochromis*. For instance, by the time young *Tilapia esculenta* leave the reed-bed " nurseries " they are too large for all but the largest individuals of any fish-eating *Haplochromis* species.

SUMMARY

(1) Sixteeen species are redescribed on the basis of new material.

(2) In addition, five new species (*Haplochromis brownae, H. nyanzae, H. bartoni, H. artaxerxes* and *H. mandibularis*) are described.

(3) With one exception (*H. brownae*) all the species are piscivorous, the principal prey being *Haplochromis*. Small cyprinids also feature in the diet, and to a slight degree, various aquatic insects. Terrestrial insects may also be eaten when, after a heavy hatch, these fall into the water.

(4) Phyletic lines and groups are less clearly defined amongst the predators than in other trophic groups. Nevertheless three major groups can be recognized.

(5) The basic predator types are considered and the anatomical specializations of the others are described.

(6) One species, *H. brownae*, is thought to represent a transitional form (in habits and anatomy) between the generalized insectivore and the basic piscivore type.

(7) Some aquarium observations on the feeding routine of *H. gowersi* are summarized.

ACKNOWLEDGEMENTS

It is with great pleasure that I thank Dr. Ethelwynn Trewavas for her advice and criticism, and for reading the manuscript of this paper. To Mr. A. C. Wheeler go my thanks for skilfully preparing the numerous radiographs on which I have based the vertebral counts. Figures 5, 9, 10, 16 and 22 are reproduced by courtesy of the Zoological Society of London.

REFERENCES

CORBET, P. S. 1961. The food of non-cichlid fishes in the Lake Victoria basin, with remarks on their evolution and adaptation to lacustrine conditions. *Proc. zool. Soc. Lond.* **136** : 1–101.

GRAHAM, M. 1929. *A Report on the Fishing Survey of Lake Victoria, 1927–1928, and Appendices.* Crown Agents, London.

GREENWOOD, P. H. 1956. A revision of the Lake Victoria *Haplochromis* species (Pisces, Cichlidae). Part I. *Bull. Br. Mus. nat. Hist., Zool.* **4** : No. 5, 223–44.

—— 1957. A revision of the Lake Victoria *Haplochromis* species (Pisces, Cichlidae). Part II. *Bull. Br. Mus. nat. Hist., Zool.* **5** : No. 4, 76–97.

—— 1959. A revision of the Lake Victoria *Haplochromis* species (Pisces, Cichlidae). Part III. *Bull. Br. Mus. nat. Hist., Zool.* **5** : No. 7, 179–218.

—— 1960. A revision of the Lake Victoria *Haplochromis* species (Pisces, Cichlidae). Part IV. *Bull. Br. Mus. nat. Hist. Zool.* **6** : No. 4, 227–81.

PLATE I

Haplochromis victorianus : Natural size. Photograph of a specimen compared with *H. victorianus* holotype.

TWO NEW SPECIES OF *HAPLOCHROMIS* (PISCES, CICHLIDAE) FROM LAKE VICTORIA

By P. H. GREENWOOD

British Museum (Natural History)

ONE of the species described below is an additional member to the group of algal-eating *Haplochromis* species in Lake Victoria. The other, besides feeding on diatoms, apparently also feeds on scales taken from the caudal fin of cichlid fishes, and is thus the first known representative of this trophic group in the lake. Several other lepidophagous cichlids are known from Lakes Nyasa and Tanganyika (see Fryer *et al.*, 1955), and the evolution of this habit is discussed.

Haplochromis phytophagus sp. nov.
[Plate XI A]

Haplochromis ishmaeli (part): Boulenger, 1915, *Cat. Afr. Fish.* **3**, 293 (one specimen, B.M.(N.H.), reg. no. 1907.5.7.70, collected by Simon from the Buddu coast of Lake Victoria).
H. cinereus (part): Regan, 1922, *Proc. zool. Soc. Lond*, 166 (the specimen noted above).
H. nuchisquamulatus (part): Boulenger, 1915, *Ibid.* 290 (one specimen, 1906.5.30.353; apparently not seen by Regan [1922]).

Holotype: an adult male, 82+21 mm. long from Kisumu, Kenya; B.M. (N.H.) reg. no. 1965.8.6.6.

Description: based on the holotype and twenty-nine paratypes, 61·0–86·0 mm. standard length.

Depth of body 36·7–42·3 (Mean, M=39·9) per cent of standard length, length of head 31·0–35·3 (M=33·4) per cent; dorsal head profile slightly curved or straight, sloping at an angle of *ca.* 40–45° to the horizontal.

Preorbital depth 12·0–18·2 (M=15·5) per cent of head length, least interorbital width 26·9–32·7 (M=30·0) per cent; snout slightly broader than long, its length 25·5–32·7 (M=30·0) per cent of head. Eye diameter 26·9–32·0 (M=29·2) per cent of head, depth of cheek 19·2–25·5 (M=22·3) per cent. In the sample studied, no character shows allometric relationship with standard length.

Caudal peduncle 13·0–16·9 (M=15·4) per cent of standard length, 1·1–1·5 (mode 1·2) times as long as deep.

Mouth horizontal or very slightly oblique, the lips not noticeably thickened. Posterior tip of the maxilla reaching the vertical through the anterior orbital margin, or slightly beyond. Lower jaw 33·7–40·8 (M=37·5) per cent of head, 1·2–1·8 (modal range 1·4–1·5) times as long as broad.

Alimentary canal: long (3–4 times standard length), the stomach large and very distensible.

Gill rakers: 8 (mode) or 9, rarely 7, on the lower limb of the first arch, the lowermost one or two reduced, the remainder moderately stout, the uppermost pair sometimes broad.

Scales: ctenoid; the lateral line with 30 (f.4), 31 (f.18) or 32 (f.8) scales (in some fishes several scales in both lateral line rows are without visible pores); 5 or 6 (modal range 5 or $5\frac{1}{2}$) scales between the upper lateral line and the dorsal fin origin, 5 or 6 between the pectoral and pelvic fin bases.

Fins: Dorsal with 24 (f.14), 25 (f.15) or 26 (f.1) rays, comprizing 15 (f.12) or 16 (f.18) spinous and 8 (f.4), 9 (f.23) or 10(f.3) branched elements. Anal with 11 (f.9) or 12 (f.20) rays, comprising 3 spinous and 8 (f.9) or 9 (f.20) branched. Caudal truncate, scaled on its basal half to two-thirds. Pectoral 27·2–31·7 (M=29·4) per cent of standard length.

Teeth: The outer row in both jaws (except for the most posterior teeth in the upper) is composed of somewhat recurved, slightly movable teeth with compressed, unequally bicuspid crowns (Fig. 1). Generally, the major cusp has an oblique edge meeting the gently curved, near vertical anterior edge at a moderately acute angle; a few teeth in all specimens have the two edges acutely apposed. The breadth (antero-posterior) of the crown is only slightly greater than that of the cylindrical neck. The minor cusp is always distinct and, generally, acute. Posteriorly in the upper jaw one to three teeth may be either unicuspid and enlarged or tricuspid and not markedly larger than the preceding teeth; less frequently these teeth are not differentiated.

There are 34–54(M=42) outer teeth in the upper jaw; despite this high variability it seems likely that the number of teeth is positively correlated with the fish's size.

In cusp form, and degree of mobility, the outer teeth of *H. phytophagus* closely resemble those of *H. nuchisquamulatus* (Hilgend.). The teeth of the two species differ in that those of *H. nuchisquamulatus* have a more slender neck (and consequently, clearer demarcation between neck and crown), are shorter, and are more closely spaced (*cf.* Figs. 1A and B). The range in number of outer teeth for *H. nuchisquamulatus* is 50–74, that for *H. phytophagus* 34–54.

Inner teeth in *H. phytophagus* are all tricuspid with the middle cusp largest, and are arranged in 3–5 (mode 3) and 2–4 (mode 3) rows in the upper and lower jaws respectively; in one specimen (79 mm. S.L.) a few bicuspid teeth occur in the first row of the inner series. A distinct space separates the inner teeth from the outer row.

Teeth in the inner rows of *H. phytophagus* are identical with those of *H. nuchisquamulatus*, but in the latter species there is a tendency for an increase in the number of inner, upper rows (4–8), and the space between these teeth and the outer row is reduced in both jaws.

Lower pharyngeal bone: relatively stout (especially for a species without enlarged pharyngeal teeth) with a well-marked bullation of its ventral surface (Fig. 2A); dentigerous surface triangular, 1·1–1·4 times broader than long, and covered by 24–30 rows of teeth.

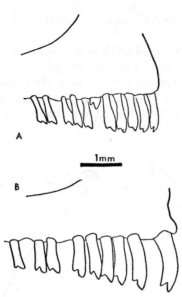

Fig. 1. Portion of the right premaxilla from: (A) *H. nuchisquamulatus*; (B) *H. phytophagus*. Both bones are from specimens of 80 mm. S.L.

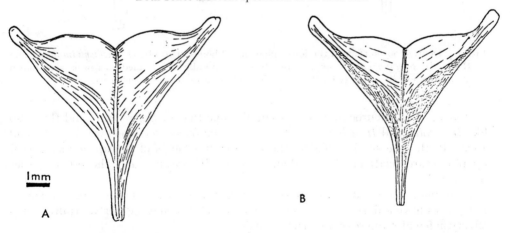

Fig. 2. Lower pharyngeal bones, in ventral view, of : (A) *H. phytophagus*, and (B) *H. nuchisquamulatus*. Note the smoothly rounded contours in *H. phytophagus*, and the well-defined crests for the insertion of the cleithro-pharyngeus muscles in *H. nuchisquamulatus*.

The teeth, although slender, are coarser and less compressed than those of other phytophagous species, and are taller than those of *H. nuchisquamulatus* (Fig. 3). All pharyngeal teeth are weakly bicuspid, those situated posteriorly having a distinctly hooked crown when viewed laterally. Some teeth in the postero-median rows may be relatively coarser.

The shape of the lower pharyngeal teeth in *H. phytophagus* and *H. nuchisquamulatus* differs quite markedly from those of *H. nigricans* (Blgr.), *H. obliquidens* Hilgend., and *H. lividus* Greenwood, the other phytophagous species in Lake Victoria. In these species the crown of the tooth is drawn out posteriorly giving a much longer occlusal surface, and and the anterior cusp is lacking in most teeth.

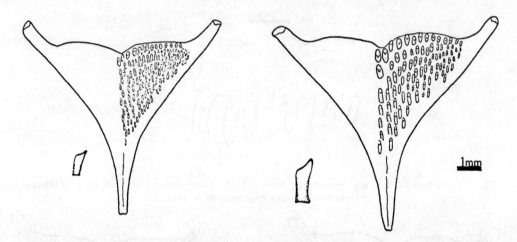

FIG. 3. Dentigerous surface of the lower pharyngeal bones in: (Left) *H. nuchisquamulatus*, and (Right) *H. phytophagus*. A single tooth from the middle of the median row is also shown (in lateral view; drawn at twice the magnification of the dentigerous surface).

Osteology: The neurocranium is essentially like that of *H. brownae* (and thus also like *H. lividus* and *H. obliquidens*, but not *H. nigricans;* see Greenwood, 1956a and 1962). In the one skull available, the basioccipital facets of the upper pharyngeal apophysis are relatively larger than those of *H. brownae* and the other species mentioned.

Coloration: No detailed information is available on the coloration of live fishes; both sexes have a dark green ground colour which becomes lighter ventrally and is silvery in females and sexually inactive males.

Preserved males (sexually active) are brown above, shading to dusky brown on the lower half of the flanks, the entire caudal peduncle, chest and belly. A broad, distinct lachrymal stripe runs from the jaw angle to the anterior orbital margin; a narrower and much fainter, rather diffuse nuchal band, and a similar interorbital stripe cross the dorsal head surface (but both these bands are absent in some specimens). A few fishes have very faint indications of six to seven vertical bars situated mid-laterally on the flank; at the level of the upper lateral line these bars may be linked by a faint longitudinal bar. Dorsal fin dusky, the melanophores most concentrated along the basal third to half of the soft part; lappets black; some dark maculæ between the last few branched rays. Caudal dark on its proximal two-thirds, dark hyaline beyond. Anal black over the spines, the rest of the fin dusky on the proximal three-quarters, hyaline distally; three greyish ocelli near the postero-ventral margin of the soft part. Pelvics dusky.

Inactive males: with a brownish ground colour, shading to silver-grey on the chest and belly; dorsal head surface dark. A faint to distinct lachrymal stripe present, but no nuchal or interorbital bars. Very faint traces of short, vertical bars or blotches may be present on the mid-flank region. Dorsal fin dusky, lappets black. Anal greyish, with 3 or 4 dead-white ocelli. Caudal dark basally (as much as the proximal half), hyaline beyond. Pelvics variable, from entirely dusky, through dusky on the outer half, to entirely hyaline.

Females and juveniles: ground colour light grey-brown shading to silver on the belly and chest, the flanks often crossed by as many as eight rather faint vertical bars which extend to the level of the upper lateral line dorsally, and ventrally to about the level of a line through the lower margin of the pelvic fin base; in some specimens the last two bars fuse into a short longitudinal blotch. All fins are yellowish hyaline, the dorsal fin sometimes faintly greyish-black; the upper half of the caudal darkly maculate.

Ecology: Habitat: The fishes described here came from four localities, all within the shallow littoral zone where the bottom is sandy, and where there are fairly dense stands of rooted aquatic plants (especially *Potamogeton*).

Food: Stomach and intestinal contents of 20 specimens (from all four localities) were examined. In all, a characteristic feature is the great bulk of finely divided fragments of phanerogam leaf tissue which fill the entire alimentary tract. Associated with the leaf tissue are large numbers of epiphytic diatoms (particularly *Suririella*, *Gomphonema*, *Navicula* and *Cyclotella*, and to a lesser degree, *Melosira*); filamentous green algæ (*Oedogonium* and *Spirogyra*) and colonial blue-green algæ are also present, but in small quantities. Little of the phanerogam tissue, or of the filamentous algæ and blue-green algae, is digested, but most diatom frustules found in the lower intestine are empty (see also Greenwood, 1956 a).

Basically, the ingested material found in *H. phytophagus* is very similar to that found in the other algal-eating species (Greenwood, *op. cit.*). However, in *H. phytophagus* the volume of macerated higher-plant tissue in the gut contents is considerably greater than was ever found in the other species.

Judging both from the nature of the dentition in *H. phytophagus*, and from this great volume of phanerogam tissue, it seems that the species obtains the nutritive portion of its diet (diatoms) not by grazing from rooted plants (as do the other species), but by browsing directly on the plants themselves.

No animal remains were recorded from the guts examined.

Breeding: No data are available on the breeding habits of *H. phytophagus*. All specimens over 80 mm. S.L. are males, the largest female being 79 mm. S.L. Females may reach sexual maturity at a standard length of 67 mm., but some larger individuals (77 mm. S.L.) are immature; the smallest sexually active male is 75 mm. S.L.

Diagnosis: Haplochromis phytophagus is very similar to *H. nuchisquamulatus*, but is distinguished by its coarser and larger lower pharyngeal bone and teeth, and by its coarser and longer outer jaw teeth (see above, pp. 304 and 305, and Figs. 1, 2 and 3). From *H. lividus* it is distinguished by the shape of the outer jaw teeth (which are also less mobile) and by the larger and stouter lower pharyngeal bone; the lower pharyngeal teeth of the two species also differ (see above, p. 306). From the other algal-grazing species of Lake Victoria (*H. obliquidens* and *H. nigricans*), *H. phytophagus* differs in the shape and number of its jaw teeth (especially so in the case of *H. obliquidens;* see Greenwood, 1956 a), and in the case of *H. nigricans*, by its straight and not decurved head profile; the stout lower pharyngeal bone of *H. phytophagus* is also diagnostic in this comparison.

In its general facies, and in dental morphology, *H. phytophagus* resembles *H. nubilus* (Blgr.), and *H. velifer* Trewavas, an endemic species from Lake Nabugabo (see Greenwood, 1965). It is readily distinguished from both these species by the greater length of intestine, and by the shape of the lower pharyngeal bone (which also lacks the somewhat enlarged median teeth found in both other species).

Distribution: The species is known only from Lake Victoria; however, some poorly preserved specimens from Lake Kioga closely resemble *H. phytophagus*, and indicate that the species (or a closely related form) may extend into that lake as well.

Discussion: The close relationship between *H. phytophagus* and *H. nuchisquamulatus* was commented upon above. Neither species differs greatly in dentition or external facies from the generalized species of *Haplochromis* as represented by, for example, *H. wingatii* and *H. nubilus*. Unlike the latter species, however, *H. phytophagus* and *H. nuchisquamulatus* are more specialized in their diet, and from this point of view may be considered as basal members of the line culminating in the specialized algal-grazing species. Unfortunately, little is known about the feeding habits of *H. nuchisquamulatus*. The gut contents recorded from the five specimens examined suggest that, unlike, *H. phytophagus*, *H. nuchisquamulatus* does not bite off pieces of rooted plant. Instead it obtains epiphytic diatoms by grazing from the stems and leaves of these plants. If this is so, then *H. nuchisquamulatus* has already reached the algal-grazing level. *Haplochromis phytophagus* perhaps represents an annectent stage between those species with feeding habits like *H. nubilus*, and those feeding in the manner of *H. nuchisquamulatus*. *Haplochromis nubilus* often feeds on small Crustacea and insect larvæ living on the submerged surfaces of rooted plants, tearing off small pieces of plant tissue in the process. The phytophagus habit probably originated in this way, and was continued and became more specialized with the evolution of a movably implanted, multiseriate dentition in certain species. Changes in tooth form, and in alimentary physiology, must have accompanied the other morphological and the behavioural changes.

The relatively stout lower pharyngeal bone, and its coarse teeth (relative to those of the true grazers) are probably adaptations associated with the maceration of plant tissue. It is interesting to recall that in *Tilapia zillii*, a species which feeds on rooted plants, the pharyngeal bones are stouter and the dentition coarser than in those species feeding on planktonic and benthic algæ (especially diatoms). The coarse outer oral teeth of *H. phytophagus*, with their slightly oblique cusps, backed by several rows of relatively robust tricuspid inner teeth seem to provide a dentition adapted for biting off pieces of plant tissue.

Study material and distribution records for *H. phytophagus*

Museum and reg. no.		Locality	Collector
		Uganda	
B.M. (N.H.)	1965.8.6.18–35	Karinya beach, near Jinja	E.A.F.F.R.O.
,,	,, 1965.8.6.16	Beach near Grant Bay, Buvuma Channel	E.A.F.F.R.O.
,,	,, 1965.8.6.17	Beach near Jinja pier	E.A.F.F.R.O.
,,	,, 1906.5.30.353	Bunjako	Degen
,,	,, 1907.5.7.70	Buddu coast	Simon
		Kenya	
,,	,, 1965.8.6.6–15	Kisumu Bay	E.A.F.F.R.O.

Haplochromis welcommei sp. nov.

[Plate XI B]

Holotype: an adult female 104·0+21·0 mm. total length, collected by E.A.F.F.R.O. from near Damba Island, Lake Victoria. B.M. (N.H.) reg. no. 1965.8.6.1. I am greatly indebted to Mr. R. Welcomme of that organization for sending this and other specimens to the B.M. (N.H.); the species is named in his honour.

Description. Although only five specimens are available, the dental characteristics of *H. welcommei* are so outstanding that I have no doubt as to its specific status. Indeed, arguments can be adduced to show that the species has attained a level of morphological differentiation equivalent to that of the endemic monotypic genera in the cichlid species flock of Lake Victoria. This point is discussed later (p. 316).

The principal morphometric characters are tabulated below, the holotype is marked with an asterisk.

S.L. (mm.)	Depth†	Head†	Preorb. %	Int. Orb. %	Snt. %	Eye %	Cheek %	L. jaw %	C.P.†
77·0	32·5	33·7	17·3	23·1	34·6	30·0	26·2	44·3	16·9
80·5	31·1	33·5	17·8	22·2	33·3	24·1	27·8	44·4	18·7
80·5	32·3	33·6	16·1	21·4	35·7	26·8	28·6	41·0	17·4
*104·0	29·8	34·6	16·6	23·6	34·8	26·4	25·0	45·8	17·3
104·5	30·6	34·0	16·9	25·3	36·6	25·3	25·3	43·3	18·2

†Per cent of standard length.
%Per cent of head length.
Caudal peduncle 1·4–1·6 times as long as deep.

The body form is relatively elongate and slender; the dorsal head profile is very gently curved, and slopes at an angle of 35°–45°, the premaxillary pedicels breaking the outline slightly or not at all.

Mouth horizontal, the lips slightly thickened; posterior tip of the maxilla extending to the vertical through the anterior orbital margin or slightly beyond this point. Jaws equal anteriorly, the lower 1·3–2·0 times as long as broad.

Gill rakers: 8 (f.4) or 9 (f.1) on the lower limb of the first arch, relatively slender except for the lowermost two or three which are reduced.

Scales: ctenoid; lateral line with 32 (f.2) or 33 (f.3) scales; cheek with 3 (f.3) or 4 (f.2) rows; $5\frac{1}{2}$ or 6 scales between the upper lateral line and the dorsal fin origin, 5 or 6 between the pectoral and pelvic fin bases.

Fins: Dorsal with 24 (f.1) or 25 (f.4) rays, comprizing 15 (f.4) or 16 (f.1) spinous and 9 (f.2) or 10 (f.3) branched elements; anal with 11 (f.1) or 12 (f.4) rays, comprizing 3 spinous and 8 (f.1) or 9 (f.4) branched rays. Caudal subtruncate, scaled over slightly more than its proximal half. Pectoral 25·5–26·8 per cent of standard length. First pelvic ray moderately elongate in both sexes (relatively longer in the adult female of a pair of equal length).

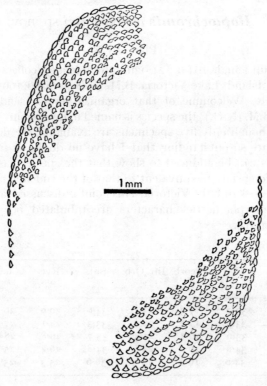

FIG. 4. *Haplochromis welcommei*. Dentigerous surface of the right premaxilla and left dentary; the teeth are seen in occlusal view. Drawn from an impression in modelling clay.

Teeth: The dental pattern, coupled with the morphology of the teeth, provide the principal diagnostic character for the species. In both jaws the teeth form broad crescentic bands not confined to the antero-lateral parts of the jaws but extending almost to the posterior limits of the dentigerous surfaces. The width of each arm decreases gradually so that the inner tooth band is multiseriate to its posterior limits in the lower jaw, and almost so in the upper jaw (Fig. 4). This arrangement contrasts strongly with that found in other *Haplochromis* having a multiseriate dental pattern, e.g. *H. obliquidens*, *H. niloticus* Greenwood, *H. nigricans* and *H. xenognathus* Greenwood. In these species the tooth bands, if continued posteriorly, are nowhere as broad as in *H. welcommei* or if broad (as in *H. xenognathus*) are confined to the anterior or anterolateral part of each dentigerous bone.

Amongst the Lake Victoria Cichlidae a dental pattern like that of *H. welcommei* is otherwise only encountered in certain monotypic genera (see below and Greenwood, 1956 b).

There is no pronounced size discrepancy between the teeth of the outermost row and those of the succeeding series, of which there are 7 to 11 in the upper jaw and 6 to 11 in the lower. The number of rows decreases posterolaterally to about 3 or 4 in the dentary, and to a single or double row in the premaxilla. On the posterior third to quarter of the latter bone, only the outer row of teeth is present. Teeth in the inner rows are slightly recurved and implanted somewhat obliquely; this arrangement combined with the space between each row, serves greatly to increase the occlusal surface.

Except for the posterior two to five teeth in the premaxilla, the outer row of teeth in both jaws is composed of recurved, movably implanted bicuspids. The crown of each tooth is expanded, compressed and clearly demarcated from the relatively slender and cylindrical neck. The occlusal edge of the major cusp slopes obliquely, but that of the minor cusp is acute or somewhat rounded (Fig. 5). In many respects these teeth may be compared with one of the common tooth types found in *H. lividus*. But, the differences include a larger size, a crown somewhat more expanded antero-posteriorly, and greater curvature of the neck. The exceptional posterior premaxillary teeth are tricuspid.

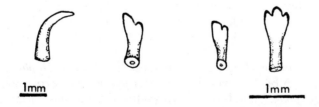

Fig. 5. *Haplochromis welcommei*. Jaw teeth. From left to right: Outer tooth from the premaxilla, viewed anteriorly to show curvature. Buccal aspect of two teeth from the premaxilla; the crown is aligned vertically. Tooth from an inner row, buccal view (drawn at twice the magnification of the outer row teeth).

Most inner teeth are tricuspid, with compressed crowns and slender necks. The middle cusp is the largest, and all three have acutely rounded tips (Fig. 5). In the outermost row of the inner series, tricuspid teeth are less common than bicuspids similar to those of the outer row but with the major cusp more protracted and spatulate. A few teeth of this type sometimes occur posteriorly in other inner rows.

Lower pharyngeal bone: slender and triangular, the dentigerous surface slightly broader than long (Fig. 6). The teeth are small and compressed, with indistinctly cuspidate and weakly hooked crowns; only the teeth forming the posterior row are enlarged.

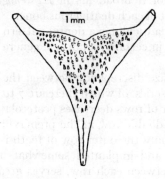

Fig. 6. *Haplochromis welcommei.* Lower pharyngeal bone, dentigerous surface.

Osteology: No complete skeleton is available but a unilateral dissection of the jaws has been made on one fish.

The premaxilla is a stout bone, with a short pedicel (about 50 per cent of premaxillary length). The ventral face is broad over the entire dentigerous area but is broadest anteriorly and anterolaterally.

In most *Haplochromis* species with multiseriate inner premaxillary tooth bands there is no appreciable widening of the dentigerous surface, the inner teeth being carried on the posterior face of the bone. An exceptional species is *H. xenognathus* (see Greenwood, 1957), which has the anterior and anterolateral dentigerous surface expanded. In this species, unlike *H. welcommei*, the expanded area slopes upwards at a sharp angle.

The condition found in *H. welcommei* is most like that of *Hoplotilapia retrodens* (see Greenwood, 1956 b); however, in the latter species, the expansion is relatively less than in *H. welcommei*, and is of uniform width over the entire length of the bone.

The dentary of *H. welcommei* also has an expanded dentigerous surface. Anteriorly and laterally it bulges outwards so as to overhang the ramus of the bone. The dental surface is virtually horizontal, and the crowns of the teeth are all at about the same level. In its gross and detailed morphology the dentary is quite unlike that of any other Lake Victoria *Haplochromis* species, including *H. xenognathus*. It also differs markedly from *Hoplotilapia retrodens*. Perhaps the closest resemblance is with the dentary of *Platytaeniodus degeni*, but even here the similarities are not great since in *P. degeni* the expanded portion has a more restricted area, is of a different shape (pyriform and not crescentic) and has a more pronounced lateral overhang.

In brief, the dentary and premaxilla of *H. welcommei* do not closely resemble these bones in any other Lake Victoria *Hapolchromis* or related species. The greatest similarity, both in the shape of the bones and their dental patterns, is with the Nyasa genus *Corematodus*, particularly *C. taeniatus;* however, the morphology of the teeth in the two genera is very different (see also p. 315).

Coloration: I am indebted to Mr. Welcomme for the following note on the colours of *live, adult males:* ground colour grey-brown shading to silver-grey with a bright red flush on the flanks, cheeks and operculum; a distinct lachrymal stripe. Bright red streaks on the soft part of the dorsal fin, and on the caudal; anal ocelli bright yellow.

Preserved specimens: adult males silvery-grey, darkest dorsally. Six rather indistinct vertical dark bars cross the flanks and extend upwards to the dorsal fin base, but only the fourth to sixth bars reach the ventral profile. An intense, dark, lachrymal stripe runs upwards from the jaw angle to below the orbit where it expands slightly; a short, dark bar runs along the upper part of the preoperculum, and extends for a short distance on to the cheek. Dorsal fin grey, the lappets black, and the soft part with elongate dusky spots. Caudal grey, darker (almost dusky) between the rays. Anal grey, with three whitish ocelli. Pelvics either entirely dusky, or dusky on the anterior three-quarters, the remainder hyaline.

Adult and juvenile females. Ground colour greyish, shading to silvery-white on the belly, chest and ventral aspects of the caudal peduncle. In the adult there are seven, ill-defined, dark, mid-lateral blotches (those on the caudal peduncle very faint), and a weak lachrymal bar. These markings are absent in the juvenile fish which has, instead, a faint longitudinal bar at the level of the lower lateral line, extending posteriorly (from a point below the last pore-scale of the upper lateral line) to the caudal fin origin. Dorsal fin greyish, the lappets dusky, and, in the juvenile, dark maculæ between the branched rays. Anal greyish, with one or two faint, elongate spots in the position of the ocelli in males. Caudal grey, weakly maculate. Pelvics hyaline.

Ecology: Habitat: The three smallest specimens were caught over a muddy bottom in the shallow inshore water of the *"Trappa* zone" of two bays near Jinja. The larger fishes were caught in about 45 feet of water over a rocky bottom near Lumya Point, Damba Island.

Food: None of the five specimens examined contains much ingested material in the stomach or intestines. The gut contents of the fishes are listed below, each preceded by the standard length of the specimen:

(i) 77 mm. Stomach (S): a few fragments of branched fin rays, and several small cycloid scales similar to those covering part of the caudal fin of cichlid fishes. Intestine (I): numerous small scales, some eroded, and a few empty diatom (*Melosira*) **frustules.**

(ii) 80·5 mm. (S): empty. (I): numerous small cycloid scales, a few *Melosira* frustules and much mucus; the diatoms and scales are from different parts of the intestine.

(iii) 80·5 mm. (S): empty. (I): much macerated phanerogam tissue (showing little evidence of digestion), the pieces of various sizes but most of approximately the same size; numerous small cycloid scales; fragments of animal epidermis (numerous melanophores visible).

(iv) 104·0 mm. (S): numerous small scales, together with large quantities of *Melosira* entangled in mucus. (I): as in the stomach.

(v) 104·5 mm. (S) empty. (I): some small fragments tentatively identified as animal epidermis; many empty *Melosira* frustules.

From these few records it seems that *H. welcommei* feeds on diatoms and on small scales scraped from the caudal fin of cichlid fishes. The fragments of well-macerated phanerogam tissue suggest that it may also browse on rooted plants. The provenance of the *Melosira* is uncertain; this diatom is principally planktonic but it also occurs epiphytically in association with rooted plants, and epilithically on rocks. Since the gill rakers and pharyngeal dentition of *H. welcommei* seem ill-adapted for collecting planktonic diatoms, the *Melosira* may be scraped from plants and rocks.

Scale-eating habits are known for seven cichlid species (see Fryer *et al.*, 1955) occurring in lakes Tanganyika and Nyasa, but this trophic specialization has not been encountered previously in Lake Victoria. The relationship of *H. welcommei* to the other scale-eating fishes is discussed below (p. 315).

Breeding: No information is available on the breeding habits of the species. The three males (80·5 (f.2) and 104·5 mm. S.L.) are all adult and sexually active; the smaller of the two females (77 mm. S.L.) is immature, and the larger (104·0 mm.) is in an advanced stage of oogenesis. Both ovaries of the latter fish are equally developed.

Diagnosis: *Haplochromis welcommei* is immediately distinguished from all other Lake Victoria *Haplochromis* species by its dental pattern (see p. 311) and tooth morphology. The same characters serve to separate the species from any of the monotypic cichlid genera in the lake.

Discussion: Two points require further comment; first, the relationship of *H. welcommei* to other scale-eating Cichlidae, and second, its taxonomic position within the Lake Victoria *Haplochromis* species-flock.

Although the studied sample of *H. welcommei* is small it does seem significant that caudal fin scales occur in four of the five guts examined (a significance enhanced by the fragments of skin and fin rays also found). On this evidence the species can at least be suspected of lepidophagy. The occurrence of diatoms and, in one fish, fragments of higher plant tissue, suggest that this species (like some other scale-eaters) is a facultative feeder. In this connection it is interesting to note that, unlike the purely algal-grazers *H. welcommei* has a short intestine (less than the body length, compared with 3 or 4 times body length).

Of the seven previously known scale-eaters, the three *Plecodus* species (Lake Tanganyika) and the monotypic *Genyochromis mento* Trewavas (Lake Nyasa) actively bite or scrape scales from the bodies of their prey (Marlier and Leleup, 1954;

Fryer, 1959, Fryer *et al.*, 1955). *Genyochromis mento* occasionally nips off pieces of fin, and algæ are also recorded from its guts (Fryer, *op. cit.*). Rather less is known about the diet of *Perissodus miolepis* another Lake Tanganyika endemic, but it too is a scale-eater, and probably has the same feeding methods as *Plecodus*.

The dentition and jaws of these three genera are different (although there are marked similarities between *Plecodus* and *Perissodus*) and none closely resembles that of *H. welcommei*, except that the dental pattern of *Genyochromis mento* is multiseriate. The mouth form in *G. mento* is most unlike that of *Plecodus* and *Perissodus*, but functionally all three genera are capable of effecting a pronounced dorso-ventral gape.

A multiseriate dentition rather like that of *H. welcommei* is found in *Corematodus shiranus* Blgr. and *C. taeniatus*, the other scale-eating species of Lake Nyasa. The jaw mechanism in these three species is also similar (weakly protusible, and a narrow gape) but there are marked intergeneric differences in the morphology of the teeth, (Boulenger, 1915; Trewavas, 1935). As Trewavas, (1947) has demonstrated, the *Corematodus* species feed by scraping small scales from the caudal fin of cichlid fishes, the fin being held between the upper and lower file-like tooth bands. When the prey pulls its caudal fin free, these teeth loosen the scales. A similar feeding method is probably employed by *H. welcommei* because the scales found in the guts are all of the cichlid caudal fin type, and the dentition is file-like.

Thus, the lepidophagous cichlids appear to employ two feeding methods, a widemouthed biting action again the prey's body in *Plecodus*, *Genyochromis* and *Perissodus*, and a filing *cum* scraping movement over two sides of a flat service in *Corematodus* and *Haplochromis welcommei*.

Fryer (1955) has suggested how the biting method could have evolved (at least for *G. mento*) from an algal-grazing one. He assumed that the basic action was that of pressing a widely opened mouth against the algal-bearing substrate, and then closing the mouth whilst it is still pressed against the substrate. This action has been observed in *Tilapia mossambica*. To achieve maximum efficiency in removing scales by this method the teeth should be at least functionally unicuspid (Fryer, 1959) and relatively immovable, a condition found in the species of *Plecodus*, *Perissodus* and *Genyochromis*.

In *Haplochromis welcommei* and *Corematodus* species, the scale-eating habit must have evolved along different lines, although still stemming from an algal-grazing ancestral habit. I have seen algal-grazing *Haplochromis* species in Lake Victoria taking a leaf in their jaws, and then swimming slowly along its length with the leaf sliding through the mouth. Such feeding behaviour, coupled with the multiseriate, movably implanted teeth of these species would seem preadapted for the feeding habits employed by *H. welcommei* and *Corematodus* species. In these fishes the teeth are relatively movable, and in *Corematodus* have oblique crowns like those in the algal-grazing *Haplochromis obliquidens* of Lake Victoria; only the outer teeth and a few inner ones in *H. welcommei* have oblique crowns, and these are of the less extreme type found in another Victoria grazer, *H. lividus*. In a scraping *cum* filing action such as these species employ, a small gape and movable teeth would be necessary to allow the substrate to slide between the jaws.

If the derivation of the lepidophagus species from algal-grazing ancestors be accepted, there is still one major step unexplained: what stimulus led the proto-lepidophages to seek food from the surface of other fishes? It has been suggested that the change could have been effected by the grazers taking small organisms (*e.g.* rotifiers) attached to the skin of fishes (Fryer *et al.*, 1955); perhaps some diatoms occur epizootically as well as epiphytically?

Returning for the moment to the possible ancestry of *Haplochromis welcommei*. The dental pattern represented a greatly expanded form of that found in most algal-grazing *Haplochromis* in Lake Victoria (Greenwood, 1956 a). The morphology of its teeth, especially the outer ones, is very like that of the teeth in *H. lividus*. In contrast, the general facies of *H. welcommei* is quiet unlike that of the algal-grazing species but resembles that of many piscivorous predators; the short gut is also characteristic of these predators. Such changes in body-form would be of value to a grazing species which has to chase a moving substrate from which to feed. The species of *Plecodus* and *Perissodus* also have the facies of piscivorous predators.

Morphologically and trophically, *H. welcommei* is but distantly related to any extant *Haplochromis* species in Lake Victoria. There are few other species in the lake which differ so distinctly from their congeners. Perhaps the most comparable example is provided by *H. xenognathus* (see Greenwood, 1957, p. 90). In this case, however, the morphological gap is less trenchant, and the species is a member of a trophic group at least four species strong. There appear, therefore, to be good grounds for thinking that *H. welcommei* has diverged sufficiently from any known group of *Haplochromis* in Lake Victoria to warrant its recognition at the generic level. I find the arguments for and against such a step less equivocal when the currently recognized endemic monotypic genera are taken into account (Greenwood, 1965b). Excepting *Paralabidochromis victoriae* Greenwood, these monotypic genera show a greater sum of divergent morphological characters, both from each other and from any *Haplochromis* species, than does *H. welcommei*. Since *Paralabidochromis victoriae* is known only from a unique specimen, judgement on that species must be reserved.

The problem posed by *Haplochromis welcommei* has wide taxonomic and philosophical ramifications when the entire species-flock of Lake Victoria is reviewed. Indeed, it is perhaps less difficult to handle than some of the problems raised by several other species and species-groups (see discussions in Greenwood, 1956 a, p. 230; and Fryer 1959, p. 239 for a similar problem in Lake Nyasa). My reasons for keeping *H. welcommei* within the genus *Haplochromis* are simply that the entire question of supraspecific grouping within cichlid species-flocks is in need of overhaul. In such a review particular reference must be given to taxa like *H. welcommei*, *H. obliquidens* and *H. xenognathus*. Preliminary investigations certainly suggest that *H. welcommei* should be given at least subgeneric status.

ACKNOWLEDGMENT

I am greatly indebted to Mr. Robin Welcomme of the East African Freshwater Fisheries Research Organization, Jinja, Uganda, who collected the specimens of *H. welcommei*, and supplied ecological and other data about the species. To him I extend my sincere thanks.

REFERENCES

BOULENGER, G. A. 1915. *Catalogue of the freshwater fishes of Africa*, **3,** British Museum (Nat. Hist.), London.

FRYER, G. 1959. The trophic interrelationship and ecology of some littoral communities of Lake Nyasa with especial reference to the fishes, and a discussion of the evolution of a group of rock-frequenting Cichlidae. *Proc. zool. Soc. Lond.* **132,** 153–281.

FRYER, G., GREENWOOD, P. H. & TREWAVAS, E. 1955. Scale eating habits of African cichlid fishes. *Nature,* **175,** 1089.

GREENWOOD, P. H. 1956 a. A revision of the Lake Victoria *Haplochromis* species (Pisces, Cichlidae). Part I. *Bull. Brit. Mus. (nat. Hist.), Zool* **4,** no. 5, 223–244.

—— 1956 b. The monotypic genera of cichlid fishes in Lake Victoria. *Ibid.* **3,** no. 7, 295–333.

—— 1957. A revision of the Lake Victoria *Haplochromis* species (Pisces, Cichlidae). Part II. *Ibid.* **5,** no. 4, 73–97.

—— 1962. A revision of the Lake Victoria *Haplochromis* species (Pisces, Cichlidae). Part V. *Ibid.* **9,** no. 4, 139–214.

—— 1965. The cichlid fishes of Lake Nabugabo, Uganda. *Ibid.* **12,** no. 9, 313–357.

MARLIER, G. & LELEUP, N. 1954. A curious ecological 'niche' among the fishes of Lake Tanganyika. *Nature,* **174,** 935.

TREWAVAS, E. 1935. A synopsis of the cichlid fishes of Lake Nyasa. *Ann. Mag. nat. Hist.* (10), **16,** 65–118.

—— 1947. An example of 'mimicry' in fishes. *Nature,* **160,** 120.

EXPLANATION OF PLATES

Plate XI.

A. *Haplochromis phytophagus;* holotype.

B. *Haplochromis welcommei;* holotype.

[Plate XI A]

[Plate XI B]

A REVISION OF THE LAKE VICTORIA *HAPLOCHROMIS* SPECIES (PISCES, CICHLIDAE) PART VI

By P. H. GREENWOOD

CONTENTS

	Page
INTRODUCTION	31
Haplochromis spekii (Blgr)	32
Haplochromis pachycephalus sp. nov.	39
Haplochromis maculipinna (Pellegrin)	43
Haplochromis boops sp. nov.	47
Haplochromis thuragnathus sp. nov.	49
Haplochromis xenostoma Regan	51
Haplochromis pseudopellegrini sp. nov.	56
Haplochromis altigenis Regan	60
Haplochromis dichrourus Regan	65
Haplochromis paraguiarti sp. nov.	69
Haplochromis acidens sp. nov.	73
Haplochromis prognathus (Pellegrin)	78
Haplochromis argenteus Regan	84
Haplochromis squamulatus Regan	87
Haplochromis barbarae sp. nov.	93
Haplochromis tridens Regan and Trewavas	97
Haplochromis orthostoma Regan	100
Haplochromis parorthostoma sp. nov.	103
Haplochromis apogonoides sp. nov.	105
DISCUSSION	108
SUMMARY	117
ACKNOWLEDGEMENTS	117
APPENDIX: *Astatotilapia nigrescens* Pellegrin, 1909	118
REFERENCES	119

INTRODUCTION

THIS is the second of two papers dealing with the piscivorous species of *Haplochromis* in Lake Victoria. In the first part (Greenwood, 1962) representatives of the principal groups of piscivores were considered, and the main morphological trends within the trophic grade were discussed. The present paper covers the remaining species which have been studied to date; undoubtedly more piscivorous *Haplochromis* species will be discovered, particularly amongst the as yet poorly sampled species of the deeper waters.

Not every species considered here is a piscivore; those of other trophic groups are included simply because the species have the morphology of a piscivore, and presumably evolved from the same stem as their fish-eating relatives.

Also included in this paper is a species apparently endemic to the Lake Kyoga system. This step was necessary because of its close relationship with a previously undescribed species from Lake Victoria.

Some of the individual species described below, and in the previous paper, would seem to be so far removed from the generality of Victoria *Haplochromis* species as to justify their elevation to generic rank. Indeed, it could be argued that even some of the species-complexes have attained this level of differentiation. However, I do not think that the question can be dealt with until the whole Lake Victoria *Haplochromis* species-flock has been described. Even then, I doubt whether it will be possible to make any such divisions, at least generically. Perhaps a number of subgeneric groups could be justified on phyletic grounds, but these will be difficult to define. The situation closely resembles that encountered by Trewavas (1964) in the genus *Serranochromis*. However, I do not believe that her solution to the *Serranochromis* problem, the recognition of a gradal genus, is applicable to the situation amongst the piscivorous *Haplochromis* of Lake Victoria, particularly because the boundary between these species and any ancestral grade (or grades) would be even more obscure and indefinite than that separating *Serranochromis* from the *Haplochromis* of central Africa. Further complications are introduced when one considers the generic status of "*Haplochromis*" species outside Lake Victoria (and this includes the Lake Nyasa species in all their complexity) relative to the possibly polygeneric *Haplochromis* species of Lake Victoria.

Haplochromis spekii (Boulenger), 1906

(Text-fig. 1)

Pelmatochromis spekii (part) Boulenger, 1906, *Ann. Mag. nat. Hist.*, (7), **17**, 440; *Idem*, 1915, *Cat. Afr. Fish*, **3**, 416, fig. 285. (Lectotype,B.M. (N.H.), reg. no. 1906.5.30.296, and probably one paralectotype, 1906.5.30.307).

Haplochromis spekii (part): Regan, 1922, *Proc. zool. Soc. Londn.*, 179 (same specimens as above).

Haplochromis serranoides Regan, 1922, *op. cit.* (Lectotype, B.M. (N.H.) 1911.3.27.17, and probably the two paralectotypes 1904.5.19.52–3).

? *Paratilapia serranus* (part): Boulenger, 1915, *Cat. Afr. Fish*, **3**, 334 (two specimens, 1904.5.19.52–3, see above).

? *Haplochromis serranus* (part): Regan, 1922, *op. cit.* 174 (paralectotype of *P. spekii* 1904.5.30.307, see above).

LECTOTYPE: a male, 191·0 mm. S.L., from Bunjako, collected by Degen; B.M. (N.H.) reg. no. 1906.5.30.296.

NOTE ON THE SYNONYMY: Certain small specimens (those indicated above with an interrogation mark) are included in the synonymy with some uncertainty. Using the diagnostic characters currently available, small preserved specimens of *H. spekii* cannot readily be separated from similar sized specimens of *H. serranus*.

Regan (1922) distinguished *H. spekii* from *H. serranoides* on two characters: the maxillary extending to below the anterior quarter of the eye (barely reaches anterior orbital margin in *H. serranoides*), and, the caudal peduncle longer than deep (as long as deep in *H. serranoides*). Additional material shows that the difference in caudal peduncle proportions is easily masked by intraspecific variability; furthermore, I am unable to confirm the marked differences in caudal peduncle proportions which Regan found in the lectotypes of the two species. The difference in the posterior extension of the maxilla is valid for the lectotypes, but it must be noted that the jaws in *H. spekii* type are somewhat distorted because of a deformed right preorbital bone. Again, more material has shown that the maxilla has a variable posterior extension which links the extremes shown by the lectotypes of the two species.

In all other characters, including the dentition and the preserved colour patterns, the two type specimens show no trenchant differences, and I consider them to be conspecific.

Because of their small size (74 and 114 mm. S.L.) the paralectotypes of *H. serranoides* have not been included in the redescription. I think it probable that these specimens are referable to the species. A similar problem is posed by three paralectotypes of *H. spekii* (B.M. [N.H.] reg. nos. 1906.5.30.301, and 1906.5.30.297–8, of standard lengths 101·0, 91·0 and 79·0 mms. respectively). The two latter may perhaps be specimens of *H. serranus*, and the former is probably referable to *H. spekii*. However, until more is known about the characteristics of smaller specimens of *H. spekii*, I consider it inadvisable to give a definite identity to these three fishes.

DESCRIPTION : based on 44 specimens (including the lectotype of the species, and the lectotype of *H. serranoides*), 128–220 mm. standard length.

Depth of body 32·8–39·8 (mean, M = 35·6) per cent of standard length, length of head 36·1–39·3 (M = 37·4) per cent. Dorsal head profile straight, sloping at an angle of 30°–35°, the premaxillary pedicels from barely to moderately prominent and interrupting the profile.

Preorbital depth 18·0–24·2 (M = 20·7) per cent of head, least interorbital width 22·0–26·0 (M = 23·3) per cent. Snout 1·2–1·3 times as long as broad, its length in fishes < 190 mm. S.L., (N = 25), 34·0–40·6 (M = 36·8) per cent of head, and in larger fishes (N = 19) 36·0–42·5 (M = 39·1) per cent. Eye diameter in fishes < 200 mm. S.L. (N = 34) 17·3–22·6 (M = 20·0), and in larger individuals (N = 10) 15·7–19·4 (M = 18·0) per cent of head ; ratio of eye/preorbital 0·8–1·3 (M = 1·0). Depth of cheek 25·7–32·9 (M = 29·5) per cent of head.

Caudal peduncle 16·7–19·8 (M = 17·9) per cent of standard length, 1·1–1·5 (modal range 1·2–1·3) times as long as deep.

Mouth horizontal or slightly oblique, jaws equal anteriorly or the lower projecting slightly, its length 49·2–61·3 (M = 53·8) per cent of head, 1·7–2·6 (modal range 1·9–2·1) times as long as broad. Mental symphysis smooth or with a slight protuberance. Premaxilla sometimes a little expanded medially but never beaked. Posterior tip of the maxilla reaching a point near the vertical through the anterior orbital margin or occasionally reaching this level (see also note on synonymy, p. 32).

Gill rakers: stout or moderately stout, the lower 1 to 3 reduced ; 8 or 9 (rarely) on the lower part of the first gill arch.

Scales : ctenoid ; lateral line with 30 (f.1), 31 (f.8), 32 (f.23), 33 (f.11) or 34 (f.1) ;

Fig. 1. *Haplochromis spekii*; lectotype; about ·77 times natural size. From Boulenger, *Fishes of the Nile*.

cheek with 3 (rare)-5 (mode 4) rows. Six to 8 (mode 7) between the upper lateral line and the dorsal fin origin, 5–9 (mode 8) between the pectoral and pelvic fin bases.

Fins: Dorsal with 24 (f.12), 25 (f.30) or 26 (f.2) rays, comprising 14 (f.1), 15 (f.29) or 16 (f.14) spinous and 8 (f.1), 9 (f.21) or 10 (f.22) branched rays. Anal with 11 (f.2), 12 (f.30) or 13 (f.9) rays, comprising 3 spinous and 8 (f.2), 9 (f.30) or 10 (f.9) branched elements. Pectoral 27·0–33·3 (M = 29·3) per cent of standard length. Pelvics with the first and second branched rays produced, slightly so in females but the first ray protracted and thread-like in males. Caudal subtruncate, scaled on its basal half.

Teeth: In all specimens examined, both the inner and outer teeth are unicuspid, those of the outer row stout and strongly curved. The smallest fish (128 mm. S.L.) shows faint indications of lateral cusps on some teeth in the inner rows.

There are 44–70 (M = 55) teeth in the outer row of the upper jaw; inner teeth in this jaw are arranged in 3–5 (usually 3 or 4) rows, and in the lower jaw in 2 or 3 (rarely 1 or 5) rows.

Osteology. The neurocranium of *H. spekii* is identical with that of *H. serranus*, that is, of the generalized predator type showing affinity with the skull of *H. guiarti* (see Greenwood, 1962).

The lower pharyngeal bone is triangular, its dentigerous surface broader than long. The lower pharyngeal teeth are relatively fine, cylindrical in cross-section and weakly bicuspid; some teeth are almost uniscuspid, with the larger cusp elongate and conical. The teeth are arranged in 22–24 rows.

Vertebral counts (precaudal and caudal) for six specimens are 13 + 16 (f.3), 13 + 17 (f.2) and 12 + 17 (f.1).

Coloration. Live coloration is unknown. *Preserved specimens*: *Males* (*adult and sexually active*): Ground colour overall dusky, including the entire head, both jaws and the branchiostegal membrane; very faint indications of a broad midlateral stripe visible behind the operculum to the beginning of the caudal peduncle where it merges with the dark general body colour. Dorsal fin dark except for the distal third to half of the soft part which is hyaline with dark spots and dashes. Caudal dark on its basal two-thirds, yellowish distally. Anal light dusky except for the distal quarter to third of the soft part which is hyaline; 4 or 5 moderately large ocelli (dead white), usually arranged in two rows or one irregular row. Pelvic fins dusky.

Adult (*but sexually quiescent males*) have a variable ground coloration which, however, is always lighter than that of sexually active fishes. The snout and jaws are darker than the flanks which vary from dusky to light golden-brown; branchiostegal membrane dark, but sometimes only in the region below the operculum. Dorsal fin dark, the lappets black, and the soft part often with close-set dark spots or dashes. Anal variable, from dusky to yellowish; ocelli whiteish-grey, 2–5 in number and arranged as in active fishes. Pelvics usually dusky but of a variable intensity; when light, the pigment concentrated over the spine and the first two branched rays.

Females (*adult and juvenile*): brownish above (and on the head and snout), shading to silvery-brown or greyish-silver on the lower flanks, belly, chest and operculum; branchiostegal membrane greyish. A faint midlateral band (of variable depth and of irregular outline) runs from behind the operculum to the caudal fin origin ; there is also a very faint upper longitudinal band running slightly above the upper lateral line visible in some specimens. All fins are brownish-yellow the soft dorsal darkly maculate. Caudal dark brown on its proximal two-thirds (because of the dense maculation in that region).

Immature males are coloured like females except that the longitudinal stripes are more distinct, and some specimens have very faint traces of 4 or 5 vertical bars crossing the longitudinal stripes on the flanks ; these bars extend from the back to a level about half way towards the ventral outline. The pelvic fins are faintly sooty.

Ecology : Habitat. *Haplochromis spekii* occurs over both hard and soft substrates, but seems to show a slight preference for the former. Few specimens were collected from nets operated over exposed beaches, most coming from gill-nets set in sheltered areas where the water was 10–30 ft. deep. Some specimens were taken from more exposed areas, but not from deeper water.

Food. Of the 42 fishes examined (from 24 localities), 22 contained food. Twenty-one of these had fed exclusively on small fishes (identified in 8 guts as *Haplochromis* species, in a further 8 as Cichlidae, and in one as a cyprinid). The exceptional fish contained unidentifiable fish remains and fragments of an ephemeropteran larva (probably *Povilla adusta*).

Breeding. All specimens < 150 mm. S.L. are immature, as is one specimen of 182 mm., but others > 150 mm. are mature. Both sexes reach the same maximum adult size.

Affinities. The close relationship between *H. spekii* and *H. serranus* has been noted already (see above p. 33). There is complete overlap in most characters but the differential growth trends shown by two characters are such that this overlap is considerably reduced in fishes more than 120 mm. S.L. The two characters are depth of preorbital, and eye diameter as proportions of head length. In *H. spekii* both are, generally, larger than in *H. serranus* when specimens of the same size are compared. However, even in these characters there is still some overlap, and, from the sample studied, it seems likely that neither is a reliable diagnostic character when fishes < 120 mm. S.L. are compared. The difference between *H. spekii* and *H. serranus* (in the size range 120–205 mm.) is perhaps best shown by the ratio of eye diameter to preorbital depth, *viz.*, 0·8–1·3 (mean 1·0) for *H. spekii*, and 1·1–1·5 (mean 1·3) for *H. serranus*.

Two other characters seem to show interspecific differences in their modal values. (i) In *H. serranus* the posterior tip of the maxilla usually lies below the eye or reaches to the vertical through the anterior orbital margin ; in *H. spekii* it rarely reaches as far posteriorly as the orbital margin (ii) *Haplochromis serranus* has a very prominent mental protuberance, but this bump is much weaker, if it is developed at all, in *H. spekii*. In many specimens of *H. serranus* the mental bump is so prominent that, in lateral view, the anterior margin of the dentary has a marked backward slope thus emphasizing the acuteness of the head profile ; in *H. spekii* the anterior margin of the dentary is, generally, almost perpendicular and so the tip of the head seems blunter than in *H. serranus*.

Unfortunately it is impossible to compare the live colours of adult males from the two species; preserved coloration is similar. This information, together with more field data on niche preferences, and small specimens of *H. spekii*, will be necessary before the precise relationships (or perhaps conspecifity) of the two species can be determined. If *H. spekii* and *H. serranus* were allopatric it would be tempting, on the information available, to consider them conspecific. However, experience with other sympatric species in Lake Victoria suggests that such slight morphological differences as are known between *H. spekii* and *H. serranus* can be the only ones manifest by biologically distinct species.

Haplochromis spekii is more easily distinguished from other members of the *H. serranus* species complex.

From *H. victorianus* it is recognizable by its larger adult size (some *H. spekii* are juvenile at a size near the upper adult limits for *H. victorianus*), larger head (36·1–39·3, M = 37·4% S.L., *cf.* 33·5–36·0, M = 34·8%), deeper cheek (25·7–32·9, M = 29·5% head, *cf.* 22·5–26·2, M = 24·6%), longer lower jaw, (49·2–61·3, M = 53·8% head, *cf.* 44·0–51·8, M = 47·1%), smaller eye in fishes < 200 mm. S.L. (17·3–22·6, M = 20·0% head, *cf.* 21·7–26·2, M = 24·6%), shorter pectoral fin (21·4–28·9, M = 25·1% S.L., *cf.* 26·2–32·7, M = 30·4%), and by having fewer and more curved outer teeth in the upper jaw (44–70, M = 55 teeth, *cf.* 64–86, M = 74).

From *H. maculipinna*, *H. spekii* differs in its larger adult size, longer head (36·1–39·3, M = 37·4% S.L., *cf.* 32·6–37·0, M = 35·5%), longer snout (34·0–40·6, M = 36·8% head, *cf.* 30·3–37·0, M = 33·7%), deeper cheek (25·7–32·9, M = 29·5% head, *cf.* 23·2–29·8, M = 25·3%), longer lower jaw (49·2–61·3, M = 53·8% head, *cf.* 43·3–52·8, M = 48·3%), and lower eye/preorbital ratio (0·8–1·3, M = 1·0, *cf.* 1·3–1·6, M = 1·5).

Although *H. spekii* resembles *H. bartoni* a little more closely in morphometric characters than it does *H. victorianus*, the species show a greater difference in neurocranial form. The neurocrania of *H. victorianus* and *H. spekii* are virtually identical, but that of *H. bartoni* is nearest the typical "*prognathus*"-group type (see p. 109). Morphometrically, *H. spekii* differs from *H. bartoni* in having a broader interorbital region (22·0–26·0, M = 23·3% of head, *cf.* 17·0–21·0, M = 18·6%), and a somewhat smaller eye (17·3–22·6, M = 20·0% head, *cf.* 20·3–24·1, M = 22·5%). Also *H. spekii* has a lower modal number of spinous dorsal fin rays (15 *cf.* 16).

From the third member of the *H. serranus* species group, *H. nyanzae*, *H. spekii* differs in its larger adult size, larger head (36·1–39·3, M = 37·4% S.L., *cf.* 33·6–36·7, M = 35·4%), deeper cheek (25·7–32·9, M = 29·5% head, *cf.* 24·4–27·6, M = 25·9%), longer lower jaw (49·2–61·3, M = 53·8% head, *cf.* 45·0–51·6, M = 48·0%), and a lower modal number of spinous dorsal fin rays (15 *cf.* 16).

Although typical specimens of *H. spekii* and *H. gowersi* are not readily confused (compare text-fig. 1 with text fig. 13 in Greenwood, 1962) there is one specimen whose appearance is such that I am unable to place it in one species or the other; it is even intermediate in the two quantifiable morphological characters (body depth and interorbital width) showing the greatest interspecific differences. *Haplochromis gowersi* and *H. spekii* differ markedly in neurocranial form, but without dissection this character cannot be checked with sufficient precision in the unique intermediate specimen. For the present, the possibility cannot be overruled that this fish is an interspecific hybrid.

Phyletically, *Haplochromis spekii* appears to be a derivative from an *H. serranus*-like ancestor, the principal difference between the species being the larger adult size attained by *H. spekii*.

Note : Gilchrist and Thompson (1917) record six specimens of *Pelmatochromis spekii* Blgr. from the Magalies river, Transvaal. I have not examined these specimens, but clearly they cannot be referred to *Haplochromis spekii* (Blgr.). Judging from their locality, it seems probable that they are specimens of *Chetia flaviventris* Trewavas. Dr. Trewavas is of a like opinion (personal communication).

STUDY MATERIAL AND DISTRIBUTION RECORDS

Museum and Reg. No.	Locality	Collector
UGANDA		
B.M. (N.H.) 1906.5.30.296 (Lectotype)	Bunjako	Degen
B.M. (N.H.) 1911.3.27.17 (Lectotype *H. serranoides*)	Victoria Nile	Melland
B.M. (N.H.) 1966.3.9.1–4, 20–21, 30–35, 39–49	Napoleon Gulf, near Jinja	E.A.F.R.O.
B.M. (N.H.) 1966.3.9.5	Beach near Nasu Point (Buvuma Channel)	E.A.F.R.O.
B.M. (N.H.) 1966.3.9.11–14	Off S. tip of Ramafuta Island (Buvuma Channel)	E.A.F.R.O.
B.M. (N.H.) 1966.3.9.28–29	Karenia, near Jinja (Napoleon Gulf)	E.A.F.R.O.
B.M. (N.H.) 1966.3.9.8–10	Pilkington Bay	E.A.F.R.O.
B.M. (N.H.) 1966.3.9.22	Thruston Bay	E.A.F.R.O.
B.M. (N.H.) 1966.3.9.27	Buka Bay (Buvuma Channel)	E.A.F.R.O.
KENYA		
B.M. (N.H.) 1966.3.9.17–19	Kisumu (Kavirondo Gulf)	E.A.F.R.O.
B.M. (N.H.) 1966.3.9.7	Naia Bay (Kavirondo Gulf)	E.A.F.R.O.
B.M. (N.H.) 1966.3.9.15–16	Sagorony (Kavirondo Gulf)	E.A.F.R.O.
B.M. (N.H.) 1928.5.24.413–5	Ulambwi Bay (Kavirondo Gulf)	Graham
TANZANIA		
B.M. (N.H.) 1966.3.9.345	Beach near Majita	E.A.F.R.O.
B.M. (N.H.) 1966.3.9.36–38	Between Ghogororo and Isanga River	E.A.F.R.O.
B.M. (N.H.) 1966.3.9.6	Mwanza (Capri Bay)	E.A.F.R.O.
LAKE VICTORIA		
B.M. (N.H.) 1966.3.9.23–26	Locality unknown	E.A.F.R.O.

Haplochromis pachycephalus sp. nov.

(Text-fig. 2)

HOLOTYPE : an adult male, 199 mm. standard length, from 40 ft. of water off Kazima island, Uganda. B.M. (N.H.) reg. no. 1966.2.21.9.

DESCRIPTION : based on the holotype and fourteen other specimens 150–232 mm. standard length.

Depth of body 36·5–42·5 (M = 39·1) per cent of standard length, length of head 35·6–39·5 (M = 36·3) per cent. Dorsal head profile straight to moderately concave, the concavity exaggerated by the prominent premaxillary pedicels ; nuchal region prominent and gently convex, prenuchal region sloping at 30°–35°. Cephalic lateral line pores large, especially on the preorbital and preopercular bones, less so on the dentary.

Preorbital depth 18·9–22·5 (M = 20·8) per cent of head, least interorbital width 24·6–31·3 (M = 27·8) per cent. Snout 1·1–1·4 (mode 1·2) times broader than long, its length 32·4–38·2 (M = 35·9) per cent of head ; eye diameter 18·8–22·2 (M = 20·6) per cent, depth of cheek 26·4–36·1 (M = 30·8) per cent.

Caudal peduncle 13·2–16·0 (M = 14·8) per cent of standard length, 1·0–1·3 (mode 1·2) times as long as deep.

Mouth oblique, sloping at an angle of 35°–45° (mode 40°). Jaws equal anteriorly or lower projecting slightly, its length 51·5–58·4 (M = 55·0) per cent of head, 1·5–1·9 (one specimen 2·2) times as long as broad. Posterior tip of maxilla reaching the vertical through the anterior orbital margin or nearly so.

Gill rakers : stout, the lower 1 or 2 sometimes reduced, the upper 3 or 4 sometimes expanded ; 8 or 9 (7 in one specimen) on the lower part of the first gill arch.

Scales ctenoid ; lateral line with 32 (f.3), 33 (f.4), 34 (f.4) or 35 (f.3), cheek with 5 or 6 (rarely 4) rows. Nine or 10 (less frequently 7, 8 or 10½) between the upper lateral line and the dorsal fin origin, 7 or 8 (less frequently 6 or 9) between the pectoral and pelvic fin bases.

Fins. Dorsal with 23 (f.1), 24 (f.6), 25 (f.6) or 26 (f.1) rays comprising 15 (f.9) or 16 (f.5) spinous and 8 (f.1), 9 (f.10) or 10 (f.3) branched rays. Anal with 11 (f.2) or 12 (f.12) rays, comprising 3 spines and 8 (f.2) or 9 (f.12) branched elements. Pectoral 21·6–30·9 (M = 24·6) per cent of standard length. Pelvics with the first branched ray produced in sexually active males, slightly so in females and quiescent males.

Teeth. The outer row in both jaws is composed of unicuspid, slender and slightly curved teeth. There are 60–80 (M = 70) teeth in the outer row of the upper jaw.

Teeth in the inner rows are small, unicuspid, curved (strongly so in the upper jaw) and implanted obliquely. There are 4 or 5 (less frequently 2 or 3) rows in the upper jaw and 2 or 3 (rarely 4) in the lower.

Osteology. No complete skeleton is available. The lower pharyngeal bone is triangular, with its dentigerous surface slightly broader than long (most markedly so in the smallest fish), or rarely, as long as broad. Lower pharyngeal teeth fairly coarse, their crowns weakly cuspidate and barely compressed ; some teeth in the two median rows are almost conical. The teeth are arranged in 18–22 rows.

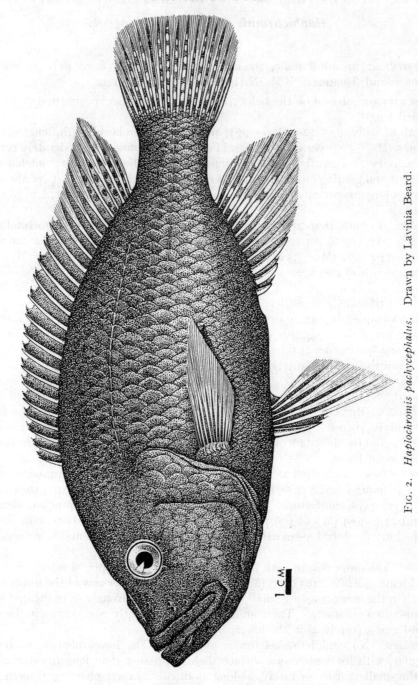

Fig. 2. *Haplochromis pachycephalus*. Drawn by Lavinia Beard.

Vertebral counts in 5 specimens are : 13 + 16 (f.3), 14 + 16 (f.1) and 14 + 17 (f.1).

Coloration : Live colours are unknown. *Preserved material : Adult males.* Ground colour variable and probably dependent on sexual state (also affected by preservation). Body greyish with black belly, chest and branchiostegal membrane, or the black replaced by a sooty-grey (in such specimens the branchiostegal membrane may be dark grey and flecked with sooty blotches) ; very faint traces of a broad midlateral stripe on the flank, originating behind the operculum and extending to the caudal fin origin. Dorsal fin greyish, dark lappets and maculae on the soft part. Caudal greyish, darkly maculate between the rays. Anal yellowish with a faintly sooty base, especially on the anterior part and around the ocelli ; the latter are dead-white, 5–9 (usually 5 or 6) in number and arranged in from 1 to 3 irregular rows. Pelvics blotched sooty to entirely black (the latter condition associated with the darkest body coloration).

Two fishes (both from the same net haul) are more sexually active than the others. Both are a very dark brown, almost uniformly so except for a black belly, chest and branchiostegal membrane. The spinous dorsal is a very dark brown (nearly black), the soft part is lighter and has a yellowish margin. Caudal light brown with lighter maculae on the basal three-quarters. Anal very faint pink, with a narrow black basal streak which expands in the region of the proximal row of ocelli and extends in amongst them ; the ocelli are whiteish.

Females (*adult but quiescent* ; based on two specimens only). Ground colour silver-grey becoming creamy on the chest and belly. Entire head (including the lower jaw) brownish with darker and irregular mottling. Body also mottled with sooty blotches, the effect being generally irregular except that on the flanks the blotches have some faint organisation into near vertical, broad bands extending from the dorsal outline to almost the ventral outline. There is some resemblance between this coloration and that of *H. cavifrons* (see Greenwood, 1962), although in *H. pachycephalus* the effect is less definitely that of freckling. Dorsal fin yellowish-grey with sooty freckling and blotching. Caudal densely and darkly blotched on its proximal third to half, greyish and darkly maculate distally. Anal greyish-yellow, with a narrow, sooty band along its base, and a duskiness over the spinous part ; both fishes have two large and distinct, dead white ocelli (an unusual feature in females). Pelvics hyaline with irregular sooty blotches.

Ecology. Habitat. The species is known from four localities ; all are some distance off-shore but close to islands. The specimens all came from nets set on a soft bottom at depths of 100–120 ft., except in one locality where the collection was made after the use of explosives. In this instance the charge was set off in about 40 ft. of water over a rock shelf with deeper water on its off-shore side.

Food. Of the 13 specimens examined (from 5 localities) six contained food in the stomach or intestines. All yielded macerated fish remains. Fragments of *Haplochromis* species were identified from three guts, a cyprinid fish in a fourth, and cichlid remains in two others.

Breeding. Little information is available about the breeding habits of *H. pachycephalus*. All specimens except the smallest (a male, 150 mm. S.L.) are mature. The two largest fishes (232 mm. and 228 mm. S.L.) are males.

Affinities. Haplochromis pachycephalus is, at least on superficial characters and those detectable on a radiograph, related to the *H. serranus* species group (see p. 109).

From *H. serranus*, *H. pachycephalus* is distinguished by its broader snout, broader interorbital (24·6–31·3, M = 27·8% head, *cf.* 20·4–26·8, M = 23·3%) and lower jaw (length/breadth ratio 1·5–1·9 *cf.* 1·8–2·5), and its smaller nuchal scales.

From *H. victorianus* it differs in its broader interorbital (24·6–31·3, M = 27·8% head, *cf.* 21·5–24·5, M = 22·6%), deeper cheek (26·4–36·1, M = 30·8% head *cf.* 22·5–26·2, M = 24·6%) and longer lower jaw (51·5–58·4, M = 55·0% head, *cf.* 44·0–51·8, M = 47·1%) ; the lower jaw is also broader in *H. pachycephalus*.

From *H. spekii* and *H. maculipinna*, the oblique mouth and broad snout of *H. pachycephalus* serve as immediately diagnostic characters, although the snout in *H. maculipinna* is broader than in other members of the " *serranus* "-group (being as much as 1·1 times broader than long, but generally as long as broad). As with other members of the group, *H. maculipinna* and *H. spekii* have a narrower interorbital region than *H. pachycephalus* ; *H. maculipinna* also has a larger eye (24·0–31·7, M = 26·3% head *cf.* 18·8–22·2, M = 20·6% in *H. pachycephalus*) but the larger adult size reached by *H. pachycephalus* may influence this character.

The same superficial characters (including the oblique mouth) serve to distinguish *H. pachycephalus* from *H. bartoni* and *H. nyanzae*, the former a member of the " *prognathus* " group, the latter a " *serranus* " group member.

Haplochromis boops and *H. thuragnathus* (both " *serranus* "-group species) closely resemble one another (see pp. 50) and *H. pachycephalus*. Both differ from *H. pachycephalus* in the following characters : a narrower interorbital, shorter snout, larger eye, and larger nuchal scales. All three species have the snout broader than it is long.

From the evidence available, *H. pachycephalus* would seem to be derived from an *H. serranus*-like ancestor, the principal morphological changes being an increase in mouth size coupled with greater obliquity of the mouth angle. The larger cephalic lateral line pores of *H. pachycephalus* are probably correlated with the deep water habitat of the species (as compared with *H. serranus* and its immediate allies).

STUDY MATERIAL AND DISTRIBUTION RECORDS

Museum and Reg. No.	Locality	Collector
	UGANDA	
B.M. (N.H.) 1966.3.9.175–177	Off S. tip of Ramafuta Island (Buvuma Channel)	E.A.F.R.O.
B.M. (N.H.) 1966.3.9.174	Deep water off Dagusi Island	E.A.F.R.O.
B.M. (N.H.) 1966.2.21.9 (Holotype)	Off Kazima Island	Uganda Fisheries Dept.
B.M. (N.H.) 1966.3.9.166–169, 171–173	Off Kazima Island	Uganda Fisheries Dept.
	TANZANIA	
B.M. (N.H.) 1966.3.9.170	Off Godziba Island	E.A.F.R.O.
	LAKE VICTORIA	
B.M. (N.H.) 1966.3.9.178	Locality unknown	E.A.F.R.O.

Haplochromis maculipinna (Pellegrin), 1913

(Text-fig. 3)

Paratilapia maculipinna Pellegrin, 1913 *Bull. Soc. Zool. France*, **37**, 311 ; *Idem*, 1914, in *Voyage de Ch. Alluaud et R. Jeannel en Afrique Occidental*, 16, Pl. 1, fig. 1, Paris.
Paratilapia prognatha (part) : Boulenger, 1915, *Cat. Afr. Fish.*, **3**, 333.
Haplochromis maculipinna : Regan, 1922, *Proc. zool. Soc. Londn.*, 177, fig. 5.

HOLOTYPE : a fish 122·0 mm. S.L. (Paris Museum No. 12–258) from Port Florence (Kavirondo Gulf) collected by Alluaud and Jeannel.

This specimen differs from all others now included in the species by its much larger eye. In other characters, however, it agrees with these specimens and differs from the few other Victoria *Haplochromis* species characterized by large eyes.

DESCRIPTION : based on 33 specimens (including the holotype), 91·5–166 mm. S.L.

Depth of body 33·3–37·0 (M = 35·9) per cent of standard length, length of head 32·6–37·0 (M = 35·5) per cent. Dorsal head profile straight or slightly concave in those fishes with prominent premaxillary pedicels, sloping at 30°–35°.

Preorbital depth 16·4–20·4 (M = 18·2) per cent of head, least interorbital width 20·7–25·5 (M = 22·8) per cent. Snout as long as broad to 1·1 times broader than long, its length 30·3–37·0 (M = 33·7) per cent of head, eye diameter 24·0–29·2 (31·7 in the type), mean 26·3 per cent, ratio of eye/preorbital 1·3–1·6 (M = 1·5) but 1·9 in the type ; depth of cheek 23·2–29·8 (M = 25·3) per cent.

Caudal peduncle 14·5–18·8 (M = 16·3) per cent of standard length, 1·2–1·8 (modal range 1·2–1·5) times as long as deep.

Mouth moderately oblique, sloping upwards at 35°–40°, lower jaw projecting slightly to strongly, its length 43·3–52·8 (M = 48·0) per cent of head, 1·6–2·3 (modal

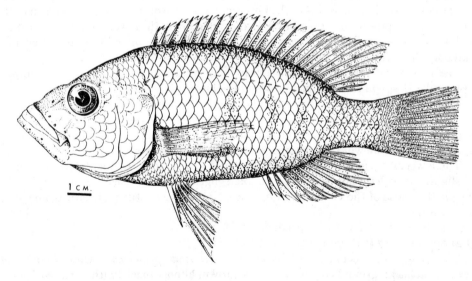

FIG. 3. *Haplochromis maculipinna*. Drawn by Barbara Williams.

range 1·7–2·1) times as long as broad. Lips not noticeably thickened, premaxilla not expanded medially. Posteriorly tip of the maxilla not quite reaching the vertical through the anterior orbital margin.

Gill rakers : moderately stout, the lower 1–3 reduced, the upper 2–5 flat and sometimes divided ; 8–11 (mode 10) on the lower part of the first arch.

Scales : ctenoid ; lateral line with 31 (f.1), 32 (f.15), 33 (f.15) or 34 (f.2), cheek with 3–5 (mode 3) rows. Five and a half (rare) to 8 (modal range 6–7) scales between the upper lateral line and the dorsal fin origin, $5\frac{1}{2}$ (rare)–9 (modal range 7–8) between the pectoral and pelvic fin bases.

Fins : Dorsal with 24 (f.5), 25 (f.26) or 26 (f.2) rays, comprising 14 (f.1), 15 (f.24) or 16 (f.8) spinous and 9 (f.11) or 10 (f.22) branched rays. Anal with 11 (f.3), 12 (f.27) or 13 (f.3) rays, comprising 3 spines and 11 (f.3), 12 (f.27) or 13 (f.3) branched elements. Pectoral 23·0–31·0 (M = 26·4) per cent of standard length. Pelvics with the first two branched rays produced in both sexes, but markedly elongate in adult males. Caudal truncate to subtruncate.

Teeth. Except in the smallest specimen (91·5 mm. S.L.), the outer teeth are unicuspid ; in the exceptional fish most teeth are weakly bicuspid but a few are unicuspids. All outer teeth are slightly curved and slender, the curvature being most marked in teeth situated laterally and posterolaterally ; teeth situated posterolaterally in the upper jaw are the smallest and finest. There are 50–80 (M = 62) teeth in the outer row of the upper jaw.

Teeth in the inner rows are more variable in form, and there is no clear-cut correlation between tooth form and the fish's size. The smallest specimen has only tricuspid inner teeth ; other and larger fishes may have only weakly tricuspids in both jaws, or tricuspid teeth predominating in both jaws but some unicuspids occurring in the lower jaw, or an admixture of tri- and unicuspids (the latter predominating) in the upper jaw and only unicuspids in the lower, or a mixture of tri- and unicuspids in both jaws, or unicuspids in the upper jaw and an admixture in the lower, or only unicuspids in both jaws. Some of the largest fishes fall in either the mixed uni- and tricuspid category or in the purely unicuspid one.

The inner teeth are arranged in 1 (rare)—4 (modes 2 and 3) rows in the upper jaw, and 2 (less commonly 1 or 3) rows in the lower.

Osteology. The neurocranium of *H. maculipinna* is virtually identical with that of *H. serranus*, differing only in having a relatively lower supraoccipital crest. The premaxilla, dentary and suspensorium are also like those of *H. serranus*, except that the dentary is somewhat deeper and shorter in *H. maculipinna*.

The lower pharyngeal bone is triangular and rather fine ; the dentigerous surface is slightly broader than long. Lower pharyngeal teeth are slender and distinctly cuspidate, those of the two median rows are the coarsest ; the teeth are arranged in 20–24 rows.

Vertebral counts for seven specimens are : 13 + 16 (f.1) ; 13 + 17 (f.3) ; 14 +16 (f.2) and 14 + 17 (f.1), giving totals of 29 to 31.

Coloration : Live colours are unknown. *Preserved specimens* : *Males* (*adult and sexually active*) : ground colour very dark brown, almost black, with a golden underlay on the flanks and operculum, and a sooty overlay on the chest. Head, including the lower jaw, dark but lips light brown. A very faint, but broad, lachrymal stripe runs from the antroventral margin of the orbit to the angle of the lower jaw. The branchiostegal membrane is black. Dorsal fin almost uniformly dark sooty, the lappets black. Caudal dark on its proximal three-quarters, dusky distally. Anal dusky, but with a black band along its base, the band expanding anteriorly to cover most of the spinous part of the fin, which is black; two or three large greyish ocelli are present. Pelvics black to dusky.

Males (adult but sexually quiescent) : ground colour dark golden brown, lightest on the anterior flanks. Head dark brown, with a distinct, narrow lachrymal stripe from orbit to angle of lower jaw, and a narrow black vertical bar on the ascending preopercular limb; branchiostegal membrane greyish-brown. Dorsal fin yellowish-grey, with black lappets, and the membrane between the branched rays dark grey-brown; the pigment often broken into discrete maculae between the last three or four branched rays. Caudal dark yellowish-grey. Anal uniformly yellowish, with two or three faint, whiteish ocelli. Pelvics dark on the anterior third, otherwise yellowish to hyaline.

Males (immature) : ground colour light brown on the flanks and belly, darker above the upper lateral line and on the dorsum of the head; the branchiostegal membrane is light brown-grey, and a faint lachrymal stripe is visible. On the flanks there is a trace of an interrupted, dark midlateral band on the anterior half of the body, and a continuous band on the posterior half, extending to the caudal origin. Dorsal fin yellowish-brown, the lappets dark, as are the maculae between the branched rays. Caudal yellowish-brown, with dark elongate blotches between the rays. Anal uniformly yellowish-brown, with two or three, distinct and dusky-grey ocelli. Pelvics yellowish with a faint dusky overlay, especially over the anterior part of the fin.

Females (adult and juvenile) : ground colour golden brown, darker on the upper half of the body, and the dorsal surface of the head; faint traces of a rather broad lachrymal stripe are often visible, the stripe generally not extending to below the level of the maxilla, but reaching the angle of the lower jaw in some specimens. Faint traces of an interrupted midlateral band on the anterior half of the body, and a continuous band on the posterior half are often visible; in some specimens no lateral band is visible, and in others the band is continuous except for a short break at about its midpoint. A few specimens show indications of a much interrupted band (really a series of 6 or 7 broad blotches) running slightly above the upper lateral line on the anterior half of the flanks, and on the lateral line posteriorly. Dorsal fin yellowish, usually darker between the posterior spines, and darkly maculate on the soft part, but uniformly yellowish with very faint maculations posteriorly in others. Anal, caudal and pelvic fins uniformly yellowish.

Ecology. Although some individuals occur over sandy, exposed and wave-washed beaches, members of this species are commoner in sheltered gulfs and bays where the water is from 10–30 ft. deep and the substrate is either soft mud or sand and shingle; a few specimens are from deeper water (35–40 ft.) near off-shore islands.

Food. Eleven of the 30 specimens examined (from 16 localities) contained food in the stomach and intestines. Seven fishes yielded fragmentary fish remains (identified as a cyprinid in one, and as *Haplochromis* species in two others), three contained fragmentary insect remains (probably larval Ephemeroptera), and one bottom debris.

Breeding. Little information is available ; most specimens less than 140 mm. S.L. are immature, as is one slightly larger individual (145 mm.). Both sexes attain the same maximum adult size.

Affinities. In both its gross and detailed morphology *H. maculipinna* shows affinity with the " *serranus* " species group, i.e. *H. serranus, H. victorianus, H. spekii,* and their deep water relatives *H. pachycephalus, H. boops* and *H. thuragnathus.* Criteria for distinguishing *H. maculipinna* from all but the first two species are considered under the descriptions of those species (see pp. 37, 42, 49 and 51 for the species respectively).

From *H. serranus, H. maculipinna* is distinguished by its larger eye (24·0–31·7, M = 26·3% head, *cf.* 20·4–26·0, M = 23·3%), shorter and more oblique lower jaw (43·3–52·8, M = 48·3% head, *cf.* 47·7–60·0, M = 54·3%), and higher eye/preorbital ratio (1·3–1·6, M = 1·5 *cf.* 1·1–1·5, M = 1·3).

From *H. victorianus,* it differs in its larger eye (24·0–31·7, M = 26·3% head *cf.* 21·7–25·5 M = 23·6%), higher eye/preorbital ratio (1·3–1·6, M = 1·5, *cf.* 1·1–1·3, M = 1·2), more oblique lower jaw (sloping at 30°–35° *cf.* horizontal or very slightly oblique) and its fewer and finer outer teeth (50–80, M = 62, *cf.* 64–86, M = 74 teeth in the upper jaw).

The close resemblance between *H. maculipinna* and these two species is obvious, and is greater than the resemblance between *H. maculipinna* and other members of the " *serranus* " group.

There are two other species, *H. nyanzae* and *H. bartoni* which, at least superficially, resemble members of the " *serranus* " group although *H. bartoni* seems to belong to a different phyletic line (see p. 109).

Haplochromis maculipinna differs from *H. nyanzae* in its larger eye (24·0–31·7, M = 26·3% head, *cf.* 19·1–24·0, M = 22·1%) and higher eye/preorbital ratio (1·3–1·6, M = 1·5, *cf.* 1·1–1·3, M = 1·2), and by its finer and less curved outer teeth.

From *H. bartoni,* it differs in having a shorter head (32·6–37·0, M = 35·5% standard length, *cf.* 36·2–39·7, M = 37·5%), broader interorbital (20·7–25·5, M = 22·8% head, *cf.* 17·0–21·0, M = 18·6%) and shorter, more oblique lower jaw (43·3–52·8, M = 48·3% head, *cf.* 50·8–57·0, M = 52·5%). Neurocranial form differs in the two species, that of *H. maculipinna* being of the " *serranus* " type, and that of *H. bartoni* being of the " *prognathus* " type (see p. 110).

Resemblances between *H. maculipinna* and *H. acidens* are discussed on p. 76. It seems unlikely that the species are closely related.

Phyletically, *H. maculipinna* was probably derived from a *H. serranus*-like ancestor.

Study Material and Distribution Records

Museum and Reg. No	Locality	Collector
	UGANDA	
B.M. (N.H.) 1906.5.30.263	Bunjako	Degen
B.M. (N.H.) 1966.3.9.145–151	Napoleon Gulf, near Jinja	E.A.F.R.O.
B.M. (N.H.) 1966.3.9.132–134	Beach near Nasu Point (Buvuma channel)	E.A.F.R.O.
B.M. (N.H.) 1966.3.9.135–144	Near Ramafuta Island (Buvuma Channel)	E.A.F.R.O.
B.M. (N.H.) 1966.3.9.129–131	Between Yempita and Busiri Island (Buvuma Channel)	E.A.F.R.O.
B.M. (N.H.) 1966.3.9. 124–6	Buka Bay	E.A.F.R.O.
B.M. (N.H.) 1966.3.9.123	Fielding Bay	E.A.F.R.O.
B.M. (N.H.) 1966.3.9.128	Kazima Island (near Entebbe)	Uganda Fisheries Dept.
B.M. (N.H.) 1966.3.9.127	Pilkington Bay	E.A.F.R.O.
	KENYA	
Paris Museum 12–258 (Holotype)	Port Florence (Kavirondo Gulf)	Alluaud & Jeannel
	LAKE VICTORIA	
B.M. (N.H.) 1966.3.9.122	Locality unknown	E.A.F.R.O.

Haplochromis boops sp. nov.

(Text-fig. 4)

HOLOTYPE: an adult male, 190 mm. standard length, from 120 ft. of water, off the southern tip of Buvuma island (Uganda). B.M. (N.H.) reg. no. 1966.2.21.7.

DESCRIPTION: based on three specimens, 179–194 mm. standard length; all are males.

Depth of body 40·5–42·3 per cent of standard length, length of head 35·3–36·1 per cent. Dorsal head profile straight, sloping steeply at 40°–50°; premaxillary pedicels not prominent. Cephalic lateral line pores enlarged, especially those on the preoperculum, preorbital and dentary.

Preorbital depth 17·8–18·6 per cent of head, least interorbital width 21·7–25·7 per cent. Snout 1·2–1·3 times as broad as long, its length 32·6–32·8 per cent of head; diameter of eye 23·9–25·7, depth of cheek 28·0–30·0 per cent.

Caudal peduncle 14·8–15·6 per cent of standard length, 1·2 times as long as deep.

Mouth somewhat oblique, sloping at 30°–35° (a horizontal line drawn through the tip of the lower jaw passes below the orbit). Jaws equal anteriorly or the lower projecting slightly, its length 50·0–52·5 per cent of head, 1·5–1·8 times as long as broad. Posterior tip of the maxilla reaching to a point below the anterior part of the eye.

Gill rakers: variable in form, from slender to stout, even in one individual; the upper 3 rakers branched in one fish. Eight or 9 on the lower part of the first gill arch.

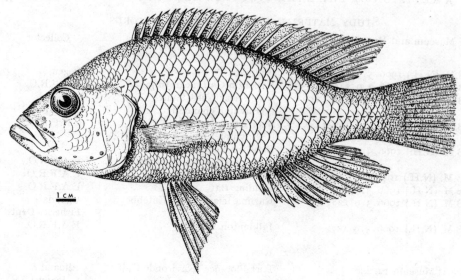

Fig. 4. *Haplochromis boops*. Drawn by Barbara Williams

Scales: ctenoid; lateral line with 33 (f.2) or 34 scales, cheek with 4 (f.2) or 5 rows. Seven or 8 scales between the dorsal fin origin and the upper lateral line, 7 or 8 between the pectoral and pelvic fin bases.

Fins: Dorsal with 15 spines and 9 (f.1) or 10 (f.2) branched rays, anal with 3 spines and 8 (f.1) or 9 (f.2) branched rays. Pectoral 25·2–32·6 per cent of standard length. Pelvics with the first branched ray produced. Caudal subtruncate.

Teeth: In the outer row of both jaws, the teeth are small, curved and slender, with about 70 in the upper jaw.

Inner teeth, arranged in three rows in both jaws, are unicuspid, small and slightly curved.

Osteology. No complete skeleton is available, but radiographs have been studied. The lower pharyngeal bone is triangular, with its dentigerous surface broader than long. Lower pharyngeal teeth are relatively stout with cylindrical necks, and compressed, weakly cuspidate crowns. The teeth are arranged in 20–22 irregular rows. The vertebral counts in three specimens are: $13 + 16$ (f.1) and $13 + 17$ (f.2).

Coloration: Live colours are unknown. *Preserved material*: *Males* (*adult and sexually active*): ground colour almost uniformly black (including the head, snout, branchiostegal membrane and belly) but with a brownish tinge. Dorsal with the spinous part blotched black on dark grey (black predominating), soft part black but with a hyaline band originating at the level of the tip of the last spine thence passing slightly downwards to end at a point about one third of the distance from the tip of the last branched ray; the dark band distal to the hyaline strip is less intense than that along the basal part of the fin. Anal black over the spines and along the basal third of the whole fin, remainder sooty; the five grey-white ocelli are arranged in two rows. Pelvic fins black.

No female specimens are available.

Ecology. Habitat. The two localities from which *H. boops* is known are in deep water (about 120 ft.) near islands and over mud substrates.

Food. Two specimens provided data on feeding habits. In both, the stomach had been everted, but fragments of macerated fish were collected from the pharynx and amongst the folds of the stomach wall. Judging from the scales and fin spines collected in this way, the fishes had fed on *Haplochromis*.

Breeding. All three specimens are adult, sexually active males.

Affinities. Haplochromis boops most closely resembles *H. thuragnathus* ; at present the species can only be distinguished by the more oblique jaw of the latter (see p. 51).

Like *H. thuragnathus*, *H. boops* appears to be a derivative of the *H. serranus* species group, probably from an ancestor resembling *H. maculipinna*. From that species *H. boops* is immediately distinguished by its broader snout (broader than long, *cf.* as long as broad), enlarged cephalic lateral line pores, and deeper body.

STUDY MATERIAL AND DISTRIBUTION RECORDS

Museum and Reg. No.	Locality	Collector
	UGANDA	
B.M. (N.H.) 1966.3.9.182	Near Dagusi Island	E.A.F.R.O.
B.M. (N.H.) 1966.2.21.7 (Holotype)	Off southern tip of Buvuma Island	E.A.F.R.O.
B.M. (N.H.) 1966.3.9.181	Off southern tip of Buvuma Island	E.A.F.R.O.

Haplochromis thuragnathus sp. nov.

HOLOTYPE : an adult male, 191 mm. standard length, from 120 ft. of water off the southern end of Buvuma island (Uganda) ; B.M. (N.H.) reg. no. 1966.2.21.8.

DESCRIPTION : based on three specimens, 191 and 200 mm. standard length. Since so few specimens are available, only ranges for morphometric characters are given.

Depth of body 39·8–41·5 per cent of standard length, length of head 34·5–35·1 per cent. Dorsal head profile slightly concave, sloping at about 30°. Cephalic lateral line pores are enlarged, especially those on the preorbital and preopercular bones.

Preorbital depth 16·4–18·8 per cent of head, least interorbital width 23·2–24·5 per cent. Snout 1·2–1·3 times as broad as long, its length 30·9–31·8 per cent of head, eye diameter 24·6–26·8, depth of cheek 27·4–29·8 per cent.

Caudal peduncle 16·0–17·7 per cent of standard length, 1·3–1·4 times as long as deep.

Mouth oblique, sloping at 40°–45°, the jaws equal anteriorly or the lower projecting slightly, length of lower jaw 53·6–56·5 per cent of head, 1·7–2·2 times as long as broad. Posterior tip of the maxilla extending to the vertical through the anterior orbital margin or to below the anterior part of the eye. A horizontal drawn anteroposteriorly through the tip of the lower jaw passes through the lower part of the eye (cf. *H. boops* where the line passes below the orbit.)

Gill rakers : stout, 9 on the lower part of the first gill arch.

Scales : ctenoid ; lateral line with 32, 33 or 34 scales, cheek with 2–4 rows. Seven or 7½ scales between the upper lateral line and the dorsal fin origin, 7 between the pectoral and pelvic fin bases.

Fins. Dorsal with 16 spines and 9 branched rays. Anal with 3 spines and 8 branched rays. Pectoral 26·0–30·0 per cent of standard length. Pelvics with the first branched ray produced, proportionately more so in males. Caudal subtruncate.

Teeth. In the outer row of both jaws, the teeth are unicuspid, small and curved ; there are 70 teeth in the upper jaw.

Inner teeth are unicuspid, small and slightly curved, and are arranged in 3 series in the upper jaw, and 2 or 3 series in the lower.

Osteology. No complete skeleton is available, but radiographs of the three specimens were examined. The lower pharyngeal bone is triangular, its dentigerous surface broader than long. The pharyngeal teeth are relatively coarse, with cylindrical necks and compressed, weakly biscuspid crowns, and are arranged in 20–22 irregular rows. Vertebral counts for all three specimens are : 13 + 17.

Coloration : Live colours are unknown. *Preserved material : Males (adult and sexually active) :* ground colour sooty over dark brown dorsally (including the head), silvery on the belly and midflank ; chest and belly darker (*i.e.* sootier), almost black There is a faint golden flush on the operculum, but it is confined to the centre of this bone and is outlined with a broad dark margin. On the flank of the lighter coloured fish are traces of a broad, dark midlateral stripe. The branchiostegal membrane is black except for its posterior and ventral margins which are greyish. Dorsal fin black except for the distal half of the soft part which is greyish. Caudal dark, but lighter towards the distal margin. Anal black on its basal half and over the spinous portion ; one large white ocellus is present in the fish with the smaller testes, but the other has 8 ocelli arranged in two irregular rows. Pelvic fins black.

Female (quiescent) : ground colour brownish, darker on back, head and snout, lighter (with silvery background) on flanks and belly ; very faint traces of a broad (three scale rows deep) interrupted midlateral band on the flanks. Operculum silvery ; a faint, dark lachrymal stripe from the orbit to behind the posterior tip of the maxilla. All fins grey-brown, the anal with 3 small, whiteish spots in the position occupied by the ocelli in males ; pelvics more grey than brown.

Ecology. Habitat. All three specimens came from nets set on the mud-bottom in water about 120 ft. deep off the southern tip of Buvuma island.

Food. Two of the three specimens examined had fragments of small *Haplochromis* species in the stomach and intestines ; the guts of the third fish were empty.

Breeding. The three specimens are adults, the two males sexually active, the female quiescent.

Affinities. Haplochromis thuragnathus is most closely related to *H. boops*. Indeed, when more material is available it may be shown that the species are not distinct. Information on the live coloration of adult males of the two species would be extremely useful in establishing their status. From *H. boops*, *H. thuragnathus* is distinguished by its more oblique lower jaw. If a horizontal line is drawn posteriorly from the tip of the lower jaw (when closed) it passes through the lower part of the eye in *H. thuragnathus*, but below the eye in *H. boops*.

A third member of this group, *H. pachycephalus*, is compared with *H. thuragnathus*. on p. 42.

It seems probable that *H. thuragnathus* was derived from an *H. maculipinna*-like ancestor, and more particularly from one like *H. boops* (assuming that the more oblique lower jaw is a derived condition). Like *H. pachycephalus* and *H. boops*, *Haplochromis thuragnathus* differs from other members of the " *serranus* " phyletic assemblage by its broad snout, and from individual members of the group by various combinations of morphometric characters (see descriptions of *H. serranus* and *H. victorianus* in Greenwood, 1962, and of *H. maculipinna* and *H. spekii* on pp. 46 and 37 above).

STUDY MATERIAL AND DISTRIBUTION RECORDS

Museum and Reg. No.	Locality	Collector
	UGANDA	
B.M. (N.H.) 1966.2.21.8	Off S. tip of Ramafuta Island	E.A.F.R.O.
B.M. (N.H.) 1966.3.9.179–180	Off S. tip of Ramafuta Island	E.A.F.R.O.

Haplochromis xenostoma Regan, 1922

(Text-figs. 5 and 6)

Paratilapia prognatha (part): Boulenger, 1915, *Cat. Afr. Fish.*, **3**, 333 (two specimens, one collected by Sir H. H. Johnston, the other from Entebbe and collected by Degen).

Haplochromis xenostoma, Regan, 1922, *Proc. zool. Soc. Londn.*, 185, fig. 10.

LECTOTYPE : an immature fish 104·0 mm. S.L. (B.M. [N.H.]reg. no. 1901.6.24.90) collected by Sir H. H. Johnston from Lake Victoria (locality unspecified).

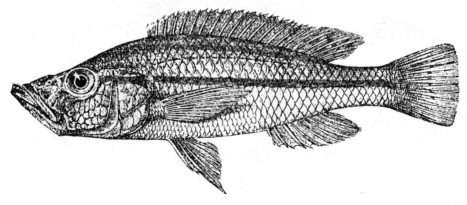

FIG. 5. *Haplochromis xenostoma*, juvenile ; lectotype, about natural size. From Regan, *Proc. zool. Soc.*

DESCRIPTION : based on 27 specimens (including the lectotype) 99–203 mm. S.L. but excluding the paralectotype. Only four specimens are less than 140 mm. S.L. (99, 104 [lectotype], 106 and 119 mm. respectively) ; on the basis of this material it appears that body depth and lower jaw length may show some slight positive allometry.

Depth of body 27·0–40·0 (M = 36·5) per cent of standard length, head length 34·6–39·7 (M = 37·3) per cent. Dorsal head profile straight or slightly convex, sloping at an angle of 20°–30°, its outline noticeably interrupted by the prominent premaxillary pedicels which give it a stepped appearance.

Preorbital depth 17·5–22·7 (M = 20·3) per cent of head, least interorbital width 20·4–27·5 (M = 24·5) per cent. Snout 1·2–1·5 times as long as broad, its length 34·2–39·2 (M = 37·4) per cent of head ; eye diameter 18·5–24·3 (M = 19·7), depth of cheek 24·3–30·8 (M = 28·1) per cent.

Caudal peduncle 14·3–18·3 (M = 15·3) per cent of standard length, 1·1–1·7 (modal range 1·3–1·4) times as long as deep ; the lectotype has an unusually shallow peduncle (ratio 1·7).

Mouth very oblique, sloping at an angle of 40°–45°, the lower jaw strongly projecting beyond the upper, its length 50·0–62·0 (M = 57·0) per cent of head and 2·0 (rarely)–3·0 times as long as broad (modal range 2·3–2·5). Posterior tip of maxilla generally not reaching the vertical through the anterior orbital margin, but reaching this point in a few specimens.

Gill rakers : short and stout, or relatively slender and elongate, the lower one or two reduced ; 8–10 (mode 9) on the lower part of the first gill arch.

Scales : ctenoid ; lateral line with 29 (f.1), 31 (f.10), 32 (f.12), 33 (f.3) or 34 (f.1), cheek with 3 (mode) or 4 rows. Six or 7 (rarely 5 or 8) scales between the upper

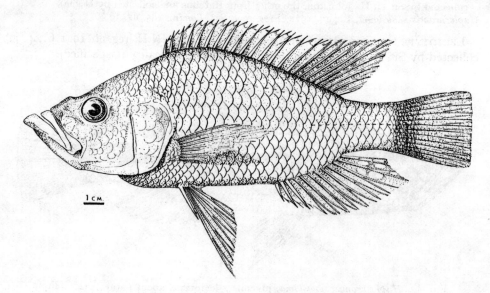

FIG. 6. *Haplochromis xenostoma* adult. Drawn by Barbara Williams.

lateral line and the dorsal fin origin, 5–7 (rarely 8) between the pectoral and pelvic fin bases.

Fins: Dorsal with 24 (f.9), 25 (f.17) or 26 (f.1) rays, comprising 14 (f.1), 15 (f.22) or 16 (f.4) spinous and 9 (f.11) or 10 (f.16) branched rays. Anal with 11 (f.11) or 12 (f.16) rays, comprising 3 spines and 8 or 9 branched elements. Pelvics with the first two branched rays produced in both sexes but proportionately more so in adult males. Pectoral 24·2–33·0 (M = 28·0) per cent of standard length. Caudal truncate, scaled on its proximal half or slightly more.

Teeth. In fishes 119 mm. S.L. and above, the outer teeth in both jaws are unicuspid and moderately stout (but occasionally slender), those in the anterior part of the jaw with a slight inward curvature, and those situated laterally and posteriorly even less curved. The lectotype (104 mm. S.L.) has an outer dentition like that of larger fishes, but in the other small fishes (88–106 mm. S.L.) the outer teeth are distinctly bicuspid anteriorly, and weakly bicuspid laterally and posterolaterally. There are 56–94 (M = 82) teeth in the outer row of the upper jaw.

Fishes less than 106 mm. S.L. have either only tricuspid teeth or a mixture of uni- and tricuspids (some weakly so) in the inner series. A specimen 119 mm. S.L. has predominantly bicuspid teeth in the upper jaw, but in the lower jaw the first row of inner teeth is composed of unicuspids, and the other rows of tri- and weakly tricuspids. In all other specimens the inner rows are composed entirely of unicuspids. Inner teeth may be implanted somewhat obliquely so as to be medially inclined. The teeth in the outermost row of the upper inner series are often noticeably larger than their congeners. There are 2–5 rows of inner teeth in the upper jaw, and 2 or 3 (rarely 4) in the lower.

Osteology. The neurocranium of *H. xenostoma* is similar to that of *H. victorianus* (see Greenwood, 1962) but has a longer preorbital face (30·3 per cent of neurocranial length *cf.* 26·0 per cent; the preorbital face being measured from the anterior tip of the vomer to the lateral ethmoid); the neurocrania of the two species also differ in that the supraoccipital crest of *H. xenostoma* is relatively higher and more pointed than in *H. victorianus*. It differs from the neurocranium of *H. serranus* (as it does from that of *H. victorianus*) in its less curved dorsal profile, and its longer preorbital face.

The very oblique and prognathous lower jaw is reflected in certain details of the suspensorium (text-fig. 7); all comparisons were made with *H. serranus*, a species

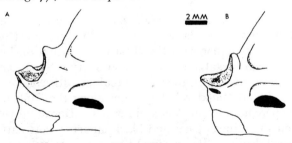

FIG. 7. Articulatory facet for the quadrate on the articular of (A) *H. xenostoma*, and (B) *H. serranus*.

with a moderately oblique jaw angle of 20°–30°. In *H. xenostoma* the articulatory surface of the articular is deeper and more nearly "U" shaped in lateral view; it lacks the posterior prolongation of its ventral border, but has a marked, near-conical eminence developed postero-medially. All these differences seem to be associated with the oblique angle of the jaw at rest, and the wide angle through which it can be abducted when the mouth is maximally protruded. The small eminence appears to function as a control for the degree of lateral movement of the dentary, particularly when that bone is dropped almost to the horizontal. The posterior vertical limb of the articular in *H. xenostoma* slopes forward at a much greater angle from the perpendicular, thus providing more space between this bone and the suspensorium.

On the basis of my material (two skeletons of *H. xenostoma*, and one each of *H. serranus* and *H. victorianus*) it appears that the horizontal length of the suspensorium (as measured in a horizontal plane from the mid-point of the hyomandibular to the articular surface of the quadrate) is greater in *H. xenostoma* than in the other two species. This could account, at least partly, for the greater prognacity of this species, whose lower jaw has the same relative length as that of *H. serranus* and *H. victorianus*. The angle of the hyomandibular relative to the perpendicular is similar in all three species.

The premaxilla has a slight median expansion of its dentigerous surface, and the pedicels are relatively shorter than those of *H. serranus*.

As compared with the dentary of *H. serranus* and *H. victorianus*, that of *H. xenostoma* is deeper and stouter, and the dentigerous surface has a more pronounced upward sweep towards the coronoid region.

The lower pharyngeal bone is triangular, and fairly stout; the dentigerous area is as long as broad or very slightly broader than long. Lower pharyngeal teeth are variable in form, usually with coarse, cylindrical necks and compressed, weakly cuspidate crowns; in some fishes, however, the crowns are distinctly cuspidate. Less commonly, the teeth are slender and compressed, with very weakly cuspidate crowns. The teeth are arranged in 22–24 rows.

Vertebral counts in 7 specimens are: $13 + 15$ (f.1), $13 + 16$ (f.5) and $13 + 17$ (f.1), giving totals of 28–30.

Coloration: *Live colours* are known only for a single *juvenile female*; ground coloration silvery, shading to yellowish-grey dorsally, the dorsal surface of the snout dark grey. Dorsal fin dark hyaline, pelvics hyaline, anal yellow, caudal yellowish-grey. *Preserved coloration*: *Males* (*adult and sexually active*) have the ground coloration generally dusky over dark brown dorsally, and silver on the flanks and belly, the latter region together with the chest often with a dusky overlay; lower jaw and entire branchiostegal membrane also sooty. Snout and preoperculum sooty, but the upper lip is dark brown; operculum with a faint golden flush. A faint but dark and broad midlateral stripe runs along the flank from the posterior margin of the preoperculum to the caudal fin origin. Dorsal fin yellow-brown, with a sooty overlay on the spinous part, and black lappets; soft part with dark spots and dashes. Caudal dusky on its proximal threequarters, yellowish-brown distally. Anal light brownish-yellow, with a thin dark line along the base, and black lappets;

3 or 4 large, dark or whitesh ocelli are present, each with a narrow black outline. The ocelli are arranged in one or two rows. Pelvics entirely dusky.

Males (*adult but quiescent*) : have essentially the same coloration as active males, but some are lighter (that is, with more silvery flanks, and greyish branchiostegal membrane).

Females (*adult, and at various degrees of sexual activity*). Greyish-brown above, shading through silvery-grey on the flanks to gold below ; snout and preorbital region dark. On the flanks, a faint dark midlateral band (as in males) may be visible, and in addition, a fainter upper band running just above the upper lateral line. Dorsal fin yellowish-grey with a sooty overlay on the spinous part, and with black lappets ; soft dorsal sometimes darkly spotted. Caudal dark on the proximal two-thirds, lighter distally. Anal yellowish with a faint sooty overlay, sometimes with ill-defined dark spots in the position of the ocelli in males. Pelvics hyaline, usually with a dusky overlay.

Ecology. Habitat. The species is apparently confined to sheltered or relatively undisturbed water, being common in bays and gulfs where the water is less than 40 ft. deep, and the bottom is of soft mud, sand or shingle. Available records suggest that the species favours a mud substrate.

Food. Of the 21 fishes examined (from 9 localities), only 6 contained food in the guts. In each case the food comprised fragmentary fish remains, unidentifiable except in one instance (a small *Haplochromis* species).

Breeding. All fishes less than 160 mm. S.L. are immature ; one larger individual (a female 163 mm. S.L.) is also immature. Males and females appear to reach the same maximum adult size.

Affinities. The very oblique mouth, marked prognacity, and relatively deep body (at least in adults) serve to distinguish *H. xenostoma* from the majority of larger *Haplochromis* species in the lake. There is some resemblance between this species and *H. macrognathus* and *H. plagiostoma*, both species with an oblique mouth, and in the case of *H. macrognathus*, a prominent lower jaw.

Haplochromis xenostoma is readily distinguished from *H. macrognathus* by its broader head (interorbital width 20·4–27·5, M = 24·5% head *cf.* 16·5–22·2, M = 18·6% ; snout 1·2–1·5 times as long as broad, *cf.* 1·5–2·2 times in *H. macrognathus*), and its more oblique mouth. The neurocranium also differs, that of *H. macrognathus* being of the "*prognathus*" type, whilst the skull of *H. xenostoma* is clearly of the "*serranus*" type (see p. 111 and discussion on pp. 109–113 ; also Greenwood, 1962).

From *H. plagiostoma*, *H. xenostoma* differs in its larger adult size, longer and narrower snout (34·2–39·2, M = 37·4% head, *cf.* 28·2–35·5, M = 32·5%), more prominent and longer lower jaw (50·0–62·0, M = 57·0% head, *cf.* 44·0–54·5, M = 49·2% ; lower jaw rarely projecting in *H. plagiostoma*) and the greater number of teeth in the outer row of the upper jaw (56–94, M = 82, *cf.* 44–68, M = 57). Neurocranial form in these two species is similar (see p. 113).

Another species with an oblique mouth is *H. cavifrons*. It is distinguished from *H. xenostoma* by its unique mottled coloration, lack of prognacity, broader snout (as long as broad or slightly broader than long, *cf.* 1·2–1·5 times as long as broad), and fewer teeth (56–74, M = 63, cf. 56–94, M = 82). The profile of the head also differs (compare fig. 6 above with fig. 20 in Greenwood, 1962).

Phylogenetically, *H. xenostoma* could be derived from a species resembling *H. plagiostoma*; its affinities seem to lie more with the " *serranus* " group than with the "*prognathus*" group to which *H. macrognathus* belongs. (See also discussion on pp. 113).

STUDY MATERIAL AND DISTRIBUTION RECORDS

Museum and Reg. No.	Locality	Collector
UGANDA		
B.M. (N.H.) 1906.5.30.257 (Paralectotype)	Entebbe	Degen
B.M. (N.H.) 1966.3.9.92–93	Napoleon Gulf, near Jinja	E.A.F.R.O.
B.M. (N.H.) 1966.3.9.88, 96–106	Ekunu Bay	E.A.F.R.O.
B.M. (N.H.) 1966.3.9.87	Pilkington Bay	E.A.F.R.O.
B.M. (N.H.) 1966.3.9.90–91	Off Ramafuta Island (Buvuma Channel)	E.A.F.R.O.
KENYA		
B.M. (N.H.) 1966.3.9.89	Naia Bay (Kavirondo Gulf)	E.A.F.R.O.
B.M. (N.H.) 1966.3.9.86	Nanga Bay (Kavirondo Gulf)	E.A.F.R.O.
B.M. (N.H.) 1966.9.9.85	Off mouth of Nzoia River	E.A.F.R.O.
LAKE VICTORIA		
B.M. (N.H.) 1901.6.24.90 (Lectotype)	Locality unknown	Sir H. H. Johnston
B.M. (N.H.) 1966.3.9.94–95	Locality unknown	E.A.F.R.O.

Haplochromis pseudopellegrini sp. nov.

(Text-fig. 8)

HOLOTYPE: an adult male 139 mm. standard length, from Pilkington Bay (Uganda). B.M. (N.H.) no. 1966.2.21.2.

Named *pseudopellegrini* because of its resemblance to *H. pellegrini* Regan.

DESCRIPTION: based on 17 specimens (including the holotype) 98 to 150 mm. standard length.

Depth of body 29·0–33·5 (M = 30·9) per cent of standard length, length of head 32·4–37·0 (M = 33·8) per cent. Dorsal head profile gently curved (rarely straight) but interrupted by the prominent premaxillary pedicels.

Preorbital depth 19·1–22·4 (M = 20·4) per cent of head length, least interorbital width 20·4–27·1 (M = 23·9) per cent. Snout a little longer than broad (1·2–1·3 times), its length 35·1–39·0 (M = 37·1) per cent of head; eye diameter 18·5–25·4 (M = 20·6), depth of cheek 24·0–29·6 (M = 26·8) per cent.

Caudal peduncle 16·3–20·8 (M = 18·3) per cent of standard length, 1·3–2·0 (modal range 1·6–1·7) times as long as deep.

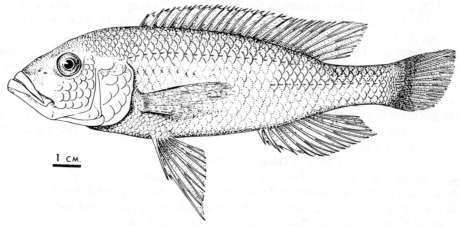

Fig. 8. *Haplochromis pseudopellegrini*; this specimen shows extreme development of an obliquely truncate caudal fin. Drawn by Barbara Williams.

Mouth slightly oblique, lower jaw projecting a little. Premaxilla slightly expanded medially. Lower jaw 46·8–53·4 (M = 49·1) per cent of head, 2·0–2·6 (modal range 2·1–2·4) times as long as broad. Posterior tip of the maxilla reaching, or almost reaching the vertical through the anterior orbital margin, extending a little beyond this point in a few specimens.

Gill rakers: of variable form, from relatively slender to moderately stout (reduced to short knobs in one specimen), the upper four sometimes flat; 9 (rarely 8 or 10) on the lower part of the first gill arch.

Scales: ctenoid; lateral line with 32 (f.4), 33 (f.7), 34 (f.4) or 35 (f.1) scales. Cheek with 4 (less commonly 3, rarely 5) rows. Six or 7 (rarely 7½) scales between the dorsal fin origin and the upper lateral line, 7 (occasionally 6, rarely 8) between the pectoral and pelvic fin bases.

Fins. Dorsal with 23 (f.1), 24 (f.4) or 25 (f.12) rays, comprising 15 (f.9) or 16 (f.8) spinous and 8 (f.1), 9 (f.12) or 10 (f.4) branched rays. Anal with 11 (f.10) or 12 (f.7) rays, comprising 3 spines and 8 (f.10) or 9 (f.7) branched elements. Pectoral 22·7–28·0 (M = 24·9) per cent of standard length. First branched ray of pelvic fin slightly produced in both sexes, proportionately more so in adult males. Caudal scaled over its proximal half; in most specimens the fin is truncate or subtruncate but in a few the lower half, or the ventro-posterior corner, slopes obliquely forward and may also be rounded.

Teeth. The outer row of teeth in both jaws is composed of slender and slightly recurved unicuspids; there are 35–52 (M = 44) teeth in the outer row of the upper jaw.

Inner teeth are also unicuspid, and are implanted at an angle varying from near vertical to almost horizontal, the latter condition being most common. There are 2 (rarely 3) rows of inner teeth in the upper jaw, and 1 or 2 rows in the lower.

Osteology. The neurocranium of *H. pseudopellegrini* shows some similarity with that of *H. mento* and other species in the "*prognathus*" group (see Greenwood, 1962, and p. 110), but at the same time it retains characteristics of the more generalized skull seen in *H. serranus* and its allies. In this respect it resembles the neurocranium of *H. prognathus*, but is slightly less "*mento*"-like. In general appearance and proportions it is similar to the neurocrania of *H. bayoni* and *H. dentex* but lacks the characteristically decurved ethmo-vomerine region of these species (see Greenwood, *op. cit.*).

The premaxilla is moderately beaked, that is, the dentigerous part of the bone is somewhat expanded medially ; the pedicels are short, being about two-thirds the length of the dentigerous arm.

The lower pharyngeal bone is triangular, its dental surface is as long as broad or slightly longer than broad. Lower pharyngeal teeth are slender and compressed, those in the two median rows are slightly coarser than their lateral congeners ; the teeth are arranged in 22–24 rows. Vertebral counts in 14 specimens are : $13 + 17$ (f.4), $13 + 18$ (f.1), $14 + 17$ (f.9), giving totals of 30 and 31.

Coloration : Live colours are known only for *adult and sexually active males*, which have the dorsal surface of the head and body dark brown, the flanks and belly golden-yellow overlain by an orange-red flush on the chest and anterior flanks, and also on the operculum. Dorsal fin dark neutral with a slight orange flush. Caudal dark neutral, with a reddish flush at the base and over the ventral quarter of the fin. Anal neutral, with yellowish-red ocelli. Pelvics are sooty.

Preserved material : Males (*adult and active*): ground colour brownish overlying silver, silvery-yellow on the chest and belly, and on the operculum ; rest of head (including the branchiostegal membrane) brownish with very faint traces of a lachrymal blotch. Dorsal fin hyaline, with a faint, narrow black band running from the tip of the eleventh spine to about the middle of the last branched ray (*i.e.* curving gently downwards). Caudal brownish to hyaline, dark on the proximal half. Anal hyaline, with faint traces of one or two whiteish-grey ocelli. Pelvics sooty, darkest on the anterior half.

Adult but quiescent males are light brown dorsally, shading to silvery below, some showing a faint but broad and dark midlateral stripe, and a fainter upper lateral band above the upper lateral line. The two lateral bands are connected by 4–6 vertical bars, which extend ventrally a little below the midlateral band ; where the lines intersect, the lateral one is diffusely expanded. A faint lachrymal blotch is present below the anterior part of the orbit. All fins are yellowish-brown, the soft dorsal and the proximal part of the caudal are often darkly maculate. Anal with 2 or 3 faint, whiteish ocelli. Pelvics variable, from yellowish-brown to sooty.

Ecology. Habitat. The species is recorded from four localities only. Two of these are shallow, sheltered bays, one is a fairly exposed, offshore and deep (90 ft.) channel, and the fourth is not fully documented except for a note that the nets were set in water about 20 ft. deep. In all, the substrate is of organic mud.

Food. Fourteen specimens were examined, and of these only two contained food, very fragmentary and generically unidentifiable fish remains.

Breeding. Little information is available on the reproductive biology of this species. The sex of the smallest specimen (98 mm. S.L.) is indeterminable ; the others (132–153 mm.) are all adults, and only two are females. The two largest fishes are males.

Affinities. In general appearance *H. pseudopellegrini* closely resembles *H. pellegrini* ; however, it reaches a much larger adult size, and the preserved coloration of adult males is much lighter (uniformly dark brown, nearly black in *H. pellegrini*, light brown over silver in *H. pseudopellegrini*). The species also differ in certain morphometric characters. *Haplochromis pseudopellegrini* has a longer snout (35·1–39·0, M = 37·1% head, *cf.* 30·8–36·0, M = 34·0%), a longer and more slender caudal peduncle (16·3–20·8, M = 18·3% standard length, *cf.* 13·2–17·8, M = 15·4% ; length/depth ratio 1·3–2·0 [modal range 1·6–1·7], *cf.* 1·1–1·5 [mode 1·3]) ; unfortunately it is not possible to determine whether, at least in part, these differences are attributable to the larger size of the *H. pseudopellegrini* specimens. There are fairly marked interspecific differences in neurocranial form, and these do not appear to be influenced by size. The neurocranium in *H. pellegrini* shows much greater departure from the generalized condition than does that of *H. pseudopellegrini* (for *H. pellegrini*, see Greenwood, 1962).

In its general facies, *H. pseudopellegrini* resembles *H. guiarti, H. altigenis, H. dentex, H. mento* and *H. gowersi,* particularly the former species.

The differences separating *H. pseudopellegrini* from *H. altigenis* are discussed on p. 64 ; *H. pseudopellegrini* could represent the ancestral morphotype from which *H. altigenis* evolved.

From *H. guiarti* it is distinguished by its different neurocranial form, and the following : slightly longer snout (35·1–39·0, M = 37·1% head, *cf.* 31·7–37·5, M = 34·4%), slightly narrower interorbital region (20·4–27·1, M = 23·9% head, *cf.* 23·4–30·2, M = 27·4%), longer and narrower lower jaw (48·6–53·4, M = 49·1% head, *cf.* 39·2–48·2, M = 44·4% ; length/breadth ratios 2·0–2·6 [modal range 2·1–2·4] *cf.* 1·5–2·3 [mode 2·0]), fewer teeth in the outer row of the upper jaw (35–52, M = 44, *cf.* 48–74, M = 62).

From *H. gowersi* it differs in head shape (and neurocranial form), and in having a much shorter head (34·2–37·0, M = 33·8% standard length, *cf.* 35·8–38·4, M = 37·0%), a broader snout, a larger eye (18·5–23·4, M = 20·6% head, *cf.* 15·5–19·3, M = 17·5%), a slightly shallower cheek (24·0–29·6, M = 26·8% head, *cf.* 27·8–33·3, M = 29·5%), and a longer caudal peduncle (16·3–20·8, M = 18·3% standard length, *cf.* 13·3–17·6, M = 14·8%).

From *H. dentex, H. pseudopellegrini* differs, superficially, by its less strongly decurved dorsal head profile. At a deeper level, there are differences in the shape of the neurocranium, that of *H. dentex* having a sharply decurved ethmoid-vomer region (see Greenwood [1962], p. 168 and fig. 25) ; but in other respects, the neurocrania of the two species are similar. Other interspecific differences lie in the more numerous and closely set teeth of *H. pseudopellegrini* (35–52, M = 44, *cf.* 32–48, M = 36), its deeper body (29·0–33·5, M = 30·9% of standard length, *cf.* 24·6–29·5, M = 26·7), and the greater posterior extension of the maxilla in this species (posterior maxillary tip reaching anterior orbital margin or to below the eye, *cf.* not reaching the orbital margin).

From *H. estor, H. pseudopellegrini* differs, principally, in having a shorter head

(32·4–37·0, M = 33·8% standard length, cf. 37·2–38·5, M = 37·8%), a shorter lower jaw (46·8–53·4, M = 49·1% head, cf. 54·2–57·5, M = 55·5%), and a longer caudal peduncle (16·3–20·8, M = 18·3% standard length, cf. 14·3–16·1, M = 15·8%). Neurocranial form in the two species differs, that of *H. estor* belonging to the "*prognathus*" group, whilst that of *H. pseudopellegrini* has stronger affinities with the "*altigenis*"-type (see p. 110).

The resemblance between *H. pseudopellegrini* and *H. mento* is probably the most distant of all. Osteologically, there is a clear-cut difference in neurocranial form (like that distinguishing *H. pseudopellegrini* and *H. estor*), and in most specimens the external head shape is distinctive (cf. fig. 12 in Greenwood [1962] with Text-fig. 8 above). Nevertheless, most cephalic morphometric characters are similar in the two species, although the snout of *H. pseudopellegrini* is broader (length/breadth ratio 1·1–1·3 cf. 1·5–1·8 in *H. mento*). The outer teeth in *H. mento* are stouter and more strongly curved than those of *H. pseudopellegrini*; the range for the number of outer upper jaw teeth overlaps in the two species, but the mean for *H. pseudopellegrini* is lower (44 cf. 52).

Phyletically, *H. pseudopellegrini* appears to be a derivative from an *H. guiarti*-like stem, and thus shows relationship with *H. bayoni* and *H. dentex*. However, unlike these species it also shows relationship with both *H. altigenis* and *H. pellegrini*. Structurally, *H. pseudopellegrini* could represent an ancestral level in the evolution of *H. altigenis*.

Museum and Reg. No.	Locality	Collector
	UGANDA	
B.M. (N.H.) 1966.3.9.299–310	Ekunu Bay	E.A.F.R.O.
B.M. (N.H.) 1966.3.9.296	Trawl in Buvuma Channel off Nasu Point	E.A.F.R.O.
B.M. (N.H.) 1966.3.9.297	Pilkington Bay	E.A.F.R.O.
B.M. (N.H.) 1966.2.21.2 (Holotype)	Pilkington Bay	E.A.F.R.O.
B.M. (N.H.) 1966.3.9.298	Sesse Islands	E.A.F.R.O.

Haplochromis altigenis Regan, 1922

(Text-figs. 9 and 10)

Paratilapia longirostris (part) : Boulenger, *Cat. Afr. Fish.*, **3**, 332.
Pelmatochromis spekii (part) : Boulenger, 1915, *op. cit., tom. cit.*, 417.
Haplochromis altigenis Regan, 1922, *Proc. zool. Soc. Londn.*, 175, Pl. 1.

LECTOTYPE : a male, 186 mm. standard length from Bunjako (Uganda), collected by Degen. B.M. (N.H.) reg. no. 1906.5.30.294.

DESCRIPTION : based on 25 specimens (including the lectotype and paralectotype), 100–202 mm. standard length.

Depth of body 28·4–34·7 (M = 31·5) per cent of standard length, length of head 36·7–39·5 (M = 38·2) per cent. Dorsal head profile gently curved, rather variable in its shape, tending to slope more steeply in large fishes which therefore have more rounded profiles; the two type specimens have the most strongly sloping head

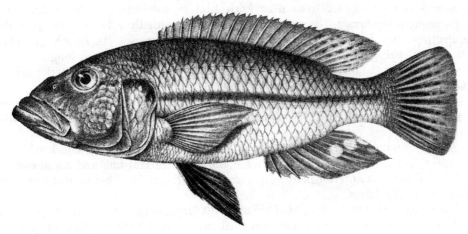

Fig. 9. *Haplochromis altigenis*; lectotype, about ·57 times natural size. From Regan, *Proc. zool. Soc.*

profiles (*ca* 40°); most other specimens lie in the range 30°–35°. The premaxillary pedicels are prominent, and clearly break the outline of the profile. Altogether, one is left with the impression of a heavy-headed fish.

Preorbital depth 17·6–23·7 (M = 19·9) per cent of head length, least interorbital width 17·4–25·0 (M = 20·2) per cent. Snout 1·1–1·3 times as long as broad, its length 36·6–42·2 (M = 39·3) per cent of head, eye diameter 16·9–21·7 (M = 19·2), depth of cheek 25·6–34·5 (M = 30·5) per cent. Cheek depth may show positive allometry with standard length in fishes > 180 mm.; the three largest fishes have the deepest cheeks.

Caudal peduncle 12·4–16·0 (M = 14·7) per cent of standard length, 1·2–1·4 (mode 1·3) times as long as deep.

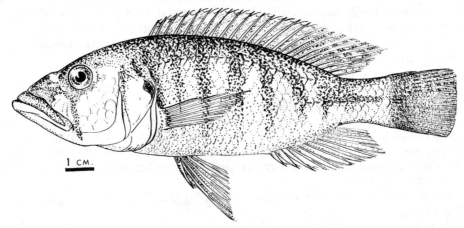

Fig. 10. *Haplochromis altigenis*, to show the usual head profile in fishes less than *ca.* 180 mm. S.L. Drawn by Barbara Williams.

Mouth slope variable, from near horizontal to moderately oblique (20°–30°). Lips slightly thickened, the premaxilla expanded medially. Lower jaw generally projecting (but jaws equal anteriorly in some specimens), its length 51·0–55·8 (M = 52·7) per cent of head, 1·8–2·4 (modal range 1·9–2·0) times as long as broad. Posterior tip of the maxilla almost reaching the vertical through the anterior orbital margin (reaching this point in a few specimens).

Gill rakers: stout (finer in the two smallest fishes), the upper 1–4 sometimes flattened and anvil-shaped; 8 or 9 (rarely 7 or 10) on the lower part of the first arch.

Scales: ctenoid; lateral line with 31 (f.7), 32 (f.13) or 33 (f.4), cheek with 4–6 (mode 5) rows. Six to 8 scales between the upper lateral line and the dorsal fin origin; chest and anterior belly scales small, 7 or 8 (rarely 9) between the pectoral and pelvic fin bases.

Fins: Dorsal with 24 (f.8), 25 (f.16) or 26 (f.1) rays, comprising 15 (f.22) or 16 (f.3) spinous and 9 (f.10) or 10 (f.15) branched rays. Anal with 11 (f.2), 12 (f.22) or 13 (f.1) rays, comprising 3 spines and 8 (f.2), 9 (f.22) or 10 (f.1) branched elements. Pectoral 20·1–26·2 (M = 22·6) per cent of standard length. Pelvics with the first branched ray produced, usually more so in males but almost as elongate in some females. Caudal subtruncate, scaled on its proximal half.

Teeth. In the size range of fishes studied, all teeth in the outer row of both jaws are stout, unicuspid and curved. There are 40–60 (M = 50) teeth in the outer row of the upper jaw.

The inner teeth of the two smallest fishes (90 and 119 mm. S.L.) are predominantly unicuspids, but some weakly tricuspid teeth occur in both jaws. In all other specimens, only unicuspids are found; these are moderately large, curved and implanted obliquely. The inner tooth rows are often irregularly arranged, with 3 (rarely 4) rows in the upper jaw, and 2 or 3 in the lower.

Osteology. The neurocranium of *H. altigenis* resembles that of *H. bayoni*, but is relatively broader in the otic region. Thus, although it shows some of the characters associated with the *H. mento*-type skull (see Greenwood, [1962], fig. 25) it still retains the curved preorbital profile, greater preorbital skull depth, and broad otic region of the more generalized neurocranium. In these characters it also resembles the neurocranium of *H. pseudopellegrini*.

The dentary and premaxillary show no outstanding characters. The former is a stout bone, and the dentigerous surface of the latter is but moderately expanded medially (less so, for example, than in *H. bayoni*).

The dentigerous surface of the lower pharyngeal bone is as broad as long but is broader than long in large fishes. Anteriorly this surface narrows rather abruptly so that the apex of the dentigerous triangle is produced into a narrow " stem ". The lower pharyngeal teeth are rather coarse, somewhat compressed and clearly but weakly cuspidate.

Vertebral counts in 7 specimens are: 13 + 16 (f.5) and 13 + 17 (f.2), giving totals of 29 and 30.

Coloration. Live colours are unknown. *Preserved material*: *Males (adult and sexually active)*: ground coloration a very dark brown, almost black on the dorsal surface and snout, and with a sooty overlay on the chest. Lower jaw and branchiostegal membrane pinkish-brown; margin of the preoperculum outlined in dark brown. Dorsal fin yellow-brown, lappets black, and with a faint black outline to the margin of the soft part. Caudal darker yellow-brown (darkest proximally). Anal yellow-brown, with 2 large, dead-white ocelli. Pelvics very dark brown, appearing black when folded. Pectorals dark brown (dark pectorals are unusual).

Males (adult but sexually quiescent) have a light brown ground coloration, darker on the dorsal surface of the snout, head and body. A prominent, but narrow, dark lachrymal stripe runs from the lower anterior border of the eye, passes almost vertically downwards behind the posterior tip of the maxilla to end on the dentary; a slightly broader dark bar lies immediately anterior to the vertical limb of the preoperculum. The lower jaw and branchiostegal membrane are very light brown; the chest is somewhat dusky. The flanks and caudal peduncle are crossed by seven very faint, moderately broad dark bars; these extend (on the flanks) from the dorsal fin base almost to the ventral body outline. Dorsal fin yellow-brown, the lappets black and the soft part darkly maculate. Caudal yellowish-brown, darkest on the proximal half. Anal dark yellow-brown. Pelvics dark brown, nearly black, on the anterior half, otherwise light yellow-brown.

Male (immature): essentially as for females (see below), but with black lappets on the dorsal fin, the caudal uniformly light but with some dark spots on the upper fifth, 3 faint whiteish ocelli on the anal, and a dusky overlay on the pelvics.

Note on the coloration of the type specimens. Both these specimens are males, but both are now a pale silver, shading to white. One specimen (that illustrated by Regan, see fig. 9 above) has a very faint midlateral stripe, which is now much fainter than is shown in the figure. Both fishes have dusky pelvics. In one, there are five, large, dead-white ocelli (arranged in three rows) on the anal fin, and in the other there are four (in two rows). The difference in coloration between these specimens and those described above is, presumably, due both to time and to the fact that the types were not fixed in formol but in alcohol.

Females (immature and adult): ground coloration light brown, darkest dorsally, and shading to yellowish-silver on the chest and belly. Lachrymal and preopercular stripes are as described for males, but in addition there are two, faint, parallel dark bands across the snout. The lower jaw and branchiostegal membrane are light yellow-brown. A faint dark midlateral band runs along the flank from slightly behind the opercular margin to the base of the caudal fin (sometimes extending onto the fin itself), and is crossed by 8 to 10, moderately broad vertical bars which extend from the dorsal fin base to about the level of the pectoral fin. Dorsal fin light greenish-brown, the soft part darkly maculate. Caudal yellow-brown on its distal quarter to third, dark brown basally. Anal greenish-brown. Pelvics light yellow-brown, somewhat darker along the anterior margin.

Three fishes (140, 148 and 149 mm. S.L.) caught on two occasions off Kisigala; Point, North Kome Island, show typical piebald coloration of black on silver; all three are immature females. These are the first known examples of piebald polychromatism amongst the piscivorous species-groups of Lake Victoria *Haplochromis*, although the phenomenon is recorded from other trophic and phyletic lines (see Greenwood, 1956, 1957, 1959, and p. 95 below).

Ecology. Habitat. The species occurs over both hard and soft substrates in sheltered bays and gulfs, as well as off-shore in places where the water is relatively undisturbed ; apparently it does not occur at depths of over 50 ft.

Food. Eleven of the 26 specimens examined (from 15 localities) contained food in the gut. In all cases this consisted of finely macerated fish remains, identifiable as cichlids in three specimens, and as *Haplochromis* species in six others.

Breeding. Haplochromis altigenis is a female mouth breeder. Fishes less than 145 mm. S.L. are immature, or, in the upper levels of the range, show early signs of maturation. The largest specimen (202 mm. S.L.) is a female, but both sexes occur in the size range 170–194 mm.

Affinities. At least superficially, four species, *H. pseudopellegrini, H. estor, H. dichrourus* and *H. gowersi*, closely resemble *H. altigenis*. A fifth species, *H. squamulatus* shows a more distant resemblance.

Similarities between *H. altigenis* and *H. dichrourus* are considered elsewhere (see p. 68) ; the resemblances between these species may indicate a fairly close phyletic relationship.

From *H. pseudopellegrini, H. altigenis* is distinguished by its longer head (36·7–39·5, M = 37·2% of standard length, *cf.* 32·4–37·0, M = 33·8%), deeper cheek (25·6–34·5, M = 30·5% head, *cf.* 24·0–29·0, M = 26·8%), longer lower jaw (51·0–55·8, M = 52·7% head, *cf.* 46·8–53·4, M = 49·1%) and its shorter and deeper caudal peduncle (12·4–16·0, M = 14·7% standard length, *cf.* 16·3–20·8, M = 18·3%, length/depth ratio 1·2–1·4 [mode 1·3] *cf.* 1·3–2·0 [modal range 1·6–1·7]). The neurocranium in the two species is generally similar. It seems that *H. altigenis* could have evolved from an *H. pseudopellegrini*-like ancestor.

From *H. estor, H. altigenis* differs in its longer snout (36·6–42·2, M = 39·3% head, *cf.* 34·5–37·0, M = 36·2%), smaller eye (16·9–21·7, M = 19·2% head, *cf.* 22·8–25·4, M = 24·5%), much deeper cheek (25·6–34·5, M = 30·5% head, *cf.* 22·8–25·4, M = 24·5%) and slightly shorter lower jaw (51·0–55·8, M = 52·7% head, *cf.* 54·2–57·5, M = 55·5%). There are fairly marked differences between the neurocrania of the two species, that of *H. estor* being of the " *prognathus* " type.

Superficially, the resemblances between *H. gowersi* and *H. altigenis* are great, and some difficulty may be experienced in separating certain specimens of the two species. However, there are distinct differences between modal specimens of the two species, and there is a clear difference in neurocranial form ; the skull of *H. gowersi* is of the " *prognathus* " type. This difference in neurocranial shape probably accounts for the more declivous snout and dorsal head profile of *H. altigenis*, and the greater prominence of the premaxillary pedicels in this species. In addition to these qualitative differences, *H. altigenis* differs from *H. gowersi* in having the interorbital width equal to or slightly greater than the preorbital depth (interorbital less than preorbital in *H. gowersi*), a broader head (greatest width, measured at about the middle of the operculum, 41·5–45·5, M = 42·7% head, *cf.* 35·6–39·0, M = 36·8) ; the greater head breadth is also reflected in the broader lower jaw of *H. altigenis*.

Haplochromis squamulatus and *H. altigenis* both have noticeably rounded head profiles, and very small chest and anterior belly scales. However, they differ in several morphometric characters. *Haplochromis altigenis* has a longer head, narrower interorbital, longer snout, deeper cheek and a longer lower jaw (see p. 87). The neurocrania show several interspecific differences (that of *H. squamulatus* showing affinities with the neurocrania of *H. martini* and *H. michaeli*) and there are marked differences in preserved coloration.

Considering the evidence available, both morphometric and osteological, it seems *H. altigenis* could have evolved from an *H. pseudopellegrini*-like stem ; certainly its neurocranial form does not favour a closer association with *H. estor* and *H. gowersi* than with *H. pseudopellegrini*.

Study material and distribution records

Museum and Reg. No.	Locality	Collector
	UGANDA	
B.M. (N.H.) 1906.5.30.294 (Lectotype)	Bunjako	Degen
B.M. (N.H.) 1906.5.30.295 (Paralectotype)	Bunjako	Degen
B.M. (N.H.) 1966.3.9.220–227	Napoleon Gulf, near Jinja	E.A.F.R.O.
B.M. (N.H.) 1966.3.9.228–235	Karenia Beach, near Jinja	E.A.F.R.O.
B.M. (N.H.) 1966.3.9.215–216	Near Ramafuta Island (Buvuma Channel)	E.A.F.R.O.
B.M. (N.H.) 1966.3.9.219	Pilkington Bay	E.A.F.R.O.
B.M. (N.H.) 1966.3.9.218	Manadu Island	E.A.F.R.O.
B.M. (N.H) 1966.3.9.213–214	Off Entebbe harbour	E.A.F.R.O.
	TANZANIA	
B.M. (N.H.) 1966.3.9.217	Mwanza Harbour	E.A.F.R.O.
	LAKE VICTORIA	
B.M. (N.H.) 1966.3.9.211–212	Locality unknown	E.A.F.R.O.

Haplochromis dichrourus Regan, 1922

(Text-fig. 11)

Paratilapia serranus (part) : Boulenger, 1915, *Cat. Afr. Fish.*, **3**, 334.
Haplochromis dichrourus Regan, 1922, *Proc. zool. Soc. Londn.*, 178, fig. 6.

HOLOTYPE : a juvenile female, 113·0 mm. S.L., from Buganga ; B.M. (N.H.) reg. no. 1906.5.30.265.

DESCRIPTION : based on eight specimens (including the holotype), 84–186 mm. S.L.

Depth of body 28·6–35·5 (M = 32·1) per cent of standard length, length of head 35·2–37·7 (M = 36·6) per cent. Dorsal head profile convex, sloping steeply (*ca* 40°) especially in the snout region ; premaxillary pedicels prominent and breaking the outline of the profile.

Preorbital depth 16·4–21·0 (M = 18·7) per cent of head, least interorbital width 18·0–24·6 (M = 21·3) per cent. Snout 1·10–1·25 times as long as broad, its length

34·1–41·0 (M = 37·4) per cent of head, eye diameter 19·4–24·6 (M = 21·1), depth of cheek 23·0–30·6 (M = 27·6) per cent.

Caudal peduncle 12·4–17·2 (M = 15·2) per cent of standard length, 1·1–1·5 times as long as deep.

Mouth slightly oblique (20°–30°), lips thickened, premaxilla slightly expanded medially. Lower jaw projecting, its length 49·0–54·0 (M = 52·7) per cent of head, 1·8–2·6 (no distinct mode) times as long as broad. Posterior tip of maxilla reaching the vertical through the anterior margin of the orbit, or to below the anterior part of the eye.

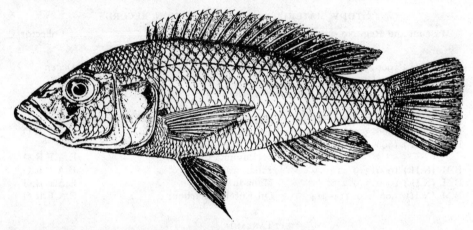

Fig. 11. *Haplochromis dichrourus*; holotype, about ·93 times natural size. From Regan, *Proc. zool. Soc.*

Gill rakers: stout, the lower 2 or 3 reduced, the upper 3 or 4 flat and expanded in some specimens; 8 or 9 (mode) on the lower part of the first gill arch.

Scales: ctenoid; lateral line with 32 (f.2), 33 (f.4) or 34 (f.2), cheek with 3 (f.1), 4 (f.1), 5 (f.5), or 6 (f.1) rows. Seven to 9 scales between the upper lateral line and the dorsal fin origin, 6–9 (mode 8) between the pectoral and pelvic fin bases; scales on the chest and belly small.

Fins: Dorsal with 25 rays, comprising 15 (f.1) or 16 (f.7) spinous and 9 (f.7) or 10 (f.1) branched rays. Anal with 11 (f.2), 12 (f.5) or 13 (f.1) rays, comprising 3 spines and 8 (f.2), 9 (f.5) or 10 (f.1) branched elements. Pectoral fin 21·2–27·3 (M = 23·7) per cent of standard length. Pelvics with the first, and to a lesser degree, the second branched rays produced, slightly so in females and markedly elongate in males. Caudal truncate or subtruncate.

Teeth. In all specimens the outer teeth (numbering 48–70, M = 58 in the upper jaw) are unicuspid, slender and very strongly curved (the tips of the anterolateral premaxillary teeth are not visible when the specimen is viewed laterally).

The inner teeth in fishes > 149 mm. S.L. are all unicuspid, curved and obliquely implanted. In the four smaller specimens, the 113 mm. fish (holotype) has mostly tricuspids in the upper jaw and an admixture of tri- and unicuspids (the latter predominating) in the lower jaw ; the 84, 101 and 121 mm. specimens have a mixture of unicuspids and weakly bicuspids in both jaws ; as in the larger fishes, these teeth are curved and obliquely implanted. The inner teeth are arranged in 1–3 series in the upper jaw, and in 2 or 3 in the lower (in a single row in one fish).

Osteology. With so few specimens available, it has not been possible to prepare a complete skeleton, but radiographs have been studied.

The lower pharyngeal bone is fine, with a triangular dentigerous surface which is equilateral or slightly broader than long. The lower pharyngeal teeth are slender, with cylindrical necks and slightly compressed weakly cuspidate crowns, and are arranged in 20–22 rows. Vertebral counts in 5 specimens are : $13 + 16$ (f.1) and $13 + 17$ (f. 4), giving totals of 29 and 30.

Ecology. Habitat. The species has a wide depth range, from 10–90 feet. It is found in both sheltered and exposed places (including beaches) over sand and shingle substrata.

Food. Of the five specimens examined (from four localities) four contained food in the gut. In three fishes this consisted solely of fish remains (*Haplochromis* sp.), and in the fourth fish-remains (a cichlid) and fragments of larval insects (one probably a dragon-fly, the other Ephemeroptera).

Breeding. Little information is available ; three of the smallest fishes (101, 113 and 121 mm. S.L.) are immature females.

An unusual feature of this species is the coloration of females, which, at least in preserved specimens, seems to be as polychromatic as that of males, and certainly more complex than the female coloration of other species (except, perhaps, *H. chromogynos*).

Coloration. The only information on live colours is provided by brief field notes made on a sexually active male caught in deep water near Soswa island. In this fish the median fins and back were described (by Mr. J. D. Kelsall) as vivid flame red, the belly as jet black.

Preserved Colours : Males. The most extreme pattern is shown by a fish which, although adult, is in an early stage of sexual activity. The upper part of the head (above the level of the lower orbital rim) brownish, snout yellowish and crossed by a thin, well-defined and dusky bar ; running parallel with the premaxillary pedicels are a pair of short, dark bars, each of which (at about the level of the pedicel tips) broadens somewhat and curves sharply at right angles to meet the orbit. The lower part of the head (cheeks, preorbital and the entire operculum) jet black. This colour extends onto the chest but does not reach above the level of the pectoral fin base. Immediately behind the pectoral fin, the margin of this black area curves ventrally but rises again above the vent so that the posterior half of the body is black on its lower half. On the caudal peduncle the dark area rises again to cover about the entire lateral aspect. Above the black areas the body is brownish. Dorsal and caudal fins are dark yellow-orange, with a narrow black crescent at the caudal base. The anal fin is a similar colour, with two ocelli, each outlined by a narrow black margin. Pelvics jet black, except for the yellowish innermost ray of each side.

In two other males (both sexually active, with convoluted testes and therefore thought to be more mature than the specimen above) the head and anterior half of the body are light orange-brown the posterior half and the caudal peduncle dusky brown. A well-defined black band crosses the snout (just above the upper lip) to the anterior margin of the preorbital bone. Another transverse black band at the level of the pedicel tips, runs from orbit to orbit. An interrupted band extends from the upper posterior margin of the orbit across the nape. A dusky, nearly vertical lachrymal stripe is present. The branchiostegal membrane and chest are black, but the belly is brown. Dorsal fin yellowish, the soft part is maculate, the spots being clear. Anal yellowish but dusky along its base, and with two, hyaline ocelli (set, in one specimen, in the dark basal zone). Caudal yellowish-grey, but with a broad, dusky band at its base. The pelvics are black except for the distal half of the membrane between the last two rays.

Females (adult and immature) : body and head dark grey-brown with a blueish hue. Branchiostegal membrane and chest dusky as are the lower parts of the operculum, preoperculum and interoperculum. Faint traces of snout and lachrymal bands (like those of males) are visible, as is a nuchal bar. Dorsal fin greyish to light sooty, the lappets dark. Anal pale orange with a single, well-defined, dead-white ocellus. Caudal with a dark base, the upper half grey and the lower orange. Pelvics light sooty, the pigment most concentrated on the anterior half of the fin.

Affinities. The anomalous coloration of female *H. dichrourus* was commented upon above. Apart from the relatively infrequent piebald females in certain species, and the occurrence of a piebald coloration as the usual one in females of *H. chromogynos* (see Greenwood, 1959), I know of no other species in Lake Victoria with colourful females.

In its gross morphology, *H. dichrourus* closely resembles *H. altigenis* although there are marked differences in the preserved coloration of the females. Both species have very small chest scales. They differ in that the maxilla of *H. dichrourus* reaches further posteriorly (rarely reaching the orbit in *H. altigenis*) and the premaxilla is not markedly expanded medially. Also, the teeth in *H. dichrourus* are more strongly curved, so that when viewed laterally, their tips are hidden ; in *H. altigenis*, the tips can always be seen.

The two smallest fishes resemble specimens of *H. pellegrini* of a similar size. They are, however, distinguished by their more slender and much more strongly recurved outer jaw teeth (in *H. pellegrini* these teeth have a gentle curvature confined to the distal part), and by having the maxilla extending further posteriorly (to below the anterior orbital margin). Also, in *H. dichrourus* the premaxillary pedicels are more prominent, and there are differences in preserved coloration.

Phyletically, *H. dichrourus* could be related to *H. altigenis* and *H. pellegrini*.

STUDY MATERIAL AND DISTRIBUTION RECORDS

Museum and Reg. No.	Locality	Collector
	UGANDA	
B.M. (N.H.) 1906.5.30.265 (type)	Buganga	Degen
B.M. (N.H.) 1966.3.9.186	Jinja, off golf course	E.A.F.R.O.
B.M. (N.H.) 1966.3.9.187	Katebo	E.A.F.R.O.
B.M. (N.H.) 1966.3.9.188–9	Karenia, Napoleon Gulf	E.A.F.R.O.

TANZANIA

B.M. (N.H.) 1966.3.9.185	Off Soswa Island . . .	J. D. Kelsall
B.M. (N.H.) 1966.3.9.183–4	Beach N. of Majita . . .	E.A.F.R.O.

Haplochromis paraguiarti sp. nov.

(Text-fig. 12)

HOLOTYPE : an adult male 130 mm. standard length, from a beach near Nasu Point, Buvuma Channel (Uganda); B.M. (N.H.) reg. no. 1966.2.21.6.

Named *paraguiarti* because of its close superficial resemblance to *H. guiarti*.

DESCRIPTION : Based on 31 specimens (including the holotype), 70–156 mm. standard length.

Depth of body 31·7–36·6 (M = 34·1) per cent of standard length, length of head 33·0–36·8 (M = 34·8) per cent. Dorsal head profile straight or very slightly curved, the premaxillary pedicels forming a slight prominence.

Preorbital depth 14·5–19·5 (M = 16·9) per cent of head length, least interorbital width 22·9–27·7 (M = 25·3) per cent. Snout as long as broad or slightly longer (1·2 times), its length 31·0–36·2 (M = 33·5) per cent of head, eye diameter 23·8–28·6 (M = 26·2), depth of cheek 20·0–26·1 (M = 23·8) per cent.

Caudal peduncle 14·7–18·6 (M = 16·6) per cent of standard length, 1·3–1·9 (modal range 1·3–1·5) times as long as deep.

Mouth horizontal or slightly oblique, the jaws equal anteriorly. Lower jaw 41·0–48·6 (M = 45·0) per cent of head, 1·7–2·4 (modal range 1·9–2·1) times as long as broad. Posterior tip of the maxilla reaching the vertical through the anterior orbital margin, occasionally extending to below the eye. Premaxilla slightly expanded medially.

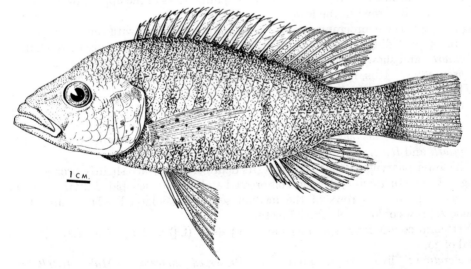

FIG. 12. *Haplochromis paraguiarti*. Drawn by Barbara Williams.

Gill rakers: moderately stout (relatively slender in fishes less than 90 mm. S.L.), the upper 4–6 flat and sometimes branched in fishes over 115 mm. S.L., the lower 1 or 2 reduced ; 8–10 (mode 9), rarely 11, on the lower part of the first gill arch.

Scales: ctenoid ; lateral line with 31 (f.4), 32 (f.12), 33 (f.12) or 34 (f.3) scales, cheek with 3 or 4 (rarely 2 or 5) rows. Five or 6 (rarely $6\frac{1}{2}$) scales between the upper lateral line and the dorsal fin origin, 6 or 7 between the pectoral and pelvic fin bases. (Because many specimens were damaged during capture by gill-nets, the last two counts are based on only 17 specimens).

Fins: Dorsal with 23 (f.1), 24 (f.8), 25 (f.19) or 26 (f.3) rays, comprising 15 (f.21) or 16 (f.10) spinous and 8 (f.1), 9 (f.15) or 10 (f.15) branched rays. Anal with 11 (f.6), 12 (f.23) or 13 (f.2) rays, comprising 3 spines and 8 (f.6), 9 (f.23) or 10 (f.2) branched elements. Pelvic fins with the first unbranched ray slightly produced in the adults of both sexes, but proportionately more so in males. Pectoral 24·3–30·0 (M=26·3) per cent of standard length. Caudal truncate or subtruncate, scaled on its basal half to two-thirds.

Teeth: The form of the teeth in the outer row of both jaws shows some correlation with standard length. Only bicuspids (with a few tricuspids posterolaterally in the upper jaw) are found in fishes 70–75 mm. S.L. Unicuspids anteriorly, with bicuspids laterally are characteristic of most fishes 84–113 mm. S.L. (and in one exceptional individual 144 mm. S.L.). In most specimens over 112 mm. S.L., only unicuspids occur, but this is also the condition in a few fishes between 99 and 112 mm. S.L. The unicuspids are moderately stout teeth, and are only slightly curved. There are 42–62 (M = 54) outer teeth in the upper jaw.

Teeth in the inner rows are relatively large ; tricuspids predominate in fishes of all sizes, but in specimens over 120 mm. S.L. weakly tricuspids are commoner than the distinctly tricuspid ones, and some unicuspid teeth also occur (especially in the outermost row of the inner series). There are 3 or 4 inner rows in the upper jaw, and 2 or 3 (less frequently 1 row) in the lower.

Osteology: The neurocranium of *H. paraguiarti* is identical with that of *H. acidens* (see p. 75). It differs somewhat from the presumed generalized piscivore skull of *H. guiarti*, and shows some of the characters found in the more specialized type of *H. prognathus*. It is, in fact, almost intermediate between the two types.

The premaxilla has a less pronounced medial expansion than in *H. acidens*, but as in that species it has long pedicels (as long as the horizontal dentigerous arms) which meet the horizontal arms at an appreciable angle. In these characters the premaxilla of *H. acidens* differs from that bone in the generalized species like *H. guiarti* and *H. serranus*.

The lower pharyngeal bone is slender, its dentigerous surface slightly broader than long. The teeth are fine, slender, compressed and clearly bicuspid, only those in the posterior one or two rows of the median series are enlarged. There are 26–30 (mode 28) rows of lower pharyneal teeth.

Vertebral counts for 6 specimens are : 13 + 17 (f.4) and 14 + 16 (f.2), giving a total of 30.

Coloration: live colours unknown. *Preserved material:* Males (*adult and sexually active*) : ground colour dark sooty, almost uniformly so except for a lighter (dusky gold) patch on the midflank, crossed by four, faint and narrow vertical stripes. Branchiostegal membrane dark (darker below the opercular series than between the jaws). A very faint, near vertical lachrymal stripe runs from the anterior orbital margin to behind the posterior tip of the maxilla. Dorsal fin dusky, lappets black, the soft dorsal maculate. Caudal dark. Anal dusky, but lighter than the dorsal except for a narrow band along the base, and over the spinous part ; two or three greyish ocelli present. Pelvics dusky, the proximal threequarters almost black, the distal quarter lighter.

Males (adult but quiescent) : ground colour dusky silver-grey, darkest dorsally ; cheek bright silver. A faint lachrymal stripe present. Dorsal fin greyish-dusky, darkly maculate between the last four spines and all the branched rays. Caudal dark. Anal dark hyaline, with two or three greyish ocelli. Pelvics sooty, darkest on the leading edge.

Females (adult) : ground colour silver grey, darker dorsally : cheek bright silver (*i.e.* like quiet males but lighter and brighter). Dorsal fin dark hyaline, with dark spots on the soft part. Caudal dark hyaline, weakly to distinctly maculate. Anal and pelvic fins hyaline.

Females (juvenile) : Two small (75 mm. S.L.) specimens have colours like those of adult females, but with traces of 7 or 8 vertical bars on the flanks and caudal peduncle ; these bars reach the dorsal outline but do not extend ventrally below the level of the pectoral fin base.

Ecology. Habitat. Most records of *H. paraguiarti* are from exposed, wave-washed beaches where the substrate is of sand, rock or shingle. There are, however, a few records from more sheltered beaches and areas, but with one exception (a mud bottom) the substrate was hard, and the depth invariably never more than 30 ft.

Food. Twenty specimens (from 12 localities) were examined. Of these, 17 contained ingested material. One individual contained fish remains (very fragmentary and unidentifiable, even to family) together with a little macerated phanerogam. tissue. Eleven others yielded either macerated plant tissue (f.4), or a mixture of plant tissue and insect fragments (larval Ephemeroptera) ; five specimens contained only fragments of larval Ephemeroptera.

Breeding. Little information is available on reproduction in this species. Fishes less than 95 mm. S.L. are immature, as is one larger specimen (a male, 100 mm. S.L.). The largest fish (156 mm. S.L.) is a male.

Affinities. Superficially, *H. paraguiarti* resembles *H. guiarti* both morphologically and trophically, although *H. guiarti* apparently includes a greater proportion of fish in its diet. Morphologically the species may be distinguished by the straight dorsal head profile of *H. paraguiarti* (gently curved in *H. guiarti*), its longer head (33·0–36·8, M = 34·8 per cent standard length *cf.* 29·5–33·8, M = 31·4 per cent), somewhat broader snout, and fewer, coarser outer teeth (42–60, M = 53, *cf.* 48–74, M = 62 is the upper jaw). Live colours are unknown for *H. paraguiarti*, but the coloration of preserved, sexually active males is noticeably darker than that of *H. guiarti* males. Osteological differences, especially in neurocranial shape, are discussed above (p. 70).

Haplochromis paraguiarti also resembles, rather closely, *H. acidens* ; the species are compared on p. 76.

Some specimens of *H. prognathus*, a rather variable species, resemble *H. paraguiarti* but there are several differences which serve to distinguish even these superficially similar individuals ; *H. paraguiarti* has a shorter head (33·0–36·8, M = 34·8 per cent of standard length, *cf.* 35·5–38·4, M = 36·9 per cent), a shallower preorbital (14·5–19·5, M = 16·9 per cent head, *cf.* 18·8–23·1, M = 20·8 per cent), shorter snout (31·0–36·2, M = 33·5 per cent head, *cf.* 33·4–39·0, M = 37·0 per cent), and a larger eye (23·8–28·6, M = 26·2 per cent head, *cf.* 20·0–25·0, M = 22·6 per cent). Differences in neurocranial form between the species are also distinctive, although the neurocranium of *H. paraguiarti* is intermediate between the more generalized *H. guiarti* type and the relatively elongate skull of *H. prognathus* (see p. 111).

When all characters are considered, *H. paraguiarti*, despite its superficial resemblance to *H. guiarti*, is probably more closely related to *H. prognathus*. It appears to possess the structural characters of the ancestral species or species group from which piscivorous predators like *H. prognathus*, *H. bartoni* and *H. mandibularis* were derived. Outside the piscivorous predator group, it shows very close relationship with *H. acidens* (see p. 76).

Study Material and Distribution Records

Museum and Reg. No.	Locality	Collector
Uganda		
B.M. (N.H.) 1966.3.9.324	Beach near Jinja	E.A.F.R.O.
B.M. (N.H.) 1966.2.21.6 (Holotype)	Beach near Nasu Point (Buvuma Channel)	E.A.F.R.O.
B.M. (N.H.) 1966.3.9.338–344	Beach near Nasu Point	E.A.F.R.O.
B.M. (N.H.) 1966.3.9.313–314	Fisherman's point near Jinja	E.A.F.R.O.
B.M. (N.H.) 1966.3.9.320–323	Near Grant Bay	E.A.F.R.O.
B.M. (B.H.) 1966.3.9.333–337	Entebbe Harbour	E.A.F.R.O.
B.M. (N.H.) 1966.3.9.317–319	Buka Bay	E.A.F.R.O.
B.M. (N.H.) 1966.3.9.312	Katebo	E.A.F.R.O.
B.M. (N.H.) 1966.3.9.325–330	Beach near Hannington Bay	E.A.F.R.O.
Kenya		
B.M. (N.H.) 1966.3.9.316	Kamaringa (Kavirondo Gulf)	E.A.F.R.O.
Tanzania		
B.M. (N.H.) 1966.3.9.311	Beach N. of Majita	E.A.F.R.O.
B.M. (N.H.) 1966.3.9.315	Majita Beach	E.A.F.R.O.
Lake Victoria		
B.M. (N.H.) 1966.3.9.331–332	Locality unknown	E.A.F.R.O.

Haplochromis acidens sp. nov.

(Text-fig. 13)

Haplochromis percoides (part) : Boulenger, 1915, *Cat. Afr. Fish.*, **3**, 296 (specimens from Kakindu, Victoria Nile, B.M. [N.H.] reg. no. 1911.3.3.82–3).
Paratilapia serranus (part) : Boulenger, 1915, *op. cit., tom. cit.*, 334 (specimens from Ripon Falls, nr. Jinja, B.M. [N.H.] reg. no. 1911.3.3.20–1).
Haplochromis nigrescens (part) : Regan, 1922, *Proc. zool. Soc. Londn.*, 172 (excluding his tentative synonymy of *Astatotilapia roberti* Pellegrin, and the synonymy of *Paratilapia parvidens* Blgr., for which see Greenwood, 1959, p. 194).

NOTES ON SYNONYMY. It is obvious that Regan's (1922) redescription of *H. nigrescens* (Pellegrin), 1909 was influenced by Pellegrin's figure of that species (Pellegrin, 1910, *Mem. soc. Zool. France*, 32, pl. 14, fig. 3). The fish depicted certainly does resemble a specimen of the species here described as *H. acidens*, but the type specimen of Pellegrin's *Astatotilapia nigrescens* more closely resembles *Haplochromis flavipinnis* (personal observations). Boulenger (1915) actually synonymized *A. nigrescens* Pellegrin with *H. percoides* Blgr., 1906 but this action was unacceptable to Regan, who resurrected the species as *Haplochromis nigrescens*. Certainly there are similarities between *A. nigrescens* type specimen and *H. percoides*, but my study of the type was revealed several differences (see appendix for a redescription of this specimen and further comments on its affinities).

In earlier papers I had accepted Regan's resurrection of *H. nigrescens*, and used specimens identified by him as a basis for comparing other Lake Victoria species with *H. nigrescens* ; thus, where *H. nigrescens* is mentioned in those papers, the name should now be corrected to read *H. acidens*.

Two of the three Lake Victoria specimens examined by Regan undoubtedly can be referred to *H. acidens*, but I am uncertain about the identity of the third fish. It is the smallest specimen, and because so little is known about small fishes of this and related groups, diagnostic characters applicable to larger individuals cannot be used with confidence. In many respects, this specimen resembles larger individuals of *H. prognathus*.

HOLOTYPE : an adult male, 108 mm. standard length, from Kisumu Kavirondo Gulf (Kenya), B.M. (N.H.) reg. no. 1966.2.21.1. Named *acidens* from its sharp, needle-like teeth.

DESCRIPTION : based on 17 specimens 67–128 mm. S.L., including the holotype and two of the specimens examined by Regan (B.M. [N.H.] reg. no. 1911.3.3.20–1).

Depth of body 34·6–40·3 (M = 38·1) per cent of standard length, length of head 34·0–36·7 (M = 35·4) per cent. Dorsal head profile gently concave (clearly so in most specimens but weakly concave in a few), sloping fairly steeply ; premaxillary pedicels not prominent.

Preorbital depth 15·2–19·2 (M = 17·1) per cent of head, least interorbital width 21·4–28·2 (M = 24·5) per cent. Snout as long as broad or 1·1–1·2 times longer, its length 31·8–37·2 (M = 34·2) per cent of head, eye diameter 23·0–29·0 (M = 26·4), depth of cheek 23·0–28·2 (M = 25·7) per cent.

Caudal peduncle 14·8–17·9 (M = 16·6) per cent of standard length, 1·1–1·5 (mode 1·3) times as long as deep.

Mouth slightly oblique, lips somewhat thickened, premaxilla with a slight median expansion. Jaws equal anteriorly, the lower 44·0–50·2 (M = 46·9) per cent of head, 1·9–2·4 (rare), mode 2·1, times as long as broad. Posterior tip of the maxilla reaching the vertical through the anterior orbital margin, occasionally not quite reaching this point.

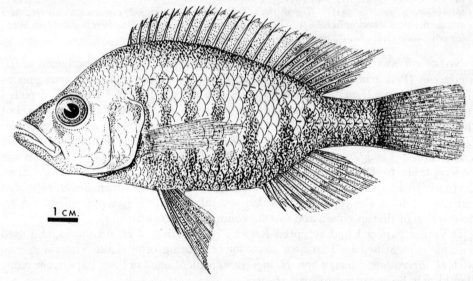

Fig. 13. *Haplochromis acidens*. Drawn by Barbara Williams.

Gill rakers: moderately slender to slender, the lower 1–4 reduced, the upper 3 or 4 flattened and lobed in some fishes; 10 (rarely 9 or 11) on the lower part of the first gill arch.

Scales: ctenoid; lateral line with 31 (f.1), 32 (f.8), 33 (f.7) or 34 (f.1), cheek with 3 (rarely 4) rows. Five and a half to 7 (mode 6) scales between the dorsal fin origin and the upper lateral line, 5 or 6 (mode), rarely 6½ or 7 between the pectoral and pelvic fin bases.

Fins: Dorsal with 23 (f.1), 24 (f.7) or 25 (f.9) rays, comprising 15 (f.12) or 16 (f.5) spinous and 7 (f.1), 9 (f.11) or 10 (f.5) branched rays. Anal with 11 (f.7) or 12 (f.10) rays comprising 3 spines and 8 or 9 branched elements. Pectoral 22·7–28·0 (M = 24·9) per cent of standard length. Pelvics with the first two branched rays slightly produced, relatively more so in adult males. Caudal truncate to subtruncate, scaled on its proximal half to two-thirds.

Teeth. In the outer row of both jaws the teeth are tall, slender and slightly curved. Fishes between 67 and 90 mm. S.L. have unicuspid teeth anteriorly, and weakly bicuspid teeth laterally and posterolaterally. Larger fishes have only unicuspids in the outer row. The anterior teeth in the upper jaw of the largest specimens may be relatively enlarged (*i.e.* stouter and longer than the lateral teeth). There are 40–64 (M = 48) teeth in the outer, upper row.

The inner rows of the upper jaw are composed of tricuspid teeth in fishes $<$ 90 mm. S.L. and a mixture of uni- and tricuspids in larger fishes; there is an increased proportion of unicuspids in specimens $>$ 115 mm. S.L. In fishes over 100 mm. S.L. the outermost row of the inner series is often composed of unicuspids and the remaining rows of tricuspids. The inner rows are widely spaced in most fishes so that the teeth form a broad band anteriorly and anterolaterally. There are 3 or 4 rows of teeth in the upper jaw.

In the lower jaw of most specimens less than 120 mm. S.L., only tricuspid teeth are found, but some unicuspids also occur in fishes of this size group. Larger individuals have a mixture of tri- and unicuspid or tri- and weakly tricuspid teeth; only rarely are all the lower, inner teeth unicuspid. There are 2 (mode) or 3 rows of inner teeth in this jaw.

Osteology. The neurocranium of *H. acidens* departs from that of *H. guiarti* towards the *H. mento* type (see Greenwood, 1962). The slope of its anterior dorsal profile (from vomer tip to the origin of the supraoccipital crest) is less steep than in *H. guiarti*, and the maximum width (across the otic region) is relatively less, as is its greatest depth (exclusive of the supraoccipital crest). In all these characters, *H. acidens* is intermediate between *H. guiarti* and *H. prognathus*, a species whose neurocranium could provide a basic " bauplan " from which the more extreme " *mento* "-type evolved.

The premaxilla is somewhat beaked, with the dentigerous surface expanded and protracted medially; the pedicels are elongate, being almost as long as the horizontal arms of the bone.

The lower pharyngeal bone is fine, and has the dentigerous surface broader than long. Lower pharyngeal teeth are arranged in 24–28 rows, and are slender, compressed and distinctly cuspidate, with those in the two median rows somewhat coarser than the others.

Vertebral counts from 12 specimens are: $13 + 16$ (f.4), $13 + 17$ (f.7) and $14 + 16$ (f.1), giving totals of 29 and 30.

Coloration. Live colours: *Males (sexually active)* have a slatey-blue ground colour, darkest dorsally. Dorsal fin smokey-grey with a pale red margin and a deep red flush on the soft part. Caudal dark grey with deep red streaks between the rays. Anal dark hyaline with a dull maroon flush, and orange-red ocelli. Pelvics black.

Females (adult and quiescent): dorsal surface yellowish-grey shading to silver on the flanks and belly. Dorsal fin hyaline but with a yellowish base. Anal and caudal dark hyaline. Pelvics hyaline.

Preserved material: Males (*adult and sexually active*): ground colour brown (darker than in females and juveniles), darkest dorsally and on chest, belly and ventral half of the caudal peduncle; six to eight vertical bands cross the flanks and caudal peduncle, each band broadening slightly below the level of the upper lateral line, and merging ventrally with the dark body coloration, but remaining discrete dorsally. A narrow, dark lachrymal stripe runs from the anterior border of the eye, sloping backwards to pass behind the posterior tip of the maxilla; a dark narrow stripe outlines the vertical limb of the preoperculum. The spinous dorsal is greyish, the soft part darker on its basal two-thirds, maculate over the posterior half. Caudal maculate on its proximal half to two-thirds. Pelvics dusky, particularly on the basal half; membrane covering the spine colourless. Anal greyish to hyaline, with two, ill-defined and dark grey to dusky ocelli.

Males (sexually quiescent) : as above but all stripes and bars are much fainter and less well-defined, and the ventral body surface is not noticeably darker than the dorsum, although still darker than the mid-flank region. Only the basal half of each pelvic fin is dusky, and the anal ocelli are almost invisible.

Immature fishes have the same coloration as females.

Ecology. Habitat. The majority of specimens is from a sheltered habitat in the Kavirondo Gulf near Kisumu. At this place the water was about 10 ft. deep, and the substrate of mud; fairly dense stands of submerged plants were common in the area. The other specimens are recorded as being from " Jinja, Ripon Falls ". I am presuming that this locality is in the Napoleon Gulf, that is, above the falls. The area, before the submergence of the Falls, was sheltered, with a hard substrate and, at least close inshore, fairly dense plant stands.

Food. All sixteen specimens examined came from the Kisumu locality, but were caught on different occasions; fifteen specimens contained ingested material in the stomach and intestines. Every gut had, as its major content, finely macerated phanerogam tissue and varying amounts of epiphytic diatoms. Some specimens also contained a flocculent, grey-green mass (as seen in preserved material) thought to be bottom debris, principally blue-green algae. In addition, one gut yielded a few fish bones, one a number of insect eggs (apparently undigested), one a mass of tissue (thought to be the foot of a gastropod) and some fragments of larval Ephemeroptera, and another, fragments of unidentifiable insects.

The phanerogam tissue is very finely divided, and a larger proportion seems to be digested than is the case in other phytophagous species examined.

The apparently vegetarian diet of *H. acidens* is at variance with the dentition and general facies which are essentially those of a piscivorous predator. The length of the gut ($1\frac{1}{2}$ to 2 times standard length) is greater than in the piscivores, but is not as long as the gut in purely phytophagous species.

Breeding. Specimens < 90 mm. S.L. are immature; both sexes reach the same maximum adult size.

Affinities. In its general appearance, *H. acidens* resembles a number of the deeper-bodied piscivorous species, and one of the specialized predators on larval and embryo fishes, *H. parvidens* (Greenwood, 1959).

Perhaps the greatest resemblance is between *H. acidens* and *H. paraguiarti*. Both species have similar neurocrania and jaw structure, but *H. acidens* has finer teeth, a deeper body (34·6–40·3, M = 38·1 per cent of standard length, *cf.* 31·7–36·6, M = 34·8%) and a concave as opposed to a straight dorsal head profile.

Also showing an overall similarity with *H. acidens* is *H. maculipinna*. The concavity of the dorsal head profile in *H. acidens* again serves as one differentiating character. Others are its less oblique jaws, fewer (40–64, M = 48 *cf.* 50–80, M = 62), more slender and longer teeth, and its larger chest scales (5–6 between pectoral and pelvic fin bases, *cf.* 6–9, mode 7 or 8, in *H. maculipinna*). The neurocrania of the two species also differ, that of *H. maculipinna* being deeper and having a steeper dorsal profile; the premaxilla in this species lacks a pronounced median expansion, and the pedicels meet the horizontal arms almost at right angles (and not at the more acute angle found in *H. acidens*).

From other species of the *H. serranus* group (*H. victorianus*, *H. serranus*, *H. spekii* and *H. nyanzae*), *H. acidens* is distinguished by its concave dorsal head profile, smaller eye, finer, longer and fewer teeth (at least as shown by the mean number, the ranges overlap), less oblique and non-prognathous lower jaw, and by the absence of a prominent mental bump at the symphysis of the dentaries. Other differences also serve to distinguish *H. acidens* from individual species of the *H. serranus* complex; for these see the species descriptions in Greenwood (1962) and p. 32 above for *H. spekii*.

At about the same level of similarity with *H. acidens* is *H. prognathus*. This species differs from *H. acidens* chiefly in its shallower body (30·1–37·1, $M = 33·3\%$ of standard length, cf. 34·6–40·3, $M = 38·1\%$), deeper preorbital (18·8–23·1, $M = 20·8\%$ of head, cf. 15·2–19·2, $M = 17·1\%$) and smaller eye (20·0–25·0, $M = 22·6\%$ of head, cf. 23·0–29·0, $M = 26·4\%$). The dentition and neurocrania of the species differ quite markedly, with *H. prognathus* having a skull nearer the *H. mento* type (see Greenwood, 1962, and p. 81) and stouter teeth.

It is difficult to assess the phyletic position of *H. acidens*. The level of anatomical specialization, especially of the syncranium and dentition, suggests affinity both with the *H. serranus* species group and with *H. prognathus*, a species not far removed from that group but probably representative of the ancestral type from which such specialized forms as *H. mento* and *H. macrognathus* evolved. Its plant diet suggests a possible trophic specialization paralleling that achieved by a species (*H. phytophagus*) related to the small, anatomically generalized, and usually insectivorous species (Greenwood, 1966).

In an earlier paper (Greenwood, 1959), I suggested that *H. parvidens* might have evolved from a species anatomically like *H. acidens* (called *H. nigrescens* in that paper); this relationship is discussed further on page 114.

Phyletically, *H. acidens* was probably derived from the same stem as *H. paraguiarti*; this stem could have been related to either an *H. guiarti*-like lineage, or to an *H. serranus*-like one.

Study Material and Distribution Records

Museum and Reg. No.	Locality	Collector
Uganda		
B.M. (N.H.) 1911.3.3.20–21	Jinja, Ripon Falls	Bayon
Kenya		
B.M. (N.H.) 1966.3.9.107–121	Kavirondo Gulf, near Kisumu	E.A.F.R.O.
B.M. (N.H.) 1966.2.21.1 (Holotype)	Kavirondo Gulf, near Kisumu	E.A.F.R.O.

Haplochromis prognathus (Pellegrin), 1904

(Text-figs. 14 and 15)

Paratilapia prognatha Pellegrin, 1904 (probably in part; the identity of the three smallest syntypes is still in doubt), *Bull. Soc. Zool. France*, **29**, 185; Idem, 1905, *Mem. Soc. Zool. France*, **17**, 181, pl. 16, fig. 4. Boulenger, 1915, *Cat. Afr. Fish.*, **3**, 333, fig. 224 (in part, the syntype B.M. [N.H.] reg. no. 1905.2.28.1, *ex* Kavirondo Gulf).
Pelmatochromis spekii (part) : Boulenger, 1915, *Cat. Afr. Fish.*, **3**, 417.
Haplochromis prognathus: Regan, 1922, *Proc. zool. Soc. London*, 177.
Haplochromis taeniatus Regan, 1922, *op. cit.*, 170, text-fig. 3.
Haplochromis macrodon Regan, 1922, *op. cit.*, 176, text-fig. 4.
Haplochromis lamprogenys Fowler, 1936 *Proc. Acad. Nat. Sci. Philad*, **88**, 330, fig 137.
Haplochromis steindachneri Lohberger, 1929. *Anz. Akad. Wiss. Wein*, no. 17, 207.
Haplochromis versluysi Lohberger, 1929, *Anz. Akad. Wiss. Wein* no. 17, 206.
Haplochromis rebeli Lohberger, 1929, *Anz. Akad. Wiss. Wein.*, no. 11, 94 (probably in part; the identity of one paratype, a female 116 mm. S.L., reg. no. 18768 of the Vienna Museum, is still doubtful).

LECTOTYPE : a fish 136·0 mm. standard length, B.M. (N.H.) reg. no. 1905.2.28.1 (presented by the Paris Museum), collected by C. Alluaud from Kavirondo Bay, Kenya. This specimen appears to be the fish figured by Pellegrin (1905), and is chosen as lectotype principally for that reason.

Note : Defining this species has proved particularly difficult, mainly because of the condsiderable intraspecific variability in head shape (see text-fig. 15). This variability, although apparent to the eye, is not readily quantifiable. Intergrades exist between the most outstanding variants, and now that a large series of specimens is available it appears that the variation is, to a large extent, size correlated. In fishes less than 110 mm. S.L., the snout seems protracted and the dorsal head profile has a marked and extended concavity above the eye. In larger fishes the elongate " face " is less noticeable, and the dorsal profile is straighter and slopes steeply (but still with a slight supraorbital concavity).

DESCRIPTION : based on 43 specimens, 70–141 mm. S.L., and including the lectotype, the syntypes of *H. taeniatus* and *H. macrodon* and the holotypes of *H. steindachneri* and *H. lamprogenys*. The types of *H. rebeli* and *H. versluysi* were examined but are not included in this redescription.

Depth of body 30·1–37·1 (M = 33·3) per cent of standard length, length of head 35·3–38·4 (M = 36·9) per cent ; head profile variable (see note above), usually with a supraorbital concavity, the dorsal outline sloping at an angle of 30°–40° with the horizontal.

Preorbital depth 18·8–23·1 (M = 20·8) per cent of head, least interorbital width 18·1–23·7 (M = 21·0) per cent. Snout length 33·4–39·6 (M = 37·0) per cent of head, $1\frac{1}{4}$ to $1\frac{1}{3}$ times as long as broad (rarely $1\frac{3}{8}$ times). Eye diameter with slight negative allometry, 20·0–25·0 (M = 22·6) per cent of head ; in fishes 70–90 mm. S.L., the eye is clearly larger than the cheek is deep but in specimens 90–110 mm. it is equal to or slightly smaller than the cheek, a relationship that holds for some larger individuals (110–125 mm. S.L.) although in most individuals in this size range the eye diameter is manifestly less than the cheek depth. Depth of cheek 20·6–28·6 (M = 24·5) per cent of head (18·9% in the smallest specimen).

Fig. 14 *Haplochromis prognathus*. Lectotype, about natural size. From Boulenger, *Fishes of the Nile*.

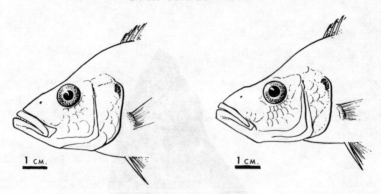

Fig. 15. *Haplochromis prognathus*, showing variability of head profile.

Caudal peduncle 14·6–19·0 (M = 16·0) per cent of standard length, 1·2 (rarely)–1·7 times is long as deep (modal range 1·3–1·5).

Mouth moderately oblique, the jaws equal anteriorly or the lower projecting slightly; lower jaw with a slight mental protuberance, its length 42·5–51·1 (M = 46·2) per cent of head, 2·0–2·8 (mode 2·4) times as long as broad. Posterior tip of the maxilla not reaching the vertical through the anterior orbital margin (except in one specimen), but usually reaching a point nearer this line than one through the nostril.

Gill rakers: 8–10 (mode 9) on the lower part of the first gill arch, the lower 1–3 rakers reduced, the upper 3 or 4 often flat or flat and lobed, the remainder slender.

Scales: ctenoid. Lateral line with 30 (f.2), 31 (f.12), 32 (f.21) or 33 (f.8) scales; some individuals lack pores in the scales of this series. Cheek with 3 (occasionally 4) rows. Five to 7 (mode 6) scales between the dorsal fin origin and the upper lateral line, 5–7 (mode 6) between the pectoral and pelvic fin bases.

Fins. Dorsal with 23 (f.6), 24 (f.26) or 25 (f.10) rays, comprising 14 (f.11), 15 (f.28) or 16 (f.3) spines and 8 (f.1), 9 (f.28) or 10 (f.13) branched rays. Anal with 11 (f.2), 12 (f.21) or 13 (f.11) rays, comprising 3 spines and 8 (f.2), 9 (f.21) or 10 (f.11) branched elements. Pectoral fin shorter than head, 22·0–29·3 (M = 25·7) per cent of standard length. First pelvic branched ray slightly produced in both sexes but proportionately more so in males. Caudal truncate or subtruncate, scaled on its basal half to two-thirds.

Teeth. In the smallest fish examined (70 mm. S.L.) the outer teeth in both jaws are mostly unequally bicuspid, but a few unicuspids are present. Fishes between this size and 105 mm. S.L. have mostly unicuspids in the outer row, but some bicuspids occur posterolaterally in the upper jaw. Fishes > 105 mm. have only unicuspid outer teeth, a condition found occasionally in individuals as small as 90 mm. S.L. The unicuspids are slightly curved inwards, and vary in form from relatively slender to moderately stout. There are 30–60 (M = 45) teeth in the outer row of the upper jaw, the number not showing any correlation with the fish's size. Regan's (1922) " key " character separating *H. taeniatus* from *H. macrodon* (outer teeth numerous, close together *cf.* outer teeth rather strong, set well apart) is hardly trenchant even when only the type specimens are compared, and is inapplicable when a large series is examined.

The inner teeth are mostly tricuspids in fishes < 110 mm. S.L., but an admixture of tri- and unicuspids (or weakly tricuspids) is found in fishes at the upper end of this size range. Some tricuspids also occur in large fishes, but in the majority of these all inner rows are composed of unicuspids. When tri- and unicuspids occur together in fishes > 120 mm. S.L. the tricuspids are confined to the innermost rows. Inner teeth are arranged in 2 (mode) or 3, rarely 1, rows in the upper jaw, and in 1 (mode) or 2 (rarely 3) rows in the lower.

Osteology. The neurocranium of *H. prognathus* is approximately intermediate between that of *H. paraguiarti* and *H. mento* (see p. 70, and Greenwood, 1962). It thus closely resembles the neurocrania of *H. bartoni, H. mandibularis, H. argenteus* and *H. longirostris* (see p. 112 and Greenwood, *op. cit*), in Lake Victoria, and *H. venator* in Lake Nabugabo (Greenwood, 1965). From the four former species it differs most noticeably in its deeper supraoccipital crest, and from *H. argenteus* also by its being somewhat broader anterior to the orbits. There are also resemblances in neurocranial form between *H. prognathus* and *H. xenostoma* (see p. 53) but the differences here are somewhat more pronounced than in the other species mentioned.

The premaxilla has an expanded median dentigerous surface which gives the bone a beaked appearance; this is a feature of those piscivorous species which I consider to be more specialized in body-form than the members of the *H. guiarti* and *H. serranus* species complexes (see Greenwood, 1962). In contrast, the dentary of *H. prognathus* is of the stout, deep-bodied type found in the latter groups.

The lower pharyngeal bone is triangular, with the dentigerous surface as long as broad or slightly broader (both conditions are of equal frequency). The lower pharyngeal teeth are fairly coarse, with cylindrical necks and compressed, weakly bicuspid crowns; these teeth are well-spaced on the bone in 18–24 rows.

Vertebral counts in 8 specimens are: $12 + 17$ (f.2), $13 + 15$ (f.1), $13 + 16$ (f.4) and $13 + 17$ (f.1) giving totals of 28–30.

Coloration. Live colours are unknown. *Preserved material: Males (adult and sexually active).* Body and head brownish above the midline, silvery-yellow below; two faint lateral bands are visible, one running midlaterally from behind the eye to the origin of the caudal fin, the other along the upper lateral line. There is a faint, ill-defined but broad lachrymal band, but it does not extend ventrally to the margin of the preorbital. Dorsal fin greyish, the dark pigment most intense basally on the spinous region, lappets dusky; soft dorsal maculate. Caudal dark grey, darkest between the rays on the proximal two-thirds of the fin. Anal greyish, lappets black, as may be the basal part of the membrane between the spines; 2 or 3 large but faint, greyish ocelli. Pelvics black.

Females (adult and juvenile). Light grey-brown above, silvery below; very faint traces of longitudinal bands as in males, the lower often visible on the posterior third of the body. All fins hyaline, but the caudal is dark on its proximal two-thirds.

Ecology. Habitat. Haplochromis prognathus is apparently confined to water less than 20 ft. deep, and to hard substrates; it occurs in both sheltered and exposed localities, including wave-washed beaches.

Food. Of the 33 specimens examined (from 11 localities), 22 contained food. Ten specimens contained only remains of fishes (predominantly small Cyprinidae, but *Haplochromis* were also identified), 3 yielded fish and insect remains (the fish unidentifiable, the insects adult Isoptera and larval Ephemeroptera), 7 contained only insects (adult Isoptera and larval Ephemeroptera), 1 contained unidentifiable fish remains and some plant fragments (undigested), and 2 specimens contained unidentifiable material (? bottom debris) and some plant remains.

Breeding. Little is known about the breeding habits of this species ; no brooding individuals have been recorded. Sexual maturity is attained at a length of about 100 mm., but larger juvenile individuals (up to 120 mm.) are known. Both sexes reach the same maximum adult size.

Affinities. There is nothing particularly outstanding about the general appearance of *H. prognathus* when it is compared with the other relatively deep-bodied predatory species of Lake Victoria. Thus, it bears a superficial resemblance to several piscivorous species, especially *H. serranus*, *H. bartoni*, *H. nyanzae* and *H. mandibularis*.

From *H. serranus*, *H. prognathus* differs in its slightly shallower body (30·1–37·1, $M = 33·3\%$ standard length *cf.* 32·7–39·2, $M = 36·0\%$), deeper preorbital (18·8–23·1, $M = 20·8\%$ head, *cf.* 14·6–20·0, $M = 17·7\%$), slightly longer snout (33·4–39·0, $M = 37·0\%$ head, *cf.* 30·8–37·0, $M = 34·0\%$), shorter lower jaw (42·5–51·1, $M = 46·2\%$ head, *cf.* 47·0–60·0, $M = 54·3\%$) and slightly larger chest scales (5–7, mode 6, between the pelvic and pectoral fin bases, *cf.* 7 or 8 [rarely 6]). The preserved coloration of the two species suggests a fairly marked difference in live colours, and there are clear-cut differences in neurocranial form (see p. 110).

From *H. bartoni* (probably its nearest extant relative), *H. prognathus* differs in having a broader and shorter lower jaw (42·5–51·1, $M = 46·2\%$ head, *cf.* 50·8–57·0, $M = 52·5\%$), fewer outer teeth in the upper jaw (30–60, $M = 45$, *cf.* 50–80, $M = 62$), the maxilla not extending so far posteriorly (reaching or almost reaching the orbit in *H. bartoni*), and its smaller chest scales. Neurocranial form in the two species is similar.

The characters separating *H. prognathus* from *H. nyanzae* are its longer snout (33·4–39·6, $M = 37·0\%$ head, *cf.* 33·4–35·8, $M = 34·5\%$), more oblique lower jaw, its maxilla not reaching the orbit, and its larger chest scales. Unfortunately it has not been possible to compare directly the neurocrania of the two species. But, from radiographs it seems probable that the skull of *H. nyanzae* is like that of *H. serranus* (see Greenwood, 1962).

Haplochromis prognathus differs from *H. mandibularis* in having a broader snout, shorter lower jaw (42·5–51·1, $M = 46·2\%$ head, *cf.* 47·3–56·8, $M = 51·5\%$), and longer caudal peduncle (14·6–19·0, $M = 16·6\%$ standard length, *cf.* 12·2–15·2, $M = 14·2\%$). Neurocranial form in the two species is similar.

Haplochromis prognathus is, apparently, closely related to *H. argenteus*. At first sight, the great variability in head shape of *H. prognathus* obscures the relationship. But, if smaller specimens of *H. prognathus* are compared with larger individuals of *H. argenteus* the likeness is striking. Morphometrically, even superficially dissimilar specimens of the two species are not readily separable. Characteristics of the lower

A REVISION OF THE LAKE VICTORIA *HAPLOCHROMIS* SPECIES 83

jaw serve to distinguish the species. The lower jaw of *H. prognathus* is broader (length/breadth ratio 2·0–2·8 [modal range 2·0–2·4], *cf.* 2·3–3·1 [modal range 2·8–3·0]), shorter (42·5–51·1, M = 46·2% of head, *cf.* 45·0–59·0, M = 50·2%), less oblique and less prognathous.

The possible phyletic relationship of *H. prognathus* within the Lake Victoria species-flock will be discussed later (p. 110) ; outside Lake Victoria, *H. prognathus* is probably related to *H. venator* of Lake Nabugabo (Greenwood, 1965). Indeed *H. venator* could well have been derived from populations of *H. prognathus* cut off when the sand bar which isolated Lake Nabugabo was formed about 4,000 years ago.

STUDY MATERIAL AND DISTRIBUTION RECORDS

Museum and Reg. No.	Locality	Collector
	UGANDA	
B.M. (N.H.) 1906.5.30.304 (Syntype *H. macrodon*)	Entebbe	Degen
B.M. (N.H.) 1906.5.30.305–6 (Syntypes *H. macrodon*)	Bugonga	Degen
B.M. (N.H.) 1906.5.30.250 (Syntype *H. taeniatus*)	Entebbe	Degen
B.M. (N.H.) 1966.3.9.50	Entebbe, Airport Beach	E.A.F.R.O.
B.M. (N.H.) 1966.3.9.64–65	Napoleon Gulf, near Jinja	E.A.F.R.O.
B.M. (N.H.) 1966.3.9.53–59, 66–67, 72–75	Beach near Nasu Point (Buvuma Channel)	E.A.F.R.O.
B.M. (N.H.) 1966.3.9.68–71	Katebo	E.A.F.R.O.
B.M. (N.H.) 1966.3.9.79–83	Bugonga, near Entebbe	E.A.F.R.O.
B.M. (N.H.) 1966.3.9.51–52	Kagera Bay, near mouth of Kagera River	E.A.F.R.O.
B.M. (N.H.) 1966.3.9.60–63	Bukafu Bay	E.A.F.R.O.
Vienna Museum 17872 (Type of *H. steindachneri*)	Sesse Islands	Rolle
Academy of Nat. Sciences Philadelphia ANSP 66131 (Holotype of *H. lamprogenys*)	Kitala	——
	KENYA	
B.M. (N.H.) 1905.2.28.1 (Lectotype)	Kavirondo Gulf	Alluaud
B.M. (N.H.) 1905.12.11.4 (Syntype *H. taeniatus*)	Kavirondo Gulf	Alluaud
	TANZANIA	
B.M. (N.H.) 1966.3.9.76–78	Mwanza Harbour	E.A.F.R.O.
B.M. (N.H.) 1966.3.9.84	Capri Bay, Mwanza	E.A.F.R.O.
	LAKE VICTORIA	
B.M. (N.H.) 1906.5.30.251–2, 256	Locality unknown	Degen
Vienna Museum 17876–7 (Types of *H. versluysi*)	Locality unknown	Rolle
Vienna Museum 18830 (Holotype of *H. rebeli*)	Locality unknown	Rolle
Vienna Museum 18769–71 (Paratypes of *H. rebeli*)	Locality unknown	Rolle

Haplochromis argenteus Regan, 1922

(Text-fig. 16)

Paratilapia longirostris (part): Boulenger, 1915, *Cat. Afr. Fish.*, **3**, 332.
Haplochromis argenteus Regan, 1922, *Proc. zool. Soc. Londn.*, 186, fig. 12.

LECTOTYPE : a juvenile female 114·0 mm. standard length, from Bunjako, Uganda (B.M. [N.H.] reg. no. 1906.5.30.266 ; the specimen figured by Regan).

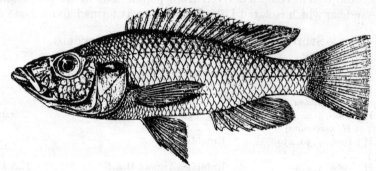

FIG. 16. *Haplochromis argenteus* ; the lectotype, a juvenile, about ·75 times natural size. From Regan, *Proc. zool. Soc.*

DESCRIPTION : based on 23 specimens (including the lecto- and paralectotype) 93·0–202·0 mm. standard length.

Depth of body 26·8–34·8 (M = 31·2) per cent of standard length, length of head 35·5–39·6 (M = 36·8) per cent. Dorsal head profile slightly concave, with prominent premaxillary pedicels, and sloping at an angle of 20°–25°.

Preorbital depth 19·6–24·0 (M = 21·7) per cent of head, least interorbital width 17·7–22·3 (M = 20·0) per cent. Snout 1·25–1·40 times as long as broad, its length 34·8–41·6 (M = 37·8) per cent of head, diameter of eye 19·4–23·5 (M = 21·5), depth of cheek 22·3–28·7 (M = 24·7) per cent.

Caudal peduncle 13·5–17·7 (M = 16·1) per cent of standard length, 1·3–1·7 (modal range 1·3–1·5) times as long as deep.

Mouth oblique (35°–40°), the lower jaw projecting moderately in some specimens, its length 45·0–59·0 (M = 50·2) per cent of head, 2·3–3·1 (modal range 2·8–3·0) times as long as broad. Lips slightly thickened. Premaxilla expanded slightly in the midline. Posterior tip of the maxilla reaching a point about midway between the orbit and the nostril, or almost reaching the orbit.

Gill rakers : generally slender but moderately stout in a few fishes ; lower 1 or 2 reduced, the upper 3 or 4 (in one fish, the upper 6) flat and lobed. Eight to 10 (mode 9) on the lower part of the first arch.

Scales : ctenoid ; lateral line with 29 (f.2), 30 (f.2), 31 (f.5), 32 (f.11), 33 (f.1), 34 (f.1) or 35 (f.1) scales. Cheek with 3 or 4 (rarely 5) rows. Five or 6 (rarely 7) scales between the dorsal fin origin and the upper lateral line, 6 or 7 (less frequently 5) between the pectoral and pelvic fin bases.

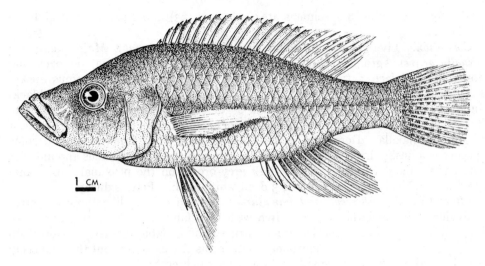

FIG. 16A. *Haplochromis argenteus*, adult. Drawn by Lavinia Beard.

Fins. Dorsal with 23 (f.2), 24 (f.12) or 25 (f.9) rays, comprising 13 (f.1), 14 (f.1), 15 (f.19) or 16 (f.2) spines and 8 (f.2), 9 (f.12) or 10 (f.9) branched rays. Anal with 11 (f.3), 12 (f.19) or 13 (f.1) rays, comprising 3 spines and 8, 9 or 10 rays. Pectoral 24·1–29·7 (M = 25·0) per cent of standard length. Pelvics with the first branched ray somewhat elongate, proportionately more so in adult males. Caudal truncate.

Teeth. Except for the smallest fish examined (the paralectotype 93 mm. S.L.) the outer teeth in both jaws are unicuspid, slender and slightly curved (those situated posterolaterally in the upper jaw more so than the others). In the smallest individual most outer teeth are like those described above but there are also some weakly bicuspid and slender teeth. There are 44–80 (M = 58) teeth in the outer row of the upper jaw, the number showing slight positive correlation with standard length.

In all except four specimens (the two types, 93 and 114 mm. S.L., and two others 88 and 112 mm. S.L.) the inner teeth are unicuspid. Of the exceptional specimens, the largest (lectotype) has an admixture of tri- and unicuspids, the 93 mm. fish (paralectotype) only tricuspids, while the 88 and 112 mm. individuals have tricuspids in the upper jaw and a mixture of tri- and some unicuspids in the lower. Inner teeth are arranged in 2–4 (rarely 5) rows in the upper jaw, and 1–3 in the lower. One exceptional fish (182 mm. S.L.) has a single, irregular row of widely separated teeth in both jaws.

Osteology. The neurocranium of *H. argenteus* is virtually identical with that of *H. longirostris* (see Greenwood, 1962), and also resembles the skull of *H. prognathus*.

The premaxilla is more beaked than that of *H. longirostris*.

The lower pharyngeal bone is narrow, but its dentigerous surface is generally broader than long; less frequently it is equilateral. The teeth are rather coarse, although they are finer and more compressed in a few fishes. These teeth are arranged in 20–24 (usually 22) rows.

Vertebral counts in 8 specimens are : 12 + 17 (f.3) and 13 + 16 (f.5) giving a total of 29.

Coloration. Live colours are unknown. *Preserved material : Males (adult and sexually active)* : ground colour grey-silver, darker (almost brown) on back and dorsal head surfaces ; tip of lower jaw and anterior part of each ramus dusky. Branchiostegal membrane sooty in the opercular region, dark (but not black) anteriorly. Sub- and interopercula with a golden-yellow flush. An ill-defined to distinct lachrymal stripe runs from the anterior orbital margin to behind the posterior tip of the maxilla. Dorsal fin greyish-yellow, with black lappets, and in some specimens a dusky, irregular banding or marbling on the entire spinous and anterior soft parts. Caudal greyish-yellow, dark grey-brown on the proximal half. Anal greyish with black lappets, and 2 or 3 dead-white ocelli. Pelvics dusky.

Males (sexually quiescent, and immature) : Ground colours like those of active individuals but somewhat lighter, often with very faint traces of 4 or 5 bars midlaterally on the flanks ; the lachrymal stripe is of variable intensity. Dorsal fin as above or without the dark marbling. Other fins also as above, but the anal ocelli may be weakly defined and small, and the pelvics lighter.

Females (adult and immature). Ground coloration as in males, but with a faint, interrupted and narrow midlateral band running from the posterior opercular margin to the basal part of the caudal fin. Even fainter traces of a more dorsal longitudinal band (following the course of the upper lateral line) are seen in some specimens ; the two bands may be linked by 3 or 4 extremely faint, short and narrow vertical bars. All fins are yellowish, the caudal dark basally.

Ecology. Habitat. Most specimens are from sheltered localities where the bottom is composed of organic mud, and at depths of less than 40 ft. Other localities are, however, more exposed and the substrate is of rock, sand or shingle, but only a few (and juvenile) fishes were obtained from nets operated over exposed beaches.

Food. Of the 20 individuals examined (from 11 localities) 8 had food in the gut. Four fishes contained only insect remains (in 2 thought to be terrestrial species, and in the others identified as larval Ephemeroptera), and 4 had fragmentary fish remains (identified as the cyprinid *Engraulicypris argenteus*).

Breeding. Fishes < 114 mm. S.L. are immature, as is one exceptional specimen of 146 mm. Both sexes reach the same maximum adult size.

Affinities. The species most like *H. argenteus* is *H. longirostris*. There are, however, a number of morphological differences which serve to separate them, including the larger adult size reached by *H. argenteus*. From *H. longirostris*, *H. argenteus* also differs in its less oblique and longer lower jaw (45·0–59·0, M = 50·2% of head, *cf.* 42·2–51·4, M = 46·0%), shorter and deeper caudal peduncle (13·5–17·7, M = 16·1% of standard length, *cf.* 17·2–22·2, M = 19·2% ; length/depth ratio 1·3–1·7 [modal range 1·3–1·5], *cf.* 1·7–2·3 [modal range 1·9–2·0]), somewhat deeper body (26·8–34·8, M = 31·2% of standard length, *cf.* 24·6–30·4, M = 27·2%) and slightly longer head (35·5–39·6, M = 36·8% standard length, *cf.* 29·2–36·2, M = 33·0%) ; also, the premaxilla of *H. argenteus* is somewhat more beaked. In an earlier paper (Greenwood, 1962) I used the ratio of eye diameter to interorbital width as a diagnostic character; more material of *H. argenteus* has shown, however, that this difference is too slight to be of value.

The similarities between *H. argenteus* and *H. prognathus* are discussed elsewhere (p. 82). When these two species and *H. longirostris* are considered together, it seems probable that *H. argenteus* and *H. longirostris* were derived from an *H. prognathus*-like ancestor, probably as distinct lines developing almost in parallel.

Study Material and Distribution Records

Museum and Reg. No.	Locality	Collector
	UGANDA	
B.M. (N.H.) 1906.5.30.266 (Lectotype)	Bunjako	Degen
B.M. (N.H.) 1906.5.30.267 (Paralectotype)	Bunjako	Degen
B.M. (N.H.) 1966.3.9.194–195	Ekunu Bay	E.A.F.R.O.
B.M. (N.H.) 1966.3.9.201–202	Pilkington Bay	E.A.F.R.O.
B.M. (N.H.) 1966.3.9.193	Beach near Nasu Point	E.A.F.R.O.
B.M. (N.H.) 1966.3.9.196–198	Ramafuta Island (Buvuma Channel)	E.A.F.R.O.
B.M. (N.H.) 1966.3.9.192	Off S. end of Buvuma Island	E.A.F.R.O.
B.M. (N.H.) 1966.3.9.208–210	Beach near Grant Bay (Buvuma Channel)	E.A.F.R.O.
B.M. (N.H.) 1966.3.9.191	Channel between Yempita and Busiri Islands	E.A.F.R.O.
B.M. (N.H.) 1966.3.9.199–200	Kazima Island	Uganda Fisheries Dept.
B.M. (N.H.) 1966.3.9.203–204	Off Entebbe Harbour	E.A.F.R.O.
B.M. (N.H.) 1966.3.9.207	Entebbe Harbour	E.A.F.R.O.
B.M. (N.H.) 1966.3.9.205–206	Bukakata Bay	E.A.F.R.O.
	LAKE VICTORIA	
B.M. (N.H.) 1901.6.24.90	Locality unknown	Sir. H. Johnston
B.M. (N.H.) 1966.3.9.190	Locality unknown	E.A.F.R.O.

Haplochromis squamulatus Regan, 1922

(Text-fig. 17)

Paratilapia pectoralis (non *Ctenochromis pectoralis* Pfeffer) Boulenger, 1911, *Ann. Mus. Genova* (3), **5**, 66, pl. 1, fig. 2. Idem, 1915, *Cat. Afr. Fish*, **3**, 339, fig. 229.
Paratilapia serranus (part): Boulenger, 1915, op.cit., tom. cit., 334.
Haplochromis guiarti (part): Regan, 1922, *Proc. zool. Soc. Londn.*, 174.
Haplochromis squamulatus Regan, 1922, *op. cit.*, 175 (*nom. nov.* for *Paratilapia pectoralis* Blgr., 1911).

HOLOTYPE: a male (probably adult), 149 mm. standard length, from Jinja (Uganda). Genoa Museum reg. no. C.E. 12977.

DESCRIPTION: based on 54 specimens (including the holotype), 66–198 mm. S.L. Depth of body 23·5–35·1 (M = 30·3) per cent of standard length, length of head 29·4–39·5 (M = 34·3) per cent. Dorsal head profile moderately to strongly sloping, the premaxillary pedicels prominent and breaking the outline of the profile, snout region sloping at an angle of 40°–50° with the horizontal.

Preorbital depth in fishes < 100 mm. S.L. (N = 11), 12·5–16·5 (M = 15·5) per cent of head, in larger fishes (N = 43) 15·6–21·6 (M = 18·9) per cent, least interorbital width 21·2–28·3 (M = 25·0) per cent. Snout as long as broad to 1·3 times longer, its length in fishes < 100 mm. S.L., 26·0–32·1 (M = 29·1) per cent of head, in larger individuals 20·6–25·9, M = 23·3 per cent. Cheek depth 19·5–29·3 (M = 25·7) per cent of head.

Caudal peduncle 14·3–22·3 (M = 17·9) per cent of standard length, 1·2 (rare)–2·0 (mode 1·7) times as long as deep.

Mouth horizontal or very slightly oblique, jaws equal anteriorly, or infrequently, lower projecting slightly. Length of lower jaw in fishes < 170 mm. S.L. (N = 46), 39·0–48·0 (M = 44·1) per cent of head, 44·2–51·0 (M = 47·5) per cent in larger fishes (N = 8); 1·5–2·5 (mode 1·8, modal range 1·5–2·0) times as long as broad in fishes of all sizes. Posterior tip of the maxilla generally reaching the vertical through the anterior orbital margin or to below the eye, but not reaching either point in a few individuals. Lips slightly thickened; premaxilla not expanded medially.

Gill rakers: short and stout, often flat, in most specimens; in fishes < 120 mm. S.L., but also in a few larger individuals, most of the rakers are relatively slender. Eight–10 (mode 9), rarely 11, on the lower part of the first arch, the lower 1–3 rakers reduced.

Scales: ctenoid. Lateral line with 32 (f.2), 33 (f.12), 34 (f.8), 35 (f.15), 36 (f.10), 37 (f.6) or 38 (f.1). Pore distribution in this series is irregular, with some pore scales occurring in the horizontal row above that in which the majority lies; also, some scales are without pores and are often smaller than those before and behind them. Cheek with 3 (f.8), 4 (f.32), 5 (f.12) or 6 (f.1) rows of scales. Scales on the nape, chest and belly are small; 6 (f.3) 6½ (f.3), 7 (f.11), 7½ (f.5), 8 (f.14) or 9 (f.16) between the

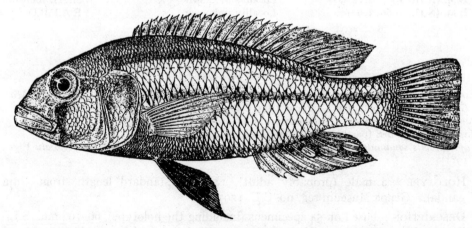

Fig. 17. *Haplochromis squamulatus*, holotype, about ·72 times natural size. From Boulenger, *Cat. Afr. Fish.*

dorsal fin origin and the upper lateral line, 6 (f.1), 7 (f.8), 8 (f.16), 9 (f.25) or 10 (f.3) between the pectoral and pelvic fin bases.

Fins. Dorsal with 24 (f.9), 25 (f.37) or 26 (f.6) rays, comprising 14 (f.2), 15 (f.26) or 16 (f.24) spines, and 8 (f.2), 9 (f.23), 10 (f.25) or 11 (f.2) branched rays. Anal with 11 (f.3), 12 (f.40) or 13 (f.5) rays, comprising 3 spines (except for one specimen with 2), and 8 (f.3), 9 (f.40) or 10 (f.5) branched elements. Pectoral 21·8–27·0 (M = 24·7) per cent of standard length. Pelvics with the first branched ray produced in both sexes but proportionately more so in adult males. Caudal subtruncate to truncate.

Teeth. Most fishes < 105 mm. S.L. have either an admixture of uni- and bicuspid teeth in the outer row of both jaws, or only bicuspids. One specimen (89 mm. S.L.) has predominantly bicuspids and a few tricuspids in the upper jaw, but mostly unicuspids in the lower. In larger fishes of the < 105 mm. group, unicuspid teeth predominate in both jaws, and in fishes > 105 mm. S.L. only unicuspids are found. All outer teeth are relatively slender and gently curved, the unicuspids more so than the bicuspids. The mean number of teeth in the outer row of the upper jaw shows some positive correlation with standard length although there is little difference in the ranges for the various groups; fishes < 120 mm. S.L. have 50–72, M = 57 teeth in this row, while larger specimens have 52–80, M = 68.

There is less obvious correlation between tooth form and the fish's size when the inner rows of teeth are considered. All fishes < 105 mm. S.L. have only tricuspids in the inner rows. Above this length many fishes have only unicuspids, but a mixture of uni- and tricuspids (or weakly tricuspids) is common even in the largest individuals. Unicuspids occur most frequently in the outer rows of the inner series.

Inner teeth are arranged in 2–6 rows (usually 3 or 4) in the upper jaw, and in 1–6 (usually 2 or 3) in the lower jaw. There is perhaps some correlation between the number of rows and the fish's length because the largest specimens (179–198 mm. S.L.) have the greatest number of inner rows. Also, in larger fishes the inner teeth are implanted very obliquely but are almost vertical in fishes < 105 mm. S.L.

Osteology. The neurocranium of *H. squamulatus* combines characteristics of both the *H. guiarti* and the *H. serranus* types (see Greenwood, 1962). The dorsal preorbital profile rises steeply and is gently curved; its line is continuous with that of the supraoccipital crest whose anterior point lies further forward than in the skulls of *H. guiarti* and *H. serranus*. In general there is a great similarity between the neurocrania of *H. squamulatus*, *H. michaeli* and *H. martini* (see Greenwood, *op. cit*).

The premaxilla is of the generalized type, and thus lacks a pronounced anterior extension of its medial dentigerous surface.

The lower pharyngeal bone has the dentigerous surface broader than long, from slightly to markedly so. The lower pharyngeal teeth are relatively slender, compressed and distinctly cuspidate; the teeth are arranged in 22–26 (mode 24) rows.

Vertebral counts in 10 specimens are: 13 + 16 (f.2), 13 + 17 (f.2), 13 + 18 (f.2), 14 + 16 (f.1), 14 + 17 (f.3) giving totals of 29–31.

Coloration. Live colours are known for females (adult) and quiescent males. *Females:* ground colour yellow-silver above shading to yellowish-white ventrally, the dorsal body and head surfaces are darkest. Two distinct longitudinal bands are invariably present, the lower running from the operculum to the caudal fin origin along the level of the lower lateral line, the upper extending from the nape to the end of the dorsal fin at a level about two scale rows below the dorsal fin insertion. Both bands have a finely zig-zagged outline; the upper band may be faint. Dorsal fin dark neutral. Caudal dark neutral with a yellow flush. Pelvic and anal fins yellow.

Quiescent males have a coloration like that of the females described above.

Preserved material: Males (*adult, sexually active but not ripe*). The general impression is one of dusky greyness, with small areas of light brown on the flanks. The entire head is dark but is lighter on the operculum and lower lip (the lower jaw is dark). A broad black band (faint in some specimens) runs from the posterior opercular margin to the caudal fin origin; it is crossed by 3 or 4 broad but faint vertical bars in the zone of lighter flank coloration. These bars merge with the dark dorsum and sooty-grey chest and belly. The ventral aspects of the caudal peduncle are also sooty-grey. Dorsal fin yellow-brown with a sooty overlay, the lappets are black, and a few ill-defined dark maculae may be visible on the posterior part of the soft dorsal fin. Caudal dark grey-brown on its proximal three-quarters, yellowish-brown distally. Anal coloured like the dorsal but greyer over the spinous part; ocelli either not clearly defined, or white with a sooty surround. Pelvic fins dusky over a yellowish ground.

Males (*adult but quiescent*). Ground colour light brown, shading to golden on the ventral surfaces which are, however, overlaid with greyish-black. Dorsal surface of the head and body dark brown. A fairly broad, distinct and almost vertical lachrymal stripe runs from the anteroventral orbital margin to the dentary. A very distinct, broad lateral stripe runs from the opercular margin to the caudal fin base; the upper margin of this band is irregularly serrate. There are 3 or 4 incomplete but broad and distinct vertical blotches on the ventral half of the flanks, each blotch originating from the midlateral band but not extending to a point more than half way between the lateral band and the ventral body outline. A very faint dark upper lateral band may be distinguished running at a level about midway between the dorsal fin base and the upper lateral line; this band is often indistinguishable from the generally dark coloration of the dorsum. A third band at the base of the dorsal fin may be visible. Dorsal fin pale brownish-yellow, with a faint sooty overlay; lappets black, dark blotches present on the soft fin. Anal pale yellowish, with a very faint sooty overlay. Pelvics yellowish with an intense sooty overlay on the anterior half. Caudal dark greyish-yellow.

Immature males are coloured like females (see below), but have a darker ground coloration (almost grey on the chest and belly) and faintly sooty pelvics.

Females (*adult and immature*). Female coloration is very distinctive because of the well-marked lateral bands. The ground coloration is a yellowish-silver, somewhat greyish above the upper lateral line and on the dorsal head surface; the cheek and operculum are yellow-silver. The midlateral band is black and has characteristically serrate upper and lower margins; the depth of this band is rather variable in any one fish, and is generally deepest above the anal fin. The upper lateral band also has irregularly serrate margins but it is generally less obvious because of its position on the dark colour of the dorsum. A third longitudinal band, narrower and interrupted, lies along the base of the dorsal fin as a series of narrow, elongate blotches. Dorsal fin yellowish with a faint sooty overlay, and sooty lappets; the soft part often has dark spots and blotches between the rays. Caudal yellowish with dark maculae on the proximal quarter to third. Anal and pelvic fins yellow.

Ecology. Habitat. The species is known from a wide variety of habitats, including sheltered bays and gulfs, exposed beaches, and from certain off-shore localities near small islands. In most places the substrate is hard (rock, sand or shingle) but *H. squamulatus* is known to occur over mud bottoms.

Food. Thirty-six specimens were examined (covering the entire size-range, and from 18 localities); of these, 19 contained food in the gut. In 16 fishes the food consisted entirely of fishes, the remains so macerated that certain identification was difficult; in two cases the remains were identified as *Haplochromis* species, in one as *Engraulicypris*, and in another as a cichlid. In two female fishes the stomach contained many small embryos of cichlid fishes; these could well be the fishes' own broods swallowed at the time of capture, although it is not known whether *H. squamulatus* is a mouth-brooder. A third female fish had several small post-larval cichlids in the stomach; judging from the gonad state of the predator it seems unlikely that the larvae were its own brood.

In addition to the sixteen piscivorous fishes, one other yielded only crustacean remains, one several insect egg-masses, and a third the remains of boring mayfly larvae (*Povilla adusta*).

Breeding. All specimens, except one, below 135 mm. S.L. are immature; the exceptional fish (93 mm.) is a ripening female. The largest fishes (180–198 mm. S.L.) are females but there is one male of 179 mm. S.L.

Affinities. The colour pattern, small chest and nuchal scales and the sharply decurved head profile of *H. squamulatus* make at least large specimens readily identifiable. But, smaller individuals are less easily distinguished from specimens of *H. martini* (although the latter are adult at a size when most *H. squamulatus* are still immature). Unfortunately, little is known about the live coloration of sexually active male *H. squamulatus* but there are several similarities in the coloration of female *H. martini* and *H. squamulatus*, particularly in the striping and the yellowish coloration. In fishes of all sizes, *H. squamulatus* differ from *H. martini* in their less strongly decurved head profile, maxilla not extending so far posteriorly (never reaching a point below the pupil as is general in *H. martini*), in having a higher modal number of lateral line scales (35 *cf.* 33) and of scales between the pectoral and pelvic fin bases (9 *cf.* 7 or 8), and in the somewhat shallower body (23·5–35·1, $M = 30·3\%$ standard length, *cf.* 30·8–38·0, $M = 34·4\%$).

When specimens over 100 mm. S.L. are compared, a number of additional morphometric differences are apparent : *H. squamulatus* has a longer snout, smaller eye and a longer lower jaw. Specimens less than 100 mm. S.L. have only one trenchant morphometric difference, the size of the eye (eye diameter in *H. squamulatus* is 25·9–30·4 [$M = 28·3$]% of head, *cf.* 29·4–37·5 [$M = 31·7$]% in *H. martini*).

Also resembling *H. squamulatus* (and *H. martini*, see Greenwood, 1962) is *Haplochromis michaeli*. The species differ in their preserved coloration (see above, and

Greenwood, *op. cit.*, p. 205), in the straighter, slightly concave, dorsal head profile of *H. michaeli*, and the greater posterior extension of the maxilla in that species (generally to below the pupil). Since specimens of *H. michaeli* over 100 mm. S.L. only are known, interspecific morphometric comparisons are restricted. These, however, show that *H. squamulatus* has a slightly shallower body (23·5–35·1, M = 30·3% of standard length, *cf.* 30·8–37·6, M = 34·3%), and a smaller eye (20·6–25·9, M = 23·3% head, *cf.* 24·0–29·1, M = 27·0% in *H. michaeli*). The nuchal scales of *H. squamulatus* are somewhat smaller (modal number of scales between the dorsal fin origin and the upper lateral line 8 or 9, *cf.* 6 or 7 in *H. michaeli*) as are those of the lateral line series (mode 35 *cf.* 33); the upper limit of the range for the lateral line scale count is also higher in *H. squamulatus* (38 *cf.* 35).

All in all, the resemblances between *H. squamulatus* and *H. martini* are greater than than those between *H. squamulatus* and *H. michaeli*, but the three species seem to form a closely related group within the more generalized piscivorous predators of Lake Victoria.

There is a certain similarity, albeit superficial, between *H. squamulatus* and *H. altigenis*. On closer examination, however, it does not seem likely that the resemblance can be construed as implying a close phyletic connection between the species. *Haplochromis altigenis* is probably a derivative of the *H. guiarti* stem.

In its general facies, *H. squamulatus* shows some similarity with *H. dichrourus*, but there is a very pronounced interspecific difference in coloration (*cf.* p. 90 with p. 67), as well as differences in some morphometric characters. *Haplochromis squamulatus* has a larger eye (20·6–25·9, M = 23·3% of head, *cf.* 19·4–22·6, M = 21·1) a shorter lower jaw (44·2–51·0, M = 47·5% of head, *cf.* 51·3–54·0, M = 52·7%) and a higher modal number of lateral line scales (35 *cf.* 33).

Phyletically, *H. squamulatus* and *H. michaeli* may represent slightly divergent developments (less so morphologically and ecologically in the former species) from an *H. martini*-like stem, an increase in adult size being a common factor in the two lines.

Study Material and Distribution Records

Museum and Reg. No.	Locality	Collector
	UGANDA	
B.M. (N.H.) 1966.3.9.262–272	Napoleon Gulf near Jinja	E.A.F.R.O.
B.M. (N.H.) 1966.3.9.282–284	Fisherman's Point near Jinja	E.A.F.R.O.
B.M. (N.H.) 1966.3.9.285–288	Beach near Nasu Point (Buvuma Channel)	E.A.F.R.O.
B.M. (N.H.) 1966.3.9.289–295	Ramafuta Island (Buvuma Channel)	E.A.F.R.O.
B.M. (N.H.) 1966.3.9.259–260	Off Buvuma Island (Buvuma Channel)	E.A.F.R.O.
B.M. (N.H.) 1966.3.9.261	Grant Bay	E.A.F.R.O.
Genoa Museum C.E. 12977 (Holotype)	Jinja (Napoleon Gulf)	Bayon
B.M. (N.H.) 1906.5.30.233–9	Nsonga	Degen
B.M. (N.H.) 1909.5.4.4–5	Sesse Islands	Bayon
B.M. (N.H.) 1966.3.9.278–281	Beach in Entebbe Harbour	E.A.F.R.O.

B.M. (N.H.) 1966.3.9.258	Entebbe, Airport Beach	E.A.F.R.O.
B.M. (N.H.) 1966.3.9.273–277	Kazima Island near Entebbe	Uganda Fisheries Dept.
B.M. (N.H.) 1966.3.9.255	Old Bukakata Bay	E.A.F.R.O.

TANZANIA

B.M. (N.H.) 1966.3.9.257	Capri Bay, Mwanza	E.A.F.R.O.
B.M. (N.H.) 1966.3.9.256	Mwanza Harbour	E.A.F.R.O.

LAKE VICTORIA

B.M. (N.H.) 1901.6.24.86	Locality unknown	Sir H. Johnston

Haplochromis barbarae sp. nov.

(Text-figs. 18 and 19)

HOLOTYPE: an adult female, 97 mm. standard length, from a beach near Nasu Point (Buvuma Channel); B.M. (N.H.) reg. no. 1966.2.21.5.

Named in honour of Mrs. Barbara Williams, whose drawings illustrate this and others of my papers.

DESCRIPTION: based on 9 specimens (including the holotype) 89·0–106·0 mm. standard length.

Depth of body 32·1–35·5 (M = 34·0) per cent of standard length, length of head 31·4–35·0 (M = 33·5) per cent. Dorsal head profile very slightly decurved or straight.

Preorbital depth 15·1–18·0 (M = 16·9) per cent of head length, least interorbital width 24·2–26·7 (M = 25·5) per cent. Snout as long as broad, its length 30·0–32·4 (M = 31·5) per cent of head, eye diameter 22·9–27·4 (M = 25·8), cheek depth 20·0–25·0 (M = 22·2) per cent.

Caudal peduncle 15·7–18·0 (M = 16·8) per cent of standard length, 1·4–1·7 times as long as deep (no well-defined mode).

Mouth very slightly oblique, the jaws equal anteriorly; lower jaw 36·6–41·2 (M = 38·0) per cent of head, 1·5–1·7 (in one fish 2·0) times as long as broad. Posterior tip of the maxilla reaching a point near the vertical through the anterior orbital margin.

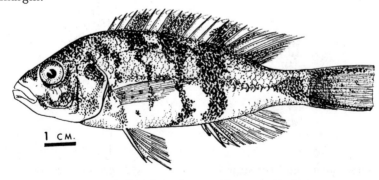

FIG. 18. *Haplochromis barbarae*. Drawn by Barbara Williams.

Gill rakers: of variable form, from moderately short and stout to slender and relatively elongate, but of constant form (except for the reduced lower rakers) in any one individual ; 8 or 9 rakers on the lower part of the first gill arch.

Scales: ctenoid ; lateral line with 33 (f.3), 34 (f.2) or 35 (f.4), cheek with 3 or 4 (rarely 2) rows. Seven (rarely $6\frac{1}{2}$ or 8) scales between the upper lateral line and the dorsal fin origin, 7 or 8 (rarely 6 or 9) between the pectoral and pelvic fin bases. Scales on the ventral aspects of the chest are noticeably smaller than those situated laterally.

Fins. Dorsal with 25 (f.7) or 26 (f.2) rays, comprising 16 spinous and 9 (f.7) or 10 (f.2) branched rays. Anal with 11 (f.2), 12 (f.6) or 13 (f.1) rays, comprising 3 spines and 8 (f.2), 9 (f.6) or 10 (f.1) branched elements. Pectoral 25·0–29·2 (M = 26·7) per cent of standard length. Caudal truncate, scaled on its proximal half.

Teeth. In the outer row of both jaws, the teeth are relatively stout, bicuspid and slightly curved ; in a few specimens some posterolateral upper teeth are unicuspid and enlarged. There are 40–65 (M = 50) teeth in the outer row of the upper jaw.

Teeth forming the inner rows are relatively large and tricuspid, and are implanted somewhat obliquely so that the crowns point inwards. There are 2 or 3 inner rows in the upper jaw, and 2 (3 in one specimen) in the lower.

Osteology. No complete skeleton is available. The lower pharyngeal bone is, compared with that of similar species, small. Its dentigerous surface is broader than long, and narrows rapidly at about the midpoint so that not only is the whole bone relatively small, but so is the area of pharyngeal teeth. The teeth are fine, compressed and distinctly cuspidate ; they are rather sparsely distributed in from 22–24 rows.

Vertebral counts in 9 specimens are : 13 + 17 (f.8) and 13 + 18 (f.1) giving totals of 30 and 31.

Coloration: Live colours are unknown. *Preserved colours*: Males (*adult but quiescent*). Ground colour dark yellowish-brown. A well-defined lachrymal stripe is present, as are faint traces of two transverse, parallel stripes across the snout. There is some darkening over the preoperculum but this is not concentrated into a

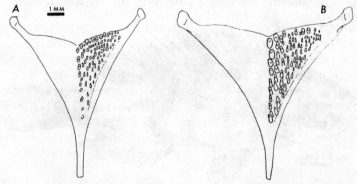

Fig. 19. Lower pharyngeal bones (in occlusal view, with dentition shown on one side only) of (A) *H. barbarae*, and (B) *H. chromogynos*.

bar or stripe. The branchiostegal membrane is dark, most intensely so below the opercular bones. On the flank is a faint midlateral stripe, and four even fainter and incomplete vertical bars. Dorsal and anal fins are yellowish with a sooty overlay, the anal being darkest along its proximal third; anal ocelli very faint. Pelvics sooty over a yellowish ground. Caudal yellowish on the margins and distal third, dark brown proximally.

Females (adult.) Three distinct colour patterns occur; two of these are probably identical with the polychrome patterns described for live *Hoplotilapia retrodens* (see Greenwood, 1956).

The colour pattern corresponding with the modal coloration in other polychromatic species has a yellow-brown ground colour with a distinct midlateral dark stripe running from the opercular margin to the caudal origin, and extending onto the caudal fin itself for about half its length; in two specimens there are four, narrow vertical bars on the flanks, extending from the dorsal fin base almost to the ventral outline. In all specimens a very faint upper line is situated slightly above the upper lateral line. All fins are yellowish-brown. Three out of the eight female specimens have this type of coloration.

The second pattern is a piebald, black on silvery-yellow (to yellow brown), the black pigment arranged in blotch-like bars of variable width and dorso-ventral extent. Some blotches extend onto the dorsal, caudal, anal and pelvic fins. The cheeks and snout may be blotched or clear. All fins are yellowish. This pattern does not differ from that shown by other species with a piebald coloration in females. Four of the eight *H. barbarae* females are piebald.

The third pattern, found only in one specimen, has an orange-yellow background peppered with fine melanophores which are, in places, aggregated into blotches (especially on the dorsum) with an irregular distribution. Some dark blotches occur on all fins (but especially the dorsal and caudal); these spots are smaller than those on the body. The ground colour of all fins is light orange-yellow.

Ecology. Habitat. The nine known specimens came from three different sites. Two of these are shallow, exposed and sandy beaches, the third, shallow water (*ca.* 10 ft. deep) over a hard substrate a short distance off-shore and near the water-lily zone fringing a papyrus swamp.

Food. Seven of the 9 specimens examined contained ingested matter in the guts. In each case this consisted of from 1–8 (mode 5) recently fertilized cichlid ova. In two specimens these could have been part of the fish's own brood swallowed during capture, because the fishes have recently spent ovaries. But, judging from the advanced stages of oogenesis shown by four other specimens, and the fact that the fifth is a male, it seems reasonable to conclude that the embryos were taken as food (see Greenwood, 1959, for a discussion of other paedophagous *Haplochromis* species).

Breeding. Nothing is known about the breeding habits of *H. barbarae*. All nine specimens are sexually mature; the sole male (102 mm. S.L.) is the second largest specimen.

Affinities. Anatomically, *H. barbarae* is very like *H. brownae*, a generalized species exhibiting many characters suggestive of affinity with anatomically generalized predators such as *H. guiarti* (Greenwood, 1962).

Haplochromis barbarae differs from *H. brownae* in having a narrower interorbital (24·2–26·7, M = 25·5% of head, *cf.* 26·0–34·0, M = 29·8%), a smaller eye (22·9–27·4, M = 25·8% of head, *cf.* 26·0–31·3, M = 28·6%), fewer gill rakers (8 or 9, *cf.* 9–12 [modes 10 and 11]), and smaller scales, especially on the ventral aspects of the chest. There is also a difference in dention in that the outer teeth of this species are stout bicuspids whereas in *H. brownae* of a similar size the teeth are slender unicuspids.

It differs from *H. guiarti* in having a shorter snout (30·0–32·4, M = 31·5% of head, *cf.* 31·7–37·5, M = 34·4) and lower jaw (36·6–41·2, M = 38·0% of head, *cf.* 39·2–48·2, M = 44·4%), fewer gill rakers (8 or 9, *cf.* 9–11, mode 10) and, at comparable sizes, the presence of bicuspid teeth anteriorly in both jaws (unicuspid in *H. guiarti*).

Because of its generalized anatomy and unspecialized dentition, *H. barbarae* closely resembles a number of other species in the Lake Victoria flock, *viz.* *H. cinereus*, *H. macrops*, *H. lacrimosus* and *H. chromogynos* (see Greenwood, 1959 for the latter species and Greenwood, 1960 for the three former). In addition to their anatomical similarities, *H. barbarae* and *H. chromogynos* also share (with several other and structurally unrelated species) the piebald coloration in females. However, *H. barbarae* differs from *H. chromogynos* in having a longer lower jaw (36·6–41·2, M = 38·0% of head, *cf.* 30·0–34·4, M = 32·5%), and, when specimens of equal size are compared, bicuspid instead of slender unicuspid teeth (specimens of *H. chromogynos* < 95 mm. S.L. have a mixed bi- and unicuspid dentition but larger individuals have only unicuspid teeth). The shape of the pharyngeal bone differs in the two species, and it is this character (see p. 94 and fig. 19) which most readily distinguishes *H. barbarae* from the other three species mentioned above, none of which is known to have piebald females or paedophagous habits.

The peculiar feeding habits of *H. barbarae* immediately suggest some affinity with the other paedophagous *Haplochromis* species, and because *H. barbarae* is anatomically unspecialized, particularly with the more " generalized " paedophages, *H. obesus* and *H. cronus*. However, in many morphological details these two differ considerably from *H. barbarae*. For instance, the peculiar dental morphology of *H. obesus* (with the small teeth deeply embedded in the gums), and the stout unicuspid teeth of *H. cronus*, serve as immediately diagnostic characters, as do the several morphometric differences between the species. It is of interest, however, to recall that piebald females occur in all three species ; but, since this character appears in other and widely different species, it is not thought to have any phyletic significance.

Any supposed close phyletic relationship between *H. barbarae* and the other members of the paedophagous species group would also be difficult to substantiate. At present all that can be suggested is that *H. barbarae* might represent an isolated line, derived from a generalized and probably insectivorous stem, paralleling trophically the paedophagous species group (itself probably of diphyletic origin). Alternatively, *H. barbarae* might represent a survivor of the stem from which such species as *H. cryptodon* and *H. microdon* arose ; more will have to be learned about the anatomy of *H. barbarae* before this possibility can be substantiated.

Study Material and Distribution Records

Museum and Reg. No.	Locality	Collector
	UGANDA	
B.M. (N.H.) 1966.3.9.243	Napoleon Gulf, off Jinja	E.A.F.R.O.
B.M. (N.H.) 1966.2.21.5 (Holotype)	Beach near Nasu Point (Buvuma Channel)	E.A.F.R.O.
	TANZANIA	
B.M. (N.H.) 1966.3.9.244–251	Beach near Majita	E.A.F.R.O.

Haplochromis tridens Regan and Trewavas, 1928
(Text-fig. 20)

Haplochromis tridens Regan & Trewavas, 1928, *Ann. Mag. nat. Hist.* (10), **2**, 226.

LECTOTYPE: a fish 116·0 mm. standard length (caudal fin damaged), B.M. (N.H.) reg. no. 1928.6.2.41, collected in Tanzanian waters at Michael Graham's station 234 (1° 4' S, 32° 13' E), at a depth of over 100 ft. (Graham, 1929).

DESCRIPTION: based on 16 specimens (including the lectotype and the paralectotype) 72–119 mm. standard length.

Depth of body 30·1–36·2 (M = 33·5) per cent of standard length, length of head 32·0–37·2 (M = 35·4) per cent. Dorsal head profile straight, but interrupted by the prominent premaxillary pedicels, sloping steeply at 40°–45°. The cephalic lateral line system with prominent pores, especially those of the preopercular and preorbital canals. These pores are probably larger than in any Lake Victoria *Haplochromis* (including *H. pachycephalus* and *H. boops*).

Preorbital depth 16·0–20·9 (M = 17·2) per cent of head, least interorbital width 15·0–19·5 (M = 16·7) per cent. Snout 1·0–1·3 (mode 1·1) times as long as broad, its length 28·0–34·8 (M = 30·4) per cent of head. Eye with an oval, horizontally aligned pupil, eye diameter 25·6–34·0 (M = 30·7) per cent of head, depth of cheek 17·5–22·8 (M = 19·2) per cent.

Caudal peduncle 16·7–19·8 (M = 17·9) per cent of standard length, 1·4–1·9 (mode 1·6) times as long as deep.

Mouth horizontal or slightly oblique, lower jaw projecting slightly and with a distinct mental projection; length of lower jaw 43·3–51·8 (M = 47·5) per cent of head, 2·0–2·8 (mode) times as long as broad. Premaxilla slightly expanded in the midline. Posterior tip of the maxilla generally reaching to below the anterior part of the eye, and occasionally to below the pupil; rarely reaching only to the vertical through the anterior margin of the orbit.

Gill rakers: relatively slender, except for the lower 1–3 which are reduced, and the upper 2–4 which are usually flat; 8–11 (mode 9) on the lower part of the first gill arch.

Scales: strongly ctenoid, especially those on the chest region. Lateral line with 31 (f.4), 32 (f.5), 33 (f.4) or 34 (f.2) scales. Cheek with 3 (mode) or 4 rows. Five and a half to 6½ (modes 6 and 6½) scales between the upper lateral line and the dorsal fin origin, 5–6½ (mode) between the pectoral and pelvic fin bases.

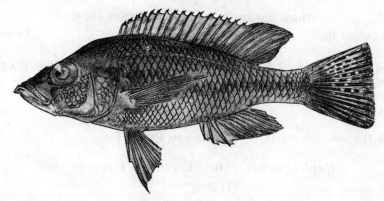

Fig. 20. *Haplochromis tridens*, paralectotype, about ·73 times natural size. Drawn by Miss M. Fasken.

Fins. Dorsal with 23 (f.7), 24 (f.7) or 25 (f.2) rays, comprising 15 (f.4) or 16 (f.12) spinous, and 7 (f.6), 8 (f.6), 9 (f.3) or 10 (f.1) branched rays. Anal with 10 (f.10), 11 (f.5) or 12 (f.1) rays, comprising 3 spines and 7 (f.10), 8 (f.5) or 9 (f.1) branched elements. (N.B. More than usual difficulty was experienced in deciding if the last dorsal and anal ray was a single, deeply divided element, or two distinct rays; this may account for the number of specimens with low (7) branched ray counts). Pectoral 27·0–33·3 (M = 29·3) per cent of standard length. Pelvics with the first branched ray produced, proportionately more so in adult males. Caudal truncate.

Teeth. The trivial name "tridens" was given to the species because the larger of the two syntypes had only tricuspid teeth in the outer tooth row of both jaws, an unusual (probably unique) feature for a *Haplochromis* species. This condition is, however, rare in the species. Uni-, bi- and tricuspid outer teeth may all occur in a single specimen. Tricuspid teeth, at least in the upper jaw, are usually found posterolaterally, the bi- and unicuspids laterally and anteriorly. This arrangement is by no means constant; in some specimens tricuspids occur anteriorly and anterolaterally. One fish (72 mm. S.L.) has only bicuspids in the upper jaw, and another (81 mm.) only unicuspids. There is no apparent correlation between the predominance of any one tooth form and the size of the fish. In three out of the sixteen specimens examined, no tricuspid outer teeth were found in either jaw.

There are 58–80 (M = 66) teeth in the outer row of the upper jaw.

Variation in the type of outer teeth occurring in the lower jaw follows the same pattern as in the upper jaw.

Teeth forming the inner rows in both jaws are predominantly tricuspids. These teeth are relatively large, are stout, and have the median cusp larger than the lateral ones. There are usually 2 inner rows (occasionally 3, rarely 4) in the upper jaw, and 2 (rarely 1) in the lower.

Osteology. No entire skeleton is available. The lower pharyngeal bone has its dentigerous surface equilateral or slightly broader than long. The lower pharyngeal teeth have cuspidate, compressed crowns, and cylindrical necks; the teeth are arranged in 22–26 rows.

Vertebral counts in 10 specimens are : 13 + 16 (f.7) and 13 + 17 (f.3), giving totals of 29 and 30.

Coloration. Live colours are unknown. *Preserved specimens : Males (adult and sexually active) :* ground colour grey-blue (gun-metal) above the upper lateral line, greyish silver on the flanks, and light dusky silver on the chest and belly. Dorsal and caudal fins hyaline. Anal hyaline except for a dusky area between the spines, and black lappets ; the dusky area extends as a fairly distinct line at the base of the spinous part, and may be expanded basally onto the soft fin. There are two, large, grey-white anal ocelli. Pelvics dusky, darkest on the lateral four-fifths of the fin.

Males (adult but quiescent) have a ground coloration more like that of females. All fins are hyaline except for a faint darkening between the anal spines, and on the pelvic fins.

Females (juvenile and adult) : greyish silver above, shading to silver below. One specimen (ovaries ripening) is darker dorsally (almost brown), and has a broad, interrupted midlateral stripe running from behind the operculum to about half the length of the caudal fin (on which the stripe narrows) ; the band is interrupted at about the middle of its length. All fins yellowish to hyaline.

Ecology. Habitat. Excepting the two types, all the material came from one trawl haul at an unknown locality (thought to be off the Kenya coast). The types came from deep water (more than 100 ft.) some distance off-shore, and from over a soft bottom. The large eyes and hypertrophied pores of the cephalic lateral line system certainly suggest adaptations to a deepwater habitat.

Food. Only 6 of the 10 specimens examined (all from the same, and unknown, locality) contained ingested material in the gut. In each, the predominant contents are undigested blue-green algae, and empty diatom frustules. Two specimens have, in addition, a few fragments of Crustacea, and some unidentifiable insect remains.

Breeding. All fourteen of the non-typical specimens are adults. The types are not well-preserved internally, but both appear to be females.

Affinities. The admixture of tri-, uni- and bicuspid outer teeth together with the hypertrophy of the cephalic lateral line pores, provide a trenchant means of distinguishing *H. tridens* from other species in the lake. Indeed, it is difficult to suggest any close relationship between this species and any other so far considered. Perhaps some relationship will become apparent when more is known about the numerous and small species of *Haplochromis* which make up the bulk of fishes caught by trawling in the deeper waters of Lake Victoria (Greenwood, unpublished). Anticipating these results, it is possible to say that the elongate body and general " predatory " facies (especially the large mouth) of *H. tridens* are not common amongst these species, and nor is the *H. tridens* dental type.

Amongst the larger species with a predatory facies, *H. tridens* most closely resembles *H. victorianus*. There are, however, many differences between the species, not least of which are their differences in ecology and the much smaller adult size of *H. tridens*. *Haplochromis victorianus* differs also in its dentition, broader interorbital region (21·5–24·5, M = 22·6% of head, *cf.* 15·0–19·5, M = 16·7%) smaller and rounder eye (21·7–25·5, M = 23·6% head, *cf.* 25·6–34·0, M = 30·7%), and deeper cheek (22·5–26·2, M = 24·6% head, cf. 17·5–22·8, M = 19·9%). It should be remembered, however, that the data for *H. victorianus* are derived from larger specimens than are available for *H. tridens*, and that the most trenchant morphometric differences are in characters most affected by allometric growth. Nevertheless, it seems very unlikely that *H. victorianus* and *H. tridens* are close relatives.

STUDY MATERIAL AND DISTRIBUTION RECORDS

Museum and Reg. No.	Locality	Collector
	TANZANIA	
B.M. (N.H.) 1928.6.2.41 (Lectotype)	1°4′ S, 32° 13′ E	M. Graham
B.M. (N.H.) 1928.6.2.42 (Paralectotype)	1°4′ S, 32° 13′ E	M. Graham
	LAKE VICTORIA	
B.M. (N.H.) 1966.3.9.152–165	Locality unknown (? Kenya coast)	E.A.F.R.O.

Haplochromis orthostoma Regan, 1922

(Text-fig. 21)

Pelmatochromis spekii (part): Boulenger, 1915, *Cat. Afr. Fish.*, **3**, 417.
Haplochromis orthostoma Regan, 1922, *Proc. zool. Soc. Londn.*, 184, fig. 9.

HOLOTYPE: a male (probably adult), 91 mm. standard length (B.M. [N.H.] reg. no. 1912.10.15.67) from Lake Salisbury (Kyoga system). The specimen is now in very poor condition, and has lost most of its scales.

This species, possibly one of the two most distinctive looking members of the Victoria-Kyoga flock, was described from a single specimen. Since then, two further specimens have been collected from Lake Salisbury, and three superficially similar specimens have been caught in Lake Victoria. The latter specimens differ from the Salisbury fishes in their dentition (and some other characters) and are thought to represent a distinct species. It is for this reason that I am redescribing the Lake Salisbury species in this paper.

The unusual physiognomy of *H. orthostoma* is well shown in text-figure 21; the snout and nuchal region meet at a distinct angle, the nuchal musculature not bulging anteriorly and laterally. Since there are only three specimens available, morphometric characters are tabulated below. The holotype is indicated with an asterisk.

S.L. (mm.)	Depth †	Head †	Preorb. %	Interorb. %	Snout %	Eye %	Cheek %	Caudal Peduncle†
67·5	37·0	36·0	19·6	20·4	32·7	24·5	24·5	19·3
83·0	35·0	36·1	20·0	23·3	30·0	21·7	30·0	17·5
*91·0	36·3	35·5	18·5	21·5	30·8	22·8	30·8	16·5

† Per cent of standard length.
% Per cent of head length.

Mouth very oblique, sloping at *ca* 50°–70°, the lower jaw projecting, its length 51·0–56·6 per cent of head length, and 2·5–3·4 times as long as broad. Lips not thickened, the lower jaw with a distinct mental protuberance. Posterior tip of the maxilla reaching a point about midway between the vertical through the anterior orbital margin and that through the nostril. Snout 1·1–1·3 times as long as broad, its dorsal surface slightly rounded, the premaxillary pedicels not prominent.

Caudal peduncle 1·5–1·6 times as long as deep.

Gill rakers: moderately stout, the lower 1–3 reduced, the upper 2 or 3 flattened (anvil-shaped in one fish); 9 or 10 on the lower part of the first gill arch (11 on one arch of a fish with 9 rakers on the other arch).

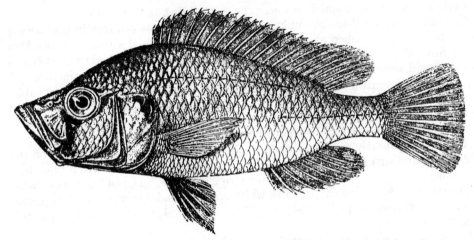

FIG. 21. *Haplochromis orthostoma*, holotype. From Regan, *Proc. zool. Soc.*, about ·86 times natural size.

Scales: ctenoid; lateral line with 30 or 31 scales (Regan gives 33 for the now scaleless type), cheek with 3 or 4 rows (4 or 5 in the type according to Regan). Six or 6½ between the upper lateral line and the dorsal fin origin (6 or 7 in the type acc. Regan), 6 or 7 between the pectoral and pelvic fin bases, the chest scales small.

Fins. Dorsal with 14 (f.1) or 15 (f.2) spines and 9 branched rays, anal with 3 spines and 8 (f.2) or 9 branched rays. First branched pelvic ray not produced in the type, but elongate in the two other specimens (male and female). Caudal subtrucate.

Teeth. In the two larger specimens (83 and 91 mm. S.L.), the outer teeth in both jaws are a most distinctive feature. These teeth are slender and unicuspid with sharply curved tips directed medially; those situated anteriorly are somewhat larger than the others. There are 66 and 61 teeth in this row for the two specimens respectively. In the lower jaw, the outer teeth, at least anteriorly, are somewhat stouter than their opposites in the upper jaw.

The smallest specimen (67·5 mm. S.L.) has outer teeth quite unlike those of the larger individuals. In the upper jaw, the anterior and lateral teeth are stout and bicuspid, those situated laterally and posteriorly are stout and tricuspid. All outer teeth in the lower jaw are bicuspid and stout, and are a little stouter than the upper jaw teeth. There are 68 teeth in the outer row of the upper jaw.

The inner teeth in the two larger fishes are unicuspid, small, slender and implanted obliquely; in the smallest fish they are small and tricuspid.

The dental arcade in all specimens is V shaped with a rounded apex; there are 2 inner tooth rows in the upper jaw, and 1 or 2 rows in the lower jaw.

Osteology. No complete skeleton is available. The lower pharyngeal bone has a triangular and equilateral dentigerous area. The teeth are relatively slender, with bicuspid and weakly compressed crowns, and are arranged in 20–22 rows; except in the smallest fish the teeth of the two median rows are coarser than the lateral rows.

Vertebral counts for 2 specimens are: $13 + 16$ (type) and $12 + 16$.

Coloration: Live colours are unknown. The type is now completely colourless; originally it was described as greyish, with a dark lachrymal stripe, and blackish pelvic fins. A juvenile female has a brown ground coloration, with very faint traces of four broad vertical bars on the flanks running from the dorsal fin origin to about the level of the ventral margin of the pectoral fin. No lachrymal stripe is visible. The lower jaw (especially over its anterior half) is rather dusky. The dorsal and caudal fins are yellowish, the former with black lappets. Anal dusky yellow. Pelvics faintly dusky, especially at their tips. Pectorals yellowish-grey.

Adult Male (probably fixed in alcohol): light brown above, shading to silver on the mid-flanks, and silvery white on the belly. Snout dark, as are the lips; branchiostegal membrane pale. A faint lachrymal stripe is visible. Dorsal fin greyish, with black lappets. Anal greyish, with two white ocelli. Caudal grey, the melanophores most concentrated along its midline. Pelvics dark along the anterior quarter, hyaline elsewhere.

Ecology. No information is available on the habitat or food of *H. orthostoma*, nor is there any information on the breeding habits of the species. The two males (91 and 83 mm. S.L.) are adult, and the female (67·5 mm.) is apparently immature.

Affinities. The peculiar head shape, large and very oblique mouth, coupled with the peculiar tooth form (at least in the larger fishes), serve to distinguish the species from all others in the Lake Victoria-Kyoga flock. The nearest relative is *H. parorthostoma* from Lake Victoria (see below). Further material, and field observations, may yet show that the two species are not distinct at that level.

Study Material and Distribution Records

Museum and Reg. No.	Locality	Collector
B.M. (N.H.) 1912.10.15.67 (Holotype)	Lake Salisbury	Presented by F. J. Jackson
B.M. (N.H.) 1958.12.5.173	Ongino, Lake Salisbury	Pitman
B.M. (N.H.) 1966.3.9.252	Lake Salisbury	E.A.F.R.O.

Haplochromis parorthostoma sp. nov.

(Text-fig. 22)

HOLOTYPE: an adult male, 117 mm. standard length, from near Zero Island (Buvuma Channel), Uganda. B.M. (N.H.) reg. no. 1966.2.21.4.

DESCRIPTION. The overall similarity between this species and *H. orthostoma* is great, particularly since both share a peculiar head profile not seen in any other *Haplochromis* species from Lakes Victoria or Kyoga.

The dorsal head profile is strongly concave, with the nuchal region meeting the snout at a noticeable but rounded angle, the junction emphasized by an anterior bulge of the cephalic epaxial body musculature. This muscular protuberance gives the fish a pronounced " forehead ", especially in the frontal plane.

Since only three specimens are available, morphometric data are tabulated below; the holotype is marked with an asterisk.

S.L. (mm.)	Depth †	Head †	Preorb. %	Interorb. %	Snout %	Eye %	Cheek %	Caudal Peduncle†
86·0	38·4	35·0	16·7	20·0	31·6	27·7	25·0	17·5
110·0	42·3	36·4	17·5	17·5	33·8	25·0	27·5	12·7
*117·0	41·0	35·0	19·5	19·5	34·9	24·4	26·8	14·5

† Per cent of standard length.
% Per cent of head length.

Mouth oblique, sloping upwards at *ca.* 40°–50°, jaws equal anteriorly, or the lower projecting slightly. Lower jaw length 48·3–53·5 per cent of head, 2·3–2·4 times as long as broad; chin with a distinct protuberance. Posterior tip of the maxilla reaching a point nearer the vertical through the anterior orbital margin than one through the nostril. Snout 1·2 times as long as broad, with a convex dorsal surface; premaxillary pedicels not prominent. Lips moderately thickened.

Caudal peduncle 1·1–1·3 times as long as deep.

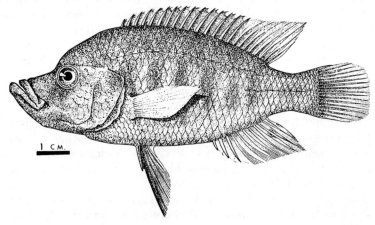

FIG. 22. *Haplochromis parorthostoma.* Drawn by Lavinia Beard.

Gill rakers: moderately stout (relatively stouter in one specimen), the lower 1–3 reduced, the upper 2 or 3 expanded and anvil-shaped; 9 (f.2) or 10 rakers on the lower part of the first gill arch.

Scales: ctenoid, lateral line with 30 or 32 (f.2) scales, cheek with 2 or 3 rows (in the former specimen, the scales not covering the ventral aspects of the cheek). Six or 7 scales between the upper lateral line and the dorsal fin origin, 6 or 7 between the pectoral and pelvic fin bases.

Fins. Dorsal with 15 (f.2) or 16 spines and 8, 9 or 10 branched rays. Anal with 3 spines and 8 or 9 (f.2) branched rays, the spines short and stout. Pectoral 24·5–26·2 per cent of standard length. First branched pelvic ray not or slightly produced (all specimens are males). Caudal almost rounded, scaled on its basal two-thirds.

Teeth. The outer row in both jaws is composed of unicuspid, slender and slightly curved teeth, the curvature being gentle and not confined to the distal part of the tooth (*cf. H. orthostoma*, p. 101); in one fish, the last three teeth in the upper jaw are larger, stouter and straighter than the anterior ones. A few bicuspid, moderately stout teeth occur posteriorly in the lower jaw of the smallest (86 mm.) fish. In no specimen are the lower jaw teeth stouter than those in the upper jaw. There are 38, 40 and 52 teeth in the outer row of the upper jaw.

Teeth forming the inner series are small, tricuspid and weakly tricuspid, and are arranged in 3 or 4 rows in the upper jaws and in 2 rows in the lower.

The dental arcade is V shaped, with the apex broadly rounded.

Osteology. No complete skeleton is available. The lower pharyngeal bone has its triangular dentigerous surface as long as broad, or slightly broader than long. The teeth, arranged in 24–30 rows, are fine, with weakly compressed bicuspid crowns; those in the two median rows are but slightly larger than the lateral teeth.

Vertebral counts in 3 specimens are: $13 + 15$ and $13 + 16$ (f.2).

Coloration: *Live colours* are known for a sexually active (but not ripe) *male*. Ground colour dark slate-grey, with faint vertical bars of a darker shade. Dorsal fin dark grey with crimson lappets on the posterior two-thirds of the spinous part, and a dark crimson margin to the entire soft part. Caudal dark grey with a crimson flush, particularly intense on its lower half. Entire anal fin, except for a dark base, crimson. Pelvics black on the anterior third, remainder dull crimson.

Preserved material: Males (*adult*). Ground colour light brown (including the branchiostegal membrane in two fishes; this membrane blackish in the third specimen); flank crossed by 5 or 6 dark but incomplete bars, each bar originating just above the upper lateral line and extending to about the level of the ventral margin of the pectoral fin. A fairly distinct vertical lachrymal stripe runs from the anteroventral margin of the orbit to the angle of the lower jaw. Dorsal fin yellowish-brown along its margin, but dark brown between the rays. Anal yellowish, sooty or dark brown between the branched rays; very faint indications of 2 or 3 whiteish ocelli. Caudal light to dark brown (almost black). Pelvics dusky on the anterior third to half, otherwise hyaline.

Ecology. One specimen came from an exposed, sandy beach, another from a rocky outcrop in about 20 ft. of water near an off-shore island, and the third from over a rocky shelf in about 40 ft. of water, also near an island.

No information is available on the food of *H. parorthostoma*. The three specimens are adult males.

Affinities. *Haplochromis parorthostoma* seems to be very closely related to *H. orthostoma* of the Kyoga system, at least in its peculiar head-shape. There is, however, a marked interspecific difference in the shape of the outer teeth. In *H. parorthostoma* these teeth are gently curved, whereas in *H. orthostoma* they have sharply recurved crowns but relatively straight necks. Furthermore, the inner teeth of *H. orthostoma* are unicuspid and arranged in one or two rows, but in *H. parorthostoma* are tricuspid and arranged in three or four rows in the upper jaw (two rows in the lower). An exception (at least with regard to outer tooth shape) is provided by the smallest specimen of *H. orthostoma* whose stout, erect and bicuspid teeth do not resemble those of its larger congeners (or, for that matter, *H. parorthostoma*). But, since the cranial morphology of this small fish is so like that of larger *H. orthostoma* individuals it is included in that species.

Haplochromis parorthostoma and *H. orthostoma* differ in characters other than dental ones. The bulging cranial epaxial muscles of *H. parorthostoma* impart a different shape to the dorsal head profile, which is further modified by the less oblique mouth. The available samples also suggest that *H. parorthostoma* has a bigger eye than *H. orthostoma* (and this despite the fact that the specimens of *H. parorthostoma* are larger) and probably a shorter and broader lower jaw.

Clearly, much more material is required before it will be possible to reach more definite conclusion about the status of the two species. Even then the decision will be complicated by the fact that they are allopatric, and their areas of distribution are physically isolated (now by the virtually impenetrable Owen Falls dam, and previously by the Ripon Falls, perhaps not a complete barrier to migration from Lake Victoria to the Kyoga system).

For the moment it is not possible to speculate on the wider relationship of these two species.

Study material and distribution records

Museum and Reg. No.	Locality	Collector
	Uganda	
B.M. (N.H.) 1966.2.21 (Holotype)	Near Zero Island (Buvuma Channel)	E.A.F.R.O.
B.M. (N.H.) 1966.3.9.253	Near Zero Island (Buvuma Channel)	E.A.F.R.O.
B.M. (N.H.) 1966.3.9.254	Near Kazima Island	Uganda Fish. Dept.

Haplochromis apogonoides sp. nov.

(Text-fig. 23)

HOLOTYPE an adult male, 118 mm. standard length, from Ekunu Bay, Uganda. B.M. (N.H.) reg. no. 1966.2.21.3.

Named "*apogonoides*" because of its fancied resemblance to the genus *Apogon*.

DESCRIPTION: based on eight specimens (including the holotype) 112–132 mm. standard length; with one exception (a quiescent female) all are males.

Depth of body 36·0–39·6 (M = 38·1) per cent of standard length, length of head 35·0–37·2 (M = 36·1) per cent. Dorsal head profile curved (strongly so in some specimens), sloping fairly steeply (*ca.* 40°), the premaxillary pedicels not prominent.

Preorbital depth 13·6–16·7 (M = 15·2) per cent of head length, least interorbital width 27·5–31·0 (M = 29·5) per cent. Snout 1·3–1·5 (mode 1·4) times as broad as long, its length 29·3–33·4 (M = 31·2) per cent of head, eye 25·3–27·5 (M = 26·6), depth of cheek 27·8–31·0 (M = 29·3) per cent.

Caudal peduncle 15·3–19·8 (M = 18·0) per cent of standard length, 1·5–1·7 times as long as deep.

Mouth slightly to moderately oblique, lips somewhat thickened, the jaws equal anteriorly. Lower jaw 45·0–51·0 (M = 47·9) per cent of head, 1·2–1·5 (modal range

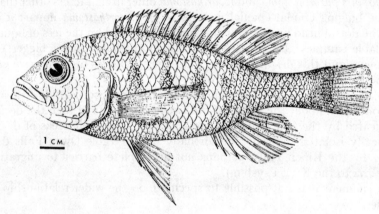

FIG. 23. *Haplochromis apogonoides*. Drawn by Barbara Williams.

1·4–1·5) times as long as broad. Posterior tip of the maxilla reaching to below the pupil. Premaxilla not expanded medially.

Gill rakers: stout, the lower 1–3 reduced, the upper 3 or 4 flat; 8–10 on the lower part of the first gill arch.

Scales: ctenoid; lateral line with 32 (f.4), 33 (f.3) or 34 (f.1) scales. Cheek with 3 or 4 (mode) rows. Six to 8 (mode 6) scales between the upper lateral line and the dorsal fin origin, 6–8 (modes 6 and 7) between the pectoral and pelvic fin bases.

Fins. Dorsal with 22 (f.1), 24 (f.6), or 25 (f.1) rays, comprising 14 (f.2) or 15 (f.6) spinous and 8 (f.1), 9 (f.5) or 10 (f.2) branched rays. Anal with 10 (f.1), 11 (f.5) or 12 (f.2) rays, comprising 3 spines and 7 (f.1), 8 (f.5) or 9 (f.2) branched elements. Pectoral 29·0–34·1 (M = 30·8) per cent of standard length. Pelvics with the first ray slightly produced. Caudal truncate to subtruncate, scaled on its basal half to two-thirds.

Teeth. The outer teeth in both jaws are a characteristic feature of the species, being unicuspid, moderately stout and with very strongly recurved tips. Such teeth are otherwise found only in the *H. sauvagei* species group (see Greenwood, 1957). There are 50–60 (M = 58) teeth in the outer row of the upper jaw.

The inner teeth are also unicuspid, large and recurved, and are implanted obliquely. There are 2 rows in the upper jaw, and 1 or 2 rows in the lower jaw. A distinct space separates the inner series from the outer row.

Osteology. No complete skeleton is available. The lower pharyngeal bone is relatively stout, the dentigerous surface noticeably broader than long ($1\frac{1}{5}$ to $1\frac{1}{4}$ times). The teeth are stout and cuspidate, and are arranged in from 16–20 rows, those of the two median rows being slightly coarser than the others.

Vertebral counts in 7 specimens are : 13 + 15 (f.1), 13 + 16 (f.5) and 14 + 16 (f.1), giving totals of 28–30.

Coloration. Live colours are unknown. *Preserved coloration : Males (adult and sexually active, but probably not ripe).* Ground colour light yellow-brown, with a silvery underlay on the flanks. Belly, chest and branchiostegal membrane dusky, the branchiostegal membrane darkest below the opercular region. On the flanks there are faint traces of 4 or 5 fairly broad vertical bars which become very faint dorsally and ventrally ; the anterior 3 or 4 bars merge ventrally with the dark belly coloration. A dark lachrymal stripe is present ; it does not reach the ventral margin of the preorbital, but appears to pass upwards across the eye. Dorsal fin yellowish, the soft part maculate in some specimens ; also in some fishes the lappets are black. Caudal yellowish marginally and on the distal third to quarter, dark yellow-brown proximally. Anal yellowish, with one large greyish to whiteish ocellus faintly outlined in black. Pelvics sooty, the colour less intense between the last two or three rays.

Female (adult and quiescent). Ground colour greyish-silver. A very faint lachrymal bar extends from below the orbit to below the posterior tip of the maxilla ; a dark spot on the upper part of the eye suggests that this bar may pass across the eye. All fins are yellowish, the dorsal with dusky lappets, and the pelvics with a faint duskiness over the anterior rays.

Ecology. Habitat. The species has been caught in two localities only. In one, a sheltered bay, the water was between 20 and 30 feet deep, and the substrate of organic mud. The second locality was at a depth of about 80 ft in the Buvuma Channel, near Buvuma Island ; again the substrate (on which the nets were set) was organic mud.

Food. Regrettably, the gut was empty in all except one specimen ; this fish contained only a little, unidentifiable sludge.

Breeding. Apart from the sexually inactive female (132 mm. S.L.), all the specimens are adult and active males.

Affinities. The stout, unicuspid outer teeth with sharply recurved crowns immediately suggest affinity with *H. sauvagei, H. prodromus* and *H. granti* (Greenwood, 1957). Like these species, the dorsal head profile of *H. apogonoides* is strongly rounded. However, unlike these species, there are fewer rows of inner teeth in *H. apogonoides*, there is a distinct interspace between the inner and outer tooth rows, and the lower jaw is much longer (45·0–51·0, M = 47·9% head cf. 30·6–37·7, M = 34·5% for *H. sauvagei*, 30·5–37·8, M = 34·3% for *H. prodromus*, and 22·2–30·6,

$M = 26·8\%$ for *H. granti*). *Haplochromis apogonoides* also differs from these species in its broader snout, and from *H. granti* in the anatomy of the jaws (see Greenwood, 1957).

Superficially, *H. apogonoides* resembles *H. cronus* (Greenwood, 1959) and, to a lesser degree, *H. empodisma* and *H. michaeli* (see Greenwood, 1960 and 1962 for the species respectively).

From *H. cronus*, it is differentiated by its dental morphology and its longer lower jaw.

From *H. empodisma*, *H. apogonoides* differs in its longer lower jaw (45·0–51·0, $M = 47·9$ per cent of head, *cf.* 39·1–48·7, $M = 43·9\%$), dental morphology (strongly *cf.* gently curved teeth), broader interorbital (27·5–31·0, $M = 29·5\%$ head, *cf.* 20·6–28·6, $M = 24·3$), broader snout, and noticeably wider dentigerous surface on the lower pharyngeal bone ($1\frac{1}{5}-1\frac{1}{4}$ times as broad as long, *cf.* longer than broad).

Haplochromis apogonoides differs from *H. michaeli* in dental morphology (see above), in its slightly deeper body, shallower preorbital, broader interorbital, markedly broader snout, and deeper cheek (27·8–31·0, $M = 29·2\%$ of head, *cf.* 22·9–27·7, $M = 25·8\%$). The lower pharyngeal bone is similar in both species, but the teeth in *H. apogonoides* are stouter.

The marked similarity between the outer jaw teeth of *H. apognoides* and those in species of the *H. sauvagei* complex (which includes *H. xenognathus*, a species not mentioned above because of its distinctive jaw morphology, tooth pattern, and head shape) suggests that *H. apogonoides* might be an off-shoot from this species group. Unfortunately, nothing is known about the feeding habits of this species; all members of the *H. sauvagei* group are snail-eaters with the unusual habit of removing the snail from its shell before ingesting it. The large mouth and not especially strong jaws of *H. apogonoides*, do not, however, suggest similar feeding methods; rather, these characters indicate piscivorous habits.

Study Material and Distribution Records

Museum and Reg. No.	Locality	Collector
	UGANDA	
B.M. (N.H.) 1966.2.21.3 (Holotype)	Ekunu Bay	E.A.F.R.O.
B.M. (N.H.) 1966.3.9.238–242	Ekunu Bay	E.A.F.R.O.
	LAKE VICTORIA	
B.M. (N.H.) 1966.3.9.236–237	Locality unknown	E.A.F.R.O.

DISCUSSION

Phylogeny

In an earlier paper (Greenwood, 1962) I outlined the various morphological trends shown by the piscivorous species, and suggested a possible ancestral morphotype (represented today by *H. brownae*) from which the different lines could have evolved. Also in that paper I indicated two major, and two minor, possibly phyletic groups of fish-eating predators. Additional information provided by the species described above does not affect the suggested trends, but does alter the phyletic picture. It now seems likely that there are three major phyletic lines, and probably three minor ones as well.

As mentioned before (Greenwood, *op. cit.*), possible phyletic lines amongst piscivorous species are less readily detected and defined than those of other trophic groups. In these latter there are dental as well as somatic characters which may be used for this purpose, but amongst the piscivores (at least when adult) the teeth are invariably unicuspid and of a very similar form. The principal " group " characters in these fishes are neurocranial shape, and body form ; the latter character often shows a greater or lesser degree of intergroup convergence. Using these two character complexes, I at first recognized two species aggregates, the " *serranus* " group, and the " *mento-macrognathus* " group. The former comprises the supposedly more generalized, broad-headed and deep-bodied species, and the latter group the more specialized, slender-bodied forms.

Information obtained from the species described in this paper suggests that my " *mento-macrognathus* " group consists of two groups, neither of which is as readily defined as the " *serranus* " group, but both being easily distinguished from that group.

One of the minor groups (that of *H. percoides*, *H. flavipinnis* and *H. cavifrons*) is now thought to be polyphyletic (see p. 113).

For convenience the groups will be referred to by the trivial epithet of a constituent species, which species, however, is not necessarily to be considered a " typical " member of the group. Indeed it is often difficult to determine just what a " typical " species would be ; at the most, the nominate species of a group is representative of a structural type found in three or four species of the group.

The three major groups are : (i) The " *serranus* " group, consisting of *H. serranus*, *H. victorianus*, *H. nyanzae*, *H. spekii*, *H. maculipinna*, *H. boops*, *H. thuragnathus* and *H. pachycephalus* : morphologically, this is a relatively homogeneous group (but, see also p. 110).

(ii) The " *altigenis* " group, comprising *H. guiarti*, *H. bayoni*, *H. dentex*, *H. pseudopellegrini*, *H. altigenis*, *H. pellegrini* and *H. dichrourus* ; a number of subgroups (some monotypic) can be recognized, and are discussed later.

(iii) The " *prognathus* " group, comprising *H. paraguiarti*, *H. acidens*, *H. prognathus*, *H. bartoni*, *H. estor*, *H. gowersi*, *H. mento*, *H. mandibularis*, *H. macrognathus*, *H. longirostris* and *H. argenteus*. This, the largest species aggregate shows several intragroup trends of which the most distinctive are the *H. longirostris*—*H. argenteus*, and the *H. mandibularis*—*H. macrognathus* subgroups.

Members of the " *serranus* " group differ from those of the other two groups in having shorter snouts* (one species out of seven with the snout $>$ 36 per cent of head length, *cf.* fourteen species out of eighteen), deeper bodies (one species out of seven with the body depth $<$ 36 per cent of standard length, *cf.* seventeen out of eighteen), and broader heads (no species with the interorbital width $<$ 22·6 per cent of head, *cf.* thirteen out of eighteen species.) Osteologically, the group is character-

* The figures given are derived from those for the mean value of a particular character in species of the groups under consideration.

ized by having a neurocranial shape nearest that of the generalized skull (see Greenwood, 1962), but with the preotic part elongate. The dorsal skull roof is straight and slopes fairly steeply, is broad both interorbitally and across the otic region, and the supraoccipital crest is high and presents a substantial area for muscle insertion (see fig. 25 in Greenwood, *op. cit.*).

The "*serranus*" group could have evolved directly from an *H. brownae*-like ancestor, the principal morphological changes being an increase in adult size, and those alterations in neurocranial proportions already mentioned. Within the group, the most differentiated species are *H. boops*, *H. thuragnathus* and *H. pachycephalus*, all three being confined to deep water (see pp. 49, 50 and 41). *Haplochromis boops* and *H. thuragnathus* were apparently derived from an *H. maculipinna*-like ancestor, whilst *H. pachycephalus* seems to show greater affinity with the *H. serranus*—*H. spekii* level of organization. *Haplochromis maculipinna* is also essentially of this affinity, but has markedly larger eyes. In turn, *H. serranus* is clearly derived from an *H. brownae*-like stem.

The "*altigenis*" and "*prognathus*" groups probably evolved from an *H. guiarti*-like ancestor or ancestors, the latter species also showing affinities with *H. brownae*. Although both the "*altigenis*" and "*prognathus*" groups have included species with a relatively deep body, the main trend shown by both groups is towards a slender, somewhat compressed body-form and a correlated head shape. It is difficult to characterize these two groups, particularly their more basic members. The most trenchant group characters are probably in neurocranial form. In lateral view there is little to differentiate the skull form in the two groups; both have the preorbital part relatively more elongate than in the "*serranus*"-type skull, the dorsal profile slopes upward at a rather slight angle (as compared with the angle in skulls of the "*serranus*" group), and the supraoccipital crest is relatively low. (These generalizations must, however, be modified somewhat for those species which seem to be structurally basal for the groups [*H. guiarti* and *H. pseudopellegrini* for the "*altigenis*" group, and *H. paraguiarti* and *H. acidens* for the "*prognathus*" group]. In these species the preorbital face is less protracted and consequently [since relative neurocranial depth varies little amongst all members of both groups] the dorsal skull roof slopes more steeply and the supraoccipital crest is higher and has a fairly extensive area). However, when the neurocranium is viewed dorsally, a difference between the groups (including their basal species) is apparent (see text-fig. 24). In members of the "*altigenis*" group, the otic region is relatively broader than in those of the "*prognathus*" group so that the outline narrows more rapidly (from a point immediately behind the orbit) than in "*prognathus*" skulls. In these the outline is that of a narrow wedge with the margins closing gradually from a point further behind the orbits. In supposedly basal members of both groups, the otic region is of about equal relative breadth but basal "*prognathus*" members nevertheless have a more gradual medial inclination of the lateral margins. Furthermore, in these species the dorsal skull profile is straighter than in the most basic "*altigenis*" group member, *H. guiarti*. Indeed, in most members of the "*altigenis*" group the profile is more curved than in species of the "*prognathus*" group.

Reasons for considering *H. guiarti* as a basic morphotype in the radiations of

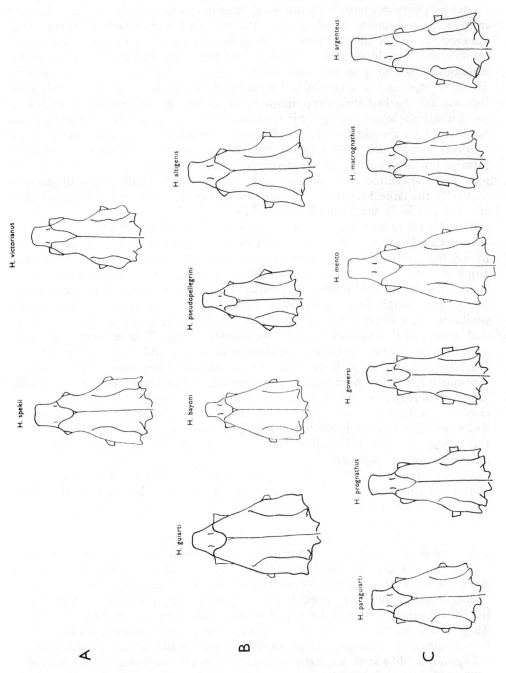

Fig. 24. Outlines of the dorsicranium in species of: line A, the "*serranus*-group", line B, the "*altigenis*-group", and line C, the "*prognathus*-group".

predatory piscivores have been discussed already (see Greenwood, 1962). Starting from a species similar to *H. guiarti*, the principal morphological changes seen amongst species of the "*altigenis*" group involve relative elongation of the preorbital face, a slight decrease in skull height (especially in the otic region) and a consequent flattening in the slope of the dorsal skull profile. In these respects *H. bayoni* represents a fairly marked departure from the basic "*guiarti*" skull form, but one less marked than that shown by *H. dentex*, *H. altigenis* or *H. pellegrini*, particularly the latter (see fig. 25 in Greenwood, 1962). A link between these forms is provided by the skull of *H. pseudopellegrini* which is intermediate between the "*bayoni*" and "*altigenis*" types. There is not a great deal of difference in body form or jaw morphology among members of the "*altigenis*" group. Perhaps the most extreme member is *H. altigenis* itself, a species with the deepest head and consequently the largest mouth. *Haplochromis pellegrini* is, because of its small adult size and relatively small mouth, atypical for the group.

There is far greater variation in body form and cranial morphology within the "*prognathus*" group. Here the basic species, *H. paraguiarti* and *H. acidens*, are morphologically similar to *H. bayoni* of the "*altigenis*" group and could be derived either from a "*bayoni*" or a "*guiarti*"-like ancestor. *Haplochromis acidens* is, of course, peculiar in that despite its predatory facies, it is apparently a phytophage (see p. 76). *Haplochromis prognathus* and *H. bartoni* are, in most respects, very similar to one another and represent the next morphological step in the evolution of such species as *H. longirostris* and *H. mandibularis* from a "*paraguiarti*"-like stem. That is to say, the neurocranium clearly shows narrowing and elongation, and there is a related refinement of body proportions. *Haplochromis estor* and *H. gowersi* continue this trend with, in addition, the development of a more oblique angle to the jaws, and in *H. gowersi* a deepening of the cheek which contributes to a larger buccal cavity. *Haplochromis mento* probably evolved from a "*prognathus*"-like ancestor, and shows many of the trends exhibited by *H. gowersi* and *H. estor*. However, in this species the mouth is almost horizontal, and the neurocranium is somewhat nearer that of *H. prognathus*.

Also apparently stemming from a "*prognathus*"-like ancestor is *H. mandibularis*. Here the trend is towards further narrowing of the skull, deepening of the cheek, increased obliquity of the jaws and lengthening of the lower jaw. This trend culminates in *H. macrognathus* (see Greenwood, 1962, pp. 180 and 186), a species which could well be a direct descendant of *H. mandibularis*.

A third derivative from a "*prognathus*" or "*bartoni*"-like ancestor is *H. longirostris*. In this line there has been little change in neurocranial shape but there is a marked increase in jaw obliquity (much greater, too, than in the "*mandibularis*"-"*macrognathus*" line), and a marked refinement in body proportions which results in one of the most slender bodies found amongst Lake Victoria *Haplochromis* species. These characters are shared by *H. argenteus*, although the elongate body-form seems less obvious in that species. *Haplochromis argenteus* could be derived either from a "*longirostris*"-like stem, or perhaps directly from a "*prognathus*"-like ancestor.

Thus, the "*prognathus*" group seems to show at least three radiations from a basal "*prognathus*"-"*bartoni*" stem, viz. the *H. estor*, *H. gowersi*, *H. mento* sub-group, and the *H. mandibularis–H. macrognathus*, and *H. longirostris–H. argenteus* species pairs (but see above for possible reservations about the last named pair). It seems reasonable to assume that the *H. prognathus* level of organization was derived from a level similar to that shown by the extant species *H. paraguiarti*. The relationships of that species are, however, less clear-cut. As was mentioned above, *H. paraguiarti* shows several structural affinities with *H. bayoni*, a species probably derived from an *H. guiarti*-like ancestor. But, it is difficult to overrule the possibility that *H. paraguiarti* evolved independently from an *H. brownae*-like stem. (The status of *H. brownae* in relation to the piscivorous predators is discussed in Greenwood, 1962).

Two species, *H. plagiostoma* and *H. xenostoma*, have not been included in the discussion so far. Both are characterized by having the mouth set at a very steep angle to the horizontal (*ca.* 40°–50°); *H. plagiostoma* is further characterized by its obliquely truncate caudal fin, and *H. xenostoma* by its pronounced prognathism. Because of these characters, especially the oblique jaws, neither species shows any obvious superficial similarity with members of the groups discussed above. The neurocranium of *H. plagiostoma* is of the " serranus " type (see also Greenwood, 1962) but also shows certain " *guiarti* "-like features. Thus, on this character complex *H. plagiostoma* could either be associated with the " serranus " group or be looked upon as an isolated offshoot of the " altigenis " group arising from an ancestor near the stem of that complex. In either eventuality, *H. plagiostoma* is not linked with the basal group by any extant, structurally intermediate species.

Haplochromis xenostoma, both in its overall organization and in its neurocranial form, seems to represent a further development from a " *plagiostoma* " stem. The chief trend involves an increase in adult size, and a relative enlargement of the mouth and jaws. The neurocranium of *H. xenostoma* reflects these changes, especially in the longer preorbital region; it is thus essentially a " serranus " group neurocranium. From the available evidence it is impossible to determine whether *H. plagiostoma* and *H. xenostoma* are part of the same phyletic lineage or whether the two species are end-points of parallel evolution from " *guiarti* " and " *serranus* "-like stems respectively.

The two smaller species complexes, the *H. michaeli*, *H. martini*, *H. squamulatus* group and the *H. percoides*, *H. flavipinnis*, *H. cavifrons* group, will now be considered.

The relationships of *H. michaeli* and *H. martini* are discussed in my paper of 1962, and there is nothing further to add. *Haplochromis squamulatus* is included with these species because of its several similarities with *H. martini*, similarities which include a basically similar coloration of a type not otherwise found in species of the Lake Victoria *Haplochromis* flock. In the " *michaeli* " group, the relationship between *H. martini* and *H. squamulatus* seems to be closer and more direct than that between either species and *H. michaeli* (see also Greenwood, 1960, pp. 245–8; *idem*, 1962, p. 206, and p. 91 above).

When discussing the relationships of the *H. percoides–H. cavifrons* group (Greenwood, 1962), I suggested that *H. percoides* could have evolved from an *H. pellegrini*-like stem, and that *H. flavipinnis* was a derivative of an *H. percoides*-like ancestor. Also in that paper I noted the less certain relationships of *H. cavifrons*, but considered it to be part of the *H. percoides* phyletic line. On revising this complex, I began to doubt my earlier conclusions about the affinities of *H. cavifrons*, which now seems to have greater relationship with the "*serranus*" group. This review provided no evidence to negative my conclusions about the interrelationships of *H. percoides* and *H. flavipinnis*, or the association of *H. percoides* with an *H. pellegrini*-like ancestor. However, I am not at all certain about the validity of my previous suggestion that the neurocrania of *H. percoides* and *H. flavipinnis* show affinity with those of *H. bartoni* and *H. longirostris* (i.e. with the "*prognathus*" group as it is now defined). With a better appreciation of neurocranial form in the piscivores as a whole, I now think that the skull of *H. percoides* is basically an "*altigenis*" group one, as is the skull of *H. pellegrini*.

Similar re-evaluation leads me to think that the syncranial organization and form shown by *H. cavifrons* links that species more closely with the "*serranus*" group than with the "*altigenis*" group and its *H. percoides*-like off-shoots. The freckled coloration of *H. cavifrons* remains unique (Greenwood, 1962), and nothing approaching it is seen in the "*serranus*" group. If *H. cavifrons* is a "*serranus*"-group derivative, then it stands in the same relationship to that group as does *H. plagiostoma*, namely as an isolated off-shoot without any extant intermediates bridging the gap.

Thus, the *H. percoides–H. cavifrons* "group", although a valid one on grounds of general similarity between the constituent species, is probably polyphyletic in origin.

Because so little material or information is available on three species described above (*H. tridens*, *H. orthostoma*, and *H. parorthostoma*), they cannot as yet be included in a discussion on phylogeny. *Haplochromis tridens* does not seem to be related to any of the piscivorous species groups; its affinities probably lie with the complex of small, bottom-living species which form the bulk of trawl catches in sheltered, mud-bottomed areas of the lake. *Haplochromis orthostoma* and *H. parorthostoma* are closely related to one another but cannot readily be associated with any other species.

Leaving for the moment those species which as adults feed on post-larval fishes, consideration will be given to species which prey on larval and embryo fishes, the paedophagous species. Only one paedophage, *H. barbarae*, is described in this paper; the others are dealt with in an earlier publication (Greenwood, 1959).

Haplochromis barbarae resembles small specimens of *H. guiarti* and adults of *H. brownae*, and does not show any close similarity in body form, jaw morphology or dentition with other paedophagous species; unfortunately it has not been possible to study its osteology in detail. The probable phylogeny of the larval and embryo fish eating species was discussed in the paper cited above; the conclusion reached was that the group had a polyphyletic origin. Little more can be added to these thoughts, except to reconsider the possible relationships existing between *H. parvidens* and *H. acidens* (previously misidentified as *H. nigrescens*; see above, p. 73). The morphology of the lower jaw in *H. parvidens* differs considerably from that of *H. acidens* (as does the diet, paedophage *cf.* herbivore); but, in other syncranial characters, and in body-form, the species are very similar. In these latter characters the two species are more similar than are *H. microdon* and *H. cryptodon*, the only known paedophages showing a morphotype which could be ancestral to that of the "*parvidens*" level. *Haplochromis cryptodon* could have evolved from an *H. brownae*-like stem (possibly one like *H. barbarae* which had already adopted paedophagous habits?), the chief morphological changes involving the dentition (reduction), jaw form (to give greater distensibility) and a differential growth of the preorbital neurocranium (also leading to greater jaw motility). The changes in neurocranial form would lead to a skull essentially like that found in basal species of the "*altigenis*" and "*prognathus*" groups, in other words one like that in *H. acidens*. Thus, the origin of the "*parvidens*" structural grade is equivocal.

Reconsidering the relationship of this grade (as represented by *H. cryptodon*, *H. parvidens* and *H. microdon*) with the other distensibly-mouthed grade of paedophages (represented by *H. obesus* and *H. maxillaris*), I can find, as before, few reasons to support a close phyletic linkage between them. Osteologically neither *H. obesus* nor *H. maxillaris* has a dentary like that occurring in the *H. parvidens* group, nor in the case of *H. obesus* is its form one from which a "*parvidens*" type might evolve. The neurocranium of *H. obesus* is most unlike that of the "*parvidens*" group, but that of *H. maxillaris* is virtually identical with the skull of *H. cryptodon*. The character which most clearly distinguishes these two species from any member of the "*parvidens*" group is the occurrence of teeth in which the crowns are curved labially (and not buccally as is usual in unicuspid teeth). This dental character is so marked, and restricted to these two species, that I am inclined to give it considerable weight when speculating on phylogenies, particularly since this tooth-form seems to have no adaptive significance. If the peculiar teeth in *H. maxillaris* and *H. obesus* do indicate a fundamental relationship between the species, then their syncranial differences would suggest an independent origin from a common stem, possibly a form like *H. cronus* (see Greenwood, 1959). Dentally, and also in its general level of organization, *H. cronus* is like those generalized *Haplochromis* species that attain a larger adult size than most members of that group; it could be derived from an *H. empodisma*-like stem (see Greenwood, 1960).

Natural History

Only broad generalizations can be made about the natural history of the piscivorous predators. To date, seventy-nine species of Lake Victoria *Haplochromis* (and related monotypic genera) have been revised. Of these, forty-two species can be classed as piscivores, thirty-four species preying on free-swimming fishes, and eight species on cichlid embryos and larvae presumably taken from the mouths of parent fishes. Anticipating results still to be obtained from those species as yet unanalyzed, it seems probable that the number of non-piscivorous species will be increased substantially. I suspect that ultimately the number of piscivorous species will be about forty per cent of the total. These figures apply to sub-adult and adult members of the species only since no data are available on the feeding habits of younger stages. The same restrictions apply to considerations on habitat preferences among the piscivores.

Piscivorous species have been found in all the localities so far sampled, but much of the deeper water in Lake Victoria remains unexplored, especially at the levels occupied by bathypelagic species. In order to give a general picture of habitats in the lake, these can be divided, rather crudely, into three types : (i) sheltered areas such as bays and the smaller gulfs. (ii) Exposed areas, especially wave-washed beaches. (iii) Relatively undisturbed, open, off-shore waters (undisturbed that is, relative to the rather turbulent conditions prevailing over exposed beaches), not enclosed in gulfs or bays. Such a subdivision is, in many respects, unsatisfactory because conditions like those in habitat (iii) do occur in the larger gulfs and bays.

More species are found in sheltered areas (habitat [i]) than in the other two habitats ; 23 species have been recorded frequently in habitat (i), 14 from habitat (ii), and 15 from habitat (iii), with, in the habitats respectively, 2, 3, and 1 species occurring infrequently. However, these various species are by no means confined to a particular habitat. For example, 11 species were found in both sheltered and exposed areas, 9 in sheltered and off-shore habitats (3 of these species also occurring over beaches) ; only 6 species are apparently confined to sheltered areas. Two species occur both over beaches and in the quieter off-shore areas, and only 4 are apparently confined to the latter habitat (but 3 of these species are known only from deep water). The nature of the substrate seems to exert a less restricting influence on the piscivores than on members of other trophic groups. Twenty-four species are recorded as occurring over hard substrata (sand, shingle or rock), and 20 over a soft substrate (organic mud) ; 10 of these species are found over both kinds of substrate, 6 are thought to be confined to a soft bottom, and 13 to a hard substrate.

The eight strictly paedophagous species (see Greenwood, 1959, and p. 114 above) are not included in the analysis above. These species appear to be rather more restricted in their distribution, particularly from the viewpoint of depth range. It seems that the paedophages are confined to the littoral and immediately sublittoral zone, and are probably restricted to sheltered bays and exposed beaches.

The depth range of the other 34 piscivores is, in general, confined to water less than 60 ft. deep (with of course, the exception of *H. boops*, *H. pachycephalus*, *H. thuragnathus* and *H. dichrourus* which have been caught in water about 120 ft. deep). Most species have a wide range within these depth limits, and few if any are restricted to purely littoral areas. However, it must be emphasized that this picture may be unduly biased by sampling limitations. The horizontal distribution of the species has not been studied critically ; in water less than 20 feet deep, the spatial distribution of piscivores caught in nets set to cover about the first five feet of water below the surface, and the five feet above the bottom, does not show any obvious horizontal stratification of the particular species. But, these observations were not tested statistically.

From the relatively few gut analyses available it would seem that the majority of piscivores prey on other *Haplochromis* species, and to a much lesser extent on small cyprinid fishes (especially *Engraulicypris argenteus*). Some species have a mixed insect-fish diet, and there are several records of otherwise exclusively piscivorous species eating insects when these are periodically and suddenly super-abundant, as for instance after a heavy termite hatch. The paedophagous species appear to feed mostly on cichlid embryos and larvae (it is presumed those of *Haplochromis* species), but insects are also recorded from the gut contents of these species.

Information on the breeding habits of piscivores is almost non-existent; in those cases where some data are available, the species are known to be female mouth brooders. The turbidity of the water in most parts of Lake Victoria has precluded field observations on the actual spawning sites of both predators and prey alike.

There has not yet been enough collecting on a lake-wide basis to establish whether or not any of the fish-eating species has a geographically restricted distribution. For eight species this possibility can definitely be overruled, (and in ten others it seems very unlikely) but for sixteen species there are suggestions of geographical restriction because they have not been caught in some regions where, on the basis of habitat and niche suitability, they should be present.

SUMMARY

(1) Ten species are redescribed on the basis of new material.

(2) Nine new species (*H. pachycephalus*, *H. boops*, *H. thuragnathus*, *H. pseudopellegrini*, *H. paraguiarti*, *H. acidens*, *H. barbarae*, *H. parorthostoma* and *H. apogonoides*) are described.

(3) Although all these species have a general facies and dentition usually associated with piscivorous habits, some do not belong to this trophic group; one species (*H. acidens*) is apparently herbivorous.

(4) The possible phyletic interrelationships of the piscivorous species are discussed. Three major morphological groups can be detected, at least on the basis of their more extreme members, but the boundaries are ill-defined. Some minor groups are also considered, as are the larval and embryo fish-eating species groups.

(5) Broad summaries of the natural history of piscivorous species are given.

(6) The holotype of *Astatotilapia nigrescens* Pellegrin is redescribed and its possible synonymy discussed.

ACKNOWLEDGEMENTS

I am deeply indebted to many people for their assistance in preparing this paper.

To my colleagues Mr. A. C. Wheeler and Dr. E. Trewavas are due my thanks for, respectively, providing numerous and excellent radiographs, and for many profitable discussions.

Through the cooperation of Dr. M. Blanc of the Paris Museum, I have been able to examine Pellegrin's type specimens and thus to settle several problems. Dr. Paul Kahsbauer of the Vienna Museum has graciously lent me Lohberger's type material; I am deeply indebted to him for this privilege.

In east Africa, Dr. M. Gee and Mr. R. Welcomme of E.A.F.F.R.O. aided my studies by proving additional material and field observations; material collected by the Uganda Fisheries Department has provided many specimens from localities which I was unable to sample personally. The cooperation of E.A.F.F.R.O. and the Uganda Fisheries Department is warmly appreciated.

APPENDIX

The disputed identity of *Astatotilapia nigrescens* Pellegrin, 1909, (*Bull. Soc. Zool. France*, **34**, 157) was mentioned on page 73. Boulenger (1915) synonymized this species with *Haplochromis percoides* Blgr. 1906. Regan (1922), however, resurrected the species as *Haplochromis nigrescens*, and included in his redescription a number of specimens which I have placed in a new species, *H. acidens*. Pellegrin's figure of *A. nigrescens* is misleading and does not convey an accurate impression of the holotype and unique specimen (Paris Museum, number 09–508).

Recently, I examined this specimen, an immature male 71 mm. standard length, collected by Alluaud from the Kavirondo Gulf, Kenya. It does not agree closely with any other specimens I have handled, but is does show affinity with both *H. percoides* Blgr., 1906 and *H. flavipinnis* (Blg.), 1906, especially the latter.

Before considering its identity further, a redescription of the holotype will be given.

	mm.	Proportional percentage
Standard length	71·0	
Depth of body	24·0	33·8 standard length
Length of head	26·0	36·6 standard length
Depth of preorbital	4·3	16·5 head length
Width of interorbital	6·0	23·1 head length
Length of snout	8·0	30·8 head length
Diameter of eye	7·0	27·0 head length
Depth of cheek	6·5	25·0 head length
Length of lower jaw	12·0	46·2 head length
Length of caudal peduncle	13·0	18·3 standard length
Length of pectoral fin	18·0	25·3 standard length

Caudal peduncle 1·4 times as long as deep.

Lower jaw slightly oblique and very slightly projecting; twice as long as broad. Posterior tip of the maxilla almost reaching the vertical through the anterior orbital margin.

Gill rakers: moderately stout, 9 on the lower part of the first gill arch.

Scales: ctenoid; lateral line with 31 scales, cheek with 4 rows (the rows short, so that the anterior part of the cheek is naked). Seven scales between the upper lateral line and the dorsal fin origin; 7 between the pectoral and pelvic fin bases. Chest, belly and nuchal scales small.

Teeth: in the outer row of both jaws relatively slender, slightly curved; about 50 in the upper jaw. Inner teeth tricuspid or weakly tricuspid, arranged in 2 and 1 rows in the upper and lower jaw respectively.

Coloration. The preserved colour pattern (in shades of brown) closely resembles that of *H. percoides* and *H. flavipinnis* (see Greenwood, 1962). The dorsal fin is marbled, and the caudal both marbled and maculate; the pelvics are dark.

DISCUSSION: In its general facies, and particularly its head shape, the holotype of *A. nigrescens* resembles both *H. percoides* and *H. flavipinnis*, especially the former because the angle of the mouth is less oblique than in most specimens of *H. flavipinnis*. However, when morphometric characters are considered a number of differences between *H. percoides* and *A. nigrescens* holotype are apparent. In six characters (body depth, preorbital depth, interorbital width, snout length, eye diameter, and caudal peduncle length) the values fall outside the known range for *H. percoides*, and the lower jaw is relatively longer than in specimens of *H. percoides* of a comparable size.

There is greater correspondence between these characters in *A. nigrescens* and *H. flavipinnis*, since only two (the shallower preorbital and larger eye) fall outside the range for *H. flavipinnis*, but there is a greater difference in head shape.

Thus, *A. nigrescens* could be an aberrant specimen of either *H. flavipinnis* (differing especially in head shape) or *H. percoides* (numerous morphometric differences).

For the time being, however, I do not think that *A. nigrescens* can be formally synonymized with either species. I would prefer to recognize it as the purely nominal species *Haplochromis nigrescens* (Pellegrin) until more is known about the range of variation in small specimens of *H. flavipinnis*, the species I think it most closely resembles.

REFERENCES

GILCHRIST, J. D. F., & THOMPSON, W. W. 1917. The freshwater fishes of South Africa. *Ann. S. Afr. Mus.*, **11** : pt. 6, 465–575.

GRAHAM, M. 1929. *A Report on the Fishing Survey of Lake Victoria, 1927–1928, and Appendices*. Crown Agents, London.

GREENWOOD, P.H. 1956. The monotypic genera of cichlid fishes in Lake Victoria. *Bull. Br. Mus. nat. Hist., Zool.* **3** : No. 7, 295–333.

—— 1957. A revision of the Lake Victoria *Haplochromis* species (Pisces, Cichlidae) Part II. *Bull. Br. Mus. nat. Hist., Zool.* **5** : No 4, 76–97.

—— 1959. A revision of the Lake Victoria *Haplochromis* species (Pisces, Cichlidae) Part III. *Bull. Br. Mus. nat. Hist., Zool.* **5** : No. 7, 179–218.

—— 1960. A revision of the Lake Victoria *Haplochromis* species (Pisces, Cichlidae) Part IV. *Bull. Br. Mus. nat. Hist., Zool.* **6** : No. 4, 227–81.

—— 1962. A revision of the Lake Victoria *Haplochromis* species (Pisces, Cichlidae) Part V. *Bull. Br. Mus. nat. Hist., Zool.* **9** : No. 4, 139–214.

—— 1965. The cichlid fishes of Lake Nabugabo, Uganda. *Bull. Br. Mus. nat. Hist., Zool.* **12** : No. 9, 313–57.

—— 1966. Two new species of *Haplochromis* (Pisces, Cichlidae) from Lake Victoria. *Ann. Mag. nat. Hist.*, (13), **8** : 303–318.

TREWAVAS, E. 1964. A revision of the genus *Serranochromis* Regan (Pisces, Cichlidae). *Ann. Mus. Roy. Afr. Cent. Tervuren, ser in 8°, Zool.* No. 125, 1–58.

A REVISION OF THE LAKE VICTORIA *HAPLOCHROMIS* SPECIES

By P. H. GREENWOOD & J. M. GEE

CONTENTS

	Page
INTRODUCTION	3
Haplochromis megalops sp. nov.	4
Haplochromis piceatus sp. nov.	7
Haplochromis paropius sp. nov.	10
Haplochromis cinctus sp. nov.	15
Haplochromis erythrocephalus sp. nov.	19
Haplochromis melichrous sp. nov.	24
Haplochromis laparogramma sp. nov.	28
Haplochromis fusiformis sp. nov.	32
Haplochromis dolichorhynchus sp. nov.	34
Haplochromis tyrianthinus sp. nov.	40
Haplochromis chlorochrous sp. nov.	44
Haplochromis cryptogramma sp. nov.	48
Haplochromis arcanus sp. nov.	52
Haplochromis decticostoma sp. nov.	55
Haplochromis gilberti sp. nov.	57
Haplochromis paraplagiostoma sp. nov.	60
ACKNOWLEDGEMENTS	63
APPENDIX	63
REFERENCES	64

INTRODUCTION

MOST of the species described in this paper were collected during experimental and exploratory trawling operations in the northern waters of Lake Victoria.

Trawling has revealed the existence of numerous undescribed, apparently benthic, species living in sublittoral habitats at depths down to more than 200 feet.

Ecological information on these fishes is still very scanty. Many of the species seem to have a wide depth range (for example from 30 to 100 feet, or for species found only at greater depths, from 70 to 200 feet), but few extend into the littoral and immediately sublittoral zones. Others seemingly have a more circumscribed depth range being confined to depths of from 50 to 100 feet.

This supposed restriction to offshore areas is inferred from the absence of " trawl species " in catches made by other fishing gear in the littoral and inshore sublittoral zones. Such reasoning has, of course, certain weaknesses. For instance, compared with a trawl, the nets used to sample the littoral and immediately sublittoral zones are highly size selective and thus might not catch small fishes. However, adult individuals of several " trawl species " are large enough to be caught by seine- and gillnets, yet none of these species has been caught, despite intensive collecting.

Conversely, only one predominantly littoral species, *Haplochromis obesus*, has been caught in deeper water. We have, however, certain reservations about the identity of these specimens which could be representatives of a deep-water species closely related to *H. obesus* (see appendix, page 63).

The geographical distribution of " trawl species " within the lake is still unknown since all the available collections are from the northern (Uganda) part of Lake Victoria. Even within this area, however, there are indications from some species of interpopulational differences in certain morphological characters. Doubtless our descriptions of the new species will have to be modified when specimens from other areas become available. Nevertheless, we are moderately confident that such additional data will not alter the specific validity of the taxa described below.

We have assumed that most, if not all, of the known " trawl species " live on or near the bottom for at least part of the day: these assumptions are based especially on the nature of the food of those species whose diet is known, and on the other ingested material found in the gut. But, we cannot overrule the possibility of fishes being caught while the net is sinking to the bottom or being hauled to the surface.

The present paper by no means covers all the new species that have been caught in trawling operations. The other species will be described in subsequent papers, where it is also proposed to discuss in more detail the relationships of the " trawl species " to each other and to the inshore species complexes.

Haplochromis megalops sp. nov.

(Text-fig. 1)

HOLOTYPE: an adult male 75·0 mm. S.L. (B.M. [N.H.] reg. no. 1968.8.30.57.) from Windy Bay, Napoleon Gulf.

The trivial name refers to the large eye.

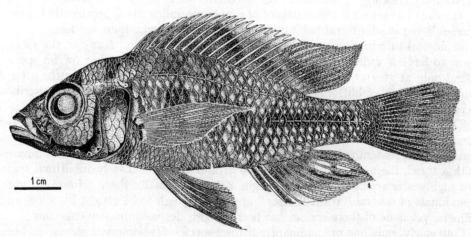

FIG. 1. *Haplochromis megalops*. Holotype. Drawn by Sharon Lesure.

DESCRIPTION: based on 27 specimens (including the holotype), 66·0–81·0 mm. S.L.

Depth of body 30·9–36·3 (mean, M = 33·4) % of standard length, length of head 32·0–35·6 (M = 33·8) %.

Dorsal head profile straight except for a slight curvature in the nuchal region, and sloping fairly steeply at 30°–40°.

Preorbital depth 11·1–15·4 (M = 12·9) % of head length, least interorbital width 22·7–28·0 (M = 24·9) %. Snout broader than long (1·1 [mode] to 1·3, rarely 1·4 times), its length 25·0–30·8 (M = 27·5) % of head; eye diameter 32·7–41·0 (M = 36·4), depth of cheek 17·3–22·6 (M = 20·0) %.

Caudal peduncle 15·3–19·2 (M = 16·3) % of standard length, 1·2 (rarely)-1·6 (mode 1·4) times as long as deep.

Mouth very slightly oblique; posterior tip of the maxilla reaching a vertical through the anterior part of the eye or, less commonly, to the anterior orbital margin. Jaws equal anteriorly or the lower jaw projecting very slightly, length of lower jaw 39·6–46·0 (M = 43·3) % of head, 1·8–2·3 (mode 2·0), rarely 1·7 or 2·4, times as long as broad.

Gill rakers: 10–12 (mode 11) on the lower part of the first gill-arch. The lower 2 or 3 rakers are reduced, and are followed by from 1 to 3 relatively slender rakers; the remainder are usually somewhat flattened, with the upper 1 or 2 often anvil-shaped.

Scales: ctenoid. Lateral line with 30 (f.3), 31 (f.6), 32 (f.12) or 33 (f.5) scales; cheek with 2 or 3 rows. Five to 6 (mode 5½) scales between the upper lateral line and the dorsal fin origin, 5–7 (mode 6) between the pectoral and pelvic fin bases.

Fins. Dorsal with 23 (f.3), 24 (f.17) or 25 (f.6) rays, comprising 14 (f.3), 15 (f.18) or 16 (f.5) spinous and 8 (f.4), 9 (f.17) or 10 (f.5) branched elements. Anal with 11 (f.8), 12 (f.16) or 13 (f.1) rays, comprising 3 spines and 8–10 branched rays. Pectoral 28·7–33·8 (M = 30·5) % of standard length. Pelvics with the first ray produced (apparently in both sexes, but only 1 female fish is available). Caudal truncate, scaled on its basal half.

Teeth. Except posteriorly in the upper jaw, the *outer* teeth in most specimens are either bicuspids, or an admixture of bi- and weakly bicuspids. The teeth are compressed, relatively stout, and slightly recurved. In some fishes most outer teeth are unicuspid and caniniform, while in other specimens unicuspids occur amongst the more numerous bicuspids; only rarely are all the outer teeth unicuspids. The outer teeth in both jaws of an individual may be similar in form, or there can be relatively more unicuspids present in the lower jaw.

The posterior outer teeth of the upper jaw are usually unicuspid and caniniform, and are often relatively large. In a few specimens, however, these teeth are similar to those occurring laterally in the jaw.

There are 48–60 (M = 52) teeth in the outer row of the upper jaw.

The *inner* teeth are generally tricuspid, but weakly so, and in a few fishes are unicuspid. All inner teeth are implanted somewhat obliquely, and are arranged in 2 rows (infrequently in 1 row) in the upper jaw, and in a single row in the lower jaw.

Osteology. The syncranium of *H. megalops* is typically that of a structurally generalized *Haplochromis* species, and as such does not depart from the type found in, for example, *H. macrops, H. nubilus, H. phytophagus* or *H. obliquidens*.

The *neurocranium* has a decurved preorbital profile and has the proportions of a generalized skull type. The openings to the cephalic lateral line canals are, however, somewhat larger than those of the species mentioned above. In contrast, the lateral line system in the dentary of *H. megalops* is not noticeably enlarged.

The lower pharyngeal bone is fine, its dentigerous surface slightly broader than long (1·1–1·2 times). The teeth are slender and cuspidate, and are arranged in 30–36 rows. In most specimens the teeth of the median rows are a little coarser than the others.

Coloration. The colours of live males are unknown; females are silvery (darker on the dorsal surfaces), with the dorsal fin hyaline, and the pelvic and anal fins pale yellow.

Preserved material: *Males (adult and sexually active)* brownish grey above the midlateral line, dusky silver below (the amount of silver visible is variable, with in extreme cases most of the ventral body half solid black save for a fine silvery sheen on the lateral aspects of the belly). A few fishes show traces of about 4 dark blotches arranged along the midlateral line of the flanks.

The snout is almost entirely black, as is the preorbital region, the lower jaw and the ventral aspects of the preoperculum; in some specimens the posterior opercular margin (otherwise silver) has a broad black margin. Two intensely black bars cross the snout, but these are only faintly discernible on the general dusky coloration of this region. A medially interrupted occipital band originates near the dorsoposterior margin of the orbit, and the nuchal region is crossed by a dark band (of variable distinctness) which originates near the opercular-preopercular junction.

Dorsal fin dusky, as are the caudal and anal fins, the latter being of variable intensity, almost black in some specimens. The anal ocelli are small, and dead white. The pelvic fins are black.

The single female available is extensively stained by rust from the metal container in which it was preserved. Thus, nothing can be said about its preserved coloration.

Ecology. Habitat. At present, the species is known from only 2 localities, one a small bay in the Napoleon Gulf near Jinja, the other in Pilkington Bay. In both places the habitat is sheltered, the water from 10–30 feet deep, and the substrate of mud or of interposed mud and sand patches.

Food. Seven of the 25 guts examined were empty. In the remainder, the predominant ingested material is macerated dipterous (?chironomid) larvae, together with small quantities of bottom mud. Chironomid pupae are also present in 8 stomachs.

Breeding. The single female examined (74 mm. S.L.) is in an advanced stage of oogenesis; both ovaries are equally developed. All the males (66–81 mm. S.L.) are adult.

Diagnosis and affinities. *Haplochromis megalops* closely resembles another new species, *H. piceatus* (see p. 7); preserved specimens of the 2 species are readily confused on superficial examination. However, *H. megalops* has a larger eye (32·7–41·0, mean 36·4% of head, *cf.* 29·0–34·0, mean 32·3% in *H. piceatus*), and a much shallower preorbital (11·1–15·4, mean 12·9% of head, *cf.* 13·6–17·8, mean 15·3%). There are other, but less trenchant differences, including dental characters (*cf.* p. 5 and p. 8).

Superficially, *H. megalops* resembles *H. cinereus*, but is distinguished from that species by its more numerous gill rakers (10–12, mode 11, *cf.* 7–9, mode 7), shallower preorbital (11·1–15·4, mean 12·9% head, *cf.* 15·0–18·0, mean 16·4%), larger eye (32·7–41·0, mean 36·4% head, *cf.* 26·2–32·0, mean 28·7%), longer lower jaw (39·6–46·0, mean 43·3% head, *cf.* 34·6–41·3, mean 37·7%) and by differences in the oral and pharyngeal dentition (see Greenwood, 1960, p. 240).

From *H. macrops*, another large-eyed species of the generalized *Haplochromis* species group (or groups), *H. megalops* is distinguished by its higher gill raker count (10–12, mode 11, *cf.* 8–11, mode 9) and its longer lower jaw (39·6–46·0, mean 43·3% head, *cf.* 38·0–42·5, mean 39·5%); the coloration of preserved specimens also differs (see Greenwood *op. cit.*, p. 236).

Haplochromis megalops is structurally and trophically a generalized species, and thus it is difficult to suggest its phyletic affinities in any more precise terms. Apart from *H. piceatus*, the species which it most closely resembles are *H. macrops*, and *H. cinereus* (see Greenwood, 1960, pp. 236–239, and 239–242).

Study Material and Distribution Records

Museum and Reg. No.	Locality	Collector
	UGANDA	
B.M. (N.H.) 1968.8.30.57 (Holotype)	Windy Bay, Napoleon Gulf	E.A.F.F.R.O.
B.M. (N.H.) 1968.8.30.58–73	Windy Bay	E.A.F.F.R.O.
B.M. (N.H.) 1968.8.30.74–88	Pilkington Bay	E.A.F.F.R.O.

Haplochromis piceatus sp. nov.

(Text-figs 2–4)

HOLOTYPE: an adult male 88·0 mm. S.L. (B.M. [N.H.] reg. no. 1968.8.30.39) from the Napoleon Gulf opposite Jinja Prison.

Named with reference to the coloration of preserved specimens (from the Latin, meaning smeared with pitch).

DESCRIPTION: based on 16 specimens (including the holotype) 67–90 mm. S.L.

Depth of body 32·0–35·3 (M = 33·3) % of standard length, length of head 32·8–35·5 (M = 34·1) %.

Dorsal head profile straight, except for slight curvature in the nuchal region, sloping at an angle of 35°–40°.

Preorbital depth 13·6–17·8 (M = 15·3) % of head, least interorbital width 21·8–25·0 (M = 23·6) %. Snout as long as broad or very slightly broader than long (1·1 times), its length 27·2–32·2 (M = 28·3) % of head; eye diameter 29·0–34·0 (M = 32·3), cheek depth 16·7–21·0 (M = 18·9) %.

Caudal peduncle 17·3–20·9 (M = 19·1) % of standard length, 1·6–2·0 (modal range 1·6–1·8) times as long as deep.

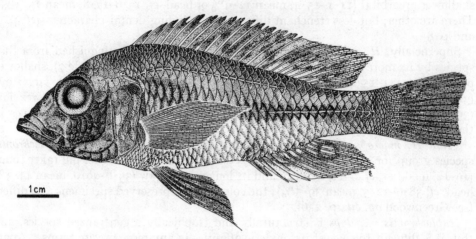

Fig. 2. *Haplochromis piceatus*. Holotype. Drawn by Sharon Lesure.

Mouth slightly oblique, the jaws equal anteriorly or, more commonly, the lower projecting slightly; length of lower jaw 41·8–45·0 (M = 43·5) % of head, 2·0 (mode)–2·6 times its breadth (in one exceptional specimen 2·8 times). Posterior tip of the maxilla reaching a vertical through the anterior part of the eye or only as far as the anterior orbital margin.

Gill rakers: 12 (mode) or 13, rarely 11 or 15, on the lower part of the first gill arch. The lower 2–5 rakers reduced, the remainder of varied form (even in one individual), usually short, or slender, or flattened and with the uppermost raker anvil-shaped in outline; some lower rakers may also be anvil-shaped or lobed.

Scales: ctenoid. Lateral line with 32 (f.7), 33 (f.7) or 34 (f.2) scales, cheek with 2 or 3 rows. Five to 6 scales between the dorsal fin origin and the upper lateral line, 5–6 (mode) between the pectoral and pelvic fin bases.

Fins. Dorsal with 23 (f.1), 24 (f.11) or 25 (f.4) rays, comprising 14 (f.2) or 15 (f.14) spinous and 9 (f.10) or 10 (f.6) branched elements. Anal with 11 (f.11) or 12 (f.5) rays, comprising 3 spines and 8 or 9 branched rays. Pectoral 27·0–32·3 (M = 29·6) % of standard length. Pelvics with the first ray produced in both sexes, but proportionately longer in males. Caudal truncate, scaled on its proximal half.

Teeth. The *outer teeth* (Text-fig. 3) situated posteriorly in the upper jaw of most specimens are unicuspid or tricuspid, those placed posterolaterally are tricuspid; in a few specimens these teeth are not, however, differentiated from the other outer teeth. The posterior teeth are deeply embedded in the gum tissue and are difficult to expose (*cf.* the situation in *H. megalops* where these teeth are unicuspid and readily exposed).

The anterolateral and anterior teeth in the upper jaw, and all outer teeth in the lower jaw, are slender, compressed and unequally bicuspid, with the major cusp produced and slender, and the minor cusp short but clearly demarcated.

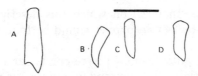

Fig. 3. *Haplochromis piceatus.* Jaw teeth. A, C and D in anterior view, B in lateral view. A: an anterior upper jaw tooth from a small individual (67 mm. S.L.); C and D: posterior upper teeth from a larger fish (85 mm. S.L.). Scale equals 0·5 mm. for A, 1 mm. for B–D.

There are 54–74 (M = 64) outer teeth in the upper jaw. The *inner* teeth are tricuspid and generally implanted somewhat obliquely; there are 2 (rarely 3) series in the upper jaw, and 1 or 2 in the lower.

Osteology. The syncranium of *H. piceatus* is indistinguishable from that of *H. megalops* (see p. 6); that is, it is of the generalized type.

The *lower pharyngeal bone* (Text-fig. 4) is fine, with its dentigerous surface slightly broader than long, and carries 36–40 rows of slender, cuspidate teeth.

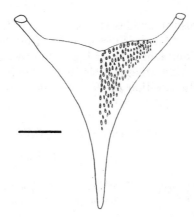

Fig. 4. *Haplochromis piceatus.* Lower pharyngeal bone; teeth indicated on left side. Scale equals 2 mm.

Coloration. The colours of live fishes are unknown. *Preserved material.* Males (adult and sexually active), are almost indistinguishable from males of *H. megalops* but differ slightly in that the demarcation between the lighter upper surfaces and the darker flanks is less obvious, the general coloration being greyer. Also, in *H. piceatus* there are no signs of any midlateral dark blotches.

Females are an almost uniform silvery-yellow, but darker dorsally; there is a very faint indication of a narrow midlateral stripe on the posterior third of the body, terminating at the base of the caudal fin. The dorsal and caudal fins are greyish, the latter weakly maculate on its upper half, and somewhat darker near the base. The anal and pelvic fins are hyaline.

Ecology. Habitat. The material on which this description is based came from a trawl haul in water 45–60 feet deep, over a mud bottom, in the relatively sheltered Napoleon Gulf.

Food. Four of the 14 guts examined were empty, and the remainder contained very little ingested material. Bottom mud and dipterous pupae (probably Chironomidae) were identified.

Breeding. All 16 specimens (67–90 mm. S.L.) are adult. One female has a few larvae in the buccal cavity, thus suggesting that the species is a mouth brooder. Three of the 4 females known have the right ovary, noticeably larger than the left; in the fourth individual (probably at an early stage of oogenesis) the ovaries are of almost equal size.

Diagnosis and affinities. Haplochromis piceatus is very similar to *H. megalops*; characters distinguishing the 2 species are given on p. 6. The principal dental differences lie in the presence of some tricuspid teeth posterolaterally in the upper jaw of most *H. piceatus* individuals, and in the absence of caniniform unicuspids anteriorly and anterolaterally in the outer tooth row of all individuals. The tendency for the posterior upper teeth of *H. piceatus* to be deeply embedded is another difference between the species.

Haplochromis piceatus also resembles *H. cinereus* and *H. macrops* (see Greenwood, 1960, pp. 236–239, and 239–242); it is distinguished from both species principally by its higher gill raker count and longer lower jaw. In other morphometric characters, however, *H. piceatus* approaches these species more closely than does *H. megalops*.

Remarks on the phyletic position of *H. megalops* (see p. 7) apply equally to *H. piceatus*.

STUDY MATERIAL AND DISTRIBUTION RECORDS

Museum and Reg. No.	Locality	Collector
	UGANDA	
B.M. (N.H.) 1968.8.30.39 (Holotype)	Napoleon Gulf, nr. Jinja prison	E.A.F.F.R.O.
B.M. (N.H.) 1968.8.30.40–56	Napoleon Gulf, nr. Jinja prison	E.A.F.F.R.O.

Haplochromis paropius sp. nov.

(Text-figs 5–7)

HOLOTYPE: an adult male 69·0 mm. S.L. (B.M. [N.H.] reg. no. 1968.8.30.89) from near Bulago island.

The trivial name (from the Greek for "eye shade") refers to the prominent lachrymal stripe.

DESCRIPTION: based on 34 specimens (including the holotype) 63·0–87·0 mm. S.L. Depth of body 35·0–40·3 (M = 37·4) % of standard length, length of head 31·5–35·8 (M = 34·0) %.

Dorsal head profile gently curved (rarely straight), sloping steeply at 40°–45°.

Preorbital depth 13·6–17·4 (M = 15·6) % of head, least interorbital width 18·3–28·0 (M = 23·1) %. Snout 1·1–1·5 (mode 1·2) times as broad as long, its length

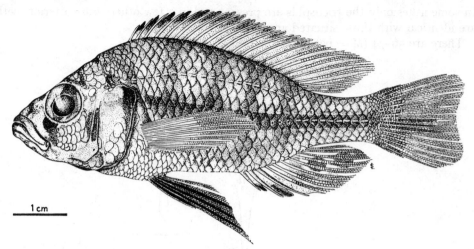

Fig. 5. *Haplochromis paropius*. Holotype. Drawn by Sharon Lesure.

25·0–31·9 (M = 29·1) % of head; eye diameter 28·0–34·8 (M = 31·9), depth of cheek 18·1–24·2 (M = 21·3) %.

Caudal peduncle 15·6–20·6 (M = 18·8) % of standard length, 1·3 (rare) to 1·9 (mode 1·6) times as long as deep.

Mouth horizontal (the usual condition) to slightly oblique (*ca* 15°); jaws equal anteriorly. Lower jaw 39·5–45·5 (M = 41·2) % of head, 1·5–1·9 (modal range 1·5–1·6) times as long as broad; a weak mental protuberance is visible in many specimens.

Posterior tip of the maxilla reaching a vertical through the anterior part of the eye or somewhat posterior to that point (rarely only reaching a vertical through the anterior orbital margin).

Gill rakers: 8–10 (mode 9), the lower 1–3 rakers reduced, the upper 2–4 flattened and club-like, branched or anvil-shaped; other rakers of various shapes, from short and stout to relatively slender.

Scales: ctenoid; lateral line with 30 (f.1), 31 (f.13), 32 (f.13) or 33 (f.3), cheek with 3 (rarely 4) rows. Five to 7 (mode 6) scales between the upper lateral line and the dorsal fin origin, 5–7 (mode 6) between the pectoral and pelvic fin bases.

Fins. Dorsal with 23 (f.8), 24 (f.23) or 25 (f.2) rays, comprising 15 (f.27) or 16 (f.6) spinous and 8 (f.12) or 9 (f.21) branched elements. Anal with 11 (f.23) or 12 (f.9) rays, comprising 3 spines and 8 or 9 rays. Pectoral fin 27·9–34·7 (M = 31·7) % of standard length. Caudal truncate scaled on its proximal half. Pelvics with the first ray produced in both sexes, but relatively longer in adult males.

Teeth. Except posteriorly, the *outer* teeth in *both jaws* are relatively stout and bicuspid (Text-fig. 6A), with compressed and slightly expanded crowns. Posteriorly and posterolaterally in the upper jaw there are, usually, 2 to 5 enlarged and near caniniform unicuspid teeth, preceded by a variable number of compressed tricuspids.

In some fishes only the tricuspids are present and in a few others the posterior teeth are identical with those situated anteriorly and laterally.

There are 56–74 (M = 62) outer teeth in the upper jaw.

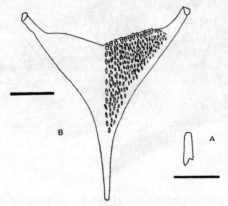

Fig. 6. *Haplochromis paropius.* A: Upper jaw tooth (anterior view). B: Lower pharyngeal bone. Scale equals 1 mm. for A, 2 mm. for B.

Inner teeth are tricuspid, compressed, and arranged in 2 or 3 (rarely 4) rows in the upper jaw and 2 (less frequently 1 or 3) rows in the lower.

Osteology. The neurocranium of *H. paropius* is typically that of the generalized *Haplochromis* type (as seen, for example, in such species as *H. nubilus*, *H. macrops* and *H. brownae*, etc.) with a gently curved preorbital dorsal profile, and relatively

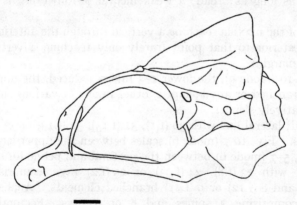

Fig. 7. *Haplochromis paropius.* Neurocranium; left lateral view. Scale equals 2 mm.

broad in the otic and interorbital regions (Text-fig. 7). The lateral line tubes and pores are not noticeably hypertrophied.

The *dentary* and *premaxilla* are also of the generalized type, although the dentary is more elongate and has slightly enlarged lateral line tubes. The lateral line tubules on the *preorbital* are also somewhat hypertrophied.

The *lower pharyngeal bone* is fine, its triangular dentigerous surface slightly broader (1·1–1·2 times) than long (Text-fig. 6B). The teeth are slender, cuspidate and fairly densely arranged in 38–46 rows, the total impression being that of a dental "felt". In some specimens, teeth in the median rows are slightly coarser than the lateral teeth.

Coloration in life. Adult, sexually active males: ground colour of body greenish, as are the snout and lips (the lower sometimes blueish); the belly is silvery. Dorsal surface of the head and body, anterior part of the belly, the opercular region and the branchiostegal membrane red (deepest red dorsally and often on the anterior ventral surfaces, otherwise orange-red). There is a pronounced, dark lachrymal stripe, a fainter transverse bar across the snout and often traces of 3 or 4 faint vertical bars on the flanks; in many individuals a faint dark midlateral longitudinal band is visible.

Dorsal fin dark green proximally, the distal part either with an overall red flush or the red pigment aggregated into spots and streaks on the soft part of the fin. Caudal green proximally, hyaline or flushed with red distally. Anal fin hyaline to white, sometimes with a dusky base; ocelli yolk-yellow. Pectoral fins in some individuals greenish, otherwise hyaline. Pelvics black.

Quiescent males are generally similar in coloration to active fishes, but the red body and head colours are much fainter, and the pelvics are dusky.

Females are dark grey (with faintly greenish undertones) dorsally, shading to silver on the flanks and belly below a distinct but interrupted dark midlateral stripe. The dorsal fin often has a red flush over the soft part, but the spinous part is hyaline. The anal fin in yellow; all other fins are hyaline.

Preserved material. Males: ground colour light brown, shading to silvery-white on the chest and belly; the flank scales in sexually active fishes have broad, dark margins. In most fishes a wide, dark midlateral band runs from the posterior opercular margin to the caudal origin; this band is sometimes broken, and it may even be entirely absent. When absent it is replaced by 5 fairly distinct vertical bars on the flanks; each bar tapers dorsally and ventrally, and none reaches the body outline. Faint to fairly distinct indications of these bars are sometimes discernible in fishes with a well-developed midlateral band.

A distinct, vertical lachrymal band runs from the orbit to immediately behind the posterior tip of the maxilla; it is always visible although of very variable intensity. In some specimens another short vertical bar is present along the vertical limb of the preoperculum. Two parallel transverse bars cross the snout, and a third, medially interrupted and broader bar crosses the occiput from orbit to orbit. This band may appear as a dorsal continuation of the lachrymal stripe.

Dorsal fin hyaline, with dusky lappets, sometimes with dark blotches between the soft rays and at the bases of the spines. Caudal hyaline, but with some darkening of the membrane between the central rays. Anal hyaline to greyish, the ocelli usually greyish-white but sometimes dead-white. Pelvic fins black.

Females have a preserved coloration very like that of males (including dark dorsal lappets and a dark base to the dorsal fin), but the ventral surfaces of the body are silvery white, the flank scales are without dark margins, and the cephalic markings are much less intense (or even absent).

Ecology. Habitat. Haplochromis paropius has been caught in several areas of northern Lake Victoria, but always over a mud substrate, in water 50–100 feet deep, and in off-shore regions.

Food. The intestine is long and coiled, the stomach large and distensible, thus suggesting a vegetarian diet. The contents of 20 guts were examined, and seem to confirm this supposition. All contain large quantities of blue-green algae and diatoms, and smaller amounts of other plant material. Eight of these guts contain, in addition, fragmentary remains of larval Diptera (probably chironomids).

Of the plant material, only the diatoms show any appreciable signs of digestion; the blue-green algae are apparently undigested.

Breeding. All the specimens examined (63–87 mm. S.L.) are adult.

Diagnosis and affinities. Structurally, *H. paropius* is a generalized species in all respects except for the long gut which is a specialization associated with essentially vegetarian feeding habits.

The coloration of adult males seems to distinguish *H. paropius* from most of the other generalized *Haplochromis* species described so far (and including those with vegetarian diets). Exceptions to this statement are *H. erythrocephalus* and *H. cinctus*, both new species described in this paper (see pp. 19 and 15 respectively).

Haplochromis erythrocephalus is distinguished by several characters (including details of coloration) and *H. cinctus* principally by colour differences; *H. paropius* is compared with these species on pp. 23 and 18. The resemblance between *H. cinctus* and *H. paropius* is very close indeed, but because the differences involve male coloration we attach great importance to them, especially since the species are sympatric. On anatomical grounds, *H. cinctus* and *H. paropius* would seem to qualify as sibling species.

Superficially, *H. paropius* resembles *H. lacrimosus* (see Greenwood, 1960). Unfortunately the live coloration of adult male *H. lacrimosus* is unknown, but there are certain similarities in the preserved coloration of the two species. However, there are also several differences (*cf.* p. 13 above with p. 231 in Greenwood *op. cit.*) and it seems likely that the differences in live colours may be fairly marked. The principal anatomical characters distinguishing the species are the longer lower jaw of *H. paropius* (39·5–45·5, M = 41·2% head, *cf.* 31·4–41·3, M = 37·1% in *H. lacrimosus*), the long, coiled gut, and the larger number of outer teeth in the upper jaw (56–74, M = 62, *cf.* 40–60, M = 50 for *H. lacrimosus*).

Another species bearing a superficial resemblance to *H. paropius* is *H. melanopus* Regan, 1922. This species is very poorly known and has not yet been revised. It is represented only by the 3 syntypes on which Regan based his description (Regan, *op. cit.*, fig. 1). *Haplochromic melanopus* differs from *H. paropius* in having more gill rakers on the lower part of the first arch (11 or 12, *cf.* 8–10, mode 9) and a much shorter lower jaw (31·0–32·6% of head, *cf.* 39·5–45·5, M = 41·2%). Until more is known about *H. melanopus* the comparison cannot be carried further.

From other generalized species with an essentially bicuspid outer dentition, *H. paropius* is distinguished by its coloration, its longer gut, and by various combinations of morphometric characters. For accounts of those species see Greenwood, 1960, and pp. 4–10 of this paper.

Certain other deeper water species resemble *H. paropius* anatomically and in some cases, in their coloration as well (for example, *H. cinctus*, see p. 17). We are studying 3 such species at present but cannot yet describe them in full detail. However, from the information we have it is clear that *H. paropius* is distinct.

Haplochromis paropius is derived from the same stem as the other anatomically generalized *Haplochromis* species in Lake Victoria; for the moment it is not possible to suggest any more precise relationships. Despite similarities in male coloration it seems unlikely that *H. paropius* is closely related to *H. erythrocephalus*, which apparently belongs to a different lineage within the complex of generalized species (see p. 23).

Notes on three specimens from near Mwama island

Three fishes (64·0, 65·5 and 69·0 mm. S.L.; all adult and sexually active males) appear to be very atypical members of this species. The specimens came from a single trawl haul made at depths ranging from 70–200 feet, over both hard (rock and sand) and soft mud substrates. The trawl was shot in a bay on the south side of Mwama island and hauled at a place some distance off-shore in the open lake.

In general facies the 3 fishes closely resemble other specimens of *H. paropius*, but in 2 specimens the eye is larger (36·4 and 39·1 % or head) and in the third the eye diameter is in the upper range known for *H. paropius* (34·8%). All 3 specimens also differ somewhat in coloration, viz., the dorsal head colour is lighter red (i.e. more orange than red) and there is no trace of red pigment on the anterior chest region or on the belly (these regions being white). Finally, the branchiostegal membrane is white and not black as in other *H. paropius* males.

With only 3 specimens available we cannot evaluate the significance of these differences. But, judging from our knowledge of other species related to *H. paropius* (including several as yet undescribed species) the 3 Mwama fishes could well represent yet another species in the "*paropius*" complex.

Study Material and Distribution Records

Museum and Reg. No.	Locality	Collector
	UGANDA	
B.M. (N.H.) 1968.8.30.89 (Holotype)	Near Bulago isl. (80–90 feet)	E.A.F.F.R.O.
B.M. (N.H.) 1968.8.30.90–117	S.W. and N.E. of Bulago isl. (55–75 feet)	E.A.F.F.R.O.
B.M. (N.H.) 1968.8.30.118–123	Near Bulago isl. (80–90 feet)	E.A.F.F.R.O.
B.M. (N.H.) 1968.8.30.124–126	Near Mwama isl. (70–200 feet)	E.A.F.F.R.O.

Haplochromis cinctus sp. nov.

(Text-fig. 8)

HOLOTYPE: an adult male 84·0 mm. S.L. (B.M. [N.H.] reg. no. 1968.8.30.13) from a trawl haul made in water 70–200 feet deep, near Mwama island.

The trivial name (from the Latin for "girded") refers to the characteristic banding seen in males.

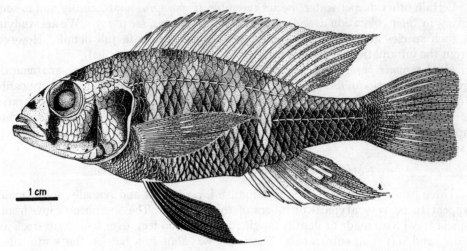

Fig. 8. *Haplochromis cinctus*. Holotype. Drawn by Sharon Lesure.

DESCRIPTION: based on 6 specimens 76·0–87·5 mm. S.L. The principal morphometric ratios are given below:

S.L.	D.†	H.†	Po. %	Io. %	Snt. %	Eye %	Ck. %	Lj %	C.P.†
76·0	36·9	33·3	15·8	25·7	30·4	31·6	19·8	39·5	18·4
81·0	38·2	33·4	14·8	25·9	29·1	33·3	21·5	40·8	17·9
84·0	36·9	32·7	16·4	27·2	31·0	34·6	20·0	43·5	15·5
84·0	37·0	32·2	18·5	25·2	33·3	36·3	21·5	41·8	19·0
86·5	38·1	34·7	16·0	24·3	30·0	30·0	20·0	40·0	16·8
87·8	36·6	35·4	16·1	25·2	29·0	30·6	21·5	41·8	19·0

† = % of standard length
% = % of head length

Dorsal head profile slightly curved, sloping at an angle of 40°–50°; in 2 specimens the snout profile slopes even more steeply than the head profile. Snout 1·1 (mode)–1·2 times broader than long.

Lower jaw horizontal or very slightly oblique, 1·5–1·8 times as long as broad. Jaws equal anteriorly. Posterior tip of the maxilla reaching a vertical through the anterior part of the eye or almost to the pupil.

Caudal peduncle 1·3–1·8 times as long as deep.

Gill rakers: 9 on the lower part of the first gill arch, the lower 1 or 2 rakers reduced, the upper 3–5 flattened, sometimes divided and anvil-shaped; intervening rakers of varied form, from relatively slender to relatively stout.

Scales: ctenoid; lateral line with 32 (f.5) or 33 (f.1) scales, cheek with 3 rows. Six and a half to 8 (mode 6½) scales between the upper lateral line and the dorsal fin origin, 6–8 (mode 7) between the pectoral and pelvic fin bases. Scales on the ventral chest region noticeably smaller than those situated laterally on the chest, or those on the belly.

Fins. Dorsal with 23 (f.1) or 24 (f.5) rays, comprising 15 (f.4) or 16 (f.2) spines and 8 (f.3) or 9 (f.3) branched rays. Anal with 11 (f.2) or 12 (f.4) rays, comprising 3 spines and 8 or 9 branched rays. Pectoral 31·0–33·1 % of standard length. Pelvics with the first and second rays produced and filamentous. Caudal slightly emarginate, scaled on its proximal half.

Teeth. The *outer* teeth in the *upper jaw* are an admixture of stout, unequally bicuspid and tricuspid teeth, with the bicuspids predominating. Posteriorly, the teeth are unicuspid and caniniform. There are 50–64 teeth.

In the *lower* jaw, the *outer* teeth are stout and unequally bicuspid anteriorly and laterally, but tricuspid posteriorly and posterolaterally.

The *inner* teeth in both jaws are tricuspid, and are arranged in 2 rows.

Osteology. No complete skeleton is available. The *lower pharyngeal bone* has a triangular dentigerous surface, 1·2 times as broad as long. The teeth are bicuspid, relatively coarse, and are arranged in 30–38 rows, those of the 2 median rows being somewhat coarser than the lateral teeth.

Coloration in life. Males: dorsal surface of head and body light grey with a distinct orange overtone; light orange on the operculum and flanks, belly dark blue-black. The flanks are crossed by 3–5 dark vertical bars which extend from the belly to the upper lateral line; these bars have turquoise highlights, as do the lateral surfaces of the caudal peduncle. The lower jaw is grey, with a turquoise sheen; the branchiostegal membrane is sooty. The ventral limb of the preoperculum has a broad, dark blotch, and 2 distinct parallel dark bands cross the snout. Above each eye there is a dark spot, the spots from each side almost meeting in the midline.

The dorsal fin is hyaline, with orange-red lappets and similarly coloured (but fainter) streaks between the spines; very distinct orange-red spots occur between the branched rays. Caudal fin is hyaline but with a reddish-orange flush. Anal with a narrow greyish area at its base but becoming faintly reddish-orange distally; the ocelli are yolk-yellow.

The coloration of *live females* is unknown.

Preserved material. Males: Dorsal surface of the head and the upper part of the body light brown. Chest, belly and almost the entire lateral aspect of the caudal peduncle have a dusky overlay; however, as the scales in these regions have light blue-grey centres, the overall coloration varies from blue-black to dark grey. Arising from the dark ventral colour of the belly and chest are 4 broad, dark stripes, lanceolate in outline, with the taper beginning just below the level of the upper lateral line; from this point dorsally, the stripe narrows rapidly and becomes much less definite until it disappears immediately below the dorsal fin base. The first stripe passes over the axil of the pectoral fin; above this point it meets a broad vertical dark stripe on the posterior margin of the operculum. The last body stripe may barely be distinguishable from the dark ground colour of the caudal peduncle. The ground colour of the body between the stripes is light, much lighter than that of the dorsal body surface.

The area of the operculum not covered by the dark posterior band is silvery. The horizontal limb of the preoperculum is blue-black or brownish, the ventral limb covered by a broad and short dark bar which expands anteriorly onto the cheek. A wide and intense lachrymal stripe extends through the eye onto the nape; the stripes from each side are narrowly separated medially above the orbit. Two distinct, narrow, parallel bands cross the snout anterior to the orbit. In some specimens the lower jaw is blackish, in others it is brownish; the colour is correlated with that of the branchiostegal membrane which may be entirely black or pale with just the posterior (i.e. opercular) part black.

The dorsal fin is yellowish to greyish darkest basally, and with black lappets. Caudal dark basally (the extent variable and its outline irregular), hyaline distally. Anal dark (almost dusky) along the proximal half, yellowish distally, the ocelli large and dead-white. Pelvic fins are black.

Female coloration is unknown.

Ecology. The 6 known specimens came from a trawl haul near Mwama island. Because this particular haul was made over both hard (rock and shingle) and soft (mud) substrates, and at depths from 70–200 feet, little can be said about the habitat of *H. cinctus*.

Only one fish contains ingested material in the gut; the stomach and intestine are filled with colonial blue-green algae, diatoms and other algaceous material. Only the diatoms show signs of digestion. Since the intestine of *H. cinctus* is long and coiled, it is reasonable to assume that the species feeds principally on plant matter.

The 6 male fishes (76·0–87·5 mm. S.L.) on which this description is based, are all adults and are sexually active.

Diagnosis and affinities. In all morphological characters, except male coloration, *H. cinctus* is indistinguishable from *H. paropius*. Yet, in both live and preserved coloration, males of the two species are immediately distinguishable; regrettably no females of *H. cinctus* are available for comparison.

Live fishes differ in obvious and subtle ways (compare p. 17 above with p. 13). Among the obvious differences may be cited the grey-orange head coloration of *H. cinctus* compared with the deep red of *H. paropius*; the clear-cut transverse barring on the flanks of *H. cinctus*, the bars arising from a deep blue-black chest and belly, compared with the silver belly and very faint (or more usually, invisible) bars in *H. paropius*; and finally, the red branchiostegal membrane of *H. paropius* compared with the dusky membrane in *H. cinctus*.

The most obvious interspecific differences in preserved coloration are the dark chest and belly of *H. cinctus*, and the distinct, dorsally incomplete bars on the flanks. No specimen of *H. paropius* has a dark chest and belly (despite the dark margin to the scales in that region) and even in those specimens with bars on the flanks, the bars are much fainter.

These differences may not seem impressive in print but are striking when specimens (especially live fishes) are compared.

Because of the close similarity between *H. cinctus* and *H. paropius*, the comparison of that species with others resembling it can be applied to *H. cinctus* as well (see p. 14). *Haplochromis cinctus* and *H. paropius* are clearly derived from the same stem, and possibly even from the same ancestral species.

Study Material and Distribution Records

Museum and Reg. No.	Locality	Collector
	Uganda	
B.M. (N.H.) 1968.8.30.13 (Holotype)	Near Mwama island	E.A.F.F.R.O.
B.M. (N.H.) 1968.8.30.14–18	Near Mwama island	E.A.F.F.R.O.

Haplochromis erythrocephalus sp. nov

(Text-figs. 9–11)

HOLOTYPE: an adult male, 74·0 mm. S.L. (B.M. [N.H.] reg. no. 1968.8.30.251) from the Buvuma channel, south of Ramafuta island.

The trivial name refers to the bright red colour of the head in adult males.

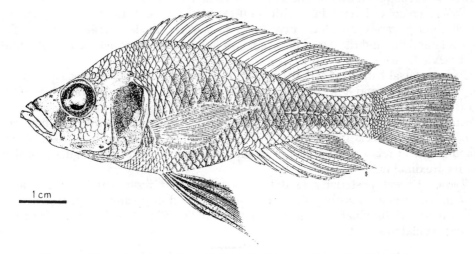

FIG. 9. *Haplochromis erythrocephalus*. Holotype. Drawn by Sharon Lesure.

DESCRIPTION: based on 41 specimens (including the holotype) 58·0–77·0 mm. S.L. Depth of body 34·5–41·5 (M = 37·1) % of standard length, length of head 30·4–35·8 (M = 33·1) %.

Dorsal head profile with a gentle concavity above the eye, sloping fairly steeply at an angle of 30°–40°.

Preorbital depth 13·6–17·4 (M = 15·8) % of head, least interorbital width 22·7–27·9 (M = 24·6) %. Snout 1·0–1·2 (mode 1·1) times as broad as long, its length 27·0–32·5 (M = 29·3) % of head, diameter of eye 28·6–35·0 (M = 31·2), depth of cheek 17·7–23·8 (M = 20·8) %.

Caudal peduncle 15·1–20·0 (M = 17·9) % of standard length, 1·3–2·0 times as long as deep. These data were obtained from specimens collected at 3 different localities, and it seems possible that there are population differences in the relative depth of the caudal peduncle. Fishes from 2 localities (Buvuma Channel south of

Ramafuta island, and off Bonga Point at the entrance to Pilkington Bay) have shallower peduncles (i.e. more are in the range 1·8–2·0 times as long as deep) than those from Pilkington Bay itself (predominantly in the range 1·5–1·6 times). It must be noted, however, that the samples from Bonga Point (N = 9) and Buvuma Channel (N = 11) are smaller than that from Pilkington Bay (N = 21).

Mouth oblique, sloping at an angle of 30°–35° (occasionally as steeply as 40°). Jaws equal anteriorly; length of lower jaw 40·2–46·8 (M = 43·6) % of head, 1·9–2·6 (mode 2·0) times as long as broad. Posterior tip of the maxilla reaching a vertical through the anterior orbital margin, occasionally a little posterior to this point.

Gill rakers: 10 (rare)–13, mode 12, on the lower part of the first arch. Lower 1–3 rakers reduced, upper 1–4 usually flattened, often branched, with the uppermost 1 or 2 anvil-shaped; intervening rakers simple, from relatively stout to slender. In a few specimens none of the upper rakers is flattened, the entire series (except the lower part) composed of unbranched, stout to slender rakers.

Scales: ctenoid; lateral line with 30 (f.1), 31 (f.10), 32 (f.20) or 33 (f.7) scales, cheek with 3 (rarely 2) rows. Five (rare) to 7 (rare), mode 6, scales between the upper lateral line and the dorsal fin origin, 5–7 (modal range 6–7) between the pectoral and pelvic fin bases.

Fins. Dorsal with 23 (f.5), 24 (f.21), 25 (f.14) or 26 (f.1) rays, comprising 15 (f.22) or 16 (f.19) spines and 8 (f.13), 9 (f.23) or 10 (f.5) branched rays. Anal with 10 (f.1), 11 (f.20), 12 (f.18) or 13 (f.2) rays, comprising 3 spinous and 7–10 branched rays. Pectoral 24·6–34·8 (M = 31·0) % of standard length. Pelvics with the first ray produced in both sexes; too few females are available to check on possible sexual dimorphism in relative elongation. Caudal truncate or slightly emarginate, scaled on its proximal half.

Teeth. Except posteriorly (and to a certain extent posterolaterally) the *outer teeth* in *both jaws* are mostly somewhat compressed and unequally bicuspid (Text-fig. 10). In some individuals the major cusp tends to be obliquely truncate (rather than almost equilateral in outline).

FIG. 10. *Haplochromis erythrocephalus*. Upper jaw teeth (anterior view), showing variation in crown shape. Scale equals 0·5 mm.

The *posterior upper* teeth are either unicuspid or tricuspid, and are markedly smaller than the anterior and lateral teeth; furthermore, these posterior teeth are deeply embedded in gum tissue and are partly hidden by a fold of the lip. The *posterior lower* teeth do not show such a marked size discrepancy, are usually bicuspid (but with cusps of almost equal size) and are exposed.

There are 42–70 (M = 56) outer teeth in the upper jaw.

All *inner teeth* are tricuspid and compressed, implanted more or less vertically, and are arranged in 2 (rarely 1) rows in the upper jaw and a single (less frequently double) row in the lower jaw.

Osteology. The *neurocranium* of *H. erythrocephalus* is essentially similar to that of *H. empodisma* (see fig. 5, in Greenwood, 1956 [the species was then wrongly identified as *H. michaeli*, see Greenwood, 1960, p. 265]). Thus it differs from the neurocranial type found in such species as *H. macrops*, *H. brownae* and *H. nubilus*, species whose syncranial architecture is thought to be of a generalized type. Parenthetically it can be noted that similar generalized neurocrania are found in species like *H. obliquidens* and *H. phytophagus* which have a specialized dentition and feeding habits (see Greenwood, 1956 and 1966).

The skull of *H. erythrocephalus* departs from the generalized type in having a flat, moderately steep dorsal profile (compared with a curved and steep profile), relatively narrower interorbital and otic regions, and a rather more elongate and narrower preorbital part of the skull.

The *dentary* also resembles the *H. empodisma* type in being relatively more slender and elongate than the generalized type.

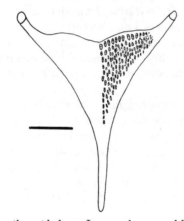

Fig. 11. *Haplochromis erythrocephalus.* Lower pharyngeal bone. Scale equals 2 mm.

The *lower pharyngeal bone* is fine, its triangular dentigerous surface $1 \cdot 1$–$1 \cdot 3$ times as broad as long (Text-fig. 11). The teeth are slender, compressed and cuspidate, and are arranged densely but irregularly in from 40–46 rows.

Coloration in life. *Adult males*: The dorsal aspects of the head are bright red, the red pigment extending onto the dorsal part of the body where it becomes fainter and takes on the appearance of a red flush. Body coloration yellowish with a silvery overlay on the flanks and belly.

The dorsal fin is pinkish-red, the caudal has a red flush on its dorsal half with the remainder yellowish to hyaline; anal fin hyaline, with 2 or 3 orange ocelli. Pelvic fins are black.

The live colours of *females* have not been recorded.

Preserved material. Adult males: ground colour light fawn to yellowish, a faint purplish-grey area extends over most of the caudal peduncle and then anteriorly as a narrow band or a series of blotches situated midlaterally on the flanks. In some specimens the belly is dusky, and scales in that region have a dark margin. The snout and preorbital region are greyish, with faint indications of 2 narrow, dusky transverse bars across the snout; a dusky, near vertical lachrymal bar extends from the lower orbital margin to the angle of the jaws.

All fins except the pelvics are hyaline, the dorsal with dusky lappets, and the central region of the caudal with a dusky zone; the anal ocelli are faint and greyish in colour. The pelvic fins vary from dusky to intense black.

Females are uniformly light straw-yellow except for a grey snout and preorbital region. Dorsal and caudal fins are greyish, the anal and pelvic fins hyaline, the caudal with a dark base.

Ecology. Habitat. The material used in this description came from 3 localities, viz. Pilkington Bay (depth 30–35 feet, over mud), Bonga Point at the entrance to Pilkington Bay (45–60 feet, over mud) and from the Buvuma Channel south of Ramafuta island (90–100 feet, over mud). Personal observations on trawl catches made in other parts of the Buvuma Channel and in the Napoleon Gulf (at depths of 30–100 feet) suggest that *H. erythrocephalus* has a wide distribution in off-shore localities in the northern parts of Lake Victoria.

Food. The guts of 26 fishes from the 3 localities were examined. In all, both the stomach and intestine are packed with colonial blue-green algae, diatoms and smaller quantities of green algae; in some fishes a few fragments of Crustacea and of insect larvae are also present.

The constitution of this ingested material closely resembles that of the organic mud substrate occurring in the areas from which the fishes were collected. Thus, we conclude that *H. erythrocephalus* feeds on the bottom mud. The ingested colonies of blue-green algae show no signs of digestion at any point in the alimentary tract. In contrast, diatoms found in the intestine are all digested, that is, only the frustules remain; diatoms in the stomach, however, are mostly intact. The crustacean and insect remains are fragmentary, irrespective of their locality in the gut, and consist only of the chitinous parts.

The stomach of *H. erythrocephalus* is large and distensible, the intestine long and coiled.

Breeding. With one possible exception (a male 64 mm. S.L.), all the specimens examined are adult (size range 58–77 mm. S.L.) and most are males. This bias is undoubtedly due to collectors selecting the brilliantly coloured male fishes: it cannot be used to draw any inferences about the distribution or relative abundance of the sexes.

One female (73 mm. S.L., from Bonga Point) with quiescent ovaries, has a few larvae in the buccal cavity, and was probably brooding young at the time of capture.

Too few females have been examined to decide whether or not there is any tendency for asymmetrical ovarian development. However, in some the right ovary is distinctly larger than the left one.

Diagnosis and affinities. Any attempt to analyse the affinities of *H. erythrocephalus* is limited by the existence of superficially similar but as yet undescribed species. For instance, we have under study at the moment four such species, and undoubtedly others will be found as more fishing is carried out in the deep, off-shore waters of the lake.

The four species mentioned above, however, can be distinguished from *H. erythrocephalus* on certain morphometric characters, and in life by differences in male coloration.

Haplochromis erythrocephalus does not seem to be closely related to any of the known in-shore species having small-sized adults with generalized, bicuspid outer teeth. From these species, *H. erythrocephalus* is distinguished by its coloration, longer lower jaw and, with respect to many species, its narrower interorbital width and higher gill raker count. The oblique lower jaw of *H. erythrocephalus* also serves as a diagnostic feature.

The oblique mouth, high gill raker count, the broader, more densely toothed lower pharyngeal bone, and the long intestine serve to separate *H. erythrocephalus* from *H. empodisma* juveniles of a comparable length. (*H. empodisma* occurs in some of the shallower localities from which *H. erythrocephalus* has been recorded.) In an earlier paper, Greenwood (1960) was referring to *H. erythrocephalus* when he wrote ". . . The nearest living relative of *H. empodisma* is a small and as yet undescribed species which occurs in the same habitat but is confined to shallow water". The latter part of this statement now requires correction (both species occur in deep water). For the moment, the first part still seems valid, particularly if the strength of the term " nearest relative " is diluted a little to read " A near relative . . . ".

Of the known (and described) species from deeper water habitats, *H. erythrocephalus* bears some superficial resemblance to *H. paropius*, including red pigment on the head of adult males. This red coloration in *H. paropius* is, however, much darker, and there are other colour differences (*cf.* p. 21 above with p. 13). Certain anatomical differences may also be noted, viz. the higher gill raker count in *H. erythrocephalus* (11–13, mode 12, *cf.* 8–10, mode 9), the concave head profile of *H. erythrocephalus* (gently convex or nearly straight in *H. paropius*) and differences in the morphology of the neurocranium (*cf.* pp. 21 and 12).

Haplochromis paropius, like *H. erythrocephalus*, is probably not an isolated species since we have evidence of an *H. paropius* species complex. But, the diagnostic characters for *H. paropius* also serve to separate the other related species from *H. erythrocephalus*.

The neurocranium of *H. erythrocephalus* is of interest. Its shape and proportions are unlike those characterizing the skulls of other species with small-sized adults, a generalized body-form and unspecialized dentition (or even specialized grazing or browsing teeth). Instead, it closely approximates to the neurocranial form found in *H. empodisma*, trophically a generalized bottom-feeding species with an unspecialized dentition but one with larger sized adults.

The significance of these differences or resemblances in neurocranial form (and correlated characters in the syncranial skeleton) are difficult to evaluate. They may suggest that *H. erythrocephalus* was derived from a different stem than that for many of the other *Haplochromis* species with small-sized adults and a generally unspecialized anatomy (including those new species described in this paper).

STUDY MATERIAL AND DISTRIBUTION RECORDS

Museum and Reg. No.	Locality	Collector
	UGANDA	
B.M. (N.H.) 1968.8.30.251 (Holotype)	Buvuma Channel, S. of Ramafuta isl. (90–100 feet)	E.A.F.F.R.O.
B.M. (N.H.) 1968.8.30.270–291	Pilkington Bay	E.A.F.F.R.O.
B.M. (N.H.) 1968.8.30.262–269	Bonga Point (mouth of Pilkington Bay)	E.A.F.F.R.O.
B.M. (N.H.) 1968.8.30.252–261	Buvuma Channel, N.W. of Vuga isl. (90–100 feet)	E.A.F.F.R.O.

Haplochromis melichrous sp. nov.
(Text-figs. 12–15)

HOLOTYPE: an adult male 98·0 mm. S.L. (B.M. [N.H.] reg. no. 1968.8.30.150) from a trawl haul made between Nsadzi island and the mainland.

The trivial name (from the Greek for honey-coloured) refers to the golden-brown coloration of female fishes.

FIG. 12. *Haplochromis melichrous*. Holotype. Drawn by Sharon Lesure.

DESCRIPTION: based on 18 specimens (including the holotype) 69·0–107·0 mm. S.L.

Depth of body 36·5–44·5 (M = 39·9) % of standard length, length of head 34·2–36·1 (M = 34·9) %.

Dorsal head profile concave (strongly so in some fishes), sloping at an angle of 30°–35°. Premaxillary pedicels not prominent. Pores of the cephalic lateral line system enlarged, the tubes and pores on the preorbital being especially prominent.

Preorbital depth 14·6–18·6 (M = 16·4) % of head, least interorbital width 18·6–23·6 (M = 22·1) %. Snout broader than long (1·1–1·3, mode 1·3, times), its length 27·0–33·3 (M = 31·1) % of head; eye diameter 24·5–29·8 (M = 27·6), depth of cheek 21·7–28·5 (M = 25·4) %.

Caudal peduncle 15·0–19·4 (M = 17·6) % of standard length, 1·1 (rare)–1·7 (mode 1·3) times as long as deep.

Mouth oblique, sloping at an angle of 35°–40°; lower jaw projecting, with the tip lying above the tip of the premaxilla (rarely with the jaws equal anteriorly). Length of lower jaw 44·7–48·8 (M = 46·9) % of head, 1·5–2·0 (modal range 1·8–1·9) times it breadth; a distinct mental protuberance is developed in most specimens. Posterior extension of the maxilla variable, from a point reaching a vertical slightly anterior to the orbit, to one passing through the anterior part of the eye.

Gill rakers: 9 or 10 on the lower part of the first arch, the lower 1–3 rakers reduced, the upper 3 or 4 flattened (some lobed, other anvil-shaped), and the intervening rakers either short and stout or relatively slender.

Scales: ctenoid. Lateral line with 30 (f.3), 31 (f.10) or 32 (f.4) scales, cheek with 3 or 4 rows. Five to 7 (mode 6½) scales between the upper lateral line and the dorsal fin origin, 5 or 6 (mode), rarely 7, between the pectoral and pelvic fin bases.

Fins. Dorsal with 23 (f.12), 24 (f.5) or 25 (f.1) rays, comprising 14 (f.2) or 15 (f.16) spines and 8 (f.11), 9 (f.5), or 10 (f.2) branched rays. Anal with 11 (f.15) or 12 (f.3) rays, comprising 3 spines and 8 or 9 branched elements. Pectoral fin 28·5–33·3 (M = 30·8) % of standard length. Caudal truncate or obliquely truncate (i.e. the lower posterior margin slopes forwards and downwards), scaled on its proximal half. Pelvics with the first 2 rays produced (the first ray longer) in both sexes but proportionately more so in adult males.

Teeth. In the *outer row* of *both jaws* there is an admixture of slender, somewhat compressed bicuspids, weakly bicuspids, and unicuspids (Text-fig. 13); the relative proportions of the different types shows great individual variability. Teeth situated laterally and posterolaterally in the premaxilla are strongly incurved, and in most specimens the posterolateral teeth are enlarged. The posterolateral and posterior teeth in the lower jaw are often tricuspid and small. There are 56–80 (M = 68) outer teeth in the upper jaw.

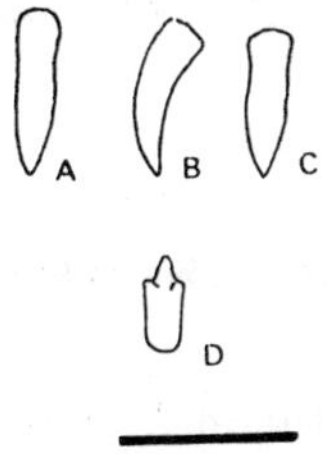

FIG. 13. *Haplochromis melichrous*. A–C upper jaw teeth. A and C: in anterior view. B: in lateral view. D: tricuspid tooth from the lower jaw. Scale equals 1 mm.

The *inner teeth* are tricuspid (sometimes weakly so) and are very obliquely implanted so as to lie almost horizontally. There are 2 (rarely 3) series of inner teeth in the upper jaw, and 2 (rarely 1) in the lower jaw.

Osteology: The *neurocranium* of *H. melichrous* (Text-fig. 14) resembles that of *H. victorianus*; in other words it is referable to the " *serranus* " group (see Greenwood, 1967, p. 109). It differs from the neurocranium of *H. victorianus*, however, in having a narrower and more acute preorbital region.

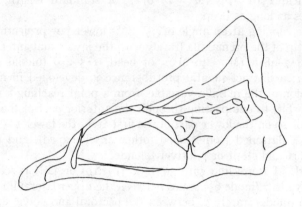

FIG. 14. *Haplochromis melichrous*. Neurocranium in left lateral view. Scale equals 2 mm.

The *premaxilla* is slightly expanded medially and anteromedially, and its lateral and posterolateral teeth are strongly incurved.

The *dentary* has a strong mental protuberance; although the openings into the lateral line tubes are enlarged, the tubes themselves are not hypertrophied.

The dentigerous surface of the *lower pharyngeal bone* is as broad as long, or slightly broader (Text-fig. 15). The teeth are fine, and arranged in from 30–34 rows; apart

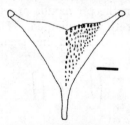

FIG. 15. *Haplochromis melichrous*. Lower pharyngeal bone. Scale equals 2 mm.

from the posterior pair of teeth in the median row, no others in this row are noticeably coarser than the lateral teeth.

Coloration in life. Males have a dark ground colour but with lighter, faintly iridescent greyish patches on the flanks, and sometimes three faint transverse bars on the anterior flanks. The dorsal fin is hyaline with dark lappets, the caudal hyaline and faintly maculate. The anal fin is dusky at its base and hyaline or faintly red distally; the ocelli are large and orange. Pelvic fins are black.

Females are dark golden-brown on the dorsal aspects of the body and on the flanks, shading to silver ventrally; a faint, dark midlateral stripe is generally visible. The dorsal fin is faintly golden-brown proximally. The pelvics are hyaline.

Preserved material. Males: have a yellow-brown ground coloration. The chest region is silvery but the scales here (and on the belly) are edged with black. The snout, lower jaw, branchiostegal membrane, lower part of the preoperculum, the entire interoperculum and most of the ventral aspects of the operculum are dusky. The ventral half to two-thirds of the body posterior to the vent is also dusky; 3–5 broad but relatively faint vertical bars originate from this dark area, and extend upwards to the level of the upper lateral line or slightly higher (but never to the dorsal body outline).

Dorsal fin dark grey to dusky, the lappets black. Caudal fin greyish to dusky, darkest basally, and maculate distally. The anal fin is dusky distally, black basally and over the spinous part; the ocelli are whiteish but very indistinct. The pelvic fins are black.

The smallest male examined (74 mm. S.L., adult but apparently quiescent) has the overall tone of the ventral body half much lighter than in the other specimens, but the margins of the scales in this region are very dark.

Females are greyish-yellow above the midlateral line, light yellow-brown below, with a faint silver overlay on the chest region. Most specimens have a faint but broad midlateral stripe, interrupted at about its mid-point for a variable distance; sometimes only a short (i.e. about 5 scale rows long) anterior part of this stripe is detectable. The tip of the lower jaw is dusky, and there is a short, rather indistinct lachrymal stripe which does not reach the ventral margin of the preoperculum. The dorsal and caudal fins are greyish, the latter sometimes maculate but always with the proximal part darkest. Anal and pelvic fins are hyaline.

Ecology. Habitat. The specimens described above came from trawl hauls made in water 70–100 feet deep, over a mud bottom between Nsadzi island and the mainland. One of us (J.M.G.) has identified *H. melichrous* in catches made south of this island over a similar bottom but at depths of from 130–160 feet.

Food. Little information could be gathered from the 10 guts examined. Most were almost empty save for small quantities of mud and fragments of crustaceans (probably *Caridina* sp); other fragments (taken from 3 guts) were tentatively identified as being remains of pupal Diptera.

Breeding. All except one of the specimens available (69–107 mm. S.L.) are adults; the smallest fish is a juvenile female, but the ovaries show early signs of oogenesis. The next smallest fish (74 mm. S.L.), a male, is adult. In all 8 adult and sexually active females, the right ovary is noticeably larger than the left one.

Diagnosis and affinities. The concave dorsal head profile, the oblique lower jaw, broad snout, and the colours of live fishes serve to distinguish *H. melichrous* from any of the deep water *Haplochromis* species so far discovered.

Haplochromis melichrous does not closely resemble any of the known inshore and shallow water species; it is readily distinguished from such oblique-mouthed species as *H. cavifrons*, *H. plagiostoma*, *H. flavipinnis* and certain extreme forms of

H. obesus (see fig. 2 and p. 183 in Greenwood, 1959) by several morphometric and dental characters.

There are few pointers to the phyletic affinities of *H. melichrous*. Judging from the dentition and syncranial architecture, the species could be associated with the "*serranus*" group (see Greenwood [1967], page 109, *et seq.*) as a rather specialized off-shoot. Equally, it could be associated with the *H. flavipinnis–H. cavifrons* group, a species complex of uncertain affinities but one probably related (at least in part) to the "*serranus*" group.

STUDY MATERIAL AND DISTRIBUTION RECORDS

Museum and Reg. No.	Locality	Collector
	UGANDA	
B.M. (N.H.) 1968.8.30.150 (Holotype)	Between Nsadzi isl. and mainland (70–100 feet)	E.A.F.F.R.O.
B.M. (N.H.) 1968.8.30.151–167	Between Nsadzi isl. and mainland (70–100 feet)	E.A.F.F.R.O.

Haplochromis laparogramma sp. nov.

(Text-figs. 16–19)

HOLOTYPE: an adult male 78·0 mm. S.L. (B.M. [N.H.] reg. no. 1968.8.30.220) from north of Nsadzi island.

The trivial name refers to the conspicuous midlateral stripe.

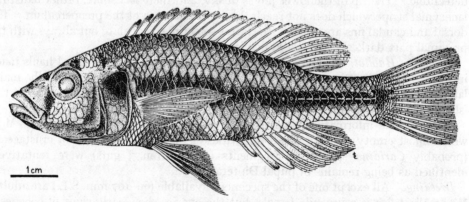

FIG. 16. *Haplochromis laparogramma*. Holotype. Drawn by Sharon Lesure.

DESCRIPTION: based on 25 specimens (including the holotype), 61·0–84·5 mm. S.L.

Depth of body 27·5–31·1 (M = 29·3) % of standard length, length of head 30·1–33·8 (M = 30·5) %.

Dorsal head profile straight or slightly curved in the nuchal region, sloping at an angle of 25°–30°.

Preorbital depth 14·6–18·8 (M = 16·3) % of head, least interorbital width 24·0–27·0 (M = 25·7) %. Snout as long as broad to 1·3 times broader than long (mode 1·1 times), its length 27·4–31·9 (M = 30·2) % of head, eye diameter 26·7–32·6 (M = 30·0), depth of cheek 15·8–21·3 (M = 19·5) %.

Caudal peduncle 17·7–22·0 (M = 20·0) % of standard length, 1·7–2·1 (modal range 1·8–1·9) times as long as deep.

Lower jaw sloping fairly steeply at an angle of 25°–35° (less frequently at *ca.* 20°), its length 40·7–45·4 (M = 42·8) % of head, and 1·8–2·6 (mode 2·2) times its breadth. Jaws equal anteriorly or the lower projecting slightly, terminating in a low mental protuberance. Posterior extension of the maxilla variable, from a point slightly anterior to the orbital margin to one below the anterior part of the eye.

Gill rakers: 11 (mode) or 12, rarely 10 or 13, on the lower part of the first gill arch. The lower 1–4 rakers are reduced, the upper 2–5 flattened, sometimes branched and often anvil-shaped; intervening rakers are relatively slender.

Scales: ctenoid; lateral line with 32 (f.4), 33 (f.10), 34 (f.6) or 35 (f.2) scales, cheek with 3 (rarely 2 or 4) rows. Five and a half or 6 scales between the upper lateral line and the dorsal fin origin, 5–6 (mode) rarely 7, between the pectoral and pelvic fin bases.

Fins. Dorsal with 24 (f.10), 25 (f.14) or 26 (f.1) rays, comprising 15 (f.18) or 16 (f.7) spines and 9 (f.16) or 10 (f.9) branched rays. Anal with 11 (f.10) or 12 (f.15) rays, comprising 3 spines and 8 or 9 branched rays. Pectoral 24·4–30·1 (M = 27·7) % of standard length. Pelvics with the first ray produced in both sexes. Caudal truncate, scaled on its basal half.

Teeth. Except posteriorly in both jaws, the *outer* teeth are compressed and unequally bicuspid (Text-fig. 17); in a few specimens there are tricuspid, compressed

Fig. 17. *Haplochromis laparogramma*. Upper jaw teeth, all from one fish, to show variation in crown shape. A: lateral view. B–D: anterior view. Scale equals 2 mm.

teeth interspersed among the bicuspids. Most individuals have weakly bicuspid or unicuspid, or (less commonly) tricuspid teeth posteriorly in the upper jaw, and tricuspids in the same position in the lower jaw. Exceptionally, there are bicuspid teeth throughout the outer row. In both jaws the posterior teeth are smaller than the anterior and lateral ones. Not infrequently the posterior quarter of the premaxilla is edentulous.

There are 46–62 (M = 54) outer teeth in the upper jaw.

The *inner teeth* are tricuspid, and are arranged in 2 (less commonly 3) series in the upper jaw and 1 or 2 series in the lower jaw.

Osteology. The *neurocranium* is close to a generalized type (see Greenwood, 1962) but shows certain specializations, such as the rather more protracted preorbital

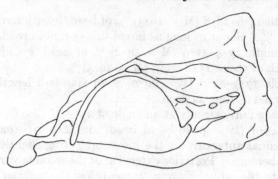

Fig. 18. *Haplochromis laparogramma*. Neurocranium in left lateral view. Scale equals 2 mm.

region which is also less steeply sloping and is straighter (Text-fig. 18). Also, the otic region is narrower than in the generalized type (greatest breadth contained *ca.* 2 times in neurocranial length *cf.* $1\frac{1}{2}$ times).

Broadly speaking, the neurocranium of *H. laparogramma* can be considered intermediate between the *H. brownae* type and the *H. serranus* type (see Greenwood, *op. cit.*).

Lateral line canals on the neurocranium are not enlarged, and neither are those on the dentary; the canals on the preorbital bone, however, are slightly enlarged.

The *dentary* departs slightly from the generalized type since it is relatively elongate. The *premaxilla* shows the development of a slight beak through the expansion of its anterior and anterolateral dentigerous surfaces.

The *lower pharyngeal bone* is slender and rather narrow (Text-fig. 19). Its dentigerous surface is as long as broad, and carries 32–38 rows of slender, cuspidate teeth.

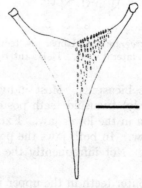

Fig. 19. *Haplochromis laparogramma*. Lower pharyngeal bone. Scale equals 2 mm.

Coloration in life. *Adult males* have the dorsal part of the body dark blue, shading to greenish-silver on the flanks, and silver on the belly. A dark midlateral stripe runs from behind the head to the origin of the caudal fin. Dorsal aspects of the head vary from brownish to yellowish-brown; the snout is sometimes crossed by 2 distinct dark bars which are weakly chevron-shaped, the apex directed orally. The opercular region and the cheek are yellowish, but the branchiostegal membrane is dusky. The dorsal and caudal fins are pale yellowish-orange, the anal is hyaline with orange ocelli. Pelvic fins are black.

Live coloration for *females* is unknown.

Preserved coloration. Males: are light to dark brown on the dorsal half of the body and caudal peduncle, silvery below, with a faint duskiness on the chest, belly and, in some specimens, the flanks as well. The brown and silver colours are distinctly demarcated by a broad midlateral stripe extending from behind the operculum to the caudal origin. This stripe is of variable thickness, being broadest over the anterior two-thirds of its length. The snout is dusky, with faint traces of 2 darker transverse bars, the upper of which is interocular in position; in some fishes the occiput is also dusky. In all specimens there is a faint but broad lachrymal stripe, and in some a faint prolongation of the midlateral band across the operculum to the vertical preopercular limb. The branchiostegal membrane is dusky.

Dorsal and anal fins are hyaline, with faintly dusky lappets; the anal ocelli are dead-white. The caudal fin is also hyaline but becomes darker near the base. The pelvic fins vary from dusky to black.

The preserved coloration of *females* is similar to that of males but lacks the ventral duskiness on the chest, belly and flanks.

Ecology. Habitat. The species has been recorded from 4 off-shore localities. In all, the bottom was of mud, and the depth varied from 50–110 feet. *Haplochromis laparogramma* has not yet been identified from catches made in deeper water.

Food. The guts of 20 individuals from 3 different localities have been examined. Apart from 5 empty guts, all contained varying amounts of fragmentary larval and pupal Diptera. In none was any bottom detritus recorded, thus suggesting that *H. laparogramma* does not feed directly on the bottom.

Breeding. All the specimens available (61–85 mm. S.L.) are adults; in the 3 females examined, the right ovary is clearly larger than the left one, the discrepancy being most marked in the single " ripe " individual.

Diagnosis and affinities. For the moment it is difficult to suggest the relationships of *H. laparogramma*, but this may become simpler when more is known about the *Haplochromis* species from deeper water habitats. Certainly *H. laparogramma* cannot be closely associated with any of the known inshore species having a similar general facies, as for example, *H. longirostris* (see Greenwood, 1962).

Among the deep water "trawl species", *H. laparogramma* may be related to *H. fusiformis* (see p. 32). Superficially *H. laparogramma* is distinguished by its coloration, slightly broader snout, larger eye (possibly correlated with the smaller adult size of the species), coarser teeth, larger scales on the nape, and its smaller adult size.

STUDY MATERIAL AND DISTRIBUTION RECORDS

Museum and Reg. No.	Locality	Collector
	UGANDA	
B.M. (N.H.) 1968.8.30.220 (Holotype)	N. of Nsadzi isl. (70–100 feet) .	E.A.F.F.R.O.
B.M. (N.H.) 1968.8.30.233–239	Buvuma Channel, W. of Vuga isl. (108 feet)	E.A.F.F.R.O.
B.M. (N.H.) 1968.8.39.240–251	Off the Bulago–Tavu bank (50–75 feet)	E.A.F.F.R.O.
B.M. (N.H.) 1968.8.30.221–232	N. of Nsandzi isl. (70–100 feet) .	E.A.F.F.R.O.

Haplochromis fusiformis sp. nov.

(Text-fig. 20)

HOLOTYPE: an adult female 110·0 mm. S.L. (B.M. [N.H.] reg. no. 1968.8.30.28) from the Buvuma Channel, west of Nienda and Vuga islands, at a depth of 108 feet. The trivial name refers to the slender, elongate body form of this species.

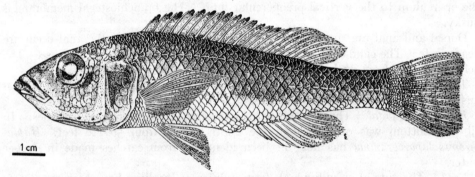

FIG. 20. *Haplochromis fusiformis*. Holotype. Drawn by Sharon Lesure.

DESCRIPTION: based on 10 specimens (including the holotype) 74·5–110·0 mm. S.L.

Depth of body 23·2–25·5 (M = 24·6) % of standard length, length of head 30·4–32·1 (M = 31·3) %. Dorsal head profile slightly curved, sloping at an angle of about 30°, its outline broken by the fairly prominent premaxillary pedicels. The openings to the cephalic lateral line system are not noticeably enlarged.

Preorbital depth 15·7–19·3 (M = 17·4) % of head, least interorbital width 21·9–26·0 (M = 24·5) %. Snout 1·0–1·2 (mode 1·1) times as long as broad, its length 29·4–34·4 (M = 31·9) % of head, eye diameter 23·4–27·7 (M = 26·3), depth of cheek 15·4–21·4 (M = 18·9) %.

Caudal peduncle 18·6–21·4 (M = 20·0) % of standard length, 1·9–2·1 (modal range 2·0–2·1) times as long as deep.

Lower jaw sloping at an angle of *ca.* 20°–30°, jaws equal anteriorly or the lower projecting slightly, its length 40·4–43·8 (M = 42·2) % of head, and 2·0–2·6 (no distinct mode) times its breadth. Posterior tip of the maxilla reaching a vertical slightly anterior to the orbital margin, or more posteriorly to one through the orbital margin.

Gill rakers: ten or 11 on the lower part of the first arch, the lower 1–3 rakers reduced, the upper 2–5 flattened and branched; intervening rakers are relatively slender.

Scales: ctenoid; lateral line with 34 (f.2), 35 (f.3), 36 (f.2) or 37 (f.1) scales, cheek with 3 or 4 rows. Seven or 8 (rarely 6) between the dorsal fin origin and the upper lateral line, 5 or 6 (mode) between the pectoral and pelvic fin bases.

Fins. Dorsal with 25 (f.2) or 26 (f.8) rays, comprising 15 (f.2), 16 (f.7) or 17 (f.1) spines and 9 (f.2), 10 (f.7) or 11 (f.1) branched rays. Anal with 11 (f.4), 12 (f.5) or 13 (f.1) rays, comprising 3 spines and 8–10 branched rays. Pectoral 23·1–27·0 (M = 25·1) % of standard length. Caudal truncate (obliquely so in 2 specimens), scaled on its proximal half to three-fifths. Pelvics with the first ray slightly produced in both sexes, but relatively more so in males.

Teeth. Slender but compressed, unequally bicuspid teeth predominate in the *outer* row of both jaws in almost all specimens, but the posterior teeth in the lower jaw are generally tricuspid or weakly tricuspid. Some weakly bicuspid, or unicuspid teeth occur anteriorly in both jaws.

Two exceptional specimens have mostly unicuspids laterally and anteriorly in both jaws, with a few bicuspids interspersed.

There are 44–64 outer teeth in the upper jaw; the 2 largest specimens have the lowest number of teeth (44 and 48), an unusual inverse correlation.

In most specimens the posterior quarter of the premaxilla is edentulous, but in a few the bone is toothed along its entire length.

The *inner rows* are composed of tricuspid or weakly tricuspid teeth arranged in 2 (mode) or 3 series in the upper jaw and 1 or 2 series in the lower jaw.

Osteology. No complete skeleton is available. The *lower pharyngeal bone* is fine, its dentigerous surface 1·1–1·2 times as broad as long, and carries 34–36 rows of slender, weakly cuspidate teeth.

Coloration of live fishes. *Males* have the dorsal part of the body brilliant purple-blue, shading to silvery-yellow on the flanks, and becoming dusky on the ventral surfaces. All fins (except the black pelvics) are dark basally and hyaline distally; the anal has 2 or 3 white ocelli.

Females have the dorsal part of the body blue-grey, shading to silver on the flanks and ventral surfaces. Fins are coloured as in the males, except that the pelvics are hyaline and there are no ocelli on the anal.

Preserved material. *Males* have a dark grey-brown ground coloration, but are sooty on the chest, belly and that area of the ventral body wall above the anterior part of the anal fin (in a few exceptional specimens the chest and belly are yellow). There are no distinct cephalic markings, although in some fishes there are traces of a medially interrupted occipito-nuchal band. The branchiostegal membrane is grey brown.

The dorsal fin is grey to sooty, the lappets black; in most specimens there is a dark but poorly defined band running along the fin base. The anal is grey, darker (nearly black) basally and over the spinous part of the fin. The pelvics are black.

Females are greyish brown dorsally, straw yellow below; the snout and dorsal surface of the head are grey. Except for the usual dark opercular spot, the head is without markings, and the branchiostegal membrane is pale. The dorsal and anal fins are greyish to lightly sooty, the anal and pelvic fins are hyaline.

Ecology. Habitat. The 10 specimens available came from two off-shore areas in the Buvuma Channel. In both places the substrate is mud, the depth 90–96 feet and about 108 feet respectively.

Food. Four of the 9 guts examined were empty. The remainder contained small quantities of unidentifiable insect fragments (probably from pupal Diptera).

Breeding. All 10 fishes are probably adults, although some doubt is felt about the sexual state of one female (93 mm. S.L.); this individual might be a juvenile. Like the three other females, the right ovary in this fish is larger than the left one.

Diagnosis and affinities. The possible relationship between *H. fusiformis* and *H. laparogramma* is mentioned on p. 31, and the features distinguishing the 2 species are listed there. Until more specimens of *H. fusiformis* are available this tentative relationship cannot be explored further. At least for the moment, the slender, elongate body-form coupled with the moderately oblique lower jaw serve to distinguish both species from the other deep-water *Haplochromis* species.

Study Material and Distribution Records

Museum and Reg. No.	Locality	Collector
	UGANDA	
B.M. (N.H.) 1968.8.30.28 (Holotype)	Buvuma Channel, W. of Nienda and Vuga isls. (108 feet)	E.A.F.F.R.O.
B.M. (N.H.) 1968.8.30.29–38	Buvuma Channel, W. of Nienda and Vuga isls. (108 feet)	E.A.F.F.R.O.

Haplochromis dolichorhynchus sp. nov.

(Text-figs. 21–24)

Haplochromis tridens (part): Greenwood, 1967, *Bull. Br. Mus. nat. Hist. Zool.*, **15**: no. 2, 97.

NOTE ON THE SYNONYMY. A single specimen (B.M. [N.H.] reg. no. 1966.3.9.165) previously identified as *H. tridens* by Greenwood (1967) is now re-identified as *H. dolichorhynchus*. This fish is the aberrantly coloured female noted on p. 99 of that paper.

HOLOTYPE: an adult male 102·0 mm. S.L. (B.M. [N.H.] reg. no. 1968.8.30.168) from Murchison Bay, at a depth of 30 feet.

The trivial name refers to the rather protracted snout of this species relative to the snout in its presumed relatives.

DESCRIPTION: based on 27 specimens (including the holotype), 67·5–119·0 mm. S.L.

Depth of body 28·6–33·6 (M = 31·9) % of standard length, length of head 33·4–36·9 (M = 35·5) %.

Dorsal head profile straight to slightly concave, sloping at an angle of 20°–30°; premaxillary pedicels fairly prominent and breaking the outline of the profile to

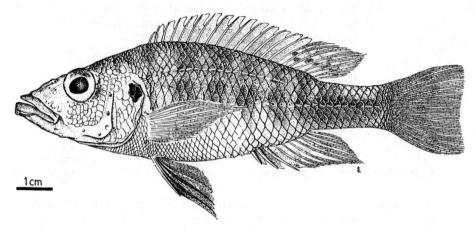

Fig. 21. *Haplochromis dolichorhynchus*. Holotype. Drawn by Sharon Lesure.

create a concavity above and a little anterior to the eye. The cephalic lateral line pores are enlarged and obvious.

Preorbital depth 16·0–19·5 (M = 17·7) % of head, least interorbital width 16·3–21·2 (M = 18·8) %. Snout 1·0–1·4 (mode 1·1) times as long as broad, its length 30·3–37·7 (M = 34·2) % of head. Eye and orbit almost circular, eye diameter 25·0–29·6 (M = 27·2) % of head, depth of cheek 16·7–23·0 (M = 19·6) %.

Caudal peduncle 17·1–21·4 (M = 19·2) % of standard length, 1·6–1·9 (modal range 1·7–1·8) times as long as deep.

Mouth somewhat oblique (*ca.* 15°–25°), lower jaw projecting slightly to strongly (the usual condition), and with a distinct mental protuberance; length of lower jaw 42·8–52·5 (M = 47·8) % of head, 2·0–2·8 (mode 2·1) times as long as broad. Premaxilla distinctly expanded medially so as to give it a beaked appearance. Posterior tip of the maxilla reaching a vertical slightly anterior to the orbital margin, occasionally a little posterior to this line.

Gill rakers: 9–11 (mode 10), rarely 8, on the lower part of the first gill arch. The lower 1–4 rakers are reduced, the upper 3 or 4 flattened and usually divided, the intervening rakers of varied form but generally either stout or slender.

Scales: ctenoid. Lateral line with 31 (f.2), 32 (f.5), 33 (f.16) or 34 (f.3) scales. Cheek with 3 (mode) or 4 rows. Five to 6½ (mode 5½) scales between the upper lateral line and the dorsal fin origin, 6 or 7 (mode) between the pectoral and pelvic fin bases.

Fins. Dorsal with 23 (f.3), 24 (f.18) or 25 (f.5) rays, comprising 14 (f.1), 15 (f.22) or 16 (f.3) spines and 8 (f.4), 9 (f.18) or 10 (f.4) branched rays. Anal with 11 (f.12) or 12 (f.14) rays, comprising 3 spines and 8 or 9 branched rays. Pectoral 24·5–30·8 (M = 27·4) % of standard length. Pelvic fins with the first ray produced, about equally so in both sexes. Caudal truncate, scaled on its proximal half.

Teeth. Except posteriorly, the *outer* teeth in the *upper jaw* are predominantly of a slender, somewhat compressed form, gently curved, with unequally bicuspid crowns (Text-fig. 22); the major cusp is noticeably produced. In many fishes there are admixtures of bi-and weakly bicuspid teeth, or of bi-and unicuspids (the latter usually situated anteriorly); rarely, some tricuspid, compressed teeth occur anteriorly and laterally in the outer row.

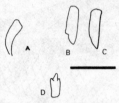

FIG. 22. *Haplochromis dolichorhynchus*. A–C upper jaw teeth. A: in lateral view. B and C: in anterior view. D: posterolateral tooth from the dentary. Scale equals 1 mm.

Posteriorly and posterolaterally it is usual to find tricuspid teeth, but unicuspids and bicuspids are sometimes found in this position, or there can be a combination of unicuspids posteriorly with tricuspids posterolaterally.

There are 50–80 (M = 70) teeth in the outer series of the upper jaw.

The *outer row* of teeth in the *lower jaw* is usually like that in the upper jaw; when tricuspid teeth occur posteriorly, they are slightly larger than their counterparts in the upper jaw.

Only tricuspid teeth occur in the obliquely implanted *inner rows*, arranged in 2 or 3 series in the upper jaw, and 1 or 2 in the lower jaw.

Osteology. The *neurocranium* of *H. dolichorhynchus* closely resembles that of *H. prognathus* (see Greenwood, 1967, pp. 81, and 109 *et seq*.) being moderately shallow with a gently sloping anterior dorsal profile, and a fairly elongate preorbital region (Text-fig. 23). The neurocranial lateral line canals and pores, however, are

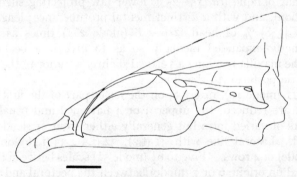

FIG. 23. *Haplochromis dolichorhynchus*. Neurocranium in left lateral view. Scale equals 4 mm.

moderately enlarged, especially when compared with the condition found in previously described species of the "*prognathus*" group (see Greenwood, 1967, *loc. cit.*).

The *premaxilla* has a pronounced beak resulting from the anterior and anterolateral expansion of its dentigerous arm; the pedicels are as long as the dentigerous arms.

The *dentary* has a marked outward flare of its upper half so that when viewed from in front, each ramus has a concave outer face. This flare is continued forward almost to the symphysis. The lower lateral faces of the symphyseal surface are produced forward as a conical mental process.

The *lower pharyngeal bone* (Text-fig. 24) has its dentigerous surface slightly longer than broad or, more frequently, a little broader than long (about 1·1–1·2 times).

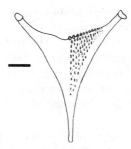

Fig. 24. *Haplochromis dolichorhynchus*. Lower pharyngeal bone. Scale equals 2 mm.

The lower pharyngeal teeth are fine and weakly cuspidate (except posteriorly), and arranged in 28–30 rows.

Coloration in life. Males are blueish-purple dorsally, becoming silvery on the flanks; the ventral surfaces are dusky. The lips and branchiostegal membrane are greenish-blue. Dorsal, caudal and anal fins are dark basally, hyaline distally, the dorsal sometimes with red maculae, and the caudal usually maculate; anal ocelli are orange-red. The pelvic fins vary from black to dusky.

Females are blue dorsally, becoming silver on the flanks and belly; the blue coloration is of variable intensity. A dark longitudinal stripe (of variable intensity) runs midlaterally along the entire body length behind the head. All fins are hyaline.

Preserved material. Males (*sexually active*) are brownish above, with the entire ventral half of the body and the lower four-fifths of the peduncle blue-black except for irregular silvery blotches on the flanks above the belly. There are traces of 5, ill-defined, narrow vertical bars stemming from the black ventral body coloration and extending to the dark streak lying along the dorsal fin base; a sixth faint bar is visible on the caudal peduncle. The cheek, branchiostegal membrane and operculum dark, the latter with a silvery sheen; the upper lip and lower jaw are greyish, the lower lip whiteish. A faint but broad band crosses the anterior part of the snout.

The dorsal fin is greyish, with sooty lappets and dark spots at the base of all the rays; the soft part of the fin is darkly maculate. The caudal is dark on its basal third, greyish-sooty distally and darkly maculate on its upper half. The anal dark grey with a deep, sooty basal band which expands in the region of the spines. Pelvic fins black on the proximal three-quarters medially, entirely black laterally.

Males (*sexually quiescent*) are dark yellow-brown dorsally, becoming darker on the ventral half of the body and caudal peduncle but with a silvery sheen on the chest and, less markedly, the belly. The snout is dark, but the rest of the head is yellowish-brown; the branchiostegal membrane is sooty. The fins are as in active males, or are much paler.

In some quiescent males faint traces of the vertical black bars are visible on the body, as are the various cephalic markings, including a postocular blotch, and a transverse bar across the snout.

Immature males have the same coloration as females (see below) except that the pelvic fins are dusky.

Females have a silvery-grey ground coloration which shades to yellowish-brown ventrally. A dark, usually interrupted midlateral band runs from behind the head to the proximal quarter of the caudal fin. The margins of this band are irregular, and the band is of variable width; the band is usually interrupted above the anterior part of the anal fin, but if it is complete, it is constricted in this region. In a few specimens the band does not appear on the anterior half of the body. A few small, irregularly placed dark blotches are often present between the upper lateral line and the dorsal fin base.

The dorsal fin is light grey, with faintly dusky lappets, and often has small dark areas at the bases of the spines. Caudal fin greyish, anal and pelvic fins hyaline.

Ecology. Habitat. Haplochromis dolichorhynchus has a wide depth range from 30 to more than 100 feet. It has been caught in open off-shore waters and in a relatively sheltered bay. In all localities, the substrate is mud.

Food. The guts of 25 specimens from three different localities were examined. Fishes from greater depths (70–108 feet) had fed on small Crustacea (especially *Caridina* sp.) and, to a lesser degree, on pupal Diptera; small quantities of bottom mud were found in these guts. One exceptional fish had at least 5 larval cichlid fishes in its stomach, and further remains of small fishes in the intestine. Since this specimen is a female with spent ovaries, the larvae could be from its own brood (assuming, of course, that the species is a mouth-brooder).

The 8 specimens from a shallower locality (Murchison Bay, at a depth of 50 feet) had ingested considerable quantities of mud; in most cases this is the only material present, but in 2 fishes fragments of pupal Diptera were identified. Possibly the high proportion of mud in the guts is unnatural and due to the fishes being, as it were, force-fed whilst the trawl was dragged along and through the near-liquid mud.

Breeding. In the sample available, all fishes less than 80 mm. S.L. are immature (all these are females); 2 larger fishes (females 101 and 96 mm. S.L.) are also immature although other fishes of the same length are adult. The largest fishes are females.

There is a tendency for the right ovary in ripe females to be larger than the left one, but a definite asymmetry in ovarian development could not be detected.

Diagnosis and affinities. Haplochromis dolichorhynchus closely resembles *H. tridens* (see note on synonymy). When *H. tridens* was redescribed (Greenwood, 1967) the existence of *H. dolichorhynchus* was unknown, and the species appeared to be very distinctive. But, since that time trawling surveys in the deeper waters of Lake Victoria have produced a number of species which, together with *H. tridens* form a "*tridens*" complex of at least 5 species, and probably a sixth. Thus some of the remarks made by Greenwood (*op. cit.*, pp. 97 and 99) about *H. tridens* (particularly with reference to the enlarged cephalic lateral line pores, the dentition, the general facies and the affinities of the species) are no longer valid.

Two species of the "*tridens*" complex are anatomically and superficially more distinct than the others, and need not be considered at this point (but see pp. 48 and 52). The more obviously similar species are *H. dolichorhynchus*, *H. tyrianthinus* (p. 40), *H. chlorochrous* (p. 44) and *H. tridens*.

As mentioned above, superficially *H. tridens* and *H. dolichorhynchus* resemble one another fairly closely. There is complete overlap in many morphometric characters, and the dentition is similar except that in *H. dolichorhynchus* the teeth are more slender and there are fewer tricuspids in the outer row of either jaw.

The species also differ in the following characters: the body is slightly deeper in *H. tridens* ($30 \cdot 1$–$36 \cdot 2$, M = $33 \cdot 5\%$ of standard length, *cf.* $28 \cdot 6$–$33 \cdot 6$, $31 \cdot 9\%$); this is probably correlated with the more steeply sloping dorsal head profile in *H. tridens* ($40°$–$45°$ *cf.* $20°$–$30°$). The orbit in *H. tridens* is noticeably elliptical (longer than deep) but is virtually circular in *H. dolichorhynchus*. The snout in *H. dolichorhynchus* is slightly longer ($30 \cdot 3$–$37 \cdot 7$, M = $34 \cdot 2\%$ of head, *cf.* $28 \cdot 0$–$34 \cdot 8$, M = $30 \cdot 4\%$ in *H. tridens*), a difference emphasised visibly by the beaked premaxilla of *H. dolichorhynchus* compared with the narrow medial part of that bone in *H. tridens*. Although the range of length/breadth ratios for the lower jaw shows complete interspecific overlap, there is a marked difference in the specific modes, with the jaw consistently narrower in *H. tridens* (mode $2 \cdot 8$ times as long as broad, *cf.* $2 \cdot 1$ times in *H. dolichorhynchus*). In *H. tridens* the posterior tip of the maxilla generally extends to a point below the eye (often to one below the pupil), but in *H. dolichorhynchus* it does not even extend to below the anterior margin of the orbit; the mouth in *H. tridens* is usually less oblique than in *H. dolichorhynchus* (horizontal to *ca.* $10°$, *cf.* $10°$–$20°$, mode *ca.* $20°$).

Unfortunately the live coloration of *H. tridens* is unknown, but the two species clearly differ in preserved coloration (*cf.* p. 37 above with p. 99 in Greenwood, 1967). *Haplochromis tridens* has no bars or longitudinal bands in either sex but all female *H. dolichorhynchus* show a longitudinal band (sometimes interrupted, sometimes faint) and vertical bars are present in males. Another difference is the very dark ventral pigmentation of adult male *H. dolichorhynchus* compared with the light grey-silver of *H. tridens*.

Haplochromis dolichorhynchus closely resembles *H. tyrianthinus*, a resemblance which extends to similarities in the live coloration of males. The principal diagnostic characters lie in the dentition and the slope of the head, but there are other, although less clear-cut, differences.

The outer teeth in *H. tyrianthinus* are slender and fine, with unicuspids predominating. The cusp in these teeth is a little compressed, but the neck and body of the tooth are distinctly cylindrical in cross-section. In contrast the commonest tooth form in *H. dolichorhynchus* is the bicuspid; these teeth are slender but compressed both at the crown and in the body. Unicuspid teeth are rare in the outer tooth row of this species, and when present are more compressed than are the unicuspids in *H. tyrianthinus*.

The dorsal head profile of *H. tyrianthinus* is more curved than in *H. dolichorhynchus*, and this feature, combined with a broader snout, gives *H. tyrianthinus* a more "heavy-headed" appearance than *H. dolichorhynchus*. Again, the distinctly beaked premaxilla of *H. dolichorhynchus* accentuates the impression of an elongate, sharp-pointed profile.

The orbit of *H. dolichorhynchus* is almost circular, that of *H. tyrianthinus* clearly longer than deep. The maxilla of *H. dolichorhynchus* does not reach the level of the anterior orbital margin, but in *H. tyrianthinus* it reaches that level or somewhat further posteriorly.

Although *H. dolichorhynchus* seems closely related to *H. chlorochrous*, superficially the two species are rather distinct. When alive they are immediately distinguishable: males of *H. dilochorhynchus* have a purple ground colour, those of *H. chlorochrous* a green one; females too are distinguishable on their coloration, being blueish-purple and lime-green in the species respectively.

Anatomically, the species may be separated by the broader snout of *H. chlorochrous* (0·9–1·0, mode 0·9 times as long as broad, *cf.* 1·0–1·4, mode 1·1 times), the more oblique lower jaw of *H. chlorochrous* and the less steeply inclined dorsal head profile of *H. dolichorhynchus*. The dentary of *H. chlorochrous* differs from that of *H. dolichorhynchus* in having the flare less pronounced and confined to the posterior and posterolateral parts of the bone (see p. 37 above). Finally, it may be noted that the neurocranium of *H. dolichorhynchus*, although essentially of the same type as that found in other species of the "*tridens*" group, differs in having the preorbital region somewhat more protracted.

The possible phyletic position of *H. dolichorhynchus* and other members of the "*tridens*" species complex will be discussed later (see p. 51).

Study material and distribution records

Museum and Reg. No.	Locality	Collector
	UGANDA	
B.M. (N.H.) 1968.8.30.168 (Holotype)	Murchison Bay (30 feet)	E.A.F.F.R.O.
B.M. (N.H.) 1968.8.30.191–196	Buvuma Channel, W. of Nienda and Vuga isls. (108 feet)	E.A.F.F.R.O.
B.M. (N.H.) 1968.8.30.177–190	North of Nsadzi isl. (70–100 feet)	E.A.F.F.R.O.
B.M. (N.H.) 1968.8.30.169–176	Murchison Bay (30 feet)	E.A.F.F.R.O.

Haplochromis tyrianthinus sp. nov.

(Text-figs. 25 and 26)

HOLOTYPE: an adult female 100·0 mm. S.L. (B.M. [N.H.] reg. no. 1968.8.30.135) from north of Nsadzi island.

The trivial name (from the Greek) refers to the predominantly purple colours of male fishes.

DESCRIPTION: based on 15 fishes (including the holotype), 85·0–105·0 mm. S.L. Depth of body 27·3–32·6 (M = 30·6) % of standard length, length of head 32·2–35·0 (M = 33·7) %.

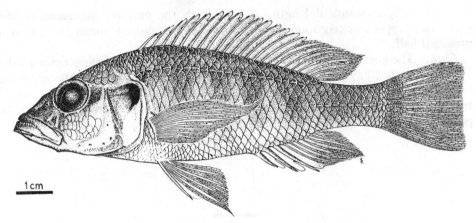

Fig. 25. *Haplochromis tyrianthinus*. Holotype. Drawn by Sharon Lesure.

Dorsal head profile slightly curved, sloping at an angle of 30°–40°; premaxillary pedicels moderately prominent. Lateral line tubes and pores of the preorbital bone prominent.

Preorbital depth 15·2–18·3 (M = 17·1) % of head, least interorbital width 15·2–18·3 (M = 17·4) %. Snout 0·9–1·0 (mode) times as long as broad, its length 29·3–35·2 (M = 32·4) % of head; snout length in this sample appears to show negative allometry with standard length. Orbit slightly eliptical (i.e. longer than deep), diameter of eye 26·1–29·3 (M = 27·7) % of head, depth of cheek 20·0–25·3 (M = 22·9) %.

Caudal peduncle 16·2–19·2 (M = 18·1) % of standard length, 1·4–1·9 (modal range 1·6–1·8) times as long as deep.

Mouth somewhat oblique, sloping at an angle of 15°–20° (mode); a horizontal line drawn from the upper anterior tip of the lower jaw passes below the orbit or less commonly through the lowermost part of the eye. Jaws equal anteriorly (rare) or the lower projecting slightly; lower jaw with a distinct but not prominent mental protuberance, its length 44·3–52·8 (M = 47·5) % of head, and 1·7–2·2 (mode 2·0) times its breadth. Premaxilla somewhat expanded medially; posterior tip of maxilla reaching a vertical through the anterior orbital margin or to slightly beyond this point.

Gill rakers: 9 (mode) or 10 on the lower part of the first arch, the lower 1–3 rakers reduced, the upper 4 or 5 flattened with at least some anvil-shaped; the intervening rakers are of varied form (from short and stout to relatively long and slender) but none is flattened.

Scales: ctenoid; lateral line with 31 (f.2), 32 (f.2), 33 (f.8) or 34 (f.3) scales, cheek with 3 or 4 rows. Five to 6 (mode 5½) scales between the dorsal fin origin and the upper lateral line, 6–7 (mode 6) between the pectoral and pelvic fin bases.

Fins. Dorsal with 24 (f.11), 25 (f.3) or 26 (f.1) rays, comprising 14 (f.1), 15 (f.11) or 16 (f.3) spines and 8 (f.1), 9 (f.10) or 10 (f.4) branched rays. Anal with 11 (f.6) or 12 (f.9) rays comprising 3 spines and 8 or 9 branched rays. Pectorals 25·7–30·0

(M = 29·3) % of standard length. Pelvics with the first ray produced in both sexes, but proportionately more so in adult males. Caudal truncate, scaled on its proximal half.

Teeth. The gently recurved *outer teeth* are slender, fine and nearly cylindrical in cross-section, but have slightly compressed crowns (Text-fig. 26). In the *upper jaw*, unicuspid, slenderly caniniform teeth predominate although slender bi-or weakly bicuspid teeth occasionally occur among the unicuspids in all parts of the row. The posterior outer teeth are sometimes tricuspid.

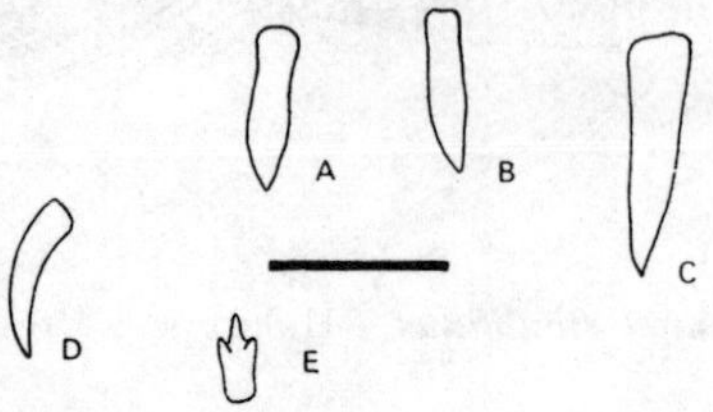

Fig. 26. *Haplochromis tyrianthinus.* A–D upper jaw teeth. A–C: in anterior view, D: in lateral view. Tooth C is from a larger individual than the other teeth. E: posterolateral tooth from the dentary. Scale equals 1 mm.

In the lower jaw, the posterior teeth are generally tricuspid and slightly larger than their counterparts in the upper jaw. The posterolateral teeth can be an admixture of tri- and bicuspids, or bicuspids alone; a few fishes have unicuspids in this position. The other lower teeth in most individuals are slender unicuspids, but some bi- and weakly bicuspid teeth are sometimes found among the unicuspids.

There are 70–86 (M = 74) teeth in the outer row of the upper jaw.

Most of the *inner teeth* are tricuspid, but in some fishes the outermost row is composed of slender unicuspids; very occasionally all the inner teeth are unicuspid. The median cusp of inner tricuspids is noticeably elongate.

All inner teeth are obliquely implanted, some lying almost horizontally; there are 2 or 3 (mode) rows in the upper jaw, and 2 (rarely 1 or 3) rows in the lower jaw.

Osteology. The *neurocranium* of *H. tyrianthinus* closely resembles that of *H. dolichorhynchus* (see p. 36) but is slightly less protracted in the preorbital region.

The dentigerous surface of the *premaxilla* is moderately expanded anteromedially but it is not so distinctly beaked as the premaxilla of *H. dolichorhynchus*; the pedicels are shorter than the dentigerous arms.

The *dentary* is flared like that of *H. dolichorhynchus* (see p. 37), and the short mental protuberance is moderately pronounced.

The toothed surface of the *lower pharyngeal bone* is slightly broader than long or, occasionally, equilateral. The teeth are cuspidate, relatively fine and usually rather widely spaced in from 24–36 rows; in some fishes, however, the teeth are finer, and more regularly and closely arranged in 30–36 rows.

Coloration in life. Detailed live colour notes are not available for this species but it was noted (J.M.G.) that adult males have a bright and intense purple coloration on the dorsal and lateral aspects of the body.

Coloration of preserved material. Adult males are brownish, lightest on the chest where a faint silver overlay is visible; the branchiostegal membrane is faintly grey. Very faint traces of 5 or 6 vertical bars are visible on the flanks and caudal peduncle, those over the body intergrade with the generally darker brown colour of the belly and ventral body region above and posterior to the anal fin. A faint lachrymal stripe is visible, being most intense immediately below the eye.

Dorsal, caudal and anal fins are grey, the lappets of the dorsal black, as is the membrane between the anal spines; the central region of the caudal fin is darker than the upper and lower parts. The pelvic fins are black.

Females are light brown above, shading to beige ventrally, and with faint traces of a silvery overlay on the chest and belly in some specimens. Very faint traces of about 5 broad vertical bars are visible on the flanks and caudal peduncle; the bars do not reach the ventral profile. In some specimens there is a faint midlateral dark stripe, most intense over the posterior half of the body. Dorsal, caudal and anal fins are greyish, the pelvics hyaline.

Ecology. Habitat. The species is known from only one locality, north of Nsadzi island. Specimens were obtained from a trawl fished over a mud bottom at a depth of 70–100 feet.

Food. One of the 10 guts examined was empty, one contained fragments of a small cichlid fish in the stomach and further fish remains in the intestine; the other 8 all yielded fragmentary remains of Crustacea (probably *Caridina* sp.). In only 2 guts was there any bottom detritus.

Breeding. Only one specimen (a female 85 mm. S.L.) is a juvenile; the smallest male (99 mm. S.L.) shows an advanced state of testicular development. There is no clear-cut indication of asymmetrical ovarian development.

Diagnosis and affinities. Within the "*tridens*" species complex, *H. tyrianthinus* most closely resembles *H. chlorochrous* and *H. dolichorhynchus*. Characters distinguishing *H. tyrianthinus* from the latter species (which it also resembles in live coloration) are discussed on p. 39. Live specimens (of either sex) are immediately distinguished from *H. chlorochrous* by their coloration which is basically purple to blue in *H. tyrianthinus* and green in *H. chlorochrous*.

Anatomically, *H. tyrianthinus* is distinguished from *H. chlorochrous* principally by dental characters and differences in head shape. The outer row of teeth in *H. tyrianthinus* is composed, mainly, of slender unicuspids, circular in cross-section over most of their length except for a slight compression of the crown. In *H. chlorochrous*, on the other hand, there is a mixture of bi- and unicuspids in this row, and the teeth although slender are relatively compressed, especially at the crown. There is also a difference in the pharyngeal teeth of the two species, but this character shows a greater degree of overlap than does the difference in jaw teeth. In most specimens of *H. tyrianthinus* the lower pharyngeal teeth are coarser, fewer and more widely spaced (particularly in the posterolateral corners of the dentigerous area) than are the teeth in *H. chlorochrous*. Small individuals of *H. tyrianthinus*, however, have a lower pharyngeal dentition like that of adult *H. chlorochrous*.

The lower jaw of *H. tyrianthinus* slopes less steeply than in *H. chlorochrous* ($15°$–$25°$, mode $20°$, *cf.* $20°$–$35°$, mode $30°$); as a result, a horizontal drawn from the upper tip of the lower jaw passes below the orbit (or through the lower orbital margin) in *H. tyrianthinus* but through the lower or middle part of the eye in *H. chlorochrous*.

The snout in *H. tyrianthinus* is somewhat narrower than in *H. chlorochrous* (0·9–1·0, mode 1·0, times as long as broad, *cf.* 0·9–1·0, mode 0·9, times in *H. chlorochrous*) and the dorsal head profile is gently curved, both factors contributing to the distinctly different physiognomy of the two species.

Osteologically, the principal interspecific differences lie in the more anteroposteriorly extensive and deeper flare of the outer face of the dentary in *H. tyrianthinus*.

Haplochromis tyrianthinus also resembles *H. tridens*, but is distinguished by its preserved coloration (especially the presence of vertical bars on the flanks) more oblique mouth, broader lower jaw (model length/breadth ratio 2·0 *cf.* 2·8 in *H. tridens*), less steeply declined and more curved upper head profile (30°–40°, *cf.* 40°–45° in *H. tridens*), and by its slender, predominantly unicuspid outer teeth.

The possible phyletic relationships of *H. tyrianthinus* are discussed on p. 51.

STUDY MATERIAL AND DISTRIBUTION RECORDS

Museum and Reg. No.	Locality	Collector
	UGANDA	
B.M. (N.H.) 1968.8.30.135 (Holotype)	North of Nsadzi isl. (70–100 feet)	E.A.F.F.R.O.
B.M. (N.H.) 1968.8.30.136–149	North of Nsadzi isl. (70–100 feet)	E.A.F.F.R.O.

Haplochromis chlorochrous sp. nov.

(Text-figs. 27–29)

HOLOTYPE: an adult male 102·5 mm. S.L. (B.M. [N.H.] reg. no. 1968.8.30.310) from water 70–100 feet deep between Nsadzi island and the mainland.

The trivial name (from the Greek) refers to the distinctive green colour of adult fishes.

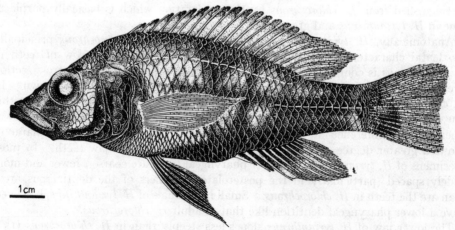

FIG. 27. *Haplochromis chlorochrous*. Holotype. Drawn by Sharon Lesure.

DESCRIPTION: based on 19 specimens (including the holotype), 70·0–120·0 mm. S.L.

Depth of body 29·5–34·5 (M = 32·0) % of standard length, length of head 32·0–35·4 (M = 34·0) %. Dorsal head profile straight or very slightly concave, sloping at an angle of 30°–40°; premaxillary pedicels only just apparent beneath the skin of the snout. Cephalic lateral line pores are prominent.

Preorbital depth 15·6–19·4 (M = 17·3) % of head, least interorbital width 15·6–20·2 (M = 17·7) %. Snout 0·9–1·0, mode 0·9 times as long as broad, its length 29·5–35·0 (M = 32·3) % of head. Orbit slightly eliptical (i.e. longer than deep), diameter of eye 25·4–29·0 (M = 18·2) % of head, depth of cheek 20·0–25·0 (M = 22·6) %.

Caudal peduncle 15·4–20·3 (M = 18·2) % of standard length, 1·3–1·9 (modal range 1·5–1·6) times as long as deep.

Mouth somewhat oblique (*ca.* 20°–35°, mode 30°); a horizontal line drawn from the upper tip of the lower jaw passes through the lower quarter of the eye or even through the centre of the eye. Lower jaw projecting, its length 44·3–51·8 (M = 47·0) % of head, 1·6–2·1 (mode 1·9) times as long as broad; mental protuberance moderate. Premaxilla moderately expanded anteromedially. Posterior tip of the maxilla reaching a vertical through the anterior orbital margin or a little more posteriorly to below the eye.

Gill rakers: Nine or 10 (rarely 8), on the lower part of the first arch. The lower 1–4 rakers are reduced, the upper 3 or 4 flattened, sometimes lobed; intervening rakers are relatively short and stout.

Scales: ctenoid; lateral line with 31 (f.1), 32 (f.7), 33 (f.9) or 34 (f.1) scales, cheek with 3 or 4, rarely 5, rows. Five and a half to 6½ (mode 6) scales between the dorsal fin origin and the upper lateral line, 6–7 (mode), rarely 5, between the pectoral and pelvic fin bases.

Fins. Dorsal with 24 (f.17) or 25 (f.2) rays, comprising 14 (f.1), 15 (f.17), or 16 (f.1) spines and 9 (f.17) or 10 (f.2) branched rays. Anal with 11 (f.6) or 12 (f.13) rays, comprising 3 spines and 8 or 9 rays. Pectoral 26·8–30·4 (M = 28·7) % of standard length. Pelvic fins with the first ray produced, not apparently showing any length correlation with sex, except that it is relatively longer in adults than in juveniles. Caudal truncate, scaled on its proximal half.

Teeth. In the *outer* tooth row of the *upper jaw* there is an admixture of slender unicuspids and slender, unequally bicuspid teeth, both types having relatively compressed crowns (Text-fig. 28); the bicuspids may be so weakly cuspidate as to appear unicuspid. In the smallest fish (70 mm. S.L.) some tricuspid teeth are

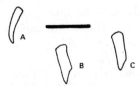

FIG. 28. *Haplochromis chlorochrous.* A–C upper jaw teeth. A: in lateral view. B and C: in anterior view. Scale equals 2 mm.

interspersed among the bicuspids. Posteriorly and posterolaterally the teeth are usually small and tricuspid, but in some specimens unicuspids or bicuspids are present in this position.

There are 66–86 (M = 74) teeth in the outer row of the upper jaw.

Generally, the *outer teeth* in the *lower jaw* are similar to those in the upper jaw, but in a few fishes all except the posterior teeth are slender bicuspids. The posterior teeth are usually tricuspid and slightly larger than their counterparts in the upper jaw.

All teeth in the *inner rows* are tricuspid, and implanted somewhat obliquely. There are 2 or 3 (mode), rarely 4, rows in the upper jaw and 2 (mode) or 3, rarely a single row, in the lower jaw.

Osteology. The neurocranium of *H. chlorochrous* is indistinguishable from that of *H. tyrianthinus*, and premaxillary shape in the two species is also identical. The *dentary*, however, differs in that the flared region is restricted to a more posterior position (see p. 42). Consequently the anterior and anterolateral face of each ramus is almost flat, only the posterior and posterolateral faces showing any concavity. The mental protuberance is stout but not especially prominent.

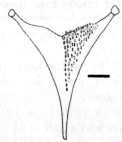

FIG. 29. *Haplochromis chlorochrous.* Lower pharyngeal bone. Scale equals 2 mm.

The dentigerous surface of the *lower pharyngeal bone* (Text-fig. 29) is as long as broad, or very slightly broader. The teeth are fine, cuspidate, and closely arranged in 30–36 regular rows.

Coloration in life. Adult males are dark green on the dorsal surfaces of the body and head, lighter green on the flanks which are crossed by 5 faint, dark, vertical bars; the ventral body surfaces are black.

Dorsal fin is greenish-yellow basally, hyaline distally, with faint traces of orange-red maculae. The caudal and pectoral fins greenish basally, hyaline distally. The anal is black basally, almost hyaline distally, the 2 or 3 ocelli are orange. Pelvic fins are black.

Females have the dorsal body surfaces lime-green, the flanks bright yellow, and the ventral body surfaces silver. There is a faint midlateral longitudinal stripe, and the flanks are crossed by 4 or 5 vertical bars. The head and upper lip are green, the lower lip yellow. All fins are yellowish-green.

Preserved coloration. Adult *males* have a yellow-brown ground coloration, the snout greyish, the lower jaw, ventral part of the preopercular region and the lower half of the operculum black. A broad and diffuse lachrymal stripe is usually visible.

The lower half of the body posterior to the vent is black or very dark brown; a narrow tongue of this dark area extends forward onto the ventral aspect of the belly as far as the base of the pelvic fins. Three to 5 dark vertical bars cross the posterior flanks; ventrally the bars merge with the black lower aspects of the body. A very faint, often incomplete, dark midlateral stripe is generally visible, extending forward from the dark posterior region almost to the opercular margin.

Dorsal fin dark grey, lappets and margin of the soft part black. The caudal fin is dark grey, indistinctly blotched with black near the base in some specimens, weakly maculate in others. Anal fin is black basally and between the spines, dark grey distally. The pelvic fins are black.

The single immature male is uniformly yellow-brown except for a darker snout region, faint lachrymal stripe, dusky lower jaw and preopercular region. The dorsal fin is grey with black lappets and margin to the soft part, the caudal and anal fins greyish, the anal dusky between the spines; the pelvics are dusky, becoming black over the anterior third.

Females have a light brown ground colour, darker dorsally and on the snout which is greyish in some individuals. A faint, narrow and interrupted dark midlateral stripe runs from behind the opercular margin to the base of the caudal fin where it is slightly expanded. The dorsal fin is greyish, the lappets black and the soft part entirely or partly maculate. Caudal fin grey, and indistinctly maculate. Anal hyaline, greyish along its margin in adults. Pelvic fins hyaline, but often greyish over the anterior third to half in adults.

Ecology. Habitat. All the specimens examined came from trawl hauls over a mud substrate, and at a depth of 70 to 100 feet, in the area between Nsadzi island and the mainland. One of us (J.M.G.) has recorded the species in several other localities, viz: south of Nsadzi island (130–160 feet, over mud), near Bulago island (55–75 feet, over mud), between the south side of Buvuma island and the northern shore of Bugaia island (105 feet, over mud) and between Mwama and Bugaia islands (200 feet, over mud).

Food. Two of the 15 guts examined were empty. Of the remainder, 2 contained only mud, and 11 mud with fragments of Crustacea (probably *Caridina* sp.).

Breeding. Three males (77, 78 and 89 mm. S.L.) are immature, as are 3 females (70, 72 and 74 mm. S.L.); all the larger fishes are adults. In the 7 adult females examined there is a distinct tendency for the right ovary to be larger than the left, irrespective of the individual's sexual state.

Diagnosis and affinities. *Haplochromis chlorochrous* is compared with *H. dolichorhynchus* on p. 40, and with *H. tyrianthinus* on p. 43; in life its green coloration serves as an immediate diagnostic character.

From *H. tridens*, the fourth member of this species complex, *H. chlorochrous* is distinguished principally by its more oblique and broader lower jaw (angle of mouth 20°–35°, mode 30°, cf. horizontal to 10° in *H. tridens*; lower jaw 1·6–2·1, mode 1·9 times as long as broad, cf. 2·0–2·8, mode 2·8 times in *H. tridens*), and by the presence of vertical bars on the flanks of preserved adult males (a longitudinal stripe in females) as well as by the very dark ventral coloration on the posterior part of the body in males.

The possible phyletic relationships of *H. chlorochrous* are discussed on p. 51.

Study material and distribution records

Museum and Reg. No.	Locality	Collector
	UGANDA	
B.M. (N.H.) 1968.8.30.310 (Holotype)	Between Nsadzi isl. and mainland (70–100 feet)	E.A.F.F.R.O.
B.M. (N.H.) 1968.8.30.311–328	Between Nsadzi isl. and mainland (70–100 feet)	E.A.F.F.R.O.

Haplochromis cryptogramma sp. nov.

(Text-figs. 30 and 31)

HOLOTYPE: an adult male 86·5 mm. S.L. (B.M. [N.H.] reg. no. 1968.8.30.197) from the Bulago–Tavu bank at a depth of 50–75 feet.

The trivial name refers to the fancied resemblance of the midlateral longitudinal stripes to symbols in the Morse code.

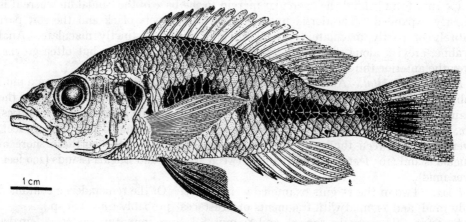

FIG. 30. *Haplochromis cryptogramma*. Holotype. Drawn by Sharon Lesure.

DESCRIPTION: based on 23 specimens (including the holotype) 55·0–94·0 mm. S.L. Depth of body 30·2–35·8 (M = 33·6) % of standard length, length of head 34·5–38·8 (M = 35·7) %.

Dorsal head profile sloping at an angle of 35°–40°, noticeably concave, the prominent premaxillary pedicels tending to exaggerate the concavity and giving the profile a characteristic outline (Text-fig. 30).

Preorbital depth 14·3–17·7 (M = 15·4) % of head, least interorbital width 20·3–23·9 (M = 22·0) %. Snout prominent in lateral view, 1·0–1·5 (mode 1·2) times as long as broad, its length 28·6–35·5 (M = 32·7) % of head; eye diameter 25·8–31·6 (M = 28·7) %, the orbit almost circular; depth of cheek 15·8–21·5 (M = 19·0) %.

A REVISION OF THE LAKE VICTORIA *HAPLOCHROMIS* SPECIES

Caudal peduncle 17·5–22·7 (M = 19·2) % of standard length, 1·5–2·0 (modal range 1·8–1·9) times as long as deep.

Mouth slightly oblique, sloping at an angle of 10°–15°. Jaws equal anteriorly or, less commonly, the lower projecting; premaxilla slightly expanded anteromedially. Lower jaw 40·0–48·5 (M = 44·8) % of head, 2·0–2·7 (modal range 2·3–2·5) times as long as broad; mental protuberance slight. Posterior tip of the maxilla generally reaching a vertical through the anterior part of the orbit, but only to the anterior orbital margin in a few specimens.

Gill rakers: Eight or 9 (rarely 7 or 10) on the lower part of the first arch, the lower 2 or 3 rakers reduced, the remainder of varied form, from short and stout through slender to flattened with some of the upper rakers anvil-shaped.

Scales: ctenoid; lateral line with 31 (f.2), 32 (f.7) or 33 (f.8) scales (several specimens have lost their scales, hence these figures are derived from only 17 fishes); cheek with 3 rows (rarely 2 or 4). Five to 6½ (mode 5½) scales between the upper lateral line and the dorsal fin origin, 6–7 (mode 7) between the pectoral and pelvic fin bases.

Fins. Dorsal with 23 (f.5), 24 (f.17) or 25 (f.1) rays, comprising 15 (f.16) or 16 (f.7) spines and 8 (f.11) or 9 (f.12) branched rays. Anal with 10 (f.1), 11 (f.21) or 12 (f.1) rays, comprising 3 spines and 7–9 branched rays. Pectoral 25·0–32·0 (M = 28·3) % of standard length. Pelvic fin with the first ray somewhat produced in both sexes, probable a little more so in adult males. Caudal truncate or slightly emarginate, scaled on its proximal half.

Teeth. The *outer row* of teeth in the *upper jaw* (except posteriorly) is composed mainly of unequally bicuspid, relatively stout teeth; however, most fishes have some tricuspid teeth interspersed amongst the bicuspids both anteriorly and laterally (Text-fig. 31). A few exceptional individuals have a predominance of tricuspid teeth in this row. Posteriorly and posterolaterally in this jaw there are either unicuspid and bicuspid teeth, or tricuspids alone; rarely, all three types of teeth occur in this position.

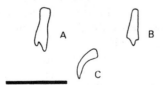

FIG. 31. *Haplochromis cryptogramma*. A–C upper jaw teeth. A and B: teeth from anterior part of outer row, in anterior view. C: tooth A in lateral view. Scale equals 2 mm.

There are 50–78 (M = 68) teeth in the outer row of the upper jaw.

In the *lower jaw*, the teeth are similar to those in the upper jaw except that there are fewer tricuspids anteriorly and anterolaterally. Posteriorly and posterolaterally most fishes have tricuspid teeth, but in a few individuals the teeth in this position are bicuspid, or there is mixture of bi- and tricuspid teeth in this position.

The *inner* series is composed of obliquely implanted tricuspid teeth arranged in 2 or 3 rows in the upper jaw, and 1 or 2 rows in the lower jaw.

Osteology. The *neurocranium* of *H. cryptogramma* is essentially of the type found in the " *tridens* " species complex (*H. tridens, H. dolichorhyncrus, H. tyrianthinus* and *H. chlorochrous*). It differs, however, in having a somewhat broader interorbital region and a lower supraoccipital crest.

The *premaxilla* is noticeably expanded medially and anteromedially, to give it a beaked appearance. The pedicels are longer than the dentigerous arms.

The *dentary* resembles that of *H. dolichorhynchus* because the lateral face is strongly flared and the resulting concavity extends forward almost to the symphyseal region. The lateral line system in the dentary shows no signs of hypertrophy.

The dentigerous surface of the *lower pharyngeal bone* is as broad as long or slightly broader than long. The teeth are relatively fine with, in some fishes, those of the median rows slightly coarser. A striking feature is the regularity with which the teeth are arranged in 26–30 rows.

Coloration in life. Adult males have the dorsal aspect of the head and body blueish-grey, shading to yellowish on the flanks, and silver ventrally. A reddish flush is usually present on the opercular and pectoral regions, and is most intense in ripe males. Two black bars cross the snout, and another, fainter, bar crosses the nuchal region behind the eyes. On the body there are 2 (but sometimes fused) black blotches at the dorsal fin base, and one on the dorsal part of the caudal peduncle. Midlaterally there is a prominent dark band interrupted in at least 2, often 3, places to give a series of short and long lines. The posterior line extends well onto the caudal fin. The dorsal fin is hyaline, with black lappets. The caudal fin is either colourless or, in ripe individuals, with a red flush. The anal is colourless or whiteish (especially in sexually active fishes) and has 2 or 3 yellow ocelli. The pelvic fins are uniformly black.

Females have a body coloration like that of males but are without the red flush; the cephalic and flank markings are present as in males, but the nuchal bar is usually less distinct. The caudal and anal fins are yellow, the other fins hyaline.

Coloration in preserved material. In *both sexes* the ground coloration is silvery white. A series of broad but elongate black blotches runs midlaterally along the flanks and caudal peduncle. The anterior (and shortest) blotch lies on the operculum, the posterior streak extends onto the caudal fin and may even reach to near the posterior border. A series of elongate blotches (sometimes confluent into 2 or even a single smudge) lies below the dorsal fin origin. In some fishes there are faint traces of 3 or 4 vertical bars on the flanks. The snout is crossed by 2 transverse bands and a faint nuchal bar is generally visible.

In *males* the dorsal fin is greyish, with black lappets, and a faintly sooty margin to the soft part; the anal is also greyish with black lappets, and faint off-white ocelli. The caudal is greyish to hyaline, somewhat darker basally and with an intense black blotch extending over at least the anterior half.

Females have all the fins hyaline but with a faint duskiness along the margin of the dorsal fin.

Ecology. Habitat. The species is known from 2 areas, namely, Namone point (at a depth of *ca.* 30 feet) and in the region of the Bulago–Tavu bank at a depth of 50–75 feet. In both areas the substrate is mud.

Food. The gut contents of 20 specimens (from one locality, near the Bulago-Tavu bank) were examined; of these, 4 were empty. The remainder all contained either pupal Diptera or adult Crustacea (especially *Caridina* sp.); less commonly, both types of food were present. In 8 guts, small quantities of bottom mud were also recorded.

Breeding. All the specimens examined (55–94 mm. S.L.) are adult. The right ovary is larger than the left in 6 of the 10 females available.

Diagnosis and affinities. The dentition and syncranial architecture of *H. cryptogramma* suggest close affinity with the " *tridens* " species complex (see p. 52). At the species level, however, *H. cryptogramma* is immediately distinguished from all other members of the group by the highly distinctive colour patterns in both live and preserved specimens.

Haplochromis cryptogramma is also distinguishable from other members of the " *tridens* " complex by its broader interorbital region.

Despite these differences we would include *H. cryptogramma* as a member of the " *tridens* " species group. Its phyletic position will, therefore, be discussed in relation to the rest of the group.

Study material and distribution records

Museum and Reg. No.	Locality	Collector
	UGANDA	
B.M. (N.H.) 1968.8.30.197 (Holotype)	Bulago–Tuva bank (50–75 feet)	E.A.F.F.R.O.
B.M. (N.H.) 1968.8.30.198–214	Bulago–Tavu bank (50–75 feet)	E.A.F.F.R.O.
B.M. (N.H.) 1968.8.30.215–219	Namone Point (Hannington Bay) (*ca.* 30 feet)	E.A.F.F.R.O.

The affinities of the *H. tridens* species group

The discovery of *H. dolichorhynchus*, *H. tyrianthinus*, *H. chlorochrous* and *H. cryptogramma* considerably modifies Greenwood's (1967) remarks about the affinities of *H. tridens*. That species was thought to occupy an isolated position among the Lake Victoria species flock, but the new species show that there is, in fact, a group of at least 5 " *tridens* "-like species.

All are characterized by their dentition, general facies and syncranial architecture (see above in the relevant sections of the species' descriptions). Although these various characters provide a means of delimiting the group as a whole, there still remains the problem of its relationships with other species complexes.

The unusual dental characteristic of the " *tridens* " group lies in the high proportion of tricuspid teeth in the outer tooth row of both jaws. The morphology of the non-tricuspid teeth is not outstanding; similar teeth occur in many other species-groups (except, of course, the most generalized) and are common types in the " *prognathus* " complex of piscivorous species (see Greenwood, 1967, p. 109).

The body-form in the " *tridens* " complex also occurs in the " *prognathus* " group, as does their type of syncranial morphology and organization. Indeed, the neurocranium of *H. dolichorhynchus* is very like that of *H. prognathus* itself, and the neurocranium in other group members does not depart greatly from this type.

These various similarities suggest to us that the "*tridens*" species complex is derived from the same stem as the "*prognathus*" group, and probably from an ancestor resembling the present-day *H. prognathus*.

The 4 species described in this paper, together with *H. tridens*, form a close-knit complex in which *H. cryptogramma* is the most deviant member, at least in its superficial appearance. But, if *H. arcanus* (see p. 54) is to be included in the "*tridens*" complex, it would qualify for that position. For the moment, however, we would prefer to leave open the question of the affinities of *H. arcanus*.

Haplochromis arcanus sp. nov.
(Text-fig. 32)

HOLOTYPE: an adult male 127·0 mm. S.L. (B.M. [N.H.] reg. no. 1968.8.30.19) from south of Nsadzi island, at a depth of between 130–160 feet.

The trivial name (from the Latin for secret) refers to our uncertainty about the relationships of this species within the Lake Victoria *Haplochromis* species-flock.

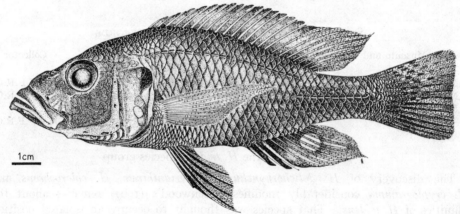

FIG. 32. *Haplochromis arcanus*. Holotype. Drawn by Sharon Lesure.

DESCRIPTION: based on 9 specimens (including the holotype) 104·0–142·0 mm. S.L. Depth of Body 30·8–33·3 (M = 31·9) % of standard length, length of head 37·3–39·4 (M = 38·1) %.

Dorsal head profile almost straight (but with a slight supraorbital depression which is intensified by the prominent premaxillary pedicels), sloping at an angle of 30°–35°.

Preorbital depth 17·3–19·8 (M = 18·8) % of head, least interorbital width 16·6–19·2 (M = 18·1) %. Snout 1·1–1·4 (mode 1·2) times as long as broad, its length 33·8–38·5 (M = 35·7) % of head, eye diameter 22·7–25·0 (M = 24·1), depth of cheek 22·2–26·0 (M = 23·8) %.

Caudal peduncle 13·7–18·3 (M = 16·0) % of standard length, 1·3–1·7 (mode 1·5) times as long as deep.

Mouth moderately oblique, sloping at an angle of 30°–35°; lower jaw with a distinct mental protuberance, projecting slightly, its length 47·0–51·8 (M = 49·3) %

of head, 1·9–2·2 (mode 2·1) times as long as broad. Premaxilla somewhat expanded anteromedially and thus slightly beaked. Posterior tip of the maxilla reaching a vertical slightly anterior to the orbital margin, or one reaching the orbit.

Gill rakers: 9–11 (mode 10) on the anterior part of the first gill arch, the lower 1–3 rakers reduced, the upper 3 or 4 flattened, expanded and often anvil-shaped; the intervening rakers are relatively stout.

Scales: ctenoid; lateral line with 31 (f.1), 32 (f.5), 33 (f.1) or 34 (f.2) scales, cheek with 4 (mode) or 5 rows. Six to 7 (mode 6) scales between the upper lateral line and the dorsal fin origin, 7–8 (mode), rarely 6, between the pectoral and pelvic fin bases.

Fins. Dorsal with 24 (f.5) or 25 (f.4) rays, comprising 14 (f.1), 15 (f.7) or 16 (f.1) spines and 9 (f.5) or 10 (f.4) branched rays. Anal with 11 (f.4), 12 (f.3) or 13 (f.2) rays, comprising 3 spines and 8–10 branched rays. Pectoral fin 26·2–31·7 ($M = 30·0$) % of standard length. Pelvics with the first ray produced. Caudal truncate, scaled on its basal half.

Teeth. Except in the smallest specimen (104 mm. S.L.), the *outer teeth* in *both jaws* are slender and unicuspid, those situated anteriorly and laterally in both jaws slightly recurved, but the posterolateral and posterior upper teeth strongly incurved.

The smallest fish has an admixture of weakly bicuspid and unicuspid teeth, and the posterolateral upper teeth are not strongly incurved.

There are 50–74 ($M = 64$) outer teeth in the upper jaw.

The *inner teeth* in most specimens are all tricuspids, but in a few individuals there is an admixture of uni-and bicuspids, with unicuspids predominating in the lower jaw. There are 2 or 3 rows of inner teeth in the upper jaw, and 2 rows in the lower jaw, all implanted at a slight angle.

Osteology. No complete skeleton of *H. arcanus* is available.

The *lower pharyngeal bone* has a triangular dentigerous surface which is as broad as, or slightly broader than, long. The teeth are fine, compressed and cuspidate, and are arranged in 26–30 rows. In some fishes the posterior teeth in the 2 median rows are noticeably coarser than their anterior congeners.

Coloration. The colours of live fishes are unknown.

Preserved material. Males (*adult*) have dark, yellowish-brown ground coloration with a faint duskiness on the chest and belly. The dorsal head surface and the snout are dusky, the branchiostegal membrane dusky in the opercular region but dark brown anteriorly. The dorsal fin is dusky, but with a darker basal region and black lappets. The caudal fin is very dark grey (almost black basally) with black maculae over most of its surface, the spots most concentrated on the upper half. Anal fin dark grey, dusky along its base and over the spinous part; ocelli large, lighter grey than the fin membrane but with a dark ring surrounding each ocellus. The pelvic fins are black.

Females (*adult*) have a greyish-yellow ground coloration dorsally, shading to light yellow-brown on the chest and belly; the snout and dorsal head surfaces dark grey. The dorsal fin is greyish with black lappets. Caudal fin greyish (darkest basally) with some dark maculae on the upper half. Anal hyaline or hyaline with a dark band basally and another slightly below the distal margin. Pelvic fins hyaline or with a faint dusky marbling.

Ecology. Habitat. The species is known from only 2 localities; one, south of Nsadzi island at a depth of 130–160 feet over a mud bottom, the other north of Nsadzi island, also over a mud bottom but at a depth of 70-100 feet.

Food. No information is available; the guts examined were empty, and several specimens had been eviscerated before preservation.

Breeding. All 9 specimens are adult.

Diagnosis and affinities. At present we can say little about the affinities of *H. arcanus*. Superficially, the species resembles certain members of the "*tridens*" complex, particularly *H. tyrianthinus* and *H. dolichorhynchus*. However, it differs from all members of the "*tridens*" group in being without any tricuspid teeth in the outer tooth row of either jaw, and in having the posterolateral upper teeth strongly incurved.

Haplochromis arcanus also differs from *H. tyrianthinus* in head shape (especially the prominent premaxillary pedicels), its longer head ($37 \cdot 3$–$39 \cdot 4$, M = $38 \cdot 1 \%$ S.L., *cf.* $32 \cdot 2$–$35 \cdot 0$, M = $33 \cdot 7 \%$), narrower snout ($1 \cdot 1$–$1 \cdot 4$, mode $1 \cdot 2$ times as long as broad, *cf.* $0 \cdot 9$–$1 \cdot 0$, mode $1 \cdot 0$ times) and in having fewer and stouter teeth (this in addition to the dental differences noted above).

From *H. dolichorhynchus*, *Haplochromis arcanus* differs in having a slightly longer head, a markedly different profile (including a more oblique lower jaw) and in various dental details (especially the unicuspid and recurved outer teeth, the relatively stouter form of all outer teeth and their greater spacing).

The dentition of *H. arcanus* resembles that of *H. argenteus*, and there is a fairly close similarity in general facies, especially when specimens of comparable size are examined (see Greenwood, 1967, p. 84). *Haplochromis arcanus* differs from *H. argenteus* in having a larger eye ($22 \cdot 7$–$25 \cdot 0$, M = $24 \cdot 1 \%$ of head, *cf.* $19 \cdot 4$–$23 \cdot 5$, M = $21 \cdot 5 \%$), broader lower jaw ($1 \cdot 9$–$2 \cdot 2$, mode $2 \cdot 1$ times as long as broad, *cf.* $2 \cdot 3$–$3 \cdot 1$, modal range $2 \cdot 8$–$3 \cdot 0$ times in *H. argenteus*), a longer pectoral fin ($26 \cdot 2$–$31 \cdot 7$, M = $30 \cdot 0 \%$ S.L., *cf.* $24 \cdot 1$–$29 \cdot 7$, M = $25 \cdot 0 \%$), and smaller chest and nape scales (see Greenwood, *op. cit.*).

On the characters we have been able to study (and these exclude osteological ones), *H. arcanus* could be related to either the *H. tridens* species complex, or to *H. argenteus*. Since both the "*tridens*" group and *H. argenteus* are probably derived from an *H. prognathus*—like ancestor (see above, p. 51, and Greenwood, *op. cit.*) these bilateral affinities of *H. arcanus* would not be unexpected if it too is descended from a similar stem.

Study material and distribution records

Museum and Reg. No.	Locality	Collector
	UGANDA	
B.M. (N.H.) 1968.8.30.19 (Holotype)	S. of Nsadzi isl. (130–160 feet)	E.A.F.F.R.O.
B.M. (N.H.) 1968.8.30.22–27	S. of Nsadzi isl. (130–160 feet)	E.A.F.F.R.O.
B.M. (N.H.) 1968.8.30.20–21	N. of Nsadzi isl. (70–100 feet)	E.A.F.F.R.O.

Haplochromis decticostoma sp. nov.

HOLOTYPE: an adult male 199·0 mm. S.L. (B.M. [N.H.] reg. no. 1968.8.30.292) from south of Nsadzi island, at a depth of 180 feet.

The trivial name (from the Greek *dektikos*: able to bite, and *stoma*: mouth) refers to the large mouth.

DESCRIPTION: based on 22 specimens (including the holotype), 129·0–229·0 mm. S.L.

Depth of body 32·9–40·1 (M = 36·6) % of standard length, length of head 36·1–42·0 (M = 38·9) %.

Dorsal head profile straight, sloping at an angle of 30°–35°, usually interrupted by the premaxillary pedicels. Pores of the cephalic lateral line system are fairly prominent, especially those on the preorbital which has a swollen appearance.

Preorbital depth 19·4–23·8 (M = 21·7) % of head length, least interorbital width 20·0–23·1 (M = 21·6) %. Snout 1·0–1·2 (mode 1·1) times as long as broad, its length 37·1–41·0 (M = 38·9) % of head. Eye diameter 17·9–22·6 (M = 20·1) % of head, eye/preorbital ratio 0·8–1·0 (mode 0·9); depth of cheek 26·4–31·3 (M = 29·1) %.

Caudal peduncle 14·4–18·7 (M = 16·4) % of standard length, 1·2–1·7 (mode 1·4) times as long as deep.

Mouth oblique, sloping at an angle of 30°–40°, lower jaw projecting slightly to strongly, its length 50·7–55·7 (M = 53·0) % of head, 1·6–2·3 (modal range 1·9–2·0) times as long as broad. A very prominent mental protuberance is developed, apparently a little longer in males than in females (but this point requires further checking on more specimens). Posterior tip of the maxilla reaching a vertical through a point slightly nearer the orbit than the nostril, or less frequently, somewhat more posteriorly, but never reaching a vertical through the anterior orbital margin.

Gill rakers: Nine or 10 (mode) on the lower part of the first gill arch, the lower 1–3 rakers reduced, the remainder short and stout, or relatively stout.

Scales: ctenoid; lateral line with 32 (f.9) or 33 (f.12) scales, cheek with 5 or 6 (mode 5) rows. Six and a half to 8 (mode 7½) scales between the upper lateral line and the dorsal fin origin, 6 (rarely) to 8 (mode) between the pectoral and pelvic fin bases.

Fins. Dorsal with 23 (f.1), 24 (f.9) or 25 (f.10) rays, comprising 14 (f.1), 15 (f.16), or 16 (f.3) spines and 9 (f.13) or 10 (f.7) branched rays. Anal with 11 (f.1), 12 (f.16) or 13 (f.3) rays, comprising 3 spines and 8–10 branched rays. Pectoral 27·4–34·8 (M = 29·7) % of standard length. Caudal fin truncate, scaled on its proximal half. Pelvics with the first ray produced, proportionately more so in males than in females.

Teeth. The *outer teeth* in *both jaws* are unicuspid, long, and relatively curved, especially over the distal half. There are 56–80 (M = 68) teeth in the outer row of the upper jaw.

The inner teeth are also unicuspid and curved, and are implanted obliquely, especially in the upper jaw where they lie almost horizontally. The inner teeth are arranged rather irregularly in 3 or 4 rows in the upper jaw and in 2 or 3 rows in the lower jaw.

Osteology. The syncranium of *H. decticostoma* is very similar to that of *H. spekii* in most details. One possible interspecific difference in the neurocranium, the height and basal extent of the supraocipital crest, cannot be substantiated without more osteological material of both species. Another probable difference is the more gently sloping preorbital profile in the skull of *H. decticostoma*.

The *lower pharyngeal bone* has a triangular dentigerous surface, as long as broad or somewhat broader (1·1–1·2 times). The lower pharyngeal teeth are fine, compressed and cuspidate, and are arranged in 24–28 irregular rows; teeth in the 2 median series are generally coarser than their lateral congeners.

Coloration in life. Males have a dark golden-brown ground coloration, the dorsal surfaces darkest and with iridescent blue highlights, shading to silver ventrally; an iridescent greenish to yellowish sheen extends over the flanks, opercular region and the cheeks.

Dorsal fin hyaline, darker (almost grey) proximally, lappets dark, soft part with orange-red streaks and spots. Anal hyaline but with a maroon flush distally, and with several orange ocelli (sometimes arranged in 2 rows). Caudal dark proximally, hyaline distally, with orange or red streaks and spots between the rays. Pelvic fins dusky.

Females. Except for the hyaline fins and absence of anal ocelli, the live coloration of females is like that of males.

Preserved material. Males have a brownish-grey coloration dorsally, becoming silver-grey on the ventral surfaces. The snout and preorbital region are dark grey, as are the lips and tip of the lower jaw (the latter almost black in some specimens). The cheek and opercular region are yellowish, the branchiostegal membrane yellowish to dead-white. In some specimens there is a very faint, narrow, but dark midlateral stripe separating the darker dorsal coloration from the lighter ventral tones; a very faint lachrymal stripe is visible in some fishes, being most intense near the angle of the jaws.

Dorsal fin greyish with dark lappets. Anal greyish basally, becoming hyaline distally, the numerous ocelli large and dead-white but sometimes with a faint grey overlay. Caudal greyish, with darker maculae. Pelvic fins dusky to black, the colour most intense on the anterior half of the fin.

Females are light yellowish-brown above, shading to yellowish-white on the chest and belly. The snout, upper lip and tip of the lower jaw are dusky grey, the cheek and opercular region yellowish. A faint dark spot situated behind the angle of the jaw seems to be comparable with the more intense part of the lachrymal stripe in males.

Dorsal fin greyish-hyaline, the lappets dusky. Anal and pelvic fins hyaline, the caudal greyish with darker maculae on its proximal half.

Ecology. Habitat. The specimens examined came from trawl hauls in 2 localities near Nsadzi island. In both places the bottom is of mud and the depth of water between 70 and 180 feet.

Food. All the 25 specimens examined had been almost completely eviscerated by pressure change during capture, and in 16 of these specimens the remaining gut was empty. In the other 9 fishes at least part of the gut was intact and contained food, namely: remains of *Caridina* sp. (Crustacea) in 5 specimens, very fragmentary and unidentifiable fish remains in 3, and a mixture of fish and crustaceans (probably *Caridina* sp.) in one other.

Breeding. All except one of the 25 specimens examined (129–229 mm. S.L.) are adult. The exceptional fish is an immature female 150 mm. S.L. (It should be noted that a female 129 mm. S.L. is adult.)

Diagnosis and affinities. In all respects, *H. decticostoma* is very similar to *H. spekii* (see Greenwood, 1967).

Unfortunately, the live colours of *H. spekii* are unknown. But, judging from *post-mortem* coloration, *H. spekii* differs from *H. decticostoma* in having a darker, more generally blue-grey ground coloration, and a dark (probably black) branchiostegal membrane. The preserved coloration of both species is similar, but again adult male *H. spekii* are darker and the branchiostegal membrane in *H. decticostoma* is lighter.

On most morphometric criteria the two species are indistinguishable. The lower jaw of *H. decticostoma* is more oblique than in *H. spekii* (30°–40°, mode 35°, *cf.* horizontal to *ca.* 15°), and the posterior tip of the maxilla does not extend quite so far posteriorly (but there is some overlap interspecifically in this character).

The modal number of gill rakers in *H. decticostoma* is higher (10 *cf.* 8), but the ranges overlap.

The least interorbital width in *H. decticostoma* is slightly narrower than in *H. spekii* (20·0–23·1, M = 21·6% head, *cf.* 22·0–26·0, M = 23·3% in *H. spekii*), and the mean eye/preorbital ratio is lower (0·8–1·0, M = 0·9, *cf.* 0·8–1·3, M = 1·0).

Haplochromis decticostoma also resembles *H. serranus* (see Greenwood, 1967, for a discussion of the *H. serranus*-*H. spekii* relationship, and Greenwood, 1962, for *H. serranus*). The resemblance includes an oblique lower jaw, but the species differ in such characters as their preserved coloration, the higher number of gill rakers in *H. decticostoma* (10, *cf.* 8 or 9 in *H. serranus*), its deeper preorbital (19·4–23·8, M = 21·7% head, *cf.* 14·6–20·0, M = 17·7% in *H. serranus*), and its lower eye/preorbital ratio (0·8–1·0, M = 0·9, *cf.* 1·1–1·5, M = 1·3).

Phyletically, *H. decticostoma* is a derivative of the *H. serranus* stem, and probably from a species anatomically indistinguishable from *H. spekii*.

STUDY MATERIAL AND DISTRIBUTION RECORDS

Museum and Reg. No.	Locality	Collector
	UGANDA	
B.M. (N.H.) 1968.8.30.292 (Holotype)	S. of Nsadzi isl. (180 feet)	E.A.F.F.R.O.
B.M. (N.H.) 1968.8.30.302–309	N. and S. of Nsadzi isl. (70–160 feet)	E.A.F.F.R.O.
B.M. (N.H.) 1968.8.30.293–301	S. of Nsadzi isl. (180 feet)	E.A.F.F.R.O.
B.M. (N.H.) 1968.8.30.330–333	S. of Nsadzi isl. (180 feet)	E.A.F.F.R.O.

Haplochromis gilberti sp. nov.

(Text-fig. 33)

HOLOTYPE: an adult male 151·0 mm. S.L. (B.M. [N.H.] reg. no. 1968.8.30.1) from near Bulago island.

The species is named in honour of Mr. Michael Gilbert, Experimental Fisheries Officer of the East African Freshwater Fisheries Research Organization. Michael Gilbert's enthusiasm and skill have added considerably to our knowledge of the Lake Victoria fishes, and especially the *Haplochromis* species from the deeper waters.

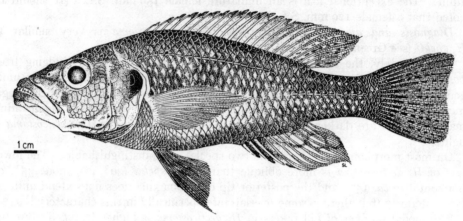

FIG. 33. *Haplochromis gilberti*. Holotype. Drawn by Sharon Lesure.

DESCRIPTION: based on 12 specimens (including the holotype) 125·0–150·0 mm. S.L. Depth of body 30·6–34·7 (M = 32·3) % of standard length, length of head 34·3–37·5 (M = 36·0) %.

Dorsal head profile straight or very slightly curved, sloping at an angle of 25°–30°, its contour interrupted by the prominent premaxillary pedicels.

Preorbital depth 16·3–19·2 (M = 17·8) % of head, least interorbital width 17·0–19·6 (M = 18·1) %. Snout 1·0 (mode)–1·1 times as long as broad, its length 32·6–36·2 (M = 34·4) % of head, diameter of eye 23·0–27·5 (M = 24·8), depth of cheek 22·2–25·3 (M = 23·7) %.

Caudal peduncle 15·4–17·3 (M = 16·6) % of standard length, 1·4–1·6 times as long as deep.

Mouth slightly oblique (*ca.* 25°), jaws equal anteriorly or the lower projecting slightly. Lower jaw length 44·3–49·1 (M = 47·5) % of head, 1·9–2·3 (mode 2·0) times as long as broad; a distinct but low mental protuberance is developed. Premaxilla with its dentigerous surface somewhat expanded medially and anteromedially giving a slightly beaked effect. Posterior tip of the maxilla reaching a vertical through the anterior orbital margin or slightly beyond, rarely not quite reaching the level of the orbit.

Gill rakers: Nine (rarely 8 or 10) on the lower part of the first gill arch, the lower 1–3 rakers reduced, the upper 3–6 (usually 4) flattened, often anvil-shaped; the intervening rakers of varied shape, from relatively slender to relatively stout.

Scales: ctenoid. Lateral line with 32 (f.2) or 33 (f.9) scales; cheek with 3 or 4 rows. Five and a half (rarely) to 7 (mode 6½) scales between the upper lateral line and the dorsal fin origin, 6 or 7 between the pectoral and pelvic fin bases.

Fins. Dorsal with 24 (f.5) or 25 (f.7) rays, comprising 15 spines and 9 or 10 branched rays. Anal with 11 (f.1) or 12 (f.11) rays, comprising 3 spines and 8 or 9 branched rays. Pectoral 24·0–29·9 (M = 27·8) % of standard length. Caudal truncate, scaled on its proximal half. Pelvics with the first ray produced (only one female is available).

Teeth. The *outer* teeth in *both jaws* are long, slender unicuspids, recurved distally. These teeth are well-spaced with 38–50 (M = 40) in the upper jaw.

The *inner teeth* are also unicuspid and curved, and are implanted obliquely, those of the lower jaw lying almost horizontally. There 3 or 4 irregular series of inner teeth in the upper jaw, and 2 or 3 series in the lower jaw.

Osteology. The *neurocranium* of *H. gilberti* is moderately elongate and slender, and except for its greater width across the interorbital region, resembles the type found in *H. tyrianthinus* and other members of the " *tridens* " species complex (see p. 51). Thus in general the skull of *H. gilberti* approaches the " *prognathus* " skull type (see Greenwood, 1967, p. 109 *et. seq.*). The cephalic lateral line tubes are moderately hypertrophied, with enlarged pores.

The *premaxilla* is moderately beaked. Together with the *dentary*, these bones can be considered typical of the condition found in many " *prognathus* "-group species.

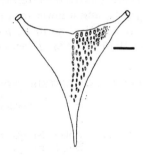

Fig. 34. *Haplochromis gilberti.* Lower pharyngeal bone. Scale equals 2 mm.

The *lower pharyngeal bone* (Text-fig. 34) is moderately slender, with its triangular dentigerous surface as long as broad. The teeth are coarse but compressed and cuspidate, and are arranged in 22–28 rows.

Coloration. The colours of live fishes are unknown.

Preserved material. Adult males are brownish-grey above the upper lateral line, on the dorsal surface of the head, the snout and the cheeks; silvery-grey below, the silver sheen most intense on the chest, belly and the operculum. The branchiostegal membrane is brownish, except for a faint dusky overlay in the opercular region of some individuals. In certain specimens there is a faint, dark lachrymal stripe, and faint traces of a dark midlateral longitudinal band.

The dorsal fin is dusky grey, the lappets black, the soft part with dark maculae. Caudal greyish, lighter along the posterior margin, the entire fin darkly maculate. Anal fin yellowish-hyaline, with a narrow, dusky base; the ocelli are large and dead-white. Pelvic fins are black.

Female (the only specimen available, an adult). Ground colour brownish grey above the upper lateral line, shading to silver below that level; a faint midlateral longitudinal stripe is visible on the flanks and caudal peduncle. The snout is greyish, the cheeks silvery, the branchiostegal membrane yellowish-white.

The dorsal fin is dark hyaline, as is the caudal which, however, is dusky at its base. The anal and pelvic fins are hyaline.

Ecology. Habitat. The species is so far known from only one locality, near Bulago island at a depth of 55–75 feet over a mud bottom.

Food. Two of the 12 guts examined were empty; of the others, 2 contained fragmentary remains of small cyprinid fishes, 1 unidentifiable fish remains, 2 fragmentary fish (cyprinid) and insect remains (chironomid pupae), 4 contained only fragments of chironomid pupae, and 1 an unidentifiable, colourless solid.

Breeding. Except for the single adult female (146 mm. S.L.), all the other specimens are adult males. The right ovary of the female is noticeably larger than the left one.

Diagnosis and affinities. Superficially (and in most aspects of its anatomy) *H. gilberti* resembles *H. paraguiarti* (see Greenwood, 1967, p. 69) and was probably derived from the same stem as that species. *Haplochromis gilberti* differs from *H. paraguiarti* principally in having a narrower interorbital ($17 \cdot 0$–$19 \cdot 6$, M = $18 \cdot 1 \%$ of head, *cf.* $22 \cdot 9$–$27 \cdot 7$, M = $25 \cdot 3 \%$) and a lower mean number of outer teeth in the upper jaw (38–50, M = 40, *cf.* 42–62, M = 54 in *H. paraguiarti*); the outer teeth in *H. gilberti* are also more slender than those of *H. paraguiarti*.

Study Material and Distribution Records

Museum and Reg. No.	Locality	Collector
	UGANDA	
B.M. (N.H.) 1968.8.30.1 (Holotype)	Near Bulago isl. (55–75 feet)	E.A.F.F.R.O.
B.M. (N.H.) 1968.8.30.2–12	Near Bulago isl. (55–75 feet)	E.A.F.F.R.O.

Haplochromis paraplagiostoma sp. nov.

(Text-fig. 35)

HOLOTYPE: an adult female $97 \cdot 0$ mm. S.L. (B.M. [N.H.] reg. no. 1968.8.30.127) from west of Bulago island, at a depth of 70 feet.

DESCRIPTION: based on 8 specimens (including the holotype) $90 \cdot 5$–$98 \cdot 0$ mm. standard length.

Depth of body $34 \cdot 1$–$38 \cdot 7$ (M = $35 \cdot 9$) % of standard length, length of head $31 \cdot 9$–$33 \cdot 4$ (M = $32 \cdot 0$) %.

Dorsal head profile sloping at an angle of 35°–40°, almost straight but gently curved in the nuchal region, and with a slight supraorbital concavity.

Preorbital depth $15 \cdot 5$–$18 \cdot 7$ (M = $17 \cdot 1$) % of head, least interorbital width $19 \cdot 0$–$22 \cdot 6$ (M = $20 \cdot 9$) %. Snout $1 \cdot 1$–$1 \cdot 3$ (mode $1 \cdot 2$) times as broad as long, its length $29 \cdot 0$–$31 \cdot 8$ (M = $30 \cdot 2$) % of head, diameter of eye $24 \cdot 1$–$29 \cdot 0$ (M = $26 \cdot 5$), depth of cheek $25 \cdot 1$–$29 \cdot 0$ (M = $27 \cdot 3$) %.

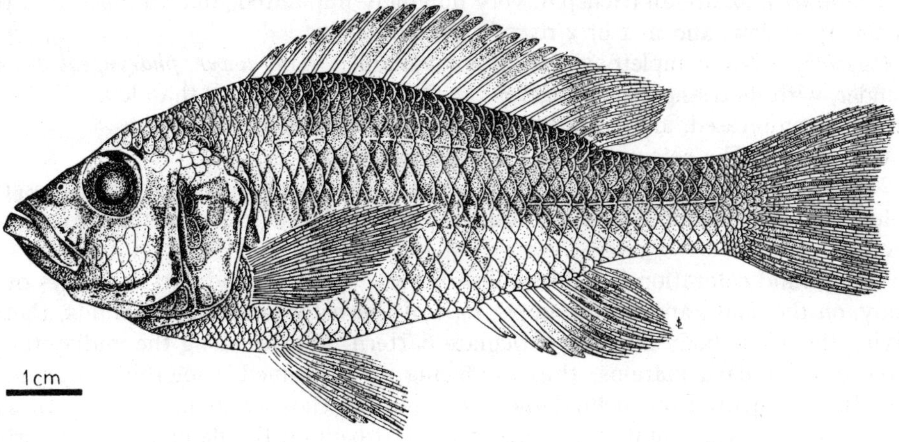

FIG. 35. *Haplochromis paraplagiostoma*. Holotype. Drawn by Sharon Lesure.

Caudal peduncle 17·6–20·4 (M = 19·0) % of standard length, 1·5–1·8 (mode 1·6) times as long as deep.

Mouth slightly oblique, sloping at an angle of 20°–35°. Lower jaw projecting slightly, and with a distinct mental protuberance, its length 41·4–46·7 (M = 44·3) % of head, 1·6–2·0 times as long as broad. Premaxilla not expanded anteromedially. Posterior tip of the maxilla reaching a vertical passing through the anterior part of the eye.

Gill rakers: Nine (mode) or 10 on the lower part of the first gill-arch, the lower 1 or 2 reduced, the upper 2–4 flattened and branched, the intervening rakers relatively slender.

Scales: ctenoid. Lateral line with 32 (f.2), 33 (f.4), 34 (f.1) or 35 (f.1) scales; cheek with 3 or 4 rows. Six to $7\frac{1}{2}$ (mode 6) scales between the upper lateral line and the dorsal fin origin, 6 or 7 (mode) between the pectoral and pelvic fin bases.

Fins. Dorsal with 24 rays, comprising 15 spines and 9 branched rays. Anal with 11 (f.5) or 12 (f.3) rays, comprising 3 spines and 8 or 9 branched rays. Pectoral 24·5–28·0 (M = 26·7) % of standard length. Caudal truncate, scaled on its proximal half to two-thirds. Pelvic fins with the first and second branched rays produced, the first more so, but without any sexual correlation.

Teeth. Outer row. In most specimens the outer teeth in both jaws are relatively compressed and unequally bicuspid, with a few weakly bicuspids and a few slender unicuspids interspersed amongst them. In the largest fish (98·0 mm. S.L.) unicuspid teeth predominate. Posteriorly in the *upper jaw* the teeth are smaller, and in some specimens are finer than the lateral and anterior teeth; unicuspid or weakly bicuspid teeth are usual in this region of the jaw. Teeth in the *lower jaw* are relatively stouter than those in the upper jaw, the discrepancy being most marked anteriorly and anterolaterally.

There are 60–72 (M = 64) outer teeth in the upper jaw.

The *inner teeth* are all tricuspid, very obliquely implanted, and arranged in 2 rows in the upper jaw, and in 1 or 2 rows in the lower jaw.

Osteology. No complete skeleton is available. The *lower pharyngeal bone* is slender, with its triangular dentigerous surface slightly broader than long. The teeth are fine, compressed, and cuspidate and are arranged in about 30 rows.

Coloration. The colours of *live fishes* are unknown.

Preserved material. There is so little sexual dimorphism apparent in the preserved coloration of the sample available that a combined description of the sexes can be given.

The ground coloration is light pinkish-brown; scales on the dorsal aspects of the body, on the flanks and on the upper part of the belly have dark margins, thereby giving the whole body a faintly reticulate pattern. Scales along the midlateral line have broader dark margins, thus producing an ill-defined, sometimes interrupted longitudinal stripe from behind the operculum to the caudal fin origin. In some fishes there are very faint traces of 5–7 vertical bars on the flanks. The preorbital region of the head is dark, as is the interopercular region, the posterior margin of the preoperculum and at least part of the branchiostegal membrane; a faint lachrymal stripe is visible in all specimens.

The dorsal and anal fins are hyaline to greyish, the former with dark lappets. In males the anal fin has two indistinct ocelli, faintly outlined anteriorly in black. Caudal greyish, faintly maculate in some individuals. Pelvic fins dark in both sexes, but more uniformly so in males.

Ecology. All 8 specimens came from the same locality near Bulago island, over a mud bottom at a depth of 70 feet.

No information was obtained on the feeding habits of *H. paraplagiostoma.*

Some difficulty was experienced in determining the sexual state of several specimens; all except one (a ripening female, 93 mm. S.L.) could be either juvenile or quiescent.

Diagnosis and affinities. There is a certain superficial similarity between *H. paraplagiostoma* and *H. plagiostoma* (see Greenwood, 1962, p. 199), but the species differ in a number of characters, including the dentition. For example *H. paraplagiostoma* has a shorter head (31·9–33·4, M = 32% standard length, *cf.* 34·0–37·5, M = 36·0%), shallower cheek (25·1–29·0, M = 27·3% head, *cf.* 28·0–36·8, M = 33·0%), shallower preorbital (15·5–18·7, M = 17·1% head, *cf.* 18·0–21·5, M = 19·8%) and a shorter lower jaw (41·4–46·7, M = 44·3% head, *cf.* 44·0–54·5, M = 49·2%); the lower jaw of *H. paraplagiostoma* also slopes less steeply than that of *H. plagiostoma.*

The dentition differs in that the outer teeth of *H. plagiostoma* at all known lengths (69–147 mm. S.L.) are unicuspid, short, and strongly curved whereas in specimens of *H. paraplagiostoma* < 96 mm. S.L. the majority of teeth are distinctly and very unequally bicuspid, and are weakly curved. Unicuspid teeth are not frequent in fishes below 96 mm. S.L., and in the larger fish the unicuspids differ in shape and relative size from those of *H. plagiostoma.*

There are clear-cut differences in the preserved coloration of the two species (*cf.* above and p. 201 of Greenwood, 1967).

Despite the superficial resemblances between *H. paraplagiostoma* and *H. plagiostoma*, we are not inclined to think that they indicate close relationship between the species.

Until more is known about *H. paraplagiostoma* the species cannot readily be linked with any of the known deep-water or inshore dwelling *Haplochromis* species. However, in many respects *H. paraplagiostoma* shows some affinity with the organizational level seen in *H. empodisma*. We do not imply that the species are closely related, but rather that *H. paraplagiostoma* could have evolved from a *H. empodisma*-like stem (see Greenwood, 1960).

Study material and distribution records

Museum and Reg. No.	Locality	Collector
	UGANDA	
B.M. (N.H.) 1968.8.30.127 (Holotype)	W. of Bulago isl. (70 feet)	E.A.F.F.R.O.
B.M. (N.H.) 1968.8.30.128–134	W. of Bulago isl. (70 feet)	E.A.F.F.R.O.

Acknowledgements

The senior author is greatly indebted to the Director and staff of the East African Freshwater Fisheries Research Organization, Jinja, Uganda, for their hospitality and assistance during two visits to their laboratory.

The junior author is especially grateful to the Freedom from Hunger Campaign for the award of a grant which enabled him to visit the British Museum (Nat. Hist.) on study leave in connection with our joint research.

Appendix

Haplochromis obesus (Blgr.) 1906

Six small fishes (67·0–71·0 mm. S.L.) caught at a depth of 80–90 feet over a mud substrate north of Nsadzi island, are tentatively identified as *H. obesus*. This species is otherwise known from littoral and immediately sublittoral habitats.

Our identification of these fishes is based on their peculiar dentition and their general facies, both of which are typical for *H. obesus* (see Greenwood, 1959). In most morphometric characters (see table below) the specimens agree with *H. obesus*, except for having a narrower interorbital (mean width 24·6% of head, *cf.* 32·2% in *H. obesus*) and a slightly longer caudal peduncle. Some of the Nsadzi fishes have a higher number of gill rakers (11) than is modal for *H. obesus* (9 or 10); however, a few *H. obesus* also have counts of 11, and one of the Nsadzi specimens has 10 gill rakers.

All six Nsadzi fishes (5 males, 1 female) are adult and sexually active. By contrast, the smallest known adult *H. obesus* from inshore and sublittoral populations is 85·0 mm. S.L. (Greenwood, 1959).

Another biological difference between the Nsadzi and the other populations concerns food. Three of the six Nsadzi fishes contained ingested material in the stomach which, in each instance, is packed with Cladocera and Copepoda. The food of individuals from inshore habitats appears to be the embryos and larvae of fishes, especially Cichlidae (Greenwood, *op. cit.*). The significance of this dietary difference is impossible to assess. In neither instance are large numbers of gut analyses available (3 from Nsadzi, 18 from other habitats), and furthermore, all the Nsadzi fishes are smaller than any of the inshore fishes examined.

This marked size difference also hampers evaluation of the observed discrepancies in certain morphometric characters (see above); only one specimen of the Nsadzi sample falls within the size range of other *H. obesus* material.

Partly for this reason, and partly because we have so few deep-water specimens, we do not feel justified in creating a new species for the Nsadzi fishes.

The principal morphological characters of the Nsadzi specimens may be summarized as follows:

S.L.	D.*	H.*	Po. %	Io. %	Snt. %	Eye %	Ck. %	Lj. %	C.P.*
67·0	34·3	31·3	14·3	23·8	28·6	33·3	23·8	47·6	17·9
69·0	34·3	33·3	13·0	23·9	28·3	30·4	21·8	43·6	18·9
69·0	35·2	33·3	13·0	26·1	28·3	28·3	21·8	43·6	18·9
70·0	34·3	31·4	15·8	22·7	28·6	31·8	25·0	45·4	17·8
70·0	37·1	32·9	15·2	26·1	30·5	30·5	23·9	43·5	17·2
71·0	34·5	33·1	17·0	25·5	28·9	29·9	23·4	46·9	18·4

* = % of standard length
% = % of head

The lower jaw is 1·4–1·9 times as long as broad; the snout 1·3–1·6 times as broad as long. The caudal peduncle is 1·4–1·6 times as long as deep.

Gill rakers: short and stout, the upper 2 or 3 flattened and anvil-shaped; 10 (f.1) or 11 (f.5) on the lower part of the first gill arch.

Teeth. The *outer row* in the *upper jaw* is composed of 30–38 (M = 36), unicuspid or weakly bicuspid, slightly recurved teeth. In the *lower jaw*, the outer teeth are unicuspid (a few weakly bicuspid in one fish), with the tips of the anterior and anterolateral teeth directed outwards.

The *inner teeth* are tricuspid in the upper jaw, unicuspid in the lower, and are arranged in 1 or 2 rows.

Fins. Dorsal with 15 (f.3) or 16 (f.3) spines, and 8 or 9 branched rays. Pectoral 27·1–31·0% of standard length.

Scales: ctenoid. Lateral line with 30 (f.2), 31 (f.2) or 32 (f.2) scales, cheek with 3 rows. Five and a half or 6½ scales between the dorsal fin origin and the upper lateral line, 6 between the pelvic and pectoral fin bases.

Register numbers: B.M. (N.H.) 1968.8.30.329; and 1968.9.3.3–7.

REFERENCES

GREENWOOD, P. H. 1956. The monotypic genera of cichlid fishes in Lake Victoria. *Bull. Br. Mus. nat. Hist.* (Zool.) **3** : No. 7, 295–333.

—— 1959. A revision of the Lake Victoria *Haplochromis* species (Pisces, Cichlidae) Part III. *Bull. Br. Mus. nat. Hist.* (Zool.) **5** : No. 7, 179–218.

GREENWOOD, P. H. 1960. A revision of the Lake Victoria *Haplochromis* species (Pisces, Cichlidae) Part IV. *Bull. Br. Mus. nat. Hist.* (Zool.) **6**: No. 4, 227–281.
—— 1962. A revision of the Lake Victoria *Haplochromis* species (Pisces, Cichlidae) Part V. *Bull. Br. Mus. nat. Hist.* (Zool.) **9** : No. 4, 139–214.
—— 1967. A revision of the Lake Victoria *Haplochromis* species (Pisces, Cichlidae) Part VI. *Bull. Br. Mus. nat. Hist.* (Zool.) **15** : No. 2, 29–119.
REGAN, C. T. 1922. The cichlid fishes of Lake Victoria. *Proc. zool. Soc. Lond.* **1922** : 157–191.

P. H. GREENWOOD, BSc.(Rand.), DSc.(Rand.)
BRITISH MUSEUM (NAT. HIST),
CROMWELL ROAD.
LONDON S.W.7

J. M. GEE, B.A. (Keele), Ph.D (Univ. Wales)
E.A.F.F.R.O.
P.O. BOX 343.
JINJA,
UGANDA.

A revision of the Lake Victoria *Haplochromis* species (Pisces, Cichlidae), Part VIII

P. H. Greenwood
Department of Zoology, British Museum (Natural History), Cromwell Road, London SW7 5BD

C. D. N. Barel
Zoologisch Laboratorium der Rijksuniversiteit Leiden, Kaiserstraat 63, Leiden, The Netherlands

Contents

Introduction	141
Haplochromis crocopeplus sp. nov.	142
Haplochromis sulphureus sp. nov.	148
Haplochromis plutonius sp. nov.	151
Comments on the *Haplochromis tridens* species complex	155
New species of the *H. serranus* group	157
Haplochromis nanoserranus sp. nov.	157
Haplochromis cassius sp. nov.	161
A new species of the *H. empodisma–H. obtusidens* group	164
Haplochromis ptistes sp. nov.	164
New species of the *H. ishmaeli–H. pharyngomylus* group	169
Haplochromis teegelaari sp. nov.	169
Haplochromis mylergates sp. nov.	174
Acknowledgements	179
Appendix: The live coloration of certain previously described *Haplochromis* species, by M. J. P. van Oijen	180
References	192

Introduction

Seven of the eight species described in this paper were collected during June 1975 by Drs G. Ch. Anker and C. D. N. Barel in the Mwanza and Speke Gulf regions (Tanzania) of Lake Victoria. The material was taken from trawl catches made, in general, over mud substrata and, for the most part, at depths of 2–10 m; one station, however, was considerably deeper (28 m); see Fig. 1. The eighth species is from much deeper water (50–60 m) in the northern (Ugandan) part of the Lake; it was collected by the senior author in 1970 during a survey cruise in the R.V. *Ibis*, then based at Jinja as part of a joint U.N.D.P.–E.A.F.F.R.O. research project into the fishery potential of Lake Victoria. All eight taxa must be considered elements of the still largely unknown offshore complex of *Haplochromis* species, now, or soon to be tapped by the developing trawl fishery on the lake.

On the basis of data from collections made in various parts of Lake Victoria, it seems likely that the Mwanza and Speke Gulf species are confined to relatively shallow and sublittoral habitats, while the Ugandan species is restricted to deeper waters. It may be of some significance that none of the new Tanzanian species has been recorded from other and similar biotopes in the lake, yet they were captured together with several species known to have a lake-wide distribution in such habitats.

The new species are of particular interest because they include three new members of the *H. tridens* species complex, two additions to the *H. ishmaeli–H. pharyngomylus* grade of mollusc crushers, a new member of the *H. empodisma–H. obtusidens* mollusc–insectivore lineage, and the first 'dwarf' member of the *H. serranus* lineage, a group of piscivorous species whose members otherwise reach some of the larger adult sizes found among the Lake Victoria species. (For

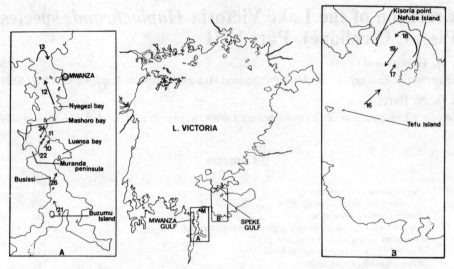

Fig. 1 Sketch map of Lake Victoria with (A) the Mwanza Gulf, and (B) the Speke Gulf, drawn to a larger scale. The numbers refer to Anker–Barel collecting stations, and the arrows to the direction in which the trawl shot at that station was fished.

details of these various groups and their presumed phyletic relationships, see Greenwood, 1974.) The eighth new taxon, also apparently a member of the *H. serranus* lineage, has a most distinctive oral and dental morphology strongly suggestive of fish-eating or other predatory habits, yet it seems to feed, at least in part, on diatoms (see p. 163).

The Anker–Barel collection has also provided a most interesting puzzle in the form of several specimens which appear to bridge the anatomical 'gap' separating the insectivorous–detritus eating *H. empodisma* from its sister species, the insectivorous–molluscivorous *H. obtusidens* (see Greenwood, 1960 & 1974). There are, however, indications that this anatomically intermediate material represents a third taxon in the lineage. Since further observations and material are needed to resolve this problem it, and the new material, will be considered in a later paper.

Drs Barel and Anker were able to gather a lot of useful information on the live coloration of various *Haplochromis* species collected during their visit to the southern end of the lake. An extensive collection of colour transparencies, together with the specimens photographed, has been deposited in the British Museum (Natural History). Many species described or redescribed in previous parts of this revision, and for which no information of live coloration was then available, are represented in the Anker–Barel photographic collection. Colour descriptions of these fishes have been prepared by Mr Martien van Oijen, a postgraduate student in the Zoology Department of Leiden University, and are published as an appendix to this paper.

The eight new taxa will be described in groups based on their presumed phyletic affinities (see Greenwood, 1974), starting with the three new members of the '*tridens*' lineage.

New species of the *H. tridens* group

Haplochromis crocopeplus sp. nov.

HOLOTYPE. An adult male, 84·0 mm standard length, from the Speke Gulf, between Nafuba and Tefu Islands at a depth of 28 m. BM(NH) reg. no. 1977.1.10:70.

The trivial name (from the Greek) refers to the basically ochrous-yellow coloration of live fishes.

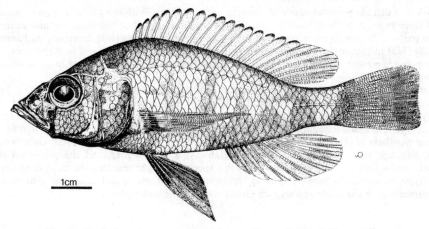

Fig. 2 *Haplochromis crocopeplus*. Holotype. Drawn by M. J. P. van Oijen.

DESCRIPTION (Figs 2–7). Based on 20 specimens (including the holotype), 71·0–100·5 mm standard length.

Depth of body 29·0–35·0% of standard length (mean, M=33·0%), length of head 34·0–38·0 (M=36·0)%.

Dorsal head profile straight, sloping at an angle of 30–35°; cephalic lateral line pores prominent, especially the pores and tubules on the preorbital bone.

Preorbital depth 15·0–19·0 (M=17·0)% of head length, least interorbital width 18·0–22·0 (M=19·5)%. Snout as broad as it is long, to slightly longer than broad (1·1 times); its length 29·0–31·0 (M=30·0)% of head. Eye and orbit slightly elliptical (i.e. longer than deep), the eye with a narrow anteroventral aphakic aperture; greatest eye diameter 28·0–33·0 (M=30·0)% of head. Cheek depth 17·0–25·0 (M=21·0)%.

Caudal peduncle 16·0–20·0 (M=18·0)% of standard length, 1·4–1·8 (modal range 1·5–1·6) times longer than deep.

Mouth inclined at an angle of 30–35° (rarely at 15–20°); posterior tip of the maxilla reaching a vertical slightly behind the anterior margin of the orbit or, less frequently, to a vertical through the anterior margin. The dentigerous arm of the premaxilla is somewhat expanded anteriorly in the midline, giving the bone a weakly beaked appearance; the dentary has a variously developed but obvious mental protuberance. Length of lower jaw 41·0–50·0 (M=47·0)% of head length, 1·7–2·4 (modal range 2·2–2·3) times its width; jaws equal anteriorly.

Gill rakers. 9–11 (rarely 8) on the lower part of the first gill arch, the lower 1–4 rakers reduced, the next 3 or 4 generally slender, the uppermost rakers flattened and often bifid or anvil-shaped.

Scales. Ctenoid; lateral line with 31 (f.2), 32 (f.6), 33 (f.8) or 34 (f.1) scales in the 17 specimens with undamaged squamation, cheek with 3 or 4 rows. Five and a half to 6½ (rarely 5) scales between the lateral line and the dorsal fin origin, 5½ to 6½ (mode 6), rarely 7, between the pectoral and pelvic fins bases.

Fins. Dorsal with 23 (f.4), 24 (f.11) or 25 (f.5) rays, comprising 14 (f.1), 15 (f.14) or 16 (f.5) spines and 7 (f.1), 8 (f.2), 9 (f.16) or 10 (f.1) branched rays. Anal with 11 (f.8), 12 (f.11) or 13 (f.1) rays, comprising 3 spinous and 8 (f.8), 9 (f.10) or 10 (f.2) branched elements. Pectoral fin 26·0–33·0 (M=30·0)% of standard length. Pelvic fins with the first branched ray produced in both sexes. Caudal truncate, scaled on its proximal third to half.

Teeth. In *both jaws* the *outer teeth* have relatively slender, near-cylindrical necks and somewhat compressed, gently recurved crowns. Three different types of crown form are present, viz. unicuspid, unequally bicuspid (sometimes weakly so) and tricuspid. It is difficult to detect any clear-cut spatially correlated arrangement of the different crown types within the outer series, or to show a definite preponderance of one type over the other. With few exceptions an admixture of all three types is present, usually with the tricuspids restricted to the posterior and posterolateral parts of the row in both jaws, and the bi- and unicuspids occurring anteriorly and anterolaterally. Not infrequently, unicuspids are found only in the upper jaw, and in 6 specimens no unicuspids are present in either jaw. Very rarely do tricuspid teeth predominate in an admixture of cusp types.

There are 70–80 (M = 74) teeth in the outer row of the upper jaw.

Teeth in the *inner series* are all small and tricuspid, implanted obliquely and arranged in 2 (rarely 1) rows in the upper jaw, and 1 or 2 (rarely 3) in the lower jaw.

Osteology. The *neurocranium* of *Haplochromis crocopeplus* (Fig. 3) closely resembles that found in other members of the *Haplochromis tridens* group (see Greenwood & Gee, 1969; Greenwood, 1974), although the supraoccipital crest is, relatively, a little higher, and the preorbital (i.e. ethmovomerine) region somewhat shorter in this species. In these features, the neurocranium of *H. crocopeplus* should be looked upon as representing a less specialized state than that seen in the neurocrania of the other species (see Greenwood, 1974; and below).

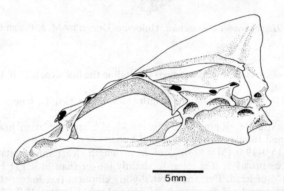

Fig. 3 *Haplochromis crocopeplus*. Neurocranium in left lateral view.

Each ramus of the *dentary* (Fig. 4) has a marked outwardly directed flare to its upper half, thus giving the lateral aspect of the bone a distinctly concave appearance (a feature best seen when the bone is viewed from the front). The concavity extends anteriorly almost to the symphysis.

The *premaxilla* has a moderately developed beak consequent upon the anterior and anterolateral expansion of its medial dentigerous surface; the pedicels (anterior ascending processes) are as long as the dentigerous arms of the bone.

The *lower pharyngeal bone* (Fig. 5) is narrow and slender, its dentigerous surface as broad as it is long (or slightly broader than long). The lower pharyngeal teeth are fine, compressed and cuspidate, and are arranged in 28–32 rows.

There are 28 (f.1), 29 (f.5), 30 (f.10) or 31 (f.1) *vertebrae* (excluding the fused PU_1 and U_1 centra) in the 17 specimens radiographed, the total comprising 12 (f.1), 13 (f.15) or 14 (f.1) abdominal and 15 (f.1), 16 (f.4) or 17 (f.12) caudal centra.

Coloration. In life a *sexually mature female* (BM(NH) reg. no. 1977.1.28:39; see Fig. 6) has the dorsum of the body and caudal peduncle grey with a yellow flush, that of the body becoming greyish-silver in the nape region. The flanks and ventral part of the caudal peduncle are ochrous becoming whitish on the belly. Traces of a very faint midlateral stripe are visible on the flanks. The dorsal surface of the head is yellowish-grey, the operculum ochrous yellow, the suboperculum somewhat silvery, the preorbital region and cheeks greyish-ochre (the latter becoming lighter ventrally) and the lips bright ochre. The branchiostegal membrane is yellowish-grey.

The dorsal fin is ochrous, with traces of red proximally, and hyaline spots distally on the soft part of the fin. The caudal is greyish-ochre proximally, yellowish distally; the anal fin is yellow with a grey margin, the pelvics are yellow, and the pectorals yellowish-hyaline.

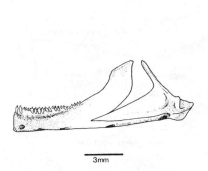

Fig. 4 *Haplochromis crocopeplus*. Left dentary and anguloarticular in lateral view.

Fig. 5 *Haplochromis crocopeplus*. Lower pharyngeal bone in occlusal view.

An *adult but sexually quiescent male* (BM(NH) reg. no. 1977.1.28:38), see Fig. 7, has the dorsum of the body dark grey, the flanks, chest belly and caudal peduncle ochrous yellow; faint traces of a dark midlateral band are visible on the flanks. The head is yellowish grey, the opercular series grey.

The dorsal fin is grey-yellow with sooty lappets, the anal reddish anteriorly, pale grey-yellow posteriorly, its egg dummies (anal ocelli) yellow. The caudal fin is yellowish with dark rays and a red flush ventrally, the pectorals are hyaline with dark rays, and the pelvics sooty with an ochrous flush.

Preserved material. Adult males have a light brown ground coloration, darkest from the dorsum to about the level of the upper lateral line; in some specimens the ventral aspects of the body, and the entire caudal peduncle, are dusky or dusky overlying silver. Very faint indications of dark vertical bars are present on the flanks and caudal peduncle, the bars never extending as far as the dorsal body outline and sometimes not to the ventral outline either; in other specimens the bars merge with the dark coloration of the belly. The lower jaw and branchiostegal membrane are sooty, the vertical limb of the preoperculum dusky silver to black. The dorsal surface of the snout is dusky, and there is a very faint lachrymal stripe extending from the orbit to the lower jaw, passing immediately behind the posterior tip of the maxilla.

The dorsal fin is sooty-grey, the lappets black; the caudal is sooty, with the pigment most intense along its middle rays. The anal fin is light sooty, the pelvics black and the pectorals greyish.

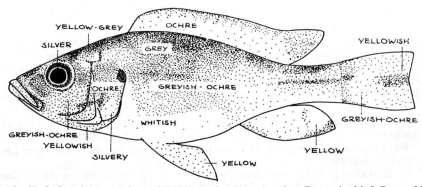

Fig. 6 *Haplochromis crocopeplus*. Adult female, showing coloration. Drawn by M. J. P. van Oijen

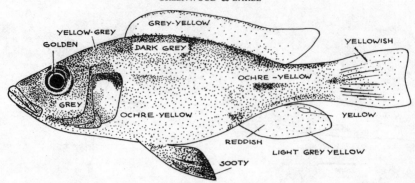

Fig. 7 *Haplochromis crocopeplus*. Adult male (sexually quiescent), showing coloration. Drawn by M. J. P. van Oijen.

Adult females have a light brown (almost fawn) ground coloration above the midlateral line, and are lighter fawn below. The dorsal aspects of the snout are faintly dusky, the operculum has a faint silvery sheen over its ventral half to two-thirds. No definite lachrymal stripe is visible, although in most specimens there is an ill-defined darkened region in the position of a typical lachrymal stripe.

The dorsal and caudal fins are greyish, the latter almost sooty between the middle few rays; the dorsal has black lappets. The anal and pelvic fins are hyaline, the pectorals greyish.

ECOLOGY. *Habitat*. Most of the specimens of *H. crocopeplus* used in this description were caught in a trawl net shot northeast of Tefu Island in the Speke Gulf. The bottom there is of soft, mostly organic mud, and the water is *c*. 28 m deep. At least one other specimen (see p. 148) was caught at the entrance to the Mwanza Gulf (sandy bottom at 6 m).

FOOD. The entire alimentary tract of all 17 specimens examined was filled with an organic sludge composed mainly of partly decomposed blue-green algae, and diatom frustules. Since such decaying organic matter is typical of the mud–water interface in Lake Victoria, we suspect that the gut contents of these fishes do not, in fact, represent normal food intake. Rather, we consider it more likely that the captive fishes ingested large quantities of the light, flocculent mud through which the net was moving.

The intestine of *H. crocopeplus*, however, is relatively long (*c*. 2·3–2·5 times the standard length) with the greater part coiled. Such a gut morphotype is often associated with a diet of detrital matter and differs from that found in the previously described and essentially carnivorous species of the '*tridens*' group (see Greenwood & Gee, 1969).

BREEDING. No information is available on the reproductive habits of *H. crocopeplus*. Only one of the 17 specimens examined is definitely immature, a female 76·5 mm S.L. A male of 71·0 mm S.L. is adult and, apparently, sexually active. There is considerable variation in the extent to which the two ovaries are developed in sexually active fishes; in some, both ovaries are of equal size, in others the left ovary is so greatly reduced (the modal condition) as to appear non-functional, while in others it is only slightly reduced. (See also Greenwood & Gee, 1969 : 38, 43, 47 & 51.)

Judging from the sample available to us, males and females reach the same maximum adult size.

DIAGNOSIS AND AFFINITIES. *Haplochromis crocopeplus* shows all the diagnostic features of an *H. tridens*-group member (see Greenwood & Gee, 1969; and p. 155 below). In addition to interspecific differences in live coloration, *H. crocopeplus* can be distinguished from other species of that group as follows:

(i) From *Haplochromis tridens* (see Greenwood, 1967 : 97, fig. 20) by its less steeply sloping dorsal head profile (30–35°, cf. 40–45°), slightly wider interorbital distance (18·0–22·0, M = 19·5%

head, cf. 15·0–19·5, M = 16·7%), more numerous teeth in the outer premaxillary series (74 cf. 66), more oblique mouth (30–35°, cf. horizontal to 10°) and its less elliptical eye.

(ii) From *Haplochromis dolichorhynchus* (see Greenwood & Gee, 1969 : 34, fig. 21) by the slightly steeper slope of its dorsal head profile (30–35°, cf. 20–30°), its shorter snout (29·0–31·0, M = 30·0% of head, cf. 30·3–38·0, M = 34·2%), slightly larger eye (28·0–33·0, M = 30·0% head, cf. 25·9–29·6, M = 27·2%), somewhat more oblique mouth (30–35°, cf. 15–25°), greater posterior extension of the maxilla and by the less marked beak-like expansion of the premaxilla.

(iii) From *Haplochromis chlorochrous* (see Greenwood & Gee, 1969 : 44, fig. 27) by its slightly longer head (34·0–38·0, M = 36·0% of standard length, cf. 32·0–35·4, M = 34·0%), very slightly shorter snout (29·0–31·0, M = 30·0% head, cf. 29·5–35·0, M = 32·3% head) and slightly larger eye (28·0–33·0, M = 30·0% head, cf. 25·4–29·0, M = 28·2%).

Both species have similarly oblique mouths and similar dorsal profiles. A minor osteological difference lies in the more concave lateral aspect of the dentary in *H. crocopeplus* (see Greenwood & Gee, 1969 : 46; and p. 144 above). *Haplochromis crocopeplus* appears to have fewer unicuspid and more bi- and tricuspid teeth in its outer tooth rows, especially that of the upper jaw, but this may be a size correlated feature since many specimens in the *H. chlorochrous* sample examined are larger than the available specimens of *H. crocopeplus*. In life, adult male coloration seems to distinguish immediately between the two species.

(iv) From *Haplochromis tyrianthinus* (see Greenwood & Gee, 1969 : 40, fig. 25) by its slightly deeper body (29·0–35·0, M = 33·0% standard length cf. 27·3–32·6, M = 30·6%), straight dorsal head profile (albeit one sloping at a similar angle), somewhat wider interorbital space (18·0–22·0, M = 19·5% head, cf. 15·2–18·3, M = 17·4%) larger eye (28·0–33·0, M = 30·0% head, cf. 26·1–29·3, M = 27·7%), and more oblique mouth (30–35° cf. 15–20°).

(v) From *Haplochromis cryptogramma* (see Greenwood & Gee, 1969 : 48, fig. 30) by the distinctive colour pattern of that species, which is retained even in preserved material, and also by the noticeably concave dorsal head profile in *H. cryptogramma*, by the slightly narrower interorbital width in *H. crocopeplus* (18·0–22·0, M = 19·5% head, cf. 20·3–23·9, M = 22·0%), by its slightly shorter snout (29·0–31·0, M = 30·0% head, cf. 28·6–35·5, M = 33·0%), deeper cheek (17·0–25·0, M = 21·0% head, cf. 15·8–21·5, M = 19·0%) slightly larger and more elliptical eye (28·0–33·0, M = 30·0% head, cf. 25·8–31·6, M = 28·7%), more steeply inclined mouth (30–35°, cf. 10–15°) and the greater number of teeth in the outer premaxillary row (70–80, M = 74, cf. 50–78, M = 68).

(vi) From *Haplochromis sulphureus* (see below, p. 148, Fig. 8), a species which it closely resembles in most morphometric features, by differences in head shape, especially its more steeply inclined mouth (30–35°, cf. 10–15°), its larger scales between the dorsal fin origin and the lateral line (5½–6, cf. 6–7½, mode 7) and its larger chest scales (5½–7, mode 6, cf. 7–8, mode 8, scales between the pectoral and pelvic fin bases).

The two species also differ in live coloration and in their habitats, with *H. sulphureus* apparently confined to deeper water (see p. 150 below).

(vii) From *Haplochromis plutonius* (see p. 151 and Fig. 11 below) by its slightly longer snout (29·0–31·0, M = 30·0% head, cf. 27·0–30·0, M = 29·0%) and lower jaw (41·0–50·0, M = 47·0% head, cf. 41·0–46·0, M = 44·0%). The two species differ markedly in the live colours of adult males (cf. pp. 144 and 153), a difference that is also reflected in the darker coloration of preserved specimens.

Little can be said about the phyletic relationships of *H. crocopeplus* within the 'tridens' species complex, except to note that its overall neurocranial morphology (see p. 144) suggests a less derived condition than that of the other species (remembering, of course, that the cranial osteology of *H. plutonius* is still unknown; see p. 153), as does the relatively slight beak on the premaxilla (see p. 144). The well-developed flare on the dentary (p. 144), however, is probably a derived condition and one shared with at least 5 species of the group (viz. *H. cryptogramma, H. dolichorhynchus, H. tridens* and *H. sulphureus*.

STUDY MATERIAL AND DISTRIBUTION RECORDS

Museum and Reg. No.	Locality	Collector
	TANZANIA	
BM(NH) 1977.1.10:70 (Holotype)	Speke Gulf, between Nafuba and Tefu Islands (28 m, mud)	Anker & Barel
BM(NH) 1977.1.10:71–87 (Paratypes)	Speke Gulf, between Nafuba and Tefu Islands (28 m, mud)	Anker & Barel
BM(NH) 1977.1.18:38 (Paratype)	Entrance of Mwanza Gulf (14m, sand)	Anker & Barel
BM(NH) 1977.1.28:39 (Paratype)	Speke Gulf, N.E. of Tefu Island (28 m, mud)	Anker & Barel

Haplochromis sulphureus sp. nov.

HOLOTYPE. An adult and sexually active male, 97·0 mm S.L., trawled over a mud bottom at a depth of *c*. 57 m in Ugandan waters at 0°45′ S, 32°38′ E. BM(NH) reg. no. 1977.1.10:106.

The trivial name (from the Latin) refers to the sulphur-yellow adult coloration in both sexes.

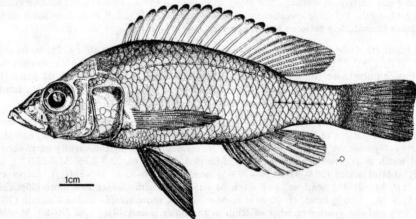

Fig. 8 *Haplochromis sulphureus*. Holotype. Drawn by M. J. P. van Oijen.

DESCRIPTION (Figs 8–10). Based on 22 specimens (including the holotype), 90·0–109·0 mm S.L.

Depth of body 29·0–35·0 (M = 32·5) % of standard length, length of head 33·0–37·0 (M = 35·0) %.

Dorsal head profile straight or slightly decurved, sloping at an angle of 30–35°; cephalic lateral line pores prominent, the pores and tubules of the preorbital bone especially so.

Preorbital depth 13·0–18·0 (M = 15·0) % of head, least interorbital width 16·0–19·0 (M = 17·0) %. Snout as long as broad to slightly longer than broad (1·1 times), its length 29·0–33·0 (M = 31·0) % of head. Orbit and eye slightly elliptical, the eye with a narrow but definite anterior and antero-ventral aphakic aperture; eye diameter 28·0–33·0 (M = 31·0) % of head. Depth of cheek 18·0–21·0 (M = 19·0) %.

Caudal peduncle 15·0–20·0 (M = 18·0) % of standard length, 1·3–1·8 (modal range 1·5–1·6) times longer than deep.

Mouth inclined at an angle of 10–15°, the posterior tip of the maxilla reaching a vertical through the anterior orbital margin or slightly beyond that level; lower jaw projecting beyond the upper anteriorly. Premaxilla with a moderately to well-developed antero-medial expansion of its dentigerous arm, giving the bone a distinctly beaked appearance; the dentary has a prominent mental protuberance. Length of lower jaw 44·0–50·0 (M = 47·0) % of head, 1·8–2·6 (modal range 2·0–2·3) times its width.

Gill rakers. 9 or 10 on the lower part of the first gill arch, the lower 1–3 rakers reduced, the next 2 or 3 relatively slender, and the remainder flattened and branched, often anvil-shaped.

Scales. Ctenoid; lateral line with 32 (f.4), 33 (f.11), 34 (f.4) or 35 (f.2) scales. Cheek with 3 (mode) or 4 rows. Six and a half (mode) or 7 scales between the lateral line and the dorsal fin origin, 7–8 between the pectoral and pelvic fin bases.

Fins. Dorsal with 23 (f.3), 24 (f.12) or 25 (f.7) rays, comprising 15 (f.17) or 16 (f.5) spines and 8 (f.4), 9 (f.15) or 10 (f.3) branched rays. Anal with 10 (f.1), 11 (f.14) or 12 (f.7) rays, comprising 3 spines and 7 (f.1), 8 (f.14) or 9 (f.7) branched rays. Pectoral fin $25 \cdot 5 – 31 \cdot 0$ (M = $29 \cdot 0$)% of standard length. Pelvic fins with the first branched ray produced in both sexes. Caudal truncate, scaled on its proximal third (rarely) to half.

Teeth. The *outer row* in *both jaws* commonly has an admixture of bicuspid (some weakly so), unicuspid and tricuspid teeth; in a few specimens no tricuspids are present, and in others unicuspids are absent. Tricuspid teeth are generally confined to the lateral and posterior parts of the tooth row, with only bi- and unicuspids occurring anteriorly. Irrespective of cusp type, the teeth have compressed and slightly recurved crowns, and near-cylindrical necks.

There are 70–82 (M = 76) teeth in the outer row of the upper jaw.

Teeth in the *inner rows* of *both jaws* are small, tricuspid, compressed and somewhat obliquely implanted; those of the upper jaw are arranged in 2 (mode) or 3 series, and those of the lower jaw in 1 or 2 (mode) series.

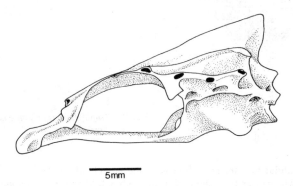

Fig. 9 *Haplochromis sulphureus.* Neurocranium in left lateral view.

Osteology. The neurocranium (Fig. 9) is of the typical '*tridens*'-group type, with a low supra-occipital crest and a relatively protracted preorbital (ethmovomerine) face.

The lateral face of the *dentary* is markedly flared, the resulting concavity extending forward to the symphysial area, the lower part of which is produced into a noticeable mental process.

The *premaxilla* is moderately beaked and its ascending processes (pedicels) are a little shorter than the dentigerous arms.

The *lower pharyngeal bone* (Fig. 10) is slender, with a narrow dentigerous surface that is slightly broader than long. When compared with the lower pharyngeal bone in other members of the '*tridens*' group, that of *H. sulphureus* appears to be relatively broader. The *lower pharyngeal teeth* are slender, compressed and cuspidate, and are arranged in from 28 to 30 rows.

There are 30 vertebrae (excluding the fused PU_1 and U_1 centra), comprising 13 abdominal and 17 caudal elements, in the 11 specimens radiographed.

Coloration. In life only slight differences exist between the colours of adult males and females, although it must be remembered that information on live coloration was obtained from fishes that had been in a trawl net for as long as half an hour before they were examined.

The ground colour is a bright sulphur yellow, shading to silvery-white on the belly, and darkening to near olivaceous on the dorsum and upper flanks. In most *males* the chest, belly, lower jaw and branchiostegal membrane are sooty.

The dorsal and caudal fins are a deep yellow, the former with black lappets and a black margin to the soft part of the fin. In *females* the anal and pelvic fins are also deep yellow, but in *males* the anal has a sooty overlay and the pelvics are black. The anal ocelli (egg dummies) of males are yolk-yellow in colour.

Preserved coloration. Adult males have a uniformly bright yellow-brown ground coloration except for some individuals in which the chest and belly are darker and may even be sooty. No trace of vertical or horizontal bars is visible on the body in most specimens but a few do show either extremely faint traces of 4 or 5 vertical bars on the flanks, or a faint, interrupted dark midlateral stripe.

The dorsal aspect of the snout is a very dark brown, as is the lower jaw and the vertical limb of the preoperculum. There is no distinct lachrymal bar, although that region of the cheek and snout shows a diffuse darkening. The branchiostegal membrane is sooty in most fishes but is a very light yellowish-brown in others; this feature, like the sooty chest and belly, is probably correlated with the degree of sexual activity.

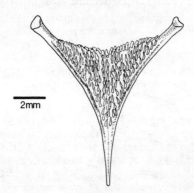

Fig. 10 *Haplochromis sulphureus*. Lower pharyngeal bone in occlusal view.

The dorsal and caudal fins are greyish-yellow, the lappets of the former are black; the anal fin is yellowish-brown (lighter than the body) but with a sooty overlay in some individuals, the colour intensifying to fully black over the spinous part of the fin. Pelvic fins sooty to black, the intensity being directly correlated with the degree of darkening manifest on the ventral aspects of the body. Pectoral fins are yellowish-hyaline.

Adult females are almost uniformly light yellowish-brown, but are slightly darker on the dorsum and much darker on the dorsal aspects of the snout. A very faint and narrow midlateral stripe is present on at least the posterior third of the body, and may extend further anteriorly.

The dorsal and caudal fins are greyish-yellow, the lappets of the dorsal black. The anal and pelvic fins are yellow, and the pectorals greyish-yellow.

ECOLOGY. *Habitat*. The five stations from which *H. sulphureus* were obtained are in the northern part of Lake Victoria (0°38′–0°50′ S); the water is from *c*. 16–20 m deep, and the bottom of soft mud.

FOOD. Of the 15 specimens examined, two contained unidentifiable sludge, one the remains of dipteran pupae and some fragments of unidentifiable crustaceans, one the remains of dipteran larvae and some fragmentary crustacean remains, another the remains of both dipteran larvae and pupae, and two only fragments of unidentifiable crustaceans.

BREEDING. No information is available on the reproductive habits of *H. sulphureus*. All the 22 specimens available are adult and most are sexually active; females appear to reach a larger adult size than do males. Sexually active females may have the left ovary much better developed and larger than the right one (the usual condition), the ovaries may be of equal size or, as in one fish, the right ovary may be larger than the left one.

DIAGNOSIS AND AFFINITIES. In addition to interspecific differences in its adult male coloration, *H. sulphureus* can be distinguished from other members of the '*tridens*' group as follows:

(i) From *H. tridens* (see Greenwood, 1967 : 97, fig. 20) by its less steeply sloping dorsal head profile (30–35°, cf. 40–45°), shallower preorbital (13·0–18·0, M=15·0% head, cf. 16·0–21·0, M=17·0%), smaller scales between pectoral and pelvic fin bases (7–8 cf. 5–6½ (mode)) and by its stouter lower jaw.

(ii) From *H. dolichorhynchus* (see Greenwood & Gee, 1969 : 34, fig. 21) by its more obviously elliptical eye, the absence in preserved female specimens of a distinct midlateral stripe, and by the smaller scales between the dorsal fin origin and the lateral line (6–7½, mode 7, cf. 5–6½, mode 5½). Although there is considerable interspecific overlap in all morphometric characters, the mean values of certain features in *H. sulphureus* indicate that this species does have a deeper caudal peduncle, a deeper preorbital, a larger eye and a shorter snout. (The means for the three latter characters, expressed as a percentage of head length, are: 15·0 cf. 18·0, 31·0 cf. 27·0 and 31·0 cf. 34·0 respectively).

(iii) From *H. chlorochrous* (see Greenwood & Gee, 1969 : 44, fig. 27) by its less oblique mouth (10–15°, mode 10°, cf. 20–35°, mode 30°), less prominent premaxillary pedicels, smaller scales between the dorsal fin origin and the lateral line (6–7½, mode 7, cf. 5½–6½, mode 6), shallower cheek (18·0–21·0, M=19·0% head, cf. 20·0–25·0, M=22·6%) and larger eye (28·0–33·0, M=31·0% head, cf. 25·4–29·0, M=28·2%).

(iv) From *H. tyrianthinus* (see Greenwood & Gee, 1969 : 40, fig. 25) by its less decurved dorsal head profile, smaller scales between the dorsal fin origin and the lateral line (6–7½, mode 7, cf. 5–6, mode 5½), shallower cheek (18·0–21·0, M=19·0% head, cf. 20·0–25·3, M=22·9%) and its larger eye (28·0–33·0, M=31·0% head, cf. 26·1–29·3, M=27·7%).

(v) From *H. cryptogramma* (see Greenwood & Gee, 1969 : 48, fig. 30) by the absence of distinctive midlateral markings of blotches and bands, by its much less prominent premaxillary pedicels, a convex or straight dorsal head profile (cf. a markedly concave one), an elliptical orbit, smaller scales between the dorsal fin origin and the lateral line (6–7½, mode 7, cf. 5–6½, mode 5½), and a narrower interorbital width (16·0–19·0, M=17·0% head, cf. 20·3–23·9 M=22·0%).

For features distinguishing *H. sulphureus* from *H. crocopeplus* and *H. plutonius* see pp. 147 and 155 for the species respectively.

The phyletic relationships of *H. sulphureus* within the *H. tridens* lineage as currently conceived cannot be determined precisely; there are some indications (especially from its coloration) that the species may be most closely related to *H. crocopeplus*.

STUDY MATERIAL AND DISTRIBUTION RECORDS

Museum and Reg. No.	Locality	Collector
	UGANDA	
BM(NH) 1977.1.10:106 (Holotype)	0°45′ S, 32°38′ E	P. H. Greenwood
BM(NH) 1977.1.10:107–108 (Paratypes)	0°44′ S, 32°30′ E	P. H. Greenwood
BM(NH) 1977.1.10:109–116 (Paratypes)	0°50′ S, 32°35′ E	P. H. Greenwood
BM(NH) 1977.1.10:117–121 (Paratypes)	0°39′ S, 32°35′ E	P. H. Greenwood

Haplochromis plutonius sp. nov.

HOLOTYPE. An adult male, 93·0 mm S.L., from a trawl made over a mud bottom in water *c.* 28 m deep, between Nafuba and Tefu Islands, Speke Gulf. BM(NH) reg. no. 1977.1.10:39.

The trivial name (from the Latin) refers to the dusky preserved coloration of adult males.

DESCRIPTION (Figs 11–13). Based on 10 specimens (including the holotype) 75·0–96·0 mm S.L.

Depth of body 29·0–33·0 (M=31·0)% of standard length, length of head 34·0–37·0 (M=35·0)%.

Dorsal profile of head straight to a point above the preoperculum, then gently decurved, sloping at an angle of *c.* 30°, the premaxillary pedicels are prominent. The cephalic lateral line pores are enlarged, the pores and tubules of the preorbital bone and dentary being especially obvious.

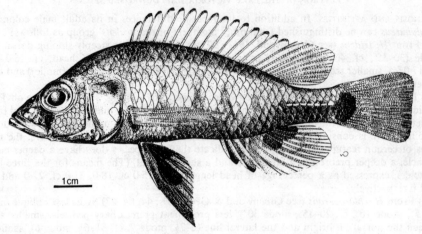

Fig. 11 *Haplochromis plutonius*. Holotype. Drawn by M. J. P. van Oijen.

Preorbital depth 15·0–17·0 (M = 16·0) % of head, least interorbital width 15·0–18·0 (M = 16·0) %. Snout as long as broad or very slightly longer than broad (1·1 times), its length 27·0–30·0 (M = 29·0) % of head. Eye and orbit distinctly elliptical, the eye with a well-developed anterior and anteroventral aphakic aperture; greatest diameter of eye 31·0–33·0 (M = 32·0) % of head. Cheek depth 16·0–21·0 (M = 19·0) %.

Caudal peduncle 1·4–1·8 (mode 1·6) times longer than deep, its length 16·0–19·0 (M = 18·0) % of standard length.

Mouth inclined at an angle of 10–15°; posterior tip of the maxilla reaching a vertical through the anterior margin of the eye or slightly beyond that level. Premaxilla with a moderately developed beak. Lower jaw projecting slightly beyond the upper, and with a moderately developed mental protuberance; length of lower jaw 41·0–46·0 (M = 44·0) % of head, 1·9–2·4 (modal range 2·0–2·1) times greater than its width.

Gill rakers. 8–10 (mode) on the lower part of the first gill arch, the lowermost 1–3 rakers reduced, the uppermost 3 or 4 flattened and usually anvil-shaped, the remaining rakers simple and moderately slender.

Scales. Ctenoid; lateral line with 32 (f.1), 33 (f.7) or 35 (f.1) scales, cheek with 3 or 4 rows. Five and a half to 7 (modal range 6–7) scales between the lateral line and the dorsal fin origin, 6–7 between the pelvic and pectoral fin bases.

Fins. Dorsal fin with 24 (f.9) or 25 (f.1) rays, comprising 15 (f.9) or 16 (f.1) spinous and 9 (f.9) or 10 (f.1) branched elements, anal with 11 (f.4) or 12 (f.6) rays, comprising 3 spines and 8 (f.4) or 9 (f.6) branched rays. Pectoral fin 26·0–30·0 (M = 29·0) % of standard length. Pelvics with the first branched ray moderately produced in both sexes. Caudal truncate, scaled on its basal half in most specimens, but not quite so extensively in a few others (only on the proximal third in one fish).

Teeth. The *outer row* in *both jaws* contains an admixture of bi- and tricuspid teeth; the teeth, irrespective of cusp shape, have compressed and slightly recurved crowns, and near-cylindrical necks. In most specimens the bicuspid teeth are situated anteriorly in the jaws, the lateral and posterolateral teeth being either all tricuspid or a mixture of tri- and bicuspids in which the tricuspids predominate. The exceptional individuals have a mixture of bi- and tricuspids anteriorly (the latter type predominating), although one specimen has only bicuspids in the outer row of both jaws.

There are 66–78 (modal range 70–74) teeth in the outer row of the upper jaw.

The *inner teeth* of *both jaws* are invariably small and tricuspid, are implanted obliquely and are arranged in 1 or 2 (mode) rows.

Osteology. Because so few specimens of *H. plutonius* are available, no complete skeleton has been prepared. Superficial dissection shows that the dentigerous surface of the *premaxilla* is moderately expanded medially, giving the bone a fairly definite beaked appearance. The dentigerous surface of the *dentary* is flared outward so that the lateral face of the bone is distinctly concave, with the concavity extending to the symphysial region of the bone.

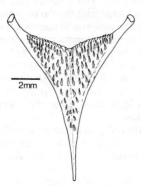

Fig. 12 *Haplochromis plutonius.* Lower pharyngeal bone in occlusal view.

The *lower pharyngeal bone* is noticeably narrow, slender and elongate (see Fig. 12); its dentigerous surface is either as broad as it is long or it may be slightly longer than broad. The *lower pharyngeal teeth* are fine, compressed and cuspidate, and are arranged in about 24 rows.

There are 29 (f.1), 30 (f.7) or 31 (f.1) vertebrae (excluding the fused PU_1 and U_1 centra) comprising 12 (f.1) or 13 (f.8) abdominal and 17 (f.8) and 18 (f.1) caudal elements.

Coloration. In *life* an *adult, sexually active male* (BM(NH) 1977.1.10:39) (see Fig. 13) has a purple ground coloration, with the ventral aspect of the flanks yellowish-grey, the chest and belly are dark grey to black, and the ventral part of the caudal peduncle a very dark grey. The head is purple except for a whitish opercular region, and a pinkish colour on the anterodorsal angle of the operculum and ventral preopercular limb; the branchiostegal membrane in sooty.

There is a faint lachrymal stripe, and three faint vertical bars on the flanks.

The dorsal fin is hyaline, with a grey base and broad, bright red streaks; the lappets are dark grey.

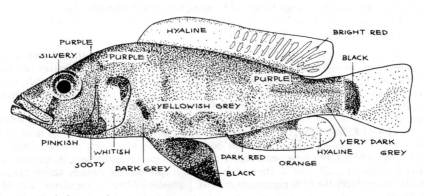

Fig. 13 *Haplochromis plutonius.* Adult male (sexually active), to show coloration. Drawn by M. J. P. van Oijen.

The anal fin has a red flush, becoming most intense on the anterior part of the fin (which may even appear to be black); the egg dummies are orange. The pelvic fins are black, the pectorals hyaline. The caudal fin is black basally, the dorsal half yellowish anteriorly but overlain by a red flush which intensifies over the ventral half of the fin.

The colours of *live females* are unknown.

Preserved coloration. Adult males. The dorsum and upper two-thirds of the flanks are dark brown, the belly and ventral aspects of the flanks dusky to sooty with a silvery-grey underlay. At least 4, sometimes 5 or 6, dark bars cross the flanks and merge with the darker ventral body colour (which is less intense than that of the bars); dorsally, the bars extend only to a level about two scale rows below the upper lateral line scale row. In some specimens there are very faint indications of an interrupted midlateral band, especially on the posterior part of the body.

The dorsal surface of the snout is a very dark brown (almost black); the branchiostegal membrane and the vertical limb of the preoperculum are black, the cheek brownish over silver, the operculum silver with a diffuse dusky overlay. A fairly distinct lachrymal stripe, of variable intensity, runs almost vertically, or with a slight anterior inclination, from the lower orbital margin to merge with the dark pigmentation of the lower jaw.

The dorsal fin is greyish to dusky, the lappets black and the soft part of the fin with dark spots and streaks between its rays. The caudal is dusky, darkest along its middle; the anal too is dusky, but with a black basal band and black pigment between the spines. The pelvic fins are black, and the pectorals faintly greyish.

Adult females. The dorsum is light brown, the remainder of the body silvery white; there are faint indications of a narrow midlateral stripe, most clearly discernible on the posterior half of the body.

The dorsal surface of the snout is very dark brown, the operculum is silvery and, save for a dark blotch anteroventrally to the orbit, there is no lachrymal stripe.

The dorsal and caudal fins are faintly sooty, the anal and pelvics hyaline, and the pectorals a faint grey.

ECOLOGY. *Habitat.* The species is known only from one locality in the Speke Gulf; the bottom is mud and the depth about 28 m. It is presumed that the specimens were caught while the trawl was fishing on the bottom.

FOOD. All 9 fishes examined had the entire alimentary tract filled with flocculent organic debris (decomposing blue-green algae). Since this type of material is typical of the mud-water interface we suspect that it was ingested while the specimens were caught in the trawl (see also pp. 146 & 168), and thus that it may not represent the natural food of *H. plutonius*.

The intestine of *H. plutonius* is relatively long (2 times standard length) and much coiled, suggesting that a certain amount of plant material may be part of the normal diet.

BREEDING. No information is available on the breeding habits of *H. plutonius*. All the 10 specimens examined were adult and sexually active. Of the two females represented in the sample, one has both ovaries equally developed, the other has only the right ovary enlarged.

DIAGNOSIS AND AFFINITIES. From *H. dolichorhynchus*, *H. tyrianthinus*, *H. chlorochrous* and *H. cryptogramma*, *Haplochromis plutonius* is differentiated by, amongst other features, its larger eye (31·0–33·0, M=32% head) and shorter snout (27·0–30·0, M=29·0% head); from *H. cryptogramma* it is also distinguished by the absence of a conspicuous broad and interrupted midlateral stripe, and by its straight, as opposed to markedly concave, dorsal head profile.

From *H. tridens*, *H. plutonius* is differentiated by its shorter lower jaw (41·0–46·0, M=44·0% head, cf. 43·3–51·8, M=47·5%), its less steeply sloping but more concave dorsal head profile (*c.* 30° cf. 40–45°) and its slightly shallower body (29·0–33·0, M=31·0% standard length, cf. 30·1–36·2, M=33·5%). The preserved coloration of the two species also differs. When males with testes in a morphologically similar state of development (presumably sexually active) are compared, *H. tridens* lacks the dark pigmentation of *H. plutonius* (*H. tridens* are silvery, with black pelvic fins).

From *H. crocopeplus*, *H. plutonius* differs in the live coloration of its males and by having a slightly shorter lower jaw (41·0–46·0, M = 44·0 % head, cf. 41·0–50·0, M = 47·0 %) and less oblique mouth (10–15°, cf. 30–35°, mode 35°).

From *H. sulphureus*, *H. plutonius* differs in its somewhat shorter snout (27·0–30·0, M = 29·0 % head, cf. 29·0–33·0, M = 31·0 %) and lower jaw (41·0–46·0, M = 44·0 % head, cf. 44·0–50·0, M = 47·0 %); live adult male coloration is also diagnostic.

The phyletic affinities of *H. plutonius* within the '*tridens*'-group cannot yet be determined.

STUDY MATERIAL AND DISTRIBUTION RECORDS

Museum and Reg. No.	Locality	Collector
	TANZANIA	
BM(NH) 1977.1.10:39 (Holotype)	Speke Gulf, N.E. of Tefu Island	Anker & Barel
BM(NH) 1977.1.10:40–48 (Paratypes)	Speke Gulf, N.E. of Tefu Island	Anker & Barel

Comments on the *Haplochromis tridens* species complex

The addition of three further species to this complex, now totalling 8 species, enables one to give a more precise definition of the lineage than hithertofore, and to review its phyletic relationships (see Greenwood, 1974 for a preliminary analysis).

One of the most diagnostic group features is, of course, the presence of several to many tricuspid teeth in the outer row of at least one and usually both jaws, these teeth not being confined to an extreme posterior position in the row. The tricuspids are sufficiently numerous, and of a size comparable with their bi- and unicuspid congeners, to exclude the possibility of their merely being teeth from the inner series that have moved outwards to fill the gaps caused by the loss of true outer row teeth.

In other Lake Victoria *Haplochromis* species a few tricuspid teeth may occur posteriorly in the outer series of the lower jaw or, less commonly, posteriorly in the upper jaw. These tricuspids are never as numerous as those in members of the '*tridens*' group, and when they occur in the lower jaw are noticeably smaller than the teeth situated laterally and anteriorly. Often the tricuspids are clearly displaced elements of the inner tooth series. Very occasionally one or two tricuspid teeth are found elsewhere in the outer rows of non-'*tridens*'-group *Haplochromis* species, but as mentioned above, these are never so numerous as are the tricuspids in '*tridens*' species.

The tricuspid outer teeth in members of the '*tridens*'-group can be considered a derived (and autapomorph) feature, although their functional significance (if any) remains unknown.

Other apomorph, but not necessarily autapomorph features shared by all members of the group are apparent in the skeleton.

The skull has slender proportions, with a low otico-occipital (brain case) region, low supraoccipital crest, and a relatively elongate ethmovomerine region, all of which give the neurocranium a characteristic appearance (see Figs 3 & 9 above, and relevant figures in Greenwood, 1974).

The preorbital bone has greatly enlarged lateral line canals and pores (Fig. 8) and, especially characteristic, a large, nearly rectangular bullation occupying almost the entire anterior portion of the bone between its margin and the first lateral line tubule. Apparently this outpocketing is associated with the relatively enlarged dorsal articular head of the maxilla, which it overlies. Enlarged preorbital lateral line tubules are, of course, found in many *Haplochromis* species, especially those inhabiting deeper or turbid waters, but the extensive preorbital bulla seems to be an autapomorphic feature of the '*tridens*'-group. (Some species belonging to other groups, e.g. *Haplochromis nanoserranus* of the *Haplochromis serranus* lineage, see p. 158 below, also have an anterior bullation of the preorbital. However, it is always relatively smaller, appears more circumscribed and is approximately circular in outline.)

The dentary in '*tridens*'-group species has a very characteristic shape (Fig. 4), low and slender but with the coronoid region rising steeply to meet the deep anguloarticular bone.

All members of the group have a narrow and slender lower pharyngeal bone (Fig. 5) with an elongate anterior blade and numerous fine, compressed and cuspidate teeth.

Ecologically, the *'tridens'* species appear to be members of the sublittoral to benthic community, the greater number of species occurring in water between 15 and 30 m deep. No clear picture has emerged yet of their feeding habits. Some species feed on pre-adult insects (especially Diptera) and adult crustaceans (see Greenwood, 1967; Greenwood & Gee, 1969; Greenwood, 1974), while others may be detritus feeders (see pp. 146 & 154 above); certain of these latter species have a relatively elongate and much coiled intestine, anatomical features often associated with that type of diet and feeding habit.

No member of the group can be considered to reach a large adult size, a standard length of 120 mm being the largest so far recorded (for *H. chlorochrous*, see Greenwood & Gee, 1969). Because of their habitat preferences and their small adult size, species of the *'tridens'* group have only been caught in small-mesh trawl nets; to the best of our knowledge none has been recorded from the catches of beach-operated seine nets or from commercially-operated set-nets.

On the basis of various derived characters shared by all known species of the *'tridens'* complex, a strong argument can be put forward for considering the group as a monophyletic assemblage within the Lake Victoria species flock. Some anatomical features (neurocranial morphology in particular) indicate close affinity with the *H. serranus* and *Haplochromis prognathus* lineages, probably as the sister group of the two latter lineages combined (see Greenwood, 1974). In his preliminary phyletic analysis of the Lake Victoria *Haplochromis* species, Greenwood (1974) also suggested the existence of a relationship between, on the one hand, the *'tridens'* group plus the lepidophagous *Haplochromis welcommei*, and on the other hand, the insectivorous–molluscivorous lineage comprising *Haplochromis riponianus*, *H. saxicola* and *H. aelocephalus* (the three lineages together forming the sister group to the combined *H. serranus* and *H. prognathus* lineages).

For the moment no further comments can be made about possible relationships between *H. welcommei* and the *'tridens'* group. However, taking into account relative specializations seen in the neurocranium of *'tridens'* species when compared with the less specialized neurocranial form of the *H. riponianus* group, and also taking into account the autapomorph features of the two groups (see Greenwood, 1974), an argument could be made against their having a recent common ancestry (but not against the *H. riponianus* group sharing more distant ancestry with both the *'tridens'* group and the *H. serranus–H. prognathus* group). In other words, the *'tridens'* group may share a more recent common ancestry with the *H. serranus–H. prognathus* lineage than with the *H. riponianus–H. aelocephalus* one.

Basically, neurocranial form in the *'tridens'* group is like that in the *H. serranus–H. prognathus* lineage, but is somewhat less specialized (see Greenwood, 1974); the lower pharyngeal bone and dentition, the form of the lower jaw, the dentition in both jaws, and the large preorbital bulla housing the enlarged maxillary dorsal articular process, however, are peculiarly *'tridens'* specializations.

Most adult fishes in the *'tridens'* lineage differ from those in the *'serranus'* group in having a narrow interorbital, a slightly to much shorter snout (*H. dolichorhynchus* is exceptional in this respect), a larger eye and a shallower cheek. Essentially the same features distinguish *'tridens'* group species from those of the *'prognathus'* line, although the intergroup differences in snout length and interorbital width are less pronounced.

Eye size and cheek depth are, in general, negatively correlated characters, and invariably eye size shows negative allometry with body length. It is thus the more unfortunate that, with few exceptions, we were unable to compare specimens of the *'tridens'* group with similar sized members of the *'serranus'* and *'prognathus'* groups. We would suggest, nevertheless, that the intergroup differences in eye and cheek proportions are probably a consequence of the very different modal adult sizes for the two groups, and that some factor controlling size at maturity may have been involved in their evolutionary histories.

We have been unable to find any features within the *'tridens'* group that can be used to establish intragroup phylogenies (a situation very familiar to the senior author amongst the more speciose lineages of Lake Victoria *Haplochromis*; see Greenwood, 1974).

It is still not possible to determine whether or not *Haplochromis arcanus* Greenwood & Gee, 1969, is a member of the *'tridens'* lineage. That none of the dental specializations found in

H. arcanus (especially the strongly incurved posterolateral teeth of the premaxilla) occurs in any of the new '*tridens*' species, seems to add further weight to the argument that *H. arcanus* is not a member of that group (see Greenwood & Gee, 1969). Its proximate relationship to the '*tridens*' lineage, through the shared common ancestry of that lineage with the *H. serranus–prognathus* line, still seems to be the most reasonable hypothesis.

When *H. dolichorhynchus*, *H. cholorochrous*, *H. tyrianthinus* and *H. cryptogramma* were first described, no radiographs could be made of the material and hence no vertebral counts were given for the species; this can now be rectified. As usual the fused PU_1 and U_1 centra are not included in the counts.

H. dolichorhynchus: 29 (f.1) or 30 (f.8) comprising 12 (f.1) or 13 (f.8) abdominal, and 16 (f.1), 17 (f.7) or 18 (f.1) caudal elements.

H. chlorochrous: 29 (f.3) or 30 (f.7), comprising 12 (f.1) or 13 (f.9) abdominal and 16 (f.3), 17 (f.6) or 18 (f.1) caudal elements.

H. tyrianthinus: 30 (f.6) or 31 (f.2), comprising 13 abdominal and 17 (f.6) or 18 (f.2) caudal elements.

H. cryptogramma: 29 (f.2), 30 (f.6) or 31 (f.2), comprising 12 (f.1) or 13 (f.9) abdominal and 16 (f.1), 17 (f.7) or 18 (f.2) caudal elements.

New species of the *H. serranus* group

Haplochromis nanoserranus sp. nov.

HOLOTYPE. An adult male 76·0, standard length, from the Mwanza Gulf, caught in a trawl shot near the eastern end of the Muranda peninsula and fished towards the northwestern point of Luansa Bay; substrate sandy mud, water depth *c*. 4–8 m. BM(NH) reg. no. 1977.1.10:54.

The trivial name is from the Latin *nanus*, a dwarf, and *serranus*, with reference to *H. serranus* (Pfeffer).

DESCRIPTION (Figs 14 & 15). Based on 6 specimens (including the holotype), 72·0–76·0 mm standard length. All specimens are adult males.

Depth of body 30·0–33·0 (M = 31·8) % of standard length, length of head 31·0–35·0 (M = 33·7) %.

Dorsal head profile straight, sloping at an angle of 30–35°; the snout profile, when viewed laterally, is noticeably acute and the premaxillary pedicels are prominent. The cephalic lateral line pores, except those of the preorbital and dentary, are not noticeably enlarged; the pre-

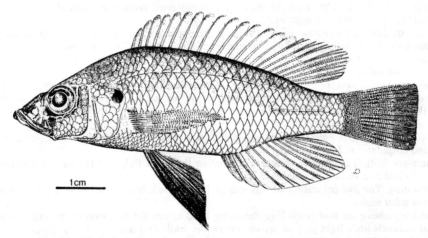

Fig. 14 *Haplochromis nanoserranus*. Holotype. Drawn by M. J. P. van Oijen.

orbital lateral line tubules are as obvious as those of species in the *H. tridens* group and there is also a small anterior bullation of that bone (see above p. 155).

Preorbital depth 16·0–18·0 (M = 17·0) % of head, least interorbital width 16·0–18·0 (M = 17·0) %. Snout varying from a little broader than long to slightly longer than broad (1·1 times), its length 29·0–31·0 (M = 30·2) % of head. The eye and orbit are noticeably elliptical, the former with a fairly well-developed anterior and anteroventral aphakic aperture; greatest diameter of eye 25·0–32·0 (M = 29·0) % of head. Cheek depth 17·0–22·0 (M = 19·3) %.

Caudal peduncle 1·6–1·9 (mode 1·6) times longer than deep, its length 18·0–21·0 (M = 19·0) % of standard length.

Mouth moderately oblique, inclined at an angle of 20–35° (mode *c.* 30°); posterior tip of the maxilla generally reaching a vertical through the anterior part of the eye, but sometimes only reaching a vertical through the anterior orbital margin. Premaxilla with its dentigerous arm somewhat expanded anteroposteriorly in the midline, giving the bone a moderately beaked appearance, jaws equal anteriorly. Dentary with a fairly prominent mental process. Lower jaw 2·3–2·8 times longer than broad, its length 45·0–52·0 (M = 47·0) % of head.

Gill rakers. 9 or 10 on the lower part of the first gill arch, the lower 1–3 rakers reduced, the remainder variously shaped but usually slender, except for the uppermost 2 or 3 which are generally flattened and either bifid or anvil-shaped.

Scales. Ctenoid; lateral line with 32 (f.2) or 33 (f.4) scales, cheek with 3 or 4 rows. Six or 6½ scales between the lateral line and the dorsal fin origin, 6–7 (mode 6½) between the pectoral and pelvic fin bases.

Fins. Dorsal with 24 (f.1) or 25 (f.5) rays, comprising 14 (f.1), 15 (f.3) or 16 (f.2) spinous and 8 (f.1), 9 (f.2) or 10 (f.3) branched elements. Anal with 12 (f.6) rays comprising 3 spines and 9 branched rays. Pectoral fin 26·0–31·0 (M = 28·5) % of standard length. Pelvics with the first branched ray moderately to strongly produced. Caudal truncate, scaled on its basal half.

Teeth. In *both jaws* the majority of teeth in the *outer row* are slender, somewhat recurved and caniniform unicuspids. A few bi- and weakly tricuspid teeth occur posteriorly and posterolaterally in the lower jaw but none was found in the upper jaw.

The occurrence of a predominantly unicuspid and caniniform outer tooth row in such small fishes is most unusual (see Greenwood, 1974 : 106); for example, in *Haplochromis pellegrini*, the only other member of the *H. serranus–H. prognathus* species complex with small adults, fishes less than 85 mm S.L. usually have a predominance of bi- and weakly bicuspid teeth in the outer row, and only a few unicuspids present anteriorly in the jaws.

The *inner teeth*, which are implanted obliquely, may all be tricuspids, or a mixture in which tricuspids predominate over unicuspid and weakly tricuspid teeth, or even one in which unicuspids predominate. The tricuspid teeth have compressed crowns but cylindrical necks, the unicuspids are somewhat compressed.

There are 2 or, rarely, 3 rows of inner teeth in the upper jaw, and 1 or 2 rows in the lower jaw.

Osteology. No complete skeleton is available, and but little information about the details of neurocranial architecture could be obtained from radiographs. The supraoccipital crest (at least as compared with that in specimens of the *H. tridens* group) is relatively high.

Superficial dissection shows that the *preorbital* bone has a small and clearly circumscribed bulla near its anterior border, and that the dentigerous surface of the dentary is flared outwards so that the lateral face of the bone is markedly concave; the concavity does not, however, extend forward to the symphysial region.

The *lower pharyngeal bone* (Fig. 15) has its dentigerous surface broader than long; its teeth are fine and cuspidate, and are arranged in about 28 rows.

There are 29 (f.1) or 30 (f.5) vertebrae (excluding the fused PU_1 and U_1 centra), comprising 13 abdominal and 16 or 17 caudal elements.

Coloration. The live coloration of this species is unknown, and *preserved colours* are known only for *adult males*.

The body above the midlateral line, the entire head except for the operculum, and the entire caudal peduncle are a light greyish-brown. Below the midlateral line (i.e. on the chest, belly and ventral flanks) the colour changes to silvery grey with, in a few specimens, a darker, almost dusky chest region. Some specimens have a broad, but faintly indicated midlateral band which is interrupted at about its midpoint and becomes broader over its posterior half. This band appears to extend onto the caudal fin (whose middle portion may be darker than the rest of the fin even in specimens lacking a midlateral stripe). The operculum is silvery (except for a typical opercular spot in its posterodorsal angle), the dorsal and anterolateral aspects of the snout are dusky, as are the median and mediolateral aspects of the upper lip, and there is a faint and relatively narrow lachrymal stripe running onto the lower jaw behind the posterior tip of the maxilla.

The dorsal and caudal fins are greyish, the membrane of the soft dorsal sometimes weakly maculate. The anal is hyaline to greyish, its ocelli (egg dummies) dead white. The pelvic fins are black, and the pectorals hyaline.

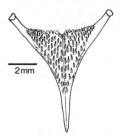

Fig. 15 *Haplochromis nanoserranus*. Lower pharyngeal bone in occlusal view.

ECOLOGY. *Habitat*. The specimens on which this description is based are all from shallow (*c.* 4–8 m) offshore waters, and were caught over a muddy sand substrate. (The senior author recalls examining specimens of a similar and probably identical taxon caught in similar habitats in the northern and eastern regions of the lake; regrettably, this material was lost in transit from east Africa to Britain.)

FOOD. One of the specimens examined had the remains of a small cichlid fish in its intestines; two others contained fragments of larval insects (in one fish larval Diptera, in the other what appeared to be the remains of a larval boring mayfly, *Povilla adusta*), and the remaining two fishes yielded only an unidentifiable sludge in both the stomach and intestines.

BREEDING. Apart from the fact that all 6 specimens are small (72–76 mm S.L.) and are sexually active males, nothing is known about the reproductive habits of this species.

DIAGNOSIS AND AFFINITIES. At first sight, specimens of *H. nanoserranus* closely resemble members of the *H. tridens* species complex. However, detailed examination shows that, unlike '*tridens*' species, *H. nanoserranus* has only unicuspid teeth in the outer series of the upper jaw, and a mixture of unicuspid and weakly bicuspid teeth in the lower jaw. Furthermore, the unicuspid teeth in *H. nanoserranus* are of the slender, near-cylindrical and caniniform type found in piscivorous predators of the *H. serranus–H. prognathus* lineage (see Greenwood, 1974), and not the more flattened, angular type characteristic of the '*tridens*' group. Also, in *H. nanoserranus* the preorbital bone has only a small and well-circumscribed, nearly circular bulla, unlike the larger and vertically more elongate bulla of the '*tridens*' type (see above, p. 155); the lower pharyngeal bone in *H. nanoserranus* (see Fig. 15) has not the slender and elongate form so characteristic of the '*tridens*' group (cf. Fig. 15 and Fig. 5).

Unfortunately, no details are available on the syncranial architecture of *H. nanoserranus*, but judging from radiographs its neurocranium has essentially the outline and proportions of an *H. serranus*-group fish rather than the lower and more elongate type found amongst members of the '*tridens*' group (see Greenwood, 1974 and p. 155 for a discussion of these neurocranial features).

Thus, at least for the moment, we are placing *H. nanoserranus* in the '*serranus*' subdivision of the *H. serranus–H. prognathus* lineage of Greenwood (1974), but noting that it does show, at least incipiently, certain features seen in members of the *H. tridens* species complex.

When making comparisons between *H. nanoserranus* and members of the *H. serranus* group we were hampered by the fact that very few small specimens of species in that complex have been described or are available for study (see Greenwood, 1962 & 1967). Consequently the small but adult specimens of *H. nanoserranus* had to be compared with much larger and often juvenile specimens of the '*serranus*' group. If, as seems most likely, some of the diagnostic features we used are subject to allometric growth, then small specimens of '*serranus*' group species may resemble *H. nanoserranus* more closely than we realize at present.

From *H. serranus* itself (see Greenwood, 1962 : 152, figs 4 & 5), *H. nanoserranus* differs in having a longer and more slender caudal peduncle (18·0–21·0, M = 19·0% standard length, cf. 13·0–19·0, M = 15·4%, and 1·6–1·9, mode 1·6, times longer than deep, cf. 1·1–1·5, mode 1·2 times), a shorter head (31·0–35·0, M = 33·7% S.L., cf. 34·8–38·7, M = 36·3%), a narrower interorbital (16·0–18·0, M = 17·0% head, cf. 20·4–26·8, M = 23·3%), a larger eye (25·0–32·0, M = 29·6% head, cf. 20·4–26·0, M = 23·3%), a shallower cheek (17·0–22·0, M = 19·3% head, cf. 22·9–31·5, M = 27·5%) and a shorter lower jaw (45·0–52·0, M = 47·0% head, cf. 48·0–60·0, M = 54·3%).

From *Haplochromis victorianus* (see Greenwood, 1962 : 156, pl. 1) it differs in its shallower body (30·0–33·0, M = 31·8% S.L., cf. 33·4–41·3, M = 37·3%), narrower interorbital (16·0–18·0, M = 17·0% head, cf. 21·5–24·5, M = 22·6%), shorter snout (29·0–31·0, M = 30·2% head, cf. 31·8–36·0, M = 34·0%), larger eye (25·0–32·0, M = 29·0% head, cf. 21·7–25·5, M = 23·6%) and a shallower cheek (17·0–22·0, M = 19·3% head, cf. 22·5–26·2, M = 24·6%).

From *Haplochromis maculipinna* (see Greenwood, 1967 : 43, fig. 3) it is differentiated by its shallower body (30·0–33·0, M = 31·8% S.L. cf. 33·3–37·0, M = 35·9%), longer and shallower caudal peduncle (18·0–21·0, M = 19·1% S.L., cf. 14·5–18·8, M = 16·3%, and 1·6–1·9, mode 1·6, times longer than deep, cf. 1·2–1·8, mode 1·1–1·2 times), narrower interorbital (16·0–18·0, M = 17·0% head, cf. 20·7–25·5, M = 22·8%), somewhat shorter snout (29·0–31·0, M = 30·2% head, cf. 30·3–37·0, M = 33·7%) and shallower cheek (17·0–22·0, M = 19·3% head, cf. 23·2–29·8, M = 25·3%).

It is interesting to note that the relative proportions of the eye diameter and lower jaw length are similar in the two species, despite the size discrepancy of the specimens examined.

From *Haplochromis boops* and *H. thuragnathus* (see Greenwood, 1967 : 47–51, fig. 4), *H. nanoserranus* differs in its much shallower body (30·0–33·0, M = 31·8% S.L., cf. 40·5–42·0 (no means given because *H. boops* and *H. thuragnathus* are known from so few specimens)), somewhat less steeply inclined dorsal head profile (30–35°, cf. 40–50°), narrower interorbital (16·0–18·0, M = 17·0% head, cf. 21·7–25·7%) and shallower cheek (17·0–22·0, M = 19·3% head, cf. 28·0–30·0%). The three species show complete overlap in the relative proportions of eye diameter and lower jaw length.

The three other previously known species of the *H. serranus* group (*Haplochromis plagiostoma*, *H. cavifrons* and *H. decticostoma*) are immediately distinguishable from *H. nanoserranus* on the basis of their gross morphology, especially their respective head shapes (compare Fig. 14 above with the figures of these species in Greenwood, 1962, and Greenwood & Gee, 1969 for *H. plagiostoma* and *H. cavifrons*, and *H. decticostoma* respectively).

From the other new and presumed member of the *H. serranus* complex, *Haplochromis cassius* (see below and Fig. 16). *H. nanoserranus* is readily distinguished by its finer, smaller and more numerous outer teeth, and by its less enlarged lips.

It is not yet possible to determine the phyletic relationships of *H. nanoserranus* within the *H. serranus* species group. The species does, however, seem to show the same morphological relationships with the congeners of its lineage as does *H. pellegrini* with its congeners in the *H. prognathus* lineage of the '*serranus–prognathus*' group (see Greenwood, 1974). In other words, it is a morphologically somewhat specialized 'dwarf' amongst a radiation of relative 'giants'.

STUDY MATERIAL AND DISTRIBUTION RECORDS

Museum and Reg. No.	Locality	Collector
	TANZANIA	
BM(NH) 1977.1.10:54 (Holotype)	East of Muranda peninsula towards the northwestern point of Luansa bay, Mwanza Gulf	Anker & Barel
BM(NH) 1977.1.10:55–59 (Paratypes)	East of Muranda peninsula towards the northwestern point of Luansa bay, Mwanza Gulf	Anker & Barel

Haplochromis cassius sp. nov.

HOLOTYPE. An adult female, 97·5 mm S.L., from the southern part of the Mwanza Gulf off Busissi, at a depth of 2 m, over a mud bottom. BM(NH) reg. no. 1977.1.10:49.

The trivial name derives from Shakespeare's 'Julius Caesar' (Act I, scene II) '... Yond Cassius has a lean and hungry look ...'.

We are well aware of the dangers inherent in describing new taxa of Lake Victoria *Haplochromis* from small and unisexual samples; but the peculiar dentition and enlarged lips of this species are so distinctive that we feel justified in our actions.

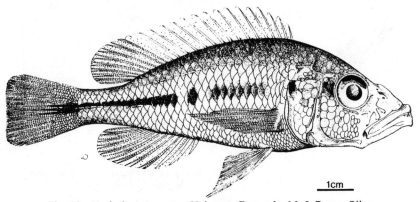

Fig. 16 *Haplochromis cassius*. Holotype. Drawn by M. J. P. van Oijen.

DESCRIPTION (Figs 16–18). Based on 5 specimens (including the holotype), 70·5–97·5 mm S.L. All are females.

Depth of body 29·0–34·0 (M = 31·0)% of standard length, length of head 36·0–40·0 (M = 37·0)%.

Dorsal head profile straight or gently decurved (its outline interrupted by the prominent premaxillary pedicels) and sloping at an angle of 30–40°. The cephalic lateral line pores, and the tubules on the preorbital bone, are not noticeably enlarged.

Preorbital depth 16·0–21·0 (M = 18·0)% of head length, interorbital width 17·0–20·0 (M = 18·0)%. Snout broader than long, its length 30·0–36·0 (M = 33·3)% of head. Eye and orbit very slightly elliptical, the eye with a well-defined anterior and anteroventral aphakic aperture; eye diameter 26·0–31·0 (M = 28·0)% of head. Cheek depth 20·0–22·0 (M = 21·0)%.

Caudal peduncle 1·7–1·9 times longer than deep, its length 17·0–19·0 (M = 18·0)% of standard length.

Mouth slightly oblique, inclined at an angle of 15–20°; jaws equal anteriorly, the posterior tip of the maxilla just reaching a vertical through the anterior margin of the eye. Premaxilla with a well-developed beak (i.e. a median anteroposterior expansion of its dentigerous arm). Both the upper and lower lips are noticeably thickened, more so than in any other species of the '*serranus*' group. The lower jaw 2·5–2·8 times longer than broad, its length 43·0–48·0 (M = 44·0)% of head length; dentary without a marked mental protuberance.

Gill rakers. 10 or 11 (mode) on the lower part of the first gill arch, the lowermost 1 or 2 (exceptionally 4) reduced, the remainder slender except for the uppermost 2 or 3 which are flattened and bi- or tri- or polyfid.

Scales. Ctenoid; lateral line with 33 (f.3) or 34 (f.2) scales, cheek with 3 (f.2) or 4 (f.3) rows. Five to $6\frac{1}{2}$ scales between the lateral line and the dorsal fin origin, 6 (mode) to 7 between the pectoral and pelvic fin bases.

Fins. Dorsal with 24 (f.1) or 25 (f.4) rays, comprising 15 (f.3) or 16 (f.2) spines and 9 (f.3) or 10 (f.2) branched rays. Anal fin with 12 rays, comprising 3 spines and 9 branched rays. Pectoral fin 26·0–28·0 (M = 27·0)% of standard length. Pelvics with the first branched ray very slightly produced. Caudal truncate, scaled on its basal third to half.

Teeth. The *outer teeth* in *both jaws* are large, somewhat recurved, caniniform unicuspids and are very widely spaced. When the jaws are closed some teeth lie outside the lip of the opposing jaw, while others seem to become embedded in the gum and lip tissues of that jaw. This unusual condition may, of course, be merely a preservation artefact, and consequent upon the thickening of gum and lip tissues in the fixative.

There are only 30–40 teeth in the outer row of the upper jaw.

The *inner teeth* are mostly small unicuspids, but some weakly tricuspid teeth also occur in these series; all are obliquely implanted, and are arranged in 1 or 2 rather irregular rows in both jaws.

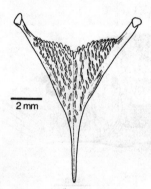

Fig. 17 *Haplochromis cassius.* Lower pharyngeal bone in occlusal view.

Osteology. No complete skeleton is available. The *lower pharyngeal bone* (Fig. 17) has its dentigerous surface very slightly longer than broad; its teeth are cuspidate and compressed, vary in form from fine to relatively robust (especially those near the midline) and are arranged in 24–26 rows. There are 30 (f.3) or 31 (f.2) vertebrae (excluding the fused PU_1 and U_1 centra), comprising 13 (f.4) or 14 (f.1) abdominal and 16 (f.1), 17 (f.2) or 18 (f.2) caudal elements.

Coloration. In life, an *adult* but *quiescent female* (BM(NH) reg. no. 1977.1.10:51, see Fig. 18) has the dorsum of the head dark grey-blue, the preorbital region, cheek, preoperculum and lips greyish, the operculum is silvery with a dark opercular blotch and the branchiostegal membrane whitish. The dorsum of the body is grey-blue anteriorly, lighter, almost silver posteriorly. The flanks are silver-grey, darkest anteriorly, with a dark midlateral stripe that is interrupted at about its midpoint. The chest, belly and caudal peduncle are silvery white, the dorsal aspect of the latter rather darker.

The dorsal and pectoral fins are hyaline, the pelvics hyaline, the anal fin grey-silver and the caudal hyaline.

Details of *preserved* coloration are available for *females* only (both immature and adult). The ground coloration is sandy-grey above the midlateral line (except the dorsum) and also on the head save for the cheeks and dorsum. Below the midlateral line the sandy-grey colour gradually becomes silvery-white. The dorsum of the head and body are dark brown, and the cheeks are silvery.

A broad and well-defined midlateral stripe (variously but narrowly interrupted) runs from behind the head to the basal part of the caudal fin. Immediately below the dark dorsum, and in places continuous with it, is an indistinct dark line which runs parallel to the dorsal outline of the body; posteriorly this line merges completely with the dark coloration of the back.

A faint, weakly V-shaped, bar crosses the snout at about the level of the lower orbital margin; in some specimens there is a short, faint and ill-defined lachrymal blotch.

All the fins are hyaline (except for a small area on the caudal where the midlateral band of the body terminates).

ECOLOGY. *Habitat*. The 5 specimens came from three different collecting stations in the Mwanza Gulf, viz. a point slightly south of the crossing between the Muranda peninsula and the opposite shore, at a depth of *c*. 4–6 m (no substrate data were recorded), another trawl haul near this area at a depth of *c*. 6–10 m over mud and, thirdly, a trawl made in the southern part of the gulf near Busissi, again over mud at a depth of only 2 m.

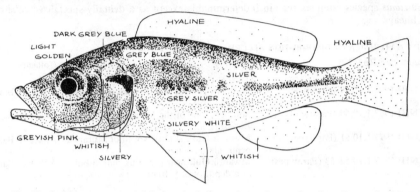

Fig. 18 *Haplochromis cassius*. Adult female (quiescent), to show coloration. Drawn by M. J. P. van Oijen.

FOOD. The feeding habits of *H. cassius* certainly cannot be determined from the small sample of guts examined, the more especially since the ingested matter is so heterogeneous. One of the 4 specimens examined was without food in any part of the gut. One fish had an empty stomach, but the remains of at least one small cichlid fish in the intestine. The other two specimens (both from the same locality and trawl haul) had the entire intestine packed with diatom frustules (mostly a colonial form resembling *Melosira*). One of these fishes had a similar diatom mass in its stomach, but the stomach of the other fish was empty.

It is difficult to account for the almost purely diatom intake of these two fishes other than by assuming that they had actively selected the diatoms as food. The nature of the gut contents certainly does not suggest that the material had been ingested while the fishes were impounded in the net and being dragged through the flocculent organic mud at the mud-water interface (see above, pp. 146 & 154) because this interface is unlikely to be composed purely of diatoms (and only one taxon at that). The organic constituents of the near-liquid mud are predominantly blue-green algae, with diatoms (and particularly the *Melosira* type of diatoms) forming but a small proportion of the whole.

Much more material of *H. cassius* will have to be examined, and more details about the substrate obtained, before this particular trophic puzzle can be solved.

The intestine of *H. cassius* is of moderate length (*c*. $1\frac{1}{2}$ times S.L.) and thus more typical of a predatory than a herbivorous species.

BREEDING. No information is available on the reproductive habits of *H. cassius*. In the one sexually active fish represented in our sample, the right ovary is much larger than the left one, although the latter does have near full-term ova present in it.

DIAGNOSIS AND AFFINITIES. *Haplochromis cassius* is readily distinguished from all other members of the *H. serranus–H. prognathus* lineage (see Greenwood, 1974), and all other species with a 'predatory' facies, by its noticeably thickened lips and by its well-spaced, caniniform teeth some of which, at least in preserved specimens, are visible when the mouth is closed. This species also differs from members of the *H. serranus–H. prognathus* complex in having a shorter lower jaw, narrower interorbital, shallower cheek, larger eye (but this possibly a correlate of its small adult size), and a higher gill raker count (modal number of rakers 11, cf. 9 for the other species, in a few of which 10 rakers have been counted in the occasional specimen).

If *H. nanoserranus* (see above, p. 159) is also a member of the *H. serranus* group, it too has a narrower interorbital, larger eye and a shallower cheek, but again the adult size of this species is much smaller than that for other members of the group. *Haplochromis nanoserranus* does, however, have a lower jaw length and a gill raker count more typical for the *H. serranus* group than does *H. cassius*.

Until more anatomical information is available for *H. cassius* its relationships within the *H. serranus* species complex remain indeterminable except as a dentally specialized offshoot of this lineage.

STUDY MATERIAL AND DISTRIBUTION RECORDS

Museum and Reg. No.	Locality	Collector
	TANZANIA	
BM(NH) 1977.1.10:49 (Holotype)	Southern part of the Mwanza Gulf near Busissi (2 m)	Anker & Barel
BM(NH) 1977.1.10:50 (Paratype)	Southern part of the Mwanza Gulf near Busissi (2 m)	Anker & Barel
BM(NH) 1977.1.10:51 (Paratype)	Mwanza Gulf, slightly south of Muranda peninsula (c. 4–6 m)	Anker & Barel
BM(NH) 1977.1.10:52–53 (Paratypes)	Mwanza Gulf near previous station but at a depth of c. 6–10 m	Anker & Barel

A new species of the *H. empodisma–H. obtusidens* group

Haplochromis ptistes sp. nov.

HOLOTYPE. An adult male 98·0 mm S.L. from the Speke Gulf northeast of Tefu Island (between Tefu and Nafuba Islands), at a depth of c. 28 m over a mud bottom. BM(NH) reg. no. 1977.1.10:60.

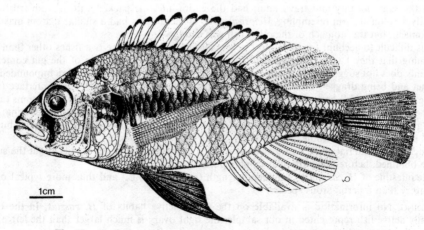

Fig. 19 *Haplochromis ptistes*. Holotype. Drawn by M. J. P. van Oijen.

The trivial name (from the Greek meaning a winnower or sheller) refers to the crushing pharyngeal mechanism of this species and the effect it has on its molluscan prey.

DESCRIPTION (Figs 19–21). Based on 10 specimens (including the holotype) 90·0–106·0 mm standard length.

Depth of body 38·6–42·0 (M = 40·0) % of standard length, length of head 34·2–37·6 (M = 36·0) %.

Dorsal head profile gently decurved and sloping at an angle of 35–40°. The cephalic lateral line pores, and the tubules on the preorbital bone, are moderately enlarged and prominent.

Preorbital depth 14·7–17·4 (M = 16·4) % of head length, least interorbital width 23·5–26·0 (M = 24·7) %. Snout broader than long, its length 29·4–32·4 (M = 30·6) % of head. Orbit and eye virtually circular, the eye with a fairly definite anterior and anteroventral aphakic aperture and, in some specimens, a more definite posterior one as well; eye diameter 26·5–32·4 (M = 30·0) % of head. Cheek depth 20·5–24·3 (M = 22·0) %.

Caudal peduncle 1·3–1·6 (modal range 1·4–1·5) times longer than deep, its length 15·0–18·0 (M = 16·4) % of standard length.

Jaws equal anteriorly, mouth almost horizontal, the posterior tip of the maxilla reaching a vertical through the anterior margin of the eye; premaxilla with a slight median anteroposterior expansion of its dentigerous arm giving it a slightly beaked appearance.

Lower jaw 1·5–1·9 (mode 1·5) times longer than broad, its length 37·3–41·2 (M = 39·0) % of head.

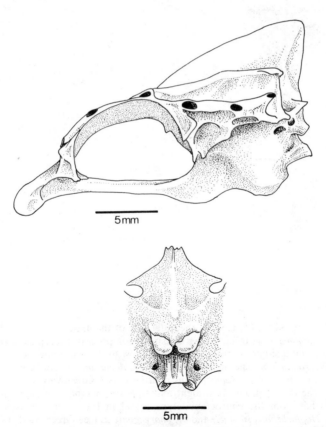

Fig. 20 *Haplochromis ptistes*. Above, neurocranium in left lateral view. Below, the apophysis for the upper pharyngeal bones.

Gill rakers. 8 (mode) or 9 on the lower part of the first gill arch, the lower 1–3 rakers reduced, the remainder short and moderately stout to stout.

Scales. Ctenoid; lateral line with 31 (f.1), 32 (f.7) or 33 (f.2) scales, cheek with 3 (mode) or 4 rows. Five to $6\frac{1}{2}$ (usually 6 or $6\frac{1}{2}$) scales between the dorsal fin origin and the lateral line, 6 or 7 (mode) between the pectoral and pelvic fin bases.

Fins. Dorsal with 23 (f.3) or 24 (f.7) rays, comprising 14 (f.2), 15 (f.4) or 16 (f.4) spines and 8 (f.5) or 9 (f.5) branched rays. Anal with 11 (f.9) or 12 (f.1) rays, comprising 3 spines and 8 (f.9) or 9 (f.1) branched rays. Pectoral 86·6–97·0 (M = 92·0)% of head length. Pelvic fins with the first branched ray produced, proportionately more so in males than in females. Caudal truncate, scaled on its basal third to half (mode).

Teeth. In most specimens less than 100 mm S.L. the *outer teeth* in the *upper jaw* are unequally bicuspid (some weakly so), relatively stout and slightly recurved; the posterior few teeth, however, are unicuspid and slightly enlarged. Specimens over 100 mm S.L. (and one fish of 90 mm S.L.) have mostly stout unicuspid teeth throughout the series. Teeth in the outer row of the *lower jaw* are similar to those in the upper jaw although some bicuspids may occur in larger specimens and a few unicuspids in smaller individuals.

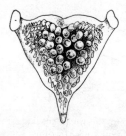

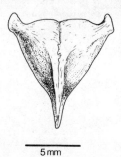

Fig. 21 *Haplochromis ptistes*. Lower pharyngeal bone in occlusal view (above), and in ventral view (below).

5 mm

There are 60–70 (mode c. 65) teeth in the outer row of the upper jaw.

The teeth of the *inner rows* in both jaws are small, compressed, tricuspids, arranged in 2 rows in the upper jaw and a single (sometimes irregular) row in the lower jaw.

Osteology. Neurocranium. The neurocranium of *Haplochromis ptistes* (Fig. 20) shows a close overall resemblance to that of *Haplochromis obtusidens* (see Greenwood, 1960 : 267, and 1974, figs 43 & 65). The dorsal profile is straight and slopes at a moderate angle, in these respects differing somewhat from the neurocranial type found in the other mollusc-crushing lineage represented by *Haplochromis ishmaeli* and *H. pharyngomylus* (see Greenwood, 1974 : 74, fig. 43). Here the orbital region is relatively high-vaulted, and consequently the preorbital profile is somewhat decurved and slopes more steeply than in *H. ptistes*.

The apophysis for the upper pharyngeal bones (Fig. 20) in this species is rather stouter and has a larger articular surface than the apophysis in *H. obtusidens*, but it is smaller and has a lesser contribution from the basioccipital than does the apophysis of *H. ishmaeli* or *H. pharyngomylus*.

The *lower pharyngeal bone* (Fig. 21) is stout, with the majority of its teeth enlarged and molariform; only those teeth contributing to the marginal row, and those situated in the posterolateral angles of the bone, are distinctly cuspidate and not particularly enlarged. The bone has a characteristic outline shape (Fig. 21) with a marked shoulder occurring a little posterior to the point where the bone narrows to form the anteriorly directed blade. In lateral view the occlusal surface is gently concave over its entire area.

When compared with the pharyngeal bones of *H. ishmaeli* and *H. pharyngomylus* (and in one of the new species described below, p. 176), that of *H. ptistes* is relatively less hypertrophied and its teeth are also less massive. However, its dentigerous area is relatively larger and there are somewhat more molariform teeth than in *H. obtusidens*.

In other words, the lower pharyngeal bone and dentition of *H. ptistes* occupy a morphologically intermediate position between those of the lineages represented by *H. ishmaeli* and *H. pharyngomylus* on the one hand, and by *H. obtusidens* on the other.

The dentary in *H. ptistes* is relatively shallow and elongate, resembling that in *H. obtusidens* rather than the dentary of *H. ishmaeli* or *H. pharyngomylus*.

There are 28 (f.2) or 29 (f.2) vertebrae (excluding the fused PU_1 and U_1 centra) in the 4 specimens radiographed, the total comprising 12 (f.2) or 13 (f.2) abdominal and 16 caudal elements.

Coloration. The *live colours* of *H. ptistes* are unknown. *Preserved coloration. Adult males.* The dorsum is yellowish-brown shading to a lighter tone on the flanks; the belly and chest are dusky. A dark, horizontally aligned blotch extends from immediately behind the eye posteriorly across the operculum where it deepens slightly and becomes confluent with, or is narrowly separated from, a broad midlateral stripe on the flank. This stripe may narrow or be interrupted at about the middle of the body; posteriorly it extends onto the caudal fin, the hind margin of which it reaches. Some specimens show traces of 3 or 4 broad vertical bars on the lower flanks and belly; very faint traces of these bars continue onto the upper flanks and back. In other specimens the bars are barely visible.

The head has two definite black bars across the snout, the upper one extending from orbit to orbit. A supraorbital stripe runs obliquely upwards from the dorsoposterior margin of the orbit almost to the midline, where it is narrowly separated from its partner of the opposite side; in most specimens the supraorbital bars are virtually rectangular in outline but in a few they are roughly triangular (but never so definitely triangular as in *Haplochromis teegelaari*, see p. 173 below).

A broad lachrymal band runs almost vertically downwards onto the anguloarticular region of the lower jaw or even further ventromedially. The branchiostegal membrane is dusky in some specimens, but pale in others.

A bar of variable intensity and completeness extends vertically upwards from a point almost at the middle of the upper opercular margin; the bar of each side meets, albeit faintly, or is narrowly separated from, its counterpart. In several specimens there is a well-defined black bar following the outline of the preoperculum, but in others it is extremely faint.

The dorsal fin is greyish, with black lappets on the spinous part and dark maculae on the soft part of the fin. The caudal is darkly maculate, especially over its upper half, and has a dark midlateral streak. The anal fin is greyish, and the pelvics are black.

Adult females have a pale yellow-brown ground colour, with the chest, belly and operculum silvery (the latter with a large dark blotch at its posterodorsal angle). There is a faint but distinct dark midlateral stripe extending from the preopercular margin to the posterior margin of the caudal fin. The lachrymal stripe is very faint and short.

The dorsal and caudal fins are greyish, the former with black lappets, the latter with a midlateral stripe, and faint maculae on its upper half. The anal and pelvic fins are hyaline.

ECOLOGY. *Habitat.* All 10 specimens came from a single trawl haul made in the Speke Gulf, between Tefu and Nafuba Islands, at a depth of *c.* 28 m over a mud bottom.

FOOD. Of the 8 specimens examined, 4 contained only flocculent organic detritus (principally blue-green algae with some diatoms and green algae) throughout the entire alimentary tract. The other 4 specimens contained, in addition to this detrital matter, fragments of mollusc shells (either of unidentifiable bivalves together with the gastropod *Melanoides tuberculata*, or of the bivalves alone).

As with the other species from this station and haul (see p. 146), the detritus may have been ingested whilst the fishes were being dragged through the mud-water interface during capture.

The intestine of *H. ptistes* is very long (c. 2½ times the standard length) and much coiled, an unusual feature for a species with the hypertrophied pharyngeal apparatus of a mollusc eater.

BREEDING. Nothing is known about the reproductive habits of *H. ptistes*. All the specimens available are adult and none is sexually active. The single female caught (90 mm S.L.) has its ovaries in an advanced stage of oogenesis, the right ovary being slightly larger than the left one.

DIAGNOSIS AND AFFINITIES. *Haplochromis ptistes* is distinguished from all previously described species with hypertrophied pharyngeal bones and teeth by the outline shape of its lower pharyngeal bone (see Fig. 21) and by the following characters for the species severally:

(i) From *H. obtusidens* (see Greenwood, 1960 : 266, fig. 18) by its more massive lower pharyngeal bone and the more extensive molarization of its lower pharyngeal dentition, its larger eye (26·5–32·4, M = 30·0% head, cf. 24·3–30·8, M = 27·2%), slightly longer snout 29·4–32·4, M = 30·6% head, cf. 26·0–31·0, M = 28·5%), shallower cheek (20·5–24·3, M = 22·0% head, cf. 21·2–30·0, M = 26·7%), longer pectoral fin (86·6–97·0, M = 92·0% head, cf. 73·5–103·0, M = 86·8%) and by the markedly different pattern of cephalic markings visible in preserved adult males. (Live colours of *H. ptistes* are unknown.)

(ii) From *H. ishmaeli* and *H. pharyngomylus* (see Greenwood, 1960 : 270–279, figs 19–21) by the presence of definite snout and supraorbital markings in preserved specimens, by its larger eye (26·5–32·4, M = 30·0% of head, cf. 23·0–31·8, M = 26·5 and 23·0–31·0, M = 27·7 for *H. pharyngomylus* and *H. ishmaeli* respectively), the greater number of teeth in the outer row of the upper jaw (modal number 65, cf. 44–52 and 36 for *H. ishmaeli* and *H. pharyngomylus* respectively), the straight preorbital profile of the neurocranium, the less massive lower pharyngeal bone, and by its higher modal number of gill rakers (8 cf. 7); *H. ptistes* is further distinguished from *H. pharyngomylus* by its longer pectoral fin (86·6–97·0, M = 92·0% head, cf. 68·5–91·0, M = 79·6%) and by the greater posterior extension of its maxilla (reaching a vertical through the anterior part of the eye in *H. ptistes*, but only to the orbital margin, or not even to that level, in *H. pharyngomylus*).

From the two newly discovered species with hypertrophied pharyngeal mills (see pp. 169–174 below), *H. ptistes* is distinguished as follows:

(i) From *H. teegelaari* (see p. 169: Figs 22–27) by its snout being broader than long, by differences in the neurocranial architecture (dorsal preorbital profile straight, compared with a more obviously vaulted and curved orbital–preorbital region; cf. Figs 20 and 24), the slightly less massive ventral apophysis, with a smaller articular area for the upper pharyngeal bones cf. Figs 20 and 25), the less markedly concave occlusal surface of the lower pharyngeal bone and by the rectangular as opposed to triangular supraorbital markings in preserved males. (The well-defined midlateral body stripe of *H. ptistes* seemingly is also diagnostic in preserved material.)

The degree of lower pharyngeal bone enlargement, and the extent to which its teeth are molarized, are similar in both species, but the bone of *H. ptistes* has a very characteristic shape when seen in occlusal view (cf. Figs 21 & 26). The two species show a virtually complete overlap in the mean values of all morphometric features except that of relative snout width (see above).

(ii) From *Haplochromis mylergates* (p. 174, Figs 29–31), *H. ptistes* is readily distinguished by its less massive lower pharyngeal bone, which also lacks the deeply concave occlusal surface seen in *H. mylergates*, by differences in skull architecture (similar to those distinguishing *H. ptistes* from *H. teegelaraai*, see above and also Figs 20 & 24), by the more gradually pointed snout as seen in dorsal view (see Fig. 23), and by differences in preserved coloration, especially the absence in *H. mylergates* of prominent cephalic markings (save for the lachrymal stripe).

On the basis of its neurocranial shape, and its relatively shallow dentary (as compared with the more generalized skull shape and deeper dentary of *H. ishmaeli*, *H. pharyngomylus* and *H.*

mylergates, and the generalized skull shape of *H. teegelaari*), *H. ptistes* is thought to be a member of the *H. empodisma–H. obtusidens* lineage of mollusc crushing species (see Greenwood, 1974), probably the derived (apomorph) sister species of *H. obtusidens*. In addition to showing certain derived morphological features, *H. ptistes* should perhaps also be considered specialized because of its relatively deeper water habitat.

STUDY MATERIAL AND DISTRIBUTION RECORDS

Museum and Reg. No.	Locality	Collector
	TANZANIA	
BM(NH) 1977.1.10:60 (Holotype)	Speke Gulf, between Tefu and Nafuba Islands, *c.* 28 m	Anker & Barel
BM(NH) 1977.1.10:61–69 (Paratypes)	Speke Gulf, between Tefu and Nafuba Islands, *c.* 28 m	Anker & Barel

New species of the *H. ishmaeli–H. pharyngomylus* group

Haplochromis teegelaari sp. nov.

HOLOTYPE. An adult male 93·0 mm standard length, from the southern part of the Mwanza Gulf near Busissi, caught over a mud bottom at a depth of *c.* 2 m. BM(NH) reg. no. 1977.1.10:16.

The species is named in honour of the late Nico Teegelaar, an outstanding Dutch biological artist whose work contributed much to the researches of the Zoology Department of Leiden University.

DESCRIPTION (Figs 22–27). Based on 23 specimens (including the holotype) 74·0–100·5 mm standard length.

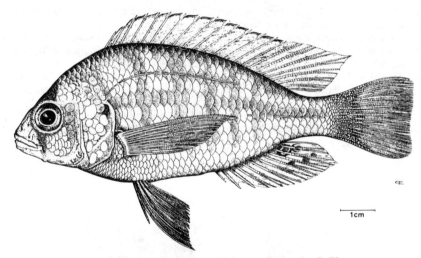

Fig. 22 *Haplochromis teegelaari*. Holotype. Drawn by C. Elzenga.

Depth of body 39·0–43·0 (M = 41·0) % of standard length, length of head 32·0–36·0 (M = 34·0) %. Dorsal head profile curved to above the eye then straight and sloping steeply downwards at an angle of 40–60° (mode 45°). Cephalic lateral line pores not enlarged, the tubules of the preorbital bone barely visible superficially. Preorbital depth 14·0–19·0 (M = 17·0) % of head, least interorbital width 25·0–30·0 (M = 27·0) %. Snout as broad as long to slightly broader than long (the modal condition), its length 27·0–31·0 (M = 29·0) % head length; when viewed from above the outline of the snout is gently and gradually rounded (see Fig. 23). Eye and orbit virtually circular, the eye with a definite anterior and anteroventral aphakic aperture; diameter of eye 27·0–33·0 M = 30·0)% of head. Depth of cheek 19·0–25·0 (M = 22·0)%.

Caudal peduncle 1·1–1·5 (mode 1·4) times longer than deep, its length 15·0–19·0 (M = 17·0)% of standard length.

Mouth very slightly oblique, inclined at an angle of 5–10° (mode 10°). Jaws equal anteriorly, the posterior tip of the maxilla reaching a vertical through the anterior orbital margin, or slightly beyond that level. Lower jaw 1·4–1·8 (modal range 1·6–1·7) times longer than broad, its length 33·0–40·0 (M = 37·0)% of head.

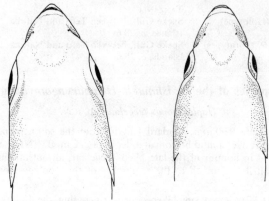

Fig. 23 Dorsal view of the snout in *H. mylergates* (left) and *H. teegelaari* (right), to show differences in outline when viewed from above.

Gill rakers. 7 or 8 (rarely 9) on the lower part of the first gill arch, the lower 1 or 2 (occasionally 3) rakers reduced, the remainder moderately stout and blunt.

Scales. Ctenoid; lateral line with 31 (f.1), 32 (f.7), 33 (f.12) or 34 (f.3) scales, cheek with 3 or 4 (mode) rows. Six to 7½ (usually 6½ or 7) scales between the dorsal fin origin and the lateral line, 6 or 7 (mode) between the pectoral and pelvic fin bases.

Fins. Dorsal with 23 (f.2), 24 (f.12), 25 (f.8) or 26 (f.1) rays, comprising 15 (f.18) or 16 (f.5) spinous and 8 (f.3), 9 (f.14) or 10 (f.6) branched elements. Anal fin with 11 (f.5), 12 (f.16) or 13 (f.2) rays, comprising 3 spines and 8 (f.6), 9 (f.15) or 10 (f.2) branched rays. Pectoral fin 84·0–103·0 (M = 91·0)% of head. Pelvic fins with the first branched ray slightly produced. Caudal truncate, scaled on its proximal half (rarely only on its proximal third) or a little further posteriorly.

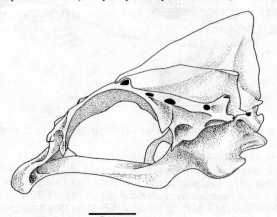

Fig. 24 *Haplochromis teegelaari.* Neurocranium in left lateral view.

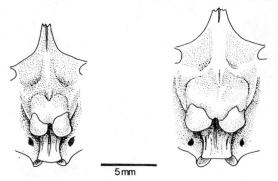

Fig. 25 Apophysis for the upper pharyngeal bones in *H. teegelaari* (left) and *H. mylergates* (right).

Teeth. In most specimens the anterior and anterolateral teeth in the *outer row* of the *upper jaw* are moderately stout bicuspids with compressed, recurved, crowns and cylindrical necks; posteriorly and sometimes posterolaterally, the teeth are unicuspid and stout, with recurved crowns. A few specimens have unicuspid teeth throughout the row, or unicuspids anteriorly, bicuspids laterally, and unicuspids posteriorly. There is no obvious correlation between a predominant tooth form and the fish's size.

In the *outer series* of the *lower jaw*, most specimens have only bicuspid teeth, although a few do have either an entirely unicuspid dentition or some unicuspids posteriorly and a mixture of bi- and unicuspids elsewhere in the jaw.

There are 38–54 (modal range 40–44) teeth in the outer row of the upper jaw.

Teeth forming the *inner series* are usually either a mixture of bi- and tricuspids or one of uni- and bicuspids; a few specimens have a mixture of all three types of teeth. There are 1 or 2 (mode) rows of inner teeth in both jaws.

Osteology. The neurocranium of *H. teegelaari* (Fig. 24) resembles that of *H. mylergates* (see p. 176 and Fig. 29 below) in having a fairly high-vaulted orbital region and a somewhat curved and relatively steeply sloping preorbital profile; the curvature and slope, however, are less marked than in the skulls of *H. pharyngomylus* and *H. ishmaeli* (see Greenwood, 1974 : 73, figs 43 & 65).

The ventral apophysis for the upper pharyngeal bones is stout, with a large articular area in which there is a substantial contribution from the basioccipitals but none from the prootics, at least in the 2 specimens examined (see Fig. 25).

The *lower pharyngeal bone* (Fig. 26) is stout and enlarged with a broad occlusal surface that is markedly concave over its entire area, a feature best seen when the bone is viewed laterally.

The *lower pharyngeal teeth*, except for those in the marginal row and a small cluster in the posterolateral angles of the bone, are enlarged and molariform; the non-molariform teeth are stout and weakly bicuspid.

The level of hypertrophy in the pharyngeal mill of *H. teegelaari* (as measured by the extent and degree of pharyngeal tooth molarization and bone enlargement) is comparable with that seen in *H. ptistes*, *H. ishmaeli* and *H. pharyngomylus*, although some specimens of the latter species do exhibit a slightly greater development of the mill.

The dentary in *H. teegelaari*, when compared with that in *H. pharyngomylus* and *H. ishmaeli*, is relatively shallower and more elongate, in these respects resembling the dentary in *H. obtusidens*, *H. ptistes* and *H. mylergates* (see p. 177 below).

There are 29 (f.11) or 30 (f.10) vertebrae (excluding the fused PU_1 and U_1 centra), comprising 13 (f.20) or 14 (f.1) abdominal and 16 (f.11) or 17 (f.10) caudal elements.

Coloration. The *live colours* of an *adult sexually active male*, see Fig. 27 (BM(NH) reg. no. 1977.1.28·41), are as follows: Body with a purplish grey dorsum, the purple colour more intense anteriorly. Flanks, chest and belly bright red, caudal peduncle yellow with a faint red overlay; traces of 6 vertical bars are visible on the flanks. Dorsum of head grey with a red flush, remainder

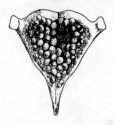

Fig. 26 *Haplochromis teegelaari*. Lower pharyngeal bone in occlusal view (above) and ventral view (below).

of head bright red except for the lower lip and branchiostegal membrane which are white. There is a faint lachrymal bar and a dark bar on the vertical preopercular limb.

Dorsal fin light grey with a faint red flush, dark grey lappets, and red maculae on the soft part of the fin. Anal light red anteriorly, greyish posteriorly; egg dummies (anal ocelli) orange to reddish. Caudal hyaline, yellowish proximally, and with red maculae and streaks. Pelvic fins mostly black, the pectorals hyaline. A second specimen (BM(NH) reg. no. 1977.1.28:40) also a sexually active male, differs slightly in having only 3 vertical bars on the flanks, a faint dark band from the opercular spot to the eye, a brownish-purple dorsum to the head, white ventral aspects of the flanks and a red flush on the otherwise black pelvic fins.

Preserved material. The coloration of *adult males* only is known. The ground colour is a light sandy brown, shading to yellowish-white on the chest and belly, the chest sometimes with a sooty overlay. Five or 6 distinct dark bars extend across the flanks from the dorsal profile almost to

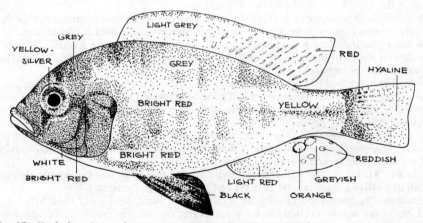

Fig. 27 *Haplochromis teegelaari*. Adult male (sexually active), to show coloration. Drawn by M. J. P. van Oijen.

the ventral outline of the body; immediately above the anal fin two or three bars generally are interconnected midlaterally by a rather ill-defined black blotch. Usually there are two vertical bars on the caudal peduncle, each somewhat broader but less well-defined than those on the flanks. Some specimens have faint indications of a dark midlateral band, especially on the anterior third of the body and again on the caudal peduncle. A faint and frequently interrupted longitudinal band is sometimes visible slightly dorsal to, but following the course of, the upper lateral line.

On the head there are two parallel and well-defined bars crossing the snout; a distinct, and relatively broad, lachrymal stripe extends in some specimens to the level of the maxillary tip, and in others further ventrally onto the lower jaw. Above the eye (and continuing the same line as the lachrymal bar) is a dark stripe which soon expands into a triangular blotch; the blotches of each side meet in the midline (cf. *H. ptistes*, p. 167). In some specimens there is a faint but dark vertical bar on the upper two-thirds of the preoperculum. The operculum itself is silvery.

The dorsal fin is yellowish with black lappets; the caudal fin is also yellowish but with a faint and ill-defined darker centre. The anal varies from hyaline to faint yellow, the pelvics are black, most intensely so over the anterior half of the fin.

ECOLOGY. *Habitat*. The specimens came from three different localities in the Mwanza Gulf (see p. 174). In all three localities, the substrate is mud; at two the depth was *c*. 2 m, and at the third *c*. 8 m.

FOOD. The guts of fishes from all three localities were examined, and gave the following results.

(i) Southern part of Mwanza Gulf near Busissi (*c*. 2 m; mud). Five specimens, all containing fragments of small, unidentifiable bivalve shells, but in 4 fishes a number of fragmented gastropod shells (*Melanoides tuberculata*) as well.

(ii) Coastal waters opposite Mashoro Bay, Mwanza Gulf (*c*. 8 m, mud). One specimen containing a few fragments of bivalve shells (specifically indeterminable).

(iii) Northeast of Buzumu Island, near the southern end of the Mwanza Gulf (*c*. 2 m; mud). Nine specimens, all except one containing a mixture of fragmentary, small and unidentifiable bivalve shells together with fragmentary gastropod shells (*Melanoides* and probably one other species); *Melanoides* remains predominate in most guts. The exceptional fish contained only *Melanoides* shell fragments.

BREEDING. Nothing is known about the breeding habits of *H. teegelaari*. Only males are available for study; all are adult and most show signs of sexual activity.

DIAGNOSIS AND AFFINITIES. The morphological characters distinguishing *H. teegelaari* from *H. ptistes* (see p. 168) are relatively slight and concerned principally with the skull and pharyngeal bones; the two species overlap in all morphometric features.

The degree of enlargement shown by the lower pharyngeal bone in both species is about equal, as is the extent to which the lower pharyngeal dentition is molarized. However, the occlusal surface of the bone is more concave in *H. teegelaari*, and the outline of the bone as seen in occlusal view lacks the small but distinct 'shoulders' immediately posterior to the blade (cf. Figs 26 & 21).

The neurocranium of *H. teegelaari* has a somewhat more vaulted orbit and thus a more steeply sloping and curved dorsal profile to the preorbital region than is the case in *H. ptistes* (cf. Figs 24 & 20).

Another anatomical feature distinguishing the two species is the much longer and more coiled intestine of *H. ptistes* (*c*. $2\frac{1}{2}$ times the standard length, cf. $1\frac{1}{2}$–2 times in *H. teegelaari*).

The most readily diagnostic feature lies in the cephalic markings of preserved specimens. In *H. teegelaari* the supraorbital blotches are clearly triangular, as opposed to rectangular in *H. ptistes*. *Haplochromis ptistes* also has a prominent midlateral stripe, a feature that is barely visible and is frequently interrupted in those specimens of *H. teegelaari* in which it is present. Regrettably, the live colours of *H. ptistes* are still unknown.

From *H. pharyngomylus* and *H. ishmaeli* (see Greenwood, 1960 : 270–279, figs 19–21), *H. teegelaari* is distinguished by the live coloration of adult males, by its shallower dentary and by its somewhat larger eye (27·0–33·0, M = 30·0% of head, cf. 23·0–31·8, M = 26·5% and 23·0–31·0, M = 27·7% for *H. pharyngomylus* and *H. ishmaeli* respectively). From *H. pharyngomylus*, *Haplochromis teegelaari* is further distinguished by its longer pectoral fin (84·0–103·0, M = 91·0% head, cf. 68·5–91·0, M = 80·0%), and from *H. ishmaeli* by usually having fewer teeth in the outer row of the upper jaw (38–54, modal range 40–44, cf. 38–66, modal range 44–52).

The complete, or almost complete, overlap of *H. teegelaari* with *H. ptistes*, *H. ishmaeli* and *H. pharyngomylus* in all morphometric and meristic characters emphasizes the difficulties encountered in taxonomic work on the Lake Victoria *Haplochromis* species flock. When live specimens are compared, the differences in adult male coloration are striking and diagnostic, and there are also subtle differences in gross morphology which cannot readily be quantified or verbalized. Together, the features of colour and shape enable one to group, quite easily, various individuals into recognizable 'taxa', an action that adds to one's conviction that these assemblages are also biologically valid species.

In most respects *H. teegelaari* seems to be related both to *H. pharyngomylus* and *H. ishmaeli*, and to *H. obtusidens* and *H. ptistes*; in particular this double relationship would seem to be manifest through the specialization expressed in the degree of pharyngeal mill hypertrophy. Similarities in neurocranial shape shared by *H. teegelaari*, *H. ishmaeli* and *H. pharyngomylus* (see above, p. 171) are probably of little value for indicating relationships because, apart from the enlarged ventral apophysis (a correlate of pharyngeal bone hypertrophy), the skull form in all three species departs little from the basic Lake Victoria *Haplochromis* type (see Greenwood, 1974). The supposedly more derived neurocranial shape of *H. ptistes* and *H. obtusidens* could, however, serve to link the two species in a phyletic lineage distinct from the lineage (or lineages) containing *H. teegelaari*, *H. ishmaeli* and *H. pharyngomylus*.

Whether or not, phylogenetically speaking, *H. teegelaari* should be associated with *H. ishmaeli* and *H. pharyngomylus* cannot be established on the basis of any derived characters shared by these three taxa alone. Likewise, there are no apomorph features shared only by *H. teegelaari* and *H. ptistes*, their common apomorph characters being shared also with *H. pharyngomylus* and *H. ishmaeli*.

Thus, for the moment, the phyletic relationships of *H. teegelaari* remain obscure, but with the probability that the species does not share an immediate common ancestor with *H. obtusidens* and *H. ptistes*.

STUDY MATERIAL AND DISTRIBUTION RECORDS

Museum and Reg. No.	Locality	Collector
	TANZANIA	
BM(NH) 1977.1.10:16 (Holotype)	Southern end of Mwanza Gulf, near Busissi (*c.* 2 m)	Anker & Barel
BM(NH) 1977.1.10:17 (Paratype)	Coastal waters opposite Mashoro Bay, Mwanza Gulf (*c.* 8 m)	Anker & Barel
BM(NH) 1977.1.10:18–26 (Paratypes)	Northeast of Buzumu Island, Mwanza Gulf (*c.* 2 m)	Anker & Barel
BM(NH) 1977.1.10:27–38 (Paratypes)	Southern part of Mwanza Gulf, near Busissi (*c.* 2 m)	Anker & Barel

Haplochromis mylergates sp. nov.

HOLOTYPE. An adult male 111·0 mm standard length, from the Speke Gulf west of Nafuba Island, at a depth of *c.* 10–12 m over a mud bottom, BM(NH) reg. no. 1977.1.10:88.

The trivial name (from the Greek, a miller) refers to the extreme hypertrophy of the pharyngeal apparatus and its effects on the molluscan prey of the species.

DESCRIPTION (Figs 28–32). Based on 18 specimens (including the holotype), 102·0–137·0 mm standard length.

Depth of body 38·0–45·0 (M = 42·0%) of standard length, length of head 33·0–37·0 (M = 34·9)%.

Dorsal head profile gently decurved or, less commonly, straight, sloping steeply at an angle of 40–45°, its outline sometimes interrupted by the fairly prominent premaxillary pedicels. The cephalic lateral line pores are enlarged, the supraorbital pore and those on the preorbital bone noticeably so; the lateral line tubules on the preorbital bone, however, are not especially prominent.

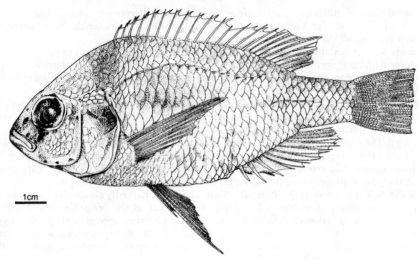

Fig. 28 *Haplochromis mylergates*. Holotype. Drawn by Gordon Howes.

Preorbital depth 12·0–20·0 (M = 16·0)% of head length, interorbital width 26·0–33·0 (M = 29·0)%. Snout as long as broad (modal condition) to 1·2 times longer than broad; when viewed from above, the outline of the snout has a characteristic appearance, narrowing abruptly to form, with the tip of the lower jaw, a relatively acute entry angle (see Fig. 23); length of snout 28·0–33·0 (M = 31·0)% of head. Eye and orbit almost circular, the eye with a definite anterior and anteroventral aphakic aperture; eye diameter 28·0–33·0 (M = 31·0)% of head. Cheek depth 20·0–29·0 (M = 23·0)%.

Caudal peduncle 1·2–1·5 (mode 1·3) times longer than deep, its length 15·0–19·0 (M = 17·0)% of standard length.

Mouth slightly oblique, inclined at an angle of 10–25°. Jaws equal anteriorly, the posterior tip of the maxilla reaching a vertical through the anterior margin of the eye or, less frequently, a little posterior to that level.

Lower jaw 1·3–1·7 (modal range 1·4–1·5) times longer than broad; its length 35·0–43·0 (M = 39·0)% of head.

Gill rakers. 8 (rarely 7 or 10) on the lower limb of the first gill arch, the lowermost 2 or 3 (rarely as many as 5) rakers reduced, the remainder relatively short and stout.

Scales. Ctenoid; lateral line with 31 (f.2), 32 (f.5), 33 (f.8) or 34 (f.2) scales, cheek with 3 (mode) or 4 rows. Six and a half to 7½ (rarely 8, mode 7) scales between the dorsal fin origin and the lateral line, 6½–7½ (rarely 8, mode 7) between the pectoral and pelvic fin bases.

Fins. Dorsal with 23 (f.2) or 24 (f.16) rays, comprising 14 (f.1), 15 (f.12) or 16 (f.5) spinous and 8 (f.5) or 9 (f.13) branched elements. Anal with 11 (f.4), 12 (f.13) or 13 (f.1) rays, comprising 3 spines (4 in one specimen) and 8 (f.5), 9 (f.12) or 10 (f.1) branched rays. The occurrence of 4 anal spines in a species of *Haplochromis* is extremely rare; it is interesting to note that he specimen with 4 anal spines also has the lowest number (14) of dorsal spines. Pectoral fin 86·0–103·0 (M = 92·0)% of head. Pelvic fins with the first branched ray noticeably produced, proportionately more so in males. Caudal truncate, scaled on its proximal half to two-thirds, rarely scaled over somewhat less than the proximal half of the fin.

Teeth. In the *outer series* of the *upper jaw* there is usually a mixture of bi- (or weakly bicuspid) and unicuspid teeth, without any positional predominance of one type over the other; in general, however, the posterior one to three teeth are unicuspid and slightly enlarged. A few specimens have only unicuspids in the outer row, but there is no obvious correlation between body size and the predominance of unicuspids, as is often the case in *Haplochromis* species. The unicuspid teeth are relatively slender but are strong and caniniform, the bicuspids have compressed cylindrical crowns; all the outer teeth are slightly recurved.

There are 44–60 (modal range 50–55) teeth in the outer series of the upper jaw.

Tooth form and arrangement in the *outer row* of the *lower jaw* are similar to those in the upper jaw, but with a tendency for bicuspids to predominate over unicuspids. Those specimens with an entirely unicuspid upper dentition also have only unicuspid teeth in the lower jaw.

In most specimens the *inner tooth rows* of both jaws have an admixture of bi-, tri- and unicuspids with, usually, tricuspids predominating; rarely are only tricuspids found in these series. There are 1 or 2 (rarely 3) rows of inner teeth in both jaws.

Osteology. The *neurocranium* (Fig. 29) of *H. mylergates* closely resembles that of *H. teegelaari*, although the orbital region is somewhat higher and consequently the preorbital dorsal profile slopes more steeply. The ventral pharyngeal apophysis is stout, with a large articulatory surface to which the prootic makes no contribution (at least in the 2 specimens examined).

The *lower pharyngeal bone* (Fig. 30) is very stout, and has a markedly concave occlusal surface, the concavity increasing to almost a broad pit in the centre of the bone. Compared with *H. pharyngomylus* and *H. ishmaeli* (the two other mollusc-crushing species with greatly hypertrophied bones), the lower and upper pharyngeal bones of *H. mylergates* have a much larger surface area (see Fig. 31); as a correlate of this feature, the ventral apophysis on the skull is, relatively speaking, also much enlarged.

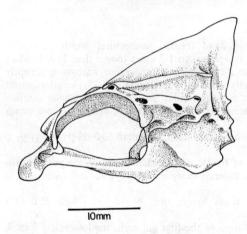

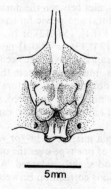

Fig. 29 *Haplochromis mylergates.* Neurocranium in left lateral view (above), and the apophysis for the upper pharyngeal bones (below).

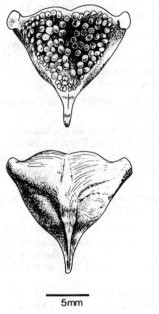

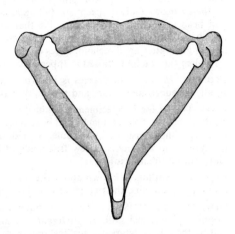

Fig. 30 *Haplochromis mylergates.* Lower pharyngeal bone in occlusal view (above) and ventral view (below).

Fig. 31 Outline of the lower pharyngeal bone (in occlusal view) of *H. mylergates* (grey shading), with that of *H. ishmaeli* (in white) superimposed on it. The bones are from a specimen of *H. ishmaeli* 105·0 mm S.L. (28·0 mm neurocranial length), and from a *H. mylergates* 109·0 mm S.L. (29·5 mm neurocranial length).

All the *lower pharyngeal teeth* (Fig. 30), except for a few in the posterolateral angles of the bone and a few in the outer row, are enlarged and molariform; the non-molariform teeth are stout and weakly cuspidate.

As in *H. teegelaari*, the dentary in *H. mylergates* is relatively shallow and elongate when compared with that bone in *H. pharyngomylus* and *H. ishmaeli*.

There are 29 (f.9) or 30 (f.8) vertebrae (excluding the fused PU_1 and U_1 centra), comprising 13 (f.13) or 14 (f.4) abdominal and 16 (f.13) or 17 (f.4) caudal elements.

Coloration. Data on *live colours* are available from an *adult and sexually active male* (BM(NH) reg. no. 1977.1.28:43), see Fig. 32. The body has a red dorsum which darkens posteriorly; the flanks and caudal peduncle are yellow to yellowish-green, becoming white ventrally; the chest and belly are a very light red. The head has a bright red dorsal surface and ethmoidal region; the preorbital region is a light reddish-grey while the cheeks and operculum are yellow with a red overlay. A faint lachrymal stripe is present. The lower jaw and branchiostegal membrane are whitish.

The dorsal fin is red anteriorly, hyaline posteriorly but with red maculae and streaks between the rays. The anal is whitish-grey, with orange egg dummies (ocelli), the caudal hyaline with red streaks and maculae. Pelvic fins are greyish posteriorly, black proximally; the pectorals hyaline with a red flush.

Preserved material. Adult males. The dorsum and the flanks to about the level of the lower lateral line are greyish-sandy to sandy; below this level the flanks, belly and chest are silvery white.

The dorsal surface of the head, excluding the snout, is sandy, the snout (both dorsally and laterally) is greyish. The cheek is silvery grey, the greater part of the opercular region silvery but the upper quarter of the operculum itself is usually darker. There are no traces of markings on the snout, but a weak and often ill-defined lachrymal stripe or blotch is present; generally this mark does not reach ventrally much below the margin of the preorbital bone but in a few specimens it extends (albeit very faintly) to a level slightly below the gape.

The dorsal, caudal and anal fins are greyish-hyaline, and are immaculate. The pelvics are black over about the anterior half of each fin, and variously sooty over the remainder.

Adult females. Only 3 specimens (all apparently spent and quiescent) are available. The body and head coloration is essentially like that of males except that there is no lachrymal bar or blotch. All the fins are hyaline, but there are very faintly sooty lappets to the spinous dorsal, and a light scattering of melanophores on the membrane between the middle few rays of the caudal fin; when the caudal is closed it appears to have a dark midlateral region.

ECOLOGY. *Habitat.* The species is known from three localities in the Speke Gulf (see p. 179). In all, the substrate is mud, and the depth between *c.* 8 and 12 m.

FOOD. Two of the 16 specimens examined were empty, the remainder all contained fragments of mollusc shells in their stomachs and intestines. The gastropod *Melanoides tuberculata* was present in all specimens, usually as the sole or predominant food organism, but in 3 fishes there were, in addition to the snails, a few fragments of bivalve shells (unfortunately too fragmentary to allow further identification).

BREEDING. Nothing is known about the reproductive habits of *H. mylergates.* All the specimens examined are adults, the two largest (128·0 and 137·0 mm S.L.) being females.

DIAGNOSIS AND AFFINITIES. From all other Lake Victoria *Haplochromis* species with a hypertrophied pharyngeal mill, *H. mylergates* is distinguished by the coloration of its adult males and by the relatively greater surface area of its pharyngeal bones; further, the lower pharyngeal bone is more concave than in any other species. The shape, in dorsal view, of the snout outline is also diagnostic (see Fig. 23).

From the 3 species with the most hypertrophied pharyngeal mills (*H. ishmaeli, H. pharyngomylus* and *H. teegelaari*), *Haplochromis mylergates* is further distinguished as follows:

(i) From *H. pharyngomylus* (see Greenwood, 1960 : 270, fig. 19) by its slightly deeper body (38·0–45·0, M=42·0% standard length, cf. 33·8–42·0, M=38·5%), larger eye, even in specimens of a comparable size or larger (28·0–33·0, M=31·0% head, cf. 23·0–31·8, M=26·5%), enlarged cephalic lateral line pores (especially those of the preorbital bone and the pore situated immediately above the eye), longer pectoral fin (86·0–103·0, M=92·0% head, cf. 68·5–91·0, M=79·6%) and more numerous teeth in the outer row of the upper jaw (44–60, modal range 50–55, cf. 30–42, mode 36).

(ii) From *H. ishmaeli* (see Greenwood, 1960 : 275, fig. 21) by its larger eye, even in specimens of a comparable size or larger (28·0–33·0, M=31·0% head, cf. 23·0–31·0, M=27·7%), by the larger scales on its chest (6½–7½, rarely 8, cf. 8 or 9, rarely 7) and by the enlarged cephalic lateral line pores (again, those on the preorbital bone and that immediately above the eye).

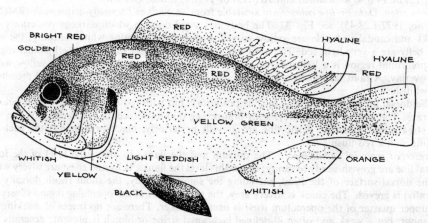

Fig. 32 *Haplochromis mylergates.* Adult male (sexually active), to show coloration. Drawn by M. J. P. van Oijen.

(iii) From *H. teegelaari* (see above, p. 169, Figs 22–27) by the absence of distinct cephalic markings (especially the large supraorbital bars or blotches), the absence of vertical bars on the body of preserved specimens, by the enlarged cephalic lateral line pores and by the presence of rather more teeth in the outer row of the upper jaw (44–60, modal range 50–55, cf. 38–54, modal range 40–44).

As with *H. teegelaari* (see above p. 174) it is difficult to determine the precise phyletic relationships of *H. mylergates*, and for the same reasons: a lack of apomorph characters that are indisputably non-convergent ones. In both species the most obvious apomorph features are connected with the hypertrophy of the pharyngeal mill. Whatever the phyletic relationships of *H. mylergates* may be (that is, either with the *H. obtusidens–H. ptistes* lineage or with the *H. ishmaeli–H. pharyngomylus* one) it must be considered to have the most highly developed pharyngeal mill of all. In terms of its habitat and depth preferences *H. mylergates* does not, however, seem to differ significantly from such species as *H. ishmaeli* and *H. pharyngomylus*, but more data on distribution and, especially, feeding habits are required before this impression is confirmed.

Since *H. mylergates* does not share the derived neurocranial features of *H. obtusidens* and *H. ptistes* it cannot be placed in that lineage. Its specialized features (pharyngeal bone shape and size, see above) in themselves do not allow its addition to the *H. ishmaeli–H. pharyngomylus* lineage with any certainty since these could well be products of convergent evolution. For the moment the species must remain in a phyletic limbo (where it joins a number of other members of the Lake Victoria *Haplochromis* flock).

STUDY MATERIAL AND DISTRIBUTION RECORDS

Museum and Reg. No.	Locality	Collector
	TANZANIA	
BM(NH) 1977.1.10:88 (Holotype)	Speke Gulf, west of Nafuba Island (*c.* 10–12 m)	Anker & Barel
BM(NH) 1977.1.10:89–90 (Paratypes)	Speke Gulf, west of Nafuba Island (*c.* 10–12 m)	Anker & Barel
BM(NH) 1977.1.10:91 (Paratype)	Speke Gulf, midway between Kisoria Point and Nafuba Island (*c.* 8 m)	Anker & Barel
BM(NH) 1977.1.10:92–105 (Paratypes)	Speke Gulf; bay north of Nafuba Island (*c.* 10 m)	Anker & Barel

Acknowledgements

Both the authors are much indebted to the many people who assisted with the field work in Tanzania; in particular we wish to thank Dr G. K. Libaba, Director of Fisheries, Tanzania, whose generous help, enthusiasm and hospitality greatly aided the work of Drs Anker and Barel. We must also thank Dr J. Okedi, Director of E.A.F.F.R.O., Jinja, and the members of the U.N.D.P. team who helped the senior author during his work on Lake Victoria in 1970 (the full results of which will be published in the next part of this revision). Dr L. B. Mkwizu (then acting principal of the Freshwater Fisheries Institute, Nyegezi) and ir. H. Bon, are to be thanked for the unstinted help they gave to Drs Anker and Barel.

We also have much pleasure in thanking our colleagues at the British Museum (Natural History) and the Zoology Department of Leiden University for their help at various stages in the preparation of this paper. In particular we thank Mr Gordon Howes (who has provided all the anatomical illustrations, and a great deal of assistance in innumerable other ways), Mr C. Elzenga who drew the figure of *H. teegelaari* and Mr M. J. P. van Oijen who made all the other illustrations of whole fishes used in this paper, except Fig. 28.

The senior author is greatly indebted to Professor Pieter Dullemeijer of Leiden University for the hospitality and facilities he has given him on numerous visits to the Zoology Department there.

Finally, the junior author would express his gratitude to the Netherlands Foundation for the Advancement for Tropical Research (WOTRO) for their generous financial aid which enabled him and Dr Anker to visit Tanzania and thus collect the data and material on which much of this paper is based.

Appendix: The live coloration of certain previously described *Haplochromis* species

M. J. P. van Oijen
Leiden University

No information was available on the live colours of several *Haplochromis* species described, or redescribed, in previous parts of this revision. Live coloration, especially that of adult male fishes, is an important diagnostic character (see Greenwood, 1974), and is often the easiest and most reliable character on which to base preliminary field identifications (especially when the worker is faced with several hundreds of specimens recently caught in a trawl or seine net); it is also an important biological feature in this closely related species flock.

For these reasons it is essential that colour descriptions should be available for all known species. With this objective in view Drs Anker and Barel took data on live colours not only from the new taxa they collected but also from those species whose coloration was previously unknown or was inadequately documented.

Live specimens were chosen from the catch and immediately photographed (using Kodachrome film) in a cuvette especially made for this purpose. The descriptions given below are based principally on the resulting colour transparencies. One set of transparencies, together with the preserved bodies of the fishes photographed, are now deposited in the British Museum (Natural History); the register number for the specimen is quoted (together with the fish's standard length, sexual state and its locality) as part of each description.

Since the colour descriptions previously published by Greenwood were also based on recently captured specimens, it is thought that emotional factors which could influence colours and colour patterns should be similar to those affecting the specimens described here. However, it should be borne in mind that Greenwood's data were derived from specimens held in air, and were taken from direct observations on the fishes and not from photographs. Another difference that should be noted is the fact that Greenwood's descriptions were compounded from several specimens (albeit ones at a similar stage of sexual development), and sometimes from fishes caught at different localities. The descriptions that follow are each based on a single specimen.

An annotated figure (the outline based on a drawing previously published with the species' description) accompanies each account of coloration. The drawing shows the pattern of body and cephalic markings, and the prominent colours for various parts of the body and fins. The pectoral fin is omitted so as not to obscure details of coloration on the anterior region of the body; notes on pectoral fin coloration are given in the description.

In the descriptions, the number of egg dummies (anal ocelli) refers to the individual described, but it must be realized that the number does show marked intraspecific variability.

Haplochromis serranus (Pfeffer), 1896
see Greenwood (1962 : 152)

Adult ♂, S.L. 182·0 mm (BM(NH) 1977.1.28:28). Fig. 33.
LOCALITY. Mwanza Gulf, depth 7 m, mud bottom.
MARKINGS. Lachrymal stripe and faint opercular blotch.
COLORATION. *Head.* Except for the whitish lips, grey-blue. *Eye.* Iris bluish, inner ring yellow. *Body.* Dorsum, chest and belly grey-blue. Flank yellow-green with three small orange spots, one just above the operculum, the others above the anal and pelvic fins. Caudal peduncle greyish dorsally and greenish ventrally. *Fins.* Dorsal, pectoral and pelvic fins grey-blue. Anal dark red, with orange egg dummies. Caudal very dark proximally and somewhat lighter grey distally, with a faint red flush and dark spots between the rays.

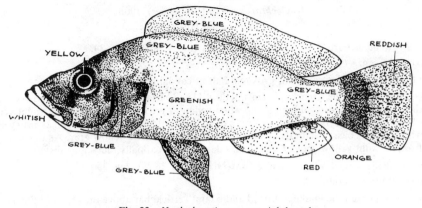

Fig. 33 *Haplochromis serranus.* Adult male.

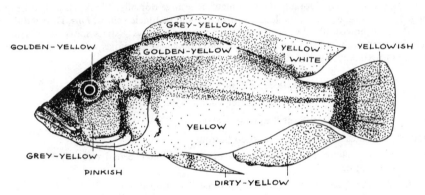

Fig. 34 *Haplochromis spekii.* Female, sexually active.

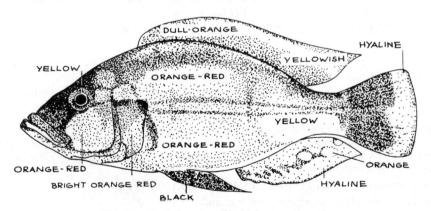

Fig. 35 *Haplochromis spekii.* Male, sexually active.

Haplochromis spekii (Boulenger), 1906
see Greenwood (1967 : 32)

Adult ♀ sexually active, S.L. 204·0 mm (BM(NH) 1977.1.28:30). Fig. 34.

LOCALITY. Mwanza Gulf.

MARKINGS. Faint midlateral band, proximal half of caudal fin very dark.

COLORATION. *Head.* Lips, preorbital area and dorsal head surface very dark; cheek and operculum greyish-yellow, branchiostegal membrane white-yellow. *Eye.* Iris dark, inner ring bright yellow. *Body.* Dorsum dark greyish-golden; ventral part of body golden yellow. *Fins.* Dorsal dirty yellow-grey with a darker base. Pectorals hyaline with dark rays. Pelvics yellow-grey. Anal yellow-grey with five very faint yellow spots. Caudal very dark grey proximally, dark grey distally.

Adult ♂ sexually active, S.L. 183·0 mm (BM(NH) 1977.1.28:29). Fig. 35.

LOCALITY. Mwanza Gulf.

MARKINGS. Faint double midlateral band and a faint vertical bar above the origin of the anal fin.

COLORATION. *Head.* Preorbital region and dorsal head surface dark. Upper lip dark red. Cheek light orange-red dorsally, white ventrally. Operculum bright orange-red. Lower lip, lower jaw, horizontal arm of the preoperculum, and branchiostegal membrane whitish. *Eye.* Iris blackish, inner ring yellow. *Body.* Dorsum dark, somewhat orange dorsally. Flank, chest and belly bright orange-red. Ventral aspect of the flank whitish. Caudal peduncle yellow. *Fins.* Dorsal dull orange, the soft part yellowish distally. Pectorals hyaline with dark rays. Pelvics sooty. Anal hyaline with a very faint red flush and some grey patches; two orange egg dummies with grey margins. Caudal dark grey proximally, hyaline distally.

Haplochromis microdon (Boulenger), 1906
see Greenwood (1959 : 200)

Adult ♂, sexually active, S.L. 88·0 mm (BM(NH) 1977.1.28:25). Fig. 36.

LOCALITY. Mwanza Gulf, depth 8 m, mud bottom.

MARKINGS. Faint lachrymal stripe, distinct opercular blotch, broad midlateral band and faint traces of a dorsal band.

COLORATION. *Head.* Dark grey, the posterior cheek margin darker still. *Eye.* Iris greyish, inner ring coppery. *Body.* Dorsum and dorsal part of the caudal peduncle yellowish grey; anterior part of the flank coppery, posterior part and caudal peduncle green. *Fins.* Dorsal grey, with a thin red margin and a green flush distally on the soft part only. Pectorals hyaline, pelvics black. Anal with a red flush, two red egg dummies. Caudal dark brownish proximally, with a purple flush distally, its margin and the distal angles red.

Adult ♂, sexually active, S.L. 92·0 mm (BM(NH) 1977.1.28:26). Fig. 37.

LOCALITY. Speke Gulf, depth 8 m, mud bottom.

MARKINGS. Very faint lachrymal stripe. Midlateral band interrupted half way; four faint vertical bars on the flank.

COLORATION. *Head.* Very dark brown except for the whitish ventral region, and the posteroventral part of the operculum which is reddish. *Eye.* Iris very dark, inner ring ivory pink. *Body.* Dorsum dark brown. Dorsal flank brown anteriorly, green posteriorly. Ventral part of the body anterior to the anal fin red. Caudal peduncle green, ventral aspect white. *Fins.* Dorsal dark grey with a brown flush. Pectorals hyaline with dark rays. Pelvics black. Anal red but greyish around the red egg dummies which have black margins. Caudal dark brown proximally, grey distally with red posterior and ventral margins.

Adult ♀, sexually active, S.L. 100·0 mm (BM(NH) 1977.1.28:27). Fig. 38.

LOCALITY. Mwanza Gulf, depth 8 m, mud bottom.

MARKINGS. Opercular blotch, five vertical bars on the flanks.

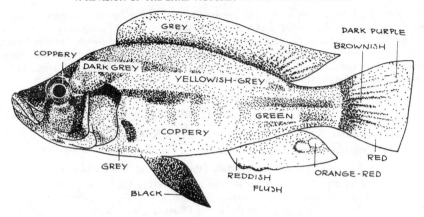

Fig. 36 *Haplochromis microdon*. Male, sexually active.

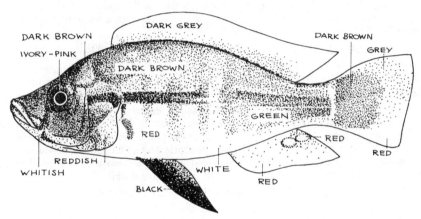

Fig. 37 *Haplochromis microdon*. Male, sexually active.

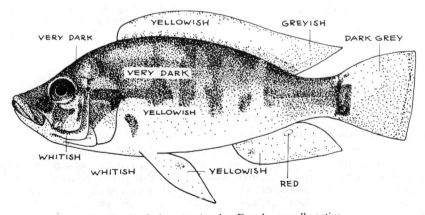

Fig. 38 *Haplochromis microdon*. Female, sexually active.

COLORATION. *Head.* Very dark except for the ventral part of the operculum, part of the cheek and the branchiostegal membrane which are whitish. *Eye.* Iris dark, inner ring light ventrally. *Body.* Dorsum and upper flank very dark, central part of flank yellowish. Whitish ventrally. *Fins.* Dorsal yellowish, faintly red between the rays; lappets and soft part greyish. Pectorals hyaline with dark rays. Pelvics yellowish. Anal yellowish with one red spot. Caudal dark grey with a dark base and dark spots between the rays.

Haplochromis macrognathus Regan, 1922
see Greenwood (1962 : 183)

Adult ♂, sexually active, S.L. 131·0 mm (BM(NH) 1977.1.28:16). Fig. 39.

LOCALITY. Mwanza Gulf, depth 2 m, mud bottom.

MARKINGS. Thin and very faint midlateral band, faint opercular blotch and lachrymal stripe.

COLORATION. *Head.* Lips, ethmoid region and the area dorsal to the eye grey. Preorbital region, cheek and operculum metallic blue-grey, somewhat pinkish ventrally. Branchiostegal membrane black. *Eye.* Iris greyish, inner ring whitish. *Body.* Dorsum grey. Flank and caudal peduncle greyish, flecked with light blue, light green and pink. Chest and belly pinkish-grey, a green sheen just ventral to the pectoral fin base. *Fins.* Dorsal grey proximally, distal part hyaline with darker patches between the rays. Pectorals hyaline. Pelvics black. Anal proximally and anteriorly dark red, distally faintly red, with two orange egg dummies. Caudal with a brownish base shading to grey distally; some fin rays black proximally.

Adult ♀, sexually active, S.L. 152·0 mm (BM(NH) 1977.1.28:18). Fig. 40.

LOCALITY. Mwanza Gulf, depth 8 m, mud bottom.

MARKINGS. Distinct midlateral band, beginning at the anterior opercular margin and terminating on the proximal part of the caudal fin; a dorsal lateral band is also present.

COLORATION. *Head.* Lips and ethmoid region dark sandy. Preorbital area, cheek and operculum dark grey-silver. Branchiostegal membrane yellowish, rest of head greyish. *Eye.* Iris sandy, inner ring golden. *Body.* Dorsum, flank and caudal peduncle yellowish-silver, becoming grey dorsally. Chest, belly and ventral aspect of caudal peduncle white. *Fins.* Dorsal, yellow proximally, hyaline distally. Pectorals hyaline. Pelvics and anal yellow, the anal with two small and light orange spots. Caudal dark yellow grey proximally and hyaline distally.

Adult ♀, sexually active, S.L. 165·8 mm (BM(NH) 1977.1.28:17). Fig. 41.

LOCALITY. Mwanza Gulf, depth 7 m, mud bottom.

MARKINGS. Broad midlateral band and narrower dorsal band, midlateral band continued faintly onto the operculum, and terminating on the proximal part of the caudal fin. Faint mental spot.

COLORATION. *Head.* Preorbital region and the area dorsal to the eye sandy; rest of head silver. *Eye.* Iris silverish, inner ring yellowish. *Body.* Silver. *Fins.* Dorsal and pectorals hyaline, pelvics and anal yellow, the anal with two small yellow spots. Caudal faintly yellow proximally, hyaline distally.

Haplochromis longirostris (Hilg.), 1888
see Greenwood (1962 : 171)

Adult ♂, sexually active, S.L. 124·0 mm (BM(NH) 1977.1.28:31). Fig. 42.

LOCALITY. Mwanza Gulf, depth 2 m, mud bottom.

MARKINGS. Faint lachrymal stripe; three faint vertical bars on the flank.

COLORATION. *Head.* Lips, preorbital region and dorsal head surface yellow. Cheek and dorsal part of the operculum silver, rest of head white. *Eye.* Iris silvery, inner ring faint yellow. *Body.* Dorsal aspect yellow, ventral aspect silvery-white. *Fins.* Dorsal yellow, posterodorsal angle of soft part hyaline. Pectorals translucent with a yellow flush. Pelvics black. Anal white with two bright red egg dummies. Caudal yellowish proximally, hyaline and maculate distally, the rays black.

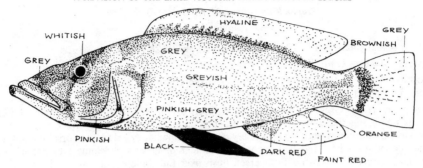

Fig. 39 *Haplochromis macrognathus*. Male, sexually active.

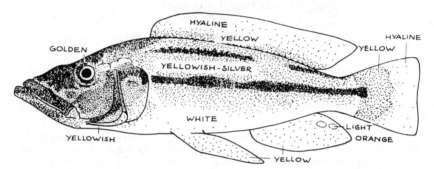

Fig. 40 *Haplochromis macrognathus*. Female, sexually active.

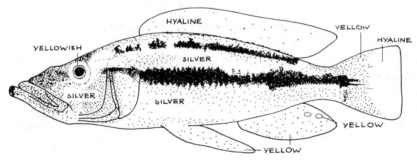

Fig. 41 *Haplochromis macrognathus*. Female, sexually active.

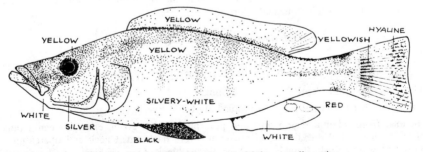

Fig. 42 *Haplochromis longirostris*. Male, sexually active.

Haplochromis paraguiarti Greenwood, 1967
see Greenwood (1967 : 69)

Adult ♂, sexually active, S.L. 107·0 mm (BM(NH) 1977.1.28:9). Fig. 43.

LOCALITY. Mwanza Gulf, depth 2 m, sand bottom.

MARKINGS. Very faint traces of a lachrymal stripe continued through the eye; an opercular blotch; and six vertical bars on the flank.

COLORATION. *Head*. Grey, darker dorsally, with an overall red flush. An iridescent blue-green stripe passes along the lower lip to the preoperculum. *Eye*. Iris black, inner ring yellowish. *Body*. Dorsum dark grey with some coppery patches immediately below the dorsal fin and dorsal to the operculum. Chest and belly sooty. Remainder of body grey. Some green spots on the flank. *Fins*. Dorsal grey with a faint green flush over the spinous part, and reddish soft rays. Pectorals with a red flush. Pelvics black. Anal reddish-grey, but sooty anteriorly; four muddy orange egg dummies with grey margins. Caudal reddish-grey.

Haplochromis percoides (Boulenger), 1915
see Greenwood (1962 : 189)

Adult ♀, sexually active, S.L. 83 mm (BM(NH) 1977.1.28:11). Fig. 44.

LOCALITY. Mwanza Gulf, depth 1 m, mud and sand bottom.

MARKINGS. Faint lachrymal stripe. Four vertical bars on the flank (two distinct, broad bands and two fainter ones).

COLORATION. *Head*. Uniformly brown except for the yellow branchiostegal membrane. *Eye*. Iris black, inner ring orange. *Body*. Dorsum orange-yellow to brown, flanks greyish-yellow. Caudal peduncle greyish-yellow, darker dorsally. *Fins*. Dorsal grey-yellow with clear maculae. Pectorals hyaline. Pelvics yellow with black streaks between some rays, and also distally. Anal yellow with one orange spot. Caudal yellow with an orange flush and clear maculae.

Haplochromis apogonoides Greenwood, 1967
see Greenwood (1967 : 105)

Adult ♂, sexually active, S.L. 151·0 mm (BM(NH) 1977.1.28:8). Fig. 45.

LOCALITY. Mwanza Gulf, depth 11 m, mud bottom.

MARKINGS. Very faint lachrymal stripe.

COLORATION. *Head*. Brown, dark spots on the branchiostegal membrane. *Eye*. Iris bluish dorsally, inner ring light. *Body*. Brown dorsum with an orange flush, but bright yellow-green flanks. Ventral aspect of body brown. Caudal peduncle brown dorsally, greyish ventrally. *Fins*. Dorsal brown with a black margin on the spinous part and bright red maculae on the soft part. Pectorals hyaline. Pelvics black except for a bright red region on the posterior margin. Anal bright red with five orange egg dummies, the area surrounding the egg dummies hyaline. Caudal dark proximally, bright red distally, maculate dorsally.

Haplochromis dichrourus Regan, 1922
see Greenwood (1967 : 65)

Adult ♂, sexually active, S.L. 115·0 mm (BM(NH) 1977.1.28:32). Fig. 46.

LOCALITY. Mwanza Gulf, depth 2 m, mud bottom.

MARKINGS. Distinct lachrymal stripe, continued above the eye.

COLORATION. *Head*. Mental region, lips and preorbital region green, ethmoid region pinkish. Dorsal head surface and dorsal part of operculum very dark red. Cheek and operculum very dark grey with a pink flush. Branchiostegal membrane black. *Eye*. Iris black, inner ring golden.

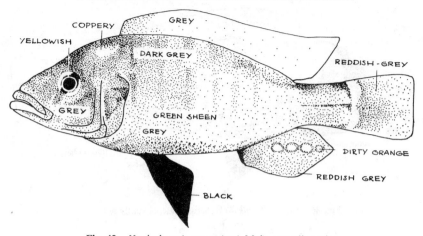

Fig. 43 *Haplochromis paraguiarti*. Male, sexually active.

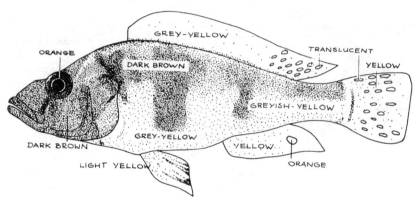

Fig. 44 *Haplochromis percoides*. Female, sexually active.

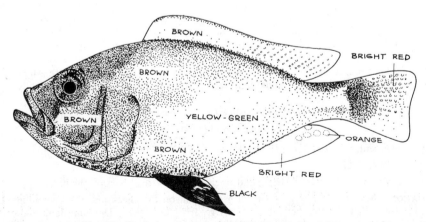

Fig. 45 *Haplochromis apogonoides*. Male, sexually active.

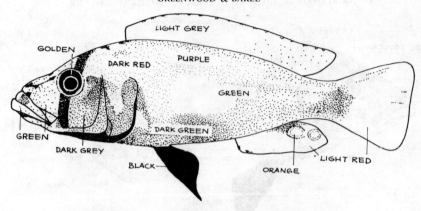

Fig. 46 *Haplochromis dichrourus*. Male, sexually active.

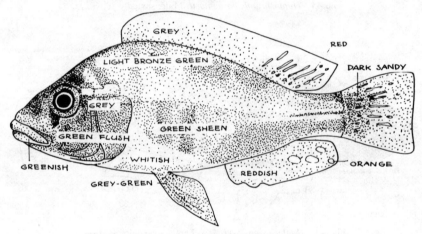

Fig. 47 *Haplochromis riponianus*. Male, sexually quiescent.

Body. Dorsum anteriorly intense dark red, purple posteriorly. Flank light iridescent green, darker posteriorly. Chest and belly dark green to black. Caudal peduncle dark green ventrally, purple dorsally. *Fins.* Dorsal light grey, with a red flush anteriorly. Pectorals hyaline. Pelvics black. Anal light red, more intensely so anteriorly. Two orange egg dummies with a light inner, and grey to black outer margin. Caudal light red ventrally, grey with a red flush dorsally.

Haplochromis riponianus (Boulenger), 1911
see Greenwood (1960 : 252)

Adult ♂, quiescent, S.L. 104·0 mm (BM(NH) 1977.1.28:22). Fig. 47.

LOCALITY. Mwanza Gulf.

MARKINGS. Very faint lachrymal stripe, and traces of two vertical bars on the flank.

COLORATION. *Head.* Uniformly grey with a green flush on the cheek and the operculum. Lips light greenish, a lighter patch behind the eye. *Eye.* Iris black. *Body.* Dorsum, flank and caudal peduncle light bronze green, the caudal peduncle darkest. A green sheen on the flank; chest and belly pinkish-white. *Fins*. Dorsal grey with a red flush and red spots between the soft rays. Pectorals hyaline. Pelvics grey-green. Anal with a dark red flush and three orange egg dummies. Caudal dark sandy with dull red spots between the rays.

Adult ♂, quiescent, S.L. 91·0 mm (BM(NH) 1977.1.28:20). Fig. 48.

LOCALITY. Mwanza Gulf, depth 2 m, mud and sand bottom.

MARKINGS. Faint lachrymal stripe continued through the eye; five faint vertical bars on the flank and a faint midlateral band caudally.

COLORATION. *Head*. Lips and preorbital region grey with a faint orange flush; remainder of head grey with a green flush which is most distinct on the cheek and the operculum. *Eye*. Iris greyish but black dorsally, inner ring yellow. *Body*. Dorsum and caudal peduncle grey, flank green-grey with a green flush over the vertical bars. Ventral body grey-white. *Fins*. Dorsal, pectorals and pelvics light grey. Anal hyaline with a very faint red flush; four orange egg dummies, some with a faint grey outline. Caudal grey with a faint brown flush and red to dark maculae.

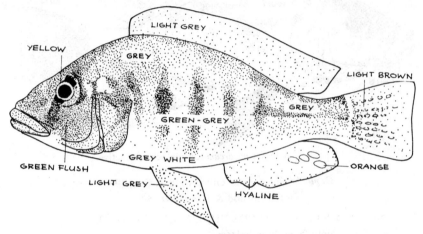

Fig. 48 *Haplochromis riponianus*. Male, sexually quiescent.

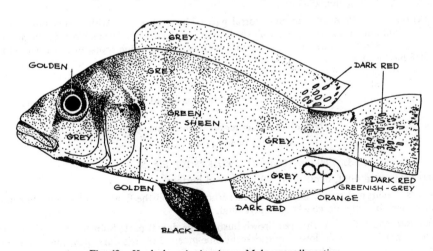

Fig. 49 *Haplochromis riponianus*. Male, sexually active.

Adult ♂, sexually active, S.L. 94·0 mm (BM(NH) 1977.1.28:21). Fig. 49.

LOCALITY. Mwanza Gulf, depth 2 m, mud and sand bottom.

MARKINGS. Faint lachrymal stripe and traces of five vertical bars on the flank.

COLORATION. *Head*. Dorsally, and posterior to the eye, dark grey with a silvery area between these regions. Rest of head grey. *Eye*. Iris black dorsally and dark grey ventrally, inner ring golden. *Body*. Grey with green iridescent scales on the flank. *Fins*. Dorsal grey, dark red basally and some red maculae on the soft part. Pectorals hyaline, with thin black streaks; pelvics black. Anal dark red anteriorly and distally, grey posteriorly; two orange egg dummies with black margins. Caudal greenish-grey proximally, becoming reddish distally; fin rays green. Part of the caudal margin is black.

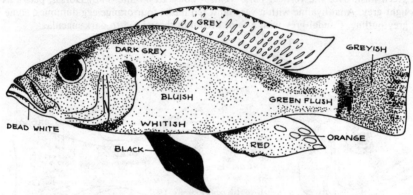

Fig. 50 *Haplochromis arcanus*. Male, sexually active.

Haplochromis arcanus Greenwood & Gee, 1969
see Greenwood & Gee (1969 : 52)

Adult ♂, sexually active, S.L. 134·0 mm (BM(NH) 1977.1.28:77). Fig. 50.

LOCALITY. Speke Gulf, depth 26 m, mud bottom.

MARKINGS. A small opercular blotch.

COLORATION. *Head*. Preoperculum and dorsal part of head dark grey with a pink flush, cheek and operculum dark grey with a green flush. Lower jaw dead white. Branchiostegal membrane blackish. *Body*. Dorsum dark grey, flank bluish with two darker blue spots, one just behind the opercular blotch and other above the anal fin; chest and belly greyish. Caudal peduncle grey with an overlying green flush. *Fins*. Dorsal grey, with black lappets and red maculae. Pectorals greyish. Pelvics black. Anal red, the anterior half of the distal margin black, hyaline around the five orange egg dummies. Caudal greyish with a black base, red maculae and a sooty distal part.

Haplochromis lacrimosus (Boulenger), 1906
see Greenwood (1960 : 230)

Adult ♂, quiescent, S.L. 79·0 mm (BM(NH) 1977.1.28:19). Fig. 51.

LOCALITY. Mwanza Gulf, depth 1 m, mud and sand bottom.

MARKINGS. Lachrymal stripe, traces of seven vertical bars on the flank and two on the caudal peduncle, dark blotches on cheek and operculum.

COLORATION. *Head*. Grey, with a red-brown flush on the dorsal part; branchiostegal membrane whitish. Lower jaw reddish, lower lip iridescent blue-green. *Eye*. Iris dark grey, inner ring light golden. *Body*. Dorsum red-brown anteriorly, yellowish-brown posteriorly. Flank and belly dull

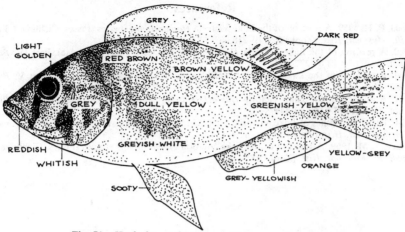

Fig. 51 *Haplochromis lacrimosus*. Male, sexually quiescent.

yellow with a green sheen. Chest and ventral aspect greyish-white. Caudal peduncle greenish-yellow. *Fins.* Dorsal grey, with a red flush and dark red maculae on the soft part. Pectorals hyaline. Pelvics sooty over yellow. Anal grey-yellowish with two orange egg dummies. Caudal yellowish-grey with a red flush proximally, dark red dorsally.

Haplochromis parvidens (Boulenger), 1911
see Greenwood (1959 : 194)

Juvenile ♀, S.L. 99·0 mm (BM(NH) 1977.1.28:35). Fig. 52.

LOCALITY. Mwanza Gulf, depth 1 m, sand bottom.

MARKINGS. Faint opercular blotch and seven more or less distinct vertical bars on the flank.

COLORATION. *Head.* Ethmoid region and frontal part of dorsal head surface dark, remainder dark golden-yellow. *Eye.* Iris dark golden-yellow, inner ring orange. *Body.* Dorsum light golden-grey, flank grey-silver with a yellow overlay. Chest and belly greyish golden. Caudal peduncle dark golden-yellow. *Fins.* Dorsal and caudal dark yellow grey. Pectorals hyaline, pelvics and anal yellow, the latter hyaline in the area around the two small orange spots.

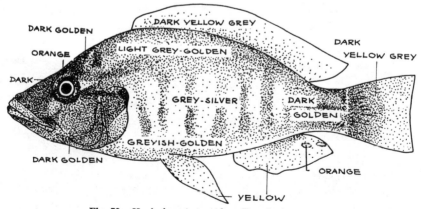

Fig. 52 *Haplochromis parvidens*. Female, immature.

485

References

Greenwood, P. H. 1959. A revision of the Lake Victoria *Haplochromis* species (Pisces, Cichlidae), Part III. *Bull. Br. Mus. nat. Hist.* (Zool.) **5** : 179–218.
—— 1960. A revision of the Lake Victoria *Haplochromis* species (Pisces, Cichlidae), Part IV. *Bull. Br. Mus. nat. Hist.* (Zool.) **6** : 227–281.
—— 1962. A revision of the Lake Victoria *Haplochromis* species (Pisces, Cichlidae), Part V. *Bull. Br. Mus. nat. Hist.* (Zool.) **9** : 139–214.
—— 1967. A revision of the Lake Victoria *Haplochromis* species (Pisces, Cichlidae), Part VI. *Bull. Br. Mus. nat. Hist.* (Zool.) **15** : 29–119.
—— 1974. The cichlid fishes of Lake Victoria, East Africa: the biology and evolution of a species flock. *Bull. Br. Mus. nat. Hist.* (Zool.) Suppl. 6 : 1–134.
—— **& Gee, J. M.** 1969. A revision of the Lake Victoria *Haplochromis* species (Pisces, Cichlidae), Part VII. *Bull. Br. Mus. nat. Hist.* (Zool.) **18** : 1–65.

THE CICHLID FISHES OF LAKE NABUGABO, UGANDA

By P. H. GREENWOOD

CONTENTS

	Page
INTRODUCTION	315
Tilapia	318
Haplochromis velifer	319
Haplochromis simpsoni sp. nov	325
Haplochromis annectidens	329
Haplochromis beadlei	335
Haplochromis venator sp. nov.	342
Haplochromis nubilus	346
Hemihaplochromis multicolor	348
Astatoreochromis alluaudi	349
DISCUSSION	351
KEY TO THE GENERA OF CICHLIDAE IN LAKE NABUGABO	355
KEY TO THE *Haplochromis* SPECIES	355
SUMMARY	356
ACKNOWLEDGEMENTS	356
REFERENCES	356

INTRODUCTION

LAKE NABUGABO is a small body of open water lying within an extensive swamp which fills a former bay on the western shore of Lake Victoria (see map, fig. 1.). Its shape is roughly pyriform, the main axis about five miles long and the width approximately three miles. Except for the western shore, the lake margin is swamp. The western shore is more varied, with, in some places, gently sloping sandy beaches and in other places forest reaching to the lake edge. The swamp margin begins with a zone of Hippo grass (*Vossia cuspidata*) whose rhizomes grow out into the open water. Behind the Hippo grass is a high "hedge" of the grass *Miscanthidium* forming a floating platform of matted roots and rhizomes jutting out into the lake. Away from the margin this platform becomes more solid through the addition and incorporation of dead and decaying vegetation. In places the platform, although still afloat, is sufficiently compacted to support the growth of trees.

Along its eastern and south-eastern lakeside boundaries the *Miscanthidium* zone is replaced by a large area of *Sphagnum* swamp, an unusual feature at this altitude in east Africa and one not encountered in the swamps around Lake Victoria.

The Bladder-wort *Utricularia* is common in the open water pools found within the eastern part of the swamp. In this region the dominant plant is the grass *Loudetia phragmitoides*, its tussocky habit allowing the development of small open pools. *Utricularia* is also common in the sheltered bays and inlets around the lake margin, and along the sheltered open shores.

Two species of water-lily (*Nymphaea lotus* and *N. caerulea*) occur in the open lake, especially in sheltered bays. Some stands of *Papyrus* are found all around the lake shore but this plant is nowhere a dominant. The relative scarcity of *Papyrus* in Nabugabo contrasts strongly with other swampy areas in the surrounding countryside and, particularly, in Lake Victoria (For a more detailed account, see Beadle and Lind, 1960).

Nowhere in Lake Nabugabo is the water more than fifteen feet deep; in most places it is between five and twelve feet. Except along the western shore the bottom does not shelve and it is in that area that the few patches of exposed sandy bottom are found. Elsewhere the sand is covered by a blanket of liquid mud (Cambridge expedition's field notes).

Very little published information is available on the hydrology of the lake. During the visit of the second Cambridge expedition (June to August, 1962) the open water was well-mixed and supersaturated with oxygen, even in the upper layers of mud. The oxygen content fell sharply in the deeper mud layers and in the water at the bottom of swamp inlets; surface water in these inlets was, however, as highly oxygenated as that of the open lake.

The water of Nabugabo is more alkaline than that of Lake Victoria (pH. of open water 8·2 *cf* 7·8 for Victoria) but in the surrounding swamps it is more acid (pH. 5·35–6·00). Perhaps the most striking hydrological feature is the very low salt concentration of the lake water; its electrical conductivity is about a quarter of that for water from Lake Victoria. Unfortunately no detailed water analyses are yet available.

The main affluents to Nabugabo are the Juma river and the Lwamunda swamp; the latter is fed by small, swampy rivers. Numerous small springs discharge along the western lake shore. The outflow of the lake is into Lake Victoria (some fifty feet lower) and is effected solely by seepage through the sand-bar which forms the eastern barrier between the two water masses; there is no surface contact.

Present day lake Nabugabo represents the greatly diminished body of open water which was gradually cut off from Lake Victoria by the formation of longshore bars across the mouth of an extensive open bay. Bishop (1959) has described the probable history of this empondment: "The landward shore of the lake was an old shoreline of Lake Victoria which consisted in the south-west of a lateritic oldland with low cliffs.

The Juma River divided the oldland into two spurs which were linked again by a series of bay bars across its mouth. Further to the north, a curving longshore bar protruded to the north-east from the mainland and finally joined a former island at Kisasa."

"At some later period another complex longshore bar commenced to grow towards the north-north-east from a point six miles south of the Juma River and finally it also reached the former island of Kisasa to complete the enclosure of Lake Nabugabo. The lake is at present approximately 50 feet above Lake Victoria and is rapidly being overgrown by swamp vegetation. The open water is now separated from the parent lake to the south-east by more than a mile of swamp and two miles of complex sand and gravel ridges comprising the longshore bar."

The age of Lake Nabugabo has been estimated at approximately 4,000 years. This figure is based on the radiocarbon dating of some rolled charcoal fragments found in a former shoreline of Lake Victoria at about the same height above the present level of the lake as is the sandbar which cuts off Nabugabo (personal communication from Dr. Bishop quoted by Beadle, 1962).

The first collections of fishes from Lake Nabugabo were made in 1930 by the

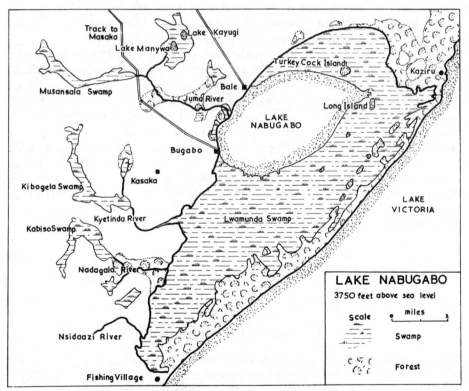

FIG. 1. Sketch map of Lake Nabugabo; after a map produced by the Cambridge Nabugabo Biological Survey.

Cambridge expedition to the East African Lakes. The Cichlidae were studied by Dr. Ethelwynn Trewavas (1933) who recognised their importance in helping to understand the evolutionary processes which had resulted in the complex *Haplochromis* species-flock of Lake Victoria. Dr. Trewavas recorded four *Haplochromis* species, of which three were described as new and endemic, and the fourth assigned to a species already known from Lake Victoria.

Since 1930 our knowledge of Lake Victoria *Haplochromis* has increased, both from the systematic and from the ecological view-points. So too has our knowledge about the geological history of the area, and there has recently been revived interest in the evolutionary problems posed by the cichlid species-flocks. Thus, it seemed desirable to revise the *Haplochromis* of Lake Nabugabo against the newly acquired information from Lake Victoria. This possibility became a reality when a group of Cambridge students offered to collect specimens from Lake Nabugabo for the British Museum (Natural History). Their collection (hereinafter referred to as the C.N.B.S. collection) has proved invaluable for several reasons. Not only did it substantially increase the number of specimens but the expedition also made detailed notes on the live colours of the fishes, and on their distribution and habitats. Observations were also made on the fishes' feeding habits and some data on breeding conditions were collected as well. With this sort of information it was possible to make a more detailed and direct comparison between the *Haplochromis* of the two lakes and thus to make a reappraisal of relationships on characters other than purely anatomical ones.

To the original four *Haplochromis* species must now be added two others and two species of *Haplochromis*-group genera, viz. *Hemihaplochromis multicolor* (Schoeller) and *Astatoreochromis alluaudi* Pellegrin. Three of the newly recorded species (*H. nubilus*, *Hh. multicolor* and *A. alluaudi*) are of fairly wide distribution in the Lake Edward-Victoria drainage basins; their occurrence in Lake Nabugabo is not surprising. Unfortunately it is not absolutely certain that one can accept their presence as natural because some introductions have been made into Lake Nabugabo since the original collections were made over thirty years ago. In 1960, the Nile Perch (*Lates*) was introduced. If the newly recorded cichlids gained access in this way it was accidental and, I would consider, unlikely if only Nile Perch were involved. *Haplochromis* are more likely to be introduced accidentally when *Tilapia* are moved from one area to another because small individuals of the two genera are easily confused.

In addition to the cichlid fishes described below, the C.N.B.S. collection contained a large number of non-cichlid species, including several new records for the lake. These fishes will be dealt with in a separate publication.

THE FISHES

I. *TILAPIA* A. Smith, 1840

Two species of *Tilapia* (*T. esculenta* Graham and *T. variabilis* Blgr.) are recorded from Lake Nabugabo; both are otherwise endemic to Lakes Victoria and Kyoga. The C.N.B.S. mentioned both species in their preliminary report but no specimens were sent to the British Museum (Nat. Hist.).

II. *HAPLOCHROMIS* Hilgendorf, 1888

Regan, C. T. 1920. The classification of the fishes of the family Cichlidae. I. The Tanganyika genera. *Ann. Mag. nat. Hist.* (9), **5** : 33–53.

Haplochromis velifer Trewavas, 1933

(Text figs. 2 and 3)

H. velifer (part) Trewavas, 1933, *J. Linn. Soc. (Zool)*, **38** : 322.

HOLOTYPE. A male 75 mm. S.L., B.M. (N.H.) Reg. No. 1933.2.23.194, collected by E. B. Worthington.

DESCRIPTION. Based on the holotype, 6 paratypes and eight additional specimens, 75–108 mm. S.L.

Depth of body 35·7–41·3 (Mean, M, = 39·1) per cent of standard length, length of head 32·3–36·0 (M = 34·7) per cent; dorsal profile of head slightly curved (but with a concavity above the orbit), sloping moderately steeply (*ca* 40°–45° to the horizontal).

Preorbital depth 13·8–18·5 (M = 16·3) per cent of head length, least interorbital width 21·8–29·0 (M = 24·1) per cent and snout length 29·1–33·4 (M = 31·3); snout slightly broader than long or as long as broad. Eye diameter shows weak negative allometry with standard length, 26·3–33·4 (M = 30·6) per cent of head; depth of cheek 21·8–26·0 (M = 23·3) per cent.

Caudal peduncle 14·5–17·5 (M = 15·9) per cent of standard length, 1·2–1·4 (mode 1·3) times as long as deep.

Mouth horizontal or very slightly oblique; jaws equal anteriorly, the lower 37·1–44·8 (M = 39·7) per cent of head, 1·4–2·0 (mode 1·5) times as long as broad. Lips often slightly thickened. Posterior tip of the maxilla reaching (the modal

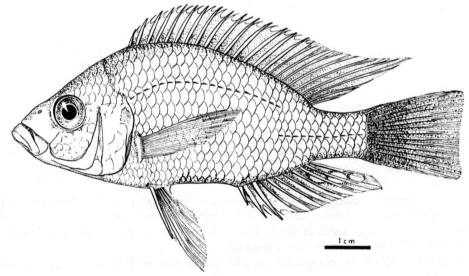

FIG. 2. *Haplochromis velifer*. Drawn by Barbara Williams.

condition) or almost reaching the vertical through the anterior orbital margin but in four specimens extending to below the anterior part of the eye.

Gillrakers short and stout, 7–9 (mode 8) on the lower limb of the first arch, the lowermost one to three rakers reduced.

Scales ctenoid; lateral line with 30 (f.4), 31 (f.7) or 32 (f.4) scales; cheek with 3 (f.14) or 4 (f.1) rows; 5–6½ (mode 6) scales between the dorsal fin origin and the upper lateral line, 5–6 (mode) between the pectoral and pelvic fin bases.

Fins. Dorsal with 23 (f.1), 24 (f.10) or 25 (f.3) rays comprising 15 (f.11) or 16 (f.3) spinous and 8 (f.3), 9 (f.9) or 10 (f.2) branched elements; anal with 11 (f.5) or 12 (f.9) rays comprising 3 spinous and 8 (f.5) or 9 (f.9) branched rays. Caudal fin truncate or subtruncate with slightly rounded distal corners (Trewavas, *op. cit.*, suggests that the degree of rounding is greatest in males), scaled on its basal half. First two branched pelvic rays produced in both sexes. Pectoral 25·4–29·8 (M = 27·6) per cent of standard length.

Teeth. The outer teeth in both jaws are slightly recurved, relatively stout and have compressed bicuspid crowns (see text-fig. 3); the postero-lateral and posterior teeth are somewhat less robust than those situated anteriorly. The acutely pointed smaller cusp is prominent and stout; the major cusp has an oblique edge which meets the nearly vertical medial aspect of the cusp at an angle of 45°–60°. There are 40–58 (mean 50) teeth in the outer row of the upper jaw.

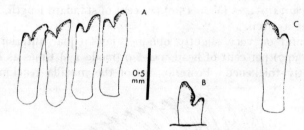

FIG. 3. *Haplochromis velifer*. Outer teeth (labial aspect) from: A, premaxilla, and C, dentary. B, newly erupted tooth (dentary).

The inner rows in both jaws are composed of small, compressed and tricuspid teeth arranged in 2–4 (mode 3) and 2 (mode) or 3 rows in the upper and lower jaws respectively. A distinct interspace separates the inner rows from the outer row.

The form of the outer teeth in *H. velifer* is one of the two generalized types found in many species of *Haplochromis*. In the other type, the major cusp is more acutely pointed.

Lower pharyngeal bone slender, the triangular dentigerous surface broader than long. The teeth are slender and cuspidate, fairly close-set and arranged in 24–28 antero-posterior rows; teeth in the two median rows are generally somewhat stouter than the others but are otherwise identical.

Vertebrae: 28 or 29, comprising 13 abdominal and 15 (f.2) or 16 (f.7) caudal elements.

Coloration in life. Males (adult but of undeterminable sexual state): ground colour dark grey-blue to olivaceous dorsally (dark indigo on head), shading on flanks to olivaceous with violet to turquoise sheen; ventral surface sooty-violet, the chest with a red flush. Males less than 65 mm. S.L. and probably juvenile have a similar coloration dorsally and laterally but the ventral surface is silvery-white and the chest lacks the red flush.

Dorsal fin dark olive (overlain with sooty in fishes >65 mm. S.L.) basally, light blue-grey distally; dull red spots and blotches between the rays; the lappets and margin to the soft part crimson. Anal fin greenish-blue basally, buff to pink distally (olivaceous in small fishes); ocelli orange with light yellow surround on a transparent area of fin membrane. Pelvics with anterior quarter sooty except for the dead-white, elongated first and second rays; remainder of fin hyaline in fishes <65 mm. S.L., pink to crimson in larger individuals. Caudal fin dark olivaceous to indigo proximally, light olive distally in small fishes, scarlet in larger ones.

Females: ground colour dull metallic grey with violet lights (particularly on the cheeks), shading through greyish-buff on the flanks to pinkish-white on the ventral surface.

Dorsal fin pale buff with a narrow basal band of crimson on that part of the fin posterior to the fifth spine. Anal with a crimson basal band followed by a broader buff band and, along the distal margin, a narrow sooty band; in some specimens there are dull orange spots in about the position of the ocelli in males. Pelvics faintly sooty.

Coloration in preserved specimens. Adult males. Body dark brown, black on belly and chest, in ripe individuals this dark coloration extending along the entire ventral surface and on the flanks as far dorsally as the upper lateral line; in some specimens there is a pearly sheen visible on the dark areas. A well-developed lachrymal stripe of varying width, two parallel transverse bars on the snout and two transverse nuchal bands are also present. The anterior nuchal band is a continuation of the lachrymal stripe; the posterior band originates at the anterior, upper angle of the operculum and also extends ventrally along the preopercular-opercular junction. On the flanks there are traces of six, moderately broad bands which reach the origin of the dorsal fin and fuse ventrally with the dark belly; these bands are much fainter than the nuchal and snout bars.

The dorsal fin in sexually quiescent fishes is hyaline but in active individuals it has dusky lappets and a solid, parabolic dusky area originating at the base of the third to sixth spinous ray from where it rises fairly steeply so that the area of membrane between the 10th and last spine is completely black; thereafter it falls rather gently so that the distal third to half of the membrane between the branched rays is hyaline but the base is black. Caudal fin is hyaline in quiescent fishes but the proximal two-thirds is black in active individuals; in some specimens darker spots are visible through the sooty ground colour. The anal has the entire interspinous membrane black; in active fishes the basal third to half of the soft part is black, the remainder hyaline (as is the entire fin in quiescent individuals); two or three, dead-white, round to oval ocelli are present in both sexes.

Certain variant patterns are fairly common; the most frequently seen being variation in the intensity and area of the dark parabola of the dorsal fin and a tendency for the posterior nuchal bar to be expanded medially into a large triangular black patch with its apex at the base of the first dorsal spine.

Females are greyish-silver, darkest dorsally; there is a fairly distinct lachrymal stripe running from the angle of the jaw through the anterior rim of the orbit. All fins are hyaline, with the soft dorsal and caudal maculate, and dark lappets to the dorsal in some specimens. Traces of six narrow vertical bars may be visible on the flanks; these do not extend to the belly and they become extremely faint near the base of the dorsal fin.

Ecology. Habitat. Haplochromis velifer has a wide distribution within the lake, being found both close inshore and at some distance out, over muddy and sandy bottoms in bays, and over the clear sandy beaches of the western shore. It does not appear to live in the isolated swamp pools, nor does it extend far up those arms of the lake which penetrate into the swamps.

Food. The C.N.B.S. notes, supplemented by a personal examination of eight additional specimens, indicate that *H. velifer* is an omnivorous bottom feeder with insect larvae (especially those of chironomids and trichopterans) providing the main source of nourishment. A considerable amount of plant material was found in the gut but it showed few signs of digestion. Since the bottom in many habitats is covered by a layer of plant matter (both algal and phanerogamic) the plant remains should probably be considered as being incidentally ingested during the search for insect larvae and other invertebrate animals. Sand grains were also recorded and in some instances these could be identified as coming from the cases of larval Trichoptera.

Breeding. No data are available. The largest specimen (108 mm. S.L.) is a female, but both sexes are found in the next largest size group, 75–85 mm. S.L.

Distribution. Known only from Lake Nabugabo.

Affinities and diagnosis. Anatomically and trophically *H. velifer* belongs to the group of generalized *Haplochromis* which are found both in the rivers and in the species-flocks of the major lakes. Comparison will be made first with the widely distributed species *H. nubilus* and *H. wingatii*, the former occurring in Lakes Nabugabo, Victoria and Edward, the latter in Lakes Edward and Albert and in the Nile.

From *H. velifer*, *H. nubilus* is immediately distinguished by the velvety black and uniform body colour of adult males, as well as by its stouter and more acutely cuspidate teeth, more obviously rounded caudal fin and its smaller eye. *Haplochromis wingatii* differs in having fewer teeth in the outer row of the upper jaw (30–40 in specimens of a size comparable with the *H. velifer* sample examined), larger and more numerous unicuspid teeth posterolaterally in this row, markedly smaller scales on the chest region, in having some blunt pharyngeal teeth and in the coloration of adult males.

Amonst the Lake Victoria endemics, *H. velifer* is perhaps nearest to *H. cinereus*, *H. macrops*, *H. lacrimosus* and *H. pallidus*. It differs from all these species in various combinations of characters (see Greenwood, 1960). In all cases the nature of the preserved coloration serves to separate males of the various species; the live coloration of male *H. macrops* (the only species of the four in which this is known) also differs markedly from that of *H. velifer*. The most obvious anatomical characters distinguishing *H. velifer* from the Victoria species are: from *H. cinereus* the dentition (in *H. cinereus* slender unicuspid or weakly bicuspid outer teeth, uni- and bicuspid, obliquely implanted inner teeth); from *H. macrops*, the absence of tricuspid teeth posterolaterally in the upper, outer tooth row of fishes < 85 mm. S.L., smaller eye (26·3–33·4, M = 30·6% of head in *H. velifer*, cf. 28·6–35·4, M = 33·0 in *H. macrops*) and deeper cheek (21·8–26·0, M = 23·3% of head cf. 17·4–24·2, M = 21·1 in *H. macrops*); from *H. lacrimosus*, the more robust less curved outer teeth with their expanded major crowns, the shorter pectoral fin (M = 81% of head in *H. velifer*, 88·5% in *H. lacrimosus*) and slightly deeper cheek (21·8–26·0, M = 23·3% of head, cf. 17·6–23·5, M = 20·5% in *H. lacrimosus*); from *H. pallidus* (which species *H. velifer* closely resembles in its oral dentition) the larger chest and nape scales, the fewer teeth in the median tooth row of the lower pharyngeal bone, and a lower modal number of gillrakers (8 cf. 9 for *H. pallidus;* the range of gillraker numbers (7–9) is identical).

In her original description of *H. velifer*, Trewavas (1933) compared the species with *H. gestri* (now a synonym of *H. obesus* (Blgr), see Greenwood 1959b). The two species are but distantly related, *H. obesus* belonging to the group of specialized larval-fish eating species which is characterized by a reduced dentition and an expansible mouth.

Amongst the Lake Edward endemics *H. velifer* shows the greatest superficial resemblance to *H. schubotzi*. It differs from this species in the following characters: its dentition, a shallower preorbital, and a shorter caudal peduncle. Small specimens of *H. schubotzi* (i.e. in the range comparable with that known for *H. velifer*) have relatively slender teeth with the minor cusp reduced and the major cusp acutely pointed; in larger specimens the teeth are, relatively, even more slender and may have the minor cusp reduced almost to vanishing point. In specimens at all sizes the inner teeth of *H. schubotzi* are tricuspid, but in specimens over 120 mm. S.L. the cuspidation is feebly manifested. Trewavas (1933), believed that the lower jaw of *H. schubotzi* is longer than in *H. velifer* but I am unable to confirm this; the lower jaw has the same relative length in both species.

The other Lake Edward species showing a superficial resemblance to *H. velifer* are *H. nigripinnis*, *H. eduardii* (including *H. vicarius* acc. Poll, 1939), *H. elegans* and *H. engystoma*.

Both *H. nigripinnis* and *H. engystoma* are distinguished from *H. velifer* by their shorter snouts, larger eyes (eye diameter about equalling snout length in *H. velifer*, much longer than snout in the two Edward species), more decurved dorsal head profile and, in *H. nigripinnis*, by the more slender and numerous gill rakers (10 or 11); also, *H. engystoma* has fewer teeth (36 in the upper, outer row of the unique holotype) with more strongly recurved cusps.

From *H. eduardii*, *H. velifer* is distinguished by its coarser teeth (and the absence of unicuspids in larger specimens), shorter, coarser and fewer gill rakers (7–9, cf.

9–12 in. *H. eduardii*), straighter dorsal head profile and shorter pectoral fin (always clearly shorter than the head in *H. velifer*, as long as the head, or nearly so, in *H. eduardii*).

Haplochromis elegans differs least of all but is nevertheless distinguished by its shallower body, somewhat shorter snout, smaller chest scales (both ventrally and laterally) and its more strongly curved dorsal head profile.

Haplochromis velifer certainly appears to have greater affinity with the Lake Victoria species discussed here than with those of Lake Edward. *Haplochromis elegans* is the only Edward species to have about the same overall degree of affinity with *H. velifer* as have the Victoria species. However, it must be remembered that these comparisons are based on fewer Edward than Victoria specimens and that less is known about their ecology and live colours.

Within Lake Nabugabo, *H. velifer* has closest affinity with the new species, *H. simpsoni* (see p.325); indeed, five paratypes of *H. velifer* are now identified as *H. simpsoni*. The species are distinguished principally by their dentition; the outer teeth of *H. simpsoni* are more slender, have an acutely pointed major cusp (the minor cusp greatly reduced or absent) and are more numerous (50–70, mean 60, in the outer, upper series, *cf.* 40–58, M = 50, for *H. velifer*). The body form of the two species is similar although the dorsal head profile of *H. simpsoni* is straighter and lacks the interorbital concavity of *H. velifer;* also, in *H. velifer* the orbit lies distinctly below the outline of the profile, whereas in *H. simpsoni* the upper margin of the orbit is generally included in the profile. In *H. velifer* the snout is broader than it is long (or at least as broad as long) but in *H. simpsoni* the snout is longer than broad (the difference becoming more pronounced in larger fishes) so that these fishes have the appearance of being thinner faced than *H. velifer*. This is reflected in the width of the lower jaw; the mean length/width ratio for *H. velifer* is 1·5 (range 1·4–1·7) and for *H. simpsoni* 2·0 (range 1·5–2·1). A difference also exists in the modal number of gill rakers (7 for *H. simpsoni*, 8 for *H. velifer*) and the lower limit for *H. velifer* is higher (7 *cf.* 6 for *H. simpsoni*). Finally, there are differences in the coloration of adult males; *H. velifer* has a red flush on the chest and crimson lappets to the spinous dorsal, this colour continuing onto the margin of the soft dorsal as well. In *H. simpsoni* the chest is sooty and the dorsal lappets black. There are also interspecific differences in the colours of the anal, caudal and pelvic fins, and the body is more definitely blue in *H. simpsoni*.

No single anatomical character can be considered diagnostic but if those mentioned above are taken in concert, the two species may be distinguished fairly readily.

STUDY MATERIAL

B.M. (N.H.) reg. no.	Collector
1933.2.23.194 (Holotype)	Worthington
1933.2.23.181–193 (Paratypes)	Worthington
1933.2.23.200–209	Worthington
1964.7.1.34–50	C.N.B.S.

Haplochromis simpsoni sp. nov.

(Text figs. 4 and 5)

H. velifer (part) Trewavas, 1933, *J. Linn. Soc. Soc.* (*Zool.*), **38** : 322. (See list of study material.)

HOLOTYPE. An adult male 88 mm. standard length (B.M.[N.H.] reg. no. 1964.7.1.12) collected by the C.N.B.S.

Named in honour of Mr. M. Simpson, one of the members of the Cambridge Nabugabo Biological Survey.

DESCRIPTION. Based on the holotype and twenty-one additional specimens, 76–114 mm. S.L.; data on dentition were also derived from eighty further specimens, collected by Capt. C. R. S. Pitman.

Depth of body 32·7–42·0 (M = 37·6) per cent of standard length, length of head 31·6–37·6 (M = 34·9) per cent. Dorsal profile of head straight (without a noticeable concavity above the eye), sloping at about 45° to the horizontal.

Preorbital depth 16·1–20·5 (M = 17·8) per cent of head, least interorbital width 18·8–24·7 (M = 22·3) per cent, length of snout 28·8–34·3 (M = 32·2) per cent; snout longer than broad. Eye diameter 25·6–32·4 (M = 29·4) per cent of head (not showing any allometry in the sample studied), depth of cheek 22·4–28·2 (M = 25·0) per cent.

Caudal peduncle 15·3–18·5 (M = 16·7) per cent of standard length, 1·1–1·6 (mode 1·4) times as long as deep.

Mouth horizontal (rarely, slightly oblique); jaws equal anteriorly, the lower 37·5–45·0 (M = 41·4) per cent of head, 1·5–2·1 (mode 2·0) times as long as broad. Posterior tip of maxilla reaching or almost reaching the vertical through the anterior orbital margin or even to slightly beyond this point.

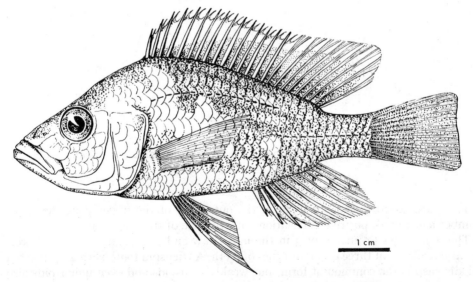

FIG. 4. *Haplochromis simpsoni.* Drawn by Barbara Williams.

Gillrakers variable, from stout to moderately slender, both extremes often occurring on the same arch; 6–9 (mode 7) on the lower part of the first arch, the lowermost 1 to 3 rakers reduced.

Scales ctenoid; lateral line with 30 (f.1), 31 (f.2), 32 (f.10) or 33 (f.8) scales; cheek with 3 (f.13) or 4 (f.9) rows; 5–7 (modal range $5\frac{1}{2}$–6) scales between the upper lateral line and the dorsal fin origin; 5 or 6 (mode) between the pectoral and pelvic fin bases.

Fins. Dorsal with 24 (f.8) or 25 (f.14) rays, comprising 15 (f.8) or 16 (f.14) spinous and 8 (f.1), 9 (f.20) or 10 (f.1) branched rays. Anal with 11 (f.12) or 12 (f.10) rays, comprising 3 spinous and 8 (f.12) or 9 (f.10) branched rays. Pectoral 26·4–33·2 (M = 27·6) per cent of standard length. First and second pelvic rays produced, proportionately more so in adult males. Caudal subtruncate.

Teeth. There are three forms of teeth in the outer row of both jaws; all are slender and recurved, and all have an acutely pointed major cusp. The commonest form has a very weakly developed minor cusp which appears as little more than a lateral spur at the base of the protracted, slender and curved major cusp (text fig. 5); a variant of this type lacks the minor cusp, either through wear or because the tooth develops without it (as can be determined from erupting teeth). The third form is relatively stouter than the other two types, has a small but distinct minor cusp and a less protracted and less acutely pointed major cusp; this form is usually restricted to a posterolateral position in the row. An admixture of all forms of tooth may occur in any fish and usually there is no difference in the proportion of the two commoner types as between upper and lower jaws. However, in some individuals the unicuspids occur in the upper jaw and the weakly cuspidate form in the lower (where there may also be a few of the stouter, more definitely bicuspid teeth). No obvious correlation exists between the sex or size of the fish and the type of dentition present.

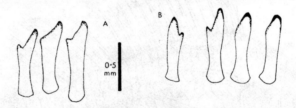

FIG. 5. *Haplochromis simpsoni.* Outer teeth (labial aspect) from:
A, dentary, and B, premaxilla.

There are 50–70 (mean 60) teeth in the outer row of the upper jaw; the tooth number has a weak positive correlation with the size of the specimen.

The inner rows (2 [mode] or 3 in the upper jaw and 1–3 [mode 2] in the lower) are also made up of three different types of teeth. A tricuspid tooth with a protracted middle cusp is the commonest form, but weakly bicuspids and even unicuspids also occur, although much less frequently and in fewer numbers.

In many specimens the teeth are coarsely disfigured by a dark brown thickening of the crown which, in many cases, almost obscures the nature of the cusp. A similar (? pathological) condition was found in *H. empodisma* of Lake Victoria (Greenwood, 1960), a species thought to be related to *H. simpsoni;* it is also seen in the very distantly related *H. obliquidens* of Lake Victoria (Greenwood, 1956).

Lower pharyngeal bone slender; its dentigerous surface triangular and slightly longer than broad. The pharyngeal teeth are slender and cuspidate, and are arranged in 26–28 rows; the median rows may contain a number of somewhat coarser teeth but in the majority of specimens examined no such differentiation was apparent.

Vertebrae: 29, comprising 12 (f.1) or 13 (f.8) abdominal and 16 (f.8) or 17 (f.1) caudal elements.

Coloration in life. Adult males: dorsal surface of body cobalt, that of the head dark umber to black; flanks with light turquoise sheen, the ventral body surface sooty except for the chest which is silvery-white with a diffuse sooty overlay.

Dorsal fin pale grey with small blotches of dull red between the spines, and darker, more irregular red blotches between the branched rays; lappets and margin to soft part black. Caudal fin dark turquoise on its proximal half, followed by a broad, pinkish vertical band and, distally, a blackish area extending to the margin; the posterior angles are sometimes outlined in red. Anal greyish-white, lappets sooty; a faint pink flush extends along the distal margin of the soft part and spreads, but less definitely, to below the black lappets of the spinous portion. Pelvics olivaceous with a sooty wash.

Females: body olive-green, with a suggestion of blue dorsally, becoming lighter on the flanks (which have a turquoise iridescence) and shading to blue-grey ventrally. Cheek and opercular region with turquoise high-lights.

Dorsal fin olive-green basally, the posterior part lightest; dull red blotches occur between the spines and give an impression of a broad red stripe along this part of the fin and even to the more distal region below the black lappets. Similar but smaller blotches occur between the soft rays so that the band is continued posteriorly where, however, it is narrower and less intense. Caudal fin with a reddish to olive-brown blotch at its base, the blotch becoming brighter red distally; the area around the blotch is olive-yellow as is the middle third of the fin; the distal third is either sooty or olive-yellow. Anal fin olive-yellow except for a dull crimson basal streak. Pelvics pale olive-yellow, the elongated rays dead white.

Coloration in preserved specimens. Adult males (*sexually active*) are light brown, with the belly, isthmus, flanks (to the level of the upper lateral line) and caudal peduncle (except dorsomedially) sooty to black; the cheeks are dark but with a pearly lustre. Two narrow transverse bars are usually visible across the snout but apparently only one nuchal band is developed; the latter originates at the anterior upper angle of the operculum and is usually interrupted medially. The lower jaw is pale but the branchiostegal membrane is black. The dorsal fin is hyaline with a sooty overlay most concentrated in the middle third of the fin and least concentrated on the distal region of the soft part. Caudal fin is dark on its proximal quarter, lighter over the remainder except along a moderately wide band outlining its ventral and posterior margins. The anal fin is faintly white to hyaline except for a narrow, intensely dark basal streak and four greyish ocelli (arranged in a single row). The pelvics are black to sooty, being lightest on the posterior third.

Sexually quiescent males are light brown, the chest, belly, branchiostegal membrane and the lower half of the caudal peduncle dusky and overlaid with a pearly sheen; the two snout- and single nuchal-bands are as in active males. Five or six vertical bars of variable intensity are visible on the flanks; these do not reach the origin of the dorsal fin and ventrally they merge with the dark ventral coloration. Dorsal fin hyaline with dark lappets, caudal dark hyaline, darkest proximally; anal dusky, darkest (almost black) in the area of the spines, two to four greyish ocelli arranged in one or two rows. Pelvic fins black to sooty.

Females are silvery-grey; a short, broad lachrymal stripe or blotch is present, as are a faint nuchal bar and an even fainter transverse bar across the snout. Six or seven faint but moderately broad bars are visible on the flanks and caudal peduncle, those on the flanks reaching neither the dorsal nor ventral body outlines. All fins are hyaline, the base of the caudal slightly darker; in one specimen there is a faint darkening between the rays of the dorsal fin (especially over the soft part) and over the proximal two-thirds of the caudal fin. In all specimens there are two dark spots on the anal in the position of the ocelli in males.

Ecology. Habitat. This species seems to have a wide distribution within the lake, being found inshore over a variety of substrata, amongst the emergent vegetation, over exposed sandy beaches and even at some distance offshore over a muddy bottom. It does not appear to inhabit isolated pools in the swamps, nor does it extend for any distance up the inlets into the swamps. Thus, it will be seen that *H. simpsoni* and *H. velifer* do not differ in their broad ecological requirements or restrictions.

Food. From the C.N.B.S. field notes, supplemented by further gut analyses on five specimens, I can find no clear-cut differences in the feeding habits of this species and *H. velifer* (see p. 322); that is, *H. simpsoni* is an omnivorous bottom feeder preying chiefly on insect larvae.

Breeding. No data are available. All the specimens examined are adult, the smallest male and female being 76 and 85 mm. S.L. respectively. The largest fish is a female (114 mm. S.L.) and the largest male is 105 mm. S.L.

Distribution.   Known only from Lake Nabugabo.

Affinities and diagnosis. The nature of the dentition, together with the relative size of the eye and snout, serves to distinguish *H. simpsoni* from the generalized species of Lake Edward (i.e. *H. eduardii, H. engystoma, H. nigripinnis, H. elegans* and *H. schubotzi*). On the basis of its dentition, *H. simpsoni* cannot be included in the same category as these generalized species (although trophically it should be considered generalized). Rather, it should be grouped with *H. empodisma* of Lake Victoria, a species with which it shows fairly close affinities.

The body form and, particularly, the dentition of *H. empodisma* and *H. simpsoni* are similar as are the feeding habits and broad ecological requirements of the two species (Greenwood, 1960). *Haplochromis simpsoni* is, anatomically, more closely related to *H. empodisma* than it is to the small, undescribed, species which I mentioned in connection with the affinities of *H. empodisma* (see Greenwood, *op. cit.*). *Haplochromis simpsoni* differs from *H. empodisma* in having somewhat fewer jaw teeth (50–70, M = 60 *cf.* 54–82, M = 70), the triangular dentigerous surface of the lower pharyngeal bone equilateral and not isoscelean, a slightly narrower head (15·0–16·7, M = 15·9 per cent of standard length, *cf.* 15·6–19·8, M = 17·4 per cent in *H. empodisma*) and a straighter dorsal head profile. The modal number of dorsal rays in *H. empodisma* (24) is lower than in *H. simpsoni* (25) although the ranges overlap; however, the range in *H. empodisma* includes a number of specimens with only 23 rays and but one fish with 25. The two species differ in the coloration of adult males, especially in the absence of red pigment on the head of *H. simpsoni*; the red head of *H. empodisma* is a characteristic feature.

Haplochromis simpsoni also resembles *H. velifer*; the diagnostic characters separating these species are discussed on page 324. In sum, it seems that *H. simpsoni* shares more characters with *H. empodisma* than with *H. velifer*.

STUDY MATERIAL

B.M. (N.H.) reg. no.	Collector
1964.7.1.12 (Holotype)	C.N.B.S.
1964.7.1.13–24 (Paratypes)	C.N.B.S.
1964.7.1.25–27 (Paratypes)	Pitman
1933.2.23.195–199 (Paratypes of *H. velifer*) Paratypes	Worthington
1964.7.1.28–33 (Paratypes)	Pitman
1935.8.23.34–63	Pitman
1935.8.23.81–100	Pitman

Haplochromis annectidens Trewavas, 1933

(Text figs. 6 and 7)

H. annectidens (part) Trewavas, 1933, *J. Linn. Soc.* (*Zool.*), **38** : 323.

HOLOTYPE. An adult male, 67 mm. S.L. from Lake Nabugabo (collected by E. B. Worthington), B.M. (N.H.) reg. no. 1933.2.23.210.

DESCRIPTION: based on the holotype, seven paratypes and thirteen additional specimens, 43–67 mm. S.L.

Depth of body 31·3–40·0 (M = 36·9) per cent of standard length, length of head 30·6–36·0 (M = 33·8) per cent. Dorsal profile of head straight, sloping moderately steeply (*ca* 40°–45° with the horizontal).

Preorbital depth 11·1–15·8 (M = 13·5) per cent of head, least interorbital width 21·0–29·2 (M = 25·3), snout length 25·0–31·6 (M = 27·9), eye diameter 30·4–37·5 (M = 33·3) and depth of cheek 12·5–21·7 (M = 18·1) per cent, the latter character showing slight positive allometry with standard length.

Caudal peduncle 14·7–18·6 (M = 16·3) per cent of standard length, 1·1–1·6 (mode 1·4) times as long as deep.

Mouth horizontal, lips sometimes slightly thickened. Jaws equal anteriorly, the lower 31·3–42·2 (M = 37·4) per cent of head length, 1·5–2·0 (modal range 1·5–1·6) times as long as broad; posterior tip of the maxilla reaching the vertical through the anterior orbital margin (the modal condition) or to below the anterior part of the eye.

Gillrakers variable, from short and stout to relatively slender, but usually of uniform shape in any one individual; 8–10 (mode 9) on the lower part of the first arch, the lowermost 1–3 rakers reduced.

Scales ctenoid; lateral line with 30 (f.1), 31 (f.6), 32 (f.12) or 33 (f.1) scales; cheek with 2 or 3 (mode) rows; 5–6 (mode) scales between the dorsal origin and the upper lateral line, 4–5½ (mode 5) between the pectoral and pelvic fin bases.

Fins. Dorsal with 23 (f.2), 24 (f.8), 25 (f.10) or 26 (f.1) rays, comprising 15 (f.8) or 16 (f.13) spinous and 8 (f.5), 9 (f.14) or 10 (f.2) branched rays. Anal with 11 (f.2), 12 (f.17) or 13 (f.2) rays, comprising 3 spinous and 8 (f.2), 9 (f.17) or 10 (f.2) branched. Caudal subtruncate or truncate. Pectoral 25·4–30·8 (M = 27·6) per cent of standard length. First two soft pelvic rays produced, proportionately more so in males.

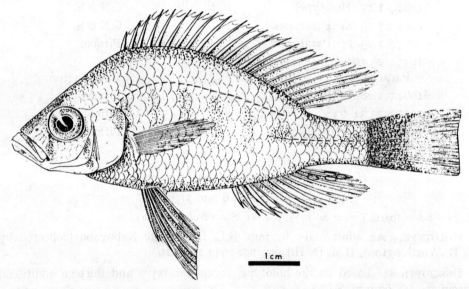

FIG. 6. *Haplochromis annectidens*. Drawn by Barbara Williams.

Teeth. The outer teeth in this species are highly characteristic. Except for a few teeth situated posteriorly in each jaw, the teeth are moveably implanted; each tooth has an elongate, slender neck and base but a flattened and expanded crown which is derived almost entirely from the enlarged major cusp (see fig. 7a, b). The occlusal margin of this cusp is obliquely truncate so that the thin occlusal surface is slightly convex and almost horizontal in position; the minor cusp is minute and acutely pointed. The anterior tip of the crown is drawn out, so that this margin of the tooth is concave; the posterior margin is curved in parallel with the anterior one (*i.e.* it is convexly arched). A few posterior teeth in the upper jaw are much smaller than their anterior congeners and are but weak replicas of them; in the dentary, the posterior teeth are tricuspid. The number of teeth in the outer, upper row shows some positive correlation with the size of the fish, *viz.* in fishes 43–50 mm. S.L. (N = 6), 38–50 (M = 45) and 48–68 (M = 56) in larger specimens (N = 15).

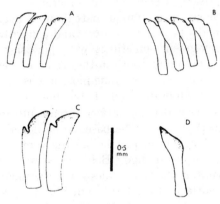

FIG. 7. A and B: *Haplochromis annectidens*, outer teeth (labial view) from, respectively, the premaxilla and the dentary. C, outer premaxillary teeth of *H. lividus* (labial view), D, outer tooth (premaxilla) in labial view, of *H. obliquidens*.

All teeth in the inner rows are small and tricuspid; in a few specimens the teeth of the outermost row may be noticeably enlarged but are still clearly tricuspid, unlike the enlarged inner teeth of *H. obliquidens* (see below p. 334) which are also obliquely cuspidate. There are 2–4 (mode) inner rows of teeth in the upper jaw and 2–4 (mode 3) in the lower; a fairly distinct interspace usually separates the inner rows from the outer row, but in some specimens it is obscured.

Lower pharyngeal bone and teeth. The lower pharyngeal bone is slender, its dentigerous surface triangular and somewhat broader than long (occasionally equilateral). The teeth are slender, weakly cuspidate, slightly curved and fairly close set in 30–36 rows. In the largest specimen examined (the holotype) a few of the posterior teeth in the median rows are slightly enlarged; in all other specimens there is no such differentiation.

Vertebrae: 28 or 29, comprising 13 abdominal and 15 (f.1) or 16 (f.5) caudal elements.

Coloration in life. The C.N.B.S. field notes on this species are extensive and cover a number of different sexual and emotional states. Overall coloration of **adult males** shows considerable variation both in intensity and in the extent of any one pigment, and is apparently correlated with the emotional state of the fish. However, the predominance of pink and red in fin coloration, irrespective of these factors, is an obvious feature, as is the amount of red on the flanks in sexually active individuals.

Quiescent males are light blue-grey above the lateral line; the flanks are orange-buff shading to light green on the ventral body surface and the lower half of the caudal peduncle. Traces of five indigo bars are visible on the mid-flank region but do not extend to the body outline. Dorsal surface of the head dark olive to reddish, cheeks and operculum greenish-yellow, sometimes with faint red blotches at the angles of the preoperculum and operculum. Lachrymal, trans-snout and nuchal stripes (see notes on preserved colours) are sometimes visible but only faintly so. Dorsal fin with a pink flush on the spinous part (lappets black), the soft part hyaline with red spots and streaks between the rays, and a narrow basal band of orange-yellow. Caudal fin grey-green, sometimes with pinkish-red streaks between the rays. Anal faintly grey, becoming sooty in the region of the spines and with a very faint pink flush over the soft part, becoming more intense at the distal margin of the anterior part of the fin; ocelli orange red. Pelvic fins dusky.

Sexually active males. Dorsal body surface intense ultramarine to purple with a faint sooty overlay especially along the base of the dorsal fin and over the nuchal region; ventral part of body, from just before the vent to the posterior tip of the caudal peduncle, dark greenish-indigo. Belly, chest and flanks below the level of the lower lateral line crimson, the belly scales with or without a narrow black margin; lower jaw and ventral surface of the head light blue to greenish-blue. Faint traces of transverse bars are sometimes visible on the mid-flank region of the body. The intensity and extent of the red flank and belly colours vary with the emotional state and may be restricted to a small triangular area on the chest extending thence dorsally to a level at about the midpoint of the operculum. The dorsal head surface is always ruddy although the intensity and the area involved are variable; however, the snout and cheeks are invariably reddish. A lachrymal stripe of variable intensity and width is always visible. Dorsal fin is pink to crimson, the colour most intense between the rays; lappets sooty. Caudal fin pink to dusky pink, the colour most concentrated proximally; posterior angles sometimes scarlet with a faint sooty overlay. Anal pink, the margins of the spinous part with a sooty overlay; ocelli orange yellow. Pelvic fins dusky with, usually, a faint pink flush.

Female coloration is also variable. Basically, the body is olive-yellow, darkest dorsally (even becoming grey-blue) and on the cheeks, lightest ventrally on the belly and jaws (almost peach-colour), the flanks sometimes with a pinkish glow. Faint traces of vertical bars are often visible on the flanks and caudal peduncle, the bars usually dark olive green and extending to the dorsal outline of the body.

In excited fishes the colours darken so that the back, dorsal head surface and the vertical flank marks appear very dark olive whilst the flanks and belly become an olivaceous yellow-green; in this condition a dark lachrymal stripe develops. Dorsal fin pale olive yellow with a faint pink flush most concentrated between the spinous and anterior branched rays. In excited individuals the pink colour darkens to scarlet and appears as blotches between the anterior soft rays. Caudal olive-yellow, darker (olive-green to sooty) proximally. Anal buff, sometimes with a faint sooty overlay in the region of the spines; spots (in the position of the ocelli in males) orange. Pelvic fins buff, the anterior half sometimes orange.

Coloration in preserved specimens. Adult males (*sexually active*) light brown, the chest, branchiostegal membrane, lower half to three-quarters of the caudal peduncle and the entire belly sooty, the latter with a silvery overlay. A broad, nearly vertical lachrymal stripe runs from the angle of the jaws to the orbit; the snout is crossed by two transverse bars and there is a medially interrupted nuchal band running upwards from the posterior margin of the orbit. The flanks are crossed by six or seven dark bars, each broadest at its midpoint and narrowing more markedly dorsally than ventrally where it merges with the dark ventral body coloration. All fins (except the pelvics) are hyaline; the lappets of the spinous dorsal are dark, as is the base of the caudal; along the base of the anal fin there is a faint, sooty crescent and near its posterior tip two or three large, circular, greyish ocelli. The pelvics are black.

Sexually quiescent adult males have a similar coloration except that the pearly-silver overlay on the belly is brighter and the lower part of the caudal peduncle is brownish rather than black. The lachrymal, nuchal and cheek stripes are as in active fishes as are the vertical flank bars except that some of the latter extend to the base of the dorsal fin; none extends to the ventral body outline. The fins are as described above but the base of the anal and caudal may be hyaline and the posterior margin of the pelvics light sooty.

Females are grey-brown above, shading to silvery white on the belly and ventral flanks. The lachrymal bar is short (not extending to the jaw angle), the nuchal stripe and snout bars very indistinct or absent. There are five or six faint vertical bars on the flanks, each bar reaching the base of the dorsal fin but not the ventral body outline. All fins are hyaline, the lappets of the spinous dorsal dark.

Ecology. Habitat. Haplochromis annectidens is an inshore species occurring mainly in the vicinity of or amongst the marginal vegetation, and only rarely over exposed sandy beaches away from rooted plants; apparently it does not penetrate deeply into the marginal swamps and is rarely recorded at the swamp ends of inlets.

Food. Data on the food of *H. annectidens* were obtained principally from fourteen preserved specimens which I examined; these observations were supplemented by notes made on four specimens by the C.N.B.S. Despite the small size of these samples the variety of organic material found in the stomach and intestines is high, suggesting that the feeding habits of the species are also varied. Perhaps the commonest gut content is a barely recognisable mush of plant debris, both algal and phanerogamic, with blue-green algae predominating. Such material often forms the flocculent "mud" which covers the bottom in inshore regions of the lake; this leads one to conclude that many fishes had fed from the bottom. This supposition gains support from the presence of dipteran larvae in the stomach contents of the same individuals. Less frequently, the guts contain fragments of plant epidermis (and sometimes the bladders of *Utricularia*) together with large quantities of epiphytic algae, especially diatoms and filamentous green algae (e.g. *Oedogonium*).

These remains suggest that the fishes had been grazing epiphytic algae off submerged plants. Sand grains and fragmentary insect remains also occur in gut contents of this type; it is impossible to tell whether the sand grains were derived from the bottom or whether they were derived from the broken-down cases of Trichoptera larvae. Likewise it is difficult to suggest the provenance of the insect larvae.

As far as could be told from the preserved guts there is little digestion of the phanerogamic material, the blue-green algae or the filamentous green algae; diatom frustules, by contrast, were always empty.

In one specimen the stomach was packed with sand grains but it also contained two larval fishes. Since the larvae were small and not Cichlidae they should, presumably, be listed amongst the food organisms of this species.

Haplochromis annectidens has the long gut ($2\frac{1}{2}$–$2\frac{3}{4}$ times the standard length) and the dentition of a herbivore. It was somewhat surprising, therefore, to find such ill-defined feeding habits. However, it may be recalled that the similar species *H. lividus* and *H. obliquidens* of Lake Victoria are also somewhat facultative in their feeding habits, although in these species there is a predominance of algal grazing over other feeding methods (Greenwood, 1956).

Breeding. One female with embryos in the mouth is recorded by the C.N.B.S. (June 1962); this specimen was not amongst those brought back to the Museum. The sex of the smallest specimen available (43 mm. S.L.) could not be determined, but a male 44·5 mm. S.L. is sexually active although another of 46 mm. is juvenile; the smallest female (50 mm. S.L.) is of undeterminable state but is probably maturing. The largest fishes examined (both 67 mm. S.L.) are of opposite sexes.

Distribution. Known only from Lake Nabugabo.

Affinities and diagnosis. The slender, obliquely cuspidate teeth of *H. annectidens* place it in the well-defined group of East African *Haplochromis* comprising the following species: *H. obliquidens* and *H. lividus* (Lake Victoria) and *H. astatodon* (Lake Kivu). Apart from their peculiar teeth and their long guts, these species have a generalized anatomy. Each differs from the others in a number of characters, including dental morphology. From *H. obliquidens*, with its invariably unicuspid anterior and anterolateral teeth, *H. annectidens* is distinguished by having bicuspid teeth in these positions, and in having teeth which are stouter and with somewhat differently shaped crowns (see fig. 7d); the interorbital is narrower in *H. annectidens* (21·0–29·2, M = 25·2%, cf. 27·8–34·7, M = 31·8 in *H. obliquidens*) and the cheek shallower (12·5–21·7, M = 18·1% of head, cf. 19·0–25·0, M = 21·5). Although the range for the lateral line scale count is identical in both species, the modal number for *H. annectidens* (32) is higher than in *H. obliquidens* (31).

The teeth in *H. lividus* are bicuspid, but crown form serves to distinguish them from those of *H. annectidens*; indeed, the condition in the latter species is almost perfectly intermediate between *H. lividus* and *H. obliquidens* (see fig. 7c). *Haplochromis annectidens* and *H. lividus* also differ in certain morphometric characters; the interorbital of *H. lividus* is broader (26·2–33·3, M = 29·7% of head, cf. 21·0–29·2, M = 25·2), the cheek is slightly deeper (17·0–24·1, M = 20·1% of head) and the range of lateral line scale counts extends to 34, although the modal number is identical in both species (32). The most pronounced difference lies in the coloration of adult males. *Haplochromis lividus* is probably unique within the genus (and certainly is unique amongst the *Haplochromis* of Lakes Victoria and Nabugabo) for the intense, almost fluorescent blue colour of the head and snout of adult males. This coloration contrasts strongly with the ruddy head tones of *H. annectidens*. (Male coloration also seems to distinguish *H. obliquidens* and *H. annectidens*; cf. p. 332 above with p. 229 of Greenwood, 1956).

The dental morphology of *H. annectidens* is very similar to that of *H. astatodon* from Lake Kivu, but there are fewer inner tooth rows in *H. annectidens*. The species also differ in head shape, the gently sloping head profile of *H. annectidens* contrasting with the declivous snout and rounded upper profile of *H. astatodon*; also, in *H. annectidens* the interorbital is markedly smaller than the eye, but in *H. astatodon* of a comparable size the two measurements are equal or the interorbital width is slightly greater.

Trewavas (1933) mentions *H. plagiodon* (a Lake Victoria species) when discussing the affinities of *H. annectidens*, but notes that the teeth of the former are "much larger and fewer", a description with which I concur (see Greenwood, 1959*b*). However, in the introduction to her paper, Trewavas suggests that *H. annectidens* "represents the stock from which *H. obliquidens* and *H. plagiodon* of Lake Victoria seem to have diverged in separate directions". Certainly, *H. annectidens* represents a dental and anatomical grade which could be ancestral to that of *H. obliquidens* but the level of its dental specialization is higher than that likely to be ancestral to the peculiar teeth found in *H. plagiodon* (see Greenwood, 1959*b*). Rather, I would support Trewavas' idea (1933, p. 324) that *H. plagiodon* evolved from a stock resembling *H. velifer* in its dental morphology. The anatomical status of *H. velifer* is that of a generalized *Haplochromis* but it does differ from many of the other generalized species in having somewhat obliquely cuspidate teeth. The *H. plagiodon*-type of tooth represents but a slight modification of the *H. velifer*-type, the *H. lividus-annectidens-obliquidens* types are much more extreme developments involving both the neck and the crown of the tooth.

STUDY MATERIAL

B.M. (N.H.) reg. no.	Collector
1933.2.23.210 (Holotype)	Worthington
1933.2.23.211–220 (Paratypes)	Worthington
1935.8.23.111–113	Pitman
1964.7.1.41–79	C.N.B.S.

Haplochromis beadlei Trewavas, 1933

(Text fig. 8)

H. beadlei Trewavas, 1933, *J. Linn. Soc. (Zool.)*, **38** : 324.

HOLOTYPE: An adult male 106·0 mm. S.L., from Lake Nabugabo (E. B. Worthington collection), B.M. (N.H.) reg. no. 1933.2.23.221.

Although *H. beadlei* is easily distinguishable from other species of *Haplochromis* in Lake Nabugabo, its systematic status is uncertain. It closely resembles *H. labiatus* of Lake Edward and *H. crassilabris* of Lake Victoria; neither of these species is sufficiently well-known to allow for a full assessment of the characters by which *H. beadlei* differs from them. As is usual amongst related *Haplochromis* species there is no single trenchant diagnostic character; since various combinations of characters (differing for large and for small specimens) seem to distinguish *H. beadlei*, it is retained as a distinct species pending a full revision of *H. crassilabris* and *H. labiatus*, especially the latter.

DESCRIPTION based on the holotype, nine paratypes and thirteen additional specimens, 72–118 mm. standard length.

Depth of body 36·0–40·3 (M = 38·1) per cent of standard length, length of head 31·4–35·8 (M = 33·8) per cent. Dorsal head profile straight or faintly concave, sloping at about 35°–40° with the horizontal; snout straight, not decurved.

Preorbital depth 14·8–18·0 (M = 15·9) per cent of head length, least interorbital width 23·2–28·0 (M = 25·0), snout length 29·6–36·0 (M = 31·7) per cent, snout slightly broader than long or, rarely, as long as broad. Eye diameter 25·0–32·0 (M = 27·8) per cent of head, depth of cheek 17·8–25·6 (M = 22·1) per cent.

Caudal peduncle 13·4–17·0 (M = 15·2) per cent of standard length, 1·1–1·6 (modal range 1·2–1·4) times as long as deep.

Mouth horizontal, both lips markedly thickened, usually to a comparable degree or with the upper lip slightly thicker; in one specimen there is an incipient median lobe developed on the upper lip. Jaws equal anteriorly in most specimens but in a few fishes > 85 mm. S.L. the lower jaw projects slightly so that the upper teeth

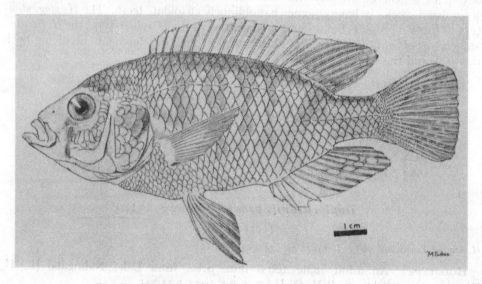

FIG. 8. *Haplochromis beadlei*. Drawn by Miss M. Fasken.

occlude behind the lower. Lower jaw 34·5–39·6 (M = 36·9) per cent of head, 1·3–1·7 (mode 1·4) times as long as broad. Posterior tip of the maxilla somewhat bullate, reaching the vertical through the anterior orbital margin in most specimens but not quite reaching this point in others (33% of the sample examined), rarely extending to below the anterior part of the eye.

Gillrakers variable (except for the lowermost one to three), from moderately slender to moderately stout, but of constant form in any one individual; 7 (f.3), 8 (f.16) or 9 (f.4) on the lower part of the first gill arch, the lowermost one to three rakers reduced.

Scales ctenoid; lateral line with 30 (f.3), 31 (f.14) or 32 (f.5) scales; cheek with 2 (f.4), 3 (f.18) or 4 (f.1) rows; 5–7 (modes 6 or 7) scales between the dorsal fin origin and the upper lateral line, 5–7 (mode 6) between the pectoral and pelvic fin bases.

Fins. Dorsal with 24 (f.11), 25 (f.11) or 26 (f.1) rays, comprising 15 (f.15) or 16 (f.8) spinous and 8 (f.1), 9 (f.16) or 10 (f.6) branched rays; anal with 10 (f.1), 11 (f.4) or 12 (f.18) rays comprising 3 spines and 7 (f.1), 8 (f.4) or 9 (f.18) branched rays. Caudal subtruncate, scaled on its basal half. First two rays of the pelvic fin barely produced in females but greatly elongate in males. Pectoral 23·3–28·0 (M = 26·0) per cent of standard length.

Teeth. The nature of the dentition changes with the size of the individual but is always characterized by the teeth being stout and slightly curved, with the anterior teeth somewhat procumbent (those of the upper jaw most obviously so). The outer, upper teeth are all bicuspid in most fishes <85 mm. S.L. (and in all individuals <75 mm.), those situated anteriorly have subcylindrical crowns but those laterally and posteriorly are more flattened. The anterior and anterolateral teeth in most specimens >85 mm. S.L. are cylindrical in section and are unicuspid or with a poorly developed minor cusp; the crown, especially in unicuspids, is slightly recurved. In exceptional individuals over 85 mm. S.L. (the holotype is one of these) the entire upper outer row is composed of distinctly bicuspid teeth.

The outer row of the lower jaw is, in general, like that of the upper except that the anterior teeth are less procumbent and are often implanted vertically; variations in tooth form follow the same size-correlated trends described above.

There are 24–32 (M = 28) teeth in the outer row of the upper jaw.

Teeth of the inner series are tricuspid and small, and are arranged in 2 or 3 (mode) rows in each jaw. In one exceptional specimen, some of the teeth in the outermost inner row of the lower jaw are enlarged and weakly tricuspid.

The shape of the dental arcade in all specimens is a broadly rounded U.

Lower pharyngeal bone moderately slender, the triangular dentigerous surface slightly broader than long. The lower pharyngeal teeth are compressed, cuspidate and slender, and are arranged in 22–26 rows. In fishes over 80 mm. S.L. the two median rows are composed of coarser and weakly cuspidate teeth. Some of the posterior teeth in neighbouring rows are also enlarged and form a small, roughly triangular zone of enlarged teeth at the posterior end of the median tooth rows; the degree of enlargement undergone by these teeth is positively correlated with the size of the individual.

Vertebrae 28 or 29, comprising 13 abdominal and 15 (f.5) or 16 (f.6) caudal elements.

Coloration in life. Sexually active male. The dorsal surface of the body is greeny-blue (rather more green than blue), the head darkest; flanks lighter, the ventral surface of the chest sooty but the lateral aspects with a red flush. Dorsal fin sooty along its base and over the whole spinous part, the soft part being pink. Anal pink, with a sooty overlay which becomes intensely black over the spinous part; ocelli orange-yellow. Pelvic fins uniformly sooty.

Females are bright olive-green, the flank scales with turquoise margins; belly blueish-green, chest golden olive to silvery white. Dorsal surface of the head dark umber, cheek with golden lights. Dorsal fin pale olive, with a sooty overlay. Anal clear olive-yellow, spots (when present occupying the place of ocelli in males) bright yellow. Caudal fin light olive with dark red maculae between the rays. Pelvics olive with a sooty overlay, the first two rays dead white.

N.B. These colour notes are based on many fewer fishes than are those for the other species; only one sexually active male is described in the C.N.B.S. field-notes.

Coloration of preserved specimens. Adult males (sexually active) are dark brown, almost black, with a pearly sheen on the belly. Lower jaw and branchiostegal membrane are greyish, the cheeks brownish and the operculum dark with a pearly overlay. The snout and dorsal aspect of the head is blackish; there is a distinct lachrymal stripe but no trace of a nuchal band; an ill-defined, blotchy stripe runs along the opercular-preopercular junction. The flanks are crossed by six vertical bars which merge dorsally and ventrally with the dark ground colour of the body. The base of the caudal fin is black, the remainder of the fin becoming progressively lighter towards its posterior margin. The entire spinous dorsal and the basal quarter of the soft dorsal are black to sooty, the remainder of the soft part darkly maculate. Anal fin hyaline, with five large, grey ocelli each outlined by a dark ring. Pelvics with anterior third black, remainder whiteish, the demarcation not well defined.

Sexually quiescent males are brownish, shading to dusky silver on the belly. Lower jaw and branchiostegal membrane are silvery white. A broad, dark lachrymal stripe originates slightly ventral to the jaw angle and continues to the anterior orbital margin; this stripe extends through the eye to form a transverse nuchal band. A weaker, posterior nuchal band is also present and originates at the junction of the operculum and body. The flanks are crossed by six, moderately broad vertical bars which extend to the base of the dorsal fin but do not reach the ventral body outline. The greater part of the dorsal fin is dusky, the soft part weakly maculate. The base of the caudal fin is dark, the rest hyaline. The anal is hyaline with three oval, dead-white ocelli. The pelvics are black on the anterior third, whiteish posteriorly, the demarcation being clear cut.

Females are light brown, becoming silvery white on the belly and ventral aspects of the flanks. The lachrymal stripe is very faint as are the nine, ventrally incomplete vertical bars on the flanks and caudal peduncle; dorsally these bars merge with the base of the dorsal fin, the latter being hyaline with a slight darkening between the posterior spines and between all the soft rays. The distal part of the soft dorsal is often darkly maculate. Caudal fin hyaline with dark maculae; anal hyaline, with three to six small, dead-white ocelli often arranged in two rows. Pelvic fins also hyaline.

Ecology. Habitat. This species is apparently confined to shallow, inshore regions of the main lake and the lakeward ends of inlets to the marginal swamps. The substrate in these places varies, and includes sand, sand with a mud-detritus overlay and deep mud.

Food. The gut contents of eight specimens containing ingested material were examined. All except one contained fragments of insect larvae (chironomid and trichopteran); in one of these fishes the gut yielded numbers of undamaged cases of Trichoptera larvae but in all the others containing identifiable trichopteran remains, no trace of the cases was found. Fragments of undigested plant epidermis were recorded in two specimens. The exceptional fish noted before contained a large amount of unidentifiable sludge.

Breeding. No data are available on the breeding habits of *H. beadlei*. One specimen (a male paratype, 77 mm. S.L.) has cichlid embryos and larvae in its mouth. Because these are at such disparate stages of development it seems unlikely that they represent a brood; rather, I suspect that the adult had snatched at young jettisoned by parents as they were captured, a not uncommon phenomenon (personal observations on the behaviour of netted *Haplochromis* in Lake Victoria).

With one exception, all the specimens examined are obviously adult; the exceptional fish, a female 72 mm. S.L., may be a juvenile or it could be a spent and quiescent adult. There is no sexual dimorphism in the size attained by fishes in this sample.

Distribution. Known only from Lake Nabugabo.

Affinities and diagnosis. The peculiar oral dentition, heavy lips and general morphology of *H. beadlei* indicate a strong affinity with *H. paucidens* (Lake Kivu), *H. crassilabris* (Lake Victoria) and *H. labiatus* (Lake Edward). Unfortunately none of these species is well-known morphologically or ecologically despite, in the case of *H. crassilabris*, intense field studies on the Lake Victoria *Haplochromis*.

The description of *H. labiatus* is based on three specimens of rather disparate sizes (60, 74 and 107 mm. S.L.); no information is available on live colours or ecology. The present concept of *H. crassilabris* stems from twelve specimens and does not agree entirely with that published by Regan (1922); my revision of *H. crassilabris* is still unpublished but will be used here as the basis for comparison with *H. beadlei*. Comparative data for *H. paucidens* were obtained from four specimens in the British Museum (N.H.).

Haplochromis beadlei differs from *H. crassilabris* in the following characters:

(i) In fishes of a comparable size, *H. beadlei* has a relatively smaller proportion of unicuspid to bicuspid outer teeth, and the teeth are more compressed and less cylindrical in cross-section (especially through the neck); also, in *H. crassilabris* some unicuspids appear anteriorly in smaller specimens.

(ii) There are more teeth in the outer row of the upper jaw (24–32, $M = 28$ *cf.* 20–30, $M = 24$ for *H. crassilabris*).

(iii) The height of the teeth in the upper jaw of *H. beadlei* is gently graded postero-anteriorly but in *H. crassilabris* the anterior teeth are markedly larger than the lateral ones.

(iv) The lower jaw in *H. beadlei* is somewhat longer (34·5–39·6, M = 36·9% of head, cf. 31·0–34·0, M = 32·4 in *H. crassilabris*).

(v) In *H. beadlei* the female has well-defined spots on the anal fin, corresponding in position with the ocelli of males; no spots are developed in *H. crassilabris*.

The orodental characters of *H. beadlei* appear less specialized than those of *H. crassilabris*. However, it is difficult to determine whether the "*beadlei*" condition represents an evolutionary stage intermediate between the generalized *Haplochromis* condition and the "*crassilabris*" level, or whether it is a slightly regressive development from a species which had already achieved the "*crassilabris*" stage. The possibility of such regressive changes must be given serious consideration because this phenomenon has been demonstrated in the cichlid *Astatoreochromis* (Greenwood, 1965). An aquarium bred specimen of *A. alluaudi*, derived from a typical Lake Victoria population, failed to develop the hypertrophied pharyngeal structures characteristic of its ancestors. The degree of hypertrophy shown by the aquarium fish resembled an intermediate stage in the development of this particular specialization. Thus, it seems possible that *H. beadlei* could have evolved from a "*crassilabris*"-like ancestor if environmental conditions in Lake Nabugabo were such that selection pressure did not demand the full expression of the specialized "*crassilabris*" dentition.

From *H. labiatus*, *H. beadlei* is distinguished by the following:

(i) The large specimen of *H. labiatus* (107 mm. S.L.) has its lower jaw shorter than the upper; in *H. beadlei* of a comparable size the jaws are equal anteriorly or the lower projects slightly.

(ii) At all sizes the eye is about equal to the interorbital width in *H. beadlei* but in *H. labiatus* the eye is distinctly larger.

(iii) In small specimens of *H. labiatus* (60 and 74 mm. S.L.) the teeth are more compressed than those in comparable sized *H. beadlei*; in larger specimens the teeth are identical.

(iv) The upper dental arcade in the large *H. labiatus* is more acutely rounded than in *H. beadlei* but the arcade in small specimens is identical.

(v) The dorsal head profile of small *H. labiatus* is more rounded than in *H. beadlei*, and the snout is more declivous (ca 60° cf. 35°–40° for *H. beadlei*); in larger fishes these differences are less marked but the profile still slopes more steeply in *H. labiatus*.

Haplochromis paucidens of Lake Kivu is very similar to *H. crassilabris* in gross morphology and dental characters. Thus, it may be distinguished from *H. beadlei* by the same characters (see above); from the few specimens available, the teeth appear to be finer than those of *H. crassilabris* and therefore stand in even greater contrast with those of *H. beadlei*.

The lower pharyngeal bone and its dentition is identical in all three species.

Live coloration is unknown for *H. labiatus* and the only record for *H. crassilabris* is from a 35 mm. colour-transparency (kindly lent to me by N. Mitton of Nairobi).

As far as I can determine, there is a general similarity between the coloration of
H. crassilabris and *H. beadlei*.

Two other species from Lake Victoria, *H. chromogynos* and *H. chilotes*, should be
considered since both species have thickened lips and a dentition obviously related
to that of *H. beadlei*. Of the two, *H. chromogynos* has the greater similarity to
H. beadlei. It is distinguished from the latter species by its fewer teeth, shorter lower
jaw (30·0–34·4, M = 32·5% of head) and the fact that all females have a piebald
black and silver coloration (Greenwood, 1959b). *Haplochromis chilotes* represents a
more extreme development of *H. chromogynos*, particularly with regard to the
hypertrophy of the lips; most specimens have both lips drawn out medially to form
large lobes, although in others there is only an incipient lobation and in a few the
lips are little more developed than in *H. beadlei*. (It will be recalled that one specimen
of *H. beadlei* has an incipient lobe developed from the upper lip.) Several morphometric characters of *H. chilotes* are correlated with the degree of lip hypertrophy,
but even specimens with poorly-developed lips may be distinguished from *H. beadlei*
by having a more acute dental arcade, finer teeth and a shorter lower jaw
(30·0–36·6, M = 33·2 cf. 34·5–39·6, M = 36·9% of head, in *H. beadlei*). On the other
hand, the lower jaw of *H. chilotes* with lobed lips is slightly longer than in *H. beadlei*
(36·0–49·0, M = 39·6% head). Finally, there are distinctive differences in the live
colours of adult males (cf. p. 338 above with p. 209, Greenwood, 1959b).

As Trewavas (1933) noted, the morphological affinities between *H. beadlei*,
H. crassilabris, *H. paucidens* and *H. labiatus* are strong; to this complex may now
be added *H. chromogynos* which, in turn, bridges the morphological gap between
this complex and the more extreme *H. chilotes*. This species complex will be
discussed again; for the moment it is only necessary to point out that *H. beadlei*
does seem most closely related to *H. labiatus* of Lake Edward and not, as might be
expected, to *H. crassilabris* of Lake Victoria. As a corollary to this paradox, the
known specimens of *H. paucidens* (Kivu) seem closer to *H. crassilabris* than to the
geographically near *H. labiatus* of Lake Edward.

Trewavas (op. cit.) also suggested that *H. beadlei* is closely related to *H. sauvagei*
of Lake Victoria, a species which, on Regan's revision of the Victoria species, is
related to *H. crassilabris*. However, recent studies (Greenwood, 1957) show that
H. sauvagei belongs to a different lineage and one not closely related to *H. crassilabris*. The jaw structure, skull architecture and dental characters of the *H. sauvagei*
line are distinctive and are not even foreshadowed in the *H. crassilabris*-*H. beadlei*
species group.

STUDY MATERIAL.

B.M. (N.H.) reg. no.	Collector
1933.2.23.221 (Holotype)	Worthington
1933.2.23.222–230 (Paratypes)	Worthington
1933.2.23.231–3	Worthington
1935.8.23.114–134	Pitman
1964.7.1.1–11	C.N.B.S.

Haplochromis venator sp. nov.

(Text fig. 9)

H. *pellegrini*: (non Regan) Trewavas, 1933, *J. Linn. Soc. Lond. (Zool.)*, **38** : 326.

Trewavas (1933) identified seven specimens of a predatory *Haplochromis* from Lake Nabugabo as *H. pellegrini* Regan on the basis of a comparison with the two syntypes of *H. pellegrini*. Since 1933 many more specimens of *H. pellegrini* have been obtained and a revised description of the species has been prepared (Greenwood, 1962). It is now clear that the Nabugabo fishes, although showing some affinity with *H. pellegrini*, should be recognised as a distinct species for which the name *venator* is proposed (Venator, Latin, a hunter).

HOLOTYPE: an adult female 158 mm. standard length (B.M. (N.H.) reg. no. 1933.2.23.240) collected by Dr. E. B. Worthington.

DESCRIPTION based on the holotype and twelve paratypes, 59–178 mm. S.L.

Depth of body 28·8–35·5 (M = 32·9) per cent of standard length, length of head 34·6–36·8 (M = 35·5) per cent; dorsal head profile straight or slightly concave (the depression increasing with size and accentuated by the prominent premaxillary pedicels), sloping at 30°–35° to the horizontal.

Preorbital depth 16·3–21·0 (M = 18·9) per cent of head length, least interorbital width 16·8–21·8 (M = 19·0), snout longer than broad, its length 33·6–38·5 (M = 35·3) per cent, probably showing slight positive allometry with standard length. Eye diameter 21·9–28·6 (M = 24·1) per cent of head, showing very slight negative allometry; cheek 19·0 (in the smallest specimen) –28·2 (M = 25·6) per cent and showing very slight positive allometry.

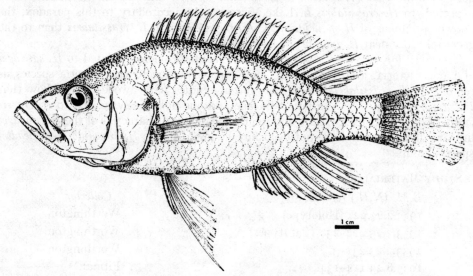

FIG. 9. *Haplochromis venator*. Drawn by Barbara Williams.

Caudal peduncle 14·3–17·5 (M = 15·8) per cent of standard length, 1·2–1·6 (modal range 1·3–1·4) times as long as deep.

Mouth oblique, sloping upwards at about 30°–35°, lips not thickened but the dentigerous surface of the premaxilla expanded antero-posteriorly in the midline. Jaws either equal anteriorly or the lower projecting slightly (both conditions equally common), its length 47·6–54·0 (M = 50·6) per cent of head length, 2·1–2·5 (mode 2·3) times as long as broad. Posterior tip of the maxilla usually not reaching the vertical through the anterior orbital margin (but nearer this point than to a vertical through the nostril), sometimes reaching that point.

Gillrakers moderately stout, the upper three or four usually flattened and anvil-shaped, the lower one to three reduced; 8–10 (mode 9) on the lower limb of the first gill arch.

Scales ctenoid; lateral line with 31 (f.1), 32 (f.9) or 33 (f.3) scales; cheek with 3 or 4 rows; 5–6 (rarely 6½) scales between the dorsal fin origin and the upper lateral line, 5–6 (mode 5) between the pectoral and pelvic fin bases.

Fins. Dorsal with 24 (f.5), 25 (f.7) or 26 (f.1) rays, comprising 14 (f.3), 15 (f.8) or 16 (f.2) spinous and 9 (f.3) or 10 (f.10) branched rays. Anal with 12 (f.9) or 13 (f.4) rays, comprising 3 spines and 9 (f.9) or 10 (f.4) branched rays. Pectoral 23·7–27·5 (M = 26·0) per cent of standard length. Caudal subtruncate, the posteroventral corner often obliquely truncate. Pelvics with the first ray produced in both sexes.

Teeth. Throughout the size-range examined, the outer row of teeth in both jaws is composed of slender unicuspids, those situated anteriorly and laterally being slightly recurved, whilst those posterolaterally are more strongly curved and are directed medially. There are 46–64 (M = 53) teeth in the outer row of the upper jaw in fishes 59–128 mm. S.L. (N = 6) and 52–80 (M = 63) in larger individuals (N = 7).

The inner rows are composed of small, unicuspid teeth in fishes > 110 mm. S.L., and of small tricuspid, weakly tricuspid and, predominantly, unicuspid teeth in individuals 59–107 mm. S.L.; the inner teeth are implanted obliquely so that their crowns may come to lie almost horizontally. Inner teeth are arranged in 2 or 3 rows in the upper jaw and 1 or 2 rows in the lower.

Lower pharyngeal bone triangular, the length of the dentigerous surface equal to its breadth or, rarely, slightly longer. The pharyngeal teeth are slender, compressed and weakly cuspidate (some almost caniniform in the median rows of larger fishes) and are arranged in 20–22 rows.

Neurocranium. The neurocranium of *H. venator* closely resembles that of *H. longirostris* and *H. mento*, being intermediate between the two (see Greenwood, 1962); in other words, it has a skull type characteristic of the group of moderately specialized predatory *Haplochromis* in Lake Victoria.

Vertebrae: 29, comprising 13 (f.4) or 12 (f.1) abdominal and 16 (f.4) or 17 (f.1) caudal elements.

Coloration in life. Adult males are bright blueish-green dorsally, shading to silvery on the lower flanks and on the ventral surfaces; operculum and cheek with a pinkish flush. Dorsal fin smokey-grey with a dull red blotch basally at the junction between the spinous and soft parts. Caudal fin sooty grey, the rays darker. Anal sooty at the base of the soft part, the remainder of the fin neutral; ocelli light vermilion-orange surrounded by a narrow white ring, each ocellus set on a transparent area of membrane. Pelvics sooty, darker along the anterior edge.

Adult females have a similar coloration to that of males but the body is a darker, more olivaceous green and the anal fin is olive with a faint salmon-pink flush over the spinous part; as many as five pale red spots occupy the position of the ocelli in males.

Preserved colours. Adult male (*sexually active*): dusky brown except on the belly and chest, which are silver with the scales outlined in black. Dorsal surface of snout dark as are the anterior and anterolateral aspects of the lips, and the branchiostegal membrane; an ill-defined, dark lachrymal blotch is present, and there are very faint traces of six dark, vertical bars on the flank and caudal peduncle. Dorsal fin grey, with black lappets and a narrow black basal stripe along the origin of the branched rays and the last few spines; the membrane between the soft rays is dark except for a narrow, whiteish band immediately above the dark basal one. Anal fin with a black basal band which is capped by a narrower, dead-white band; the ocelli are very faint. Caudal fin greyish to sooty. Pelvics sooty, darkest along the anterior quarter.

Females: greyish-brown to silvery brown (depending on fixative used, the former for formol, the latter for alcohol), shading to silvery white on chest and belly. Snout, lips and dorsal head surface dark grey or brown. A broad mid-lateral band visible on the flanks in formol fixed specimens only; the band is faint and appears interrupted at about its mid-point.

All fins are hyaline, the anal with faint spots in the position of the ocelli in males; the membrane between the caudal rays is maculate, the spots darkest proximally.

Ecology. Habitat. Haplochromis venator is widely distributed in the lake but appears to be more abundant in open, off-shore areas than in other regions. It has been caught in surface gill-nets set in the middle of the lake over a deep mud bottom (water depth 10 ft.) but is not recorded from bottom nets in the same places, nor from the inlets to swampy areas. Specimens caught in beach operated seines are smaller (30–90 mm. S.L.) than those caught off-shore but this may be a reflection of the type of gear used in the two places.

Food. In addition to data from the C.N.B.S. field-notes, I have examined seven preserved specimens. From these records it is clear that *H. venator* is predominantly a piscivorous predator, although insects (especially adult Ephemeroptera) are also eaten. Only two specimens of the fourteen for which detailed gut analyses are available contained other ingested material, in both cases a few fragments of plant tissue. *Haplochromis* (of *ca* 20–40 mm. total length) and small *Barbus* species seem to be the commonest prey species but in some cases the fish remains were too fragmentary to allow for further identification. One fish (a juvenile 59 mm. S.L.) contained twelve larval cichlids of various sizes but all were within the size range at which larvae are carried by the parent; it is impossible to tell whether such small individuals represent "normal" prey for *H. venator* or whether they were jettisoned young swallowed whilst the seine net was being brought to the shore.

Breeding. No data are available. The two smallest specimens (59 and 82 mm. S.L.) are juveniles; all others are adults (and predominantly females) in differing states of sexual activity.

Distribution. Known only from Lake Nabugabo.

Diagnosis and affinites. The resemblance between *H. venator* and *H. pellegrini* of Lake Victoria has been noted already (p. 342). Several small differences serve to distinguish the two species. In addition, *H. venator* reaches a much larger adult size than does *H. pellegrini* (178 mm. S.L. *cf.* 104 mm.); considering the ecological differences obtaining in the two lakes, this difference is difficult to evaluate. This size difference may underlie a number of the observed morphometric differences between the species. *Haplochromis venator* has a longer lower jaw than *H. pellegrini* (47·6–54·0, M = 50·6%.of head, *cf.* 42·3–51·5, M = 46·8 in *H. pellegrini*), a longer pectoral fin (23·7–27·5, M = 26·0% of standard length, *cf.* 19·4–25·3, M = 21·3), and a slightly narrower head as measured by interorbital width (16·8–21·8, M = 19·0% of head, *cf.* 18·2–24·0, M = 21·0 in *H. pellegrini*). In dental morphology the species hardly differ except that unicuspids occur in the inner rows of *H. venator* at all sizes. No inner unicuspids are found in *H. pellegrini*, although in some specimens a few inner teeth may be only weakly tricuspid. The pharyngeal dentition is similar in both species.

A most marked difference is the coloration; live female *H. pellegrini* are a dark chocolate-brown (shading to light brown ventrally) and have greyish-black fins (see Greenwood, 1962). The body of *H. venator* females is olivaceous green shading to silver, the fins are olivaceous to sooty and the anal has a pink flush. Live colours of male *H. pellegrini* are unknown but the dark brown coloration of preserved specimens contrasts with the lighter colours of preserved male *H. venator*.

Neurocranial architecture differs in the two species, the neurocranium of *H. pellegrini* being of a rather distinctive type (see Greenwood, *op. cit.*), whereas that of *H. venator* is closely allied to the *H. mento* and *H. longirostris* types.

The overall morphology of *H. venator* is not closely similar to that of *H. longirostris* from which species it is distinguished by its deeper body, longer and less oblique lower jaw, medially expanded premaxilla and, most obviously, by its shorter and deeper caudal peduncle (14·3–17·5, M = 15·8% of standard length, length/depth ratio 1·2–1·6, modal range 1·3–1·4, *cf.* 17·2–22·2, M = 19·2, 1·9–2·0 modal range, for *H. longirostris*). There are also interspecific differences in dental morphology, especially the finer and more numerous outer teeth of *H. longirostris*.

Haplochromis mento is so obviously distinct from *H. venator* that no detailed comparison is required (see Greenwood, *op. cit.*).

Some resemblance exists between *H. venator* and the group of Lake Victoria *Haplochromis* comprising the "species" *H. macrodon*, *H. taeniatus* and *H. lamprogenys*. This group is under revision, hence the uncertainty as to the specific status of its members. *Haplochromis venator* differs in having more numerous teeth, a somewhat larger eye and narrower interorbital (eye diameter equals interorbital width in the *H. macrodon* group but is larger than the interorbital in *H. venator*) and a longer and more oblique jaw; there are also differences in preserved coloration. Looked at in relation to the morphological groupings of the Lake Victoria predatory *Haplochromis*, *H. venator* belongs to a more specialized grade than does the *H. macrodon* group.

STUDY MATERIAL

B.M. (N.H.) reg. no.	Collector
1933.2.23.240 (Holotype)	Worthington
1933.2.23.234–239 (Paratypes)	Worthington
1964.7.1.80–85 (Paratypes)	C.N.B.S.

Haplochromis nubilus (Blgr.) 1906

(Text fig. 10)

Tilapia nubila Blgr., 1906, *Ann. Mag. nat. Hist.*, (7), **17** : 450.

Haplochromis nubilus (part): Regan, 1922, *Proc. Zool. Soc.*, 164 (excluding *Paratilapia victoriana* Pellegrin, 1903, for which see Greenwood, 1962).

Haplochromis annectidens (part) Trewavas, 1933, *J. Linn. Soc. Lond. (Zool.)*, **38** : 323.

This synonymy is by no means definitive since a revision of the species is still incomplete; it is given here especially to include the two paratypical specimens of *H. annectidens* which are now identified as *H. nubilus*.

From the information I already have on this rather widely distributed east African species there are indications that the various geographical groups may be differentiable on certain anatomical and morphometric characters (see also Trewavas, 1933). Thus, a brief description and tabulation of these characters is given for the five specimens from Lake Nabugabo. The paratypes of *H. annectidens* are indicated with an asterisk.

S.L.	Depth†	Head†	Po.%	Io.%	Snt.%	Eye%	Cheek%	␣l.j.%	C.P.†
62·5*	39·2	33·6	14·3	28·6	28·0	28·6	23·8	38·1	14·4
65·0*	35·4	32·3	14·6	26·2	31·0	29·2	23·8	38·4	16·8
72·0	37·5	34·7	16·0	24·0	30·0	28·0	20·0	40·0	13·9
86·0	37·2	35·0	16·7	24·4	31·7	26·7	21·7	40·0	15·2
86·0	34·9	35·5	14·7	23·0	31·1	29·6	19·7	39·3	14·0

† = per cent of standard length.
% = per cent of head length.

Caudal peduncle 1·1–1·6 times as long as deep.

Dorsal head profile straight but usually with a marked concavity above the orbit.

Mouth horizontal or very slightly oblique, the lips somewhat thickened; posterior tip of the maxilla reaching the vertical through the anterior margin of the orbit. Jaws equal anteriorly, the lower 1·6–2·0 times as long as broad.

Gillrakers: 8 or 9 (7 in one specimen), relatively stout, on the lower part of the first gill arch, the lowermost 3 or 4 rakers reduced.

Scales ctenoid; lateral line with 31 (f.4) or 32 (f.1) scales, cheek with 3 rows; 6 scales between the dorsal fin origin and the upper lateral line, 5 (f.3) or 6 (f.2) between the pectoral and pelvic fin bases.

Fins. Dorsal with 24 (f.2) or 25 (f.3) rays, comprising 15 (f.3) or 16 (f.2) spinous and 9 (f.4) or 10 (f.1) branched rays. Anal with 12 (f.3) or 13 (f.2) rays comprising 3 spines and 9 (f.3) or 10 (f.2) branched rays. Pectoral 23·6–28·8 per cent of standard length. Caudal distinctly subtruncate or rounded. Pelvics with the first branched ray slightly produced in both sexes.

FIG. 10. *Haplochromis nubilus*. From Boulenger, *Fishes of the Nile:* a Lake Victoria specimen is depicted.

Teeth in the outer row of both jaws relatively stout, immovably implanted and unequally bicuspid; the major cusp slopes somewhat obliquely (*cf. H. velifer*). The posterior three or four teeth on either side of the upper jaw are stouter than the anterior teeth and are unicuspid. The number of upper, outer teeth in the five specimens is 40, 42 and 46 (f.3).

The inner rows in both jaws are composed of small tricuspid teeth arranged in 2, 3, 4 or 5 series in the upper jaw and 2 or 3 series in the lower jaw; with one exception, the inner series are separated from the outer row by a distinct interspace; in the exceptional specimen the gap is obscured by the irregular arrangement of the inner series.

Lower pharyngeal bone slender and triangular, its dentigerous surface broader than long. The pharyngeal teeth are slender, bicuspid and laterally compressed, those of the two median rows somewhat coarser. The teeth are fairly close-set (especially in the upper corners of the bone) and are arranged in 24–26 rows.

Coloration. Judging from the few notes prepared by the C.N.B.S. there does not seem to be any noticeable difference in the colours of Nabugabo fishes. Adult males have an overall velvety black colour, with a bright scarlet margin to the entire dorsal fin, scarlet maculae on the soft part, a scarlet distal half of the caudal fin and a similar colour spread over the entire anal fin; the anal ocelli are orange-yellow. The pelvic fins are black.

Ecology. Because so few specimens of *H. nubilus* were recorded by the C.N.B.S. little is known about the habits of this species in Lake Nabugabo. The one locality at which *H. nubilus* was caught is in shallow water (about 3 ft. deep) over a sand and mud bottom and a few feet away from a swamp shore.

No data are available on the food of the Nabugabo population, nor is anything known about their breeding biology except that all five specimens described above are adults (4 males, 1 female).

III. *HEMIHAPLOCHROMIS* Wickler, 1963

See Wickler (1963) for a full discussion of this genus.

Hemihaplochromis multicolor (Schoeller) 1903

(Text fig. 11)

Chromis multicolor Schoeller, 1903, *Bl. Aq. Terrk.*, **14** : 185.
Paratilapia multicolor: Hilgendorf, 1903, *Sitzber. Ges. Naturf. Fr. Berlin*, 429–32.
Haplochromis multicolor: Regan, 1922, *Ann. Mag. nat. Hist.* (9), **10** : 249–64.
Hemihaplochromis multicolor: Wickler, 1963, *Senk. biol.*, **44** : 83–96.

As yet, insufficient revisional work has been done on this widespread species to determine the relationships of the Nabugabo population. The C.N.B.S. field notes on coloration certainly do not suggest that the Nabugabo fishes differ in this important character.

Hemihaplochromis multicolor is recorded from the White Nile, Lower Nile, Bahr el Jebel, Lake Albert, the Semliki river, Lakes Victoria and Kyoga, the Malawa and

FIG. 11. *Hemihaplochromis multicolor*. From Boulenger, *Fishes of the Nile*: a Lake Victoria specimen is depicted.

Aswa rivers (Uganda) and the small, swampy lakes Kachira, Kijanebalola and Nakavali which lie between Lakes Edward and Victoria. The C.N.B.S.'s material constitutes the first record of the species from Lake Nabugabo.

In Nabugabo, *H. multicolor* was collected from a range of habitats; indeed, it is probably the only *Haplochromis*-group species to occur in the isolated pools of the floating border swamp and in the sand-bar swamp at the eastern lake edge. It is found frequently amongst the rooted vegetation fringing the main lake, and there are records of it from the deep inlets, which penetrate into the swamps. A few specimens were caught amongst flooded tree roots at the mouth of the Juma river. There are no records from the open lake but limitations imposed by the gear used there, coupled with the small size of the fishes, would reduce the chances of their being captured.

The C.N.B.S. concluded that *H. multicolor* are most common in areas where the water is relatively sheltered and where there is a good growth of water-weed, particularly *Utricularia* and *Ceratophyllum*. Of all the species in Lake Nabugabo, *H. multicolor* seems to be peculiar in exploiting *Utricularia* as a food source; presumably it is the animals trapped by the plant, and not the plants, which provide the nourishment.

The feeding habits of *H. multicolor* are diverse. Data from the C.N.B.S. notes indicate that many individuals had fed exclusively on other fishes, especially the cyprinodont *Aplocheilichthys pumilus* and species of the cyprinid *Barbus*. Insects (particularly chironomid and ephemeropteran larvae) and Crustacea (*Cyclops* and Ostracoda) are also recorded, as are some specimens in which the entire gut is packed with filamentous and blue-green algae, or with a mixture of algae and the remains of *Utricularia*. In these latter fishes remains of small Crustacea are also found, suggesting that they were derived from the *Utricularia* bladders since similar animals are trapped by these plants. A few specimens contained, in addition to the plant matter, a number of small oligochaet worms.

No breeding females were recorded by the C.N.B.S., but several females had ovaries in an advanced stage of oogenesis.

IV. *ASTATOREOCHROMIS* Pellegrin, 1903

See Greenwood (1959a) for a complete generic synonymy and a discussion of generic characters; also, see Greenwood (1965) for further comments on the generic diagnosis.

Astatoreochromis alluaudi Pellegrin, 1903
(Text fig. 12)

For synonymy see Greenwood (1959a and 1964).

A single specimen of this species was collected by the C.N.B.S. and it represents the first record from Lake Nabugabo. Unfortunately it is impossible to be certain that *A. alluaudi* forms a natural element in the Nabugabo fauna. In recent years this species has been introduced into several ichthyo-faunal regions of east Africa as a biological control agent against snails. However, I can find no definite record of introduction to Lake Nabugabo and since it does occur naturally in other small lakes (Greenwood, 1959a) and because snail control in Nabugabo would not be necessary (snails are reputedly rare) the evidence does seem to favour the consideration of *A. alluaudi* as a natural element.

The specimen available is an adult male 85 mm. S.L. (85 + 22 mm. total length) and does not differ in any morphometric characters from specimens described before (Greenwood, 1959a).

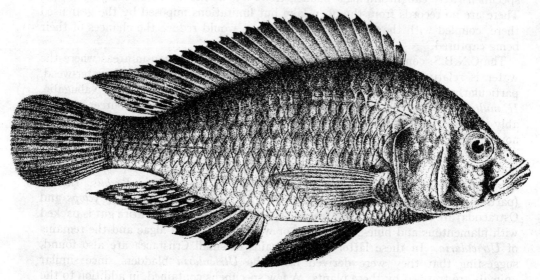

FIG. 12. *Astatoreochromis alluaudi*. From Boulenger. *Fishes of the Nile*.

The oral dentition is likewise typical. The pharyngeal bones and teeth are directly comparable with the greatly reduced type characterizing fishes from Lakes Edward, George, Nakavali and Kachira (*i.e.* of the populations formerly described as the subspecies *A.a.occidentalis*; but see Greenwood, 1965). The lower pharyngeal bone is weak and only slightly enlarged but the upper pharyngeal bones are relatively stouter. The two median rows of lower pharyngeal teeth are composed of slightly enlarged teeth still with remnants of the smaller cusp (*i.e.*, not of the broad-crowned unicuspid type found in specimens from Lake Victoria and in some from Lake Edward). The remaining teeth are small, compressed and weakly bicuspid with the major cusp very prominent. Some enlarged teeth occur on the upper pharyngeal bones and these are relatively stouter than their counterparts on the lower bone; however, they are distinctly bicuspid.

The neurocranial apophysis for the upper pharyngeal bones is markedly reduced in size, with proportionately much smaller basioccipital facets than are found in any other natural population of the species. Indeed, the morphology of the apophysis is comparable with that of an aquarium raised specimen which was fed on a snail-free diet (see Greenwood, 1965).

Unfortunately, the gut of the unique Nabugabo specimen is empty so no data are available on its feeding habits. The extreme scarcity of snails in the lake suggests, however, that the food of these fishes would not be at all like the predominantly snail diet of the Lake Victoria population, and it probably contains even fewer snails than does the diet of populations from Lake Edward.

It is clear from the C.N.B.S. notes that the coloration of the Nabugabo fish is identical with that of specimens from other areas. Like males from Lake Kachira, the Nabugabo fish is adult at a smaller size than has been recorded from Lake Victoria. No data are available on the minimum sizes of adult males from other localities.

DISCUSSION

Before discussing the evolutionary aspects of the Lake Nabugabo *Haplochromis*, it is necessary to consider the status of the five endemic "species", particularly their status *vis à vis* related species in Lake Victoria.

But for one exceptional species pair, the endemic *Haplochromis* are easily distinguished from each other both anatomically and with regard to male breeding coloration. No intermediate specimens are encountered and there is no *prima facie* evidence to suggest that interspecific crossing takes place.

The exceptional pair is *H. simpsoni* and *H. velifer*. There, the overall level of morphological differentiation, although appreciable, is slight and concentrated in dental characteristics. Two specimens out of a total of seventy have a dentition which could be considered a mixed *"velifer-simpsoni"* type; however, each fish has a greater ratio of *"simpsoni"* to *"velifer"* characters. In other respects the morphology of these fishes is *"simpsoni"*-like, but it must be remembered that it is in dental characters that specimens of the two species show the greatest divergence. With only this evidence on which to work, one cannot decide if specimens with apparently interspecific dental characters are of hybrid origin. Whatever their origin, their frequency is very low.

The real problem posed by the Nabugabo species flock is to estimate the degree of biological separation existing between the endemic Nabugabo species and those *Haplochromis* endemic to Lake Victoria which are morphologically closest to them and from which the Nabugabo species were probably derived. Phenotypically, as we have seen, there is every reason to consider them specifically distinct. Indeed, the morphological gaps between any one Nabugabo species and its counterpart in Victoria are as great as those between any two related *Haplochromis* in Lake Victoria. Similar differences are apparent when Nabugabo species are compared with species from Lake Kivu or Lake Edward.

I believe that the marked differences in male coloration between the anatomically nearest species of Nabugabo and Victoria suggest that members of each pair or group would behave as biological species should they ever become sympatric. Ethological studies (Baerends and Baerends van Roon, 1950; Wickler, 1963) all stress the importance of male coloration in species discrimination amongst cichlids; indirect evidence from the multispecific *Haplochromis* flock of Lake Victoria adds support to the experimental studies because one repeatedly finds distinctive male colours characterising otherwise phenotypically similar species.

This evidence is inferential with regard to the extralacustrine validity of the Nabugabo species but it seems as well-grounded as any such evidence can be.

The history of Lake Nabugabo (see page 316) makes it seem unlikely that its fishes were derived from any source other than Lake Victoria, a conclusion amply supported by the non-cichlid fishes.

If this source is accepted, it then remains to consider the stage of speciation which the Victoria *Haplochromis* had reached 4,000 years ago. Allowing for some inevitable genetical differences, it seems likely that there was little major difference between the species then and now. The reasons for my thinking this are based mainly on the difficulty of understanding how any significant number of species could have evolved in Lake Victoria during that short period. If it be argued that some degree of habitudinal isolation would allow for the differentiation of species, then one is faced with the problem of how the species then became widely distributed around the extensive shores of Victoria. Also militating against the concept of habitudinal isolation is the fact that no morphologically distinct populations are recognisable in any present-day Victoria species. This does not, of course, deny the possibility of there being genetically distinguishable populations (particularly since Victoria *Haplochromis* species are usually restricted in their habitat preferences) but the present argument is concerned with the morphological expression of the genotype and the recognition of species in that way.

If habitudinal segregation be ruled out, what of isolation through changes in the topography of the Lake basin? Again there is nothing to support the hypothesis; all the geological and climatological evidence strongly suggests that there has not been any great change in the form of Lake Victoria during the past four thousand years. The problem of intralacustrine distribution after speciation is also relevant.

Thus, it seems reasonable to conclude that at the time of its formation Lake Nabugabo would have been populated by *Haplochromis* similar to those inhabiting a comparable bay in the Lake Victoria of today.

Trewavas (1933) implies that the Nabugabo endemics are relict species. Her views were influenced by the fact that the extreme youth of Nabugabo had not then been fully appreciated, and also by the lack of knowledge about the *Haplochromis* species-flock of Lake Victoria. For instance, it was not realised that there are species in Victoria which, anatomically speaking, are of stock status when compared with the more specialized species existing alongside them, often in the same habitat. This ancestor-descendant relationship is found in all the trophic groups of Victoria *Haplochromis* but is seen especially well amongst the algal grazers, mollusc crushers, mollusc shellers and the piscivorous predators. There is no reason to consider *H. annectidens* anatomically more "basal" than its relative in Lake Victoria, *Haplochromis lividus* (see above, p. 334); *Haplochromis venator* belongs to the Lake Victoria species group containing some of the more advanced piscivorous predators, and there is nothing about *H. simpsoni* or *H. velifer* either more or less specialized than in similar species of Lake Victoria.

The case of *H. beadlei* is equivocal (see p. 340) since although the dental characters of this species seem less specialized than those of *H. crassilabris* in Victoria, they could be interpreted as having evolved, through regression, from a *"crassilabris"*-like condition.

If the endemic *Haplochromis* of Lake Nabugabo were evolved from Lake Victoria *Haplochromis* virtually identical with those of the present day, the evolutionary process was simply one of speciation. It did not involve the development of new adaptive lines from the point of view of feeding mechanisms; indeed, this could hardly have been possible since the Victoria flock had already exploited this field to saturation. Adaptation seems more likely in those physiological characters concerned with respiration and osmoregulation, because the hydrology of Lake Nabugabo certainly differs from that of Victoria.

Historically, Nabugabo is a cut-off bay of Lake Victoria. If, on this basis, it is compared with a bay of comparable size in Victoria, one of the more remarkable features is the reduction in the number of *Haplochromis* and *Haplochromis*-group species: eight species compared with at least thirty in Victoria (i.e. *ca.* 25 per cent of species one might expect to find). There is also a reduction in the number of trophic types amongst the species (four compared with six in Lake Victoria). The broad specializations not represented in Lake Nabugabo are the embryo and larval-fish eaters and the mollusc shellers; there is, also, a reduction in the number of intra-group specializations represented in the Nabugabo flock. Another difference is in the proportions of the various trophic types; the number of insectivorous species in Nabugabo is proportionately much higher than would be found in a bay of Lake Victoria. There, piscivorous species would be most numerous, followed by phytophagous, mollusc-eating and insectivorous species in about equal proportions. These proportions are based on adult and subadult fishes because the feeding habits of immediately post-larval individuals are still unknown.

The non-cichlids of Lake Nabugabo, on the other hand, show much less depauperization, with about seventy-three per cent of the Victoria species also occurring in Nabugabo (figures based on the C.N.B.S. collections now deposited in the B.M. [N.H.].).

There is no indication of the fate of those Lake Victoria trophic groups which are not represented in Lake Nabugabo. Some may have retreated from the embryo lake before it was completely sealed-off and others may have been unable to survive the hydrological changes which took place once the bar was completed, hydrological changes directly affecting the water chemistry or acting indirectly through altered food chains.

The history and ultimate differentiation of the surviving isolated populations provide a text-book case of geographical isolation resulting in speciation. The rate at which speciation occurred shows how rapidly a genetic revolution (*sensu* Mayr, 1963) can be achieved. In relation to the species of Lake Victoria, those of Nabugabo must be considered inferential but it is significant that the degree of morphological differentiation achieved is as great as that existing between related sympatric species in Lake Victoria. The marked differences in male coloration between similar species in the two lakes is striking and, I believe, an important element in providing interspecific barriers. In Lake Nabugabo there could have been no selection in favour of strengthening such potential interspecific barriers because the species evolved there were derived from fully differentiated Victoria species. In such circumstances differences in male coloration must be looked upon as coincidental results of the genetical revolution undergone by the isolates, possibly even byproducts of selection acting on other components of the genotype directly concerned with adaptation to altered and altering environmental conditions.

If these suppositions are accepted, they throw some light on the rate at which genetical isolating mechanisms could evolve in isolated populations of *Haplochromis*. This is one of the principal problems involved in any attempt to explain the evolution of such multispecific flocks as that of Lake Victoria.

Recently, Hubbs (1961) has criticized the emphasis I placed on the role of spatial isolation (i.e. geographical and physical) in accounting for the history of the Lake Victoria *Haplochromis* flock. Lake Nabugabo seems to provide a "pertinent indication" (Hubbs' phrase) of speciation through geographical isolation which Hubbs did not consider when arguing against this concept in favour of essentially sympatric speciation within cichlid flocks. Hubbs believes that "in general, shallow ponds are not scenes of extensive speciation . . . ." and that "No great diversity has arisen within small lakes and ponds over Africa". Perhaps Hubbs was placing much emphasis on the qualifying words "extensive" and "great"; neither, I agree, applies to Nabugabo. However, the fact that speciation seems to have occurred in Nabugabo should be considered and the evidence for it is as good as it is for any allopatric species; it could hardly be more pertinent to the problem of cichlid speciation in Lake Victoria (see Greenwood, 1959c). The question of diversity is another issue if the word is taken to cover more extensive evolution than just the multiplication of species. Because of the highly differentiated species-flock from which it arose one would not expect to find great new diversity amongst the Lake Nabugabo species. Nabugabo provides evidence for phylogenesis and not anagenesis. The latter will always be a function of the adaptive levels attained by the ancestral stock or stocks and the environmental conditions obtaining during the evolutionary period under consideration.

The effects of ecological segregation (and with it in many cases spatial but not physically insuperable segregation) on the species of Lake Victoria has certainly been much less marked than has the effect of physical isolation on the ancestors of the present-day Nabugabo species. Lake Nabugabo was cut off about four thousand years ago and in that period the physiography of Lake Victoria has remained unchanged. The Victoria *Haplochromis* species must then have had virtually the same intralacustrine distribution and habitat preferences as at present, yet none has shown any differentiation comparable with that undergone by the populations isolated in the bay that was to become Lake Nabugabo. Any argument that the present-day Victoria species-flock has also evolved and reached the present pattern of species distribution (no species shows a geographically restricted intralacustrine distribution) in the last four thousand years, seems inconceivable in the light of evidence we have on the *Haplochromis* of Lakes Victoria and Edward (particularly the species common to both lakes), and the geological history of the area (Greenwood, 1951, 1959a and c).

Key to the genera of *CICHLIDAE* in Lake Nabugabo

	Scales ctenoid	1
	Scales cycloid . . . . **TILAPIA**	
1	Anal fin with three spines	2
	Anal fin with more than three spines . . **ASTATOREOCHROMIS**	
2	Many scales in lateral line series without pores; no ocelli on anal fin in adult males but posterior tip of fin with pigmented spot . . **HEMIHAPLOCHROMIS**	
	All scales in the lateral line series with pores; ocelli on anal fin in adult males, no pigment spot on posterior tip . . . . . **HAPLOCHROMIS**	

Key to the Species of *HAPLOCHROMIS*

	Lower jaw more than 47 per cent of head length; teeth unicuspid and moderately stout, more than 40 in the upper jaw . . . . . **H. venator**	
	Lower jaw less than 47 per cent of head length; teeth generally bicuspid or if unicuspid, either slender and numerous or stout and fewer than 40 in upper jaw . . .	1
1	Caudal truncate or weakly subtruncate	2
	Caudal distinctly subtruncate or, more usually, rounded; adult males jet black **H. nubilus**	
2	Teeth bicuspid or unicuspid (or a mixture of both); if predominantly unicuspid, then slender and more than 30 in upper jaw; lips not markedly thickened . . .	3
	Teeth unicuspid or bicuspid, stout and procumbent, less than 34 in upper jaw; lips markedly thickened (upper sometimes thicker than lower). . **H. beadlei**	
3	Teeth distinctly bicuspid	4
	Teeth weakly bicuspid or unicuspid or a mixture of both types, slender; more than 50 in upper jaw . . . . **H. simpsoni**	
4	Teeth moderately stout, distinctly bicuspid, immovably implanted; major cusp not protracted towards symphysis, and occlusal surface not horizontally aligned **H. velifer**	
	Teeth with slender shafts, weakly bicuspid, major cusp expanded, obliquely truncated so that occlusal surface is almost horizontal and produced towards the symphysis; moveably implanted . . . . **H. annectidens**	

SUMMARY

1. Lake Nabugabo, a small swampy lake, is separated from Lake Victoria by a relatively narrow sand-bar and swamp. The sand-bar is estimated to be about 4,000 years old; prior to that date the lake was a bay of Lake Victoria. A short description of the lake and its history is given.

2. The cichlid fishes are reviewed, mainly on the basis of new material collected by the Cambridge Nabugabo Biological Survey of 1962.

3. Six species of *Haplochromis* (five endemic to the lake) are now recorded; of these, two are new (*Haplochromis simpsoni* and *H. venator*; the latter was previously confused with *H. pellegrini* of Lake Victoria) and one (*H. nubilus*) is recorded for the first time.

4. Two other new records are: *Hemihaplochromis multicolor* and *Astatoreochromis alluaudi*. The latter species shows certain interesting osteological differences when compared with specimens from Lakes Victoria and Edward.

5. The evolutionary history of the Nabugabo *Haplochromis* is discussed. The evidence strongly suggests that the endemic species were derived from Lake Victoria species similar to, if not identical with species still extant in Lake Victoria. The significance of this rapid speciation in understanding speciation in Lake Victoria *Haplochromis* is considered.

ACKNOWLEDGEMENTS

In many respects this paper should be regarded as a joint effort between the members of the Cambridge Nabugabo Biological Survey and myself. I have drawn heavily upon their extensive and detailed field-notes as well as on the numerous coloured drawings they made of live fishes. Without their painstaking attention to detail and deep appreciation of the *Haplochromis* problem it would have been almost impossible to gain a full impression of the Nabugabo species. To the members of the C.N.B.S., Alan Roberts, Barney Hopkins, Michael Simpson and Robin Sturdy, I express my warmest thanks.

My thanks are also due to my colleagues, Dr. Ethelwynn Trewavas for the numerous discussions we have had about these fishes, and Mr. A. C. Wheeler for making several radiographs.

REFERENCES

BAERENDS, G. P. & BAERENDS VAN ROON, J. M. 1950. An introduction to the study of the ethology of cichlid fishes. *Behaviour*, Supplement I: 1–242. Leiden.

BEADLE, L. C. 1962. The evolution of species in the lakes of East Africa. *Uganda J.* **26**: 44–54.

BEADLE, L. C. & LIND, E. M. 1960. Research on the swamps of Uganda. *Uganda J.*, **24**: 84–98.

BISHOP, W. W. 1959. Raised swamps of Lake Victoria. *Rec. Geol. Survey Uganda* (1955–56), 33–42.

Cambridge Nabugabo Biological Survey, 1962. Preliminary Report (mimeographed), 21 pp.

Greenwood, P. H. 1951. Evolution of the African cichlid fishes; the *Haplochromis* species flock in Lake Victoria. *Nature*, London, **167** : 19.

—— 1956. A revision of the Lake Victoria *Haplochromis* species (Pisces, Cichlidae) Part I. *Bull. Br. Mus. nat. Hist., Zool.*, **4** : No. 5, 223–44.

—— 1957. A revision of the Lake Victoria *Haplochromis* species (Pisces, Cichlidae) Part II. *Bull. Br. Mus. nat. Hist., Zool.*, **5** : No. 4, 73–97.

—— 1959a. The monotypic genera of cichlid fishes in Lake Victoria. Part II. *Bull. Br. Mus. nat. Hist., Zool.*, **5** : No. 7, 163–177.

—— 1959b. A revision of the Lake Victoria *Haplochromis* species (Pisces, Cichlidae) Part III. *Bull. Br. Mus. nat. Hist., Zool.*, **5** : No. 7, 179–218.

—— 1959c. Evolution and speciation in the *Haplochromis* fauna (Pisces, Cichlidae) of Lake Victoria (preprint 1958). *Proc. XVth. Intern. Cong. Zool., London*, 147–150.

—— 1960. A revision of the Lake Victoria *Haplochromis* species (Pisces, Cichlidae) Part IV. *Bull. Br. Mus. nat. Hist., Zool.*, **6** : No. 4, 227–281.

—— 1962. A revision of the Lake Victoria *Haplochromis* species (Pisces, Cichlidae) Part V. *Bull. Br. Mus. nat. Hist., Zool.*, **9** : No. 4, 139–214.

—— 1965. Environmental effects on the pharyngeal mill of the cichlid fish *Astatoreochromis alluaudi*. *Proc. Linn. Soc. Lond.*, **176** : 1. 1–10.

Hubbs, C. L. 1961. Isolating mechanisms in the speciation of fishes: in *Vertebrate Speciation*. University of Texas Press, Austin.

Mayr, E. 1963. *Animal species and evolution*. Harvard University Press, Cambridge, Mass.

Poll, M. 1939. Poissons. *Explor. Parc. Nat. Albert. Mission H. Damas* (1935–6). Fasc. 6 : 1–73.

Regan, C. T. 1922. The cichlid fishes of Lake Victoria. *Proc. Zool. Soc. London*, 1922 : 157–191.

Trewavas, E. 1933. Scientific results of the Cambridge expedition to the East African lakes, 1930–1931. II. The cichlid fishes. *J. Linn. Soc. (Zool.)*, **38** : 309–341.

Wickler, W. 1963. Zur Klassifikation der Cichlidae, am Beispiel der Gattungen *Tropheus, Petrochromis, Haplochromis* und *Hemihaplochromis* n. gen. **Senck. biol. 44** : 83–96.

A REVISION OF THE *HAPLOCHROMIS* AND RELATED SPECIES (PISCES : CICHLIDAE) FROM LAKE GEORGE, UGANDA

By P. H. GREENWOOD

CONTENTS

	Page
INTRODUCTION	141
MATERIALS AND METHODS	144
Haplochromis elegans Trewavas	145
Haplochromis aeneocolor sp. nov.	150
Haplochromis nigripinnis Regan	155
Haplochromis oregosoma sp. nov.	159
Haplochromis macropsoides sp. nov.	162
Haplochromis limax Trewavas	167
Haplochromis mylodon sp. nov.	172
Haplochromis angustifrons Blgr.	177
Haplochromis schubotzi Blgr.	183
Haplochromis schubotziellus sp. nov.	189
Haplochromis taurinus Trewavas	192
Haplochromis labiatus Trewavas	196
Haplochromis pappenheimi (Blgr.)	199
Haplochromis squamipinnis Regan	204
Haplochromis petronius sp. nov.	209
Haplochromis eduardianus (Blgr.)	215
NON-ENDEMIC *Haplochromis* AND *Haplochromis*-group SPECIES	221
Haplochromis nubilus (Blgr.)	221
Astatoreochromis Pellegrin	224
Astatoreochromis alluaudi Pellegrin	224
Hemihaplochromis Wickler	225
Hemihaplochromis multicolor (Schoeller)	226
DISCUSSION	227
Biology of the Lake George *Haplochromis* species flock	227
The relationships and history of the Lake George *Haplochromis* species	230
ACKNOWLEDGEMENTS	237
APPENDICES : I	238
II	238
BIBLIOGRAPHY	239
GUIDE TO THE IDENTIFICATION OF THE *Haplochromis* SPECIES	241

INTRODUCTION

LAKE GEORGE, smallest of the African ' Great Lakes ' (text-fig. 1), occupies a virtually square basin of about 270 km² area in the western Rift valley (0°55′ N to 0°05′ S, and 30°02′ E to 30°18′ E). Water depth over much of the lake rarely exceeds 2·5 m, although there are some circumscribed areas with depths of up to 4·0 m.

Most of the lake is bordered by flat savannah-bush, but to the north and east there are areas of papyrus swamp extending, albeit as narrow fingers, for some 15 km from the lake edge. The principal affluent rivers enter Lake George through these papyrus swamps. The Rivers Sibwe, Nsonge and Mubuka arise in the Ruwenzori mountains; the Mpanga, however, is a westward flowing tributary of the Katonga, a river which also flows eastward to enter Lake Victoria. The shared headwater of this river is a swamp divide (Doornkamp & Temple 1966), apparently impenetrable to all but air-breathing fishes.

Much of the shore line is simple, but there are a few deeply indented bays and one steep-sided bay formed from a volcanic crater.

Considerable areas of the lake bottom are covered by thick (ca 3 m) deposits of flocculent organic ooze, overlying a firm clay substrate. In some places, both in- and offshore, a sandy substrate is exposed or is but thinly overlaid by mud.

Two large islands (Kankurunga and Akika) lie close to the western lake shore; a third (Irangara Island), on the north-western shore, almost occludes the entrance to the lake's largest and most sheltered bay, Hamukunga Bay. The island shorelines are varied and include slightly indented muddy bays, short stretches of sandy beach and extensive but narrow fringes of papyrus.

For a more detailed description of the lake and a brief outline of its limnological features, reference should be made to Dunn, Burgis, Ganf, McGowan & Viner (1969).

In addition to its small size and extreme shallowness, Lake George also differs from the other 'Great Lakes' in being directly linked with another water body, Lake Edward. Connection between Lakes Edward and George is effected through the Kazinga Channel, a 36 km long, river-like passage uninterrupted by swamps or rapids. There is a definite net outflow of water from Lake George into Lake Edward but the current is slight, and on occasion, undergoes wind-induced reversal of flow, at least in the upper layers of water.

To what extent the Kazinga Channel allows an actual exchange of fishes between the lakes (or of gene flow between populations of fishes in the lakes) has yet to be determined. Certainly many species of cichlid and non-cichlid fishes are present in both lakes, and the apparent endemism of some Lake George *Haplochromis* species may well be just a reflection of inadequate collecting in Lake Edward. Nevertheless, some habitats in Lake Edward are not represented in Lake George, and it is almost certain that a few Edward species are absent from Lake George. These absentees include not only species from deep-water habitats but also several from inshore habitats as well. Their absence from Lake George is hardly attributable to inadequate sampling because that lake has been intensively collected during the past six years.

Fairly comprehensive fish collections have been made recently along the whole length of the Kazinga Channel (*see* Appendix II). These collections indicate that the *Haplochromis* species of the channel are exclusively those common to both Lake Edward and Lake George. Surprisingly, even as close to Lake Edward as the Mweya Peninsula none of the inshore-living and apparently endemic Lake Edward species was found in the channel. Clearly, detailed ecological studies will have to be made (particularly at the Lake Edward end of the channel) before this situation is understood. For the moment, however, there seem to be good *a priori* grounds for believing that, for many species, there is continuity of populations between the lakes.

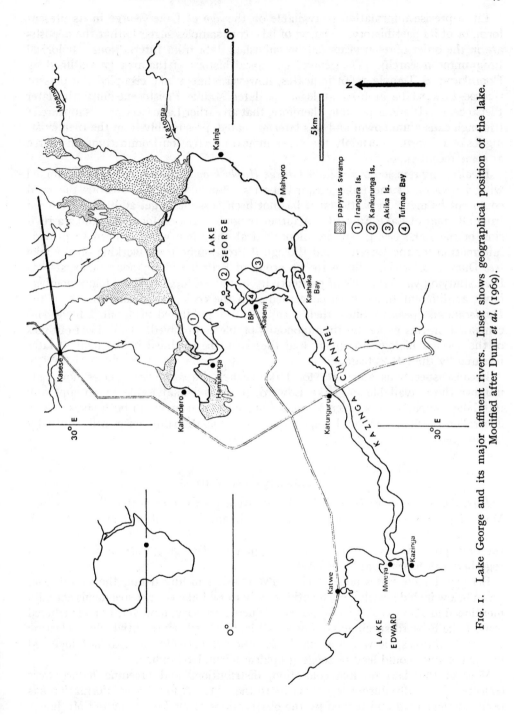

FIG. 1. Lake George and its major affluent rivers. Inset shows geographical position of the lake. Modified after Dunn et al. (1969).

Little precise information is available on the age of Lake George in its present form, or of its past history. Dating of lake core samples suggests that the deposits are in the order of 3500 years old (unpublished data from International Biological Programme research). The general geological history of the area (as outlined by Doornkamp & Temple 1966) indicates, however, that a lake occupied the present George–Edward basin from at least the later Middle Pleistocene until the later Pleistocene. It seems possible, therefore, that an earlier Lake George ' disappeared ' (through causes unknown) and was later recreated, presumably from the then existing Lake Edward. Certainly the fishes indicate derivation from a common source at some recent time.

Ideally, any revision of the Lake George *Haplochromis* species should be combined with a revision of the Lake Edward species. For a variety of reasons the ideal could not be met; in particular it has not been possible to get additional material from all areas of Lake Edward. A rather pragmatic reason for undertaking a revision of the Lake George species alone and at this time was the need to provide information for the International Biological Programme team working on the lake (*see* Dunn *et al*. 1969). Since 1967 a group of British and Ugandan biologists has been studying various levels of productivity in Lake George. Three team members have, at different times, been concerned with the ecology and distribution of the *Haplochromis* species. Since their results will be published in detail, I have concentrated, in this paper, on the taxonomic problems involved. Only brief outlines of the species' biology are given, and these may be modified in the light of later research by the I.B.P. team.

Because specimens and data for Lake George *Haplochromis* species now outnumber those available for Lake Edward, it seemed inadvisable to attempt any interlake comparisons between samples from species occurring in both lakes. As a general impression, however, I suspect that interpopulation differences will eventually be detected.

MATERIALS AND METHODS

Haplochromis species from Lake George were poorly represented in the British Museum (Natural History) collections. Consequently, most of the material on which this paper is based is that obtained during the I.B.P. investigation of the lake. In the only previous revision (Trewavas 1933), Lake George specimens were treated together with fishes from Lake Edward.

Because I have not studied Lake Edward fishes in any detail, the synonymies given below include, with few exceptions, only those Lake George specimens actually mentioned in Trewavas' (*op. cit.*) paper. Where necessary, however, I have included some Lake Edward specimens. For example, this has been essential when selecting certain lectotypes, or where a misidentification is corrected and, if not included in a synonymy, could lead to zoogeographical misunderstanding.

Most of the data on live coloration, distributions and breeding habits were collected personally during several visits to the lake. A lot of this information has been supplemented and refined by the observations of Dr Ian Dunn and Mr James

Gwahaba, the fish biologists of the I.B.P. team. Both these workers have given unstintingly of their time and information, and I am extremely indebted to them.

Measurements used in describing the species are those I have employed in other papers on *Haplochromis* species, viz. :

Standard length : measured directly[1] from the snout tip (including the premaxilla) to the posterior margin of the hypural bones (located by bending the caudal fin at right angles to the body's long axis).

Head length : measured directly[1] across the head from snout tip to the most posterior point on the opercular bone.

Preorbital depth : is the greatest depth of the first infraorbital bone (= lachrymal bone).

Interorbital width : is the least distance between the bony (frontal) margins of the orbit.

Snout length : measured directly[1] from the snout tip (i.e. the premaxillary symphysis) to the anterior orbital margin.

Eye diameter : is the greatest diameter of the bony orbit in the horizontal plane.

Cheek depth : is the greatest depth of the muscular part of the cheek (even when this extends below the scale rows) and is measured vertically.

Lower jaw length : is measured directly[1] from the dentary symphysis to the posterior margin of the articular bone (located by opening the lower jaw and finding its point of articulation).

Upper jaw length : is measured directly[1] from the premaxillary symphysis to the posterior margin of the maxilla.

Caudal peduncle length : is taken from the posterior margin of the hypurals to a vertical projected from the insertion of the last anal ray. *Peduncle depth* is the least depth.

A character I have used for the first time concerns the so-called *pseudorakers* on the first gill arch. These structures lie on the anterior (i.e. upper) face of the arch, between the inner and outer rows of true gill rakers. Pseudorakers are localized thickenings of the tissue covering the arch. In gross appearance they resemble true gill rakers, but unlike those structures they lack a bony central core.

Vertebral counts do not include the fused first preural and ural vertebrae (which support the parhypural and hypurals).

Haplochromis elegans Trewavas, 1933

(Text-figs. 2 & 3)

Haplochromis nubilus (part) : Trewavas, 1933, *J. Linn. Soc. (Zool.)*, **38** : 329 (specimens BMNH reg. nos. 1933.2.23 : 288–295 from Lake George).

H. elegans (part) Trewavas, *op. cit.* : 332 (3 paralectotypes, reg. nos. 1933.2.23 : 387–389, supposedly from Lake George, and 4 other specimens, 1933.2.23 : 390–393, also from that lake).

NOTE ON THE SYNONYMY. According to Trewavas (1933), three *H. elegans* syntypes (all females) are from Lake George, the other syntypical material being from Lake

---

[1] In a direct measurement, one tip of the dividers or calipers is placed at one of the points specified and the other tip is placed on the second point; the distance measured may thus run across the long axis of the fish (as, for example, in snout and head lengths).

Edward. I have examined the three syntypes (reg. nos. 1933.2.23 : 387–389) and agree with Trewavas' identification. However, the bottle label, and the Museum register, give the locality for these fishes as Lake Edward and not Lake George.

Trewavas (*op. cit.*) also refers seven specimens from the hypodigm (reg. nos. 1933.2.23 : 390–395) to this species, giving their localities as Worthington's (1932) stations 522 (Lake Edward) and 613 (Lake George). Six fishes (reg. nos. 1933.2.23 : 390–395) are in a bottle now labelled ' Lake George ' but without any station number quoted. I take these to be the fishes from station 613. Of these specimens, three are referable to *H. elegans*.

The *lectotype*, an adult male 65·5 mm standard length (BMNH 1933.2.23 : 381), is from Lake Edward.

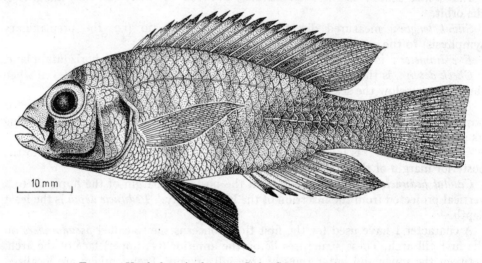

FIG. 2. *Haplochromis elegans*. Lake George specimen ; a male.

DESCRIPTION. Based on 34 specimens, 58·0–72·5 mm standard length (but not including the lectotype since it is from Lake Edward).

Depth of body 35·7–40·8 (mean, M = 37·6) per cent of standard length, length of head 32·2–35·4 (M = 33·7) per cent. Dorsal profile of head gently decurved or straight, sloping at about 35° to the horizontal ; dorsal margin of eye not entering the line of the profile, but clearly below it.

Preorbital depth 11·8–16·5 (M = 14·5) per cent of head, showing very slight positive allometry. Least interorbital width 21·9–27·3 (M = 24·2) per cent of head, length of snout 25·6–32·5 (M = 28·4) per cent, 0·8–0·9 of its breadth. Eye diameter 28·0–37·0 (M = 33·5) per cent of head (not showing clear-cut allometry in the size range examined), depth of cheek 18·2–24·4 (M = 20·8) per cent.

Caudal peduncle 13·8–18·7 (M = 16·2) per cent of standard length, 1·2–1·5 (modal range 1·2–1·3) times as long as deep.

Mouth horizontal or very slightly oblique ; lips somewhat thickened. Length of upper jaw 28·6–34·0 (M = 30·3) per cent of head, length of lower jaw 35·0–40·2

(M = 37·9) per cent, 1·3–1·8 (mode 1·4) times as long as broad. Posterior tip of maxilla reaching the vertical through the anterior orbital margin, but not quite reaching this level in a few specimens.

Gill rakers variable in form but usually rather stout, the lower 1 or 2 greatly reduced ; 8 or 9 rakers in the outer row on the lower part of the first gill arch. No pseudorakers are developed between the inner and outer rows of gill rakers on this arch.

Scales. Ctenoid ; lateral line with 30 (f.2), 31 (f.15), 32 (f.15), 33 (f.1) or 34 (f.1) scales ; cheek with 2 or 3 (mode) rows. Five to $6\frac{1}{2}$ (mode $5\frac{1}{2}$) scales between the upper lateral line series and the dorsal fin origin, 6–8 (mode 6) between the pectoral and pelvic fin bases.

Fins. Dorsal with 14 (f.2), 15 (f.24) or 16 (f.8) spinous and 8 (f.1), 9 (f.21) or 10 (f.12) branched rays. Caudal subtruncate, scaled on its basal half, or a little beyond. Pectoral 26·6–33·6 (M = 30·3) per cent of standard length, 80·0–97·8 (M = 90·0) per cent of head. Pelvics with the first two rays produced, especially in adult males.

Teeth. The *outer row* of teeth in both jaws (text-fig. 3) is composed principally of relatively stout, well-spaced, unequally bicuspid teeth ; anteriorly in the lower jaw, the teeth are implanted so as to slope forward at a slight angle.

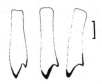

FIG. 3. *H. elegans.* (Left). Premaxillary teeth, left side, viewed from a point slightly anterior of lateral. The teeth are from an anterolateral position in the jaw. (Right). Dentary teeth (right side), lateral in position. Viewed laterally. Scale = 0·25 mm.

The major cusp is isoscelene in outline (*see* text-fig. 3) and very slightly incurved ; the neck of the tooth is slightly flattened in cross-section. Some teeth in each jaw have one margin of the major cusp partly flattened from below the tip so that it appears as a narrow step-like flange adjacent to the minor cusp (*cf. H. aeneocolor* where the flange is present on most teeth and is more obvious).

Posteriorly in the upper jaw there may be from 1 to 5 unicuspid and dagger-shaped teeth ; less often these posterior teeth are tricuspid.

There are 34–42 (mean = 38) teeth in the outer premaxillary row.

The inner rows (usually 2, less commonly 3 in the upper jaw, and 2 or 3 in the lower) are composed of small tricuspid teeth, often irregularly arranged (particularly in the upper jaw).

OSTEOLOGY. The *neurocranium* of *H. elegans* is typically that of a generalized *Haplochromis* species (*see* Greenwood 1962), although the preorbital profile is a little less decurved.

The lower *pharyngeal bone* is fairly stout, with its dentigerous area 1·1–1·2 times broader than long. The teeth are fine, cuspidate and compressed and are arranged in 24–26 rows. Teeth in the two median rows are somewhat coarser than the others.

Vertebral counts for the 30 specimens radiographed are : 28 (f.3), 29 (f.22) or 30 (f.5), comprising 12 (f.5), 13 (f.24) or 14 (f.1) abdominal and 15 (f.2), 16 (f.20) or 17 (f.8) caudal elements.

COLORATION IN LIFE. *Adult males* : ground colour smokey grey overlying bluish-silver. Snout, lips and cheek with a livid iridescence. Belly and branchiostegal membrane dark cinder grey. Dorsal fin with the spinous part sooty, the inter-spinous membrane generally darkest ; lappets black but with a narrow red streak (or spot) at the tip. Soft dorsal with maroon streaks between the rays. Caudal fin with maroon spots and blotches between the rays, and a suffuse maroon flush around the fin margin. Anal dark hyaline (or faintly grey) often dusky at its base and with a pinkish-maroon border. The pelvics are black.

Females : ground colour sandy green shading to silvery white on the belly and lower flanks. All fins yellowish-green. Because female *H. elegans* are not immediately identifiable in the field, these ' live ' colours are in fact ' post-mortem ' colours and should not be considered at all precise.

COLORATION IN PRESERVED SPECIMENS. *Adult males* : ground colour variable but basically grey-brown ; belly and chest dusky, as are, sometimes, the flanks. The flanks are generally crossed by 3–6 faint vertical bars. The branchiostegal membrane is black. Cephalic markings comprise a distinct lachrymal stripe, two bars across the snout and a broader bar immediately behind the orbits ; in many specimens there is an even broader, but fainter, bar or blotch transversely across the nape. The lower part of the cheek and the vertical limb of the preoperculum are sometimes dusky.

The dorsal fin is dusky, with darker streaks between the spines and rays, or the latter region maculate. Anal fin dusky or indistinctly maculate. Caudal with a dark central area and a light marginal zone. Pelvics are black, and the pectorals hyaline.

Females have a greyish-yellow ground coloration, and sometimes very faint traces of 3–6 vertical bars on the flanks. On the head there are slight indications (sometimes just a darker area) of two bars across the snout, and a lachrymal stripe. All the fins are hyaline, the dorsal usually darker than the others ; the caudal is often maculate.

ECOLOGY. *Habitat*. *Haplochromis elegans* is essentially a species of the inshore regions of the lake, especially near papyrus shores or where the bottom is sandy. It rarely occurs in open-water localities or in shallow places where the substrate is mud.

Food. Mostly chironomid larvae, although emergent aquatic Diptera are also eaten when available.

Breeding. *Haplochromis elegans* is a female mouth-brooder. All specimens, of both sexes, within the size range studied are adult ; females appear to reach a larger size than do males. In the 15 sexually active females examined, 11 have the right ovary noticeably larger than the left, 2 have the ovaries equally developed, and 2 have the right ovary a little larger than the left one.

Distribution. Lakes Edward and George, and the Kazinga Channel.

DIAGNOSIS AND AFFINITIES. Within Lake George, *H. elegans* most closely resembles *H. aeneocolor* (see p. 150), both morphologically and in its habitat preferences. Adult males of the two species are readily distinguished by their coloration, but females and preserved specimens are differentiated chiefly by the fewer and somewhat stouter teeth of *H. elegans* (32–42, M = 38, *cf.* 40–58, M = 48 in *H. aeneocolor*), the slightly procumbent anterior dentary teeth of *H. elegans*, the well-developed flange on the major cusp of most teeth in *H. aeneocolor* (see p. 151), the shorter upper jaw of *H. elegans* (28·6–34·0, M = 30·3 per cent of head, *cf.* 30·0–37·8, M = 34·9 per cent) and the shorter lower jaw of *H. elegans* (35·0–40·2, M = 37·9 per cent of head, *cf.* 38·0–44·0, M = 41·0 per cent in *H. aeneocolor*). In life the lips of *H. elegans* appear thicker than those of *H. aeneocolor*, but this distinction is less obvious in preserved material.

Haplochromis elegans shows few specializations in its dental or cranial anatomy, and must be ranked amongst the 'generalized' *Haplochromis* species. Outside the Lake Edward–Lake George species complex it resembles *H. pallidus* (Blgr.) of Lake Victoria (see Greenwood 1960). From *H. pallidus*, *H. elegans* differs in its adult male coloration, some morphometric characters (e.g. having a shorter snout) and in its overall morphology. The significance of this apparent resemblance will be discussed elsewhere (p. 230); however, it should be noted that the resemblance between *H. elegans* and *H. pallidus* cannot be shown to be more significant than that existing between it and species of the *H. bloyeti* complex (see Greenwood 1971).

Trewavas (1933) compared *H. elegans* with *H. cinereus* (Blgr.) of Lake Victoria, but this comparison is no longer valid now that we have a clearer concept of *H. cinereus* (see Greenwood 1960). In fact, *H. cinereus* shows some specialized characters (its dentition for one). These specializations would not be apparent in 1933, because at that time '*H. cinereus*' was a dumping ground for several of the generalized Lake Victoria species.

Resemblances which I noted between *H. elegans* and *H. velifer* Trewavas of Lake Nabugabo (Greenwood 1965b) are somewhat diluted by the greater amount of information now available on *H. elegans*. For example, the teeth of *H. elegans* (at least in Lake George populations) have a more acutely pointed cusp, and there are fewer teeth in the outer premaxillary row. There is, of course, a marked difference in the male breeding coloration of the two species.

Diagnostic problems arising in connection with *H. elegans* and species at present known only from Lake Edward (and then very imperfectly known) are virtually identical with those discussed in relation to *H. aeneocolor* on page 153.

STUDY MATERIAL

Register number BMNH	Locality: Lake George
1972.6.2 : 166–167	N.E. lake shore (papyrus)
1972.6.2 : 168–171	Papyrus shore off I.B.P. Laboratory
1972.6.2 : 172–177	Kankurunga Island

1972.6.2 : 178–179	Kankurunga Island
1972.6.2 : 180–182	Kankurunga Island
1972.6.2 : 183–206	Kankurunga Island
1972.6.2 : 207–224	Kankurunga Island
1972.6.2 : 225–230	Kankurunga Island
1972.6.2 : 231–236	Kankurunga Island
1972.6.2 : 237–238	Akika Island
1972.6.2 : 239–243	Akika Island
1972.6.2 : 244–260	Akika Island
1972.6.2 : 261–272	I.B.P. Jetty
1972.6.2 : 273–285	Kashaka Bay
1972.6.2 : 286–291	Tufmac Bay
1972.6.2 : 292	Close to shore (muddy)

Haplochromis aeneocolor sp. nov.
(Text-figs. 4 & 5)

Haplochromis nubilus (part) : Trewavas, 1933, *J. Linn. Soc. (Zool.)* **38** : 329 (4 specimens, BMNH reg. nos. 1933.2.23 : 296–299).

HOLOTYPE. A male, 68·0 mm standard length, BMNH reg. no. 1972.6.2 : 43.
The specific name refers to the brassy appearance of adult males.

DESCRIPTION. Based on 36 specimens (including the holotype), 58·0–75·0 mm standard length.

Depth of body 35·7–41·1 (M = 37·7) per cent of standard length, length of head 32·0–36·8 (M = 34·5) per cent. Dorsal profile of head straight or slightly concave, sloping fairly steeply at *ca* 35°–40° with the horizontal; dorsal margin of orbit not entering the line of the profile but distinctly below it.

Preorbital depth 12·0–18·2 (M = 14·6) per cent of head (not showing any clear-cut allometry), least interorbital width 22·7–29·3 (M = 25·5) per cent, snout length 26·7–31·8 (M = 28·8) per cent, 0·8–1·0 (mode 0·9) of its breadth. Eye diameter 28·6–35·0 (M = 31·4) per cent of head (showing no obvious allometry), depth of cheek 19·0–25·0 (M = 22·8) per cent.

Caudal peduncle 12·9–17·4 (M = 15·3) per cent of standard length, 1·2–1·5 (modal range 1·2–1·3) times as long as deep.

Mouth angle ranging from horizontal to slightly oblique; lips somewhat thickened. Length of upper jaw 30·0–37·8 (M = 34·9) per cent of head, lower jaw 38·0–44·0 (M = 41·0) per cent of head, 1·5–2·1 (modal range 1·6–1·8) times as long as broad. Posterior tip of the maxilla reaching a vertical through the anterior part of the eye or even to a vertical through the anterior margin of the pupil.

Gill rakers of various shapes, from short and stout to relatively slender; the lower 1 or 2 rakers on the first gill arch are greatly reduced, the upper 2 or 3 often flattened. There are 8 or 9 rakers on the lower part of the first arch.

Pseudorakers are poorly developed.

Scales. Ctenoid; lateral line with 30 (f.11), 31 (f.18), 32 (f.4) or 33 (f.1) scales, cheek with 3 (rarely 2) rows. Five to 6½ (mode 5½) scales between the lateral line and the dorsal fin origin, 6 (mode) or 7, rarely 5½ or 5, between the pectoral and pelvic fin bases.

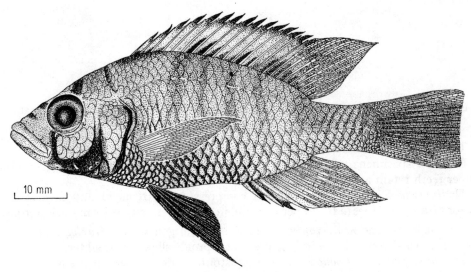

FIG. 4. *Haplochromis aeneocolor*. Holotype.

Fins. Dorsal with 14 (f.3), 15 (f.31) or 16 (f.2) spinous and 8 (f.3), 9 (f.25) or 10 (f.8) branched rays. Anal with 3 spines and 7 (f.1), 8 (f.9), 9 (f.20) or 10 (f.6) branched rays. Caudal subtruncate, scaled on its basal half. Pectoral 27·0–31·7 (M = 28·9) per cent of standard length, 73·8–93·5 (M = 84·5) per cent of head. Pelvics with the first and second rays produced (the first markedly so), and relatively longer in adult males than in females.

Teeth. The *outer teeth*, although basically of the generalized, unequally bicuspid type, are nevertheless rather distinctive. This is due to the presence of a well-developed, thin flange on that margin of the outer cusp which is adjacent to the minor cusp (*see* text-fig. 5). Few individuals fail to show flange development on at least the majority of anterior and lateral teeth in both jaws. The flange can be so well developed that the tooth seems to have an expanded and obliquely sloping major cusp (i.e. to be like the teeth of *H. limax*, see p. 168). Usually the flange is thin and almost transparent, so that there appears to be a dividing line between it and the more substantial body of the cusp itself. Although the flange may be continuous with the occlusal (i.e. distal) part of the cusp (thereby simulating an *H. limax*-like tooth) it is generally confined to the proximal half or two-thirds of the cusp. In this way a distinct step is developed between the flange and the occlusal tip of the cusp.

Apart from the flange, outer teeth in *H. aeneocolor* are typical bicuspids, with the major cusp having the outline of an isosceles triangle rather than of an equilateral one. The minor cusp is well developed, but its tip is not very acute. The crown of an outer tooth has virtually no incurvature, and the neck is a slightly compressed cylinder.

The posterior 1–4 upper teeth are either compressed tricuspids or are unicuspid and caniniform.

FIG. 5. *H. aeneocolor*. Dentary teeth (left side), lateral in position. Viewed laterally. Scale = 0·25 mm.

There are 40–56 (mean 48) teeth in the outer row of the upper jaw.

In a few specimens, all the outer teeth in the lower jaw are unicuspid, but the upper teeth retain a typical bicuspid form.

The *inner teeth* in both jaws are small, compressed and tricuspid, and are arranged in 2 or 3 (rarely 4) series in the upper jaw, and in 2 (rarely 1 or 3) series in the lower jaw.

OSTEOLOGY. The *neurocranium* is of the typical generalized *Haplochromis* type (see Greenwood 1962), but with the preorbital profile slightly straighter.

The *lower pharyngeal bone* is moderately stout, its dentigerous area equilateral or slightly broader than long. The teeth are fine, compressed and cuspidate, and are arranged in *ca* 24–26 rows; the median teeth are not noticeably larger or coarser than those of the lateral rows.

Vertebral counts in the 16 fishes examined are 27 (f.1), 28 (f.7) and 29 (f.8), comprising 12 (f.6) or 13 (f.10) abdominal and 15 (f.4), 16 (f.11) or 17 (f.1) caudal elements.

COLORATION IN LIFE. *Adult males*: the flanks, lateral aspect of the chest and belly, lower part of the head, the branchiostegal membrane and the lips are dark sulphurous yellow, with an orange overlay on the operculum. The rest of the flank (i.e. the posterior part) and the caudal peduncle are yellowish-green with a faint bluish overlay, and the ventral aspect of the chest is sooty. The dorsal body surface is dull bronze posteriorly, becoming purple above the flanks, and crimson anteriorly. The snout and anterior dorsum of the head are puce.

The overall colour impression gained from a newly caught male is one of brassyness, despite the various colour elements described above.

The dorsal fin is dark hyaline on the spinous part (the lappets black), but lighter on the soft part where the margin is crimson. The anal fin is hyaline over the basal third of the soft part, but with the spines and distal two-thirds of the soft part pinkish-crimson; the ocelli are orange-yellow. The caudal fin is pinkish to red, the colour intensified on the ventral third of the fin and at its posterior angle. The pelvic fins are black, the pectorals hyaline.

Male coloration is difficult to describe adequately because the intensity of the various colours is variable and changes rapidly after the fish is removed from water. Some fishes, for example, appear almost black a short while after capture.

PRESERVED COLORATION. *Males*: the ground colour is essentially like that described for *H. elegans* (*see* p. 148), but in *H. aeneocolor* the dark ventral pigment is more extensive; in some individuals it covers the entire caudal peduncle and the flanks to a level just below the upper lateral line. Cephalic markings are identical in both species.

The dorsal fin is dusky to black; if dusky, the pigment is often concentrated basally so that this region of the fin is almost black. The caudal is more or less uniformly dark, except for hyaline areas on the ventral and posteroventral margin. The anal varies from grey to dusky; the area over the spines is generally black. The pelvic fins are black, the pectorals hyaline.

Females have a greyish-silver to greyish-yellow ground colour; the head shows very faint traces of two transverse bars across the snout and an ill-defined, short, lachrymal stripe or streak. The dorsal fin has dark streaks between the rays, especially on the spinous part. The caudal is maculate, usually weakly so, and with the spots most obvious on the centre of the fin; a few specimens have intense maculae distributed over most of the fin. The anal is hyaline as are the pelvics (which may be faintly dusky).

ECOLOGY. *Habitat.* This species is particularly common near papyrus shores, and is rare elsewhere in the inshore region. Apparently it never occurs offshore.

Food. *Haplochromis aeneocolor* seems to be a detritus feeder since plant fragments and insect larvae are predominant elements of its gut contents. Adult insects are, however, also eaten.

Breeding. *Haplochromic aeneocolor* is a female mouth brooder. Of the 10 adult females examined, the right ovary is much larger than the left in 6 individuals, slightly larger in 3 and of equal size in 1 fish.

All specimens within the size range studied are adult and there is apparently no sexual dimorphism in the maximum size attained.

Distribution. Lake George and the Kazinga Channel.

DIAGNOSIS AND AFFINITIES. Until more is known about most Lake Edward species (and especially those not recorded from Lake George) an adequate diagnosis for *H. aeneocolor* is difficult to compile. For example, *H. eduardii* Regan superficially resembles *H. aeneocolor* but appears to differ in having stouter, non-flanged and less acutely cuspidate teeth, a shallower body and more rounded (i.e. decurved) head profile. *Haplochromis engystoma* Trewavas (known only from the holotype and one other specimen) has dental characteristics more like those of *H. aeneocolor*, but differs in several morphometric characters, especially in its higher (2·0) eye/cheek ratio, shorter lower jaw (34·8 per cent of head) and its strongly decurved head profile. *Haplochromis vicarius* Trewavas (at least as restricted to the holotype) has an overall superficial resemblance, but differs in having obliquely cuspidate outer teeth, more rows of inner teeth and a larger eye (36·0 per cent of head); Poll (1939) has synonymized *H. vicarius* with *H. eduardii* but I doubt the correctness of this decision (*see* Appendix I).

Regrettably, in none of these comparisons could life colours be taken into account because these are unavailable for the Lake Edward species.

Considering now those species which also occur in Lake George, *H. elegans* has the closest overall and detailed resemblances with *H. aeneocolor* (see p. 149). Male coloration is, however, very different in the two species, and there are dental and morphometric differences as well (again, see p. 147). Distinguishing between females of the species is especially difficult, although in the field the more rounded head profile of *H. elegans* does give some guidance for preliminary sorting.

Haplochromis limax Trewavas shows a fairly marked resemblance to *H. aeneocolor* in its superficial morphology and there is also a certain convergence in dental morphology. This arises from the peculiar flange developed on the major cusp of outer teeth in *H. aeneocolor* (see p. 152). If this flange is hypertrophied, it increases the area of the major cusp and imparts to it an oblique cutting edge. However, if such teeth in *H. aeneocolor* are closely examined, the junction between flange and main body is apparent, as is a slight indentation on the cutting edge. Furthermore, the flange is much thinner (nearly transparent) than the corresponding margin of a tooth in *H. limax*. Another distinguishing feature of *H. limax* is the broader array of inner teeth, and their larger size. Again, male breeding coloration is very different in the two species.

Beyond the confines of Lakes Edward and George, *H. aeneocolor*, like *H. elegans*, resembles the generalized *Haplochromis* species of Lake Victoria, in particular *H. pallidus*. But, in the absence of any clearly defined specializations in the species involved, little significance can be attached to these resemblances. The peculiar flange formation on the outer teeth of *H. aeneocolor* is an unusual feature for *Haplochromis* but its recognition as a specialization remains to be confirmed. Certainly it is rarely manifest among Lake Victoria species, but it does occur more frequently (if only as an individual variant) amongst the species of Lakes Edward and George.

Finally, and as if to reinforce the generalized nature of *H. aeneocolor* anatomy, the resemblance between this species and *H. nubilus* (Blgr.) should be noted. *Haplochromis nubilus* is one of the anatomically and ecologically most generalized species occurring in the Victoria–Edward drainage basin (see p. 221), and in turn shows close affinity with the fluviatile species of east Africa. On all morphometric characters *H. nubilus* and *H. aeneocolor* cannot be separated, but male coloration is markedly distinct, the caudal of *H. nubilus* has a nearly round distal outline, flanged teeth are not found (the teeth are unicuspid in large fishes) and the dorsal head profile is more concave than in *H. aeneocolor*.

STUDY MATERIAL

Register number BMNH	Locality: Lake George
1972.6.2 : 43 (Holotype)	N.E. shore near River Mpanga mouth
1972.6.2 : 44–50 (Paratypes)	Kankurunga Island
1972.6.2 : 52–54 (Paratypes)	Kashaka Crater
1972.6.2 : 55–63 (Paratypes)	N.E. shore
1972.6.2 : 64–67 (Paratypes)	N.E. shore
1972.6.2 : 68–72 (Paratypes)	Kashaka Bay
1972.6.2 : 73–79 (Paratypes)	Kankurunga Island
1972.6.2 : 81–84	Papyrus shore off I.B.P. Laboratory

1972.6.2 : 85–103 N.E. shore
1972.6.2 : 104–111 Kankurunga Island

Haplochromis nigripinnis Regan, 1921
(Text-figs. 6 & 7)

Haplochromis nigripinnis Regan, 1921, *Ann. Mag. nat. Hist.* (9) **8** : 635.
Haplochromis nigripinnis : Trewavas, 1933, *J. Linn. Soc. (Zool.)*, **38** : 330 (refers to Lake Edward fishes only).
Haplochromis guiarti (part) : Trewavas, 1933, *op. cit.* : 340 (1 of the 3 small fishes from Worthington's (1932) stations 613 and 618, Lake George, viz. BMNH reg. no. 1933.2.23 : 476.

HOLOTYPE. A male (probably adult), 64·0 mm standard length from Lake Edward, BMNH reg. no. 1914.4.8 : 14.

DESCRIPTION. Based on 36 specimens (excluding the holotype), 50·0–68·0 mm standard length.

Depth of body 32·3–41·5 (M = 36·8) per cent of standard length, length of head 31·5–35·3 (M = 33·6) per cent. Dorsal head profile straight or gently curved, sloping at *ca* 35°–40° to the horizontal ; dorsal margin of orbit barely entering the line of the head profile.

Preorbital depth 10·0–14·6 (M = 13·0) per cent of head, least interorbital width 20·7–27·5 (M = 24·2) per cent, ratio of interorbital width to eye diameter 1·28–1·75 (M = 1·49). Snout length 24·0–30·3 (M = 27·4) per cent of head, 0·8–0·9 (rarely 1·0) its breadth ; eye diameter 33·3–40·0 (M = 35·8) per cent, with no detectable allometry ; depth of cheek 16·3–22·9 (M = 19·8) per cent.

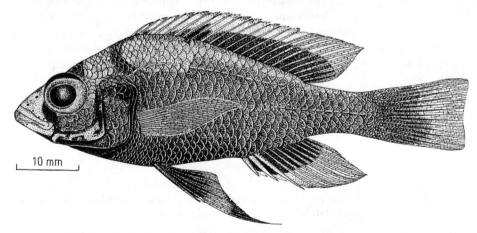

FIG. 6. *Haplochromis nigripinnis*. Lake George specimen ; a male.

Caudal peduncle 15·8–19·8 (M = 17·4) per cent of standard length, 1·3–1·8 (modal range 1·5–1·7) times as long as deep.

Mouth horizontal, lips not noticeably thickened. Length of upper jaw 30·6–36·9 (M = 34·0) per cent of head, length of lower jaw 39·0–47·5 (M = 43·6) per cent,

1·8–2·2 (modal range 1·9–2·0) times as long as broad. Posterior tip of the maxilla reaching a vertical through the anterior part of the eye or even to one through the anterior margin of the pupil.

A noticeable feature of the snout in *H. nigripinnis* is the size of the anterior opening to the nasal laterosensory canal. The opening is as large as (or almost as large as) the nostril. In most other *Haplochromis* species from Lake George (and apparently Lake Edward also) the opening to this canal is much smaller than the nostril, and is often difficult to locate.

The intestine in *H. nigripinnis* is long (*ca* 2–2½ times total length) and much coiled on itself.

Gill rakers on the first arch are, except for the reduced lower 1–3 and the occasional flattened and anvil-shaped upper 1–3, slender and relatively elongate. There are 8–10 (mode 9), rarely 11, rakers on the lower part of this arch. No pseudorakers are developed (see p. 145).

Scales. Ctenoid; lateral line with 30 (f.6), 31 (f.13), 32 (f.14) or 33 (f.3) scales, cheek with 2 or 3 (mode) rows. Five to 6½ (mode 5½) scales between the upper lateral line and the dorsal fin origin, 6 or 7 (mode), rarely 5, between the pectoral and pelvic fin bases.

Fins. Dorsal with 14 (f.1), 15 (f.28) or 16 (f.7) spinous and 8 (f.7), 9 (f.24) or 10 (f.5) branched rays. Caudal generally truncate but weakly emarginate in some fishes; scaled on its basal half. Pectorals 25·8–31·3 (M = 29·7) per cent of standard length, 79·5–94·5 (M = 87·3) per cent of head. Pelvics with the first ray slightly produced.

Teeth. The *outer teeth* in both jaws (except posteriorly in the upper) are slender, compressed and unequally bicuspid (text-fig. 7). The outline of the major cusp varies from equilateral to isoscelene; all intergrades may occur in one individual or one type of cusp outline may predominate. The crown is slightly incurved. Posterior teeth in the upper jaw are often either unicuspid and slender, or small, compressed and tricuspid. There are 40–60 (M = 52) teeth in the outer premaxillary row.

Fig. 7. *H. nigripinnis.* Premaxillary teeth (left), anterolateral in position. Viewed from an anterolateral position. Scale = 0·25 mm.

Teeth of the *inner rows* are small, compressed and tricuspid, and are arranged in 1 or 2 rows (rarely in 3) in the upper jaw, and 1 or 2 in the lower jaw. The serial arrangement of these teeth is often rather irregular.

OSTEOLOGY. The *neurocranium* of *H. nigripinnis* is identical with that of *H. elegans* and *H. aeneocolor*, that is, of a generalized type.

The *lower pharyngeal bone* gives an impression of being long, slender and fine (especially when compared with the bone in *H. elegans* or *H. aeneocolor*). Its dentigerous surface, however, is about 1·2 times broader than long. The teeth on this bone are fine, slender and cuspidate, and are arranged in *ca* 34 rows; teeth situated on the posterolateral angles of the bone are more densely crowded than elsewhere.

Vertebral counts in the 13 specimens radiographed are 28 (f.1), 29 (f.9), 30 (f.2) or 31 (f.1), and comprise 12 (f.2), 13 (f.10) or 14 (f.1) abdominal and 16 (f.9) or 17 (f.4) caudal elements.

COLORATION IN LIFE. *Adult males*: ground colour (including that of the head) is a dark malachite green with a silvery underlay; the branchiostegal membrane is black. On the head there is a prominent lachrymal stripe and a less intense interorbital bar; both marks are intensified after death. The dorsal fin is dark grey, with black lappets, and a darker irregular line along the base; a pink suffusion is visible over the soft part of the fin. The caudal has an overall pink flush except basally, where the membrane is dark hyaline. The anal is black over the spinous part, pinkish elsewhere (the colour intensifying distally); ocelli orange-yellow. Pelvic fins are black.

Adult females have an overall greyish-silver coloration above a midlateral line, and are chalky white below that level. The upper half of the caudal fin is hyaline but the basal area and lower half of the fin are suffused with pale lemon-yellow; this pigmentation sometimes extends over the entire fin but even then is most intense on the lower half. Some individuals show dark rather elongate spots along the middle of the caudal. The anal has, distally, similar yellow colour to that of the caudal but it is hyaline basally. The dorsal and pelvic fins both are hyaline.

PRESERVED COLORATION. *Adult males*: the ground colour is dark brown to black below the midlateral line, and to varying degrees above that level as well. When dark pigment does extend dorsally it is generally less intense than on the ventral body. In some fishes up to seven dark, fairly narrow vertical bars extend across the light brown of the upper body; sometimes there is a longitudinal dark bar extending for a variable length along the upper lateral line scale row.

The ventral half of the head is dark brown, the branchiostegal membrane black. A fairly distinct lachrymal stripe is usually visible through the general dark ground coloration of the snout. Two bars (of equal thickness) cross the snout, and often there is a small median blotch above the posterodorsal margin of the orbit. A larger dark blotch crosses the nape anterior to the dorsal fin origin; this mark seems to be a medial continuation of the first vertical bar of the flank.

The dorsal fin is dark, black on its proximal half and dusky beyond; the lappets are black. The caudal fin has a dark central area basally but otherwise it is greyish. The anal is black on its basal half, and dusky or hyaline distally. The pelvics are uniformly black or blotched black and dusky, the outer half of the fin being the darker part. The pectorals are hyaline.

Females have a greyish-silver to greyish-brown ground colour, and are lighter ventrally. In some few specimens very faint indications of vertical bars are visible on the flanks; such marks are confined to the central flank region and do not extend as far as the dorsal or ventral body outline. All fins are hyaline, the dorsal sometimes greyish with dark lappets.

ECOLOGY. *Habitat*. Although predominantly a species of offshore areas and the open central part of the lake, *H. nigripinnis* is sometimes found within a few feet of the shoreline, especially where the substrate is sandy.

Food. As the long and coiled intestine suggests, *H. nigripinnis* is a vegetarian species. It feeds principally on suspended phytoplankton; there is no indication from the gut contents of any bottom feeding habits. Like many *Haplochromis* species, *H. nigripinnis* is an opportunistic feeder; insect remains (of both larvae and pupae) are recorded from the gut.

Breeding. Female mouth brooding is practised by *H. nigripinnis*. Of the 8 adult females examined, 5 have the right ovary noticeably larger than the left, 1 has the right ovary slightly the larger and 2 show equal ovarian development.

Distribution. Lakes Edward and George and the Kazinga Channel.

DIAGNOSIS AND AFFINITIES. Trewavas (1933) compared *H. nigripinnis* with *H. cinereus* (Blgr.) of Lake Victoria. The invalidity of this comparison has been commented upon above (p. 149). It is due entirely to there being, at that time, insufficient material of either species to permit of precise comparison.

Surprisingly, neither Trewavas (*op. cit.*) nor Regan (1921) compared *H. nigripinnis* with any other species in the Edward–George complex. *Haplochromis nigripinnis* is indeed a distinctive species, especially when its coloration, fine dentition, long gut, fine gill rakers and its feeding habits are considered. But, preserved specimens (or live females) have a great similarity with specimens of *H. macropsoides*, a new taxon described on p. 162.

Haplochromis nigripinnis differs from *H. macropsoides* in having finer outer teeth in both jaws, fewer rows of inner jaw teeth, a longer and more slender caudal peduncle, and in having a relatively larger opening to the nasal laterosensory canal (*see above*, p. 156).

The same character combination (and especially the dental ones) serves to distinguish *H. nigripinnis* from such species as *H. elegans*, *H. aeneocolor* and *H. limax*. In all instances, of course, male breeding coloration provides the most outstanding interspecific difference.

Outside Lakes Edward and George, the greatest morpho-anatomical (and ecological) resemblances are with *H. erythrocephalus* Greenwood & Gee, of Lake Victoria (Greenwood & Gee 1969). Both species have, besides a similar gross morphology, a diet of phytoplankton, fine and numerous teeth, slender gill rakers and a long, coiled intestine; all, of course, correlated characters within each species. Male coloration is particularly different. Male *H. erythrocephalus* have a bright red head, while the head in *H. nigripinnis* is dark malachite green (*cf.* p. 157 above with p. 21 in Greenwood & Gee, *op. cit.*). There are several other interspecific differences, particularly in the pharyngeal dentition and the neurocranial shape. Skull form in *H. erythrocephalus* is more like that in the moderately specialized *Haplochromis* species, and the pharyngeal teeth are finer, more numerous and more densely arranged than in *H. nigripinnis*. *Haplochromis erythrocephalus* also has relatively longer and more slender gill rakers. In other words, *H. erythrocephalus* shows greater specialization for phytoplankton feeding than does *H. nigripinnis*.

STUDY MATERIAL

Register number BMNH	Locality : Lake George
1972.6.2 : 549–554	Tufmac Bay (trawl)
1972.6.2 : 598–601	Small island north of Kankurunga Island
1972.6.2 : 602–605	Small island north of Kankurunga Island
1972.6.2 : 606	Kashaka Bay
1972.6.2 : 607–611	Kashaka Bay
1972.6.2 : 648–654	East side of Akika Island
1972.6.2 : 636–647	Mid-lake *ca* 5 miles east of Kankurunga Island
1972.6.2 : 805 (figured specimen)	Small island north of Akika Island (trawl)

Haplochromis oregosoma sp. nov.

(Text-figs. 8 & 9)

HOLOTYPE. A female, 66·5 mm standard length, BMNH reg. no. 1972.6.2 : 141.

The specific name (from the Greek *orego* to stretch and *soma* the body) alludes to the rather elongate form of this species.

DESCRIPTION. Based on 20 specimens, including the holotype, 48·0–72·5 mm standard length.

Depth of body 30·3–34·3 (M = 32·1) per cent of standard length, length of head 32·0–36·0 (M = 33·8) per cent. Dorsal head profile gently curved, less commonly straight, sloping at an angle of *ca* 35° to the horizontal ; dorsal margin of eye entering the profile or extending slightly above it.

Preorbital depth 10·5–15·2 (M = 13·3) per cent of head (showing ill-defined positive allometry with standard length), least interorbital width 20·0–25·0 (M = 22·3) per cent, ratio of interorbital width to eye diameter 1·5–1·8 (M = 1·7). Snout length 23·5–29·2 (M = 26·1) per cent of head, 0·7–0·9 (rarely 1·0) times broader than long ; eye diameter 33·4–41·2 (M = 38·2) per cent, depth of cheek 15·2–20·8 (M = 17·9) per cent.

Caudal peduncle 15·9–21·1 (M = 17·9) per cent of standard length, 1·4–2·0 (modal range 1·6–1·7) times as long as deep.

Mouth horizontal, lips not thickened. Length of upper jaw 28·6–34·7 (M = 31·7) per cent of head, length of lower jaw 38·1–45·6 (M = 41·9) per cent and 1·6–2·3 (mode 2·0) times as long as broad. Posterior tip of the maxilla reaching a vertical through the anterior margin of the eye.

Intestine about $1\frac{1}{2}$ times the total length.

Gill rakers. The lower 1–3 on the first arch are reduced, the remainder either all slender and elongate or, less commonly, with the upper 2–4 flattened and branched. There are 9–11 (mode 10) rakers on the lower part of the first arch. No pseudorakers are developed.

Scales. Ctenoid ; lateral line with 30 (f.1), 31 (f.3), 32 (f.10) or 33 (f.5), cheek with 2 or 3 rows. Five to 6 (bimodal at $5\frac{1}{2}$ and 6) scales between the upper lateral line and dorsal fin origin, 5 or 6 (mode) between the pectoral and pelvic fin bases.

Fins. Dorsal with 15 (f.11) or 16 (f.9) spines, and 8 (f.5) 9 (f.9) or 10 (f.6) branched rays. Anal with 3 spines and 7 (f.1), 8 (f.8) or 9 (f.11) branched rays. Caudal slightly emarginate, scaled on its basal half or a little further posteriorly. Pectoral fin 25·6–30·0 (M = 28·1) per cent of standard length, 75·0–88·0 (M = 82·7) per cent of head. Pelvics with the first ray slightly produced.

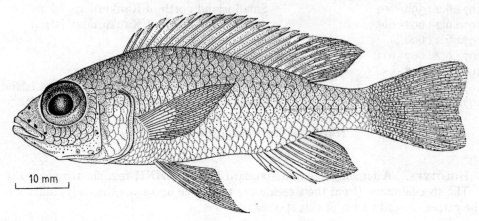

Fig. 8. *Haplochromis oregosoma.* Holotype.

Teeth. The predominant tooth form in the *outer row* of both jaws is a moderately slender, compressed and unequally bicuspid tooth (text-fig. 9). The major cusp in such teeth is produced, isoscelene in outline and slightly incurved. A distinct, step-like flange is sometimes developed on that margin of the cusp adjacent to the minor cusp. Some teeth may have the minor cusp greatly reduced in size. Posterolateral teeth in the upper jaw may be bicuspid like the others, slender unicuspids or compressed tricuspids. Tricuspid teeth are occasionally intercalated amongst the anteriorly situated bicuspids in either or both jaws. There are 42–60 (M = 50) outer teeth in the premaxilla.

Fig. 9. *H. oregosoma.* Premaxillary teeth (left side), anterior in position. Viewed anteriorly. Scale = 0·25 mm.

Inner tooth rows in both jaws are composed of small tricuspid and compressed teeth, arranged in a single (rarely double) series.

OSTEOLOGY. Basically, the *neurocranium* of *H. oregosoma* is of the generalized *Haplochromis* type, but differs in having a low supraoccipital crest and a relatively shorter ethmo-vomerine region.

The *lower pharyngeal bone* is fine, its dentigerous area slightly broader than long (1·1–1·2 times). The teeth are slender and cuspidate, and are arranged in *ca* 26–30 rows.

Vertebral counts in the 12 specimens radiographed are 29 (f.3) or 30 (f.9), comprising 13 (f.5) or 14 (f.7) abdominal and 15 (f.1), 16 (f.8) or 17 (f.3) caudal centra.

COLORATION IN LIFE. *Adult males*: ground colour metallic purple above the midlateral line, shading through iridescent turquoise to silvery on the belly and ventral flanks; a sooty overlay spreads across the chest and belly. The dorsal fin is a dark sooty colour, as is most of the anal except for its scarlet tip; the anal ocelli are large, near circular in outline, and orange-yellow in colour. The entire caudal fin is scarlet, but the basal fifth may be sooty or solid black. The pelvic fins are uniformly black.

Adult females are an overall greyish-silver. The dorsal fin is hyaline, the anal pale yellow, the caudal dark hyaline and the pelvics greyish.

PRESERVED COLORATION. *Adult males*: ground coloration black or intensely dusky, with a silvery underlay; the chest and midventral aspects of the belly are light dusky. No distinct markings are visible on the head (probably they are obscured by the general dark coloration). Dorsal fin dusky to black, the pigment most intense between the rays. Anal black over its distal two-thirds, dusky beyond. Caudal with variable coloration but always dark over the proximal third; distally the fin is usually yellowish with a dusky overlay that intensifies between the middle rays. The pelvic fins are black, the pectorals hyaline.

Adult females are silvery-grey, shading to silver on the chest and belly; the dorsum and snout are dark grey. No cephalic markings are visible. All fins are greyish-hyaline.

BIOLOGY. Very little is known about the biology of *H. oregosoma*. Apparently the species is confined to offshore areas of the lake, and it does not occur close to papyrus or other shores. Specimens have been caught over both sand and mud substrata.

The breeding habits are unknown; of 9 adult females examined, 4 have the right ovary considerably larger than the left one, 4 have the ovaries equally developed and 1 has the left ovary larger than the right one. Individuals less than 55 mm standard length are immature, although males of 56 mm standard length are ripening. Fishes, of both sexes, are fully adult at 60 mm standard length.

Haplochromis oregosoma seems to feed on phytoplankton, but as yet too few specimens have been examined to establish whether the food is taken in suspension or from bottom deposits.

Distribution. Lake George and the Kazinga Channel.

DIAGNOSIS AND AFFINITIES. From all other *Haplochromis* species in Lake George, *H. oregosoma* is distinguished by the following character combination: large eye, numerous and slender gill rakers, slender and elongate body form. Perhaps the species showing most superficial similarity with *H. oregosoma* is *H. nigripinnis*; the characters listed above, together with a difference in eye/interorbital ratio (1·5–1·8, mean 1·7, *cf*. 1·3–1·8, mean 1·5 for *H. nigripinnis*) serve to distinguish the species.

The totality of characters (including skull and jaw form) suggest that *H. nigripinnis* and *H. oregosoma* are probably not very closely related.

Among the Lake Edward species not recorded from Lake George, there is some resemblance between *H. oregosoma* and *H. engystoma* Trewavas. Unfortunately, *H. engystoma* is known only from the holotype (now in a poor state of preservation) and another, much smaller specimen which may not be a member of the species. Comparing *H. engystoma* holotype with *H. oregosoma*, the latter differs in having more gill rakers (10 or 11, *cf.* 8), straighter teeth, longer lower jaw (38·1–45·6, M = 41·9 per cent head, *cf.* 34·8 per cent in *H. engystoma*), and a slightly larger eye (33·4–41·2, M = 38·2 per cent head, *cf.* 34·8 per cent).

No known species from Lake Victoria shows any close resemblance to *H. oregosoma*.

STUDY MATERIAL

Register number BMNH	Locality : Lake George
1972.6.2 : 141 (Holotype)	North end of Kankurunga Island
1972.6.2 : 142–146 (Paratypes)	North end of Kankurunga Island
1972.6.2 : 147 (Paratype)	Northern tip Kankurunga Island
1972.6.2 : 148–152 (Paratypes)	Northern tip Kankurunga Island
1972.6.2 : 153 (Paratype)	Kankurunga Island
1972.6.2 : 154 (Paratype)	Tufmac Bay
1972.6.2 : 165 (Paratype)	In sandy shallows
1972.6.2 : 155–160	Tufmac Bay
1972.6.2 : 161–164	Tufmac Bay

Haplochromis macropsoides sp. nov.

(Text-figs. 10–12)

H. vicarius (part) Trewavas, 1933, *J. Linn. Soc.* (*Zool.*), **38** : 330–331 (1 of the paratypes, BMNH reg. no. 1933.2.23 : 353 from Worthington's station 613, Lake George. Trewavas incorrectly lists this station as '. . . East shore of Lake Edward ', but *vide* Worthington 1932).

HOLOTYPE. A male, 76·0 mm standard length, BMNH reg. no. 1972.6.2 : 718.

The trivial name refers to the overall similarity between this species and *H. macrops* (Blgr.) of Lake Victoria (and, so it was once thought, of Lake Edward as well).

DESCRIPTION. Based on 30 specimens (including the holotype, and the paratype of *H. vicarius*, see above), 59·0–77·0 mm standard length.

Depth of body 34·9–41·0 (M = 36·5) per cent of standard length, length of head 32·2–37·2 (M = 34·3) per cent. Dorsal head profile straight or very slightly curved, sloping at an angle of *ca* 35°–40° to the horizontal ; dorsal margin of the eye entering the line of the profile or extending slightly above it.

Preorbital depth 12·0–15·2 (M = 13·5) per cent of head, least interorbital width 20·5–25·0 (M = 23·1) per cent. Snout length 25·0–30·5 (M = 27·6) per cent, 0·8–0·9 times broader than long, eye diameter 33·3–39·1 (M = 36·0) per cent, showing very slight negative allometry with standard length, depth of cheek 17·5–23·8 (M = 21·4) per cent.

Caudal peduncle 13·6–19·3 (M = 15·8) per cent of standard length, 1·0–1·5 (modal range 1·2–1·3) times as long as deep.

FIG. 10. *Haplochromis macropsoides*. Holotype.

Mouth horizontal, lips not thickened. Length of upper jaw 31·3–41·5 (M = 34·2) per cent of head, length of lower jaw 39·2–45·4 (M = 41·0) per cent, 1·7–2·2 (modal range 1·7–1·9) times as long as broad. Posterior tip of the maxilla reaching a vertical through the anterior margin of the eye or somewhat behind that level.

The anterior opening to the nasal laterosensory canal is much smaller than the nostril (*cf. H. nigripinnis*), and indeed, can be difficult to locate.

The intestine is $1\frac{1}{3}$–$1\frac{1}{2}$ times the total length.

Gill rakers. Except for the reduced lower 1 or 2 rakers, others on the first gill arch are relatively slender, although some of the uppermost ones may be flat and some lower ones stout. There are 8–10 (mode 9) rakers on the lower part of this arch. Pseudorakers are present, but are poorly developed and small.

Scales. Ctenoid; lateral line with 30 (f.2), 31 (f.9), 32 (f.15) or 33 (f.4) scales, cheek with 2 or 3 (bimodal) rows. Five to $6\frac{1}{2}$ (mode 6) scales between the upper lateral line and the dorsal fin origin, 6 or 7 (rarely 5) between the pectoral and pelvic fin bases.

Fins. Dorsal with 14 (f.4), 15 (f.20) or 16 (f.6) spinous and 8 (f.4), 9 (f.10) or 10 (f.16) branched rays. Anal with 3 spines and 8 (f.7), 9 (f.22) or 10 (f.1) branched rays. Caudal truncate or, less frequently, weakly emarginate; scaled on its basal half. Pectorals 25·7–32·3 (M = 29·4) per cent of standard length, 73·0–95·5 (M = 85·0) per cent of head. Pelvics with the first ray slightly produced.

Teeth. The majority of the *outer teeth* in both jaws are relatively stout, compressed and very unequally bicuspid; the larger cusp is isoscelene in outline and the crown slightly incurved (text-fig. 11). Posteriorly in the upper jaw the last few teeth are often slender and unicuspid. Some tricuspid teeth may be intercalated amongst the bicuspids anteriorly in either or both jaws.

Fig. 11. *H. macropsoides*. Dentary teeth (right side), anterolateral in position. Viewed anteriorly. Scale = 0·5 mm.

As described for *H. elegans* (see p. 147), some bicuspids can have one margin of the major cusp produced into a narrow flange.

There are 42–60 (M = 52) teeth in the outer premaxillary series.

The *inner tooth rows* (2 or 3 in both jaws) are composed of tricuspid, compressed teeth.

OSTEOLOGY. The *neurocranium* of *H. macropsoides*, identical with that of *H. elegans*, *H. aeneocolor* and *H. nigripinnis*, is of a generalized *Haplochromis* type.

The *lower pharyngeal bone* (text-fig. 12) is moderately fine, with its dentigerous surface *ca* 1·2 times broader than long. The pharyngeal teeth are slender, compressed and cuspidate, and are arranged in *ca* 32–36 rows. Teeth in the two median rows are not noticeably coarser than their lateral congeners; teeth situated in the posterolateral corners of the bone are more closely set than elsewhere.

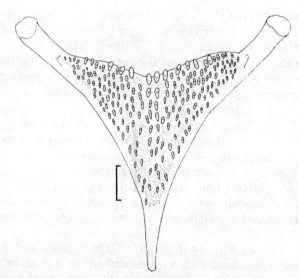

Fig. 12. *H. macropsoides*. Lower pharyngeal bone, occlusal view. Scale = 1·0 mm.

Vertebral counts in the 7 specimens examined are: 29 (f.6) or 30 (f.1), comprising 13 abdominal and 16 (f.6) or 17 (f.1) caudal centra.

COLORATION IN LIFE. *Adult males*: ground colour smokey grey overlying bluish-silver; snout, cheeks and lips a muted iridescent blue-green. Shortly after death, six faint vertical bars appear on the flanks; the first and second bars are separated by a greater distance than that between any of the succeeding bars. Dorsal fin with the spinous part sooty, the interspinous membrane generally darkest in colour; lappets black but with a narrow red streak or spot at the tip. Soft part of dorsal with maroon streaks between the rays. Caudal fin with maroon spots and blotches between the rays and a maroon flush around the margin. Anal dark hyaline (or faintly grey), dusky at the base but with a pinkish border. Pelvics are black.

In every major detail the adult, sexually active male coloration of *H. macropsoides* is identical with that of *H. elegans*. The sole colour difference, and that a subtle one, lies in the less brilliantly iridescent blue colour of the snout, cheeks and lips; in *H. macropsoides* the colour is more blue-green. More vertical stripes (9) appear after death in *H. elegans*.

I have been able to compare live fishes, of both species, in the same advanced state of sexual activity (i.e. ripe-running) and from the same locality. Except for the slight differences noted, I would consider the coloration to be identical. This is a most unusual situation amongst syntopic *Haplochromis* species and has not been recorded from the species flock of Lake Victoria. Further comment is reserved until p. 229.

Adult females have a golden-silver ground coloration shading to white on the belly. The dorsal fin has sooty lappets and an overall dark coloration except for reddish vertical stripes between the spines. The caudal fin is fairly dusky, with dark red streaks between the middle rays, and traces of red on the upper posterior margin. The anal is pale yellow, with a prominent orange spot occupying the position of an ocellus in a male fish. The pelvics are pale yellow.

PRESERVED COLORATION. *Adult males* have a generally dusky ground colour, darkest on the ventral aspects of the flanks and belly, but greyish-silver on the chest. At least 5 dark vertical bars are visible on the flanks; each bar extends from the dorsal to the ventral body outline. The head is dusky overall, with a distinct lachrymal stripe. Paired trans-snout bars are rarely visible, and then but faintly. No other cephalic markings can be detected (cf. *H. elegans*, p. 148). The branchiostegal membrane is black. The dorsal fin is dark grey to dusky, sometimes with a narrow black band along the entire base; the lappets are black, and the soft part of the fin is darkly maculate. Caudal fin greyish (darkest basally) and maculate but sometimes only weakly so. The anal is black or dark grey along its basal half, hyaline to light grey distally. The pelvics are dusky to black, the pectorals hyaline.

Adult females are greyish-silver, the ventral half of the body more silver than grey. No distinct cephalic markings are present. The dorsal and caudal fins are greyish hyaline, the margin of the dorsal often dusky. The caudal is immaculate, but has dark, ill-defined streaks between its rays. All other fins are hyaline.

ECOLOGY. *Habitat*. Not a great deal of information is available for this species, probably because in the field it is easily confused with *H. nigripinnis* and *H. elegans*. The specimens of *H. macropsoides* at my disposal are either from lake areas close to the fringing papyrus of islands (especially Kankurunga and Akika) or from more exposed areas offshore from these islands. In all localities the substrate is either sand with a thin mud overlay or organic mud.

Food. Only 6 of the specimens I have examined contained ingested material in the guts ; in all, this comprised dipteran larvae and pupae. As no other material was present (and particularly no phytoplankton or sand grains) the fishes may have been feeding away from the bottom.

Breeding. *Haplochromis macropsoides* is a female mouth brooder. All 7 of the adult, sexually active females examined have the right ovary much larger than the left one.

Distribution. Definitely recorded from Lake George and the Kazinga Channel (where it is scarce). The species probably occurs in Lake Edward as well since some specimens in the British Museum (Natural History), misidentified as *H. macrops* (Blgr.), a Lake Victoria endemic, are probably referable to *H. macropsoides*.

DIAGNOSIS AND AFFINITIES. As noted above (p. 165) the male reproductive coloration of *H. macropsoides* is virtually identical with that of *H. elegans*. However, the species clearly differ in a number of characters, including dentition (more outer teeth and more inner tooth rows in *H. macropsoides*) and the larger eye and longer upper and lower jaws of *H. macropsoides*.

From *H. nigripinnis*, *H. macropsoides* is distinguished by its deeper caudal peduncle (length/depth ratio 1·0–1·5, modal range 1·2–1·3, *cf.* 1·3–1·8, modal range 1·5–1·7 for *H. nigripinnis*), broader and more numerous inner tooth rows (3 in both jaws, *cf.* 1 or 2) and in having the anterior opening to the nasal laterosensory canal much smaller than the nostril (equal to it in *H. nigripinnis*). The species also differ in male coloration and, apparently, in their feeding habits (the intestine of *H. nigripinnis* is much longer and more coiled than that of *H. macropsoides*, and the former species is known to be a specialized phytoplankton eater).

From *H. oregosoma*, another large-eyed Lake George species, *H. macropsoides* differs in body form (depth 30·3–34·3, M = 32·1 per cent of standard length in *H. oregosoma*, *cf.* 34·9–41·0, M = 36·5 per cent ; also the caudal peduncle is more slender in *H. oregosoma*, namely 1·4–2·0, modal range 1·6–1·7 times longer than deep, *cf.* 1·0–1·5 modal range 1·2–1·3), in dentition (2–3 inner tooth rows in *H. macropsoides cf.* 1 in *H. oregosoma*), in neurocranial shape and in adult male coloration.

At first glance, *H. macropsoides* resembles *H. macrops* (Blgr.) of Lake Victoria. Closer inspection shows that the species differ in several characters. For example, the interorbital width is less in *H. macropsoides* (20·5–25·0, M = 23·1 per cent of head, *cf.* 26·6–32·2, M = 29·7 per cent in *H. macrops*), the eye is larger (33·3–39·1, M = 36·0 per cent of head, *cf.* 28·6–35·4, M = 33·0 per cent in *H. macrops*), the lower jaw is longer (39·2–45·4, M = 41·0 per cent head, *cf.* 38·0–42·5, M = 39·5) and the outer jaw teeth are fewer (42–60, M = 52, *cf.* 46–66, M = 60 in *H. macrops*). Male breeding coloration differs (*cf.* p. 164 above with p. 237 in Greenwood 1960). The supposed occurrence of *H. macrops* in Lakes Edward and George is discussed on p. 232.

Haplochromis velifer Trewavas of Lake Nabugabo resembles *H. macropsoides* in several respects. However, the species may be differentiated by the shallower preorbital of *H. macropsoides* (12·0–15·2, M = 13·5 per cent head, *cf.* 13·8–18·5, M = 16·3 per cent), the shorter snout (25·0–30·5, M = 27·6 per cent head, *cf.*

29·1–33·4, M = 31·3 per cent for *H. velifer*) and larger eye (33·3–39·1, M = 36·0 per cent head, *cf.* 26·3–33·4, M = 30·6 per cent); male coloration is also quite different in the two species (compare p. 164 above with pp. 321–322 in Greenwood, 1965b).

STUDY MATERIAL

Register number BMNH	Locality: Lake George
1972.6.2 : 718 (Holotype)	Kankurunga Island
1972.6.2 : 723–728 (Paratypes)	Small island north of Kankurunga Island
1972.6.2 : 734–738 (Paratypes)	East side of Akika Island
1972.6.2 : 739–744 (Paratypes)	East side of Akika Island
1972.6.2 : 746–752 (Paratypes)	East side of Akika Island
1972.6.2 : 753 (Paratype)	Kankurunga Island
1972.6.2 : 754 (Paratype)	Kankurunga Island
1972.6.2 : 802–803 (Paratypes)	Kankurunga Island
1933.2.23 : 354 (Paratype)	Worthington collection
1972.6.2 : 719–722	Small island north of Kankurunga Island
1972.6.2 : 729–733	Small island north of Kankurunga Island
1972.6.2 : 755–756	Kankurunga Island
1972.6.2 : 757–765	East side of Akika Island
1972.6.2 : 766–772	Small island north of Kankurunga Island
1972.6.2 : 773–777	Small island north of Kankurunga Island
1972.6.2 : 778–783	Small island north of Kankurunga Island

Haplochromis limax Trewavas, 1933

(Text-figs. 13 & 14)

H. elegans (part) Trewavas, 1933, *J. Linn. Soc. (Zool.)*, **38** : 332 (1 specimen, BMNH reg. no. 1933.2.23 : 395, collected by Worthington from Lake George [no other locality data given]).
? *H. nubilus* (part) : Trewavas, 1933, *op. cit.* (2 specimens BMNH reg. nos. 1933.2.23 : 301–302 from Lake George, are tentatively referred to *H. limax*).

HOLOTYPE. A male 80 mm standard length (BMNH reg. no. 1933.2.23 : 243) from Lake Edward.

DESCRIPTION. Based on 22 specimens (excluding the holotype), 61·0–84·0 mm standard length.

Depth of body 35·4–40·3 (M = 37·9) per cent of standard length, length of head 31·8–34·4 (M = 33·0) per cent. Dorsal head profile straight (rarely with a slight concavity or a slight convexity), sloping fairly steeply at an angle of *ca* 40°–45° with the horizontal.

Preorbital depth 13·6–18·2 (M = 15·4) per cent of head, least interorbital width 23·3–30·5 (M = 26·0) per cent. Snout 26·5–31·8 (M = 29·0) per cent, 0·8–0·9 (rarely 1·0) times broader than long; eye diameter 28·2–34·1 (M = 31·6) per cent, cheek depth 21·1–26·2 (M = 23·9) per cent. Caudal peduncle 13·6–18·1 (M = 15·3) per cent of standard length, 1·0–1·4 (mode 1·1) times as long as deep.

Mouth slightly oblique, lips a little thickened. Length of upper jaw 29·2–36·0 (M = 33·1) per cent of head (showing slight positive allometry with standard

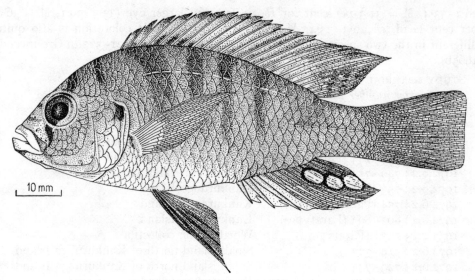

Fig. 13. *Haplochromis limax*. Lake George specimen; a male.

length), length of lower jaw 36·0–40·9 (M = 38·3) per cent, 1·3–1·9 (modal range 1·4–1·6) times as long as broad. Posterior tip of maxilla reaching or almost reaching a vertical through the anterior margin of the eye.

Intestine long, ca $1\frac{1}{2}$ to $1\frac{3}{4}$ times total length, and much coiled.

Gill rakers. Lower 1 or 2 rakers reduced, the remainder relatively slender or with 1 or 2 lower rakers short and stout. (One individual has no reduced rakers, but in this fish the total count is only 7 rakers.) There are 7 (rare) to 10 (mode 9) rakers on the lower part of the first gill arch.

The pseudorakers are well developed and are directed medially so that they overlie the true gill rakers of the inner row.

Scales. Ctenoid; lateral line with 29 (f.1), 30 (f.9), 31 (f.8) or 32 (f.4) scales; cheek with 2 or 3 (mode) rows. Five to 7 (rarely), mode $5\frac{1}{2}$, scales between the upper lateral line and the dorsal fin origin, $5–6\frac{1}{2}$ (mode 6) between the pectoral and pelvic fin bases.

Fins. Dorsal with 14 (f.4), 15 (f.15) or 16 (f.3) spinous and 8 (f.1), 9 (f.12) or 10 (f.9) branched rays. Anal with 3 spines and 8 (f.1), 9 (f.18) or 10 (f.3) rays. Caudal subtruncate, scaled on its basal half. Pelvics with the first ray produced. Pectoral 26·4–31·0 (M = 29·3) per cent of standard length, 80·0–98·0 (M = 89·1) per cent of head.

Teeth. Although basically the form of *outer row* jaw teeth is that of an obliquely cuspidate bicuspid, there is some individual variability, especially in the upper jaw (text-fig. 14). This variability concerns the angle of the cutting edge to the major cusp. In all specimens examined, this edge is most acute in teeth situated posterolaterally on the premaxilla; teeth more anteriorly placed sometimes have the cusp so obliquely truncate that the cutting edge is almost horizontal. The modal condition, however, is one where the edge is at an angle of about 60° with the vertical. It may be noted that, in this respect, the Lake George fishes differ from the holotype whose teeth are of the more acute type.

FIG. 14. *H. limax*. Dentary teeth (left), anterolateral in position. Viewed from a slightly ventrolateral position. Scale = 0·5 mm.

In the lower jaw, teeth tend to be more uniform with regard to cusp shape, and are like the modal upper jaw teeth already described. The minor cusp, in teeth of both jaws, is very small irrespective of major cusp shape. Posterior premaxillary teeth (usually the last 1–4 of the series) are generally unicuspid or are of the generalized bicuspid type, that is, with an acutely pointed major cusp.

There are 30–54 (M = 46) teeth in the outer premaxillary row.

Teeth of the *inner series* are invariably tricuspid and compressed. There are 4 or 5 (rarely 3) rows anteriorly in the premaxilla, and 3 or 4 (mode) anteriorly in the dentary; laterally and posteriorly the number of rows, in both jaws, decreases to 1. A very distinct interspace separates the outermost row of the inner series from the outer row.

OSTEOLOGY. The *neurocranium* is of a generalized *Haplochromis* type but with the preorbital region more noticeably decurved than in *H. macropsoides* and *H. elegans*. Also, when compared with these fishes, the premaxilla of *H. limax* is more robust; the dentary, however, is similar in all three species.

Comparison of *H. limax* syncranium with that in Lake Victoria species of similar feeding habits (i.e. scraping epilithic and epiphytic algae) shows that *H. limax* is more like *H. obliquidens* Hildg. and *H. lividus* Greenwood than *H. nigricans* (Blgr.). In the latter species the preorbital face of the skull is more strongly decurved and the dentary is deeper and more robust.

The lower *pharyngeal bone* in *H. limax* is moderately stout; the dentigerous area is *ca* 1·2 times broader than long. The pharyngeal teeth are fine, compressed and cuspidate, and are arranged in *ca* 28–30 rows.

Vertebral counts in the 6 specimens radiographed are: 28 (f.5) and 29 (f.1), comprising 12 (f.1) or 13 (f.5) abdominal and 15 (f.4) or 16 (f.2) caudal elements.

COLORATION IN LIFE. *Adult males*: ground colour greyish, with a faint overlay of lime on the caudal peduncle and ventrally on the flanks as far forward as the anal fin. Laterally on the flanks and ventrally on the belly there is a scarlet flush, the intensity and area of which vary with sexual state. In quiescent fishes the flush is the colour of dried blood but it is bright scarlet in sexually active individuals. A similarly coloured flush is developed on the operculum and cheek.

The dorsal fin is dark hyaline with deep scarlet streaks between the spines; between the branched rays the streaks are more precisely demarcated as scarlet lines. Anal fin is hyaline with a pink flush; the ocelli are small and yolk-yellow in colour. The pelvic fins are dusky overall.

Adult females have a golden-grey ground colour, shading through silver to white on the lower flanks, belly and chest. The dorsal fin is hyaline but is somewhat dusky along its base. Caudal greyish-yellow, the yellow predominating basally. Anal fin also pale grey-yellow, with small, deep yellow spots in the position of ocelli in males. Pelvic and pectoral fins are hyaline.

PRESERVED COLORATION. *Adult males*: the ground coloration is silvery grey. The flanks and caudal peduncle are crossed by up to seven faint vertical bars, none of which extends ventrally below the level of the pectoral fin insertion. In some individuals a faint midlateral band is visible on the caudal peduncle, and extending forward to about a vertical through the origin of the soft dorsal fin. The chest, in some fishes, is sooty; the branchiostegal membrane, in all, is black. A pair of parallel bars (of variable intensity) cross the snout; a faint transverse bar extends across the head immediately behind the orbit, but in most specimens only that part of the bar immediately above the orbit is at all intense and discrete.

The dorsal fin is always dark, sometimes sooty, sometimes almost solid black between the rays; the lappets are black. The caudal fin is greyish, becoming yellow on its ventral third; in a few specimens there are concentrations of melanophores between the middle rays. The anal is greyish basally, yellowish distally. The pelvics are dusky to black on the outer (i.e. anterior) half, but yellowish elsewhere.

Adult females have a coloration similar to that of males, but the ground coloration is somewhat lighter and the pelvic fins are hyaline.

ECOLOGY. *Habitat.* The distribution of *H. limax* is closely correlated with the presence of emergent rooted vegetation, or of other places suitable for the growth of aufwuchs. *Haplochromis limax* has never been recorded far from the shore line, but the substrata over which it occurs are varied.

Food. Aufwuchs, its associated microfauna and macerated phanerogam tissue are the commonest types of ingested matter recorded from the gut. Little of the higher plant tissue is digested; its occurrence in the gut is probably accidental and associated with the plant-scraping feeding habits of the species. Since sand grains and other inorganic bottom material are sometimes found in the gut, it is presumed that *H. limax* also feeds by scraping suitable food items from the lake bottom.

Breeding. Haplochromis limax is a female mouth brooder. Of the 6 adult females examined, 5 have the right ovary much larger than the left and 1 has the ovaries equally developed. One of the 2 smallest fishes available (both 61·0 mm standard length) is a juvenile, the other is a male with indications of early sexual development. At a standard length of 64 mm, fishes of both sexes are adult.

Distribution. Lakes Edward and George. The absence of this species from samples made in apparently suitable areas of the Kazinga Channel is noteworthy and inexplicable (*see* Appendix II).

DIAGNOSIS AND AFFINITIES. No other *Haplochromis* species in Lake George shows the dental characteristics of *H. limax*; the male coloration is also highly diagnostic. In Lake Edward, on the other hand, there are two species, *H. serridens*

Regan and *H. fuscus* Regan, both with multiseriate inner tooth rows, and obliquely cuspidate outer teeth. *Haplochromis limax* is distinguished from *H. serridens* by its straighter dorsal head profile (distinctly curved in *H. serridens*), the fewer rows of inner teeth anteriorly in the jaws (3–5, *cf.* 5–8 in *H. serridens*) and by the presence of a distinct space between the outer tooth rows and the inner series of teeth. From *H. fuscus*, *H. limax* is distinguished primarily by the outer teeth having a broader and more obliquely truncate cusp, by the smaller size of the minor cusp on these teeth and by having a truncate (as opposed to rounded) caudal fin. In addition, preserved male *H. fuscus* are uniformly dark (nearly black) whereas *H. limax* males are silvery grey.

With so few specimens of *H. fuscus* and *H. serridens* available for comparison with *H. limax* it is impossible to evaluate the apparent interspecific differences in some morphometric characters. Data on live colours are not available for *H. serridens* or *H. fuscus*.

From the little information available, it seems reasonable to consider the three species closely related, with *H. serridens* the most specialized (at least in its oral dentition).

Trewavas (1933) noted similarities between *H. limax* and *H. vicarius*. Certainly the outer teeth in many of the *H. vicarius* specimens available do resemble the *H. limax* type. But, they are equally like those of *H. fuscus* (see above). It will be necessary to examine further samples of *H. vicarius*, and get information on live male coloration, before more definite conclusions can be reached on this possible interspecific relationship.

I have compared *H. limax* with those Lake Victoria species having a similar diet and dental specializations (viz. *H. lividus*, *H. nigricans* and *H. obliquidens*). All three species can be distinguished from *H. limax* by various characters or character combinations.

Tooth form and dental pattern in *H. limax* is most like that of *H. lividus*, but it is by no means identical. The teeth of *H. lividus* are more slender, their crowns are relatively less expanded, have curved not straight vertical margins, and are more movably implanted. On these characters, *H. lividus* would seem more specialized than *H. limax*.

The subequally, or almost subequally, bicuspid teeth of *H. nigricans*, coupled with the strongly decurved preorbital skull profile, and the relatively massive dentary of this species, all suggest that it belongs to a different lineage from that of the other Lake Victoria algal grazers. These same characters also serve to distinguish *H. nigricans* from *H. limax*.

The extreme modification of the teeth in *H. obliquidens* (see Greenwood 1956b) immediately distinguishes this species from *H. limax* but does not necessarily rule out a fairly close relationship between the species. The teeth in *H. obliquidens* seem to be the ultimate expression of a specialization already apparent in *H. lividus* and *H. limax* (see Greenwood, *op. cit.*, and above).

A fourth Lake Victoria species, *H. nuchisquamulatus* (Hildg.), has feeding habits similar to those discussed above. However, its teeth retain the basic, unequally bicuspid crown, and are not closely like those of *H. limax* (see Greenwood, *op. cit.*, and above).

Except for *H. nuchisquamulatus*, where it is not known, the breeding coloration of these species is clearly different.

Finally, comparison should be made with *H. annectidens* Trewavas of Lake Nabugabo. This species has about the same degree of resemblance to *H. limax* as does *H. lividus*. The same can probably be said of *H. astatodon* Regan of Lake Kivu, but far less is known about intraspecific morphological and dental variability in this species.

In brief, *H. lividus*, *H. limax*, *H. astatodon* and, despite its highly specialized teeth, *H. obliquidens* could well be members of a phyletic lineage.

STUDY MATERIAL

Registered number BMNH	Locality
1933.2.23 : 243 (Holotype)	Lake Edward (collected by Worthington)
1933.2.23 : 395	Lake George (collected by Worthington)
1972.6.2 : 112–118	Lake George, various localities
1972.6.2 : 119–123	Lake George, various localities
1972.6.2 : 124	Lake George, Akika Island
1972.6.2 : 125	Lake George, papyrus edge
1972.6.2 : 126–128	Lake George, Kankurunga Island
1972.6.2 : 129–135	Lake George, Kankurunga Island
1972.6.2 : 136–139	Lake George, Kankurunga Island
1972.6.2 : 140	Lake George, no locality
1972.6.2 : 808 (Figured specimen)	Lake George, Kankurunga Island

Haplochromis mylodon sp. nov.
(Text-figs. 15 & 16)

Haplochromis ishmaeli (*non* Boulenger) : Trewavas, 1933, *J. Linn. Soc.* (*Zool.*), **38** : 334 (both specimens identified as *H. ishmaeli* are from Lake Edward).

HOLOTYPE. A male, 85·0 mm standard length, BMNH reg. no. 1972.6.2 : 656.

The trivial name, from the Greek, refers to the mill-like crushing dentition of the pharyngeal bones.

DESCRIPTION. Based on 21 specimens (including the holotype), 68·0–115·0 mm standard length.

Depth of body 36·0–40·5 (M = 38·6) per cent of standard length, length of head 31·0–35·8 (M = 33·1) per cent. Dorsal head profile with some size correlated variation in outline, being more decurved in larger individuals and almost straight in smaller fishes ; sloping at an angle of 40°–45° with the horizontal at all sizes.

Preorbital depth 13·3–20·8 (M = 15·9) per cent of head, least interorbital width 24·1–28·6 (M = 26·6) per cent. Snout length 28·0–32·2 (M = 30·1) per cent of head, 0·8–0·9 (rarely 1·0) times its breadth, eye diameter 26·0–34·0 (M = 29·8) per cent, cheek depth 20·7–26·3 (M = 22·8) per cent. Caudal peduncle 14·7–18·4 (M = 17·9) per cent of standard length, 1·1–1·5 (mode 1·3) times as long as deep.

Mouth horizontal, lips not thickened. Length of lower jaw 35·2–40·5 (M = 37·6) per cent of head, 1·2–1·8 (modal range 1·5–1·6) times as long as broad. Posterior tip of premaxilla reaching a vertical through the anterior margin of the orbit or a little further posteriorly.

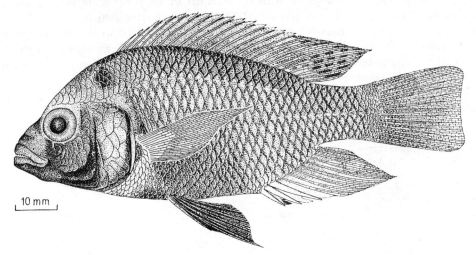

Fig. 15. *Haplochromis mylodon*. Holotype.

Gill rakers short and stout, the lower 1 or 2 reduced ; 7–9 (mode 7) rakers on the lower part of the first gill arch. The median row of pseudorakers on the first arch is well developed but individual pseudorakers are low.

Scales. Ctenoid ; lateral line with 31 (f.11), 32 (f.9) or 34 (f.1) scales, cheek with 3 (rarely 2) rows. Five to 7 (mode $5\frac{1}{2}$) scales between the upper lateral line and the dorsal fin origin, 6–8 (mode 7) between the pectoral and pelvic fin bases.

Fins. Dorsal with 14 (f.2), 15 (f.10), 16 (f.8) or 17 (f.1) spinous and 8 (f.5), 9 (f.13) or 10 (f.3) branched rays. Caudal subtruncate, scaled on its basal half. Pectoral 27·6–33·3 (M = 29·5) per cent of standard length, 82·0–96·5 (M = 84·1) per cent of head. Pelvics with the first ray slightly produced.

Teeth. Except for 1 to 3 unicuspids posteriorly in the upper jaw of most fishes, the *outer teeth* in both jaws are stout and unequally bicuspid ; the major cusp is almost equilateral in outline, moderately protracted and barely incurved. There are 32–46 (M = 40) teeth in the outer premaxillary row.

The *inner* teeth are small and tricuspid, and are arranged in 1 or 2 rows in the upper jaw, and a single (rarely double) row in the lower one.

Osteology. The neurocranium of *H. mylodon* is virtually identical with that of *H. ishmaeli* (or *H. pharyngomylus*) of Lake Victoria (*see* Greenwood 1960). The shape and size of the facet for the upper pharyngeal bones is strictly comparable in all three species, as is the relative contribution to this facet of the basioccipital and parasphenoid bones.

The *lower pharyngeal bone* is a massive structure (text-fig. 16). Compared with this bone in *H. ishmaeli* and *H. pharyngomylus*, that of *H. mylodon* is slightly less massive. The difference is not nearly so marked, however, as that between the bone in Lake Victoria and Lake George populations of *Astatoreochromis alluaudi* (*see* Greenwood 1959a, 1965b). The dentigerous area of the lower pharyngeal bone in *H. mylodon* is slightly smaller than in *H. ishmaeli* or *H. pharyngomylus*. Except in the outer rows, and in the posterolateral part of the toothed area, the lower pharyngeal teeth are all massive, stout and molariform (at least in the size range of specimens examined). Even the non-molariform teeth are stout, and are but weakly cuspidate. The extent of 'molarization' in *H. mylodon* is thus comparable with that found in *H. ishmaeli* and *H. pharyngomylus*.

Vertebral counts in the 8 specimens radiographed are : 28 (f.1), 29 (f.6) or 30 (f.1), comprising 13 (f.8) abdominal and 15 (f.1), 16 (f.6) or 17 (f.1) caudal centra.

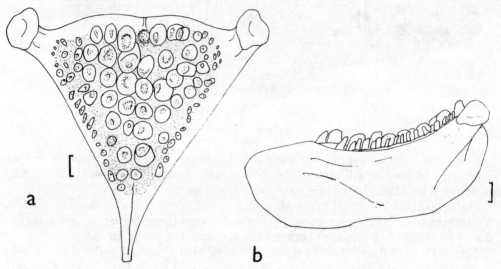

Fig. 16. *H. mylodon.* Lower pharyngeal bone. (a) In occlusal view. (b) In lateral view. From a specimen 110 mm standard length. Scale = 1·0 mm.

COLORATION IN LIFE. *Sexually active males* have a blue-grey ground colour with an iridescent turquoise sheen that is particularly concentrated around the margin of flank scales ; the belly is silvery grey, the chest and branchiostegal membrane are charcoal-grey. The head is blue-grey but the interorbital region is iridescent blue and is crossed by 2 dark bars. A prominent and dark lachrymal stripe continues to above the eye, where it expands to form a dark blotch. The dorsal fin is greyish-hyaline, the lappets are dusky, the soft dorsal has an orange-red to red margin, and there are dark red streaks between the rays. The caudal is dark over its proximal third, the remainder being rather dusky but with a red (crimson-lake) margin ; this red colour tends to extend forward onto the lower half of the fin, resulting in a dusky-pink coloration. The anal varies from dark hyaline to dusky pink on the spinous part (the lappets are black) ; the soft part is dark hyaline proximally, crimson to pink distally (the darker shades found nearest the middle of the fin). The anal ocelli are deep yellow. The pelvic fins are uniformly dusky.

In general, all colours are more intense in the sexually more active individuals.

Sexually quiescent and starting males have a light brassy ground colour with a pale-lime overlay, and are white on the chest and belly. The dorsal fin is dark hyaline, the soft part the darker. The caudal is yellowish basally, hyaline on the distal half, and has a bright orange-red margin that widens on the ventral half of the fin. The anal is pinkish, the pelvics dusky.

Adult females are silvery, shading to white on the belly. The dorsal fin is hyaline, as is the caudal which, however, has a yellowish flush on the ventral half and a bright scarlet posterior margin. The distal part of the ventral half of the caudal is also light scarlet. The anal fin is yellowish with a pinkish to scarlet overlay that is particularly intense along the margin of its soft part ; two orange spots occur in the position of the ocelli in males. The pelvic fins are hyaline or faintly yellow.

PRESERVED COLORATION. *Adult males* are brown above the midlateral line, shading to dark brown (almost bitter chocolate) below. Up to 8 rather narrow and fairly faint vertical bars are visible across the lighter part of the flank in some specimens ; ventrally the bars merge with the darker ventral flank coloration. The chest is dusky silver, the branchiostegal membrane dusky below the operculum but lighter between the jaws. The head, except for the operculum, is dark yellow-brown ; the operculum is even darker (i.e. it is comparable with the ventral flanks). There is a well-defined and broad lachrymal stripe, and two narrow transverse bars across the snout ; a broad, rather ill-defined band crosses the head behind the posterodorsal margin of the orbits, while another broad but more diffuse band or blotch extends across the nape immediately before the dorsal fin origin. The dorsal fin is dark grey, with short black blotches or streaks between the spines (at least basally). The soft part of this fin is darkly maculate. The caudal is greyish, becoming darker over its basal half. Proximally the anal is dark grey to dusky, the dark area becoming more extensive posteriorly and may occupy as much as the basal half of the fin. The rest of the fin is light yellowish-brown. The pelvics are dusky to black, the pectorals are hyaline.

Adult females are yellowish-brown to light greyish-brown dorsally, shading to silvery yellow on the belly and lower flanks. The dorsal and caudal fins are greyish, the anal, pelvics and pectorals are yellowish-hyaline.

ECOLOGY. *Habitat.* *Haplochromis mylodon* occurs near the shoreline over mud and mud–sand substrata. A few specimens have been caught in more open-water localities over a sandy bottom.

Food. Within the size range sampled, the diet of *H. mylodon* seems to consist mainly of gastropods, particularly *Melanoides tuberculata*. Chironomid larvae are also eaten.

Breeding. *Haplochromis mylodon* is a female mouth brooder. Individuals, of both sexes, less than 75 mm standard length are immature, and a few larger fishes (80 mm standard length) also show no signs of gonadial activity. Females may reach a larger size than the males ; the largest male recorded is 90 mm standard length, whereas the largest female is 115 mm standard length.

Distribution. Lakes Edward and George and the Kazinga Channel.

DIAGNOSIS AND AFFINITIES. Amongst the *Haplochromis* species of Lake George (and, apparently, also of Lake Edward), *H. mylodon* is immediately recognizable by its massive pharyngeal bones and dentition.

On the basis of purely anatomical characters, Trewavas (1933) very reasonably identified Lake Edward specimens of *H. mylodon* as *H. ishmaeli*, a species otherwise known only from Lake Victoria. Certainly on such characters it is difficult to distinguish between the two species. However, in life the coloration of adult, sexually mature males is very different (compare p. 174 above with p. 277 in Greenwood 1960).

When Lake George specimens are compared with *H. ishmaeli* from Lake Victoria (*see* Greenwood 1960) there are, in fact, some slight anatomical differences as well. For example, the cheek is a little shallower in *H. mylodon*, there are fewer teeth in the outer premaxillary tooth row than in *H. ishmaeli* and the chest scales are smaller.

A shallower cheek but larger chest scales and a longer pectoral fin distinguish *H. mylodon* from *H. pharyngomylus*, the other Lake Victoria species with a similar crushing pharyngeal dentition. Once again, adult male coloration provides a ready interspecific difference when live fishes are compared. But, in this instance the coloration is rather less different than in the case of *H. ishmaeli* and *H. mylodon*.

In brief, *H. mylodon*, *H. ishmaeli* and *H. pharyngomylus* are alike in nearly all morphometric characters and in most anatomical ones as well, but each species has a characteristic male coloration. It is chiefly because of the differences in coloration that I place *H. mylodon* in a distinct species (and do not include it with *H. pharyngomylus*). The importance of male coloration in cichlid courtship and species recognition is such that it would be biologically unsound to consider *H. mylodon* as anything other than specifically distinct.

Poll (1959) described three species (one from Lake Edward and two from nearby localities) with enlarged pharyngeal bones and molariform pharyngeal teeth. Of these species, one, *H. malacophagus* (from Lake Kibuga, *ca* 50 km south of Lake Edward), need not be considered in detail. Its pharyngeal bones and dentition are but slightly enlarged and there are other characters which differentiate it from *H. mylodon*.

The second species, *H. placodus* (from the Molindi River, near Lake Kibuga), has a greatly enlarged lower pharyngeal bone and an almost completely molariform pharyngeal dentition ; in both characters it is comparable with *H. mylodon*. In overall appearance, too, *H. placodus* is rather like *H. mylodon*. The holotype and only specimen (101 mm standard length) differs from *H. mylodon* in its larger pectoral and nuchal scales ($4\frac{1}{2}$ between upper lateral line and dorsal origin, 5 between pectoral and pelvic fin bases, *cf.* 5–7 [mode $5\frac{1}{2}$] and 6–8 [mode 7] in *H. mylodon*), in having a much smaller eye (22·2 per cent of head, *cf.* 26–34·0, mean = 29·8 per cent) and a shorter pectoral fin (72·3 per cent head length, *cf.* 82·0–96·5, mean 84·1 per cent). When more specimens of *H. placodus* are available, its relationships with *H. mylodon* can be reviewed more critically. But, unless *H. placodus* holotype is an aberrant individual, it seems unlikely that the two species will prove to be conspecific.

The third species, *H. pharyngalis*, is from the western shore of Lake Edward, at Bugazia. In two of the three syntypes, the pharyngeal mill exhibits a degree of development almost comparable with that of *H. mylodon* and *H. placodus*. In the third specimen, however, the bones are not greatly enlarged and only the median rows of the lower pharyngeal teeth are molariform. Body form in *H. pharyngalis* is unlike that of *H. mylodon,* being elongate and slender. The thickened lips and slightly shorter lower jaw of *H. pharyngalis* give to the face a most distinctive appearance. Several other characters serve to distinguish this species from *H. mylodon* (and the other species considered here). Outstanding among these diagnostic features are the minute nuchal and chest scales. I count, in the three *H. pharyngalis* syntypes, 8 or 9 scales between the upper lateral line and the dorsal fin origin, and about the same number between the pelvic and pectoral fin bases ; furthermore, scales lower on the chest are so small and thin that, at first sight, this area seems naked. Other diagnostic characters are the low number of gill rakers (5 or 6), the longer snout (34·5 and 35·0 per cent head, *cf.* 28·0–32·2, mean 30·1 per cent in *H. mylodon*) and deeper cheek (26·7 and 29·3 per cent of head, *cf.* 20·7–26·3 per cent, mean = 22·8 per cent in *H. mylodon*). Because the two *H. pharyngalis* syntypes measured (82·5 and 88·0 mm standard length) are within the size range of the *H. mylodon* sample, these morphometric differences are unlikely to be the results of allometric growth. (The third syntype is rather distorted and was, therefore, not measured.)

There do not seem to be any grounds for assuming a close or even distant relationship between *H. pharyngalis* and *H. mylodon* (or, indeed, between that species and *H. placodus* or *H. malacophagus*). The peculiarly small nuchal and thoracic squamation of *H. pharyngalis* is, however, characteristic of a species recently discovered in Lake George (*see* p. 209). The two species also have a similar body form and physiognomy, but the new Lake George species does not have a hypertrophied pharyngeal mill ; their possible relationship is considered below (p. 213).

STUDY MATERIAL

Register number BMNH	Locality : Lake George
1972.6.2 : 656 (Holotype)	Small island north of Kankurunga Island
1972.6.2 : 655 (Paratype)	N.E. corner of the lake
1972.6.2 : 657 (Paratype)	Over sandy shallows
1972.6.2 : 661–667 (Paratypes)	Various localities
1972.6.2 : 668–676 (Paratypes)	Various localities
1972.6.2 : 799–801 (Paratypes)	Small island north of Kankurunga Island
1972.6.2 : 658	Sandy shallows
1972.6.2 : 677–678	Locality unknown
1972.6.2 : 679–680	50–70 m from bush shore
1972.6.2 : 681	Locality unknown
1972.6.2 : 682	Locality unknown

Haplochromis angustifrons Blgr., 1914

(Text-figs. 17–19)

SYNONYMY. Trewavas (1933) gives a full synonymy for *H. angustifrons*, a species which Regan (1921) had previously synonymized with *H. schubotzi* Blgr.

Trewavas' redescription of the species (and her synonymy for it) was based entirely on Lake Edward specimens in the B.M.(N.H.) collections, which include 5 of Boulenger's syntypes (*see* Boulenger 1914 and 1915). One of the latter specimens was referred to *H. schubotzi* by Trewavas (*op. cit.*), and only 3 of the remaining 4 syntypes were included in her redescription of *H. angustifrons*. I have examined the fourth and neglected specimen, and can confirm its identity as *H. angustifrons*. The 4 syntypical specimens have the B.M.(N.H.) register numbers, 1914.4.8 : 25–28.

Through the courtesy of Dr K. Deckert (Berlin Museum) I was able to examine 48 syntypes of this species (including the specimen figured in Boulenger 1914 and 1915). It should be noted that there are apparently 53 syntypes in existence, although Boulenger (1914) originally recorded 56 specimens.

As far as I can tell without detailed knowledge of *H. angustifrons* in Lake Edward, all except 2 of the Berlin syntypes can be referred to this species. I do, however, have some reservations about the identity of a few small specimens in this series.

The 2 specimens which I do not consider to be *H. angustifrons* provide something of a puzzle that may only be solved when a large-scale revision of the Lake Edward *Haplochromis* species is carried out. Both these fishes are from the Berlin Museum lot number 19778. One, a female 71 mm standard length, appears to be of an *H. elegans*-like species. The other, a female, 86·0 mm standard length, I am tentatively referring to a new species described below (p. 188).

In his original description of *H. angustifrons*, Boulenger (1914) mentions some females as having '...einem breite, dunklen, braunen Seitenband vom kiemendeckel zur Schwanzflosse,...'. The 86 mm female mentioned above is the only syntype I examined with such a midlateral band. *Haplochromis angustifrons* females do not exhibit this colour pattern which is, however, a characteristic of the new species to which this syntypical fish is now tentatively referred (*see* p. 190 below).

As *lectotype* of *H. angustifrons* I have chosen the figured specimen, a male 91·0 mm standard length, Berlin Museum number 19118, collected by Schubotz from Lake Edward.

At least with respect to Lake George populations of *H. angustifrons*, the lectotype is unusual in being a male of such large size. In Lake George, adult males are generally much smaller than females. Despite this size discrepancy, the lectotype is a modal *H. angustifrons* in all morphological characters.

Trewavas' (1933) synonymy of *H. angustifrons* must now be expanded to include :

H. elegans (part) Trewavas, 1933, *J. Linn. Soc.* (*Zool.*), **38** : 333 (1 specimen from Lake George, BMNH reg. no. 1933.2.23 : 394).
H. vicarius (part) Trewavas, 1933, *op. cit.* : 331 (1 of the 2 paratypes from Lake George, [Worthington's (1932) station 613], BMNH reg. no. 1933.2.23 : 353).
H. schubotzi (part) : Trewavas, 1933, *op. cit.* : 337 (the 5 small specimens from Lake George, BMNH reg. nos. 1933.2.23 : 409–413, from Worthington's stations 613 and 627).
H. nubilus (part) : Trewavas, 1933, *op. cit.* : 329 (2 specimens, BMNH reg. nos. 1933.2.23 : 287 and 300, from Worthington's stations 613 and 618, Lake George).
H. guiarti (part) : Trewavas, 1933, *op. cit.* : 339 (1 specimen, BMNH reg. no. 1933.2.23 : 477, collected by Worthington from Lake George but no station number was given).

DESCRIPTION. Based on 41 specimens, 40·0–90·0 mm standard length, all from Lake George.

Depth of body 34·3–40·5 (M = 36·2) per cent of standard length, length of head 34·5–38·3 (M = 36·4) per cent. Dorsal head profile straight or very weakly convex, sloping at an angle of *ca* 35°–40° to the horizontal; premaxillary pedicels prominent and breaking the dorsal head outline to give the fish a very characteristic 'Roman nose' profile. The upper margin of the orbit just enters the line of the head profile.

Preorbital depth 12·5–18·9 (M = 16·6) per cent of head, least interorbital width 17·3–24·0 (M = 20·5) per cent, snout length 25·0–34·0 (M = 29·5); all three proportions show slight positive allometry with standard length. The snout varies from slightly broader than long to a little longer than broad (0·8–1·1), but modally is as long as broad.

The eye diameter and the cheek depth both show marked allometry with standard length, the former negatively allometric, the latter positively so. Thus, for these measurements two figures are given, first for fishes < 60 mm standard length (N = 12), and second for larger fishes (N = 29). Eye 32·3–37·6 (M = 35·1) per cent head, and 27·8–33·4 (M = 31·0) per cent; cheek 15·3–23·7 (M = 21·3) per cent, and 22·7–29·6 (M = 25·7) per cent.

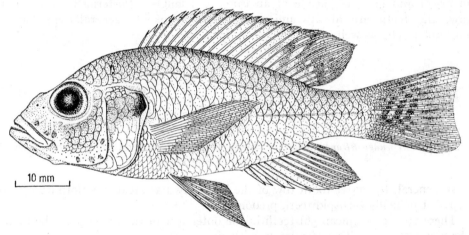

FIG. 17. *Haplochromis angustifrons*. Lake George specimen; a female.

Caudal peduncle 15·6–20·0 (M = 17·5) per cent of standard length, 1·2–1·8 (modal range 1·3–1·5) times as long as deep.

Mouth slightly oblique, or horizontal; lips not thickened. Length of upper jaw 30·2–37·5 (M = 34·4) per cent of head, length of lower jaw 38·8–45·8 (M = 42·5) per cent, 1·5–2·3 (modal range 2·0–2·2) times as long as broad. Posterior tip of the maxilla reaching a vertical through the anterior part of the eye or even through the pupil.

Gill rakers. The lower 1 or 2 rakers are reduced, the remainder relatively slender and elongate, although the rakers immediately above the reduced ones may be short and stout. There are 7 or 8 (rarely 10) rakers on the lower part of the first gill arch.

Pseudorakers are barely developed; the tissue between the inner and outer rows of true rakers is slightly thickened and thrown into low and barely discrete projections.

Scales. Ctenoid; lateral line with 30 (f.13), 31 (f.22) or 32 (f.6) scales, the cheek with 2 (rare)–4 (mode 3) rows. Four and a half to $5\frac{1}{2}$ (rarely $6\frac{1}{2}$), mode 5, scales between the upper lateral line and the dorsal origin, 5–7 (rarely 8), mode 6, between the pectoral and pelvic fin bases.

Fins. Dorsal with 14 (f.15) or 15 (f.26) spinous and 8 (f.3), 9 (f.29) or 10 (f.9) branched rays. Anal with 3 spines and 7 (f.2), 8 (f.29) or 9 (f.10) branched rays. Caudal truncate to very weakly emarginate, scaled on its basal half or a little further posteriorly. Pectoral 27·4–33·7 (M = 30·4) per cent of standard length, 74·0–89·9 (M = 83·1) per cent of head. Pelvics with the first and second rays somewhat produced.

Teeth. The *outer row* in the upper jaw usually is composed of both unicuspid and bicuspid teeth (text-fig. 18) the latter sometimes showing every gradation from fully and unequally bicuspid to weakly bicuspid (with the minor cusp virtually absent). As far as I can ascertain, this variability is not size correlated. All bicuspids, like the unicuspids, are slender and compressed; the major cusp is protracted, slightly incurved and has the outline of an isosceles triangle. Posteriorly in the upper jaw, the teeth are always unicuspid, and unicuspids generally predominate posterolaterally as well.

Fig. 18. *H. angustifrons.* Dentary teeth (left), anterolateral in position. Viewed laterally. Scale = 0·5 mm.

In general, the outer tooth row of the lower jaw has greater uniformity of tooth type. Unequally bicuspid teeth predominate.

There are 44–66 (mean 56) teeth in the outer row of the upper jaw, the number showing a slight positive correlation with size.

Tricuspid teeth predominate in the inner tooth series of both jaws, but many individuals have an admixture of tricuspid and weakly tricuspid teeth, or of unicuspids and weakly tricuspids. There are 1 or 2 (rarely 3) rows in both jaws.

OSTEOLOGY. The *neurocranium* closely resembles that in *H. schubotzi* and the new taxon, *H. schubotziellus* (see p. 190). It represents a somewhat specialized departure from the basic *Haplochromis* type; the preorbital region is relatively elongate and gently sloping, and there is an overall reduction in neurocranial width.

The *lower pharyngeal bone* is fine, its outline noticeably elongate and narrow (text-fig. 19), especially in comparison with the pharyngeal bone of other species in the Lake George flock. The dentigerous area is a little longer than broad (*ca* 1·1 times), the teeth fine, compressed and cuspidate and are arranged in *ca* 24–28 rows. Some teeth in the median rows are a little stouter than those situated laterally.

Vertebral counts for the 10 specimens radiographed are 28 (f.4) or 29 (f.6), comprising 12 (f.9) or 13 (f.1) abdominal and 16 (f.5) or 17 (f.5) caudal elements.

COLORATION IN LIFE. *Adult males*: the dorsum of the head and body is an iridescent violet which shades to turquoise on the midflank and greenish-golden on the belly. Chest and lower jaw are sooty, the branchiostegal membrane dusky to black, and the cheeks greenish-turquoise. Cephalic markings are not always visible, but when developed consist of a prominent, saddle-shaped nuchal bar and two parallel stripes across the snout.

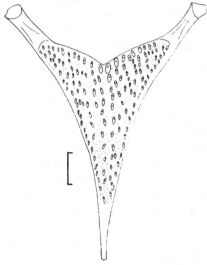

FIG. 19. *H. angustifrons*. Lower pharyngeal bone, in occlusal view. Scale = 1·0 mm.

The dorsal fin is dark hyaline with a sinuous black band running the entire length of the fin at a level about one-third of the distance between margin and base. Caudal fin dark hyaline, as is the anal which, however, is black basally and along its margin, and may show a faint pink flush; the anal ocelli are orange. The pelvic fins are black.

Shortly after death, traces of 7–10 vertical bars may appear on the flanks and caudal peduncle.

Adult females are metallic grey dorsally, shading through silver on the flanks to whitish on the belly. All fins are hyaline or faint yellow; the caudal is densely and clearly maculate, the spots dark grey and very obvious. (Indeed, this feature is diagnostic for the species in Lake George.)

PRESERVED COLORATION. *Males*: the ground colour is dark brown to black, the dark pigment most concentrated on the snout, cheeks, operculum, belly and lower half of the flanks. The lips are usually lighter than the cheeks, and the thoracic region is lighter than the flanks and belly. Traces of up to 6 narrow, fairly close-set vertical bars are often visible on the flanks. The bars are most distinct dorsally because ventrally they merge with the overall dark coloration for that region. Cephalic markings are not always visible, but when present consist of 2 bars across the snout, a lachrymal stripe, a large posteriorly directed triangular blotch on the posterior interorbital region, and a broad band across the nape.

The dorsal, pelvic and anal fins are dusky to black, the dorsal lappets intensely black; the pigment on the dorsal and anal fins may be concentrated along the fin base. The caudal is lighter than the others, and has its pigment concentrated between the middle rays.

Females are silvery brown, some with faint traces of about 6 ill-defined vertical bars on the flanks. All fins are hyaline, but narrow dark streaks occur between the spines of the dorsal fin, and the soft part of that fin is sometimes weakly maculate. The caudal is invariably maculate, and distinctly so even if the spots are rather pale. (As in live fishes, this feature is a diagnostic one.)

ECOLOGY. *Habitat.* *Haplochromis angustifrons* is essentially an offshore species, and is rarely captured near any type of shore. In its habitat it occurs over both mud and sand substrates, but it seems to prefer the latter.

Food. Both planktonic and benthic animals are eaten, of which, respectively, chaoborid and chironomid larvae are the dominant food organisms.

Breeding. At least in the Lake George populations there is a very marked sexual dimorphism in the adult size attained. Males are noticeably smaller than females, individuals more than 65 mm standard length are rare, and the smallest male fish examined (40 mm standard length) was sexually active. Females, on the other hand, only reach sexual maturity at a length of about 63–65 mm, and attain a maximum adult size of at least 90 mm standard length. Some males do reach this size (one is known from Lake George, and the holotype, from Lake Edward, is 91 mm standard length) but are rare.

In addition to this sexually correlated size disparity, there also appears to be a marked imbalance in sex ratio at all times, but especially during daylight hours when males are particularly scarce (about 1 in 20 adult fishes). This problem of diurnal sex ratio change (with the concomitant problem of male 'migration'), and the apparently real predominance of females at all times and all places, is under active research by the I.B.P. team on Lake George. No further comments can be made at this time.

Of the 25 sexually active females sampled, 12 have the right ovary considerably larger than the left one, 6 have the left slightly larger and 7 have both ovaries equally developed.

Distribution. Lakes Edward and George, and the Kazinga Channel.

DIAGNOSIS AND AFFINITIES. *Haplochromis angustifrons* is immediately distinguishable from other Lake George *Haplochromis* species by its deep body, distinctive 'Roman nose' profile and, at least in females, by the clearly maculate caudal fin. The distinctive coloration and small adult size of males are further diagnostic features.

Haplochromis angustifrons does not appear to be closely related to any other species in Lake George, nor, as far as can be estimated from known collections, to any species in Lake Edward. The dentition, narrow and elongate lower pharyngeal bone and the body form (especially head shape) distinguish this species from, on the one hand, the *H. elegans–macropsoides* complex, and on the other hand, from *H. schubotzi* and related species.

These same characters give *H. angustifrons* a superficial resemblance to *H. empodisma* Greenwood of Lake Victoria (*see* Greenwood 1960), and to *H. simpsoni* Greenwood of Lake Nabugabo (Greenwood 1965b). These two species are, however, distinguished from *H. angustifrons* by several morphometric and colour differences.

STUDY MATERIAL

Register number BMNH	Locality: Lake George
1933.2.23 : 287	Collected by Worthington, station no. 618
1972.6.2 : 412	Over sandy shallows
1972.6.2 : 414	Over sandy shallows
1972.6.2 : 420–428	Various localities
1972.6.2 : 432–433	Tufmac Bay
1972.6.2 : 434–437	Tufmac Bay
1972.6.2 : 438–440	Tufmac Bay
1972.6.2 : 441–500	Tufmac Bay
1972.6.2 : 512–515	Between Akika and Kankurunga Islands
1972.6.2 : 516–518	Kankurunga Island
1972.6.2 : 519–526	Small island north of Kankurunga Island
1972.6.2 : 531	Small island north of Kankurunga Island
1972.6.2 : 542–548	Small island north of Kankurunga Island
1972.6.2 : 804 (figured specimen)	Between Akika and Kankurunga Islands

Haplochromis schubotzi Blgr., 1914

(Text-figs. 20 & 21)

Haplochromis schubotzi (part) : Trewavas, 1933, *J. Linn. Soc.* (*Zool.*), **38** : 337 (the 2 specimens 107 and 110 mm standard length, collected by Worthington. These specimens were not previously registered in the B.M. [N.H.] collection and are now given the reg. nos. 1972.2.24 : 1–2).

The five small specimens (52–80 mm standard length) from Lake George which Trewavas (*op. cit.*) mentions in her description of *H. schubotzi* are now identified as *H. angustifrons* (*see above*, p. 178).

The type series of *H. schubotzi* consists of 5 large males, all from Lake Edward (*see* Boulenger 1914). One of these fishes (reg. no. 1914.4.8 : 18) is in the collections of the British Museum (Nat. Hist.), the others are in the Berlin Museum. Through the courtesy of Dr K. Deckert I have been able to examine these specimens and thus to select a lectotype for the species.

LECTOTYPE. A male, 118·0 mm standard length, collected by Schubotz from Lake Edward (Berlin Museum number 19116). The three paralectotypes from that museum (also *ex* Lake Edward) have the lot number 22699.

DESCRIPTION. Based on 30 specimens from Lake George, 69·0–125·0 mm standard length.

Depth of body 33·7–39·3 (M = 36·4) per cent of standard length, length of head 32·5–37·8 (M = 34·9) per cent. Dorsal head profile straight or, less commonly, gently curved, sloping at an angle of *ca* 35°–40° with the horizontal.

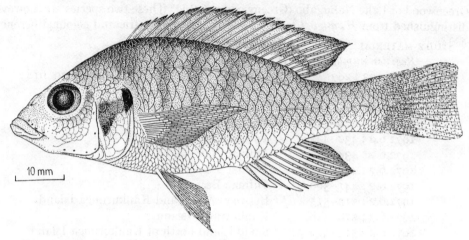

Fig. 20. *Haplochromis schubotzi*. Lake George specimen; a juvenile female.

Preorbital depth 16·4–20·9 (M = 19·0) per cent of head, least interorbital width 19·2–25·0 (M = 22·4) per cent; snout length 30·4–39·6 (M = 33·4) per cent, its breadth equal to (mode) or slightly greater than its length. Eye diameter 25·5–32·0 (M = 28·4) per cent of head, depth of cheek 20·7–26·0 (M = 22·8) per cent.

Fig. 21. *Haplochromis schubotzi*. A Lake Edward specimen (the lectotype). From Boulenger, *Cat. Afr. Fishes* (1915).

Caudal peduncle 16·2–19·2 (M = 17·7) per cent of standard length, 1·2–1·7 (modal range 1·3–1·5) times as long as deep.

Mouth horizontal, lips slightly thickened. Length of upper jaw 27·0–33·3 (M = 30·5) per cent of head, length of lower jaw 35·8–44·0 (M = 39·2) per cent, 1·4–2·0 (modal range 1·6–1·7) times as long as broad. Posterior tip of the maxilla usually reaching a vertical slightly anterior to the orbital margin, but reaching that level in a few fishes.

Gill rakers. A characteristic feature of the first gill arch in *H. schubotzi* is the well-developed papillose area of tissue immediately preceding the first (i.e. lowermost) gill raker. This raker, and usually the next one, is reduced; the others are well developed and range in form from relatively stout to relatively slender. The pseudo-rakers on this arch are especially well developed.

There are 7 or 8 (mode) rarely 9 gill rakers on the lower part of the first arch.

Scales. Ctenoid; lateral line with 31 (f.10), 32 (f.14), 33 (f.5) or 34 (f.1) scales; cheek with 3 rows (rarely 2 or 4), the scales deeply embedded in the skin. Five to $6\frac{1}{2}$ (mode $5\frac{1}{2}$) scales between the upper lateral line and the dorsal fin origin; 6 or 7 (deeply embedded) scales between the pectoral and pelvic fin bases.

Fins. Dorsal with 13 (f.1), 15 (f.17) or 16 (f.11) spinous and 8 (f.8), 9 (f.19) or 10 (f.2) branched rays; anal with 3 spines and 8 (f.15) or 9 (f.15) branched rays. Caudal weakly emarginate, scaled on its basal half. Pectoral 25·6–30·0 (M = 27·6) per cent of standard length, 73·0–87·0 (M = 79·4) per cent of head. Pelvics with the first ray produced in adults, proportionately longer in males.

Teeth. Tooth form is loosely correlated with body size. In fishes less than 71 mm standard length, the *outer teeth* are clearly, but unequally bicuspid, the major cusp is produced and isoscelene in outline. In larger individuals the disparity in cusp size is more marked so that the inner cusp is virtually invisible; the major cusp seems to be even more protracted and slender. At all sizes, the outer teeth are slender and compressed, with the crown slightly incurved. Irrespective of the fish's size the posterior and some posterolateral teeth in the upper jaw are relatively more slender than the others and are usually unicuspid or very weakly bicuspid. Elsewhere in this jaw (and particularly in larger fishes) there is usually an admixture of clearly bicuspid teeth, weakly bicuspids and, to a lesser extent, unicuspids.

Tooth form in the lower jaw is, on the whole, more uniform.

There are 46–62 (M = 52) teeth in the outer series of the premaxilla; the number of teeth does not show any clear-cut correlation with standard length, but the two largest fishes (123 and 125 mm standard length) do have the two highest numbers of teeth recorded (60 and 62 for the fishes respectively).

The *inner series* in most fishes less than 90 mm standard length are composed of slender tricuspids, but in larger individuals there may be an admixture of tricuspids, weakly tricuspids and weakly bicuspids. All inner teeth are slender, and are generally implanted so as to lie horizontally. There are 3 rows (less frequently 2 or 4) in the upper jaw and 2 or 3 in the lower jaw.

OSTEOLOGY. The *neurocranium* of *H. schubotzi* shows many of the characters seen in the skull of *H. riponianus* (Blgr.) from Lake Victoria (*see* Greenwood 1960). In other words, it is a slightly specialized derivative of the generalized skull-type seen in, for example, *H. elegans*. The principal differences lie in the more elongate preorbital region of *H. schubotzi* skull, and in its straighter and less steeply sloping dorsal profile.

The *lower pharyngeal bone* is relatively slender, with its dentigerous area almost equilateral. The teeth are fine, compressed and cuspidate, and are arranged in from 26 to 30 rows. Some fishes have the teeth in the median rows slightly coarser than the others.

Vertebral counts in the 6 fishes radiographed are 29 (f.5) or 30 (f.1) comprising 12 (f.2) or 13 (f.4) abdominal and 16 (f.3) or 17 (f.3) caudal elements.

COLORATION IN LIFE. *Adult males* : the dorsum of the head, snout and body, the operculum and the anterolateral aspects of the flanks have a pinkish to orange-red flush ; the remainder of the body is blue-grey except for the sooty chest, and yellow tinge on the upper part of the caudal peduncle. Lips, lower margin of the preoperculum, the lower jaw and the lateral aspects of the snout are bright iridescent blue, or the cheek may be orange-red. Branchiostegal membrane is black but with traces of iridescent blue over its anterior half.

Dorsal fin dark but with reddish to orange streaks between the rays and a faint overall reddish-orange flush on the soft part ; the lappets are black. The caudal is hyaline with a faint red tinge between the rays, especially noticeable on the upper half of the fin. Anal bluish to dusky, the ocelli yolk-yellow. Pelvics dusky to black.

Immature males are basically silver-grey, with a faint rose flush on the operculum and anterior flanks, and some iridescent blue on the cheek and lips. Dorsal and anal fins are hyaline, the former with reddish streaks between the spines and rays ; caudal hyaline with a faint pink flush on its ventral half.

Adult females have a silvery-grey ground colour, shading to white on the belly. The dorsal, caudal and anal fins are hyaline with a faint yellowish to yellowish-grey flush, the pigment being most concentrated basally. The pelvic fins are hyaline.

PRESERVED COLORATION. *Adult males* : ground colour greyish-brown or greyish-silver above the midlateral line, becoming dusky silver ventrally. Lateral aspects of the belly and the entire thoracic region are dusky. The snout, cheeks and most of the opercular region are dusky or at least darker than the dorsum. Cephalic markings are of variable intensity depending on the basic tone of the head coloration. The lachrymal stripe is generally intense ; the snout is crossed by 2 bars, the upper of which is the wider and is often interrupted medially. On either side of the midline behind the level of the orbits is a dark, near-triangular blotch extending ventrally to the upper orbital margin ; an ill-defined black blotch crosses the nape, anterior to the dorsal fin origin.

The dorsal and anal fins are dark grey to greyish-brown, the margin of the soft part pale, the lappets of the dorsal black. The caudal is greyish, either darkest over its basal third or almost uniformly grey-brown. The pelvic fins are black or dusky, the pectorals hyaline.

Females are greyish to brown over a silvery underlay, silvery on the lower flanks and belly. Some individuals have very faint traces of transverse barring on the upper part of the body. Cephalic markings apparently are not developed save for a very faint lachrymal blotch. The dorsal and caudal fins are greyish, the soft dorsal sometimes maculate. The anal is hyaline to yellowish, the pelvics and pectorals are hyaline.

ECOLOGY. Virtually nothing is known about the feeding and breeding habits of *H. schubotzi*, and little is known of its distribution within Lake George.

Most of the specimens described above came from offshore localities, over sand or muddy sand substrates. Some localities are exposed, others relatively protected. Certainly the species is rarely caught in nets set close to a papyrus margin or close to other emergent aquatic plants; nevertheless it does sometimes occur in such habitats. Apparently the species is absent from the open waters of the centre lake.

The few available records of gut contents suggest that *H. schubotzi* is insectivorous, but the extent of its dependence on this food source requires confirmation.

It is still not known whether or not *H. schubotzi* is a mouth brooder. Fishes, of both sexes, less than 75 mm standard length are immature, as are some larger fishes (up to 80 mm standard length). Of the 8 adult females studied, 5 have the right ovary considerably larger than the left one and 3 have both ovaries equally developed.

Distribution. Lakes Edward and George and the Kazinga Channel.

DIAGNOSIS AND AFFINITIES. Probably the species from Lake George most like *H. schubotzi* is the new taxon *H. schubotziellus*. This species is described on p. 188, and its relationship with *H. schubotzi* is discussed on p. 192.

Superficially, *H. schubotzi* also resembles *H. mylodon* but is readily distinguished from that species by its unmodified pharyngeal bones, narrower interorbital (19·2–25·0, M = 22·4 per cent head, *cf.* 24·1–28·6, M = 26·6 per cent for *H. mylodon*), and by the presence of the extensive papillose area on the lower part of the first gill arch. In life, male breeding coloration is distinctive.

Rather less similar in its overall morphology is *H. angustifrons*, although small individuals of *H. schubotzi* could be confused with members of that species. *Haplochromis schubotzi* differs from *H. angustifrons* in having the inner jaw teeth horizontally aligned, in its shorter upper jaw (27·0–33·3, M = 30·5 per cent head, *cf.* 30·2–37·5, M = 34·4 per cent head), in possessing a papillose area preceding the first gill raker (*see above*), and in the failure of the posterior tip of the maxilla to reach the anterior orbital margin. Again, male coloration is diagnostic.

Considering species from outside Lakes Edward and George, *H. cinereus* (Blgr.) of Lake Victoria shares several characteristics with *H. schubotzi*. As pointed out before (Greenwood 1960), *H. cinereus* is, in fact, not the generalized species it was once thought to be by many workers. Its dentition and skull are relatively specialized when compared with the generalized *Haplochromis* type (Greenwood *op. cit.*). In these particular characters *H. cinereus* resembles *H. schubotzi*, as it also does in having well-developed pseudorakers and a papillose area before the lower gill raker (characters not previously recorded for *H. cinereus*). *Haplochromis cinereus* differs from *H. schubotzi* principally in having rather more unicuspid teeth in the jaws (at least when equal-sized fishes are compared) and in having the median teeth of the lower pharyngeal bone noticeably enlarged. The possible relationship of these two species will be considered again later in this paper (p. 233).

Gross morphology, neurocranial shape and dental characters are also similar in *H. schubotzi* and two other Lake Victoria species, *H. riponianus* (Blgr.) and *H. saxicola* Greenwood (*see* Greenwood 1960). The slender lower pharyngeal bone (lacking enlarged median teeth) of *H. schubotzi* is more like that of *H. saxicola* than that of *H. riponianus*. As usual, the coloration of adult males is different and there are morphometric characters distinguishing *H. schubotzi* from the two Lake Victoria species.

STUDY MATERIAL

Register number BMNH	Locality: Lake George
1972.6.2 : 683–685	Bay at the north end of Akika Island
1972.6.2 : 687–692	Bay at the north end of Akika Island
1972.6.2 : 697–698	Bay at the north end of Akika Island
1972.6.2 : 699	Bay at the north end of Akika Island
1972.6.2 : 700–701	Bay at the north end of Akika Island
1972.6.2 : 702–703	Off east shore of Akika Island
1972.6.2 : 704–705	Off papyrus edge of Akika Island
1972.6.2 : 706–707	Locality unknown
1972.6.2 : 708	Locality unknown
1972.6.2 : 709	1 m offshore from papyrus edge
1972.6.2 : 710	Over sandy shoal
1972.6.2 : 711–712	Locality unknown
1972.6.2 : 713	Northeast of Kankurunga Island
1972.6.2 : 714–715	Off Kankurunga Island
1972.6.2 : 716	Tufmac Bay
1972.6.2 : 717	Tufmac Bay
1972.6.2 : 807 (figured specimen)	Tufmac Bay

Haplochromis schubotziellus sp. nov.
(Text-figs. 22 & 23)

? *H. angustifrons* (part) Boulenger, 1914, in *Wiss. Ergebn. Deuts. Zentral-Afrika Exped., 1907–1908, Zool.* **3** : 256–257 (1 of the paralectotypes in the Berlin Museum [no. 19778], a female 86 mm standard length from Lake Edward ; see also under synonymy of *H. angustifrons* on p. 178).

HOLOTYPE. A female, 76·0 mm standard length, BMNH reg. no. 1972.6.2 : 351.
The trivial name (a diminutive) is given because, in the field, specimens of this species are often confused with small specimens of *H. schubotzi*.

DESCRIPTION. Based on 28 specimens (including the holotype) 45·0–79·0 mm standard length, all from Lake George. The syntype of *H. angustifrons* is not included because its identification as *H. schubotziellus* is tentative, and it comes from Lake Edward.

Depth of body 33·3–38·0 (M = 35·7) per cent of standard length, length of head 32·7–37·3 (M = 35·1) per cent. Dorsal head profile straight or gently curved, sloping at an angle of *ca* 35°–40° with the horizontal.

Preorbital depth 12·5–18·5 (M = 15·6) per cent of head, showing slight positive allometry with standard length; least interorbital width 17·9–20·8 (M = 19·5) per cent, length of snout 25·0–33·3 (M = 29·2) per cent, 0·8–0·9 times its breadth. Eye diameter 28·8–34·7 (M = 31·8) per cent of head, cheek depth 17·6–24·1 (M = 22·0) per cent, showing very slight positive allometry.

Caudal peduncle 13·3–18·5 (M = 16·5) per cent of standard length, 1·1–1·6 (modal range 1·3–1·4) times as long as deep.

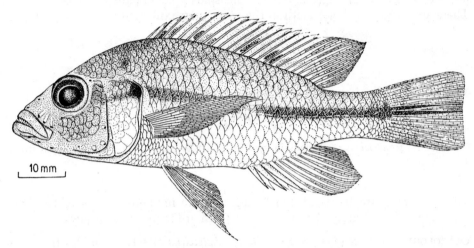

FIG. 22. *Haplochromis schubotziellus*. Holotype.

Mouth horizontal, lips very slightly thickened. Length of upper jaw, showing slight positive allometry with standard length, 28·2–37·0 (M = 34·2) per cent of head, length of lower jaw 36·7–45·0 (M = 41·6) per cent, 1·4–2·3 (modal range 1·7–1·9) times as long as broad. Posterior tip of the maxilla reaching a vertical through the anterior part of the orbit or a little further posteriorly.

Gill rakers. The tissue immediately anterior to the first gill arch is but slightly thickened, and is thrown into low, rather ill-defined folds (thus contrasting with *H. schubotzi* where this area is distinctly papillose and markedly pachydermatous). The first raker, and sometimes the 1 or 2 succeeding it, is reduced; the other rakers are short and relatively stout, with 7–9 (mode 8) on the lower part of the first arch. Pseudorakers are present but are small and sometimes ill defined.

Scales. Ctenoid; lateral line with 30 (f.1), 31 (f.9), 32 (f.15) or 33 (f.1) scales, cheek with 3 (rarely 2 or 4), rows, the scales not deeply embedded. Five to 6 (no distinct mode) scales between the upper lateral line and dorsal origin, 6 (mode) or 7, rarely 5 between the pectoral and pelvic fin bases, the scales not deeply embedded.

Fins. Dorsal with 14 (f.1), 15 (f.17) or 16 (f.10) spinous and 8 (f.1), 9 (f.18) or 10 (f.9) branched rays. Anal with 3 spines and 8 (f.17) or 9 (f.11) branched rays. Caudal weakly emarginate, scaled on its basal half. Pectoral fin 27·7–34·1 (M = 31·1) per cent of standard length, 78·0–94·5 (M = 88·9) per cent of head. First ray of pelvic fin produced, especially so in adult males.

Teeth. The predominant tooth type of the *outer row* in both jaws is a slender, very unequally bicuspid, with the major cusp produced, isoscelene to subequilateral in outline, and fairly strongly incurved (text-fig. 23). Slender unicuspids, and tricuspids, also occur in the outer row, and some fishes have all three types of teeth. A nearly constant feature is the presence of at least 1, usually 3, unicuspids at the posterior end of the premaxillary tooth row. Tooth form is less variable in larger fishes, where slender, strongly incurved unicuspids predominate.

There are 40–56 (M = 50) teeth in the outer premaxillary row.

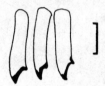

Fig. 23. *H. schubotziellus.* Premaxillary teeth (left), anterior in position. Viewed from anterior. Scale = 0·5 mm.

The *inner teeth* are tricuspid and broad, and are implanted so as to lie almost horizontally. Two (rarely 3) rows of inner teeth are found in both jaws.

OSTEOLOGY. The *neurocranium* of *H. schubotziellus* closely resembles that of *H. schubotzi*, but the dorsal preotic profile (especially anterior to the midpoint of the orbit) is somewhat more decurved in *H. schubotziellus*.

The *lower pharyngeal bone* is moderately stout and has an equilateral dentigerous area. The teeth are relatively fine, compressed and cuspidate, and are arranged in *ca* 20–24 rows. Teeth in the two median rows (especially those in the posterior third of the rows) are a little coarser than the others.

Vertebral counts in the 8 specimens radiographed are : 28 (f.2), 29 (f.4) or 30 (f.2), comprising 13 (f.5) or 14 (f.3) abdominal and 15 (f.3) or 16 (f.5) caudal centra.

COLORATION IN LIFE. *Adult males* : the ground coloration is greyish-silver with a faint iridescent blue-green sheen, particularly on the midflank region. The belly and ventral body surfaces are whitish. A fairly distinct, deep blue-black stripe extends midlaterally from the caudal fin base to the posterior opercular margin.

The dorsal fin is faintly sooty, with short black blotches along its base ; the lappets are black, the margin of the soft part is red and there are deep red spots between the branched rays. The caudal fin has similar red streaks on its proximal half and a pinkish-red flush distally. The anal is faintly dusky, with a slight pink flush ; the ocelli are orange-yellow. The pelvic fins are black.

Females are silver. A prominent black stripe runs midlaterally along the body and onto the caudal fin where it extends nearly to the midpoint. The band is of almost constant depth along the body but tapers somewhat on the fin. The dorsal fin, proximal half of the caudal and the distal part of the anal are hyaline with a suffusion of pale yellow.

PRESERVED COLORATION. *Adult males* : the ground colour is brownish above the midlateral line, greyish to sooty below ; the thoracic region is greyish to tarnished silver. The lower jaw is yellowish-brown, the branchiostegal membrane black.

Body markings are variable, but there is usually a complete dark midlateral band from the opercular margin to the caudal origin or else a band from about the middle of the body to the caudal base (sometimes this band is restricted to the posterior third of the body). Occasionally, a second longitudinal band is present, and follows approximately the course of the upper lateral line. Five to 7 rather faint but broad vertical bars are present on the flanks, and extend from the dorsal fin origin to about the level of the ventral margin of the pectoral fin. Cephalic markings are generally present (but faint) and comprise a lachrymal bar or blotch, a small blotch above and in contact with the posterodorsal margin of the orbit, and 2 faint, narrow bars across the nape.

The dorsal fin is greyish to dusky, the soft part generally maculate, the lappets dark or black. The caudal fin varies but usually is dark grey with lighter posterior and ventral margins ; otherwise the entire fin is light except for a central grey basal area. The anal is grey to dusky, particularly over the spinous part and along its distal margin. The pelvics are dusky to black, the pectorals hyaline.

Females have a light brown ground coloration shading to silver on the lower flanks and belly. A prominent and broad, dark midlateral band runs from the posterior opercular margin onto the basal part of the caudal fin ; in some specimens it extends to the posterior margin of the fin. This band is generally broken at about its midpoint, or at least is much thinner in that region. A second, but far less definite band runs a little above and parallel to the upper lateral line. The dorsal fin is greyish, often with dark lappets and sometimes with several concentrations of dark pigment along its base ; each blotch extends for a short distance upwards onto the fin membrane. The caudal fin is greyish (and has a continuation of the midlateral body stripe). All other fins are hyaline.

ECOLOGY. *Habitat.* The species is widely distributed in Lake George and occurs in most habitats. It is particularly common in muddy bays and near papyrus-fringed shorelines, but is rarely encountered in the open waters of the midlake region.

Food. Very little information is available on the food or feeding habits of *H. schubotziellus*. The presence in the gut of plant and other organic debris, together with dipteran larvae, suggests bottom feeding, possibly insectivorous habits.

Breeding. Almost no data are available on breeding habits. The size range of individuals available for analysis is such that one cannot tell precisely at what length sexual maturity is attained. The three smallest fishes examined (45–48 mm standard length) are immature ; the next smallest fish (66 mm standard length) and all others are adult and sexually active.

Of the 6 adult females studied, 4 have the right ovary much larger than the left and 2 have the gonads equally developed.

Distribution. Lake George and the Kazinga Channel (and probably Lake Edward as well).

DIAGNOSIS AND AFFINITIES. The close resemblance between *H. schubotziellus* and *H. schubotzi* has been noted already (p. 187). However, the species are immediately distinguishable on their coloration, even when preserved. The prominent midlateral band (especially in females) is diagnostic, and also serves to distinguish *H. schubotziellus* from all other species in Lake George (and probably Lake Edward as well). Compared with *H. schubotzi*, *H. schubotziellus* has a shallower preorbital (12·5–18·4, M = 15·6 per cent head, *cf.* 16·4–20·9, M = 19·0 per cent) a shorter snout (25·0–33·3, M = 29·2 per cent head, *cf.* 30·4–39·6, M = 33·4 per cent) a longer upper jaw (28·2–37·0, M = 34·2 per cent head, *cf.* 27·0–33·3, M = 30·5 per cent) and a longer pectoral fin (78·0–94·5, M = 88·9 per cent head, *cf.* 73·0–87·0, M = 79·4 per cent). There are also slight differences in the shape of the outer teeth, and in the relative stoutness of the lower pharyngeal bone (*H. schubotziellus* having a coarser bone with, usually, some teeth in the median rows noticeably coarser than the others).

This overall resemblance between the species means that *H. schubotziellus* also resembles the same Lake Victoria species as does *H. schubotzi* (*see above* p. 187). Indeed, the stouter lower pharyngeal bone and somewhat coarser median teeth in *H. schubotziellus* enhance its resemblance to *H. riponianus*, although in the latter species the lower pharyngeal dentition is rather more specialized (*see* Greenwood 1960). In many respects *H. schubotziellus* bears the same phenetic relationships to *H. schubotzi* as does *H. riponianus* to *H. saxicola* (Greenwood, *op. cit.*).

The relationship between *H. schubotzi* and *H. schubotziellus* could well be a truly phyletic one.

STUDY MATERIAL

Register number BMNH	Locality : Lake George
1972.6.2 : 351 (Holotype)	Kankurunga Island
1972.6.2 : 352 (Paratype)	Tufmac Bay
1972.6.2 : 353–355 (Paratypes)	Kankurunga Island
1972.6.2 : 356–358 (Paratypes)	Kankurunga Island
1972.6.2 : 359–366 (Paratypes)	Kankurunga Island
1972.6.2 : 367–372 (Paratypes)	Kankurunga Island
1972.6.2 : 373 (Paratype)	Kankurunga Island
1972.6.2 : 376–377 (Paratypes)	I.B.P. Jetty
1972.6.2 : 378 (Paratype)	I.B.P. Jetty
1972.6.2 : 374–375	Papyrus fringe of shore

Haplochromis taurinus Trewavas, 1933
(Text-figs. 24 & 25)

Haplochromis taurinus Trewavas, 1933, *J. Linn. Soc. (Zool.)*, **38** : 336 (description based on Lake Edward fishes only).

HOLOTYPE. A female, 135·0 mm standard length, BMNH reg. no. 1933.2.23 : 406 from Lake Edward.

DESCRIPTION. Based on 12 specimens, 72·0–140·0 mm standard length, all from Lake George.

Depth of body 30·4–38·5 (M = 36·2) per cent of standard length, length of head 27·9–32·0 (M = 30·6) per cent. Dorsal head profile variable but usually concave, sloping at an angle of *ca* 40°–45° with the horizontal.

Preorbital depth 13·6–18·2 (M = 15·2) per cent of head, least interorbital width 22·2–29·6 (M = 25·4) per cent, length of snout 27·4–33·3 (M = 30·6) per cent, 0·7–0·9 of its breadth. Eye diameter 27·8–33·3 (M = 30·6) per cent of head, depth of cheek 23·7–30·0 (M = 27·1) per cent.

Caudal peduncle 15·7–19·0 (M = 17·0) per cent of standard length, 1·2–1·5 (modal range, 1·2–1·3) times as long as deep.

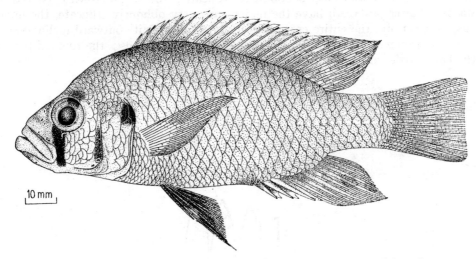

FIG. 24. *Haplochromis taurinus*. Lake George specimen; an adult male.

Mouth somewhat oblique, lips thickened; jaws equal anteriorly or the lower projecting a little. Upper jaw 38·0–42·3 (M = 41·0) per cent of head, lower jaw 43·3–56·0 (M = 46·8) per cent, 1·4–2·0 (modal range 1·6–1·8) times longer than broad. Posterior tip of the maxilla mostly exposed, reaching a vertical through the anterior part of the eye or one through the anterior margin of the pupil.

Gill rakers of variable form, from short and relatively stout to moderately long and slender; the lower 1 or 2 rakers are reduced, the upper 2 or 3 often flattened and anvil-shaped. There are 8–11 (mode 9) rakers on the lower part of the first gill arch. No clearly defined pseudorakers are present on this arch, but the tissue between the inner and outer rows of gill rakers is raised into a distinct ridge with slight but circumscribed thickenings in the position usually occupied by pseudorakers.

Scales. Ctenoid; lateral line with 31 (f.6) or 32 (f.6) scales, cheek with 3 (mode) or 4 rows. Five to 7 (mode 6) scales between the upper lateral line and the dorsal fin origin, 6 or 7 (rarely 8) between the pectoral and pelvic fin bases.

Fins. Dorsal with 15 (f.8) or 16 (f.4) spinous and 9 (f.7) or 10 (f.5) branched rays; anal with 3 spines and 8 (f.4) or 9 (f.8) branched rays. Caudal subtruncate, scaled on its basal half or a little more. Pectoral fin 23·0–29·8 (M = 27·4) per cent of standard length, 81·5–100·0 (M = 89·3) per cent of head. Pelvics with the first ray noticeably prolonged.

Teeth. The *outer teeth* in *H. taurinus* show the form (text-fig. 25) which, in Lake Victoria *Haplochromis* species, is associated with paedophagus habits (Greenwood 1959b). Also, as in those species, the teeth of *H. taurinus* are deeply embedded in the mucosa of the jaws. In both jaws the basic tooth form is similar, namely a cylindrical neck and lower crown, but with a markedly compressed, chisel-like bicuspid upper crown. Upper jaw teeth have a crown in which the minor cusp is distinct and the major cusp is obliquely truncate; the entire crown is curved inwards. Lower jaw teeth have the major cusp very obliquely truncate, the minor cusp distinct and the entire crown has a slight but definite outward inclination.

The posterior third to half of the premaxilla is edentulous; the toothed part of the bone carries 32–48 (M = 36) teeth.

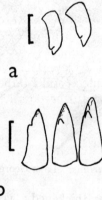

Fig. 25. *H. taurinus*. Teeth. (a) Premaxillary teeth (right), anterior in position. Viewed laterally. (b) Dentary teeth (right), anterolateral in position. Viewed from a point slightly anterior of lateral. Scale = 0·5 mm.

The *inner teeth* in both jaws are small and tricuspid, and are arranged in 1 or 2 rows. Like the outer teeth, those of the inner rows are deeply embedded in the mucosa.

OSTEOLOGY. The *neurocranium* of *H. taurinus* is of the generalized *Haplochromis* type. The premaxilla and dentary are also basically of a generalized type, the premaxilla not therefore showing the relative elongation of its ascending process (as occurs in some Lake Victoria paedophages, e.g. *H. parvidens*). Thus, in all syncranial features *H. taurinus* is comparable with the less specialized embryo and larval fish-eating species of Lake Victoria, viz. *H. maxillaris* and *H. obesus*.

The *lower pharyngeal bone* is relatively fine, its dentigerous area a little broader than long (ca 1·1 times). The teeth are slender, compressed and cuspidate, with those in the two median rows coarser than the others; there are ca 20–22 rows of teeth.

Vertebral counts in the 7 specimens radiographed are 28 (f.1), 29 (f.5) or 30 (f.1), comprising 13 abdominal and 15 (f.1), 16 (f.5) or 17 (f.1) caudal centra.

COLORATION IN LIFE. *Males*: the live colours of sexually active males are unknown. A *juvenile male* had similar coloration to that of a female (*see below*) except that there were faint traces of a rosy flush on the operculum and anterior parts of the flanks (especially intense above the pectoral fin insertion). Other differences noted were that the lower limb of the preoperculum, the cheek and the lower lip were a pale iridescent blue. The dorsal fin had red streaks between the rays, as had the caudal fin where the colour was most intense on the middle of the fin. Two well-defined yolk-yellow ocelli were present on the otherwise hyaline anal fin.

Females, both adult and juvenile, have a silvery-grey ground colour shading to white on the belly, and a faint, yellowish overlay on the flanks (more intense in adults than in juveniles). The dorsal fin is hyaline but faintly yellow along its insertion. The caudal fin is yellowish-green over its basal half, pale yellow-green distally. The anal is faintly yellow, and the pelvics are hyaline.

PRESERVED COLORATION. *Adult males* are brownish above, shading to silvery grey (with faint dusky overtones) on the flanks and belly; the chest is dusky silver, the branchiostegal membrane dusky grey. The head has a well-defined and intense lachrymal stripe continued through the eye and terminating as a blotch above and slightly behind the dorsal margin of the orbit. Other cephalic markings include a dark vertical arm of the preoperculum, and 2 rather faint transverse bars across the snout. A dark area just anterior to the dorsal fin origin is faintly visible. The dorsal fin is greyish, the caudal yellow-brown with traces of dark pigment between the rays (especially those in the middle of the fin). The anal fin is yellowish with a faint dusky overlay, the pelvics are black and the pectorals hyaline.

Females are light brownish-yellow above, shading to silvery yellow ventrally. The snout is grey-brown and there is an ill-defined and faint lachrymal stripe which does not extend through the eye to the dorsum. The dorsal and caudal fins are greyish (the former slightly the darker); all other fins are hyaline.

BIOLOGY. So few specimens of *H. taurinus* have been caught that it is impossible to generalize on the biology of the species. Apparently it is confined to inshore regions of the lake, where it has been taken off the papyrus fringe and also over sandy beaches in sheltered areas.

Judging from the dentition (*see above* p. 194) and the widely distensible mouth, *H. taurinus*, like similarly adapted species in Lake Victoria, feeds on the embryos and larvae of other cichlid fishes (*see* Greenwood 1959b). This supposition is borne out by the only two guts that yielded food remains. In these there were fragments of larval cichlids, bones of small fishes (of a size compatible with their being from larval fishes) and a fatty, yellow fluid closely resembling yolk.

Little information has been collected on the breeding habits of *H. taurinus*. The two available fishes less than 80 mm standard length are both immature; all specimens of 83 mm standard length and longer are sexually active. Only 2 of the adult females examined have ovaries in an advanced stage of oogenesis; in both fishes the right ovary is slightly larger than the left one.

Distribution. Lakes Edward and George and the Kazinga Channel.

DIAGNOSIS AND AFFINITIES. Among the *Haplochromis* species of Lakes George and Edward, *H. taurinus* is immediately recognizable by its dentition (*see above* p. 194) and by its broad and laterally distensible mouth.

Outside these lakes, *H. taurinus* bears a close resemblance in both general morphology and in its dentition to *H. maxillaris* Trewavas of Lake Victoria (*see* Greenwood 1959b). Morphometric differences between the species are slight, with the jaws of *H. maxillaris* being somewhat larger and thus the gape in this species being a little greater than in *H. taurinus*. As far as can be told from the colours of juvenile male *H. taurinus* (compared with both adult and juvenile *H. maxillaris*) there is also a difference in this character. Taking all anatomical characters into consideration, *H. maxillaris* is the more specialized species of the two.

STUDY MATERIAL

Register number BMNH	Locality : Lake George
1972.6.2 : 29	Sandy shoal
1972.6.2 : 30	Kankurunga Island
1972.6.2 : 31–34	Papyrus fringe opposite I.B.P. Laboratory
1972.6.2 : 35	Locality unknown
1972.6.2 : 36–37	Locality unknown
1972.6.2 : 38–42	Locality unknown
1972.6.2 : 806 (figured specimen)	Kankurunga Island

Haplochromis labiatus Trewavas, 1933

(Text-figs. 26 & 27)

Haplochromis labiatus Trewavas, 1933, *J. Linn. Soc.* (*Zool.*), **38** : 335 (holotype and only specimen; from Lake Edward).

Haplochromis labiatus was described from the holotype alone, although two smaller fishes were also mentioned (Trewavas 1933); all 3 specimens are from Lake Edward.

Only 1 specimen identifiable as *H. labiatus* has been caught in Lake George (from a locality close to the northern shore of Akika Island, in shallow water ca $1-1\frac{1}{2}$ m deep, over mud, and near sparse stands of the reed *Phragmites*).

The Lake George fish is 87·0 mm standard length and is of indeterminable sex.

Depth of body 41·4 per cent of standard length, length of head 32·2 per cent. Dorsal head profile concave above the orbit, sloping steeply at ca 45° to the horizontal.

Preorbital depth 16·1 per cent of head, least interorbital width 24·3 per cent, length of snout 0·8 its breadth and 28·6 per cent of head length. Eye diameter 28·6 per cent of head, depth of cheek 25·0 per cent.

Caudal peduncle 14·1 per cent of standard length, 1·1 times its depth.

Mouth horizontal, lips noticeably thickened (but not produced into lobes medially) posterior tip of the maxilla reaching a vertical through the anterior part of the eye. Upper jaw 32·2 per cent of head, lower jaw 35·7 per cent, 1·4 times longer than broad.

Gill rakers. On the lower part of the first gill arch there are 8 rakers of which the lowermost is reduced, the following 2 are stout and the remaining 5 are relatively slender. Pseudorakers are present but weakly developed.

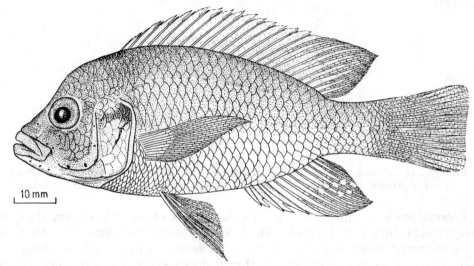

Fig. 26. *Haplochromis labiatus.* Lake George specimen.

Scales. Ctenoid, lateral line with 30 scales, cheek with 3 rows. There are $5\frac{1}{2}$ scales between the upper lateral line and the dorsal fin origin, 7 between the pectoral and the pelvic fin bases. Scales on the thoracic region are small.

Fins. Dorsal with 15 spines and 10 branched rays, anal with 3 spines and 9 branched rays. Pectoral 31·0 per cent of standard length, 96·5 per cent of the head. Caudal scaled on its basal half; the distal margin is frayed but was apparently truncate when intact. Pelvics with the first ray barely produced.

Teeth of the *outer row* in both jaws are stout and somewhat compressed (text-fig. 27). In the upper jaw the two posterior teeth on each side are unicuspid, but the remainder are unequally bicuspid, with the crown vertically orientated. There are 32 teeth in this row.

All teeth in the lower jaw are bicuspid, but those located posteriorly are smaller than the more anterior teeth. Anteriorly, the lower jaw teeth are slightly procumbent, each tooth lying forward at an angle of about 80° to the horizontal.

The *inner teeth* in both jaws are small and tricuspid, and are arranged in 3 rows in the upper jaw and 2 in the lower.

The *lower pharyngeal bone* is moderately stout and its dentigerous area is slightly broader than long (1·1 times). The teeth are compressed and cuspidate, with the posterior teeth of the median rows coarser than the others.

The *vertebral count* is 13 abdominal + 16 caudal centra. (The type has 13 + 17 centra.)

Only *preserved coloration* is known. The body is brownish above and shades to silvery on the ventral flanks, chest and belly; the entire head is brownish. All fins, except the pelvics, are hyaline-greyish, the soft dorsal is maculate posteriorly. The pelvics are dusky on the anterior half but hyaline posteriorly.

Distribution. Lakes Edward and George ; not yet recorded from the Kazinga Channel.

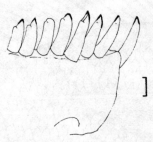

FIG. 27. *H. labiatus.* Dentary (anterior portion) with anterior and anterolateral teeth *in situ* (right side, viewed laterally); from the Lake George specimen. Scale = 0·5 mm.

COMPARISON WITH THE HOLOTYPE. The holotype is a larger fish (109 mm standard length) and is from Lake Edward. The Lake George specimen differs from the holotype in several minor ways, but in most details and in its overall morphology it resembles that specimen more closely than it does specimens of any other species.

The principal morphometric difference is in the longer jaws of the Lake George fish (35·7 and 32·2 per cent of head, *cf.* 32·3 and 28·0 per cent for the upper and lower jaws respectively); such a difference, however, is well within the range of variation for these characters in other *Haplochromis* species.

The lips of the holotype are clearly much better developed than are those of the Lake George specimen, and its teeth are predominantly unicuspid, not bicuspid as in the Lake George fish. However, some teeth in the holotype do show indications of a very small lateral cusp remnant. Both these differences could be attributable to the larger size of the holotype. Certainly the difference in lip development is well within the range of variation encountered in other species with hypertrophied lips and is not necessarily size-correlated. A further dental difference lies in the more clearly procumbent anterior teeth of the holotype. I am unable to comment on the sigficance of this character.

DIAGNOSIS AND AFFINITIES. As only two specimens are available (and those from different lakes) it is difficult to provide a precise diagnosis.

With so few specimens studied doubts might well be raised as to the validity of the species. However, if various dental and morphometric characters are combined, it seems most likely that *H. labiatus* is a valid species.

The teeth, thick lips and short lower jaw (*ca* 35 per cent of head) in *H. labiatus* together with its strongly concave profile distinguish the species from *H. limax* (teeth with obliquely cuspidate major cusps, compressed and relatively slender ; lower jaw 36·0–40·9, mean 38·5 per cent).

The concave profile of *H. labiatus* is an immediately obvious difference when a comparison is made with *H. elegans* ; the species also differ in that *H. elegans* has a shallower cheek (18·2–24·5, mean 20·8 per cent head, *cf.* 23·5 per cent), and has obviously bicuspid teeth, even in large specimens.

Dental differences like these noted above distinguish *H. labiatus* from *H. aeneocolor* (which also has more teeth in the premaxillary, viz. 40–56, mean 48, *cf.* 32), a longer lower jaw (38·0–44·0, mean 41·0 per cent head, *cf.* about 35·0 per cent) and a straight or convex dorsal head profile.

Similarly, the dentition and gross morphology serve to separate *H. labiatus* from other Lake George species. Within this lake the appearance of *H. labiatus* (and, basically, its dentition) is most like that of *H. elegans* and *H. aeneocolor*, but the resemblances are less close than are those with *H. beadlei* Trewavas of Lake Nabugabo, and with species of the *H. crassilabris* species complex in Lake Victoria. Unfortunately I have still to resolve satisfactorily the *H. crassilabris* problem (see Greenwood 1965b). Nevertheless, species of this complex can each be distinguished from *H. labiatus*. Characters separating *H. beadlei* from *H. labiatus* holotype were detailed in Greenwood (1965b). These will be found less trenchant when the Lake George fish is taken into account. More specimens of *H. labiatus* must be studied before the relationships, both phyletic and phenetic, of the two species can be settled.

STUDY MATERIAL

Register number BMNH *Locality* : Lake George
1972.6.2 : 809   East side of Akika Island

Haplochromis pappenheimi (Blgr.), 1914
(Text-figs. 28–30)

Tilapia pappenheimi (part) Boulenger, 1914, in *Wiss. Ergebn. Deuts. Zentral-Afrika Exped.*, *1907–1908, Zool.* **3** : 254–255 (10 of the 32 syntypes, all from Lake Edward, see note below).

T. pappenheimi (part) Boulenger, 1915, *Cat. Afr. Fishes*, **3** : 232–233 (4 of the 6 specimens listed, all from Lake Edward ; the figured specimen is not in the B.M. (N.H.) collections but is in Berlin [see below]. The skeleton listed cannot be identified with certainty, but probably it is not this species).

Haplochromis pappenheimi : Regan, 1921, *Ann. Mag. nat. Hist.* (9), **8** : 634–635 (all Lake Edward fishes).

Haplochromis pappenheimi : Trewavas, 1933, *J. Linn. Soc. (Zool.)*, **38** : 334 (all Lake Edward fishes).

NOTE ON THE TYPE SERIES. Six of the syntypes were deposited in the British Museum (Natural History) ; the other 26 specimens were retained by the Berlin Museum.

Regan (1921) reviewed the B.M. (N.H.) material and referred 2 specimens to the new species *H. nigripinnis* and *H. eduardii* described in that paper. Regan did not examine the Berlin syntypes, and no lectotype was chosen. The 4 remaining syntypes were considered conspecific by Regan, a conclusion with which I concur.

Through the kindness and cooperation of Dr Deckert, I have been able to study the 26 syntypes (including the figured specimen) from the Berlin Museum collections. This series proved to be polyspecific and cannot be fully evaluated until the Lake Edward *Haplochromis* species are revised. For the moment, however, it should be noted that 12 syntypical specimens are conspecific and are considered to be *H.*

pappenheimi (*see below*). The other specimens, in part, are probably referable to *H. nigripinnis* (8 specimens Z.M. Berlin, nos. 22693 and 22698), *H. eduardii* (2 specimens, Z.M. Berlin, no. 22692) and 2 specimens to species as yet undescribed (Z.M., nos. 22695 and 22696).

Boulenger's (1914) original description of *H. pappenheimi* is quite inadequate by current standards, and thus it is impossible to determine from it the morphological limits of his species. The reason for my deciding that certain specimens are '*H. pappenheimi*' is essentially an attempt to avoid unnecessary nomenclatural change. There is certainly a biologically and morphologically valid taxon, occurring in both Lakes Edward and George, whose characteristics are recognizable in 16 of the *H. pappenheimi* syntypes (4 specimens from the B.M. [N.H.] and 12 from the Berlin Museum (Z.M.B. lot nos. 22689 and 22697). The four B.M. [N.H.] fishes are those on which Regan [1921] based his redescription of the species).

It is to these 16 specimens that I have decided the name '*pappenheimi*' should be restricted and from which the lectotype should be chosen. If any of the other syntypes was chosen as lectotype, then the name '*pappenheimi*' would either fall into synonymy or would replace the name of an already established taxon. Either way, a new name would have to be found for the taxon here considered to be *Haplochromis pappenheimi*.

Regrettably, the specimen illustrated in Boulenger (1914, 1915) cannot be referred to the taxon *H. pappenheimi* as recognized by Regan (*op. cit.*) or myself. It is a fish of 66·0 mm standard length, probably a female, ZMB no. 22692. As far as I can determine this fish is a specimen of *H. eduardii* Regan.

To avoid the nomenclatural changes that would follow the choice of this fish as lectotype, I have selected for that purpose a specimen from the 16 syntypes showing the diagnostic features of *H. pappenheimi*, *sensu* Regan (1921). This fish, a female 73·0 mm standard length (ZMB no. 19110) has a characteristically elongate and slender body, and also clearly shows the dental and gill raker characters of the species (text-fig. 28).

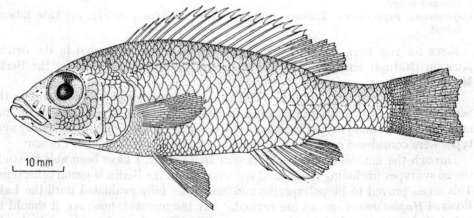

FIG. 28. *Haplochromis pappenheimi*. Lectotype; a Lake Edward specimen.

DESCRIPTION. Based on 20 specimens, 38·5–61·0 mm standard length, all from Lake George.

Depth of body 26·6–30·4 (M = 29·9) per cent of standard length, length of head 30·5–34·4 (M = 32·3) per cent.

Dorsal head profile straight, sloping gently at an angle of *ca* 20°–25° with the horizontal.

Preorbital depth 12·5–16·8 (M = 14·8) per cent of head, least interorbital width 23·3–29·2 (M = 26·4) per cent, length of snout 1·0–1·1 times its breadth, and 24·0–30·8 (M = 28·1) per cent of head. Eye diameter 31·5–37·0 (M = 33·9) per cent (not showing any allometry in this size range), depth of cheek 13·4–19·5 (M = 17·0) per cent.

Caudal peduncle 19·5–24·5 (M = 22·2) per cent of standard length, 1·5–2·0 (modal range 1·7–1·8) times as long as deep.

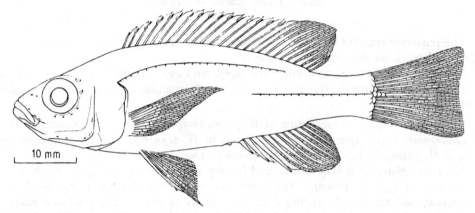

FIG. 29. *H. pappenheimi*. Outline drawing of a Lake George fish.

Mouth slightly oblique, jaws equal anteriorly. Posterior tip of the maxilla reaching a vertical slightly anterior to the orbital margin. Upper jaw 25·0–29·2 (M = 27·5) per cent of head, lower jaw 35·7–41·5 (M = 39·0) per cent, 1·7–2·5 (modal range 2·3–2·5) times longer than broad.

Gill rakers. The lower 1 or 2 rakers (rarely the first 4) are reduced, the others are long and relatively slender. Occasionally some of the uppermost rakers are flattened, or flattened and bifid. There are 10(rare)–13, mode 11, rakers on the lower part of the first gill arch. No pseudorakers are present.

Scales. Ctenoid; lateral line with 33 (f.7) or 34 (f.12) scales, cheek with 2 or 3 rows. Five (rarely) to 7 (mode 6) scales between the upper lateral line and the dorsal fin origin, 5–7 (mode 6) between the pectoral and pelvic fin bases.

Fins. Dorsal with 14 (f.1), 15 (f.10) or 16 (f.9) spinous and 8 (f.4), 9 (f.11) or 10 (f.5) branched rays. Anal with 3 spines and 7 (f.2), 8 (f.7) or 9 (f.11) branched rays. Caudal slightly emarginate, scaled on its basal half or slightly more. Pectoral fin 22·0–27·5 (M = 25·0) per cent of standard length, 71·0–84·5 (M = 77·1) per cent of head. Pelvics with the first ray slightly produced.

Teeth. The *outer teeth* in both jaws are unequally bicuspid, with the major cusp obliquely truncate in most fishes, but especially so in larger individuals (text-fig. 30); both the crown and the neck of the tooth are compressed. Occasionally, the teeth situated posteriorly and posterolaterally in the premaxilla are tricuspid or unicuspid; this difference may be size correlated, with unicuspids commoner in larger fishes (i.e. > 55 mm standard length).

FIG. 30. *H. pappenheimi.* Dentary teeth (right), lateral and anterolateral in position. Viewed laterally. Scale = 0·25 mm.

The posterior quarter to third of the premaxilla is edentulous; there are 28–38 (M = 34) teeth on the rest of the bone.

Inner teeth are small, invariably tricuspid, and are arranged in a single row in both jaws. Sometimes in the upper jaw the row is irregular and gives the impression of being double.

OSTEOLOGY. The *neurocranium* of *H. pappenheimi* departs from the generalized *Haplochromis* type, approaching that found in *H. schubotzi* and *H. schubotziellus*, and in *H. guiarti* of Lake Victoria (*see above* p. 186, and Greenwood 1962).

The *lower pharyngeal bone* is fine, and its dentigerous surface is distinctly broader than long (1·25–1·30 times). The teeth are very fine, slender, compressed and cuspidate, and are closely arranged in from 30 to 34 rows. The posterior margin of the bone is noticeably concave in outline when viewed from above, having the shape of a shallow V.

Vertebral counts in the 8 specimens examined are 30 (f.3), 31 (f.4) or 32 (f.1), comprising 14 (f.6) or 15 (f.2) abdominal and 16 (f.4) or 17 (f.4) caudal centra.

COLORATION IN LIFE. There appears to be no marked sexual dimorphism in coloration, although the possibility of this occurring cannot be overruled because no sexually active males have yet been examined. However, large and adult males were caught in the Kazinga Channel and these did not differ from females, except in being slightly darker.

The ground colour in both sexes is silver, shot with green iridescence above the midlateral line, and whitish on the belly. All fins are hyaline, but in males the lappets of the dorsal fin are black, as are the pelvic fins. The anal fin of males carries from 1 to 3 yolk-yellow ocelli.

PRESERVED COLORATION. Ground coloration is greyish-silver, darker (i.e. greyer) on the dorsum and flanks to about the midlateral line. All fins are hyaline to greyish, the lappets of the spinous dorsal black in males, as are the pelvic fins.

Distribution. Lakes Edward and George and the Kazinga Channel.

ECOLOGY. *Habitat.* The slender body form and pelagic habits of this species have resulted in relatively few specimens being caught in the nets used by the I.B.P. team. Thus, not a great deal is known about the habitat preferences of *H. pappenheimi*. The species is apparently confined to upper water levels in offshore regions of the lake (both in bays and in the open lake). An intensive study of the species is now in progress. The use of purse-seine nets and small-mesh trawls should result in many more samples being taken.

Food. The few guts examined contained only zooplankton, particularly copepods and cladocerans.

Breeding. Most of the specimens on which this description is based are sexually immature, but 2 fishes (females, 60 and 61 mm standard length) show signs of ovarian activity. Obviously, adults of *H. pappenheimi* in Lake George must reach a larger size (as they do in Lake Edward and the Kazinga Channel). There is, of course, the possibility that adult individuals move out of Lake George, and that the breeding sites are in the Kazinga Channel. To date the only evidence (and that rather flimsy) supporting this hypothesis is the capture, in May 1972, of numerous large adults in the channel, when, using the same gear (a purse seine), only juveniles were collected in the Lake itself.

One of the paralectotypes from Lake Edward, a female (72 mm standard length) is brooding young in the buccal cavity, as are 2 females (72·0 and 69·0 mm standard length) in the paralectotypical series of the British Museum (Natural History). Unfortunately it is impossible to sex the other paralectotypes with any certainty. All are apparently female, the largest a fish 92·5 mm standard length.

DIAGNOSIS AND AFFINITIES. The species is immediately distinguishable from other Lake George *Haplochromis* species by its slender body form, head shape, by its long and slender caudal peduncle, and by the shape and distribution of its outer jaw teeth. In life the silvery blue-green coloration is also diagnostic.

There is an overall similarity between *H. pappenheimi* and *H. guiarti*, a species endemic to Lake Victoria but which was once thought to occur in Lake Edward (Trewavas 1933; but *see below* p. 232).

Haplochromis pappenheimi differs from *H. guiarti* in several characters; for example, the dentition (obliquely cuspidate teeth, contrasted with the usual unicuspid and unequally but acutely bicuspid teeth of *H. guiarti*; the fewer teeth in *H. pappenheimi*, and the fully toothed premaxilla of *H. guiarti* compared with the posteriorly edentulous bone of *H. pappenheimi*), a shallower preorbital (12·5–16·8, M = 14·8 per cent head, *cf.* 16·3–21·5, M = 18·3 per cent in *H. guiarti*), shorter snout (24·0–30·8, M = 28·1 per cent head, *cf.* 31·7–37·5, M = 34·4 per cent), larger eye (31·5–37·0, M = 33·4 per cent head, *cf.* 23·6–29·8, M = 26·5 per cent in *H. guiarti* of the same size), and a longer caudal peduncle (19·8–24·5, M = 22·2 per cent standard length, *cf.* 16·2–20·8, M = 18·9 per cent).

Among the Lake Victoria *Haplochromis* species there is only one other, *H. fusiformis* Greenwood & Gee, that resembles *H. pappenheimi*. *Haplochromis fusiformis* is confined to the deeper waters (90–100 ft) of the lake and is benthic in habits. Like *H. pappenheimi* it has a partly edentulous premaxilla and is slender bodied. Interspecific differences, however, are well marked and include a shallower body, deeper preorbital, smaller eye and longer lower jaw in *H. fusiformis*, as well as the retention of typical bicuspid teeth in that species.

STUDY MATERIAL

Register number BMNH	Locality : Lake George
1972.6.2 : 297-298	Kankurunga Island
1972.6.2 : 317-320	Small island north of Kankurunga Island
1972.6.2 : 328-332	Small island north of Kankurunga Island
1972.6.2 : 333-336	Small island north of Kankurunga Island
1972.6.2 : 337-342	Tufmac Bay

Haplochromis squamipinnis Regan, 1921
(Text-fig. 31)

Haplochromis squamipinnis Regan, 1921, *Ann. Mag. nat. Hist.* (9), **8** : 636 (a single specimen, the holotype, BMNH reg. no. 1914.4.8 : 32, from Lake Edward).

Haplochromis squamipinnis : Trewavas, 1933, *J. Linn. Soc. (Zool.)*, **38** : 338-339 (mostly Lake Edward fishes, but 1 from Lake George).

NOTE ON *H. mentatus* Regan 1925, A PUTATIVE SYNONYM OF *H. squamipinnis*. Trewavas (1933) noted that the holotype and only specimen of *H. mentatus* (a Lake Edward species) closely resembled *H. squamipinnis*. She was unable to examine the holotype and thus did not see fit to formally synonymize the species with *H. squamipinnis*.

During a recent visit to the Museum of Comparative Zoology, Harvard University, I was able to examine *H. mentatus* holotype (MCZ no. 31523), a small fish 94·0 mm standard length, of indeterminable sex. In all morphometric characters and in most anatomical features this specimen is, as Trewavas suggested, like a young *H. squamipinnis*. However, I could not find any trace of the minute scales which are closely adherent to the bases of the anal and dorsal fin rays of *H. squamipinnis*, irrespective of the individual's size (*see below* p. 206).

These scales, which extend for a short distance onto the fins, are delicate and easily dislodged. Furthermore, there is considerable individual variability with regard to the number and position of the fin rays with which the scale rows are associated. Despite this variation, however, I have yet to examine a specimen of *H. squamipinnis* in which there is absolutely no trace of fin scales.

Thus, it is difficult to assess the significance of their total absence in *H. mentatus* holotype, especially since in all other trenchant characters the specimen agrees with comparable-sized *H. squamipinnis*.

Personally, I would be inclined to consider it either a young *H. squamipinnis* in which all traces of fin scales rows are lost, or an aberrant member of the species in which these scales failed to develop.

DESCRIPTION OF *H. squamipinnis*. Based on 34 specimens 34·0-202·0 mm standard length, from Lake George (including the specimens collected by Worthington in 1931). Because most characters show some allometry with standard length, the sample has been divided into two groups, viz. : (a) fishes < 120 mm standard length and (b) fishes > 129 mm standard length. Ranges and means for the various morphometric characters are given accordingly.

Depth of body (a) 32·1–37·0 (M = 34·7), (b) 34·7–41·0 (M = 37·4) per cent of standard length, length of head (a) 33·5–37·6 (M = 35·4), (b) 33·9–36·2 (M = 35·0) per cent.

Dorsal head profile straight but broken by the prominent premaxillary pedicels, sloping at an angle of *ca* 40° with the horizontal.

Preorbital depth (a) 15·0–19·0 (M = 17·0), (b) 17·8–20·4 (M = 21·9) per cent of head, least interorbital width (a) 19·2–23·0 (M = 21·9), (b) 21·3–25·4 (M = 23·0) per cent; length of snout equal to or slightly greater than its breadth in fishes of both size groups, but in (a) 30·8–33·8 (M = 32·3) per cent of head and in (b) 31·5–37·1 (M = 34·9) per cent.

Eye diameter (a) 23·0–33·0 (M = 27·5), (b) 21·2–25·5 (M = 23·0) per cent of head; depth of cheek (a) 18·3–27·6 (M = 24·5), (b) 27·8–31·7 (M = 29·3) per cent. For the cheek depth and eye diameter the lowest and highest values respectively relate to the smallest specimen examined.

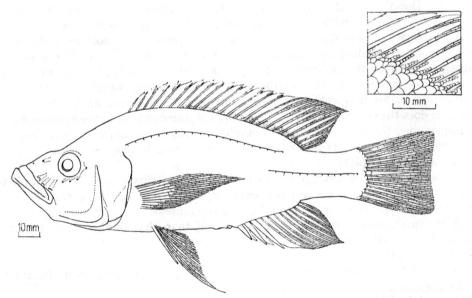

FIG. 31. *Haplochromis squamipinnis*. Lake George specimen. Inset shows squamation on dorsal fin.

Caudal peduncle not showing allometric growth, its length 14·7–20·6 (M = 16·8) per cent of standard length, 1·1–1·5 (modal range, 1·2–1·4) times as long as deep.

Mouth distinctly oblique, sloping upwards at an angle of *ca* 35–45 degrees to the horizontal. Jaws equal anteriorly or the lower projecting slightly (the usual condition). Length of lower jaw in (a) 41·8–53·8 (M = 47·5), (b) 47·3–56·6 (M = 51·3) per cent of head, length of upper jaw in (a) 33·3–40·0 (M = 37·0), in (b) 39·3–45·5

(M = 42·4) per cent, 1·1–1·5 (modal range 1·2–1·4) times longer than broad in both size groups. The smallest specimen examined (34·0 mm standard length) has, proportionately, the shortest jaws.

Posterior tip of the maxilla reaching a vertical through the orbital margin in most fishes, not quite reaching this level in a few, and extending beyond it to below the eye in others; this variation does not seem to be size correlated.

Gill rakers. The lower 1–3 rakers are reduced, the remainder are relatively slender. In some specimens the uppermost 1 or 2 rakers may be flat and anvil-shaped, and in some individuals all the rakers are short and relatively stout. There are 8–11 (rare), usually 9, rakers on the lower part of the first gill arch.

Pseudorakers are poorly developed: the tissue lying between the inner and outer gill raker rows is thickened and slightly produced into ill-defined, low projections.

Scales. Ctenoid; lateral line with 31 (f.1), 32 (f.17) or 33 (f.12) scales, cheek with 3, 4 (mode) or 5 rows. Five and a half to 7 (rare), usually 6, scales between the upper lateral line and the dorsal fin origin, 5–7 (mode 6) between the pectoral and pelvic fin bases.

Fins. Dorsal with 14 (f.1), 15 (f.21) or 16 (f.9) spinous and 9 (f.19) or 10 (f.12) branched rays. Anal with 3 spines and 9 (f.22) or 10 (f.9) branched rays.

A peculiarity of both dorsal and anal fins is the presence of short vertical rows of minute scales extending from the body onto the fin (*see* text-fig. 31). The scales are closely applied to the fin rays, both spinous and branched. Not all rays have associated scale rows, and the distribution of the rows shows considerable individual variability. Most frequently, the entire soft part of the dorsal fin has basal scale rows, as does the entire soft anal. Generally the anterior part of the spinous dorsal is asquamous, the scale rows only beginning at about the sixth or seventh spine. However, the rows may begin as far forward as the fourth spine, or may not appear until the eleventh spine.

The scales are easily dislodged and part of this variability may be attributable to damage sustained when the fish is caught, especially if it has been trapped in a gill-net. No specimen was seen in which all the fin scale rows are missing, and the anal fin squamation is usually better preserved than that of the dorsal (another reason for thinking that damage during capture may account for some of the observed variability).

The caudal fin is truncate to weakly subtruncate, and is scaled on its basal half or a little more.

Pectoral fin length shows no marked allometry with standard length; however, the relatively shortest fin is found in the smallest fish (34·0 mm standard length). Length of pectoral 23·5–30·0 (M = 28·1) per cent of standard length, 66·6–87·0 (M = 79·9) per cent of head.

Pelvic fins have the first ray produced and relatively more elongate in adult males than in females.

Teeth. Except for the smallest specimen (34·0 mm standard length), the predominant tooth form in the *outer row* of either jaw in fishes of 49·0–202·0 mm standard length is a slender but strong unicuspid with a slightly to strongly incurved crown. Some fishes in the size range 75–85 mm standard length have a few weakly and unequally bicuspid teeth interspersed among the unicuspids anteriorly and anterolaterally in both jaws; in one fish, most outer teeth in the lower jaw are bicuspid, although some typical unicuspids occur anteriorly. In fish of all sizes the posterior premaxillary teeth are unicuspid.

The 34 mm standard length fish has typically unicuspid teeth posteriorly and posterolaterally in the upper jaw, but unequally bicuspid teeth (the major cusp long and slender) anteriorly; all the lower jaw teeth in this fish are unicuspid and typical.

The number of outer premaxillary teeth is positively correlated with standard length; in fishes < 118 mm standard length there are 34 (in the smallest specimen) to 70 (M = 48) teeth, and in fishes > 130 mm standard length, 46–80 (M = 60) teeth.

Fishes > 90 mm standard length have the *inner tooth rows* composed of slender unicuspids, although in fishes as long as 100 mm some weakly tricuspid teeth may also occur. At lengths between 70 and 90 mm, there is usually an admixture of tri- and unicuspid teeth, but in some fishes only tri- and weakly tricuspids are found. In smaller specimens, tricuspids predominate.

OSTEOLOGY. The *neurocranium* in *H. squamipinnis* is of a relatively specialized type, both with respect to the basic *Haplochromis* neurocranial type, and also to the presumed basic skull form in piscivorous species (such as is shown by *H. guiarti*, *H. victorianus* and *H. serranus* of Lake Victoria, see Greenwood 1962, 1967). It compares closely with the neurocranial type found in Lake Victoria species like *H. longirostris* and *H. mento*, species that have deviated from the near basic *H. serranus* grade (Greenwood, *op. cit.*).

The *lower pharyngeal bone* is relatively fine and has an equilateral dentigerous area. The teeth are compressed, but strong, with weakly developed cusps and are somewhat sparsely distributed on the bone in *ca* 18–20 rows.

Vertebral counts in the 11 specimens radiographed are 29 (f.3) and 30 (f.8), comprising 13 (f.11) abdominal and 16 (f.3) or 17 (f.8) caudal elements.

COLORATION IN LIFE. *Adult males*: the ground colour is greyish with a turquoise to blue-green sheen covering most of the flank and caudal peduncle. The dorsal head and body surfaces are dark grey, the chest and belly greyish to greyish-sooty, the branchiostegal membrane sooty.

Dorsal fin dusky, darkest along its basal third, the upper margin of this dark area with a gently undulating outline; the lappets are black but there is a narrow, pinkish margin to the soft part of the fin. Deep orange-red spots occur between the posterior rays of the soft dorsal. Caudal fin dark hyaline, with deep red spots and a light red (almost pink) flush over the ventral half of the fin and at its upper and lower distal angles. Anal fin almost completely pink, but with a hyaline area around the yolk-yellow ocelli. The pelvic fins are black.

Females and immature males are golden-silver dorsally, shading to yellowish-green on the flanks, and to white on the belly and ventral flanks. Dorsal fin yellowish-green, becoming hyaline dorsally. The caudal is similarly coloured, but has dark spots over its proximal two-thirds. Anal fin dark hyaline basally, yellow-green distally; females often have yellow spots in the position of the ocelli in males, but young males may lack any indications of such markings. The pelvics are hyaline or yellowish, becoming dusky in near-adult males.

PRESERVED COLORATION. *Adult males*: the ground colour is brownish-grey above, shading to silvery grey or dusky silver on the lower flanks and belly; the chest is silvery or dusky silver. Upper surface of the snout and the lips (at least anteriorly) are dark grey, the lower jaw and branchiostegal membrane dusky grey. The only cephalic marking is an ill-defined lachrymal stripe.

The dorsal fin is grey, the lappets black. Basally there is a darker band (in places almost black) which widens over the soft part of the fin so that the basal third to half of the fin is black or very dark grey in colour. The distal part of the soft dorsal is densely and darkly maculate. The caudal is dark grey, most intensely so over its proximal third to half. The anal is grey to dusky grey, with the basal third noticeably darker. The pelvic fins are black.

Females and juvenile males are light brown above, shading to silvery yellow on the flanks and belly. The lips, snout and lower jaw are coloured as in adult males, but are slightly paler in some individuals. The lachrymal blotch is poorly defined but is generally visible. The coloration of the dorsal, caudal and anal fins is like that of adult males, but is more variable in its intensity. The pelvics are hyaline or a little dusky (more so in juvenile males) over the distal part of the first 3 rays.

ECOLOGY. *Habitat*. At least when adult, individuals of *H. squamipinnis* are found in all habitats, but are especially common in the offshore open-water areas of the lake. Juveniles may have a more restricted distribution to areas nearer the shoreline.

Food. Fishes predominate in the diet of *H. squamipinnis* at all the growth stages investigated (70-200 mm standard length), and appear to be the sole food source of fishes more than 150 mm standard length. In small fishes insects contribute substantially to the diet.

As far as could be determined, the principal prey species are other *Haplochromis*, but the macerated nature of the gut contents in *H. squamipinnis* generally precludes accurate identification.

Breeding. *Haplochromis squamipinnis* is a female mouth brooder. All examined individuals less than 120 mm standard length are immature; the first indications of sexual activity are found in fishes (of both sexes) in the size range 125-130 mm standard length.

Because all specimens larger than 155 mm standard length are females, there may be sexual dimorphism in the maximum adult size attained.

Of the 13 sexually active females examined, 5 have the right ovary larger than the left one, 2 have the right ovary slightly larger and 6 have the ovaries equally developed.

Distribution. Lakes Edward and George and the Kazinga Channel.

DIAGNOSIS AND AFFINITIES. The overall appearance of *H. squamipinnis* with its long and obliquely sloping jaws, its unicuspid teeth, and the peculiar scale rows on the median fins, immediately distinguishes this species from all known *Haplochromis* species in Lake George. The same criteria would distinguish it from all known *Haplochromis* species in Lake Edward, but a rather similar species is now known to occur in that lake (Greenwood, unpublished information).

Haplochromis squamipinnis resembles a number of piscivorous *Haplochromis* species in Lake Victoria, but none of these has scale rows on the fins.

Taken in concert, the cranial and dental characters of *H. squamipinnis* place it near *H. victorianus* and *H. serranus* (i.e. the '*serranus*' group of Greenwood 1967). However, in certain characters, especially skull form, *H. squamipinnis* approaches the rather more specialized structural grade seen in *H. mento* and *H. longirostris* (i.e. of the '*prognathus*' group as defined by Greenwood, *op. cit.*).

Despite the general and often particular resemblances between *H. squamipinnis* and these various Lake Victoria species, it can be distinguished from them by various combinations of morphometric characters. It may be significant, in phylogenetic terms, that the male coloration of *H. squamipinnis* is not markedly different from that of *H. victorianus* (*see* Greenwood 1962, p. 157), and that several Lake Victoria species of the '*serranus*' group show basically similar male breeding coloration.

Haplochromis squamipinnis also resembles *H. venator* Greenwood, the sole piscivorous *Haplochromis* species of Lake Nabugabo (Greenwood 1965b). Skull morphology in *H. venator* is rather more like that of *H. mento* than of *H. squamipinnis*, and there are some differences in gross body form as well as some specific morphometric ones (especially the deeper body and wider interorbital of *H. squamipinnis*). *Haplochromis venator*, of course, lacks the scale rows on its dorsal and anal fins. Certainly the similarities between this species and *H. squamipinnis* are no closer than those between *H. squamipinnis* and the Lake Victoria species considered above.

STUDY MATERIAL

Register number BMNH	Locality: Lake George
1972.6.2 : 379	I.B.P. Jetty
1972.6.2 : 380	I.B.P. Jetty
1972.6.2 : 381–386	Kashaka Bay
1972.6.2 : 387–395	Various localities
1933.2.23 : 449	Collected by Worthington
1933.2.23 : 439–443	Collected by Worthington
1933.2.23 : 450–451	Collected by Worthington
1933.2.23 : 452–455	Collected by Worthington
1933.2.23 : 444–448	Collected by Worthington

Haplochromis petronius sp. nov.

(Text-figs. 32–34)

HOLOTYPE. A male 85·5 mm standard length, BMNH reg. no. 1972.6.2 : 1, from Kashaka Crater, Lake George.

The trivial name, from the Latin, meaning 'of, or pertaining to rocks', refers to the usual habitat of this species in Lake George.

DESCRIPTION. Based on 25 specimens, 67·0–88·0 mm standard length (including the holotype), all from Lake George.

Depth of body 33·2–38·3 (M = 35·8) per cent of standard length, length of head 31·8–35·8 (M = 33·8) per cent.

Dorsal head profile straight, sloping steeply at an angle of *ca* 50°–55° with the horizontal.

Preorbital depth 14·5–19·3 (M = 17·6) per cent of head, least interorbital width 19·2–25·0 (M = 22·1) per cent. Snout length 0·7 – 0·9 (mode 0·8) of its breadth, 28·0–34·6 (M = 31·2) per cent of head, diameter of eye 25·0–30·8 (M = 28·0), depth of cheek 24·5–30·2 (M = 27·2) per cent.

Caudal peduncle 12·6–16·8 (M = 15·7) per cent of standard length, 1·1–1·4 (modal range 1·2–1·4) times as long as deep.

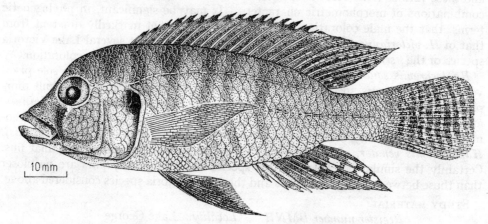

FIG. 32. *Haplochromis petronius*. Holotype.

Jaws equal anteriorly, mouth horizontal, the lips thickened. Posterior tip of the maxilla reaching a vertical through the anterior margin of the pupil, rarely not reaching so far posteriorly. Length of upper jaw 34·2–41·6 (M = 36·8) per cent of head, length of lower jaw 34·8–41·5 (M = 37·8) per cent, 1·3–1·7 (modal range 1·3–1·5) times its breadth.

Gill rakers. The lower 1–3 rakers of the first gill arch are reduced, the others are usually short and stout, but rather slender in some fishes. There are 7 or 8 (mode) rarely 9, rakers in the outer row on this arch.

The pseudorakers are very well developed and are transversely aligned so as to link the inner and outer rows of true gill rakers on the first arch.

Scales. Ctenoid; lateral line with 30 (f.3), 31 (f.4), 32 (f.15) or 33 (f.2) scales, cheek with 3 or 4 rows. Scales on the nape and chest are small, the latter rather deeply embedded. The transition between the larger ventral belly scales and the much smaller thoracic scales is abrupt, and occurs at about the level of the pectoral fin insertion. This abrupt type of size transition is unusual in *Haplochromis* (see Greenwood 1971). Another unusual character of the squamation in *H. petronius* is the presence of a small naked area (equivalent to about 4 or 5 scales) immediately anterior to the insertion of the first dorsal fin spine.

There are 5–7 (mode 6) scales between the upper lateral line and the dorsal fin origin, and 7–10 (modal range 8–9) between the pectoral and pelvic fin bases.

Fins. Dorsal with 15 (f.3), 16 (f.21) or 17 (f.1) spinous and 8 (f.1), 9 (f.21) or 10 (f.3) branched rays. Caudal strongly subtruncate, almost rounded ; scaled on its basal third to half.

Pectoral fin 26·5–32·4 (M = 28·1) per cent of standard length, 75·0–97·8 (M = 82·0) per cent of head length. Pelvics with the first ray somewhat produced in both sexes.

Teeth. In the *outer row*, the teeth posteriorly in the premaxilla are unicuspid and caniniform ; anteriorly in this jaw and throughout the lower jaw the teeth are stout, unequally bicuspid (often weakly so), and have the crown incurved (text-fig. 33). These teeth are cylindrical in cross-section, with the upper part of the crown compressed. Occasionally, a few unicuspids occur anteriorly in the upper jaw, interspersed among the usual bicuspids. Also, in a few fishes the posterior premaxillary teeth are slender and bicuspid, not unicuspid as is usual for teeth in that position.

There are 36–50 (M = 42) teeth in the outer premaxillary row.

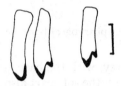

FIG. 33. *H. petronius.* Premaxillary teeth (left), anterolateral in position. Viewed laterally. Scale = 0·5 mm.

Inner teeth are usually tricuspids in fishes < 75 mm standard length (and in a few larger individuals) but most fishes > 75 mm long have an admixture of tricuspid, weakly tricuspid and unicuspid teeth. There are 2 or 3 rows of inner teeth in the upper jaw, and 2 (less frequently 3) in the lower jaw.

OSTEOLOGY. The *neurocranium* is of the generalized *Haplochromis* type (see p. 147) with a moderately decurved preorbital profile.

The *lower pharyngeal bone* is relatively stout, its dentigerous surface equilateral. The teeth are cuspidate and compressed, with those of the median rows noticeably coarser (text-fig. 34). In a few specimens some or all of the median teeth are submolariform. There are *ca* 20–24 rows of teeth.

Vertebral counts in the 12 specimens radiographed are 29 (f.10) and 30 (f.2), comprising 12 (f.1), 13 (f.10) or 14 (f.1) abdominal and 15 (f.1), 16 (f.8) or 17 (f.3) caudal centra.

COLORATION IN LIFE. *Males, adult but not sexually active* : the flanks are greenish-yellow, tinged with blue, the blue concentrated along the scale margins ; dorsally the colour changes to greenish-violet. The belly and chest are white, the branchiostegal membrane light grey. The lips are turquoise, and there are strong tinges of turquoise on the basically grey snout and cheeks.

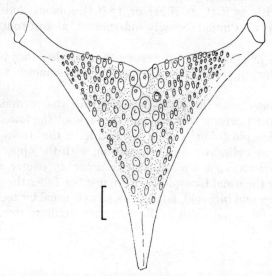

Fig. 34. *H. petronius*. Lower pharyngeal bone, occlusal view. Scale = 1·0 mm.

The dorsal fin is light sooty-grey, with the base greenish-blue, and the lappets red (as is the margin of the soft part) ; the soft part also has a scattering of well-marked, elongate, deep red streaks and spots. The anal fin is also sooty-grey (sometimes flushed with pink or red, perhaps a correlate of sexual activity) with a light-blue area along its base and a narrow scarlet outline to its margins ; there are as many as eight bright orange ocelli on the posterior part of this fin. The caudal is dark grey, outlined in red, this marginal band expanding at the posterodorsal and ventral angles of the fin ; deep maroon streaks occur between the rays. The pelvic fins are black or sooty, the spine and first ray are, however, bluish-white.

Females : no live females have been observed.

PRESERVED COLORATION. Only males are available, all are adult. The ground colour is uniformly light grey-brown ; in some specimens there are traces of up to 7 dark but faint vertical bars across the flanks and caudal peduncle, those on the flanks extending from the dorsal body outline almost to the ventral body margin. The snout is dark grey to dusky, the lower jaw dusky-grey and the lips pale. The ventral aspects of the cheek and operculum are of variable duskiness, almost black in some fishes but only a dark brown in others. The branchiostegal membrane is greyish-sooty. Cephalic markings comprise a usually distinct and intense lachrymal stripe, and 2 transverse bars, of variable intensity, across the snout.

The dorsal fin is dark grey to sooty, the membrane between the last few branched rays maculate. The caudal is grey, maculate distally and dark grey, almost black basally. The anal fin is greyish to sooty, the pelvics variable, from dusky to black but with the spine and first ray much lighter.

ECOLOGY. *Habitat*. *Haplochromis petronius* is the only Lake George *Haplochromis* species that appears to have a clearly circumscribed habitat. With a few exceptions (*see below*) the species has been found only in a rocky bay situated immediately behind the village of Kashaka. This bay is an old volcanic crater, one wall of which has collapsed and thus connected the crater with the lake. The bay is roughly ovoid in outline, its greatest and least axes being about 1·3 and 0·8 km. At its centre the water is some 6 m deep, but around the margins it is between 1 and 3 m. This marginal area has a rough substrate composed of rocks and stones derived from the crater walls. Some plant debris (including dead trees) lies among the rocks which, in places, are also covered lightly by a thin slick of organic mud. The bay is sheltered from all directions and has a relatively narrow entrance.

Haplochromis petronius is found only over the marginal rocky area of Kashaka Bay; nets set in the deeper central area, either at the bottom or floating at the surface, caught no *H. petronius*, although other fishes (both cichlid and non-cichlid), including species found inshore, were caught.

The species is common in catches from its habitat; other *Haplochromis* species inhabiting the same region include *H. angustifrons*, *H. elegans*, *H. aeneocolor* and *H. schubotziellus*, but none occurs in such abundance as does *H. petronius*.

Outside Kashaka bay, *H. petronius* is rarely encountered; it is probably significant in this connection that no other areas of the lake have a similar rock-boulder substrate. The very few *H. petronius* caught in the main lake are either from near the papyrus fringe over a mud-bottom or from over a sandy substrate on an exposed shore facing the lake centre.

Food. Insects, both larval and emergent, seem to be the commonest food organisms in the diet of *H. petronius*, but the sample I examined was small (20 specimens).

Breeding. The specimens I have examined are all males, and indeed I have been unable to catch any females, despite intensive fishing in the area. Thus, it would seem that there is a definite segregation of the sexes, and probably at all times of the year, itself an unusual phenomenon amongst *Haplochromis* species.

The smallest fish examined (67 mm standard length) is a ripening male, although another specimen 69 mm standard length is probably immature.

Distribution. Known only from Lake George.

DIAGNOSIS AND AFFINITIES. Taken in its totality, the appearance of *H. petronius* is highly characteristic, and readily distinguishes this species from other *Haplochromis* in Lake George. The small chest and nape scales, the abrupt size transition between the thoracic and ventral body squamation, and the small scaleless area before the first dorsal fin spine are trenchant diagnostic features.

However, these particular characters are also found in a species known from Lake Edward, viz. *H. pharyngalis* Poll (*see also* p. 177). The overall morphology of *H. pharyngalis* is also very like that of *H. petronius*, especially the steep head profile, the horizontal mouth with its thickened lips and the near-rounded caudal fin outline.

The principal interspecific differences are twofold. First, in *H. pharyngalis* the lower pharyngeal bone is greatly enlarged (massive in 1 of the 3 known specimens), and there are several rows of enlarged molariform or near-molariform teeth (*see* Poll 1939a, p. 46, fig. 26). Second, the nuchal and thoracic scales (particularly the former) are relatively smaller in *H. pharyngalis*. In addition, there are fewer (i.e. 6) gill rakers in *H. pharyngalis*, and there is a naked area, about 1 scale row deep, on the ventral margin of the cheek.

With only 3 specimens of *H. pharyngalis* available for comparison and with no information on their live colours, it is difficult to assess the precise relationship between the species. Certainly the differences in pharyngeal bone development (and correlated dental differences) and in the size and shape of the apophysis for the upper pharyngeal bones seem well marked. But, these are of similar nature to those distinguishing the George–Edward populations of *Astatoreochromis alluaudi* from those of Lake Victoria (*see* Greenwood 1959a, and especially 1965a). That *H. pharyngalis* shows a range of pharyngeal bone enlargement (even in only 3 specimens), and that *H. petronius* exhibits incipient molarization of the pharyngeal teeth, adds to the impression of close relationship between the species. The question can only be pursued when larger samples of *H. pharyngalis* are available.

An even closer resemblance exists between *H. petronius* and *H. wingatii* (Blgr.), a species of the Nile and Lake Albert (*see* Greenwood 1971 for a revision of this often misidentified species). Again, comparisons are hampered by the small number of specimens available. Only 2 well-preserved and 1 poorly preserved specimen of *H. wingatii* are known, all much smaller than the smallest *H. petronius* available, and there are no data on their live colours.

Morphometrically, *H. wingatii* and *H. petronius* are indistinguishable except for the longer pectoral fin in *H. petronius* (75·0–97·8, M = 86·4 per cent head length, *cf.* 65·0–66·5 per cent for *H. wingatii*) and a higher vertebral count (29 or 30, *cf.* 28). Two slight interspecific differences are the absence of a naked predorsal area in *H. wingatii* and the existence of a naked strip below the cheek scales in that species. There is also a difference in the dentition. In the holotype of *H. wingatii* (53 mm standard length) the majority of the outer teeth in both jaws are slender, strongly recurved unicuspids. The outer teeth in *H. petronius* are relatively coarser and, except in the largest fishes, are all bicuspid. Even in large individuals the predominant tooth form is the bicuspid. The two other *H. wingatii* specimens are smaller and do have bicuspid teeth (as is usual for small individuals of species with a unicuspid definitive dentition). The bicuspids in these specimens are slender (like the few bicuspids of the holotype) and thus are unlike the bicuspids of *H. petronius*.

Resemblances between *H. wingatii*, *H. petronius* and *H. pharyngalis* are striking, particularly since they involve, principally, details of squamation not found in other species of Lakes Edward, George or Victoria, or even in the fluviatile *Haplochromis* species of east Africa. Reduced pectoral and nuchal squamation is known from other *Haplochromis* and *Haplochromis*-like species, but in all these fishes it is associated with rheophilic habits (Greenwood 1954; Thys van den Audenaerde 1963). The habitats of the 3 species under consideration are far removed from the torrential, and thus it is difficult to attribute the resemblances in squamation to environmentally induced parallelism. If, therefore, the similarities are a reflection of phyletic affinity, then *H. wingatii*, *H. petronius* and *H. pharyngalis* would appear to be of a lineage distinct from all other *Haplochromis* in Lakes Victoria, Edward and George. The implications of this conclusion will be considered later (p. 235).

STUDY MATERIAL

Register number BMNH	Locality : Lake George
1972.6.2. : 1 (Holotype)	Kashaka Crater Bay
1972.6.2 : 2–10 (Paratypes)	Kashaka Crater Bay
1972.6.2 : 11–21 (Paratypes)	Kashaka Crater Bay
1972.6.2 : 22 (Paratype)	North end of Kankurunga Island
1972.6.2 : 23 (Paratype)	Caught over sandy shallows
1972.6.2 : 788 (Paratype)	Kashaka Crater Bay
1972.6.2 : 811 (Paratype)	Kashaka Crater Bay
1972.6.2 : 293	Sandy shoal

Haplochromis eduardianus (Blgr.), 1914
(Text-figs. 35–37)

Schubotzia eduardiana Boulenger, 1914, in *Wiss. Ergebn. Deuts. Zentral-Afrika Exped., 1907–1908, Zool.* **3** : 258–259 (Lake Edward).
Schubotzia eduardiana : Regan, 1921, *Ann. Mag. nat. Hist.* (9), **8** : 639.
Schubotzia eduardiana : Trewavas, 1933, *J. Linn. Soc. (Zool.)*, **38** : 340.

All these references are to specimens from Lake Edward; the species was not discovered in Lake George until recently.

NOTE ON THE ALTERED GENERIC STATUS OF THE SPECIES. Boulenger (1914) defined the monotypic genus *Schubotzia* solely on the basis of its dental morphology.

The outer jaw teeth are unusual (text-fig. 36). The crown is somewhat expanded relative to the cylindrical neck, has the tip distinctly rounded and strongly incurved so as to lie almost horizontally. In outline (i.e. as a flattened object) the tooth is paddle-shaped.

The inner teeth, by contrast, are typical tricuspids but are restricted to 1 or 2 rows in each jaw.

In other characters, *Schubotzia* does not differ from *Haplochromis*. The lower jaw, especially the dentary, is deep and stout, and the premaxilla is rather inflated. However, both these characteristics can be seen in other *Haplochromis* species (e.g. *H. nigricans* and *H. obesus* of Lake Victoria), although the condition could not be described as a common one; it is usually associated with a specialized dentition. The lower jaw of *Schubotzia* is slightly overhung by the upper jaw, again an unusual condition but one found in some *Haplochromis* species (e.g. *H. xenognathus* of Lake Victoria, a species with an unusual dentition, albeit one quite unlike that of *Schubotzia eduardiana*; see Greenwood 1957).

The teeth of *Schubotzia* are outstandingly different when compared with those of other Lakes Edward and George cichlids (text-fig. 36) and especially with those of the

Haplochromis species known to Boulenger in 1914. But, when the *Schubotzia* tooth type is seen against the wide range of tooth morphology found within the genus *Haplochromis* as a whole (or just a segment like the species of Lake Victoria), then it does not seem to be so unusual.

In my opinion, the morphological differences separating *Schubotzia* from *Haplochromis* are relatively slight, and certainly less than those distinguishing *Haplochromis* from the Lake Victoria monotypic genera *Hoplotilapia* and *Platytaeniodus* (or those genera from one another).

To retain *Schubotzia eduardiana* in a separate and monotypic genus serves only to hide its close phyletic relationship with *Haplochromis*. Thus, I would favour classifying this taxon with its closest relatives, that is in the genus *Haplochromis*.

Rosen and Bailey (1963, p. 6) have succinctly stated the pragmatic and theoretical difficulties associated with the generic concept. Particularly they stress the often phyletically misleading results of undue emphasis placed on one or two outstanding morphological differences as generic criteria. I am fully in agreement with these authors' support for a wider use of the subgenus to indicate morphological divergence without losing sight of phyletic relationships. Reducing the genus *Schubotzia* to subgeneric status could well meet the requirements of this particular case. However, there is a great need for new and careful consideration of the generic and infrageneric classification of the ' genus ' *Haplochromis* as currently defined. Such a study must be based on phyletic principles and must test the phyletic integrity of what might well be a polyphyletic taxon.

It is my intention to undertake just such a study ; for the moment, I prefer not to establish *Schubotzia* formally as a subgenus of *Haplochromis*.

LECTOTYPE. A specimen 71·0 mm standard length, BMNH reg. no. 1914.4.8 : 35, from Lake Edward. This specimen is eviscerated and its sex cannot be determined ; the absence of dark pigment on the body suggests that it is a female (*see below* p. 220).

Three paralectotypes are in the collections of the Berlin Museum ; all are from Lake Edward.

DESCRIPTION. Based on 20 specimens, 50·5–79·0 mm standard length, from Lake George. A smaller fish, 35 mm standard length is not included in the morphometric section, since it is distorted, but certain features of its dentition are considered on p. 218.

Depth of body 30·5–36·7 (M = 33·5) per cent of standard length, length of head 31·5–34·3 (M = 33·1) per cent.

Dorsal head profile gently decurved, but sometimes straight, sloping at an angle of 30°–35° with the horizontal.

Preorbital depth 11·2–15·4 (M = 13·3) per cent of head, least interorbital width 23·0–28·3 (M = 25·9) per cent. Length of snout 0·8–0·9 of its width, 23·0–28·3 (M = 25·9) per cent of head, eye diameter 29·6–36·0 (M = 32·8) per cent, depth of cheek 16·7–22·2 (M = 20·0) per cent.

Caudal peduncle 14·7–19·5 (M = 16·9) per cent of standard length, 1·2–1·7 (modal range 1·3–1·5) times longer than deep.

Mouth horizontal, lips slightly thickened; lower jaw shorter than the upper when the mouth is closed. Length of upper jaw 30·8–36·0 (M = 32·8) per cent of head, length of lower jaw 29·2–35·1 (M = 33·1) per cent, 1·0–1·4 (mode 1·1) times longer than broad. Posterior tip of the maxilla reaching a vertical through the anterior margin of the eye, rarely not quite reaching that level.

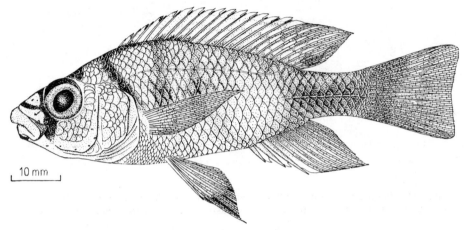

FIG. 35. *Haplochromis eduardianus*. Lake George specimen; a male.

Gill rakers. The lowermost 1 or 2 rakers are reduced, the others are relatively slender although as many as 3 of the lower rakers may be relatively stout. The pseudorakers are well developed and prominent, but are short and stout.

There are 8 or 9 (rarely 10) gill rakers in the outer series on the lower part of the first gill arch.

Scales. Ctenoid; lateral line with 31 (f.8), 32 (f.11) or 33 (f.1) scales, cheek with 2 or 3 (mode), rarely 4 rows. Five to 7 (modes at 5 and $5\frac{1}{2}$) scales between the upper lateral line and the dorsal fin origin, 6–8 (mode 7) between the pectoral and pelvic fin bases.

Fins. Dorsal with 14 (f.1), 15 (f.11) or 16 (f.8) spinous and 8 (f.6), 9 (f.13) or 10 (f.1) branched rays. Anal with 3 spines and 7 (f.2), 8 (f.16) or 9 (f.2) branched rays. Pectoral fin 26·5–32·4 (M = 28·1) per cent of standard length, 80·5–98·0 (M = 82·0) per cent of head. Caudal subtruncate, scaled on its basal half. Pelvics with the first ray slightly elongate.

Teeth. With one exception in the material studied, the shape of the *outer teeth* in both jaws is remarkably uniform. Each tooth is unicuspid, with a flattened crown that is almost half the total length of the tooth. The whole crown is strongly incurved (especially in teeth situated anteriorly and anterolaterally), and its tip is broadly rounded (text-figs. 36 & 37). The neck of the tooth is cylindrical and, compared with that of a typical bicuspid tooth, much stouter. If a tooth were straightened out so that the crown and neck are in one plane, then it would have the outline of a paddle.

The exceptional specimen mentioned earlier (a fish 50·5 mm standard length) differs from the others only in having a few small tricuspid teeth intercalated among the typical teeth posterolaterally in the upper jaw.

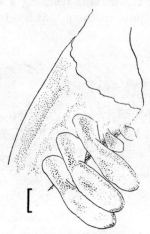

FIG. 36. *H. eduardianus*. Premaxillary teeth (right) anterior in position. Viewed from below and a little laterally, to show labial aspect of the teeth. Scale = 0·25 mm.

In the smallest available specimen (35 mm standard length), most teeth are like those of larger fishes, but in both jaws there are a few teeth with traces of a small lateral cusp. This minor cusp is not separated from the major one by a distinct gap, as in typical bicuspids. Instead, the demarcation between cusps is more in the nature of a narrow, V-shaped groove. Because a minor cusp is present, the outline of these teeth is not rounded but is rather bluntly oblique. Posteriorly in the lower jaw of this small fish there are a few tricuspid teeth.

There are 40–52 (M = 48) teeth in the outer row of the upper jaw; the number not showing any correlation with the fish's size.

Inner teeth (text-fig. 37) in both jaws are tricuspid, with the crown compressed and strongly incurved; arranged in 2 or rarely 3 rows in the upper jaw and a single (rarely double) row in the lower jaw. Virtually no interspace exists between the outer tooth row and the first row of inner teeth.

FIG. 37. *H. eduardianus*. Dentary teeth (left), anterolateral in position. Viewed from medial aspect to show lingual face of teeth. Scale = 0·25 mm.

OSTEOLOGY. The *neurocranium* is of the generalized *Haplochromis* type with a moderately decurved preorbital profile ; that is, a skull of the type found in *H. limax* and *H. petronius* rather than of the type in *H. macropsoides* and *H. elegans*. The premaxilla is a stout and inflated bone (especially the dentigerous arm) and resembles a more extreme form of the premaxilla found in *H. limax*. Stoutness and compactness also characterize the lower jaw. Both dentary and articular are stout, deep bones, and elements of the suspensorium are so arranged and proportioned that the dentary tip lies behind the vertical through the premaxillary symphysis. Presumably the general stoutness of the jaws is at least partly correlated with the presence of stout teeth.

The *lower pharyngeal bone* is relatively stout, although its teeth are fine (except for a few coarser posterior teeth in the middle tooth rows of larger fishes). There are *ca* 24–26 rows of teeth arranged over a dentigerous area about 1·3 times broader than it is long.

Vertebral counts in the 10 specimens radiographed are 29 (f.7) and 30 (f.3), comprising 13 (f.7) or 14 (f.3) abdominal and 15 (f.2), 16 (f.6) or 17 (f.2) caudal elements.

COLORATION IN LIFE. *Adult males* have a silvery blue-grey ground coloration, shading to white on the thoracic region ; many flank scales have deep red-brown centres, the intensity of the colour and the number of reddened scales correlated with the degree of sexual activity. Dark cephalic markings are always well developed. A thin stripe crosses the snout at the level of the lower orbital margin, and a much broader band runs between the upper orbital margins. Often this upper band is interrupted medially. A broad lachrymal stripe is continued above the eye as a wide blotch, the blotches from each side usually meeting medially. Behind this mark is another, this time in the form of a broad band crossing the nape from an origin at about the level of the upper opercular margin's midpoint.

The dorsal fin is hyaline with bright scarlet lappets and a diffuse scarlet flush between the spines ; this flush becomes concentrated into discrete blotches and spots between the branched rays. The anal is dusky on its proximal half, pinkish to scarlet distally ; the almost round ocelli (1 or 2) are very large and are yolk-yellow in colour. The caudal is entirely suffused with bright scarlet, although its ground colour is greyish-green. The pelvic fins are dusky but have scarlet streaks between the rays posteriorly on the proximal half of the fin.

Females (at all stages of sexual activity) : the ground colour is dark silver-grey shading to white on the belly and chest. Cephalic markings are usually absent, but if visible are a faint replica of those in the male (*see above*). The dorsal fin is hyaline, with a narrow scarlet marginal band on the soft part, and thin red streaks on the lappets of the spinous part ; in some individuals red streaks are present between the spines and rays. The caudal is hyaline with an overall pink flush, sometimes intensified to scarlet streaks between the rays and along the posterior margin of the fin. The anal is very pale yellow, with 2 deeper yellow spots (not ocelli) in the position of ocelli in males. The pelvics are whitish-hyaline.

PRESERVED COLORATION. *Adult males* : the ground colour is silver-grey, the flanks crossed by up to 10 vertical bars of variable intensity and definition ; the first 2 bars lie immediately behind the opercular margin, the last on the caudal peduncle. Dorsally, the bars extend to the body outline and at least anteriorly may extend onto the dorsal fin membrane; ventrally the bars do not reach much below the level of the pelvic fin insertion.

Cephalic markings are very intense. Across the snout are 2 parallel bars, the upper generally twice the width of the lower. A broad lachrymal stripe is present, and there is a well-defined bar or triangular blotch extending towards the midline immediately behind the orbit. Posterior to this mark is another, but strap-shaped one which lies anterior to the dorsal fin origin. This bar appears to be a medial extension of the first vertical bar on the flank.

All fins, except the pectorals and pelvics, are yellowish, the basal region of the caudal often dusky. The pectorals are hyaline, and the pelvics black on the outer third, otherwise dusky.

Females are silvery grey or yellowish, with all fins yellow or hyaline. No cephalic markings are visible except, in some specimens, for a faint darkening below the eye, that is, in the position of a lachrymal stripe.

ECOLOGY. *Habitat*. The species is found in most inshore areas of the lake, over sand and mud bottoms, and in both exposed and sheltered localities. It has also been caught in the near-shore areas of the open lake but not further than about 100 m from the nearest land (in this case a small, reed-fringed island). No specimens have been collected in midlake or other distinctly offshore regions.

Food. The diet of *H. eduardianus* is still unknown. Most specimens examined had nothing recognizable except for a few sand grains and a few macrophyte fragments in any part of the gut.

The highly specialized dentition would suggest equally specialized feeding habits. By analogy with the similarly shaped teeth of *Plecodus* spp., one might suspect a similar diet of fish scales (Marlier & Leloup 1954).

Since the intestine of *H. eduardianus* is short (about half the length of the body) a vegetarian diet is almost certainly ruled out.

As so many (*ca* 90 per cent) of the specimens examined have nothing in the guts it seems probable that, whatever the food, it is rapidly digested.

Breeding. *Haplochromis eduardianus* is a female mouth brooder. Sexual maturity is reached at a standard length of *ca* 55 mm, and both sexes attain the same maximum length.

All of the 9 sexually active females examined have the right ovary much larger than the left one, and in some individuals only the right ovary is developed.

Distribution. Lakes Edward and George and the Kazinga Channel.

DIAGNOSIS AND AFFINITIES. *Haplochromis eduardianus* is distinguished from all other *Haplochromis* species in Lakes George and Edward by the morphology of its teeth. This peculiar tooth form led Boulenger (1914) to place the species in a distinct and monotypic genus. Reasons for not accepting Boulenger's classification are discussed above (p. 215). In essence, I argue that to place *H. eduardianus* in a monotypic genus (i.e. *Schubotzia*) is to obscure its phyletic relationships. That the dentition (both oral and pharyngeal) of *Haplochromis* group cichlids is, in an evolutionary sense, easily modified can be seen readily amongst the component species of the Lake Victoria *Haplochromis* species flock (*see* Greenwood 1965c for summary and further references). Indeed, in at least one instance the morphological sequence leading from generalized to highly specialized dentitions is still preserved among the extant species of that lake (Greenwood 1957). There are also examples where intraspecific variability is such that if only the extreme condition was known, and the criteria for generic status were based solely on morphological ' gaps ', then the species would have to be accorded generic status (*see* Greenwood, *op. cit.*, p. 96, with reference to *H. xenognathus*).

The reality of a morphological gap like that between the *H. eduardianus* tooth form and the shape of the teeth in any other *Haplochromis* species cannot be denied ; the difficulty lies in attempting to interpret the significance of the gap. The existence within a single species flock of such examples as *H. xenognathus* (and others where different species bridge a morphological gap) warns against hasty action that could obscure the essence of a phyletic classification ; that is, the demonstration of relationships as well as divergence. Is there any reason to suppose that, phylogenetically, the current evolutionary end-point seen in ' *Schubotzia* ' *eduardiana* and *Platytaeniodus degeni* or *Hoplotilapia retrodens* (both Lake Victoria monotypic genera ; *see* Greenwood 1956a) is any different from that represented by *H. xenognathus?* Certainly all three monotypic genera have as their nearest relatives a species of *Haplochromis*. That such a species can be found for *H. xenognathus* (Greenwood 1957) but not for *Schubotzia, Playtaeniodus* or *Hoplotilapia* may be more a reflection of a past epigenetic situation than of phyletic history.

STUDY MATERIAL

Register number BMNH	Locality : Lake George
1972.6.5 : 1–3	Tufmac Bay
1972.6.5 : 4	Kashaka Crater
1972.6.5 : 5–7	Small island north of Kankurunga Island
1972.6.5 : 8–13	Various localities
1972.6.5 : 14–20	Various localities

NON ENDEMIC *HAPLOCHROMIS* AND *HAPLOCHROMIS*-GROUP SPECIES IN LAKE GEORGE

Haplochromis nubilus (Blgr.), 1906

(Text-fig. 38)

Haplochromis nubilus (part) : Trewavas, 1933, *J. Linn. Soc. (Zool.)*, **38** : 329.

Most of the Lake George specimens referred to *H. nubilus* by Trewavas (1933) were misidentified, but two specimens (BMNH reg. nos. 1933.2.23 : 285–286) from Worthington's (1932) station 618 appear to be of this species. They are not included in the description given below.

Four specimens collected by the I.B.P. team can be identified with certainty since all are sexually active males and their live colours were recorded. These fishes came from a catch made close inshore, near reeds, in a bay of Akika Island.

A summary of morphometric characters is given below. It is based on 3 specimens only because the fourth is distorted and damaged. The latter specimen is, however, used in the description of the teeth and for fin ray and scale counts.

S.L.	Depth †	Head †	PO %	IO %	Snt %	Eye %	Cheek %	Lj %	Uj %	CP †
67·0	37·3	34·3	17·4	21·7	28·3	30·3	21·8	34·8	34·8	16·4
70·5	39·3	34·1	16·7	27·0	31·3	29·2	22·9	37·5	37·5	16·5
74·5	37·5	34·2	15·6	24·7	28·4	27·4	23·5	39·1	35·3	16·1

† = per cent of standard length.
% = per cent of head.
Lj = lower jaw, Uj = upper jaw.

Caudal peduncle 1·3–1·4 times longer than deep.

Dorsal head profile sloping steeply at *ca* 40°–45° with the horizontal, its outline straight except for a marked concavity above the orbital region. Upper margin of the orbit distinctly below the level of the dorsal profile.

Mouth horizontal, lips somewhat thickened; posterior tip of the maxilla reaching a vertical through the anterior margin of the orbit. Jaws equal anteriorly, the lower *ca* 1·3 times longer than broad. Snout slightly broader than long.

FIG. 38. *Haplochromis nubilus*. A Lake Victoria specimen. From Boulenger, *Fishes of the Nile*.

Gill rakers. Eight or 9 on the lower part of the first arch; the lower 2 rakers are reduced, the remainder are relatively stout. Pseudorakers are present and discrete, but are rather low and small.

Scales. Ctenoid; lateral line with 31 (f.2) or 32 (f.2) scales, cheek with 3 or 4 rows. Five or 5½ scales between the upper lateral line and the dorsal fin origin, 6 between the pectoral and pelvic fin bases.

Fins. Dorsal with 15 (f.4) spinous and 9 (f.2) or 10 (f.2) branched rays, anal with 3 spines and 8 (f.2) or 9 (f.2) branched rays. Pectoral 26·2–28·2 per cent of standard length, 76·5–83·3 per cent of head. Pelvics with the first ray variably produced, markedly elongate in one fish. Caudal distinctly subtruncate, almost rounded; scaled on its basal third to half.

Teeth. Posteriorly in the upper jaw the *outer teeth* are large and unicuspid. Elsewhere in this jaw, and throughout the lower jaw, the teeth are unequally bicuspid and moderately stout, or are an admixture of such teeth with slender unicuspids. Bicuspid teeth have the major cusps equilateral in outline and barely incurved; some teeth have faint indications of a flange on one aspect of the cusp (see p. 151).

There are 40–44 teeth in the outer premaxillary row.

Inner teeth in both jaws are small and tricuspid, are arranged in 2 or 3 rows in the upper jaw, and in 1 or 2 rows in the lower.

OSTEOLOGY. The *syncranium* of *H. nubilus* is typically that of a generalized *Haplochromis* species.

The *lower pharyngeal bone* is relatively slender, its teeth compressed and bicuspid, with those of the median rows slightly coarser (especially posteriorly). There are ca 24–28 rows of teeth arranged on a dentigerous area that is ca 1·2–1·3 times broader than long.

Vertebral count is 29 (comprising 13 abdominal and 16 caudal centra) in the 4 fishes radiographed.

COLORATION IN LIFE. *Adult males* have a highly characteristic velvety-black coloration that is virtually uniform over the whole body. The dorsal fin also is black except for a narrow scarlet margin along its entire length, and some scarlet spots and streaks on the soft part. A scarlet flush covers most of the anal fin, although the spinous part may be a little dusky; the ocelli are yolk-yellow. Proximally the caudal fin is black, and this dark colour may extend along the centre of the fin almost to its margin. The margin is bright scarlet, the colour expanding at the posteroventral margin of the fin. The pelvics are jet black.

The live colours of *females* from Lake George are unknown; in Lake Victoria the body is deep olive-green and all the fins are greyish-green.

Preserved males are either uniformly black on the body or the dorsum (above the midlateral line) may be lighter (i.e. a deep brown) and crossed by 4 or 5 rather faint vertical bars. The dorsal fin is black except for a pale (yellowish-orange) margin. Almost the entire anal is pale yellow, although there is a faint and narrow dark band basally. The caudal is dark proximally and between the middle rays, but otherwise it is yellowish. The pelvics are black.

ECOLOGY. Little can be said about *H. nubilus* in Lake George. The specimens I examined came from a shallow inshore area close to emergent vegetation. A similar habitat seems to be the preferred one for *H. nubilus* in Lakes Victoria and Nabugabo (Greenwood 1965b).

No data are available on the feeding and breeding habits of the species in Lake George. Lake Victoria populations are female mouth brooders, and have a rather omnivorous diet in whichv laral insects and small Crustacea predominate.

Distribution. Lakes Victoria, Kyoga, Nabugabo, Edward and George, and in many rivers and streams connected with these lakes. To date no specimens have been caught in the Kazinga Channel.

DIAGNOSIS. Morphologically, *H. nubilus* closely resembles *H. aeneocolor* (see p. 154) in nearly all characters, especially morphometric ones. Pseudorakers are present in *H. nubilus* but not in *H. aeneocolor*, and the caudal fin is virtually rounded in *H. nubilus* (but truncate in *H. aeneocolor*).

There are some dental differences between the species but these are not particularly trenchant. However, in *H. nubilus* flange development on the major cusp is certainly less common than in *H. aeneocolor*, and the flange, when developed, is less prominent. Also, judging from Lake Victoria *H. nubilus*, it seems probable that unicuspid outer teeth occur more frequently in this species than in *H. aeneocolor*.

In life, male breeding coloration is certainly diagnostic.

ASTATOREOCHROMIS Pellegrin, 1903

This genus is readily distinguished from *Haplochromis* by the higher number of anal fin spines : 4 or more, usually 5.

For a full diagnosis of the genus *see* Greenwood 1959a and 1965a, b.

Astatoreochromis alluaudi Pellegrin, 1903
(Text-fig. 39)

Astatoreochromis alluaudi : Trewavas, 1933, *J. Linn. Soc. (Zool.)*, **38** : 321 (1 specimen from Lake George).
Astatoreochromis alluaudi occidentalis Greenwood, 1959, *Bull. Br. Mus. nat. Hist. (Zool.)*, **5** : 174–175.

Very few specimens of *A. alluaudi* have been collected from Lake George by the I.B.P. team, despite an intensive fishing effort. Four specimens (45–104 mm standard length) are available for study.

Because *A. alluaudi* is easily recognized by its having 4 or more anal spines and greatly enlarged pharyngeal bones, and because the Lake George fishes do not differ greatly from those described elsewhere (Greenwood 1959a) no full description is necessary.

The live colour of Lake George fishes (previously unknown) is identical with that described for the Lake Victoria populations (*see* Greenwood, *op. cit.*, p. 172).

The lower pharyngeal bone and dentition of *A. alluaudi* from Lakes Edward and George are much less massive than those from comparable-sized fishes in Lake Victoria. This led me to describe a subspecies, *A. a. occidentalis*, for these western lakes (Greenwood 1959a). Subsequent research, however, strongly suggests that the degree of pharyngeal bone development (and of tooth molarization) is probably under direct environmental control (Greenwood 1965a). In other words, the differences between Lake Victoria and Lake Edward–George fishes is not genetically determined. Thus it seems inadvisable to continue recognizing two subspecies.

Lower pharyngeal bones and teeth in the 4 Lake George fishes fit broadly into the reductional pattern described for fishes from lakes other than Victoria and Kyoga

(Greenwood 1959a). However, the Lake George fishes seem to have rather more massive bones (and greater molarization of the teeth) than do fishes from Lake Nakavali. In the two smallest Lake George specimens (45 and 46 mm standard length) the bone is only a little less developed than in a comparable-sized specimen from Lake Victoria, and there is equal molarization of the teeth (*see* fig. 3 top right, in Greenwood 1959a). The 80 mm standard length Lake George fish has a bone comparable with that of the 123 mm standard length Lake Nakavali specimen figured (*op. cit.*), but the 104 mm standard length fish has a relatively less massive bone which is comparable with the same specimen from Lake Nakavali.

FIG. 39. *Astatoreochromis alluaudi*. A Lake Victoria specimen. From Boulenger, *Fishes of the Nile*.

The ratio of head length to pharyngeal bone width (measured from tip to tip of the upper arms) is in the range 2·9–3·0 (*see* Greenwood, *op. cit.*).

Vertebral counts for the 4 specimens are 13 + 16 (f.2), 13 + 15 (f.1) and 14 + 15 (f.1).

ECOLOGY. Little more can be added to our knowledge of this species in Lake George. The 4 specimens came from different areas of the lake, but a common feature for each locality is its proximity to the shore and the presence of rooted aquatic vegetation in the area.

Only fragments of gastropod shells were found in the intestines of the larger fishes (80 and 104 mm standard length); the 2 smaller individuals (45 and 46 mm standard length) yielded fragmentary remains of chironomid larvae.

HEMIHAPLOCHROMIS Wickler, 1963

The primary generic distinction of this superficially *Haplochromis*-like taxon is its reproductive biology (*see* Wickler 1963).

The two morphological characters that separate it from *Haplochromis* (at least those species occurring in Uganda) are :

(i) The typical elongate and raised cover to the lateral line canal opening in each pore scale on the body is absent from many of these scales in *Hemihaplochromis*. Instead there is either a simple pore or seemingly no opening at all; scales in the posterior part of the upper lateral line series and those of the entire lower line are those most often missing a cover.

(ii) In males, instead of there being well-developed ocelli on the anal fin, there is a bright orange spot at the posteroventral angle (or tip) of the fin. This character is, of course, associated with the different reproductive behaviour of species in this genus (Wickler, *op. cit.*).

Hemihaplochromis multicolor (Schoeller), 1903
(Text-fig. 40)

For a full synonymy of this species, *see* Greenwood 1965b.

Only 1 specimen has been collected by the I.B.P. team. This apparent scarcity of *H. multicolor* in collections from the lake is probably a reflection of the small adult size attained and the habitats occupied, rather than a true indication of its abundance. Trewavas (1933) does not record *H. multicolor* from either Lake George or Lake Edward, and the only record from the Kazinga Channel is a few specimens I caught (by dip-netting amongst reeds) near Katungura (unpublished information).

The Lake George fish (BMNH reg. no. 1972.6.5 : 21) is 32 mm standard length and was caught near Busatu Island; its sex is indeterminable.

Depth of body 35·9 per cent of standard length, length of head 35·9 per cent.

Dorsal head profile very gently curved, sloping at an angle of *ca* 40° with the horizontal.

Preorbital depth 17·4 per cent of head, least interorbital width 30·3 per cent, snout length 0·8 of its breadth and 26·0 per cent of head; eye diameter 30·3 per cent of head, cheek depth 26·0 per cent.

Caudal peduncle as long as deep, 15·6 per cent of standard length.

FIG. 40. *Hemihaplochromis multicolor*. A Lake Victoria specimen. From Boulenger, *Fishes of the Nile*.

Mouth slightly oblique, lips not thickened. Upper jaw 30·3 per cent of head, lower jaw 34·8 per cent. Posterior tip of the maxilla not quite reaching a vertical through the anterior orbital margin.

Gill rakers all short and stout, the lower 3 a little shorter than the others ; 7 rakers on the lower part of the first gill arch.

Scales. Ctenoid ; 29 in the lateral line series. Many of these scales (particularly in the lower series and posteriorly in the upper series) lack the longitudinal, arched cover to the pore opening, which is represented by a simple pit ; other scales are without any visible opening at all.

Cheek with 2 rows of scales ; 5 scales between the upper lateral line and the dorsal fin origin, 3 between the pectoral and pelvic fin bases.

Fins. Dorsal with 14 spines and 10 rays, anal with 3 and 9. Pectoral 23·4 per cent of standard length, 65·2 per cent of head length. Caudal fin rounded, scaled over slightly more than its proximal third.

Teeth. Posteriorly in the upper jaw, the *outer teeth* are very slender and unicuspid. Elsewhere in this jaw, and in the entire outer row of the lower jaw, the teeth are stout and unequally bicuspid. The major cusp is somewhat oblique in outline. There are 34 teeth in the upper jaw outer series.

Inner teeth are slender, slightly curved and unicuspid, and are arranged in a single row in each jaw.

Vertebral count. Thirteen abdominal and 16 caudal vertebrae.

ECOLOGY. Nothing is known about the bionomics of this species in Lake George. It seems probable (by analogy with its behaviour elsewhere) that in Lake George *H. multicolor* lives among submerged plants and in open-water areas within the margin of papyrus swamps (*see* Greenwood 1965b).

Hemihaplochromis multicolor is a female mouth brooder ; adults more than 55 mm standard length are uncommon.

Distribution. Widespread in the lakes, rivers, swamps and streams of Uganda, and probably in other regions of eastern Africa as well. The species also occurs in the Nile.

DISCUSSION

Biology of the Lake George *Haplochromis* species flock

Detailed biological studies of the fishes are being carried out by members of the International Biological Programme team. Some of their results will be available shortly and it is thus appropriate now merely to make some general observations.

Unlike the Lake Victoria *Haplochromis* species, those of Lake George are less closely associated with a particular habitat or substrate type. Nevertheless, any one kind of habitat, for example, offshore open water, the papyrus fringe, protected bays or an exposed sandy shore, has a definable assemblage of species in which a few are numerically dominant.

Open offshore waters have the fewest species, and of these only a few are found also in other habitats, and then never as the dominant elements.

Species with clearly restricted habitat preferences are : the mollusc-eating *H. mylodon*, the supposed larval and embryo cichlid-eating *H. taurinus* and the grazer on epiphytic and epilithic algae, *H. limax*. Species caught infrequently (despite widespread sampling in many habitats) like *H. schubotzi*, *H. labiatus*, *H. nubilus*, *Hemihaplochromis multicolor* and *Astatoreochromis alluaudi*, may be presumed to have limited habitat preferences.

Only *Haplochromis petronius* can be said to have a truly restricted habitat since it is virtually confined to a rocky crater bay (the only one of its kind in the lake). Even in this example, however, there are exceptions since three specimens of *H. petronius* have been collected in the main lake (over sand and close inshore).

As with the *Haplochromis* species of Lake Victoria, the most readily observed adaptations seen in the Lake George fishes are those associated with feeding habits. Differences in tooth form and number, pharyngeal bone shape and dentition, and in jaw size and arrangement are the more obvious characters involved in this adaptive radiation. The overall impression gained is of a small-scale Victoria flock. In other words, one in which the major trophic adaptations are developed but to a slightly lower degree of specialization, and with fewer species occupying one trophic niche or showing the same level of specialization.

Lakes George and Edward harbour one trophic specialization not found among the *Haplochromis* of Lake Victoria, namely a species (*H. pappenheimi*) feeding, as an adult, almost exclusively on pelagic zooplankton. Possibly the absence of an *Engraulicypris* species from Lakes Edward and George has enabled a *Haplochromis* to fill this niche.

Two *Haplochromis* species in Lake Edward (*H. mylodon* and *H. pharyngalis*) have the enlarged pharyngeal bones and teeth associated with a diet of molluscs. Only *H. mylodon* occurs in Lake George, but, as in Lake Edward, another mollusc crushing cichlid, *Astatoreochromis alluaudi*, is present.

Specialized mollusc shellers (as opposed to crushers) have not evolved in the Edward-George flock. This specialization is well-represented in the *H. sauvagei-Macropleurodus bicolor* series of Lake Victoria (Greenwood 1957, 1965c). Possibly snails are a less abundant food source in Lakes Edward and George than in Lake Victoria; qualitative sampling in all three lakes certainly suggests that this is so (personal observations).

Another sharp contrast between the flocks of Lake Victoria and George is the presence in Lake George of only one truly piscivorous species (*H. squamipinnis*). Among the inshore species of *Haplochromis* from Lake Victoria, about 30 per cent are piscivores. The situation in Lake Edward is different again because there are at least three piscivorous species among the known but yet undescribed species from that lake. However, the proportion of piscivores to non-piscivores in Edward is still much lower than in Victoria, probably about one in ten species.

A discrepancy is also noticed in the absolute number of larval and embryo fish-eating species. Only one paedophagus species, *H. taurinus*, is known from Lake George (and Edward) whereas at least seven species occur in Victoria. The relative proportion of paedophages to all other species (from all habitats and trophic groups) is, however, less disparate, being *ca* 6 per cent in Lake George and *ca* 4 per cent in Lake Victoria.

Although there is only one epiphytic and epilithic algal grazer in Lake George (*H. limax*) as compared with at least four such species in Lake Victoria (Greenwood

1956b), the total for Lakes Edward and George together is three, proportionately a much higher number than in Lake Victoria.

It is difficult to generalize about the insectivores and detritus feeders, except to say that in both flocks species belonging to these trophic groups are numerous (Greenwood 1965c).

At least one species in Lake Victoria (*H. erythrocephalus*) and one in Lakes Edward and George (*H. nigripinnis*) is a specialized feeder on suspended phytoplankton.

In Lake George there is a very rapid extinction of incident light so that over much of the lake the photic zone extends merely to a depth of about 60 cm. Most *Haplochromis* species live below this zone or at least spend a great deal of the day in very poorly lit waters. This raises several interesting questions regarding the fishes' adaptations and behavioural responses to such photic conditions. In particular there are problems associated with breeding behaviour; there would seem to be insufficient light for potential mates to recognize one another visually. Another difficulty associated with breeding is the nature of the substrate in Lake George. Over much of the lake the bottom is composed of soft, near liquid, organic ooze. This would appear to provide a most unsuitable substrate on which to spawn in the typical *Haplochromis* fashion. There are, of course, places in the lake where the bottom is hard and where there is reasonably good light penetration. But, no evidence has been collected to suggest that these areas are the only ones used as spawning sites. Brooding females have been found in all habitats, although this does not necessarily imply that the spawning site was near by. It does, however, indicate that for the species involved there are no definite ' nursery ' zones.

Field conditions in Lake George virtually preclude direct observations on the fishes' spawning behaviour, and what information we may get in the future must therefore be indirect, i.e. from aquarium studies. A comparative study of the Lake George species with those from the deep, near-aphotic waters of Lake Victoria would be particularly valuable, especially since the latter species are also faced with a generally soft substrate.

Whether or not vision plays an important part in courtship and species recognition, the Lake George *Haplochromis* species, like those from other and clearer lakes, show distinctive and species-specific male coloration. The Lake George species do, however, differ in one respect. Many females have large, pigmented spots in the same position and of the same colour as the ocelli (or egg dummies, *see* Wickler 1962) in males. These pigment spots lack the clear border that, in males, makes the spot an ocellus. Unfortunately I do not know if female ' egg spots ' are so well developed, or if they are present at all, in Lake Edward populations of the same species.

As a footnote to these remarks on coloration it may be noted that no Lake George (or, as so far recorded, no Lake Edward) *Haplochromis* species exhibits a colour polymorphism like that found in several Lake Victoria species. In these fishes a certain percentage of females has an outstanding coloration in the form of piebald black on silver or black on yellow, or a peppered black, red and orange on a yellowish background. Sex-limited polychromatism is now recorded in at least eight species from Lake Victoria, but none has been found in any Lake George species from among the thousands of *Haplochromis* specimens examined by the I.B.P. team.

This absence of polychromatism in the Edward–George flock is rather surprising, not only in view of the flock's obvious relationship to that of Victoria, but because polychromatism is found in species of *Haplochromis* from Lake Kivu (*see* Poll 1939b). (The Kivu *Haplochromis* are related to those of Lake George in probably much the same way as are those of Lake Victoria.)   No explanation is immediately apparent.

Species living in such poorly lit waters as those of Lake George might be expected to show certain compensatory hyperdevelopment of various sensory organs, especially the eyes and acustico-lateralis systems. Only the gross morphology of these organs has been investigated so far, and the conclusions reached are equivocal. The cephalic lateral line canals and their openings are not noticeably enlarged. In general, the eye (as measured by its diameter relative to head length) of Lake George species does not seem to be greatly enlarged. A comparison was made between the mean eye diameter in Lake George *Haplochromis* and their ecological counterpart species from Lake Victoria (the comparisons confined to individuals of the same size, and from species with the same maximum adult size). In the Lake Victoria species examined, the range of mean eye diameter is from 27·0–31·0 per cent of head length, whereas in the Lake George species it is from 28·0–38·0 per cent (the modal range being 31–35 per cent). *Haplochromis squamipinnis* is excluded from these figures because individuals attain a larger adult size, and eye diameter proportions generally show a strong negative allometry with body length. When a comparison is made between *H. squamipinnis* and similar-sized individuals of Lake Victoria piscivores, no noticeable difference was noted in eye size (mean eye diameter for *H. squamipinnis* 23·0 per cent of head, *cf.* 20·0–26·0 per cent for the Victoria species).

These comparisons were extended to include species from the deeper waters of Lake Victoria (*see* Greenwood & Gee 1969), where light values are probably similar to those found below the upper 30 cm of water in Lake George. The deep-water Victoria fishes have a range of eye diameter between 25·0 and 36·0 per cent of head length, with a modal value at about 31·0 per cent. This compares with values of 27·0–38·0 per cent (modal range 31·0–35·0 per cent) for the Lake George species, a suggestive similarity.

A comparison of the cephalic lateral line canals in these two species-groups was most inconclusive, mainly because of the difficulty in quantifying the characters involved.

The relationships and history of the Lake George *Haplochromis* species

From both zoogeographical and historical standpoints, the fishes of Lake George should be considered in conjunction with those of Lake Edward. The two lakes are now interconnected by the Kazinga Channel, they share many otherwise endemic cichlid species, and there is no evidence to suggest that Lake George has ever been in direct connection with any other major water body.

Regrettably, it was neither possible to effect a full revision of the Lake Edward *Haplochromis* species, nor was it feasible to collect in parts of the lake never previously sampled. There is no doubt that many species remain to be discovered in the deeper (i.e. western) parts of the lake (as, for example, was the case in the deep waters of Lake Victoria). The few recent collections made in Lake Edward (by Dr Dunn of the I.B.P. team), coupled with a brief re-examination of existing collections, show that there are definitely several undescribed species from inshore habitats.

Despite the drawback of having to exclude Lake Edward in detail, the material examined, together with that from Lake George enables one to reconsider currently held views on the origin of the Lake Edward–George *Haplochromis* species flock. Such reflection is very necessary, both in view of the more detailed geological and palaeontological knowledge now available (Greenwood 1959c; Doornkamp & Temple 1966; Bishop 1969) and because of the rather different conclusions I have reached on the interrelationships of the Lake George and Lake Victoria *Haplochromis* species (themselves extensively revised since Trewavas' [1933] pioneer work on the Edward–George species).

That the *Haplochromis* species flocks of Lakes Victoria and Edward–George have a close phyletic as well as a phenetic relationship is beyond doubt. What has still to be determined is whether the Edward–George flock was derived directly from part of the Victoria species assemblage, or whether the two flocks evolved independently, but in parallel, from common ancestral species.

Trewavas (1933) believed that Lake Edward '...received its Cichlidae, or their not very remote ancestors, from Lake Victoria,...'. This concept has been basic to thinking on the subject ever since (Brooks 1951; Greenwood 1959c; Temple 1969). Trewavas' views were influenced mainly by the overall similarity of the *Haplochromis* species in the two lakes, and by the fact that three otherwise endemic Victoria species were thought to be present in both lakes (*see below*). At the time of Trewavas' paper there was little geological evidence available to suggest either the nature or the duration of the route through which the faunal exchange might have taken place. The Rivers Katonga and Ruizi (now with a drainage via swamp divides into both the Victoria and Edward–George basins) suggested a possible passage way, particularly if, in earlier times, the swampy areas were readily passable. Later, Wayland's (1934) geological and palaeoclimatic hypotheses seemed to support the idea of an aquatic connection between the lake basins (Greenwood 1951, 1959c).

The ichthyological evidence once used in support of a Victoria–Edward (and George) interconnection will be reviewed first.

On Trewavas' reckoning there were six cichlid species shared between the lakes, viz. *Hemihaplochromis multicolor*, *Astatoreochromis alluaudi*, *Haplochromis nubilus*, *H. guiarti*, *H. macrops* and *H. ishmaeli*. Furthermore, every endemic *Haplochromis* species from Lake Edward–George was, in her opinion, closely related to a species from Lake Victoria (the endemic Edwardian monotypic genus *Schubotzia eduardianus* providing the only clear-cut exception [but *see above* p. 215]).

As noted earlier, the idea of a close overall relationship between the Victoria and Edward–George *Haplochromis* is still valid (and in many instances is reinforced by new information). I would find it difficult, however, to establish a direct phyletic relationship of an ancestor-descendant kind between each Edward–George species and its Victoria counterpart (the supposed *H. guiarti* of Lake Edward and *H. mylodon* excepted).

It is my opinion that *H. ishmaeli*, *H. guiarti* and *H. macrops* are not present in Lake Edward or Lake George. The fishes once identified as *H. ishmaeli* are now placed in a new taxon (*H. mylodon*, see p. 172) and the specimens thought to be *H. macrops* do not conform with the revised definition of that species (Greenwood 1960), nor are they conspecific with any other endemic Victoria species (*see below*). The status of the supposed *H. guiarti* from Lake Edward is difficult to determine without a full revision of the Lake Edward *Haplochromis*; no similar species occurs in Lake George. For the moment I can only say that *H. guiarti* might be the sole example of an otherwise endemic Victoria species occurring in Lake Edward. The importance of determining the identity of Edward '*H. guiarti*' needs no further emphasis.

Two specimens identified by Boulenger (1914) as *H. macrops* were kindly lent to me by the Berlin Museum. A detailed morphometric and morphological study shows that both specimens differ from *H. macrops* (*see* Greenwood 1960) in dental and certain proportional characters. One specimen (a female 70 mm standard length) can be identified as a specimen of *H. nigripinnis*. The other (64 mm standard length, probably a female) is of *H. macropsoides* (*see above* p. 162). A third specimen (in the British Museum [Natural History], reg. no. 1933.2.23 : 397), identified by Trewavas (1933), has outer jaw teeth with markedly oblique major cusps, quite unlike the acute cusps of *H. macrops* (*see* Greenwood, *op. cit.*). This specimen also differs from *H. macrops* in several morphometric characters. In all these divergent characteristics, and especially in its dentition, the B.M. (N.H.) fish agrees closely with the type (and some paratypes) of *H. vicarius* Trewavas, a Lake Edward endemic (*see* Appendix I, p. 238, for a discussion on the status of this species).

Thus, all three Lake Edward fishes formerly identified as *H. macrops* are now referred to endemic Edward–George species.

The identity of Edward–George specimens previously identified as *H. ishmaeli* is discussed on p. 176. All the specimens are now included in a new and endemic species from Lakes Edward and George, *H. mylodon*. Anatomically, *H. mylodon* is very like *H. ishmaeli* and *H. pharyngomylus* of Lake Victoria. The main interspecific difference lies in the coloration of the adult males. In this respect, *H. mylodon* bears the same relationship to its Victoria counterparts as do certain endemic *Haplochromis* species of Lake Nabugabo to their counterparts in Lake Victoria (*see* Greenwood 1965b). It could, therefore, be argued that *H. mylodon* represents an instance of direct speciation from an *H. ishmaeli* or *H. pharyngomylus*-like ancestor that invaded the Edward basin at some time past.

Material collected by Worthington from Lake Edward and subsequently identified by Trewavas as *H. guiarti* is polyspecific. In fact, only a small part of it can be confused with *H. guiarti* as currently defined (*see* Greenwood 1962). Of the remaining specimens, one resembles *H. squamulatus* of Lake Victoria, and the others show characters of the *H. victorianus*–*H. serranus* species complex in that lake. It must be stressed that none of these specimens is referable to its Lake Victoria counterpart. Preliminary work suggests that in Lake Edward there are, in addition to an *H. guiarti*-like species, two other piscivorous species endemic to Lake Edward.

On the basis of preserved material alone, it is difficult to separate the *H. guiarti*-like specimens from the true *H. guiarti*. When specimens are placed side by side, the Lake Edward fishes are distinguishable on the basis of their total morphology, especially head shape. The situation here is quite comparable with that existing between *H. mylodon* and *H. ishmaeli* (or *H. pharyngomylus*) but without the benefit of information on live male coloration.

Turning for the moment to the cichlid species which are definitely shared by the lakes. *Hemihaplochromis multicolor* has such a wide distribution in eastern Africa (including the Nile) that it is irrelevant to this discussion. Its absence from Lakes Edward and George would be of greater significance than its presence.

Haplochromis nubilus has a somewhat more restricted range and can definitely be categorized as a species of the Victoria drainage basin. *Astatoreochromis alluaudi* can also be categorized in this way. Both species, unlike other Victorian *Haplochromis* and related genera, are common in streams and rivers entering the lake, and both penetrate for some distance into papyrus swamps.

Taken in its entirety, the ichthyological evidence does not really seem to provide a strong argument in favour of a strictly Victorian derivation for the Edward-George cichlid species. In particular it does not support the idea of derivation from a developed, or partly developed *Haplochromis* species flock, an idea that I had previously espoused (Greenwood 1959c ; also Temple 1969).

The degree of anatomical differentiation between most known Edward-George species and their morphological counterparts in Lake Victoria is sufficiently well marked to suggest that one is observing the results of parallel evolution and not direct speciation in Edward-George from an already specialized invader species. Since both *Astatoreochromis alluaudi* and *Haplochromis nubilus* are relatively eurytopic, their presence in both lake basins could mean that they were components of the cichlid complex inhabiting the area prior to lake formation. Possibly, but less likely on ecological grounds, the two species could have gained access to Lake Edward-George via the Katonga-Mpanga River system.

The distribution of the extant non-cichlid fishes in the area contributes little of value to this discussion (*see* Greenwood 1959c). Only the occurrence of *Barbus altianalis* in both Victoria and Edward-George argues strongly for some past connection between the basins (as it does for a connection with Lake Kivu ; *see below*). Otherwise, the non-cichlid fishes of these lakes have little in common ; the number of endemic Victoria fishes contrasts with the depauperate but clearly Nilo-Albertine nature of the Edward-George non-cichlid species assemblage (Greenwood, *op. cit.*).

At this point brief mention should be made of Lake Kivu and its small *Haplochromis* flock. Historically, Lake Kivu was derived from a river that once flowed northwards into what is now the Edward-George basin. This river was dammed by the formation of the Bufumbiro volcanic chain, probably during the late Pleistocene. As the embryo Kivu gradually filled, it found a new outlet, now the Ruzizi, which drained into Lake Tanganyika. Rapids in the Ruzizi seem to block the passage of fishes (at least northwards) between Lake Kivu and Lake Tanganyika, although certain non-cichlids (e.g. *Barilius moorii* and *Barbus pellegrini*) are found in both lakes, perhaps as relicts of an earlier, unimpeded river connection.

The *Haplochromis* of Lake Kivu definitely show no relationships with those of Lake Tanganyika, but are distinctly of the Victoria–Edward type. There has been no recent revision of the Kivu *Haplochromis* species, and data on their live coloration are unavailable; furthermore, an examination of the type series of two species (personal observations) strongly hints of more species than are currently recognized (Poll 1939a, b).

Comparing the Edward–George *Haplochromis* with the Kivu species, on a purely morphological basis, suggests that the Kivu fishes are quite distinct, although showing affinity with Lake Edward–George species (or, in one case, a Lake Victoria species).

Of the Kivu species I have studied in detail, *H. astatodon* Regan resembles *H. serridens* of Lake Edward, *H. graueri* Blgr. (at least, that is, one of the types) resembles fairly closely *H. schubotzi*, and *H. paucidens* Regan has the general orodental specializations of *H. labiatus* but in many features is more like members of the *H. crassilabris* species complex in Lake Victoria (*see* Greenwood 1965b). *Haplochromis vittatus* (Blgr.), too, shows most phenetic affinity with a Victoria species group (especially *H. gowersi*, a member of the '*prognathus*' group in that lake; *see* Greenwood 1967); it does not closely resemble *H. squamipinnis* of Lakes Edward and George.

The remaining Kivu *Haplochromis* species (and those still undescribed) I feel less able to comment upon. *Haplochromis wittei* Poll and *H. schoutedeni* Poll could be related to either *H. elegans* or *H. aeneocolor* of Lake Edward–George, especially the former species, while *Haplochromis adolphifrederici* (Blgr.), if it is distinct from *H. graueri*, has superficial resemblances to *H. schubotzi* and *H. schubotziellus* of Edward and George.

As noted earlier (p. 230) sex-limited female polychromatism occurs in at least two Kivu species (*H. wittei* and *H. adolphifrederici*) but has not been recorded from any of the Edward–George species.

A detailed revisionary study of Lake Kivu *Haplochromis* species may throw more light on their phylogeny. This would be of great interest because the ancestors of these fishes could have been derived from the proto-George–Edward flock (before the Bufumbiro dam was formed) or could have evolved after that time, from ancestors living in the river before it was dammed. Since this river originated in the Ruanda Highlands it might well have been populated by different species from those in the westward flowing rivers of the Kenya Highlands which populated the embryo Lakes Victoria and Edward–George.

Modern geological studies on the Pleistocene sequence in Uganda also seem to support the idea of parallel evolution in the cichlid species flocks of the Victoria and Edward basins. (*See* summaries in Doornkamp & Temple 1966; Bishop 1969). Older ideas and temporal sequences based on Wayland's pluvial hypothesis (1934) are no longer tenable.

The formation of Lake Victoria is currently dated at about the later mid-Pleistocene, and is thought to be consequent upon the reversal and ponding-back of rivers that flowed across its present basin into the western Rift lake system (i.e. into a proto-Lake Edward–George and Albert). For a summary of the evidence, *see* particularly Doornkamp & Temple (1966).

River reversal was initiated by local uplift along a line nearer the western Rift than the developing Victoria basin. As a result of this uplift the formerly westward-flowing rivers drained both to the east and to the west, an anomalous situation still persisting. Extensive swamps developed over the watershed, and today these provide an effective barrier to fish dispersal along the rivers.

If one accepts the geological evidence, then one must conclude that a lake existed in the western Rift some time before Lake Victoria started to develop as a series of small lakes in the eastern sections of the reversed rivers. There is good palaeontological evidence for the existence of the western Rift lake or lakes from at least Kaiso Formation times (earlier Pleistocene) onwards (Greenwood 1959c). Essentially, this fossil record is one of non-cichlid fishes so it throws little direct light on the question of *Haplochromis* relationships.

Judging from the reconstructed topography of western Uganda in the earlier Pleistocene (Doornkamp & Temple, *op. cit.*) there was a steep escarpment bordering the eastern shoreline of proto-Lake Edward–George. It seems unlikely, therefore, that *Haplochromis* species could enter this western lake after the formation of Victoria. Furthermore, if the species that evolved in the developing Lake Victoria were as stenotopic (i.e. lacustrine) as are their present derivatives, it is highly improbable that they would spread along the inter-lake rivers (even assuming that such a passage was physically possible).

Thus the conclusion seems inevitable that, for all of their histories as lakes, Victoria and Edward–George have been effectively isolated from each other, and that Lake Edward–George is older than Lake Victoria. Since both basins were filled from the same river systems (the old east–west drainage) it is reasonable to assume that their initial fish colonizers were the same. In other words, their present-day *Haplochromis* species flocks were derived from common ancestral species, presumably of the generalized type now represented by *H. bloyeti* (*see* Greenwood 1971).

One Lake George species, *H. petronius* (*see* p. 209), does not fit this picture of a close phyletic relationship between the flocks of Lakes Victoria and Edward–George. Nor does it seem to be related to the *H. bloyeti* stock. As discussed in greater detail above (p. 213), *H. petronius* shows marked affinities with *H. wingatii*, a species known from the Nile and Lake Albert (Greenwood 1971). The characters relating these two species (and also *H. pharyngalis* of Lake Edward ; *see* p. 214) are not present in any Lake Victoria *Haplochromis* species. It is unlikely, too, that these characters are products of convergent evolution.

To me, the implication is that *H. petronius* was derived from a different lineage than that of the other species. It cannot, of course, be told if that lineage occurred in the Victoria basin but failed to survive there. Certainly there is no indication of *H. wingatii*-like species in any of the streams and rivers flowing into Lake Victoria today.

That related species appear to have persisted in Lake Albert and the Nile, and also in Lake George, suggests that the ancestor of *H. petronius* entered that lake from a source other than the old westward draining rivers. The nearest living relative of *H. wingatii* is probably *H. desfontainesi*, a species now restricted to North Africa (*see* Greenwood 1971). Perhaps the ancestor of *H. petronius* (and *H. pharyngalis*) was a northern rather than an east–west river species, that gained access to Lake Edward–George from the Nile before the lake was isolated from that river by the Semliki rapids (*see* Greenwood 1959c).

The distribution of *Haplochromis nubilus* and *Astatoreochromis alluaudi* in Lakes Victoria and Edward–George, as well as in the small lakes lying between these basins (Trewavas 1933) suggests that the species are remnants of the original species complex inhabiting the old east–west river systems. Trewavas (*op. cit.*) interpreted the presence of *H. nubilus* and *A. alluaudi* in Lakes Nakavali, Kachira and Kijanebalola as possible evidence of the route through which the postulated Victoria to Edward faunal exchange took place. It now seems more likely that the species are fluviatile relicts in those lakes. The absence of other and more typically Lake Victoria or Lake Edward species from these small lakes puzzled Trewavas (*op. cit.*, p. 311). Probably the explanation is simply that these species or their immediate ancestors were never in that area.

Elsewhere I have argued (Greenwood 1965c) that the *Haplochromis* species flock in Lake Victoria represents the amalgamation of several smaller flocks, each evolved in isolation from a common ancestor or, later in the lake's history, a few common ancestral species. The isolation I envisaged was essentially one of small lakes lying within the area of what is now the basin of a single large lake. The present fauna of Lake Edward–George could be looked upon as another of these isolates but one which, because its basin retained its physical identity, has been given the status of a separate species flock. Phyletically speaking it is perhaps wrong to do so. Rather, one should refer to it as the Edward–George subflock.

As matters stand, there is insufficient knowledge of the physical and ecological factors involved in the processes of speciation and adaptive radiation within the Edward–George subflock. Lake George has now been sufficiently well sampled for one to be almost certain that some species occurring in Lake Edward are absent from Lake George. Likewise it is clear that there are many more species in Lake Edward than are currently recorded. (Personal observations on recent collections from Lake Edward.) Collections from the Kazinga Channel show that its cichlid fauna is virtually identical with that of Lake George. That is, the Edward species not recorded from George are also absent from the Channel (*see* Appendix II). It seems, therefore, that the channel is at least partially a differential species filter between the lakes. The factors inhibiting occupation by certain species (and these do include some from Lake George) have not been discovered. This question is yet another whose solution will depend upon learning more about the ecology of the fishes, especially those from Lake Edward.

The unusually complete fossil record for the fishes of Lake Edward shows that throughout the Pleistocene, and well into the Holocene, the non-cichlid fishes were more diverse than at present (Greenwood 1959c). The genera *Lates* and *Synodontis*, now absent, were present until local Mesolithic times, and another present-day absentee, *Polypterus*, persisted into the early Holocene (de Heinzelin's level N.F.P.R. at Ishango is now dated at *ca* 8000–10 000 years B.P.).

Depauperization of this Nilo-Albertine fauna was sudden and of a relatively recent date (*see* Greenwood, *op. cit.*, p. 73). Localized vulcanicity polluting the water (especially of inflowing streams) may have been a major factor in this process. The differential adaptability of species to these altered conditions could account for the fact that some survived while others were wiped out.

If the arguments presented above on the origin of the Lake Edward–George cichlids are sound, then these fishes must have survived the environmental hiatus that exterminated several non-cichlid species. There is no evidence that the cichlids or their ancestors reinvaded the lake after the volcanic period, although the time elapsed could have been sufficient for the flock to evolve (*see* Greenwood 1965b).

Assuming that the Edward–George *Haplochromis* evolved from mid-Pleistocene fluviatile colonizers implies that speciation and adaptive radiation took place in the presence of such predators as *Lates* and *Hydrocynus*. Worthington's ideas on the inhibitory effects of *Lates* and *Hydrocynus* on these processes are well known and well argued over (*see* Fryer & Iles, 1972, for a comprehensive summary of various viewpoints in this discussion). The history of the Lake Edward–George *Haplochromis* species flock now seems to provide an even stronger counter-argument to the Worthington hypothesis than the one presented in my 1959c paper. There, I had assumed that the flock was derived from an at least partly differentiated one (at the species and adaptational levels) invading from Lake Victoria.

ACKNOWLEDGEMENTS

This paper stemmed from the need to provide the Royal Society–International Biological Programme team investigating Lake George, with a means of identifying the *Haplochromis* species found there. Thus, I am particularly indebted and grateful to the Royal Society *ad hoc* Committee for I.B.P./Lake George who so generously approved the expenditure that enabled me to visit Lake George on several occasions. To the team members themselves I am equally indebted. Every member has at some time or other supplied me with data, helped make collections and given me access to their unpublished research; at all times they have been the most magnificent hosts. Even at the risk of seeming invidious, I must thank especially Dr Ian Dunn and Mr James Gwahaba, the two workers principally concerned with fish research, whose unstinted help has proved invaluable; and also Dr Dunn, in his capacity as team field leader, for much help. Later Dr George Ganf, as team leader, did much to help me both in the field and with local organization.

In the Museum, I owe a great many thanks to my assistant Mr Gordon Howes who has undertaken many jobs to my gain (not least the excellent radiography he has produced), and to Mrs Sharon Chambers for her excellent drawings.

To Dr K. Deckert and Dr C. Karrer of the Zoologisches Museum der Humboldt-Universität, Berlin, I am much indebted for lending me several type specimens. Likewise I am indebted to Dr Max Poll of the Musée Royal de l'Afrique Centrale, Tervuren. To them collectively and individually go my sincere thanks. My thanks go also to Mrs M. M. Dick of the Museum of Comparative Zoology, Harvard University, for the gracious help she has given whenever I have made use of the M.C.Z.'s collections.

Finally, it is a great pleasure to thank Professor W. Banage, Zoology Department of Makerere University, Kampala, who, with his staff, has provided me with hospitality and facilities whenever I visited the University.

APPENDIX I

The status of *Haplochromis vicarius* Trewavas, 1933

Poll (1939), synonymized *H. vicarius* with *H. eduardii* Regan, 1921 on the grounds that the large collection of specimens available to him bridged the morphological gap distinguishing the species. However, Poll seems only to have considered as specifically trenchant the posterior extent of the maxilla, and does not mention the dental characteristics of either taxon. My experience with various *Haplochromis* leads me to place little importance on the maxillary character, but considerable value on the form of the teeth.

The question raised by Poll's proposed synonymy is complicated by the fact that the type series for *H. vicarius* is very probably polyspecific. One specimen is labelled 'Holotype' although no holotype was formally designated (Trewavas 1933). This fish and at least two paratypes have a distinctive cusp shape to the outer teeth of both jaws (*see* p. 171), a cusp type that does not occur in the teeth of *H. eduardii* holotype. Furthermore, I can find no reasons to believe that the tooth shape in *H. vicarius* holotype represents an extreme variant of the *H. eduardii* tooth-type (or vice versa).

Thus, I would suggest that *H. vicarius* is specifically distinct from *H. eduardii*. When the Lake Edward *Haplochromis* species are fully revised I suspect that other characters will be found to support this separation.

APPENDIX II

Kazinga Channel fishes

During May and June, 1972, collections were made at several places in the Kazinga Channel, particularly in the neighbourhood of Katungura (approximately the midpoint of the channel). Other regions sampled were near the Lake George end of the channel and at Mweya, near the opening into Lake Edward. Small-mesh gill nets and a purse seine were used, and sites near the shoreline and in midchannel were sampled.

A list of the species collected in the area around Katungura, with notes on the region of the channel in which they most frequently occur, and a subjective evaluation of their abundance, is given below.

Haplochromis elegans: common inshore, especially near reed beds; also caught offshore, but is less abundant there.

H. aeneocolor : inshore near reeds ; not very abundant.
H. nigripinnis : only in midchannel ; rare.
H. oregosoma : inshore ; rare.
H. macropsoides : inshore and midchannel ; rare.
H. mylodon : inshore ; very rare.
H. angustifrons : mostly from midchannel where it is fairly abundant ; occurs inshore but is rare.
H. schubotzi : inshore ; rare.
H. schubotziellus : rare in midchannel, even rarer inshore.
H. taurinus : midchannel only and then infrequently.
H. pappenheimi : abundant everywhere, particularly inshore. Unlike catches of this species in Lake George, those from the channel contained large (110-130 mm standard length) and sexually active individuals of both sexes.
H. squamipinnis : ubiquitous, but in small numbers.
H. eduardianus : infrequently caught, and then only by dip-netting among the reeds.

Collections made near the Mweya landing were hampered by technical difficulties ; only gill nets, set inshore and in midchannel, were used and then on but one occasion. These yielded specimens of *H. elegans*, *H. aeneocolor*, *H. angustifrons*, *H. taurinus*, *H. pappenheimi* and *H. squamipinnis*, all in small numbers.

Because of inadequacies in the sampling methods used at Mweya landing, and since only one collection was made there, this list must be incomplete.

It is surprising that the well-sampled Katungura area did not produce any specimens of *Haplochromis limax*, *H. nubilus* or *Astatoreochromis alluaudi*. All three species were found in similar habitats in Lake George. There is, of course, a noticeable water flow in the channel, but this alone could hardly be the cause of these particular species' absence. More probably, their ' absence ' is a reflection of the sampling methods used (and the time available for sampling).

BIBLIOGRAPHY

BISHOP, W. W. 1969. Pleistocene stratigraphy in Uganda. *Mem. geol. Surv. Uganda*, no. 10.
BOULENGER, G. A. 1914. Family Cichlidae. In *Wiss. Ergebn. Deuts. Zentral-Afrika Exped., 1907-1908, Zool.* **3** : 253-259.
—— 1915. *Catalogue of the Freshwater Fishes of Africa*, **3**. London.
BROOKS, J. L. 1950. Speciation in ancient lakes. *Q. Rev. Biol.* **25** : 30-60 & 131-176.
DUNN, I. G., BURGIS, M. J., GANF, G. G., McGOWAN, L. M. & VINER, A. B. 1969. Lake George, Uganda : a limnological survey. *Verh. int. Verein. theor. angew. Limnol.* **17** : 284-288.
DOORNKAMP, J. C. & TEMPLE, P. H. 1966. Surface, drainage and tectonic instability in part of southern Uganda. *Geogrl. J.* **132** : 238-252.
FRYER, G. & ILES, T. D. 1972. *The Cichlid Fishes of the Great Lakes of Africa. Their Biology and Evolution.* Oliver and Boyd. Edinburgh.
GREENWOOD, P. H. 1951. Evolution of the African cichlid fishes : the *Haplochromis* species-flock in Lake Victoria. *Nature, Lond.* **167** : 19-20.
—— 1954. On two species of cichlid fishes from the Malagarazi River (Tanganyika), with notes on the pharyngeal apophysis in species of the *Haplochromis* group. *Ann. Mag. nat. Hist.* (12), **7** : 401-414.

GREENWOOD, P. H. 1956a. The monotypic genera of cichlid fishes in Lake Victoria. *Bull. Br. Mus. nat. Hist.* (Zool.), **3** : 295-333.
—— 1956b. A revision of the Lake Victoria *Haplochromis* species (Pisces, Cichlidae), Part I. *Bull. Br. Mus. nat. Hist.* (Zool.), **4** : 223-244.
—— 1957. A revision of the Lake Victoria *Haplochromis* species (Pisces, Cichlidae), Part II. *Bull. Br. Mus. nat. Hist.* (Zool.), **5** : 73-97.
—— 1959a. The monotypic genera of cichlid fishes in Lake Victoria, Part II. *Bull. Br. Mus. nat. Hist.* (Zool.), **5** : 163-177.
—— 1959b. A revision of the Lake Victoria *Haplochromis* species (Pisces, Cichlidae), Part III. *Bull. Br. Mus. nat. Hist.* (Zool.), **5** : 179-218.
—— 1959c. Quarternary fish fossils. *Explor. Parc. natn. Albert Miss. J. de Heinzelin de Braucourt*, **4** : 1-80.
—— 1960. A revision of the Lake Victoria *Haplochromis* species (Pisces, Cichlidae), Part IV. *Bull. Br. Mus. nat. Hist.* (Zool.), **6** : 227-281.
—— 1962. A revision of the Lake Victoria *Haplochromis* species (Pisces, Cichlidae), Part V. *Bull. Br. Mus. nat. Hist.* (Zool.), **9** : 139-214.
—— 1965a. Environmental effects on the pharyngeal mill of a cichlid fish, *Astatoreochromis alluaudi*, and their taxonomic implications. *Proc. Linn. Soc. Lond.* **176** : 1-10.
—— 1965b. On the cichlid fishes of Lake Nabugabo, Uganda. *Bull. Br. Mus. nat. Hist.* (Zool.), **12** : 313-357.
—— 1965c. Explosive speciation in African lakes. *Proc. R. Instn Gt Br.* **40** : 256-269.
—— 1967. A revision of the Lake Victoria *Haplochromis* species (Pisces, Cichlidae), Part VI. *Bull. Br. Mus. nat. Hist.* (Zool.), **15** : 29-119.
—— 1971. On the cichlid fish *Haplochromis wingatii* (Blgr.), and a new species from the Nile and Lake Albert. *Revue Zool. Bot. afr.* **84**, 3-4 : 344-365.
—— & GEE, J. M. 1969. A revision of the Lake Victoria *Haplochromis* species (Pisces, Cichlidae), Part VII. *Bull. Br. Mus. nat. Hist.* (Zool.), **18** : 1-65.
MARLIER, G. & LELEUP, N. 1954. A curious ecological ' niche ' among the fishes of Lake Tanganyika. *Nature, Lond.* **174** : 935-936.
POLL, M. 1939a. Poissons. *Explor. Parc. natn. Albert Miss. H. Damas* (1935-1936), **6** : 1-73.
—— 1939b. Poissons. *Explor. Parc. natn. Albert Miss. G. F. de Witte* (1933-1935), **24** : 1-81.
REGAN, C. T. 1921. The cichlid fishes of Lakes Albert Edward and Kivu. *Ann. Mag. nat. Hist.* (9), **8** : 632-639.
ROSEN, D. E. & BAILEY, R. M. 1963. The poeciliid fishes (Cyprinodontiformes), their structure, zoogeography, and systematics. *Bull. Am. Mus. nat. Hist.* **126** : 1-176.
TEMPLE, P. H. 1969. Some biological implications of a revised geological history for Lake Victoria. *Biol. J. Linn. Soc.* **1** (4) : 363-373.
THYS VAN DEN AUDENAERDE, D. F. E. 1963. Description d'une espèce nouvelle d'*Haplochromis* (Pisces, Cichlidae) avec observations sur les *Haplochromis* rhéophiles du Congo oriental. *Revue Zool. Bot. afr.* **68**, 1-2 : 140-152.
TREWAVAS, E. 1933. Scientific results of the Cambridge expedition to the East African lakes, 1930-1. 11. The cichlid fishes. *J. Linn. Soc.* (*Zool.*), **38** : 309-341.
WAYLAND, E. J. 1934. Rifts, rivers, rains and early man in Uganda. *Jl R. anthrop. Inst.* **64** : 333-352.
WORTHINGTON, E. B. 1932. Scientific results of the Cambridge expedition to the East African lakes, 1930-1. 1. General introduction and station list. *J. Linn. Soc.* (*Zool.*), **38** : 99-119.
WICKLER, W. 1962. Zur Stammesgeschichte functionell korrelierter Organ- und Verhaltensmerkmale : Ei-Attrappen und Maulbrüten bei afrikanischen Cichliden. *Z. Tierpsychol.* **19** : 129-164.
—— 1963. Zur Klassification der Cichlidae, am Beispiel der Gattungen *Tropheus, Petrochromis, Haplochromis* und *Hemihaplochromis* n. gen. (Pisces, Perciformes). *Senckenberg. biol.* **44** : 83-96.

A GUIDE TO THE IDENTIFICATION OF THE *HAPLOCHROMIS* SPECIES FROM LAKE GEORGE

A simple dichotomous key cannot be compiled for these fishes. Intraspecific variability is high and few species can be diagnosed on the basis of single characters. Thus, this ' key ' should be used as a general guide rather than as a means of identifying a taxon without recourse to other discriminating characters included in the full species descriptions. It is based on adult and subadult specimens.

Haplochromis labiatus is not included here because only one Lake George specimen is known. A specimen of *H. labiatus* would probably key out to *H. elegans*, but its dental characters (see p. 198) should prove diagnostic.

Morphometric characters are defined on p. 145. Unless otherwise specified, ' teeth ' refers to the outer row of teeth in both jaws.

The Lake George species[1]

Teeth spatulate, the upper half of each tooth strongly incurved (see text-fig. 37); lower jaw shorter than the upper . . . . **H. eduardianus**

Teeth bicuspid or unicuspid and caniniform, sometimes a mixture of both, and occasionally with some tricuspids intercalated; lower jaw not shorter than upper . *A*

A Scales on the chest very small and deeply embedded (difficult to detect), especially when compared with those on the belly; a small scaleless area immediately anterior to the first dorsal spine . . . . **H. petronius**

Scales on chest not deeply embedded, not disproportionately smaller than those on the belly, and readily visible; no naked area at base of the first dorsal spine . *B*

B A vertical row of small scales extending onto the fin membrane along the basal part of many (if not all) dorsal and anal fin rays and spines. Lower jaw long (42–57, mode *ca* 50 per cent of head length, showing positive allometry) and oblique; teeth usually unicuspid. Adults reach a large size (> 150 mm) . **H. squamipinnis**

No small scales extending onto the dorsal and anal fins. Adults rarely more than 115 mm long, modally *ca* 80 mm . . . . *C*

C Lower pharyngeal bone massive (see text-fig. 16), most of its teeth strong and molariform . . . . **H. mylodon**

Lower pharyngeal bone not massive, if molariform pharyngeal teeth present, few in number, small, confined to middle row . . . . *D*

D Teeth few in number (32–48, mean 36 in upper jaw), stout, deeply embedded in jaw tissue (difficult to see) and, although bicuspid, of characteristic shape (see text-fig. 25); gape of mouth manifestly large, lower jaw 43–56, mean 47 per cent of head; dorsal profile of head concave. Adults reach a standard length of 140 mm
H. taurinus

Teeth otherwise than above; mean lower jaw length less than 45 per cent of head usually less than 40 per cent . . . . *E*

E Depth of body less than 35 per cent of standard length (mean = 31 per cent); modal number of gill rakers 10 or 11 (but as many as 13), the rakers slender . . *F*

Depth of body usually more than 35 per cent of standard length; modal number of gill rakers less than 10 (usually 8 or 9) . . . . *G*

F At least the posterior third of the premaxilla without teeth; teeth small, flat and of a characteristic shape (see text-fig. 30), 28–38 (mean = 32) in upper jaw. Body colour uniformly silver in both sexes. Body fusiform, its depth 27–31 (mean = 30) per cent of standard length . . . . **H. pappenheimi**

[1] The two other *Haplochromis*-group species are identified as follows:
More than 3 (usually 4 or 5) anal spines: *Astatoreochromis alluaudi*.
Many scales of the lateral line series without pores: *Hemihaplochromis multicolor*.

Entire length of premaxilla toothed; teeth relatively slender (see text-fig. 9), 42–60 (mean = 50) in upper jaw. Males dark, females greyish-silver. Body depth 30–34 (mean = 32) per cent of standard length . . . . **H. oregosoma**

G Teeth in outer row of both jaws with an obliquely truncate cusp (see text-fig. 14), long and movably implanted; 4 or 5 (rarely 3) rows of inner teeth in the upper jaw . . . . **H. limax**

Teeth otherwise than above, and only 2 or 3 inner rows (often only 1 row) . . H

H Usually less than 40 teeth in the upper jaw (34–42, mean = 38); teeth bicuspid, most without a well-developed flange on the major cusp (see text-fig. 3, and cf. text-fig. 5). Upper jaw 28–34 (mean = 30) per cent of head (i.e. equal to or less than the eye diameter) . . . . **H. elegans**

More than 40 teeth in the upper jaw (40–60, mean = 50) I

I Distinct and prominent midlateral dark band running from behind operculum onto the caudal fin; snout length 31–40 (mean = 33) per cent of head **H. schubotziellus**

No distinct midlateral band (or, if a series of short midlateral streaks present, the last not extending onto caudal fin); snout length usually less than 30 per cent of head length . . . . J

J When fish is viewed laterally, the upper margin of the orbit is seen to be continuous with the dorsal profile, or the eye appears to extend above this line . . . K

The upper margin of the orbit lies below the dorsal profile of the head . . . L

K Dorsal head profile sloping smoothly (not obviously interrupted by prominent premaxillary pedicels). Preorbital depth 12–15 (mean = 14) per cent of head. Outline of toothed area on lower pharyngeal bone broader than long (see text-fig. 12). Caudal fin not distinctly maculate . . . . **H. macropsoides**

Slope of dorsal head profile interrupted by the prominent premaxillary pedicels. Preorbital depth 13–19 (mean = 17) per cent of head. Outline of toothed area on lower pharyngeal bone noticeably longer than broad (bone appears long and narrow, see text-fig. 19). Caudal fin very distinctly maculate . . **H. angustifrons**

L Thickened and papillose area of tissue preceding first gill raker of first gill arch; pseudorakers between inner and outer row of gill rakers especially well developed and prominent. Snout length 31–40 (mean = 33·4) per cent of head . **H. schubotzi**

No manifestly thickened and papillose area preceding first gill raker (or if tissue in that region slightly thickened, definitely not papillose); pseudorakers absent or poorly developed. Snout length usually less than 30 per cent of head . . M

M Caudal fin with an almost rounded distal margin . . . . . **H. nubilus**

Caudal fin with truncate or weakly subtruncate distal margin. Two species, viz.:

(i) Most teeth with a well-developed flange on the major cusp (see text-fig. 5). Upper jaw 30–38 (mean 35) per cent of head. Eye diameter 28–35 (mean = 31·4) per cent of head. Lips slightly thickened. Nostril opening much larger than the anterior opening to the nasal lateral line canal. Intestine ca 1½ times total body length . . . . **H. aeneocolor**

(ii) Few teeth with a flange on the major cusp (see text-fig. 7). Eye diameter 33–40 (mean = 36) per cent of head. Lips not noticeably thickened. Opening to nostril of equal size to that of nasal lateral line canal. Intestine long (ca 2–2½ times total body length) and much coiled . . . . **H. nigripinnis**

P. H. GREENWOOD, D.Sc.
Department of Zoology
BRITISH MUSEUM (NATURAL HISTORY)
CROMWELL ROAD
LONDON SW7 5BD

THE *HAPLOCHROMIS* SPECIES (PISCES : CICHLIDAE) OF LAKE RUDOLF, EAST AFRICA

By P. H. GREENWOOD

CONTENTS

	Page
INTRODUCTION	141
Haplochromis rudolfianus	142
Haplochromis turkanae sp. nov.	150
Haplochromis macconneli sp. nov.	154
DISCUSSION	161
ACKNOWLEDGEMENTS	164
REFERENCES	164

INTRODUCTION

LAKE RUDOLF is outstanding amongst the Rift Valley Great Lakes of Africa for the paucity of its endemic cichlid species, and particularly for the absence of a *Haplochromis* species flock (Trewavas, 1933). Whereas all other large Rift Valley lakes have a well-defined flock of endemic *Haplochromis* species (even if, like Lake Albert, the flock comprises only a few species), Lake Rudolf was thought to possess but a single *Haplochromis* species, the endemic *H. rudolfianus* Trewavas, 1933. The lake's one positive ichthyological peculiarity, the presence of an endemic species belonging to the west African genus *Pelmatochromis*, has now been shown to stem from a misidentification. The Rudolf *Pelmatochromis* is, in fact, a specimen of the widespread taxon *Hemichromis bimaculatus* Gill (see Trewavas, 1973).

Endemicity amongst the non-cichlid species is also at a low level (see Worthington & Ricardo, 1936; personal observations on collections recently made in the lake).

Various ideas have been advanced to explain the absence of a *Haplochromis* species flock in Lake Rudolf (see summaries and comment in Fryer and Iles, 1972). Of these, the most likely would seem to be the relative youth of the present lake fauna, an invasion from the Nile in post-Middle Pleistocene times. Coupled with this factor are the shape and recent history of the lake basin, neither of which would provide opportunities for the isolation (and subsequent speciation) of populations living in the lake. That until now the only *Haplochromis* known to inhabit the lake was of a structurally and ecologically generalized type similar to the fluviatile species of eastern Africa, would seem to agree with such a postulated lake history.

Recently, however, a second and anatomically specialized species has been discovered in the deeper waters of the lake. A few specimens of a third species (similar to *H. rudolfianus*, see p. 150) have also been collected, and there are indications of yet another taxon (see p. 149).

These and other specimens, made available through the efforts of the Lake Rudolf Fishery Research Project, have provided sufficient material to describe the new deep-water taxon, the new *Haplochromis rudolfianus*-like fish, and to redescribe in greater detail *H. rudolfianus* itself. Unfortunately the putative fourth species is represented by so few and distorted specimens that I consider it inadvisable to describe it at present. Nevertheless, sufficient information has been gathered from these few specimens to show that this species too is like *H. rudolfianus*.

As part of a planned review of phyletic relationships within the *Haplochromis*-group cichlids, some consideration is also given to the relationships of the Lake Rudolf species with those of other lakes, especially Lakes Albert and Victoria.

Haplochromis rudolfianus Trewavas, 1933
(Text-figs. 1-4)

Haplochromis rudolfianus Trewavas, 1933, *J. Linn. Soc.* (*Zool.*), **38** : 321-322.

LECTOTYPE. A specimen, 51·0 mm standard length (BMNH reg. no. 1933.2.23 : 163), from a weedy lagoon on the east shore of Lake Rudolf, near Mt El Moitat (station number 285, *see* Worthington, 1932). The specimen has been eviscerated, but judging from its preserved coloration it is probably an adult male.

PARALECTOTYPES. Three specimens (BMNH reg. nos. 1933.2.23 : 164-166), 39·0-42·5 mm S.L., from the same locality as the lectotype (all are eviscerated), and one other specimen (BMNH reg. no. 1933.2.23 : 167), 45·0 mm S.L. from Central Island (Worthington's station no. 264). This latter fish is also eviscerated but is probably an adult male.

Comment on the original description of H. rudolfianus

Trewavas' (1933) original description was based on 5 syntypes, but 25 other specimens were also examined although not included in the description. I have re-examined these fishes and would confirm their identification as *H. rudolfianus*. Dr Trewavas also included in this species, but with certain reservation, a larger specimen, 61·0 mm S.L. and 80 mm total length (BMNH reg. no. 1933.2.23 : 169). The locality label for this fish reads '? Lake Rudolf', the uncertainty stemming from the collector's notes on the provenance of the specimen. A number of differences between this fish and other specimens of *H. rudolfianus* were noted by Trewavas. I can confirm these differences and would add others. The size discrepancy once existing between this specimen and others of *H. rudolfianus* is virtually obliterated by the larger specimens of the latter species now in the Museum's collections. Thus it seems very unlikely that the various differences listed by Trewavas are, as she then suggested, size correlated ones.

The specimen in question departs from *H. rudolfianus* in having a deeper body (37·4 per cent of standard length), somewhat deeper preorbital (17·7 per cent of head length), wider interorbital distance (25·6 per cent of head), a deeper cheek (25·6 per cent of head), a smaller eye diameter (30·0 per cent of head) and a markedly shorter caudal peduncle (14·7 per cent of standard length, *cf.* 16·5-19·7, mean 18·0 per cent in *H. rudolfianus*). The fish also differs from specimens of *H. rudolfianus* in having fewer gill rakers (7 *cf.* 8 or 9), larger scales on the nape and in the lateral line series (30), in lacking the characteristic dark vertical bars on the body and caudal peduncle (*see* p. 147), in having no melanic pigment in the ovarian walls (*see* p. 146) and in having stouter and somewhat obliquely cuspidate outer jaw teeth. The lower pharyngeal bone is finer than in *H. rudolfianus*, and the median lower pharyngeal teeth are also finer (*see* p. 146).

In my opinion, this specimen cannot be identified as a member of *H. rudolfianus*, nor can it be placed in either of the other *Haplochromis* species from Lake Rudolf. All the evidence certainly suggests that it in fact belongs to the fauna of another lake. On this assumption, the specimen was compared with *Haplochromis* species from all the lakes sampled by the Cambridge University Expedition of 1930–31. It cannot be identified with any known Lake Albert *Haplochromis* species, nor does it closely resemble any of the described (or known but undescribed) species of Lakes Edward and George; *see* Greenwood, 1973. It does, however, closely agree in all morphometric and anatomical features (especially the dentition) with *H. velifer* Trewavas of Lake Nabugabo, Uganda (*see* Greenwood, 1965). Since extensive collections were made in this lake by the Cambridge Expedition, I would identify the fish as a specimen of *H. velifer* and suggest that the locality label be altered to read 'Lake Nabugabo'.

The redescription of *H. rudolfianus* given below is based on the 5 type specimens (39·0–51·0 mm S.L.) and 20 additional fishes (30·5–58·0 mm S.L.) collected in 1972 by the Lake Rudolf Fisheries Research Project team at Topi point, Allia Bay, and Ferguson's Gulf. Coloration and certain anatomical details were also checked on the other specimens from which meristic and morphometric data were not taken (BMNH reg. nos. 1973.11.13 : 151–170).

All counts and measurements used in this description are those defined by Greenwood (1973).

Dorsal head profile straight or gently curved, sloping at an angle of 25–30 degrees with the horizontal; premaxillary pedicels rarely breaking the smooth outline of the profile.

Length of head 31·0–35·3 (mean, $M = 33·5$) per cent of standard length, depth of body 29·4–33·7 ($M = 32·1$) per cent.

Preorbital depth 12·5–17·9 ($M = 15·0$) per cent of head, least interorbital width 20·0–27·4 ($M = 22·7$) per cent, neither dimension showing allometry with standard length. Length of snout 26·6–32·3 ($M = 29·9$) per cent of head, 0·7–1·1 (mode 0·8) times its breadth. Eye diameter 30·5–35·7 ($M = 32·9$) per cent of head (showing ill-defined negative allometry), depth of cheek 17·4–25·8 ($M = 21·7$) per cent.

Caudal peduncle 16·5–19·7 ($M = 18·0$) per cent of standard length, 1·2–1·7 (modal range 1·3–1·5) times as long as deep.

Mouth horizontal or slightly oblique. Length of upper jaw 30·8–38·0 ($M = 33·7$) per cent of head, length of lower jaw 35·7–48·3 ($M = 39·7$) per cent, 1·4–1·8 (modal range 1·5–1·7) times as long as broad. Posterior tip of maxilla reaching a vertical through the anterior margin of the eye.

Gill rakers. Relatively stout, with the lower 1 or 2 on the first gill arch reduced; 9 (less frequently 8, rarely 10) on the lower limb of that arch. Well-developed 'pseudorakers' (fleshy protuberances between the inner and outer row of true rakers) are present on the first arch.

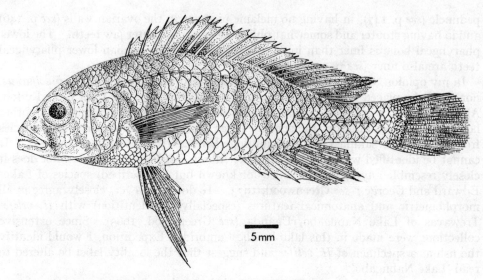

Fig. 1. *Haplochromis rudolfianus*. Lectotype. Drawn by Gordon Howes.

Scales. Strongly ctenoid, those on the anteroventral aspects of the thoracic region very small, with an abrupt size transition between them and the posterior scales of this region. Immediately anterior to the first dorsal fin spine there is a small naked area (about $1\frac{1}{2}$ or 2 scales in area), and often there is also a narrow naked strip below the ventral horizontal row of cheek scales. Lateral line with 30 (f2), 31 (f10), 32 (f12), or 33 (f1) scales, cheek with 3 (rarely 2 or 4) rows. Six or 7 scales between the dorsal fin origin and the upper lateral line, 6 or 7 (rarely 5) between the pectoral and pelvic fin bases.

Fins. Dorsal with 24 (f12), 25 (f11) or 26 (f2) rays, comprising 14 (f3), 15 (f21) or 16 (f1) spinous and 9 (f9), 10 (f15) or 11 (f1) branched rays. Anal with 3 spinous and 8 (f3), 9 (f20), 10 (f1) or 11 (f1) branched rays.

First branched ray of the pelvic fin very slightly produced in both sexes, but relatively more so in adult males.

Caudal subtruncate, scaled on its basal quarter to third.

Teeth. The majority of teeth in the *outer row* of *both jaws* are unequally bicuspid and moderately stout (Text-fig. 2) ; the major cusp is acutely pointed and equilateral in outline. A few unicuspid teeth sometimes occur posteriorly in the upper jaw, but more often the posterior teeth are tricuspid ; rarely are these teeth bicuspid. A noteworthy feature of the samples examined is the pronounced wear pattern seen on the outer teeth. Worn teeth have the major cusp either obliquely truncate or the wear may be so great that all demarcation between major and minor cusps has disappeared and the crown is spatulate.

There are 28–48 teeth in the outer premaxillary row, the number showing some positive correlation with the fish's length.

All *inner row* teeth are small tricuspids, and are arranged in 2 or 3 (mode), rarely 4, rows in the upper jaw and in 2 (mode) or 3 rows, rarely a single row, in the lower jaw.

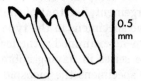

Fig. 2. *Haplochromis rudolfianus.* Outer row jaw teeth (in labial view) from left dentary, anterolateral in position.

OSTEOLOGY. The *neurocranium* (Text-fig. 3A) of *H. rudolfianus* is of the generalized '*H. bloyeti*'-type (see Greenwood, 1974), with a moderately decurved profile and a relatively short preotic skull length (ca 62·5 per cent of total neurocranial length). None of the cephalic laterosensory canals or pores, nor any of the bones carrying the canals, is at all hypertrophied (*cf.* p. 155; Text-fig. 3B).

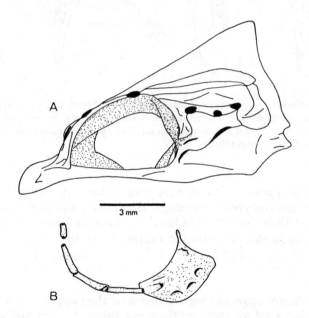

Fig. 3. *Haplochromis rudolfianus.* A: Neurocranium, left lateral view.
B: Bones in the infraorbital series of the right side.

The *lower pharyngeal bone* (Text-fig. 4) is triangular in outline, with the dentigerous area a little broader than long (ca 1·1 times). In all specimens examined the bone is noticeably stout, especially compared with that of the other Rudolf species or that of the generalized species in other lakes and in the east African rivers. The degree of enlargement, however, varies between individuals.

The two median tooth rows are composed of teeth clearly stouter than their lateral congeners; the degree of enlargement, like that of the bone itself, shows considerable individual variability. All five specimens in the type series have the stoutest bones and dentition of all the specimens examined; the lectotype (51 mm S.L.) is exceptional even amongst the type series in having submolariform teeth posteriorly in the median rows (in all other specimens the bicuspid crown is still retained). The lectotype also has the relatively most massive lower pharyngeal bone (Text-figs. 4A and B).

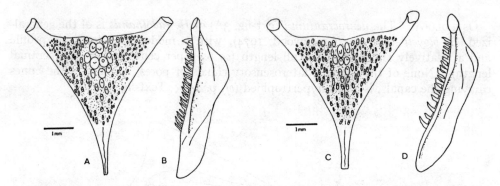

FIG. 4. *Haplochromis rudolfianus*. Lower pharyngeal bones to show variation in size of median teeth. A and B: Bone from one of the paratypes, in occlusal and left lateral views respectively. C and D: Bone from another specimen of the same size, in occlusal and left lateral views respectively.

Lower pharyngeal bone and tooth enlargement does not seem to be size correlated because some of the syntypes are amongst the smaller fishes examined, and a specimen 7 mm longer than the lectotype has a less massive bone and dentition.

Vertebral counts in the 30 specimens radiographed are: 27 (f1), 28 (f3), 29 (f18) or 30 (f8), comprising 11 (f1), 12 (f4), 13 (f23) or 14 (f2) abdominal and 16 (f23) or 17 (f7) caudal vertebrae. (The fused first ural and preural centra are not included in these figures.)

Only two specimens show any sign of fusion or close apposition between elements in the hypural series (*cf.* p. 158); in these two fishes the first and second hypurals seem to be fused.

VISCERA. The intestine is from $1\frac{1}{3}$ to $1\frac{1}{2}$ times the standard length; the ovaries are of unequal size with the right ovary noticeably larger than the left in most specimens or, rarely, it alone shows signs of oogenesis. A very characteristic feature is the intensely melanic tissue of the ovary wall; the testes, however, are but slightly pigmented. The entire peritoneum is also densely melanic. This extreme visceral melanism is probably correlated with the intense sunlight to which fishes living in the shallows of Lake Rudolf would be subjected.

COLORATION. In freshly killed specimens of *adult male H. rudolfianus* the ground colour is greenish-yellow (but whitish on the chest). Dark vertical bars cross the flank and caudal peduncle (*see* notes on preserved colours below). Each scale on the flanks has an opalescent centre in which yellow, blue and green colours can be detected; yellowish pigment predominates on the anterior body scales, especially those around the base of each pectoral fin. Posteriorly on the body the dominant colour in the scale centres is electric blue, and is especially noticeable around the base of the anal fin. The head, like the anterior part of the body, is an opalescent greenish-yellow and blue, the operculum marked with a golden yellow area on its lower part. The branchiostegal membrane is a delicate pale yellow except for a clearly demarcated black area anteriorly.

The dorsal fin membrane is dark golden-yellow, the lappets are scarlet and the soft part of the fin has a bright yellow basal streak and light yellow dots dorsally, the intervening areas melanic. Black spots also occur on the spinous part of the fin. Two dark basal blotches occur on the spinous dorsal; the first lies between the sixth to ninth spines, the second between the last two or three spines. The anal fin is yellow but with scattered black and red chromatophores, and with two or three light golden-yellow ocelli. The caudal is a marbled yellowish-green, with red and black spots, the red colour being most intense in the dorsal and ventral angles of the distal margin. The pelvic fins are dusky.

Adult females have a similar but more subdued coloration, with the whitish ventral areas more extensive and without the black anterior region to the branchiostegal membrane. The pelvic fins are hyaline tinged with yellow, and the red lappets and margin to the dorsal and anal fins respectively are barely discernible. On the anal fin there are pale yellow spots in the same position as the ocelli of males.

I am indebted to Mr and Mrs Hopson for supplying the notes on which this description is based.

Preserved coloration is virtually identical in both sexes, except that the pelvics in females are clear and not dusky, and the ground coloration is lighter than in males.

The ground coloration is a pale brown (fawn) shading to greyish-charcoal on the chest and belly. The flanks and caudal peduncle are crossed by 5-7 (rarely 8-10) dark and clearly defined vertical bars; the bars on the flanks reach almost to the ventral profile, but those on the caudal peduncle rarely extend to below the level of the midlateral line where they merge with a short and faint horizontal bar extending the length of the peduncle. This bar is of variable intensity and is barely visible in some specimens. Anteriorly there is a broad, sometimes ill-defined dark bar overlying the cleithrum and following the outline of that bone; dorsally the bar joins a dark, saddle-shaped blotch on the nape. A fainter bar is sometimes visible along the vertical limb of the preoperculum. The snout is crossed by a pair of parallel dark bars, and there is an intense and clearly demarcated lachrymal stripe.

All fins are a greyish-hyaline, the soft dorsal and the entire caudal are maculate, the dorsal with two distinct dark blotches basally (*see above*), and with dark lappets. Spotting on the caudal fin is most distinct proximally, the spots often arranged so as to produce two to four dark vertical bands on that part of the fin. The pelvics are dusky in males, greyish in females.

ECOLOGY. *Haplochromis rudolfianus* is apparently confined to the shallow and protected inshore areas of the lake, although there is a population inhabiting a crater lake on Central Island (*see below*). No data are yet available on the feeding habits of this species, nor whether it shows any particular substrate preferences.

Males appear to reach a larger size than do females, and are adult at a standard length of 45-47 mm. Females, however, mature at a smaller size, namely *ca* 30 mm. No data are available on the breeding habits or seasons of the species.

THE CENTRAL ISLAND POPULATION. One of the paralectotypes (*see* p. 142) and 12 additional specimens collected in 1965 by Dr R. L. Welcomme are from Central Island. The paralectotype is apparently from the shore of the island (station no. 264, *see* Worthington, 1932) but the other fishes are from one of the crater lakes in that island (*see* Beadle, 1932, for details of the lakes).

The Central Island fishes, particularly those from the crater itself, are of considerable interest since they are apparently isolated from other populations inhabiting the mainland shores of the lake. The apparent absence of *H. rudolfianus* from open-water localities suggests that it does not leave (or at least not frequently) its shallow-water inshore habitats.

A population of *Sarotherodon* living in Crater Lake A of Central Island is sufficiently distinctive for it to have been referred to a new species (*Tilapia vulcani*, Trewavas, 1933, *see also* Trewavas, 1973, for distinguishing features of the genera *Tilapia* and *Sarotherodon*). A population from the neighbouring Crater Lake C, however, did not differ significantly from the populations of *Sarotherodon niloticus* living in the main lake (Trewavas, 1933). The explanation for this seemingly anomalous situation appears to be that Lake C is, on occasion, connected with the main lake. (Personal communication from Dr K. E. Banister, based on information he was given by members of the Fishery Research Team on Lake Rudolf; the last interconnection was in 1972.) Lake A, on the other hand, is completely isolated from the main lake by a crater wall at least 10 m above current lake level.

It is the more regrettable then that the *H. rudolfianus* collected in 1965 bear no more precise locality data than 'Crater Lake, Central Island', and that all are distorted and poorly preserved. Allowing for the difficulty of measuring distorted specimens, I can find no meristic or morphometric differences between the Central Island fishes and those from the main lake. Several of the Island fishes are, however, larger than any recorded from the lake (maximum size 72 mm, *cf*. 58 mm for lake fishes). The largest Island fish has four or five unicuspid teeth situated posterolaterally in the upper jaw, but this could well be a size-correlated phenomenon (*see* Greenwood, 1974). Most of the Island fishes are much darker than are the lake fishes and consequently the vertical barring on the body is less obvious; in a few paler individuals, however, the bars are quite distinct. Like their main lake congeners, female Crater Lake fishes have only the right ovary well developed, but unlike the latter populations the ovarian wall in these fishes is but faintly melanized and then only on the dorsal and lateral aspects. These observations were paralleled by those I was able to make on specimens and colour photographs obtained by Dr Banister during his recent visit to the lake. Once again, the specimens are not well preserved and add little anatomico-morphological information to that already available.

However, the similarity in coloration between Dr Banister's fishes from Crater Lake A and those from the Welcomme sample strongly indicate that the latter are from the same source.

The dark coloration of Central Island *H. rudolfianus* seems to parallel that of '*Tilapia vulcani*' from Crater Lake A on the island (*see* Trewavas, 1933). In other characteristics the parallelism is not clearly apparent. That is to say, the *H. rudolfianus* specimens do not show the leanness, larger eyes, larger heads, longer dorsal fin spines and broader bands of inner teeth that Trewavas noted in the *Tilapia* specimens (Trewavas, 1933).

In contrast to the Central Island Crater Lake A specimens of *H. rudolfianus*, the paralectotype from the shore of the island does not show any marked darkening of the ground coloration. Unfortunately, it is eviscerated so no check can be made on the melanization of its gonads. The peritoneum is very dark.

Without more precise locality data, more material from different localities on the island and more details on live coloration, little can be said about the taxonomic status of Central Island populations of *H. rudolfianus*. It does, however, seem probable that, like the Crater Lake A population of *Sarotherodon niloticus* (with which species '*T. vulcani*' should now be synonymized; Dr Trewavas, personal communication), the Crater Lake A *Haplochromis* show some ecophenotypic response to their peculiar environment. The relative melanism shown by Crate Lake A *Haplochromis rudolfianus* populations is of particular interest because male coloration seems to be an important species recognition character (*see* Greenwood, 1974). Altered male coloration apparently is one of the first morphological differences seen in recently speciated *Haplochromis* (Greenwood, 1965).

DIAGNOSIS AND AFFINITIES OF *H. rudolfianus*. The distinctively barred colour pattern of this species immediately serves to distinguish it from all other *Haplochromis* species in Lake Rudolf (*see* footnote, below). From *H. macconneli* described on p. 154, *H. rudolfianus* is further distinguished by the absence of hypertrophied laterosensory canals and pores on the head, by several morphometric characters (*see* p. 155) and in having few, if any tricuspid teeth in the outer series of either jaw. In addition to the hypertrophied canal system, the neurocranium in *H. macconneli* is of a more derived type than that of *H. rudolfianus*. The nearest living relatives of *H. rudolfianus* are probably *H. turkanae* (see p. 153), and the fourth but as yet undescribed *Haplochromis* species in the lake.[1]

The validity of these relationships can only be tested when more material of the putative relatives is available.

---

[1] Four specimens 37·0–64·0 mm standard length (and a fifth prepared as a skeleton) may represent a fourth *Haplochromis* species. Morphometrically these fishes are not distinguishable from *H. rudolfianus*, and the dentition is similar except for there being only a single row of inner teeth in each jaw. The principal 'interspecific' difference seems to be in the coloration, both that observed by Mr Hopson (*in litt.*) when the fishes were alive and that remaining in the preserved material. There is also an observable (but non-quantifiable) difference in head shape, and these fishes have a colourless (not black) peritoneum. The gonads show no sign of melanization, unlike those of *H. rudolfianus*.

All five specimens are from water between 15 and 35 m deep; that is, from somewhat greater depths than *H. rudolfianus*, but within the range of *H. turkanae*, from which species they differ in the same morphometric characters as does *H. rudolfianus*.

Since only five specimens are available and because all are in some way damaged or distorted, I would consider it inadvisable to describe a new taxon on this material.

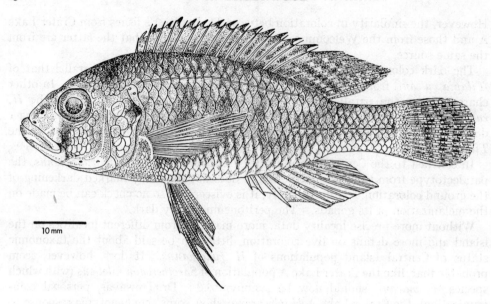

Fig. 5. *Haplochromis turkanae*. Holotype. Drawn by Gordon Howes.

In its gross morphology and in the details of its pectoral and predorsal squamation, *H. rudolfianus* resembles *H. albertianus* Regan, a species endemic to Lake Albert (see Trewavas, 1938). The two species differ in their preserved colour patterns, and in adult *H. albertianus* (of the same size as *H. rudolfianus*) having unicuspid outer jaw teeth and a more massive lower pharyngeal bone with a greater number of enlarged, submolariform teeth. In other words, *H. albertianus* shows a greater degree of specialization in those characters that are already somewhat specialized in *H. rudolfianus*.

STUDY MATERIAL

Register number BMNH	Locality : Lake Rudolf
1973.11.13 : 151–154	Ferguson's Spit (shallow water, inshore)
1973.11.13 : 155–170	Topi Point, Allia Bay

Haplochromis turkanae sp. nov.
(Text-figs. 5 and 6)

HOLOTYPE. An adult male, 73·0 mm standard length (BMNH reg. no. 1973.11.20 : 1), caught in a bottom trawl fished at a depth of 16 m, over a mud and rock bottom, 5·6 km north-west of Porr.

PARATYPES. Three adult males (BMNH reg. nos. 1973.11.20 : 2–4), 77·5–86·0 mm S.L. from the same locality and trawl as the holotype.

DESCRIPTION. Based on these four fishes, 73·0–86·0 mm S.L. With so few specimens available most of the morphometric data can be presented most conveniently in tabular form.

S.L.	Depth*	Head*	PO %	IO %	Snt %	Eye %	Cheek %	Lj %	Uj %	CP*
73·0	35·0	33·5	20·5	20·5	30·6	28·6	28·6	45·0	41·0	17·5
77·5	35·0	32·3	18·0	22·0	32·0	30·0	32·0	44·0	40·0	17·4
81·0	34·5	32·7	18·9	20·8	34·0	34·0	26·4	45·3	41·5	18·5
86·0	33·2	31·4	18·5	22·0	29·6	27·9	27·0	44·5	40·8	16·9

* = per cent of standard length. % = per cent of head length.
PO = preorbital depth; IO = least interorbital width; Snt = snout length; Lj = lower jaw, and Uj = upper jaw length. CP = caudal peduncle length.

Caudal peduncle 1·4–1·5 times longer than deep. Dorsal profile of head straight or gently curved dorsally, but straight anteriorly, sloping at an angle of 30–35 degrees with the horizontal. The ascending processes of the premaxillae barely interrupt the outline of the profile.

Mouth slightly (but noticeably) oblique, the lips a little thickened; posterior tip of the maxilla reaching a vertical through the anterior part of the eye. Jaws equal anteriorly, the lower 1·6–1·7 times longer than broad. Snout broader than long, its anterior profile, when viewed from above, smoothly rounded. None of the cephalic laterosensory canals (or their pores) is noticeably enlarged.

Gill rakers. Relatively stout, 8 (f3) or 9 (f1) on the lower part of the first arch, the lowermost one or two rakers reduced in size. Pseudorakers (*see* p. 143) are developed between the inner and outer rows of gill rakers, but are not conspicuous.

Scales. Ctenoid; lateral line with 31 (f3) or 32 (f1) scales, cheek with 4 (f1) or 5 (f3) imbricating rows. Six or 7 scales between the upper lateral line and the dorsal fin origin; a naked area (about 1½ scales in extent) immediately before the first dorsal fin spine. Nine or 10 scales between the pectoral and pelvic fin bases; the scales of the chest are also very small and grade abruptly with the larger scales of the post-pectoral region.

Fins. Dorsal with 14 (f2) or 15 (f2) spinous rays and 9 (f4) branched rays, anal with 3 spines and 7 (f2) or 8 (f2) branched rays. The first pelvic ray is produced, very noticeably so in two specimens where it extends as far as the second branched anal ray. The caudal fin is strongly subtruncate, almost rounded, and scaled on its proximal quarter to half. Pectoral fin 25·6–28·7 per cent of standard length, 81·5–88·0 per cent of head length.

Teeth. In all four specimens the posterior three or four teeth in the *outer premaxillary row* are strong, dagger-like unicuspids. In two fishes the other teeth in this row are a mixture of caniniform unicuspids and weakly bicuspids, while in the other two specimens distinctly bicuspid teeth predominate, although a few unicuspids occur anterolaterally. There are 42–52 teeth in this row. There is a similar difference in the predominant tooth type of the outer row in the *lower jaw*. The first two specimens have mainly unicuspid teeth with a few bicuspids, the second

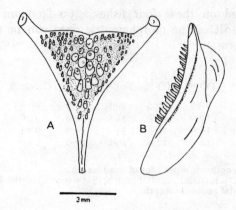

Fig. 6. *Haplochromis turkanae*. Lower pharyngeal bone of holotype in A: occlusal view; B: left lateral view.

pair have a predominance of bicuspids. Bicuspid teeth in the lower jaw have their crowns more distinctly incurved than do the upper jaw teeth.

Bicuspid teeth in both jaws are of the typical generalized *Haplochromis* type, and closely resemble the teeth of *H. rudolfianus* (see p. 144; Text-fig. 2).

The *inner series* of teeth in both jaws are composed of small tricuspids arranged in 3 rows in the upper and 2 (or irregularly 2) rows in the lower jaw.

OSTEOLOGY. With only four specimens available, no skeletal preparations were possible. Vertebral counts made from radiographs are: 28 (f3) and 29 (f1), comprising 13 abdominal and 15 or 16 caudal vertebrae (the fused first preural and ural centra not included). No specimen has any fused elements in the caudal fin skeleton, but in two fishes hypurals 1 and 2 are closely apposed.

COLORATION. No information is available on the live colours of this species. The four adult (but not sexually active) male specimens (fixed in formalin) have a light grey to yellowish-grey ground colour that extends ventrally to a clearly demarcated horizontal line on the body and caudal peduncle, at which level it becomes pearly white. The line is at about the horizontal level of the lowermost insertion of the pelvic fin; thus in lateral view little of the caudal peduncle shows the white ventral coloration. On the flanks, most dorsal, lateral and ventrolateral scales have a narrow margin of dark pigment. The lower jaw, branchiostegal membrane, sub-operculum and the lower part of the cheek are also pearly-white.

About five very faint vertical dark bars are visible on the flanks, and another may also occur posteriorly on the caudal peduncle; these bars are narrower dorsally than ventrally, and do not extend onto the white ventral coloration of the body. The vertical limb of the preoperculum is faintly to clearly dusky and there is a broad, intensely black lachrymal bar that, at about the level of the posterior maxillary tip, narrows abruptly and then continues ventrally and a little medially onto the lower jaw.

The entire soft dorsal fin and the posterior half of the spinous dorsal are densely and distinctly spotted, with the spots arranged in from four horizontal rows anteriorly to six or more rows posteriorly on the fin. On the anterior part of the spinous dorsal the spots are confluent and form vertically aligned, dark, interspinous streaks. The lappets of the dorsal fin are black. The entire caudal fin is covered with dark and discrete spots so arranged as to form wavy, vertical bars when the fin is not fully opened. The anal is hyaline except for three rather pale ocelli. The pelvics have the anterior half dusky, the posterior half hyaline; the elongate first pelvic ray is dead-white.

ECOLOGY. Virtually nothing is known about the biology of this species. The four specimens came from deeper water than is usual for *H. rudolfianus*, and shallower water than is usual for *H. macconneli* (see pp. 148 and 159).

All four specimens have the stomach and intestine packed with fragments of ostracod shells.

It is interesting to note that, unlike *H. rudolfianus* but like *H. macconneli*, there is no trace of dark pigment in the peritoneum and neither is there any on the gonads (*cf.* p. 146).

All four specimens are adult males, but judging from the size and shape of the testes, none is sexually active.

DIAGNOSIS AND AFFINITIES. *Haplochromis turkanae* is immediately distinguishable from *H. macconneli* because of its non-hypertrophied cephalic laterosensory canal system. There are also morphometric differences between the species, differences in their dentition and, apparently, in the preserved coloration of adult males (*see below*, pp. 156–159).

In its general appearance, its dentition and in several morphometric characters, *H. turkanae* closely resembles *H. rudolfianus*. It differs principally in having more rows of scales on the cheek (4 or 5, *cf.* 3), smaller scales between the pectoral and pelvic fin insertions (9 or 10, *cf.* 6 or 7), a deeper cheek (26·4–32·0 per cent of head, *cf.* 17·4–25·8, M = 21·7 per cent), longer upper jaw (40·8–41·5, *cf.* 30·8–38·0, M = 33·7 per cent of head), and a somewhat longer lower jaw (44·0–45·3, *cf.* 35·7–48·3, M = 39·7 per cent of head). *Haplochromis turkanae* also differs in having a more elongate first pelvic ray (which is distinctively coloured) and in its overall coloration. For example, it lacks the prominent vertical bars on the body, has more intensely and densely maculate dorsal and caudal fins, and has the peculiar dribble-like extension of the lachrymal bar onto the lower jaw. The lower pharyngeal bone and dentition of *H. turkanae* are somewhat more massive than those of most *H. rudolfianus*, but are quite comparable with those in the type series of that species (*see above*, p. 146, and *cf.* figs. 4 and 6).

Without more osteological information on *H. turkanae* it is difficult to say much about its affinities with *H. rudolfianus*. From what is known, however, the two species would seem to be very closely related phyletically. Indeed, the resemblance parallels that seen between many pairs of *Haplochromis* species in Lake Victoria. As in many of the Victoria pairs there is also an apparent ecological replacement involved, with *H. rudolfianus* being the inshore, shallow-water species and *H. turkanae* the species of deeper waters (*see* Greenwood, 1974).

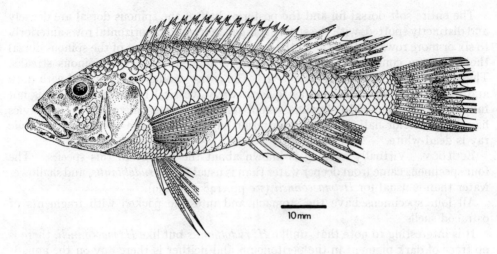

Fig. 7. *Haplochromis macconneli*. Holotype. Drawn by Gordon Howes.

Haplochromis macconneli sp. nov.

(Text-figs. 7–10)

This most distinctive species was discovered by Mr and Mrs A. J. Hopson when the Lake Rudolf Research Project team began trawling operations in the deeper waters of the lake. Apparently it is confined to water more than 20 m deep and has been collected from depths down to 75 m.

Besides its various anatomical peculiarities, *H. macconneli* is also noteworthy for its marked sexual dimorphism in adult size. No males larger than 35·5 mm standard length have yet been recorded. Indeed, despite intensive efforts, only two males have so far been recognized amongst the several hundreds of specimens examined. Females, on the other hand, are common components of deep-water trawl catches.

The taxon is named for Mr R. B. McConnel, Officer in Charge of the Fisheries Department at Lake Rudolf, in grateful recognition of the assistance he has unstintingly given to Mr Hopson and his research team.

HOLOTYPE. A female, 77·0 mm standard length (BMNH reg. no. 1973.11.13 : 37), caught near the bottom in 50–64 m of water 3 miles north-west of Central Island.

PARATYPES. (i) Twenty-one specimens (BMNH reg. nos. 1973.11.13 : 38–58), 51·0–72·0 mm S.L., all females and from the same locality as the holotype ; (ii) 6 juveniles (of indeterminable sex), 22·0–38·0 mm S.L., caught in a bottom trawl at a depth of 20 m 1·6 km off Ferguson's Spit (BMNH reg. nos. 1973.11.13 : 59–64 ; (iii) 2 adult males, 31·0 and 35·5 mm S.L., caught in a bottom trawl at a depth of 30 m, 2·4 km east of North Island (BMNH reg. nos. 1973.11.13 : 65–66).

DESCRIPTION. Based on the holotype and 29 paratypes, 22·0–77·0 mm S.L. Various characters have been checked on the 78 additional specimens (size range 15·0–59·0 mm S.L.) from the Ferguson's Spit station and one other station 3·2 km off-shore at a depth of 75 m. None of this extra material has, however, been included in the morphometric counts and measurements.

Because most of the proportional measurements used in this description show allometry with the fish's size, the material has been divided into two size groups, one of fishes 51–77 mm S.L. and the other of fishes 22–38 mm S.L. Ranges and means for the larger specimens (N = 22) are given first, followed in parentheses by those for the smaller fishes (N = 8).

Length of head 33·3–39·9, M = 35·3 (31·6–39·4, M = 36·0) per cent of standard length, depth of body 31·7–35·1, M = 33·3 (25·3–32·3, M = 29·7) per cent.

Dorsal profile of head clearly but gently concave above the eye, sloping at an angle of *ca* 35–40 degrees to the horizontal. The profile of the snout varies from straight to somewhat decurved, with the premaxillary pedicels always breaking the outline. The anterior tip of the lower jaw is usually produced into a symphysial knob, which is most obvious in fishes over 65 mm S.L.

An outstanding feature of the head (in specimens of all sizes) is the greatly enlarged openings to all the laterosensory canals (Text-figs. 7 and 9). Particularly obvious are those on the preorbital and preopercular bones. The underlying canals are hypertrophied, with the result that the bones involved have a distinctly inflated appearance.

Laterosensory canals on the neurocranium, especially the temporal canal of the pterotic bone, are also inflated, particularly in comparison with those in *H. rudolfianus* and *H. turkanae*, and indeed with those of the generality of *Haplochromis* species (*see below*, p. 158). Canals, and their openings, in the extrascapular and supracleithrum are equally affected by this trend, as are the nasals although the latter are relatively the least cavernous of the cephalic laterosensory canal bones. In well-preserved specimens the canal openings are occluded by a thin membrane.

Depth of preorbital 16·6–22·2, M = 20·1 (12·5–16·7, M = 15·4) per cent of head, least interorbital width 18·8–24·2, M = 21·8 (14·7–25·0, M = 21·2) per cent. Snout slightly broader than long (rarely as long as broad) in specimens of all sizes, its length 27·8–38·1, M = 33·8 (26·1–33·3, M = 29·6) per cent of head for the size groups respectively. Diameter of eye 23·7–29·3, M = 26·8 (25·0–33·3, M = 28·7) and depth of cheek 22·8–29·3, M = 26·7 (18·1–25·0, M = 21·2) per cent of head.

Caudal peduncle 16·5–21·5, M = 18·6 (17·8–22·6, M = 20·0) per cent of standard length, 1·5–2·2 times as long as deep in fishes of all sizes.

Lower jaw with a distinct symphysial knob in fishes of all sizes, the protuberance most marked in individuals of > 65 mm S.L. Length of lower jaw 40·8–51·0, M = 45·6 (37·6–46·2, M = 42·0) per cent of head, 1·6–2·0 (modal range 1·8–2·0) times longer than broad in specimens of all sizes. Length of upper jaw 31·8–44·0, M = 38·4 (30·6–37·6, M = 33·3) per cent of head.

Mouth slightly oblique, the posterior tip of the maxilla reaching a vertical through the anterior margin of the orbit, or to a point slightly posterior to that line.

Gill rakers. Of variable form, but usually with the upper 2 or 3 rakers on the first gill arch flattened and branched, those on the middle section of the arch relatively slender, and the lowermost 1 or 2 rakers reduced; no pseudorakers are developed

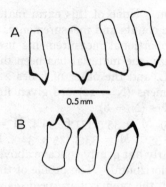

Fig. 8. *Haplochromis macconneli*. Dentition. A: Outer teeth from the premaxilla. B: Outer teeth from the dentary. In both, the teeth are from an anterolateral position in the jaw and are viewed labially.

(see p. 143). There are 7 (f1), 8 (f8) or 9 (f21) rakers on the lower limb of the first gill arch.

Scales. Ctenoid, those on the chest very small but grading in size with those on the subpectoral region (that is, the demarcation between small and large scales in this region is less abrupt than in *H. rudolfianus*; see p. 154).

Five and a half to 7 (mode 6) scales between the dorsal fin origin and the lateral line, 7–9 (modes 8 and 9), rarely 5, between the pectoral and pelvic fin bases. Cheek with 3 or 4 rows of imbricating scales. In most specimens there is a small naked area (about the area of one scale) in front of the first dorsal fin spine.

Fins. Dorsal with 22 (f6), 23 (f16) or 24 (f8) rays, comprising 13 (f2), 14 (f16), 15 (f11) or 16 (f1) spinous and 8 (f11), 9 (f18) or 10 (f1) branched rays. Anal with 3 spines and 7 (f2), 8 (f25) or 9 (f3) branched rays. First ray of the pelvic fin slightly produced, more so in larger fishes. Caudal truncate, scaled on its proximal quarter (mode) to third. Pectoral fin 25·8–34·5, M = 28·6 (23·7–28·3, M = 26·1) per cent of standard length, and 73·0–92·5, M = 81·1 (63·5–77·0, M = 72·3) per cent of head.

Teeth. In fishes more than 50 mm S.L. the *outer row* of *premaxillary* teeth is composed of unequally bicuspid teeth anteriorly, but of tricuspids laterally and posteriorly (Text-fig. 8); the median cusp of the tricuspids, and the major cusp of the bicuspids is slightly incurved. Many specimens have some tricuspids intercalated amongst the biscuspids anteriorly, and in a few fishes almost the entire outer row is composed of tricuspid teeth. When there is a mixture of bi- and tricuspids the latter predominate. At all positions in the tooth row the bi- and tricuspids are of equal size.

Tooth morphology and arrangement in the *lower jaw* are similar to those in the upper, although more individuals have only tricuspid teeth present; a predominantly bicuspid outer row is rarely encountered.

Fishes in the size range 20–40 mm S.L. have mainly bicuspid teeth in the upper jaw, with those tricuspids present restricted to a posterolateral position in the row.

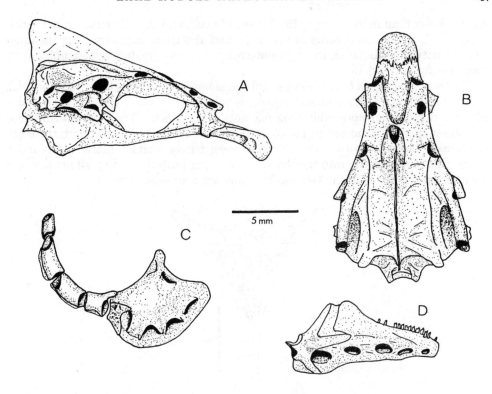

Fig. 9. *Haplochromis macconneli*. A : Neurocranium in right lateral view. B : Neurocranium, dorsal surface. C : Bones of the infraorbital series, right side. D : Right lower jaw, viewed from a slightly ventrolateral position.

Bicuspid teeth also predominate in the lower jaw, but some unicuspid (and slender) teeth may occur posterolaterally. Tooth form is like that of the larger fishes.

There are 41–64 (M = 52) outer premaxillary teeth in fishes 50–77 mm S.L., and 24–42 in specimens 22–39 mm long.

Irrespective of a fish's size, the *inner teeth* in both jaws are all tricuspid and small, and are usually arranged in a single series but double rows are encountered occasionally.

OSTEOLOGY. The characteristic hypertrophy in the cephalic laterosensory canal system has been commented upon above (*see also* Text-figs. 7 and 9). Canal bones in the pectoral skeleton are also affected, and the otic region of the skull is noticeably inflated (*see* Text-fig. 9).

The *neurocranium* (Text-fig. 9) departs from the generalized *Haplochromis* type (as seen in *H. rudolfianus*, Text-fig. 3) and clearly approaches that found in *H. saxicola* and allied species in the Lake Victoria species flock (Greenwood, 1974). In other words, the preotic region of the skull is slightly more elongate than in the generalized type, and associated with this and the shallower braincase, the preotic skull roof is straighter and slopes upwards at a smaller angle. The supraoccipital crest is lower than in most generalized types of skull, and the otic region is narrower.

Although all the canal bones of the skull (and also those bones encasing the inner ear) are noticeably inflated, the dermopterotics show the greatest hypertrophy of all (*see* Text-fig. 9A and B).

The *lower pharyngeal bone* (Text-fig. 10) is triangular in outline (length and breadth of the dentigerous surface almost equal), is relatively slender, and has an anterior blade that is neither noticeably long nor noticeably short. The teeth on this bone are rather sparsely arranged in 16–20 irregular rows. Without exception, the teeth are weakly bicuspid, with a low, blunt or even barely visible anterior cusp and a crown that slopes gently into the sharper and larger posterior cusp; all are fine and compressed but those in the two median rows are somewhat coarser.

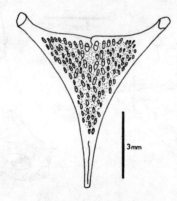

Fig. 10. *Haplochromis macconneli*. Lower pharyngeal bone in occlusal view.

Vertebrae and the caudal fin skeleton. Vertebral counts for the 24 specimens radiographed are : 27 (f1), 28 (f8), 29 (f14) or 30 (f1), comprising 12 (f16) or 13 (f8) abdominal and 15 (f1), 16 (f15) or 17 (f8) caudal centra. The fused first ural and first preural centra are excluded from these counts.

The caudal fin skeleton is unusual in showing a high degree of fusion between various hypurals (or if not fusion, such extremely close apposition as to be indistinguishable from fusion on radiographed specimens). The extent of fusion or apposition is generally complete, but in a few fishes there are short unfused sections between the otherwise conjoined elements.

About 77 per cent of the 26 specimens radiographed showed some degree of fusion between various hypurals. Only 6 specimens (*ca* 23 per cent of the sample) had all 5 hypurals completely free from each other. Most frequently (i.e. in 20 fishes) hypurals 1 and 2 are fused. In the upper part of the skeleton the commonest pattern of fusion is to have hypurals 3 and 4 fused, but hypural 5 free (11 specimens); only 4 specimens had all 3 upper hypurals fused, and 5 fishes showed no fusion between any of the 3 upper hypurals. In *H. rudolfianus*, it will be recalled, only 2 out of 30 specimens (i.e. about 7 per cent) had any fused hypurals; in both these fishes fusion was between hypurals 1 and 2 (*see* p. 146).

For comparison with the situation in Lake Rudolf, the caudal skeleton in several endemic *Haplochromis* species from other lakes was examined. Of 174 specimens (representing 12 species) from Lake George, Uganda, about 8 per cent showed some (but never complete) fusion between elements in both the upper and the lower parts of the skeleton. The frequency of fusion seems equally distributed amongst the species examined.

In Lake Victoria, too, hypural fusion is relatively rare. One hundred and seventy-eight specimens representing 22 species (with a modal sample size of 10 specimens per species) were examined. Of these, 20 specimens (i.e. about 13 per cent) had fused hypurals. The commonest pattern here is of fusion between hypurals 3 and 4, less frequently it occurs between hypurals 1 and 2, and only once was it recorded, with certainty, between hypurals 1 and 2, and 3 and 4 in the same individual (although two other individuals may show this pattern). The Lake Victoria species examined are from several phyletic lineages within the *Haplochromis* flock of that lake and cover a wide range of body forms.

Far fewer specimens are available of the four endemic *Haplochromis* species from Lake Albert. Of these, *H. bullatus* Trewavas has 3 out of 18 fishes (i.e. about 17 per cent) with hypurals 1 and 2 fused (or very closely apposed), while *H. avium* Regan (9 specimens) and *H. albertianus* Regan (15 specimens) have none. Both the latter species, however, have some individuals in which the hypurals are closely apposed.

Thus, even allowing for the small sample sizes involved in this survey, it does seem that the frequency of hypural fusion in *H. macconneli* is exceptionally high. The significance of this phenomenon remains unexplained.

VISCERA. Because of poor preservation it is impossible to measure precisely the length of the gut in *H. macconneli*; I would estimate the intestine to be about $1\frac{1}{2}$ times longer than the standard length. In strong contrast to *H. rudolfianus* there is no trace of melanin in the peritoneal tissue and neither are the gonadial walls pigmented (*see* p. 146 above).

COLORATION. In life, *adult females* are a pale greenish-fawn with traces of greenish iridescence on the flanks. All fins are colourless except for three conspicuous and bright yellow spots on the anal.

Live colours for *males* have not been recorded.

Preserved colours. I suspect that the material I have examined is somewhat bleached and thus the coloration is probably lighter than it might otherwise have been.

There is apparently little sexual dichromatism. In both sexes the ground colour of the body and head is a pale yellowish-fawn with no tonal variation between dorsum and venter. All fins are hyaline with, in both sexes, dark maculations on the proximal third to half of the caudal fin. The males examined have the dorsal fin somewhat darker than that of the females, and there is a fairly dense aggregation of melanophores on the anterior third of the pelvic fins. (These specimens may, however, be less bleached than are the females.)

ECOLOGY. Judging from Mr Hopson's records, *H. macconneli* is confined to water more than 20 m deep, and is probably most abundant in deeper water (i.e. at depths of 50-70 m). Some of the smallest specimens collected (15-20 mm S.L.) were caught at a depth of 75 m, and none, of any size, has been recorded from littoral habitats. The pale coloration of this species and the absence of melanic pigments in the peritoneum and gonads may well be correlated with this deep-water distribution.

No data are available yet on the feeding habits of the species.

Breeding biology. The marked sexual dimorphism in adult size has been noted already (p. 154). Despite a very thorough search, Mr and Mrs Hopson were at first unable to find any males amongst the several hundreds of *H. macconneli* they examined. Eventually two males, 31·0 and 35·5 mm S.L., were identified in a catch made some 2·4 km east of North Island at a depth of 30 m. Both these fishes appear to be sexually adult; one is probably in an advanced stage of ripening, the other at a slightly earlier stage of development. Females are certainly adult at a length of *ca* 50 mm and it seems likely that they may mature at a smaller size (Hopson *in litt*.). Be that as it may, females seem to grow to a much greater length than do males.

In a sample of 21 adult females examined, only 2 have both ovaries equally developed. Eleven fishes have the right ovary much larger than the left one (at all stages of oogenesis), and 8 have only the right ovary developed. Unlike *H. rudolfianus*, there is no trace of melanization in the ovarian wall of *H. macconneli*.

No details are available on the breeding habits or seasons of the species.

DIAGNOSIS AND AFFINITIES. *Haplochromis macconneli* is readily distinguished from the other Lake Rudolf species by the hypertrophy of its cephalic laterosensory canal system, its dentition (*see* p. 156), and in specimens > 35 mm S.L., from *H. rudolfianus* by several morphometric characters (deeper cheek and preorbital, longer snout, smaller eye and longer jaws).

Indeed, *H. macconneli* is so distinct morphologically from *H. rudolfianus* and *H. turkanae* (and the putative fourth species) that it is difficult to establish its phyletic affinities with these other taxa. In addition to the characters listed above, *H. macconneli* also differs from *H. rudolfianus* (and from the undescribed species too) in the shape of its neurocranium, which is of a more specialized type (*see above*, p. 157 and Greenwood, 1974). Almost certainly the hypertrophy of the cephalic lateral line canals is an adaptation associated with the deep-water habitat of *H. macconneli*, and strongly suggests that the species evolved within the lake. No fluviatile *Haplochromis* exhibits this specialization.

It is interesting to note that the cephalic lateral line canal hypertrophy seen in *H. macconneli* is much greater than that found in any known *Haplochromis* species living at comparable depths in Lake Victoria (Greenwood & Gee, 1969). Amongst *Haplochromis*-group species an equivalent hypertrophy is seen only in *H. bullatus* of Lake Albert, and in species of the endemic Lake Malawi genus *Trematocranus* (Trewavas, 1935). Species of another Malawi endemic, *Aulonocara*, show greater development in certain parts of the system (the infraorbital series for example), but otherwise exhibit a level of hypertrophy comparable with that of *H. macconneli*.

Trematocranus and *Aulonocara* are manifestly more closely related to each other and to other taxa from Lake Malawi (Trewavas, 1935; personal observations) than to any species occurring outside the lake. Neither need be considered further in the possible phylogeny of *H. macconneli*.

On purely morphological grounds *H. bullatus* of Lake Albert could be considered the nearest living relative of *H. macconneli*. Both species share the specialization of enlarged laterosensory canals on the head, and both share (with other species from Lakes Albert and Rudolf) certain peculiarities in the predorsal and thoracic squamation patterns (*see above*, p. 144). The latter character is, however, difficult to assess with respect to its being a primitive or a derived one. At present all that can be said is that the pattern is not encountered amongst the *Haplochromis* species of Lake Victoria nor is it seen in the fluviatile *Haplochromis* of Kenya, Uganda and Tanzania. It is rarely encountered in the *Haplochromis* species flock of Lakes Edward and George (where it is known from two species, *H. pharyngalis* Poll and *H. petronius* Greenwood; *see* Greenwood, 1973), but it does seem to characterize the *Haplochromis* of the River Nile, Lake Albert, Lake Rudolf and the River Zaire (personal observations; also Greenwood, 1971).

Haplochromis macconneli differs from *H. bullatus* in the shape of its neurocranium (*see* p. 157 above) which is like that found in the moderately specialized insectivore–piscivore radiation in Lake Victoria (*see* Greenwood, 1974, pp. 80–93). This difference would not, of course, debar *H. bullatus* from consideration as the living plesiomorph sister species of *H. macconneli*. But the fact that the Lake Rudolf and Lake Albert basins have never been interconnected (and if there had been some riverine connection, the probability that any presumed common ancestor of the two species would itself have been abyssal in habits) seems to rule out any such close phyletic relationship. Interspecific similarities in laterosensory canal hypertrophy are thus to be interpreted as the product of parallel evolution. The shared peculiarities in squamation patterns may well reflect a common ancestry but this is likely to be a relatively distant one (*see below*, p. 162).

It seems probable, therefore, that the relationships of *H. macconneli* should be sought amongst the species of Lake Rudolf. Two interpretations seem possible. First, *H. macconneli* may be an immediate derivative of an *H. rudolfianus*-like ancestor (i.e. *H. rudolfianus* and *H. macconneli* may be true sister species). Alternatively, *H. macconneli* might be the apomorph survivor of another lineage the relatively plesiomorph, that is *H. rudolfianus*-like members of which have become extinct (unlike the situation in Lake Victoria, for example, where it is possible to follow, from species still extant, the specialization of a lineage; see Greenwood, 1974).

DISCUSSION

Even with the discovery of two and possibly three new *Haplochromis* species, the total fish fauna of Lake Rudolf still stands low on the scale of endemicity in African lakes. Its *Haplochromis* species flock also shows a low level of adaptive radiation, probably lower than that of Lake Albert (*see* Trewavas, 1938; Greenwood, 1971) where an anatomically specialized mollusc-eater, a specialized grazer on epiphytes and a species adapted for life in deep water have evolved.

This comparison must, however, be interpreted with care. For one thing, *H. mahagiensis* David & Poll (the mollusc-crushing species) of Lake Albert may well belong to a different lineage from that of the other species in the lake. It could be the local representative of a fluviatile mollusc-crushing species represented elsewhere by *H. straeleni* Poll and *H. vanderhorsti* Greenwood (*see* Greenwood, 1954 & 1959a, for discussion). Furthermore, a temporal element is probably involved. There is little evidence of Lake Albert having dried out at any time in its history, but Lake Rudolf probably was severely reduced, or even completely desiccated, during the middle part of the Pleistocene (*see* Fryer & Iles, 1972, for review). Refilling of the Rudolf basin appears to have been through what is now the River Sobat at some time within the later Pleistocene. Subsequently the connection was broken and has never been re-established. Such an historical background has two consequences, namely that Lake Rudolf is to be considered a relatively young lake, and that its colonizers (or, perhaps more accurately, its recolonizers) were species of Nilotic origin.

The relative youth of Lake Rudolf, coupled with the nature of its basin may, as Fryer & Iles (1972) suggest, account for the paucity of endemic species and, I would also suggest, for the muted adaptive radiation seen amongst the three or four *Haplochromis* species that evolved there.

With two exceptions, all the *Haplochromis* species of Lake Albert are apparently confined to the basin of that lake. The two more widely distributed species, *H. wingatii* (Blgr.) and *H. loati* Greenwood, both have dental specializations that are not shared with any Lake Rudolf taxa (Greenwood, 1971). Thus, it seems unlikely that an extant Albertine *Haplochromis* species was the original recolonizer of Lake Rudolf in later Pleistocene times. Nor does it seem probable that the recolonizer closely resembled any species from the Nile (amongst which must be considered *H. wingatii* and *H. loati* or their ancestors). In this situation the only conclusion that can be drawn is that some fluviatile species, now extinct, provided the founder population for the Lake Rudolf microflock.

It is, of course, possible that the present-day Rudolf species (or some part of them) are descendants from the relicts of a previous flock, possibly a more complex one, that inhabited the early Pleistocene lake and which survived the subsequent period of desiccation. The neurocranial differences between *H. rudolfianus* and *H. macconneli* might be explained in this way. If this was the history of the present-day species then it follows that their ancestors were derived not from the Nile (which did not then exist in its present form; *see* Berry & Whiteman, 1968), but from a river that originated in the eastern highlands and emptied into the developing Nile system in the region of the present River Sobat.

Shared peculiarities in the squamation pattern of the thoracic and predorsal regions hint at a common ancestry for the Lake Rudolf and Lake Albert *Haplochromis* species (*see above*, p. 161). This character complex is not found in the *H. bloyeti*-like species group that is widespread in the rivers of Uganda, Kenya and Tanzania. Members of this species complex are thought to be close relatives of the ancestral species which gave rise to the sister species flocks in Lake Victoria and Lakes Edward and George (*see* Greenwood, 1973 & 1974). Fishes in these flocks, with one exception in Lake George and one in Lake Edward, all lack the Albert–Rudolf scale pattern. The exceptional species, *H. petronius* and *H. pharyngalis*, resemble the Nilotic *H. wingatii* in several features as well as the one of scale pattern, and they may represent an exotic element amongst the otherwise *H. bloyeti*-like derivatives inhabiting these lakes (Greenwood, 1974). The possible phyletic relationship of *Haplochromis* species from Lakes Rudolf and Albert raises some interesting points of zoogeography. The lakes are several hundred kilometres apart and any form of past interconnection they may have had would have been of an indirect nature (*see above*, p. 162). In contrast, Lake Albert is close to Lake Edward and the lakes are in direct contact through the River Semliki. Yet, their faunas, both cichlid and non-cichlid, are quite distinct (Greenwood, 1959b). The present barrier to faunal interchange, principally the Semliki rapids, is clearly an effective one.

I have argued elsewhere for a close phyletic relationship between the *Haplochromis* species of Lakes Edward and Victoria and for their derivation in parallel from a common ancestor that once inhabited the westward flowing rivers of eastern Africa during the Pleistocene (Greenwood, 1973). It seems now that perhaps this concept should be qualified by postulating an ancestral species that lived in some but not all of those rivers. The reasoning behind this qualification is, of course, the presence of species in Lake Albert that would seem to be derived from a different lineage, a lineage that also gave rise to the species of the Nile and, possibly, Lake Rudolf as well (*see above*, p. 161). Furthermore, it is possible that the *Haplochromis* species of the River Zaire may share ancestry with these species (p. 161).

The evidence upon which these postulated phylogenies are based, a shared scale pattern, is admittedly tenuous, particularly since it is not yet possible to determine which of the two types is to be considered the primitive condition. Further research is planned to investigate the phylogeny of *Haplochromis*-group species and I would certainly not consider the ideas expressed here as more than a working hypothesis.

The presence of two species with Albert–Rudolf scale patterns (*H. petronius* and *H. pharyngalis*) in Lakes George and Edward demands explanation. Three possible explanations can be considered. First, the ancestor of these species made its way into the Lake Edward basin in fairly recent times and *via* the River Semliki. The likelihood of this, however, is reduced by the fact that no reciprocal exchange of *Haplochromis* seems to have taken place (although the two lakes share another cichlid *Sarotherodon leucostictus* [Trewavas]). Second, the prerift rivers each may have contained species of both squamation types. The absence of species with the Albert–Rudolf pattern from most rivers in eastern Africa (and probably from Lake Victoria as well) would seem unlikely if both types had been represented there previously. The third possibility is one based on the assumption of there having been a single basin in which the proto-Lakes Albert and Edward developed, probably as a series of partly interconnected small and swampy lakes. The northern region of this basin (the future Lake Albert) could have been fed by rivers in which the *Haplochromis* species had an Albert–Rudolf facies, while the southern end (future Lakes Edward and George) was fed by rivers with *Haplochromis* of a Victoria–Edward facies. The next assumption would be that only a limited exchange of species took place between the two regions before their continuity was broken. Victoria-type species, if any penetrated to the north, were, presumably, unsuccessful in that environment or in competition with the Albert–Rudolf types. Such a general contact between the early lakes may also account for the similarity in their Pleistocene fish faunas (*see* Greenwood, 1959b; also new and unpublished observations), although one can equally argue that a fauna of this type was widespread in the prerift river systems.

Whatever the explanation, it does seem that both Lake Albert and Lake Rudolf have, since their inceptions, been relatively isolated from each other and from other water bodies in eastern Africa.

ACKNOWLEDGEMENTS

I am particularly grateful to Mr and Mrs A. J. Hopson for, in the first place, sending me the specimens on which this paper is based, and then for the great pains they took to obtain extra information and specimens. To my colleague Mr Gordon Howes goes my gratitude for all the help he has given in the preparation of the paper, for his draughtsmanship displayed in Text-figs. 5, 7 and 8, and for his skill in radiographing several hundred specimens. Finally, I thank my colleague Dr Keith Banister whose visit to Lake Rudolf (made since this paper went to press) led to his providing me with valuable extra information on Crater Lake species, and on the scarcity of *H. macconneli* males; despite diligent searches he found none of the latter.

REFERENCES

BEADLE, L. C. 1932. Scientific results of the Cambridge expedition to the East African lakes, 1930-1. 4. The waters of some East African lakes in relation to their fauna and flora. *J. Linn. Soc. (Zool.)*, **38**: 157-211.

BERRY, L. & WHITEMAN, A. J. 1968. The Nile in the Sudan. *Geogrl. J.* **134**: 1-37.

FRYER, G. & ILES, T. D. 1972. *The Cichlid Fishes of the Great Lakes of Africa. Their Biology and Evolution.* Oliver and Boyd. Edinburgh.

GREENWOOD, P. H. 1954. On two species of cichlid fishes from the Malagarazi River (Tanganyika), with notes on the pharyngeal apophysis in species of the *Haplochromis* group. *Ann. Mag. nat. Hist.* (12), **7**: 401-414.

—— 1959a. The monotypic genera of cichlid fishes in Lake Victoria, Part II. *Bull. Br. Mus. nat. Hist. (Zool.)*, **5**: 163-177.

—— 1959b. Quaternary fish fossils. *Explor. Parc. natn. Albert Miss. J. de Heinzelin de Braucourt*, **4**: 1-80.

—— 1965. On the cichlid fishes of Lake Nabugabo, Uganda. *Bull. Br. Mus. nat. Hist. (Zool.)*, **12**: 313-357.

—— 1971. On the cichlid fish *Haplochromis wingatii* (Blgr.) and a new species from the Nile and Lake Albert. *Revue Zool. Bot. afr.* **84**: 344-365.

—— 1973. A revision of the *Haplochromis* and related species (Pisces: Cichlidae) from Lake George, Uganda. *Bull. Br. Mus. nat. Hist. (Zool.)*, **25**: 139-242.

—— 1974. The cichlid fishes of Lake Victoria, east Africa: the biology and evolution of a species flock. *Bull. Br. Mus. nat. Hist. (Zool.)*, Suppl. No. 6: 1-134.

—— & GEE, J. M. 1969. A revision of the Lake Victoria *Haplochromis* species (Pisces, Cichlidae), Part VII. *Bull. Br. Mus. nat. Hist. (Zool.)*, **18**: 1-65.

TREWAVAS, E. 1933. Scientific results of the Cambridge expedition to the East African lakes, 1930-1. 11. The cichlid fishes. *J. Linn. Soc. (Zool.)*, **38**: 309-341.

TREWAVAS, E. 1935. A synopsis of the cichlid fishes of Lake Nyasa. *Ann. Mag. nat. Hist.* (10), **16** : 65–118.
—— 1938. Lake Albert fishes of the genus *Haplochromis*. *Ann. Mag. nat. Hist.* (11), **1** : 435–449.
—— 1973. On the cichlid fishes of the genus *Pelmatochromis* with proposal of a new genus for *P. congicus*; on the relationship between *Pelmatochromis* and *Tilapia* and the recognition of *Sarotherodon* as a distinct genus. *Bull. Br. Mus. nat. Hist.* (Zool.), **25** : 1–26.
WORTHINGTON, E. B. 1932. Scientific results of the Cambridge expedition to the East African lakes, 1930–1. I. General introduction and station list. *J. Linn. Soc.* (Zool.), **38** : 99–119.
—— & RICARDO, C. K. 1936. Scientific results of the Cambridge expedition to the East African lakes, 1930–1. 15. The fish of Lake Rudolf and Lake Baringo. *J. Linn. Soc.* (Zool.), **39** : 353–389.

P. H. GREENWOOD, D.Sc.
Department of Zoology
BRITISH MUSEUM (NATURAL HISTORY)
CROMWELL ROAD
LONDON SW7 5BD

Towards a phyletic classification of the 'genus' *Haplochromis* (Pisces, Cichlidae) and related taxa. Part I.

Peter Humphry Greenwood
Department of Zoology, British Museum (Natural History), Cromwell Road, London SW7 5BD

Contents

Introduction	266
Methods and materials	269
Methods	269
Materials	276
Classification	277
Section I	278
Haplochromis Hilgendorf	278
Description	278
Contained species	280
Diagnosis and discussion	281
Astatotilapia Pellegrin	281
Description	281
Contained species	283
Diagnosis and discussion	284
Astatoreochromis Pellegrin	285
Discussion	285
Contained species	286
Ctenochromis Pfeffer	287
Description	287
Contained species	289
Diagnosis and discussion	289
Thoracochromis gen. nov.	290
Description	291
Contained species	293
Diagnosis and discussion	294
Orthochromis Greenwood	295
Synonymy	295
Description	296
Contained species	297
Diagnosis and discussion	297
Section II	299
Serranochromis Regan	299
Subgenus *Serranochromis* Regan	299
Description	299
Contained species	302
Subgenus *Sargochromis* Regan	303
Description	303
Contained species	304
Diagnosis and discussion	306
Chetia Trewavas	307
Description	307
Contained species	308
Diagnosis and discussion	308

Pharyngochromis gen. nov.	310
Description	310
Contained species	311
Diagnosis and discussion	311
The Angolan *Haplochromis* species	312
Summary and conclusions	313
Key to the genera	315
Appendix 1	
A replacement 'generic' name for the Lake Malawi '*Haplochromis*' species	317
Appendix 2	
The taxonomic status of the genus *Limnotilapia* Regan, 1920	317
Acknowledgements	319
References	319
Index	321

Introduction

As currently recognized, the genus *Haplochromis* Hilgendorf encompasses over 300 species, some doubtless nominal but the majority of apparent biological validity (see Fryer & Iles, 1972; Greenwood, 1974a). It is the most speciose African taxon in the family Cichlidae and, next to the genus *Sarotherodon* has the widest distribution in the continent, extending from Tunisia in the north to Namibia (South West Africa) in the south. It is, however, virtually absent from west Africa, being represented there by only one or two species from Nigeria.

Amongst its numbers, indeed contributing the greatest number of species, are the well-known *Haplochromis* flocks of Lakes Victoria and Malawi (Trewavas, 1935; Fryer & Iles, 1972; Greenwood, 1974a), together with the smaller and less studied flocks of Lakes Edward, George, Turkana (Rudolf), Albert and Kivu (see Regan, 1921a; Poll, 1932; Trewavas, 1933; Trewavas, 1938; Greenwood, 1973, 1974b).

It is amongst the species of these various lacustrine flocks that one encounters the great range of anatomical, dental and morphological differentiation usually associated with the genus. The fluviatile species appear to be less diversified, but even here there is more diversity than is realized at first.

With this wide range of anatomical and morphological variation it is not surprising that the present concept of the genus, both in morphological and in phyletic terms, is very ill-defined. Indeed, the concept of *Haplochromis* seems to be based entirely on some intuitive appreciation of 'overall similarity' amongst its constituent species. There has been, so far, no real attempt to test the validity of the implicit monophyly of these species. The recognizable and often noted intrageneric variability in *Haplochromis* has, until recently, not been seen as an analytical taxonomic tool because thinking amongst systematists working on cichlids has been dominated by a 'size of the morphological gap' approach to supraspecific classification.

A reappraisal of the situation with this variation seen in terms of derived (apomorph) and primitive (pleisomorph) character states has not been applied to the genus as a whole (but see Greenwood, 1974a, for the Lake Victoria species). It is this basically Hennigian approach (Hennig, 1966) that I have attempted to apply to the problem. Its use, I believe, does allow one to produce a more realistic classification of the species now lumped together in *Haplochromis*, or separated from that genus because of their showing an extreme manifestation of features already indicated in species still retained in *Haplochromis*.

The taxon *Haplochromis* was first introduced by Hilgendorf (1888), as a subgenus of *Chromis*, for his new species aptly named '*obliquidens*' (see Greenwood, 1956a). The fine, closely packed and multiseriate teeth of '*obliquidens*', with their protracted and obliquely truncate crowns (see Fig. 7B), provided the diagnostic features for Hilgendorf's subgenus.

No further species were added to *Haplochromis* until Boulenger (1906) elevated the taxon to generic rank and included in it six new species from Lake Victoria and the Victoria Nile. Boulenger gave no reasons for raising Hilgendorf's subgenus to a full genus, nor did he attempt to define *Haplochromis* so as to accommodate the new species, none of which had teeth like those of

H. obliquidens. A footnote to the paper (Boulenger, 1906 : 443) might be interpreted as a generic definition, but it is completely inadequate and rather confusing, merely noting that '... in addition to the character of the dentition, intermediate between *Paratilapia* and *Tilapia*, the fishes of this genus differ from the latter in usually having a considerable portion of the maxillary bone exposed when the mouth is fully closed'.

The following year Boulenger (1907 : 495) did provide a formal definition of *Haplochromis*, in which genus he then synonymized Pffefer's (1893) genus *Ctenochromis* and Pellegrin's (1903) *Astatoreochromis* (now recognized as a distinct genus, see Greenwood, 1959a; Poll, 1974 and p. 285 below). This definition is, however, very vague and so worded that it is impossible to distinguish Boulenger's concept of *Haplochromis* from that of his redefined *Paratilapia* Bleeker.

The situation remained virtually unchanged, except for the addition of several more species, with the publication of the third volume in Boulenger's *Catalogue of African Freshwater Fishes* (1915) in which he again comments that some *Haplochromis* species '... vary to such an extent in their dentition that [they] might be referred to *Tilapia* and others to *Paratilapia*'.

Regan's (1920, 1922a) fundamental studies on the osteology of African Cichlidae, and his consequent revision of Boulenger's genera, resulted in many more species being included in *Haplochromis* (which then became the '... largest African genus', Regan, 1920 : 45). In his 1920 paper Regan also defined (in a footnote) several genera which, although apparently related to *Haplochromis*, differed from that genus in various dental features, both oral and pharyngeal.

Surprisingly, in the light of these other generic definitions, Regan was content to include in *Haplochromis* a majority of species whose dental characters were quite unlike those of the type species. In effect, Regan's redefinition of *Haplochromis* in these and subsequent papers (especially those of 1921a & b and 1922a & b) was only a slight improvement of that provided by Boulenger. *Haplochromis* remained a polymorphous assemblage of species showing a wide range of dental and other anatomical peculiarities, only united by having a particular kind of cranial apophysis for the upper pharyngeal bones. Since a similar apophysis occurs in other taxa defined by Regan, the monophyletic origin of *Hapolchromis* was not established.

That Regan was aware of his system's shortcomings is shown by remarks in his papers on the cichlids of Lakes Malawi (1921b) and Victoria (1922b). For example, regarding the *Haplochromis* of Lake Victoria he wrote (Regan, 1922b : 158): 'The species of *Haplochromis* exhibit almost as great a diversity as in Nyassa, yet there are certain features which enable one to say almost at a glance to which lake a species belongs', and on page 160: 'From what has been said above as to the evolution and relationships of the Cichlidae of Victoria, it will be evident that I do not regard the classification here proposed as entirely satisfactory'.

Regarding the species of Lake Malawi, Regan (1921b : 686) has this to say: '... the absence of evident relationship to species found elsewhere leads to the conclusion that the Nyassa species are a natural group and may, perhaps, have evolved in the lake from a single ancestral form'. Regrettably, Regan does not elaborate on his remark about the absence of evident relationship to species found elsewhere, particularly since a year later he was to place the majority of Lake Victoria species in the same genus.

When revising the Lake Victoria species, Regan (1922b) divided the *Haplochromis* into five subgenera, *Neochromis* for *H. nigricans* and *H. nuchisquamulatus*, *Bayonia* for *H. xenodon* (now considered a synonym of *Macropleurodus*, see Greenwood, 1956b), *Haplochromis* for *H. obliquidens*, and *Ctenochromis* for the remaining 42 species. These latter were characterized by their having conical or bicuspid teeth separated by an interspace from the smaller inner teeth, the other subgenera having variously specialized crown forms to the teeth. Regan disregarded, or perhaps failed to appreciate the principal diagnostic feature which Pfeffer (1893) used to diagnose *Ctenochromis*, namely the very small scales on the thoracic region. Both Pfeffer and Regan overlooked other diagnostic features in *Ctenochromis pectoralis* (type species of the genus), none of which is found in any of the 42 Victoria species placed in Regan's *Ctenochromis* subdivision of that flock (see p. 287 below). Recent research (summarized in Greenwood, 1974a) also indicates that these 42 species, and about an equal number described since Regan's 1922 revision, can be subdivided into several distinct groups.

Similar arguments can be marshalled against Regan's (1922a : 253) statement that '... the species (of *Haplochromis*) not peculiar to the Great Lakes all belong to the subgenus *Ctenochromis*, Pfeffer ...'; this aspect of the problem will be discussed later.

Since Regan's time, no real attempts have been made to subdivide the genus (which now contains almost double the number of species known to Regan). Some species have been separated off as mono- or oligotypic genera, but these actions have in no way simplified the problem either taxonomically or phylogenetically, and the genus has still not been shown to be a monophyletic unit.

Clearly, to test the phylogenetic integrity of such a large, ill-defined taxon will require much detailed and critical analysis. The present paper must be looked upon as a tentative first step in that direction. I shall limit my detailed analysis to those *Haplochromis* species which I have studied in some depth, viz. the species flock of Lake Victoria (which contains the type species, *H. obliquidens*) and those of Lakes Turkana, Albert, Edward and George, together with the few *Haplochromis* occurring in Lake Tanganyika, and the purely fluviatile species from Africa and the Middle East. Also included are the *Haplochromis*-like riverine genera *Orthochromis* Greenwood, *Serranochromis* Regan and *Rheohaplochromis* Thys van den Audenaerde, and the partly lacustrine *Astatoreochromis* Pellegrin. Unfortunately, I have been unable, through lack of firsthand knowledge, to include the Lake Malawi *Haplochromis* flock. However, I trust that the results of my analysis of these other *Haplochromis* species will enable workers on the Malawi fishes to review the species of that lake in a new light.

My review of anatomical, osteological and morphological features, including details of secondary sexual markings and coloration, has yielded one particularly significant (but not surprising) result; there is, apparently, not one derived feature shared only by the 190 species examined.

The commonly occurring tooth form, an unequally bicuspid tooth, is found in several other genera, as is the unicuspid and caniniform type. Even some of the specialized dental types seem to have evolved independently in other genera, these genera, and those in which bi- and unicuspid teeth also occur, each being recognizable on the basis of derived features not shared by *Haplochromis*.

The structure of the cranial apophysis for the upper pharyngeal bones (see Regan, 1920) is probably a derived feature (see Greenwood, 1978), but again it is a feature widely distributed amongst several genera whose close affinity with *Haplochromis* cannot be established. At best the pharyngeal apophysis can be used as an indicator of relationship at a more distant level than the 'generic' one (see Greenwood, 1978).

No derived features of the anatomy or the squamation are universally shared amongst all the species although, as with various other characters, distinct groups can be defined within *Haplochromis* on the basis of shared derived features.

The anal fin markings found in adult male *Haplochromis*, the so-called anal ocelli or eggdummies (see Wickler, 1962a & b; Trewavas, 1973), have been considered a unique feature of the genus. Trewavas (1973 : 34) expressed the generally held view on these markings when she wrote '... within their endless diversity the species of *Haplochromis* have almost universally in common a feature of the colour-pattern, the well-known *ocellar spots* on the anal fin of the male' (italics mine). Certainly such ocellar markings are present in all the described species of *Haplochromis* from Lake Victoria, Edward, George and Kivu, and probably in those from Lake Turkana as well. But, true ocelli (i.e. a central coloured spot with a clear surround) are not found in the species of Lake Albert, in the majority of species occurring in the rivers, nor even in many of the Lake Malawi species (see figs in Axelrod & Burgess, 1977). Coloured markings do occur on the anal fins of these fishes (sometimes in both sexes), but are in the form of spots without a clear surround, often smaller than the true ocellar type, sometimes more numerous and covering the greater part of the fin, sometimes only as one or two spots, or, less commonly, similar in number (3–5) and linear arrangements to the true ocellar type.

Clearly, the presence of ocellar anal markings cannot be considered a character of *Haplochromis* as that genus is currently conceived, and the value of anal markings *per se* as an indicator of phyletic relationship must be reassessed.

Although a monophyletic origin for the 'genus' *Haplochromis* cannot be established, it is possible to recognize several seemingly monophyletic lineages (reconstructed on the basis of synapomorphic characters) amongst the species of Lake Victoria (Greenwood, 1974a and unpublished). None of the six major lineages recognized in that lake, however, could be interrelated on a sister-group basis (although sister-groups could be recognized within five of the lineages themselves). In other words, the synapomorphic features of each lineage are superimposed on a basic, plesiomorphic 'bauplan' shared by all*.

A similar picture emerges when the fluviatile species, and those from Lakes Albert, Turkana and Tanganyika are examined closely. That is, one can postulate a number of lineages (some containing both fluviatile and lacustrine members), but none can be further interrelated on the basis of synapomorphic features.

With the possible exception of their occurrence in two species (one from Lake Victoria, the other from Lake George), none of the apomorph features used to delineate these lineages has been observed amongst the '*Haplochromis*' species of Lakes Victoria, Edward, George and Kivu.

Although no apomorph character has been found to unite all the species of Lakes Victoria, Edward, George and Kivu, and thus suggest their common ancestry, species from the different lakes can be grouped into common lineages each of presumed monophyletic origin. For that reason the '*Haplochromis*' of Lakes Edward, Kivu and George will be treated together with those of Victoria in a forthcoming paper (except for those species which are now referred to the redefined genus *Haplochromis*, see p. 280).

To summarize, the so-called *Haplochromis* species of Africa (excepting those of Lake Malawi which are not included in this review) can be split into a number of major lineages. Most of these lineages are characterized by derived features unique to its members.

The different lineages cannot be interrelated on a sister-group basis for want of ascertainable synapomorphic features which would permit the recognition of their sister-group status. I use the qualification 'most of these' because one of the groups cannot be defined on the basis of even a single shared apomorph character. This is the group in which must be placed the widespread *H. bloyeti* species complex of east Africa (see Greenwood, 1971, 1974a) and, probably, certain of the generalized endemic species of Lake Victoria, Edward and Kivu; it is recognized merely on the overall similarity (and plesiomorphy) of its constituent species.

Wherever breeding habits are known, members of the various lineages described in this paper are female mouth brooders, and all have a '*Haplochromis*'-type cranial apophysis for the upper pharyngeal bones (Greenwood, 1978), features shared with the '*Haplochromis*' and several seemingly related species in Lake Malawi (Trewavas, 1935; Greenwood, 1978). Oral brooding and its associated spawning behaviour, as compared with substrate spawning and brood-care, is a derived condition; the '*Haplochromis*'-type apophysis would also seem to be a derived feature. One may therefore hypothesize a shared common ancestry, at some point, both for the lineages described below and for those which eventually will be recognized amongst the Lake Malawi haplochromine species (i.e. those with a '*Haplochromis*'-type pharyngeal apophysis and, probably, species with a '*Tropheus*'-type apophysis as well; see Greenwood, 1978).

For the moment, however, and until it is possible to interrelate dichotomously the various lineages on a sister-group basis, one is faced with a series of unresolved dichotomies (see, for example, the problem discussed on p. 313). In classifying this assemblage I have followed the convention suggested by Nelson (1972), namely that the taxa (i.e. the individual lineages) be given equal rank. At this stage in our knowledge of supraspecific relationships amongst African cichlids, generic rank would seem to be the most appropriate.

Methods and materials

Methods

In essence I have attempted to break up the 'genus' *Haplochromis* into a number of monophyletic

---

* It has been assumed (on the basis of overall morphological similarity between the least specialized members of each lineage) that the endemic *Haplochromis* species of Lake Victoria are of monophyletic origin (Greenwood, 1974a). Since no apomorph feature unique to the Victoria species has yet been found, that hypothesis is without formal support.

lineages, the members of each lineage being related by their relative recency of common ancestry. Recency of common ancestry, in turn, is recognized by members of a lineage possessing derived (apomorph) characters which are not shared with other species.

Determining the primitive (plesiomorph) or derived status of characters in the Cichlidae is at present a very difficult task. No guidance is available from the entirely inadequate fossil record, and the family's nearest living relatives have yet to be recognized. Comparisons between different character states (outgroup comparison, see Hecht & Edwards, 1977) ideally should be carried out across the whole family. As there are well over 600 nominal species in Africa and America, few of which have been studied in the detail necessary for proper phyletic analysis, the level of outgroup comparisons employed in this paper is, perforce, a low one.

All comparisons have been restricted to African taxa, in particular to species and lineages within the group having a '*Haplochromis*'-type of pharyngeal apophysis. This decision was made on the assumption that all such taxa were derived from a common ancestor, albeit a distant one, and that the '*Haplochromis*'-type apophysis, relative to the '*Tilapia*'-type, is itself a derived character. The most detailed comparisons, of course, have been those made between species comprising the lineages discussed in this paper.

Outgroup comparisons have also been made with species having a '*Tilapia*'-type apophysis, in particular the lineages represented by the genera *Sarotherodon* and *Tilapia*.

When comparisons were made with *Haplochromis* from Lake Victoria, the Victorian lineages were those discussed in Greenwood (1974a). Since no such breakdown is available for the endemic *Haplochromis* of Lake Malawi or for the endemic genera with a '*Haplochromis*'-type apophysis in Lake Tanganyika, these various taxa were not involved in the analysis.

The particular characters and character transformations studied are those which, after a preliminary survey of the taxa involved*, seemed to be most likely to yield information on their derived or primitive states within the material available and within the limits of the tests which could be applied to the conclusions reached.

As might be expected, the principal test was that of the distribution of a character state amongst the species compared. The state having the widest occurrence is assumed to be the most primitive one, that with the most circumscribed distribution the derived one (the so-called commonality principle of Schaeffer, Hecht & Eldredge, 1972).

The characters finally selected, and a few others that deserve comment, can now be discussed.

(*i*) *Squamation*. All *Haplochromis* have the scales on the chest region (the area anterior to a line through the pelvic and pectoral fin insertions, and ventral to a horizontal line through the ventral part of the pectoral fin insertion) smaller than those on the ventral and ventrolateral parts of the body. The common condition is that in which the size change between the scales of the two regions is a gradual one, see Fig. 1; even when, as in *H. squamulatus* of Lake Victoria, the chest scales are noticeably small, the size change is still gradual (see fig. 17 in Greenwood, 1967). The less frequent condition is that in which the size transition (usually along the line between pectoral and pelvic fin insertions, but sometimes a little further posteriorly) is abrupt; since in these fishes the chest scales are generally small and numerous, the chest squamation is noticeably distinct from that of the belly and ventral flank regions (Figs 2 & 3).

A totally scaled chest, irrespective of squamation pattern, is the usual condition; circumscribed, bilaterally symmetrical naked patches are uncommon and are confined to species showing an abrupt size transition in thoracic-abdominal scale sizes. A completely naked chest is the most uncommon condition and would seem to be the end point in the apomorphic morphocline: abrupt size change ➤ bilateral naked patches ➤ completely naked chest.

Although the ventral body scales extending posteriorly from the pelvic fin insertions to the anus are smaller than those on the lateral and ventrolateral aspects of the flanks, the size gradation between the two fields is generally gradual. However, in a few species the ventral (belly) scales are much reduced in size and thus are clearly demarcated from the flank scales above them. This

---

* The *Haplochromis* species of Lakes Victoria, Albert, Turkana, Tanganyika, Edward, George and Kivu, of the African rivers and those of Syria and Israel, and the species of *Serranochromis*, *Rheohaplochromis*, *Orthochromis*, *Astatoreochromis*, *Macropleurodus*, *Platytaeniodus* and *Hoplotilapia*, a total of some 390 species.

condition is correlated with an equally marked and abrupt size reduction in the scales on the chest, so that the tiny abdominal scales appear as a posterior and ventrolateral extension of those on the chest (Fig. 3). Such an arrangement is also considered to be a derived condition.

Surprisingly, in a group of species where most morphological features appear as elements in a continuum of differentiation, the various scale patterns discussed above are very trenchantly separated from one another. The few intermediate specimens I have observed are clearly individual rather than populational or specific variants.

As with the chest, a completely scaled cheek is the common condition, the scales being arranged in three or four horizontal rows. Reduced squamation is encountered infrequently, but ranges from a narrow naked band (one or two rows deep) along the ventral margin, to an almost completely naked cheek with only the suborbital row, or part of that row, persisting.

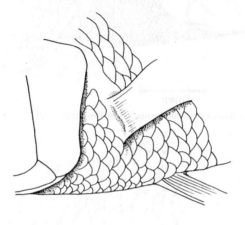

Fig. 1 Thoracic-abdominal scale transition in *Astatotilapia nubila*; left lateral view.

On the principle of commonality (Schaeffer *et al*, 1972), strongly ctenoid body scales should be looked upon as the primitive condition, and an increase in the area of the body covered by cycloid or reduced ctenoid scales should be considered the derived one*.

Some uncertainty about this conclusion could be raised by the situation in *Hemichromis*, also a '*Haplochromis*' group species (see Regan, 1922a). Here the scales are mostly cycloid with a few weakly ctenoid ones confined to the anterior part of the body; that is, a presumably derived condition. But *Hemichromis* species are substrate spawners and brood guarders, a presumed primitive condition amongst African cichlids. Since certain other characters in *Hemichromis* are apparently derived ones (the unicuspid outer teeth, the number of inner tooth rows (one or none), and the form of the upper jaw), the cycloid scales may have evolved independently in the lineage. On the other hand, the presence of cycloid scales in '*Tilapia*' group species (see footnote), some of which are also substrate spawners and all of which have an apparently plesiomorph type of pharyngeal apophysis (see Greenwood, 1978), would appear to strengthen the argument for considering cycloid scales as primitive features. In the face of such contradictory observations it would seem advisable not to use this type of scale ornamentation in phyletic analysis.

All the *Haplochromis* and *Haplochromis* group species used in this review (see footnote p. 270) have less than the proximal two-thirds of the caudal fin covered by small scales; usually only the

* Most taxa in the '*Tilapia*' group, as defined by apophyseal structure (see Greenwood, 1978), have cycloid scales, although some have a few weakly ctenoid scales on the anterior part of the body).

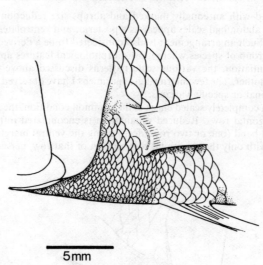

Fig. 2 Thoracic-abdominal scale transition in *Thoracochromis wingatii*; left lateral view.

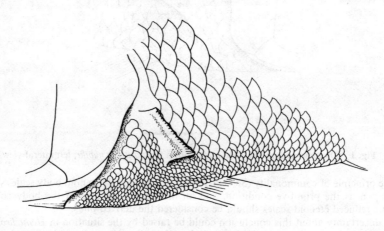

Fig. 3 Thoracic–abdominal scale transition in *Orthochromis polyacanthus*; left lateral view.

proximal half is covered. In contrast, all the endemic lacustrine species from Lake Malawi have the entire fin densely scaled (Trewavas, 1935). The partly scaled condition is assumed to be the plesiomorph one.

(*ii*) *Dentition*. Outer tooth row in both jaws. The most frequently occurring tooth form (Fig. 4) is that with an unequally bicuspid crown, moderately distinct neck and relatively stout body firmly attached to the underlying bone. Neither cusp is strongly compressed, their tips are acute or subacute and lie in or but slightly outside a vertical drawn through the corresponding outer margin of the tooth's body. Such teeth, apart from providing the definitive dental form in many species, also precede the definitive tooth type in species having unicuspid teeth in adult fishes, and also, in at least some species, precede the definitive types when these are much modified versions of the basic bicuspid (e.g. in *H. obliquidens*). Unfortunately, ontogenetic data on tooth replacement are not available for many species, so the generality of the latter observation is unknown.

Because of its common occurrence and its primary position in the ontogenetic sequence of tooth replacement, the unequally bicuspid tooth is taken to be the plesiomorph dental type. Bicuspid teeth in which there is a differential growth of one cusp (usually the larger one) or equal development of both cusps are considered to be derived features, as are unicuspid teeth.

As mentioned above, most taxa having a definitive outer row dentition composed of unicuspid teeth also have an ontogenetically earlier one of bicuspids (usually persisting until an individual fish is between 80 and 100 mm standard length). Any shift forward in the time or body size at which the definitive unicuspid teeth appear can therefore be interpreted as being a derived condition.

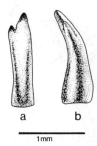

Fig. 4 Outer row jaw teeth (premaxillary) of *Astatotilapia flaviijosephi*. A. Labial view. B. Lateral view (posterior aspect).

From one to six (rarely as many as twelve) enlarged and unicuspid teeth occur posteriorly on the premaxilla, even when the other teeth on that bone are bicuspids. The replacement of these unicuspids by teeth similar to those on the rest of the premaxilla must be considered a derived condition.

Inner row teeth. Here, on the grounds of common occurrence, small, tricuspid teeth must represent the primitive condition. As with the outer teeth, there can be an ontogenetic succession of teeth types, tricuspids or a mixture of tri- and bicuspids preceding unicuspids. Occasionally some or all inner rows are composed of highly modified bicuspid types resembling, albeit on a smaller scale, those of the outer row (e.g. *Haplochromis obliquidens* and *Macropleurodus bicolor*; see Greenwood, 1974a).

The presence of inner teeth other than tricuspids is a derived condition. Since most commonly there are from 2 to 3 rows of inner teeth, any increase or decrease in the number of rows must also indicate an apomorph condition.

(*iii*) *The lower pharyngeal bone and its dentition.* The most commonly encountered form of lower pharyngeal bone has an approximately equilateral, triangular dentigerous surface, is not noticeably thickened or robust, and has its anterior blade-like portion neither noticeably elongate nor short (Fig. 5).

The teeth are arranged anteroposteriorly in about 30 to 50 rows, with those in the two median and in the posterior transverse row stouter than the others but, like them, retaining an unequally bicuspid crown in which the minor cusp is a near horizontal shoulder and the major one is weakly falciform and vertically aligned.

Apomorphic derivations from this basic type include changes in overall outline shape of the dentigerous area (Fig. 14), elongation of the anterior blade, increase or decrease in the number of tooth rows, an increase or, less commonly, a decrease in the number of rows of coarser teeth

(Figs 20 & 8), and changes in crown morphology of the teeth (generally a process of molarization associated with a general coarsening of tooth form; see Fig. 18B).

(*iv*) *Neurocranial morphology*. Modal neurocranial form (and thus the presumed plesiomorph condition) is best appreciated from a drawing (Fig. 6).

Salient features are the moderately high supraoccipital crest (*c*. three-quarters of the depth of the otic skull region measured from roof to ventral parasphenoidal face, but excluding the pharyngeal apophysis); the preorbital skull profile (from vomerine tip to the anterior point of the supraoccipital crest) rising at an angle of $c.45°$, its outline gently curved and its ethmovomerine region sloping forwards and downwards at a slight angle; the preotic part of the skull (measured from the vomerine tip to the anterior vertical wall of the prootic bone) comprising some 55–60% of the total length of the neurocranium, and the otic region of the skull not inflated. The pharyngeal apophysis is not enlarged, and the prootic does not contribute to the articular surface (Greenwood, 1978).

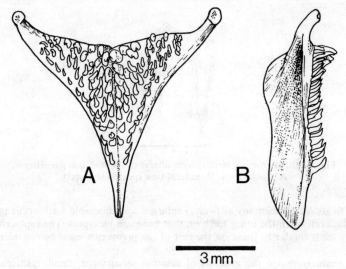

Fig. 5 Lower pharyngeal bone of *Astatotilapia bloyeti*. A. Occlusal view. B. Right lateral view (bone aligned vertically).

Derivative conditions include elongation of the preotic part of the skull (to about 70% of the neurocranial length) correlated with a flattening of the preorbital skull profile (Fig. 13); narrowing of the otic region, and in some variants a relative lowering of the supraoccipital crest; the retention of a basic skull form in the otic region but a marked increase in the slope of the ethmovomerine region (in some species almost to the vertical) and a correlated increase in the slope and curvature of the preorbital skull profile; the retention of basic otic and ethmovomerine regions but the elevation of the preorbital skull roof so that the neurocranium becomes higher and more angular in outline (see Greenwood, 1974*a*, for further analysis and figures).

Departure from the plesiomorph condition for the pharyngeal apophysis is always associated with an hypertrophy of the upper and lower pharyngeal bones and their dentition (see Greenwood, 1965*a*, 1974*a*, 1978).

(*v*) *Anal fin markings (egg-dummies) in male fishes*. Reference has already been made (p. 268) to the variety of these markings in *Haplochromis*. (See Wickler, 1962*a* & *b*, 1963 for a discussion of their importance in the breeding biology of these fishes).

Regrettably there is little information about these markings in live fishes, and what has been recorded is often insufficiently detailed to be of value. For instance, it is important to know if the markings are merely coloured spots, whether each spot has a contrasting border, or whether it is truly an ocellus with a wide and translucent surround. The number and distribution of the markings are also important data. My own observations on live fishes from different parts of Africa, and on preserved material as well, all suggest that the anal markings (or their absence) may be of considerable value in helping to define lineages. But, because of a paucity of information for many species considered below it has proved impossible to use the character fully in this study.

It seems reasonable to assume that the egg-dummy markings (using that term in its widest sense and not just for true ocelli) were derived from coloured streaks and spots like those that are an almost universal feature on the dorsal fins of cichlids (see Wickler, 1962a). The first steps in the evolution of egg-dummies from a maculate colour pattern would involve a slight reduction in the number of spots and a consequent increase in the space between them, and the intensification or alteration of their colour so as to differentiate the anal spots from those in the dorsal fin. The end point in this process of differentiation seemingly would be reached with the development of ocellar spots.

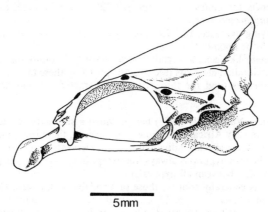

Fig. 6 Neurocranium of *Astatotilapia bloyeti*; left lateral view.

In species with true ocellar egg-dummies (e.g. the *Haplochromis* of Lake Victoria; see Greenwood, 1974a, especially plate 1) the number of spots is reduced to modes of 3 or 4, the coloured centre of each spot is ovoid in outline, generally has a narrow black or dark border and is surrounded by a clear zone of fin membrane; other spots and markings on the fin are suppressed, although the greater part of the fin may have a coloured flush. The result is a most distinctive mark, with an illusion of three dimensionality, that is readily distinguished from any other fin or body markings.

There are, of course, other kinds of anal marking which, in their appearance, size and distribution on the fin, are intermediate between the supposed plesiomorph type (numerous, non-ocellate spots) and the presumed apomorph kind described in the last paragraph. Their possible phylogenetic importance will become apparent when more information is available not only on their appearance in live animals, but also on their functional role in mate recognition, courtship and spawning.

For the moment one can assume that the ocellar spots represent the apomorph condition and that the multiple spot type of anal marking is the plesiomorph one. Those *Haplochromis*-like species apparently without any spatially or chromatically differentiated anal markings (e.g. *Orthochromis malagaraziensis*; Greenwood, 1954) provide a particular problem because we know nothing of their reproductive behaviour. Thus the absence of 'egg-dummies' cannot necessarily be construed as representing a plesiomorph condition in these species.

(vi) *Vertebral numbers.* The modal range of total vertebral counts (excluding the fused $PU_1 + U_1$ urostylar element) is 27–29 (comprising 12–14 abdominal and 15 or 16 caudal centra). On the grounds of its being the modal number, it is taken to be the plesiomorph condition.

Apomorphic deviations occur amongst the *Haplochromis* species of Lake Victoria where a few species show a higher modal count (30–32); these are all long-bodied piscivorous predators and the increase in the number of vertebrae occurs in the caudal section of the column. Parenthetically, it may be noted that Lake Malawi *Haplochromis* species with more than 32 vertebrae also show an increase in the number of caudal elements.

The most marked increase in vertebral numbers (apart from that in some Malawi species) is found in the genus *Serranochromis*. Here the modal counts are 33 and 34 (abdominal modes 16–17, caudal modes also 16 and 17), with an increase in the number of abdominal rather than the caudal elements as was the situation in the Victoria and Malawi *Haplochromis* species.

A similar increase in the number of abdominal vertebrae is also found in the seven Zambezi species of so-called *Haplochromis* revised by Bell-Cross (1975). In these species there is, however, a reduction in the number of caudal vertebrae as compared both with *Serranochromis* and with the plesiomorphic *Haplochromis* condition (14 and 15 in the Zambezi species, 16 and 17 in the others). In consequence, the modal total count (29–32) for the Zambezi fishes overlaps that of the plesiomorph *Haplochromis* type.

For want of falsifying evidence, the *Serranochromis* and 'Zambezi *Haplochromis*' conditions are both considered to be apomorphic ones.

In some *Haplochromis*-like genera, certain species have low counts for caudal vertebrae, but modal ones for the abdominal elements (see p. 290); probably these taxa, in the phyletic context of their particular lineages (and for this particular character), should be considered derived (i.e. autapomorphic).

(vii) *Caudal fin skeleton.* Vandewalle (1973) has provided a summary of the caudal fin skeleton in 108 cichlid species, mostly African. He shows that, overall, there is a remarkable constancy in this feature but that in some species individual hypural elements may fuse (especially hypural 1 with 2, and 3 with 4; the fifth hypural is always free except in one species (see p. 292) – Vandewalle finds no instance of fusion between all hypurals).

My own observations generally confirm those of Vandewalle, but strongly indicate that, with one possibly exceptional lineage (see p. 297), hypural fusion is an individual and not a specific or lineage trait. It is thus of very restricted value as an indicator of phyletic affinities, as is the organization of the whole caudal fin skeleton amongst the taxa examined.

(viii) *Number of dorsal and anal fin rays.* Amongst the *Haplochromis* species and related taxa reviewed (see footnote p. 270) the modal numbers of dorsal fin rays are 15 and 16, and of branched rays 9 and 10. Except for obvious individual variants, and two species of *Astatoreochromis*, all these taxa have 3 spinous rays in the anal fin and, modally, 8 or 9 branched rays.

If, on the principle of commonality, these numbers are taken to be the basic (i.e. plesiomorph) counts, then modal ray counts for either fin that are higher or lower should be considered derived features.

(ix) *Gill rakers.* There is a fairly narrow range of both gill raker numbers and shapes (counts and observations restricted to the outer row of gill rakers on the lower part of the first gill arch). In most species the rakers are relatively robust, simple structures (with sometimes the upper 2 or 3 of the series flattened and bi-, tri- or even polyfid), of moderate length and numbering from 7 to 12 (modal counts 8 and 9).

Materials

All the BMNH material (spirit specimens, alizarin preparations, dry skeletons and radiographs) of all the taxa named in this paper has been examined, as have the Museum's collections of *Haplochromis* species from Lakes Victoria, Edward, George and Kivu, and selected specimens from the collection of Lake Malawi *Haplochromis* species and related genera.

In addition, the following specimens, borrowed from other institutions, have also been studied (and radiographed).

Haplochromis albolabris	(Holotype)	H1784
Haplochromis angusticeps		ANSP 54369–76
Haplochromis bakongo	(Paratypes)	RMAC 16945–947
Haplochromis buysi	(Holotype)	WH P1219
Haplochromis darlingi		AM/P2461
Haplochromis fasciatus		RMAC 48407–415
Haplochromis giardi	(Holotype)	MHN A2754
Haplochromis luluae	(Paratypes)	ANSP 51759–62
Haplochromis oligacanthus		RMAC 167930–931
Haplochromis polli	(Paratypes)	RMAC 99403–404
Haplochromis stappersi		RMAC uncatalogued
Haplochromis thysi	(Holotype)	RMAC 163991
Haplochromis toddi	(Holotype)	RMAC 1346
Haplochromis torrenticola		IRSN 1809–1960
Chetia brevis	(Holotype)	AM/P951
Chetia brevis	(Paratypes)	AM/P952
Chetia brevis		AM/P1425–6
Chetia flaviventris		AM/P1298
Ctenochromis pectoralis	(Lectotype)	ZMH402
Ctenochromis pectoralis	(Paratypes)	ZMH403

AM/P, Albany Museum; ANSP, Academy of Natural Sciences of Philadelphia; IRSN, Institut Royal des Sciences Naturelles de Belgique; H, Zoologisches Museum, Hamburg; MNH, Museum National d'Histoire Naturelle, Paris; RMAC, Musee Royal de l'Afrique Centrale, Tervuren; WH, Windhoek Museum; ZMH, Zoologisches Museum, Berlin.

Classification

Applying the methodology and reasoning discussed above and in the Introduction, nine lineages, here given generic rank, may be recognized amongst the taxa studied. With the exception only of *Chetia*, each of these genera now contains species that were previously placed in *Haplochromis*.

Unless indicated otherwise, all the genera have a cranial apophysis for the upper pharyngeal bones formed from the parasphenoid and basioccipital (see '*Haplochromis*'-type apophysis in Greenwood, 1978) and a caudal fin scaled on its proximal half or less.

All vertebral counts quoted exclude the fused PU_1 and U_1 centra, and may thus be lower than those used by some other authors; abdominal vertebrae are identified as those bearing pleural ribs (including, of course, the first two vertebrae that have no ribs), and the caudal centra as those without ribs but, except occasionally the first (anterior) centrum, with a haemal arch.

Since the genus *Haplochromis* is now restricted to five species (see p. 280), difficulties arise when reference is made either to species formerly included in that genus, but which have not yet been assigned to other genera, or to the former concept of the genus *Haplochromis*. To avoid confusion, I have adopted the convention proposed and used by Patterson & Rosen (1977 : 163) for dealing with such situations. Namely, to prefix the species name with its former generic name cited between quotation marks, i.e. '*Haplochromis*' *nigricans* or '*H*' *nigricans*. When reference is made to the former concept of the genus the generic name alone, but in quotation marks, is used.

The Lake Victoria '*Haplochromis*' species will be reviewed in a paper now in preparation; until its publication these species can be referred to by using the old generic name in quotes. As a temporary expedient for general use until such times as the Lake Malawi '*Haplochromis*' are revised, a purely formal generic name for these species is proposed on p. 317.

Species mentioned in this paper are listed in the index on p. 321 under their former generic names (usually *Haplochromis*), with a reference first to the page on which they are listed in their new generic grouping, and secondly to the page on which that genus is described.

In the generic descriptions, presumed apomorph (i.e. derived) character states are italicized.

The generic revision which follows is arranged in two parts. After redefining the genus *Haplochromis* Hilgendorf, 1888, the first section will deal with '*Haplochromis*' species from Lakes Turkana, Albert, Tanganyika and Mweru, and with those from the Nile and Zaire river drainage systems, and the rivers of Kenya, Uganda and Tanzania.

The genera *Orthochromis* Greenwood, 1954, and *Astatoreochromis* Pellegrin, 1903, will also be considered in this section of the paper.

The second section (p. 299) will be concerned with the genera *Serranochromis* Regan 1920 and *Chetia* Trewavas, 1961, together with those '*Haplochromis*' species from the Zambesi, Limpopo and Angola river systems, which were thought to be related to *Serranochromis* and *Chetia* (see Trewavas, 1964).

Section I

HAPLOCHROMIS Hilgendorf, 1888

TYPE SPECIES: *Chromis* (*Haplochromis*) *obliquidens* Hilgendorf, 1888 (type specimens in the Humboldt Museum, Berlin).

Description

Body relatively deep (depth 35–40 % of standard length).

Squamation. Scales on the body below the lateral line, and behind a line through the pectoral and pelvic fin insertions, are ctenoid; those above the upper lateral line and on the head and chest are cycloid.

The small scales on the chest grade imperceptibly in size with those on the ventrolateral and ventral aspects of the flanks (p. 270).

Cheek and chest fully scaled.

Lateral line with 29–34 scales (modal range 30–32); all but the last 3 or 4 scales of the upper lateral line are separated from the dorsal fin base by at least two scales of approximately equal size.

Neurocranium. The skull is of a generalized type (see p. 274; Fig. 6, and Greenwood, 1974*a*), its ethmovomerine region having only a slight downward slope, the dorsal surface of the vomer sloping in the same plane and at the same angle as the anterior part of the skull roof; the preotic part of the skull comprises some 55–60 % of the total neurocranial length.

Vertebral numbers: 28–30 (modes 28 and 29), comprising 12–14 (mode 13) abdominal and 15 or 16 caudal elements.

Dentition. The outer teeth in both jaws are weakly bicuspid or unicuspid, *the crown of the tooth compressed and noticeably expanded relative to its slender, cylindrical neck and body* (Fig. 7). *The major cusp in bicuspid teeth is very much larger than the minor one, which is often little more than a slight, obliquely truncated basal point on the posterior margin of the anteriorly protracted and slightly incurved (i.e. buccally directed) major cusp. The compressed, anteriorly protracted and dorsoventrally expanded major cusp gives to the tooth, be it bi- or unicuspid, the appearance of having an obliquely truncated crown. The tip of this cusp lies outside the vertical formed by the anterior margin of the tooth's body.*

All outer teeth, save in some species for a few posterior teeth on the premaxilla, are moveably attached to the underlying bone.

In some species the posterior one to six teeth on the premaxilla are unicuspid or acutely bicuspid, and are stouter and larger than the others in that series; *these posterior teeth in other species of the genus closely resemble, in size and cusp morphology, their anterior congeners.*

Teeth forming the inner rows in both jaws mostly are small and tricuspid, but *in some species the anterior and anterolateral teeth in the outermost row may be identical with those of the outer row.* The inner teeth are arranged so as to form a tooth band that is wide anteriorly and anterolaterally, but narrow posterolaterally. This is effected by an *increase in the tooth rows from the primitive state of two or three to the derived condition of from 4 to 6.*

Lower jaw is relatively slender in lateral aspect and not noticeably deepened posteriorly; the length of its dentigerous surface is equal or about equal to the greatest depth of the dentary (as

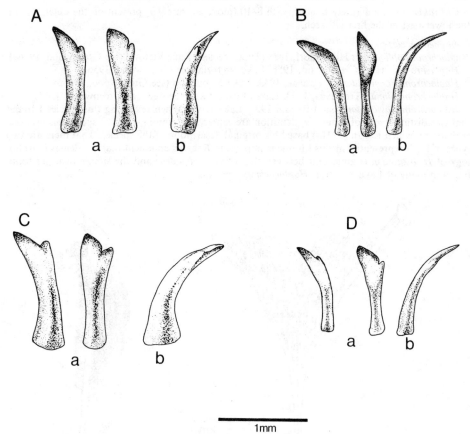

Fig. 7 Outer row jaw teeth (premaxillary) in various *Haplochromis* species; a. labial view, b. lateral (posterior) view. A. *H. annectidens*. B. *H. obliquidens*. C. *H. limax*. D. *H. astatodon*.

measured vertically from the posterior tip of its ascending coronoid process to the ventral margin of the bone).

Lower pharyngeal bone and dentition. The dentigerous surface of the bone is triangular and approximately equilateral in outline (Fig. 8). The teeth, except for those forming the posterior transverse row, are slender with the greater part of the cusp protracted and curved posteriorly so that most of the occlusal surface lies nearly parallel with the surface of the pharyngeal bone. (*The absence of a distinctly coarser median series of teeth probably is a derived feature*; see p. 273.)

Dorsal fin with 14–16 (modes 15 and 16) spinous and 8–10 (mode 9) branched rays.

Anal fin with 3 spinous and 7–10 (mode 9) rays.

Caudal fin skeleton without fusion between any of the hypural elements; none of the species reviewed here was examined by Vandewalle (1973).

Caudal fin truncate or subtruncate, the posterior margin straight or weakly emarginate.

Pelvic fins with the first branched ray the longest.

Anal fin markings. True ocellar egg dummies (see p. 274), *usually 3 or 4 in a single row, are present in adult males.* Some females may have a similar number of small, non-ocellate spots present in the same position on the fin.

Gill rakers are moderately slender, with 8–10 (mode 9), rarely 7, present on the outer row on the lower part of the first gill arch.

Contained species

Haplochromis obliquidens Hilgendorf, 1888 (Type species). Lake Victoria (see Greenwood, 1956a).
 Haplochromis lividus Greenwood, 1956. Lake Victoria (see Greenwood, 1956a).
 Haplochromis annectidens Trewavas, 1933. Lake Nabugabo (see Greenwood, 1965b).
 Haplochromis limax Trewavas, 1933. Lakes Edward and George (see Greenwood, 1973).
 Haplochromis astatodon (part) Regan, 1921. Lake Kivu. When reviewing this species I found that two distinct types of outer jaw dentition are represented amongst the 13 specimens and one skeleton on which Regan (1921a) based his original description of the species. That there are two types of teeth represented in this sample is implicit in Regan's comment that the dental morphology of *H. astatodon* is annectant between that of *H. obliquidens* and the simple bicuspid tooth found in many of Lake Victoria '*Haplochromis*' species.

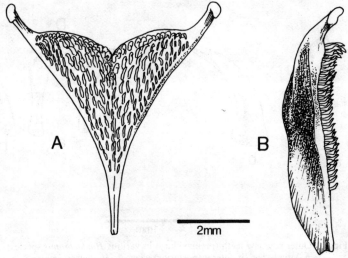

Fig. 8 Lower pharyngeal bone of *H. lividus*. A. Occlusal view.
B. Right lateral view (bone aligned vertically).

Six of the syntypical specimens (BMNH reg. nos 1906.9.6 : 124–129) have outer teeth remarkably like those of *H. obliquidens*. That is, the crown is obliquely truncate and greatly produced (Fig. 7D), the neck and body fine, and the teeth are moveably implanted in the gum tissue. However, in these specimens a minute second cusp is present on many of the anterior and anterolateral teeth, a condition seen only amongst some of the posterior teeth in *H. obliquidens*. These particular specimens of *H. astatodon* also resemble *H. obliquidens* in having many teeth in the outermost of the multiseriate inner rows enlarged and often obliquely cuspidate; the inner teeth, like those in *H. obliquidens*, being small and tricuspid.

The pharyngeal dentition in those specimens is also like that in *H. obliquidens*, but the two species differ in certain morphometric features and thus would seem to represent distinct species.

The eight other syntypical specimens of *H. astatodon* (BMNH reg. nos 1906.9.6:130–132; 1977.5.2:1–4 (the material was originally under-registered, hence the lot registered in 1977)), and a skeleton (BMNH reg. no. 1906.9.6:133) have distinctly bicuspid teeth in which the major cusp, although somewhat obliquely truncate, is neither protracted nor expanded, nor is it markedly compressed; the minor cusp is not so greatly reduced as it is in the other syntypical specimens of *H. astatodon*, or in *H. limax*, *H. annectidens* and *H. obliquidens*.

The tooth form and dental arrangement in these aberrant syntypes does, however, approach closely that found in '*Haplochromis*' *nuchisquamulatus* of Lake Victoria and '*H*'. *aeneocolor* of Lake George (see Greenwood, 1956a and 1973 for the species respectively).

Because of this marked difference in dental morphology I am restricting the name '*astatodon*' to the six syntypes (BMNH reg. nos 1906.9.6:124–129) with *H. obliquidens*-like teeth, and have selected specimen 1906.9.6:125 as the lectotype. The eight specimens with '*nuchisquamulatus*'-like teeth will be dealt with in a subsequent paper.

Diagnosis and discussion

The lineage here recognized as the genus *Haplochromis* is characterized by having obliquely truncate and protracted crowns to the outer jaw teeth, an increase in the number of rows of inner teeth in both jaws (some of these teeth also having obliquely truncate crowns), and all jaw teeth moveably articulated with the underlying bone. The species are also characterized by their fine lower pharyngeal teeth, and the presence of truly ocellar egg-dummies on the anal fin of adult males. In their gross morphology, and in most aspects of their osteology and anatomy (the greatly elongate intestine excepted), the species retain a generally unspecialized level of organization.

Dental features similar to those of *Haplochromis* do occur in other African Cichlidae, for example *Cyathochromis obliquidens* Trewavas (1935) of Lake Malawi, but are associated with characters which indicate that these other taxa do not share a recent common ancestry with *Haplochromis*. Since all the taxa are grazers on epilithic or epiphytic algae their dental similarities must be considered the results of convergent evolutionary trends.

Within the genus *Haplochromis*, *H. obliquidens* and *H. astatodon* have what appears to be the most specialized (i.e. derived) form of teeth, and *H. limax* the least specialized teeth (see Fig. 7C). The common ancestor could well have been a species with a '*limax*'-like dentition.

In a preliminary phyletic analysis of the Lake Victoria '*Haplochromis*' species flock (Greenwood, 1974a), '*H*'. *nigricans* was considered to be the sister species of *H. lividus* and *H. obliquidens*, and '*H*'. *nuchisquamulatus* the sister species of the other three species combined. My recent investigations now suggest, however, that '*H.*' *nigricans* (together with species from Lakes Edward, George and Kivu) form a distinct monophyletic lineage which cannot be related to *Haplochromis* (as here defined) on the basis of shared derived characters. Neither, on those grounds, can '*H.*' *nuchisquamulatus* be considered the sister group of *Haplochromis*. But, it cannot be denied that '*H.*' *nuchisquamulatus* (and at least two other species, including the atypical *H. astatodon* noted above) do have a tooth morphology approaching that of *Haplochromis* (in particular *H. limax*). Tooth form in these species is what might be expected as an early stage of differentiation from the primitive bicuspid type in a morphocline leading to the '*obliquidens*' type. More research is needed before any possible relationship between the five *Haplochromis* species and the '*H.*' *nuchisquamulatus* complex can be clarified.

ASTATOTILAPIA Pellegrin, 1903

TYPE SPECIES: *Labrus desfontainii* Lacépède, 1803 (type specimens, once in the Paris Museum but now apparently lost).

Description

Body relatively deep (35–40% of standard length).

Squamation. Over most of the body the scales are ctenoid, the ctenii generally strong and extending around the greater part of the scale's free margin; scales above the upper lateral line may be less strongly ctenoid than those below it, and in three species are partly or entirely cycloid. Scales on the nape and cheek are cycloid, as may be those on the chest, which is always completely scaled.

The chest scales show a gradual size transition with those on the ventral and ventrolateral aspects of the flanks (see Fig. 1); in some species the chest scales are not noticeably small, but in *A. flaviijosephi*, *A. dolorosa* and *A. desfontainesi* these scales are distinctly smaller (and in *A. calliptera* somewhat smaller) than are the chest scales in other species of the genus.

The cheek is fully scaled, usually with 3 (less commonly 2 or 4) horizontal scale rows.

The lateral line has from 28 to 30 scales (31–33 in *A. desfontainesi*), all but the last 1 to 4 scales of the upper lateral line separated from the dorsal fin base by two or more scales of approximately equal size. *Astatotilapia swynnertoni* is exceptional in having the last 6–8 scales of the upper lateral line separated by less than two scales of equal size.

Neurocranium. The skull is of the generalized haplochromine type (see Greenwood, 1974a : 58–59, and p. 274 above). The preotic portion of the skull is not protracted (comprising some 55–60% of the neurocranial length), and the ethmovomerine region is short, sloping upwards at a slight angle. The dorsal skull roof may be straight or very slightly convex anterodorsal to the orbit. The supraoccipital crest is not reduced in length; its anterior border continues the line formed by the dorsal surface of the ethmovomerine region and the orbital part of the skull roof.

The ventral apophysis for the upper pharyngeal bones (see p. 274) is slightly enlarged in one species (*A. flaviijosephi*). The otic region of the skull is relatively deep, and is not inflated.

Vertebral numbers: 27–30, rarely 26 (modes 28 and 29), comprising 12–14 (mode 13) abdominal and 14–16 (mode 15) caudal elements. The apophysis for the *retractor dorsalis* pharyngeal muscles is small, and situated on the third vertebra.

Dentition. In fishes less than 70 mm standard length unequally bicuspid teeth (Fig. 4) predominate in the outer row of both jaws. The crown in these teeth is not noticeably compressed, neither is it clearly demarcated from the neck of the tooth (cf. *Haplochromis*, p. 278). Except in two species, the cusps are acutely pointed and both lie within the verticals formed by the anterior and posterior margins of the tooth's body; in *A. calliptera* and *A. swynnertoni*, however, at least the major cusp is somewhat obliquely truncate. Unlike *Haplochromis*, the outer teeth in *Astatotilapia* are firmly attached to the bone.

Even in specimens less than 70 mm SL some weakly bicuspid teeth and some unicuspids are found in both jaws. In larger specimens the unicuspid type predominates. The proportion of uni- to bicuspid teeth increases with the size of the individual; *A. dolorosa*, known only from the holotype, a fish 95 mm SL, has only unicuspid teeth.

All species, and specimens of all sizes, have the posterior 3–12 teeth in the premaxilla enlarged and unicuspid.

Teeth forming the inner rows are generally tricuspid and small; some weakly bicuspid or even unicuspid teeth may occur in these rows, particularly in larger fishes. There are 1–3 (usually 2) rows of teeth anteriorly and anterolaterally in both jaws, a single row posteriorly and posterolaterally.

Lower jaw is not foreshortened, nor is it noticeably deepened posteriorly.

Lower pharyngeal bone and dentition. The dentigerous surface is apparently equilateral in all species. Except in *A. flaviijosephi*, all the teeth are compressed, slender and cuspidate, with only those teeth in the two median and the posterior transverse row somewhat coarser than the others. Cusp form is essentially similar to that in *Haplochromis* but the crown is not so markedly produced as in that genus (cf. p. 279 above). In *A. flaviijosephi*, the teeth forming the median rows are enlarged, with molariform or submolariform crowns; some teeth in the lateral rows also are somewhat enlarged and have submolariform crowns. The lower pharyngeal bone, as compared with that in the other species, is stouter.

Dorsal fin with 14–16 (mode 15), rarely 13, spinous rays and 8–11 (modes 9 and 10) branched rays.

Anal fin with 3 spinous and 7–10 (modes 8 and 9) branched rays; specimens of *A. desfontainesi* with 4 spines have been recorded.

Caudal fin skeleton. *Astatotilapia burtoni* (7 specimens radiographed) and *A. nubila* (9 specimens) show no hypural fusion, but the other 7 species all yielded certain individuals with some degree of fusion in either the upper or the lower set of hypurals, or far less frequently, in both sets. It must be stressed, however, that these observations were made principally from radiographs and that these can be difficult to interpret with accuracy if the hypurals are closely apposed to one another. Vandewalle (1973) reports no fusion in the two specimens of *A. burtoni* he examined. (It is not possible to tell from his paper whether these were radiographed or dissected specimens.)

Caudal fin strongly subtruncate to rounded.

Pelvic fin with the first branched ray longest.

Anal fin markings in male fishes. True ocellar markings are present, usually 2–4 in number (but as many as 9 in large specimens of some species) arranged in one or less commonly in two rows, the number of rows positively correlated with the number of ocelli and the size of the specimen. The row or rows of ocelli run along a line situated approximately midway between the base and the distal margin of the fin. This linear arrangement (and that in *Haplochromis*) contrasts strongly with the near random arrangement of the anal spots in *Serranochromis* and *Chetia* (see pp. 302 and 308 respectively).

No information is available on the occurrence of anal spots (not ocelli, see p. 275) in the females of *Astatotilapia* species, except that 3 or 4 linearly arranged spots do occur in *A. nubila* from Lakes Victoria and George. Trewavas (1973), however, records seeing ripe female *A. nubila* and *A. bloyeti* with male-type ocelli on the anal fin, a phenomenon I never encountered when working with these species in the field.

Where breeding habits are known, *Astatotilapia* species are female mouthbrooders. Indeed, *A. burtoni* is the species on which Wickler (1962b) developed his dummy-egg theory to explain the function of anal ocelli in spawning.

Gill rakers of various shapes, from short and stout to moderately long and slender; 8 or 9, less commonly 7, in the outer row on the lower part of the first gill arch.

Contained species

Astatotilapia desfontainesi (Lacép.) 1803 (Type species)*; as restricted by Regan (1922a) to specimens from Tunisia and Algeria.

Astatotilapia flaviijosephi (Lortet), 1883. Israel and Syria. See Trewavas (1942) and Werner (1976) for redescription and biology.

Astatotilapia bloyeti (Sauv.), 1883. The type specimens are from Kandoa (Great Ruaha system), Tanzania. Regan (1922a) included three other species in the synonymy of *A. bloyeti*, viz. *Ctenochromis strigigena* Pfeffer, 1893, *Tilapia sparsidens* Hilgendorf, 1903 and *Paratilapia kilossana* Steindachner, 1916. For the purposes of this revision I have accepted Regan's synonymy, but much more research is needed into the alpha level taxonomy of fluviatile haplochromine cichlids in eastern and southern Africa before the species can be defined adequately. I have examined material from many localities in Kenya, Uganda and Tanzania, and also from Lake Chad and the Upper Niger. This material would seem to represent a taxon closely similar, if not identical, to *A. bloyeti*, at least on anatomical and morphometric characters. However, judging from colour notes made on certain specimens from Tanzania there are strong indications that at least some populations are distinguishable on the basis of male coloration.

An apparently undescribed species recently distributed by the aquarium trade under the name 'Nigerian mouthbrooder' or 'Nigerian *H. burtoni*' probably should be included in the *A. bloyeti* complex, as should an undescribed species from the Malagarasi river (personal observations).

Until the necessary revision of these 'species' and populations has been carried out, the different taxa involved can be referred to as the '*A. bloyeti* complex' (see Greenwood, 1971). Its distribution includes rivers, streams and certain lakes in Kenya, Uganda, Tanzania, Nigeria and, probably the Nile.

Astatotilapia nubila (Blgr.), 1906. Lakes Victoria, Kioga, Edward, George, Nabugabo, Kachira, Nakavali and Kijanebalola, and river systems in Uganda which are connected with these lakes (see Trewavas, 1933; Greenwood, 1965b and 1973). The species has been widely distributed in Uganda (and possibly Kenya and Tanzania as well) as a result of fishfarming and dam stocking activities; its natural distribution is that listed in the first sentence.

Astatotilapia dolorosa (Trewavas), 1933. Known only from the holotype, a specimen collected from the Chambura river which flows into the Kazinga channel connecting Lakes Edward and George.

---

* The emended spelling of the trivial name, '*desfontainesi*', was first used by Boulenger (1899). Since Lacépède intended that the species be named for M. Desfontaines, Boulenger's emendation, although not explained, would seem to be justified.

Astatotilapia burtoni (Günther), 1893. Lake Tanganyika and rivers associated with that lake (see Poll, 1956). The specimens recorded from Lake Kivu by Boulenger (1915) were misidentifications (see Regan, 1921a).

Astatotilapia stappersi (Poll), 1943. Rivers associated with Lake Tanganyika, see Poll (1956).

Astatotilapia swynnertoni (Blgr.) 1907. Lower Buzi river, Mozambique.

Astatotilapia calliptera (Günther), 1893. Lakes Malawi and Chilwa; '. . . coastal rivers as far as the Save river, Mocambique' (Bell-Cross, 1976); Busi and lower Sabi-Lundi systems; Lower Zambezi and Pungwe systems.

Jubb (1967a) treats *A. swynnertoni* as a synonym of this species, but gives no reason for so doing. Judging from the material I have examined, I would consider the two species to be distinct.

Incertae sedis: *Chetia brevis* Jubb, 1968. The presence of 3 or 4 true ocellar markings on the anal fin would seem to exclude this species from the genus *Chetia* (see p. 307 below, and p. 274 above), as would the predominance of ctenoid over cycloid scales on the body, and the retention of bicuspid teeth as the predominant tooth form in specimens as large as 90 mm standard length (see p. 273).

The inclusion of *Chetia brevis* in *Astatotilapia* is, however, very tentative and may well be altered when the phylogeny and systematics of the 'Angolan *Haplochromis*' species are revised (see p. 312).

Diagnosis and discussion

The genus *Astatotilapia* is distinguished from the other fluviatile '*Haplochromis*' group species by the following combination of characters: male anal fin markings are true ocelli, large and usually numbering from 3 to 6, and are arranged in a single or, less frequently, a double row (the number of ocelli and hence the number of rows correlated positively with the size of the fish); scales on the chest region not sharply size-demarcated from those on the ventrolateral and ventral aspects of the body; chest and cheek fully scaled; most body scales are ctenoid, the ctenii on each scale not restricted to a narrow median arc on the scale's free margin but distributed along almost the entire free margin; most teeth in the outer row of both jaws are bicuspid, the cusps of unequal size, but the minor one never minute. The major cusp is acute or, rarely, somewhat obliquely truncate but not protracted (cf. Figs 4 and 7). A few stout unicuspid teeth occur posteriorly in the premaxilla of fishes at all sizes, and some may also be present anteriorly, in both jaws, of fishes > 80 mm SL; the inner teeth usually are tricuspid (occasionally some are weakly bicuspid or unicuspid) and are arranged in two rows anteriorly and anterolaterally; 27–30 (rarely 26) vertebrae, of which 12–14 (mode 13) are abdominal, and 14–16 (mode 15) are caudal elements; pelvic fin with the first branched ray the longest.

Apart from the ocellar anal fin markings, none of these characters can be considered derived, and the anal ocelli are an apomorphic feature shared with *Haplochromis* and most, if not all '*Haplochromis*' species from Lakes Victoria, Edward, George and Kivu, and some species from Lake Malawi as well. Thus, the possibility cannot be overruled that *Astatotilapia* is a non-monophyletic assemblage.

The absence of other synapomorphic features shared with the genera described in this paper at least indicates that no members of *Astatotilapia* are closely related to any one of those lineages (as was implied when, hitherto, most were placed in the genus *Haplochromis*).

The relationship of *Astatotilapia* to *Haplochromis* as now redefined is obscure. Both lineages share the apomorphic feature of anal ocelli, suggesting that both share a more recent common ancestry than either lineage does with any taxa not having this feature. Uncertainty also exists about the relationship between *Astatotilapia* and the anal ocelli-bearing lineages of Lakes Victoria, Edward, George and Kivu, and for that matter some of the anatomically generalized '*Haplochromis*' species of Lake Malawi.

Any member of the *Astatotilapia* line with acutely bicuspid teeth (except, because of its specialized pharyngeal mill, *A. flaviijosephi*) could, on purely morphological grounds, be taken to represent the ancestral species for many lineages within the Victoria–Edward–Kivu species flock (see Greenwood, 1974a). It is, indeed, likely that a number of generalized but endemic species from that flock will have to be included in *Astatotilapia*, as may some from Lake Malawi.

Relationships within the *Astatotilapia* lineage cannot be indicated at present, partly because no intragroup synapomorphies are apparent and partly because the species are as yet poorly defined and understood (see p. 283).

Astatotliapia, like *Thoracochromis* (see p. 290) has a wide geographical distribution (one, indeed that extends beyond Africa into the Middle East). Both genera occur in north Africa, although *Astatotilapia* does not apparently occur in the Nile drainage (except in Lakes Victoria and Kioga); *Thoracochromis*, on the other hand, is widely distributed in the Nile system but is poorly represented, if at all, in Lake Victoria, and does not extend so far into southern Africa as does *Astatotilapia*.

In general, it could be said that *Astatotilapia* is a lineage of eastern and southern Africa, with outliers in the northeast (*A. flaviijosephi*) and northwest (Algeria and Tunisia), and *Thoracochromis* a lineage of north, central and western Africa. Since the phyletic integrity of both lineages is uncertain (see above, and p. 294), and because large parts of the Zaire system are poorly known, this difference may be more artefactual than real.

ASTATOREOCHROMIS Pellegrin, 1903

TYPE SPECIES: *Astatoreochromis alluaudi* Pellegrin, 1903 (type specimens in the Paris Musuem). For synonymy see Greenwood (1959*a*) and discussion below.

Discussion

Pellegrin (1903) distinguished *Astatoreochromis* (then monotypic) from similar '*Haplochromis*'-group species and genera principally on its having 5 or 6 anal and 18 or 19 dorsal fin spines. The type species is from Lake Victoria, but later, specimens were collected from Lakes Edward and George, the Victoria Nile and Lakes Kioga, Nakavali and Kachira (see Greenwood, 1959*a*).

Redescriptions based on this enhanced material added to the number of diagnostic features, at least with respect to the '*Haplochromis*' species of Lakes Edward and Victoria (Greenwood, 1959*a*). Amongst those features are the rounded caudal fin, the high number and multiserial arrangement (3 or 4 rows) of the anal ocelli in male fishes, the unusual coloration (golden overlain with olivaceous green, the median fins olive-yellow, flushed with maroon and margined with black), and the lack of sexual dimorphism in basic body and fin colours. This material also extended the known range of dorsal fin spine numbers (16–20) as well as those of the anal fin (4–6).

The species *Astatoreochromis alluaudi* is further characterized by its strongly hypertrophied crushing pharyngeal dentition and bones (with a correlated hypertrophy of the cranial apophysis for the upper pharyngeal bones; see Greenwood, 1959*a* and 1965*a*). A similar degree of pharyngeal hypertrophy does, of course, occur in at least five other species from Lakes Victoria, Edward and George (Greenwood, 1960 : 270–279; 1973 : 172–177; Greenwood & Barel, 1978 : 164–179), but these species differ from *A. alluaudi* in several features, all of which suggest that *A. alluaudi* represents a distinct phyletic lineage.

The principal diagnostic characters for *Astatoreochromis* are not easily assessed on a basis of their apo- or plesiomorphy. The enlarged pharyngeal mill clearly is a derived feature, but is one that has evolved independently in at least two '*Haplochromis*' lineages (see Greenwood, 1974*a*, and p. 279 above); it is thus of little value in assessing relationships at the level with which we are here concerned.

As yet too little is known about the evolution and phyletic distribution of anal ocelli (and other anal fin markings) to say whether the increased number and multiserial arrangement in *Astatoreochromis* is a derived feature. The basis for comparison here is with the fewer ocelli and their uni- or biserial arrangement in *Haplochromis*, *Astatotilapia* and the Lake Victoria '*Haplochromis*' species.

The absence of sexually dimorphic coloration in *Astatoreochromis* is a most unusual feature amongst '*Haplochromis*'-like taxa, but would seem, *a priori*, to be a primitive rather than a derived feature (although its correlation with the increased number of anal ocelli and thus, possibly its degree of relative importance in breeding behaviour, cannot be interpreted without appropriate ethological studies).

Only the increased number of anal and dorsal fin spines (but, it should be noted, not the total number of rays in these fins) would seem to be derived features.

In the absence of other and synapomorphic characters, however, it is impossible to use fin spines numbers to suggest any possible close phyletic affinities for *Astatoreochromis*. For example, the squamation pattern in the genus is of the supposedly plesiomorph type, and this would seem to rule out any possible relationship with the *Orthochromis* lineage (see p. 295) in which there is also a marked trend towards increased numbers of dorsal and anal fin spines. *Orthochromis*, it may be added, does not have anal ocelli or, apparently, any other anal fin markings in the males (the breeding habits of no *Orthochromis* species are recorded).

That *Astatoreochromis* may be related (possibly as the derived sister-group) to some of the '*Haplochromis*' species with enlarged pharyngeal mills, cannot be completely discounted; but, equally there is little unequivocal evidence to support such an hypothesis (see above).

For the moment, then, *Astatoreochromis* is maintained as a distinct lineage because of its various distinctive features, taken in combination, and because its sister-group relationship to any other lineage cannot be hypothesised on the basis of uniquely shared derived characters.

Recently, Poll (1974) added a second species (*Haplochromis straeleni* Poll, 1944, from the Lukuga and Ruzizi rivers, Zaire) to the genus *Astatoreochromis*. This step was taken because some specimens of *straeleni* have 4 anal spines (i.e. 2 of the 7 specimens known), because of close similarities in overall coloration, in the pattern and number of anal ocelli, and because *straeleni* has a relatively enlarged lower pharyngeal bone with some molariform teeth (see fig. 1, Poll, 1974). The species also has, as compared with *Astatotilapia* and the Lake Victoria '*Haplochromis*' species, more dorsal fin spines (17 or 18) but the same number of branched rays in that fin (8 or 9); in other words, the *Astatoreochromis* condition (see above). Furthermore, according to Poll's account, there is no sexually dimorphic coloration in '*H.*' *straeleni*.

There is another '*Haplochromis*' species, '*H*'. *vanderhorsti* Greenwood, 1954 (Malagarasi river, Tanzania) which closely resembles *straeleni* in all the characters under consideration, differing only in its slightly lower dorsal fin spine count (16 or 17), and in none of the 54 specimens examined having 4 anal fin spines; its lower pharyngeal bone and dentition are more massive than those of *straeleni* (i.e. like the condition in *A. alluaudi* of the same size). The resemblances between '*H.*' *vanderhorsti* and *A. alluaudi*, and those between '*H.*' *straeleni* and '*H.*' *vanderhorsti* have been noted already (Greenwood, 1954 : 405–407; 1959a : 166–167), but were not analysed in terms of their apo- or plesiomorphy, and no conclusion was reached about the interrelationships of the species or their formal taxonomic status.

Apart from the increased number of dorsal fin spines (and the four-spined individuals of '*H.*' *straeleni*), the only other shared, and probably derived, characteristic common to the three species is the enlarged lower pharyngeal bone and its at least partly molariform teeth (again a trend character, least developed in *straeleni*, most developed in *Astatoreochromis alluaudi*, and one which is known to have evolved independently in several haplochromine lineages). But, taking into account the virtually identical, non-sexually dichromatic coloration of the three species (and the ubiquity of sexual dichromatism alongst fluviatile haplochromines) the most parsimonious solution would be to consider *alluaudi*, *straeleni* and *vanderhorsti* as being more closely related to one another than any one of them is to any other lineage.

On these grounds I would agree with Poll's (1974) inclusion of '*H.*' *straeleni* in *Astatoreochromis* and would now include '*H.*' *vanderhorsti* in that genus as well.

Contained species
Astatoreochromis alluaudi Pellegrin, 1903 (Type species). Lakes Victoria, Kioga, Edward, George, Nabugabo, Kachira, and Nakavali; rivers and streams associated with these lakes. The species has been widely distributed in Kenya, Uganda and Tanzania as a biological control agent against snails (McMahon, Highton & Marshall, 1977). For a full description of the species see Greenwood (1959a), and for evidence invalidating the two subspecies described in that paper see Greenwood (1965a).

Astatoreochromis straeleni (Poll), 1944. Lukuga and Ruzizi rivers, Lake Tanganyika drainage. See Poll (1974) for a redescription of the species.

Astatoreochromis vanderhorsti (Greenwood) 1954. Malagarasi river and swamps, Lake Tanganyika drainage.

CTENOCHROMIS Pfeffer, 1893

TYPE SPECIES: *Ctenochromis pectoralis* Pfeffer, 1893 (type specimens in Hamburg Museum and BMNH).

Description
Body relatively deep to relatively slender (depth 30–40% of standard length).

Squamation. Scales on the body below the upper lateral line are strongly to moderately ctenoid (weakly ctenoid in one species, *C. horii*), becoming cycloid over the posterior half of the body. Scales above the upper lateral line show the same range of ctenoidy or are all cycloid, the kind of scale being constant intraspecifically and positively correlated with those below the lateral line. Scales on the head and, when present, on the cheek, are cycloid.

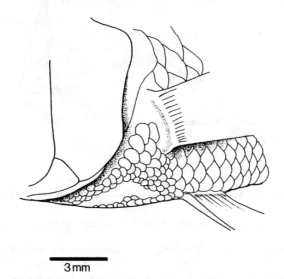

Fig. 9 Thoracic–abdominal scale transition in *Ctenochromis pectoralis*. Left lateral view.

The cheek always has a naked area along its ventral margin; in one species this area is less than a scale row in depth, in two others it is from 1 to 3 horizontal scale rows deep, and in a fourth virtually the whole cheek is naked, a few scales remaining immediately below and, or, behind the orbit.

Along or slightly behind a line joining the pectoral and pelvic fin bases there is an abrupt size transition between the very small scales on the chest and the much larger scales on the lateral and ventrolateral aspects of the body (Fig. 9).

The chest has a well circumscribed naked patch on each side of the body (Fig. 9), the two patches joined in some species by a ventral naked area. The size of the naked patch shows some interspecific variability, from a small and ventrolaterally situated area, to one covering most of the lateral and ventrolateral (but not the medial) aspects of the chest.

There are 27–33 scales in the lateral line series (modal numbers 30 and 31 for the two species from Tanzania, and 28–30 for the three Zaire river system species (see p. 290 below)); *the last 8–12 (usually 8 or 9) pored scales in the upper lateral line are separated from the dorsal fin base by less than two scales of approximately equal size.*

Neurocranium. The neurocranium is apparently of the generalized type (see Fig. 6), but in at least one species (*C. horii*; Fig. 10) its preotic region is more elongate (*c.* 68–70% of neurocranial length) and in others the preorbital region is slightly vaulted. Since little skeletal material is available these remarks are based mainly on radiographs and should be checked on actual skeletons.

Vertebral numbers: 25–29 (modal range 27–28), comprising 12 or 13 abdominal and 13–17 caudal elements (see p. 290 below).

Dentition. The outer teeth are unequally bicuspid, or, in two species, subequally bicuspid, are relatively stout and firmly attached to the underlying bone. Some posterior premaxillary teeth (as many as 16 on each side in *C. horii*) are unicuspid, caniniform and relatively larger than the preceding bicuspids. The crowns of the bicuspid teeth are not noticeably compressed, nor are they sharply demarcated from the shaft of the tooth; the cusps are acutely pointed.

The inner teeth are small and tricuspid, and arranged in 2 or 3 series anteriorly and antero-laterally, but a single series posteriorly.

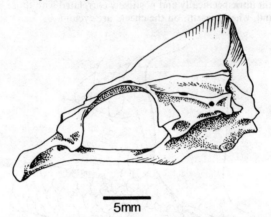

Fig. 10 Neurocranium of *Ctenochromis horii*; left lateral view.

Lower jaw. The dentary is relatively slender in lateral aspect, and not greatly deepened posteriorly.

Lower pharyngeal bone and teeth. The dentigerous surface is either triangular and subequilateral (slightly broader than long) in outline, or noticeably broader than long (c. 1½ times; Fig. 11). The teeth are cuspidate and compressed, those in the median and the posterior transverse row somewhat stouter than the others (the degree of stoutness, especially of teeth in the two median rows, shows a positive correlation with the fish's size).

Dorsal fin with 14–17 (modes 15 and 16) spinous and 8–10 (mode 9) branched rays. The holotype of *C. oligacanthus* (Regan) has only 12 spines, but all other specimens have 15.

Caudal fin skeleton. Because few dry skeletons or alizarin preparations are available, information on the caudal skeleton has been obtained mainly from radiographs. The difficulty of differentiating between fused and closely apposed hypural elements as seen in radiographs makes these observations of limited value.

Ctenochromis pectoralis (10 specimens radiographed) has all five hypurals free.

C. horii. Seven specimens (radiographed) have hypurals 1 and 2, and 3 and 4 fused, as does the dry skeleton examined. Two other specimens (radiographed) have hypurals 1 and 2 free, but 3 and 4 fused. Vandewalle (1973) found no fused hypurals in the two specimens he examined.

C. polli (2 specimens radiographed) has all hypurals free.

C. oligacanthus. Of the three specimens radiographed, one (the holotype) has all hypurals free, one has hypurals 1 and 2 free but 3 and 4 fused, and the third has hypurals 1 and 2, and 3 and 4 fused.

Caudal fin is markedly subtruncate, almost rounded in some species.

Pelvic fin with the first branched ray the longest.

Anal fin markings in male fishes. Where known (3 of the 5 species) from preserved and, or, living specimens, these are in the form of one or two (rarely three) brilliant white or yellow spots, without a dark margin and without a clear surround (cf. *Haplochromis* and *Astatotilapia*, p. 279 and

p. 283 respectively). The spot or spots may be on the anterior or the posterior part of the soft fin with, apparently, their position constant intraspecifically. For a colour picture of *C. polli* see Voss (1977 : 74).

Gill rakers short and stout in all species except *C. horii* where they are long and slender; there are 7–9 rakers in the outer row on the first gill arch, except in *C. horii* where there are 10–13.

Contained species
Ctenochromis pectoralis Pfeffer, 1893 (Type species). Known only from streams in south eastern Tanzania, near Korogwe.

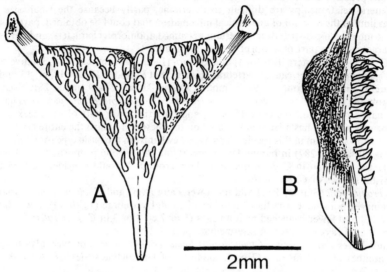

Fig. 11 Lower pharyngeal bone of *Ctenochromis polli*. A. Occlusal view. B. Right lateral view (bone aligned vertically).

Ctenochromis horii (Günther), 1893. Lake Tanganyika and the coastal reaches of associated rivers. See Poll (1956) for a detailed description and notes on biology.

Ctenochromis polli (Thys van den Audenaerde), 1964. Pool Molebo (Stanley Pool) and the lower Zaire river. For coloured plate see Voss (1977). For notes regarding specimens formerly identified as *H. fasciatus* see p. 293.

Ctenochromis oligacanthus (Regan) 1922. Ubangi river, an affluent of the Zaire river. The holotype and sole specimen available to Regan has only 12 dorsal fin spines and 9 branched rays; the two additional specimens I have examined have 15 spines and 9 branched rays.

Ctenochromis luluae (Fowler), 1931. Lulua river, Kasai drainage system, lower Zaire. I have examined four paratypes, and on the basis of that material would consider the species to be distinct from *C. polli* and *C. oligacanthus* (the other two *Ctenochromis* from the lower Zaire drainage), and from *C. pectoralis* and *C. horii*.

Diagnosis and discussion
Members of the genus *Ctenochromis* are characterized by the abrupt size transition between the very small chest scales and the larger scales on the ventrolateral aspects of the anterior flanks, by a naked area on either side of the chest, and by a failure of the cheek squamation to reach the ventral margin of the cheek. (In one species, *C. horii*, the entire suborbital region of the cheek is scaleless in some individuals and in others there is, at most, no more than a single scale row below and behind the orbit.) In three of the five species the anal fin markings of male fishes are in the form of one or two (rarely three), relatively small and simple, non-ocellate spots; no information is available on the other two species.

Virtually nothing is known about the biology of *Ctenochromis* species, which are, with the exception of *C. horii*, very poorly represented in study collections. Four of the five species are fluviatile and reach a small adult size (80 mm standard length); their feeding and breeding habits are unknown. The fifth species, *C. horii*, is essentially lacustrine, reaches a larger adult size (185 mm SL) and is at least partly piscivorous.

The geographical range of *Ctenochromis* extends from Tanzania in the east to the Zaire drainage (including Lake Tanganyika) in the west and lies between the latitudinal limits of c. 3° N and 8° S.

Various apomorph features shown by *Ctenochromis* suggest its relationships with two other genera; this problem will be discussed later (see p. 313).

Intrageneric relationships are difficult to determine, partly because the small size of some specimens limited the amount of anatomical information that could be obtained, partly because of the mosaic interspecific distribution of certain presumed apomorph characters, and partly because other characters form part of a morphoclinal continuum.

Both *C. pectoralis* (eastern Tanzania) and *C. horii* (Lake Tanganyika) have, relative to *C. polli* and *C. oligacanthus*, higher caudal vertebral counts (15–17, modes 16 and 17, cf. 13 and 14) and higher lateral line scale counts (29–33, modes 30 and 31, cf. 27–29, mode 28); these features should probably be considered plesiomorph ones (see p.276). *Ctenochromis pectoralis* has relatively small naked patches on the chest (Fig. 9), and only the lower part of the cheek is scaleless. In comparison, *C. horii* has a large naked area on the chest and almost the entire cheek is without scales; the neurocranium in this species departs somewhat from the basic type found in *C. pectoralis* (see Fig. 10 and p. 287) in having larger preorbital and preotic proportions, and individuals reach a larger size than in *C. pectoralis*. On these grounds I would consider *C. horii* to be the derived sister species of *C. pectoralis*.

Ctenochromis polli (Stanley Pool and the lower Zaire river) and *C. oligacanthus* (Ubangi river, Zaire drainage) both have a reduced number of caudal vertebrae (13 or 14), fewer lateral line scales (27–29) and fewer branched anal fin rays (6 or 7 cf. 8 or 9 in *C. pectoralis*; *C. horii* with 7 (mode) or 8 anal rays occupies an intermediate position).

The third lower Zaire species, *Ctenochromis luluae* (Lulua river), however, also has an intermediate number of anal fin rays (7 or 8, mode 8), of lateral line scales (28–30) and of caudal vertebrae (15).

The outer jaw teeth in *C. polli*, *C. luluae* and *C. oligacanthus* are similar and differ from the basic, unequally bicuspid type found in *C. horii* and *C. pectoralis* in having the cusps subequal in size, with the tip of the smaller cusp directed away from the near vertical larger cusp. In both *C. polli* and *C. oligacanthus* the dentigerous surface of the lower pharyngeal bone is noticeably broader than long, whereas in *C. pectoralis* and *C. horii* its length and breadth are approximately equal (see Fig. 11); the bone in *C. luluae* has proportions that are intermediate between these two types. Finally, the naked area of the chest in *C. oligacanthus* is much larger than in *C. polli*, but a greater area of the cheek is scaled in the former species. Most of the cheek is scaled in *C. luluae*, and the naked chest area is intermediate between that of *C. polli* and *C. oligacanthus*.

On the basis of their dental morphology I would suggest that *C. polli*, *C. luluae* and *C. oligacanthus* together form the sister group to *C. pectoralis* and *C. horii*; *C. pectoralis* would seem to be the least derived taxon of the lineage.

As a postscript to this discussion it may be mentioned (with the reservations noted on p. 282) that hypural fusions in *Ctenochromis* species are relatively common, and certainly commoner than in *Haplochromis*, *Astatotilapia* and the '*Haplochromis*' species of Lake Victoria (see p. 276 above, and Greenwood, 1974*b* : 159).

THORACOCHROMIS gen. nov.

TYPE SPECIES: *Paratilapia wingatii* Boulenger, 1902 (see Greenwood, 1971 for a redescription of the species).

ETYMOLOGY. The name is derived from the latinized Greek word for a breastplate + *chromis*, a name when used in such a combination now associated with many genera of African Cichlidae;

it refers to the small and clearly size-demarcated scales on the thoracic region of species in this lineage.

Description
Body form ranging from relatively deep to relatively slender (depth 30–40% of standard length).

Squamation. In the majority of species, the scales on the body above and below the upper lateral line, and behind a line through the pectoral and pelvic fin insertions, are ctenoid. A few species have cycloid scales above the upper lateral line, and weakly ctenoid scales below it. Scales on the cheek, head and chest are cycloid.

The scales on the chest are small to very small and meet, with an abrupt change in size, the larger scales on the lateral and ventrolateral aspects of the flanks (Fig. 2). Generally the line of this abrupt size change lies approximately between the insertions of the pectoral and pelvic fins, but may be a little behind or, less frequently, a little before that level. The chest is always completely scaled, although in two species the scales are so small and deeply embedded that the area appears to be naked.

The cheek is completely or almost completely scaled (in two species there is a very narrow, horizontal naked strip along the ventral margin, and in several other species there is a naked embayment at the anteroventral angle of the cheek squamation).

There are 29–32 (modal range 30–32) scales in the lateral line series; *about the last eight pore-bearing scales of the upper lateral line are separated from the dorsal fin base by not more than one large and one much smaller scale.*

Neurocranium. Most *Thoracochromis* species have a skull form that departs but slightly from the type found in *Ctenochromis* (see p. 287). That is, a generalized type (see p. 274) in which the preotic region of the neurocranium comprises some 65–70% of the total neurocranial length.

The most marked departure from this skull form is seen in two species, *Th. bullatus* (Lake Albert) which has a greatly inflated otic capsule and somewhat enlarged lateral line sensory canals, and *Th. macconneli* (Lake Turkana) where the sensory canals are hypertrophied and the braincase is shallower.

Thoracochromis demeusii (Zaire) deviates in a different way; here the supraoccipital crest is deepened, extends further anteriorly than in the other species and has a steeper slope to its anterior margin. These features may all be associated with the pronounced dermal hump developed in the nuchal region of this species.

Vertebral numbers: 26–31 (modes 28 and 29), comprising 12–14 (modes 12 and 13) abdominal and 13–17 (modes 14 and 17) caudal elements. With one exception (*Th. moeruensis*) the lower modal counts for caudal vertebrae are found in fluviatile species from the Zaire river drainage system, the higher ones in species from Lakes Turkana, Albert, George and Mweru.

Dentition. Unequally bicuspid or unicuspid, caniniform outer teeth are the the most frequently occurring types. The crown in bicuspids is not noticeably compressed, and the cusps are acutely pointed. Unicuspid teeth may be recurved or almost straight; where the material covers a sufficiently wide size range of specimens it shows that the unicuspid dentition is preceded by a bicuspid one. Fishes in all species with a bicuspid definitive dentition have a few (1–6) unicuspids posteriorly on the premaxilla, these teeth generally being larger than the anterior bicuspids.

In two species (*Th. fasciatus* and *Th. loati*) the outer teeth, although unequally bicuspid, have the major cusp obliquely truncate, somewhat protracted and relatively compressed; the minor cusp is much reduced and is also obliquely truncate. Thus there is a close resemblance between these teeth and those in *Haplochromis lividus* and *H. limax* (see Fig. 7C and p. 281 above; also Greenwood, 1971 : 360, fig. 5).

The inner teeth generally are tricuspid and in both jaws are arranged in 2 or 3 series anteriorly and laterally, but in a single series posteriorly. In those species with unicuspid outer teeth at least the outermost row of the inner series contains some unicuspids, a mixture of tri- and unicuspids, or it may be composed entirely of unicuspids.

Lower jaw relatively slender in lateral outline and not obviously deepened posteriorly.

Lower pharyngeal bone and teeth. With respect to the outline shape of the dentigerous area, two fairly distinct types of pharyngeal bone occur in this genus. One type (found in species of the Zaire drainage, including *Th. moeruensis* but excluding *Th. demeusii*) has the surface clearly broader than long (Fig. 12). The second type (in species from Lakes Turkana, Albert and George) has its dentigerous surface only slightly broader than long, i.e. about $1\frac{1}{5}$ times.

In all species the median tooth rows are noticeably coarser than their lateral congeners, and are even coarser than those in the posterior transverse row. Some species have stout and molariform or submolariform teeth in the median rows, and in three other species (*Th. albertianus*, *Th. mahagiensis* and *Th. pharyngalis*) teeth lateral to the median rows are also enlarged and submolariform to molariform, (see Trewavas, 1938 : 441 and 444; Poll, 1939 : 47; Greenwood, 1973 : 213). Associated with this enlargement of the dentition (especially in *Th. mahagiensis* and *Th. pharyngalis*) the lower pharyngeal bone is markedly thickened.

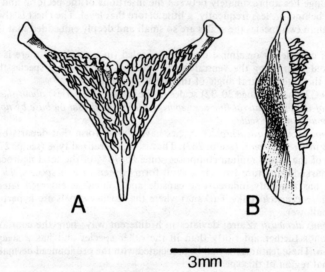

Fig. 12 Lower pharyngeal bone of *Thoracochromis bakongo*. A. Occlusal view. B. Right lateral view (bone aligned vertically).

Teeth other than the enlarged ones are compressed and cuspidate in all species.

Dorsal fin with 13–16 (modal range 14–16), rarely 17, spinous and 8–10 (modes 9 and 10), rarely 11, branched rays.

Anal fin with 3 spines and 6–10 (modal range 7–9) branched rays.

Caudal fin skeleton: the distribution of fused hypural elements amongst *Thoracochromis* species, as determined mainly from radiographs, is as follows:

(i) Lake Albert species, and those from the Nile and Lake George: no fusion in *Th. wingatii* (holotype), *Th. albertianus* (6 specimens), *Th. mahagiensis* (1), *Th. avium* (holotype), *Th. bullatus* (13), *Th. loati* (16) and *Th. petronius* (12); some specimens in most species have certain hypurals very closely apposed.

(ii) Lake Turkana. *Th. rudolfianus* has all hypurals free in 28 specimens examined, but hypurals 1 and 2 are closely apposed or perhaps fused in two others; *Th. turkanae* (4 specimens) has all hypurals free; in *Th. macconneli* 6 of the 26 specimens examined have all hypurals free, 4 have hypurals 1 and 2, and 4 and 5 fused, 11 have hypurals 1 and 2, and 3 and 4 fused, but hypural 5 free, and 5 have only hypurals 1 and 2 fused.

(iii) Lower Zaire drainage, and Lake Mweru. *Th. moeruensis* has hypurals 1 and 2, and 3 and 4 fused (6 specimens). *Th. demeusii* (holotype) has all free; *Th. fasciatis* (6 syntypes) has all free, as does *Th. bakongo* (3 specimens); *Th. stigmatogenys* has 2 specimens with hypurals 3 and 4 fused, and 4 others with all hypurals free. Vandewalle (1973) reports no fusion in the two species he examined, namely, *Th. bakongo* and *Th. fasciatus*.

Caudal fin subtruncate to almost rounded, most species being in the latter category.

Pelvic fin with the first branched ray the longest.

Anal fin markings. Regrettably little information is available on this character, either from live or preserved specimens. True ocellar spots (3–8 in number) are present in *Th. petronius* (Lake George) and apparently in the three species from Lake Turkana, although the clear surround in these latter species is much narrower than in the ocelli of *Haplochromis* and *Astatotilapia*. The four species from Lake Albert and the Nile for which data are available (i.e. *Th. albertianus, Th. bullatus, Th. wingatii* and *Th. loati*) do not appear to have any clear area around the small, coloured or white spots, which are relatively large, well defined and number from 1 to 3. However, these observations were made on preserved material only and, since the whole fin is somewhat pigmented, a narrow hyaline surround could well be overlooked.

No information is available for the remaining species (Lake Mweru and the Zaire river drainage).

Gill rakers are of various shapes, with 6–12 (modal range 7–9) in the outer row on the first gill arch.

Contained species

Thoracochromis wingatii (Blgr.), 1902, type species; see Greenwood (1971) for a redescription and definition of the species. Upper Nile (Bahr-el-Jebel) and probably Lake Albert also.

Thoracochromis loati (Greenwood), 1971. Upper Nile (Bahr-el-Jebel) and Lake Albert.

Thoracochromis rudolfianus (Trewavas), 1933. Lake Turkana.

Thoracochromis turkanae (Greenwood), 1974. Lake Turkana.

Thoracochromis macconneli (Greenwood), 1974. Lake Turkana.

See Greenwood (1974b) for notes on the biology of the last three species, and for comments on their relationships.

Thoracochromis albertianus (Regan), 1929. Lake Albert.

Thoracochromis mahagiensis (David & Poll), 1937. Lake Albert (see also Greenwood, 1971 : 356).

Thoracochromis avium (Regan), 1929. Lake Albert. I follow Trewavas (1938) in considering *Haplochromis lanceolatus* David & Poll, 1937, a synonym of this species.

Thoracochromis petronius (Greenwood), 1973. Lake George, Uganda.

Thoracochromis pharyngalis (Poll), 1939. Lake Edward. See Greenwood (1973 : 213) for a discussion on the relationship of this species.

Thoracochromis moeruensis (Blgr.), 1899. Lake Mweru.

Thoracochromis demeusii (Blgr.), 1899. Lower Zaire river. See Thys van den Audenaerde (1964) for a redescription of the species and for other data; this author considers that the presumed type locality (Bangala country, Upper Congo) is in error.

Thoracochromis bakongo (Thys van den Audenaerde), 1964. Kasai drainage to the lower Zaire river.

Thoracochromis fasciatus (Perugia), 1892. Lower Zaire drainage at Vivi (5°38′ S, 13°30′ E; see Thys van den Audenaerde, 1964). At present I am restricting the concept of this species to the six syntypical specimens in the BMNH collections (reg. nos 1898.12.12:1–6). Certain other specimens in the BMNH collections identified as *fasciatus*, and at least part of the material on which Thys van den Audenaerde (1964) based his redescription of the species, are referable to one or possibly two other species. These, or this, species differ from the syntypes of *fasciatus* in dental characters and in having a graded rather than an abrupt size change between the chest and ventrolateral flank scales (i.e. they cannot be referred to the genus *Thoracochromis*; see p. 291 above).

Three further specimens in the BMNH collections (reg. nos 1899.9.6:2–4, ex Stanley Pool) which were included in *fasciatus* by Boulenger (1915 : 215–216) do, however, show an abrupt size change in the scales of this region. All 3 specimens are now in a very poor state of preservation but, judging from certain morphometric characters and also from their dental morphology, it seems that they should be identified as *Ctenochromis polli* (see p. 289).

Regan (1922a) tentatively included *Paratilapia toddi* Blgr., 1905 (Kasai river, Zaire drainage) in the synonymy of *fasciatus*. Regan's concept of *fasciatus* was essentially that of Boulenger (1915) since his study material included the misidentified specimens noted above. The only known specimen of *Paratilapia toddi*, the holotype, is considerably larger (127·0 mm SL) than any member of a known *Thoracochromis* species, and has the general facies and external cranial morphology of a *Serranochromis*-like fish (see Trewavas (1964) who, indeed, thought that *P. toddi* might be related to *Serranochromis*). However, the holotype of *P. toddi* does show an abrupt size transition between the scales of the chest and flanks, and it also has a low number of caudal vertebrae (13) and a low branched anal fin ray count (7), features shared with some species of *Thoracochromis* (see p. 291 above). On the other hand, there are several features of *P. toddi* that are not encountered in any member of that genus.

Until more specimens are available it would seem best to treat *P. toddi* as a taxon *incertae sedis*; it certainly cannot be considered a synonym of *Thoracochromis fasciatus*.

Thys van den Audenaerde (1964) considers Boulenger's (1899) *Chromis monteiri* (from Boma) to be a synonym of *fasciatus*. Regrettably the holotype (and unique) specimen of *C. monteiri* cannot now be found and so a comparison between it and the syntypes of *Th. fasciatus* could not be made. But, judging from Boulenger's original description and figure, it seems highly improbable that the specimens are from the same species. The possibility that *monteiri* holotype and some of the misidentified *fasciatus* material are conspecific cannot, however, be overlooked.

Diagnosis and discussion
Members of the genus *Thoracochromis* are characterized by the abrupt size transition between the small chest scales and the much larger scales on the ventrolateral and ventral aspects of the flanks. This is, apparently, the only derived character shared by all members of the lineage. Several other derived features are, however, found in member species. For example, the obliquely truncate tooth cusps in *Th. loati* and *Th. fasciatus*, the enlarged pharyngeal mills in *Th. mahagiensis*, *Th. albertianus* and *Th. pharyngalis*, the bullate otic region in *Th. bullatus*, the hypertrophied cephalic lateral line canals in *Th. macconneli* (and probably its near-dwarf males as well; see Greenwood, 1974b) and, finally, the short but broad lower pharyngeal bone in most species from the lower Zaire system. But, the restricted distribution of these apomorphic characters amongst the species obliges one to rank them either as autapomorphies or as low level synapomorphies suggesting possible intrageneric relationship (if, in the latter case, it can be shown that the characters have not evolved independently, an impossible task when there are no other features on which to establish intrageneric relationships).

Even the single synapomorphy used to define the lineage as a whole, the abrupt thoracic-flank scale size transition, is shared with *Ctenochromis* (p. 287) and *Orthochromis* (p. 296 below). The two latter lineages, however, have each their own derived features which can be interpreted as indicators of their monophyletic origin and thus their phyletic distinctiveness.

The possible interrelationships of *Thoracochromis*, *Ctenochromis* and *Orthochromis* are considered later (p. 313).

Thoracochromis has a wide but disjunct geographical distribution. In the north there are three species endemic to Lake Turkana, and two others in the Nile, both of which also occur with the three endemic species in Lake Albert. There is one species in Lake George, another in Lake Edward (with the possibility that the George species also occurs in Edward) and possibly a third in Lake Victoria.

Much further south (and a little to the west) there is one species in Lake Mweru, and a group of five species apparently confined to the lower Zaire drainage.

From an historical viewpoint (Greenwood, 1974b) the species of the Nile, Lake Turkana and Lake Albert could well be closely related and could also be related to the species from Lakes Edward and George. Indeed, *Th. mahagiensis* (Lake Albert) and *Th. pharyngalis* (Lake Edward) have three apparently derived characters in common, viz. hypertrophied pharyngeal mills, a low number of gill rakers (as compared with other species from Lake Albert) and a reduced cheek squamation; and again, *Th. bullatus* (Lake Albert) and *Th. macconneli* (Lake Turkana) both have hypertrophied cranial lateral line systems, although in this instance, since both species live in deep waters, the resemblance could be the result of parallel evolution.

The lower Zaire species, with the exception of *Th. demeusii*, have distinctly broad and short lower pharyngeal bones (that of *Th. demeusii* is but slightly broader than long and resembles the bone found in all other *Thoracochromis* species). *Thoracochromis moeruensis*, a geographically isolated Zairean species from Lake Mweru, also has a short and broad lower pharyngeal bone, suggesting its possible relationship with the lower Zaire species group (perhaps, geographically speaking, through some past linkage via the Kasai drainage system).

More collecting in the Zaire river system, especially its middle reaches, and more information about the northern (i.e. Nile, Turkana, Albert) species is needed before any of these suggested intralineage groups can be developed further, and indeed before the phyletic integrity of the whole lineage can be tested adequately. Data on live coloration, anal fin markings and cranial osteology are particularly needed.

The absence, save for two or possibly three species of *Thoracochromis* from the Lake Victoria–Edward–George–Kivu cichlid flock (totalling some 200 species), is of particular zoogeographical interest, especially when it is recalled that in Lakes Turkana and Albert species of *Thoracochromis* are the only '*Haplochromis*'-group taxa represented. Likewise one may note the predominance of *Thoracochromis*, *Ctenochromis* and *Orthochromis* species in the Zaire river system.

ORTHOCHROMIS Greenwood, 1954

TYPE SPECIES: *Haplochromis malagaraziensis* David, 1937 (type specimens in the Musee Royal de l'Afrique Centrale, Tervuren).

Synonymy
Rheohaplochromis Thys van den Audenaerde, D. F. E. (1963), *Revue Zool. Bot. afr*. 68, 1–2 : 145 (as a subgenus of *Haplochromis*); idem (1964), *Revue Zool. Bot. afr*. 70, 1–2 : 169 (raised to generic rank). No type species by original designation.

When discussing the affinities of *Rheohaplochromis*, Thys van den Audenaerde (1964 : 169) mentions my observations (*in litt*.) that the genus showed strong affinities with *Orthochromis*. Although agreeing with my remarks, Thys van den Audenaerde considered that '. . . l'écaillure nuchale et ventrale vraiment miniscule des *Rheohaplochromis* (*polyacanthus* et *torrenticola*) nous semble un caractère suffisamment important pour maintenir ces espèces dans un genre séparé . . .'. I would argue that the suite of derived characters shared by these species and the two other species discussed below (including *O. malagaraziensis*) are a stronger argument in favour of their inclusion in a single lineage of presumed monophyletic origin (i.e. within the scope of this revision, a genus see p. 269 above).

Thys van den Audenaerde's supplementary argument for placing the species *polyacanthus* and *torrenticola* together in a separate genus (because of their overlap in distribution as compared with the allopatric distribution of the other Zaire haplochromine species) might well be used to explain the presence of derived features shared only by *polyacanthus* and *torrenticola*, but it seems to have little bearing on the problem of determining their overall phyletic relationships.

In an earlier paper, Thys van den Audenaerde (1963) considered that the small ventral and nuchal scales, and the rounded pelvic fins, of *polyacanthus* and *torrenticola* could be ecophenotypic features associated with their rheophilic habits. In support of his contention he mentions similar features in *Steatocranus*, an unrelated taxon (see Greenwood, 1978). This argument of ecophenotypically evolved characters could also be used to explain the similar scale and fin characters in the two other species I would include in the same lineage as *polyacanthus* and *torrenticola*. But, to me, it would seem more parsimonious to conclude that, although the features possibly have selective advantage in a torrential habitat, their association in a number of species sharing other derived features is more likely to be indicative of common ancestry than of repeated parallel evolution. Since *Steatocranus* may well be a member of a much more distantly related branching within the African Cichlidae (see Greenwood, 1978), the similarity in scale and fin organization in that instance would, I agree, be the result of convergence (and thus indicative of the characters having adaptive value in that type of habitat).

Description
Body elongate and slender (*its depth 25-30% of standard length*); *dorsal head profile strongly decurved, eyes generally suprolateral in position*, giving the fish a somewhat goby-like appearance.

Squamation. Scales on the head and on the body above the upper lateral line are cycloid or weakly ctenoid, or cycloid over the anterior third of the upper body and ctenoid over the posterior two-thirds. Scales below the upper lateral line are ctenoid except on the chest and belly, where they are cycloid.

The chest is naked or scaled (if the latter there is sometimes a small naked area on one or both sides of the body); *the chest scales, when present, are very small, as are the scales on the ventral and ventrolateral body surface as far posteriorly as the anus* (Fig. 3).

The small ventral body scales have an abrupt size transition with the moderatley larger scales on the ventrolateral aspects of the flanks. When the entire chest is scaled, the small scales of that region extend posteriorly beyond a line joining the pectoral and pelvic fin insertions (Fig. 3). There is also a sharply defined size difference between the larger ventrolateral body scales and the small thoracic ones, the line of size demarcation curving gently in a posteroventral direction to merge with the demarcation line separating the belly and ventrolateral flank scales (Fig. 3). In effect, the corslet of small scales covering the chest trails backwards to the anus (cf. *Ctenochromis* and *Thoracochromis* where the corslet is confined to an area anterior to the pelvic–pectoral fin insertions). Even when the major part of the chest is naked, there is a patch of small scales between and somewhat posterior to a line through the pelvic and pectoral fin insertions; as in the other species, these small scales are sharply demarcated from the larger ones on the flank.

The nuchal scales in two species (*O. polyacanthus* and *O. torrenticola*) are very small and deeply embedded.

There are 30–35 (modal range 30–32) scales in the lateral line, *all the pore-bearing scales of the upper lateral line being separated from the dorsal fin base by not more than one large and one small scale* (cf. *Ctenochromis* and *Thoracochromis* where only the last few scales of the upper lateral line are separated from the dorsal fin by less than two scales of equal size).

The cheek is naked or, if scaled, has a distinct naked area along its entire ventral border; in some individuals with otherwise naked cheeks, a few irregularly arranged scales may occur posterodorsally.

Neurocranium. The skull in *Orthochromis* differs from the generalized type *in having a relatively low and short supraoccipital crest, and in having the skull roof anterior to the supraoccipital crest gently rounded* (not concave or flat as is the generalized skull); *the entire neurocranium is relatively narrow, most noticeably in the interorbital region, and the preorbital skull profile slopes downwards at a steep angle*.

Vertebral numbers: 27–30 (modal range 28–30), comprising 12 or 13 (mode 13) abdominal and 14–17 (mode 17) caudal elements.

Dentition. The outer teeth in both jaws are either bicuspids (generally with the shaft of the tooth curved buccally) or slender unicuspids (in which case small fish have bicuspid teeth). Some slender unicuspids are present posteriorly in the premaxilla of all species.

Inner row teeth are small and tricuspid (with some unicuspids present when the teeth of the outer row are predominantly unicuspids), are arranged in 2 or 3 series anteriorly and laterally, and in a single row posteriorly.

Lower jaw appears foreshortened in lateral view because its posterior region (*angulo-articular bone and the coronoid process of the dentary*) *are deepened* relative to the generalized condition seen, for example in *Astatotilapia*.

Lower pharyngeal bone and dentition. The dentigerous surface of the lower pharyngeal bone is somewhat broader than it is long (*c.* $1\frac{1}{2}$ times), but is not sufficiently broad to give the bone an overall short and broad appearance. The teeth are compressed and cuspidate, those of the two median rows showing some interspecific variation in form, from not as coarse or slightly coarser than the lateral teeth, to being markedly coarser; teeth forming the posterior transverse row are coarse (but cuspidate) in all species.

Dorsal fin with *16–20* (*modes 17 and 18*) spinous and 9–11 (modes 9 and 10) branched rays.

Anal fin with 3 or, in one species, 4 spines, and 7–10 branched rays.

Caudal fin skeleton. The occurrence of fused hypural elements (as determined from radiographs, and in the case of *O. malagaraziensis* an alizarin preparation) is as follows: in *O. malagaraziensis* (7 specimens, including 1 paratype), *O. polyacanthus* (11) and *O. machadoi* (2), hypurals 1 and 2, and 3 and 4 are fused, but in *O. torrenticola* (2) none is fused although all are closely apposed to one another in each half of the skeleton. Vandewalle (1973) records no fusion in the specimen of *O. torrenticola* he examined.

Caudal fin is moderately to strongly subtruncate (almost rounded).

Pelvic fin with the second, or the second and third branched rays the longest, thus giving the fin a rounded rather than an acute distal margin.

Anal fin markings in male fishes. No discrete, egg-dummy-like markings have been described for any *Orthochromis* species, nor are any visible in the preserved material examined; certainly none was visible in the live specimens of *O. malagaraziensis* I examined (Greenwood, 1954). In some species the fin is without any form of maculate colour pattern so that if egg-dummies were present they should be visible. *Orthochromis torrenticola* does have a maculate anal fin (the spots arranged in oblique rows) and Thys van den Audenaerde (1963) reports that males have more densely spotted fins that do females.

Observations on live *O. malagaraziensis* suggest that sexually dimorphic coloration in that species may be confined to differences in the colour of the lips, anal fin, and branchiostegal membrane (Greenwood, 1954).

Gill rakers relatively slender but short, 6–9 (modes 7 and 8) in the outer row on the lower part of the first gill arch.

Contained species
Orthochromis malagaraziensis (David), 1937. Malagarasi river (Burundi and Tanzania); see Greenwood (1954) for a redescription of the species and notes on its biology.

Orthochromis polyacanthus (Blgr.), 1899. Lake Mweru, Upper Zaire river (Stanley Falls and Stanleyville, and certain affluent rivers (see Thys van den Audenaerde, 1963)). I have, for the moment, accepted Regan's (1922a) synonymy of Boulenger's (1902) *Tilapia stormsi* with this species; however, a review of material in the BMNH suggests that Regan's opinion may not be correct.

Orthochromis torrenticola (Thys van den Audenaerde), 1963. Lufira river (Upper Zaire river drainage).

Orthochromis machadoi (Poll), 1967. Cunene river, Angola.

Diagnosis and discussion
Members of the genus *Orthochromis* are characterized, principally, by the abrupt size change between the large scales on the ventrolateral aspects of the flanks and the small scales of the chest and belly, by the curved and posteroventrally directed line of size demarcation between these scales, and the union of this line with that separating the very small scales on the belly from the larger scales on the flanks, see Fig. 3. The very small belly scales, extending backwards to the anus, are another characteristic feature. Also characteristic (when taken in combination with those characters listed above) is the absence or extensive reduction of the cheek squamation, the posteriorly deepened lower jaw, the increased number of spinous rays in the dorsal fin (without a corresponding reduction in the number of branched rays, this comparison being based on the modal counts for branched rays in *Ctenochromis* and *Thoracochromis*), the elongate second or second and third branched rays in the pelvic fin and, apparently, the absence of egg-dummy-like markings on the anal fin of adult males.

Other diagnostic features are reviewed on pp. 295–296 above; the high frequency of hypural fusion, affecting both the upper and lower halves of the caudal fin skeleton, is particularly noteworthy but requires confirmation from larger samples and the use of skeletal rather than radiographed material.

The absence of egg-dummy-like markings on the anal fin also requires confirmation from observations made on live specimens (their absence in *O. malagaraziensis*, however, seems certain; Greenwood (1954)). This is a most unusual feature amongst '*Haplochromis*'-group species, and may imply that the courtship and breeding habits of *Orthochromis* species are also unusual for the group. Until something is known about these habits in *Orthochromis* it is impossible to determine whether the absence of egg-dummies is to be considered a primitive or a derived feature for the genus.

I am unable to demonstrate any clear-cut interspecific relationships within the *Orthochromis* lineage. *Orthochromis machadoi* (Cunene river) is probably the least derived member. It has a partly scaled cheek, the chest is either entirely scaled or, as in one specimen, it can have a small scaleless area unilaterally, the ventral (belly) body scales are relatively large and, finally, in its general facies the species has not fully achieved the elongate goby-like body form seen in the other species.

In his original description of *O. machadoi*, Poll (1967) argues that the species is closely related to *Pseudocrenilabrus philander* (Weber) an opinion I cannot accept (especially since Poll's views are, it seems, largely based on supposed similarities in coloration). Anatomically, and with regard to their squamation patterns, the taxa are quite distinct.

The preserved colours of *O. machadoi*, on the other hand, are like those of *O. malagaraziensis*. Both species have all the body scales (except on the chest and belly) narrowly outlined in black, giving the body an overall 'diamond-mesh' pattern; they also have a distinctive and vertically elongate dark blotch at the base of the caudal fin.

In *O. torrenticola* this diamond-mesh pattern is very faint but general over the body, whereas in *O. polyacanthus* it is restricted to a pair of narrow bands, one situated midlaterally, the other following the upper lateral line. *Orthochromis torrenticola* retains the caudal spot which is lost in *O. polyacanthus*. Both species have the body crossed by several closely spaced vertical bands. The apo- or plesiomorph states of these colour patterns cannot be determined.

Orthochromis torrenticola and *O. polyacanthus* have minute scales on the dorsal surface of the head and nuchal region (in *O. machadoi* and *O. malagaraziensis* these scales are only slightly smaller than those on the dorsal body surface), and the scales on the thoracic region are relatively smaller than in the other two species, especially *O. machadoi*. In other words, *O. torrenticola* and *O. polyacanthus* share derived features in their squamation.

If these various characters can be taken as indicators of relationship, then *O. machadoi* and *O. malagaraziensis* would be sister species, as would *O. torrenticola* and *O. polyacanthus*. But, one must set against these similarities the fact that the chest and cheek are naked (or largely naked) in *O. malagaraziensis* and *O. torrenticola*, and that both species have similar general facies (sharply decurved anterior head profile, elongate body and a supralateral eye), all features which would appear to be derived rather than plesiomorph ones. The four anal spines in *O. torrenticola* must be considered an autapomorphic feature and as such cannot be used to assess relationships.

Orthochromis, *Ctenochromis* and *Thoracochromis* share one derived feature, the abrupt size transition between chest and body scales, and thus are presumed to be derived from a common ancestor also possessing this feature. However, no synapomorph character can be found to indicate which two of the three genera are more closely related to one another.

Since *Thoracochromis* has only one apomorph feature (chest–body scale size transition), a character shared by all three taxa, it can on that basis be considered to represent the least derived member of the group.

Ctenochromis and *Orthochromis* both exhibit, but do not share, a number of derived features which must, therefore, be considered autapomorphic for the lineage in which they occur (and define). If one were to consider 'trend' characters, for example a tendency to reduce cheek and chest squamation, then *Ctenochromis* and *Orthochromis* could be said to share some derived features not shared with *Thoracochromis*. But, I can find no trenchant synapomorphic character that would allow one to establish an unequivocal sister-group relationship between the two taxa. It is for this reason that I have given each lineage in this ultimately monophyletic assemblage the status of a genus (see p. 269) rather than ranking *Orthochromis* and *Ctenochromis* as subgenera (i.e. implicit sister-groups) on the grounds of their having shared and presumed apomorph 'trend' characters.

Section II

Although several of the species dealt with in this section have previously been referred to the genus *Haplochromis* (see Bell-Cross, 1975), at least one author (Trewavas, 1964) has suggested that these same species, together with the genera *Serranochromis* and *Chetia*, are more closely related to one another than to any of the species already accounted for. In part I would agree with Trewavas' groupings, but the available evidence does not allow one to substantiate, in their entirety, the relationships indicated in her phyletic diagram (Trewavas, 1964 : fig. 1), nor is it possible to determine the relationships of these 'southern' taxa with the more northern '*Haplochromis*'-group genera considered in Section I.

SERRANOCHROMIS Regan, 1920

TYPE SPECIES. *Chromys thumbergi* Castelnau, 1861 (neotype, designated by Trewavas (1964), in BMNH collections).

I have united several species (those previously placed in this genus by Trewavas (1964) and others placed in *Haplochromis* by Bell-Cross (1975)) into one lineage (=genus) because all share the following apparently derived features: (i) *A high number of abdominal vertebrae, 16–18, rarely 15 or 19 (modal numbers 16 and 17)*. (ii) *A large number of gill rakers, 9–15 (modal range 10–13) in the outer row on the lower part of the first gill arch.* (iii) *A high number of branched fin rays in the dorsal fin.*

In addition, members of this lineage reach a large adult size, all have cycloid or a mixture of cycloid and weakly ctenoid scales (the ctenii confined to a small median sector on the scale's posterior margin) in which the cycloid kind predominate, and the anal fin markings (egg-dummies) in males are numerous, small and non-ocellate (in some species differing little in size, shape or colour from the spots on the soft part of the dorsal fin). It is not, however, possible to assess the primitive or derived states of these features which, therefore, are of no direct value in assessing phylogenetic affinities, (but see p. 274 regarding egg-dummies.)

Two sublineages, each based on shared derived features common to their constituent species, can be recognized within the genus *Serranochromis*, and these are given subgeneric rank.

Subgenus *SERRANOCHROMIS* Regan, 1920

TYPE SPECIES. *Chromys thumbergi* Castelnau, 1861.

Description
The body form varies from deep to moderately slender (body depth 30–45% of standard length).

Squamation. The scales on the head, chest, cheek and above the upper lateral line are cycloid, those elsewhere on the body mostly cycloid. When ctenoid scales are present these are weakly ctenoid, with the ctenii confined to a short median sector on the free margin of the scale.

The scales on the chest (which may be relatively small) show a gentle size gradation with those on the lateral and ventrolateral aspects of the flanks; the chest is always fully scaled.

The cheek is fully scaled, *with from 3 (rare) to* 11 *horizontal rows of scales (usually 5–9 rows)*.

There are 35–41, *rarely* 34 *scales in the lateral line*, all but the last 2 or 3 pore-bearing scales of the upper lateral line are separated from the dorsal fin origin by two scales of approximately the same size.

Neurocranium. The skull has a protracted preotic region (comprising some 65–70% *of the total neurocranial length), especially noticeable in the ethmovomerine region which comprises c.* 27–33% *of the total neurocranial length. The ethmovomerine part of the skull is almost horizontally aligned, its dorsal surface sloping at a small angle (Fig. 13).* The supraoccipital crest is variously developed, high in some species, relatively lower in others but never shallow relative to the total skull proportions.

Vertebral numbers and apophysis for the dorsal retractor muscles of the upper pharyngeal bones. There are 31–36 vertebrae, comprising 16–18, rarely 15 or 19 (modes 16 and 17) abdominal and 16–18, rarely 15 (modes 16 and 17) caudal elements. Such a high number of both caudal and abdominal vertebrae is rarely encountered amongst '*Haplochromis*'-group cichlids, and is unique amongst the fluviatile taxa (see, also p. 313 below).

An apophysis for the origin of the dorsal retractor muscles of the upper pharyngeal bones is developed on the ventral face of either the 3rd or 4th abdominal vertebra; although the apophysis does occur on the 4th vertebra in other '*Haplochromis*'-group taxa, it is usually confined to the 3rd centrum (see **Trewavas**, 1964 for comments on this feature).

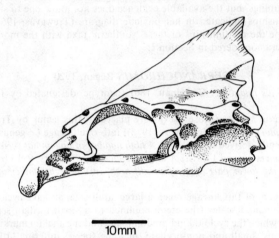

Fig. 13 Neurocranium of *Serranochromis* (*Serranochromis*) *robustus*; left lateral view.

Dentition. The teeth in both the inner and the outer rows of the jaws are unicuspid, even in the smallest (29 mm SL) specimens examined. Other '*Haplochromis*'-group species in which the adult dentition is a unicuspid one (and for which specimens less than 100 mm SL are available), have bicuspid outer teeth, and usually tri- or bicuspid inner teeth, in fishes less than 80–100 mm standard length.

The inner teeth are arranged in a single (rarely a double) row; when two rows are present, these are confined to the anteromedial part of the jaw, the series continuing posteriorly as a single row. In many species the inner teeth of the lower jaw are confined to a narrow, anteromedial arc. The majority of '*Haplochromis*'-group taxa have a more extensive inner dental pattern, with the teeth arranged in at least 2 (and usually 3) rows over the anteromedial and anterolateral parts of the jaw bones.

Lower pharyngeal bone and teeth. With regard to its outline shape when viewed occlusally, two kinds of pharyngeal bone can be recognized (Fig. 14). *In one, the commoner type, the bone is long and narrow, the dentigerous surface having the outline of an isosceles triangle* (Fig. 14A & B). The second type (found in two species) is relatively broader and its dentigerous surface, although still slightly broader than long (c. $1\frac{1}{3}$ times) is more nearly equilateral (Fig. 14C).

Irrespective of the bone's outline shape, the teeth (except those in the two median rows) are fine and either simply pointed or with a weakly developed shoulder anterior to the pointed cusp. The two median and the posterior transverse rows are made up of stouter teeth, those in the median rows are relatively the stouter and have the shoulder more clearly demarcated than it is in the outer teeth.

Jaws. The lower jaw is relatively slender in lateral view (Fig. 15B) and is not noticeably deepened posteriorly (angulo-articular region). *The premaxillae have, in most species, long ascending pro-*

A REVISION OF THE *HAPLOCHROMIS* GENERIC CONCEPT 301

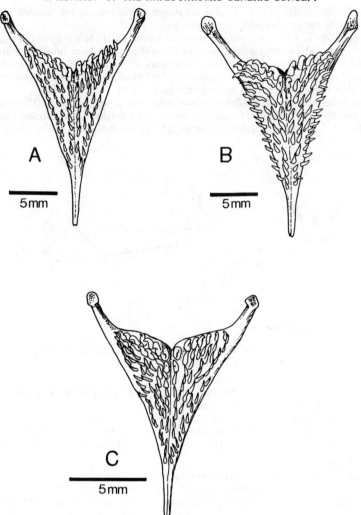

Fig. 14 Lower pharyngeal bones of various *Serranochromis* (*Serranochromis*) species, seen in occlusal view. A. *Serranochromis* (*S.*) *thumbergi*. B. *S.* (*S.*) *robustus*. C. *S.* (*S.*) *macrocephalus*.

cesses (Fig. 15A) *which, in the entire fish, extend to the level of the midpoint of the dorsal orbital margin or even further dorsoposteriorly*.

 Dorsal fin: with 13–18 (modes 15 and 16) spinous, and 13–16 (*usually* 14–16) *branched rays* (a high branched ray count when compared with that in other fluviatile haplochromine taxa).

 Anal fin: with 3 spines and *9–13* (*modes 10 and 11*) *branched rays* (again, a high branched ray count).

 Caudal fin skeleton. No hypural fusion was noted in any of the radiographed material examined, i.e. *S. macrocephalus* (12 specimens), *S. spei* (1), *S. robustus* (14), *S. longimanus* (4), *S. angusticeps* (17), *S. stappersi* (1), *S. meridionalis* (1). No fusion was reported by Vandewalle (1973) in the *S. macrocephalus* (1) or *S. robustus* (1) he examined.

Caudal fin: subtruncate (slightly emarginate in one species) to weakly rounded.
Pelvic fin: with the first branched ray the longest.

Anal fin markings in male fishes. Most species have many small, generally circular spots without a clear surround and covering a large area of the soft anal fin, sometimes extending onto the spinous part as well. In their size and shape these spots are similar to those on the soft part of the dorsal fin and on the caudal fin. An exception to these generalizations is *S. spei* which has fewer and larger (but non-ocellate) spots covering the greater part of the soft fin.

From the little information available on live coloration it seems possible that the anal spots may differ slightly from the dorsal fin spots in colour and intensity, but this requires confirmation. (For coloured illustrations, see Jubb (1967a, pls 41–44) and Bell-Cross (1976: pls 26–28).

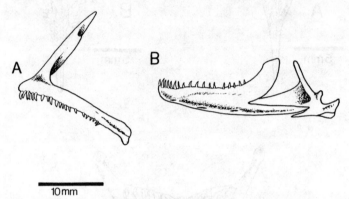

Fig. 15 Jaws of *Serranochromis (S.) robustus*; left lateral view. A. Premaxilla, B. Lower jaw.

Trewavas (1964) reports that similar spots are present on the anal fin of females, but are grey rather than red, yellow or orange as in males.

Where breeding habits are known, the species are female mouthbrooders.

Gill rakers are of various forms, from short and stout to moderately slender; there are 9–13 (modal range 10–12) rarely 8 rakers in the outer row on the lower part of the first gill arch.

Contained species

See Trewavas (1964) for detailed descriptions, figures, etc.

Serranochromis (S.) robustus (Günther), 1864. Lake Malawi, Upper Shire river; Mossamedes; Okavango; Upper Zambesi; Kafue river; Luangwa system (tributary of the Middle Zambesi); Bangweulu region; Luembe river, Kasai system (see Poll, 1967); possibly also in Lake Mweru and the Lualaba system.

Serranochromis (S.) thumbergi (Castelnau), 1861 (Type species). Mossamedes; Okavango river and Lake Ngami; rivers Kafue and Luansemfwa (Luangwa system); Bangweulu region; Upemba basin.

Serranochromis (S.) macrocephalus (Blgr.), 1899. Mossamedes; Okavango river; Lake Cameia (on an Angolan tributary of the Upper Zambesi); Upper Zambesi; Kafue river; Luansemfwa river, Luangwa system; Luapula river; Lake Mweru; Lulua river; Angolan Kasai.

Serranochromis (S.) angusticeps (Blgr.), 1861. Mossamedes; Okavango river and Lake Ngami region, Upper Zambesi; Kafue river; Bangweulu region; Luapula river; possibly Lake Mweru (see also Poll, 1967).

Serranochromis (S.) longimanus (Blgr.), 1911. Okavango river and the Upper Zambesi.

Serranochromis (S.) stappersi Trewavas, 1964. Lake Mweru and the lower Luapula river.

Serranochromis (S.) spei Trewavas, 1964. Lake Kafakumba (23°40′ E, 9°40′ S) on a tributary of the Kasai system; Lake Kabongo in the Lake Upemba depression.

Serranochromis (*S.*) *janus* Trewavas, 1964. Malagarasi swamps (Malagarasi river), Tanzania.

Serranochromis (*S.*) *meridionalis* Jubb, 1967. Incomati river system, Transvaal, South Africa (see Jubb, 1967b).

Dr Trewavas (1964) has discussed the possible affinities of these species (except *S. meridionalis*) at what should now be considered an intra-subgeneric level. Until more material is available for anatomical studies no further comment would be worthwhile. The relationships of the subgenus with its sister-group (*Sargochromis*), and of the genus as a whole, will be considered below (p. 306).

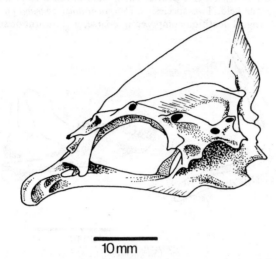

Fig. 16 Neurocranium of *Serranochromis* (*Sargochromis*) *codringtoni*; left lateral view.

Subgenus *SARGOCHROMIS* Regan, 1920.

TYPE SPECIES: *Paratilapia codringtoni* Blgr., 1908 (type specimen in the collections of the BMNH, see Bell-Cross, 1975).

Description

Body deep and stout (depth 35–50% of standard length).

Sargochromis differs from the nominate subgenus in the following characters:

Squamation. There are fewer lateral line scales (28–34, modes 30 and 31, cf. 35–41 rarely 34).

Neurocranium. Although basically of the same type as that in *Serranochromis*, most *Sargochromis* species have a somewhat shorter ethmoid region (but similar preotic skull proportions), a deeper otico-occipital region and, in some species, a more robust apophysis for the upper pharyngeal bones (Fig. 16). This latter character is positively correlated with the degree of enlargement of the pharyngeal bones and the extent to which their dentition is molarized (see Greenwood, 1965a and 1978). The more massive the pharyngeal bones the greater is the relative contribution of the basioccipital to the articular surface of the apophysis, and in those species with the largest bones the prootic also contributes to that surface.

Vertebral numbers. There are fewer caudal vertebrae (12–16, modal numbers 14 and 15), and hence a lower total count (28–32, mode 31). *The number of abdominal vertebrae, however, is high* in both subgenera.

Jaws. The dentary differs from that in *Serranochromis* in being relatively more foreshortened and thus deeper (Fig. 17). The premaxillary ascending processes do not extend beyond about the midpoint of the anterior orbital margin (beyond that point in most *Serranochromis*).

Dentition. Unlike small specimens of *Serranochromis*, small *Sargochromis* do have some bicuspid inner and outer teeth (at least some specimens < 10–15 cm, depending on the species, have predominantly bicuspid outer teeth). The dental pattern is similar in both subgenera, save that *S. (Sargochromis) thysi* has 4 inner series in both jaws.

Lower pharyngeal bone and teeth. The bone shows some interspecific variation in outline shape (Fig. 18) but is always relatively broader than in *Serranochromis*, and thus the dentigerous surface more closely approximates to the equilateral. In only one species, *S. (Sargochromis) greenwoodi* (Fig. 18A), are there no markedly enlarged median teeth. Most of the other species have some enlarged, often submolariform, teeth in addition to those forming the two median rows. Generally these enlarged teeth are restricted to a central patch, several tooth rows wide, in the posterior (oesophageal) dentigerous field. Two species, *S. (Sargochromis) codringtoni* and *S. (Sa.) giardi*, have most of the pharyngeal dentition composed of coarse, molariform or submolariform teeth.

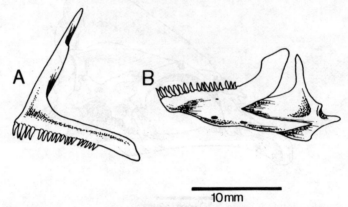

Fig. 17 Jaws of *Serranochromis (Sargochromis) codringtoni*; left lateral view. A. Premaxilla, B. Lower jaw.

Bell-Cross (1975: fig. 1) described intrapopulational differences in the extent to which the pharyngeal dentition is enlarged.

Most *Sargochromis* species have the lower pharyngeal bone coarser than it is in the nominate subgenus, and markedly so when the pharyngeal dentition is hypertrophied.

Dorsal fin has a lower modal branched ray count (12 or 13 cf. 14–16 in *Serranochromis*) but there is an extensive overlap in the total ranges (11–16 cf. 13–16). There is also a broad overlap in spinous ray counts, although the higher numbers (17 and 18) recorded for the nominate subgenus have not been reported for *Sargochromis*.

Anal fin markings are essentially the same in both subgenera. For colour illustrations see Jubb (1967a: pls 40 and 45); Bell-Cross (1976: pls 17 and 18).

Caudal fin skeleton. No hypural fusion was seen in the radiographs of *S. (Sa.) coulteri*, *S. (Sa.) greenwoodi*, and *S. (Sa.) codringtoni* (1 specimen each); in *S. carlottae* one specimen has hypurals 3 and 4 fused but two other fishes show no fusion. Of the two *S. (Sa.) mellandi* examined, one has hypurals 3 and 4 fused, but the other has none fused. Vandewalle (1973) records *S. (Sa.) mellandi* as having either no fusion (4 specimens) or hypurals 3 and 4 fused (2 specimens).

Caudal fin strongly subtruncate to virtually rounded.

Gill rakers are more numerous in *Sargochromis* (9–15, modal numbers 12 and 13).

Contained species
For a systematic review and notes on the ecology and distribution of the first seven species listed below see Bell-Cross (1975).

Serranochromis (Sa.) greenwoodi (Bell-Cross), 1975. Upper Zambesi; Kafue system; Okavango system.

Serranochromis (Sa.) coulteri (Bell-Cross), 1975. Upper Cunene system.

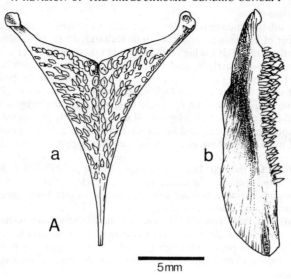

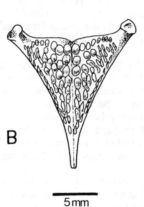

Fig. 18 Lower pharyngeal bone of : A. *Serranochromis* (*Sargochromis*) *greenwoodi* in a. Occlusal, and b. Right lateral view (bone aligned vertically). B. *S.* (*Sa.*) *codringtoni*; occlusal view.

Serranochromis (*Sa.*) *mortimeri* (Bell-Cross), 1975. Upper Zambesi; Kafue river (above the Lufwanyama–Kafue confluence); part of the Mulungishi river (a Middle Zambesi tributary).

Serranochromis (*Sa.*) *mellandi* (Blgr.), 1905. Chambesi river and Lake Bangweulu; the Luapula river and Lake Mweru; Lake Calundo, Angola (see Poll, 1967).

Serranochromis (*Sa.*) *carlottae* (Blgr.), 1905. Upper Zambesi; Okavango and Kafue systems.

Serranochromis (*Sa.*) *codringtoni* (Blgr.), 1905 (Type species of the subgenus). Upper and Middle Zambesi rivers (including the larger tributaries); Kafue and Okavango river systems.

Serranochromis (*Sa.*) *giardi* (Pellegrin), 1904. Middle and lower regions of the Upper Zambesi; the Okavango and the plateau section of the Kafue river; one record from the Cunene system.

Serranochromis (*Sa.*) *thysi* (Poll), 1967. Luembe river, Angola.

Diagnosis and discussion

Members of both subgenera comprising the genus *Serranochromis* are characterized by having a high modal number of abdominal vertebrae (16 or 17, rarely 15 or 19, modes 16–18) and thus a high total vertebral count (29–36), a high number of gill rakers in the outer row on the lower part of the first gill arch (10–15, rarely 9, modal range 10–13), a generally high number of branched dorsal fin rays (11–16, modal range 12–16), mostly cycloid scales on the body (if some ctenoid scales are present they are weakly so), a high number of lateral line scales (28–41) as compared with other fluviatile '*Haplochromis*'-group species and, at least in one subgenus, a greater number of scale rows on the cheek (5–9), in having a skull with a relatively protracted preotic region and relatively high supraoccipital crest, and in having numerous, small non-ocellate coloured spots on the anal fin of male fishes (these markings differing but slightly in size and colour from those on the dorsal fin).

Diagnostic features for the two subgenera are detailed on pp. 299–301 & 303–304, and in the key (p. 316). They involve, chiefly, the higher number of caudal vertebrae and branched dorsal fin rays in *Serranochromis* (*Serranochromis*) species, and the stouter lower pharyngeal bones and dentition in *Serranochromis* (*Sargochromis*) species.

Trewavas (1964) and Bell-Cross (1975) have considered intragroup relationships within the subgenera *Serranochromis* and *Sargochromis* respectively. A reconsideration of their conclusions is beyond the scope of this paper, although Trewavas' (1964 : fig. 1 and p. 10) grouping of the *Serranochromis* (*Serranochromis*) species would seem, on the basis of the characters used, to be a sound hypothesis.

Trewavas (1964) also made an extensive analysis of the intergroup (i.e. intergeneric) relationships of what I am treating as the subgenus *Serranochromis* (treated by Trewavas as a genus). She brought into these considerations the 'genus' *Sargochromis* (*S. codringtoni* only) and three '*Haplochromis*' species (*mellandi, frederici* and *carlottae*) which are now referred to *Sargochromis* (as a subgenus of *Serranochromis*). In discussing Trewavas' ideas, unless quoting directly, I shall use the terms '*Serranochromis*', '*Haplochromis*' and '*Sargochromis*' to cover her concept of these taxa.

In Trewavas' view (1964 : also fig. 1, p. 8) '*Serranochromis*' is '. . . a gradal genus rather than a clade', of diphyletic origin from '. . . a small species-flock of *Haplochromis*' (i.e. the four *Haplochromis* of Angola, '*H*'. *lucullae*, '*H*'. *humilis*, '*H*'. *acuticeps* and '*H*'. *angolensis*, plus '*H*'. *darlingi* of the Zambesi (see below, pp. 310–313).

'A cladal grouping,' Trewavas continues, 'would recognize *Chetia, S. robustus* and *S. thumbergi* on the one hand, and *H. welwitschii, S. macrocephalus* and the other species of *Serranochromis* on the other, but definitions would be almost impossible. . . . The broken line' (referring to fig. 1) 'at the *Haplochromis–Sargochromis* transition reflects the absence here too of a clear generic division'.

Because all '*Serranochromis*' species share a high caudal vertebral count and other apparently derived features (see p. 299), I cannot accept Trewavas' concept of that taxon having a diphyletic origin, nor can I accept, without considerable qualification, the inclusion of *Chetia* (i.e. *C. flaviventris*) and '*H.*' *welwitschii* in one cladal grouping. Neither *Chetia* nor '*H.*' *welwitschii* has the high caudal vertebrae count of '*Serranochromis*' (i.e. the nominate subgenus recognized above) and, although these two species together with certain other endemic Angolan '*Haplochromis*' and '*H.*' *darlingi* do share some features with *Serranochromis*, these are not of the kind that would suggest a close cladistic relationship.

The question of possible relationships between the Angolan species, '*H.*' *darlingi* and *Chetia flaviventris* will be considered on pp. 312–313.

Trewavas (1964 : fig. 1, p. 9) recognizes the phyletic affinity between *Serranochromis* and *Sargochromis* (the latter now of course broadened to include the three '*Haplochromis*' (see p. 305) species which she indicated as being more closely related to '*Sargochromis*' than '*Serranochromis*'). We would differ, however, in our interpretation of the relationship between *Sargochromis* and '*Haplochromis*' *darlingi*. Trewavas (1964 : 9) writes of 'The evolutionary line which leads from *H. darlingi* to *Sargochromis . . .*' But I can find only one derived character (the enlarged pharyngeal mill) that might link '*darlingi*' more closely with *Sargochromis* than with *Serranochromis*, and none of the synapomorph characters shared by *Serranochromis* and *Sargochromis* alone. In the absence of these characters from '*H.*' *darlingi*, and because an enlarged pharyngeal mill has apparently evolved independently in several haplochromine lineages, I consider that the affinities between this species and *Serranochromis* (including *Sargochromis*) are not as close as those implicit in Trewavas' proposed ancestor–descendant relationship.

In my view, *Serranochromis* and *Sargochromis* shared a recent common ancestry not shared with '*H.*' *darlingi* (the common ancestor for the former taxa could well have resembled *S.* (*Sa*) *greenwoodi* in its anatomical, morphological and meristic features; see description in Bell-Cross, 1975).

Any relationship between the genus *Serranochromis* and '*H.*' *darlingi* would be at a more distant level because these two taxa share fewer derived features than do *Serranochromis* and *Sargochromis*.

Finally, comment must be made on the superficially close resemblance between members of the subgenus *Serranochromis* (*Serranochromis*) and certain '*Haplochromis*' species of Lake Victoria (the *spekii–serranus* species complex, see Greenwood, 1967 : 109, and 1974a : 80 *et seq.*; also Trewavas, 1964 : 6). That the resemblance is the result of convergent evolutionary trends towards the production of an adaptive morphotype (piscivorous predator) and not one of close phyletic relationship seems evident from the several features in which the two taxa differ from one another. For example, the predominantly cycloid and weakly ctenoid scales of *Serranochromis* compared with the strongly ctenoid scales of the '*Haplochromis*' species, the few and fully ocellate egg-dummies of the latter as contrasted with the numerous, small and non-ocellate anal spots in *Serranochromis*, and the more numerous gill rakers, branched fin rays and, particularly, the high number of abdominal vertebrae in the latter taxon.

Certainly it would seem more parsimonious to suppose that *Serranochromis* and the Lake Victoria *Haplochromis* were derived from different lineages, rather than to suggest a common ancestry from some widespread lineage of fluviatile, piscivorous predators (an idea I had entertained previously when considering the phyletic history of the Lake Victoria species flock).

It would seem possible, too, that there is no close phyletic relationship between *Serranochromis* and certain '*Haplochromis*' species in Lake Malawi (see Trewavas, 1964 : 6), but more research is required on the Malwai species before this idea can be tested adequately.

CHETIA Trewavas, 1961.

TYPE SPECIES. *Chetia flaviventris* Trewavas, 1961 (Holotype and paratypes in the BMNH, 3 paratypes in the Transvaal Museum, Pretoria).

NOTE. The species *Chetia brevis* Jubb, 1968 is excluded from this genus because in adult males the anal fin markings are large, true ocelli and few in number (3 or 4). Also, unequally bicuspid outer jaw teeth are still present in specimens of a size (86–89 mm SL) when, in *Chetia flaviventris*, the outer row is comprised mainly of unicuspid and caniniform teeth; the few bicuspid teeth present in *C. flaviventris* of that size are different from those in *C. brevis* since the minor 'cusp' is a shoulder and not a point.

Description

The body form is moderately slender (depth of body 29–35% of standard length).

Squamation. The scales on the head, chest, cheek and body above the upper lateral line are cycloid, and cycloid scales predominate on the body below that level as well; a few weakly ctenoid scales may be present anteriorly on the body, the ctenii on these scales being confined to a short median arc on the scale's free margin. It seems possible that a higher proportion of ctenoid scales is present in smaller than in larger individuals; the largest specimen examined has only cycloid scales on all parts of the body and head (see also Trewavas, 1961).

The cheek is completely scaled (5 or 6 *horizontal rows*). The chest scales show a gentle size gradation with those on the belly and ventrolateral aspects of the flanks.

There are 34 *or* 35 *scales in the lateral line series*, with only the last one or two pore-bearing scales of the upper lateral line separated from the dorsal fin base by less than two scales of almost equal size.

Neurocranium. The skull has a moderately produced preotic region (c. 68–70% of total neurocranial length). The ethmovomerine region is not noticeably extended, *and slopes at a slight angle*. In its proportions and general shape, the neurocranium in *Chetia* approaches that in the subgenus *Serranochromis* (*Serranochromis*), but has a less elongate ethmovomerine region.

Vertebral numbers and apophysis for the dorsal retractor muscles of the upper pharyngeal bones. There are 30–32 (mode 31) vertebrae, comprising 14 or 15 (mode 15) abdominal and 15–17 (modes 16 and 17) caudal elements.

Trewavas (1961) reports an absence of any bony apophysis for the origin of the pharyngeal muscles; from the radiographs I have examined (i.e. of the holo- and 4 paratypes) the structure is visible in one specimen. Trewavas (1961) implies that the apophysis serves principally for the attachment of the swimbladder. That organ certainly is attached to the posterior face of the apophysis in all cichlids I have examined, but the greater surface area of the apophysis serves as a point of origin for the pharyngeal retractor muscles.

Dentition. Unicuspid teeth predominate (or are the only kind of teeth present) in specimens more than 30 mm standard length; the few bicuspid teeth present have a much reduced, shoulder-like minor cusp, and are mostly replaced by unicuspids in specimens > 35 *mm S.L.* There is, however, a size correlated change in the kind of unicuspid teeth present. Fishes < 35 mm long have rather flattened, almost spear-shaped unicuspids whereas in larger fishes the teeth are caniniform.

Unicuspid teeth also predominate in the inner rows of fishes at all sizes, although a few weakly bicuspid teeth are present in specimens less than 40 mm SL. *There are one, or, less commonly, two rows of inner teeth anteriorly in both jaws, and a single series laterally.*

Lower and upper jaws. The lower jaw has the appearance and proportions of that in *Serranochromis* (*Serranochromis*) species, but in the upper jaw the ascending premaxillary process does not reach to between the orbits as it does in many of the latter species; it reaches only to about the midpoint of the anterior orbital margin.

Lower pharyngeal bone and teeth. The bone is not thickened, has an almost equilateral dentigerous surface, and its teeth are slender and weakly cuspidate (Fig. 19). Those teeth forming the two median rows and the posterior transverse row are slightly coarser than their congeners.

Dorsal fin with 14 or 15 spinous and 11 or 12 branched rays.

Anal fin with 3 spines and 9 or 10 branched rays.

Caudal fin skeleton. All the hypurals are free in the five specimens radiographed (the type series).

Caudal fin is subtruncate.

Pelvic fin has the first branched ray the longest.

Anal fin markings in male fishes. As in *Serranochromis* (see p. 302) there are numerous, small and non-ocellate spots covering a large area of the soft anal fin, the spots resembling in size and coloration those on the soft part of the dorsal fin.

According to Du Plessis & Groenewald (1953) the anal spots in *C. flaviventris* are more plentiful in males than in females, and the species is a female mouthbrooder.

Gill rakers are moderately short and slender, with 9 or 10 rakers in the outer row on the lower part of the first gill arch.

Contained species
Chetia flaviventris Trewavas, 1961 (Type species). Tributaries of the Limpopo and Incomati rivers, Transvaal, South Africa.

Diagnosis and discussion
The single species in this genus is distinguished from the other fluviatile '*Haplochromis*'-group species, except *Serranochromis*, by the nature of the anal fin markings in adult males, which are numerous, small and non-ocellate (and which barely differ from those in females). In addition, *Chetia* is distinguished from *Ctenochromis, Orthochromis* and *Thoracochromis* by the nature of

the scale pattern in the thoracic–abdominal region (a gradual as compared with an abrupt size change in the scales of the two body regions).

From *Serranochromis*, *Chetia* is distinguished mainly by having fewer (14 or 15) abdominal vertebrae (cf. 16–18, rarely 15, in *Serranochromis*) and by having bicuspid teeth in specimens of a larger size. In meristic characters, other than vertebral numbers, the two genera have a comparable overlap, but for each feature the modal values are distinct, those for *Serranochromis* being the higher.

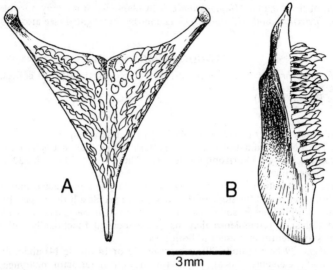

Fig. 19 Lower pharyngeal bone of *Chetia flaviventris*. A. Occlusal view. B. Right lateral view (bone aligned vertically).

I can detect no apomorph features which would suggest that *Chetia* might be related to any of the three lineages showing an abrupt size change in the thoracic–abdominal squamation, and nor can I find apomorph characters to associate it with *Haplochromis*, *Astatotilapia* or *Astatoreochromis*.

That *Chetia* and *Serranochromis* share a similar kind of anal fin marking does not necessarily imply a close relationship between them either, since it seems likely that this is a primitive (plesimorph) feature for '*Haplochromis*'-group species (see above, p. 275). I can detect no unequivocally synapomorphic features common to *Chetia* and *Serranochromis* and thus, despite their superficial similarities, cannot place the taxa in the same genus.

Likewise, the almost identical anal fin markings in *Chetia* and '*Haplochromis*' *darlingi* (see below, p. 310) cannot be taken to indicate a close relationship. It is for this reason, as well as their lack of uniquely shared apomorph features and the presence of autapomorph features in each species, that has led me to place *Chetia flaviventris* and '*H.*' *darlingi* in separate lineages, and thus to give the latter taxon generic rank (see p. 312).

Trewavas (1964 : 10) has remarked on the similarity between *Chetia* and certain *Serranochromis* species, a similarity which led her to consider *Chetia* an offshoot from a lineage that also contains *S.* (*Serranochromis*) *thumbergi* and *S.* (*S.*) *robustus*. That there are similarities between the three species is undeniable, but *Chetia* does not share with the two *Serranochromis* species (and with other species of the genus) the derived feature of a high number of abdominal vertebrae. It does, of course, share with all *Serranochromis* (*Serranochromis*) species the early ontogenetic appearance of unicuspid outer and inner jaw teeth (see p. 300), an apomorph feature which *Serranochromis* (*Sargochromis*) does not share with the nominate subgenus.

Thus at present, one cannot find a totality of shared apomorph features which would indicate a clear-cut sister group relationship for *Chetia*. For that reason I would consider that *Chetia* is best represented as a monotypic lineage (genus) of uncertain affinities. Intuitively one suspects that *Chetia* is related either to *Serranochromis* (especially the nominate subgenus of that taxon) or to '*Haplochromis*' *darlingi*. But, the evidence to propose formally one or other of these relationships is not available if the classification adopted is to reflect phyletic relationships.

Superficially, *Chetia* also resembles one of the Angolan '*Haplochromis*' species, '*H.*' *welwitschii* Blgr., a taxon known only from its now poorly preserved holotype. Until more and better documented material of the Angolan '*Haplochromis*' is available for study, any possible relationship between *Chetia flaviventris* and '*H.*' *welwitschii* cannot be investigated (see also p. 312 below).

PHARYNGOCHROMIS gen. nov.

TYPE SPECIES. *Pelmatochromis darlingi* Blgr., 1911. (Holotype in the BMNH collections.)

SYNONYMY. See Regan (1922a).

Description
Body form moderately slender (body depth 30–33% of standard length).

Squamation. The body squamation type and pattern is like that in *Chetia* (see p. 307). The cheek is fully covered by 4 or 5 horizontal scale rows. There are 32–34 (modes 32 and 33), rarely 31, scales in the lateral line.

Neurocranium. The preotic portion of the skull is slightly less protracted that in *Chetia*, the brain case is a little higher and the slope of the dorsal skull profile a little steeper. In other words, the overall skull morphology is somewhat more like that in *Serranochromis* (*Sargochromis*) species than in *Chetia*, a resemblance that may be associated functionally with the enlarged pharyngeal bones and dentition present in both taxa.

Vertebral numbers: 29 or 30 (mode 29), comprising 13 or 14 (mode 14) abdominal and 15 or 16 (mode 15) caudal elements. An apophysis for the dorsal retractor pharyngeal muscles is present on the third centrum.

Dentition. There is a predominance of unicuspid, caniniform teeth in the outer row of both jaws in fishes over 60 mm standard length, but even in the largest specimens examined (90 mm SL) many unequally bicuspid teeth persist (and, occasionally, may be the predominant form). Unicuspids also predominate in the inner tooth rows, the other teeth being bi- or weakly bicuspid.

The inner rows of both jaws are arranged in two series anteromedially and a single row laterally and posteriorly.

Jaws. The lower jaw is somewhat shorter and deeper than in *Chetia*, but the premaxilla is similar in both genera.

Lower pharyngeal bone and dentition. The dentigerous surface is equilateral or almost so, and the bone itself is somewhat thickened medially (noticeably so when compared with that in *Chetia*). The two median tooth rows are composed of coarse, stout and molariform or submolariform teeth (Fig. 20), the latter retaining traces of a small, near-central point on the occlusal surface. The teeth in the row, or the two rows on either side of the median series, are markedly coarser than those in the lateral rows (which are also clearly cuspidate), and may have submolariform crowns.

Dorsal fin with 14 or 15 (mode 14), rarely 13, spines and 10–12 (mode 11) branched rays.

Anal fin with 3 spines and 7–8 branched rays.

Caudal fin skeleton. All hypurals are free in the 4 specimens (including the holotype) radiographed.

Caudal fin: strongly truncate to virtually rounded.

Pelvic fin: with the first branched ray the longest.

Anal fin markings in male fishes. As in *Chetia* and *Serranochromis*, there are numerous (up to 18, according to Bell-Cross, 1976), small orange spots on the soft part of the fin, and sometimes extending onto the membrane between the spines as well. *Pharyngochromis darlingi* is a female mouthbrooder (Bell-Cross, 1976). For coloured illustrations see Jubb (1967a : pl. 46); Bell-Cross (1976 : pl. 16).

Gill rakers are short and stout, with 9 or 10 (less commonly 7 or 8) in the outer row on the lower part of the first gill arch.

Contained species
Pharyngochromis darlingi (Blgr.), 1911. Type species. Widely distributed in the Zambesi river system and southwards to the Limpopo.

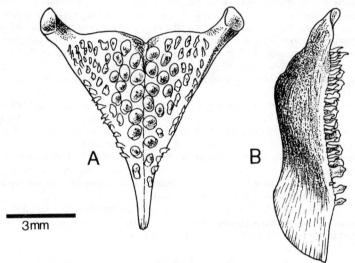

Fig. 20   Lower pharyngeal bone of *Pharyngochromis darlingi*. A. Occlusal view. B. Right lateral view (bone aligned vertically).

Poll (1967) recorded this species (as *Haplochromis darlingi*) from Lake Calundo (Zambesi drainage), Angola, and also redetermined, as *H. darlingi*, specimens from the Cubango river which Pellegrin (1936) had identified as *Haplochromis welwitschii*. However, judging from the nature of the anal fin markings in male specimens, I suspect that these specimens should not be referred to *P. darlingi*. Their identity should be more firmly established (possibly as a yet undescribed species) when a thorough revision of the Angolan '*Haplochromis*' species is carried out (see p. 312).

The same features distinguishing *Chetia* from the other fluviatile '*Haplochromis*'-group genera, including *Serranochromis*, also serve to distinguish *Pharyngochromis*.

From *Chetia* itself, *Pharyngochromis* is distinguished, chiefly, by its stouter lower pharyngeal bone and its partly molarized dentition, by having lower modal numbers of abdominal and caudal vertebrae, fewer lateral line scales, and by having a larger proportion of bicuspid teeth in the outer tooth row of both jaws in fishes more than 40 mm standard length.

My reason for not treating *Chetia flaviventris* and *Pharyngochromis darlingi* as members of the same genus is their lack of shared derived features (see above, p. 310). The same reasons led me to exclude *Pharyngochromis* from the *Sargochromis* division of the *Serranochromis* lineage.

Presumably it is the presence of enlarged pharyngeal teeth, as well as overall similarity in body form and oral dentition, that led Trewavas (1964 : fig. 1 and p. 9) to place *P. darlingi* at the base of a lineage leading to *Sargochromis* (then restricted to the type species, *S. codringtoni*). Derivatives from, and members of, this lineage also included a number of Zambesi '*Haplochromis*' species which I now place in *Sargochromis*.

Since the subgenera *Serranochromis* and *Sargochromis* share certain derived features (especially an increased number of abdominal vertebrae) not found in *Pharyngochromis* it would seem more parsimonious to consider that the two former taxa share a recent common ancestry and that any relationship they may have with *Pharyngochromis* is a more distant one.

The alternative classification implicit in Trewavas' (1964) phyletic diagram, that *Chetia* and *Serranochromis* (i.e. my subgenus *Serranochromis* (*Serranochromis*)) are sister-groups, and that *Sargochromis* (i.e. my subgenus *Serranochromis* (*Sargochromis*)) plus *Pharyngochromis* is the sister-group of *Chetia* and *Serranochromis* combined, is not supported by the distribution of derived characters amongst the taxa involved. Such an arrangement would also imply that the derived features shared by *Sargochromis* and *Serranochromis* were evolved independently. Admittedly in my scheme one specialized feature (the enlarged pharyngeal bones and dentition of *Pharyngochromis* and most *Sargochromis* species) would have to be evolved independently. But, evidence from haplochromine lineages in Lake Victoria (see Greenwood, 1974a) and from *Thoracochromis* (e.g. *Th. mahagiensis* and *Th. pharyngalis*) seems to indicate that the independent evolution of an enlarged pharyngeal mill is not uncommon amongst '*Haplochromis*'-group cichlids.

To summarize: the relationships amongst those '*Haplochromis*'-group taxa with non-ocellate and numerous anal fin spots (a group essentially of the Zambesi, Limpopo, and Angolan rivers) cannot clearly be recognized at present. Two lineages with a presumed recent common ancestry (*Serranochromis* and *Sargochromis*) are treated as sister-groups and given subgeneric rank; the other two lineages cannot be related unequivocally (on the basis of shared derived characters) with either the *Serranochromis–Sargochromis* lineages or with one another; each therefore is treated as a monotypic and monophyletic assemblage (on the basis of autapomorphic features) and given generic rank (*Chetia* and *Pharyngochromis*).

The further resolution of relationships amongst these taxa awaits more detailed studies of their contained species (and of the Angolan '*Haplochromis*' species), and an understanding of the phyletic importance which can be attached to anal fin markings.

The Angolan '*Haplochromis*' species

Several references have been made to these little-known and poorly represented taxa. The last comprehensive revision of the Angolan *Haplochromis* was that of Regan (1922a) who recognized three species, *H. humilis* (Steindachner), 1866, *H. acuticeps* (Steindachner) 1866 and *H. multiocellatus* (Blgr.) 1913. For some reason not stated (but probably because Boulenger (1915) included both species in *Pelmatochromis*), Regan omitted Steindachner's (1865) *Hemichromis angolensis* and Boulenger's (1898) *Pelmatochromis welwitschii*. Both species, however, would have fallen into Regan's definition of *Haplochromis*.

All the specimens representing these species (and others synonymized therein by Regan, 1922a) are poorly preserved, mostly represented by a single individual (or at best 4 or 5 syntypes), and often without precise locality data; the type of *P. angolensis* is now lost (see Bell-Cross, 1975 : 427).

In recent years Poll (1967) has added three species (*H. thysi*, *H. machadoi* and *H. schwetzi*; see pp. 305 and 297 above for the first two species respectively), Trewavas and Thys van den Audenaerde (1964) a fourth (*H. albolabris*) and Penrith (1970) a fifth (*H. buysi*).

My revision of this material indicates that probably several lineages are represented within it, and that Regan's (1922a) re-definition of *H. acuticeps* (Steindachner) embraces at least two species. Some of the Angolan species seem to show affinity with the genus *Astatotilapia*, others with *Chetia* and some may represent lineages yet unrecognized. But, until the species can be revised and reviewed on the basis of more extensive, better preserved and better documented collections I believe that it is inadvisable to place any species in the genera recognized in this paper. It is certainly impossible to demonstrate that the Angolan species are more closely related to one another than to any other lineage, although intuitively one recognizes, in at least some species, an 'Angolan facies' and feels that this overall appearance suggests relationships with the Zambesi–Limpopo genera.

Trewavas' (1964) phyletic diagram illustrating the possible relationships of '*Serranochromis*' and '*Sargochromis*' indicates that a number of Angolan '*Haplochromis*' species are related to '*Serranochromis*' (i.e. the subgenus *Serranochromis* (*Serranochromis*) as defined on p. 299). I can find no synapomorph characters to support this supposition. The very faint traces of anal fin markings left on the holotype of *H. welwitschii* suggest that they may be of the *Serranchromis–Chetia–Pharyngochromis* pattern but this is probably a plesiomorph feature. Where anal fin markings are detectable in the other species mentioned by Trewavas (1964), for example *H. lucullae*, they appear to be of the true ocellar type and thus a derived feature not represented in *Serranochromis*.

The number of vertebrae (especially the abdominal elements) in *H. welwitschii* and the other Angolan species is lower than that in *Serranochromis* (although within the range for *Chetia*) and again represents a plesiomorph condition.

Other characters and character states are equally lacking in shared apomorph features, or represent autapomorphies characterizing the Angolan taxa alone.

As a temporary expedient I can only suggest that the Angolan species be given no formal generic status and that they should be referred to under the informal epithet '*Haplochromis*', whose use in no way implies a close relationship with the species of *Haplochromis* (or, indeed, the majority of species previously referred to that genus).

Summary and conclusions

I am well aware of the shortcomings in this preliminary attempt to clarify the phylogenetic relationships of '*Haplochromis*'-group cichlids, and in particular members of that manifestly polyphyletic 'genus' *Haplochromis*.

Two major difficulties were encountered, and although one has been overcome to a greater or lesser degree, the other still stands in the way of a fully phylogenetic classification. The first difficulty lies in determining morphocline polarity amongst the characters available for research of this kind (p. 270). Then, when plesiomorph and apomorph features are recognized, there is the problem caused by an apparent absence of synapomorphic features at the various levels of relationship necessary to construct a truly cladistic classification.

In other words, one can identify fairly readily what appear to be monophyletic lineages, but the difficulties arise when one attempts to interrelate the different lineages on a sister-group basis.

The problem is well exemplified by the genera *Ctenochromis*, *Thoracochromis* and *Orthochromis*. All three taxa share the presumably derived feature of an abrupt size-change between the scales on the thoracic and ventrolateral flank regions of the body (see p. 270), and thus are assumed to share, at some level, a common ancestry. Both *Orthochromis* and *Ctenochromis*, but not *Thoracochromis* exhibit derived features that are unique for each genus (i.e. autapomorphies) but there are no synapomorphic characters that would indicate which two of the three genera are more closely related to one another. Since *Thoracochromis* shows only one apomorph feature common to all its species it is, presumably, the least derived member of the trio.

Similar difficulties arise with *Haplochromis*, *Astatotilapia* and *Astatoreochromis*, taxa which appear to be interrelated (along with the components of the Victoria–Edward–Kivu species flock) only on the basis of their possessing true ocellar spots on the anal fin of male fishes (p. 274). *Haplochromis* and *Astatoreochromis* (and each of the major lineages in the Victoria flock) have clear-cut autapomorphic features; *Astatotilapia*, apparently, has none. Again one is left with an unresolved polychotomy, but in this case, because the lake flocks are involved, a far more complex one.

Finally, but in a rather different category, since no unifying synapomorphic characters have been detected, are the genera *Chetia*, *Pharyngochromis* and *Serranochromis*. Intuitively the taxa would seem to be interrelated (as they have been assumed to be by other workers, e.g. Trewavas, 1964), probably because of their similar overall morphology, coloration, and the repeated occurrence of enlarged pharyngeal mills amongst their constituent species; in addition, the species form a well-defined, Zambesian geographical group. Yet, I have failed to substantiate their presumed relationship because there are no apparently derived features common to all three genera (see p. 312). Of course, my interpretation of one shared feature (the non-ocellate, and very numerous anal spots) as a primitive condition may be incorrect (see p. 275); only further research, especially comparative ethological research, can clarify that point.

Further research is also needed to test the phylogenetic homogeneity of the speciose lineage *Thoracochromis* (see above and p. 294). As yet no way has been found to test the possibility, indicated by certain morphological features, that there are three infragroups represented in the lineage, viz. one in Lake Turkana, another in the Nile and Lake Albert (including also the outlier species from Lake Edward and possibly Lake Victoria), and a third from the Zaire river system.

Despite these limitations I believe that the classification suggested here is a more efficient one than that existing at present (the term 'efficient' used *sensu* Patterson & Rosen (1977 : 158–159) to denote a classification from which a '... theory of relationships is recoverable ... without loss of information'). Clearly its efficiency can be improved, but that must await the phyletic analysis of '*Haplochromis*'-group species in the Great Lakes, especially those of Lakes Malawi and Victoria.

It has been generally assumed (see Regan 1921b; Trewavas, 1935; Fryer & Iles, 1972) that the Malawi '*Haplochromis*'-group species were derived from an anatomically generalized fluviatile '*Haplochromis*' (i.e. *Astatotilapia*) species. I now suspect, however, that the story is far more complex, that the Malawi flock is probably of polyphyletic origin and that lineages related to *Thoracochromis* as well as to *Astatotilapia* and even to *Serranochromis* and *Chetia* may have contributed to the flock. Possibly some of the ideas put forward in this paper may contribute to the elucidation of that problem.

Likewise the assumed monophyly of the Lake Victoria '*Haplochromis*' species flock (Greenwood, 1974a) must be thrown into doubt, because no characters have been found to support this concept (see p. 269). As compared with Lake Malawi, however, it does seem more likely that fewer and phyletically more closely related lineages were involved, and that most are related to the *Astatotilapia* lineage.

Geographically, the different lineages dealt with in this paper have interesting patterns of distribution.

Thoracochromis is essentially a Nilotic–Zairean taxon (see p. 294). Unlike the others with Zairean representatives (see below), it is best represented in the lower reaches of that river, since only one species (*Th. moeruensis*) is recorded from the upper Zaire system (see p. 293).

The virtual absence of *Thoracochromis* from Lakes Victoria, Edward and Kivu is, on the basis of its overall distribution, rather surprising. Possibly this is attributable to the relatively recent association between these lakes and the Nile system (see Beadle, 1974 : 139–146; Greenwood, 1974b and 1976; Berry, 1976; Livingstone, 1976; Rzóska, 1976a & b : 2–29). Lake Turkana and, as far as can be told, Lake Albert as well, have only ever had major and direct interconnections with the Nile system; the geologically recent riverine connection between Lakes Albert and Victoria probably is made impassable to fishes by the presence of the Murchison Falls (now Kabalega Falls), and the connection between Lakes Edward and Albert via the Semliki river also seems to be impassable for most fishes (see discussions in Greenwood, 1959b, 1973 and 1976; also Rzóska, 1976c : 197–202).

Astatotilapia, apart from its outliers in North Africa (*A. desfontainesi*), Syria and Israel (*A. flaviijosephi*) and possibly in Nigeria (see p. 283), is essentially a lineage of the eastern Rift Valley (except Lake Turkana) and the rivers of eastern Africa; it is represented in the Zaire drainage only by its species in Lake Tanganyika (see p. 284). The absence of *Astatotilapia* from Lake Turkana probably is to be explained through the history of that lake (see above). Far more puzzling is the occurrence of two *Astatotilapia* species north of the Sahara (Tunisia, Algeria, Syria and Israel), and the possibility of one or two other species in Nigeria. This disjunct distribution may, of course, be the result of incorrectly assessing the phyletic relationships of the outlier species. On currently available evidence, however, there is nothing to suggest how else these outlier species might be interrelated.

Ctenochromis, with one exceptional species (*C. pectoralis*) from southeastern Tanzania (Indian Ocean drainage), is totally Zairean (including Lake Tanganyika) in its distribution (see p. 289). *Orthochromis* too is an essentially Zairean lineage and, like *Ctenochromis*, is confined to the upper parts of that system; it has a representative in the Malagarasi river system of Tanzania, but historically that river should be considered part of the upper Zaire drainage (Poll, 1956). The only outlier species, *Orthochromis machadoi*, occurs in the Cunene river, Angola, a river whose ichthyofaunal affinities are closer to those of the Zambesi than the Zaire (Poll, 1967).

Serranochromis is widely distributed (see p. 302), having representatives in the Zaire and Zambesi systems, as well as in the Limpopo and certain Angolan rivers (including the Cunene). Both its Zairean and Zambesi components are confined to the upper portions of their respective systems (cf. the distribution of *Thoracochromis*).

Chetia and *Pharyngochromis* have, geographically speaking, the most restricted distributions of all the fluviatile species considered in this paper (apart from *Astatoreochromis* which occurs only in parts of the Malagarasi and Lukuga rivers, in Lakes Edward and Victoria and in some physiographically related water bodies, see p. 286). *Chetia*, a monotypic genus, is confined to the Limpopo drainage system, and *Pharyngochromis*, also monotypic, to the Upper and Middle Zambesi, the Sabi–Lundi system and the Limpopo system.

The only truly lacustrine lineage discussed in this paper, *Haplochromis*, is confined to Lakes Victoria, Edward, George and Kivu. The significance of this distribution, and the possible relationships of the genus, will be discussed in a forthcoming revision of the lineages from those lakes.

Key to the genera

Notes (i) When citing the range for meristic characters, values rarely encountered are given in square brackets and precede or follow, respectively, the most frequently recorded low and high values for that character.

(ii) Modal values (or modal ranges) are in bold type and enclosed in round brackets.

(iii) Gill raker counts are for the outer row of rakers on the lower part of the arch, and do not include the raker (if such is present) on the epi-ceratobranchial articulation.

(iv) For further notes, and definitions of the characters used see pp. 270–276.

Key

A gradual change in size between the scales on the chest (i.e. ventral and ventrolateral body region anterior to the insertions of the pectoral and pelvic fin bases) and those on the ventral and ventrolateral aspects of the flanks and belly (see Fig. 1) 1

An abrupt size change between the small scales on the chest and the larger scales on the ventrolateral and ventral aspects of the body, the size demarcation line usually running between the pectoral and pelvic bases (but sometimes a little before or behind that level; see Figs 2, 3 & 9) . . . . 2 (p. 316)

1) (a) Anal fin in adult males with 3–9 (**3 or 4**) ocelli (coloured spots each with a clear or translucent area surrounding it) arranged in one or two lines and lying about midway between the base and the distal margin of that fin. (Most females and juvenile males with 3 or 4 non-ocellate spots in the same position, or fin without spots.) Scales below (and often those above) the upper lateral line ctenoid, the ctenii arranged along almost the entire free margin of the scale. Anal fin with 3 spines (individuals with 4 spines are so rare that this number can be considered as an individual abnormality) Dorsal fin rarely with more than 16 spines. Marked sexual dimorphism in adult coloration (males colorful, females drab) . . . . A

(b) Anal fin in adult males with 6–20 ocelli arranged in 3–5 regular rows and thus occupying a large area on the soft part of the fin; females with a similar pattern if spots (non-ocellate) are present. Anal fin with 3–6 spines, dorsal fin with 16–20 (**17–19**) spines. Caudal fin rounded. Lower pharyngeal bone thickened (strongly so in two species), its dentition partially or completely molarized. No marked sexual dimorphism in coloration; body colour yellow–green, fins with a maroon flush. Other features as in 1(a) above

Astatoreochromis (p. 285)

(c) Anal fin in both sexes with numerous (18–40) small spots, none with a clear or translucent surround, not arranged in regular rows but covering most of the area of the soft anal fin; similar spots on the soft dorsal and the caudal fins. Scales below the upper lateral line are cycloid or predominantly cycloid; when ctenoid scales are present the ctenii are weak and confined to a small median sector on the free margin of the scale. . . . B

A) (i) Jaw teeth in the outer row (and sometimes the inner rows as well) with obliquely cuspidate compressed crowns, the major cusp drawn out beyond the tooth's vertical axis (see Fig. 7), the minor cusp reduced or absent. 12–14 (**13**) abdominal and 15 or 16 caudal vertebrae (total 28–30; **28** and **29**). Dorsal fin with 14–16 (**15** and **16**) spines and 8–10 (**9**) branched rays. Anal fin with 3 spines and 7–10 (**9**) branched rays. Lateral line with 29–34 (**30–32**) scales. Cheek with 3 [4] horizontal rows of scales. Caudal fin truncate or weakly subtruncate. Lower pharyngeal bone without any noticeably enlarged or coarse teeth in the two median rows. Gill rakers [7] 8–10 (**9**) . . . . *Haplochromis* (p. 278)

(ii) Jaw teeth in the outer row unequally bicuspid or unicuspid, the crown neither compressed nor obliquely truncate, its tip lying within the tooth's vertical axis (see Fig. 4). Outer teeth mostly bicuspids in fishes <70 mm SL; an admixture of bi- and unicuspids in larger fishes, with unicuspids predominating in specimens >100 mm SL. Inner teeth predominantly tricuspid, small. 12–14 (**13**) abdominal and 14–16 (**15**) caudal vertebrae (total 27–30; **28** and **29**). Dorsal fin with 14–16 (**15**) spines and 8–11 (**9** and **10**) branched rays. Lateral line with 28–30 (in one species 31–34) scales, cheek with [2], **3**, [4] horizontal rows of scales. Caudal fin rounded or slightly subtruncate. Lower pharyngeal bone with at least the two median rows composed of coarser (sometimes molariform) teeth. Gill rakers [7] 8 or 9 . *Astatotilapia* (p. 281)

B) (i) Abdominal vertebrae 13 or 14 (**14**), caudal vertebrae 15 or 16 (**15**), total number of vertebrae 29 or 30 (**29**). Dorsal fin with 14 or 15 (**14**) spines and 10–12 (**11**) branched rays. Lateral line with 32–34 (**32** and **33**) scales, cheek with 4 or 5 horizontal rows. Outer row of jaw teeth composed of unequally bicuspids in fishes <60 mm SL, unicuspids present and becoming commoner in larger individuals. Lower pharyngeal bone thickened, at least the two median rows composed of enlarged and molariform teeth (see Fig. 20). Gill rakers [7 or 8] **9** or **10**. Anal fin with up to 20 spots . . . . *Pharyngochromis* (p. 310)

(ii) Abdominal vertebrae 14 or 15, caudal 15–17 (**16** and **17**), total number of vertebrae 30–32 (**31**). Dorsal fin with 14 or 15 spines and 11 or 12 branched rays. Lateral line with 34 or 35 scales, cheek with 5 or 6 horizontal rows of scales. Outer row of jaw teeth mainly unicuspids in fishes >30 mm SL, some weakly bicuspids (the minor cusp a shoulder rather than a point) present in smaller individuals. Lower pharyngeal bone not thickened, without molariform or submolariform teeth (see Fig. 19). Gill rakers 9 or 10 . . . . *Chetia* (p. 307)

(iii) Abdominal vertebrae [15] 16–18 [19] (**16** and **17**), caudal vertebrae 12–16 (**14** and **15**), total number of vertebrae 28–32 (**31**). Dorsal fin with 13–16 (**15** and **16**) spines and 11–16 (**12** and **13**) branched rays. Lateral line with 28–34 (**30** and **31**) scales, cheek with 3–6 (**3 – 5**) horizontal rows of scales. Outer jaw teeth mostly unequally bicuspids in fishes <150 mm SL, predominantly unicuspids in larger individuals. Lower pharyngeal bone thickened in all but one species, and in all but that species with at least the two median tooth rows composed of enlarged and molariform teeth (see Fig. 18); the exceptional species has coarse and slightly enlarged, but cuspidate, teeth in the median rows (see Fig. 18A), Gill rakers 9–15 (**12** and **13**). Anal fin with up to 40 spots . . . . . . *Serranochromis (Sargochromis)*; p. 303)

(iv) Abdominal vertebrae [15] 16–18 [19] (**16** and **17**), caudal vertebrae [15] 16–18 (**16** and **17**), total number of vertebrae 31–36. Dorsal fin with 13–18 (**15** and **16**) spines and 13–16 (**14–16**) branched rays. Lateral line with [34] 35–41 scales, cheek with 3–11 (**5–9**) horizontal rows of scales. Outer jaw teeth predominantly or entirely unicuspids in fishes >30 mm SL. Lower pharyngeal bone not thickened, either elongate and narrow (see Fig. 14A & B) or its dentigerous surface almost equilateral in outline (see Fig. 14C); no teeth molariform, even the median row teeth only slightly coarser than the others. Gill rakers [8] 9–13 (**10–12**). Anal fin with up to 40 spots
Serranochromis (Serranochromis); p. 299)

2) (a) Pelvic fin with the first branched ray the longest. Scales on ventral body surface behind pelvic fins not markedly reduced in size (see Fig. 9) 2A

(b) Pelvic fin with the second or third branched ray the longest. Scales on ventral body surface and on ventrolateral aspects of flanks small to minute (see Fig. 3). Cheek naked or, if scaled, with a definite naked area along its entire ventral (preopercular) margin. Chest completely scaled, or partly scaled, or naked. Dorsal fin with 16–20 (**17** and **18**) spines and 9–11 (**9** or **10**) branched rays. Anal fin with 3 or 4 spines and 7–10 branched rays. Lateral line with 30–35 (**30** and **31**) scales. Head profile strongly decurved, eyes supro-lateral in most species . . . . ***Orthochromis*** (p. 295)

2A) (i) Chest with a naked patch or extensive naked area on each side of the body (see Fig. 9). At least the ventral part of the cheek scaleless (almost the entire cheek naked in one species). Anal fin with 3 spines and 6–9 (**6–8**) branched rays. Lateral line with 27–33 (**28** and **30** or **31**) scales . . . . ***Ctenochromis*** (p. 287)

(ii) Chest completely scaled. Cheek completely or almost completely scaled (i.e. one horizontal row absent ventrally). Dorsal fin with 13–16 [17] (**14–16**) spines and 8–10 [11] (**9** and **10**) branched rays. Anal fin with 3 spines and 6–10 (**7–9**) branched rays. Lateral line with 29–32 (**30–32**) scales . . . ***Thoracochromis*** (p. 290)

Appendix 1

A replacement 'generic' name for the Lake Malawi '*Haplochromis*' species

Since the genus *Haplochromis* is now restricted to five species, all members of the Lakes Victoria, Edward, George and Kivu species flock (p. 280), the Lake Malawi species formerly referred to *Haplochromis* are without a generic name. Because it is obvious that the '*Haplochromis*' of Lake Malawi are a polyphyletic group, any generic placement at the present time must be considered merely a formal nomeclatural action unrelated to the phyletic affinities of the species.

Two generic names would appear to be available for this purpose (see Trewavas, 1935), namely *Cyrtocara* Boulenger (1902) and *Champsochromis* Boulenger (1915). A third name, *Otopharynx*, Regan (1920), apparently is also available, but it is junior to the others and there are anatomical grounds for regarding its contained species as representing a lineage distinct from that to which many Malawi '*Haplochromis*' belong (Greenwood, 1978).

Cyrtocara (type species *C. moori*) has a pharyngeal apophysis of the typical '*Haplochromis*'-type (Trewavas, 1935) and its oral dentition is composed of slender unicuspid outer teeth and mixed uni- and tricuspid inner teeth.

Although at least some members of the type species have a moderately developed hump in the frontal region of the head, I can see no morphological grounds for not accepting *Cyrtocara* as a temporary formal name for the '*Haplochromis*' species of Lake Malawi. I thus propose that it be used in that capacity until the Malawi species are revised. This action by no means implies that I consider many of these species to have a true phyletic relationship with *Cyrtocara moori*.

Appendix 2

The taxonomic status of the genus *Limnotilapia* Regan, 1920

In a recent paper (Greenwood, 1978) I treated the genus *Limnotilapia* Regan (1920) as a synonym of *Simochromis* Boulenger, 1898, thus unintentionally anticipating the publication of a paper giving detailed reasons for this nomenclatural change. Since publication of the paper in which the two 'genera' are to be discussed is likely to be delayed further, the reasons for synonymizing *Limnotilapia* with *Simochromis* are dealt with below.

A comparison of Regan's (1920) description for *Limnotilapia* with his redescription of *Simochromis* reveals that the taxa apparently are differentiable only on the former having a rather small, terminal mouth, and the latter having the mouth subterminal and rather wide.

When the type species of the genera, *Limnotilapia dardennii* (Blgr.) and *Simochromis diagramma* (Günth.), are compared, these differences can be translated into more substantial osteological ones involving the morphology of the premaxilla and dentary.

Viewed from below (i.e. occlusally), the premaxillary outline in *L. dardennii* is gently curved and relatively narrow; in other words, it has an outline approximating to that of a Norman arch. The premaxillary outline in *S. diagramma*, by contrast, has a virtually straight and wide anterior margin, with the short posterior dentigerous arms meeting it almost at right angles; the outline of the bone is thus more nearly that of a hollow square. The posterior dentigerous arms of the premaxilla in *S. diagramma* are slightly bullate, whereas in *L. dardennii* they are slender and compressed.

There are, of course, comparable interspecific differences in the occlusal outline of the dentary. In *Limnotilapia dardennii* the lateral arms of the dentary are protracted relative to the transversely directed anterior part of the bone, and the outline of the whole bone is similar to that of the premaxilla. In *Simochromis diagramma* the dentary, like the premaxilla, is foreshortened, with the short lateral dentigerous arms (about equal in length to the transverse part) meeting the slightly curved transverse portion at almost a right angle. The dentary in *S. diagramma* also differs from that of *L. dardennii* in having virtually no upward sweep to its coronoid portion; in *L. dardennii* this region slopes upward at a gentle but noticeable angle.

Seen in these terms, the osteological 'morphological gap' separating the taxa would appear to be a more substantial one than that expressed in Regan's (1920) key and generic synopsis. But, the 'gap' is bridged when one examines the premaxilla and dentary of *Limnotilapia loocki* Poll, 1949 (see Poll, 1956 : 62, fig. 10 for an expanded description of the species, and illustrations of the jaws and dentition).

The morphology of both these bones in *L. loocki* is virtually intermediate between those in *L. dardennii* and *S. diagramma*. Thus, it is impossible to differentiate the 'genera' on the osteological features characterizing the jaws of the type species. Furthermore, the external oral characters used by Regan (1920) also intergrade when growth-series of the type species are examined, and I have been unable to detect other characters that might serve to distinguish the taxa (it being understood that the 'genera' are being interpreted here, as they were by Regan, merely on the presence of a discrete morphological gap that is 'greater' than one which might be used to characterize species). There would, therefore, seem to be no grounds for treating *Limnotilapia* and *Simochromis* as distinct genera, the more so when one considers the various (and apparently synapomorphic) features that are shared by all but one of their included species.

Limnotilapia loocki (like *L. dardennii*, *Simochromis diagramma*, *S. babaulti* Pellegrin, *S. curvifrons* Poll and *S. marginatus* Poll) has, in both jaws, slender-shafted, recurved, outer teeth with markedly compressed and expanded, obliquely bicuspid crowns, a greatly reduced (or absent) interspace between the numerous inner and single outer tooth rows in both jaws, a densely toothed lower pharyngeal bone (the teeth fine and compressed) and a strongly decurved anterior profile to the neurocranium (where, in some species, the ethmovomerine region is almost vertically inclined) In all these species, too, the chest scales are small, deeply embedded and have an abrupt size demarcation with the larger scales on the anterior abdominal region of the body.

For the moment it is these apparently apomorphic features which should be used to define the genus *Simochromis* Blgr., 1898 (with which is now included, as a junior synonym, the genus *Limnotilapia* Regan, 1920).

The one species not included in the character analysis given above is *Limnotilapia trematocephala* (Blgr., 1901), a taxon known only from its holotype. I have not, of course, been able to examine all the relevant osteological features in this specimen, but its relatively sparsely toothed lower pharyngeal bone, the morphology of its outer row jaw teeth (which are without noticeably compressed, expanded and obliquely bicuspid crowns, and which are not strongly recurved), and its relatively large pectoral scales, all suggest that the species probably belongs to a different lineage and should not, therefore, be included in the genus *Simochromis*.

For the moment it is impossible to indicate the phyletic relationships of the genus *Simochromis*, either within or without the cichlid flocks of Lake Tanganyika. Much further research will be required before this can be achieved (and will also be needed before a generic placement of '*Limnotilapia*' *trematocephala* can be effected).

As was noted in my paper on the pharyngeal apophysis in African cichlids (Greenwood, 1978), *Simochromis dardennii* has a near-typical *Tilapia*-type of apophyseal structure. *Simochromis loocki*, on the other hand, has an apophysis of the modified *Tropheus*-type; the basioccipital is inflated and bullate, with its ventral tip almost reaching the level of the parasphenoidal facets but not contributing in any way to the articular surface provided by these facets. In its general organization, the apophysis in *S. loocki* is intermediate between the *Tilapia* and *Tropheus* types (see Greenwood, 1978), but differs from the modal condition of both types in having the basioccipital noticeably inflated.

Since *S. dardennii* (with a *Tilapia*-type apophysis) has the least specialized premaxillary and dentary of any *Simochromis* species, and since *S. babaulti* and *S. diagramma* have the most derived jaws (the species having, respectively, *Tropheus* and near *Haplochromis* type apophyses; see Greenwood, 1978), it is tempting to conclude that in this lineage the *Tilapia*-type apophysis is the plesiomorph one. That *S. loocki* (whose jaw morphology is intermediate between that of *S. dardennii* and those of the other *Simochromis* species) has an apophysis intermediate between the *Tilapia* and *Tropheus* types, would also seem to support this hypothesis.

Acknowledgements

I am deeply indebted to my colleague, Gordon Howes, for all the assistance he has given me in the preparation of this paper, and in particular for his skill and patience in preparing the illustrations. For the loan and gifts of specimens used in this work, I gratefully acknowledge the cooperation of the Curator of Fishes, Museum for Naturkunde, Humboldt-Universität, Berlin (D.D.R.); Dr R. A. Jubb and Mr P. S. Skelton, Albany Museum, South Africa; Dr D. F. E. Thys van den Audenaerde, Musee Royal de l'Afrique Centrale, Tervuren; Dr J.-P. Gosse, Institut Royal des Sciences Naturelles de Belgique, Brussels; Dr M. L. Bauchot, Museum National d'Histoire Naturelle, Paris; Dr H. Wilkens, Zoologisches Institüt und Zoologische Museum, University of Hamburg; Dr J. E. Böhlke, Academy of Natural Sciences, Philadelphia; and Dr M. J. Penrith, State Museum, Windhoek.

Finally, it is my pleasure to thank my colleagues of the freshwater fish section (then including Dr Richard Vari, a NATO postdoctoral research fellow) for the innumerable arguments we have had on the subject of phylogenetic systematics.

References

Axelrod, H. R. & Burgess, W. E. 1977. *African cichlids of Lakes Malawi and Tanganyika*. New Jersey.
Beadle, L. C. 1974. *The inland waters of tropical Africa*. London.
Bell-Cross, G. 1975. A revision of certain *Haplochromis* species (Pisces : Cichlidae) of Central Africa. *Occ. Pap. natn. Mus. Rhod. ser* B. **5** (7) : 405–464.
—— 1976. *The fishes of Rhodesia*. Salisbury.
Berry, L. 1976. The Nile in the Sudan, geomorphological history. *Monographiae biol.* **29** : 11–19.
Boulenger, G. A. 1899. A revision of the African and Syrian fishes of the family Cichlidae. Part II. *Proc. zool. Soc. Lond.* **1899** : 98–143.
—— 1902. List of the fishes collected by Mr W. L. S. Loat at Gondokoro. *Ann. Mag. nat. Hist.* (7) **10** : 260–264.
—— 1906. Descriptions of new fishes discovered by Mr E. Degen in Lake Victoria. *Ann. Mag. nat. Hist.* (7) **17** : 433–452.
—— 1907. *Fishes of the Nile*. London.
—— 1915. *Catalogue of the fresh-water fishes of Africa* 3. London.
Du Plessis, S. S. & Groenewald, A. A. 1953. The kurper of Transvaal. *Fauna Flora Pretoria* **3** : 35–43.
Fryer, G. & Iles, T. D. 1972. *The cichlid fishes of the Great Lakes of Africa. Their biology and evolution*. Edinburgh.
Greenwood, P. H. 1954. On two species of cichlid fishes from the Malagarazi river (Tanganyika), with notes on the pharyngeal apophysis in species of the *Haplochromis* group. *Ann. Mag. nat. Hist.* (12) **7** : 401–414.
—— 1956a. A revision of the Lake Victoria *Haplochromis* species (Pisces, Cichlidae), Part I. *Bull. Br. Mus. nat. Hist.* (Zool.) **4** : 223–244.
—— 1956b. The monotypic genera of cichlid fishes in Lake Victoria. *Bull. Br. Mus. nat. Hist.* (Zool.) **3** : 295–333.
—— 1959a. The monotypic genera of cichlid fishes in Lake Victoria. Part II. *Bull. Br. Mus. nat. Hist.* (Zool.) **5** : 163–177.
—— 1959b. Quaternary fish fossils. *Explor. Parc. natn. Albert Miss. J. de Heinzelin de Braucourt* **4** : 1–80.
—— 1960. A revision of the Lake Victoria *Haplochromis* species (Pisces, Cichlidae), Part IV. *Bull. Br. Mus. nat. Hist.* (Zool.) **6** : 227–281.

—— 1965a. Environmental effects on the pharyngeal mill of a cichlid fish, *Astatoreochromis alluaudi*, and their taxonomic implications. *Proc. Linn. Soc. Lond.* **176** : 1–10.
—— 1965b. The cichlid fishes of Lake Nabugabo, Uganda. *Bull. Br. Mus. nat. Hist.* (Zool.) **12** : 315–357.
—— 1967. A revision of the Lake Victoria *Haplochromis* species (Pisces, Cichlidae), Part VI. *Bull. Br. Mus. nat. Hist.* (Zool.) **15** : 29–119.
—— 1971. On the cichlid fish *Haplochromis wingatii* (Blgr.), and a new species from the Nile and Lake Albert. *Revue Zool. Bot. afr.* **84** (3–4) : 344–365.
—— 1973. A revision of the *Haplochromis* and related species (Pisces, Cichlidae) from Lake George, Uganda. *Bull. Br. Mus. nat. Hist.* (Zool.) **25** : 139–242.
—— 1974a. Cichlid fishes of Lake Victoria, east Africa: the biology and evolution of a species flock. *Bull. Br. Mus. nat. Hist.* (Zool.) Suppl. **6** : 1–134.
—— 1974b. The *Haplochromis* species (Pisces : Cichlidae) of Lake Rudolf, east Africa. *Bull. Br. Mus. nat. Hist.* (Zool.) **27** : 139–165.
—— 1976. Fish fauna of the Nile. *Monographiae biol.* **29** : 127–141.
—— 1978. A review of the pharyngeal apophysis and its significance in the classification of African cichlid fishes. *Bull. Br. Mus. nat. Hist.* (Zool.) **33** : 297–323.
—— & Barel, C.D.N. (1978) A revision of the Lake Victoria *Haplochromis* species (Pisces : Cichlidae), Part VIII. *Bull. Br. Mus. nat. Hist.* (Zool.) **33** : 141–192.
Hecht, M. K. & Edwards, J. L. 1977. Phylogenetic inference above the species level. *NATO Advanced Study Inst. ser.A.* **14** : 3–51.
Hennig, W. 1966. *Phylogenetic systematics.* Urbana.
Hilgendorf, E. 1888. Fische aus dem Victoria-Nyanza (Ukerewe-See). *Sber. Ges. naturf. Freunde Berl.* **1888** : 75–79.
Jubb, R. A. 1967a. *Freshwater fishes of southern Africa.* Cape Town.
—— 1967b. A new *Serranochromis* (Pisces, Cichlidae) from the Incomati river system, eastern Transvaal, South Africa. *Ann. Cape prov. Mus.* **6** (5) : 55–62.
—— 1968. A new *Chetia* (Pisces, Cichlidae) from the Incomati river system, eastern Transvaal, South Africa. *Ann. Cape prov. Mus.* **6** (7) : 71–76.
Lacépède, B. G. E. (1803). *Histoire naturelle des poissons* **4**. Paris.
Livingstone, D. A. 1976. The Nile – palaeolimnology of headwaters. *Monographiae biol.* **29** : 21–30.
McMahon, J. P., Highton, R. B. & Marshall, T. F. de C. 1977. Studies on biological control of intermediate hosts of schistosomiasis in western Kenya. *Envir. Conserv.* **4** (4) : 285–289.
Nelson, G. J. 1972. Phylogenetic relationship and classification. *Syst. Zool.* **21** (2) : 227–231.
Patterson, C. & Rosen, D. E. 1977. Review of ichthyodectiform and other Mesozoic teleost fishes and the theory and practice of classifying fossils. *Bull. Am. Mus. nat. Hist.* **158** : 81–172.
Pellegrin, J. 1903. Contribution à l'ètude anatomique, biologique et taxonomique des poissons de la famille des cichlidés. *Mém. Soc. zool. Fr.* **16** : 41–402.
—— 1936. Contribution à l'ichthyologie de l'Angola. *Archos. Mus. Bocage* **7** : 45–62.
Penrith, M-L. 1970. Report on a small collection of fishes from the Kunene river mouth. *Cimbebasia* ser A. **1** : 165–176.
Pfeffer, G. 1893. Ostafrikanische Fische gesammelt von Herrn Dr F. Stuhlmann. *Jb. hamb. wiss. Anst.* **10** : 131–177.
Poll, M. 1932. Contribution à la faune des Cichlidae du lac Kivu (Congo Belge). *Revue Zool. Bot. afr.* **23** (1) : 29–35.
—— 1939. Poissons. *Explor. Parc. natn. Albert Miss. H. Damas* (1935–1936) **6** : 1–73.
—— 1956. Poissons Cichlidae. *Résult. scient. Explor. hydrobiol. lac Tanganika* (1946–1947), **3**, fasc. 5b : 1–619.
—— 1967. Contribution à la faune ichthyologique de l'Angola. *Publicoes cult. Co. Diam. Angola* no. 75 : 1–381.
—— 1974. Contribution à la faune ichthyologique du lac Tanganika, d'après les récoltes de P. Burchard. *Revue Zool. afr.* **88** (1) : 99–110.
Regan, C. T. 1920. The classification of the fishes of the family Cichlidae – I. The Tanganyika genera. *Ann. Mag. nat. Hist.* (9) **5** : 33–53.
—— 1921a. The cichlid fishes of Lakes Albert, Edward and Kivu. *Ann. Mag. nat. Hist.* (9) **8** : 632–639.
—— 1921b. The cichlid fishes of Lake Nyasa. *Proc. zool. Soc. Lond.* **1921** : 675–727.
—— 1922a. The classification of the fishes of the family Cichlidae. – II. On African and Syrian genera not restricted to the Great Lakes. *Ann. Mag. nat. Hist.* (9) **10** : 249–264.
—— 1922b. The cichlid fishes of Lake Victoria. *Proc. zool. Soc. Lond.* **1922** : 157–191.
Rzóska, J. 1976a. The geological evolution of the river Nile in Egypt. *Monographiae biol.* **29** : 2–4.

—— 1976b. Pleistocene history of the Nile in Nubia. *Monographiae biol.* **29** : 5–9.
—— 1976c. Descent to the Sudan plains. *Monographiae biol.* **29** : 197–214.
Schaeffer, B., Hecht, M. K. & Eldredge, N. 1972. Phylogeny and paleontology. *Evolut. Biol.* **6** : 31–57.
Thys van den Audenaerde, D. F. E. 1963. Descriptions d'une espèce nouvelle d'*Haplochromis* (Pisces, Cichlidae) avec observations sur les *Haplochromis* rhéophiles du Congo oriental. *Revue Zool. Bot. afr.* **68** (1–2) : 140–152.
—— 1964. Les *Haplochromis* du Bas-Congo. *Revue Zool. afr.* **70** (1–2) : 154–173.
Trewavas, E. 1933. Scientific results of the Cambridge expedition to the East African lakes, 1930–1. II. The cichlid fishes. *J. Linn. Soc. (Zool.)* **38** : 309–341.
—— 1935. A synopsis of the cichlid fishes of Lake Nyasa. *Ann. Mag. nat. Hist.* (10) **16** : 65–118.
—— 1938. Lake Albert fishes of the genus *Haplochromis*. *Ann. Mag. nat. Hist.* (11) **1** : 435–449.
—— 1942. The cichlid fishes of Syria and Palestine. *Ann. Mag. nat. Hist.* (11) **9** : 526–536.
—— 1961. A new cichlid fish in the Limpopo basin. *Ann. S. Mus.* **46** (5) : 53–56.
—— 1964. A revision of the genus *Serranochromis* Regan (Pisces, Cichlidae). *Annls. Mus. r. Congo belge.* Ser. 8vo, Zool. no. 125 : 1–58.
—— 1973. II. A new species of cichlid fish of rivers Quanza and Bengo, Angola, with a list of the known Cichlidae of these rivers and a note on *Pseudocrenilabrus natalensis* Fowler. *Bull. Br. Mus. nat. Hist.* (Zool.) 25 : 27–37.
—— & **Thys van den Audenaerde, D. F. E.** 1969. A new Angolan species of *Haplochromis* (Pisces, Cichlidae). *Mitt. zool. StInst. Hamb.* **66** : 237–239.
Vandewalle, P. 1973. Osteologie caudale des Cichlidae (Pisces, Teleostei). *Bull. Biol. Fr. Belg.* **107** (4) : 275–289.
Voss, J. 1977. Les livrées ou patrons de coloration chez les poissons cichlidés Africaines. *Revue franc. Aquariol.* **4** (2) : 33–81.
Werner, Y. L. 1976. Notes on reproduction in the mouth-brooding fish *Haplochromis flaviijosephi* (Teleostei : Cichlidae) in the aquarium. *J. nat. Hist.* **10** : 669–680.
Wickler, W. 1962a. Ei-Attrappen und Maulbrüten bei afrikanischen Cichliden. *Zeit. Tierpsychol.* **19** (2) : 129–164.
—— 1962b. Egg-dummies as natural releasers in mouth-breeding cichlids. *Nature* **194** : 1092–1093.
—— 1963. Zur Klassification der Cichlidae, am Beispiel der Gattungen *Tropheus, Petrochromis, Haplochromis* und *Hemihaplochromis* n. gen. (Pisces, Perciformes). *Senckenberg. biol.* **44** : 83–96.

Index

The species discussed in this paper are here indexed under their former generic names. Each entry is followed by two numbers; the first refers to the page on which the species is listed in its new generic group and the second to the page on which that genus or subgenus is described.

Astatoreochromis alluaudi   286   285
Chetia brevis   284   281
C. flaviventris   307   307
Chromis monteiri   294   294
Haplochromis acuticeps   312   312
H. albertianus   293   290
H. albolabris   312   312
H. angolensis   312   312
H. annectidens   280   278
H. astatodon   280   278
H. avium   293   290
H. bakongo   293   290
H. bloyeti   283   281
H. burtoni   284   281
H. buysi   312   312
H. callipterus   284   281
H. carlottae   305   303
H. codringtoni   305   303
H. coulteri   304   303
H. darlingi   311   310
H. demeusii   293   290

H. desfontainesi   283   281
H. dolorosus   283   281
H. fasciatus   293   290
H. flaviijosephi   283   281
H. giardi   305   303
H. greenwoodi   304   303
H. horii   289   287
H. humilis   312   312
H. limax   280   278
H. lividus   280   278
H. loati   293   290
H. luluae   289   287
H. macconneli   293   290
H. machadoi   297   295
H. mahagiensis   293   290
H. mellandi   305   303
H. moeruensis   293   290
H. mortimeri   305   303
H. multiocellatus   312   312
H. nubilus   283   281
H. obliquidens   280   278

H. oligacanthus 289 287
H. pectoralis 289 287
H. petronius 293 290
H. pharyngalis 293 290
H. polli 289 287
H. polyacanthus 297 295
H. rudolfianus 293 290
H. stappersi 284 281
H. straeleni 286 285
H. swynnertoni 284 281
H. thysi 305 303
H. torrenticola 297 295
H. turkanae 293 290
H. vanderhorsti 286 285
H. welwitschii 312 312

H. wingatii 293 290
Orthochromis malagaraziensis 297 295
Paratilapia toddi 294 294
Rheohaplochromis torrenticola 297 295
Sargochromis codringtoni 305 303
S. mellandi 305 303
Serranochromis angusticeps 302 299
S. janus 303 299
S. longimanus 302 299
S. macrocephalus 302 299
S. meridionalis 303 299
S. robustus 302 299
S. spei 302 299
S. stappersi 302 299
S. thumbergi 302 299

Manuscript accepted for publication 20 April 1978

Towards a phyletic classification of the 'genus' *Haplochromis* (Pisces, Cichlidae) and related taxa. Part II; the species from Lakes Victoria, Nabugabo, Edward, George and Kivu.

Peter Humphry Greenwood
Department of Zoology, British Museum (Natural History), Cromwell Road, London SW7 5BD

Contents

Introduction	2
Methods and materials	4
Methods	4
Materials	6
Classification	6
Astatotilapia Pellegrin	6
Harpagochromis gen. nov.	10
Prognathochromis gen. nov.	14
subgenus *Prognathochromis* nov.	19
subgenus *Tridontochromis* nov.	20
Yssichromis gen. nov.	22
Pyxichromis gen. nov.	24
Lipochromis Regan	26
subgenus *Lipochromis* Regan	29
subgenus *Cleptochromis* nov.	31
Gaurochromis gen. nov.	32
subgenus *Gaurochromis* nov.	36
subgenus *Mylacochromis* nov.	36
Labrochromis Regan	37
Enterochromis gen. nov.	43
Xystichromis gen. nov.	46
Neochromis Regan	49
Haplochromis Hilgendorf	53
Psammochromis gen. nov.	53
Allochromis gen. nov.	57
Ptyochromis gen. nov.	60
Paralabidochromis Greenwood	67
Hoplotilapia Hilgendorf	72
Platytaeniodus Boulenger	75
Macropleurodus Regan	80
Schubotzia Boulenger	85
Lake Victoria haplochromines of uncertain generic relationship.	88
'*Haplochromis*' *cronus* Greenwood	88
'*Haplochromis*' *apogonoides* Greenwood	89
'*Haplochromis*' *theliodon* Greenwood	89
Lake Kivu haplochromines of uncertain generic relationship	90
'*Haplochromis*' *schoutedeni* Poll	90
'*Haplochromis*' *wittei* Poll	90
Summary and conclusions	91
Acknowledgements	95
Guide to the identification of haplochromine genera in Lakes Victoria, Kioga, Nabugabo, Edward, George and Kivu	96

Bull. Br. Mus. nat. Hist (Zool.) 39(1): 1–101

Issued 30 October 1980

References 98
Index 100

Introduction

In the first part of this paper (Greenwood, 1979) I dealt at length with the philosophical and taxonomic problems posed by the polyspecific 'genus' *Haplochromis*. Applying a basically Hennigian approach to the problem (Hennig, 1966), I attempted to sort the fluviatile species of Africa and Israel into a number of monophyletic lineages. Since the interrelationships of these lineages could not be resolved cladistically, each was treated as having equal rank and classified as a genus (see Nelson, 1972).

The nominate genus, *Haplochromis*, was restricted to five species sharing a particular and highly derived form of dentition. All five are essentially lacustrine in habitat, and are confined to Lakes Victoria, Kioga, Edward, George, Nabugabo and Kivu (see Greenwood, 1979 : 278–281).

As in the first section of this review, the second part is concerned with an attempted arrangement of haplochromine taxa into presumed monophyletic lineages. Each lineage, as before, is recognized by its member species possessing uniquely derived features (synapomorphies). Where possible, the sister-groups of these lineages were determined but are not, at this stage of the revision, given any formal taxonomic status (save that of subgenus when that rank seemed the most appropriate one to indicate apparent relationship).

The taxa considered here are from the east African Lakes Victoria, Kioga, Edward, George, Nabugabo and Kivu. These lakes are well-known for the high levels of endemicity exhibited by their haplochromine species, and are interrelated through their hydrographical and geographical histories (see Greenwood, 1965, 1973 & 1974; and Fryer & Iles, 1972 for discussions and relevant literature).

A close taxonomic relationship between the '*Haplochromis*' species of these lakes, despite the high level of intralake endemicity, has long been acknowledged. But, this factor was, perhaps, overshadowed by the size and ecological complexity of the so-called *Haplochromis* 'species-flocks' in Lakes Victoria and Edward-George. Either explicitly, or implicitly, the flocks were assumed to be of mono- or at most oligophyletic origin (see Greenwood, 1974 : 19–20). In Lake Victoria it was thought that '... The existing species are certainly more closely related to one another than to any species outside the lake, and it seems justifiable to refer to the assemblage as a species flock' (Greenwood, 1974).

Research embodied in this paper readily falsifies that statement. Many lineages described below (*ie* groups of taxa sharing a recent common ancestry and therefore more closely related to one another than to members of another lineage) are composed of species from at least four of the lakes. Furthermore, it is rarely possible to establish, within a lineage, that the species from one lake are in fact each other's closest relatives. There are, of course, a few lineages known only from one lake (eg *Macropleurodus* and *Hoplotilapia* in Lake Victoria, and *Schubotzia* in Lake Edward–George), but the overall picture is one of a super-flock comprised of several lineages whose members cut across the boundaries imposed by the present-day lake shores.

Is it correct then, to continue using the term 'species flock' (= species swarm of Mayr, 1963) if the component species cannot be shown to stem from a single and fairly recent ancestor? Strictly the answer is 'no'; in future the term should be used sparingly and informally (whether to describe the haplochromines of one lake or of the lakes combined) and it should be employed more in an ecological than a taxonomic context.

Regrettably, no new data have emerged which might establish a monophyletic origin for the haplochromine species of Victoria–Edward–Kivu[1] (see also Greenwood, 1979 : 269). In

---

[1] Since Lake Kioga is an extension of Lake Victoria, albeit one now isolated by the Owen Falls Dam at Jinja, Lake Nabugabo a cut-off bay of Lake Victoria (Greenwood, 1965 & 1974), and because Lakes Edward and George are in direct contact with one another (see Greenwood, 1973), the entire system can be referred to by the shorthand name Victoria–Edward–Kivu.

other words, no uniquely derived character (or characters) has been found amongst all or the majority of haplochromines in these lakes.

Anatomically, and particularly in their squamation patterns, the Victoria–Edward–Kivu species are of a basically *Astatotilapia* type, as they are in the occurrence of true ocelli on the anal fin of adult male fishes (see Greenwood, 1979 : 274–5 & 281–3). They far exceed *Astatotilapia*, however, in their range of body form, dental morphology, gill-raker shapes and number, and syncranial architecture (see Greenwood, 1973 & 1974). They also differ from fluviatile *Astatotilapia* in having a higher modal range of lateral line pore scales (31–33 **cf.** 28–30), the range extending to 36 (**cf.** 30, rarely 33 in riverine *Astatotilapia*), and in having a higher modal number of caudal vertebrae (16 **cf.** 15) and thus a higher count for the total number of vertebrae.

The ranges for both these counts, it should be noted, do overlap, although the lower values seen in fluviatile *Astatotilapia* species are rarely encountered in the Victoria–Edward–Kivu lineages (including species referred to *Astatotilapia* itself; see p. 8).

The significance of these differences is difficult to assess. A shift in modal values and in the end points of a range could, for such meristic characters, be genetically determined or might be the result of environmental factors acting during ontogeny. For the moment there are insufficient data on which to develop either argument further. This is, however, a problem that could have bearings on the possible monophyletic origin of the lake's super-flock (see above).

An earlier attempt to resolve phyletic relationships within the Lake Victoria *Haplochromis* flock (Greenwood, 1974) is now seen to be quite inadequate, and in many respects misleading when taxa from the historically related lakes Edward and Kivu are taken into account. The new arrangement, apart from suggesting a different grouping for certain taxa, also indicates that fewer species and lineages can be interrelated on a sister-group basis.

Outside lakes Victoria–Edward–Kivu, too, no sister-group relationships can be established for the lineages from those lakes. This situation may, however, change when the Lake Malawi haplochromines are studied more critically and within a cladistic framework.

At present, the former genus *Haplochromis* can be resolved into a number of apparently monophyletic lineages; the search for characters uniting these lineages through various levels of common ancestry must continue. More discriminating anatomical studies may reveal linkages so far undiscovered, and there would seem to be great scope for developing biochemical techniques directed towards that end.

In some respects it might seem that the application of a cladistic approach to the *Haplochromis* problem has failed, particularly when it appears impossible to construct dichotomously branching phylogenies at either the inter- or the intralineage levels (**cf.** for example, Vari's (1978) work on the teraponids which did achieve these goals).

It must be accepted that within any such polyspecific assemblage as the former taxon *Haplochromis*, some species will, through descent from a shared common ancestor, be more closely related to one another than to other species or groups of species. So far it has been possible only to make a first step towards discovering these relationships.

Yet, I believe that even this step could not have been made without the application of a basically Hennigian philosophy to the problem. In that way any morphological gaps (*ie* apomorph characters) have been used as positive characters to construct testable hypotheses (*ie* lineages) rather than as a means of ranking lineages through the subjectively estimated 'size' of the gap.

The creation of over twenty-five 'genera' where there was previously only one, has, I know, upset some of my colleagues who look upon my actions as those of a splitter run wild. But that indeed may have been the way in which the haplochromines (particularly the lacustrine species) evolved. To lump these lineages in a single genus (even as subgenera), whose monophyletic origin has not been established (and whose artificial and polyphyletic nature is strongly hinted at), is to hide the phylogenetic element of a classification. Regrettably, the constraints imposed by the Linnaean system of classification oblige one to use the formal rank of genus for the different lineages if one is not to be forced into implying a relationship that may not exist.

Part of the difficulty encountered in classifying the Victoria–Edward–Kivu haplochromines at both the inter- and intragroup level may be a consequence of their recent and explosive speciation (see Greenwood, 1974). Take, for example, the large number of *Astatotilapia* species in these lakes (page 8 below). All are remarkably similar, differing only in male coloration and minor (at any rate to the taxonomist) morphometric features. Any one of these anatomically and ecologically unspecialized species could be the nearest living relative of the ancestor to a lineage recognized by its one or, at most, its few derived features. In other words, one is dealing with a situation where the first phenotypically manifest 'marker' apomorphies have evolved only recently. It is thus impossible to recognize the plesiomorph sister group (or species) since the characters it shares with the derived sister species are primitive ones (*ie* symplesiomorphies).

Methods and materials

Methods

Little more need be added to the points discussed in the 'methods' section in the first part of this paper (see Greenwood, 1979 : 269–276), except to define certain characters not employed in that paper. These are concerned either with describing neurocranial shape or with the morphology of the jaw skeleton.

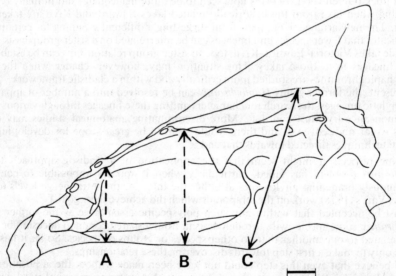

Fig. 1 Diagram to show reference points for various neurocranial measurements. A. Preorbital depth. B. Depth of orbit. C. Depth of otic region. (Solid line = direct measurement. Dashed line = to horizontal from parasphenoid margin.) Neurocranial outline from a skull of *Prognathochromis argenteus*. Scale = 3 mm.

Neurocranial length. The length of the skull measured directly[1] from the anterior tip of the vomer to the posterior point on the rim of the basioccipital facet for articulation with the first vertebra.

---

[1] That is, with the tips of the dividers actually contacting the points concerned, and the measurement then made between these points.

Preorbital depth. The depth of the skull through the anterior part of the orbit; measured in the vertical from the highest point on the horizontal surface of the frontal (*ie* disregarding the frontal wing which rises to meet the anterior tip of the supraoccipital) immediately above the point where the posterior margin of the lateral ethmoid meets the frontal, to the level of a horizontal line extended from the ventral face of the parasphenoid (see Fig. 1).

Depth of the otic region. The greatest depth measured directly[1] from the highest point on the supraoccipital base (*ie* excluding the crest), downwards and somewhat forwards to the lowest point on the parasphenoid posterior to the orbit (usually situated below the lateral commissure); see Fig. 1. It was necessary to choose the latter reference point as the ventral one because of considerable intra- and interspecific variability in the depth of curvature between that part of the parasphenoid and the articular surface of the pharyngeal apophysis.

Greatest width across the otic region. The maximum width of the skull as measured directly[1] across the pterotics.

Preotic skull length. Measured directly[1] from the anterior point of the vomer to the junction between the prootic and the ascending wing of the parasphenoid (*ie* near the base of the lateral commissure).

Depth of orbit. Measured vertically from the highest point on the curve of the frontal margin forming the dorsal rim of the orbit (ignoring the lateral line tubule if that should coincide) to a horizontal extended from the ventral margin of the parasphenoid in that vertical (Fig. 1).

Height of premaxillary ascending processes. Measured directly[1] from the bony distal tip of the processes to a point on the anterior face of the dentigerous arm (*ie* the beak or peak) level with the upper margin of its horizontal posterior prolongation (Fig. 2).

Length of premaxillary dentigerous arm. Measured directly[1] from the premaxillary symphysis to the posterior point on the horizontal arm.

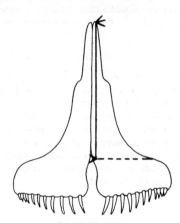

Fig. 2 Diagram showing reference points for measuring height of premaxillary ascending processes.

Length of lower jaw. Measured directly[1] from the symphysis to the posterior face of the anguloarticular bone immediately below the articular facet for the quadrate. All lower jaw measurements quoted were taken from whole specimens and, since some soft tissue is interposed between bone and divider point, these are fractionally greater than measurements taken from a skeleton.

The shape and proportions of jaw elements (dentary, anguloarticular bone, premaxilla and maxilla) in *Astatotilapia* species are taken to represent the plesiomorph (*ie* primitive) condition amongst haplochromines. This conclusion is based on the principle of commonality (see discussion in Greenwood, 1979 : 270). Departure from the *Astatotilapia* condition is, therefore, interpreted as being a derived (*ie* apomorph) state.

Such departures include bullation of the dentary in the region where it divides into ascending (coronoid) and horizontal arms; a change in various of its proportions so that the dentary no longer appears to be a slender, elongate bone but is relatively deeper and stronger, thereby assuming a stout, foreshortened appearance; and, finally, extensive lateral development of the dentigerous region so that its alveolar surface projects outwards as a distinct shelf overhanging the lateral wall of the dentary. Primitively, the symphysial region of the dentary is vertically aligned and narrow, and there is no distinct mental protuberance developed at its anteroventral angle.

Derived conditions in the anguloarticular are usually associated with a stout and foreshortened dentary; they include a general thickening of the bone, a reduction in its relative height (*ie* of the coronoid arm, the so-called primordial process of Barel *et al.*, 1976), and a change in the shape of its anteroventral projection (*ie* the arm which underlies the dentary), especially its anterior angle (which, primitively is produced and acute, but deepened and rounded or rectangular in its derived form).

Derived characteristics in the premaxilla include a lengthening of its ascending processes relative to the dentigerous arms, inflation of the dentigerous arms which thus become nearly cylindrical or distinctly ovoid in cross-section, and an increase in the width of their alveolar surfaces (a change generally but not invariably correlated with inflation of the arm itself). An expansion of the dentigerous arms anteriorly and anterolaterally, in the region below, and in front of the ascending processes, is also considered to be a derived condition. This gives the bone a 'beaked' or 'peaked' appearance.

The maxilla shows fewer changes in its morphology; chief amongst the derived conditions recognized are those in which the posterior part is relatively deepened, or is bullate, or when the articulatory head is turned medially at a marked angle (sometimes almost to form a right angle with the shaft of the bone). These apomorph conditions may occur together, in various combinations, or singly.

Materials

In addition to the specimens noted in Part I of this paper, all the BM(NH) haplochromine material from the five lakes has been involved in this review. It includes spirit specimens, dry skeletons (many prepared for this paper) alizarin transparencies, and radiographs.

Descriptions and comments are generally based on adult or subadult specimens. This is partly because the specific identity of juvenile and post-larval fishes cannot always be determined, and partly because some morphological features change during ontogeny (the dentition and some syncranial features are particularly liable to be affected in this way). However, where ontogenetic modifications appear to throw some light on problems of phylogeny, these have been taken into account.

Classification

In the generic diagnoses which follow, presumed apomorph (*ie* derived) features or conditions of a character-complex are italicized.

The serial listing of the genera should not be interpreted as having any particular phylogenetic significance. However, the most generalized taxon is taken first, and some of the more specialized lineages are dealt with later in the paper.

ASTATOTILAPIA Pellegrin, 1903

Several endemic species from Lakes Victoria, Kioga, Edward, George, Nabugabo and Kivu must now be added to this genus (see Greenwood, 1979 : 281 *et seq* for redescription and basic diagnosis of the genus). A typical *Astatotilapia* is shown in Fig. 3.

Unfortunately, this additional material in no way clarifies the status of the taxon, nor does it help to establish its monophyly. *Astatotilapia* remains, as before, a genus based on plesiomorph characters widely distributed amongst haplochromine cichlids; the one probable apomorphy (anal ocelli in male fishes) is shared by at least three other lineages (see Greenwood, 1979 : 268 & 284–5).

Fig. 3 *Astatotilapia elegans*. Lake George. About natural size.

The endemic *Astatotilapia* species from Victoria–Edward–Kivu differ from their fluviatile congeners in two major features: a higher modal number of lateral line pore scales (32–33 **cf.** 28–30) and a higher modal number of caudal vertebrae (16 **cf.** 15). In addition, not all the lacustrine species have several enlarged unicuspid teeth posteriorly in the outer premaxillary row (see Greenwood, 1979 : 282), although one or two such teeth are present in the majority.

There is an overlap in the total range of lateral line scale counts for fluviatile and lacustrine species (28–30, rarely 33, compared with 30–34 in the latter), but no lake species has a scale count in the lower part of the fluviatile range. Likewise, there is an overlap in the ranges for the number of caudal vertebrae, but again higher counts are confined to lake species (14–16 caudal vertebrae in fluviatile species, 15–17 in lake species).

Astatotilapia desfontainesi, from Tunisia and Algeria, is exceptional amongst the fluviatile taxa in having a high lateral line scale count (31–33), but its vertebral counts are typically those of fluviatile *Astatotilapia* species.

In most other respects the redescription of *Astatotilapia*, based on fluviatile taxa, given in Greenwood (1979 : 281–5) covers the species now included in the genus. Note may be made, however, that the range of gill raker numbers is increased to 13 (but the mode remains at 8 or 9), and that one of the newly included species (*A. oregosoma*) has a more slender body form than its congeners (30–34% SL **cf.** 35–40%).

The maximum adult size range is from 70 to 100 mm standard length.

Certain syncranial characters not discussed in Greenwood (1979) were found to be of value when dealing with the Victoria–Edward–Kivu haplochromines. These may be noted briefly, and are applicable both to the fluviatile and the lacustrine representatives of *Astatotilapia*.

Jaws. Over its posterior half, the lateral wall of each dentary is flared outwards. Consequently the alveolar surface is carried on a narrow, laterally produced shelf projecting slightly beyond the body of the ramus. The length of the entire lower jaw (dentary and anguloarticular) ranges from 31–46% of head length (modal range 40–43%).

The ascending processes of the premaxilla are shorter than the dentigerous arms, usually much shorter (about half the length) but in a few species only fractionally shorter. In all but one species (*A. velifer*, Lake Nabugabo) the dentigerous arms are compressed in cross section; *A. velifer* has the arms very slightly inflated and ovoid in section. No *Astatotilapia* species has the anterior and anterolateral aspects of the dentigerous arms drawn out into a beak- or shelf-like projection.

The maxilla is relatively short and deep, its articulatory head inclined medially at a distinct angle.

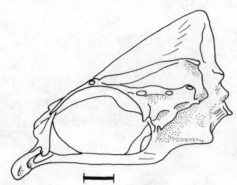

Fig. 4 Neurocranium (left lateral view) of *Astatotilapia macrops*. Scale = 3 mm.

Neurocranium (Fig. 4). Although the overall skull outline, especially the slope of the preorbital face, shows little intralineage variation, there is some variability in the profile of the supraoccipital crest. Most species have a relatively low crest with a distinctly wedge-shaped outline; in a few species, however, the crest is deeper and consequently its profile approaches the near-pyramidical shape found in members of certain other lineages (*A. bloyeti* typifies this condition in *Astatotilapia*; see Greenwood, 1979: fig. 6).

Preorbital skull depth in *Astatotilapia* ranges from 25–30% of neurocranial length, skull width from 50–61% (modal range 56–60%) and depth of the otic region from 43–50% (modal range 47–50%).

Contained species
Lakes Victoria, Kioga, Edward, George, Nabugabo and Kivu endemics only are listed; for the fluviatile species and those occurring in other lakes see Greenwood (1979 : 283–4).

Astatotilapia aeneocolor (Greenwood), 1973. Lake George; see Greenwood (1973 : 150–154).

Astatotilapia barbarae (Greenwood), 1967. Lake Victoria; see Greenwood (1967 : 93–97). Because of its partially paedophagous habits this species was previously associated with the major paedophage lineage *Lipochromis* (see p. 32 below and Greenwood, 1974). Anatomically, however, it shows none of the derived features characterizing *Lipochromis* (see Greenwood, 1967 : 96), and thus cannot be included in that lineage. Amongst *Astatotilapia*, *A. barbarae* is outstanding for its low otic skull depth (*ca* 43% of neurocranial length).

Astatotilapia brownae (Greenwood), 1962. Lake Victoria; see Greenwood (1962 : 142–9).

Astatotilapia cinerea (Blgr.), 1906. Lake Victoria; see Greenwood (1960 : 239–242).

Astatotilapia eduardi (Regan), 1921. Lake Edward; see Trewavas (1933 : 332). Poll (1932 : 42) considers *Haplochromis vicarius* Trewavas (1933) to be a synonym of this species.

Astatotilapia elegans (Trewavas), 1933. Lakes Edward and George; see Greenwood (1973 : 145–9), and Barel *et al.* (1976).

Astatotilapia engyostoma (Trewavas), 1933. Lake Edward; see Trewavas (1933 : 331–2). A poorly known species represented only by the holotype, a specimen now in very poor condition (see Greenwood, 1973 : 162).

Astatotilapia lacrimosa (Blgr.), 1906. Lake Victoria; see Greenwood (1960 : 230–233).

Astatotilapia latifasciata (Regan), 1929. Lake Kioga; see Regan (1929 : 390).

Astatotilapia macrops (Blgr.), 1911. Lake Victoria and possibly the Victoria Nile; see Greenwood (1960 : 236–9).

Astatotilapia macropsoides (Greenwood), 1973. Lakes Edward and George; see Greenwood (1973 : 162–7).

Astatotilapia martini (Blgr.), 1906. Lake Victoria; see Greenwood (1960 : 245–8). The sharply decurved head profile, and the golden-yellow body coloration in both sexes, make this an outstanding species amongst the *Astatotilapia* of Victoria–Edward–Kivu. Skull morphology is also atypical since the preorbital and orbital depths are above modal, and the preorbital skull profile is decurved and slopes steeply. Dentally and in other osteological features, however, it does not depart from the usual *Astatotilapia* condition. The relationship of *A. martini* can be reviewed when more is known about '*H*' *cronus* and '*H*' *apogonoides* (see pp. 88 & 89 respectively).

Astatotilapia megalops (Greenwood & Gee), 1969. Lake Victoria; see Greenwood & Gee (1969 : 4–7).

Astatotilapia melanopus (Regan), 1922. Lake Victoria; see Regan (1922 : 165–6). A taxon of uncertain status in the Lake Victoria flock.

Astatotilapia oregosoma (Greenwood), 1973. Lake George; see Greenwood (1973: 159–162). An unusual species amongst the *Astatotilapia* complex because of its shallow body (depth 30–34% SL, mean 32%) and elongate habitus.

Astatotilapia pallida (Blgr.), 1911. Lake Victoria and possibly the Victoria Nile; see Greenwood (1960 : 233–6).

Astatotilapia piceata (Greenwood & Gee), 1969. Lake Victoria; see Greenwood & Gee (1969 : 7–10).

Astatotilapia schubotziella (Greenwood), 1973. Lake George and probably Lake Edward; see Greenwood (1973 : 188–192). The previously suggested close relationship between this species and *Psammochromis schubotzi* (see p. 56 below, and Greenwood, 1973 : 192) was not corroborated by more detailed anatomical knowledge.

Astatotilapia velifer (Trewavas) 1933. Lake Nabugabo; see Greenwood (1965 : 319–324).

DISCUSSION

Nothing much can be added to the comments already made on this supposed lineage (Greenwood, 1979 : 269, remarks on the *A. bloyeti* species complex, and also pp 284–5). Even information derived from the additional taxa now referred to *Astatotilapia* has failed to isolate a single derived feature uniquely common to all, or even the majority, of *Astatotilapia* species.

Species endemic to the lakes do, however, differ from their fluviatile congeners in having higher modal numbers of lateral line scales and caudal vertebrae (see p. 7 above). Phylogenetically speaking, the significance of those features is difficult to assess. Members of all other lineages in the lakes have modes and ranges for these two features comparable with those of lake *Astatotilapia* (or, in some genera, even higher). But, there is, currently, no way of telling whether this should be identified as an ecophenotypic response, or the result of a distant shared common ancestry.

Taken in their entirety, the lacustrine *Astatotilapia* show a little more diversity in their syncranial and dental morphology than do the fluviatile species.

Amongst the species of Lakes Victoria, Edward and George there are some whose dental and pharyngeal jaw morphology, and, or, their feeding habits, vaguely foreshadow the definitive characteristics of certain other lineages. For example, the slightly enlarged lower pharyngeal bone (with its enlarged median teeth) in *A. pallida* suggests affinity with *Labrochromis* (see p. 37 below, and Greenwood, 1960 : 234); skull morphology, and a partially paedophagous diet in *A. barbarae* suggests *Lipochromis* affinities (see Greenwood, 1974); the unicuspid jaw teeth and partially piscivorous habits of large *A. brownae* hint at a relationship with *Harpagochromis* (see p. 10 below, and Greenwood, 1974); the slender body-form of *A. oregosoma* resembles that in *Yssichromis* (see p. 23 below), and

A. schubotziella has certain features that suggest it might be related to *Gaurochromis* (see p. 32 below, and Greenwood, 1973 : 192).

However, in none of these examples is the resemblance either sufficiently clear-cut, or reflected in unequivocally derived morphological features, for it to be used as a reliable indicator of phyletic relationship. Hence, all these taxa are included in *Astatotilapia* and not in the lineage with some or all of whose members they seem to bear some resemblance.

It has not proved possible to demonstrate intralineage relationships between the numerous *Astatotilapia* species because the synapomorphies that would permit such an analysis were not discovered.

HARPAGOCHROMIS gen. nov.

TYPE SPECIES: *Hemichromis serranus* Pfeffer, 1896 (Type specimens in the Humboldt Museum, Berlin); see Greenwood (1962 : 152–6).

ETYMOLOGY. The name is derived from the Greek *harpage*, meaning a robber + *chromis*, a word, when used in such a combination, is now associated with many genera of African Cichlidae; it refers to the predatory habits of species in this lineage.

DIAGNOSIS. Robust and deep to relatively deep-bodied haplochromines (body depth 30–42% of standard length, modal range 34–36%), *reaching a large maximum adult size (146–200 mm SL)*. Mouth generally horizontal or slightly oblique, but sometimes distinctly oblique; *lower jaw long (43–61% of head length, modal range 47–54%)* and with a prominent mental protuberance. Anterior and anterolateral regions of the premaxilla not produced to form a distinct beak or peak.

Neurocranium essentially of the generalized type, *but with a shallower otic region (40–44% neurocranial length* **cf.** *47–50% in the generalized type)* and a higher supraoccipital crest which is generally near pyramidical in outline. Preorbital skull depth 23–28% neurocranial length, mean 25%, skull width 54–60% (no distinct mode), greatest orbital depth 25–31% (modal range 25–28%).

Outer jaw teeth strong and recurved, unequally bicuspid and a few unicuspids in fishes <90 mm SL, the proportion of unicuspids increasing in larger fishes until, in specimens >120 mm SL, only unicuspids are present; outer row in premaxilla with 48–80 teeth (modal range 60–70). One or 2 (less commonly 3, rarely up to 5) inner rows of teeth in each jaw.

Cheek fully scaled, generally with 4 or 5 rows of scales (less frequently with 2, 3 or 6 rows).

DESCRIPTION
Habitus (Fig. 5). Most members of this lineage have a deep or moderately deep body (30–42% SL) which is never manifestly compressed. Head shape, and particularly the angle of the mouth, show some intraspecific variation, the mouth angle varying from almost horizontal (*Harpagochromis victorianus, H. serranus, H. michaeli*) to distinctly oblique (*H. cavifrons, H. plagiostoma*, see Fig. 5B).

Anatomically, osteologically and in their dentition, *Harpagochromis* species depart but slightly from the generalized condition typified by species of *Astatotilapia*. All, however, reach a much greater maximum adult size (146–200 mm SL cf. 60–100 mm), and most only attain sexual maturity at a length which is never reached by an *Astatotilapia* (ie at a standard length of more than 100 mm in *Harpagochromis*, modally between 120 and 140 mm, compared with lengths of 50–80 mm in *Astatotilapia*). The adult size reached by species of *Harpagochromis* is, with few exceptions, also greater than that attained by members of most other lineages (but see p. 14 below).

In some morphometric features, especially in having a relatively longer lower jaw and, less noticeably, a relatively smaller eye and deeper cheek (two characters generally correlated), *Harpagochromis* species differ from *Astatotilapia*, at least in modal values. The lower jaw in *Harpagochromis* is 43–61% of head length (modal mean values 47–54%), eye diameter 17–29% (modal mean values 20–24%) and cheek depth 22–37% (modal mean values 27–32%). There is a slight overlap in the ranges of these ratios when large specimens of

Fig. 5 A. *Harpagochromis spekii*. Lake Victoria. About half natural size. B. *Harpagochromis cavifrons*. Lake Victoria. About three-quarters natural size.

Astatotilapia species are compared with equal-sized *Harpagochromis* individuals. Since the characters in question invariably show either a positive or a negative allometry with body length, this overlap is to be expected. However, the modal value for lower jaw length is always higher for *Harpagochromis*. This might suggest that increased jaw size in *Harpagochromis* is not entirely attributable to allometric relationships with overall body size. In other words, it is in itself a derived feature, which is further emphasised through that allometric relationship.

Squamation. The cheek is fully scaled, usually with 4 or 5 rows of scales (less frequently 2, 3 or 6 rows); there are 30–34 (modal range 32–33) scales in the lateral line series.

Neurocranium (Fig. 6). The neurocranium in *Harpagochromis* is similar to that in *Astatotilapia*, except that the otic region is shallower (40–44% of neurocranial length cf. 47–50%) and the supraoccipital crest is higher and pyramidical in outline. Because the skull is deep preorbitally and orbitally, and since it is wide in the otic region, it is less obviously streamlined than is the skull form characterizing the other major piscivore lineage, *Prognathochromis* (see p. 16). Also, as compared with that lineage, the supraoccipital crest in *Harpagochromis* is taller relative to skull length, and slopes downward and forward more steeply.

Dentition. Fishes over 90 mm SL mostly have strong, somewhat recurved unicuspid teeth in the outer row of both jaws. Smaller specimens have predominantly bicuspid outer teeth, the cusps of unequal size, and the crown and upper neck slightly recurved. There are 48–80 outer teeth in the premaxillary series (modal range 60–70).

Inner row teeth are either tricuspid, unicuspid or a mixture of both types, arranged in 1 or 2 (less frequently 3) rows anteriorly and anterolaterally.

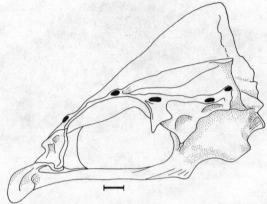

Fig. 6. Neurocranium (left lateral view) of *Harpagochromis maculipinna*; scale = 3 mm.

Jaws. The proportionally long lower jaw has been noted already. In all other respects it barely departs from the *Astatotilapia* type, apart from having a well-developed mental protuberance at the base of the symphysis. In some species the crown of the coronoid process (the ascending dentary limb) has a distinct lateral deflection.

The anterior and immediately lateral regions of the premaxilla are but slightly produced forward (Fig. 7), so that in lateral view it is hardly beaked (**cf.***Prognathochromis*, p. 18). The dentigerous arms are very slightly inflated, and have a compressed ovoid cross-section.

Vertebral numbers: 28–31 (mode 30), comprising 12–14 (mode 13) abdominal, and 15–18 (modes 16 or 17) caudal elements (the fused PU_1 and U_1 centra excluded).

Caudal fin. Most species have the posterior margin truncate or weakly subtruncate, but in one species (*H. plagiostoma*) it is obliquely truncate, that is, sloping forward and downward to meet the upwardly curved ventral region.

Contained species

The taxa are arranged in groups approximating to their degree of morphological departure from the generalized '*Astatotilapia*' bauplan.

Harpagochromis serranus (Pfeffer), 1896. Lake Victoria; see Greenwood (1962 : 152–5).
Harpagochromis victorianus (Pellegrin), 1904. Lake Victoria; see Greenwood (1962 : 156–8; plate I).

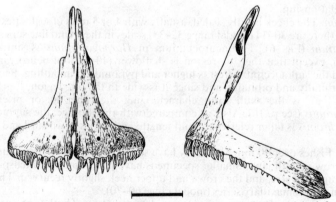

Fig. 7 Premaxilla of *Harpagochromis maculipinna*, seen anteriorly and in left lateral view. Scale = 5 mm.

Harpagochromis nyanzae (Greenwood), 1962. Lake Victoria; see Greenwood (1962 : 159–161).
Harpagochromis spekii (Blgr.), 1906. Lake Victoria and the Victoria Nile; see Greenwood (1967 : 32–38).
Harpagochromis maculipinna (Pellegrin), 1913. Lake Victoria; see Greenwood (1967 : 43–47).

Harpagochromis squamipinnis (Regan), 1921. Lake Edward and George; see Greenwood (1973 : 204–9).
   This species differs from all its congeners in Lake Victoria (and all known haplochromine species) in having short rows of small scales extending from the body onto the membrane of the dorsal and anal fins. The scales are closely applied to the fin rays, both spinous and branched (see fig. 31 in Greenwood, 1973 : 206). In its gross morphology *H. squamipinnis* closely resembles *H. serranus* and *H. victorianus* of Lake Victoria. At present no other *Harpagochromis* species are known from Lakes Edward and George.

Harpagochromis boops (Greenwood), 1967. Lake Victoria; see Greenwood (1967 : 47–49).
Harpagochromis pachycephalus (Greenwood), 1967. Lake Victoria; see Greenwood (1967 : 39–42).
Harpagochromis thuragnathus (Greenwood), 1967. Lake Victoria; see Greenwood (1967 : 49–51).

Harpagochromis guiarti (Pellegrin), 1904. Lake Victoria; see Greenwood (1962 : 145–9). The distribution recorded in that paper included Lake Edward; this was based on earlier identifications of material from Lake Edward which most probably does not belong to this species (see Greenwood, 1973 : 232). The identity of the Edward material, and the true distribution of *H. guiarti*, must await the outcome of further research on the haplochromines of Lakes Edward and George.
Harpagochromis artaxerxes (Greenwood), 1962. Lake Victoria; see Greenwood (1962 : 170).

Harpagochromis altigenis (Regan), 1922. Lake Victoria, see Greenwood (1967 : 60–65).

Harpagochromis pectoralis (Blgr.), 1911. Lake Victoria. This species was given the replacement name '*squamulatus*' by Regan (1922), since at that time '*pectoralis*' would have been a junior homonym of *Haplochromis pectoralis* (Pfeffer), 1893. Because Pfeffer's species is now placed in the genus *Ctenochromis*, for which it is indeed the type species (Greenwood, 1979 : 287), the original trivial name for Boulenger's 1911 species can be resurrected.
   Harpagochromis pectoralis differs from other members of the lineage in a number of features, but especially in having a more strongly decurved preorbital skull profile. Its relationships with (or within) the *Harpagochromis* lineage may have to be reconsidered when several undescribed and '*pectoralis*'-like species from Lake Victoria have been studied.
   The species is redescribed in Greenwood (1967 : 60–65).

Harpagochromis plagiostoma (Regan), 1922. Lake Victoria; see Greenwood (1962 : 199–202).
Harpagochromis cavifrons (Hilgendorf), 1888. Lake Victoria and possibly the Victoria Nile; see Greenwood (1962 : 196–9).

Harpagochromis michaeli (Trewavas), 1928. Lake Victoria; see Greenwood (1962 : 203–6).

Incertae sedis
'*Haplochromis*' *diplotaenia* Regan & Trewavas, 1928, Lake Victoria. This species is known from very few specimens, and no skeletal material is available. Judging from radiographs its

neurocranial architecture is of the *Harpagochromis* type, but the difficulties associated with obtaining accurate measurements from radiographs preclude definite conclusions on that point.

'*Haplochromis*' *paraplagiostoma* Greenwood & Gee, 1969. Lake Victoria. Again, a shortage of osteological material has made it impossible to check several critical features. Although superficially this species does resemble other members of the *Harpagochromis* lineage, it seems advisable to keep the generic placement of '*H*'. *paraplagiostoma* as *incertae sedis* until more specimens are available for study.

'*Haplochromis*' *worthingtoni* Regan, 1929. Lake Kioga. Known only from the holotype (141·0 mm SL), this species closely resembles species of the *Harpagochromis serranus—maculipinna* group in all visible anatomical features. From a radiograph its neurocranium also seems to be of a *Harpagochromis* type.

Discussion

The phyletic integrity of this presumed lineage depends on a single synapomorphy, the attainment of a large adult size (with which is, of course, associated a number of other characters, see p. 11 above).

There are indications that members of this lineage, as compared with species of the genus *Astatotilapia*, do have a basically longer lower jaw over and above the relative increase in jaw length effected through its positively allometric growth pattern.

Anatomically, most *Harpagochromis* species retain a generalized level of organization like that in *Astatotilapia*. Only two species, *H. plagiostoma* and *H. cavifrons*, with their very oblique mouths, show some departure from that condition, while a third (*H. pectoralis*) has a slightly derived type of skull architecture.

The possible relationship between *Harpagochromis* and *Prognathochromis* (as suggested in Greenwood, 1974) is discussed on p. 22.

Although the list of *Harpagochromis* species (p. 12) is arranged so that morphologically similar species are grouped together (in order of increasing departure from the *Astatotilapia* level of organization), this should not be taken to imply true phyletic relationships. Such intralineage relationships cannot be determined on the basis of data currently available.

PROGNATHOCHROMIS gen. nov.

Type species: *Paratilapia prognatha* Pellegrin, 1904; see Greenwood (1967 : 78).

Etymology. The name is derived from that of the type species.

Diagnosis. Body form variable, from shallow to deep (24–45% standard length) but most species relatively slender (body depth 30–34% SL) and none is markedly compressed. *Lower jaw long (41–62% head length, modal range 45–53%)* and with a prominent mental protuberance; premaxilla distinctly beaked or peaked.

Overall habitus one of a slender, streamlined fish with a large, often oblique mouth, and the head with a relatively acute entry angle. Maximum adult size extending over a wide range of lengths (70–230 mm SL), but most specimens reaching 140–200 mm SL.

Neurocranium (the principal diagnostic feature, particularly for differentiating *Prognathochromis* from *Harpagochromis*), *elongate, slender and shallow, with a low, supraoccipital crest*, wedge-shaped in lateral outline. Preorbital skull depth 18·6–23·% of neurocranial length (mode 21%), greatest orbital depth 22–28% (modal range 22–23%), depth of otic region 31–42% (no distinct mode), skull width 42–55% (modal range 47–50%), all expressed as ratios of neurocranial length.

Teeth in outer row of both jaws strong and recurved, mostly unicuspid in fishes > 90 mm SL, unequally bicuspid with some unicuspids in smaller fishes. Thirty-four to 94 outer teeth in the premaxilla, modal range in the nominate subgenus 50–60, but 66–74 in the other subgenus. In that subgenus, whose members have a small adult size (95–120 mm SL), *tricuspid teeth occur ... terally and anterolaterally, interspersed amongst the predominantly unicuspid outer teeth.* The inner teeth in both subgenera are uni- or tricuspid, and are arranged in 2 or 3 (rarely 1, 5 or 6) rows anteriorly and anterolaterally.

Cheek fully scaled, usually with 3 or 4 rows but sometimes 5 or 6, rarely with only 2.

DESCRIPTION

Habitus (Fig. 8). Body form is variable, the depth varying from shallow to relatively deep (24–45% SL) but moderately shallow in the majority of species (modal range 30–34% SL). Except in a few species, the body is not noticeably compressed.

With few exceptions, members of this genus have a 'typical' predator facies (see Fig. 8; and figs 13–15 in Greenwood, 1974), *ie* a slender streamlined body, large mouth and a predominantly unicuspid dentition. In all these features, *Prognathochromis* represents a marked development of the habitus-type beginning to appear in the *Harpagochromis* lineage. No

Fig. 8 A. *Prognathochromis (P.) prognathus.* Lake Victoria. About four-fifths natural size. B. *Prognathochromis (P.) mento.* Lake Victoria. About half natural size. C. *Prognathochromis (Tridontochromis) sulphureus.* Lake Victoria. Scale = 1 cm.

member of the *Prognathochromis* line retains the near-*Astatotilapia* habitus which characterizes so many species of *Harpagochromis*.

Nine species attain only a small adult size (70–120 mm SL), but for the others the maximum size lies in the range 140–230 mm SL.

As would be expected, those species reaching a small maximum size have, as compared with their larger congeners, relatively larger eyes and shallower cheeks (24–34% head length, modal range 27–30%, and 16–29%, modal range 19–22% head length, for the characters respectively). In these features they depart but slightly from the generalized condition. Species reaching greater maximum sizes (modally 150–200 mm SL) have relatively smaller eyes (17–26% head, modal range 19–22%) and somewhat deeper cheeks (18–33% head, modal range 24–26%).

The angle of the mouth varies from near horizontal to markedly oblique, and the lower jaw is long (41–62% head, modal range 45–53%). As in *Harpagochromis*, the relative lower jaw length is greatest in larger individuals, but even in those species with a small adult size (*ie* < 110 mm SL), the mean lower jaw length is distinctly greater than that in comparable sized *Astatotilapia* species. It is also greater than in most species of other, but non-piscivorous, groups.

Squamation. The cheek is fully covered by 3 or 4 (less commonly 2, 5 or 6) rows of scales, the lateral line has 30–34 scales (modes 32 or 33), rarely 35.

Neurocranium (Fig. 9). A slender, shallow and generally streamlined neurocranium is found in all species of *Prognathochromis* (see Fig. 9; and fig. 69 in Greenwood, 1974).

Preorbital skull depth is from 18·6–23·1% of neurocranial length (mode 21%), greatest orbital depth 21·7–28·0% (modal range 22–23%), and the greatest width across the otic region 42–55% (modal range 47–50%). Preotic skull length varies from 63–70% (mode 66%) of neurocranial length, and is thus virtually identical with that in *Harpagochromis*.

In most species the supraoccipital crest is low (particularly in relation to that in *Harpagochromis*), and it slopes gently downwards and forwards. The outline of the crest is thus more nearly wedge-shaped than is the pyramidical crest in *Harpagochromis* (and in a few *Prognathochromis* species, eg *P. xenostoma* and *P. flavipinnis*; however, even in these taxa the crest is relatively low and the rest of the skull has proportions typical for the lineage). A wedge-shaped crest characterizes the presumed generalized haplochromine skull (eg *Astatotilapia bloyeti*, see fig. 6 in Greenwood, 1979), but it is relatively lower in *Prognathochromis*.

Within *Prognathochromis* there is a graded range of skull forms (see Greenwood, 1962 : 208, fig. 25; 1967 : 108–115; and 1974 : 98, fig. 69) but even in those species with the least derived skull form (eg *P. melichrous, P. dichrourus, P. arcanus*), gross neurocranial morphology is quite distinct from that in other lineages.

There is also some variation in the form of the ethmovomerine region. Modally, this part of the skull appears as an almost uninterrupted anterior prolongation of the dorsal skull outline, and the vomer tip lies slightly below the level of the parasphenoid. Exceptionally, the ethmovomerine region is decurved, projecting well below the level of the parasphenoid, and the skull has a more convex dorsal profile than that of the modal type. Examples (Fig. 9B) are seen in *P. dentex, P. bayoni* and *P. vittatus.* Intermediate types, however, link the extreme with the modal form.

In the nominate subgenus of *Prognathochromis* (see p. 19), the posterior (*ie* orbital) face of the lateral ethmoid slopes backwards at an angle of 45°–60° with the horizontal; in the other subgenus it is more nearly vertical (75°–80°), the common condition amongst haplochromine lineages. In that subgenus too, the lateral aspects of the bone are more expansive and anteriorly protracted than in the nominate subgenus; again, this is the more usual haplochromine condition.

Characters contributing to the low, slender and streamlined appearance of the skull in *Prognathochromis* (*ie* the shallow preorbital depth, low orbital and otic regions, the low, gently sloping supraoccipital crest, and the narrow otic region) are all derived features, and constitute the principal synapomorphies defining the lineage.

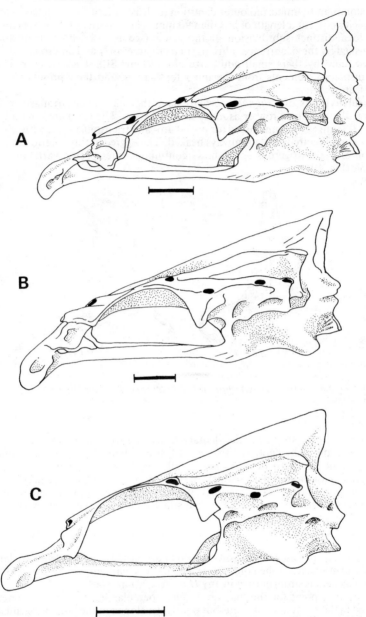

Fig. 9 Neurocranium (in left lateral view) of: A. *Prognathochromis (P.) mento*. B. *Prognathochromis (P.) dentex*. C. *Prognathochromis (T.) sulphureus*. Scale = 5 mm.

Dentition. Specimens over 90 mm SL in most *Prognathochromis* species have a preponderance of strong, recurved, unicuspid teeth in the outer row of both jaws. Some unequally bicuspid teeth may also be present in specimens between 90 and 110 mm SL, but are in a minority, only predominating in fishes less than 80 mm long.

A predominantly unicuspid outer dentition (at least anteriorly in the jaws) also occurs, in some species only, at a length of less than 90 mm. The majority of these species (8 out of 11) belong to a distinct subdivision of the genus (see p. 20) whose members are further distinguished by the occurrence of tricuspid teeth anteriorly and anterolaterally in the outer tooth row, and by their small adult size (96–120 mm SL). These tricuspid teeth are not simply displaced elements from the inner tooth rows, and their presence is considered a derived feature.

Both subdivisions of *Prognathochromis* have the inner teeth arranged in 2 or 3 rows, exceptionally in a single row or as many as 5 or 6 rows. The teeth are generally tricuspid in fishes < 100 mm SL and unicuspid (or mixed uni- and tricuspids) in larger individuals.

Upper jaw. The premaxilla is always beaked. The degree of this anterior and anterolateral extension of the dentigerous arms shows continuous interspecific variation, but is always noticeable (Fig. 10).

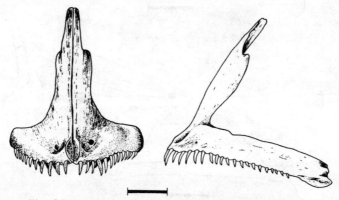

Fig. 10 Premaxilla of *Prognathochromis (P.) argenteus*, seen anteriorly and in left lateral view. Scale = 5 mm.

Lower jaw. In no species is the dentary foreshortened, but there is considerable interspecific variation in the depth of this bone. Most members of the nominate subgenus have a relatively stout (*ie* deep) dentary, but in all species of the other subgenus the bone is noticeably shallower and more slender. In this group too, the lateral aspect of the dentary is deeply concave; when viewed anteriorly, it has a pronounced upward and outward flare and the alveolar surface projects laterally as a distinct shelf (see Greenwood & Barel, 1978 : 155; and Fig. 11).

A prominent mental protuberance at the ventral end of the symphysis is present in both subgenera.

The crown of the ascending (coronoid) dentary arm is always deflected laterally, most obviously so in species of the nominate subgenus, less markedly in the others. The deflection in both subgenera is stronger than in any *Harpagochromis* species.

The insertion point for the mandibulo-interopercular ligament is prominent and well-developed in the nominate subgenus but poorly developed in the other subgenus.

Lower pharyngeal bone and teeth. The dentigerous surface of this bone is triangular, usually as broad as it is long, but sometimes slightly broader than long.

In one subgenus (see p. 20) the lower pharyngeal bone is relatively narrow and its dentigerous surface has an anteroposteriorly attenuate appearance (Fig. 12; and figs 5, 10 and 12 in Greenwood & Barel, 1978). The pharyngeal teeth in this subgenus are finer and more compressed than those of the nominate subgenus, and the teeth in the two median and the posterior transverse row are, relatively speaking, less enlarged.

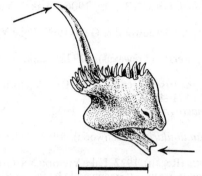

Fig. 11 Anterior view of right dentary in *Prognathochromis (T). dolichorynchus* to show overhang of alveolar shelf, and 'flare' of the dentary's outer wall. Arrows point to: ascending (coronoid) arm of the ramus (upper arrow), and the anterior part of the anguloarticular (lower arrow). Scale = 3 mm.

Vertebral numbers: 28–32 (mode 30), comprising 12–14 (mode 13) abdominal and 15–18 (modes 17 and 18) caudal elements excluding the fused PU_1 and U_1 centra.

Caudal fin. An obliquely truncate fin (see p. 12 above) occurs in some specimens of one species (*P. pseudopellegrini*), but otherwise the fin is truncate or weakly subtruncate.

The two subgenera recognized are:

Subgenus **PROGNATHOCHROMIS** nov.

TYPE SPECIES: *Paratilapia prognatha* Pellegrin, 1904; see Greenwood (1967 : 78).

DIAGNOSIS. *Prognathochromis* without tricuspid teeth in the outer tooth row of either jaw (except, very rarely, as obviously displaced elements from the inner series), usually reaching a large maximum adult size (140–230 mm SL) but only to a length of 93–105 mm in 3 species. First infraorbital (lachrymal) bone with a very slight anterior bullation preceding the anterior infraorbital lateral line tubule, the bullation barely visible without dissection. *Maximum orbital depth of the skull 22–25% neurocranial length (modal range 22–23%). Lateral ethmoid relatively narrow, its posterior face sloping backwards at an angle of 45°–60° to the horizontal.* Lower jaw (especially the dentary) not noticeably shallow, and with a moderately pronounced alveolar shelf visible when the bone is viewed frontally. Lower pharyngeal teeth coarser than those in the other subgenus, the bone itself not distinctly attenuated.

Contained species
Since no intragroup relationships can be determined, the species are listed alphabetically.

Prognathochromis (Prognathochromis) arcanus (Greenwood & Gee), 1969. Lake Victoria; see Greenwood & Gee (1969 : 52–4).
Prognathochromis (P.) argenteus (Regan), 1922. Lake Victoria; see Greenwood (1967 : 84–7).
Prognathochromis (P.) bartoni (Greenwood), 1962. Lake Victoria; see Greenwood (1962 : 161–4).
Prognathochromis (P.) bayoni (Blgr.), 1909. Lake Victoria; see Greenwood (1962 : 149–52).
Prognathochromis (P.) decticostoma (Greenwood & Gee), 1969. Lake Victoria; see Greenwood & Gee (1969 : 55–7).
Prognathochromis (P.) dentex (Regan), 1922. Lake Victoria; see Greenwood (1962 : 167–9)
Prognathochromis (P.) dichrourus (Regan), 1922. Lake Victoria; see Greenwood (1967 : 65–9).
Prognathochromis (P.) estor (Regan), 1929. Lake Victoria; see Greenwood (1962 : 164–7).

Prognathochromis (*P.*) *flavipinnis* (Blgr.), 1906. Lake Victoria; see Greenwood (1962 : 192–5).
Prognathochromis (*P.*) *gilberti* (Greenwood & Gee), 1969. Lake Victoria; see Greenwood & Gee (1969 : 57–60).
Prognathochromis (*P.*) *gowersi* (Trewavas), 1928. Lake Victoria; see Greenwood (1962 : 180–3).
Prognathochromis (*P.*) *longirostris* (Hilgend.), 1888. Lake Victoria and possibly the Victoria Nile; see Greenwood (1962 : 171–4).
Prognathochromis (*P.*) *macrognathus* (Regan), 1922. Lake Victoria; see Greenwood (1962 : 183–6).
Prognathochromis (*P.*) *mandibularis* (Greenwood), 1962. Lake Victoria; see Greenwood (1962 : 178–80).
Prognathochromis (*P.*) *mento* (Regan), 1922. Lake Victoria; see Greenwood (1962 : 174–8).
Prognathochromis (*P.*) *nanoserranus* (Greenwood & Barel), 1978. Lake Victoria; see Greenwood & Barel (1978 : 157–61).
Prognathochromis (*P.*) *paraguiarti* (Greenwood), 1967. Lake Victoria; see Greenwood (1967 : 69–72).
Prognathochromis (*P.*) *pellegrini* (Regan), 1922. Lake Victoria; see Greenwood (1962 : 186–9).
Prognathochromis (*P.*) *percoides* (Blgr.), 1915. Lake Victoria; see Greenwood (1962 : 189–91).
Prognathochromis (*P.*) *prognathus* (Pellegrin), 1904. Lake Victoria; see Greenwood (1967 : 78–83).
Prognathochromis (*P.*) *pseudopellegrini* (Greenwood), 1967. Lake Victoria; see Greenwood (1967 : 56–60).
Prognathochromis (*P.*) *venator* (Greenwood), 1965. Lake Nabugabo; see Greenwood (1965 : 342–6).
Prognathochromis (*P.*) *vittatus* (Blgr.), 1901. Lake Kivu; see Regan (1921 : 638).
Prognathochromis (*P.*) *xenostoma* (Regan), 1922. Lake Victoria; see Greenwood (1967 : 51–6).

Incertae sedis
Astatotilapia nigrescens Pellegrin, 1909 (Lake Victoria). This taxon, known only from the holotype, has a close superficial resemblance to both *P.* (*P.*) *flavipinnis* and *P.* (*P.*) *percoides* (see Greenwood, 1967 : 118–19). It is for this reason alone that I am including it, tentatively, as a member of this subgenus.

Subgenus **TRIDONTOCHROMIS** nov.

TYPE SPECIES: *Haplochromis tridens* Regan & Trewavas, 1928 (see Greenwood, 1967 : 97). Lake Victoria.

ETYMOLOGY. The name alludes to the tricuspid teeth which are a feature of the outer tooth row in both jaws.

DIAGNOSIS. *Prognathochromis* species in which *tricuspid teeth occur anteriorly and anterolaterally* (as well as posteriorly) *in the outer tooth rows of, generally, both jaws;* the size and number of these teeth, together with their inevitable presence, militate against their merely being displaced elements from the inner tooth series. *The lachrymal bone (1st infraorbital) has, in 8 of the 9 species known, an enlarged, ovoid to rectangular bullation occupying the greater part of the bone anterior to the first lateral line tubule, the bulla visible without dissection.* Maximum orbital depth is 23–28% of the neurocranial length (modal range 26–27%), the lateral ethmoid is more expansive than in species of the nominate subgenus, and its posterior face is aligned almost vertically or at an angle of 70°–80° with the horizontal. *The lower pharyngeal bone is narrow, its dentigerous surface having an anteroposteriorly attenuate appearance* (see Fig. 12 and figs 5, 10 & 12 in Greenwood & Barel,

1978); teeth on this bone are fine and compressed. *The lower jaw is shallow, its lateral face having a pronounced upward and outward flare so that the alveolar surface is carried as a prominent shelf overhanging the body of the bone* (see fig. 4 in Greenwood & Barel, 1978).

Members of this subgenus reach a small maximum adult size (95–120 mm SL), becoming sexually mature at a standard length of between *ca* 55–85 mm.

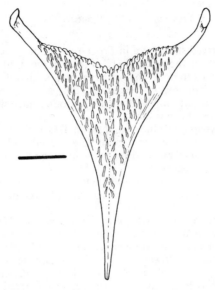

Fig. 12 Lower pharyngeal bone of *Prognathochromis (T.) crocopeplus* in occlusal view. Scale = 2 mm.

Contained species
No intragroup relationships can be determined, and the species are therefore listed alphabetically.

Prognathochromis (Tridontochromis) chlorochrous (Greenwood & Gee), 1969. Lake Victoria; see Greenwood & Gee (1969 : 44–8).
Prognathochromis (T.) crocopeplus (Greenwood & Barel), 1978. Lake Victoria; see Greenwood & Barel (1978 : 142–8).
Prognathochromis (T.) cryptogramma (Greenwood & Gee), 1969. Lake Victoria; see Greenwood & Gee (1969 : 48–51).
Prognathochromis (T.) dolichorhynchus (Greenwood & Gee), 1969. Lake Victoria; see Greenwood & Gee (1969 : 34–40).
Prognathochromis (T.) melichrous (Greenwood & Gee), 1969. Lake Victoria; see Greenwood & Gee (1969 : 24–8).
Prognathochromis (T.) plutonius (Greenwood & Barel), 1978. Lake Victoria; see Greenwood & Barel (1978 : 151–5).
Prognathochromis (T.) sulphureus (Greenwood & Barel), 1978. Lake Victoria; see Greenwood & Barel (1978 : 148–51).
Prognathochromis (T.) tridens (Regan & Trewavas), 1928. Lake Victoria; see Greenwood (1967 : 97–100).
Prognathochromis (T.) tyrianthinus (Greenwood & Gee), 1969. Lake Victoria; see Greenwood & Gee (1969 : 40–4).

Incertae sedis
Haplochromis eutaenia Regan & Trewavas, 1928. The type and only specimen of this species is now rather damaged, particularly about the jaws. As a result the dentary is broken and there are almost no outer series teeth in either jaw. The remaining teeth are bicuspids, but the morphology of the dentary, the overall proportions of the specimen, and what little I can learn about its neurocranial architecture, all suggest that the species could be referred to this subgenus of *Prognathochromis*.

Lake Victoria; see Regan & Trewavas (1928 : 225–6).

DISCUSSION

The characteristic and derived skull form in *Prognathochromis* (see above, p. 16) seems to provide a strong argument for the monophyly of the lineage. Certainly it would be more parsimonious to consider this to be so than to argue that such a distinctive skull form had evolved independently and on several occasions. But, as is so often the case with the Victoria–Edward–Kivu haplochromines, there are few other unequivocally synapomorphic features to back-up the single, diagnostic one.

Again, an absence of synapomorphic characters makes it difficult to identify the sister group of *Prognathochromis*. In an earlier attempt (Greenwood, 1974), certain paedophagous species were tentatively identified as the sister taxon of *Prognathochromis* (then represented by what is now the nominate subgenus).

A more critical analysis of the features on which that suggestion was based, shows that it is no longer tenable; some of the characters involved proved to be plesiomorphies, and others to be autapomorphies.

I have also not been able to find new features that would corroborate my suggestion that *Prognathochromis* and *Harpagochromis* are closely related (Greenwood, 1974). *Prognathochromis* could, on available anatomical evidence, be derived from an *Harpagochromis* or an *Astatotilapia*-like ancestor. However, since in certain respects the skull form in *Harpagochromis* does depart from the generalized *Astatotilapia* type towards that of *Prognathochromis*, there may be grounds for suspecting some relationship between the two lineages. Unfortunately, since there are no other features to support (or negate) this idea, it must remain as no more than a suggestion.

In the same tentative phylogeny (Greenwood, 1974 : fig. 70), the *Tridontochromis* division of *Prognathochromis* was thought to have a rather distant relationship with the nominate subgenus. It was, indeed, allied with a taxon now accorded the status of a monotypic genus, namely *Allochromis welcommei* (see p. 57); the two taxa were, at that time, considered to be the sister group of the *Haplochromis riponianus* complex here included in the genus *Psammochromis*, see p. 53.

First doubts about these proposed relationships were expressed by Greenwood & Barel (1978 : 156), and are confirmed by the research embodied in this paper (see below, p. 60).

For the moment, all that can be established on the basis of synapomorphic characters is the sister-group relationship between the two divisions of *Prognathochromis* itself; their affinities with the other haplochromine lineages from Victoria–Edward–Kivu have still to be discovered.

Intralineage relationships remain indetectable at the level of investigation employed so far. Each subgenus has its morphologically outstanding taxa, but the majority differ from one another only in such features as male coloration and certain morphometric characters.

Within the subgenus *Tridontochromis*, however, *Prognathochromis (T.) melichrous* stands apart because of the greater number of plesiomorph features it displays, and the skull architecture in all species is less derived than in species of the nominate subgenus. The intralineage dichotomy would seem to have occurred early in the history of the genus.

YSSICHROMIS gen. nov.

TYPE SPECIES: *Haplochromis fusiformis* Greenwood & Gee, 1969. Lake Victoria.

ETYMOLOGY. From the Greek *yssos*, javelin, + *chromis*, alluding to the slender, elongate body form.

DIAGNOSIS. *Shallow bodied, elongate haplochromines (body depth 23–30% SL, modal range 27–29%, caudal peduncle 17–25% of standard length, modal range 19–22%, its depth contained 1·7–2·1 times (modally 1·8–2·0) in its length)*, reaching a small maximum adult size (85–110 mm SL).

Neurocranium of the generalized type with a low, wedge-shaped supraoccipital crest. Premaxilla not beaked anteriorly; *edentulous over the posterior $\frac{1}{4}$–$\frac{1}{3}$ of its dentigerous arms*.

Teeth in the outer premaxillary row compressed and unequally bicuspid, those in the dentary similar but with a few tricuspids posteriorly and laterally; 28–64 teeth (no distinct modal number) in the outer row of the premaxilla.

Lower pharyngeal bone slender and elongate, all its teeth fine and compressed.

Lateral line with 32–37 scales (modal range 33–35); cheek fully scaled, with 3 or 4 (rarely 2) rows.

DESCRIPTION

Habitus (Fig. 13). The body is shallow and elongate, the head profile in lateral view moderately acute. Maximum adult size recorded for each of the three constituent species is 110 mm, 93 mm and 85 mm SL respectively.

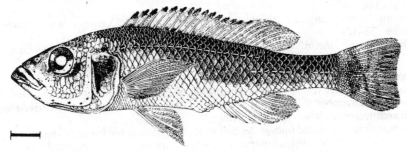

Fig. 13 *Yssichromis fusiformis*. Lake Victoria. Scale = 1 cm.

Superficially, members of this genus resemble those of the *Prognathochromis* lineage, especially members of the subgenus *Tridontochromis*. However, *Yssichromis* species retain several generalized features in the syncranium (see below), and the lower jaw length is shorter, although some overlap does occur (*viz* 35·7–43·8% head length, modal range 40–42% **cf.** 41–62%, modal range 45–53%).

Neurocranium. Neurocranial architecture in this genus is basically of the generalised type (see fig. 18 in Greenwood & Gee, 1969). In one species (*Y. pappenheimi*), however, the preorbital and orbital depths, and the maximum otic width, are reduced and approach the condition found in *Prognathochromis* (which skull-type that of *Y. pappenheimi* also resembles in having a straight rather than a gently curved preorbital skull profile).

The supraoccipital crest in all species is relatively low and wedge-shaped in profile, but it is not as low as that in *Prognathochromis*.

Upper jaw. The premaxilla is not produced anteriorly into a definite beak. Posteriorly over about its last $\frac{1}{4}$–$\frac{1}{3}$ each horizontal dentigerous arm is edentulous in all known specimens of *Y. pappenheimi*, and, apart from the rare exception, is edentulous in the other two species as well.

Lower jaw. The mouth is slightly to moderately oblique (20°–35°), with the tip of the lower jaw not, or but marginally projecting beyond the upper jaw.

The dentary departs but little from the generalized type (and is thus relatively deeper than in *Tridontochromis* species). It does, however, have a well-defined upward and outward flare to the lateral walls when viewed frontally, and in this feature closely approximates to the dentary in *Tridontochromis*. Two of the three species (*Y. pappenheimi* is the exception) have a poorly defined mental protuberance at the symphysis, which is visible in skeletal material but barely detectable in whole fishes.

Dentition. Most outer row teeth in both jaws are compressed and unequally bicuspid; posteriorly in the lower jaw there are often some tricuspid teeth, and tricuspids or unicuspids are sometimes present posteriorly in the upper jaw. Posterior teeth in both jaws are either smaller than those situated anteriorly or may be of approximately the same size, even when unicuspid. (The generalized condition, as seen for example in *Astatotilapia* and some other genera, is for the posterior few teeth to be enlarged.)

Teeth forming the inner row or rows are tricuspid and small.

Lower pharyngeal bone and teeth. The bone is narrow and slender, its dentigerous surface slightly broader than it is long. Two species (*Y. pappenheimi* and *Y. fusiformis*) have the transverse posterior margin of the bone deeply indented so that it is acutely 'V' shaped rather than broadly 'V' shaped (the usual condition in all genera except those with hypertrophied pharyngeal bones, and in the third *Yssichromis* species, *Y. laparogramma*).

All lower pharyngeal teeth are fine, laterally compressed, and weakly cuspidate.

Contained species

Yssichromis fusiformis (Greenwood & Gee), 1969. Lake Victoria; see Greenwood & Gee (1969 : 32–34).

Yssichromis laparogramma (Greenwood & Gee), 1969. Lake Victoria; see Greenwood & Gee (1969 : 28–32).

Yssichromis pappenheimi (Blgr.), 1914. Lakes Edward and George; see Greenwood (1973 : 199–204).

DISCUSSION

Although superficially resembling certain species of *Prognathochromis*, *Yssichromis* cannot be considered a member of that lineage because it does not share with it any derived features in skull architecture or lower jaw proportions.

Yssichromis is an isolated lineage defined by its autapomorphic features (shallow, elongate body, and posteriorly edentulous premaxilla), but otherwise is of a generalized type.

Within the genus, *Y. pappenheimi* from Lakes Edward and George is apparently the most derived species, judging from its skull shape, and *Y. laparogramma* (Lake Victoria) the most plesiomorphic one; *Y. fusiformis*, also from Lake Victoria, occupies an intermediate position in this morphocline.

PYXICHROMIS gen. nov.

TYPE SPECIES: *Haplochromis parorthostoma* Greenwood, 1967. Lake Victoria.

ETYMOLOGY. From the Greek *pyx* (later form of *pyge*), the rump, an allusion to the rump-like protuberance of the nuchal musculature, and, punningly, to the gnome-like physiognomy of the known species.

DIAGNOSIS. Small relatively deep-bodied and compressed haplochromines (body depth 35–42% SL; maximum adult size 117 mm SL), *with a very oblique lower jaw (sloping upwards at 50°–70° to the horizontal), a sharply concave dorsal head profile, and the dorsal surface of the snout virtually horizontal.* The very characteristic profile of these fishes is contributed to by the bulging anterior portion of the cephalic epaxial musculature (see Fig. 14).

The anatomy of the upper jaw is distinctive (see p. 25).

DESCRIPTION

Habitus (Fig. 14). The external features of *Pyxichromis* are highly characteristic. Considering the small adult size attained, the eye diameter (22–28% head length) is small, especially in comparison with that of certain piscivorous groups (eg *Prognathochromis*

(*Tridontochromis*) species) and in *Astatotilapia* and *Haplochromis*; in contrast, the cheek depth in *Pyxichromis* (24–32% head length) is greater.

Members of the genus are sexually mature at a standard length of *ca* 70 mm, and are not known to exceed a length of 117 mm.

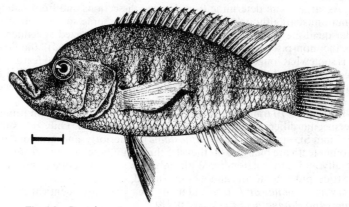

Fig. 14 *Pyxichromis parorthostoma.* Lake Victoria. Scale = 1 cm.

Anatomy. Unfortunately, very few *Pyxichromis* specimens are available and consequently knowledge of its anatomy and osteology is confined to information gleaned from partial dissections and from radiographs.

Neurocranium. Skull form is essentially of the *Astatotilapia* type except that the supraoccipital crest is relatively more expansive, a probable correlate of the somewhat hypertrophied nuchal muscle mass.

Upper jaw. An unusual feature of *Pyxichromis* is the near-horizontal alignment of the dorsal snout surface. Amongst other members of the Victoria–Edward–Kivu haplochromine complex the snout profile slopes downwards and forwards, albeit at various angles, but it is never horizontal. The angle at which the snout descends in these other species virtually parallels the slope of the underlying ethmovomerine region of the skull.

In *Pyxichromis* the ethmovomerine region slopes at almost the same angle as it does in the skulls of *Harpagochromis* and *Prognathochromis* species. That the upper snout profile is, nevertheless, horizontal in *Pyxichromis* can be explained by the hypertrophy of certain articulatory menisci and other surfaces associated with the maxillae and premaxillae.

For example, the median rostral cartilage is much deeper than it is in other taxa, and its ventral face (which is apposed to the sloping ethmovomer) is angled so that its dorsal surface (in contact with the premaxillary process) lies horizontally, not sloping forward and downward as it would otherwise do if the anterior part of the cartilage were not deeper than the posterior part. The premaxillary processes are thus elevated above the ethmovomerine surface and, since the upper surface of the cartilage is almost horizontal, held horizontally as well.

As a result of this arrangement there is a considerable gap anteriorly between the processes and the rostral part of the dorsal ethmoid surface. The premaxillary processes are supported in this region by an hypertrophy of the membrane and cartilage cushion surrounding the condyle of the medially directed posterior process on the maxillary head (the neurocranial process of Barel *et al.*, 1976). More support is derived from the enlarged cushion of tissue capping the anteroventral process of the maxilla (Barel *et al*'s 'premaxilliad wing').

Thus, when the mouth is closed, the premaxilla is supported, and held away from the ethmovomer, by enlarged articulatory points at three places: posteriorly by the rostral cartilage, near its midpoint by the neurocranial process, and anteriorly (where the ascending processes join the body of the bone) by a pad of tissue on the premaxilliad wing of the maxilla.

The alignment and relative enlargement of these three surfaces is such that the premaxillary processes lie almost horizontally, despite the forward and downward slope of the ethmoid and vomer against which two of them articulate.

Lower jaw. As far as I can determine from limited dissections and from radiographs, the lower jaw is not unusual (although I suspect that there are some specialized features in the anguloarticular-quadrate joint). When compared with similar-sized specimens of species belonging to the non-piscivorous lineages (especially *Astatotilapia*), the lower jaw in *Pyxichromis* is somewhat longer (48–57% head length). This is a derived feature shared with both *Harpagochromis* and *Prognathochromis* (see pp. 10 & 16). The jaw is narrow (its maximum width contained more than twice in the length), a correlate of the generally compressed body-form.

Dentition. In fishes >70 mm SL, the outer teeth are mostly slender and unicuspid, but there are interspecific differences in tooth shape and orientation. Only one smaller fish is known (a 67·5 mm SL specimen of *P. orthostoma*); anteriorly and anterolaterally in the upper jaw its outer teeth are bicuspid (as they are throughout the lower jaw) but are short and tricuspid laterally and posterolaterally. With only one small specimen known, the significance of these tricuspid outer teeth cannot be evaluated.

Lower pharyngeal bone and teeth. These are virtually identical with those of *Harpagochromis* and *Prognathochromis* (see p. 18).

Caudal fin. One species (*P. orthostoma*, Lake Kioga system) has a truncate fin, the other species (*P. parorthostoma*; Lake Victoria) has the fin strongly subtruncate, almost rounded.

Contained species

Pyxichromis orthostoma (Regan), 1922. Lake Salisbury, Kioga system; see Greenwood (1967 : 100–2).

Pyxichromis parorthostoma (Greenwood), 1967. Lake Victoria; see Greenwood (1967 : 103–5).

DISCUSSION

Because of its peculiar autapomorphic features, and within the limits imposed by inadequate anatomical and osteological data, it is particularly difficult to assess the affinities of *Pyxichromis*.

Its high relative jaw length, especially in a species with a small maximum adult size, suggests a possible relationship with *Harpagochromis* and *Prognathochromis*. As far as can be told, neurocranial form in *Pyxichromis* is of the near-generalized type and like that found in *Harpagochromis* (see p. 11). Except for the narrow otic region, the neurocranium shows none of the derived features characterizing the skull of *Prognathochromis* (*ie* low preorbital skull depth, low orbital depth, and a relatively shallow, gently sloping supraoccipital crest).

Pyxichromis does, however, share with *Prognathochromis* the derived features of a unicuspid dentition in small individuals, the presence of some tricuspid outer teeth laterally in the jaw (at least in one specimen, the smallest known) and, of course, a relatively, long lower jaw.

LIPOCHROMIS Regan, 1920

TYPE SPECIES: *Paratilapia obesus* Boulenger, 1906. Lake Victoria; for details of synonymy *etc*, see Greenwood (1959b : 182–3).

DIAGNOSIS. Haplochromine fishes with an adult size range of 130–170 mm SL, and a varied body form (see Figs 15A & B). All are characterized by having a thick-lipped, *widely distensible and protractile mouth, and small teeth deeply embedded in the oral mucosa (often invisible without dissection)*.

The nominate subgenus is characterized by many of its outer row jaw teeth having the crowns reflected labially (not buccally as is usual), *and by its broadly rounded lower jaw*. The *other subgenus is recognizable by its peculiarly boat-shaped lower jaw which narrows abruptly over about its anterior third so that this part of the jaw closes within the upper jaw;* no outer row jaw teeth have their crowns curved labially, and are either erect or recurved.

The neurocranium is essentially of the generalized type, but does have a relatively tall and expansive supraoccipital crest which is near-pyramidical in lateral outline.

DESCRIPTION

Habitus (Fig. 15). Body form is variable, especially with regard to the head and snout profiles; these range from heavy and blunt ('pug-headed'), to slender and subacute (see Figs 15A & B respectively). Body depth ranges from 27–47% of standard length, with some species (as currently defined) showing considerable intraspecific variation (eg 33–47% in *L. obesus,* and 32–43% in *L. maxillaris*). In general, deep-bodied species are pug-headed, and slender bodied ones have a more refined profile.

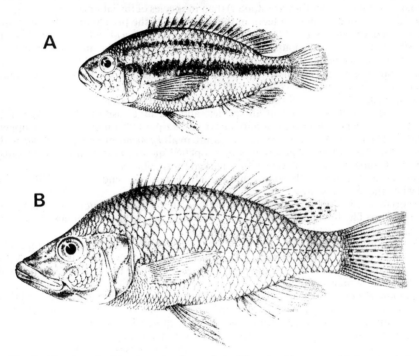

Fig. 15 A. *Lipochromis (Lipochromis) obesus.* Lake Victoria. About two-thirds natural size. B. *Lipochromis (Cleptochromis) parvidens.* Lake Victoria. About two-thirds natural size.

Maximum adult size ranges from 130–170 mm SL, with sexual maturity reached at between 85 and 105 mm SL depending on the maximum adult size for a particular species.

Neurocranium. Basically, the skull is of the near-generalized type found in species of the *Harpagochromis* group (see p. 11) but it retains the deeper otic region of the more generalized skull (see Greenwood, 1974 : fig. 45, excluding skull of *H. barbarae*).

As in *Harpagochromis,* the supraoccipital crest in *Lipochromis* species is high, with a near-pyramidical profile, but it is more expansive than in the majority of *Harpagochromis* species.

Greatest departure from the modal *Lipochromis* skull type is seen in *L. obesus,* where the orbital depth is greater and consequently the preorbital skull profile slopes at a greater angle.

Dentition. All jaw teeth are deeply embedded in the thickened oral mucosa so that, at most, only the crowns of the teeth are visible; often the inner rows are completely buried.

Two distinct forms of outer teeth are present. In one, the tip of the crown is inclined anteriorly or laterally (see below) whilst in the other it is either vertical or fairly strongly recurved (*ie* directed buccally; see below p. 31).

Most outer row teeth in fishes between 70 and 100 mm SL are weakly bicuspid (with, in certain species, some unicuspids and weakly bicuspids also present); above that size the majority of teeth are unicuspid, although in one species (*L. taurinus*) bicuspids predominate even in the largest individuals (see Greenwood, 1973 : 194).

The inner rows, usually 1 or 2, are composed of tricuspid teeth (with a few unicuspids) in fishes <80–100 mm SL, and predominantly of unicuspids in larger individuals.

Compared with the teeth in equal-sized specimens from other lineages, those of *Lipochromis* are shorter (as little as half the height of teeth in members of the piscivorous lineages *Harpagochromis* and *Prognathochromis*); teeth in *Lipochromis* are also often finer (although in some species they are stouter) than in species of the latter genera.

Modally, the total number of teeth in the outer row of the premaxilla is less than in comparable-sized specimens of *Harpagochromis* and *Prognathochromis*, the number ranging from 30–62, but generally about 40. A comparable reduction in the number of outer teeth in the dentary is also noted in *Lipochromis*.

Most individuals in certain species have almost the posterior third of the premaxilla devoid of teeth; when premaxillary teeth are present posteriorly in these taxa, they are widely spaced, as they are in those species with a completely toothed premaxilla.

Lip tissue is well-developed in all species, and the inner aspect of the upper lip is so arranged that it generally covers, or partly covers, the tips of the outer teeth in the upper jaw.

Mouth. The mouth is a very distinctive feature in all *Lipochromis* species, in particular the wide lateral gape of the upper jaw in a fully-opened mouth; in one subgenus this distensibility of the upper jaw is combined with a marked protrusibility.

In all species the lips are thickened, but the bullate posterior end of the maxilla is obvious, even when the mouth is closed.

The orientation of the mouth is slightly oblique in most species, more obviously so in one (*L. microdon*). The lower jaw may project a little way beyond the upper jaw, particularly in those species with an acute head profile.

Relative lower jaw length in *Lipochromis* (38–56% of head length, modal range 42–48%) overlaps that in *Harpagochromis* and *Prognathochromis*, but modally it is shorter. Similarly, its length range overlaps that of most other lineages but in these instances the mode for *Lipochromis* is somewhat higher.

Upper jaw. Posteriorly the maxilla is markedly bullate, its lateral face convex and the inner face concave. The ventral margin over the entire bone, except in the bullate region, is distinctly thickened. Some variation exists in the degree to which the anterior half of the bone is incurved relative to the posterior portion; in most species the curvature is very noticeable, especially so in members of the nominate subgenus (see below, p. 30).

A prominent feature of the premaxilla is the stoutness of its dentigerous arms, which are almost cylindrical in cross-section over the greater part of their lengths. The ascending processes are either as long as the dentigerous arms or are distinctly shorter, a feature positively correlated with the degree of mouth protrusibility.

Lower jaw. The dentary is bullate in the region surrounding its bifurcation into ascending (coronoid) and horizontal arms. This horizontally directed, dorsoventrally compressed swelling is produced forward for almost half the anterior length of each ramus as a thick, shelf-like lateral projection.

The tip of the ascending process has a definite, but interspecifically variable, deflection laterally. A similar deflection occurs in *Prognathochromis*, particularly in members of its nominate subgenus, but in no species is it so noticeable as it is in most *Lipochromis* species. The insertion for the mandibulo-interopercular ligament is prominent and well-developed (to a level comparable with that in *Prognathochromis* (*Prognathochromis*) species).

Two extreme forms of lower jaw morphology are found amongst *Lipochromis* species (see Fig. 16). In one (the *obesus* type) the anterior margin of the jaw is broadly rounded (most clearly so in *L. obesus*); when viewed from below, the alveolar surface appears as a broad shelf projecting above and beyond the ventral half of each ramus, the lateral wall of which is sharply angled towards the midline.

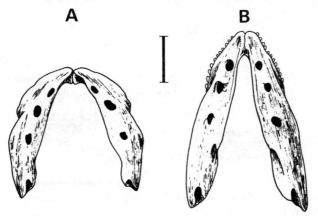

Fig. 16 Dentary in ventral view, of: A. *Lipochromis (L.) obesus*, and B. *Lipochromis (C.) parvidens*. Scale = 5 mm.

In the second, or *parvidens* type, the entire forward portion of the lower jaw anterior to the dentary bifurcation narrows rapidly; the alveolar 'shelf' is thus much less obvious and the anterior jaw outline is more acute (Fig. 16B). Also, since the ventral half of each ramus is, as it were, pinched medially, this anterior narrowing is emphasized and the ventral contours of the rami are rounded. In section, the jaw is rather boat-shaped, especially when viewed from in front. (In nautical terms, the shape is like that of a pram seen bow-on.)

Although the extreme conditions, as represented by *L. obesus* and *L. parvidens*, are very distinctive, some species have a jaw shape that almost bridges this morphological gap (see, for example, the lower jaws in *L. maxillaris* and *L. cryptodon*; Greenwood, 1959 : 189, 192 & 198–200).

Lower pharyngeal bone and dentition. The bone is broad and short, its dentigerous surface triangular and as broad as, or more often, broader than it is long. The teeth are fine and weakly bicuspid, with only those of the posterior transverse row, and some posteriorly in the two median rows, coarser than the others.

Taking into account the differences in dentition and in jaw morphology, it would seem that intragroup relationships of *Lipochromis* are best expressed by recognizing two subgenera. This action reflects and corrects an earlier view (Greenwood, 1974) that the paedophage trophic radiation (here represented by the genus *Lipochromis*) was of diphyletic origin. Further consideration of that idea has led me to give greater phylogenetic emphasis than before to the derived features shared by all *Lipochromis* species (*viz* buried teeth reduced in size and number, coupled with great distensibility and protrusibility of the upper jaw), and to the lack of characters suggesting an alternate relationship for any or all of the species involved. Hence, the recognition of one lineage comprising two subdivisions.

Subgenus *LIPOCHROMIS* Regan, 1920

TYPE SPECIES: *Pelmatochromis obesus* Blgr., 1906 (see Greenwood, 1959*b* : 182–8).

Members of this subgenus are characterized by the presence, in *the outer tooth row, of stout uni- or bicuspid teeth whose crowns are inclined labially (*ie *anteriorly or laterally depending* *on their position in the jaw).* Such teeth (Fig. 17) usually are the predominant type in the lower jaw; if present in the upper jaw they are intercalated amongst the more numerous recurved (*ie* buccally directed) or erect and conical teeth. Bicuspid teeth are the commonest type in fishes < 100 mm SL, but in one species (*L. taurinus* from Lakes Edward and George) most teeth are bicuspid even in specimens 140 mm long. Compared with the teeth in members of the other subgenus, those in *Lipochromis* are stouter and shorter.

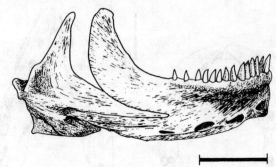

Fig. 17 Dentary (right) of *Lipochromis (L.) obesus* to show labial curvature of certain outer row teeth. Scale = 5 mm.

The lower jaw, save in one species, is of the *obesus* type (see above, and Fig. 16A), although in the majority of species it is not as broad anteriorly as it is in *L. (L.) obesus* itself. With the same exceptional species, the lower jaw does not close within the upper. The exception, *L. (L.) melanopterus*, was known only from the holotype; however, material collected recently in Lake Victoria indicates that it is not, as once was thought, an aberrant specimen (see Greenwood, 1959b : 192–4). As far as could be determined from superficial examination, radiographs and limited dissection on the holotype, the dentary in *L. (L.) melanopterus* is essentially of the *parvidens* type (see above p. 29) a conclusion confirmed from the examination of new material. In its dentition, head shape and oral features, however, *L. (L.) melanopterus* is typically a member of the nominate subgenus.

Regan's (1920) original description of *Lipochromis* (based solely on the species *obesus*) gave as the only diagnostic feature '... Lower jaw shutting within the upper'. The type specimen of *Pelmatochromis obesus*, the only specimen available at that time, is very atypical and also rather poorly preserved. Certainly the lower jaw does seem to shut within the upper, but in my view this is more likely to be a preservation artefact than the natural condition (see also Greenwood, 1959b : 183). In none of the 47 other specimens on which my redescription of the species was based does the lower jaw fail to occlude with the upper.

Species of *Lipochromis (Lipochromis)* have a 'pug-headed' morphotype unlike the more elegant head form in all known species of the second subgenus.

The mouth is moderately protractile but is markedly distensible laterally. The premaxillary ascending processes are much shorter than the dentigerous arms of the bone (even in *L. (L.) melanopterus*), hence, presumably, the comparatively restricted protusibility of the upper jaw.

Contained species
The taxa are listed in order of their apparently increasing level of derivation.
Lipochromis (Lipochromis) taurinus (Trewavas), 1933. Lakes Edward and George; see Greenwood (1973 : 192–6).
Lipochromis (L.) maxillaris (Trewavas), 1928. Lake Victoria; see Greenwood (1959b : 189–192).

Lipochromis (*L.*) *obesus* (Blgr.), 1906. Lakes Victoria and Kwania (Uganda); see Greenwood (1959*b* : 182–8).
Lipochromis (*L.*) *melanopterus* (Trewavas), 1928. Lake Victoria; see Greenwood (1959*b* : 192–4).

Subgenus *CLEPTOCHROMIS* nov.

TYPE SPECIES: *Paratilapia parvidens* Blgr., 1911. Lake Victoria (see Greenwood, 1959*b* : 194–8).

ETYMOLOGY. From the Greek *kleptes*, a thief, + *chromis*, with reference to the paedophagous habits of its member species.

Species of this subgenus are characterized by the unusual form of the dentary, and by the outer row of jaw teeth being mainly slender, recurved, often strongly recurved unicuspids in specimens over *ca* 100 mm SL, and weakly recurved bicuspids in smaller fishes; in no species are there any teeth with anteriorly or laterally directed crowns (**cf.** subgenus *Lipochromis*).

The lower jaw, at least anteriorly, closes within the upper, and has a boat-shaped dentary of the 'parvidens' type (see above p. 29 and Fig. 16B). The mouth is both markedly distensible and protractile.

In all known species the ascending premaxillary processes are as long as, or longer than the dentigerous arms of that bone.

Contained species
The taxa are arranged in their apparently increasing order of derivation.
Lipochromis (*Cleptochromis*) *cryptodon* (Greenwood), 1959. Lake Victoria; see Greenwood (1959*b* : 198–200).
Lipochromis (*C.*) *microdon* (Blgr.), 1906. Lake Victoria; see Greenwood (1959*b* : 200–3).
Lipochromis (*C.*) *parvidens* (Blgr.), 1911. Lake Victoria; see Greenwood (1959*b* : 194–8).

DISCUSSION

The genus *Lipochromis* comprises two groups of embryonic and larval cichlid-eating haplochromines (the paedophages) discussed in Greenwood (1959*b* & 1974 for the species of Lake Victoria) and 1973 for the Edward–George species. They are now united in a single lineage because of their presumed synapomorphies, namely: jaw teeth deeply embedded in the thickened oral mucosa, the teeth reduced in size (relative to those in comparable sized specimens from other lineages) and often absent from the posterior part of the premaxilla, the mouth widely distensible (through a mechanism effecting a marked lateral displacement of the upper jaw moieties when the mouth is opened), a highly protrusible premaxilla, the pronounced bullation of the posterior maxillary arm, and the thickened ventral margin of the maxilla.

There are two other derived characters shared by all members of this lineage: the premaxillary dentigerous arms are inflated, and the dentary is greatly swollen posteriorly in the region of its bifurcation into ascending and horizontal arms. However, there is evidence indicating that these features could be the result of convergent trends associated with the evolution of a jaw that is much involved in the handling of prey objects. Probably in the case of *Lipochromis* those jaw features are truly synapomorphic, but since the possibility of their convergent evolution in other genera exists, they are unreliable indicators of any relationship between *Lipochromis* and lineages showing the same features (see discussions on pp. 52, 71 & 75).

Previously (Greenwood, 1974) I postulated a diphyletic origin for the paedophage trophic radiation. Some of the cranial characters on which that argument was based are now seen to be plesiomorphous, and the dental features used are probably associated with the large adult size attained by the paedophage species. Thus, there seem to be no adequate grounds for suggesting that *Lipochromis* might share a recent common ancestry with *Harpagochromis* and *Prognathochromis* (the lineages in which most of the species linked with the paedophages in my earlier analysis are now placed; see Greenwood, 1974, fig. 70 and discussions in the text). Indeed, I can find no unequivocally apomorph features that would allow one reasonably to identify the sister group of *Lipochromis*.

Two paedophage (or partly paedophagous) species, '*H*' *cronus* and '*H*' *barbarae* were formerly associated with, respectively, the '*obesus*' and '*parvidens*' lineages recognized by Greenwood (1974). Since neither of these species has any of the synapomorphic features characterizing *Lipochromis*, neither is currently included in *Lipochromis* (see p. 8 and p. 88 for '*barbarae*' and '*cronus*' respectively; also Barel, Witte & van Oijen (1976) for a comparative anatomical study of the palate in '*H*' *barbarae* and various *Lipochromis* species).

Within the *Lipochromis* lineage it is only possible to note that, for the species so far described, *L. (L.) taurinus* from Lake Edward and George has the least derived dental morphology for taxa in its subgenus, and that *L. (C.) parvidens* shows the greatest jaw distensibility and tooth reduction amongst members of its subgenus.

When the several species recently collected in Lake Victoria are studied further it may be possible to produce a more satisfactory indication of intralineage relationships (personal observations based on material collected by the Leiden University research team).

GAUROCHROMIS gen. nov.

TYPE SPECIES: *Haplochromis empodisma* Greenwood, 1960. Lake Victoria.

ETYMOLOGY. From the Greek '*gauros*', haughty + *chromis*, alluding to the physiognomy in at least four of the member species.

DIAGNOSIS. Deep to relatively deep-bodied haplochromines (body depth 30–44% SL, modal range 38–39%), with a straight or slightly concave but steeply sloping dorsal head profile interrupted by the prominent ascending process of the premaxilla. Mouth horizontal or slightly oblique, lips not thickened, and the teeth small, fine and numerous but rarely contiguous.

Premaxilla with compressed (*ie* not inflated) dentigerous arms that are longer than the ascending processes, and which are not produced anteriorly into a beak or shelf.

Outer teeth in both jaws (particularly when compared with those in Labrochromis and Astatotilapia) finer, more compressed, shorter and closer set, with 44–82 (modal range 60–70) in the premaxillary outer row. Fishes < 90 mm SL have unequally bicuspid teeth (the major cusp acutely pointed but sometimes obscured by a dark brown accretion), *the crown barely broader than the neck.* Larger fishes have a mixed dentition of weakly bicuspid and unicuspid teeth, the latter predominating in the largest specimens. An exclusively unicuspid outer dentition has never been recorded.

Inner row teeth small, generally tricuspid, and arranged in 1 or 2 (less frequently 3) rows anteriorly and laterally in both jaws; separated from the outer row by a distinct space.

Two kinds of lower pharyngeal bone occur. One is slender, narrow and relatively elongate, with all or the majority of its teeth fine and compressed. A few teeth in the two median rows may be coarser than the others, but none is molariform or submolariform.

The second type of lower pharyngeal bone is distinctly hypertrophied and stout, and has thick, short, articular horns. The two median tooth rows are composed of enlarged and molariform teeth, and usually there are several other rows containing enlarged and submolariform (or even molariform) teeth. Fine, compressed and distinctly unicuspid teeth are virtually confined to the posterolateral angles of the dentigerous surface.

DESCRIPTION
Habitus (Fig. 18). Although not particularly deep-bodied (depth 30–44% SL, modal range 38–39%) the total impression gained from a specimen of *Gaurochromis* is one of a deep-bodied fish. The premaxillary ascending processes are prominent and break the outline of steeply sloping dorsal head profile, giving the fish a distinctive 'Roman nose'. The lips are not thickened, and the mouth is horizontal or slightly oblique.

The maximum adult size range is 90–117 mm SL.

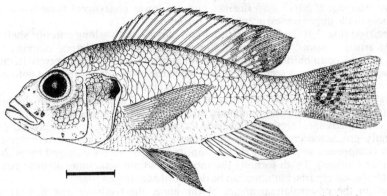

Fig. 18 *Gaurochromis (Gaurochromis) angustifrons.* Lake George. Scale = 1 cm.

Neurocranium Skull architecture is of the generalized type, with a straight and moderately sloping dorsal profile; preorbital depth ranges from 19–27% neurocranial length (being lowest in *G. angustifrons* from Lakes Edward and George), the mode lying in the lower part of the range for taxa with a generalized skull type. The height of the supraoccipital crest varies from low to relatively low, the bone being wedge-shaped in lateral profile.

In some species the neurocranial apophysis for the upper pharyngeal bones is hypertrophied (see p. 36).

Dentition. Teeth in both jaws are fine, slender and compressed. Those in the outer row are slightly recurved and, especially in the anteroposterior dimensions of the crown, noticeably finer than the teeth in similar-sized specimens from other taxa; in general these teeth are also shorter.

Most outer teeth in fishes < 90 mm SL have the generalized type of unequally bicuspid crown which is marginally broader than its subcylindrical neck; the major cusp is acutely pointed. Some slender, compressed unicuspids may be present in fishes less than 90 mm long, and weakly bicuspid teeth may also occur. The latter are the predominant teeth in larger fishes. An entirely or even a mainly unicuspid outer dentition has not been recorded in large individuals (*ie* > 100 mm SL) of any *Gaurochromis* species (see Greenwood, 1960 : 264; 1965 : 326; 1973 : 180).

The outer teeth are close-set but rarely contiguous, with 44–82 (modal range 60–70) in the premaxilla. Although the range overlaps that of *Astatotilapia, viz* 34–74 (and thus the presumed primitive numbers), the modal range is higher in *Gaurochromis* (60–70 **cf.** 48–54).

Inner row teeth are small and usually tricuspid, but some weakly tricuspid and even unicuspid teeth may be present in fishes > 100 mm SL. The teeth in both jaws are arranged in 1 or 2 (less commonly 3) rows anteriorly and anterolaterally, and a single row posteriorly.

Upper jaw. The maxilla is slender and elongate, its articular head with a fairly marked medial curvature relative to the shaft of the bone.

The dentigerous arms of the premaxilla are compressed, and are longer than the ascending processes. Anteriorly, the bone is not produced into a shelf or beak.

Lower jaw. In two species (*G. empodisma* and *G. simpsoni*) the dentary is slender and relatively shallow, its alveolar surface produced laterally into a narrow but distinct shelf; in the third species (*G. angustifrons*), the shelf is less obvious but is nevertheless clearly defined, more so than in the generalized condition.

The lower jaw length is 38–49% head (modal range 41–44%) thus overlapping the range in the generalized taxon *Astatotilapia* and in other but trophically specialized non-piscivorous lineages; however, the upper part of its range exceeds that in those taxa, and the modal range is also higher.

Lower pharyngeal bone. Two distinct types of lower pharyngeal bone (with correlated differences in the upper bones) occur in *Gaurochromis.*

In one type (Fig. 19) the bone is slender and narrow, with a long anterior shaft and fine, elongate articular horns. Its dentigerous area is attenuated, the lateral margins narrowing rapidly to produce an outline which is nearer that of an isosceles than an equilateral triangle. This overall attenuation is seen most clearly when the bone is superimposed onto one from a similar-sized specimen of some other lineage, for example, any species of *Astatotilapia.*

There is some intrageneric variation in the degree of attenuation, with the bone in *G. angustifrons* being the narrowest and most attenuated (see Greenwood, 1973 : fig. 19).

The lower pharyngeal teeth are fine, slender, and compressed, with weakly cuspidate and not greatly protracted crowns; even those teeth forming the posterior transverse row are strongly compressed so that they do not appear to be relatively enlarged (as is the case in most other lineages). Teeth forming the median rows are sometimes slightly coarser than their lateral counterparts, but none can be described as enlarged.

Except in the posterolateral angles of the bone, the teeth are not densely arranged; consequently the occlusal surface does not have the appearance of a fine dental felt.

The other type of lower pharyngeal bone (and dentition) differs markedly from the slender, attenuate bone described above (Fig. 20). It is moderately hypertrophied and stout (almost massive in some specimens), thus in many respects resembling the type of bone found in the genus *Labrochromis* (see p. 40).

At least the two median rows (and often the two rows lateral to them as well) are composed of enlarged and molariform teeth. In *Gaurochromis obtusidens* nearly all the other teeth are somewhat enlarged, with molariform or submolariform crowns, only those in the posterolateral angles of the bone being distinctly finer and cuspidate.

There appears to be a related species (currently undescribed) in which the bone is less massive and has fewer enlarged and molariform teeth outside the median series.

Neither *Gaurochromis obtusidens,* nor the undescribed species have the entire bone or its dentigerous surface so characteristically attenuated as it is in the other species. Yet, when compared with *Labrochromis,* the bone is narrower relative to its length (Fig. 21) and the dentigerous area is more nearly isoscelene than equilateral. Also, the articular horns, though stouter than in the other *Gaurochromis,* are neither as short nor as massive as those in *Labrochromis.*

Based on these intralineage differences in pharyngeal bone morphology and dentition, two divisions of *Gaurochromis* are recognised, each characterized by its autapomorphic features.

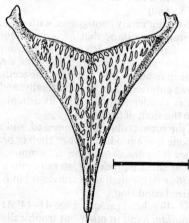

Fig. 19 Lower pharyngeal bone of *Gaurochromis (Gaurochromis) simpsoni,* in occlusal view. Scale = 3 mm.

A REVISION OF THE *HAPLOCHROMIS* GENERIC CONCEPT

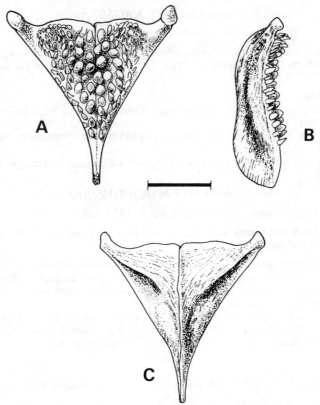

Fig. 20 Lower pharyngeal bone of *Gaurochromis (Mylacochromis) obtusidens* in. A. Occlusal. B. Right lateral. C. Ventral view. Scale = 5 mm.

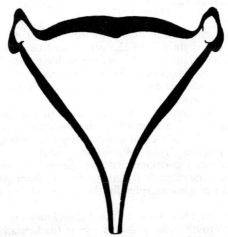

Fig. 21 Outline of lower pharyngeal bone, in occlusal view, of *Gaurochromis (Mylacochromis) obtusidens* (white) superimposed on that of *Labrochromis ishmaeli* (in black). Both bones are from adult specimens of the same standard length.

Subgenus *GAUROCHROMIS* nov.

TYPE SPECIES: *Haplochromis empodisma* Greenwood, 1960. Lake Victoria.

DIAGNOSIS. *Gaurochromis* species with a *slender, attenuated and fine lower pharyngeal bone* without molariform lower pharyngeal teeth. The dentigerous area of the bone is narrower than it is long.

Contained species
Gaurochromis (*Gaurochromis*) *empodisma* (Greenwood), 1960. Lake Victoria; see Greenwood (1960 : 262–6).
Gaurochromis (*G.*) *simpsoni* (Greenwood), 1965. Lake Nabugabo; see Greenwood (1965 : 325–9).
Gaurochromis (*G.*) *angustifrons* (Blgr.), 1914. Lakes Edward and George; see Greenwood (1973 : 177–83).

Subgenus *MYLACOCHROMIS* nov.

TYPE SPECIES: *Haplochromis obtusidens* Trewavas, 1928. Lake Victoria.

ETYMOLOGY. From the Greek *mylakris,* a millstone, + *chromis,* referring to the enlarged pharyngeal jaws and dentition.

DIAGNOSIS. *Gaurochromis* in which *the lower pharyngeal bone is enlarged and stout, with at least the two median tooth rows composed of enlarged and molariform teeth;* often with many additional teeth enlarged and molariform or submolariform, and others coarse and somewhat enlarged but still cuspidate.

The neurocranial apophysis for the upper pharyngeal bones has an expanded articular surface (both the parasphenoid and the basioccipital facets contributing to this enlargement), and strengthened, bullate lateral walls (especially the prootic component).

Contained species
Gaurochromis (*Mylacochromis*) *obtusidens* (Trewavas), 1928. Lake Victoria; see Greenwood (1960 : 266–9).

Research in progress at the University of Leiden, Netherlands, indicates that there is probably a second species of *Mylacochromis* in Lake Victoria (Dr C. D. N. Barel, *pers. comm.* and personal observations).

DISCUSSION
In its overall level of anatomical differentiation (except for its larger adult size), *Gaurochromis* departs but slightly from the generalized condition seen in *Astatotilapia* species. Even in its oral dentition (p. 33) *Gaurochromis* has retained the basic bicuspid tooth form and the basic dental pattern. Its derived dental features involve a reduction in tooth size (the teeth becoming finer, shorter and more slender) combined with an increase in the number of outer row teeth in both jaws.

Departure from the generalized haplochromine condition is also seen in the attenuated and fine lower pharyngeal bone in members of the nominate subgenus (p. 34) and, in the opposite direction, by the still relatively attenuate but greatly enlarged pharyngeal bone and molariform teeth in the subgenus *Mylacochromis* (see above).

For these various reasons I conclude that the species here grouped under the name *Gaurochromis* represent a monophyletic lineage distinct from *Astatotilapia,* with which genus it shares only features plesiomorphic for the Victoria–Edward–Kivu haplochromines in general.

When compared with the pharyngeal mill in *Labrochromis* (p. 40 and Fig. 21), that of *Gaurochromis* (*Mylacochromis*) species is less robust (particularly so in the undescribed taxon). This difference is apparent in specimens of all sizes, but it is more especially obvious in larger individuals.

Enlargement of the pharyngeal bones and dentition in *Gaurochromis* (*M.*) *obtusidens* prompted an earlier suggestion (Greenwood, 1960 : 265 & 268) of a possible relationship between this species and the genus *Labrochromis* (then known only from two species, *Haplochromis ishmaeli* and *H. pharyngomylus*). In turn, the resemblance between *Gaurochromis* (*M.*) *obtusidens* and *Gaurochromis* (*G.*) *empodisma* was thought to indicate a direct phyletic linkage between *Labrochromis* and the generalized *Astatotilapia* species of the Lake Victoria flock (Greenwood, 1960 : 265 & 268).

Later studies (Greenwood, 1974; and p. 42 below), indicated, however, that *Gaurochromis* (*G.*) *empodisma* and *G.* (*M.*) *obtusidens* belong to a distinct lineage and were unlikely to be linked with '*Haplochromis*' *ishmaeli* and '*H.*' *pharyngomylus* through recent common ancestry. This conclusion is apparently corroborated by the peculiar dental features of *Gaurochromis* species belonging to both subdivisions of the genus (see p. 33 above).

These features (and the form of the lower pharyngeal bone in *Gaurochromis* (*Gaurochromis*) species) also negate the idea that '*Haplochromis*' *erythrocephalus* (see p. 46) has a recent shared common ancestry with *Gaurochromis* (*G.*) *empodisma*, and also that '*Haplochromis*' *acidens* could be included, as a sister group, in the same major lineage (see Greenwood, 1974 : fig. 70). Incidentally, '*Haplochromis*' *erythrocephalus* is the '... small undescribed species' that was considered to be the nearest living relative of *G.* (*G.*) *empodisma* when that taxon was first described (Greenwood, 1960 : 265).

To summarize, *Gaurochromis* appears to be an independent lineage whose sister-group relationships cannot yet be determined. Its once supposed relationships with *Labrochromis* are no longer supported because the shared specialization, an enlarged pharyngeal mill, is more parsimoniously explained as the result of convergent evolution.

Within the nominate subgenus *Gaurochromis, G.* (*G.*) *angustifrons,* from Lakes Edward and George, has the most derived pharyngeal bone morphology (see Greenwood, 1973 : fig. 19) and also differs from its congeners, probably in an apomorphic way, in having a marked sexual dimorphism in adult size range, males being much smaller than females (Greenwood, 1973 : 182).

LABROCHROMIS Regan, 1920

TYPE SPECIES: *Haplochromis ishmaeli* Blgr., 1906 (not *Tilapia pallida* Blgr., 1911, as cited by Regan 1920, p. 45, footnote). Lake Victoria.

Regan (1920) apparently defined *Labrochromis* on the basis of a single specimen (BMNH 1911.3.3 : 132), a skeleton prepared from one of the paratypical series of *Tilapia pallida* Blgr., 1911. This specimen and one other paratype were misidentified by Boulenger (see Greenwood, 1960 : 275); both are clearly referrable to *Haplochromis ishmaeli* Blgr., 1906, a fact implicitly recognized by Regan in 1922. In that paper Regan remarks apropos of *H. ishmaeli*, '... The remarkable pharyngeal dentition might well be held to justify the genus *Labrochromis* (Regan, 1920) were it not that in all other characters the species is nearly identical with *H. cinereus*' (Regan, 1922 : 170).

Amongst the species synonymized with *H. ishmaeli*, Regan (1922 : 169) includes '... *Tilapia pallida* (part) Blgr., Cat. Afr. Fish. 3 : 231', but does not state specifically whether the skeletal preparation in question was included in that 'part'. However, judging from Regan's (1922) comments on *Labrochromis* quoted above, it seems reasonable to conclude that the skeleton was indeed included in the '*pallida*' material reidentified as '*ishmaeli*'.

The holotype of Boulenger's *Tilapia pallida* represents a quite distinct taxon (see Greenwood, 1960 : 233–6; and p. 43 below), one showing none of the diagnostic features for *Labrochromis* mentioned by Regan, nor any of those to be considered below.

DIAGNOSIS. Haplochromines characterized by *a massive hypertrophy of the pharyngeal mill (especially the lower pharyngeal bone and its dentition),* and having stout but generalized jaw teeth.

The lower pharyngeal bone is massive, relatively short and broad, the dentigerous surface concave, and the articular horns short and stout. Its dentition is composed almost entirely of stout molariform teeth; a few smaller submolariform or cuspidate teeth sometimes occur in the posterolateral angles of the dentigerous field, or as the teeth forming the perimeter of that field.

The neurocranium is of the generalized type but has a somewhat more decurved preorbital profile and a relatively higher supraoccipital crest whose outline is nearer pyramidical than wedge-shaped. *The apophysis for the upper pharyngeal bones is enlarged and stout, its expansive articular surface almost square in outline.* As compared with the generalized type of apophysis, the basioccipital facets make a much larger contribution to the articular area, and the walls of the apophysis (particularly the prootic part) are manifestly strengthened.

Outer jaw teeth, as compared with those in *Gaurochromis*, are coarser and less numerous (30–70, modal range 36–50, in the premaxilla).

Differences in the morphology and number of the jaw teeth, and the presence of a broader, more massive and more extensively molarized pharyngeal mill, are the features most readily distinguishing *Labrochromis* from *Gaurochromis*.

DESCRIPTION

Habitus (Fig. 22). In their overall appearance, members of the genus *Labrochromis* have a typically generalized facies, but most species do have a rather 'heavy headed' appearance.

Maximum adult size ranges from 90–140 mm SL.

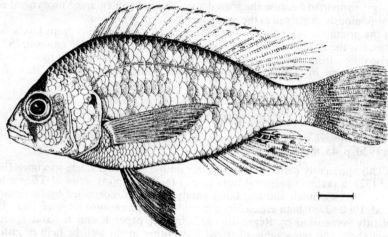

Fig. 22 *Labrochromis teegelaari.* Lake Victoria. Scale = 1 cm.

Neurocranium. The skull departs slightly from the generalized *Astatotilapia* type in having the preorbital profile gently curved (rather than straight), and the supraoccipital crest relatively high and near-pyramidical rather than wedge-shaped in profile. However, in one species (*L. ptistes*), the skull is more like the generalized kind in these features (see Greenwood & Barel, 1978 : fig. 20).

All *Labrochromis* have a stout and well-developed ventral articular apophysis for the upper pharyngeal bones, which exhibits relatively little interspecific variation in its form. Particularly noticeable are the expanded articular surface (almost square in outline), the enlarged parasphenoidal and basioccipital facets (Fig. 23), and the strengthened, somewhat bullate lateral walls (especially that part contributed by the prootics).

As the individual grows so the apophysis becomes relatively more massive. However, even in the smallest specimens examined, the basioccipital facets are larger, and the total area of the apophysis greater than in a similar-sized specimen from any other lineage (including

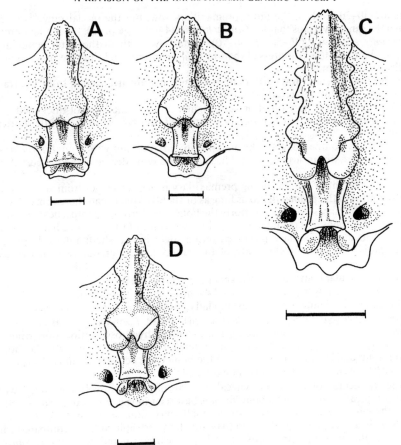

Fig. 23 Structure of the neurocranial apophysis for the upper pharyngeal bones (the pharyngeal apophysis) in various genera. A. *Gaurochromis (G.) empodisma* (typical of condition found in most haplochromine species). B. *Gaurochromis (M.) obtusidens* (see p. 36). C. *Labrochromis humilior* (the *Labrochromis* species with the least hypertrophied pharyngeal jaws in that lineage). D. *Labrochromis ishmaeli* (massive pharyngeal jaws and dentition). Scale = 3 mm.

Gaurochromis [p. 36 above], but excluding *Astatoreochromis*, see Greenwood, 1979b : 285–6 and also 1959a : 165–176; 1974 : fig. 44). This ontogenetic change in the apophysis is correlated with the size-related hypertrophy of the upper and lower pharyngeal bones and dentition.

Closest approximation to the *Labrochromis* apophyseal type is found in one species of the subgenus *Gaurochromis (Mylacochromis)*, see above, p. 36, but the differences, even if less well-marked, are nevertheless apparent.

Dentition. Teeth in the outer row of both jaws are mostly of the basic bicuspid type, moderately stout to stout, slightly recurved, the neck subcylindrical, and the crown not markedly compressed. The minor cusp is small and the major one equilateral in outline. Some unicuspid teeth (otherwise similar in their morphology to the bicuspids) occur in specimens of all species at a length of *ca* 70–80 mm, the proportion increasing with the fish's length. However, even in fishes ⩾ 100 mm SL, an exclusively unicuspid jaw dentition is uncommon.

There are 30–70 teeth in the outer premaxillary row, but the modal range is *ca* 36–50. Although the range overlaps that for *Gaurochromis* (44–82, see p. 33), the modal numbers in *Labrochromis* are lower (36–50 **cf**. 60–70). Outer jaw teeth in the latter genus are also stouter, stockier and less compressed than in *Gaurochromis*, and thus are more akin to the generalized tooth form.

Teeth forming the inner rows are small, tricuspid, and are arranged in from 1–3 (rarely 4) rows anteriorly and anterolaterally in both jaws.

Mouth. The mouth is horizontal or very slightly oblique, the lips not thickened, and the jaws equal anteriorly except in *L. humilior* where the lower jaw is usually a little shorter than the upper.

Upper jaw. As compared with the maxilla in *Gaurochromis*, that in *Labrochromis* is shorter and deeper, but its articular head has about the same degree of medial curvature (see p. 33).

The relative height of the ascending premaxillary processes ranges from shorter than the dentigerous arms of the premaxilla to as long as or slightly larger than those arms. The dentigerous arms are compressed, and anteriorly the bone is not produced into a beak or shelf.

Lower jaw. The dentary is slender and shallow, with almost the posterior half (sometimes a little less) of its alveolar surface produced into a slight lateral shelf; anteriorly there is no shelf-like projection because the body of the ramus merges gradually with the alveolar surface.

The anguloarticular complex is of the generalized type (see p. 6).

The length of the lower jaw is from 34–44% head length (modal range 37–40%), that is, within the generalized range and, at least modally, shorter than in *Gaurochromis*.

Lower pharyngeal bone and teeth (Fig. 24). In all *Labrochromis* species the lower pharyngeal bone is massive and strong, the extensive hyperossification imparting to the ventral surface a characteristic bulbous appearance (see Fig. 24C). Its articular horns are short and stout, and the posterior margin of the bone lying between them is always strongly convex, save for a slight median depression (Fig. 24C).

Because the degree of lower pharyngeal hypertrophy is positively correlated with the fish's size, 'typical' *Labrochromis* bone-form is best seen in specimens over 80 mm SL. But, even in the one species reaching only a small maximum adult size (*L. humilior*, *ca* 90 mm SL), the lower pharyngeal is seen to be much hypertrophied when compared with the bone from similar-sized specimens in other lineages (including *Gaurochromis* (*M.*) *obtusidens*; see p. 34). It also shows the characteristic bulbous ventral profile, and the short, stout articular horns characteristic of larger specimens (see Greenwood, 1960 : fig. 11; and 1974 : fig. 5C).

In some *Labrochromis* species the broadly triangular dentigerous surface is barely concave but in others it is markedly so, with a deep and extensive central pit (see Greenwood & Barel, 1978 : figs 26 & 30). The outline of the toothed surface is, relatively speaking, wider overall than in *Gaurochromis* (*Mylacochromis*) and, when viewed occlusally, it narrows gradually rather than rapidly from its maximum posterior width to the narrow anterior angle (Fig. 21). Generally the toothed area is as long as it is broad, but sometimes it is broader than long. In a few species, the surface, after beginning to narrow, actually broadens slightly at a point about two-thirds of the way along its antero-posterior length before it narrows again near the base of the short and deep anterior keel (see Greenwood & Barel, 1978 : figs 21 & 30).

Two outstanding features of the lower pharyngeal dentition in all *Labrochromis* species are the large size of the molariform teeth and, except in *L. humilior*, the extent to which pharyngeal molarization has proceeded. (These latter remarks, of course, are based on larger individuals; molarization is less pronounced in fishes < 60 mm SL).

Apart from a few compressed and bi- or unicuspid teeth situated in the posterolateral angles of the dental field, and the teeth comprising the perimeter series (*ie* the outermost one or two teeth in each transverse row), all the remaining teeth are molariform. Teeth within the central area are the most enlarged, and rarely show any trace of the low cusp that usually is

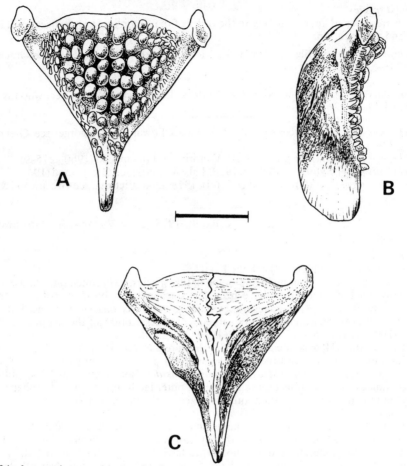

Fig. 24 Lower pharyngeal bone of *Labrochromis ishmaeli* in: A. Occlusal, B. Right lateral. C. Ventral view. Scale = 3 mm.

present on smaller molariform teeth. Loss of this cusp is at least partly attributable to wear, but even in newly erupted teeth it is insignificant.

Labrochromis humilior has proportionately fewer enlarged and molariform teeth, and these are restricted to the median rows. But, it must be stressed, these teeth are greatly enlarged, particularly when compared with the coarser teeth occurring in the median rows of the bone in other lineages (including comparable-sized specimens of *Gaurochromis* (*Mylacochromis*) *obtusidens*).

Specimens of *Labrochromis humilior* reach a maximum adult size of only *ca* 90 mm SL. When compared with like-sized individuals from other *Labrochromis* species the pharyngeal dentition is virtually identical (as is the degree to which the lower pharyngeal bone is hypertrophied). It is for these reasons (and because of its oral dentition) that I have included '*humilior*' in *Labrochromis* and not *Gaurochromis*.

All *Labrochromis* species have a correlated hypertrophy and molarization of the upper and lower pharyngeal elements.

Contained species

The taxa are grouped approximately in the order of their increasing pharyngeal mill hypertrophy and molarization.

Labrochromis humilior (Blgr.), 1911. Lake Victoria and the Victoria Nile; see Greenwood (1960 : 248–52).

Labrochromis ptistes (Greenwood & Barel), 1978. Lake Victoria; see Greenwood & Barel (1978 : 164–9).

Labrochromis mylodon (Greenwood), 1973. Lakes Edward and George; see Greenwood (1973 : 172–7).
Labrochromis ishmaeli (Blgr.), 1906. Lake Victoria; see Greenwood (1960 : 275–9).
Labrochromis pharyngomylus (Regan), 1929. Lake Victoria; see Greenwood (1960 : 270–5).
Labrochromis teegelaari (Greenwood & Barel), 1978. Lake Victoria; see Greenwood & Barel (1978 : 169–74).

Labrochromis mylergates (Greenwood & Barel), 1978. Lake Victoria; see Greenwood & Barel (1978 : 174–9).

Incertae sedis
Tilapia adolphifrederici Blgr., 1914. Lake Kivu.

I have been unable to examine the holotype of this species, a specimen once housed in the Berlin Museum but which may have been lost during the 1939–45 war. In his original description, Boulenger (1914) makes no reference to the lower pharyngeal bone of this fish, and it has not been mentioned in any subsequent redescription of the taxon (see Regan, 1921 : 637; Poll & David, 1937 : 259; Poll, 1939 : 9).

In the three BM(NH) specimens (one a skeleton), reg. nos: 1935.8.26 : 18–20, the pharyngeal mill is hypertrophied and the lower pharyngeal dentition is molarized (to an extent equaling that in *Gaurochromis* (*Mylacochromis*) *obtusidens* specimens of a comparable size). The morphology of the oral teeth, the number of outer teeth in both jaws, the shape of the dentary, and the proportions of the toothed surface on the lower pharyngeal bone are, however, of the *Labrochromis* and not the *Gaurochromis* types.

Judging from various published comments on this taxon (especially those of Regan, 1921; and Poll & David, 1937), it seems likely that at least two taxa have been confused under one name. Until more material, and the holotype, have been examined in detail, it seems inadvisable to place *Tilapia adolphifrederici* formally in *Labrochromis*. Nevertheless, the three BM(NH) specimens noted here can be referred to that genus.

Haplochromis placodus Poll, 1939, from the river Molindi, near Lake Kibuga, Zaire (Lake Edward drainage basin).

This species is known only from the holotype, and thus little detailed information is available on its anatomy. Considering the greatly enlarged and extensively molarized lower pharyngeal bone, and the nature of the oral dentition, the species probably should be included in *Labrochromis* (see also Greenwood, 1973 : 176).

DISCUSSION

Apart from the hypertrophied pharyngeal mill (and correlated modifications to the pharyngeal apophysis on the skull base) members of this lineage share no other derived features indicative of their monophyletic origin. Some doubt can even be cast in this instance on the hypertrophied pharyngeal mill being a true synapomorphy.

An enlarged lower pharyngeal bone, coupled with some degree of dental molarization, occurs in other lineages amongst the Victoria–Edward–Kivu haplochromines, and amongst haplochromine lineages from other areas as well. *Gaurochromis* (*Mylacochromis*) *obtusidens* is an example from the Victoria area, whilst *Astatoreochromis* species furnish examples from that region and beyond. Within the polyspecific lineage *Thoracochromis*, *Th. pharyngalis* and *Th. mahagiensis* are examples from Lakes Edward and Albert, whilst species of the *Serranochromis* subgenus *Sargochromis*, together with the monotypic *Pharyngochromis darlingi*, are examples from the more southerly parts of Africa (see Greenwood, 1979).

Because these lineages do not appear to be more closely related to one another (or to *Labrochromis*) than they are to any other lineage, and since in some cases (eg in *Thoracochromis*) the species with hypertrophied mills are related to others without that specialization, the evolution of this character must have occurred independently on a number of occasions.

Thus, in the absence of unifying synapomorphies uniquely shared by all *Labrochromis* species, one cannot consider the presence of an hypertrophied pharyngeal mill (and various correlated characters) as unequivocal indicators of monophyly for the genus.

Labrochromis (in particular the species *ishmaeli* and *pharyngomylus*), has, in the past, been considered a derived relative of *Gaurochromis* (*M.*) *obtusidens* (see Greenwood, 1954 : 412–13; in that discussion, for *H. michaeli* read *H. empodisma*, see Greenwood, 1960 : 262, 266 & 269).

Later (Greenwood, 1974 : 72–4), it was suggested that *Labrochromis* (as represented by *ishmaeli* and *pharyngomylus*) and *Gaurochromis* (represented by *empodisma* and *obtusidens*) probably belonged to separate lineages. This suggestion is apparently borne out by the dental and pharyngeal differences discussed above (p. 40), differences which are apomorphic features serving to distinguish all *Gaurochromis* from every *Labrochromis* species. This general situation would still hold even if *Labrochromis* proves to be a non-monophyletic assemblage (see above).

Amongst *Labrochromis* species, *L. humilior* (Lake Victoria), with its small adult size and moderate degree of pharyngeal development, seems to be the least derived taxon. *Labrochromis ptistes*, *L. ishmaeli*, *L. pharyngomylus*, *L. teegelaari* (all from Lake Victoria) and *L. mylodon* (Lakes Edward and George) are at approximately the same level of anatomical derivation, whilst *L. mylergates* (Lake Victoria) appears to be the most derived species in the genus (see Greenwood & Barel, 1978 : 176–7).

In a previous analysis of the Lake Victoria flock (Greenwood, 1974 : fig. 70), *Haplochromis pallidus* (now *Astatotilapia pallida*, p. 9) was considered to be the plesiomorph sister taxon of three species now included in *Labrochromis*. This supposed relationship was based on *pallida* having somewhat enlarged median teeth on its slightly enlarged lower pharyngeal bone, and on the overall dental and syncranial similarities shared with the other species. It is apparent that the latter features are symplesiomorphies (and thus of no value as phyletic indicators), and that the pharyngeal characters are of equivocal significance. Several species have pharyngeal features like those of *A. pallida*, but in none (including *A. pallida*) is the bone so hypertrophied, nor its teeth so extensively molarized as in *Labrochromis humilior*, the least derived member of that genus. In other words, there are no clear-cut synapomorphies allowing one to postulate a recently shared common ancestry between *Astatotilapia pallida* and *Labrochromis*; at best the available evidence is but faintly suggestive of such a relationship.

ENTEROCHROMIS gen. nov.

TYPE SPECIES: *Haplochromis erythrocephalus* Greenwood & Gee, 1969. Lake Victoria.

ETYMOLOGY. From the Greek *enteron*, the bowel, + *chromis*, referring to the long intestine in members of this lineage.

DIAGNOSIS. Small haplochromines (maximum adult size range 68–88 mm SL), with a generalized body form, head shape and syncranial skeleton, but *with a long, much coiled intestine that is at least 3 or 4 times longer than the standard length.*

From other haplochromines with a long intestine, *Enterochromis* is distinguished as follows:

From *Xystichromis* (p. 46), by its narrow bands of inner jaw teeth (1–3 rows) separated from the outer row by a distinct interspace, the crowns of the outer teeth distinctly broader than the neck of the tooth, and by having *the anterior opening to the nasal lateral line canal as large as the nostril.*

From *Neochromis* (p. 49) by its straight and sloping dorsal head profile (compared with a strongly decurved one), its narrow bands of inner teeth separated from the outer series by a distinct interspace, by its unequally bicuspid, and not equally or subequally bicuspid teeth, by its elongate and not foreshortened and laterally bullate dentary, its compressed and not inflated premaxillary dentigerous arms, by its gently sloping and not near-vertically aligned ethmovomerine skull region, and by having the opening to the nasal lateral line tubule as large as or larger than the nostril.

From *Haplochromis* (Greenwood, 1979 : 278–81), by not having the major cusp in the outer teeth drawn-out, compressed, expanded, and disproportionately larger than the minor cusp. It also differs in having no elements of the inner tooth rows similar in size or cusp form to teeth in the outer series, and in its nasal opening as large as, or larger than the nostril.

DESCRIPTION

Habitus and anatomy (Fig. 25). In most respects *Enterochromis* closely resembles *Astatotilapia*, and only those features distinguishing the two taxa (or which are developed to a different degree in *Enterochromis*) will be noted.

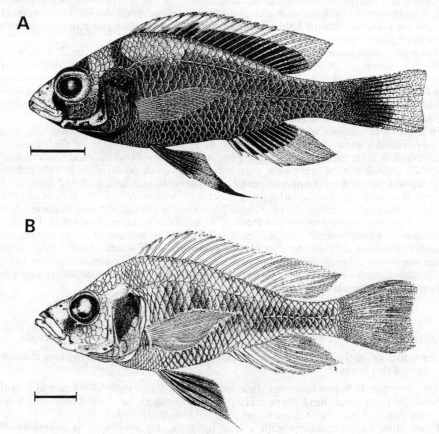

Fig. 25 A. *Enterochromis nigripinnis.* Lake George. B. *Enterochromis erythrocephalus.* Lake Victoria. Scale = 1 cm.

Neurocranium. One species (*E. erythrocephalus*) has a low preorbital skull depth (ca 24% neurocranial length), and the skull is narrow (otic width 50–51% neurocranial length); in the other three taxa these particular measurements are modal for the generalized skull type. (It should be noted that the skull is also narrow [ca 50% neurocranial width] in the otherwise generalized skull of some *Astatotilapia* species, but none has such a shallow preorbital depth as has *E. erythrocephalus*.)

Dentition. Compared with the modal condition in *Astatotilapia*, the teeth in *E. erythrocephalus* and *E. nigripinnis* are somewhat finer. Teeth in the other *Enterochromis* species, however, are of the typical *Astatotilapia* type. One or more outer teeth posteriorly in the premaxilla are enlarged and, generally, unicuspid, as they are in most *Astatotilapia* species. Also, as in that genus, some of the posterior outer teeth in both the premaxilla and the dentary are tricuspid.

No unicuspid teeth have been recorded from an anterior or anterolateral position in either jaw; possibly this is a consequence of the small adult size reached by members of the various species (see Greenwood, 1974 : 106).

There are 40–74 (modal range 50–56) teeth in the outer premaxillary row; the modal range for *Enterochromis* lies in the upper half of that for *Astatotilapia*, and the upper extremes of the *Enterochromis* range are rarely encountered in that genus.

Jaws. One *Enterochromis* species (*E. erythrocephalus*) has an oblique mouth, sloping upwards at an angle of 30°–35° (occasionally 40°) to the horizontal; the mouth in the remaining species is almost horizontally aligned.

Cephalic lateral line pores. All *Enterochromis* species, as far as I can determine, are outstanding amongst at least the Victoria–Edward–Kivu haplochromines in having the anterior opening to the nasal lateral line tubule as large (or almost as large) as the nostril. In other lineages the canal opening is much smaller.

Gut. The intestine in *Enterochromis* is long (ca 3–4 times the standard length) and much coiled. The folding is mostly in a horizontal plane, with 3 or 4 coils arranged below the elongate, greatly distensible stomach; posteriorly the intestine appears to be thrown into at least one vertical loop.

Lower pharyngeal bone and dentition (see also Greenwood & Gee, 1969 : 12–13, 21; and Greenwood, 1973 : 157). The bone is slender, with a triangular and equilateral dentigerous surface. Except for one species, all the teeth are fine, strongly compressed and of approximately the same size; even the posterior transverse row is made up of teeth only a little stouter than the others. In the exceptional species (*E. cinctus*), the teeth are somewhat coarser, and those in the two median rows are slightly stouter than the others. But, even in *E. cinctus* all the pharyngeal teeth are relatively finer and more compressed than those in any *Astatotilapia* species.

Again with the exception of *E. cinctus*, the teeth are numerous and close set, producing a coarse dental felt (coarse that is, compared with the dental felt in most tilapiine species of the genera *Tilapia* and *Sarotherodon*, but fine in comparison with the majority of haplochromine species).

In *E. cinctus* the teeth are more widely spaced and the dental felt is, as a result, coarser.

Contained species
The taxa are listed in approximately the order of their increasing derivation from the generalized condition.

Enterochromis cinctus (Greenwood & Gee), 1969. Lake Victoria; see Greenwood & Gee (1969 : 15–19).

Enterochromis paropius (Greenwood & Gee), 1969. Lake Victoria; see Greenwood & Gee (1969 : 10–15).

Enterochromis nigripinnis (Regan), 1921. Lakes Edward and George; see Greenwood (1973 : 151–9).

Enterochromis erythrocephalus (Greenwood & Gee), 1969. Lake Victoria; see Greenwood & Gee (1969 : 19–24).

DISCUSSION

The long, much coiled intestine of *Enterochromis* may indicate a somewhat distant common ancestry with the other phytophagous genera, namely, *Haplochromis, Xystichromis* and *Neochromis*. This question is discussed on p. 48.

Earlier attempts to relate *E. erythrocephalus* with the *Gaurochromis* lineage, in particular with *G. (G.) empodisma* (see Greenwood, 1974 : 66–7 and fig. 70; also Greenwood & Gee, 1969 : 23) can no longer be substantiated. The two taxa share no unequivocally derived features, and each has its own apomorph features which suggest relationships with other lineages (see p. 37).

Within the genus, *E. erythrocephalus* is the most derived species. Its narrow and preorbitally shallow skull, the fine and densely toothed lower pharyngeal bone, and the oblique mouth, are all characters contributing to that status, as are the high number (12) and fine shape of the gill-rakers (see Greenwood & Gee, 1969 : 20). Otherwise, little else can be said about intragroup relationships.

XYSTICHROMIS gen. nov.

TYPE SPECIES: *Chromis nuchisquamulatus* Hilgend., 1888. Lake Victoria; see Greenwood (1956*b* : 241).

ETYMOLOGY. From the Greek '*xyster*', one who scrapes, + *chromis*, alluding to the grazing habits of its member species.

DIAGNOSIS. Small haplochromines with a maximum adult size range of 85–105 mm SL, *a much coiled and long intestine (ca 3–4 times SL), and the broad bands (4–6 rows deep) of inner teeth anteriorly and anterolaterally in both jaws, narrowly, if at all separated from the outer tooth row.*

Neurocranium of the generalized type except that the *preorbital skull profile slopes more steeply and the supraoccipital crest is deeper and more pyramidical in shape.*

Teeth in the outer row of each jaw very close set (usually contiguous), moveably implanted, tall, and slender but strong, showing only a slight antero-posterior decline in their height and size. All (except for a few unicuspids posteriorly in the upper jaw) are unequally bicuspid, the minor cusp prominent but clearly smaller than the major one; the crown is not distinctly broader than the neck.

Inner row teeth are tricuspid, those of the outermost one or two rows almost as large as their counterparts in the outer row.

Lower pharyngeal bone without molariform or submolariform teeth; in some specimens the teeth of the median rows are enlarged and coarser than those of the lateral rows.

From other genera with long and coiled intestines, *Xystichromis* is distinguished as follows:

From *Neochromis*, by its gently sloping, not strongly decurved dorsal head profile, its unequally as opposed to equally or subequally bicuspid teeth, its elongate and not foreshortened dentary (which also is not bullate laterally), and by the ethmovomerine region of the skull sloping at an angle of 40°–50° to the horizontal and not almost vertically aligned.

From *Haplochromis*, particularly by its unequally bicuspid teeth, as compared with the very unequally bicuspid teeth in which the major cusp is protracted and compressed, and the minor cusp is virtually or entirely suppressed.

From *Enterochromis* it is distinguished by the characters listed on p. 43.

DESCRIPTION

Habitus (Fig. 26). There is little to differentiate members of this genus from *Astatotilapia* species. Maximum adult size range is 86–105 mm SL; no information is available on the size at which sexual maturity is reached.

Neurocranium. Skull form in *Xystichromis* is essentially like that in *Astatotilapia* except that the preorbital skull region slopes more steeply, and the supraoccipital crest is relatively deeper and more pyramidical in outline.

A REVISION OF THE *HAPLOCHROMIS* GENERIC CONCEPT

Dentition. It is the dentition of *Xystichromis* which provides the greatest number of derived features and, indeed, the synapomorphies uniting members of the lineage.

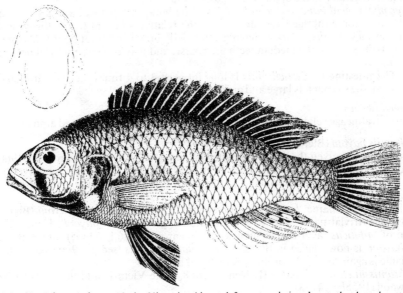

Fig. 26 *Xystichromis bayoni* Lake Victoria. About 1·3× natural size. Inset: the dental pattern of the upper and lower jaws (about 2·5× natural size).

Outer row teeth in both jaws are very close-set (contiguous or partially continguous), moveably implanted and tall, with only a slight antero-posterior decline in their height and overall size. Apart from one to three enlarged caniniform teeth posteriorly in the premaxilla, and the occasional intercalation of a tricuspid tooth posterolaterally, the teeth are all unequally bicuspids. The acutely pointed minor cusp is prominent but distinctly smaller than the major cusp, which has a somewhat obliquely slanting posterior margin and a broadly acute tip. Many teeth have this posterior margin produced into a low flange over part of its length; such flanges are known from members of other lineages as well (*eg Astatotilapia macropsoides, A. elegans* and *A. aeneocolor,* from Lake George (Greenwood, 1973), and *Gaurochromis empodisma* in Lake Victoria); their significance is not known.

Although relatively slender, the outer teeth in *Xystichromis* are robust and but slightly recurved. They also differ from the generalized bicuspid tooth in not having the crown distinctly broader than the neck and body of the tooth; as a result, the anterior and posterior margins of the entire tooth are almost parallel. The crown is also somewhat more compressed than in the generalized bicuspid tooth, although it could not be described as flattened.

There are 36–70 (modal range 50–65) teeth in the outer premaxillary series.

Because the insertion line of the outer teeth is lower than that of the inner teeth, the crowns of the teeth in both series are, effectively, at the same level. Presumably this is a feature associated with the algal-grazing habits of the known species.

Although the modal number of inner tooth rows in any one species is elevated in comparison with the generalized condition, the lower end of the total range (2–8) does overlap that for *Astatotilapia*, but *Xystichromis* individuals with only 2 or 3 inner rows are uncommon.

Upper jaw. Compared with the generalized type of premaxilla, that in *Xystichromis* has a broader alveolar surface and the dentigerous arms are slightly inflated anteriorly and anterolaterally.

Lower jaw. The dentary is not deep and foreshortened (as in *Neochromis,* see p. 51), but neither is it as slender as the dentary in *Astatotilapia.*

Lower pharyngeal bone and dentition. Both the bone and its teeth are of the generalized type. There is some intrageneric variation in the relative width of the dentigerous surface, with one species, *X. phytophagus,* having a distinctly broad and stout bone (see Greenwood, 1966 : 304–6). Some of the median teeth are coarse, and may even be enlarged posteriorly in *X. bayoni.*

Gut. The intestine in *Xystichromis* is long (at least 3 to 4 times the standard length) and much coiled; the stomach is large and greatly distensible.

Contained species

Since no intralineage relationships can be determined the species are listed alphabetically.

Xystichromis bayoni (Blgr.), 1911. Victoria Nile.

This species was given the replacement trivial name '*niloticus*' by Greenwood (1960 : 243) who, disagreeing with Regan's (1922 : 169) idea that the taxon was synonymous with *Haplochromis humilior,* resurrected it to full specific status within the genus *Haplochromis* as then defined.

At that time the name '*bayoni*' was preoccupied by *Haplochromis bayoni* (Blgr.), 1909, and so a new trivial name was required for Boulenger's (1911) '*bayoni*'. Hence the introduction of '*niloticus*' as a replacement (see Greenwood, 1960 : 243–5). Since Boulenger's 1909 '*bayoni*' is now placed in the genus *Prognathochromis* (see p. 19 above), Boulenger's 1911 name is again available for the species listed here.

Xystichromis nuchisquamulatus (Hilgend.), 1888. Lake Victoria and the Victoria Nile; see Greenwood (1956*b* : 241–3).

Xystichromis phytophagus (Greenwood), 1966. Lake Victoria; see Greenwood (1966 : 303–9).

Discussion

Apart from its dental specializations and long, coiled intestine, *Xystichromis* is, anatomically speaking, a generalized haplochromine.

However, the dental pattern, the tall and slender but robust teeth, and the enlarged, broadbanded inner teeth do resemble those of *Neochromis,* a lineage with which *Xystichromis* also shares the derived feature of an elongate and much coiled gut (see p. 52).

It is possible, therefore, that these two genera share a common ancestor in which such dental and alimentary features are present, and that they should be ranked as sister taxa (but see also p. 52).

A long coiled gut is also present in *Haplochromis* and in *Enterochromis* (see p. 45) but neither genus shares all the dental synapomorphies common to *Xystichromis* and *Neochromis.* *Haplochromis* has a uniquely derived crown form in its outer jaw teeth (see Greenwood, 1979 : 278–9), but also has broad bands of teeth anteriorly and anterolaterally in the jaw. In that latter feature it shares an apomorph character with both *Neochromis* and *Xystichromis.* *Enterochromis* has an essentially plesiomorph haplochromine dentition (see p. 45).

If the shared apomorphy of a long, much coiled gut really does indicate a common ancestry for all four genera, then the various dental specializations suggest that *Xystichromis, Neochromis* and *Haplochromis* are more closely related to one another than any one is to *Enterochromis.* In other words, *Haplochromis,* because of its greater dental specializations, is the sister taxon of *Xystichromis* and *Neochromis* combined, and *Enterochromis,* because of its relatively generalized dentition is the plesiomorph sister group to all three.

An acceptance of this solution is complicated by various derived syncranial features which *Neochromis* alone shares with a number of other lineages, and which might therefore indicate its relationship to them (the dental features being then taken as parallelisms). However, the first set of relationships proposed above is the more parsimonious since its resolution involves fewer and less profound dental and anatomical changes (see discussion p. 52).

Intrageneric species grouping within *Xystichromis* is not possible on the basis of available data; no two species appear more closely related to one another than either does to the third member of the lineage.

NEOCHROMIS Regan, 1920

TYPE SPECIES: *Tilapia simotes* Blgr., 1911 (see Regan, 1920 : 45), now considered a junior subjective synonym of *Tilapia nigricans* Blgr., 1906 (see Greenwood, 1956*b* : 237).

DIAGNOSIS. Small haplochromines (maximum adult size 95 mm SL), with *a very strongly decurved dorsal head profile (sloping at 70°–80° to the horizontal), a long, much coiled intestine (ca 3–4 times SL), broad bands of inner teeth anteriorly and anterolaterally in both jaws, not separated from the outer row, and equally or subequally bicuspid outer teeth.*

Neurocranium with a strongly decurved preorbital face, the ethmovomerine region almost vertically aligned.

Dentary markedly foreshortened, deep and stout, its anterior margin strongly curved medially so that the anterior outline of the lower jaw is almost rectangular. The region of the dentary surrounding its posterior division into coronoid and horizontal limbs is markedly bullate. Length of lower jaw 30–38% head length, modal range 34–36%.

The anguloarticular complex of the lower jaw is stout, *the anterior point of its antero-ventral arm blunt or rectangular* (never acute).

Premaxilla with noticeably inflated dentigerous arms, almost cylindrical in cross section, the alveolar surfaces broad; its ascending processes as long as the dentigerous arms.

Outer jaw teeth tall, slender but robust, and without any marked antero-posterior decline in height; moveably implanted and very close set (contiguous). Cusp form characteristic, the minor (*ie* posterior) cusp well-developed and only a little smaller than the major (*ie* anterior) cusp, from which it is separated by a narrow notch; the points of each cusp are almost spatulate. The crown is compressed relative to the cylindrical neck and body of the tooth (but is not flattened), and is broader than the body.

Inner tooth rows numerous (3–8, usually 5 or 6), the teeth in a row contiguous, and the rows close set; there is no discrete interspace between the outermost row and the outer row of teeth. Inner teeth tricuspid (with all cusps of about the same height), the teeth in the two outermost rows distinctly larger than those of the innermost rows.

Teeth in the median rows on the lower pharyngeal bone relatively stout in two species, not so in the remainder.

Features distinguishing *Neochromis* from the other taxa with long, coiled, intestines are listed on p. 44 for *Enterochromis*, and p. 46 for *Xystichromis*.

DESCRIPTION

Habitus (Fig. 27). The strongly decurved and steep dorsal head profile, and the near-horizontal mouth, combine to give the pug-headed appearance which is so characteristic a feature of all *Neochromis* species. None reaches a large adult size (*ca* 95 mm SL).

Neurocranium. The preorbital face of the skull is very strongly decurved (sloping at an angle of *ca* 70°–80° to the horizontal), the ethmovomerine region is aligned almost vertically and its tip extends ventrally to a point below a horizontal through the anterior part of the parasphenoid (see Fig. 28). The supraoccipital crest is of variable outline and relative height, but it is always somewhat deeper and less wedge-shaped than in a generalized skull.

Dentition. Teeth in the outer row of both jaws are tall (and without any marked antero-posterior decline in height), slender but strong, are moveably implanted and so close-set as to be contiguous. There are 40–70 teeth (modal range 50–56) in the outer premaxillary row.

One species (*N. nigricans*) has the last, or sometimes the last two teeth in the premaxillary row enlarged and unicuspid, but usually in that species, and in the other two *Neochromis* species, the posterior teeth are like the others.

Fig. 27 *Neochromis nigricans.* Lake Victoria. Scale = 1 cm.

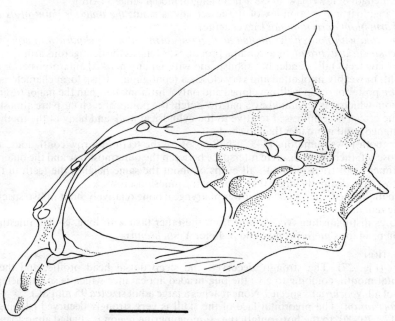

Fig. 28 Neurocranium of *Neochromis nigricans*; left lateral view. Scale = 3 mm.

Cusp form in *Neochromis* is very characteristic (Fig. 29). The inner (*ie* posterior) cusp is well-developed, often only a little smaller than the major (*ie* anterior cusp), and is directed obliquely backwards (rather than vertically upwards). The points of both cusps are spatulate or somewhat acutely spatulate (but never acute). The crown is compressed (relative to the cylindrical neck and body of the tooth) but is by no means flattened; it is also broader than the neck so that the margins of the tooth are not parallel (**cf.** *Xystichromis* p. 47).

Fig. 29 Outer row teeth from the dentary of *Neochromis nigricans*, viewed labially. Scale = 1 mm.

Inner tooth rows are composed of moveably implanted tricupsid teeth in which the two lateral cusps are of almost the same height and width as the median one. Those teeth forming the two outermost rows in each jaw are enlarged, and even those in the remaining rows are relatively larger than their counterparts in other lineages except *Xystichromis*; all, however, are shorter than the outer row teeth. There are 3–8 (modes 5 or 6) rows of inner teeth anteriorly and anterolaterally in each jaw, and one or two rows laterally. Teeth in these rows are contiguous, and the rows themselves are close set so that only a very narrow interspace separates them; the interspace between the outermost row of the inner series and the outer tooth row is barely discernible. Thus, as compared with *Xystichromis*, although the area covered by the inner tooth rows is almost the same, there are more rows of teeth in *Neochromis*, and the rows are set much closer together (see p. 47). As in *Xystichromis*, tooth insertion levels are such that the crowns of the inner and outer row teeth are at the same level despite the inner teeth being slightly shorter.

Mouth. The mouth is horizontal and the jaws equal anteriorly; the lower jaw is broad and its anterior outline, when viewed from below, is almost rectangular.

Upper jaw. The premaxilla has notably inflated dentigerous arms, oval to near-circular in cross section anteriorly and anterolaterally, but somewhat more compressed posteriorly. Compared with the generalized premaxilla, that in *Neochromis* has a broader alveolar surface, and the ascending processes are longer (as long as the dentigerous arms).

Lower jaw. The dentary is deep relative to its length, and has the appearance of being a stout, foreshortened bone (Fig. 30). Anterolaterally, each ramus curves inwards rather abruptly so that the anterior margin of the entire jaw is rectangular.

The anguloarticular complex is stout, with the anterior point of its anteroventral arm blunt or rectangular in outline.

The crown of the coronoid process (the ascending arm) on the dentary has a slight but definite medial inflection.

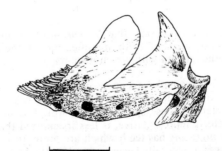

Fig. 30 Dentary of *Neochromis nigricans*, left lateral view. Scale = 3 mm.

Lower jaw length in *Neochromis* (30–38% head length, modal range 34–36%) broadly overlaps that in the majority of non-piscivore lineages with small-sized adults (*ie* <110 mm SL), although the modal length in several of these species is higher (40–44%) than in *Neochromis*. Lower jaw width in *Neochromis* (1·0–1·4, mode 1·2 times longer than broad) also overlaps that in lineages with a generalized syncranial morphology, but modally it is distinctly greater in *Neochromis*.

Lower pharyngeal bone and dentition. The dentigerous surface is triangular and as broad as it is long. In *Neochromis nigricans* and *N. serridens* the teeth are fine and compressed bicuspids, densely arranged on the bone, with only the posterior transverse row composed of stouter teeth; *N. fuscus*, however, has all the pharyngeal teeth relatively coarser, less closely set and with the two median rows composed of somewhat enlarged (but cuspidate) teeth.

Gut. The intestine is long (at least 3–4 times longer than the standard length) and much coiled; the stomach is large and distensible.

Contained species
For the possible interrelationships of these taxa, see below.

Neochromis nigricans (Blgr.), 1906. Lake Victoria and the Victoria Nile (*Tilapia simotes* Blgr., 1911, the type species of *Neochromis*, is currently considered to be a junior synonym of this species; see Greenwood, 1956*b* : 237–40).
Neochromis serridens (Regan), 1925. Lake Edward; see Trewavas (1933 : 327–8).
Neochromis fuscus (Regan), 1925. Lake Edward; see Trewavas (1933 : 329).

DISCUSSION

In all *Neochromis* species certain derived syncranial features, such as the strongly decurved preorbital skull, the deep, foreshortened dentary, the stout anguloarticular complex, the laterally bullate dentary, and the inflation of the premaxillary arms, are shared either *in toto*, in part or in varying degrees of expression, with several lineages (*Macropleurodus, Ptyochromis, Hoplotilapia, Platytaeniodus, Paralabidochromis, Lipochromis, Schubotzia* and *Allochromis*).

It is difficult to assess the significance of this situation. Other synapomorphies occurring in these taxa, but not in *Neochromis*, point to further groupings that can be made amongst them (see discussions on pp. 92–94) and suggest that the syncranial features shared with *Neochromis* are, at most, an indication of distant (rather than recent) common ancestry. Alternatively, the syncranial synapomorphies could be parallelisms associated with the independent evolution of strong jaws and dentition (often multiseriate), or of a multiseriate dentition alone.

The latter interpretation must be invoked if the hypothesized relationship between *Neochromis, Haplochromis, Enterochromis* and *Xystichromis*, put forward on p. 48, is accepted. It should be stressed that this relationship, unlike that associating *Neochromis* with *Macropleurodus, Ptyochromis* etc, does not require the independent evolution of a long gut (and presumably associated physiological changes) in *Neochromis*, nor the unique development of its dental type within a 'lineage' having totally different dental specializations and a simple gut form.

Thus, it is more parsimonious to propose that *Neochromis* is related to the other species with long, coiled intestines and bicuspid teeth, modified though the teeth may be in some species, than to the eight other genera with which it shares some syncranial specializations (see pp 48–49).

Unfortunately there are few available specimens of *Neochromis serridens* and *N. fuscus*, the Lake Edward representatives of the genus. As a result, little is known about the range of variation in critical characters in these species, but it would seem that the dentition of *N. nigricans*, the Lake Victoria representative, is less specialized than that in the Edward species. Since *Neochromis nigricans* has teeth which are more unequally bicuspid than in the others, and its inner rows are generally fewer in number, the Edward taxa would seem to be more closely interrelated than either is to *N. nigricans*.

HAPLOCHROMIS Hilgendorf, 1888

TYPE SPECIES: *Chromis (Haplochromis) obliquidens* Hilgendorf, 1888. This now much impoverished genus is redescribed in the first part of this paper (see Greenwood, 1979 : 278–81).

Contained species
The taxa are grouped and listed in order of their increasing derivation.

Haplochromis limax Trewavas, 1933. Lakes Edward and George; see Greenwood (1973 : 167–72).

Haplochromis annectidens Trewavas, 1933. Lake Nabugabo; see Greenwood (1965 : 329–35).
Haplochromis lividus Greenwood, 1956. Lake Victoria; see Greenwood (1956*b* : 232–7).

Haplochromis astatodon (part) Regan, 1921. Lake Kivu; see Greenwood (1979 : 280).
Haplochromis obliquidens Hilgendorf, 1888. Lake Victoria; see Greenwood (1956*b* : 226–32).

PSAMMOCHROMIS gen. nov.

TYPE SPECIES: *Pelmatochromis riponianus* Blgr., 1911 (as redefined by Greenwood, 1960 : 252–6). Lake Victoria.

ETYMOLOGY. From the Greek '*psammos*', sand, + *chromis*, referring to the sandy substrata seemingly preferred by most members of the genus.

DIAGNOSIS. Haplochromines reaching a maximum adult size of 100–123 mm SL, the body relatively slender to moderately deep (31–43% SL, modal range 36–38%); lips thickened in all species, the lower lobate in one.

Neurocranium of a near-generalized type but *shallower in the otico-occipital region (40–46% neurocranial length)*.

Outer jaw teeth tall and slender, their crowns recurved and either compressed or finely acuminate and cylindrical in cross-section. Very unequally bicuspid teeth present in specimens of all sizes, but unicuspids predominate in fishes >80–90 mm SL; 24–68 teeth in the outer premaxillary row.

Inner teeth tall and slender, tri- or unicuspid, and usually implanted so as to lie almost horizontally; commonly arranged in 2–4 rows anteriorly.

Premaxilla with a definite anterior beak, its ascending processes longer than the dentigerous arms, which have a slight ventral decurvature over the posterior half (more marked in some species than in others).

Dentary with a very distinctive form, each ramus noticeably inflated anteriorly and antero-laterally, this circumscribed swelling extending almost to the bone's ventral profile. Over this region (and slightly behind it) the narrow outer margin of the alveolar surface dips distinctly downward so that the outer tooth row also has a ventral inflection (see Fig. 32).

Lower pharyngeal bone in some species moderately stout, its median teeth enlarged and submolariform, but the bone slender and without enlarged teeth in others.

DESCRIPTION
Habitus (Fig. 31). There are few outstanding features in the habitus of most *Psammochromis* species. The body varies from relatively slender to moderately deep (31–43% SL, modal range 36–38%), the dorsal head profile is straight or gently curved, and slopes fairly steeply. All species have thickened lips, and the lower lip may be lobate in *P. aelocephalus*, which species also has a highly variable snout form (noticeably protracted in some individuals; see Greenwood, 1959*b* : 214–17).

Maximum adult size ranges from 100–123 mm SL; individuals reach sexual maturity at a length of 80–85 mm in those species attaining the larger maximum sizes.

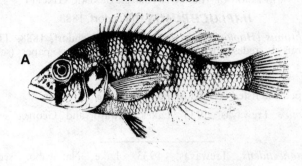

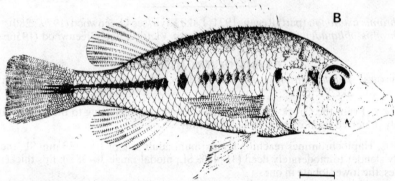

Fig. 31 A. *Psammochromis riponianus*. Lake Victoria. About two-thirds natural size. B. *Psammochromis cassius*. Lake Victoria. Scale = 1 cm.

Neurocranium. The skull is of a near-generalized type but has a lower otico-occipital region (depth 40–46% neurocranial length) so that the preotic dorsal profile slopes downwards and forwards fairly gently; the supraoccipital crest is relatively low and is wedge-shaped in profile.

Although in one species (*P. riponianus*) some individuals have the pharyngeal mill enlarged to a degree comparable with that in *Gaurochromis* (*Mylacochromis*) *obtusidens* (see p. 34), there is no corresponding enlargement of the pharyngeal apophysis on the skull base. As compared with those congeneric species not having an enlarged pharyngeal mill, the parasphenoid contribution to the apophysis in *P. riponianus* is slightly more expansive, but the basioccipital facets are barely larger.

Dentition. The outer row jaw teeth are slender and tall, the body and recurved crowns either relatively compressed (only the lower part of the tooth cylindrical in cross-section) or the crown is finely acuminate and the whole tooth cylindrical in cross-section and very slender. Species with the latter type of teeth have the teeth widely spaced, especially in the lower jaw.

When bicuspid, the coarser type of tooth has the minor cusp greatly reduced, the major cusp vertically protracted and pointed; bicuspid forms of the finely acuminate unicuspids are as yet unknown.

Some bicuspid teeth are present in most specimens of all other species, but predominate only in fishes <90 mm SL; above that size, unicuspids and, or, weakly bicuspid teeth are more frequent. *Psammochromis cassius* is unusual in having only unicuspid teeth present in specimens as small as 70 mm SL, and, apparently, in having some lower teeth lying outside, or even penetrating into, the upper lips (see Greenwood & Barel, 1978 : 162).

A characteristic feature of two species from Lake Victoria (*P. riponianus* and *P. saxicola*), is the very abraded crowns on most outer teeth, which then appear bluntly incisiform.

Inner teeth are also slender and tall, tricuspid in small individuals but unicuspid in larger fish, implanted so as to lie almost horizontally, and generally embedded deeply in the oral mucosa (Greenwood, 1960 : 254). There may be as many as 5 rows of inner teeth anteriorly in each jaw; the modal numbers are, however, 2–4.

Mouth. Lips are clearly and equally thickened, but in one species (*P. aelocephalus*) the lower lip may be produced anteriorly into a small but definite mental lobe. The mouth is horizontal or but slightly oblique; this, combined with the thickened lips and particular head profile, impart to the members of this lineage a very characteristic but indefineable physiognomy (see Fig. 31; also figs 13 & 14 in Greenwood, 1960; and figs 20 & 21 in Greenwood, 1973).

Upper jaw. The premaxilla is somewhat expanded and protracted anteriorly and antero-medially into a definite beak or peak. Its ascending processes are longer than the dentigerous arms, which have a slight but distinct ventral curvature over their posterior halves (the curve more marked in some species than in others).

The maxilla is elongate and slender in *P. saxicola* but relatively foreshortened and deep in the other species. In none is the medial face of the posterior arm strongly concave (and thus the lateral aspect is but slightly bullate), and none has a marked medial curvature of its articular head.

Lower jaw. The most trenchant and diagnostic synapomorphy linking members of this lineage lies in the morphology of the dentary (Fig. 32).

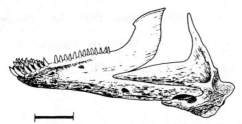

Fig. 32 Dentary of *Psammochromis saxicola* in left lateral view. Scale = 3 mm.

Anteriorly and anterolaterally the bone immediately below the alveolar surface is distinctly inflated, the well-circumscribed swelling extending almost to the ventral margin of each ramus. Over this swollen region, and a little behind it as well, the narrow outer margin of the alveolar surface dips downwards so that the line of outer teeth also dips ventrally in that region of the jaw. Consequently the tips of the outer teeth are on a level with those of the much smaller inner teeth.

The dentary also departs from the generalized type in being relatively shallower. Thus, although the length of the entire lower jaw (dentary + anguloarticular) is within the 'generalized' range (33–49% head length) it gives the impression of being much more slender and attenuated.

No mental protuberance is developed at the dentary symphysis. Indeed, the ventral symphysial profile slopes backwards so that the jaw appears 'chinless', except for a slight vertically directed ventral projection at the symphysial base.

Lower pharyngeal bone and teeth. There is considerable inter- and some intraspecific variation in the stoutness of the bone, and there are correlated differences in the nature of its dentition (see Greenwood, 1959 : 216; 1960 : 254 & 258, and figs 4 & 5).

All or some of the teeth in the four median rows may be enlarged and are often molariform or submolariform. The remaining teeth, and sometimes those of the median rows as well, are weakly cuspidate and compressed (except, as is usual, for the robust teeth in the posterior transverse row).

Squamation. Except in two species, the scales anteriorly and ventrally on the chest region are distinctly smaller than those on the ventral flanks and belly, and appear to be more deeply embedded. There is, however, no abrupt size change between the scales of the two regions, which grade imperceptibly into one another (see Greenwood, 1979 : 270–2).

In the exceptional taxa (*P. acidens* and *P. cassius*) the chest scales are not obviously smaller than the belly scales, and do not give the appearance of being deeply embedded.

Contained species

The taxa are listed in order of their increasing apomorphy.

Psammochromis graueri (Blgr.), 1914. Lake Kivu.

Amongst the BM(NH) material identified as *graueri*, only three specimens (BMNH reg. nos: 1914.4.8 : 16, 19 & 20) are apparently conspecific. The concept of *graueri* used in this paper is thus based on those specimens, of which one, reg. no. 1914.4.8 : 20, a fish 99·0 mm SL, is chosen as the lectotype of the species.

Psammochromis schubotzi (Blgr.), 1914. Lakes Edward and George; see Greenwood (1973 : 183–8).

The possible relationship between *P. schubotzi* and *A. schubotziella* (see p. 9) suggested in that paper can no longer be upheld since the latter taxon shares none of the derived features shown by *P. schubotzi*.

Psammochromis riponianus (Blgr.), Lake Victoria and probably the Victoria Nile; see Greenwood (1960 : 252–6)

Psammochromis saxicola (Greenwood), 1960. Lake Victoria, and probably the Victoria Nile; see Greenwood (1960 : 256–9).

Psammochromis aelocephalus (Greenwood), 1959. Lake Victoria; see Greenwood (1959b : 214–17).

Psammochromis acidens (Greenwood), 1967. Lake Victoria, and probably the Victoria Nile; see Greenwood (1967 : 73–7).

Psammochromis cassius (Greenwood & Barel), 1978. Lake Victoria; see Greenwood & Barel (1978 : 161–4).

DISCUSSION

Previously (Greenwood, 1974 : fig. 70), most of the species included in *Psammochromis* were thought to be related to a lineage comprising, amongst other taxa, those now divided between the genera *Harpagochromis* and *Prognathochromis* (including its subgenus *Tridontochromis*); the *Psammochromis* species were considered to be most closely related to the latter group (the so-called *tridens* complex). In turn, the '*tridens* complex' and *Psammochromis* (then the '*riponianus* complex') were ranked as the sister group of a species (*H. welcommei*) here classified as the monotypic genus *Allochromis* (see p. 57).

Anatomical and osteological information now available for all these various taxa renders that hypothesis, based chiefly on neurocranial architecture, untenable in its entirety but not in part.

Skull form in *Psammochromis* is of a near-generalized type, and its derived dental features are unlike those in either subgenus of *Prognathochromis*.

The form of the dentary in *Psammochromis* exhibits apomorphies not shared with *Prognathochromis*, and there are no obvious synapomorphies linking *Psammochromis* with the dentally specialized *Allochromis* (the former *H. welcommei*).

However, there are certain derived features in the morphology of the dentary in *Psammochromis* which might still indicate its relationship with *Allochromis*, an argument which is taken up later (p. 60). These same features may also indicate a shared common ancestry with *Macropleurodus*, *Paralabidochromis* and *Ptyochromis* (see p. 66). In other words, *Psammochromis* and *Allochromis* together may constitute the sister group of the three other genera (see also pp 92–94).

Psammochromis acidens previously was given, tentatively, the status of sister group to the '*empodisma–obtusidens*' lineage, that is, the genus *Gaurochromis*; p. 32 (Greenwood, 1974).

Now that skeletal material of *P. acidens* is available it is clear that the lower jaw morphology in this species is far removed from that in *Gaurochromis*. Likewise, a possible relationship of *P. cassius* with the '*serranus* group' (that is, *Harpagochromis*), as suggested by Greenwood & Barel (1978 : 164), is not supported by the peculiar morphology of its lower jaw; both *P. acidens* and *P. cassius* have the distinctive and derived type of dentary characterizing the genus. *Psammochromis cassius* and *P. acidens* do, however, depart from other members of the lineage in having both a very different tooth form (see p. 54), and chest scales which are not noticeably smaller than those on the ventrolateral flanks and belly. The latter feature must be ranked as plesiomorphic, the dental one as derived. Possibly the two species together represent a subgroup within the lineage, but more material must be studied before they are formally recognized as such.

ALLOCHROMIS gen. nov.

TYPE SPECIES: *Haplochromis welcommei* Greenwood, 1966. Lake Victoria.

ETYMOLOGY. From the Greek '*allos*', different, strange + *chromis*, alluding to the unusual tooth shape and dental pattern, as well as to the lepidophagous habits of the type species.

DIAGNOSIS. Haplochromine fishes having an adult size range of *ca* 80–105 mm SL, a shallow, streamlined body (depth 30–33% SL), a gently decurved dorsal head profile (sloping at *ca* 35°–40°), a horizontal mouth and slightly thickened lips (Fig. 33). In general, the habitus is like that of many *Prognathochromis* species. *Allochromis* is, however, immediately distinguished by its dental morphology and the wide, broadly crescentic bands of fine teeth.

Teeth in the outer row of both jaws are close set and *have a very slender, tall and near-cylindrical neck which expands abruptly into a compressed, bicuspid crown which is about twice as broad as the neck; the crown and upper third of the neck are strongly recurved and lie almost at right angles to the rest of the tooth* (see Fig. 34). Inner row teeth are mostly tricuspid, the cusps of approximately equal size, and the whole crown strongly recurved.

Both jaws have the teeth arranged in a broad, almost crescentic band extending nearly to the posterior limits of the dentigerous surfaces involved (Fig. 35).

DESCRIPTION
Habitus (Fig. 33). A shallow, streamlined body combined with the gently curved and sloping head profile give this taxon a very *Prognathochromis*-like appearance (**cf.** Fig. 8). The mouth is horizontal and the lips are slightly thickened.

Adult size range for the few specimens known is *ca* 80–105 mm SL.

Neurocranium. Overall skull shape closely approaches that in species of the *Psammochromis* lineage; that is, a near-generalized neurocranium but with somewhat shallower otico-occipital and pre-orbital regions (*ca* 25% and 45% of neurocranial length, respectively). The supraoccipital crest is relatively low and wedge-shaped in profile.

Fig. 33 *Allochromis welcommei*. Lake Victoria. Scale = 1 cm.

Dentition. Outer row teeth have a very characteristic shape and cusp form (see Fig. 34). In lateral view the tooth has a very slender, nearly cylindrical and tall neck which expands rather abruptly, but equally on either side, into a compressed bicuspid crown. The crown is almost twice as wide as the body of the tooth, its outline resembling a stylized drawing of a tulip. Both the crown and the upper third of the neck are strongly recurved.

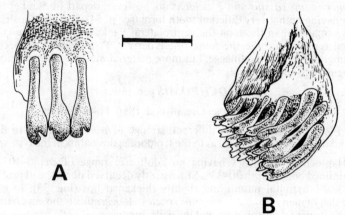

Fig. 34 Teeth from the left premaxilla of *Allochromis welcommei*. A. Labial aspect, viewed anteriorly. B. Ventromedial aspect viewed somewhat anteriorly. Scale = 1 mm.

Most outer teeth are unequally bicuspid, but there are a few tricuspids posteriorly in the row. The major cusp has a sub-acuminate, almost rounded distal margin; the minor cusp is more acute but is by no means pointed. In life these teeth, and those of the inner series, are moveably attached to the jaw.

Teeth in the inner series have the same overall shape as the outer ones but the crown is generally tricuspid, although bicuspids do occur frequently in the outermost row. There is little difference in the size of the three cusps, but the middle one is slightly higher and broader; their distal margins are sub-acuminate. All inner teeth are strongly recurved.

A very gradual size gradient exists across the inner rows; teeth in the outermost row are almost as tall as those in the outer series.

There are 70–80 close-set teeth in the outer premaxillary row, the margins of their cusps contiguous or slightly overlapping.

The dental pattern in both jaws is of a highly derived and distinctive type (see Fig. 35). The teeth are set out in broad crescentic bands which are not confined to the anterolateral parts of the jaw, but extend almost to the posterior limits of their respective dentigerous surfaces. Each arm of the crescent decreases gradually in width so that the inner tooth bands are multiseriate to their posterior limits in the dentary, and almost to those limits in the premaxilla (Fig. 35).

At its broadest point there are 7–11 rows in the upper, and 6–11 in the lower jaw, the numbers decreasing posterolaterally to *ca* 3 or 4 in the latter and a single or double row in the former. The rows are very closely spaced and there is no gap between the inner and outer tooth rows in either jaw.

Upper jaw. The premaxilla has the greater part of its dentigerous arms inflated and ovoid in cross section; the posterior quarter of each arm, however, is less enlarged and is more compressed.

The ascending processes are only about half as long as the horizontal dentigerous arms, which are produced anteriorly and anterolaterally into a slight but broad beak.

The maxilla is elongate and relatively shallow, its articular head with only a slight medial curvature.

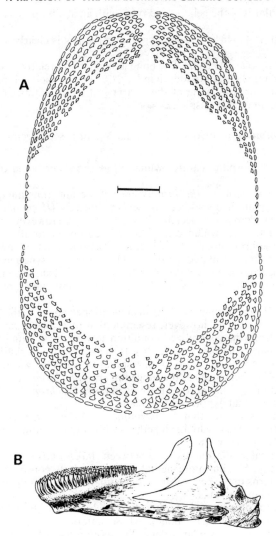

Fig. 35 A. Dental pattern of *Allochromis welcommei*. Scale = 1 mm. B. Dentary of *Allochromis welcommei*, in lateral view. Scale = 3 mm.

Lower jaw. The dentary is elongate and relatively shallow. Its alveolar surface, however, is inflated and broad, forming a deep bullation extending beyond, and overhanging, the lateral wall of each ramus. Anterolaterally, the inflated region extends almost to the ventral margin of the ramus.

Each alveolar surface has a decided anteroventral inclination, sloping forwards and downwards from a point immediately in front of the ascending dentary arm to the superficially shallow symphysial area. The symphysis itself extends vertically through the bullation so that it is both deep vertically and, since it incorporates the bullate part of the bone, also wide dorsally.

The anguloarticular is of the generalized type, with an acute tip to its horizontally aligned ventral limb.

Lower jaw length is 41–46% of head length, and the jaw is clearly longer (1·3–2·0 times) than broad.

Lower pharyngeal bone and teeth. The bone is slender and relatively elongate, its triangular dentigerous surface slightly broader than long (see Greenwood, 1966 : fig. 6). Apart from the transverse posterior row, none of the pharyngeal teeth is noticeably enlarged; all are small, compressed and weakly cuspidate.

Contained species
Allochromis welcommei (Greenwood), 1966. Lake Victoria; see Greenwood (1966 : 309–18).

DISCUSSION

There are no synapomorphic features which allow the sister taxon of *Allochromis* to be identified precisely.

The outer row jaw teeth in *Allochromis* are unique and thus autapomorphic, but the multiseriate dental pattern does occur in several lineages (*Hoplotilapia, Platytaeniodus, Neochromis* and *Ptyochromis,* especially *P. xenognathus*). However, *Allochromis* does not exhibit other characteristics which can be considered unequivocally synapomorphic with those in any of these taxa, and each, including *Allochromis,* has its own distinctive dental pattern when these are compared in detail. One apparent synapomorphy, the inflated premaxillary dentigerous arms, would seem to be a parallelism associated either with the presence of enlarged teeth, a multiseriate dentition, or a combination of both (see discussion on p. 52).

Allochromis also differs from most of the taxa mentioned above in the gross morphology of its neurocranium, which does, however, resemble that in *Psammochromis.* In both genera the skull deviates from the generalized condition towards that found in the least derived species of the *Prognathochromis* lineage. It was this similarity in skull architecture which previously led me (Greenwood, 1974) to suggest that *Allochromis* was the sister group of the '*tridens* complex' (now recognized as the subgenus *Tridontochromis* of *Prognathochromis;* p. 20). A reconsideration of other derived features in both *Allochromis* and *Prognathochromis, sensu lato,* now renders that hypothesis untenable.

In that paper, I also suggested that *A. welcommei* (plus the '*tridens* complex') might be the sister group of three taxa which, together with others, now constitute the genus *Psammochromis*; see above, p. 56.

Certain neurocranial similarities existing between *Allochromis* and *Psammochromis* have already been noted (p. 57), but more significant (particularly considering the very different tooth form and patterns in the genera) are their similarities in lower jaw morphology, which is undeniably derived in both taxa.

Both genera have a pronounced anteroventral inclination to the outer tooth row of the dentary (itself a slender, elongate bone), whose anterior and anterolateral aspects are markedly inflated below the alveolar surface. The tooth rows in *Psammochromis* are neither as numerous nor as spatially extensive as they are in *Allochromis* so that those shared derived features cannot be ascribed to that cause, and thus be dismissed as parallelisms.

If these apparent synapomorphies in the morphology of the dentary can be accepted as truly synapomorphic, *Allochromis* could be the derived sister taxon of *Psammochromis.* If, in turn, one can accept the arguments put forward for a common ancestry shared by *Psammochromis* on the one hand and *Paralabidochromis, Ptyochromis* and *Macropleurodus* on the other (see p. 66), then *Psammochromis* and *Allochromis* together should comprise the sister group of the other three genera combined. Another possible member of this lineage *sensu lato, Schubotzia eduardiana,* is discussed on pp. 87–88 & 94.

PTYOCHROMIS gen. nov.

TYPE SPECIES: *Ctenochromis sauvagei* Pfeffer, 1896. Lake Victoria (see Greenwood, 1957 : 76–81, plate 4, upper figure).

A REVISION OF THE *HAPLOCHROMIS* GENERIC CONCEPT

ETYMOLOGY. From the Greek '*pyto*', to spit out, +*chromis*, alluding to the way in which these fishes crush mollusc shells orally and then spit out the fragments.

DIAGNOSIS. Haplochromine fishes with a maximum adult size range of 105–130 mm SL, a dorsal head profile ranging from straight and steeply sloping to strongly decurved, a small,

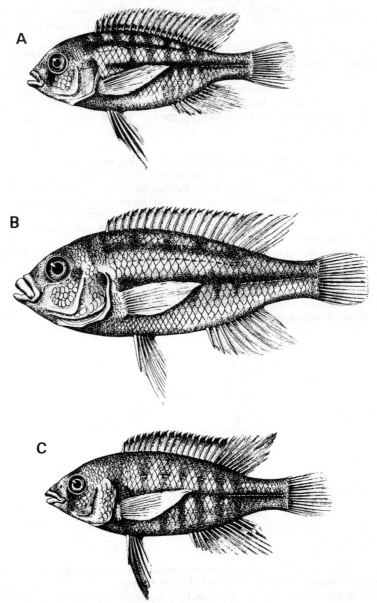

Fig. 36 A. *Ptyochromis sauvagei*. Lake Victoria. About natural size. B. *Ptyochromis granti*. Lake Victoria. About two-thirds natural size. C. *Ptyochromis xenognathus*. Lake Victoria. About natural size.

horizontal mouth with thickened lips, and a *lower jaw* that is *usually shorter than the upper. The slender teeth are very strongly recurved, those of the inner series arranged in a broad band across the anterior part of each jaw.*

Neurocranium with the preorbital face sloping fairly steeply (ca. 60°–65°, but 70°–75° in one species), *its preorbital depth 30–33% of neurocranial length.*

Premaxilla with somewhat inflated dentigerous arms which, anteriorly, are produced into a broad, shelf-like 'beak'. *Twenty-six to 56 teeth in the outer premaxillary row (modal range 40–44).*

Dentary deep posteriorly but shallowing rapidly over the anterior two-thirds of its length, the lateral walls curving abruptly mediad from a level immediately below the alveolar surface. The outer margin of this surface, over its anterior half, dips downwards and slightly outwards so that the insertions of the outer row of teeth lie below those of the inner series.

Lower jaw length 22–38% of head length (modal range 34–35%).

Lower pharyngeal bone stout and broad, the median rows with coarser teeth, but none is submolariform or molariform.

DESCRIPTION

Habitus (Fig. 36). The dorsal head profile is variable, both inter- and intraspecifically, and ranges from strongly decurved to straight but steeply sloping. The mouth is horizontal, or less commonly, slightly oblique. There is a tendency for the lower jaw to be slightly shorter than the upper one; the lips are thickened, more so in some species than in others.

Body form shows no outstanding features, and is of the generalized type.

Maximum adult size ranges from 105–130 mm SL; sexual maturity is reached at lengths between 70 and 100 mm.

Neurocranium. Skull form in this genus combines features seen in *Macropleurodus* and *Paralabidochromis* (see pp. 81 & 68) but with most species and individuals approximating more closely to the latter condition (Fig. 37). Generally, the preorbital face slopes fairly steeply (*ca* 60°–65° to the horizontal) but in one species (*P. annectens*) the slope may reach 70°–75°; preorbital skull depth ranges from 30–33% of neurocranial length. The tip of the vomer lies distinctly below the level of the parasphenoid.

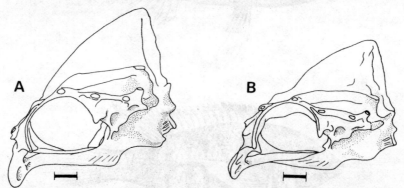

Fig. 37 Neurocranium (in left lateral view) of A. *Ptyochromis annectens*, and B. *Ptyochromis xenognathus*. Scale = 3 mm.

Supraoccipital crest outline varies from near pyramidical to a deep wedge-shape, with a corresponding variation in the height of its posterior margin.

Dentition. In both jaws the teeth in the outer row are slender but strong, with markedly recurved crowns (Fig. 38); the angle formed between the buccal face of the crown and the neck is *ca* 130°. The crown is neither expanded nor noticeably compressed, and joins imperceptibly the cylindrical neck.

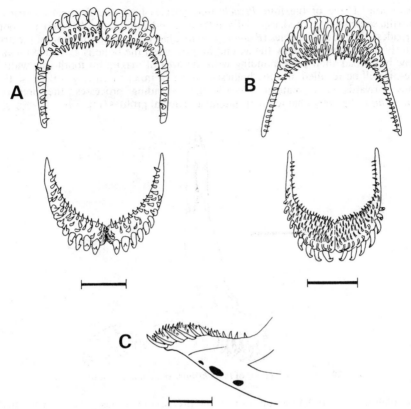

Fig. 38 Dental pattern in: A. *Ptyochromis annectens*, B. & C. *P. xenognathus* (entire pattern, and as seen laterally in dentary respectively). Scale = 3 mm.

Unequally bicuspid crowns predominate in the denition of fishes < 80 mm SL; a mixture of bi- and unicuspids occurs in fishes between 80 and 100 mm long (with unicuspids first appearing anteriorly in the jaws). In larger individuals the teeth are all or predominantly unicuspid.

Anteriorly and anterolaterally in the lower jaw the teeth are implanted almost procumbently and below the level of those situated laterally (see also the description of the dentary, p. 64). As a result, their crowns are on about the same level as those of the inner teeth. Tooth insertion on the premaxilla is nearly vertical or very slightly procumbent.

There are 26–56 (modal range 40–42) teeth in the outer premaxillary row.

Teeth forming the inner rows are small and arranged in a characteristic pattern, namely a wide anteromedial band lying transversely across the front of each jaw, but narrowing abruptly to a single or double row laterally, and a single row posterolaterally (see Fig. 38A). Modally, there are 4 or 5 rows in the upper jaw and 3 or 4 in the lower; *Ptyochromis xenognathus* (Fig. 38B) is exceptional in having modes of 7 and 5 rows in the jaws respectively. The total range of tooth row numbers is from 3–9 (rarely 2) in the premaxilla and 2–9 in the dentary (see Greenwood, 1957 : 83 for comments on the aberrant tooth pattern in the holotype of *P. annectens*; also Regan, 1922 : fig. 14).

Most inner teeth in fishes less than 80 mm SL are tricuspid, but are predominantly unicuspid in fishes > 90 mm SL.

Upper jaw. Three of the four *Ptyochromis* species have the dentigerous arms of the premaxilla somewhat inflated, especially anteriorly and anterolaterally where the tooth rows are broadest. The fourth species (*P. xenognathus*), has little or no inflation of the arms, but anteriorly (in the region below the ascending processes) the bone is extended forward as a shallow but broad shelf corresponding with the area of maximum tooth row width. (This species, it will be recalled, has the highest number of inner tooth rows.) Because this shelf extends forwards as a plateau beneath the ascending processes, the premaxilla in *P. xenognathus* has very characteristic lateral and dorsal profiles (Fig. 39).

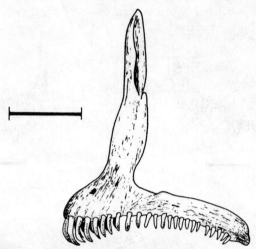

Fig. 39 Premaxilla (left) of *Ptyochromis xenognathus*. Scale = 3 mm.

Anterolaterally, each dentigerous arm of the premaxilla curves rather sharply mediad so that in occlusal view the entire structure has the outline of a broad-based U; the base is narrowest in *P. xenognathus*.

Premaxillary ascending processes are shorter than the dentigerous arms in all species except *P. xenognathus*, where they are equal.

Each maxilla is foreshortened, its posterior arm relatively deep, with a strongly concave median face (and a corresponding bullation of its lateral face). There is some variation in the extent to which the articulatory head is curved medially; curvature is strong in *P. granti* and *P. annectens*, but only moderate in *P. sauvagei* and *P. xenognathus*.

Dentary. There are several outstanding, and derived, features in this bone. In all species it is deep posteriorly but shallows rapidly forward from the region near the point where the ascending (coronoid) arm begins to rise. Consequently, in lateral view the ventral profile of the dentary appears to slope steeply upwards into a shallow symphysial region (Fig. 40). Anteriorly, when compared with the condition seen in both generalized and differently derived dentaries, the side wall of the dentary in *Ptyochromis* species does not descend vertically for some distance before it begins to curve inwards. Instead, its medially directed curvature begins only a short distance below the alveolar surface, and the curvature is unusually abrupt. This pattern of curvature, coupled with the relatively deep coronoid region of the bone, gives the dentary a very characteristic appearance (Fig. 38C). The appearance is most extreme in *P. xenognathus* and is least marked in *P. sauvagei* and *P. granti*. Parenthetically it should be noted here that the dentary of *P. annectens* figured in Greenwood (1974 : 70, fig. 40, captioned *Haplochromis prodromus*, but see p. 66 below for nomenclature) was in fact drawn from a misidentified skeleton of a *Paralabidochromis* species (see p. 67), probably *Paralabidochromis crassilabris*.

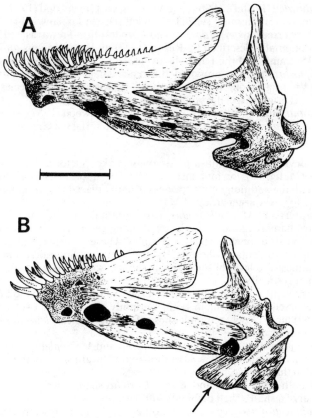

Fig. 40 Left dentary of: A. *Ptyochromis sauvagei* seen in lateral view. B. *Ptyochromis xenognathus*, viewed somewhat ventrolaterally, from the left, to show the shape of the anterior margin to the anguloarticular (arrowed). Scale = 3 mm.

Anteriorly and anteromedially the alveolar surface of the dentary has a noticeable fore-and-aft expansion (corresponding, as in the premaxilla, to the greatest width of the inner tooth rows), and a correlated antero-posterior lengthening of the symphysial surface.

About halfway along its length the alveolar surface for the outer tooth row in the dentary dips downwards and forward to occupy a position below the alveolar surface for the inner tooth rows. Immediately below the somewhat ventrolaterally displaced anterior outer teeth, the dentary is thickened and a little bullate. A similar swelling, associated with a displaced outer tooth row, occurs in *Psammochromis* (see p. 55 and further discussion on p. 66).

The anguloarticular is deep and stout. Its anteroventral arm, which barely underlies the posteroventral edge of the dentary, has a blunt anterior margin (Fig. 40).

Lower jaw length is 22–38% head length (modal range 34–35%) and is thus shorter than in a generalized syncranium but comparable with that in some derived lineages (eg *Paralabidochromis* and *Macropleurodus*). Usually the lower jaw is longer than broad (1·3–1·4 times longer).

Contained species
The taxa are listed in order of their increasing derivation.

Ptyochromis sauvagei (Pfeffer), 1896. Lake Victoria; see Greenwood (1957 : 76–81).
Ptyochromis annectens (Regan), 1922. Lake Victoria; see Greenwood (1957 : 82–6) where the species is referred to as *Haplochromis prodromus* Trewavas, 1935. *Ptyochromis annectens* was originally described by Regan (1922) as *Haplochromis annectens* but when Trewavas (1935) transferred the Lake Malawi species *Cyrtocara annectens* Regan, 1921 to the genus *Haplochromis,* she proposed the replacement name *prodromus* for the then homonymous Victoria taxon. With the removal of Regan's 1922 species from the genus *Haplochromis,* his original specific epithet again becomes available.
Ptyochromis granti (Blgr.), 1906. Lake Victoria; see Greenwood (1957 : 86–90).
Ptyochromis xenognathus (Greenwood), 1957. Lake Victoria; see Greenwood (1957 : 90–5).

Discussion

In an earlier review of relationships amongst Lake Victoria haplochromine species (Greenwood, 1974, fig. 70), the four species now referred to *Ptyochromis* were also grouped together, and with the addition of two species currently placed in *Paralabidochromis* (p. 71), were considered to be the sister group of *Macropleurodus.*

Both *Ptyochromis* and *Macropleurodus* share certain derived features in the morphology of the lower jaw, namely an outer tooth row dipping ventrally over the anterior half of its length, a deep and stout anguloarticular complex whose anteroventral arm has an obtuse anterior margin, and a dentary whose lateral walls, at least anteriorly, are abruptly curved inwards from almost the level of the alveolar surface, so that they are aligned more nearly horizontally than vertically.

Dentally, there is little in common between the two lineages (**cf.** pp. 62 & 82), except for the strongly recurved tooth form in both. Certainly the teeth in two *Paralabidochromis* species (see p. 69) more closely resemble those of juvenile *Macropleurodus* than do the teeth in any *Ptyochromis* species.

All the derived jaw features shared by *Ptyochromis* and *Macropleurodus* are also present in *Paralabidochromis* (see pp. 64, 70 & 83) although the inward slope of the dentary wall is less well-developed in that lineage.

Trophically, both *Ptyochromis* and *Macropleurodus* share a derived feeding habit, that is, the oral removal of a snail's shell before its soft parts are ingested (see Greenwood, 1974 : 69 *et seq.*). *Paralabidochromis* species, on the other hand, are all insectivores with some taxa known to be specialized in their ability to remove insect larvae and pupae from burrows in wood or rock (Greenwood, 1959*b* : 210).

In brief, *Ptyochromis, Macropleurodus* and *Paralabidochromis* all share a number of derived features in the jaws and detailed morphology of their jaw bones (especially the lower jaw), and in the way the outer row of teeth is inserted on the dentary. All have a similar and derived neurocranial form. The teeth and the dental pattern are derived in all three genera, with those of *Ptyochromis* and *Paralabidochromis* more alike than either is to *Macropleurodus* which has the most derived tooth morphology and dental pattern (see p. 82).

From the evidence available it is impossible to indicate precise interrelationships between the three genera, although the original concept of their being more closely related to one another than to any other lineage (Greenwood, 1974) does seem to be corroborated by the characters discussed above and on p. 84. (See also p. 95).

The peculiar alignment of the outer tooth row in the dentary in these genera resembles closely that of *Psammochromis* (see p. 55). There is a further resemblance in the way the anterior wall of the dentary immediately below the displaced section of the tooth row is thickened and bullate. In other respects, however, the morphology of the dentary (and its anguloarticular bone) in *Psammochromis* is near the elongate and shallow generalized type, and not like the relatively deep, short and stout bone in *Macropleurodus, Paralabidochromis* and some *Ptyochromis* species. But, in two *Ptyochromis* (*P. sauvagei* and *P. xenognathus*) the proportions of the dentary are intermediate between those of *Psammochromis* and the dentary in the other *Ptyochromis* species. In turn, these other *Ptyochromis* species intergrade with the *Paralabidochromis* and *Macropleurodus* conditions, so that there is, in effect, a morphocline in dentary proportions running from *Psammochromis* to *Macropleurodus* (as representatives of the two extreme conditions).

The slender but strong, tall and recurved outer jaw teeth in *Psammochromis* must be ranked as derived in relation to the unequally bicuspid, slightly recurved and stout teeth characterizing many lineages. Basically, the *Psammochromis* tooth-form aproaches that of *Ptyochromis*, albeit one less strongly recurved and with the crown more obviously compressed. They could, however, be considered the plesiomorph 'sister form' of the *Ptyochromis* type.

Neurocranial shape in *Psammochromis* differs quite markedly from that in the other three genera under consideration (see p. 54); it represents a slight departure from the generalized type towards that of the predatory piscivorous lineage *Prognathochromis*. That is, a slight overall elongation and streamlining of the basic form, as compared with the foreshortening and elevation of that type manifest in *Ptyochromis*, *Paralabidochromis* and *Macropleurodus*.

Considering the various synapomorphies discussed above, their range of expression and their differences, *Psammochromis* could be included as a member of an assemblage forming the sister group of *Ptyochromis*, *Paralabidochromis* and *Macropleurodus* combined. In most respects the level of derivation (from their hypothetical common ancestor) which *Psammochromis* has reached, appears to be much less marked than that shown by any other member of the total assemblage.

PARALABIDOCHROMIS Greenwood, 1956

TYPE SPECIES: *Paralabidochromis victoriae* Greenwood, 1956. Lake Victoria.

Note. Regan (1920:45) erected the genus *Clinodon* for the species *Hemitilapia bayoni* Blgr, 1908. The paratype of that species, which Regan examined, was later chosen as the holotype for *Haplochromis plagiodon* Regan & Trewavas (1928:224–5), a species which I now include in the genus *Paralabidochromis*. Since the holotype of *Hemitilapia bayoni* Blgr., (see Boulenger, 1915: 491, fig. 340) the designated type species of Regan's *Clinodon* is, however, referrable to *Haplochromis obliquidens* (see Greenwood, 1956*b*: 226–232, and 1979:278), the older name *Clinodon* is not available for the taxon here called *Paralabidochromis*.

DIAGNOSIS. Haplochromines with a maximum adult size range of 70–150 mm SL, *a forceps-like dentition (lower teeth implanted procumbently)*, lips somewhat thickened (hypertrophied and lobate in one species), mouth horizontal, dorsal head profile straight, or slightly concave, and sloping fairly steeply.

Neurocranium with a deep preorbital region (33–37% neurocranial length **cf.** *25–30%, modal range 26–27% in the generalized skull):* entire preorbital gently curved and sloping at an angle of 45°–50°. Supraoccipital crest of variable outline, from near pyramidical to deeply wedge-shaped.

Dentary foreshortened and deep, with a marked lateral bullation in the region of its division into coronoid and horizontal arms. *Profile of the symphysial region with a pronounced posteroventral slope giving the jaw a distinctly chinless appearance. Lower jaw length 30–49% head length (modal range 33–35%).*

Premaxilla with slightly inflated dentigerous arms (oval in cross-section), its ascending processes as long as, or slightly longer than dentigerous arms. *Maxilla foreshortened,* its posterior arm deep but not markedly bullate.

Outer jaw teeth strong, slender, recurved and cylindrical in cross-section, the crown somewhat compressed when bicuspid, otherwise cylindrical. Teeth anteriorly and anterolaterally in the lower jaw implanted procumbently, sloping forwards and upward at an angle of ca 45°–50°. Upper jaw teeth implanted almost vertically, but when the premaxilla is *in situ* they are inclined forwards to form, with the procumbent lower teeth, a forceps-like dentition. *Relatively few outer teeth in both jaws*, 16–48 (modal range 30–35) in the premaxillary outer row. Cusp form variable but usually bicuspid in fishes <65–70 mm SL, and unicuspid, near conical in larger individuals.

Inner teeth arranged in 2 or 3 (rarely 1 or 4) rows, separated from the outer row by a distinct interspace.

Lower pharyngeal bone short and broad, the median row teeth coarse but rarely submolariform.

DESCRIPTION

Habitus (Fig. 41). Body form departs but slightly from the generalized *Astatotilapia* shape, with the dorsal head profile straight or slightly concave and sloping fairly steeply. The lips are somewhat thickened in all species and are hypertrophied, even lobate, in *P. chilotes*.

Maximum adult size range is from *ca* 70–150 mm SL.

Neurocranium. Skull form departs somewhat from the generalized condition in that the preorbital region is relatively deeper (33–37% neurocranial length, **cf.** 25–30%, modal range 26–27%) and in consequence the preorbital profile slopes more steeply (*ca* 45°–50°) and is slightly curved (Fig. 42). The supraoccipital crest is a little higher relative to the generalized condition, and varies in profile from pyramidical to deeply wedge-shaped.

Dentition. The outer jaw teeth are strong and robustly slender, recurved, and cylindrical in cross-section, the crown slightly compressed when bicuspid. Teeth anteriorly in the lower jaw are implanted procumbently so that they slope forward and upward at an angle of *ca* 45°–50° to the horizontal. All species have the anterior and immediately anterolateral

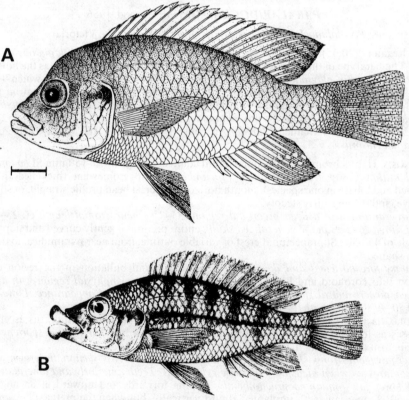

Fig. 41 A. *Paralabidochromis labiatus*. Lake George. Scale = 1 cm. B. *Paralabidochromis chilotes*. Lake Victoria. About two-thirds natural size.

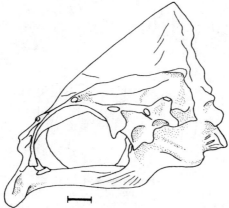

Fig. 42 Neurocranium (left lateral view) of *Paralabidochromis chilotes*. Scale = 3 mm.

teeth longer than the lateral ones; the size discrepancy being more marked in some species than in others. Since anteriorly the line of insertion for these outer teeth has a distinctly ventral direction, their tips lie at about the same level as those of the inner rows.

The upper teeth are inserted almost vertically on the premaxilla, but when that bone is *in situ* it slopes at an angle such that the anterior upper teeth are directed procumbently, and they occlude with their counterparts in the lower jaw to provide a forceps-like dental arrangement.

In most species the dental arcade in both jaws, but particularly that in the premaxilla, narrows anteriorly to give it a narrow-based 'U'-shaped outline.

There are relatively few teeth in the outer row of either jaw, the premaxillary series having 16–48 (modal range 30–35).

Cusp form is interspecifically variable, but is predominantly bicuspid in most specimens of all species at a standard length of less than 65–70 mm.

Species in which the definitive outer dentition is unicuspid have a near-conical tip to the crown, which is thus more robust than in the typical unicuspid teeth in such lineages as *Harpagochromis* and *Prognathochromis*.

In one species, *P. victoriae*, the anterior teeth in both jaws are more slender and are relatively longer than those in other members of the genus, thus enhancing the forceps-like nature of the dentition (see Greenwood, 1956a : 328, and fig. 10).

Two types of bicuspid teeth occur in *Paralabidochromis*. The commoner is close to the generalized kind but has the cusps of markedly unequal size, the major one approximately equilateral in outline and less compressed than in the generalized tooth; the minor cusp may be relatively smaller in *Paralabidochromis* teeth. The second cusp type has so far been recorded, as the predominant form, in only one species, *P. plagiodon*. Here the minor cusp is aligned at a slight angle so that it resembles a weak spur, and the posterior margin of the major cusp slopes obliquely forward to meet its near vertical anterior margin at a somewhat obtuse angle; the entire cusp is also more compressed than in the other type of tooth (Fig. 43). In other words, cusp form in *P. plagiodon* resembles that in *Haplochromis lividus* (Greenwood, 1956b : fig. 2B; 1959 : 206; and 1979 : 278), although it is stouter overall and the anterior angle of the major cusp lies in the same vertical as the neck of the tooth (not outside it as in *Haplochromis*).

Unlike the generalized bicuspid, where the crown is expanded relative to the neck, in neither type of *Paralabidochromis* tooth is the crown much wider than the neck.

Inner teeth are tricuspid (sometimes weakly so) or, in specimens > 90 mm SL, a mixture of tri- and unicuspids. They are arranged in 2 or 3 rows (rarely in 1 or 4 rows), and are always

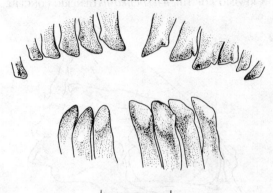

Fig. 43 Outer jaw teeth of *Paralabidochromis plagiodon*, viewed *in situ* and from in front, to show the nature of the crowns. Since the lips are not fully reflected, the bases of the teeth are not visible. Scale = 1 mm.

separated from the outer teeth by a distinct interspace; in some individuals the outermost inner row is composed, anteriorly, of teeth slightly larger than those of the other inner rows, but still distinctly smaller than the outer row teeth.

Mouth. The lips are thickened in all *Paralabidochromis* species and are hypertrophied in *P. chilotes*; some individuals of that species have both lips produced into prominent medial lobes, but in others the lobes may be represented only by a slight bulbous swelling (see Greenwood, 1959b : 208, and fig. 11). The mouth is horizontal, and both jaws are equal anteriorly.

Upper jaw. The dentigerous arms of the premaxillae are somewhat inflated, especially anteriorly and anterolaterally (where the bone is a compressed oval in cross section). Inflation is more marked in some species than in others. The ascending processes are as long as, or slightly longer than the dentigerous arms.

The maxilla is foreshortened and its posterior arm is relatively deep; the medial face of the latter is not markedly concave and thus there is no strong bullation of its lateral aspect. The articular head of the bone curves gently mediad.

Lower jaw. The dentary is short and deep, giving the entire lower jaw a foreshortened appearance (Fig. 44). This impression is enhanced by the bullation of each ramus near its

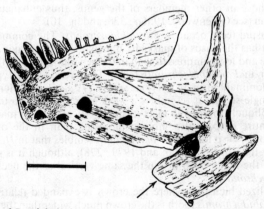

Fig. 44 Lower jaw of *Paralabidochromis crassilabris*, seen ventrolaterally from the left, to show the anterior margin of the anguloarticular (arrowed). Scale = 2 mm.

division into ascending (coronoid) and horizontal limbs. In these respects the dentary resembles that found in some other lineages (eg *Lipochromis, Neochromis* and *Macropleurodus*).

Its anterior profile, both in lateral and in ventral views, is most characteristic. The symphysial region of the bone has a pronounced posteroventral slope which gives it a rather 'chinless' look in lateral view. This area is also slightly expanded horizontally and the bone hereabouts is noticeably thickened. Anteriorly, each ramus has only a gently curved mediad inclination so that the entire jaw, when viewed from above, has a bluntly apexed V-shaped outline. In all these respects the dentary of *Paralabidochromis* is unlike that in *Lipochromis* and *Neochromis*, and in most respects that of *Macropleurodus* as well; see above.

In three species (*P. crassilabris, P. plagiodon* and *P. beadlei*) the lateral wall of each ramus has, anteriorly, a fairly marked mediad curvature so that it slopes inwards rather than almost vertically downwards (see also *Ptyochromis* and *Macropleurodus*, pp. 64 & 84 respectively).

The anguloarticular is deep and robust, the anterior point of its anteroventral arm blunt or rectangular in outline (Fig. 44).

Lower jaw length ranges from 30–49% of head length (modal range 33–35%); the jaw is always narrower than it is long. In both these features *Paralabidochromis* broadly overlaps the majority of non-piscivorous lineages whose adults have a maximum length of less than 115 mm, but its modal lower jaw length is slightly lower.

Lower pharyngeal bone and dentition. The bone is short and broad, its triangular dentigerous area always broader than it is long. All but two species have the median tooth rows composed of coarse teeth, and in some individuals of certain species these teeth (especially the posterior few) have submolariform crowns; otherwise the crown is bicuspid and compressed.

Contained species
The taxa are grouped, approximately, in order of their related and increasing derivation.

Paralabidochromis beadlei (Trewavas), 1933. Lake Nabugabo; see Greenwood (1965: 335–41).
Paralabidochromis paucidens (Regan), 1921. Lake Kivu; see Regan (1921: 638).
Paralabidochromis crassilabris (Blgr.), 1906. Lake Victoria; I have not yet published a revised description of this species; the information used here is derived from the specimens used by Regan (1922: 167–8) in his previous revision, supplemented by new material and osteological preparations.

Paralabidochromis labiatus (Trewavas), 1933. Lakes Edward and George; see Greenwood (1973: 196–9).

Paralabidochromis plagiodon (Regan & Trewavas), 1928. Lake Victoria; see Greenwood (1959*b*: 205–7); and note on p. 67 above.

Paralabidochromis chromogynos (Greenwood), 1959. Lake Victoria; see Greenwood (1959*b*: 212–4).
Paralabidochromis chilotes (Blgr.), 1911. Lake Victoria and probably, the Victoria Nile; see Greenwood (1959*b*: 207–12).

Paralabidochromis victoriae Greenwood, 1956. Lake Victoria; see Greenwood (1956*a*: 328–9).

DISCUSSION
Paralabidochromis victoriae has the most derived and forceps-like dentition of all the species, yet morphologically speaking, the various stages in its evolution are represented in

other members of the lineage. The least derived conditions are those seen in *P. beadlei* (Lake Nabugabo), *P. paucidens* (Lake Kivu) and *P. labiatus* (Lake Edward and George) and *P. crassilabris* (Lake Victoria). Two Lake Victoria species *P. chromogynos* and *P. chilotes*, in that order, most closely approach *P. victoriae*.

Paralabidochromis plagiodon, although retaining a basically plesiomorph tooth form for a member of this lineage (but a distinctly apomorph one relative to the basic bicuspid tooth) does exhibit certain autapomorphic features in the crown shape of these teeth (see p. 69 above). The expansive, obliquely margined major cusp, the small spur-like minor cusp, and the somewhat buccally orientated crown (Fig. 43) set *P. plagiodon* apart from other members of the lineage (with which it nevertheless shares several synapomorphies).

Its distinctive dental features, however, do approach those seen in small specimens (< 90 mm SL) of *Macropleurodus*, a genus which differs from *Paralabidochromis* in several autapomorphic dental characters and some osteological ones as well. The *Macropleurodus*-like features in the dentition of *P. plagiodon*, together with certain derived features in the morphology of the lower jaw shared by *Macropleurodus* and all species of *Paralabidochromis*, suggest that the lineages could have shared a relatively recent common ancestry. This, and other possible interrelationships of *Paralabidochromis*, are discussed further on pages 66, 84 & 93.

HOPLOTILAPIA Hilgendorf, 1888

TYPE SPECIES (*Paratilapia*?) *retrodens* Hilgendorf, 1888. Lake Victoria; see Greenwood (1956a: 319 & 321) for detailed synonymy.

DIAGNOSIS. Haplochromines with an adult size range of *ca* 96–145 mm SL, characterized by a number of dental and syncranial specializations.

The dentary has an almost square anterior outline, is very shallow over most of its length, is 'chinless' and has a marked lateral bullation of the area surrounding its bifurcation into ascending (coronoid) and horizontal arms. In the entire fish it has a very shovel-like appearance.

Premaxilla with very strongly inflated dentigerous arms, the broad alveolar surface extending almost to their posterior tips, and virtually circular in cross-section.

Teeth in both jaws are arranged in broad bands (5–10 rows deep) of almost uniform width over their entire length; those of the outer row not separated by a distinct interspace from the inner rows, and, at least in fishes > 75 mm SL, continuing almost to the crown of the coronoid process (and often accompanied by one or more inner rows). Unicuspid teeth predominate in both the inner and outer rows of specimens in the known size range (ca 55–145 mm SL), but some bi- and tricuspid teeth occur amongst the inner rows in fishes < 100 mm SL.

DESCRIPTION

Habitus (Fig. 45). The straight or weakly concave, steeply sloping dorsal head profile, coupled with the broad, horizontal mouth and shallow, square-ended lower jaw give *Hoplotilapia* a very characteristic appearance. In other respects it is a typical, moderately deep-bodied haplochromine (depth of body 31–42% of SL, mean 38%).

The adult size range if from 96–145 mm SL.

Neurocranium. The skull has a moderately steep (50°–55°) and straight or very gently curved preorbital profile (see Greenwood, 1956a: fig. 8A; and 1974: fig. 76). Its preorbital depth is 34–35% of neurocranial length (*ie* deeper than the generalized type).

The supraoccipital crest is relatively high and expansive, with a near-pyramidical outline.

Dentition (See also Greenwood, 1956a: 322, fig. 8B; and 1974: fig. 74). In both jaws the outer row teeth are slender but strong, cylindrical in cross-section, strongly recurved, and with the crown but slightly, if at all, compressed.

Outer teeth in the lower jaw (Fig. 46) are implanted procumbently (the neck almost horizontal); those situated anteriorly in the upper jaw are effectively procumbent because the

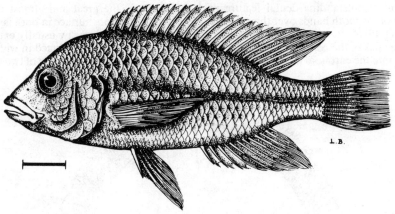

Fig. 45 *Hoplotilapia retrodens*. Lake Victoria. Scale = 2 cm.

premaxillary alveolar surface slopes forward and upward at an appreciable angle. Because of their strong recurvature, the crowns of the anterior teeth in the lower jaw are directed almost vertically, despite the near-horizontal alignment of their necks.

Most fishes >75 mm SL have the outer row of dentary teeth continuing onto the coronoid process of the bone, and ending near the coronoid crest (see below).

Unicuspid teeth predominate in the outer row of most specimens in the size range available for study (*ie* 55–144 mm SL), but a few lateral and posterolateral teeth may show faint traces of a minor cusp.

There are 40–68 teeth in the outer premaxillary row.

Unicuspid teeth also predominate in the inner tooth rows, with some bi- and tricuspid teeth present, especially in smaller fishes.

This predominance of unicuspid teeth in fishes < 90 mm SL can be considered a derived characteristic.

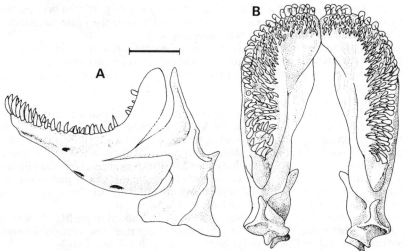

Fig. 46 Lower jaw of *Hoplotilapia retrodens* in: A. Lateral view, and B. occlusal view. Scale = 3 mm.

The most outstanding dental feature in *Hoplotilapia* is the great and almost uniform breadth of the tooth bands over the entire length of the dentigerous surface in both jaws (Figs 46 & 47). It will be recalled that the outer row of teeth in the lower jaw usually extends to near the apex of the coronoid process; the inner rows (here somewhat reduced in width) also extend onto the coronoid process, but stop short of the outer row by a distance of two or three outer teeth.

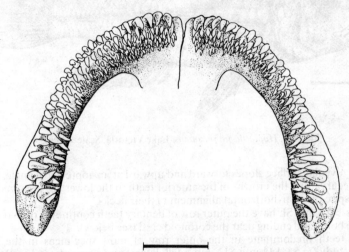

Fig. 47 Dentigerous surface of the premaxilla in *Hoplotilapia retrodens*, to show dental pattern. Scale = 3 mm.

Teeth forming the inner rows are not particularly close set, but are inserted across the entire width of the alveolar surfaces involved. There are 5–8 rows of teeth anteriorly, and 4–5 rows posteriorly on the premaxilla, with 5–10 anteriorly and 3–5 posteriorly on the dentary (see also Greenwood, 1956a: 322).

Upper jaw. The premaxilla is a robust bone with very strongly inflated dentigerous arms that are virtually cylindrical in cross-section, each widest over the posterior quarter of its length. The posterior fifth of each arm is bulbous (Fig. 47).

The ascending processes are much shorter than the dentigerous arms (half to two-thirds the length), and the articular processes are stout and anteroposteriorly expanded.

The maxilla is somewhat foreshortened and relatively deep, with a deeply concave medial face to its posterior arm (and a correspondingly bullate lateral face). Its articular head is strongly curved mediad.

Lower jaw. The dentary is a most distinctive bone, unlikely to be confused with the dentary in any other haplochromine from the Victoria–Edward–Kivu flock.

It is stout, with an almost vertical ascending (coronoid) arm which is, relative to the depth of the ramus, markedly elevated (Fig. 46A). This apparent shallowness of the dentigerous ramus is attributable to the way in which its lateral wall curves abruptly mediad from a level just below the alveolar surface. As a result of this curvature, the greater part of the ramus is almost horizontally (and not vertically) aligned and underlies the very broad alveolar surface (see Fig. 46B) beyond which it extends medially.

In lateral view the dentary is devoid of a 'chin', its anterior profile sloping steeply backwards and downwards. Whereas in other haplochromines the symphysial surface is either vertical or is inclined posteriorly, in *Hoplotilapia* it lies at an angle of *ca* 45° to the horizontal. The actual articular surface of the symphysis is relatively narrow in the vertical plane, but it is extensive in the near horizontal plane.

The region surrounding the dentary's bifurcation into ascending and horizontal arms is inflated, the lateral bullation emphasized by the shallowness of the ramus in that area.

About the anterior half of each ramus curves strongly mediad so that the anterior margin of the entire lower jaw is subrectangular when viewed dorsally.

The anguloarticular is deep and stout, the anterior margin of its anteroventral arm deep and rectangular in outline.

Lower jaw length is 34–41% of head length; the length/breadth ratio of the jaw varies intraspecifically from broader than long to 1·3 times longer than broad.

Lower pharyngeal bone and dentition. The bone is short and broad, its triangular dentigerous area equilateral in outline. All the teeth are cuspidate and relatively coarse, with those of the median rows stouter than the others (except for the posterior transverse row).

Contained species
Hoplotilapia retrodens Hilgendorf, 1888. Lake Victoria and, probably, the Victoria Nile; see Greenwood (1956a: 319–326).

DISCUSSION
Although many of the derived features characterizing *Hoplotilapia* are autapomorphies (eg. the gross morphology of the lower jaw, the form of the premaxilla, and the dental pattern in both jaws), most would seem to be foreshadowed in the *Ptyochromis* lineage, especially in *Ptyochromis xenognathus* and *P. annectens* (see pp. 62–65 above).

For example, in *Pytochromis* the tooth bands are broad in both jaws (but particularly in the dentary), there is a distinct tendency for the lateral wall of the dentary to curve sharply mediad, the teeth in the lower jaw are procumbent, the outer teeth in both jaws of larger specimens are strong but slender unicuspids and have markedly recurved crowns.

There are other resemblances too; for instance in overall neurocranial architecture, the robust lower jaw, the morphology of the anguloarticular bone, and the inflated dentigerous arms of the premaxilla. These could, however, be the products of convergence associated with the evolution of strong jaws since some or all of the features occur in *Neochromis* and *Lipochromis*, lineages that share no other derived characters with *Hoplotilapia* (see p. 52).

Those characters apart, the other synapomorphies do seem to suggest that *Hoplotilapia* may share a common ancestry with *Pytochromis*, and by an extension of that relationship, with *Paralabidochromis* and *Macropleurodus* as well (see pp. 66–67). This argument will be elaborated and evaluated when the possible relationships of *Hoplotilapia* and *Platyaeniodus* have been considered (see p. 80 below).

PLATYTAENIODUS Boulenger, 1906

TYPE SPECIES: *Platytaeniodus degeni* Blgr., 1906. Lake Victoria; see Greenwood (1956a: 312 & 315) for detailed synonymy.

DIAGNOSIS. Haplochromine fishes with an adult size range of *ca* 70–155 mm SL, readily diagnosed by their peculiar dental pattern.

Teeth in the dentary are grouped into two broad pyriform patches, contiguous anteriorly. In the premaxilla the teeth are arranged in a broad, inverted U-shaped band, whose arms and base are of almost uniform width in fishes < 100 mm SL, but in larger fishes the posterior parts of the arms are expanded medially so that they approach one another closely in the midline (Fig. 51). *There are corresponding modifications to the shape of the premaxilla and dentary, the latter having a near-circular outline when viewed occlusally* (Fig. 50).

DESCRIPTION
Habitus (Fig. 48). The gently curved, moderately steeply sloping dorsal head profile, coupled with the thickened, broad, beak-like projection of the premaxilla and the rounded anterior margin of the dentary, give to *Platytaeniodus* a most unusual physiognomy. This distinction is enhanced by the thickened lips (the lower often lying outside the upper lip), and by the almost completely hidden maxilla.

Fig. 48 *Platytaeniodus degeni.* Lake Victoria. Scale = 2 cm.

Adult size range is from 71–154 mm SL.

Neurocranium. Skull morphology in *Platytaeniodus* is close to that of *Paralabidochromis* (Fig. 49); that is, an anteriorly deepened variant of the generalized form (preorbital skull depth 36–37% of neurocranial length), with the preorbital region fairly strongly decurved and sloping at an angle of *ca* 55°–60° to the horizontal. The supraoccipital crest is moderately high, its outline near-pyramidical, and the anterior margin straight or gently curved.

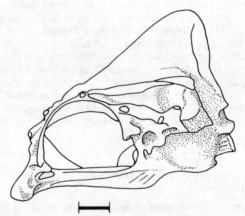

Fig. 49 Neurocranium (left lateral view) of *Platytaeniodus degeni.* Scale = 3 mm.

Dentition (See also Greenwood, 1956a: 312–3, 316 & fig. 6; 1974: fig. 73). The outer row teeth in both jaws are moderately slender but strong, are implanted vertically and are close-set. An admixture of uni- and unequally bicuspid teeth occurs in fishes of all sizes (smallest seen 67 mm SL), with unicuspids predominating in fishes > 90 mm SL. Unicuspid teeth have slender, conical crowns that are slightly broader than the near-cylindrical neck of the tooth; bicuspids have the crown a little compressed. In neither form is the crown strongly recurved; usually it is straight.

There are 36–50 teeth in the outer row of the premaxilla.

Inner teeth are either unicuspid or tricuspid, or there may be an admixture of both kinds with tricuspids predominating in fishes < 100 mm SL (particularly in the innermost rows).

There is a very gentle size-gradient between the outer row teeth and those of the inner rows. Virtually no interspace separates the two series. Like the outer teeth, those of the inner series are tall, slender and strong. Teeth in the innermost 2 or 3 rows are more compressed than the others. Implantation is vertical, and the crowns are slightly recurved.

It is in the arrangement of its jaw teeth that *Platytaeniodus* departs most markedly from the other lineages (Figs 50 & 51).

In both jaws the teeth are arranged in broad bands composed of 5–9 rows in the premaxilla and 5–7 in the dentary. There is no obvious anteroposterior decrease in the number of premaxillary rows but there is a size-correlated change in the dental pattern (Figs 50 & 51).

Small fishes (< 100 mm SL) have the bands either of uniform width over the entire length of the premaxillary alveolar surface, or there is a slight medial expansion of the rows on the posterior third of that surface (Fig. 50A). In larger fishes this medial expansion is continued

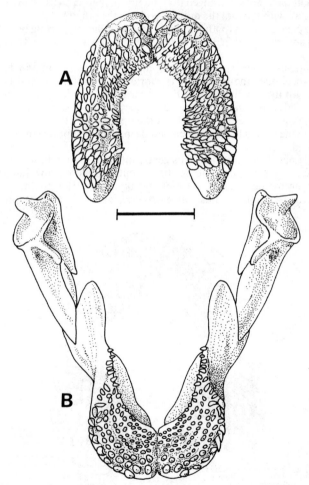

Fig. 50 *Platytaeniodus degeni.* A. Occlusal view of premaxillary dental surface. B. Lower jaw, in occlusal view.
From a small specimen (80 mm SL) to show dental pattern of premaxilla (cf. Fig. 51). Scale = 3 mm.

until, in fishes > 120 mm SL, the inner aspects of each arm are narrowly separated from one another (Fig. 51).

It should be stressed that this local expansion of the alveolar surface is not correlated with any increase in the number of tooth rows carried on it; these are always equal in number, or at most slightly fewer than those on the anterior and anterolateral regions of the bone.

Tooth bands on the lower jaw are less subject to variation in shape with body size. The teeth are confined to the anterior and anterolateral regions of the jaw, and are grouped into two broad patches roughly pyriform in outline and contiguous at the symphysis. Posteriorly on each side there is a short, single row of up to seven teeth lying between the main dental concentration and the base of the ascending arm of the dentary; these teeth are apparently a posterior extension of the outer tooth row.

Each pyriform patch is broadest anteriorly, narrowing rather abruptly over about its posterior sixth (Fig. 50B); before that point there is no decrease in the number of tooth rows.

The teeth in both jaws are so arranged that the tips of their crowns all lie in the same plane (and not, as is usual, with those of the outer row above the others).

As might be expected with a tooth pattern of this sort, the supporting bones, especially the dentary, are considerably modified even when compared with those in *Macropleurodus* and *Hoplotilapia*.

Mouth. The unusual, almost 'duck-billed' appearance of the mouth has already been noted. In most specimens the lower jaw is shorter than the upper, and laterally its well-developed, deep, lip usually lies outside the upper lip, hiding its lower margin posteriorly. The lips are not only thickened, but their free margins are also produced so as to extend above, or as in the case of the upper jaw, below the tips of the teeth.

Upper jaw. The maxilla is foreshortened and deep, with no appreciable mediad curvature of its articular head.

The premaxilla has its dentigerous arms gently inflated, even anterior to the origin of the ascending processes. Here the bone is drawn out into a narrow but deep, shelf-like expansion. The entire dentigerous part of the bone posterior to the shelf is virtually cylindrical in cross-section; it ends, on each side, as a short blunt projection formed

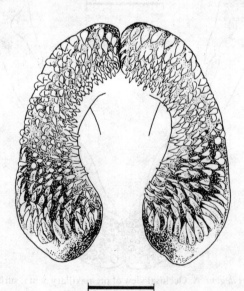

Fig. 51 *Platytaeniodus degeni.* Premaxillary dental surface (in occlusal view); from a specimen 120 mm SL. Scale = 3 mm.

immediately behind the last teeth (Fig. 51). Each arm of the bone has a marked downward curvature over almost the posterior third of its length (Fig. 52).

Both the ascending and the articular processes of the premaxilla are moderately stout; the length of the former is greater than that of the dentigerous arms.

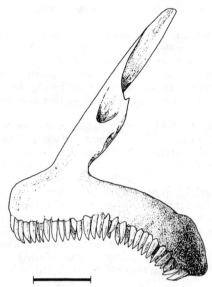

Fig. 52 *Platytaeniodus degeni.* Premaxilla in left lateral view; from a specimen 120 mm SL. Scale = 3 mm.

Lower jaw. In addition to being stout and foreshortened, the dentary has a most extraordinary overall apperance. Posteriorly, in the region of its division into ascending and horizontal arms, it is distinctly inflated, but the greatest departure from the generalized condition is seen anteriorly. The very broad and pyriform alveolar surface, and the bone supporting it, are produced outwards (almost to the ventral margin of the ramus) as an inflated, ovoid swelling. This projects laterally well beyond the rest of the bone (Fig. 50B). Below the swollen area, the ramus continues ventrally in almost the vertical plane; posteriorly, however, it curves sharply mediad, thus further emphasizing the swollen anterior region of the jaw.

As a result of this anterior swelling, the symphysial area on each ramus is both deep and wide anteroposteriorly, and the dentary has a very characteristic bulbous semicircular anterior outline.

The anguloarticular bone is deep and stout, the anterior margin of its anteroventral arm deep and subrectangular in outline.

Lower jaw length is from 32–40% of head length (mean 37%); the jaw is clearly longer than it is broad (1·2–1·7 times longer).

Lower pharyngeal bone and dentition. The bone is short and broad, its triangular dentigerous surface equilateral in outline. The teeth are coarse and cuspidate, those of the two median rows and the posterior transverse row stouter than the others.

Contained species
Platytaeniodus degeni Blgr., 1906. Lake Victoria; see Greenwood (1956a: 312–318).

DISCUSSION
Most of the derived features characterizing *Platytaeniodus* are autapomorphies (eg the

peculiar dental pattern and shape of the lower jaw; the medial expansion of the premaxillary alveolar surfaces and the overall morphology of that bone). As such they give no indication of the taxon's interrelationships with other haplochromines in or outside the lakes under review.

Only the uniformly broad premaxillary tooth bands extending along the entire length of each premaxillary arm appear at first sight to be a synapomorphic feature shared with *Hoplotilapia* (see p. 74).

In other derived dental features, *Hoplotilapia* and *Platytaeniodus* have little in common, unless it be argued that fundamentally both taxa do have uniformly broad bands of teeth in the lower jaw as well. Those in *Hoplotilapia*, however, extend far onto the ascending process of the dentary, whereas in *Platytaeniodus* they are confined to the horizontal part of the bone, and have a very different arrangement anteriorly (cf Figs 46 & 50B). If the basic pattern is synapomorphic, then each taxon has departed from that basic condition along differently derived pathways.

Another possible synapomorphy lies in the nature of the dentary immediately below the anterior and anterolateral portions of its alveolar surface. In *Platytaeniodus* this region is hypertrophied to form the characteristically bulbous bow of the dentary. In *Hoplotilapia* this region of the dentary is inflated to form a deepened shelf of bone overhanging, laterally, the anteroventral aspects of the ramus; the shelf so formed resembles, albeit in an embryonic way, the peculiar stage-like development of the jaw in *Platytaeniodus*.

The sum of the various apparent synapomorphies shared by *Platytaeniodus* and *Hoplotilapia* would suggest that each taxon is the other's nearest living relative. That *Hoplotilapia* shows some synapomorphic similarities with, ultimately, the *Pytochromis* lineage, could indicate a shared common ancestry for that lineage (that is *Paralabidochromis*, *Macropleurodus*, *Hoplotilapia* and *Platytaeniodus*). The precise sequence of dichotomies interrelating the different taxa has, however, still to be resolved (see also p. 93).

MACROPLEURODUS Regan, 1922

TYPE SPECIES: *Haplochromis bicolor* Blgr., 1906 (type specimen only; see Greenwood, 1956a: 299–301). Lake Victoria.

DIAGNOSIS. Haplochromine fishes reaching a maximum adult size of *ca* 150 mm SL, usually with a very strongly decurved dorsal head profile (dorsum of snout sloping at *ca* 70°–80° to the horizontal), a small mouth and thickened lips, *the upper lip displaced laterally by the hypertrophied outer premaxillary teeth which, consequently, are exposed when the mouth is shut.*

Teeth in the outer row of both jaws are stout, with an inwardly directed, strongly recurved major cusp (lying at almost right angles to the neck) and a greatly reduced minor cusp (often merely a slight protuberance on the crown). The minor cusp is vertical and, because of the extreme curvature of the major cusp, lies labially to the tip of that cusp. Fishes > 80 mm SL have, laterally on the premaxilla, one or more inner tooth rows composed of enlarged teeth morphologically similar to those in the outer series.

DESCRIPTION

Habitus (Fig. 53). *Macropleurodus* has a very distinctive head and mouth, at least in specimens over 60 mm SL (smaller individuals are unknown). The dorsal head profile is generally very strongly decurved, the dorsum of the snout sloping steeply at an angle of 70°–80°; less commonly the profile is slightly curved (and may be straight). Annectant forms link the two extremes (see Greenwood, 1956a: 304–5, fig. 3).

The mouth appears to be relatively small, and has fairly well-developed lips. On one or both sides the upper lip is displaced dorsally by the hyperdeveloped lateral dentition. As a result, the teeth in that region are exposed, and the fish could be described as having a permanent leer on one or both sides of its face.

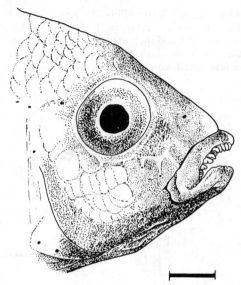

Fig. 53 Head of *Macropleurodus bicolor* to show naturally displaced upper lip and exposed dentition. Scale = 5 mm.

Adult size range for the only species, *M. bicolor,* is 80–150 mm SL; most specimens are sexually mature at lengths of between 90 and 100 mm.

Neurocranium. Despite some variation in the superficial dorsal outline of the head, the underlying preorbital profile of the skull is invariably steep and curved, the ethmovomerine region sloping at an angle of 80°–85°, its tip clearly reaching a level below that of the parasphenoid (Fig. 54; also Greenwood, 1956a: fig. 4A; and 1974: figs 66 & 76).

As would be expected in a skull of this shape, the preorbital depth of the neurocranium is high (34–37% of neurocranial length), particularly when compared with a generalized skull (24–30%, modal range 26–27%).

The supraoccipital crest is tall and expansive, with a near-pyramidical lateral profile.

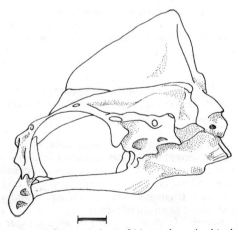

Fig. 54 Neurocranium (left lateral view) of *Macropleurodus bicolor.* Scale = 3 mm.

Dentition (see also Greenwood, 1956a: 299–301, and 304–5; 1974, fig. 75). The most derived features of *Macropleurodus* are to be found in its dentition, especially in the morphology of the upper jaw teeth and their arrangement on the premaxilla.

In both jaws, the outer teeth are stout, with a cylindrical to subcylindrical neck which merges gradually into a protracted, conical major cusp aligned almost at right angles to the neck.

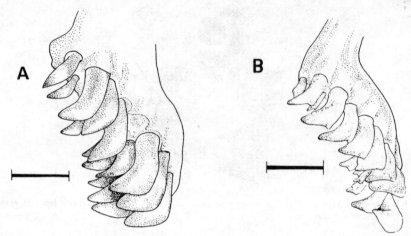

Fig. 55 *Macropleurodus bicolor*. Left premaxilla; the occlusal surface viewed medially and from somewhat below. A. Large fish (115 mm SL); the minor cusp is absent from all teeth; and B. Smaller individual (88 mm SL), to show the prominent minor cusp on all teeth (**cf.** Figs 43 & 57).

Superficially, the teeth in fishes > 100 mm SL appear to be unicuspid, but closer examination usually discloses a low bump on one side of the crown near the region of its juncture with the neck (Fig. 55A). In smaller fishes this swelling is discernible as a discrete, spur-like cusp inclined labially (Figs 55B & 57).

The major cusp is directed buccally, the arrangement of the teeth being such that in those situated laterally in the jaws, the crown lies at right angles to the anteroposterior axis of the row.

There is a clear-cut anteroposterior size gradient amongst the outer teeth, better marked in the lower than the upper jaw. Posteriorly in the lower jaw a few teeth barely show the characteristic morphology of the rostral elements in the row, and are little more than stout, incurved, uni- or weakly bicuspid teeth.

Anteriorly and anterolaterally, the outer dentary teeth are implanted somewhat procumbently, with the result that the morphologically 'outer' aspect of the crown forms a sloping occlusal surface. This unusual condition is emphasized because the line of tooth insertion dips distinctly ventrad over the anterior half of the dentary.

Premaxillary outer row teeth are, in contrast, vertically implanted; there are 24–40 (modally 34–36) teeth in this row.

Two kinds of inner teeth are present in the premaxilla, at least in fishes > 90 mm SL. Those in the first, or first and second inner rows, are stout, and in shape closely approximate to their counterparts in the outer row. Anterolaterally and laterally these teeth are larger than elsewhere in the row and are crowded together. Teeth forming the innermost row or rows of the series are weakly tricuspid (almost unicuspid) and small (but coarser than equivalent-sized tricuspids of the generalized type).

Such coarse, weakly tricuspid teeth are present in all the inner rows of fishes less than 90 mm SL.

There are 2–4 inner tooth rows anteriorly in the premaxilla, and 1 or 2 posterolaterally. Most of the specimens examined have the dextral tooth band a little wider than the sinistral one, and composed of slightly larger teeth. Symmetrical or sinistral hypertrophy has, however, been recorded.

A somewhat similar morphological differentiation is apparent amongst the inner teeth of the dentary. Here the outermost row is composed of relatively large, strongly recurved and uni- or bicuspid teeth that are both smaller and more refined than their counterparts of the outer row, but have essentially similar gross morphology. Teeth in the second and third rows are small, compressed tricuspids (occasionally bicuspids). The serial arrangement of inner row dentary teeth is regular, with 2 or 3 (sometimes 4) rows anteriorly and a single row posteriorly.

An infrequent variant of the basic tricuspid form in *Macropleurodus* is known. The lateral cusps are displaced behind and slightly medial to the central cusp, so that the tooth has a triangular crown with a cusp in each angle. Such teeth generally occur only in the outermost row of the inner premaxillary series.

What little information there is on the ontogeny of tooth form and pattern shows that buccal larvae (*ca* 9 mm total length) have slender, conical outer teeth indistinguishable from those in like-sized larvae of *Astatotilapia* (Greenwood: 1956a: 308). During later ontogenetic stages these teeth must be replaced by others having the characteristic *Macropleurodus* form and pattern. At a standard length of *ca* 85–95 mm it is known that certain lateral inner teeth on the premaxilla are replaced by enlarged teeth closely resembling their counterparts in the outer row.

Upper jaw. The dentigerous arms of the premaxilla are greatly inflated (especially anterolaterally), are cylindrical or nearly cylindrical in cross section, and have an expansive alveolar surface over the greater part of their length. This arm of the bone is distinctly arched at about its midpoint, the degree of curvature being greater on that side of the maxilla with the most enlarged teeth; see above (Greenwood, 1956a: fig. 4C).

The ascending premaxillary processes, as compared with the dentigerous arm, are slender and as long or longer than those arms.

The maxilla is foreshortened, its posterior portion relatively deep and with a pronounced median concavity (and thus a correspondingly strong bullation of the lateral face). Its articular head has a pronounced medial curvature.

Lower jaw (Fig. 56). The robust lower jaw (dentary + anguloarticular) is deep relative to its length, thus appearing foreshortened and massive.

Each ramus of the dentary is much inflated in the region surrounding its division into ascending and horizontal arms. Anteriorly the bone has a pronounced median curvature so that the entire jaw, viewed from above, has the outline of a broad-based U.

Anteriorly, the upper part of the dentary is greatly thickened, and has a broad superficial surface extending from the symphysis anteriorly to the base of the ascending arm posteriorly.

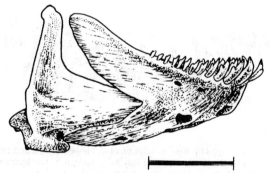

Fig. 56. *Macropleurodus bicolor.* Right dentary in lateral view. Scale = 3 mm.

Only about the outer half of this surface is dentigerous. Laterally, the ramus wall curves abruptly towards the midline so that it lies more nearly horizontally than vertically.

Each aspect of the symphysial surface is wide anteroposteriorly, and runs almost vertically (**cf.** the backwardly sloping symphysis and resultant 'chinless' outline in *Paralabidochromis* p. 71).

The anguloarticular is massive and carries a broad and deep articular surface for the head of the quadrate. Its anteroventral prolongation has a near rectangular anterior margin and is so aligned that it runs downward in parallel with the posterior margin of the dentary (not ventral to that margin as it does in most other lineages).

Lower jaw length is from 28–36% of head length (mean 32%), that is somewhat shorter than in the generalized syncranium, but similar to that in most species of the *Paralabidochromis* and *Ptyochromis* lineages. Usually the lower jaw is as broad as it is long, but in some specimens it may be slightly longer than broad.

Lower pharyngeal bone and dentition. The relatively stout bone is short and broad, its triangular dentigerous surface broader than long. The teeth are cuspidate and compressed, but not noticeably fine; those in the two median series usually are coarser than the others (except those in the transverse posterior row).

Contained species
Macropleurodus bicolor (Blgr.), 1906. Lake Victoria; see Greenwood (1956a : 299–312)

DISCUSSION
Some of the derived features seen in *Macropleurodus* appear to be an intensification or elaboration of those found in *Paralabidochromis*. This is particularly apparent when the jaw skeletons and dentition of the two genera are compared. The skeletal modifications could, at least in part, be associated with the more massive teeth and expansive dental pattern characterizing *Macropleurodus*.

Neurocranial form in *Macropleurodus* also seems to represent a modification of the derivative trend manifest in the majority of *Paralabidochromis* species. Namely, an increase

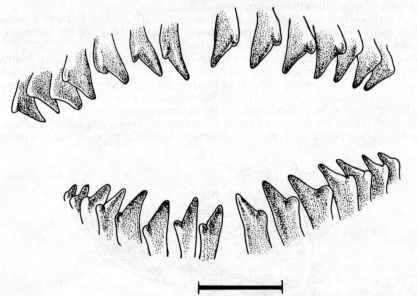

Fig. 57 Dentition of a juvenile *Macropleurodus bicolor*, viewed *in situ* and from in front, to show the nature of the crowns. Since the lips are not fully reflected, the bases of the teeth are not visible. (For comparison with Fig. 43.) Scale = 1 mm.

in the slope of the preorbital skull face, especially in the ethmovomerine region (cf the neurocranial descriptions for the two taxa, p. 81 and p. 68 respectively).

An apparently similar trend of apomorphic intensification could be invoked to explain the evolution of the peculiar outer tooth form in *Macropleurodus*. That is to say, from the unequally bicuspid teeth of a type seen in *P. beadlei* or *P. plagiodon*, the *Macropleurodus* type could develop by an increase in stoutness, the differential elongation of the major cusp, and an increase in its curvature towards the midline (cf. Figs 43 & 57). As noted before (p. 72), the outer teeth in small *Macropleurodus* do resemble those of adult *P. plagiodon*. It is also known that in *Macropleurodus* the peculiar inner teeth are developed relatively late in ontogeny, when the fish is between 85 and 95 mm long. Smaller individuals have inner teeth like those of comparable-sized, and larger, *Paralabidochromis*.

In the absence of other and more clear-cut synapomorphies it is possible to reach only a tentative conclusion about the relationships of *Macropleurodus*; namely, that *Macropleurodus* could share a fairly recent common ancestor with *Paralabidochromis*. The possible further relationships of *Paralabidochromis* are discussed on p. 93.

SCHUBOTZIA Boulenger, 1914

TYPE SPECIES: *Schubotzia eduardiana* Blgr., 1914. Lake Edward; for synonymy see Greenwood (1973 : 215–21).

Note. In my paper on the haplochromine species from Lake George (Greenwood, 1973), I placed this monotypic taxon in the genus *Haplochromis* (as then recognized), mainly on the grounds that '... To retain *Schubotzia eduardiana* in a separate and monotypic genus serves only to hide its close phyletic relationship with *Haplochromis*'. That sentiment was certainly valid when the genus *Haplochromis* had such a broad and non-monophyletic interpretation. However, the redefinition of *Haplochromis* (*sensu stricto*) and other elements of the old *Haplochromis* concept (Greenwood, 1979), negates my action and the reasons I gave for it. Following the methodology applied in the latter paper, this species must be given generic rank, and thus it returns to its former name, *Schubotzia*.

DIAGNOSIS. Small haplochromines (adult size range *ca* 55–80 mm SL), with a relatively shallow body (30–37% SL, mean 34%), thickened lips, a horizontal mouth in which the *lower jaw (29–35% head length, mean 33%) is shorter than the upper, and jaw teeth of a distinctive type.*

These teeth have an expanded crown that is markedly compressed and strongly recurved, and which constitutes almost half the length of the tooth; the overall appearance is one of a paddle with its blade bent at right angles to the shaft. Most teeth are unicuspid and the distal margin of the crown is rounded.

All outer teeth are moveably implanted and close-set, those of the dentary extending almost to the crest of the low ascending arm (coronoid) of the bone.

Inner teeth are small and tricuspid, with strongly compressed and recurved crowns; virtually no interspace separates the outer row of teeth from the 2 or 3 inner series in the upper jaw, or the single (rarely double) series in the lower jaw.

DESCRIPTION
Habitus (Fig. 58). The body is relatively slender (30–37% SL, mean 34%), the dorsal head profile straight or gently curved, sloping at an angle of 30°–35°. The thickened lips and the shorter lower jaw give the head a distinctive profile.

Neurocranium. Apart from its moderately decurved preorbital profile and more steeply sloping ethmovomerine region (which projects distinctly below the level of the parasphenoid), the skull is of a generalized type. Its preorbital depth is *ca* 30% of the neurocranial length, and the supraoccipital crest is low and wedge-shaped in outline.

Dentition. Outer teeth in both jaws have the crown greatly expanded relative to the slender subcylindrical neck and body, the much compressed crown constituting about half the height of the tooth (Fig. 59). Except for a few weakly bicuspid teeth in the smallest specimen

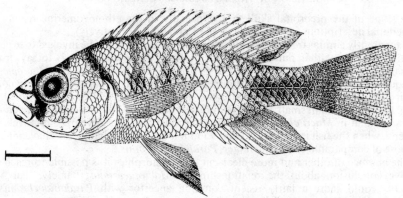

Fig. 58 *Schubotzia eduardiana*. Lake George. Scale = 1 cm.

examined (35 mm SL), the crown is unicuspid and its distal margin rounded. If the tooth were flattened out, it would have the shape of a paddle.

In bicuspid teeth, the minor cusp is not separated from the major one by a distinct gap, but merely by a narrow, V-shaped, groove.

The teeth are implanted vertically, but the strong buccal curvature of the crown results in that part of the tooth being aligned almost horizontally. In the lower jaw, the outer row is continued posteriorly almost to the tip of the low coronoid arm of the dentary. Rarely, a few small tricuspid teeth are intercalated, posteriorly, amongst the unicuspid outer teeth of the lower jaw. There is a slight anteroposterior size gradient in the height of the premaxillary teeth, and a more marked gradient in the lower jaw.

Both jaws have the teeth fairly close-set (but never with the crowns contiguous); in fresh material the teeth are moveably implanted.

There are 40–52 (mean 48) teeth in the outer premaxillary row. Teeth forming the inner rows are tricuspid and small, and have strongly compressed, recurved crowns. Almost no

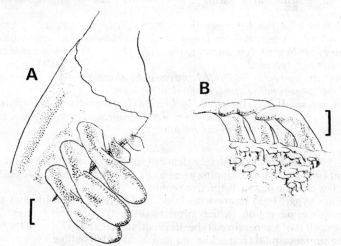

Fig. 59 *Schubotzia eduardiana*. A. Right premaxilla (in part), viewed medially and from below to show lateral aspect of the outer teeth. B. Teeth from the left dentary (anterolateral in position) viewed medially to show lingual aspect of the teeth. Scale = 0·25 mm.

space separates the outer teeth from those of the inner rows, of which there are 2 (rarely) or 3 in the upper jaw and a single (rarely double) row in the lower jaw.

Upper jaw. The maxilla is moderately foreshortened and deep; its articular head curved gently mediad.

The premaxilla is inflated and almost cylindrical in cross-section; its ascending processes are almost half the length of the dentigerous arms which, when viewed from below, have a broadly rounded U-shaped outline.

Lower jaw. The dentary is stout, deep (especially posteriorly) and foreshortened in appearance, with a low ascending (coronoid) arm; it is not noticeably inflated in the region around its bifurcation into ascending and horizontal arms (Fig. 60). The alveolar surface, at

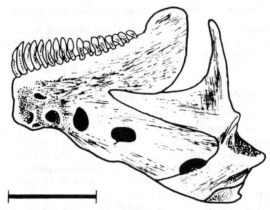

Fig. 60 *Schubotzia eduardiana.* Left ramus of lower jaw, seen from a somewhat ventrolateral position. Scale = 3 mm.

least anteriorly and laterally, is somewhat broadened, the inflation extending a little below the alveolar surface as a shallow dorsal bullation of the ramus wall. This bullation, like the overlying alveolar surface, clearly slopes forward and downward from a point near the upward curve leading into the coronoid arm (onto the greater part of which the outer tooth row is continued posteriorly and upwards).

The anguloarticular is a little stouter than in the generalized type of jaw but, as in the latter, its anteroventral arm has an acute anterior angle; the arm itself lies almost horizontally beneath the posterior part of the dentary's ventral margin.

Length of the lower jaw is 29–35% (mean 33%) of head length; the length/breadth ratio of the jaw ranges from unity to 1·4 times longer than broad (mode 1·1 times).

Lower pharyngeal bone and teeth. The bone is relatively elongate and narrow; none of its teeth is enlarged (except, as usual, those in the transverse posterior row and, in larger fishes, a few posterior teeth in the median rows as well). The triangular dentigerous surface is about 1·3 times longer than it is broad.

Contained species
Schubotzia eduardiana Blgr., 1914. Lakes Edward and George; see Greenwood (1973 : 215–21).

DISCUSSION
Like the monotypic genera in Lake Victoria, *Schubotzia* is readily distinguished by its autapomorphic characters. But, it is extremely difficult to find its sister group amongst potentially related taxa because few of its derived features are shared ones.

The subalveolar bullation of the dentary in *Schubotzia* suggests a relationship with the Victoria lineages *Psammochromis* and *Allochromis* (see pp 55 & 59), as does the anteroventral slope to the outer tooth row on that bone; but, neither of the latter genera has the foreshortened and deep dentary of *Schubotzia*, and the actual dentition is distinctive in the three taxa (each being derived in its own way). The gross and detailed morphology of the dentary in *Schubotzia* also shares some features with that bone in the *Ptyochromis-Macropleurodus* assemblage, the possible sister group of *Psammochromis* and *Allochromis*. Thus there is a suggestion that *Schubotzia* is related to this lineage in its broadest sense, but it cannot be placed with certainty in either of the supposed sister divisions (see p. 94). On balance, the nature or degree of expression of its apparently synapomorphic features indicate closer relationship with the *Ptyochromis* assemblage.

Lake Victoria haplochromines of uncertain generic relationship

The status of the three species considered below will probably be resolved when studies on several newly discovered taxa are completed. Only one of the species treated here is represented by adequate study material, another factor hampering their full inclusion in the present revision.

'*HAPLOCHROMIS*' *CRONUS* Greenwood, 1959

When first described (Greenwood, 1959*b* : 180–2), and in a later publication (Greenwood, 1974), '*H*' *cronus* was thought to be a relative of species now placed in the nominate subgenus of *Lipochromis* (see p. 29 above). This assignment was based on the paedophagous habits of the few *cronus* specimens with food remains in the gut, and on some osteological features of the jaws. Superficially at least, and in its dentition, *cronus* appears to be more like the generalized species now placed in *Astatotilapia* (see p. 6) although it reaches a larger adult size (135 mm SL) than any member of that genus.

In one feature, an almost completely scaled caudal fin, *cronus* differs from all known haplochromines in the Victoria–Edward–Kivu assemblage; a completely scaled caudal is, indeed, generally considered to be the 'hall-mark' of haplochromines from Lake Malawi (see Regan, 1922 : 158). The value of this character as a phylogenetic marker, however, has still to be tested, and for the moment I shall disregard its presence in *cronus*.

Since the original description of *cronus* was prepared I have been able to examine one rather damaged and incomplete cranial skeleton, and thus gain more data on its osteological characteristics.

The morphology of the premaxilla and maxilla is very like that in members of the nominate subgenus of *Lipochromis*, as is the morphology of the lower jaw (dentary and anguloarticular bones), see p. 30. Other elements of the syncranium too are like those in *Lipochromis*, but all the resemblances involve either plesiomorph (*ie* non-derived) characters or those suspected of repeated and convergent origins.

The outer row jaw teeth in *cronus* are unlike those in either subgenus of *Lipochromis*, being large, relatively stout and caniniform unicuspids with recurved tips; none is embedded in the oral mucosa, which is not noticeably thickened.

In recently dead specimens the jaws are neither so markedly distensible nor so protractile as those in *Lipochromis*, even when compared with species of the nominate subgenus (see p. 30 above).

Thus, there are no clear-cut apomorphies shared by *cronus* and *Lipochromis*; the few apparent synapomorphies in jaw morphology also occur in lineages not thought to share a recent common ancestry with *Lipochromis* (see discussion on p. 52).

Dentally, '*H*.' *cronus* resembles members of the genus *Harpagochromis*, but its inclusion in that lineage is precluded by the morphology of its skull and jaws, features which, together with its large adult size, also exclude it from *Astatotilapia*.

Amongst the haplochromine material recently collected by members of the Leiden University research team there is at least one undescribed species which appears to resemble *cronus* both superficially and anatomically. When this taxon has been studied (especially in relation to other newly discovered paedophages) it may be possible to reconsider the status of '*H.*' *cronus.*

Until that time and until more material of '*H.*' *cronus* is available, little purpose would be served by giving the species any formal supraspecific grouping.

'*HAPLOCHROMIS*' *APOGONOIDES* Greenwood, 1967

I suggested originally (Greenwood, 1967 : 108), on the basis of its dentition and strongly decurved snout, that '*H.*' *apogonoides* could be related to species now placed in the genus *Ptyochromis* (see p. 60).

The dental resemblance, however, is actually confined to the morphology of the teeth (stout, very strongly recurved unicuspids), and does not include the dental pattern or other details.

In *apogonoides,* unlike *Ptyochromis,* the anterior outer teeth of the lower jaw are not implanted procumbently, there is no pronounced anteroventral dip in their line of insertion, there is a decided gap between the inner and outer series of both jaws, and the number of tooth rows anteriorly and anterolaterally is not increased above the generalized number (hence the tooth bands, compared with those in *Ptyochomis*, are narrower).

'*Haplochromis*' *apogonoides* also lacks the foreshortened and deep dentary and the stout, anteriorly obtuse ventral arm to the angulo-articular, derived features found in *Ptyochromis.* In fact the lower jaw of *apogonoides* is, in one respect, derived along the opposite morphocline from that manifest in *Ptyochromis.* It is, relatively speaking, longer than in the generalized type and much longer than that in *Ptyochromis* (45–51% of head length in *apogonoides,* **cf.** 22–38%, mode 35%, in *Ptyochromis*).

The two taxa do share certain apomorph jaw features (posteriorly bullate dentary, inflated dentigerous arms in the premaxilla) but, as noted for '*H.*' *cronus,* these are of doubtful validity when establishing interlineage relationships.

Thus there seem to be no grounds for assuming that '*H.*' *apogonoides* and *Ptyochromis* shared a recent common ancestor. Nor is it yet possible to find other characters indicating a close relationship of *apogonoides* with any other lineage in the Victoria–Edward–Kivu flock.

The extreme recurvature of the teeth in *apogonoides,* coupled with the morphology of its dentary, would seem to preclude any close relationship with *Astatotilapia, Harpagochromis, Gaurochromis* or *Lipochromis* (see pp. 7, 10, 32, & 27 respectively), some of whose species do have a superficial likeness to *apogonoides* (see Greenwood, 1967). Furthermore, *apogonoides* shares no apomorph features with any of these taxa, except perhaps with *Harpagochromis* an increased relative lower jaw length.

As with '*H.*' *cronus,* the situation may be clarified when certain newly discovered species have been studied more closely.

'*HAPLOCHROMIS*' *THELIODON* Greenwood, 1960

The few specimens of '*H.*' *theliodon* available for study (seven in all) are remarkably uniform in appearance (see Greenwood, 1960 : fig. 15) and all have two features which, when taken in combination, are very distinctive. Namely, a stout lower pharyngeal bone with numerous molariform teeth, and very small, deeply embedded scales on the thoracic region of the body. (The upper pharyngeal bones of the specimen prepared as an alizarin transparency are enlarged, but have only a few molariform teeth.)

The extent to which the lower pharyngeal bone and its dentition are hypertrophied is comparable with that in one species of *Labrochromis* (*L. humilior*) and that in *Gaurochromis* (*Mylacochromis*) *obtusidens*; the bone's outline is nearer that of *L. humilior.*

Rather surprisingly, the neurocranial apophysis for the upper pharyngeal bones in *theliodon* is not developed to a degree comparable with the apophysis in *L. humilior* or *Gaurochromis* (*M.*) *obtusidens.* It has only a slight expansion of the parasphenoid facet, and the basioccipital facets are no larger than those in the generalized type.

From both *L. humilior* and *G.* (*Mylacochromis*) *obtusidens*, '*H.*' *theliodon* is immediately distinguished by the very small chest scales, a derived feature found in the lineage *Thoracochromis* (see Greenwood, 1979 : 290–5). In its hypertrophied pharyngeal mill, *theliodon* closely resembles *Thoracochromis pharyngalis* (Poll, 1939) of the Lake Edward drainage system. However, the small chest scales in '*H.*' *theliodon* do not have an abrupt size demarcation with the scales on the lateral and anterolateral aspects of the body (as is the case in all *Thoracochromis* species). Again, unlike *Thoracochromis*, there are more scales between the posterior part of the upper lateral line and the dorsal fin insertion (see Greenwood, 1979 : 291). Thus there are no grounds for placing *theliodon* in the *Thoracochromis* lineage, especially since it also lacks another *Thoracochromis* apomorphy, an incompletely scaled cheek.

Returning for the moment to those haplochromine lineages in Lake Victoria with which '*H.*' *theliodon* shares some derived features, *Labrochromis* and *Gaurochromis*.

The lower pharyngeal bone, as noted before, is like that in *Labrochromis humilior* but the apomorph chest scale pattern in *theliodon*, coupled with its lower, more streamlined skull, caution against supposing a sister-group relationship between these taxa on the basis of the pharyngeal mill alone. Also, an enlarged pharyngeal mill has evolved independently in lineages other than *Labrochromis* (eg in *Gaurochromis*, *Astatoreochromis* and *Thoracochromis*), and so its value as a phyletic marker in this context must be ranked rather low (see also p. 42).

The fine and numerous jaw teeth in all *Gaurochromis* are a derived feature (see p. 32), and are not shared with '*H.*' *theliodon*. An hypertrophied pharyngeal mill is also a derived feature but it is not possible to suggest that *theliodon* is the plesiomorph sister group to *Gaurochromis* on that basis because only one taxon in the *Gaurochromis* lineage has enlarged pharyngeal bones and teeth. The other species have fine pharyngeal bones and teeth, and since in other lineages this is the plesiomorph condition, it seems unlikely that the trend would be reversed in *Gaurochromis* (and that apart from the problems of convergence already noted).

Certain species of the genus *Psammochromis* do have small chest scales and a slight hypertrophy of the lower pharyngeal bone and dentition (see pp 54 & 56). But, all *Psammochromis* share derived features in the morphology of the lower jaw and its dental arrangement, (see p. 55). None of these features is present in '*H.*' *theliodon*, and the greater hypertrophy of its pharyngeal mill would also preclude it from plesiomorph sister-group status with *Psammochromis*. Likewise, since not all *Psammochromis* species have small chest scales, one cannot use that feature to suggest a close relationship with '*H.*' *theliodon*.

Similar arguments can be ranged against any attempted pairing of '*H.*' *theliodon* with other lineages from Lakes Victoria, Edward and Kivu. The species would, therefore, seem to be the sole representative of a distinct lineage and thus coordinate with those already accorded generic rank (see pp. 55 & 56). However, I hesitate to give '*H.*' *theliodon* that status until more specimens from a wider range of localities are available, and more is known of its syncranial anatomy.

Lake Kivu haplochromines of uncertain generic relationship

'*HAPLOCHROMIS*' *SCHOUTEDENI* Poll, 1932

This Lake Kivu species probably should be referred to *Paralabidochromis* (see p. 67), but I have not been able to see enough specimens, nor to obtain relevant osteological data, to reach a definite conclusion; see Poll (1932).

'*HAPLOCHROMIS*' *WITTEI* Poll, 1939

As with '*H.*' *schoutedeni*, a lack of material renders it difficult to place this Kivu species. Again, it would seem to be a member of the *Paralabidochromis* lineage.

Summary and conclusions

Some points of a general nature arising from this revision have been discussed in the introduction (p. 2), and others have been treated in an earlier paper (Greenwood, 1979 : 313–14). They need no repetition, except perhaps to reiterate that no progress has been made in resolving the problem of whether or not the Victoria–Edward–Kivu flock is of mono- or polyphyletic origin.

No unique apomorphy is shared by members of that flock and one cannot therefore erect a satisfactory hypothesis for its monophyletic origin. Equally, one cannot as yet find sister-groups outside the lake area for any of the endemic lineages described here. Thus, any hypothesized polyphyletic origin is also without adequate foundations.

A lack of differentially shared apomorphies has also rendered it virtually impossible to establish sister-group relationships for lineages within the flock. Since, therefore, a complete and sequentially dichotomous cladogram for the flock as a whole cannot be established either, another line of evidence, in this case one indicating a monophyletic origin, is not available.

I say virtually impossible because some lineages do seem to be relatable on the basis of one, or at best a few, synapomorphic features. But, in these instances all that can be achieved is the formulation of the broad hypothesis '... These various lineages are more closely related to one another than to any other lineage (or group of lineages) because they alone share this (or these) synapomorph feature or features'. The 'others' still remain as phyletic isolates.

Again, within a lineage it has proved almost impossible to provide a precise and sequentially ordered series of dichotomies. A few dichotomies can be resolved at a high level of generality for the lineage (or group of related lineages) but beyond that one is usually left with the all too familiar unresolved tri- or polychotomy.

Brundin (1972) has laid great emphasis on the importance of the 'search for the sister-group' when constructing a cladistic phylogeny. In that respect my analysis of the Victoria–Edward–Kivu haplochromines has failed. The search for sister groups will have to be continued.

It cannot be argued, at the interlineage (*ie* intergeneric) level of universality, that the Victoria–Edward–Kivu flock is too young (at most 1my; see Greenwood, 1974) for the synapomorphies identifying sister groups to have evolved (see Brundin, 1972 : 110). The lineages exist and are presumed to be monophyletic on the basis of apomorphies unique to each. Any apparent synapomorphies developed in the future would not be true synapomorphies but convergences. Clearly, if synapomorphies exist they must be present now and are either indeterminable by the techniques applied to the problem, or I have failed to recognize them, or the sister-groups exist elsewhere.

Possibly the sister lineages may yet be identified outside the geographical area encompassing the Victoria–Edward–Kivu flock, for example amongst the '*Haplochromis*' species of Lake Malawi. They certainly cannot be recognized amongst the fluviatile haplochromines of east and central Africa (see Greenwood, 1979).

On the other hand, it has been suggested (Greenwood, 1974) that the Victoria–Edward–Kivu flock is but distantly related to that of Malawi. It would then have evolved from one or even several species more closely related to those now inhabiting the local river systems than to those associated with Lake Malawi. If that is the history of the two flocks, then I can foresee the greatest difficulties in ever constructing a fully sequential cladogram for the Victoria–Edward–Kivu flock.

The morphological equivalents of any hypothetical common ancestor (or ancestors) must be of an *Astatotilapia* type since none of the Victoria–Edward–Kivu taxa shows the derived features characterizing the other fluviatile lineages (see Greenwood, 1979). Because present-day *Astatotilapia* species cannot be sorted into sister species on the basis of anatomical synapomorphies, it is reasonable to suppose that the situation was no different in Pleistocene and pre-lake times. The synapomorphic features identifying a present-day lineage would,

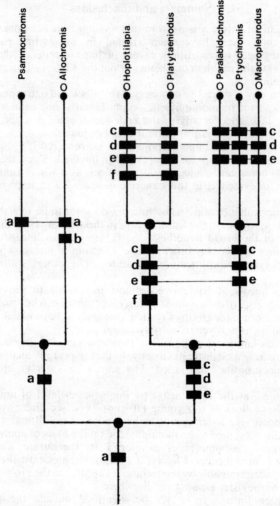

Fig. 61 Tentative cladogram for the *Psammochromis-Macropleurodus* super-lineage; see text pp. 93–94.
- ● Hypothetical common ancestor
- ○ Various autapomorphies (see text. pp. 57, 72, 75, 67, 62, & 80 for the taxa respectively).

a. Dentary slender, with a strong ventrad dip in the line of outer row tooth insertions anteriorly and anterolaterally, a marked lateral bullation of the ramus below that region.
b. Teeth in both jaws arranged in broad, crescentic bands over most but not all of the alveolar surfaces on the dentary and premaxilla.
c. Dentary foreshortened and deep, its lateral walls curving abruptly mediad a short distance below the alveolar surface of the bone.
d. Anterior outer row teeth of the dentary implanted procumbently (except in *Platytaeniodus*; see text p. 79).
e. Anterior margin to anteroventral limb of anguloarticular rectangular or bluntly rounded.
f. Teeth in broad bands on entire alveolar surface of the dentary and premaxilla.

therefore, be the apomorphic features evolved at the first (speciational) dichotomy leading to the origin of that lineage, and thereafter carried in it through successive speciation events. Under such circumstances it would be impossible to relate the two taxa resulting from that speciation event because the plesiomorph taxon would not show the diagnostic apomorph features of its derived sister species (and, subsequently, its descendent taxa as well).

Should the flock be of monophyletic origin, then a similar situation would exist if the plesiomorph daughter species of each dichotomy remained, in terms of derived features, indistinguishable from the mother species. In other words, one would have, through time, an anatomically identical line of stem taxa from which a series of primary side branches were split off. Each branch would be characterized by a different set of apomorph characters superimposed on the plesiomorph 'bauplan' common to every branch and to the successive stem species as well.

Clearly, if sister groups are to be identified, no matter how the Victoria–Edward–Kivu flock evolved, there is need for further and more detailed examination of the resultant taxa and, I would suspect, the use of characters other than strictly anatomical ones.

As mentioned earlier, some sister groups have been recognized amongst the lineages of that flock, with the result that two 'super-lineages' can be reconstructed.

The assemblage of species in the genera *Psammochromis, Allochromis, Paralabidochromis, Macropleurodus, Platytaeniodus* and *Hoplotilapia* (with possibly, *Schubotzia* as well) is the larger of the two super-lineages so far delimited (Fig. 61).

These, in many respects dentally diverse taxa are, with one exception, united on the basis of a single synapomorphy, the marked ventrad inclination, anteriorly and anterolaterally, of the outer tooth row in the dentary (see Fig. 61;a). The exceptional taxon, *Hoplotilapia*, has a very highly modified dentary, modified in such a way as to obscure the primary nature of its outer tooth implantation. *Hoplotilapia* is included in the group because of other synapomorphies which are shared with certain taxa having the group synapomorphy (see below).

The first dichotomy within this major assemblage (Fig. 61) is based on the nature of the overall morphology of the dentary.

In one of these primary divisions (*Psammochromis* and *Allochromis*) the dentary retains the slender, elongate facies of a kind not greatly different from the generalized condition, although it is somewhat inflated and bullate on either side of the symphysis. Dental pattern and tooth shape in *Psammochromis* and *Allochromis* are very different, with those of *Allochromis* showing a much derived condition, and those in *Psammochromis* retaining a more nearly generalized one (see pp. 54 & 59).

The other primary division (Fig. 61) is characterized by two, or possibly three, major synapomorphies. The first of these is shared by all members, namely a foreshortened, deep dentary, usually with the rami especially robust in that region occupied by the anterior bullation in *Psammochromis* and *Allochromis*. The second synapomorphy is shared by all but one genus (*Platytaeniodus*, see below) and involves the anterior dentary teeth being implanted procumbently (almost horizontally in two taxa). The third synapomorphy may be only a correlated character associated with the foreshortening, deepening and strengthening of the lower jaw; it concerns the shape and proportions of the anteroventral arm of the anguloarticular bone, which is deep and has a rectangular or rounded anterior margin (shallow and acute in the plesiomorph condition).

Other apparently derived features in the dentary involve the manner in which its lateral walls incline steeply, and in varying degrees abruptly, mediad, a condition foreshadowed in *Allochromis* of the other primary division (see above).

Hoplotilapia (see above) is included with the other taxa because of its deep, stout and foreshortened dentary, the shape of its anguloarticular, and because it has horizontally implanted dentary teeth which, in their derived gross morphology, resemble those in two other group members. *Platytaeniodus* (see above) is included, despite the vertical implantation of its teeth, because of the overall shape of its dentary and the anguloarticular.

Within this second primary division there are two further dichotomies, one of which terminates in an unresolved trichotomy (Fig. 61).

Members of one subdivision are characterized by having their oral teeth arranged in broad bands that extend over almost the entire dentigerous surface of both the premaxilla and the dentary. Its constituent taxa, both currently monotypic, are *Hoplotilapia* and *Platytaeniodus*; each is readily distinguished by various autapomorphic features.

Species belonging to the second subdivision (Fig. 61) are characterized by their teeth not occurring in broad bands over the entire dental surface of the jaws. In one genus (*Ptyochromis*) the inner tooth rows are wide anteriorly in the jaw, or even anteriorly and anterolaterally (*P. xenognathus*), but are reduced to a single row laterally and posterolaterally.

Three genera go to make up this subdivision, namely *Paralabidochromis*, *Ptyochromis* and *Macropleurodus*. Each is recognizable on the basis of its autapomorphies, but it is not possible to determine which two of the three taxa is the more closely related. Hence, the existence here of an unresolved trichotomy (Fig. 61).

Schubotzia, as was mentioned earlier, is a possible member of this super-lineage, but it typifies in many ways some of the difficulties encountered when one attempts to interrelate different lineages on a simple dichotomous basis.

The dentary in *Schubotzia* has a typical, downward sloping anterior deflection to the outer tooth row, thus suggesting that the fundamental relationships of the genus lie with the super-lineage. Its overall dentary form is nearest that in members of the *Ptyochromis-Paralabidochromis-Hoplotilapia-etc* primary division (Fig. 61) since it is relatively deep and foreshortened (not elongate and slender) although it does have the generalized type of anguloarticular seen in the *Psammochromis-Allochromis* primary division. The outer jaw teeth in *Schubotzia* have some autapomorphic features (see p. 85) but in their gross morphology (especially the broad spatulate crown strongly recurved on its fine and narrow neck) and in their vertical insertion on the dentary, these teeth closely resemble those of *Allochromis* and thus are of a shape not found elsewhere in the Victoria–Edward–Kivu flock. The dental pattern, however, is of a generalized type.

In brief, *Schubotzia* shares derived features with members in both the primary divisions of this super-lineage. It could be placed in one division or the other only if particular weight was given to a specific apomorphy (ie either the jaw shape or the tooth shape), and there are no grounds on which such an action could be based.

The problem created by *Schubotzia*, together with the uncertainties I entertain about the validity of the tooth-line character uniting the two primary divisions, are some of the reasons why I am treating this super-lineage as an informal assemblage of possibly doubtful phyletic significance. Likewise, in the current state of knowledge, I would hesitate to recognize formally the aggregated taxa in any of the three secondary dichotomies (ie *Psammochromis-Allochromis*, *Paralabidochromis-Ptyochromis-Macropleurodus*, and *Platytaeniodus-Hoplotilapia*).

A second super-lineage brings together four genera, *Enterochromis*, *Xystichromis*, *Neochromis* and *Haplochromis*. Here the sole apomorphy uniting the taxa is a long and much coiled intestine (all member species are phytophagous).

The primary dichotomy, based on dental apomorphies, results in one genus, *Enterochromis*, becoming the sister group of the other three genera combined.

Enterochromis retains an essentially underived dental morphotype and pattern, whereas its sister group shows various kinds of derived cusp form, and a trend towards an increase in the number of inner tooth rows with the resulting elimination of the gap between the inner and outer tooth series.

Within the tri-generic sister group, a secondary dichtotomy would separate *Xystichromis* as the plesiomorph sister taxon; the specializations seen in the dental morphology of *Haplochromis* and *Neochromis* are very disparate and thus for the moment it is probably best to consider the trio as an unresolved trichotomy.

Apart from the members of the two super-lineages discussed above, none of the other genera currently recognized can be interrelated. Consequently the general phylogenetic picture remains much as shown in the 'wheel diagram' used in my earlier analysis of the Victoria flock, in which, of course, the lineages were treated as subdivisions of a single genus (Greenwood, 1974).

The detailed picture, on the contrary, is different, with some groupings enlarged and others reduced as a result of redefined specific interrelationships, as well as the incorporation of species from lakes other than Victoria. Also, there would now be many more 'spokes' to the wheel, the result of a more critical assessment of the presumed apomorph or plesiomorph status of the characters (chiefly cranial ones) used to construct that provisional phylogeny.

A noticeable change involves the paedophagous species, which were considered to be a trophic group (or grade) of diphyletic origin (see Greenwood, 1974 : fig. 7, and discussions in the text). These species (with two exceptions, '*H.*' *cronus* [p. 88] and *Astatotilapia barbarae* [p. 32]) are now treated as a single phyletic lineage with two subdivisions (*ie* subgenera); see discussion, p. 31.

This arrangement, too, may come to be altered when a number of newly discovered paedophagous species from Lake Victoria has been studied in greater detail. Preliminary studies indicate that the present nominate subgenus (*Lipochromis*) may have to be given lineage (*ie* generic) status (and itself be subdivided). The phyletic integrity of the total lineage as now conceived, however, will be unaltered, and the paedophages will then become the third super-lineage.

More profound anatomical investigation of the two piscivorous lineages *Harpagochromis* and *Prognathochromis*, particulary the latter, may also lead to rearrangements within the genera and perhaps help to clarify their interrelationships with other members of the flock.

Naturally, any further research on the Victoria–Edward–Kivu flock will help towards clarifying relationships, but any equally important step in that direction must involve the haplochromines of Lake Malawi, if only to establish on more adequate grounds the suggestion that their relationship with the Victoria–Edward–Kivu assemblage is a distant one (Greenwood, 1974 : 99).

Almost 60 years ago, Regan (1922 : 160) concluded his revision of the Lake Victoria Cichlidae by writing '... it will be evident that I do not regard the classification here proposed as entirely satisfactory... at present I am not in a position to improve this arrangement'.

I can but appropriate his statement as an epilogue for my own work.

Acknowledgements

Since, in many respects this paper incorporates all my various '*Haplochromis*' studies, I can thank again all those people who helped me then, and many of whom have helped me again in the preparation of this paper.

In particular I am much indebted to my colleagues in the Fish Section of the British Museum (Natural History), Drs Ethelwynn Trewavas and Keith Banister, Margaret Clarke, Jim Chambers and Gordon Howes. They have contributed in so many ways: as audiences for my ideas, as critics, and as the source of ideas and information (not least in the difficult task of finding suitable names for the new taxa). To them all I proffer my warmest thanks. Especially is it a pleasure to thank Gordon Howes who has done so much for me, particularly in preparing most of the figures illustrating the paper (some taken from earlier work, others newly drawn) and in radiographing innumerable specimens (a job in which he was ably assisted by Margaret Clarke).

From amongst these many people I am, however, particularly beholden to Dr Ethelwynn Trewavas. She it was who introduced me to the fascinations of cichlid taxonomy some thirty years ago, and who has ever since been a source of inspiration and help (as she has been to so many other ichthyologists and fishery workers throughout the world). Thus, I would dedicate this paper to her on the occasion of her 80th birthday in November this year, with affection and gratitude.

A guide to the identification of haplochromine genera in Lakes Victoria, Kioga, Nabugabo, Edward, George and Kivu

Because few of the haplochromine genera from these lakes are recognizable on superficial characters alone, it is impossible to construct a simple key for their identification.

Instead, I have tried to produce a set of introductory pointers as a guide to generic identification using only features that can be ascertained readily. Any tentative identification made on this basis can then be checked against the relevant generic diagnoses, descriptions and figures. That step will be particularly important in those cases where the characters used in the guide are more applicable to groups of genera than to a single genus.

All haplochromines so far recorded from these lakes have a gradual size transition between the scales of the chest and those covering the ventral and ventrolateral aspects of the flanks; also, in all, adult males have true ocelli on the anal fin (see Greenwood, 1979 : 270, 281–2 and 274–6 for figures and discussions of these characters).

The five species treated as *incertae sedis* (pp 88–90) are not included in the guide.

Finally it must be emphasized that the guide is based on adult or near adult specimens because some of the characters utilized may not have reached their definitive expression in smaller fishes. The term 'adult size' refers to the size range of specimens which are either sexually mature or nearly so.

For a description of the various measurements used, see pp 4–6.

Anal fin with 4 or more spines . . . . ***Astatoreochromis***
(see Greenwood, 1979 : 285)
Anal fin with 3 spines . . . . A

A (i) Intestine at least 3–4 times standard length, and much coiled[1].
Fishes with a small adult size (80–105 mm SL). Usually the teeth fine, close-set and arranged in broad bands; in most taxa the inner series not separated from the outer row by a distinct gap   B
(ii) Intestine less than 2½ times standard length and with few coils.
A wide range of adult sizes. Many different kinds of teeth represented amongst the numerous taxa, but generally the teeth are robust. Some taxa have the teeth in broad bands, but in most the inner teeth are in 2 or 3 series, and distinct from the outer row C

B (i) Outer teeth with a protracted and compressed major cusp, giving the crown an obliquely truncated appearance; the tip of the major cusp clearly lies beyond the neck of the tooth . . . . ***Haplochromis*** (p. 53)
(see also, Greenwood, 1979 : 278–81)
(ii) α. Outer teeth unequally or subequally bicuspid, when unequally cuspid the major cusp clearly larger but not protracted; inner teeth in broad bands (4–6 rows). Dorsal head profile sometimes strongly decurved . . . . (a)
β. Outer teeth as above; inner teeth not in broad bands (1–3 rows); separated from the outer row by a distinct space. Outer teeth with the crown broader than the neck, close-spaced but not contiguous . . . . ***Enterochromis*** (p. 43)

(a) (i) Outer teeth unequally bicuspid, the crown not much, if at all wider than the neck; very close-set (often contiguous). Inner teeth in broad bands (usually 4–6 rows, sometimes up to 8) anteriorly and anterolaterally in the jaws, not obviously separated from the outer row by a distinct gap. Dorsal head profile not strongly decurved, usually straight . . ***Xystichromis*** (p. 46)
(ii) Outer teeth equally or almost equally bicuspid, close-set (usually contiguous); inner teeth in broad bands (usually 5 or 6 rows) anteriorly and anterolaterally, not separated from the outer row by a distinct gap. Dorsal head profile very strongly decurved. Mouth appears small . . . . ***Neochromis*** (p. 49)

---

[1] If it is impossible to examine the gut, three of the four genera included under B can be recognized by their close-set, compressed, outer jaw teeth, which are moveably implanted, and by the rather wide bands of close-set inner teeth (generally not separated from the outer row by a distinct space). The fourth genus (*Enterochromis*) is not separable on dental characters from *Astatotilapia* (cf. full descriptions for other diagnostic features, especially the relative size of the nostril and the nasal lateral line canal opening).

C (i) Lower pharyngeal bone manifestly enlarged and stout, with more than the two median tooth rows composed of enlarged, molariform teeth . . . ***Labrochromis*** (p. 37)
(see also **Gaurochromis (Mylacochromis)** (p. 36)
(ii) Lower pharyngeal bone not manifestly enlarged and stout (it may be slightly thickened), the median tooth rows not composed of molariform teeth, although often with slightly larger or coarser teeth (but these never have molariform crowns) . . . . . D

D (i) Jaw teeth exposed, not buried in the oral mucosa, and of varied form in the different taxa. Mouth with varying degrees of protrusibility and distensibility but never very markedly distensible laterally and never with the anterior part of the lower jaw closing within the upper (Lower jaw may, however, be shorter than the upper). . . . . E
(ii) Jaw teeth small, deeply embedded in the oral mucosa, usually with only the tips of the outer teeth visible; in some species the tips of these teeth curve outwards. Inner teeth often completely hidden. Mouth large and protrusile, markedly distensible laterally. In some species the lower jaw is boat-shaped (narrow anteriorly), and its anterior part closes within the upper jaw; in others the lower jaw is broadly rounded anteriorly and does not close within the upper . . . . ***Lipochromis*** (p. 26)

E (i) Inner tooth rows in both jaws arranged anteriorly in 2 or 3 series (very exceptionally as many as 5 or 6), narrowing gradually to a single row posterolaterally; always separated from the outer row by a distinct gap . . . . F
(ii) Inner tooth rows in both jaws arranged in bands of equal or almost equal width over the entire dentigerous surface of the jaws, not separated from the outer row by a distinct gap.
  (a) Inner tooth bands wide (5–10 rows), outer teeth neither enlarged nor stout . . (α)
  (b) Inner tooth bands narrow (2 or 3 rows), outer and some inner teeth enlarged and stout; crown of outer teeth with a minute minor cusp and a strongly recurved and inwardly directed major cusp . . . . (β)

(α) (*i*) In both jaws the inner tooth bands (5–10 rows) are in the form of a broad-based U (outer and some inner rows in lower jaw usually continued onto the ascending process of dentary) Teeth unicuspid even in specimens < 70 mm SL. Lower jaw flat and shovel-like, its anterior margin rectangular . . . . ***Hoplotilapia*** (p. 72)
(*ii*) Wide tooth bands (6–10 rows) in the upper jaw either in the shape of a broad-based U (*ie* in fishes < 100 mm SL) or, in larger fishes, also U-shaped but with the posterior part of each arm greatly expanded medially so as almost to touch in the midline. Tooth bands of the lower jaw in the shape of two pyriform patches, contiguous medially. Anterior region of lower jaw rounded and bullate . . . . ***Platytaeniodus*** (p. 75)
(*iii*) Wide tooth bands (6–11 rows) in both jaws in the form of broad crescents which abruptly taper posteriorly, a single row (the outer) continuing for a short distance beyond the crescent. The bicuspid teeth in the outer row have the crown markedly expanded and much broader than the slender neck. Body slender (30–33% of SL), and elongate ***Allochromis*** (p. 57)

(β) Outer row of jaw teeth (and usually the first inner row) composed of stout bicuspids in which the minor cusp is greatly reduced and lies labially to the strong, elongate and bucally orientated major cusp; major cusp lying almost at right angles to the body of the tooth. Laterally on one or both sides the upper lip is reflected to expose the teeth, even when the jaws are closed . . . . ***Macropleurodus*** (p. 80)

F (i) Outer row of jaw teeth simple bi- and/or unicuspids (sometimes a few tricuspids as well); all inner row teeth smaller than those of the outer rows, and tri- and unicuspid . . G
(ii) Outer row jaw teeth with a long, markedly expanded and compressed, spatulate crown that is about half the length of the tooth (which is thus paddle-shaped in outline) and is strongly recurved (almost at right angles). In the lower jaw the outer row teeth extend onto the ascending arm of the dentary. Lower jaw shorter than the upper . . ***Schubotzia*** (p. 85)

G (i) Dorsal profile of head sharply concave; dorsal surface of snout almost horizontal; lower jaw sloping obliquely upwards at a marked angle (50–70° to the horizontal); cephalic portions of the epaxial musculature prominent. Head and body relatively compressed. Maximum adult length 117 mm . . . . . ***Pyxichromis*** (p. 24)
(ii) Dorsal head profile and overall head shape otherwise than in G (i); when mouth is oblique, the snout profile always slopes downwards and forwards (*ie* never horizontal or almost horizontal). When the dorsal profile is concave, the concavity is a gentle one . . . H

H (i) Body form slender and elongate (body depth 23–30% of SL, modal range 27–29%). Posterior ¼–¼ of the premaxilla edentulous. A small maximum adult size (85–110 mm SL)   ***Yssichromis*** (p. 22)   also cf. ***Prognathochromis*** (p. 14)

Body form other than H (i); save for exceptional individuals, the body is deeper (30% (rarely)–47% of SL). Premaxilla fully toothed (edentulous posteriorly in some *Lipochromis* for which see back, D (ii)). Wide range of maximum adult sizes (from 50–230 mm SL).   . . .   I

I   (i) Outer jaw teeth in both jaws slender, with strongly recurved crowns; 26–56 (usually 40–44) teeth in the outer row of the premaxilla. Inner teeth, also strongly recurved, arranged in broad bands (3–9 rows) anteriorly and anterolaterally in the jaws, narrowing to a single or double row laterally; separated from outer teeth by a distinct space. Lips usually thickened but not lobate. Lower jaw length 22–38% of head length (modal range 34–35%); lower jaw usually shorter than the upper. Dorsal head profile straight and steeply sloping (gently sloping in one species), or strongly decurved   . . . . . .   ***Ptyochromis*** (p. 60)

(ii) Outer jaw teeth strong but slender, moderately recurved; those situated anteriorly in the lower jaw inserted somewhat procumbently so as to form with the upper teeth a forceps-like bite. Rather few teeth in the outer premaxillary row (16–48, modal range 30–35). Inner teeth not arranged in broad bands. Lips thickened, lobate in one species. Dorsal head profile straight and sloping steeply   . . . . . .   ***Paralabidochromis*** (p. 67)

(iii) Dental patterns and gross morphology of the teeth, taken in combination, not as in I (i) & (ii) above: the genera, *Gaurochromis, Harpagochromis, Prognathochromis, Astatotilapia* and *Psammochromis*.

These genera are defined, principally, on osteological features. For their identification reference must be made to the full descriptions. As a rough guide, the following comments may be useful.

Gaurochromis (p. 32), has short, fine, compressed and close-set outer teeth, and in all but one species, a narrow, elongate lower pharyngeal bone with fine teeth. The exceptional species has a moderately enlarged lower pharyngeal bone with some molariform teeth; it is easily confused with *Labrochromis* (see p. 37).

Harpagochromis (p. 10) species reach a large maximum adult size (>200 mm SL) and have large jaws; there is, however, a strong superficial resemblance to *Astatotilapia*, especially in small specimens.

Prognathochromis (p. 14) species also have large jaws, but are more streamlined and shallower-bodied than species of *Harpagochromis*; they have what is generally thought of as a 'typical' predatory facies. There is a wide range of maximum adult body sizes, some species exceeding 220 mm SL, while others reach only *ca* 100 mm SL.

Astatotilapia (p. 6) species have a very generalized anatomy, pharyngeal and oral dentition, and few outstanding features in their gross appearance. The maximum adult size range is between *ca* 70–100 mm SL.

Psammochromis (p. 53) species, in their gross appearance, are intermediate between the slender-bodied, streamlined, *Prognathochromis* and the deeper, stouter-bodied *Astatotilapia* species. The lips are thickened (lower lip lobate in one species), and in some species the lower pharyngeal bone is slightly hypertrophied (with some enlarged, submolariform teeth).

References

Barel, C. D. N., Witte, F. & van Oijen, M. J. P. 1976. The shape of the skeletal elements in the head of a generalized *Haplochromis* species: *H. elegans* Trewavas 1933 (Pisces, Cichlidae). *Neth. J. Zool.* **26** (2) : 163–265.

Boulenger, G. A. 1914. Fische, in *Wiss. Ergebn. Deuts. Zentral-Afrika Exped., 1907–1908, Zool.* **3** : 354.

—— 1915. *Catalogue of the fresh-water fishes of Africa.* **3.** London.

Brundin, L. 1972. Evolution, causal biology and classification. *Zoologica Scr.* **1** (3–4) : 107–120.

Fryer, G. & Iles, T. D. 1972. *The cichlid fishes of the Great Lakes of Africa. Their biology and evolution.* Edinburgh.

Greenwood, P. H. 1954. On two species of cichlid fishes from the Malagarazi river (Tanganyika), with notes on the pharyngeal apophysis in species of the *Haplochromis* group. *Ann. Mag. nat. Hist.* (12) **7** : 401–414.

—— 1956a. The monotypic genera of cichlid fishes in Lake Victoria. *Bull. Br. Mus. nat. Hist.* (Zool.) **3** : 295-333.
—— 1956b. A revision of the Lake Victoria *Haplochromis* species (Pisces, Cichlidae), Part I. *Bull. Br. Mus. nat. Hist.* (Zool.) **4** : 223-244.
—— 1957. A revision of the Lake Victoria *Haplochromis* species (Pisces, Cichlidae), Part II. *Bull. Br. Mus. nat. Hist.* (Zool.) **5** : 73-97.
—— 1959a. The monotypic genera of cichlid fishes in Lake Victoria, Part II. *Bull. Br. Mus. nat. Hist.* (Zool.) **5** : 163-177.
—— 1959b. A revision of the Lake Victoria *Haplochromis* species (Pisces, Cichlidae), Part III. *Bull. Br. Mus. nat. Hist.* (Zool.) **5** : 179-218.
—— 1960. A revision of the Lake Victoria *Haplochromis* species (Pisces, Cichlidae), Part IV. *Bull. Br. Mus. nat. Hist.* (Zool.) **6** : 227-281.
—— 1962. A revision of the Lake Victoria *Haplochromis* species (Pisces, Cichlidae), Part V. *Bull. Br. Mus. nat. Hist.* (Zool.) **9** : 139-214.
—— 1965. The cichlid fishes of Lake Nabugabo, Uganda. *Bull. Br. Mus. nat. Hist.* (Zool.) **12** : 315-357.
—— 1966. Two new species of *Haplochromis* (Pisces, Cichlidae) from Lake Victoria. *Ann. Mag. nat. Hist.* (13) **8** : 303-318.
—— 1967. A revision of the Lake Victoria *Haplochromis* species (Pisces, Cichlidae), Part VI. *Bull. Br. Mus. nat. Hist.* (Zool.) **15** : 29-119.
—— 1973. A revision of the *Haplochromis* and related species (Pisces, Cichlidae) from Lake George, Uganda. *Bull. Br. Mus. nat. Hist.* (Zool.) **25** : 139-242.
—— 1974. Cichlid fishes of Lake Victoria, east Africa; the biology and evolution of a species flock. *Bull. Br. Mus. nat. Hist.* (Zool.) Suppl. **6** : 1-134.
—— 1979. Towards a phyletic classification of the 'genus' *Haplochromis* (Pisces, Cichlidae) and related taxa. *Bull. Br. Mus. nat. Hist.* (Zool.) **35** : 265-322.
—— & **Barel, C. D. N.** 1978. A revision of the Lake Victoria *Haplochromis* species (Pisces, Cichlidae), Part VIII. *Bull. Br. Mus. nat. Hist.* (Zool.) **33** : 141-192.
—— & **Gee, J. M.** 1969. A revision of the Lake Victoria *Haplochromis* species (Pisces, Cichlidae), Part VII. *Bull. Br. Mus. nat. Hist.* (Zool.) **18** : 1-65.
Hennig, W. 1966. *Phylogenetic systematics.* Urbana.
Mayr, E. 1963. *Animal species and evolution.* Harvard.
Nelson, G. J. 1972. Phylogenetic relationship and classification. *Syst. Zool.* **21** (2) : 227-231.
Pellegrin, J. 1903. Contribution à l'ètude anatomique, biologique et taxonomique des poissons de la famille des cichlidés. *Mém. Soc. zool. Fr.* **16** : 41-402.
Poll, M. 1932. Contributions à la faune des Cichlidae du lac Kivu (Congo Belge). *Revue Zool., Bot., afr.* **23** (1) : 29-35.
—— 1939. Poissons. *Explor. Parc. natn. Albert Miss. H. Damas* (1935-1936) **6** : 1-73.
—— & **David, L.** 1937. Contribution à la faune ichthyologique du Congo Belge; collections du Dr H. Schouteden (1924-1926) et d'autres récolteurs. *Annls. Mus. r. Congo Belge, Zool.,* ser. 1 **3** (5) : 189-294.
Regan, C. T. 1920. The classification of the fishes of the family Cichlidae-I. The Tanganyika genera. *Ann. Mag. nat. Hist.* (9) **5** : 33-53.
—— 1921. The cichlid fishes of Lakes Albert, Edward and Kivu. *Ann. Mag. nat. Hist.* (9) **8** : 632-639.
—— 1922. The cichlid fishes of Lake Victoria. *Proc. zool. Soc. Lond.* **1922** : 157-191.
—— 1929. New cichlid fishes from Lakes Victoria, Kioga and Albert. *Ann. Mag. nat. Hist.* (10) **3** : 388-392.
—— & **Trewavas, E.** 1928. Four new cichlid fishes from Lake Victoria. *Ann. Mag. nat. Hist.* (10) **2** : 224-226.
Trewavas, E. 1933. Scientific results of the Cambridge expedition to the East African lakes, 1930-1. 11. The cichlid fishes. *J. Linn. Soc. (Zool.)* **38** : 309-341.
—— 1935. A synopsis of the cichlid fishes of Lake Nyasa. *Ann. Mag. nat. Hist.* (10) **16** : 65-118.
Vari, R. P. 1978. The terapon perches (Percoidei, Teraponidae). A cladistic analysis and taxonomic revision. *Bull. Am. Mus. nat. Hist.* **159** (5) : 174-340.

Manuscript accepted for publication 19 February 1980

Index

Certain Victoria–Edward–Kivu species which are essentially fluviatile in habitat, but which do occur in the lakes, are dealt with in Greenwood (1979). They are listed here but marked with an asterisk, as are the lacustrine species dealt with in that paper.

Former binomen	Current generic placement	Page
Astatoreochromis alluaudi *	*Astatoreochromis*	..
Astatotilapia nigrescens	?*Prognathochromis*	20
Haplochromis acidens	*Psammochromis*	56
H. adolphifrederici	?*Labrochromis*	42
H. aelocephalus	*Psammochromis*	56
H. aeneocolor	*Astatotilapia*	8
H. altigenis	*Harpagochromis*	13
H. angustifrons	*Gaurochromis*	36
H. annectidens	*Haplochromis*	53
H. apogonoides	*Incertae sedis*	89
H. arcanus	*Prognathochromis*	19
H. argenteus	*Prognathochromis*	19
H. artaxerxes	*Harpagochromis*	13
H. astatodon	*Haplochromis*	53
H. barbarae	*Astatotilapia*	8
H. bartoni	*Prognathochromis*	19
H. bayoni	*Prognathochromis*	19
H. beadlei	*Paralabidochromis*	71
H. boops	*Harpagochromis*	13
H. brownae	*Astatotilapia*	8
H. cassius	*Psammochromis*	56
H. cavifrons	*Harpagochromis*	13
H. chilotes	*Paralabidochromis*	71
H. chlorochrous	*Prognathochromis*	21
H. chromogynos	*Paralabidochromis*	71
H. cinctus	*Enterochromis*	45
H. cinereus	*Astatotilapia*	8
H. crassilabris	*Paralabidochromis*	71
H. crocopeplus	*Prognathochromis*	21
H. cronus	*Incertae sedis*	88
H. cryptodon	*Lipochromis*	31
H. cryptogramma	*Prognathochromis*	21
H. decticostoma	*Prognathochromis*	19
H. dentex	*Prognathochromis*	19
H. dichrourus	*Prognathochromis*	19
H. diplotaenia	?*Harpagochromis*	13
H. dolichorhynchus	*Prognathochromis*	21
H. dolorosus *	*Astatotilapia*	..
H. eduardiana	*Schubotzia*	87
H. eduardi	*Astatotilapia*	8
H. elegans	*Astatotilapia*	8
H. empodisma	*Gaurochromis*	36
H. engyostoma	*Astatotilapia*	8
H. erythrocephalus	*Enterochromis*	45
H. estor	*Prognathochromis*	19
H. eutaenia	*Prognathochromis*	22
H. flavipinnis	*Prognathochromis*	20
H. fuscus	*Neochromis*	52
H. fusiformis	*Yssichromis*	24
H. gilberti	*Prognathochromis*	20
H. gowersi	*Prognathochromis*	20
H. granti	*Ptyochromis*	66
H. graueri	*Psammochromis*	56
H. guiarti	*Harpagochromis*	13
H. humilior	*Labrochromis*	42
H. ishmaeli	*Labrochromis*	42
H. labiatus	*Paralabidochromis*	71
H. lacrimosus	*Astatotilapia*	8
H. laparogramma	*Yssichromis*	24
H. latifasciatus	*Astatotilapia*	8
H. limax	*Haplochromis*	53
H. lividus	*Haplochromis*	53
H. longirostris	*Prognathochromis*	20
H. macrognathus	*Prognathochromis*	20
H. macrops	*Astatotilapia*	9
H. macropsoides	*Astatotilapia*	9
H. maculipinna	*Harpagochromis*	13
H. mandibularis	*Prognathochromis*	20
H. martini	*Astatotilapia*	9
H. maxillaris	*Lipochromis*	30
H. megalops	*Astatotilapia*	9
H. melanopterus	*Lipochromis*	31
H. melanopus	*Astatotilapia*	9
H. melichrous	*Prognathochromis*	21
H. mentatus	*Harpagochromis* (see *H. squamipinnis*)	13
H. mento	*Prognathochromis*	20
H. michaeli	*Harpagochromis*	13
H. microdon	*Lipochromis*	31
H. mylergates	*Labrochromis*	42
H. mylodon	*Labrochromis*	42
H. nanoserranus	*Prognathochromis*	20
H. nigricans	*Neochromis*	52
H. nigripinnis	*Enterochromis*	45
H. niloticus	*Xystichromis* (see *X. bayoni*)	48
H. nubilus *	*Astatotilapia*	..
H. nuchisquamulatus	*Xystichromis*	48
H. nyanzae	*Harpagochromis*	13
H. obesus	*Lipochromis*	31
H. obliquidens	*Haplochromis*	53
H. obtusidens	*Gaurochromis*	36
H. oregosoma	*Astatotilapia*	9
H. orthostoma	*Pyxichromis*	26
H. pachycephalus	*Harpagochromis*	13
H. pallidus	*Astatotilapia*	9
H. pappenheimi	*Yssichromis*	24
H. paraguiarti	*Prognathochromis*	20
H. paraplagiostoma	?*Harpagochromis*	14
H. paropius	*Enterochromis*	45
H. parorthostoma	*Pyxichromis*	26

H. parvidens	*Lipochromis*	31	*H. squamulatus*	*Harpagochromis*	
H. paucidens	*Paralabidochromis*	71		(see *H. pectoralis*)	13
H. pectoralis	*Harpagochromis*	31	*H. sulphureus*	*Prognathochromis*	21
H. pellegrini	*Prognathochromis*	20	*H. taurinus*	*Lipochromis*	30
H. percoides	*Prognathochromis*	20	*H. teegelaari*	*Labrochromis*	42
H. pharyngomylus	*Labrochromis*	42	*H. theliodon*	*Incertae sedis*	89
H. phytophagus	*Xystichromis*	48	*H. thuragnathus*	*Harpagochromis*	13
H. piceatus	*Astatotilapia*	9	*H. tridens*	*Prognathochromis*	21
H. placodus	*Labrochromis*	42	*H. tyrianthinus*	*Prognathochromis*	21
H. plagiodon	*Paralabidochromis*	71	*H. velifer*	*Astatotilapia*	9
H. plagiostoma	*Harpagochromis*	13	*H. venator*	*Prognathochromis*	20
H. plutonius	*Prognathochromis*	21	*H. vicarius*	*Astatotilapia*	
H. prodromus	*Ptyochromis*			(see *A. eduardi*)	8
	(see *P. annectens*)	66	*H. victorianus*	*Harpagochromis*	12
H. prognathus	*Prognathochromis*	20	*H. vittatus*	*Prognathochromis*	20
H. pseudopellegrini	*Prognathochromis*	20	*H. welcommei*	*Allochromis*	60
H. ptistes	*Labrochromis*	42	*H. wittei*	*Incertae sedis*	90
H. riponianus	*Psammochromis*	56	*H. worthingtoni*	?*Harpagochromis*	14
H. sauvagei	*Ptyochromis*	66	*H. xenognathus*	*Ptyochromis*	66
H. saxicola	*Psammochromis*	56	*H. xenostoma*	*Prognathochromis*	20
H. schubotzi	*Psammochromis*	56	*Hoplotilapia*	*Hoplotilapia*	75
H. schubotziellus	*Astatotilapia*	9	retrodens		
H. schoutedeni	*Incertae sedis*	90	*Macropleurodus*	*Macropleurodus*	84
H. serranus	*Harpagochromis*	12	bicolor		
H. serridens	*Neochromis*	52	*Paralabidochromis*	*Paralabidochromis*	71
H. simpsoni	*Gaurochromis*	36	victoriae		
H. spekii	*Harpagochromis*	13	*Platytaeniodus*	*Platytaeniodus*	79
H. squamipinnis	*Harpagochromis*	13	degeni		

Species-flocks and explosive evolution

P. H. Greenwood

The species-rich, narrowly endemic, and ecologically diverse flocks of cichlid fishes from certain east and central African lakes are sometimes described as the products of 'explosive speciation' and 'explosive evolution' (Mayr, 1976, pp. 168–70; also Fryer & Iles, 1972 and Greenwood, 1974).

Such graphic imagery is perhaps justified when one considers the relatively short time-scales involved (750,000 years to 1.5 or 2 million years), the wide range of morphological and ecological diversity produced, and the fact that each species flock has evolved within one lake basin (see Fig. 6.1).

That explosive evolution in these lakes is confined virtually to one family, the perch-like Cichlidae, seems further to highlight its unusual nature. Are there, then, unusual evolutionary processes involved (as has been suggested), and are there special characteristics of cichlid fishes not shared by other families with which they coexist in each lake? There is also the question of whether explosive speciation could have played a significant role in evolution generally.

These are some of the problems that have intrigued students of evolution and ichthyology for almost seventy years.

Equally intriguing are the ecological consequences of lacustrine explosive evolution. Frequently this has led to the coexistence in one niche of two or more species (often close relatives) with what appear to be identical demands on that niche. In other words, this appears to be an apparent negation of the competitive exclusion principle. Such close coexistence may, of course, be more apparent than real, the consequence of insufficiently detailed research. Be that as it may, even apparent total interspecific overlap in such fundamental requirements as food, breeding times, and spawning sites raises the question of whether interspecific competition is inevitably, or even commonly, a major factor in evolution.

Cichlid fishes, the subjects used in this essay to investigate certain questions relating to species-flocks, are perch-like teleosts. These may be familiar to readers since many species are kept in aquaria. Cichlids are widely distributed in the freshwaters of Central and South America, Africa, Syria, Madagascar and in certain brackish waters of the Indian subcontinent. There are at least 1000 species but well over half of these are found in Africa and especially in the Great Lakes where endemism reaches a very high level.

The Great Lakes of tropical Africa, in which the cichlid species flocks occur, lie in or between the eastern and western Rift Valleys, and owe their existence to earth movements and surface changes associated with rift formation. The actual ages of the lakes are still uncertain, but Lake Tanganyika is probably the oldest (c. 3 million years BP) and Lake Victoria the youngest (c. 750 000 years BP). Since the various lakes differ in age and developmental history, and because most were populated from different river systems, it is not surprising that each has its distinctive hydrological and geomorphological features, and a characteristic fish fauna.

But, despite these differences all the larger lakes, with the two exceptions of Lakes Turkana and Albert, share one prominent feature: the taxonomic and ecological dominance of their fish-faunas by numerous species belonging to one family – the Cichlidae. The cichlids have entered into and exploited most of the many habitats and ecological niches available in the lakes, sometimes coexisting with species from other families, sometimes as the sole inhabitants. Even in Lakes Turkana and Albert, where the predominating fishes belong to other families, small cichlid species flocks have evolved. These too show a surprising range of ecological diversity (Greenwood, 1979a).

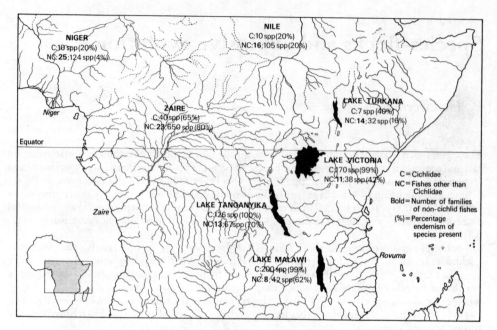

Figure 6.1 Map showing the number of cichlid and non-cichlid fishes in the major lakes and rivers of Africa. Abbreviations: C = number of cichlid species present, NC = non-cichlid fishes; the figure in **bold** type is the number of non-cichlid families present, and is followed by the number of species in those families. The percentage of endemic species in a lake or river is given in parentheses.

The larger flocks inhabiting lakes stand in marked contrast to the relatively few cichlid species inhabiting the rivers of Africa. Although there is a fairly high level of endemism for a particular river system, there has been far less speciation (Fig. 6.1) and adaptive radiation amongst the fluviatile species. Clearly, environmental conditions within a lake, coupled with the developmental history of the lake, provide a more suitable background for cichlid radiation than does a riverine environment in the course of its history. By contrast, fishes from other families show a somewhat higher level of major anatomical diversity, and a much higher level of speciation, in rivers than in lakes (see Fig. 6.1).

When comparing the number of endemic cichlid species in the three major centres of explosive speciation (Lakes Malawi, Tanganyika and Victoria) one finds only a slight difference in the total number of species for each lake, and no great difference in their degree of endemicity. In contrast, many more endemic genera have been described from Lakes Tanganyika and Malawi than in Lake Victoria, a fact which reflects the greater morphological differentiation within the cichlids of the two former lakes. In Lake Victoria most of the endemic species have been referred to a single, non-endemic genus, *Haplochromis*, a genus that also looms large in the cichlid fauna of Lake Malawi. Recent research (Greenwood, 1979a, 1980), however, indicates that a more precise phylogenetic picture is presented if the 'genus' *Haplochromis* is divided into a number of distinct lineages (or genera). It is also clear that there is little close phyletic relationship between the so-called *Haplochromis* species of Lake Malawi and those of Lake Victoria (Greenwood, 1979a). That until now more genera have been described from Lakes Tanganyika and Malawi is, nevertheless, a valid reflection of the relatively muted morphological diversity seen amongst the cichlids of Lake Victoria. Why there should be this pronounced difference is still a debated question (see Fryer & Iles, 1972).

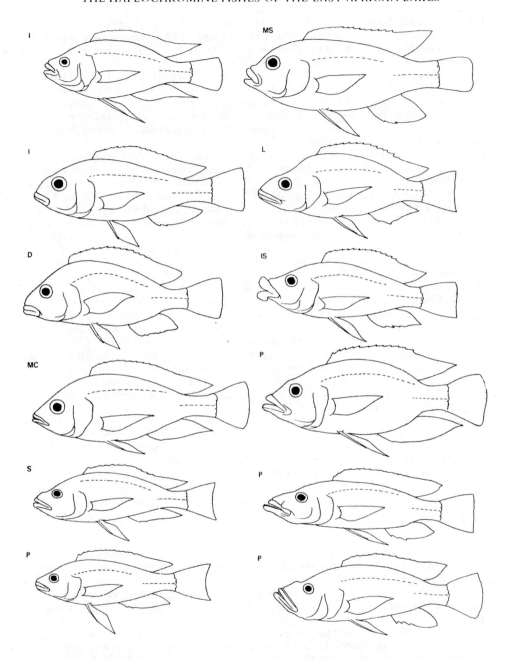

Figure 6.2 The range of body form found amongst *Haplochromis*-group cichlids in Lake Victoria. The feeding groups of the species illustrated are: D–Detritus eater, I–Insectivore, IS–Specialised insectivore (removes larvae and pupae from burrows), L–Paedophage, MC–Mollusc eater (pharyngeal crusher), MS–Mollusc eater (oral sheller), P–Piscivore, S–Scale eater. The drawings are not to the same scale.

Lake Victoria has about 170 known *Haplochromis*-like species but only 38 species of non-cichlid fishes and can be taken as an example of a cichlid dominated lake. As compared with Lakes Tanganyika and Malawi it is shallow (100 metres, *cf*. 1470 and 704 metres) and has a much greater surface area (69 000 km^2, *cf*. 34 000 and 29 604 km^2); and its indented shoreline, shallow offshore region and deep central region provide a variety of macro- and microhabitats, all of which are inhabited by cichlid and non-cichlid species. The Lake's relative youth (750 000 years, compared with 1.5 to 3 million years for Malawi and Tanganyika), the moderate level of morphological differentiation shown by its species flock, and our knowledge of its geological history, all make Lake Victoria a particularly informative example of lacustrine explosive evolution (but still an inperfectly studied one; see Greenwood, 1974, 1980).

Despite the low level of superficial diversity among the Lake Victoria cichlids (Fig. 6.2), there is a wide spectrum of smaller differences, especially in dental characters and in cranial anatomy. These differences, brought together in various combinations (and doubtless with certain physiological specialisations too) have produced a species flock that differs little from those of Malawi and Tanganyika in its range of trophic specialisations and habitat exploitation.

The evolution of many different trophic specialisations is apparently a key element in the ecological success of the lacustrine cichlids, and Lake Victoria is no exception in this respect. As in the other lakes, the cichlids of Victoria encompass the entire range of feeding habits (Fig. 6.3) practised by species in other families, and in addition they show specialisations not represented among those fishes. The term specialisation, it must be emphasised, does not mean that a species feeds exclusively on one food; it is used in this context to indicate its usual diet.

Adults of most Lake Victoria *Haplochromis*-group species are rarely more than 100–130 mm long, but some piscivorous species reach lengths of between 150–250 mm. There is a marked sexual colour difference in all species (but virtually no other dimorphism), with males brightly coloured and females drably so; no species are known to have identical breeding liveries.

Figure 6.2 shows the relatively narrow range of variation in body form encountered amongst the species so far discovered, a uniformity that is even more obvious when compared with the diversity found in Lake Tanganyika (see Fryer & Iles, 1972). The most strikingly different body shape is that of certain piscivores (Fig. 6.2), but even in this group many species have what may be described as the modal form for the flock.

It is only when the anatomy of the head is examined closely, in particular skull and jaw morphology and the shape and arrangement of the teeth (Fig. 6.4 & 6.7), that one encounters any real diversity. Important elements in the cranial anatomy are the upper and lower pharyngeal bones and teeth. These modified segments of the gill arches (Fig. 6.5) provide, in effect, a second pair of jaws situated immediately in front of the oesophagus.

The basic oral dentition, found in over 50 Lake Victoria species, comprises an outer row of moderately stout bicuspid teeth, backed by two or three rows of small, tricuspid teeth. Both cusps of the outer teeth are acutely pointed, the anterior one distinctly larger than the posterior one (Fig. 6.4). Individuals in most species having outer teeth like these rarely exceed an adult size of more than 100 mm; specimens from the upper end of this size range generally have some unicuspid teeth interspersed amongst the bicuspids. In other words there is a change in the type of replacement teeth correlated with body size, a feature of some importance in the evolution of certain trophic types.

The basic pharyngeal dentition (shown on the lower pharyngeal bone in Fig. 6.5) is also composed of bicuspid teeth, with those in the two median rows slightly stouter than the lateral teeth.

Species in Lake Victoria with this type of oral and pharyngeal dentition are mostly insectivorous, feeding chiefly on the benthic larvae of dipterous insects. They closely resemble many of the *Haplochromis* that inhabit the rivers of East Africa. Since the developing Rift Valley lakes were first populated from these rivers, it seems reasonable to assume that it was from such fishes that the more specialised species of the lake evolved.

Slight departures from the generalised type are seen in five species that feed on phytoplankton and on organic bottom ooze (mostly moribund phytoplankton). Jaw and pharyngeal teeth are finer and more numerous in these species, and the intestine is two or three times longer than in the

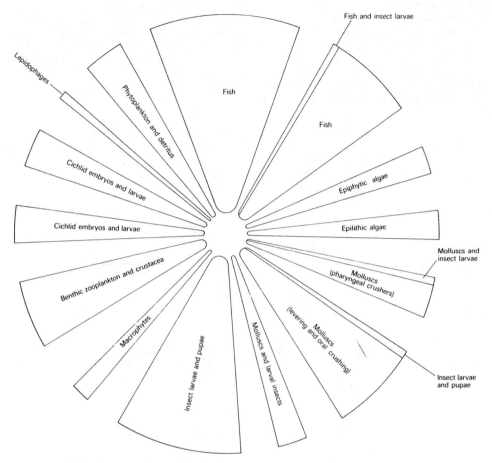

Figure 6.3 Diagram showing the variety of feeding habits amongst species of the *Haplochromis* group from Lake Victoria.
The area of each segment is proportional to the number of species utilising a particular food source.

insectivores. The two species that browse on macrophytic vegetation also have a lengthened gut but otherwise barely depart from the modal type.

Somewhat greater dental deviation occurs in a group of at least six species which graze algae from rocks and rooted plants. Here the teeth are longer, finer, and are moveably attached to the jaw bones. There is also some change in cusp form, with the crowns having an oblique rather than an acute cutting edge; in a few species the posterior cusp is almost as large as the anterior one. The inner teeth are arranged in several rows (5–8, compared with 2 or 3 in the generalised type), but retain a tricuspid crown.

The skull departs from the generalised shape in having the snout region noticeably down-curved (Fig. 6.7). The lower jaw is short and stout, relative to the modal type, and the bones of the upper jaw are also strengthened.

A second group of algal grazing species retains a generalised type of skull and jaw shape, but shows greater departure in tooth form (Fig. 6.4). The anterior cusp is drawn out and the posterior cusp is much reduced or even suppressed completely.

Unfortunately the significance of the dental and cranial differences between the two lineages is not really understood. Apparently the same algal species are consumed, but some field observations suggest that each group of algal grazers has different feeding methods. The species with stout jaws, decurved snouts and teeth with cusps of subequal size seem to graze algae directly from a hard, unyielding substrate. Members of the other group take a leaf in the mouth and, holding it loosely, then swim along its length, scraping off the epiphytes as the leaf passes between the several rows of teeth in each jaw. The shape of the teeth would thus seem to be associated with the method of scraping employed.

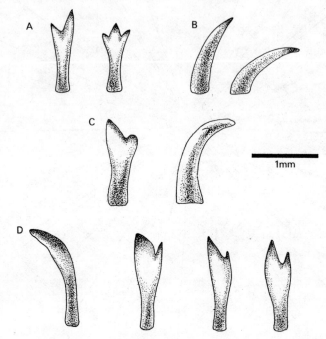

Figure 6.4 Tooth form in various Lake Victoria *Haplochromis*-group species. A–Outer (left) and inner (right) teeth of a generalised type; taken from an insectivore. B–Unicuspid, canine-like teeth, seen in side view to show variation in the degree of curvature; taken from piscivorous species. C–Stout bicuspid form (frontal view left, side view right) from a scale eating species. D–Teeth from four different species to show the gradual change in cusp form; on the right is a generalised bicuspid type, and on the left an extreme obliquely cuspid type associated with algal scraping. The second and third teeth in the row are also from algal scraping species.

Of particular interest is the fact that when the five species of the second group are compared, they can be arranged in a graded series according to cusp shape in the outer teeth. At the lower end of the series is a species in which the cusp is close to the generalised type (but with some protraction of the major cusp and its oblique cutting edge already apparent). The series terminates with a species in which the minor cusp is suppressed and the major cusp greatly protracted (Fig. 6.4).

A third type of scraping dentition has evolved in a species feeding on scales rasped from the tail fin of other *Haplochromis*. As in the algal grazers there are broad bands of inner teeth but, unlike those species, the lepidophage has stout, strongly recurved and acutely bicuspid teeth in the outer rows. This species also differs in being more slender and streamlined, in this way closely resembling many of the piscivorous predators. Presumably its feeding habits require a greater turn of speed when approaching and leaving its prey than is demanded of the algal grazers.

Molluscs, both gastropods and bivalves, form an important or even exclusive element in the food of at least twenty Lake Victoria *Haplochromis* species. Two quite distinct methods of handling this hard-shelled prey have been evolved. One method involves crushing the entire mollusc between the upper and lower pharyngeal 'jaws'. The other involves removing the mollusc from its shell, either by crushing the shell between the fish's jaws or by actually levering out the body and leaving the shell intact.

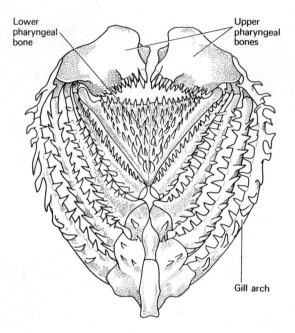

Figure 6.5 'Prey's-eye' view of the gill arches and the pharyngeal bones in a *Haplochromis*-group fish.

In the pharyngeal crushers the oral dentition is of a generalised type but the pharyngeal bones, especially the lower one, are much enlarged and some or almost all the teeth are strong and molar-like (Fig. 6.6). Skull shape in these species hardly differs from the generalised type.

Within the species of pharyngeal crushers one can observe a series of intergrading steps involving increasing hypertrophy of the pharyngeal bones and a correlated molarisation of their dentition (see Fig. 6.6). Species with slightly enlarged and molarised pharyngeals tend to have a mixed mollusc-insect diet, those with hypertrophied pharyngeal mills feed, at least when adult, almost exclusively on snails.

Those species that lever out the snail's body, or crush molluscs orally, show, as one might expect, modifications to the jaws and oral dentition. In several species there is also some departure from the generalised skull type, involving a strong downward flexure of the snout region. The latter modification allows the upper jaw to be protracted straight downwards, rather than forward and downward as in the generalised skull, thus enabling the fish to grab its prey from above rather than having to scoop it into the mouth from below.

The jaws are stout and the tooth-bearing surfaces are broad, markedly so in a few species. The teeth are strong, sharply pointed, and have a pronounced curvature towards the buccal cavity.

Functionally, the teeth, their disposition in the jaws and the shape of the jaws, all combine to form a strong, vice-like mechanism. With this vice, the body of the snail is held whilst the fish, with lever-like movements of its body, wrenches the snail from its shell. Alternatively, the shell is crushed free from the body (the usual process when bivalves are eaten). In contrast to the jaws, neither the pharyngeal bones nor their dentition is modified.

As with the algal grazers and the pharyngeal snail crushers, one can observe intergrading changes in tooth form, tooth pattern and skull type amongst the 'winkle-picking' and oral crushing species. These changes involve various combinations of characters which, starting with

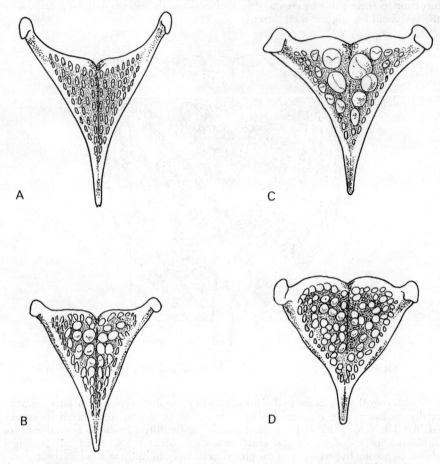

Figure 6.6 The toothed surface of the lower pharyngeal bones from four *Haplochromis*-group species in Lake Victoria. A–Detritus eater; B–D–Mollusc crushing species. In B the diet is a mixed mollusc-insect one, in C small gastropods and bivalves, and in D large, hard-shelled gastropods. Drawn to different scales, C greatly enlarged.

a species that differs only slightly from the generalised type, end with species so modified that systematists have placed them in genera distinct from one another and from other members of the trophic group.

A similar sequence in morphology can be seen in a lineage of specialised insectivores which are able, with their forceps-like dentition, to remove the larvae of burrowing insects from rocks and wood.

All species in this group have the lower jaw foreshortened and robust. The upper jaws are strengthened and capable of being protruded downwards, and there are correlated changes in the preorbital skull region like those seen in the mollusc eaters described before. The strong jaw teeth are reduced in number, uni- or weakly bicuspid, have sharply pointed crowns and protrude forwards. The unusual angle of tooth implantation, combined with an anteriorly narrowed dental arcade, imparts to the biting action of the jaws a forceps-like action as well. Because the teeth protrude, the fishes are able to insert what is virtually an extension of the jaws into the larval burrows.

Finally, we must consider what is probably the largest trophic group in Lake Victoria, the piscivorous predators, to which about 30–40 per cent of the known *Haplochromis* species of that lake belong. The principal food of these species is other *Haplochromis*, although a few non-cichlid fishes are also eaten.

Predatory *Haplochromis*, unlike their non-cichlid counterparts in the lake, do not bolt their prey whole, but macerate it between the upper and lower pharyngeal teeth. Thus the jaws and oral teeth serve to capture the prey and then to hold it during the lengthy process of pharyngeal mastication.

In the light of what we know about other trophic groups, it is not surprising to find an almost complete morphological sequence in adaptive characters amongst the piscivores (of which there are at least two phylogenetically distinct lineages). Most obvious are an increase in size, improved streamlining of the body, lengthening and increased protrusibility of the jaws, and a change from bicuspid to sharply pointed, unicuspid jaw teeth. Many predatory species also show a deepening of the pharyngeal and buccal cavities, and show a well-marked trend towards an increase in skull length coupled with a decrease in its height (Fig. 6.7). The decrease in skull height (affecting as it does that part of the skull over which the toothed part of the upper jaw slides during its protrusion), correlated with a deepening of the pharyngeal and buccal cavities, are important factors in developing an efficient prey catching device.

Modifications to the pharyngeal dentition are relatively slight, incorporating some reduction in tooth number, coarsening of the teeth, and the tendency for the crowns to develop a strong cutting edge.

The mouth-brooding habits of female cichlids have been exploited as a source of food by at least eight piscivorous species, the so-called paedophages, whose diet is principally the embryos and larvae obtained from the mouths of brooding *Haplochromis* females.

Two distinct lineages have adopted this bizarre feeding strategy. Both lines share certain oral modifications, namely a widely distensible and protrusible mouth, a reduction in the number of teeth and an hypertrophy of the oral mucosa so that the teeth are deeply embedded in soft tissue. Aquarium observations suggest that these modifications are associated with the paedophages' feeding methods, that is, engulfing the snout of the brooding female and forcing her to jettison the brood into the predator's mouth.

Although trends of increasing specialisation are seen within both paedophage lineages, these species do seem to differ from other specialist feeders in showing a more trenchant morphological gap between their least derived members and those of other trophic groups (including the true piscivores with which they share certain features).

There are a few lineages amongst the Lake Victoria *Haplochromis* (for example the small, benthic feeding zooplankton eaters) in which these seemingly orthoselective trends (Grant, 1977) cannot be detected because the member species all show a comparable level of specialisation. Within the flock as a whole, however, lineages having a distinct trend in morphological specialisation are commoner than those which do not. Nevertheless, it must be stressed that even in lineages showing clear-cut morphological sequences one invariably finds two or more species at any one level of specialisation.

This latter phenomenon is particularly obvious amongst the two major piscivore lines (but not the paedophages), the mollusc crushers (both oral and pharyngeal types), the specialist insectivores, and the algal grazers. Thus, from an anatomical point of view, it seems that two kinds of speciation may be recognised: an apomorphic kind in which one or more anatomical characters show further development in daughter taxa, and a stasimorphic one in which there is only a multiplication of species without further anatomical differentiation. These latter, then, may

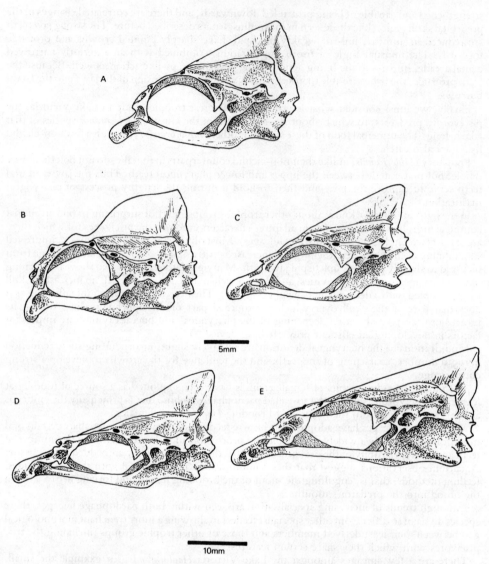

Figure 6.7 Skull shape in some Lake Victoria *Haplochromis*-group species. A–The generalised type (an insectivore). B–Skull with strongly decurved snout region (from an algal grazer). C, D and E–Skulls from piscivorous predators. Note the gradual change in proportions, particularly the relative elongation of the snout region and the overall increase in streamline form of the skull; compare with Figure 6.2.

display characteristics seen in sibling species-groups. The adjectival terms apomorphic and stasimorphic refer to extreme points in a continuum of the changes in external characteristics that are associated with speciation. Apomorphic is that which shows greater deviation from what is presumed to be the primitive condition.

Such a brief summary of morphological features in a trophically multiradiate species flock must, perforce, be an oversimplification, but it does serve to underline certain aspects which are of importance in attempting to understand this example of explosive evolution.

On analysis it seems that the anatomical features that underlie the trophic specialisations originated and developed through simple morphological transformations. The skeletal transformations (especially in the head) involved only slight alterations in the differential growth patterns of the various skull regions, coupled with similar changes in the jaws and the suspensorium. Likewise, taking the generalised bicuspid tooth as a starting point, the observed dental changes seem to stem from altered differential growth rates, this time in cusp sizes, curvature of the tooth, its length and robustness. Other dental changes involved an increase, or decrease, in the number of teeth or tooth rows. Changes in pharyngeal bone shape and in the form of the pharyngeal teeth also seem attributable to the same processes.

Finally, there are some modifications, especially important among piscivorous lineages, linked with an increase in adult size and possibly an enhanced growth rate. Particularly obvious are shape and proportional changes associated with allometric growth patterns, and the ontogenetic changes from bi- and unicuspid teeth with body growth beyond a certain size.

The simplicity of these structural changes, individually or in combination, may help to explain not only the wide range of trophic specialisations that were produced but also the speed with which this happened (that is, in about 750 000 years).

It would seem unnecessary to postulate a genetical revolution to produce morphological changes of this magnitude, and that too may account for the rapidity of change.

Regrettably, nothing is known about the genetical basis for the various transformations. Their phenotypic manifestation, however, suggests that genes with a regulatory or epigenetic function could have been more important than those with structural effects (see Stansfield, 1977, p. 331 for discussion of epigenesis).

In many respects (and with particular regard to the way in which speciation probably occurred in Lake Victoria) the origin of this flock seems to accord well with Grant's concept of quantum speciation, except that the genetical changes may not be of the revolutionary kind which he associates with that phenomenon (see Grant, 1977).

Parenthetically, it should be stressed that such seemingly simple, genetical and anatomical transformations are not confined to the species flock in Lake Victoria. They would seem equally applicable to the superficially more complex cichlid flocks of Tanganyika and Malawi.

That it is the cichlid species and not members of other families which have undergone explosive evolution and adaptive radiation probably is due in large measure to the suitability of the cichlid 'bauplan' as a substrate for the type of anatomical and dental specialisations discussed above. Most cichlids, unlike members of other families in the lakes, have a 'bauplan' that is neither too generalised nor too specialised for such simple changes to be effective. Perhaps, over a greater time-span, species from other families could have produced the necessary modifications to rival those of the cichlids. From their performance elsewhere the Cyprinidae (minnows) and Characidae might well have succeeded in so doing, particularly if the African characid species were of a dentally less specialised kind. To produce the range of dental and cranial specialisations seen in the Characidae as a whole (the only basis for comparison with a lake cichlid flock) from even the most generalised African species would require very considerable remodelling of the skull and dentition. It could not be achieved, as apparently it was in the cichlids, by simple changes in differential growth patterns. The cyprinids lack jaw teeth, and have only the lower pharyngeal bones toothed, two inhibiting factors that are compensated for only by the development of extensive modifications to mouth and lip form, and to the morphology and pattern of the pharyngeal teeth. Thus in this family too it would seem reasonable to hypothesise a greater level of genetical reorganisation than was required in the cichlids. The cichlids have two further advantages over the non-cichlids. Firstly, their ability to breed throughout the year and, secondly, the fact that most species have short generation times. Thus, compared with the annually breeding, slower maturing non-cichlids, the cichlids have far greater opportunity for genetical reshuffling, so providing more raw material for the production of novel morphotypes.

No matter what epigenetic mechanisms control the morphological differences underlying the trophic specialisations in a cichlid species flock, each flock owes its origin to repeated acts of speciation, many of which still persist in the contemporary lake. In other words, the Lake Victoria species flock, in its origins, consequent results and the evolutionary potential contained in those results, would seem to fit closely Grant's concept of speciational evolution, a thesis further developed in Gould & Eldredge's model of punctuational evolution (see Grant, 1977; Gould & Eldredge, 1977; Stanley, 1975).

Gould & Eldredge believe that the history of life has been dominated by concentrated outbursts of rapid speciation (followed by the differential success of certain species), rather than by slow directional transformations within a lineage (that is, evolution by accelerated cladogenesis rather than through phyletic gradualism).

Admittedly, within lineages of the Lake Victoria flock one can see many examples of gradual change in a particular character or suite of characters. But, each point in the grade is a species, and the different species are contemporaneous, not successive elements in a temporal and phyletic continuum as they would be in a case of true phyletic gradualism. Perhaps one should call this phenomenon 'cladistic gradualism'?

Unless truly quantum jumps occur, cladistic gradualism must be the norm in punctuational evolutionary histories, but it is doubtless a difficult feature to recognise when only fossils are available.

The concept of cladistic gradualism raises some interesting questions, especially when, as in Lake Victoria, it invokes characters associated with trophic specialisations. Take, for example, the phyletic lineages whose members are, respectively, mollusc eaters, piscivores, or algal grazers. From the data currently available there seem to be no qualitative differences between the diet of most species in a lineage (nor between members of different lineages with similar feeding habits). What factors, one asks, could have initiated development towards further levels of character derivation when, apparently, the 'lower' levels provide an equally effective degree of trophic specialisation? And, how is it that the latter species still remain in successful coexistence with their derived relatives if the concept of competitive exclusion has general applicability?

Paradoxically, within a moderately wide range of anatomical specialisation a trophically specialised species can also be a more generalised one. That is, it retains the ability to utilise the food sources tapped by its ancestors, and also has the capabilities to exploit sources not open to the ancestors because these lacked the dental and other necessary specialisations. Thus, in Lake Victoria today specialised algal grazers can, and do, feed on insects, but the insectivores are incapable of grazing algae.

Questions relating to the origin and further differentiation of derived features are relevant to the continuing debate on the role of natural selection in creating evolutionary novelties (see review by Rosen, 1978, pp. 371–3).

The slight and seemingly orthoselective differences between some species in a lineage might be interpreted as the sort of 'fine tuning' (Avise, 1977) that could be produced by natural selection (that is, the differential survival and ultimate fixation of certain alleles within a genotype). Such an interpretation would imply that the species are competing for a limited food resource, and that their morphological differences bring about more effective resource partitioning. Data currently available would not appear to corroborate the 'selection' hypothesis since there is apparently complete interspecific overlap in environmental requirement. However, more refined analyses are needed before these ideas can be tested fully.

Also uncorroborated is an alternative hypothesis, that such slight interspecific differences have no 'selective' value, and are stochastic in nature, the consequence of genetic sampling inevitably associated with a speciation event (that is, the so-called founder effect of Mayr). The high level of intraspecific variability in the characters under discussion, and the interspecific overlap in ecological requirements, however, seem to add support to the 'non-selectionist' viewpoint.

The major morphological differences, the obvious evolutionary novelties, which characterise each of the various lineages appear to be changes associated with altered growth patterns. These features, at least superficially, seem to be of greater magnitude than the intralineage differences discussed above. They certainly exceed the scale of change that could be labelled 'fine tuning'. The origin of such features is impossible to explain realistically on the basis of either true Darwinian or neo-Darwinian selection. The alternative is to hypothesise an origin stemming from the chance mutations established in a population at the time of the speciation event. The products of such changes would be, effectively, Goldschmidtian 'hopeful monsters' (although their origin, in the case of Lake Victoria cichlids, would not seem to involve the large-scale systemic mutations postulated by Goldschmidt; for further discussion of Goldschmidt's ideas, the reader is referred to Stansfield [1977]). In other words, the major lineage traits could have evolved in essentially the same non-selective way as the minor, intralineage traits.

Whatever evolutionary factors are involved, most lineages in the contemporary cichlid species flock of Lake Victoria clearly display the phenomenon of cladistic gradualism in several morphological traits. In this respect they retain a more complete account of their history than do the flocks of Lakes Tanganyika and Malawi, especially the former. The reasons for these differences are difficult to determine. The relative youth of Victoria, and its mode of origin, could both be contributory factors, as could differences in the diversity of the riverine forms that first colonised the developing lakes.

One can virtually be certain that cichlids were not the only fishes to populate a developing lake. If the riverine fish faunas of east Africa during the Pleistocene were like those of today, then both in the number of their species and in their trophic diversity, the cichlids were very much a minority group; the little fossil evidence available supports this conjecture. Thus, any suggested origin of a cichlid flock through multiple invasions of the lake can be discounted as an important element in its history, but the early non-cichlid invaders could well have been important in shaping its early development.

If multiple invasions are ruled out, then speciation within the lake basin, at all stages of its ontogeny, would seem to be the only way the cichlid flocks could have evolved.

Much debate has centred around the actual way in which speciation has taken place in the African lakes, in particular whether or not the cichlid flocks provide a *prima facie* case for sympatric as opposed to allopatric speciation (see Fryer & Iles, 1972; Greenwood, 1974 for references).

Supporters of sympatric speciation in Lake Victoria have generally suggested habitudinal segregation as the means whereby the evolving species are isolated from one another. It is difficult to conceive how such segregation alone could have led to the origin of 170 species, especially when the existing species are segregated more by differences of feeding habits than by differences in habitat, and when there is no or little appreciable spatial segregation of preferred food organisms within a habitat. Also it could be argued that if the original cichlid invaders did, by some means or other, become habitudinally segregated, then any subsequent speciation would, in fact, conform to the allopatric model. Indeed, it is difficult to know at what spatial cut-off point one should draw the line between so-called allo- and sympatric speciation (see discussion in White, 1978, p. 146). Perhaps only stasipatric speciation, speciation through disruptive selection and speciation through polyploidy in plants should qualify as truly sympatric models.

As we have no genetical or karyotypic information for the Lake Victoria cichlids (and very little for other cichlids either) one cannot usefully speculate on the possibility of stasipatric speciation playing some part in the origin of the flock.

Evidence from an American cichlid does, however, suggest a way in which disruptive selection could have been involved (Sage & Selander, 1975). In a Mexican lake studied by these authors, the supposedly single species present exists in a state of balanced polymorphism. The three morphs show distinct anatomical differences associated with the oral and pharyngeal dentition, and there are correlated differences in feeding habits (algae and detritus, molluscs, and fishes respectively). If the genes controlling these features were to become linked with genes affecting male breeding coloration (or other reproductive features), then, through the effects of assortative mating, the morphs could become true species (a state which the Mexican species seemingly has not yet reached). In other words, this model proposes that the primary isolating mechanism is selective mating, affected through differences in breeding livery. Unfortunately, there is no biological information to support Sage & Selander's suggestion that balanced polymorphism could explain the situation in the African lake flocks (see Greenwood, 1974), although it might have played a part in their origins.

The geomorphological and hydrographical history of Lake Victoria seems to suggest that allopatric speciation, through actual geographical isolation, played the prime role in the establishment and diversification of its species flock.

When the lake first formed (during the mid-Pleistocene c. 750 000 years BP) its future basin was crossed by four or more westward flowing rivers. These drained the eastern highlands of present-day Kenya and emptied into the Zaire river system. A gradual but large-scale surface warping in the west led to a reversal of river flow and a consequent backponding in the western reaches of the rivers. As the shallow western river valleys filled with water each became an expansive, dendritic lake. Eventually these lakes overflowed, joining their neighbours to form a large but shallow water body occupying an area considerably greater than that of the present lake. Later, the lake basin was subject to further periods of tectonic instability which considerably modified the margins of the lake. Such changes would, at times, result in the isolation of water bodies on the periphery, and at other times their reunion with the main lake. Still later, and when the lake had assumed its present form, local climatic changes led to alterations in the lake level. These changes again resulted in the formation of small peripheral lakes and their ultimate reunion with the lake (see Fryer & Iles, 1972; Greenwood, 1974 for references).

In effect then, present-day Lake Victoria can be considered the product of repeated fractioning and reunion; that is, an amalgam of several lakes which developed in a closed drainage basin.

Initial differentiation, through acts of speciation, of the main phyletic lines within the cichlid species flock, and the concomitant origin of what are now seen as the various trophic specialisations, must have taken place in the shallow lakes formed when the rivers were first reversed and ponded back. Potentially each lake could be the cradle of a species. Later, as the number of species was increased, each peripheral isolate could serve as the cradle for as many new species as there were existing species cut off in it, a kind of exponential species growth.

Lake Nabugabo, a small (30 km^2) lake now separated from Victoria by a narrow sand bar, is a good example of this process, and also of the rate at which certain cichlids can speciate (Greenwood, 1965). Nabugabo was cut off about 4000 years ago; yet five of its seven *Haplochromis* species are endemic. Each endemic species closely resembles a species in Lake Victoria, but differs both in male coloration and, less markedly, in some morphometric characters (an example of stasimorphic speciation).

A model of alternating allopatry and sympatry may also help to explain a notable feature of the Lake Victoria flock. That is, the occurrence in many lineages of more than one species at a particular level in a trend of trophic specialisation. That is to say, these species are the products of speciation events that did not involve the evolution of any change in anatomical or physiological characters associated with feeding habits, but simply the evolution, in different peripheral lakes, of only specific reproductive barriers between daughter species derived from segments of the same parental species.

Although, ultimately, the cichlid flock in Lake Victoria came to resemble those of Lakes Tanganyika and Malawi, the history of the lakes, and the phylogeny of their cichlid flocks, differ is several respects (see Fryer & Iles, 1972; Greenwood, 1974, 1979a). Each lake, in its own way, contributes to our understanding of explosive evolution, a multifaceted phenomenon.

In this context it is relevant to compare the cichlid species flocks with certain other, and better publicised, examples of adaptively multiradiate and geographically restricted species groups, the Galapagos finches, the Hawaiian honeycreepers and the Hawaiian fruitflies (see Grant, 1977; White, 1978; Dobzhansky *et al.*, 1977). As in the lakes, these island faunas are, so to speak, dominated by one group of related animals, a feature further emphasised in the islands by the absence of many components of the mainland faunas from which they were first populated.

There are many broad similarities in the ways in which these otherwise dissimilar organisms have differentiated. In all three island flocks (as in the lake cichlids) differentiation has been primarily towards trophic specialisation. All share, as contributory features in their evolution, the historical elements of geographical (or at least spatial) isolation and the availability of new habitats and niches for exploitation. A high level of endemicity is a further common feature, as is the fact that the island birds and fruitflies, like the lake faunas, show a greater range of morphological diversity than is encountered in their close relatives from other regions. Except for the fruitflies, however, far fewer species are involved in the islands flocks (for example 13 finches and

22 honeycreepers, but as many as 160 species of fruitflies).

The manner in which the spatial isolation necessary for speciation was achieved may differ in the different island flocks (as indeed it probably did in the African lakes). For example, the pattern amongst the Galapagos finches, of evolution in isolation followed by inter-island migration and the ultimate occurrence of several species coexisting on any one island, closely parallels the pattern proposed in the Lake Victoria fishes. There, the first speciation events occurred in isolated lakes, whilst the later union of the separate small lakes would be equivalent to the inter-island migration phases of evolution in the Galapagos flock. Isolation through the development of ecological barriers within an island (for example, lava flows surrounding patches of vegetation) is thought to be the principal factor in speciation amongst Hawaiian fruitflies. The parallel here would be with the way in which part of the Lake Malawi cichlid flock probably evolved, with sandy beaches and rocky outcrops providing the ecological barriers (see discussion in Fryer, 1977; Fryer & Iles, 1972).

The fish, bird and insect flocks are similar, too, in that their adaptational successes have been effected through simple anatomical changes. The complexity, or otherwise, of the underlying genetic changes involved are unknown for the birds and the fishes. Those concerned in the fruitfly radiation would not appear to be revolutionary ones (see White, 1978).

It is generally believed that one factor stimulating the evolution of trophic adaptations in the Galapagos finches and the Hawaiian honeycreepers was an absence of competitors on the newly created islands. To a large extent this hypothesis seems inapplicable to the cichlid fishes. The original cichlid invaders would be in the company of other species from other families which, in total, would seem capable of occupying all the trophic and ecological niches provided by the developing lake (and, from their present-day habits, those of the mature lake as well). These interfaunal reactions (especially in the exploitation of micro-niches), rather than their absence, may have played an important role in the early shaping of the cichlid flocks. Nevertheless, the creation of increasing numbers of niches by members of the flock itself could well have been an important influence on later stages of differentiation within the species flocks of the island birds and the fruitflies, as it undoubtedly was amongst the lake fishes.

In general then, there would seem to be greater levels of similarity than dissimilarity in the way these various radiations have evolved. White (1978, p. 232), on the contrary, believes that '. . . there is probably not much resemblance between the patterns and modes of speciation in ancient lakes and on oceanic islands'. His views are based essentially on supposedly differing degrees of isolation and ecological diversity obtaining in the two sitations, a viewpoint that is not substantiated by information from the African lakes, nor from Lake Baikal (see Greenwood, 1974; Fryer & Iles, 1972; Kozkov, 1963). White (1978) would, however, concede a greater level of similarity between the situation in Lake Victoria and that in the Galapagos and the Hawaiian islands.

The species flocks and other localised, sometimes explosive, radiations discussed above are by no means the only ones known amongst animals and plants (see White, 1978). From what we can reconstruct of their histories, and learn from the living animals, it seems unnecessary to invoke any special kind of evolutionary phenomenon to explain their origin and development. Rather, each appears to represent a combination, and temporal condensation, of many factors thought to have been generally operative in the evolution of bisexual organisms. In particular, these flocks focus attention on the primary factor in evolution – the act of speciation (Stanley, 1975; Greenwood, 1979*b*). It is the elements of temporal and spatial condensation, coupled with the persistence of all or nearly all stages in the development of different anatomical specialisations, that make species flocks and explosive evolution so conspicuous; on studying a species flock one experiences the feeling of looking into a factory where prototypes are still in production alongside the latest models.

The question of whether or not explosive speciation has contributed to the origin of major animal and plant groups is unlikely to be answered by direct evidence and certainly not by fossil evidence. But, since the origin of every major group lies in a single speciation event, the origin of numerous and diverse species in a circumscribed area, and over a geologically short time-span, could well be of importance in providing the raw materials for further evolutionary development and diversification (see also Greenwood, 1979*b*; Gould & Eldredge, 1977; Rensch, 1959 for further discussion).

References

Avise, J. C. 1977. Is evolution gradual or rectangular? Evidence from living fishes. *Proceedings of the National Academy of Sciences of the United States of America* **74**: 5083–7.
Dobzhansky, T., Ayala, F. J., Stebbins, G. C. & Valentine, J. W. 1977. *Evolution*, 572 pp. San Francisco: Freeman.
Fryer, G. 1977. Evolution of species flocks of cichlid fishes in African Lakes. *Zeitschrift für Zoologische Systematik und Evolutionsforschung* **15**: 141–65.
Fryer, G. & Iles, T. D. 1972. *The cichlid fishes of the Great Lakes of Africa. Their biology and evolution*. 641 pp. Edinburgh: Oliver & Boyd.
Gould, S. J. & Eldredge, N. 1977. Punctuated equilibria: the tempo and mode of evolution reconsidered. *Paleobiology* **3**: 115–51.
Grant, V. 1977. *Organismic evolution*. 418 pp. San Francisco: Freeman.
Greenwood, P. H. 1965. The cichlid fishes of Lake Nabugabo, Uganda. *Bulletin of the British Museum (Natural History), Zoology* **12**: 315–57.
—— 1974. Cichlid fishes of Lake Victoria, East Africa: the biology and evolution of a species flock. *Bulletin of the British Museum (Natural History), Zoology* Supplement 6: 1–134.
—— 1979a. Towards a phyletic classification of the 'genus' *Haplochromis* and related taxa. Part I. *Bulletin of the British Museum (Natural History), Zoology* **35**: 265–322.
—— 1979b. Macroevolution – myth or reality? *Biological Journal of the Linnean Society of London* **12**: 293–304.
—— 1980. Towards a phyletic classification of the 'genus'*Haplochromis* and related taxa. Part II. *Bulletin of the British Museum (Natural History), Zoology* **39**: 1–101.
Kozkov, M. 1963. Lake Baikal and its life. *Monographiae Biologicae* **11**: 1–344.
Mayr, E. 1976. *Evolution and the diversity of life*, 721 pp. Harvard: Belknap Press.
Rensch, B. 1959. *Evolution above the species level*, 419 pp. London: Methuen.
Rosen, D. E. 1978. Darwin's demon (A review of *Introduction to natural selection*, by C. Johnson [1976]), *Systematic Zoology* **27**: 370–3.
Sage, R. D. & Selander, R. K. 1975. Trophic radiation through polymorphism in cichlid fishes. *Proceedings of the National Academy of Sciences of the United States of America* **72**: 4669–73.
Stanley, S. M. 1975. A theory of evolution above the species level. *Proceedings of the National Academy of Sciences of the United States of America* **72**: 646–50.
Stansfield, W. D. 1977. *The science of evolution*. 614 pp. London: Collier Macmillan.
White, M. J. D. 1978. *Modes of speciation*. 455 pp. San Francisco: Freeman.

Indexes

Some remarks about the indexing system are necessary, especially since the principal haplochromine genus, *Haplochromis*, has been divided recently into a number of different genera.

i) For simplicity's sake, the species dealt with in these volumes are still entered under the name *Haplochromis* if they were previously placed in that genus.

The current generic name for every former *Haplochromis* species is, however, listed alongside it (together with the page on which that genus is described or redescribed).

ii) All page numbers cited in the indexes refer to the pagination adopted for this reprint edition (centred at the foot of the page) and *not* to the original pagination of the papers which, except for two pages which were re-set, appears in the upper left or right hand corner of a page.

iii) The page reference (or, occasionally, references) relates to the place where a genus or species is fully described. Other references to a taxon, unless they are of particular taxonomic importance, are not cited.

Index to species

Former binomen	Page	Current generic placement	Page
Astatoreochromis alluaudi	83	*Astatoreochromis*	679; 83
Astatotilapia nigrescens	368; 736	?*Prognathochromis*	736
Chetia brevis	678	?*Astatotilapia*	678
C. flaviventris	701	*Chetia*	701
Chromis monteiri	688	*Thoracochromis fasciatus*	688
Haplochromis acidens	323	*Psammochromis*	769
H. acuticeps	706	'*Haplochromis*'	706
H. adolphifrederici	758	?*Labrochromis*	758
H. aelocephalus	132	*Psammochromis*	769
H. aeneocolor	540	*Astatotilapia*	675; 722
H. albertianus	309	*Thoracochromis*	684
H. albolabris	706	'*Haplochromis*'	707
H. altigenis	310	*Harpagochromis*	726
H. angolensis	706	'*Haplochromis*'	707
H. angustifrons	567	*Gaurochromis*	748
H. annectidens	501	*Haplochromis*	672
H. apogonoides	355	*Incertae sedis*	805
H. arcanus	420	*Prognathochromis*	730
H. argenteus	334	*Prognathochromis*	730
H. artaxerxes	220	*Harpagochromis*	726
H. astatodon	674	*Haplochromis*	672
H. avium	687	*Thoracochromis*	684
H. bakongo	687	*Thoracochromis*	684
H. barbarae	343; 724	*Astatotilapia*	675; 722
H. bartoni	211	*Prognathochromis*	730
H. bayoni	199	*Prognathochromis*	730
H. beadlei	507	*Paralabidochromis*	783
H. bloyeti	677	*Astatotilapia*	675
H. boops	297	*Harpagochromis*	726
H. brownae	192	*Astatotilapia*	675; 722
H. burtoni	678	*Astatotilapia*	675; 722
H. buysi	706	'*Haplochromis*'	706
H. callipterus	678	*Astatotilapia*	675
H. carlottae	699	*Serranochromis* (*Sargochromis*)	697

H. cassius	455	Psammochromis	769
H. cavifrons	246	Harpagochromis	726
H. chilotes	125	Paralabidochromis	783
H. chlorochrous	412	Prognathochromis	730
H. chromogynos	130	Paralabidochromis	783
H. cinctus	383	Enterochromis	759
H. cinereus	147	Astatotilapia	675; 722
H. codringtoni	699	Serranochromis (Sargochromis)	697
H. coulteri	698	Serranochromis (Sargochromis)	697
H. crassilabris	787	Paralabidochromis	783
H. crocopeplus	436	Prognathochromis	730
H. cronus	98	Incertae sedis	805
H. cryptodon	116	Lipochromis	742; 747
H. cryptogramma	416	Prognathochromis	730
H. darlingi	704	Pharyngochromis	704
H. decticostoma	423	Prognathochromis	730
H. demeusii	309	Thoracochromis	684
H. dentex	217	Prognathochromis	730
H. desfontainesii	677	Astatotilapia	675; 722
H. dichrourus	315	Prognathochromis	730
H. diplotaenia	729	?Harpagochromis	726; 729
H. dolichorhynchus	402	Prognathochromis	730
H. dolorosus	677	Astatotilapia	675; 722
H. eduardi	724	Astatotilapia	722
H. eduardiana	605	Schubotzia	801
H. elegans	535	Astatotilapia	675; 722
H. empodisma	170	Gaurochromis	748
H. engyostoma	724	Astatotilapia	675; 722
H. erythrocephalus	387	Enterochromis	759
H. estor	214	Prognathochromis	730
H. eutaenia	738	Prognathochromis	730
H. fasciatus	687	Thoracochromis	684
H. flavijosephii	677	Astatotilapia	675; 722
H. flavipinnis	242	Prognathochromis	730
H. fuscus	768	Neochromis	765
H. fusiformis	400	Yssichromis	738
H. giardi	699	Serranochromis (Sargochromis)	697
H. gilberti	425	Prognathochromis	730
H. gowersi	230	Prognathochromis	730
H. granti	70	Ptyochromis	776
H. graueri	772	Psammochromis	769
H. greenwoodi	698	Serranochromis (Sargochromis)	697
H. horii	683	Ctenochromis	681
H. humilior	156	Labrochromis	753
H. humilis	706	'Haplochromis'	706
H. ishmaeli	183	Labrochromis	753
H. labiatus	586	Paralabidochromis	783
H. lacrimosus	138	Astatotilapia	675; 722
H. laparogramma	396	Yssichromis	738
H. latifasciatus	724	Astatotilapia	675; 722
H. limax	557	Haplochromis	672; 769
H. lividus	46	Haplochromis	672; 769
H. loati	687	Thoracochromis	684
H. longirostris	221	Prognathochromis	730

H. luluae	683	Ctenochromis	681
H. macconneli	646	Thoracochromis	684
H. machadoi	691	Orthochromis	689
H. macrognathus	233	Prognathochromis	730
H. macrops	144	Astatotilapia	675; 722
H. macropsoides	552	Astatotilapia	675; 722
H. maculipinna	293	Harpagochromis	726
H. mahagiensis	687	Thoracochromis	684
H. mandibularis	228	Prognathochromis	730
H. martini	153	Astatotilapia	675; 722
H. maxillaris	107	Lipochromis	742
H. megalops	372	Astatotilapia	675; 722
H. melanopterus	110	Lipochromis	742
H. melanopus	725	Astatotilapia	675; 722
H. melichrous	392	Prognathochromis	730
H. mellandi	699	Serranochromis (Sargochromis)	697
H. mentatus	594	Harpagochromis (see H. squamipinnis)	726
H. mento	224	Prognathochromis	730
H. michaeli	253	Harpagochromis	726
H. microdon	118	Lipochromis	742
H. moeruensis	687	Thoracochromis	684
H. mortimeri	699	Serranochromis (Sargochromis)	697
H. multiocellatus	706	'Haplochromis'	706
H. mylergates	468	Labrochromis	753
H. mylodon	562	Labrochromis	753
H. nanoserranus	451	Prognathochromis	730
H. nigricans	51	Neochromis	765
H. nigripinnis	545	Enterochromis	759
H. niloticus	151	Xystichromis (see H. bayoni)	762; 764
H. nubilus	518; 611	Astatotilapia	675; 722
H. nuchisquamulatus	55	Xystichromis	762
H. nyanzae	209	Harpagochromis	726
H. obesus	100	Lipochromis	742
H. obliquidens	40	Haplochromis	672
H. obtusidens	174	Gaurochromis	748
H. oligacanthus	683	Ctenochromis	681
H. oregosoma	549	Astatotilapia	675; 722
H. orthostoma	350	Pyxichromis	740
H. pachycephalus	289	Harpagochromis	726
H. pallidus	141	Astatotilapia	675; 722
H. pappenheimi	589	Yssichromis	738
H. paraguiarti	319	Prognathochromis	730
H. paraplagiostoma	428	?Harpagochromis	726; 730
H. paropius	378	Enterochromis	759
H. parorthostoma	353	Pyxichromis	740
H. parvidens	112	Lipochromis	742
H. paucidens	787	Paralabidochromis	783
H. pectoralis	683	Ctenochromis	681
H. pectoralis	337	Harpagochromis	726
H. pellegrini	236	Prognathochromis	730
H. percoides	239	Prognathochromis	730
H. petronius	599	Thoracochromis	684
H. pharyngalis	687; 603	Thoracochromis	684
H. pharyngomylus	178	Labrochromis	753

H. phytophagus	265	Xystichromis	762
H. piceatus	375	Astatotilapia	675; 722
H. placodus	758; 566	Labrochromis	753
H. plagiodon	123	Paralabidochromis	783
H. plagiostoma	249	Harpagochromis	726
H. plutonius	445	Prognathochromis	730
H. polli	683	Ctenochromis	681
H. polyacanthus	313	Orthochromis	689
H. prodromus	66	Ptyochromis (see P. annectens)	776
H. prognathus	328	Prognathochromis	730
H. pseudopellegrini	240	Prognathochromis	730
H. ptistes	458	Labrochromis	753
H. riponianus	160	Psammochromis	769
H. rudolfianus	634	Thoracochromis	684
H. sauvagei	60	Ptyochromis	776
H. saxicola	164	Psammochromis	769
H. schoutedeni	806	Incertae sedis	806
H. schubotzi	573	Psammochromis	769
H. schubotziellus	578	Astatotilapia	675; 722
H. serranus	202	Harpagochromis	726
H. serridens	768	Neochromis	765
H. simpsoni	497	Gaurochromis	748
H. spekii	282	Harpagochromis	726
H. squamipinnis	594	Harpagochromis	726
H. squamulatus	337	Harpagochromis (see Har. pectoralis)	726
H. stappersi	678	Astatotilapia	675; 722
H. straeleni	680	Astatoreochromis	679; 83
H. sulphureus	442	Prognathochromis	730
H. swynnertoni	678	Astatotilapia	675; 722
H. taurinus	582	Lipochromis	742; 745
H. teegelaari	463	Labrochromis	753
H. theliodon	168	Incertae sedis	805
H. thuragnathus	299	Harpagochromis	726
H. thysi	699	Serranochromis (Sargochromis)	697
H. tridens	347	Prognathochromis	730
H. turkanae	642	Thoracochromis	684
H. tyrianthinus	408	Prognathochromis	730
H. vanderhorsti	680	Astatoreochromis	679; 83
H. velifer	491	Astatotilapia	675; 722
H. venator	514	Prognathochromis	730
H. vicarius	724	Astatotilapia (see A. eduardi)	675; 722
H. victorianus	206	Harpagochromis	726
H. vittatus	736	Prognathochromis	730
H. welcommei	271	Allochromis	773
H. welwitschii	706	'Haplochromis'	706
H. wingatii	687	Thoracochromis	684
H. wittei	806	Incertae sedis	806
H. worthingtoni	730	?Harpagochromis	726
H. xenognathus	74	Ptyochromis	776
H. xenostoma	301	Prognathochromis	730
Hoplotilapia retrodens	25	Hoplotilapia	788; 23
Macropleurodus bicolor	5	Macropleurodus	796; 3
Orthochromis malagaraziensis	691	Orthochromis	689
Paralabidochromis victoriae	32	Paralabidochromis	783; 31

Paratilapia toddi	688	*Incertae sedis*	688
Platytaeniodus degeni	19	*Platytaeniodus*	791; 16
Rheohaplochromis torrenticola	691; 692	*Orthochromis*	689
Sargochromis codringtoni	699	*Serranochromis (Sargochromis)*	697
S. mellandi	699	*Serranochromis (Sargochromis)*	697
Serranochromis angusticeps	693	*Serranochromis (Serranochromis)*	693
S. janus	697	*Serranochromis (Serranochromis)*	693
S. longimanus	696	*Serranochromis (Serranochromis)*	693
S. macrocephalus	696	*Serranochromis (Serranochromis)*	693
S. meridionalis	697	*Serranochromis (Serranochromis)*	693
S. robustus	696	*Serranochromis (Serranochromis)*	693
S. spei	696	*Serranochromis (Serranochromis)*	693
S. stappersi	696	*Serranochromis (Serranochromis)*	693
S. thumbergi	696	*Serranochromis (Serranochromis)*	693

Generic and subgeneric index

Allochromis	773	*Neochromis*	765
Astatoreochromis	83; 679	*Orthochromis*	689
Astatotilapia	675; 722	*Paralabidochromis*	783
Chetia	701	*Pharyngochromis*	704
Cleptochromis	747	*Platytaeniodus*	16; 791
Ctenochromis	681	*Prognathochromis*	730
Enterochromis	759	*Psammochromis*	760
Gaurochromis	748	*Ptyochromis*	776
Haplochromis	672	*Pyxichromis*	740
'Haplochromis'	707	*Schubotzia*	801
Harpagochromis	726	*Sargochromis*	697
Hoplotilapia	23; 788	*Serranochromis*	693
Labrochromis	753	*Thoracochromis*	684
Lipochromis	742	*Tridontochromis*	736
Macropleurodus	3; 796	*Xystichromis*	762
Mylacochromis	752	*Yssichromis*	738

M01 40006 96550

RENNER LEARNING RESOURCE CENTER
Elgin Community College
Elgin, IL 60123